PROCEEDINGS OF THE INTERNATIONAL CONFERENCE ON

ALGEBRA 2010

Advances in Algebraic Structures

Gadjah Mada University, Indonesia   7 – 10 October 2010

Wanida Hemakul
Chulalongkorn University, Thailand

Sri Wahyuni
Gadjah Mada University, Indonesia

Polly W Sy
University of the Philippines-Diliman, Philippines

editors

NEW JERSEY • LONDON • SINGAPORE • BEIJING • SHANGHAI • HONG KONG • TAIPEI • CHENNAI

Published by

World Scientific Publishing Co. Pte. Ltd.
5 Toh Tuck Link, Singapore 596224
USA office: 27 Warren Street, Suite 401-402, Hackensack, NJ 07601
UK office: 57 Shelton Street, Covent Garden, London WC2H 9HE

British Library Cataloguing-in-Publication Data
A catalogue record for this book is available from the British Library.

PROCEEDINGS OF THE INTERNATIONAL CONFERENCE ON ALGEBRA 2010
ADVANCES IN ALGEBRAIC STRUCTURES

ISBN-13 978-981-4366-30-4
ISBN-10 981-4366-30-7

Printed in Singapore by World Scientific Printers.

Preface

This volume contains proceedings of the International Conference on Algebra held in honor of Professor Shum Kar-Ping in the occasion of his 70th birthday. It took place in the Faculty of Mathematics and Natural Sciences of the Gadjah Mada University, Yogjakarta, Indonesia on 7–10 October 2010. More than a hundred participants from all over the world attended this Conference.

As a consequence of the wide coverage of his research interests and work, this volume contains 54 research articles which describe the latest research and development, and address a variety of issues and methods in semigroups, groups, rings, modules, lattices, combinatorics and Hopf algebra. The book also includes five well-written expository survey articles on the structure of finite groups, new results of Gröbner-Shirshov basis, polygroups and their properties, main results on abstract characterizations of algebras of n-place functions obtained in the last 40 years, as well as inverse semigroups and their generalizations. All manuscripts are original and have been carefully peer-reviewed. In view of this, we would like to express our gratitude to all authors for their contributions and the referees for their time and the high quality job.

We would like to sincerely thank all plenary and invited speakers who warmly accepted our invitation to come to the Conference and the paper contributors for their overwhelming response to our call for short presentations. Moreover, we are very grateful for the financial assistance and support that we received from the Gadjah Mada University, the Southeast Asian Mathematical Society, and the Indonesian Mathematical Society. Finally, we thank all members of the International Organizing Committee chaired by Professor Tan Eng Chye, International Steering Committee chaired by Professor Ann-Chi Kim, International Program Committee chaired by Professor Efim Zelmanov, and the Local Organizing Committee chaired by Professor Sri Wahyuni for contributing to the success of this Conference.

The Editors
July 31, 2011

Conference Program

October 7 (Thursday)

08:30 – 09:30 **Opening Ceremony**

09:30 – 10:20 **Victoria Gould** (University of York)
Restriction and Ehresmann semigroups

10:40 – 11:30 **Leonid Bokut** (Russian Academy of Sciences)
Gröbner and Gröbner-Shirshov bases theory and some applications for algebras, groups, and semigroups

11:30 – 12:20 **Ngai Ching Wong** (National Sun Yat-Sen University)
Kaplansky type theorems for completely regular spaces

13:00 – 13:50 **Xiuyun Guo** (Shanghai University)
On power automorphisms and induced automorphisms in finite groups

13:50 – 14:40 **Laszlo Marki** (Hungarian Institute of Sciences)
Divisibility theory in commutative rings: Bezout monoids

14:40 – 15:10 **Maxim Goncharov** (Sobolev Institute of Mathematics)
Structures of Malcev bialgebras on the simple non-Lie Malcev algebra

15:30 – 16:00 **Ivan Kaygorodov** (Sobolev Institute of Mathematics)
δ-superderivations on semisimple Jordan superalgebras

October 8 (Friday)

08:00 – 9:30 Parallel Session Invited Talks

10:00 – 11:20 Parallel Session 1 of Contributed Talks

13:00 – 15:00 Parallel Session 2 of Contributed Talks

15:20 – 17:00 Parallel Session 3 of Contributed Talks

October 9 (Saturday)

08:00 – 08:50 **Nguyen Van Sanh** (Mahidol University)
A generalization of Hopkins -Levitzki theorem

08:50 – 09:40 **Ivan Chajda** (Palacky University Olomouc)
Homomorphisms of relational systems and the corresponding groupoids

10:00 – 10:50 **Yuqun Chen** (South China Normal University)
Gröbner-Shirshov bases for Lie algebras over a commutative algebra

10:50 – 11:50 **Closing Ceremony**

Parallel Sessions of Invited Talks
October 8, 2010, Friday, 08:00 – 09:30

Parallel Session 1

Time	Title	Invited speaker
08:00– 08:30	Polygroups and their properties	Bijan Davvaz (Yazd University)
08:30– 09:00	A new type of shadow problem	Gyula O.H. Katona (Hungarian Academy of Sciences)
09:00– 09:30	Notes on infinite dimensional Lie algebras	Ki-Bong Nam (University of Wisconsin-Whitewater)

Parallel Session 2

Time	Title	Invited speaker
08:00 - 08:30	The development of the theory of almost distributive lattices	G. C. Rao (Andhra University)
08:30 – 09:00	Extension of certain fuzzy ideals of semirings on strong lattice of semirings	T. K. Dutta (University of Calcutta)
09:00 – 09:30	Lecture on pseudo-complemented almost distributive lattices	G. Nanji Rao (Andhra University)

Parallel Session 3

Time	Title	Invited speaker
08:00 – 08.30	Some properties of semirings	T. Vasanthi (Yoji Vemana University)
08:30 – 09:00	K-restricted duplication closure of languages	Masami Ito (Kyoto Sangyo University)
09:00 – 09:30	Sheaves over Boolean space	U M Swamy (Andhra University)

Contents

A Short Biography of Professor Shum Kar-Ping

Wanida Hemakul,* Sri Wahyuni† and Polly W. Sy‡

Professor Shum Kar-Ping grew up in Ping Shek, a small town in the northern part of Guangdong Province of China where he was born on January 2, 1941. He received his primary education in Guangzhou and later his junior high school education in Macau. In 1955, he followed his parents to Hong Kong and studied in Pui Kiu Middle School where he finished his high school in 1958.

Due to his excellent performance in the Hong Kong School Certification Public Examination, he was awarded a five-year Hong Kong Government Scholarship. In 1962, he obtained a Diploma course in Mathematics and Physics at Hong Kong Baptist College (now the Hong Kong Baptist University). During this period, he also obtained a Higher Certificate in Electrical Engineering from the Hong Kong Technical College (now, the Hong Kong Polytechnic University).

A year later, he finished a special one-year course from the Hong Kong Northcote College of Education. He then worked as a high school teacher of mathematics and science for the next two years. From 1964–67, while teaching in high school, he continued to study Refrigeration, as a part-time student, and consequently obtained certificates in AC and DC Machinery from the City and Guilds of London Institute via external examinations.

In 1968, Shum received his MS in Mathematics from the University of Leeds, England. In 1971, he obtained his PhD in Mathematics from the University of Alberta, Canada. Soon after, he joined the faculty of Chung Chi College of the Chinese University of Hong Kong (CUHK) as a Lecturer of Mathematics in October 1971, and subsequently rose through the academic ranks from Lecturership to full Professorship. During these years, he was the Chairman of the Department of Mathematics in 1976 for one term. He served CUHK for more than 30 years as an educator and a researcher.

*Department of Mathematics, Chulalongkorn University, Bangkok, Thailand
†Department of Mathematics, Gadjah Mada University, Yogyakarta, Indonesia
‡Institute of Mathematics, University of the Philippines, Diliman, Quezon City, Philippines

In January 2008, he retired from CUHK and was appointed an Honorary Professor of the University of Hong Kong for three years. At present, he is an Honorary Director of the Institute of Mathematics of Yunnan University, China.

During his active years, Shum had visited and given lectures in many prestigious universities and academic institutions in Southeast Asia, Asia, North America, Canada as well as Europe. Some countries he visited were Russia, India, Pakistan, Iran, Japan, Bangladesh, Mongolia, Australia, South Korea, New Zealand, USA, and mainland China. Moreover, he was a Visiting Professor/Honorary Professor/Consulting Professor of Mathematics at the University of Alberta, University of Manitoba, Wuhan University, Shanghai University, Southwest Jiaotung University, South China University of Technology, the University of the Philippines-Diliman, Chulalongkorn University, Mahidol University, South China Normal University, and Yunnan University. In April 2010, he was awarded an Honorary Doctorate degree at the 80th anniversary ceremony of Gomel University in Russia.

Academically, Shum has authored more than 300 publications in well-known international mathematical journals. He also co-authored at least 7 books. His research work is mainly in algebraic structures and his research interests include semi-groups, groups, ring theory, module theory, universal algebra, and fuzzy algebras. He has collaborated with many mathematicians around the world. Among his main collaborators are Guo Yuqi, Leonid Bokut, M. K. Sen, Nguyen Van Sanh, Xu Yonghua, Guo Xiaojiang, Guo Wenbin, and his students Guo Xiuyun, Ren Xueming, Chen Yuqun, Ng Siuhung, Kam Shiuman, Lok Tsangming, Lee Sukyee, and Leung Chikwan.

Aside from his dynamic role as an educator, mentor, and researcher, he was actively involved in the Hong Kong Mathematical Society in which he served as its President for four consecutive terms (in 1985/86, 86/87, 87/88 and 88/89). He is also a former President of the Southeast Asian Mathematical Society (SEAMS) in 1988/89 and 2002/2003 for two terms. As SEAMS President, he organized and sponsored mathematics symposia, meetings of mathematicians, and other academic activities among SEAMS members.

Since 1991, he has been the Chief Editor of the Southeast Asian (SEA) Bulletin of Mathematics, a publication made possible through his determined effort to find sponsors to sustain the expenses incurred in its editing, printing, and distribution. This has been considered as one of his greatest contributions to SEAMS.

He is a founding Editor of Algebra Colloquium which first appeared in 1993. Moreover, he is also a founding Editor of Asian European Journal of

Mathematics, established in 2007/2008. These two journals together with SEA Bulletin aim to reflect the latest developments in mathematics and promote international academic exchanges. They publish original research articles of high level in mathematics.

Shum was the first to conceptualize and organize several important international conferences in Hong Kong, Southeast Asia, and Asia. One of them is the first Asian Mathematical Conference (AMC) held in Hong Kong in 1990. One main objective of AMC is to gather mathematicians in Asia to exchange research results and to foster friendship among them. The 5th AMC was held in Kuala Lumpur, Malaysia in June 2009, while the 6th AMC will be held in Pusan, South Korea in June 2013.

The first International Congress in Algebras and Combinatorics (ICAC) held in CUHK in 1997 and the first International Conference on Algebras and Number Theory held in CUHK in 1989 that gathered mathematicians in algebras, combinatorics and number theory were organized by Shum. These conferences and others such as the International Congress of Mathematicians (ICM) Satellite Conference in Algebras held in CUHK in 2002 organized by him had great impact on the development of algebras in the SEA region.

In addition to his various academic activities, Shum's extension services include the Chairmanship of the Federation of Hong Kong Higher Education Staff Associations held since 1996, in which he has fought very hard to safeguard and improve the welfare and interests of the staff of higher education institutions in Hong Kong; the Chairmanship of the Hong Kong International Mathematical Olympiad Committee held since 1985; and Chairmanship of the Popularization Mathematics Committee in Hong Kong, a position he held since 2007. It is noteworthy to mention that he also chaired the Organizing Committee of the 35th International Mathematical Olympiad held in Hong Kong in 1994.

We would like to congratulate Professor Shum Kar-Ping on his 70th birthday and take this opportunity to thank him for his invaluable contribution to the Southeast Asia mathematical community. Indeed, we deeply appreciate his selflessness to share his expertise and the time he spent towards the development of mathematics in Southeast Asia. Moreover, we are always inspired by his vigorous energy and determination, stimulated by his brilliant ideas, fascinated by his extraordinary experiences, and delighted by his enormous sense of humor. The International Conference on Algebra is an occasion for us to celebrate his achievements and generosity with the hope that his legacy and example will encourage and influence the succeeding generations of researchers, teachers and students in this part of the world. We wish that he will celebrate more birthdays.

Interval-Valued Bifuzzy Graphs

Muhammad Akram[1] and Karamat H. Dar[2]

1. Punjab University College of Information Technology,
University of the Punjab, Old Campus, Lahore-54000, Pakistan.
E-mail: m.akram@pucit.edu.pk, makrammath@yahoo.com
2. Department of Mathematics, G. C. University Lahore,
P.O. Box 54000, Pakistan
E-mail: prof_khdar@yahoo.com

Abstract

In this article, we introduce the notion of interval-valued bifuzzy graphs and describe various methods of their construction. We also present the concept of interval-valued bifuzzy regular graphs.

Keywords: Graph, Interval-valued bifuzzy graphs, Interval-valued bifuzzy regular graphs

Mathematics Subject Classification 2000: 05C99

1 Introduction

In 1736, Euler first introduced the notion of graph theory. In the history of mathematics, the solution given by Euler of the well known Königsberg bridge problem is considered to be the first theorem of graph theory. This has now become a subject generally regarded as a branch of combinatorics. The theory of graph is an extremely useful tool for solving combinatorial problems in different areas such as geometry, algebra, number theory, topology, operations research, optimization and computer science.

In 1975, Rosenfeld [8] introduced the concept of fuzzy graphs. The fuzzy relations between fuzzy sets were also considered by Rosenfeld and he developed the structure of fuzzy graphs, obtaining analogs of several graph theoretical concepts. Mordeson and Peng [4] introduced some operations on fuzzy graphs. Shannon and Atanassov [7] introduced the concept of intuitionistic fuzzy relations and intuitionistic fuzzy graphs, and investigated some of their properties. Recently, Akram and Dudek [1] have studied some properties of interval-valued fuzzy graphs. n this article, we introduce the notion of interval-valued bifuzzy graphs and describe various methods of their construction. We also present the concept of interval-valued bifuzzy regular graphs.

2 Preliminaries

A *fuzzy graph* $G = (\mu, \nu)$ is a non-empty set V together with a pair of functions $\mu : V \to [0,1]$ and $\nu : V \times V \to [0,1]$ such that $\nu(\{x,y\}) \leq \min(\mu(x), \mu(y))$ for all x, $y \in V$.
Fuzzy graph is a graph consists of pairs of vertex and edge that have degree of membership containing closed interval of real number [0,1] on each edge and vertex.
In 1975, Zadeh [9] introduced the notion of interval-valued fuzzy sets as an extension of fuzzy sets [10] in which the values of the membership degrees are intervals of numbers instead of the numbers.
An *interval number* D is an interval $[a^-, a^+]$ with $0 \leq a^- \leq a^+ \leq 1$. The interval $[a,a]$ is identified with the number $a \in [0,1]$. $D[0,1]$ denotes the set of all interval numbers.

For interval numbers $D_1 = [a_1^-, b_1^+]$ and $D_2 = [a_2^-, b_2^+]$, we define

- $\text{rmin}(D_1, D_2) = \text{rmin}([a_1^-, b_1^+], [a_2^-, b_2^+]) = [\min\{a_1^-, a_2^-\}, \min\{b_1^+, b_2^+\}]$,
- $\text{rmax}(D_1, D_2) = \text{rmax}([a_1^-, b_1^+], [a_2^-, b_2^+]) = [\max\{a_1^-, a_2^-\}, \max\{b_1^+, b_2^+\}]$,
- $D_1 + D_2 = [a_1^- + a_2^- - a_1^- \cdot a_2^-, b_1^+ + b_2^+ - b_1^+ \cdot b_2^+]$,

The *interval-valued fuzzy set* A in V is defined by

$$A = \{(x, [\mu_A^-(x), \mu_A^+(x)]) : x \in V\},$$

where $\mu_A^-(x)$ and $\mu_A^+(x)$ are fuzzy subsets of V such that $\mu_A^-(x) \leq \mu_A^+(x)$ for all $x \in V$.

Definition 2.1. By an *interval-valued fuzzy graph* of a graph $G^* = (V, E)$ we mean a pair $G = (A, B)$, where $A = [\mu_A^-, \mu_A^+]$ is an interval-valued fuzzy set on V and $B = [\mu_B^-, \mu_B^+]$ is an interval-valued fuzzy relation on E.

The notion of interval-valued intuitionistic fuzzy sets was introduced by Atanassov and Gargov in 1989 as a generalization of both interval-valued fuzzy sets and intuitionistic fuzzy sets. Gerstenkorn and Mańko [6] re-named the intuitionistic fuzzy sets as bifuzzy sets in 1995.

Definition 2.2. For a nonempty set G, we call a mapping $A = (\mu_A, \nu_A) : G \to D[0,1] \times D[0,1]$ an *interval-valued bifuzzy set* in G if $\mu_A^+(x) + \nu_A^+(x) \leq 1$ and $\mu_A^-(x) + \nu_A^-(x) \leq 1$ for all $x \in G$, where the mappings $\mu_A(x) = [\mu_A^-(x), \mu_A^+(x)] : G \to D[0,1]$ and $\nu_A(x) = [\nu_A^-(x), \nu_A^+(x)] : G \to D[0,1]$ are the *degree of membership* functions and the *degree of non-membership functions*, respectively. We use $\widetilde{0}$ to denote the *interval-valued fuzzy empty set* and $\widetilde{1}$ to denote the *interval-valued fuzzy whole set* in a set G, and we define $\widetilde{0}(x) = [0,0]$ and $\widetilde{1}(x) = [1,1]$, for all $x \in G$.

Definition 2.3. If $G^* = (V, E)$ is a graph, then by an *interval-valued bifuzzy relation* B on a set E we mean an interval-valued bifuzzy set such that

$$\mu_B(xy) \leq \mathrm{rmin}(\mu_A(x), \mu_A(y)),$$

$$\nu_B(xy) \geq \mathrm{rmax}(\nu_A(x), \nu_A(y))$$

for all $xy \in E$.

3 Interval-valued Bifuzzy Graphs

Definition 3.1. An *interval-valued bifuzzy graph* with underlying set V is defined to be a pair $G = (A, B)$ where

(i) the functions $\mu_A : V \to D[0,1]$ and $\nu_A : V \to D[0,1]$ denote the degree of membership and nonmembership of the element $x \in V$, respectively such that $\widetilde{0} \leq \mu_A(x) + \nu_A(x) \leq \widetilde{1}$ for all $x \in V$,

(ii) the functions $\mu_B : E \subseteq V \times V \to D[0,1]$ and $\nu_B : E \subseteq V \times V \to D[0,1]$ are defined by

$$\mu_B(\{x,y\}) \leq \mathrm{rmin}(\mu_A(x), \mu_A(y)) \quad \text{and} \quad \nu_B(\{x,y\}) \geq \mathrm{rmax}(\nu_A(x), \nu_A(y))$$

such that $\widetilde{0} \leq \mu_B(\{x,y\}) + \nu_B(\{x,y\}) \leq \widetilde{1}$ for all $\{x,y\} \in E$.

We call A the *interval-valued bifuzzy vertex set* of V, B the *interval-valued bifuzzy edge set* of G, respectively. Note that B is a symmetric interval-valued bifuzzy relation on A. We use the notation xy for an element of E. Thus, $G = (A, B)$ is an interval-valued bifuzzy graph of $G^* = (V, E)$ if

$$\mu_B(xy) \leq \mathrm{rmin}(\mu_A(x), \mu_A(y)) \quad \text{and} \quad \nu_B(xy) \geq \mathrm{rmax}(\nu_A(x), \nu_A(y)) \quad \text{for all} \quad xy \in E.$$

Example 3.2. Consider a graph $G^* = (V, E)$ such that $V = \{x, y, z\}$, $E = \{xy, yz, zx\}$. Let A be an interval-valued bifuzzy set of V and let B be an interval-valued bifuzzy set of $E \subseteq V \times V$ defined by

$$A = < (\frac{x}{[0.1, 0.4]}, \frac{y}{[0.1, 0.2]}, \frac{z}{[0.3, 0.4]}), (\frac{x}{[0.2, 0.5]}, \frac{y}{[0.3, 0.5]}, \frac{z}{[0.1, 0.5]}) >,$$

$$B = < (\frac{xy}{[0.06, 0.1]}, \frac{yz}{[0.05, 0.09]}, \frac{zx}{[0.06, 0.3]}), (\frac{xy}{[0.1, 0.4]}, \frac{yz}{[0.07, 0.4]}, \frac{zx}{[0.07, 0.4]}) > .$$

By routine computations, it is easy to see that $G = (A, B)$ is an interval-valued bifuzzy graph of G^*.

Definition 3.3. Let $A = (\mu_A, \nu_A)$ and $A' = (\mu'_A, \nu'_A)$ be interval-valued bifuzzy subsets of V_1 and V_2 and let $B = (\mu_B, \nu_B)$ and $B' = (\mu'_B, \nu'_B)$ be interval-valued bifuzzy subsets of E_1 and E_2, respectively. The Cartesian product of two interval-valued bifuzzy graphs G_1 and G_2 of the graphs G_1^* and G_2^* is denoted by $G_1 \times G_2 = (A \times A', B \times B')$ and is defined as follows:

(i) $(\mu_A \times \mu'_A)(x_1, x_2) = \text{rmin}(\mu_A(x_1), \mu'_A(x_2))$
$(\nu_A \times \nu'_A)(x_1, x_2) = \text{rmax}(\nu_A(x_1), \nu'_A(x_2))$ for all $(x_1, x_2) \in V$,

(ii) $(\mu_B \times \mu'_B)((x, x_2)(x, y_2)) = \text{rmin}(\mu_A(x), \mu'_B(x_2y_2))$,
$(\nu_B \times \nu'_B)((x, x_2)(x, y_2)) = \text{rmax}(\nu_A(x), \nu'_B(x_2y_2))$ for all $x \in V_1$, for all $x_2y_2 \in E_2$,

(iii) $(\mu_B \times \mu'_B)((x_1, z)(y_1, z)) = \text{rmin}(\mu_B(x_1y_1), \mu'_A(z))$
$(\nu_B \times \nu'_B)((x_1, z)(y_1, z)) = \text{rmax}(\nu_B(x_1y_1), \nu'_A(z))$ for all $z \in V_2$, for all $x_1y_1 \in E_1$.

Proposition 3.4. *Let G_1 and G_2 be the interval-valued bifuzzy graphs. Then Cartesian product $G_1 \times G_2$ is an interval-valued bifuzzy graph.*

Proof. Let $x \in V_1$, $x_2y_2 \in E_2$. Then

$$\begin{aligned}
(\mu_B \times \mu'_B)((x, x_2)(x, y_2)) &= \text{rmin}(\mu_A(x), \mu'_B(x_2y_2)) \\
&\leq \text{rmin}(\mu_A(x), \text{rmin}(\mu'_A(x_2), \mu'_A(y_2)) \\
&= \text{rmin}(\text{rmin}(\mu_A(x), \mu'_A(x_2)), \text{rmin}(\mu_A(x), \mu'_A(y_2))) \\
&= \text{rmin}((\mu_A \times \mu'_A)(x, x_2), (\mu_A \times \mu'_A)(x, y_2)), \\
(\nu_B \times \nu'_B)((x, x_2)(x, y_2)) &= \text{rmax}(\nu_A(x), \nu'_B(x_2y_2)) \\
&\geq \text{rmax}(\nu_A(x), \text{rmax}(\nu'_A(x_2), \nu'_A(y_2)) \\
&= \text{rmax}(\text{rmax}(\nu_A(x), \nu'_A(x_2)), \text{rmax}(\nu_A(x), \nu'_A(y_2))) \\
&= \text{rmax}((\nu_A \times \nu'_A)(x, x_2), (\nu_A \times \nu'_A)(x, y_2)).
\end{aligned}$$

Let $z \in V_2$, $x_1y_1 \in E_1$. Then

$$\begin{aligned}
(\mu_B \times \mu'_B)((x_1,z)(y_1,z)) &= \text{rmin}(\mu_B(x_1y_1), \mu'_A(z)) \\
&\leq \text{rmin}(\text{rmin}(\mu_A(x_1), \mu_A(y_1)), \mu'_A(z)) \\
&= \text{rmin}(\text{rmin}(\mu_A(x), \mu'_A(z)), \text{rmin}(\mu_A(y_1), \mu'_A(z))) \\
&= \text{rmin}((\mu_A \times \mu'_A)(x_1,z), (\mu_A \times \mu'_A)(y_1,z)), \\
(\nu_B \times \nu'_B)((x_1,z)(y_1,z)) &= \text{rmax}(\nu_B(x_1y_1), \nu'_A(z)) \\
&\geq \text{rmax}(\text{rmax}(\nu_A(x_1), \nu_A(y_1)), \nu'_A(z)) \\
&= \text{rmax}(\text{rmax}(\nu_A(x), \nu'_A(z)), \text{rmax}(\nu_A(y_1), \nu'_A(z))) \\
&= \text{rmax}((\nu_A \times \nu'_A)(x_1,z), (\nu_A \times \nu'_A)(y_1,z)).
\end{aligned}$$

This completes the proof. □

Definition 3.5. Let $A = (\mu_A, \nu_A)$ and $A' = (\mu'_A, \nu'_A)$ be interval-valued bifuzzy subsets of V_1 and V_2 and let $B = (\mu_B, \nu_B)$ and $B' = (\mu'_B, \nu'_B)$ be interval-valued bifuzzy subsets of E_1 and E_2, respectively. The composition of two interval-valued bifuzzy graphs G_1 and G_2 is denoted by $G_1[G_2] = (A \circ A', B \circ B')$ and is defined as follows:

(i) $(\mu_A \circ \mu'_A)(x_1,x_2) = \text{rmin}(\mu_A(x_1), \mu'_A(x_2))$
$(\nu_A \circ \nu'_A)(x_1,x_2) = \text{rmax}(\nu_A(x_1), \nu'_A(x_2))$ for all $(x_1,x_2) \in V$,

(ii) $(\mu_B \circ \mu'_B)((x,x_2)(x,y_2)) = \text{rmin}(\mu_A(x), \mu'_B(x_2y_2))$,
$(\nu_B \circ \nu'_B)((x,x_2)(x,y_2)) = \text{rmax}(\nu_A(x), \nu'_B(x_2y_2))$ for all $x \in V_1$, for all $x_2y_2 \in E_2$,

(iii) $(\mu_B \circ \mu'_B)((x_1,z)(y_1,z)) = \text{rmin}(\mu_B(x_1y_1), \mu'_A(z))$
$(\nu_B \circ \nu'_B)((x_1,z)(y_1,z)) = \text{rmax}(\nu_B(x_1y_1), \nu'_A(z))$ for all $z \in V_2$, for all $x_1y_1 \in E_1$,

(iv) $(\mu_B \circ \mu'_B)((x_1,x_2)(y_1,y_2)) = \text{rmin}(\mu'_A(x_2), \mu'_A(y_2), \mu_B(x_1y_1))$,
$(\nu_B \circ \nu'_B)((x_1,x_2)(y_1,y_2)) = \text{rmax}(\nu'_A(x_2), \nu'_A(y_2), \nu_B(x_1y_1))$ for all $(x_1,x_2)(y_1,y_2) \in E^0 - E$.

Proposition 3.6. *Let G_1 and G_2 be two interval-valued bifuzzy graphs. Then the composition $G_1[G_2]$ is an interval-valued bifuzzy graph.*

Proof. Let $x \in V_1$ and $x_2y_2 \in E_2$. Then

$$\begin{aligned}
(\mu_B \circ \mu'_B)((x,x_2)(x,y_2)) &= \text{rmin}(\mu_A(x), \mu'_B(x_2y_2)) \\
&\leq \text{rmin}(\mu_A(x), \text{rmin}(\mu'_A(x_2), \mu'_A(y_2)) \\
&= \text{rmin}(\text{rmin}(\mu_A(x), \mu'_A(x_2)), \text{rmin}(\mu_A(x), \mu'_A(y_2))) \\
&= \text{rmin}((\mu_A \circ \mu'_A)(x,x_2), (\mu_A \circ \mu'_A)(x,y_2)), \\
(\nu_B \circ \nu'_B)((x,x_2)(x,y_2)) &= \text{rmax}(\nu_A(x), \nu'_B(x_2y_2)) \\
&\geq \text{rmax}(\nu_A(x), \text{rmax}(\nu'_A(x_2), \nu'_A(y_2)) \\
&= \text{rmax}(\text{rmax}(\nu_A(x), \nu'_A(x_2)), \text{rmax}(\nu_A(x), \nu'_A(y_2))) \\
&= \text{rmax}((\nu_A \circ \nu'_A)(x,x_2), (\nu_A \circ \nu'_A)(x,y_2)).
\end{aligned}$$

Let $z \in V_2$ and $x_1y_1 \in E_1$. Then

$$\begin{aligned}(\mu_B \circ \mu'_B)((x_1,z)(y_1,z)) &= \text{rmin}(\mu_B(x_1y_1), \mu'_A(z)) \\ &\leq \text{rmin}(\text{rmin}(\mu_A(x_1), \mu_A(y_1)), \mu'_A(z)) \\ &= \text{rmin}(\text{rmin}(\mu_A(x), \mu'_A(z)), \text{rmin}(\mu_A(y_1), \mu'_A(z))) \\ &= \text{rmin}((\mu_A \circ \mu'_A)(x_1,z), (\mu_A \circ \mu'_A)(y_1,z)), \\ (\nu_B \times \nu'_B)((x_1,z)(y_1,z)) &= \text{rmax}(\nu_B(x_1y_1), \nu'_A(z)) \\ &\geq \text{rmax}(\text{rmax}(\nu_A(x_1), \nu_A(y_1)), \nu'_A(z)) \\ &= \text{rmax}(\text{rmax}(\nu_A(x), \nu'_A(z)), \text{rmax}(\nu_A(y_1), \nu'_A(z))) \\ &= \text{rmax}((\nu_A \circ \nu'_A)(x_1,z), (\nu_A \circ \nu'_A)(y_1,z)).\end{aligned}$$

Let $(x_1,x_2)(y_1,y_2) \in E^0 - E$, so $x_1y_1 \in E_1$, $x_2 \neq y_2$. Then

$$\begin{aligned}(\mu_B \circ \mu'_B)((x_1,x_2)(y_1,y_2)) &= \text{rmin}(\mu'_A(x_2), \mu'_A(y_2), \mu_B(x_1y_1)) \\ &\leq \text{rmin}(\mu'_A(x_2), \mu'_A(y_2), \text{rmin}(\mu_A(x_1), \mu_A(y_1))) \\ &= \text{rmin}(\text{rmin}(\mu_A(x_1), \mu'_A(x_2)), \text{rmin}(\mu_A(y_1), \mu'_A(y_2))) \\ &= \text{rmin}((\mu_A \circ \mu'_A)(x_1,x_2), (\mu_A \circ \mu'_A)(y_1,y_2)), \\ (\nu_B \circ \nu'_B)((x_1,x_2)(y_1,y_2)) &= \text{rmax}(\nu'_A(x_2), \nu'_A(y_2), \nu_B(x_1y_1)) \\ &\geq \text{rmax}(\nu'_A(x_2), \nu'_A(y_2), \text{rmax}(\nu_A(x_1), \nu_A(y_1))) \\ &= \text{rmax}(\text{rmax}(\nu_A(x_1), \nu'_A(x_2)), \text{rmax}(\nu_A(y_1), \nu'_A(y_2))) \\ &= \text{rmax}((\nu_A \circ \nu'_A)(x_1,x_2), (\nu_A \circ \nu'_A)(y_1,y_2)).\end{aligned}$$

This completes the proof. □

Definition 3.7. Let $A = (\mu_A, \nu_A)$ and $A' = (\mu'_A, \nu'_A)$ be interval-valued bifuzzy subsets of V_1 and V_2 and let $B = (\mu_B, \nu_B)$ and $B' = (\mu'_B, \nu'_B)$ be interval-valued bifuzzy subsets of E_1 and E_2, respectively. The union of two interval-valued bifuzzy graphs G_1 and G_2 is denoted by $G_1 \cup G_2 = (A \cup A', B \cup B')$ and is defined as follows:

(A) $(\mu_A \cup \mu'_A)(x) = \mu_A(x)$ if $x \in V_1 \cap \overline{V_2}$,
$(\mu_A \cup \mu'_A)(x) = \mu'_A(x)$ if $x \in V_2 \cap \overline{V_1}$,
$(\mu_A \cup \mu'_A)(x) = \text{rmax}(\mu_A(x), \mu'_A(x))$ if $x \in V_1 \cap V_2$.

(B) $(\nu_A \cap \nu'_A)(x) = \nu_A(x)$ if $x \in V_1 \cap \overline{V_2}$,
$(\nu_A \cap \nu'_A)(x) = \nu'_A(x)$ if $x \in V_2 \cap \overline{V_1}$,
$(\nu_A \cap \nu'_A)(x) = \text{rmin}(\nu_A(x), \nu'_A(x))$ if $x \in V_1 \cap V_2$.

(C) $(\mu_B \cup \mu'_B)(xy) = \mu_B(xy)$ if $xy \in E_1 \cap \overline{E_2}$,
$(\mu_B \cup \mu'_B)(xy) = \mu'_B(xy)$ if $xy \in E_2 \cap \overline{E_1}$,
$(\mu_B \cup \mu'_B)(xy) = \text{rmax}(\mu_B(xy), \mu'_B(xy))$ if $xy \in E_1 \cap E_2$.

(D) $(\nu_B \cap \nu'_B)(xy) = \nu_B(xy)$ if $xy \in E_1 \cap \overline{E_2}$,
$(\nu_B \cap \nu'_B)(xy) = \nu'_B(xy)$ if $xy \in E_2 \cap \overline{E_1}$,
$(\nu_A \cap \nu'_B)(xy) = \mathrm{rmin}(\nu_B(xy), \nu'_B(xy))$ if $xy \in E_1 \cap E_2$.

Proposition 3.8. *Let G_1 and G_2 be two interval-valued bifuzzy graphs. Then the union $G_1 \cup G_2$ is an interval-valued bifuzzy graph.*

Proof. Let $xy \in E_1 \cap E_2$. Then

$$\begin{aligned}
\mu_B \cup \mu'_B)(xy) &= \mathrm{rmax}(\mu_B(xy), \mu'_B(xy)) \\
&\leq \mathrm{rmax}(\mathrm{rmin}(\mu_A(x), \mu_A(y)), \mathrm{rmin}(\mu'_A(x), \mu'_A(y))) \\
&= \mathrm{rmin}(\mathrm{rmax}(\mu_A(x), \mu'_A(x)), \mathrm{rmax}(\mu_A(y), \mu'_A(y))) \\
&= \mathrm{rmin}((\mu_A \cup \mu'_A)(x), (\mu_A \cup \mu'_A)(y)), \\
(\nu_B \cap \nu'_B)(xy) &= \mathrm{rmin}(\nu_B(xy), \nu'_B(xy)) \\
&\geq \mathrm{rmin}(\mathrm{rmax}(\nu_A(x)\nu_A(y)), \mathrm{rmax}(\nu'_A(x), \nu'_A(y))) \\
&= \mathrm{rmax}(\mathrm{rmin}(\nu_A(x)\nu'_A(x)), \mathrm{rmin}(\nu_A(y), \nu'_A(y))), \\
&= \mathrm{rmax}((\nu_A(x) \cap \nu'_A)(x), (\nu_A(y) \cap \nu'_A)(y)).
\end{aligned}$$

Similarly, we can show that if $xy \in E_1 \cap \overline{E_2}$, then

$$(\mu_A \cup \mu'_A)(xy) \leq \mathrm{rmin}((\mu_A \cup \mu'_A)(x), (\mu_A \cup \mu'_A)(y)),$$

$$(\nu_A \cap \nu'_A)(xy) \geq \mathrm{rmax}((\nu_A \cap \nu'_A)(x), (\nu_A \cap \nu'_A)(y)).$$

If $xy \in E_2 \cap \overline{E_1}$, then

$$(\mu_B \cup \mu'_B)(xy) \leq \mathrm{rmin}((\mu_A \cup \mu'_A)(x), (\mu_A \cup \mu'_A)(y)),$$

$$(\nu_B \cap \nu'_B)(xy) \geq \mathrm{rmax}((\nu_A \cap \nu'_A)(x), (\nu_A \cap \nu'_A)(y)).$$

This completes the proof. □

Definition 3.9. Let $A = (\mu_A, \nu_A)$ and $A' = (\mu'_A, \nu'_A)$ be interval-valued bifuzzy subsets of V_1 and V_2 and let $B = (\mu_B, \nu_B)$ and $B' = (\mu'_B, \nu'_B)$ be interval-valued bifuzzy subsets of E_1 and E_2, respectively. The join of two interval-valued bifuzzy graphs G_1 and G_2 is denoted by $G_1 + G_2 = (A + A', B + B')$ and is defined as follows:

(i) $(\mu_A + \mu'_A)(x) = (\mu_A + \mu'_A)(x)$ if $x \in V_1 \cup V_2$,
$(\nu_A + \nu'_A)(x) = (\nu_A + \nu'_A)(x)$ if $x \in V_1 \cup V_2$,

(ii) $(\mu_B + \mu'_B)(xy) = (\mu_B \cup \mu'_B)(xy) = \mu_B(xy)$
$(\nu_B + \nu'_B)(xy) = (\nu_B \cap \nu'_B)(xy) = \nu_B(xy)$ if $xy \in E_1 \cup E_2$,

(iii) $(\mu_B + \mu'_B)(xy) = \text{rmin}(\mu_A(x), \mu'_A(y))$
$(\nu_B + \nu'_B)(xy) = \text{rmax}(\nu_A(x), \nu'_A(y))$ if $xy \in E'$.

Proposition 3.10. *Let G_1 and G_2 be two interval-valued bifuzzy graphs. Then the join $G_1 + G_2$ is an interval-valued bifuzzy graph.*

Proof. Let $xy \in E'$. Then

$$\begin{aligned}
(\mu_B + \mu'_B)(xy) &= \text{rmin}(\mu_A(x), \mu'_A(y)) \\
&\leq \text{rmin}((\mu_A \cup \mu'_A)(x), (\mu_A \cup \mu'_A)(y)) \\
&= \text{rmin}((\mu_A + \mu'_A)(x), (\mu_A + \mu'_A)(y)), \\
(\nu_B + \nu'_B)(xy) &= \text{rmax}(\nu_A(x), \nu'_A(y)) \\
&\geq \text{rmax}((\nu_A \cap \nu'_A)(x), (\nu_A \cap \nu'_A)(y)) \\
&= \text{rmax}((\nu_A + \nu'_A)(x), (\nu_A + \nu'_A)(y)).
\end{aligned}$$

Let $xy \in E_1 \cup E_2$. Then the result follows from Proposition 3.8. This completes the proof. □

We formulate the following characterizations.

Proposition 3.11. *Let G_1 and G_2 be the interval-valued bifuzzy graphs and let $V_1 \cap V_2 = \emptyset$. Then union $G_1 \cup G_2$ is an interval-valued bifuzzy graph if and only if G_1 and G_2 are interval-valued bifuzzy graphs.*

Proposition 3.12. *Let G_1 and G_2 be two interval-valued bifuzzy graphs and let $V_1 \cap V_2 = \emptyset$. Then the join $G_1 + G_2$ is an interval-valued bifuzzy graph if and only if G_1 and G_2 are interval-valued bifuzzy graphs.*

Definition 3.13. Let $G = (A, B)$ be an interval-valued bifuzzy graph. If each vertex has same degree n, then G is called an *interval-valued bifuzzy n-regular graph.*

Definition 3.14. Let $G = (A, B)$ be an interval-valued bifuzzy graph. The total degree of a vertex x is defined by $T\deg(x) = \sum_{xy \in E} \mu_B(xy) + \mu_A(x) = \deg(x) + \mu_A(x)$, $T\deg(x) = \sum_{xy \in E} \nu_B(xy) + \nu_A(x) = \deg(x) + \nu_A(x)$. If each vertex of G has total degree n, then G is called an *interval-valued bifuzzy n- totally regular graph.*

Proposition 3.15. *If an interval-valued bifuzzy graph G is both regular and totally regular, then A is constant.*

Proof. Let G be a m-regular and interval-valued bifuzzy n-totally regular graphs. Then $\deg(x) = m$ and $T\deg(x) = n$ for all $x \in V$. Now

$$\begin{aligned} T\deg(x) &= n \\ \deg(x) + \mu_A(x) &= n \\ m + \mu_A(x) &= n \\ \mu_A(x) &= n - m \\ \mu_A(x) &= constant. \end{aligned}$$

Likewise, $\nu_A(x)$= constant. Hence A is a constant function. □

We state a characterization without proof.

Proposition 3.16. *Let G be an interval-valued bifuzzy graph. Then A is a constant function if and only if the following conditions are equivalent:*

(1) *G is an interval-valued bifuzzy regular graph*

(2) *G is an interval-valued bifuzzy totally regular graph.*

References

[1] M. Akram and W.A. Dudek, *Interval-valued fuzzy graphs*, Computers and Mathematics with Applications, **61** (2011) 289-299.

[2] M. Akram and K.H. Dar, *Generalized fuzzy K-algebras*, VDM Verlag, 2010, pp.288, ISBN 978-3-639-27095-2.

[3] K.T. Atanassov, *Intuitionistic fuzzy fets: Theory and applications, Studies in fuzziness and soft computing*, Heidelberg, New York, Physica-Verl., 1999.

[4] J.N. Mordeson and C.S. Peng, *Operations on fuzzy graphs*, Information Sciences **79** (1994) 159-170.

[5] J.N. Mordeson and P.S. Nair, *Fuzzy graphs and fuzzy hypergraphs*, Physica Verlag, Heidelberg 1998; Second Edition 2001.

[6] T. Gerstenkorn and J. Mańko, *Bifuzzy probabilistic sets*, Fuzzy Sets and Systems **71** (1995), 207-214.

[7] A. Shannon, K.T. Atanassov, *A first step to a theory of the intuitionistic fuzzy graphs*, Proceeding of FUBEST (D. Lakov, Ed.), Sofia, (1994) 59-61.

[8] A. Rosenfeld, *Fuzzy graphs*, Fuzzy Sets and their Applications(L.A.Zadeh, K.S.Fu, M.Shimura, Eds.), Academic Press, New York, (1975) 77-95.

[9] L.A. Zadeh, *Similarity relations and fuzzy orderings*, Information Sciences **3**(2)(1971)177-200.

[10] L.A. Zadeh, *The concept of a lingusistic variable and its application to approximate reasoning*, Information Sci. **8** (1975)199-249.

Injective Envelope

Yunita Septriana Anwar
Mathematics Graduate Student, Gadjah Mada University,
Yogyakarta, Indonesia
E-mail: na2_math@yahoo.com

Indah Emilia Wijayanti
Department of Mathematics, Gadjah Mada University,
Yogyakarta, Indonesia
E-mail: ind_wijayanti@yahoo.com

Based on the fact that every module M is contained in an injective module, it can be found a minimal injective module which contains M as a submodule which is called injective envelope. In this paper, we will investigate the characterizations of injective envelopes that can be used to determine a cogenerator in $\sigma[M]$.

Keywords: $\sigma[M]$, Cogenerator, Injective Module, Injective Envelope

Introduction

Injective module first appeared in the context of Abelian Groups. L.Zippin observed in 1935 that an Abelian group is divisible if and only if it is a direct summand of any larger group containing it as a subgroup. The general notion of an injective module over an arbitrary ring was first investigated by R.Baer in the paper *Abelian groups that are direct summands of every containing Abelian group.* R.Baer worked with injective modules over a ring R, namely modules N such that every homomorphism from a one-sided ideal of R to N extends to a homomorphism from R to N. R.Baer proved that every module is a submodule of a injective module, and that a module is injective if only if it is a direct summand of every module that contains it. Injective envelope of a module is an injective minimal module contains that module as a submodule.

In Theorem 1.1 we show that every module in category $R-MOD$ and subcategory $\sigma[M]$ have an injective envelope. In a general way injective envelope in $R - MOD$ is different from injective envelope in $\sigma[M]$.

Module Q is said to be cogenerator in $\sigma[M]$ if Q cogenerates every module in $\sigma[M]$. In particular, if Q is injective module in $\sigma[M]$, we get in Proposition 1.5 that Q is cogenerator in $\sigma[M]$ if and only if Q cogenerates every simple module in $\sigma[M]$. More generally, Q is cogenerator in $\sigma[M]$ if and only if Q contains all simple module in $\sigma[M]$ with their M-injective envelopes as submodules. In this paper we will investigate properties and characterization of injective envelope in $\sigma[M]$, and also characterization of cogenerator in $\sigma[M]$.

1. Preliminaries

We assume throughout the paper that R is an associative ring with identity, and M is a fixed left R-module.

1.1. *The Category of $\sigma[M]$*

The left R-module N is said to be M-*generated* if there is an R-epimorphism from a direct sum of copies of M onto N. The category $\sigma[M]$ is defined to be the full subcategory of $R - MOD$ that contains all submodules N such that N is isomorphic to a submodule of an M-generated module. The subcategory $\sigma[M]$ coincides with $R-MOD$ if and only if R belongs to $\sigma[M]$. It is shown in [1] that $\sigma[M]$ is closed under taking homomorphic images, submodules, and direct sums. For example, since $\bigoplus_{\mathbb{N}} \mathbb{Z}_n$ is a generator in $\mathbb{Z} - MOD$, we have $\sigma[\bigoplus_{\mathbb{N}} \mathbb{Z}_n]$ is subcategory of the torsion modules in $\mathbb{Z} - MOD$.

1.2. *Cogenerator*

Let $\mathcal{U}$ be a class of modules. A module N is *cogenerated* by $\mathcal{U}$ or $\mathcal{U}$ cogenerates N if there is an indexed set $\{U_\lambda\}_{\lambda \in \wedge}$ in $\mathcal{U}$ and a monomorphism

$$N \to \prod_{\wedge} U_\lambda.$$

If $\mathcal{U} = \{U\}$ is a singleton, then we simply say that U cogenerates N and there is a monomorphism $N \to U^\wedge$ for some set $\wedge$. $\mathcal{U}$ is said to be a class of cogenerator for subcategory $\sigma[M]$ if every module in $\sigma[M]$ is cogenerated by $\mathcal{U}$. For example, module $\mathbb{Q}/\mathbb{Z}$ over $\mathbb{Z}$ is cogenerator for module $\mathbb{Z}_n$ over $\mathbb{Z}$, since there is a monomorphism $f\colon \mathbb{Z}_n \to \mathbb{Q}/\mathbb{Z}$ with $f(z + \mathbb{Z}n) = \frac{z}{n} + \mathbb{Z}$.

1.3. *Cocyclic Module*

Let N be any left R-module. Module N is called *cocyclic* if there is an $n_0 \in N$ with the property: every morphism $g\colon N \to M$ with $n_0 \notin Ker\ g$ is monomorphism. For example, every simple modules are cocyclic. Suppose that left R-module N be a non-zero simple module. Let $0 \neq n \in N$ with $n \notin Ker\ g$ and $g\colon N \to M$ be an arbitrary morphism. Since $Ker\ g \subseteq N$ and N is simple module, we obtain $Ker\ g = 0$ or $Ker\ g = N$. But $n \notin Ker\ g$ and $0 \neq n \in N$, we get $Ker\ g = 0$. That means g is a monomorphism. Thus, N is a cocyclic module.

The following proposition yields some characterizations of cocyclic modules.

Proposition 1.1. *For a non-zero left R-module N the following assertions are equivalent:*

(1) N is cocyclic;
(2) N has a simple submodule K which is contained in every non-zero submodule of N;
(3) the intersection of all non-zero submodules of N is non-zero;
(4) N is subdirectly irreducible;
(5) for any monomorphism $\varphi\colon N \to \prod_{\Lambda} U_\lambda$ in $R-MOD$, there is a $\lambda_0 \in \Lambda$ for which $\pi_{\lambda_0}\varphi\colon N \to U_{\lambda_0}$ is monomorphism;
(6) N is an essential extension of a simple module.

The next result shows connection between cogenerator and cocyclic module.

Proposition 1.2. *Every module is cogenerated by its cocyclic factor modules.*

Proof. Let N be a non-zero left R-module and $0 \neq n \in N$. Consider the set $\mathcal{F}$ of all submodules U of N with the property that $n \neq U$, i.e.,

$$\mathcal{F} = \{U \subseteq N | n \notin U\}.$$

It is a partially ordered set with respect to the relation of subset inclusion. By Zorn's Lemma, there exist a maximal element in $\mathcal{F}$. Let $U_n \subset N$ be maximal in $\mathcal{F}$ and L/U_n submodule of N/U_n with $U_n \subset L \subset N$. Since U_n is a maximal element in $\mathcal{F}$ with $n \notin U_n$, we have $n \in L$. Therefore $(Rn+U_n)/U_n$ is contained in every non-zero submodule of N/U_n, i.e. N/U_n is cocyclic. The canonical morphisms $\varphi_n\colon N \to N/U_n$ with $0 \neq n \in N$, yield a morphism $\varphi\colon N \to \prod_{N\setminus 0} N/U_n$ with $Ker\,\varphi = \bigcap_{N\setminus 0} Ker\,\varphi_n = \bigcap_{N\setminus 0} U_n = 0$. Thus, N is cogenerated by N/U_n. $\square$

1.4. *M-Injective Modules*

A left R-module U is called $M-injective$ if for any monomorphism $\varphi\colon K \to M$ and for any homomorphism $\psi\colon K \to U$ there exist a homomorphism $h\colon M \to U$ such that $\psi = h\varphi$. This means that any diagram of the form

$$\begin{array}{ccc} 0 \to K & \xrightarrow{\varphi} & M \\ \downarrow & & \\ U & & \end{array}$$

with the top row exact can be extended commutatively by a morphism $h\colon M \to U$. This property is equivalent to the condition that the map

$$Hom_R(\varphi, U)\colon Hom_R(M, U) \to Hom_R(K, U)$$

is surjective for every monomorphism $\varphi\colon K \to M$. Since the functor $Hom_R(-, U)$ is left exact, we have U is M-injective if and only if $Hom_R(-, U)$ is exact with respect to all exact sequences $0 \to K \to M \to N \to 0$. Module U is called *quasi injective* if it is U-injective.

The following proposition shows characterization of M-injective module (see subsection 16.3 [1], 16 [3], 10.5 [2]).

Proposition 1.3. *For a left R-modules U and M the following are equivalent:*

(1) U is M-injective;
(2) U is N-injective for every (finitely generated, cyclic) submodule N of M;
(3) U is N-injective for any $N \in \sigma[M]$;
(4) the functor $Hom(-, U)\colon \sigma[M] \to AB$ is exact.
If U belongs to $\sigma[M]$, then (1) until (4) are also equivalent to:
(5) every exact sequence $0 \to U \to L \to N \to 0$ in $\sigma[M]$ splits;
(6) every exact sequence $0 \to U \to L \to N \to 0$ in $\sigma[M]$, in which N is a factor module of M splits.

For $R-MOD$ we get from Proposition 1.3 the following proposition.

Proposition 1.4 (Baer's Criterion). *For a left R-module U the following properties are equivalent:*

(1) U is injective module in $R-MOD$;

(2) U is R-injective;

(3) for every left ideal $I \subset R$ and every morphism $h: I \to U$, there exists $u \in U$ with $h(a) = au$ for all $a \in I$.

Baer's criterion shows that R-injective modules are actually injective in the category $R - MOD$. Concept cogenerator in $\sigma[M]$ have special characterization if cogenerator be an injective module in $\sigma[M]$ which is given by the following proposition.

Proposition 1.5. *An injective module Q in $\sigma[M]$ is a cogenerator in $\sigma[M]$ if and only if it cogenerates every simple module in $\sigma[M]$.*

For example, module $\mathbb{Q}/\mathbb{Z}$ and $\mathbb{R}/\mathbb{Z}$ over $\mathbb{Z}$ are injective cogenerators in $\mathbb{Z} - MOD$. Injective cogenerator in $R - MOD$ and $\sigma[M]$ is given by the following proposition.

Proposition 1.6. *Let M be a left R-module.*

(1) ${}_R Hom_{\mathbb{Z}}(R, \mathbb{Q}/\mathbb{Z})$ is an injective cogenerator in $R - MOD$.

(2) If Q is an injective module in $R - MOD$, then $Tr(M, Q)$ is an injective module in $\sigma[M]$. If Q is an injective cogenerator in $R - MOD$, then $Tr(M, Q)$ is an injective cogenerator in $\sigma[M]$.

Injective module first appeared in the context of Abelian groups, in particular, divisible group. Recall that an additive Abelian groups G is said to be divisible if for any $g \in G$ and any non-zero $n \in \mathbb{Z}$ there exists $g' \in G$ with $ng' = g$. Since every Abelian groups can be considered as a $\mathbb{Z}$-module, therefore we can say that $\mathbb{Z}$-module M is divisible if $nM = M$ for every non-zero $n \in \mathbb{Z}$. For arbitrary ring, a left R-module N is called divisible if for every $s \in R$ which is not zero divisor and every $n \in N$, there exist $m \in N$ with $sm = n$.

The field of rational numbers $\mathbb{Q}$ as a $\mathbb{Z}$-module is a divisible module. It is easy to show that direct product and direct sum of divisible group is a divisible group and that a quotient group of a divisible group is also divisible (see subsection 5 [4]).

Proposition 1.7. *A $\mathbb{Z}$-module N is injective if and only if it is divisible.*

Let R be a ring and let D be an Abelian group. Since any ring R can be considered as a left $\mathbb{Z}$-module, the Abelian group $Hom_{\mathbb{Z}}(R, D)$ can be made into left R-module if we set $(rf)(x) = f(rx)$ for any $r, x \in R$ and $f \in Hom_{\mathbb{Z}}(R, D)$. In particular, if D is divisible group we get $Hom_{\mathbb{Z}}(R, D)$ is an injective module which is given by the following lemma.

Lemma 1.1. *If R is a ring and D is a divisible $\mathbb{Z}$-module, then $Hom_{\mathbb{Z}}(R, D)$ is an injective left R-module.*

The next result shows that any module can be embedded into an injective module.

Theorem 1.1 (Baer's Theorem). *Every module is a submodule of an injective module.*

Proof. Let K be a left R-module. It can also be considered as a left $\mathbb{Z}$-module. Since any module is isomorphic to a quotient module of a free module, we can write $K \simeq F/L$ where $F = \bigoplus_{\wedge} \mathbb{Z}$ is a free $\mathbb{Z}$-module and L is submodule of F. Let $E = \bigoplus_{\wedge} \mathbb{Q}$ is a $\mathbb{Z}$-module, then $F \subset E$ and $F/L \subset E/L$. Since E/L is divisible groups, then by Proposition 1.7 E/L is an injective $\mathbb{Z}$-module. So we have $K \subset E/L$, where E/L is an injective $\mathbb{Z}$-module. Write $E/L = D$, then we have an exact sequence of $\mathbb{Z}$-modules

$$0 \to K \to D \to D/K \to 0$$

and the sequence

$$0 \to Hom_{\mathbb{Z}}(R, K) \to Hom_{\mathbb{Z}}(R, D) \to Hom_{\mathbb{Z}}(R, D/K)$$

is also exact. So we have an inclusion

$$Hom_{\mathbb{Z}}(R, K) \subseteq Hom_{\mathbb{Z}}(R, D)$$

where $Hom_{\mathbb{Z}}(R, D)$ is an injective R-module by Proposition 1.1.

For any $m \in K$ there exist a group homomorphism $f_m \colon R \to K$ given by $f_m(r) = rm$ for any $r \in R$. Let $h \colon R \to K$ be an arbitrary R-homomorphism, then there is an element $m \in K$ such that $h(1) = m$ and $h(r) = rm$ for any $r \in R$. Therefore we can consider a map $\varphi \colon Hom_R(R, K) \to Hom_{\mathbb{Z}}(R, K)$ given by $\varphi(h)(r) = f_m(r)$ for any $h \in Hom_R(R, K)$ and $r \in R$. Obviously, it is an R-homomorphism. We shall show that φ is a monomorphism. Suppose, $\varphi(h)(r) = 0 = rm$ for any $r \in R$. Since $h(1) = m$ for any $r \in R$, we have $h(r) = rm = 0$. Hence $h = 0$, i.e., φ is a monomorphism. Finally, since there exist a natural isomorphism of R-modules K and $Hom_R(R, K)$ given by $m \mapsto f_m$, where $f_m(1) = m$, we have a sequence of inclusions of R-modules

$$K \simeq Hom_R(R, K) \subseteq Hom_{\mathbb{Z}}(R, K) \subseteq Hom_{\mathbb{Z}}(R, D)$$

with an injective R-module $Hom_{\mathbb{Z}}(R, D)$. So we obtain an exact sequence $0 \to K \to Hom_{\mathbb{Z}}(R, D)$. The proposition is proved. □

2. Injective Envelope

2.1. *Essential Extensions*

In the previous section it was shown that any module can be embedded into an injective module. There may be many such injective modules for a given module M. The goal of this section is to show that among them there exist a minimal one. We shall prove that every module M has such a minimal injective module and we show that it is an essential extension of M, which is unique up to isomorphism.

Definition 2.1. A submodule K of a left R-module M is called *essential* or *large in* M if for every non-zero submodule $L \subset M$, we have $K \cap L \neq 0$.

Then M is called an *essential extension of* K and we write $K \trianglelefteq M$. A monomorphism $f\colon L \to M$ is said to be *essential* if *Im* f is an essential submodule of M. For example, any module is always an essential extension of itself. The field of all rational numbers $\mathbb{Q}$ as a $\mathbb{Z}$-module is an essential extension of the integers $\mathbb{Z}$.

The next simple lemma gives a very useful test for essential extensions.

Lemma 2.1. *An R-module M is an essential extension of a left R-module N if and only if for any $0 \neq x \in M$ there exist $r \in R$ such that $0 \neq rx \in N$.*

Proof. Let M be an essential extension of N and $0 \neq x \in M$, then $Rx \cap N \neq 0$, that means there exists $r \in R$ such that $0 \neq rx \in N$. Conversely, let $X \subseteq M$ and $0 \neq x \in M$. By hypothesis there exist $r \in R$ such that $0 \neq rx \in N$. Then $0 \neq rx \in N \cap X$, i.e., N is essential in M. □

An interesting characterization of essential monomorphisms is given by the following proposition.

Proposition 2.1. *A monomorphism $f\colon L \to M$ in $R-MOD$ is essential if only if for every morphism $h\colon M \to N$ in $R-MOD$, hf monomorphism implies that h is monomorphism.*

Proof. Suppose f be essential monomorphism and hf monomorphism. Let $x \in Ker\ h \bigcap Im\ f$. Then $x \in Ker\ h$ and $x \in Im\ f$. Hence, there exist $y \in L$ such that $f(y) = x$. We have $hf(y) = h(f(y)) = h(x) = 0$. Since hf monomorphism, we obtain $y = 0$. Therefore $Ker\ h \bigcap Im\ f = 0$. Since f is essential monomorphism, it follows that $Ker\ h = 0$. Thus, h is a monomorphism.

Conversely, assume that f has the given property and $K \subset M$ is a submodule with $Im\ f \bigcap K = 0$. Let $p\colon M \to M/K$ be a canonical projection and $x \in Ker\ pf$. Then $pf(x) = p(f(x)) = 0$. Therefore, $f(x) \in Im\ f \bigcap Ker\ p$. Since $Im\ f \bigcap Ker\ p = 0$, we get $f(x) = 0$. Since f monomorphism, it follows that pf monomorphism. By assumption, p is monomorphism. Therefore $Ker\ p = K = 0$. □

The next proposition shows properties of essential extensions.

Proposition 2.2. *Let K, L and M be left R-modules.*

(1) Two monomorphism $f\colon K \to L$, $g\colon L \to M$ are essential if and only if gf is essential.
(2) If $K \subset L \subset M$, then $K \trianglelefteq M$ if and only if $K \trianglelefteq L \trianglelefteq M$.
(3) If $h\colon K \to M$ is a morphism and $L \trianglelefteq M$, then $(L)h^{-1} \trianglelefteq K$.
(4) If $K_1 \trianglelefteq L_1 \subset M$ and $K_1 \cap K_2 \trianglelefteq L_1 \cap L_2$.
(5) The intersection of two (finitely many) essential submodules is an essential submodule in M.

Proof.

(1) Assume f, g are essential monomorphism. Let $h\colon M \to M'$ is a monomorphism with hgf monomorphism. Since f and g are essential monomorphism, by Proposition 2.1 we obtain h monomorphism. Therefore, gf is essential monomorphism by Proposition 2.1.
Conversely, suppose that gf be essential monomorphism. Let $h\colon M \to M'$ is a morphism with hg monomorphism. Then hgf is monomorphism. Since gf essential monomorphism, by Proposition 2.1 we obtain h is monomorphism. Hence g is essential monomorphism. We are going to show f is essential monomorphism. For any $k\colon L \to L'$ with kf monomorphism, we form the pushout diagram

$$\begin{array}{ccccc} K & \xrightarrow{f} & L & \xrightarrow{g} & M \\ & & k \downarrow & & \downarrow^{p_2} \\ & & L' & \xrightarrow{p_1} & P \end{array}$$

with $P = Cokernel\ q^*$ where $q^*: L \to L' \oplus M$. By properties of pushout (see subsection 10 [1]), g monomorphism implies that p_1 is a monomorphism. Since kf monomorphism, we have $p_1 kf = p_2 gf$ is monomorphism and since gf is essential monomorphism, it follows that p_2 is monomorphism. By diagram, $p_2 g = p_1 k$ implies that k is monomorphism. Thus, f is essential monomorphism.

(2) It follows from (2) applied to the inclusion $K \to L$, $L \to M$.
(3) Assume $h\colon K \to M$ is a morphism and $L \trianglelefteq M$. Let $U \subset K$. If $h(U) = 0$ then $U \subset Ke\ h \subset h^{-1}(L)$. Hence $Ker\ h \cap h^{-1}(L) \neq 0$ and it follows that $h^{-1}(L) \trianglelefteq K$. In case $h(U) \neq 0$, since $h(U) \subset M$ and $L \trianglelefteq M$, then $h(U) \cap L \neq 0$. Hence, there exist a non-zero $u \in U$ with $h(u) \in L$. Therefore, $u \in U \cap h^{-1}(L)$. Thus, $h^{-1}(L) \trianglelefteq K$.
(4) Suppose $K_1 \trianglelefteq L_1 \subset M$ and $K_2 \trianglelefteq L_2 \subset M$. Let $0 \neq X \subset (L_1 \cap L_2)$. Then $X \subset L_1$ and $X \subset L_2$. Since $K_1 \trianglelefteq L_1$, we get $X \cap K_1 \neq 0$ and from $X \subset L_2$, it follows that $0 \neq (X \cap K_1) \subset L_2$. Since $K_2 \trianglelefteq L_2$, we obtain $(X \cap K_1) \cap K_2 = X \cap (K_1 \cap K_2) \neq 0$. Thus, $K_1 \cap K_2 \trianglelefteq L_1 \cap L_2$.
(5) It follows immediate from (4) by induction. □

By Proposition 2.2(5), intersection of a family of essential submodules of M need not be essential in M. For example, in $\mathbb{Z}$ we have $Zn \trianglelefteq \mathbb{Z}$ for all $n \in \mathbb{N}$, but $\bigcap_{n\in\mathbb{N}} \mathbb{Z}n = 0$ and hence is not essential in $\mathbb{Z}$. The next result shows direct sums of essential submodules.

Proposition 2.3. *Let $\{K_\lambda\}_\wedge$ and $\{L_\lambda\}_\wedge$ be families of submodules of the R-module M. If $\{K_\lambda\}_\wedge$ is an independent family of submodules in M and $K_\lambda \trianglelefteq L_\lambda$ for all $\lambda \in \wedge$, then $\{L_\lambda\}_\wedge$ also is an independent family and $\bigoplus_\wedge K_\lambda \trianglelefteq \bigoplus_\wedge L_\lambda$.*

Proof. In case $\wedge = 1, 2$, let $K_1 \trianglelefteq L_1$ and $K_2 \trianglelefteq L_2$ be a submodules of M with $K_1 \cap K_2 = 0$. By Proposition 2.2(4), $K_1 \cap K_2 \trianglelefteq L_1 \cap L_2$. Since $K_1 \cap K_2 = 0$, we get $L_1 \cap L_2 = 0$. Hence $\{L_1, L_2\}$ is an independent family. We form the projections $p_1\colon L_1 \oplus L_2 \to L_1$ and $p_2\colon L_1 \oplus L_2 \to L_2$ with $K_1 \trianglelefteq L_1$ and $K_2 \trianglelefteq L_2$. By Proposition 2.2(3), $p_1^{-1}(K_1) \trianglelefteq L_1 \oplus L_2$ and $p_2^{-1}(K_2) \trianglelefteq L_1 \oplus L_2$. Since $p_1^{-1}(K_1) = K_1 \oplus L_2$ and $p_2^{-1}(K_2) \trianglelefteq L_1 \oplus K_2$, then $K_1 \oplus L_2 \trianglelefteq L_1 \oplus L_2$ and $L_1 \oplus K_2 \trianglelefteq L_1 \oplus L_2$. Suppose $x \in K_1 \oplus K_2$, then $x = (k_1, k_2)$ where $k_1 \in K_1$ and $k_2 \in K_2$. Since $K_2 \trianglelefteq L_2$, we obtain $x \in K_1 \oplus L_2$, and since $K_1 \trianglelefteq L_1$, it follows that $x \in L_1 \oplus K_2$. Thus $x \in (K_1 \oplus L_2) \cap (L_1 \oplus K_2)$, that means $(K_1 \oplus K_2) \subset (K_1 \oplus L_2) \cap (L_1 \oplus K_2)$.

Conversely, suppose $y \in (K_1 \oplus L_2) \cap (L_1 \oplus K_2)$. Then $y \in (K_1 \oplus L_2)$ and $y \in (L_1 \oplus K_2)$. Hence $y \in K_1 \oplus K_2$. Thus, $(K_1 \oplus L_2) \cap (L_1 \oplus K_2) \subset (K_1 \oplus K_2)$, i.e., $(K_1 \oplus K_2) = (K_1 \oplus L_2) \cap (L_1 \oplus K_2)$. By Proposition 2.2 (4) we obtain

$$K_1 \oplus K_2 = (K_1 \oplus L_2) \cap (L_1 \oplus K_2) \trianglelefteq L_1 \oplus L_2.$$

For an arbitrary index set $\wedge$, a family $\{L_\lambda\}_\wedge$ is independent if every finite subfamily is independent. For any non-zero $m \in \bigoplus_\wedge L_\lambda$, we have $m \in \bigoplus_E L_\lambda$ for some finite subset $E \subset \wedge$. Since $\bigoplus_E K_\lambda \trianglelefteq \bigoplus_E L_\lambda$, we get

$0 \neq (Rm \bigcap \bigoplus_E K_\lambda) \subset (Rm \bigcap \bigoplus_\Lambda K_\lambda)$. Hence the intersection of a non-zero submodule of $\bigoplus_\Lambda L_\lambda$ with $\bigoplus_\Lambda K_\lambda$ is again non-zero, i.e., $\bigoplus_\Lambda K_\lambda \trianglelefteq \bigoplus_\Lambda L_\lambda$. □

The connection between injectivity and essential extensions is given by the following theorem.

Theorem 2.1. *A left R-module N is M-injective if and only if N has no proper essential extensions.*

Proof. Let N be M-injective module and E be an essential extension of N. By Proposition 1.3, N is a direct summand of E, i.e., $E = N \oplus K$, where $N \cap K = 0$. If $K \neq 0$, then E is not an essential extension. Hence $K = 0$ and $E = N$. Thus, N has no proper essential extensions.

Conversely, suppose N has no proper essential extensions. By Baer's Theorem, there exist an injective module Q containing N. Consider the set $\mathcal{W}$ of all submodules S of Q with the property that $S \cap N = 0$. This set is not empty because $0 \in \mathcal{W}$. It is a partially ordered set with respect to the relation of subset inclusion. By Zorn's Lemma, there exist a maximal element in this set. Let $L \subset Q$ be a maximal in the set $\mathcal{W}$. Then $N \cap L = 0$ and $N + L \subseteq Q$. We shall show that $Q = N + L$. Suppose $N + L \neq Q$, then $(N + L)/L \subset Q/L$ and $(N + L)/L \neq Q/L$. Consider a non-zero submodule $0 \neq X/L \subset Q/L$. Then $L \subset X$ and $L \neq X$. Since L is a maximal element in $\mathcal{W}$, we have $N \cap X \neq 0$. Now taking into account that $N \cap L = 0$ we obtain $N \cap X \nsubseteq L$. Therefore $L \subset X \cap (N + L)$. This means that $X/L \cap (N + L)/L \neq 0$ and so Q/L is an essential extension of $(N + L)/L$. Since $(N + L)/L \simeq N/(N \cap L) \simeq N$ implies N is essential in Q/L. By hypothesis, N has no proper essential extension, $Q/L = (N + L)/L$ and this implies $Q = N + L$. Since $N \cap L = 0$, we have $Q = N \oplus L$. Hence by Proposition 1.3, N is M-injective module. □

2.2. *Injective Envelope*

Let N be a left R-module in $\sigma[M]$ and M in $R{-}MOD$. An injective module E in $\sigma[M]$ (or $R{-}MOD$) together with an essential monomorphism $\varepsilon\colon N \to E$ is called an *injective envelope (hull)* of N in $\sigma[M]$ (resp. $R - MOD$).

The injective envelope of N in $\sigma[M]$ is also called M-injective envelope of N and is usually denoted by $\widehat{N}$. With this terminology the injective envelope of N in $R - MOD$ is the R-injective envelope and denoted by $E(N)$. In general $\widehat{N} \neq E(N)$. The existence of injective envelopes is given by the next proposition.

Proposition 2.4. *Let M be an R-module.*

(1) Every module in $\sigma[M]$ has an injective envelope in $\sigma[M]$.
(2) The injective envelopes of a module are unique up to isomorphism.
(3) Every module N in $R-MOD$ has an injective envelope $E(N)$ in $R-MOD$. If $N \in \sigma[M]$, then $\widehat{N} \simeq Tr(M, E(N))$.

Proof.

(1) Let N be a left R-module in $\sigma[M]$. By Baer's Theorem there is an injective module Q containing a given module N. Consider the set $\mathcal{W}$ of all essential extensions of N contained in the module Q, i.e.,

$$\mathcal{W} = \{K \subset Q | N \trianglelefteq K\}.$$

This set is not empty because $N \in \mathcal{W}$ and it is a partially ordered set with respect to subset inclusion. We shall show that any increasing chain of modules contained in the set $\mathcal{W}$ has an upper bound in $\mathcal{W}$. Let

$$N \subseteq E_1 \subseteq E_2 \subseteq \ldots \subseteq E_n \subseteq \ldots \subseteq Q$$

be a chain of modules $E_\lambda \in \mathcal{W}$. Let $E^* = \bigcup_{\lambda \in \wedge} E_\lambda$, then $N \subseteq E^* \subseteq Q$. Let $0 \neq X \subset E^*$, then there is $\lambda \in \wedge$ such that $X \cap E_\lambda \neq 0$. Therefore $(X \cap E_\lambda) \cap N \neq 0$. Hence $X \cap N \neq 0$ and E^* is an essential extension of N, i.e., $E^* \in \mathcal{W}$ and it is an upper bound of all E_λ for $\lambda \in \wedge$. Therefore we can apply Zorn's Lemma to the set $\mathcal{W}$ and conclude that there exist a maximal element E in $\mathcal{W}$. We are going to show that E is an injective module. Let L be an essential extension of E. Since $N \subseteq E \subseteq L$ and $N \trianglelefteq E$, $E \trianglelefteq L$, by Proposition 2.2(1) we get $N \trianglelefteq L$. Therefore $L \in \mathcal{F}$ and owing to maximality of E we obtain that $E = L$. Thus, E has no proper essential extension. By Theorem 2.1, E is an M-injective module, i.e., E is an injective envelope of N in $\sigma[M]$.

(2) Let $\widehat{N}$ and $\widehat{N}_1$ be two injective envelopes of R-module N. Since $\widehat{N}$ and $\widehat{N}_1$ are injective modules, there exist a homomorphism $\tau\colon \widehat{N} \to \widehat{N}_1$ such that the diagram

$$\begin{array}{cccc} 0 \to & N & \to \widehat{N} \\ & \|_{id_N} & \downarrow^{\tau} \\ 0 \to & N & \to \widehat{N}_1 \end{array}$$

with exact top and bottom rows is commutative. Since $Ker\ \tau \subseteq \widehat{N}$ and $N \trianglelefteq \widehat{N}$ we get $Ker\ \tau \cap N = 0$. Since $\widehat{N}$ is an essential extension of N, $Ker\ \tau = 0$, i.e., τ is a monomorphism. So $\widehat{N} \simeq Im\ \tau \subseteq \widehat{N}_1$ and from the injectivity of $\widehat{N}$, we get $\widehat{N}_1 = \widehat{N} \oplus L$, where $L \subseteq \widehat{N}_1$. Since $N \subseteq Im\ \tau$, we have $N \cap L = 0$ and since $\widehat{N}_1$ is an essential extension of N, it follows that $L = 0$. Thus $Im\ \tau \simeq \widehat{N}_1$, that means τ is an epimorphism and therefore τ is an isomorphism extending the identity id_N of N.

(3) For $M = R$ we obtain the injective envelope in $R - MOD$ from (1). For $N \in \sigma[M]$, let $E(N)$ be an injective envelope of N in $R - MOD$. Since $Tr(M, E(N))$ is the largest submodule of $E(N)$ generated by M, then $N \subset Tr(M, E(N)) \subset E(N)$. Hence $N \trianglelefteq Tr(M, E(N))$. Since $E(N)$ is injective module, by Proposition 1.6 we obtain $Tr(M, E(N))$ is M-injective. Therefore $Tr(M, E(N))$ is an injective envelope of N in $\sigma[M]$. It follow from (2) that $\widehat{N} \simeq Tr(M, E(N))$. □

The next proposition shows that $\widehat{L}$, injective envelope of L, is a maximal essential extension of a module L if no module properly containing $\widehat{L}$ can be an essential extensions of L and $\widehat{L}$ is a minimal M-injective module if no module properly contained in $\widehat{L}$ and properly containing L can be M-injective.

Proposition 2.5. *Let M be an R-module and L, N in $\sigma[M]$ with M-injective envelopes $\widehat{L}$, $\widehat{N}$, respectively.*

(1) If $L \trianglelefteq N$, then $\widehat{L} \simeq \widehat{N}$. In particular, N is isomorphic to a submodule of $\widehat{L}$.

(2) If $L \subset N$ and N is M-injective, then $\widehat{L}$ is isomorphic to a direct summand of N.

(3) If for a family $\{N_\lambda\}_\Lambda$ of modules in $\sigma[M]$, the direct sum $\bigoplus_\Lambda N_\lambda$ is M-injective, then $\bigoplus_\Lambda \widehat{N}_\lambda$ is an M-injective envelope of $\bigoplus_\Lambda N_\lambda$.

Proof.

(1) Suppose $L \trianglelefteq N$. Then $L \subset N \subset \widehat{N}$. By Proposition 2.2 we have $L \trianglelefteq \widehat{N}$. Thus, $\widehat{N}$ is an M-injective envelope of L. By Proposition 2.4(2) we get $\widehat{L} \simeq \widehat{N}$.

(2) Suppose we have $L \subset N$ and N is M-injective. Consider a diagram

$$\begin{array}{ccc} 0 \to & L & \xrightarrow{i} \widehat{L} \\ & {\scriptstyle f}\downarrow & \\ & N & \end{array}$$

with the top row exact. Since N is M-injective module there exist a homomorphism $g\colon \widehat{L} \to N$ extending i which makes the diagram commutative. Assume g is not a monomorphism, i.e., $Ker\ g \neq 0$. Then $Ker\ g \cap L = 0$, since $gi = f$ and i, f are monomorphism. But this contradicts the fact that $\widehat{L}$ is an essential extension of L. So, we obtain that g is a monomorphism. Then $g(\widehat{L}) \simeq \widehat{L} \subset N$. Since $\widehat{L}$ is M-injective module, by Proposition 1.3 we obtain that $\widehat{L}$ is isomorphic to a direct summand of N.

(3) Let $\bigoplus_{\wedge} \widehat{N}_{\lambda}$ is M-injective. Since $N_{\lambda} \trianglelefteq \widehat{N}_{\lambda}$, by Proposition 2.3 we have $\bigoplus_{\wedge} N_{\lambda} \trianglelefteq \bigoplus_{\wedge} \widehat{N}_{\lambda}$. Thus, $\bigoplus_{\wedge} \widehat{N}_{\lambda}$ is an M-injective envelope of $\bigoplus_{\wedge} N_{\lambda}$. □

In particular, for $R - MOD$ we get from Proposition 2.5:

Corollary 2.1. *In* $R - MOD$

(1) If $N \trianglelefteq L$, *then* $E(N) \simeq E(L)$.
(2) If $K \subseteq Q$, *with* Q *injective module, then* $Q = E(K) \oplus K'$.
(3) If $\bigoplus_{\wedge} E(M_{\lambda})$ *is injective then* $E(\bigoplus_{\wedge} M_{\lambda}) = \bigoplus_{\wedge} E(M_{\lambda})$.

By Proposition 1.5, an M-injective module Q is a cogenerator in $\sigma[M]$ if and only if Q contains all simple modules in $\sigma[M]$. The next result shows that Q is a cogenerator in $\sigma[M]$ if and only if Q contains all simple modules with their injective envelopes.

Proposition 2.6. *Let* M *be a non-zero* R*-module,* $\{E_{\lambda}\}_{\wedge}$ *a minimal representing set of the simple modules in* $\sigma[M]$ *and* $\widehat{E}_{\lambda}$ *the* M*-injective envelope of* E_{λ}, $\lambda \in \wedge$.

(1) A module $Q \in \sigma[M]$ *is a cogenerator in* $\sigma[M]$ *if and only if it contains, for every* $\lambda \in \wedge$ *a submodule isomorphic to* $\widehat{E}_{\lambda}$.
(2) $\{\widehat{E}_{\lambda}\}_{\wedge}$ *is a set of cogenerators in* $\sigma[M]$.
(3) Every cogenerator in $\sigma[M]$ *contains a submodule isomorphic to* $\bigoplus_{\wedge} \widehat{E}_{\lambda}$ *which is a minimal cogenerator.*

Proof.

(1) Assume $Q \in \sigma[M]$ be a cogenerator and $\widehat{E}_{\lambda}$ be an M-injective envelope of E_{λ}. Since $\widehat{E}_{\lambda} \in \sigma[M]$, there exist monomorphism $\psi\colon \widehat{E}_{\lambda} \to Q^{\wedge}$. First we show $\widehat{E}_{\lambda}$ are cocyclic modules. Since $\widehat{E}_{\lambda}$ are essential extension of E_{λ} and E_{λ} is a simple module, then every non-zero submodules of $\widehat{E}_{\lambda}$ contains E_{λ}. By Proposition 1.1, we get $\widehat{E}_{\lambda}$ are cocyclic module. Therefore $\psi_{\lambda}\colon \widehat{E}_{\lambda} \to Q$ is a monomorphism for every $\lambda \in \wedge$.
Conversely, assume module $Q \in \sigma[M]$ contains all $\widehat{E}_{\lambda}$ as submodules. By Proposition 1.1, every cocyclic module is an essential extension of

some E_λ, hence it is a submodule of $\widehat{E}_\lambda$. But every module is cogenerated by its cocyclic factor modules (see Proposition 1.2) and hence is cogenerated by Q. Thus, Q cogenerator in $\sigma[M]$.

(2) Let $\{E_\lambda\}_\wedge$ be a minimal representing set of simple modules in $\sigma[M]$, i.e., $E_{\lambda_1} \neq E_{\lambda_1}$ for every $\lambda_1, \lambda_2 \in \wedge$ with $\lambda_1 \neq \lambda_2$. Since $E_\lambda \subset \widehat{E}_\lambda$ and E_λ is cocyclic factor modules in $\sigma[M]$, we get $\widehat{E}_\lambda$ is a cogenerator in $\sigma[M]$. Hence $\{\widehat{E}_\lambda\}_\wedge$ is a set of cogenerators in $\sigma[M]$.

(3) Let module Q be a cogenerator in $\sigma[M]$. From (1), $\widehat{E}_\lambda \subset Q$ for every $\lambda \in \wedge$, then $\sum_\wedge \widehat{E}_\lambda \subset Q$. For every $\lambda \in \wedge$, we will show that $\widehat{E}_\lambda \cap \sum_{\lambda \neq \lambda'} = 0$. Suppose $E_\lambda \subset \sum_{\lambda \neq \lambda'} \widehat{E}_{\lambda'}$. Then $E_\lambda \subset \widehat{E}_{\lambda'}$ for some $\lambda' \neq \lambda$. This contradicts the fact that $\{E_\lambda\}_\wedge$ is a minimal representing set of simple modules in $\sigma[M]$.Therefore $\{\widehat{E}_\lambda\}_\wedge$ is an independent family and $\bigoplus_\wedge \widehat{E}_\lambda = \sum_\wedge \widehat{E}_\lambda \subset Q$. □

Some example of injective envelope is given by the following proposition.

Proposition 2.7. *For any prime number $p \in \mathbb{N}$, $\mathbb{Z}_{p^\infty}$ is an injective envelope of $\mathbb{Z}_p$, where $\mathbb{Z}_{p^\infty}$ is p-component of $\mathbb{Q}/\mathbb{Z}$ and $\mathbb{Z}_p = \mathbb{Z}/p\mathbb{Z}$.*

Proof. First we show that every torsion module M over $\mathbb{Z}$ is a direct sum of its p-components, i.e., $M = \bigoplus\{p(M) \mid p \textit{ a prime number}\}$ where $p(M) = \{a \in M | p^k a = 0 \textit{ for some } k \in \mathbb{N}\}$ is p-components of M. Let $a \in M$. Since M is torsion module, there exist $n \in \mathbb{N}$ with $na = 0$. Hence $n \in Ann_{\mathbb{Z}}(a) = n\mathbb{Z}$ with $n = p_1^{k_1} \ldots p_r^{k_r}$ for different prime numbers $p_1 \ldots p_r$. Suppose $n_i = n/p_i^{k_i}$, $i = 1, 2, \ldots, r$. Then n_i have greatest common divisor 1, $i = 1, 2, \ldots, r$. Hence there are $\alpha_1, \ldots, \alpha_r \in \mathbb{Z}$ with $\sum \alpha_i n_i = 1$. Therefore $a = \alpha_1 n_1 a + \ldots + \alpha_r n_r a$. Since for every $i = 1, 2, \ldots, r$, $\alpha_i n_i a p_i^{k_i} = \alpha_i (n/p_i^{k_i}) a p_i^{k_i} = \alpha_i n a = 0$, we get $\alpha_i n_i a \in p_i(M)$. Then $a \in \sum_i^r p_i(M)$ and hence $M = \sum\{p(M) | p \textit{ a prime number}\}$. Suppose that $a \in p_1(M) \cap (p_2(M) + \ldots + p_l(M))$ for different prime numbers $p_1, \ldots, p_l$. Then $a \in p_1(M)$ and $a \in p_2(M) + \ldots + p_l(M)$. Hence $p_1^{k_1} a = 0$ and $a = a_2 + \ldots + a_l$ where $a_i \in p_i(M)$, $i = 2, 3, \ldots, l$. Therefore $p_2^{k_2} \ldots p_l^{k_l} a = 0$ for some $k_i \in \mathbb{N}$. Since $p_1^{k_1}$ and $p_2^{k_2} \ldots p_l^{k_l}$ are relatively prime, there exist $\beta_1, \beta_2 \in \mathbb{Z}$ with $\beta_1 p_1^{k_1} + \beta_2 (p_2^{k_2} \ldots p_l^{k_l}) = 1$, this implies $a = 1a = (\beta_1 p_1^{k_1} + \beta_2 (p_2^{k_2} \ldots p_l^{k_l}))a = \beta_1 p_1^{k_1} a + \beta_2 (p_2^{k_2} \ldots p_l^{k_l}) a = 0$. Thus, $\{p(M) | p \textit{ a prime number}\}$ is an independent family of submodules. So $M = \bigoplus\{p(M) \mid p \textit{ a prime number}\}$, that means every torsion module over $\mathbb{Z}$ is a direct sum of its p-components.

Since module $\mathbb{Q}/\mathbb{Z}$ over $\mathbb{Z}$ is torsion module, then $\mathbb{Q}/\mathbb{Z} = \bigoplus\{\mathbb{Z}_{p^\infty} | p \textit{ a prime number}\}$ with $\mathbb{Z}_{p^\infty}$ is p-component of $\mathbb{Q}/\mathbb{Z}$. Since module $\mathbb{Q}/\mathbb{Z}$ is divisible module, we get $\mathbb{Z}_{p^\infty}$ is divisible $\mathbb{Z}$-module. Therefore $\mathbb{Z}_{p^\infty}$ is injective $\mathbb{Z}$-module. We are going to show $\mathbb{Z}_{p^\infty}$ is essential extension of $\mathbb{Z}_p$. Since $\mathbb{Z}_p \simeq \{\frac{z}{p} + \mathbb{Z} | z \in \mathbb{Z}\}$ is a submodule of $\mathbb{Z}_{p^\infty}$, suppose K be a proper submodule of $\mathbb{Z}_{p^\infty}$, and choose $n \in \mathbb{N}$ such that

$$\frac{1}{p^n} + \mathbb{Z} \in K$$

but $\frac{1}{p^{n+1}} + \mathbb{Z} \notin K$. For any element $\frac{k}{p^m} + \mathbb{Z} \in K$ with $k \in \mathbb{Z}$, p not dividing k, and $m \in \mathbb{N}$, we can find $r, s \in \mathbb{Z}$ with $kr + p^m s = 1$. This yields

$$\frac{1}{p^m} + \mathbb{Z} = \frac{kr + p^m s}{p^m} + \mathbb{Z} = r(\frac{k}{p^m} + \mathbb{Z}) \in K.$$

By the choice of n, this means $m \leq n$ and $K = \mathbb{Z}(\frac{1}{p^n} + \mathbb{Z})$. Thus, $\mathbb{Z}_p \subset K$, that means $\mathbb{Z}_{p^\infty}$ is an essential extension of $\mathbb{Z}_p$. So, $\mathbb{Z}_{p^\infty}$ is an injective envelope of $\mathbb{Z}_p$. □

Acknowledgment

The results of this paper were obtained during my M.Sc studies at Gadjah Mada University and are also contained in my thesis. I would like to express deep gratitude to my supervisor Dr. Indah Emilia Wijayanti and Prof. Sri Wahyuni whose guidance and support were crucial for the successful completion of this project.

References

1. R. Wisbauer, *Foundation of Module and Ring Theory* (Gordon and Breach Science Publishers, Philadelphia, 1999).
2. D. S. Dummit and R. M. Foote, *Abstract Algebra* (John Wiley & Sons, New York, 2004).
3. F. W. Anderson and K. R. Fuller, *Rings and Categories of Modules* (Springer-Verlag, New York, 1992).
4. M. Hazewinkel, *Algebras, Rings and Modules* (Kluwer Academics Publishers, New York, 2004).

Cover and Avoidance Properties and the Structure of Finite Groups

A. Ballester-Bolinches[*] R. Esteban-Romero[†] Yangming Li[‡]

Dedicated to Professor K. P. Shum
on the occasion of his seventieth birthday

Abstract

If a subgroup H of a finite group G covers or avoids every chief factor of G, then we say that H has the cover and avoidance property in G or H is a CAP-subgroup of G. If H just covers or avoids the chief factors of a given chief series of G, then H is called partial CAP-subgroup of G. These subgroup embedding properties are extremely useful to clear up the group structure. The main aim of this survey is to present some recent results about the cover and avoidance properties in the context of the old ones.

Mathematics Subject Classification (2010): 20D10, 20D20, 20D30

Keywords: Finite groups, Maximal subgroups, Minimal subgroups, cover and avoidance property, Saturated formation, Supersoluble group

1 Introduction

All groups considered in this paper will be finite. One fruitful topic in group theory is the study of embedding properties of subgroups and their influence in

[*]Departament d'Àlgebra, Universitat de València, Dr. Moliner, 50, E-46100 Burjassot, València, Spain, email: Adolfo.Ballester@uv.es

[†]Institut Universitari de Matemàtica Pura i Aplicada, Universitat Politècnica de València, Camí de Vera, s/n, E-46022 València, Spain, email: resteban@mat.upv.es

[‡]Department of Mathematics, Guangdong University of Education, Guangzhou, 510310, People's Republic of China, email: liyangming@gdei.edu.cn

the structure of the group. One of the most important subgroup embedding properties extending normality is the covering and avoidance property.

Definition 1.1 *Let A be a subgroup of a group G, and let H and K be subgroups of G such that K is normal in H. We say that*

1. *A* covers *the section H/K if $H \leq KA$, and*

2. *A* avoids *the section H/K if $H \cap A \leq K$.*

Definition 1.2 ([9, A, 10.8]) *Let A be a subgroup of a group G. Then we say that A has the* cover and avoidance property *in G, or that A is a* CAP-subgroup *of G, if A either covers or avoids every chief factor of G.*

Unlike other embedding properties, the cover and avoidance property is not persistent in intermediate subgroups, that is, if A is a CAP-subgroup of G and H is a subgroup of G containing A, it does not follow necessarily that A is a CAP-subgroup of H, as the following example shows.

Example 1.3 ([6, Example 1.3]) *Let $G = \mathrm{Sym}(6)$ be the symmetric group of degree 6. We consider in the group G the subgroups $A = \langle x, a, b\rangle$ with $x = (1,2,3)$, $a = (1,2)(3,4)$, and $b = (1,4)(2,3)$, and $C = \langle c\rangle$ with $c = (5,6)$. We observe that $A = \langle x, a, b \mid x^3 = a^2 = b^2 = 1 = [a,b], a^x = b, b^x = ab\rangle$ is isomorphic to the alternating group of degree 4 and C is a cyclic group of order 2, $C \cap A = 1$, and $C \leq \mathrm{C}_G(A)$. Let K be the direct product $K = A \times C$. Obviously, K is soluble.*

The element $ac = (1,2)(3,4)(5,6)$ generates a cyclic subgroup of order 2, $H = \langle ac\rangle$, and, since ac is odd, we have that $G = HN$ for $N = \mathrm{Alt}(6)$, that is, H is a CAP-subgroup of G. The Sylow 2-subgroup of K is $P = \langle a, b, c\rangle$. Note that P is isomorphic to an elementary abelian group of order 2^3 and P is normal in K. The subgroup H is not a CAP-subgroup of K: for the chief factor P/C of K, we have that $C < HC = \langle c, a\rangle < P$.

Fan, Guo, and Shum introduced in [11] the following generalisation of the cover and avoidance property, which turns out to be persistent in intermediate subgroups:

Definition 1.4 *Let A be a subgroup of G. Then we say that A has the* partial cover and avoidance property *or that A is a* partial CAP-subgroup *(or* semi-CAP-subgroup *or* SCAP-subgroup*) of G if there exists a chief series Γ_A of G such that A either covers or avoids each factor of Γ_A.*

It is clear that CAP-subgroups are partial CAP-subgroups, but the converse does not hold in general.

Example 1.5 *In Example 1.3, the subgroup H is a partial CAP-subgroup of K, but H is not a CAP-subgroup of K. In the chief series*

$$1 < V < P < K$$

of K, $H \cap V = 1$, that is, H avoids $V/1$, $P = HV$, that is, H covers P/V, and H avoids K/P, because $H \leq P$.

The purpose of this survey is to present results on CAP-subgroups and partial CAP-subgroups which show the fundamental role played by these subgroup in the structural study of the groups.

2 Cover and avoidance property

The most typical examples of CAP-subgroups are the normal ones. In the soluble universe, many relevant families of subgroups enjoy this property: every maximal subgroup and every Hall subgroup of a soluble group are CAP-subgroups. In fact, if we impose the partial cover and avoidance property on the maximal subgroups and Sylow subgroups we get solubility as it is proved in [15], [23]. Prefrattini subgroups, introduced by W. Gaschütz in [12], and in general any kind of subgroups of prefrattini type are other interesting types of CAP-subgroups in soluble groups (see [5, Section 4.3]).

In addition to maximal subgroups and Sylow subgroups, the so-called 2-maximal subgroups can be used to get solubility in the presence of the cover and avoidance property. Recall, a subgroup U of a group G is called a 2-*maximal subgroup* of G if U is a maximal subgroup of a maximal subgroup of G.

Theorem 2.1 ([14], [15]) *Let G be a group.*

1. *If every 2-maximal subgroup of G is a (partial) CAP-subgroup of G, then G is soluble.*

2. *G is soluble if and only if G has a 2-maximal subgroup which is a soluble (partial) CAP-subgroup of G.*

The first statement of Theorem 2.1 is a consequence of the following result ([22, Theorem 2]).

Theorem 2.2 *Let H/K be a chief factor of a group G. Then H/K is abelian if one of the following conditions holds:*

1. *If every 2-maximal subgroup of G covers or avoids H/K.*
2. *There exists a soluble maximal subgroup of G that covers or avoids H/K.*
3. *Every maximal subgroup of every Sylow p-subgroup of G covers or avoids H/K for the smallest prime p dividing the order of H/K.*

It is well-known that the class of all supersoluble groups is characterised as the class of all groups in which every subgroup has the cover and avoidance property. One natural question is whether supersolubility can be deduced if we assume that the members of smaller interesting families of subgroups satisfy the cover and avoidance property. An affirmative answer was obtained by Ezquerro in [10].

Theorem 2.3 *Let p be a prime, let G be a p-soluble group, and let H be a normal subgroup of G such that G/H is p-supersoluble. Suppose that all maximal subgroups of the Sylow p-subgroups of H are CAP-subgroups of G. Then G is p-supersoluble.*

Recall that a group is said to be *p-supersoluble* for a prime p if every chief factor of order divisible by p is cyclic. Moreover, a group is supersoluble if and only if it is p-supersoluble for all primes p.

In particular, *a soluble group G is supersoluble if and only if all maximal subgroups of the Sylow subgroups are CAP-subgroups of G.*

In fact, the solubility hypothesis can be removed from the above result. This is a consequence of the following result proved by Ezquerro in [10, Theorems C and D].

Theorem 2.4 *Let G a group with a normal subgroup H such that G/H is supersoluble. If either all maximal subgroups of the Sylow subgroups of H are CAP-subgroups of G or H is soluble and all maximal subgroups of the Sylow subgroups of $\mathrm{F}(H)$ are CAP-subgroups of G, then G is supersoluble.*

The above characterisations of soluble and supersoluble groups are the starting point for a research project consisting in characterising some formations by the cover and avoidance property of some relevant families of subgroups like maximal subgroups of Sylow subgroups, second maximal subgroups of Sylow subgroups, minimal subgroups, or second minimal subgroups of Sylow subgroups.

Recall that a *formation* is a class of groups $\mathfrak{F}$ which is closed under epimorphic images and such that if G is a group with two normal subgroups N_1 and N_2 such that G/N_1 and G/N_2 are $\mathfrak{F}$-groups, then $G/(N_1 \cap N_2) \in \mathfrak{F}$. If, in addition, $G \in \mathfrak{F}$ when $G/\Phi(G) \in \mathfrak{F}$, we say that $\mathfrak{F}$ is saturated. Given a non-empty formation $\mathfrak{F}$, each group G has a smallest normal subgroup whose quotient belongs to $\mathfrak{F}$. This subgroup is called the $\mathfrak{F}$-*residual* of G and is denoted by $G^{\mathfrak{F}}$. It is clear that $G^{\mathfrak{F}}$ is a characteristic subgroup of G (see [5, 9] for details).

The classes of all p-supersoluble groups and supersoluble groups, denoted by $\mathfrak{U}_p$ and $\mathfrak{U}$ respectively, are typical examples of saturated formations.

Let $\mathfrak{H}$ be a non-empty class of groups. According to [5, 1.2.9, 2.3.18], a chief factor H/K of a group G is said to be $\mathfrak{H}$-*central* if $[K/H] * G$ belongs to $\mathfrak{H}$, where $[H/K] * G$ is the semidirect product $[H/K]\big(G/\mathrm{C}_G(H/K)\big)$ if H/K is abelian and $G/\mathrm{C}_G(H/K)$ if H/K is not abelian. A normal subgroup N of a group G is said to be $\mathfrak{H}$-*hypercentral* in G if every chief factor of G below N is $\mathfrak{H}$-central in G.

The generalised Jordan-Hölder theorem [5, 1.2.36] ensures that the product of all $\mathfrak{H}$-hypercentral normal subgroups of a group G is again $\mathfrak{H}$-hypercentral in G. Thus every group G possesses a unique maximal normal $\mathfrak{H}$-hypercentral subgroup called the $\mathfrak{H}$-*hypercentre* of G and denoted by $\mathrm{Z}_{\mathfrak{H}}(G)$. By the generalised Jordan-Hölder theorem, every chief factor of G below the $\mathrm{Z}_{\mathfrak{H}}(G)$ is $\mathfrak{H}$-central in G. For the class $\mathfrak{N}$ of all nilpotent groups, we obtain that $\mathrm{Z}_{\mathfrak{N}}(G) = \mathrm{Z}_\infty(G)$, the hypercentre of G, which coincides with the limit of the upper central series.

A subgroup U of a group G is said to be *hypercentrally embedded* in G if $U/\mathrm{Core}_G(U)$ is contained in $\mathrm{Z}_\infty\big(G/\mathrm{Core}_G(U)\big)$. U is called *hypercyclically embedded* in G if $U/\mathrm{Core}_G(U)$ is contained in $\mathrm{Z}_{\mathfrak{U}}\big(G/\mathrm{Core}_G(U)\big)$, the supersoluble hypercentre of $G/\mathrm{Core}_G(U)$.

If U is either hypercentrally embedded or hypercyclically embedded in G, then every chief factor of G between $\langle U^G \rangle$, the normal closure of U in G, and $\mathrm{Core}_G(U)$ is cyclic of prime order and so U covers or avoids such chief factors. Applying [23, Proposition 13], it follows that U is a CAP-subgroup of G. Since every permutable subgroup is hypercentrally embedded ([24, Theorem 5.2.3], [2, Corollary 1.5.6]) and every modular subgroup is hypercyclically embedded ([24, Theorem 5.2.5]), we have:

Proposition 2.5 ([23]) *Permutable and modular subgroups have the cover and avoidance property.*

The converse of the above proposition is not true since there are soluble groups with Sylow subgroups which are neither permutable nor modular.

Asaad [1] extended Theorem 2.4 to saturated formations containing the class of all supersoluble groups:

Theorem 2.6 *Let $\mathfrak{F}$ be a saturated formation containing $\mathfrak{U}$ and let G be a group such that G/H belongs to $\mathfrak{F}$. If either all maximal subgroups of the Sylow subgroups of H are CAP-subgroups of G or G is soluble and all maximal subgroups of the Sylow subgroups of $\mathrm{F}(H)$ are CAP-subgroups of G, then G belongs to $\mathfrak{F}$.*

In [20], the authors extended the above result to non-soluble groups replacing the Fitting subgroup by the generalised Fitting subgroup. More precisely they proved:

Theorem 2.7 *Let $\mathfrak{F}$ be a saturated formation containing $\mathfrak{U}$. Suppose that E is a normal subgroup of G such that $G/E \in \mathfrak{F}$. If all maximal subgroups of the Sylow subgroups of generalised Fitting subgroup $\mathrm{F}^*(E)$ of E are CAP subgroups of G, then $G \in \mathfrak{F}$.*

In [18] the third author proved the following remarkable common extension of Theorem 2.7 and the main result of [7]:

Theorem 2.8 *Let $\mathfrak{F}$ be a saturated formation containing the class $\mathfrak{U}$ of all supersoluble groups. Let E be a normal subgroup of a group G such that $G/E \in \mathfrak{F}$. Suppose that every non-cyclic Sylow subgroup P of $\mathrm{F}^*(E)$ has a subgroup D such that $1 < |D| < |P|$ and all subgroups H of P with order $|H| = |D|$ or order $|H| = 2|D|$ (if P is a non-abelian 2-group and $|P : D| > 2$) satisfy the cover and avoidance property in G. Then $G \in \mathfrak{F}$.*

For the saturated formation of all supersoluble groups we have:

Corollary 2.9 *A group G is supersoluble if and only if every non-cyclic Sylow subgroup P of $\mathrm{F}^*(G)$ has a subgroup D such that $1 < |D| < |P|$ and all subgroups H of P with order $|H| = |D|$ and with order $2|D|$ (if P is a non-abelian 2-group and $|P : D| > 2$) are CAP-subgroups of G.*

3 Partial cover and avoidance property

As a point of departure, we recall that supersolubility can be also characterised by the partial cover and avoidance property and so we can also wonder if we can weaken the requirements and impose this property on some relevant families of subgroups. In particular, bearing in mind the results of the above section, the following question naturally arises:

Question 3.1 *Can we replace the cover and avoidance property by the partial cover and avoidance property in Theorem 2.8?*

The answer to Question 3.1 is negative, as the following example (see [4]) shows:

Example 3.2 *Consider an elementary abelian group*

$$H = \langle a, b \mid a^5 = b^5 = 1, ab = ba \rangle$$

of order 25 *and let* α *be an automorphism of* H *of order* 3 *satisfying that* $a^\alpha = b$, $b^\alpha = a^{-1}b^{-1}$. *Let* $H_1 = H$, $H_2 = \langle a', b' \rangle$ *be a copy of* H_1 *and* $G = [H_1 \times H_2]\langle \alpha \rangle$. *For any subgroup* A *of* G *of order* 25, *there exists a minimal normal subgroup* N *such that* $A \cap N = 1$. *Then* A *covers or avoids the factors of the chief series of* G

$$1 < N < AN < G.$$

In other words, A *is a partial CAP-subgroup of* G. *However,* G *is not supersoluble.*

However, on the positive side, we can give very interesting and useful information about the structure of groups in which some distinguished families of subgroups satisfy the partial cover and avoidance property.

For our study, we follow a local method as in [6]. By "local" we mean that the results are generalised in a form depending on a prime p, usually by considering p-elements or p-chief factors. The reason for choosing this local method is to discover new situations. Roughly speaking, the global hypothesis, referring to all primes, force the solubility and some non-soluble cases not to appear. This leads us to the interesting question of how the global properties can be obtained as the conjunction of the local ones for all primes.

First for a fixed prime p, we will concentrate ourselves with the family of all maximal subgroups of all Sylow p-subgroups of the group. There are two classes of groups in which obviously every maximal subgroup of each Sylow p-subgroup is a partial CAP-subgroup.

1. The class of p-supersoluble groups, and
2. The class of all groups whose Sylow p-subgroups are cyclic groups of order p.

As we have noted before, the class of p-supersoluble groups is a well-known saturated formation. The groups in the second class can be described as follows.

Proposition 3.3 **([6])** *Let G be a group whose Sylow p-subgroups are cyclic groups of order p. Then:*

1. *Either G is a p-soluble (and then p-supersoluble) group, or*
2. *$G/\mathrm{O}_{p'}(G)$ is a quasi-simple group and*

$$\mathrm{Soc}\big(G/\mathrm{O}_{p'}(G)\big) = \mathrm{O}^{p'}\big(G/\mathrm{O}_{p'}(G)\big)$$

is a simple group whose Sylow p-subgroups are of order p.

The following result proved in [6, Theorem 3.2] shows that apart from the groups described in the above paragraph, no other group satisfies that all maximal subgroups of all Sylow p-subgroups of the group are partial CAP-subgroups.

Theorem 3.4 *Let p be a prime dividing the order of a group G. Suppose that all maximal subgroups of every Sylow p-subgroup of G are partial CAP-subgroups of G. Then, either G is a group whose Sylow p-subgroups are cyclic groups of order p, or G is a p-supersoluble group.*

The global result easily follows from the above theorem and appears in the seminal paper [11].

Corollary 3.5 *Suppose that all maximal subgroups of every Sylow subgroup of a group G are partial CAP-subgroups of G. Then G is a supersoluble group.*

The following extension of Corollary 3.5 was proved in [19, Theorem 3.6].

Corollary 3.6 *Let $\mathfrak{F}$ is a saturated formation containing all supersoluble groups. Let G be a group with a normal subgroup N such that $G/N \in \mathfrak{F}$. If every maximal subgroup of every Sylow subgroup of N is a partial subgroup of G, then G belongs to $\mathfrak{F}$.*

For odd primes p, the following p-nilpotency criterion was proved in [13, Theorem 3.5].

Theorem 3.7 *Let p be an odd prime dividing the order of a group G and P a Sylow p-subgroup of G. Suppose that every maximal subgroup of P is a partial subgroup of G. Then G is p-nilpotent if and only if $N_G(P)$ is p-nilpotent.*

Related results were proved in [26].

Now we focus our attention on the family of all maximal subgroups of all Sylow subgroup of the generalised p-Fitting subgroup.

If p is a prime and G is a group, we denote by $\mathrm{F}_p^*(G)$ the set of all elements of G which induce and inner automorphism of every chief factor of G whose order is divisible by p. It is known that $\mathrm{F}^*(G)$, the generalised Fitting subgroup of G, is the intersection of all $\mathrm{F}_p^*(G)$ for all primes p (see [17, X.13.9]).

Clearly, for every group X, the subgroup $\mathrm{F}_p^*(X)$ is the normal subgroup of X such that $\mathrm{F}_p^*(X)/\mathrm{O}_{p'}(X) = \mathrm{F}^*\big(X/\mathrm{O}_{p'}(X)\big)$.

Theorem 3.8 ([6, Theorem 3.6]) *Suppose that $\mathfrak{F}$ is a saturated formation containing all p-supersoluble groups such that $\mathfrak{E}_{p'}\mathfrak{F} = \mathfrak{F}$. Let G be a group and let p be a prime dividing the order of G. Consider a normal subgroup N of G such that such that $G/N \in \mathfrak{F}$. If every maximal subgroup of every p-Sylow subgroup of $\mathrm{F}^*(N)$ is a partial CAP-subgroup of G, then either G belongs to $\mathfrak{F}$ or $\mathrm{F}^*(N)/\mathrm{O}_{p'}(N)$ is a non-abelian simple group whose Sylow p-subgroups are of order p.*

Note that Theorem 3.8 generalises [16, Theorem 3.6].

For the saturated formation of all p-supersoluble groups we have:

Corollary 3.9 *Let G be a group and let p be a prime dividing the order of G. If every maximal subgroup of every Sylow p-subgroup of $\mathrm{F}_p^*(G)$ is a partial CAP-subgroup of G, then either G is p-supersoluble or $\mathrm{F}^*(N)/\mathrm{O}_{p'}(N)$ is a non-abelian simple group whose Sylow p-subgroups are of order p.*

With the same arguments to those used in the proof of Theorem 3.8, the following result follows.

Theorem 3.10 ([6, Theorem 3.8], [19, Theorem 3.10]) *Let $\mathfrak{F}$ be a saturated formation containing the class $\mathfrak{U}$ of all supersoluble groups and let E be a normal subgroup of G such that $G/E \in \mathfrak{F}$. Suppose that every maximal subgroup of every Sylow subgroup of $\mathrm{F}^*(E)$ is a partial CAP-subgroup of G. Then G belongs to $\mathfrak{F}$.*

Our next goal is to describe the groups G enjoying the following property:

> (†) *Every 2-maximal subgroup of every Sylow p-subgroup of G is a partial CAP-subgroup of G.*

This is the main result of the paper [8]. Clearly a 2-maximal subgroup of a p-group of order p^n has order p^{n-2}. Therefore we have to think about groups G whose order is a multiple of p^2. Let us remark on some key points of this classification (see [8]):

1. Trivially, a group whose Sylow p-subgroups have order p^2 has property (†). The class of all p-supersoluble groups is composed of groups enjoying property (†) as well. Thus, important steps of our research will be the description of the structure of all groups whose Sylow p-subgroups have order p^2 and the characterisation of all p-soluble groups with property (†) which are not p-supersoluble.

2. Note that in a quaternion group of order 8, the unique 2-maximal subgroup is the centre. By the celebrated Brauer-Suzuki Theorem, if the Sylow 2-subgroups of a group G are isomorphic to a quaternion group of order 8, then $\mathrm{Z}(G/\mathrm{O}_{2'}(G))$ has even order (see [25, pages 306–315]). This is to say that every group whose Sylow 2-subgroups are quaternion groups of order 8 has property (†).

Theorem 3.11 **([8, Theorem 7])** *Let G be a group whose Sylow p-subgroups have order p^2. Consider the quotient group $G^+ = G/\mathrm{O}_{p'}(G)$ and denote $S = \mathrm{Soc}(G^+)$ and $F = \mathrm{O}^{p'}(G^+)$. Then one of the following holds:*

1. *S is a chief factor of G. In this case, one of the following holds:*
   - *(a) S is isomorphic to a cyclic group of order p, G is p-supersoluble and F is isomorphic to a cyclic group of order p^2.*
   - *(b) G^+ is a primitive group of type 1, G is p-soluble and $F = S$ is isomorphic to an elementary abelian group of order p^2.*
   - *(c) G^+ is a primitive group of type 2 and $F = S$ and either S is a product of two copies of a non-abelian simple group whose Sylow p-subgroups have order p or S is a simple group whose Sylow p-subgroups have order p^2.*
   - *(d) G^+ is a primitive group of type 2 and $S < F$ and G^+ is an almost-simple group such that S is a non-abelian simple group with Sylow p-subgroups of order p and G^+/S is a soluble group with Sylow p-subgroups of order p.*
2. *S is the direct product of two distinct minimal normal subgroups of G^+, N_1 and N_2 say. In this case, $F = S$ and N_1 and N_2 are simple groups with cyclic Sylow p-subgroups of order p.*

Theorem 3.12 **([8, Theorem 17])** *Let p be a prime number and let G be a group such that p^2 divides the order of G. Then all 2-maximal subgroups of the Sylow p-subgroups of G are partial CAP-subgroups of G if and only if one of the following statements holds:*

1. *G is a p-supersoluble group;*
2. *G is a p-soluble group such that if P is a Sylow p-subgroup of G and Q is a 2-maximal subgroup of P, then $\Phi\big(G/\mathrm{O}_{p'}(G)\big) \leq Q\mathrm{O}_{p'}(G)/\mathrm{O}_{p'}(G)$;*
   - *(a) if $\Phi\big(G/\mathrm{O}_{p'}(G)\big) = Q\mathrm{O}_{p'}(G)/\mathrm{O}_{p'}(G)$, then every chief series of the group G has exactly one complemented p-chief factor; moreover, this p-chief factor has order p^2;*
   - *(b) if $\Phi\big(G/\mathrm{O}_{p'}(G)\big) < Q\mathrm{O}_{p'}(G)/\mathrm{O}_{p'}(G)$, then all complemented p-chief factors of G are G-isomorphic to a 2-dimensional irreducible G-module V which is not an absolutely irreducible G-module;*

3. *G is a non-p-soluble group whose Sylow p-subgroups have order p^2;*

4. *$p = 2$ and G is a non-2-soluble group whose Sylow 2-subgroups are isomorphic to the quaternion group of order 8.*

Note that, according to Theorem 2.1 and Example 3.2, the partial cover and avoidance property of all second maximal subgroups of all Sylow subgroups implies solubility but not supersolubility in general.

A natural further step in our analysis of the influence of the partial cover and avoidance property on the structure of a groups is to study the effect of imposing this property on all subgroups of prime order or order 4, which can be regarded in some sense as dual subgroups of the maximal subgroups of the Sylow subgroups. The structural implications were studied in [4], where the following results were proved.

Theorem 3.13 *Let p be a prime and let P be a normal p-subgroup of a group G. If every subgroup of P of order p or 4 (if $p = 2$) is a partial CAP-subgroup of G, then P is contained in the supersoluble hypercentre $Z_{\mathfrak{U}}(G)$ of G.*

Theorem 3.14 *Suppose that p is a prime dividing the order of a group G. If every cyclic subgroup of order p or order 4 (when $p = 2$) is a partial CAP-subgroup of G, then G is p-supersoluble.*

Applying the above results, we prove:

Theorem 3.15 *Let $\mathfrak{F}$ be a saturated formation containing the class $\mathfrak{U}$ of all supersoluble groups, and let G be a group with a normal subgroup H such that $G/H \in \mathfrak{F}$. Then $G \in \mathfrak{F}$ if every cyclic subgroup of $F^*(H)$ of prime order or order 4 is a partial CAP-subgroup of G.*

Therefore a *group G is supersoluble if and only if every subgroup of prime order or order 4 is a partial CAP-subgroup of G.*

As a consequence of Theorem 3.15, we get the corresponding local result.

Theorem 3.16 *Assume that p is a prime and $\mathfrak{F}$ is a saturated formation containing all p-supersoluble groups such that $\mathfrak{E}_{p'}\mathfrak{F} = \mathfrak{F}$. Suppose that G is a group with a normal subgroup N such that G/N belongs to $\mathfrak{F}$. If every cyclic subgroup of $\mathrm{F}_p^*(N)$ of order p or 4 (when $p = 2$) is a partial CAP-subgroup of G, then G belongs to $\mathfrak{F}$.*

We have seen in Example 3.2 that the partial cover and avoidance property of second minimal subgroups of the Sylow subgroups (that is, subgroups of order p^2 for a prime p) does not imply supersolubility. However, we can give a fairly accurate description of groups whose subgroups of order p^2, p a prime, satisfy the partial cover and avoidance property and, in some sense, these groups are not very far from being p-supersoluble. This is the aim of [3]. This property can be seen as the dual of property (†).

Theorem 3.17 *Let p be a prime number and let G be a group. Let $G^+ = G/\mathrm{O}_{p'}(G)$. Then every subgroup of G of order p^2 is a partial CAP-subgroup of G if and only if one of the following statements holds:*

1. *the order of the Sylow p-subgroups of G is at most p;*
2. *G is a p-supersoluble group;*
3. *$\Phi(G^+) = 1$ and if P is a Sylow p-subgroup of G, then $P^+ = \mathrm{Soc}(G^+)$ is the direct product of minimal normal subgroups of G^+ which are all G-isomorphic to a 2-dimensional irreducible G-module V over $\mathrm{GF}(p)$, the finite field of p-elements. Furthermore, V is not an absolutely irreducible G-module when $r > 1$.*

Applying Theorem 3.17, it follows that a *group in which the second minimal subgroups of the Sylow subgroups are partial CAP-subgroups must be soluble.*

We bring this section to a close with the following

Open question 3.18 *Let p be a prime. What can be said about the structure of a group G in which the subgroups of order p^m are partial CAP-subgroups of G for $2 < m < n - 2$, where p^n is the order of a Sylow p-subgroup of G?*

4 Postscript

In a recent paper [21], Liu, Guo, and Li introduced an interesting variation of the cover and avoidance property and studied the structural implications of this subgroup embedding property.

They called a subgroup H of a group G a *CAP*-subgroup* if H either covers or avoids every non-Frattini chief factor of G. The authors show that this property is inherited in epimorphic images, that is, if N is a normal subgroup of G and H is a CAP*-subgroup of G, then HN/N is a CAP*-subgroup of G/N (Lemma 2.3.1 of [21]), but it is not persistent in intermediate subgroups (see Example 1.3).

Of notable feature are the following results.

Theorem 4.1 *Let G be a group. Then the following statements are equivalent:*

1. *G is soluble;*
2. *every Hall subgroup of G is a CAP*-subgroup of G;*
3. *every Sylow subgroup of G is a CAP*-subgroup of G;*
4. *every maximal subgroup of G is a CAP*-subgroup of G.*

Theorem 4.2 *Let G be a group. The following statements are equivalent:*

1. *G is soluble;*
2. *there exists a maximal subgroup M of G such that M is a soluble CAP*-subgroup of G;*
3. *there exists a maximal subgroup M of G such that every Sylow subgroup of M is a CAP*-subgroup of G.*

Theorem 4.3 *Let G be a group.*

1. *If every 2-maximal subgroup of G is a CAP*-subgroup of G, then G is soluble.*
2. *If there is a 2-maximal subgroup H of G such that H is a soluble CAP*-subgroup of G, then G is soluble.*

Acknowledgement

The work of the two first authors has been supported by the grant MTM-2010-19938-C03-01 from the *Ministerio de Ciencia e Innovacin* (Spanish Government).

References

[1] M. Asaad. On maximal subgroups of Sylow subgroups of finite groups. *Comm. Algebra*, 26(11):3647–3652, 1998.

[2] A. Ballester-Bolinches, R. Esteban-Romero, and M. Asaad. *Products of finite groups*, volume 53 of *de Gruyter Expositions in Mathematics.* Walter de Gruyter, 2010.

[3] A. Ballester-Bolinches, R. Esteban-Romero, and Y. Li. On second minimal subgroups of sylow subgroups of finite groups. Preprint.

[4] A. Ballester-Bolinches, R. Esteban-Romero, and Y. Li. A question on partial CAP-subgroups of finite groups. Preprint.

[5] A. Ballester-Bolinches and L. M. Ezquerro. *Classes of Finite Groups*, volume 584 of *Mathematics and its Applications.* Springer, New York, 2006.

[6] A. Ballester-Bolinches, L. M. Ezquerro, and A. N. Skiba. Local embeddings of some families of subgroups of finite groups. *Acta Math. Sinica*, 25:869–882, 2009.

[7] A. Ballester-Bolinches, L. M. Ezquerro, and A. N. Skiba. Subgroups of finite groups with strong cover-avoidance property. *Bull. Austral. Math. Soc.*, 79(3):499–506, 2009.

[8] A. Ballester-Bolinches, L. M. Ezquerro, and A. N. Skiba. On second maximal subgroups of Sylow subgroups of finite groups. *J. Pure Appl. Algebra*, 215:705–714, 2011.

[9] K. Doerk and T. Hawkes. *Finite Soluble Groups*, volume 4 of *De Gruyter Expositions in Mathematics.* Walter de Gruyter, Berlin, New York, 1992.

[10] L. M. Ezquerro. A contribution to the theory of finite supersolvable groups. *Rend. Sem. Mat. Uni. Padova*, 89:161–170, 1993.

[11] Y. Fan, X. Guo, and K. P. Shum. Remarks on two generalizations of normality of subgroups. *Chinese Ann. Math.*, 27A(2):169–176, 2006. Chinese.

[12] W. Gaschütz. Praefrattinigruppen. *Arch. Math.*, 13:418–426, 1962.

[13] X. Guo, P. Guo, and K. P. Shum. On semi cover-avoiding subgroups of finite groups. *J. Pure Appl. Algebra*, 209:151–158, 2007.

[14] X. Guo and K. P. Shum. Cover-avoidance properties and the structure of finite groups. *J. Pure Appl. Algebra*, 181:297–308, 2003.

[15] X. Guo, J. Wang, and K. P. Shum. On semi-cover-avoiding maximal subgroups and solvability of finite groups. *Comm. Algebra*, 34:3235–3244, 2006.

[16] X. Guo and L. L. Wang. On finite groups with some semi cover-avoiding subgroups. *Acta Math. Sin. (Engl. Ser.)*, 23(9):1689–1696, 2007.

[17] B. Huppert and N. Blackburn. *Finite groups III*, volume 243 of *Grundlehren der Mathematischen Wissenschaften [Fundamental Principles of Mathematical Sciences]*. Springer-Verlag, Berlin, 1982.

[18] Y. Li. On cover-avoiding subgroups of Sylow subgroups of finite groups. *Rend. Sem. Mat. Uni. Padova*, 123:249–258, 2010.

[19] Y. Li, L. Miao, and Y. Wang. On semi cover-avoiding maximal subgroups of Sylow subgroups of finite groups. *Comm. Algebra*, 37(4):1160–1169, 2009.

[20] Y. Li and G. Wang. A note on maximal CAP subgroups of finite groups. *Algebras Groups Geom.*, 23:285–290, 2006.

[21] J. Liu, X. Guo, and S. Li. The influence of CAP*-subgroups on the solvability of finite groups. To appear in *Bull. Malaysian Math. Sci. Soc.*

[22] X. Liu and N. Ding. On chief factors of finite groups. *J. Pure Appl. Algebra*, 210(3):789–796, 2007.

[23] J. Petrillo. The embedding of CAP-subgroups in finite groups. *Ricerche mat.*, 59:97–107, 2010.

[24] R. Schmidt. *Subgroup lattices of groups*, volume 14 of *De Gruyter Expositions in Mathematics*. Walter de Gruyter, Berlin, 1994.

[25] M. Suzuki. *Group theory. II*, volume 248 of *Grundlehren der Mathematischen Wissenschaften (Fundamental Principles of Mathematical Sciences)*. Springer-Verlag, New York, 1986.

[26] L. Wang and G. Chen. Some properties of finite groups with some (semi-p-)cover-avoiding subgroups. *J. Pure Appl. Algebra*, 213:686–689, 2009.

Semilattices of Archimedean Semigroups*

S. BOGDANOVIĆ and Ž. POPOVIĆ
University of Niš, Faculty of Economics
Trg kralja Aleksandra 11, P.O.Box 121, 18000 Niš, Serbia
E-mail: sbogdan@eknfak.ni.ac.rs
E-mail: zpopovic@eknfak.ni.ac.rs
www.eknfak.ni.ac.rs

M. ĆIRIĆ
University of Niš, Faculty of Science, Department of Mathematics
Višegradska 33, P.O.Box 224, 18000 Niš, Serbia
E-mail: ciricm@bankerinter.net
www.pmf.ni.ac.rs

Dedicated to Professor Kar-Ping Shum on the occasion of his 70th birthday

M. S. Putcha, in 1973, gave the first complete description of semilattices of Archimedean semigroups. Other characterizations of this class of semigroups have been given by T. Tamura in 1972, by S. Bogdanović and M. Ćirić in 1992 and by M. Ćirić and S. Bogdanović in 1993. In this paper, for $m, n \in \mathbf{N}$, we define a relation $\rho_{(m,n)}$ and prove that it is a congruence relation on an arbitrary semigroup. Using the congruence relation $\rho_{(m,n)}$ we will give some new characterizations of semilattices of Archimedean semigroups. At the end we describe hereditary properties of semilattices of k-Archimedean semigroups.

Keywords: band congruence, semilatice decomposition, semilatice of Archimedean semigroups, semilatice of k-Archimedean semigroups, semilatice of hereditary k-Archimedean semigroups

1. Introduction and preliminaries

Semilattice decompositions of semigroups were defined and studied by A. H. Clifford [7], 1941. After that, several authors have worked on this very important topic. The existence of the greatest semilattice decompo-

*Research supported by Ministry of Science and Environmental Protection, Republic of Serbia, Grant No. 174013

sition of a semigroup was established by M. Yamada [19], 1955, and by T. Tamura and N. Kimura [18], 1955. Every semigroup has the greatest semilatice decomposition. So, very interesting and very important are decompositions in which all its components are Archimedean semigroups. T. Tamura and N. Kimura [17], 1954, showed that every commutative semigroup is a semilattice of Archimedean semigroups. This well known result has since been generalized by many authors. Semilattices of Archimedean semigroups are described by M. S. Putcha [13], 1973, by T. Tamura [15], 1972, by S. Bogdanović and M. Ćirić [1], 1992, [3], 1993, and by M. Ćirić and S. Bogdanović [6], 1993. Various other results concerning semilattices of Archimedean semigroups can be found in paper by S. Bogdanović and M. Ćirić [3], M. Mitrović [11] and by S. Bogdanović, Ž. Popović and M. Ćirić [5].

By $\mathbf{N}$ we denote the set of all positive integers. Starting from the division relation on a semigroup M. S. Putcha [13] proved that a semigroup S is a semilattice of Archimedean semigroups if and only if $a \mid b \Longrightarrow (\exists n \in \mathbf{N})\ a^2 \mid b^n$. P. Protić [12], and S. Bogdanović and M. Ćirić [1], have given some equivalent statements for Putcha's theorem. In this paper, using divisibility on a semigroup, for $m, n \in \mathbf{N}$, we define a relation $\rho_{(m,n)}$ and by it we give some new characterizations of a semilattices of Archimedean semigroups. The results obtained generalize some results by previously mentioned authors. Also, we give some new characterizations of a semilattices of hereditary k-Archimedean semigroups.

By S^1 we denote a semigroup S with identity 1. A semigroup in which all elements are idempotents is a *band.* A commutative band is a *semilattice.* By $\mathcal{S}$ we denote the class of all semilattices.

Let a be an element of a semigroup S. By $\langle a \rangle$ we denote a subsemiogroup of S generated by a.

Let ξ be a binary relation on a non-empty set X. For $n \in \mathbf{N}$, ξ^n will denote the n-th power of ξ in the semigroup of binary relations on X, and ξ^0 will denote the equality relation on X.

Let ϱ be an arbitrary relation on a semigroup S. Then the *radical* $R(\varrho)$ of ϱ is a relation on S defined by:

$$(a, b) \in R(\varrho) \Leftrightarrow (\exists p, q \in \mathbf{N})\,(a^p, b^q) \in \varrho.$$

The radical $R(\varrho)$ was introduced by L. N. Shevrin in [14].

An equivalence relation ξ is a *left* (*right*) *congruence* if for all $a, b \in S$, $a\,\xi\,b$ implies $ca\,\xi\,cb$ ($ac\,\xi\,bc$). An equivalence ξ is a congruence if it is both left and right congruence. A congruence relation ξ is a band congruence on

S if S/ξ is a band, i.e. if $a\,\xi\,a^2$, for all $a \in S$.

Let ξ be an equivalence on a semigroup S. By $\xi^\flat$ we define the largest congruence relation on S contained in ξ. It is well-known that

$$\xi^\flat = \{(a,b) \in S \times S \mid (\forall x, y \in S^1)\ (xay, xby) \in \xi\}.$$

Let a and b be elements of a semigroup S. Then:

$$a \mid b \Leftrightarrow b \in S^1 a S^1, \quad \text{and} \quad a \longrightarrow b \Leftrightarrow (\exists n \in \mathbf{N})\ a \mid b^n.$$

Using previous relations we will define the following relation:

$$\text{---} = \longrightarrow \cap (\longrightarrow)^{-1}.$$

A semigroup S is *Archimedean*, if $a \longrightarrow b$, for all $a, b \in S$. The class of all Archimedean semigroups we denote by $\mathcal{A}$.

Let $k \in \mathbf{N}$ be a fixed positive integer. A semigroup S is *k-Archimedean* if $b^k \in S^1 a S^1$, for all $a, b \in S$. The class of all k-Archimedean semigroups we denote by $k\mathcal{A}$. A semigroup S is *intra k-regular* if $a^k \in S a^{2k} S$, for all $a \in S$. The class of all intra k-regular semigroups we denote by $\mathcal{I}k\mathcal{R}$.

Let $k \in \mathbf{N}$ be a fixed positive integer and let a and b be elements of a semigroup S. Then:

$$a \uparrow_k b \Leftrightarrow b^k \in \langle a, b \rangle a \langle a, b \rangle.$$

A semigroup S is *hereditary k-Archimedean* if $a \uparrow_k b$, for all $a, b \in S$. The class of all hereditary k-Archimedean semigroups we denote by $Her(k\mathcal{A})$.

Let a be an element of a semigroup S and let $n \in \mathbf{N}$. We will use the following notation:

$$\Sigma_n(a) = \{x \in S \mid a \longrightarrow^n x\}.$$

We introduce the following equivalence on a semigroup S:

$$a\ \sigma_n\ b \Leftrightarrow \Sigma_n(a) = \Sigma_n(b).$$

Let ξ be an arbitrary relation on a semigroup S. A semigroup S is *ξ-simple*, if $(a, b) \in \xi$, for all $a, b \in S$. It is evident that every Archimedean semigroup is a σ_1-simple semigroup.

For two classes $\mathcal{X}_1$ and $\mathcal{X}_2$ of semigroups, $\mathcal{X}_1 \circ \mathcal{X}_2$ will denote the *Mal'cev product* of $\mathcal{X}_1$ and $\mathcal{X}_2$, i.e. the class of all semigroups S on which there exists a congruence ϱ such that S/ϱ belongs to $\mathcal{X}_2$ and each ϱ-class of S which is a subsemigroup of S belongs to $\mathcal{X}_1$. If $\mathcal{X}_2$ is a subclass of $\mathcal{S}$, then $\mathcal{X}_1 \circ \mathcal{X}_2$ is the class of all semigroups having a semilattice decomposition whose related factors belongs to $\mathcal{X}_2$ and the components belong to $\mathcal{X}_1$. Such decompositions will be called *$\mathcal{X}_1 \circ \mathcal{X}_2$-decompositions.*

For undefined notions and notations we refer to [2], [8] and [9].

2. The main results

Let $m, n \in \mathbf{N}$. On a semigroup S we define a relation $\rho_{(m,n)}$ by

$$(a,b) \in \rho_{(m,n)} \Leftrightarrow (\forall x \in S^m)(\forall y \in S^n)\ xay \text{---} xby,$$

i.e.

$$a\rho_{(m,n)}b \Leftrightarrow (\forall x \in S^m)(\forall y \in S^n)(\exists i, j \in \mathbf{N})(xay)^i \in SxbyS \wedge (xby)^j \in SxayS.$$

The relation $\rho_{(1,1)}$ we simply denote by ρ.

If instead of the relation — we assume the equality relation, then we obtain the relation which was introduced and discussed by S. J. L. Kopamu in [10], 1995. So, the relation $\rho_{(m,n)}$ is a generalization of Kopamu's relation.

By the following theorem we give a very important characteristic of the $\rho_{(m,n)}$ relation.

Theorem 2.1. *Let $m, n \in \mathbf{N}$. On a semigroup S the relation $\rho_{(m,n)}$ is a congruence relation.*

Proof. It is evident that $\rho_{(m,n)}$ is a reflexive and symmetric relation on S.

Assume $a, b, c \in S$ such that $a\rho_{(m,n)}b$ and $b\rho_{(m,n)}c$. Then

$$a\rho_{(m,n)}b \Leftrightarrow (\forall x \in S^m)(\forall y \in S^n)(\exists i, j \in \mathbf{N})\ (xay)^i \in SxbyS \wedge (xby)^j \in SxayS,$$

$$b\rho_{(m,n)}c \Leftrightarrow (\forall x \in S^m)(\forall y \in S^n)(\exists p, q \in \mathbf{N})\ (xby)^p \in SxcyS \wedge (xcy)^q \in SxbyS.$$

So, $(xay)^i = uxbyv$ and $(xcy)^q = wxbyz$, for some $u, v, w, z \in S$. Since $b\rho_{(m,n)}c$, then for $x \in S^m$ and $yvu \in S^{n+2} \subseteq S^n$ we have that there exists $t \in \mathbf{N}$ such that $(xbyvu)^t \in SxcyvuS$ and

$$((xay)^i)^{t+1} = (uxbyv)^{t+1} = u(xbyvu)^t xbyv \in uSxcyvuSxbyv \subseteq SxcyS.$$

Thus $(xay)^{i(t+1)} \in SxcyS$.

Similarly, we prove that $(xcy)^k \in SxayS$, for some $k \in \mathbf{N}$. Hence, $a\rho_{(m,n)}c$. Therefore, $\rho_{(m,n)}$ is a transitive relation on S.

Now, assume $a, b, c \in S$ are such that $a\rho_{(m,n)}b$. Then for $x \in S^m$, $y \in S^n$ we have $cy \in S^{n+1} \subseteq S^n$, so, there exist $p, q \in \mathbf{N}$ such that

$$(x(ac)y)^p = (xa(cy))^p \in Sxb(cy)S = Sx(bc)yS,$$

$$(x(bc)y)^q = (xb(cy))^q \in Sxa(cy)S = Sx(ac)yS.$$

Hence $ac\rho_{(m,n)}bc$. Similarly, we prove that $ca\rho_{(m,n)}cb$. Thus, $\rho_{(m,n)}$ is a congruence relation on S. $\square$

Remark 2.1. Let μ be an equivalence relation on a semigroup S and let $m, n \in \mathbf{N}$. Then a relation $\mu_{(m,n)}$ defined on S by

$$(a,b) \in \mu_{(m,n)} \Leftrightarrow (\forall x \in S^m)(\forall y \in S^n)\ (xay, xby) \in \mu$$

is a congruence relation on S. But, there exists a relation μ which is not equivalence, for example $\mu = \longrightarrow$, for which the relation $\mu_{(m,n)}$ is a congruence on S.

The following two lemmas were proved by S. Bogdanović, Ž. Popović and M. Ćirić in [4] and we will use them in the further work.

Lemma 2.1. *Let ξ be an equivalence on a semigroup S. Then ξ is a congruence relation on S if and only if $\xi = \xi^\flat$.*

Lemma 2.2. *Let ξ be an equivalence relation on a semigroup S. Then $\xi^\flat$ is a band congruence if and only if*

$$(\forall a \in S)(\forall x, y \in S^1)\ (xay, xa^2y) \in \xi.$$

Now we are ready to prove the main result of this paper. It gives some new characterizations of a semilattices of Archimedean semigroups.

Theorem 2.2. *Let $m, n \in \mathbf{N}$. The following conditions on a semigroup S are equivalent:*

(i) $\rho_{(m,n)}$ *is a band congruence;*

(ii) $(\forall a \in S)(\forall x \in S^m)(\forall y \in S^n)\ xay \longrightarrow xa^2y$;

(iii) $S \in \mathcal{A} \circ \mathcal{S}$;

(iv) $R(\rho_{(m,n)}) = \rho_{(m,n)}$;

(v) $\rho^\flat_{(m,n)}$ *is a band congruence;*

(vi) $(\forall a \in S)(\forall u, v \in S^1)\ (uav, ua^2v) \in \rho_{(m,n)}$;

(vii) ρ *is a band congruence.*

Proof. (i) $\Longrightarrow$ (ii) This implication follows immediately.

(ii) $\Longrightarrow$ (iii) Let (ii) hold. Then for every $a, b \in S$, if $x = (ab)^l$ and $y = (ba)^k b$, for some $k, l \in \mathbf{N}$, $l \geq m$ and $k \geq n$, there exists $i \in \mathbf{N}$ such that

$$((ab)^l a (ba)^k b)^i \in S(ab)^l a^2 (ba)^k b S \subseteq Sa^2S,$$

i.e.

$$(ab)^{(2k+l+1)i} \in Sa^2S.$$

Thus, by Theorem 1 [6] S is a semilattice of Archimedean semigroups.

(iii) $\Longrightarrow$ (i) Let S be a semilattice Y of Archimedean semigroups S_α, $\alpha \in Y$ and let $m, n \in \mathbf{N}$ be fixed elements. By Theorem 2.1 $\rho_{(m,n)}$ is a congruence relation on S. It remains to prove that $\rho_{(m,n)}$ is a band congruence on S. Assume $a \in S$, $x \in S^m$ and $y \in S^n$. Then $xay, xa^2y \in S_\alpha$, for some $\alpha \in Y$. Since S_α, $\alpha \in Y$, is Archimedean, then there exist $p, q \in \mathbf{N}$ such that

$$(xay)^p \in S_\alpha xa^2yS_\alpha \subseteq Sxa^2yS,$$

$$(xa^2y)^q \in S_\alpha xayS_\alpha \subseteq SxayS,$$

hence $a\rho_{(m,n)}a^2$, i.e. $\rho_{(m,n)}$ is a band congruence on S. Thus (i) holds

(i) $\Longrightarrow$ (iv) The inclusion $\rho_{(m,n)} \subseteq R(\rho_{(m,n)})$ always holds, so it remains to prove the opposite inclusion. Since $\rho_{(m,n)}$ is a band congruence on S, then we have that

$$(\forall a \in S)(\forall k \in \mathbf{N}) \;\; a\rho_{(m,n)}a^k.$$

Now assume $a, b \in S$ such that $aR(\rho_{(m,n)})b$. Then $a^i\rho_{(m,n)}b^j$, for some $i, j \in \mathbf{N}$, and by previous statement we have that $a\rho_{(m,n)}a^i\rho_{(m,n)}b^j\rho_{(m,n)}b$. Thus $a\rho_{(m,n)}b$. So $R(\rho_{(m,n)}) \subseteq \rho_{(m,n)}$. Therefore, (iv) holds.

(iv) $\Longrightarrow$ (i) Since $\rho_{(m,n)}$ is reflexive, then by hypothesis for every $a \in S$ we have that

$$a^2\rho_{(m,n)}a^2 \Leftrightarrow (a^1)^2\rho_{(m,n)}(a^2)^1 \Leftrightarrow aR(\rho_{(m,n)})a^2 \Leftrightarrow a\rho_{(m,n)}a^2.$$

Thus, (i) holds.

(i) $\Longrightarrow$ (v) This implication follows by Lemma 2.1.

(v) $\Longrightarrow$ (vi) This implication follows by Lemma 2.2.

(vi) $\Longrightarrow$ (i) Let (vi) hold. By Theorem 2.1 $\rho_{(m,n)}$ is a congruence and by (vi) for $u = v = 1$ we obtain that $(a, a^2) \in \rho_{(m,n)}$, for every $a \in S$, i.e. $\rho_{(m,n)}$ is a band congruence. Thus, (i) holds.

(i)$\Longleftrightarrow$(vii) This equivalence follows immediately by the equivalence (i)$\Longleftrightarrow$(iii). □

The following result gives connections between relations ρ and $\sigma_1^\flat$.

Theorem 2.3. *Let S be an arbitrary semigroup. Then $\rho = \sigma_1^\flat$.*

Proof. Assume $a, b \in S$ such that $a\rho b$. If $c \in \Sigma_1(a)$, then $c^k = uav$, for some $u, v \in S$ and some $k \in \mathbf{N}$. Since $a\rho b$ then we obtain that $(uav)^i \in SubvS \subseteq SbS$, for some $i \in \mathbf{N}$. Thus

$$c^{ki} = (c^k)^i = (uav)^i \in SbS,$$

whence $c \in \Sigma_1(b)$. So, we proved that $\Sigma_1(a) \subseteq \Sigma_1(b)$. Similarly we prove that $\Sigma_1(b) \subseteq \Sigma_1(a)$. Therefore, $\Sigma_1(a) = \Sigma_1(b)$, i.e. $a\sigma_1 b$. Thus, $\rho \subseteq \sigma_1$.

Let ξ be an arbitrary congruence relation on S contained in σ_1 and let $a, b \in S$ be elements such that $a\xi b$. Since ξ is a congruence, then for every $x, y \in S$ we have that

$$(xay, xby) \in \xi \subseteq \sigma_1 \subseteq \text{—},$$

so it follows that $(\forall x, y \in S)\ xay$—$xby$, i.e. $a\rho b$. Therefore, $\xi \subseteq \rho$. Since $\sigma_1^\flat$ is the greatest congruence contained in σ_1, then by the previous statement is evident that $\rho = \sigma_1^\flat$. □

On an arbitrary semigroup S, it is clear that the following inclusion holds

$$\rho^\flat = \rho = \sigma_1^\flat \subseteq \sigma_1.$$

Theorem 2.4. *Let $k \in \mathbf{N}$. A semigroup S is a semilattice of k-Archimedean semigroups if and only if*

$$(\forall a, b \in S)\ (ab)^k \in Sa^2S \quad \& \quad S \in \mathcal{I}k\mathcal{R}.$$

Proof. Let S be a semilattice Y of k-Archimedean semigroups S_α, $\alpha \in Y$. For $a, b \in S$ there exists $\alpha \in Y$ such that $ab, a^2b \in S_\alpha$, whence $(ab)^k \in S_\alpha a^2 b S_\alpha \subseteq Sa^2S$. By Theorem 3.1 [4] we have that $S \in \mathcal{I}k\mathcal{R}$.

Conversely, from the first condition by Theorem 2.2 we have that S is a semilattice of Archimedean semigroups and since S is intra k-regular we have by Theorem 3.1 [4] that the assertion follows. □

T. Tamura [16] proved that the class of all semigroups which are semilattices of Archimedean semigroups is not subsemigroup closed. By the following theorem we determine the greatest subsemigroup closed subclass of the class of all semigroups which are semilattices of k-Archimedean semigroups.

Theorem 2.5. *Let $k \in \mathbf{N}$. Then $k\mathcal{A} \circ \mathcal{S}$ is a subsemigroup closed if and only if*

$$(\forall a, b \in S)\ (ab)^k \in \langle a, b\rangle a^2 \langle a, b\rangle \quad \& \quad a^k \in \langle a, b\rangle a^{2k} \langle a, b\rangle.$$

Proof. Assume $a, b \in S$ and $T = \langle a, b\rangle$. Since T is a semilattice of k-Archimedean semigroups then by Theorem 2.4 we obtain

$$(ab)^k \in Ta^2T = \langle a, b\rangle a^2 \langle a, b\rangle,$$

and

$$a^k \in Ta^{2k}T = \langle a, b\rangle a^{2k}\langle a, b\rangle.$$

Conversely, let T be an arbitrary subsemigroup of S. Assume $a, b \in T$. By the hypothesis we have that

$$(ab)^k \in \langle a, b\rangle a^2\langle a, b\rangle \subseteq Ta^2T,$$

so by Theorem 2.2 or Theorem 1 [6] T is a semilattice of Archimedean semigroups. Also, by the second part of hypothesis we have that

$$a^k \in \langle a, b\rangle a^{2k}\langle a, b\rangle \subseteq Ta^{2k}T,$$

thus T is intra k-regular semigroup. Therefore, by Theorem 2.4 T is a semilattice of k-Archimedean semigroups. □

Theorem 2.6. *Let $k \in \mathbf{N}$. Then $S \in Her(k\mathcal{A})$ if and only if every its subsemigroup is k-Archimedean.*

Proof. Let S be a hereditary k-Archimedean semigroup and let T be a subsemigroup of S. Assume $a, b \in T$, then $\langle a, b\rangle \subseteq T$, and also by hypothesis we have that

$$b^k \in \langle a, b\rangle b\langle a, b\rangle \subseteq TaT.$$

Thus, T is k-Archimedean.

Conversely, assume $a, b \in S$. Then $a, b \in \langle a, b\rangle$ and since $\langle a, b\rangle$ is k-Archimedean, we obtain that

$$b^k \in \langle a, b\rangle a\langle a, b\rangle.$$

Thus, S is hereditary k-Archimedean. □

By the following theorem we describe semilettices of hereditary k-Archimedean semigroups which are subsemigroup closed.

Theorem 2.7. *Let $k \in \mathbf{N}$. The following conditions on a semigroup S are equivalent:*

(i) $S \in Her(k\mathcal{A}) \circ \mathcal{S}$;

(ii) $(\forall a, b \in S)\ a \longrightarrow b \implies a^2 \uparrow_k b$;

(iii) $(\forall a, b \in S)\ a \longrightarrow c\ \&\ b \longrightarrow c \implies ab \uparrow_k c$;

(iv) $(\forall a, b, c \in S)\ a \longrightarrow b\ \&\ b \longrightarrow c \implies a \uparrow_k c$;

(v) the class $Her(k\mathcal{A}) \circ \mathcal{S}$ is subsemigroup closed.

Proof. (i) $\Longrightarrow$ (ii) Let S be a semilattice Y of hereditary k-Archimedean semigroups S_α, $\alpha \in Y$. Let $a, b \in S$, such that $a \longrightarrow b$. Then $b, a^2b \in S_\alpha$, for some $\alpha \in Y$ and by the hypothesis we have that

$$b^k \in \langle b, a^2b\rangle a^2b\langle b, a^2b\rangle \subseteq \langle a^2, b\rangle a^2\langle a^2, b\rangle,$$

i.e. $a^2 \uparrow_k b$. So, (ii) holds.

(ii) $\Longrightarrow$ (i) By Theorem 2.2 S is a semilattice Y of Archimedean semigroups S_α, $\alpha \in Y$. Assume $a, b \in S_\alpha$, $\alpha \in Y$. Then $a \longrightarrow b$ and by (ii) we have that $a^2 \uparrow_k b$, whence

$$b^k \in \langle a^2, b\rangle a^2\langle a^2, b\rangle \subseteq \langle a, b\rangle a\langle a, b\rangle.$$

Hence, $a \uparrow_k b$ in S_α, i.e. S_α, $\alpha \in Y$ is a k-Archimedean semigroup.

(ii) $\Longrightarrow$ (iii) Assume $a, b, c \in S$ such that $a \longrightarrow c$ and $b \longrightarrow c$. Then by (i)$\Leftrightarrow$(ii) and by Theorem 1.1 [3] or Theorem 3 [5] we have that $ab \longrightarrow c$. Now, by the hypothesis we have that $(ab)^2 \uparrow_k c$, whence $ab \uparrow_k c$.

(iii) $\Longrightarrow$ (iv) By Theorem 3 [5] we have that $\longrightarrow$ is transitive. Assume $a, b, c \in S$ such that $a \longrightarrow b$ and $b \longrightarrow c$. Then $a^2 \uparrow_k c$, whence $a \uparrow_k c$.

(iv) $\Longrightarrow$ (i) Since $\longrightarrow$ is transitive, then by Theorem 3 [5] we have that S is a semilattice Y of Archimedean semigroups S_α, $\alpha \in Y$. Let $a, b \in S_\alpha$, $\alpha \in Y$. Then $a \longrightarrow b$ and $b \longrightarrow b$ and by (iv) we have that $a \uparrow_k b$. Hence, (i) holds.

(ii) $\Longrightarrow$ (v) Let T be a subsemigroup of S and let $a, b \in T$ such that $a \longrightarrow b$ in T. By (ii), $a^2 \uparrow_k b$, i.e.

$$b^k \in \langle a^2, b\rangle a^2\langle a^2, b\rangle \subseteq Ta^2T.$$

Hence, $a^2 \uparrow_k b$ in T. By (i)$\Leftrightarrow$(ii) we have that T is a semilattice of hereditary k-Archimedean semigroups.

(v) $\Longrightarrow$ (i) This implication is obvious. □

References

1. S. Bogdanović and M. Ćirić, *Semigroups in which the radical of every ideal is a subsemigroup*, Zbornik radova Fil. fak. Niš, Ser. Mat. **6** (1992), 129–135.
2. S. Bogdanović and M. Ćirić, *Semigroups*, Prosveta, Niš, 1993 (in Serbian).
3. S. Bogdanović and M. Ćirić, *Semilattices of Archimedean semigroups and (completely) π-regular semigroups* I (*A survey*), Filomat (Niš) **7** (1993), 1–40.
4. S. Bogdanović, Ž. Popović and M. Ćirić, *Bands of k-Archimedean semigroups*, Semigroup Forum **80** (2010), 426–439.
5. S. Bogdanović, Ž. Popović and M. Ćirić, *Band decompositions of semigroups* (*A-survey*), Mathematica Macedonica (submitted for publication).

6. M. Ćirić and S. Bogdanović, *Decompositions of semigroups induced by identities*, Semigroup Forum **46** (1993), 329–346.
7. A. H. Clifford, *Semigroups admitting relative inverses*, Ann. of Math. **42** No. 2 (1941), 1037–1049.
8. P. M. Higgins, *Techniques in semigroup theory*, Oxford press, 1992.
9. J. M. Howie, *An introduction to semigroup theory*, Acad. Press. New York, 1976.
10. S. J. L. Kopamu, *On semigroup species*, Communications in Algebra, **23** (1995), 5513-5537.
11. M. Mitrović, *On semilattice of Archimedean semigroups* (*A-survey*), Proc. of the Workshop, Semigroups and Languages, Lisboa, Portugal, 2002 (I. M. Aranjo, M. J. J. Branco, V. H. Fruandes, G. M. S. Gomes, Eds), World Scientific Publishing Co. Pte. Ltd. 2004. ISBN 981-238-917-2
12. P. Protić, *A new proof of Putcha's theorem*, PU.M.A. Ser. A, **2/3-4** (1991), 281–284.
13. M. S. Putcha, *Semilattice decompositions of semigroups*, Semigroup Forum **6** (1973), 12–34.
14. L. N. Shevrin, *Theory of epigroups* I, Mat. Sbornik **185** No. 8 (1994), 129-160 (in Russian).
15. T. Tamura, *On Putcha's theorem concerning semilattice of Archimedean semigroups*, Semigroup Forum **4** (1972), 83–86.
16. T. Tamura, *Quasi-orders, generalized Archimedeaness, semilattice decompositions*, Math. Nachr. **68** (1975), 201–220.
17. T. Tamura and N. Kimura, *On decomposition of a commutative semigroup*, Kodai Math. Sem. Rep. **4** (1954), 109–112.
18. T. Tamura and N. Kimura, *Existence of greatest decomposition of a semigroup*, Kodai Math. Sem. Rep. **7** (1955), 83–84.
19. M. Yamada, *On the greatest semilattice decomposition of a semigroup*, Kodai Mat. Sem. Rep. **7** (1955), 59–62.

Some New Results on Gröbner-Shirshov Bases*

L. A. Bokut[†]
School of Mathematical Sciences, South China Normal University
Guangzhou 510631, P. R. China
Sobolev Institute of Mathematics, Russian Academy of Sciences
Siberian Branch, Novosibirsk 630090, Russia
bokut@math.nsc.ru

Yuqun Chen[‡]
School of Mathematical Sciences, South China Normal University
Guangzhou 510631, P. R. China
yqchen@scnu.edu.cn

K. P. Shum
Department of Mathematics, The University of Hong Kong
Hong Kong, P. R. China
kpshum@maths.hku.hk

Abstract

In this survey article, we report some new results of Gröbner-Shirshov bases, including new Composition-Diamond lemmas and some applications of some known Composition-Diamond lemmas.

Key words: Gröbner-Shirshov basis; Composition-Diamond lemma; normal form; Lie algebra; right-symmetry algebra; Rota-Baxter algebra; L-algebra; conformal algebra; metabelian Lie algebra, dendriform algebra; category; group; braid group; inverse semigroup; plactic monoid; quantum group; operad.

AMS 2000 Subject Classification: 16S15, 13P10, 13N10, 16-xx, 17-xx, 17B01, 17B66, 18Axx, 20F36, 20M18, 17B37, 18D50

---

*Supported by the NNSF of China (Nos. 10771077, 10911120389).

[†]Supported by RFBR 09-01-00157, LSS–3669.2010.1 and SB RAS Integration grant No. 2009.97 (Russia) and Federal Target Grant Scientific and educational personnel of innovation Russia for 2009-2013 (government contract No. 02.740.11.5191).

[‡]Corresponding author.

1 Introduction

We continue our previous survey [12] on papers given mostly by participants of an Algebra Seminar at South China Normal University, Guangzhou.

This survey consists of two blocks:

(I) *New Composition-Diamond (CD-) lemmas.*

(II) *Applications of known CD-lemmas.*

In [82], A. I. Shirshov established the theory of one-relator Lie algebras $Lie(X|s = 0)$ and provided the algorithmic decidality of the word problem for any one-relator Lie algebra. In order to proceed his ideas, he first created the so called Gröbner-Shirshov bases theory for Lie algebras $Lie(X|S)$ which are presented by generators and defining relations. The main notion of Shirshov's theory was a notion of composition $(f,g)_w$ of two Lie polynomials, $f,g \in Lie(X)$ relative to some associative word w. The following lemma was proved by Shirshov [82].

Shirshov Lemma. *Let $Lie(X) \subset k\langle X\rangle$ be a free Lie algebra over a field k which is regarded as an algebra of the Lie polynomials in the free algebra $k\langle X\rangle$, and let S be a Gröbner-Shirshov basis in $Lie(X)$. Then $f \in Id(S) \Rightarrow \bar{f} = u\bar{s}v$, where $s \in S^c$, $u,v \in X^*$, $\bar{f}, \bar{s}$ are the leading associative words of the corresponding Lie polynomials f, s with respect to the deg-lex order on X^*.*

Nowadays, Shirshov's lemma is named the "Composition-Diamond lemma" for Lie and associative algebras.

We explain what it means of "CD-lemma" for a class (variety or category) $\mathcal{M}$ of linear Ω-algebras over a field k (here Ω is a set of linear operations on $\mathcal{M}$) with free objects.

$\mathcal{M}$-CD-lemma Let $\mathcal{M}$ be a class of (in general, non-associative) Ω-algebras, $\mathcal{M}(X)$ a free Ω-algebra in $\mathcal{M}$ generated by X with a linear basis consisting of "normal (non-associative Ω-) words" $[u]$, $S \subset \mathcal{M}(X)$ a monic subset, $<$ a monomial order on normal words and $Id(S)$ the ideal of $\mathcal{M}(X)$ generated by S. Let S be a Gröbner-Shirshov basis (this means that any composition of elements in S is trivial). Then

(a) If $f \in Id(S)$, then $[\bar{f}] = [a\bar{s}b]$, where $[\bar{f}]$ is the leading word of f and $[asb]$ is a normal S-word.

(b) $Irr(S) = \{[u] | [u] \neq [a\bar{s}b], s \in S, [asb]$ *is a normal* $S-$ *word*$\}$ is a linear basis of the quotient algebra $\mathcal{M}(X|S) = \mathcal{M}(X)/Id(S)$.

In many cases, each of conditions (a) and (b) is equivalent to the condition that S is a Gröbner-Shirshov basis in $\mathcal{M}(X)$. But in some of our "new CD-lemmas", this is not the case.

How to establish a CD-lemma for the free algebra $\mathcal{M}(X)$? Following the idea of Shirshov (see, for example, [11, 17, 82]), one needs

1) to define appropriate linear basis (normal words) N of $\mathcal{M}(X)$;

2) to find a monomial order on N;

3) to find normal S-words;

4) to define compositions of elements in S (they may be compositions of intersection, inclusion and left (right) multiplication, or may be else);

5) to prove two key lemmas:

Key Lemma 1. Let S be a Gröbner-Shirshov basis (any composition of polynomials from S is trivial). Then any element of $Id(S)$ is a linear combination of normal S-words.

Key Lemma 2. Let S be a Gröbner-Shirshov basis, $[a_1s_1b_1]$ and $w = [a_2s_2b_2]$ normal S-words, $s_1, s_2 \in S$. If $\overline{[a_1s_1b_1]} = \overline{[a_2s_2b_2]}$, then $[a_1s_1b_1] - [a_2s_2b_2]$ is trivial modulo (S, w).

Let the normal words N of the free algebra $\mathcal{M}(X)$ be a well-ordered set and $0 \neq f \in \mathcal{M}(X)$. Denote by $\bar{f}$ the leading word of f. f is called monic if the coefficient of $\bar{f}$ is 1.

A well ordering on N is monomial if for any $u, v, w \in N$,

$$u > v \Longrightarrow \overline{w|_u} > \overline{w|_v},$$

where $w|_u = w|_{x\mapsto u}$ and $w|_v = w|_{x\mapsto v}$.

For example, let X be a well-ordered set and X^* the free monoid generated by X. We define the deg-lex order on X^*: to compare two words first by length and then lexicographically. Then, such an order is monomial on X^*.

Let $S \subset \mathcal{M}(X)$, $s \in S$, $u \in N$. Then, roughly speaking, $u|_s = u|_{x_i \mapsto s}$, where x_i is the individuality occurrence of the letter $x_i \in X$ in u, is called an S-word (or s-word). More precisely, let $\star \notin X$ be a new letter and $N(\star)$ the normal words of the free algebra $\mathcal{M}(X \cup \{\star\})$. Then by a $\star$-word we mean any expression in $N(\star)$ with only one occurrence of $\star$. Now, an s-word means $u|_s = u|_{\star\mapsto s}$, where u is a $\star$-word. Therefore, the ideal $Id(S)$ of $\mathcal{M}(X)$ is the set of linear combination of S-words.

An S-word $u|_s$ is normal if $\overline{u|_s} = u|_{\bar{s}}$.

Definition 1.0.1 *Given a monic subset $S \subset \mathcal{M}(X)$ and $w \in N$, an intersection (inclusion) composition h is called trivial modulo (S, w) if h can be presented as a linear combination of normal S-words with leading words less than w; a left (right) multiplication composition (or another kinds composition) h is called trivial modulo (S) if h can be presented as a linear combination of normal S-words with leading words less than or equal to $\bar{h}$.*

The set S is a Gröbner-Shirshov basis in $\mathcal{M}(X)$ if all the possible compositions of elements in S are trivial modulo S and corresponding w.

If a subset S of $\mathcal{M}(X)$ is not a Gröbner-Shirshov basis then one can add all nontrivial compositions of polynomials of S to S. Continuing this process repeatedly, we finally obtain a Gröbner-Shirshov basis S^C that contains S. Such a process is called Shirshov algorithm. S^C is called Gröbner-Shirshov complement of S.

We establish new CD-lemmas for the following classes of algebras:

2.1 Lie algebras over commutative algebras (L. A. Bokut, Yuqun Chen, Yongshan Chen [14]).

2.2 Associative algebras with multiple operations (L. A. Bokut, Yuqun Chen, Jianjun Qiu [20]).

2.3 Rota-Baxter algebras (L. A. Bokut, Yuqun Chen, Xueming Deng [15]).

2.4 Right-symmetric (or pre-Lie) algebras (L. A. Bokut, Yuqun Chen, Yu Li [16]).

2.5 L-algebras (Yuqun Chen, Jiapeng Huang [34]).

2.6 Differential algebras (Yuqun Chen, Yongshan Chen, Yu Li [31]).

2.7 Non-associative algebras over commutative algebras (Yuqun Chen, Jing Li, Mingjun Zeng [36]).

2.8 n-conformal algebras (L. A. Bokut, Yuqun Chen, Guangliang Zhang [21]).

2.9 λ-differential associative algebras with multiple operators (Jianjun Qiu, Yuqun Chen [73]).

2.10 Categories (L. A. Bokut, Yuqun Chen, Yu Li [17]).

2.11 Metabelian Lie algebras (Yuqun Chen, Yongshan Chen [30]).

2.12 S-act algebras (Xia Zhang [87]).

A new CD-lemma is given in:

2.13 Operads (V. Dotsenko, A. Khoroshkin [45], see also V. Dotsenko, M. V. Johansson [46]).

Some new applications of known CD-lemmas are given in:

3.1 Artin-Markov normal form for braid groups in Artin-Burau generators (L. A. Bokut, V. V. Chainikov, K. P. Shum [10] and Yuqun Chen, Qiuhui Mo [37]).

3.2 Braid groups in Artin-Garside generators (L. A. Bokut [8]).

3.3 Braid groups in Birman-Ko-Lee generators (L. A. Bokut [9]).

3.4 Braid groups in Adyan-Thurston generators (Yuqun Chen, Chanyan Zhong [42]).

3.5 GSB and normal forms for free inverse semigroups (L. A. Bokut, Yuqun Chen, Xiangui Zhao [22]).

3.6 Embeddings of algebras (L. A. Bokut, Yuqun Chen, Qiuhui Mo [19]).

3.7 Word problem for Novikov's and Boone's group (Yuqun Chen, Wenshu Chen, Runai Luo [32]).

3.8 PBW basis of $U_q(A_N)$ (L. A. Bokut, P. Malcolmson [26] and Yuqun Chen, Hongshan Shao, K. P. Shum [40]).

3.9 GSB for free dendriform algebras (Yuqun Chen, Bin Wang [41]).

3.10 Anti-commutative GSB of a free Lie algebra relative to Lyndon-Shirshov words (L. A. Bokut, Yuqun Chen, Yu Li [18]).

3.11 Free partially commutative groups, associative algebras and Lie algebras (Yuqun Chen, Qiuhui Mo [39]).

3.12 Plactic monoids in row generators (Yuqun Chen, Jing Li [35]).

Some applications of CD-lemmas for associative and Lie algebras are given in:

3.13 Plactic algebras in standard generators (Lukasz Kubat, Jan Okniński [64]).

3.14 Filtrations and distortion in infinite-dimensional algebras (Y. Bahturin, A. Olshanskii [2]).

3.15 Sufficiency conditions for Bokut' normal forms (K. Kalorkoti [60]).

3.16 Quantum groups of type G_2 and D_4 (Yanhua Ren, Abdukadir Obul [74], Gulshadam Yunus, Abdukadir Obul [86]).

3.17 An embedding of recursively presented Lie algebras (E. Chibrikov [43]).

3.18 GSB for some monoids (Canan Kocapinar, Firat Ates, A. Sinan Cevik [62]).

Notations

CD-lemma: Composition-Diamond lemma.
GSB: Gröbner-Shirshov basis.
X^*: the free monoid generated by the set X.
X^{**}: the set of all non-associative words (u) in X.
k: a field.
K: a commutative ring with unit.
$Id(S)$: the ideal generated by the set S.
$k\langle X\rangle$: the free associative algebra over k generated by X.
$k[X]$: the polynomial algebra over k generated by X.
$Lie(X)$: the free Lie algebra over k generated by X.
$Lie_K(X)$: the free Lie algebra over a commutative ring K generated by X.

2 New CD-lemmas

2.1 Lie algebras over a commutative algebra

A. A. Mikhalev, A. A. Zolotykh [70] prove the CD-lemma for $k[Y] \otimes k\langle X\rangle$, a tensor product of a free algebra and a polynomial algebra. L. A. Bokut, Yuqun Chen, Yongshan Chen [13] prove the CD-lemma for $k\langle Y\rangle \otimes k\langle X\rangle$, a tensor product of two free algebras. Yuqun Chen, Jing Li, Mingjun Zeng [36] (see §2.7) prove the CD-lemma for $k[Y] \otimes k(X)$, a tensor product of a non-associative algebra and a polynomial algebra.

In this subsection, we introduce the CD-lemma for $k[Y] \otimes Lie(X)$, Lie algebra $Lie(X)$ over a polynomial algebra $k[Y]$, which is established in [14]. It provides a Gröbner-Shirshov bases theory for Lie algebras over a commutative algebra.

Let K be a commutative associative k-algebra with unit and $\mathcal{L}$ a Lie K-algebra. Then, $\mathcal{L}$ can be presented as K-algebra by generators X and some defining relations S,

$$\mathcal{L} = Lie_K(X|S) = Lie_K(X)/Id(S).$$

In order to define a Gröbner-Shirshov basis for $\mathcal{L}$, we first present K in a form

$$K = k[Y|R] = k[Y]/Id(R),$$

where $R \subset k[Y]$. Then the Lie K-algebra $\mathcal{L}$ has the following presentation as a $k[Y]$-algebra

$$\mathcal{L} = Lie_{k[Y]}(X|S, Rx, \ x \in X).$$

Let the set X be well-ordered, and let $<$ and $\prec_X$ be the lex order and the deg-lex order on X^* respectively, for example, $ab < a, \ a \prec_X ab, \ a, b \in X$.

A word $w \in X^* \setminus \{1\}$ is an associative Lyndon-Shirshov word (ALSW for short) if

$$(\forall u, v \in X^*, u, v \neq 1) \ w = uv \Rightarrow w > vu.$$

A nonassociative word $(u) \in X^{**}$ is a non-associative Lyndon-Shirshov word (NLSW for short), denoted by $[u]$, if

(i) u is an ALSW;

(ii) if $[u] = [(u_1)(u_2)]$ then both (u_1) and (u_2) are NLSW's (from (i) it then follows that $u_1 > u_2$);

(iii) if $[u] = [[[u_{11}][u_{12}]][u_2]]$ then $u_{12} \leq u_2$.

We denote the set of all NLSW's (ALSW's) on X by $NLSW(X)$ $(ALSW(X))$.

Let $Y = \{y_j | j \in J\}$ be a well-ordered set and $[Y] = \{y_{j_1} y_{j_2} \cdots y_{j_l} | y_{j_1} \leq y_{j_2} \leq \cdots \leq y_{j_l}, l \geq 0\}$ the free commutative monoid generated by Y. Then $[Y]$ is a k-linear basis of the polynomial algebra $k[Y]$.

Let $Lie_{k[Y]}(X)$ be the "double" free Lie algebra, i.e., the free Lie algebra over the polynomial algebra $k[Y]$ with generating set X.

From now on we regard $Lie_{k[Y]}(X) \cong k[Y] \otimes Lie_k(X)$ as the Lie subalgebra of $k[Y]\langle X\rangle \cong \mathbf{k}[Y] \otimes k\langle X\rangle$ the free associative algebra over polynomial algebra $k[Y]$, which is generated by X under the Lie bracket $[u, v] = uv - vu$.

For an $ALSW$ t on X, there is a unique bracketing, denoted by $[t]$, such that $[t]$ is $NLSW$:

$$[t] = t \in X, \ [t] = [[v][w]],$$

where $t = vw$ and w is the longest $ALSW$ proper end of t (then v is also an ALSW).

Let

$$T_A = \{u = u^Y u^X | u^Y \in [Y], \ u^X \in ALSW(X)\}$$

and

$$T_N = \{[u] = u^Y [u^X] | u^Y \in [Y], \ [u^X] \in NLSW(X)\}.$$

Let kT_N be the linear space spanned by T_N over k. For any $[u], [v] \in T_N$, define

$$[u][v] = \sum \alpha_i u^Y v^Y [w_i^X]$$

where $\alpha_i \in k$, $[w_i^X]$'s are NLSW's and $[u^X][v^X] = \sum \alpha_i [w_i^X]$ in $Lie_k(X)$.

Then $k[Y] \otimes Lie_k(X) \cong kT_N$ as k-algebra and T_N is a k-basis of $k[Y] \otimes Lie_k(X)$.

We define the deg-lex order $\succ$ on

$$[Y]X^* = \{u^Y u^X | u^Y \in [Y], u^X \in X^*\}$$

by the following: for any $u, v \in [Y]X^*$,

$$u \succ v \ \text{ if } \ (u^X \succ_X v^X) \ \text{ or } \ (u^X = v^X \ \text{ and } \ u^Y \succ_Y v^Y),$$

where $\succ_Y$ and $\succ_X$ are the deg-lex order on $[Y]$ and X^* respectively.

Lemma 2.1.1 *(Shirshov [80, 83]) Suppose that $w = aub$ where $w, u \in T_A$ and $a, b \in X^*$. Then $[w] = [a[uc]d]$, where $[uc] \in T_N$ and $b = cd$.*

Represent c in a form $c = c_1 c_2 \ldots c_k$, where $c_1, \ldots, c_n \in ALSW(X)$ and $c_1 \leq c_2 \leq \ldots \leq c_n$. Denote by

$$[w]_u = [a[\cdots[[[u][c_1]][c_2]]\ldots[c_n]]d].$$

Then, $\overline{[w]_u} = w$.

Definition 2.1.2 *Let $S \subset Lie_{k[Y]}(X)$ be a k-monic subset, $a, b \in X^*$ and $s \in S$. If $a\bar{s}b \in T_A$, then by Lemma 2.1.1 we have the special bracketing $[a\bar{s}b]_{\bar{s}}$ of $a\bar{s}b$ relative to $\bar{s}$. We define $[asb]_{\bar{s}} = [a\bar{s}b]_{\bar{s}}|_{[\bar{s}]\mapsto s}$ to be a normal s-word (or normal S-word).*

It is proved that each element in $Id(S)$ can be expressed as a linear combinations of normal S-words (see [14]). The proof is not easy.

There are four kinds of compositions.

Definition 2.1.3 *Let f, g be two k-monic polynomials of $Lie_{k[Y]}(X)$. Denote the least common multiple of $\bar{f}^Y$ and $\bar{g}^Y$ in $[Y]$ by $L = lcm(\bar{f}^Y, \bar{g}^Y)$.*

If $\bar{g}^X$ is a subword of $\bar{f}^X$, i.e., $\bar{f}^X = a\bar{g}^X b$ for some $a, b \in X^*$, then the polynomial

$$C_1\langle f, g\rangle_w = \frac{L}{\bar{f}^Y} f - \frac{L}{\bar{g}^Y}[agb]_{\bar{g}}$$

is called the inclusion composition of f *and* g *with respect to* w*, where* $w = L\bar{f}^X = La\bar{g}^X b$.

If a proper prefix of $\bar{g}^X$ *is a proper suffix of* $\bar{f}^X$*, i.e.,* $\bar{f}^X = aa_0$, $\bar{g}^X = a_0 b$, $a, b, a_0 \neq 1$, *then the polynomial*

$$C_2\langle f, g\rangle_w = \frac{L}{\bar{f}^Y}[fb]_{\bar{f}} - \frac{L}{\bar{g}^Y}[ag]_{\bar{g}}$$

is called the intersection composition of f *and* g *with respect to* w*, where* $w = L\bar{f}^X b = La\bar{g}^X$.

If the greatest common divisor of $\bar{f}^Y$ *and* $\bar{g}^Y$ *in* $[Y]$ *is non-empty, then for any* $a, b, c \in X^*$ *such that* $w = La\bar{f}^X b\bar{g}^X c \in T_A$*, the polynomial*

$$C_3\langle f, g\rangle_w = \frac{L}{\bar{f}^Y}[afb\bar{g}^X c]_{\bar{f}} - \frac{L}{\bar{g}^Y}[a\bar{f}^X bgc]_{\bar{g}}$$

is called the external composition of f *and* g *with respect to* w.

If $\bar{f}^Y \neq 1$*, then for any normal* f*-word* $[afb]_{\bar{f}}$, $a, b \in X^*$*, the polynomial*

$$C_4\langle f\rangle_w = [a\bar{f}^X b][afb]_{\bar{f}}$$

is called the multiplication composition of f *with respect to* w*, where* $w = a\bar{f}^X ba\bar{f}b$.

Theorem 2.1.4 ([14], CD-lemma for Lie algebras over a commutative algebra) *Let* $S \subset Lie_{k[Y]}(X)$ *be nonempty set of* k*-monic polynomials and* $Id(S)$ *be the* $k[Y]$*-ideal of* $Lie_{k[Y]}(X)$ *generated by* S*. Then the following statements are equivalent.*

(i) S *is a Gröbner-Shirshov basis in* $Lie_{k[Y]}(X)$.

(ii) $f \in Id(S) \Rightarrow \bar{f} = \beta a\bar{s}b \in T_A$ *for some* $s \in S$, $\beta \in [Y]$ *and* $a, b \in X^*$.

(iii) $Irr(S) = \{[u] \mid [u] \in T_N,\ u \neq \beta a\bar{s}b,$ *for any* $s \in S$, $\beta \in [Y]$, $a, b \in X^*\}$ *is a* k*-basis for* $Lie_{k[Y]}(X|S) = Lie_{k[Y]}(X)/Id(S)$.

Now, we give some applications of Theorem 2.1.4.

A Lie algebra is called special if it can be embedded into its universal enveloping associative algebra.

Only a few of non-special Lie algebras were known.

Example 2.1.5 *(Shirshov [79, 83]) Let the field* $k = GF(2)$ *and* $K = k[Y|R]$*, where*

$$Y = \{y_i, i = 0, 1, 2, 3\},\ R = \{y_0 y_i = y_i\ (i = 0, 1, 2, 3),\ y_i y_j = 0\ (i, j \neq 0)\}.$$

Let $\mathcal{L} = Lie_K(X|S_1, S_2)$*, where* $X = \{x_i, 1 \leq i \leq 13\}$*,* S_1 *consist of the following relations*

$$[x_2, x_1] = x_{11},\ [x_3, x_1] = x_{13},\ [x_3, x_2] = x_{12},$$
$$[x_5, x_3] = [x_6, x_2] = [x_8, x_1] = x_{10},$$
$$[x_i, x_j] = 0 \quad (\textit{for any other } i > j),$$

and S_2 consist of the following relations

$$
\begin{aligned}
&y_0x_i = x_i \ (i = 1, 2, \ldots, 13),\\
&x_4 = y_1x_1,\ x_5 = y_2x_1,\ x_5 = y_1x_2,\ x_6 = y_3x_1,\ x_6 = y_1x_3,\\
&x_7 = y_2x_2,\ x_8 = y_3x_2,\ x_8 = y_2x_3,\ x_9 = y_3x_3,\\
&y_3x_{11} = x_{10},\ y_1x_{12} = x_{10},\ y_2x_{13} = x_{10},\\
&y_1x_k = 0\ (k = 4, 5, \ldots, 11, 13),\ y_2x_t = 0\ (t = 4, 5, \ldots, 12),\ y_3x_l = 0\ (l = 4, 5, \ldots, 10, 12, 13).
\end{aligned}
$$

Then $S = S_1 \cup S_2 \cup RX \cup \{y_1x_2 = y_2x_1,\ y_1x_3 = y_3x_1,\ y_2x_3 = y_3x_2\}$ is a GSB in $Lie_{\mathbf{k}[Y]}(X)$. By Theorem 2.1.4, $x_{10} \neq 0$ in $\mathcal{L}$. But from the CD-lemma for $k[Y] \otimes k\langle X\rangle$ in A. A. Mikhalev, A. A. Zolotykh [70], it easily follows that $x_{10} = 0$ in the universal enveloping associative algebra $U_K(\mathcal{L}) = K\langle X|S_1^{(-)}, S_2\rangle$. Hence, $\mathcal{L}$ is non-special as a K-algebra.

Example 2.1.6 *(Cartier [29]) Let $k = GF(2)$, $K = k[y_1, y_2, y_3|y_i^2 = 0,\ i = 1, 2, 3]$ and $\mathcal{L} = Lie_K(X|S)$, where $X = \{x_{ij}, 1 \leq i \leq j \leq 3\}$ and*

$$S = \{[x_{ii}, x_{jj}] = x_{ji}\ (i > j), [x_{ij}, x_{kl}] = 0\ (others),\ y_3x_{33} = y_2x_{22} + y_1x_{11}\}.$$

Then $S' = S \cup \{y_i^2x_{kl} = 0\ (\forall i, k, l)\} \cup S_1$ is a GSB in $Lie_{k[Y]}(X)$, where S_1 consists of the following relations

$$
\begin{aligned}
&y_3x_{23} = y_1x_{12},\ y_3x_{13} = y_2x_{12},\ y_2x_{23} = y_1x_{13},\ y_3y_2x_{22} = y_3y_1x_{11},\\
&y_3y_1x_{12} = 0,\ y_3y_2x_{12} = 0,\ y_3y_2y_1x_{11} = 0,\ y_2y_1x_{13} = 0.
\end{aligned}
$$

The universal enveloping algebra of $\mathcal{L}$ has a presentation:

$$U_K(\mathcal{L}) = K\langle X|S^{(-)}\rangle \cong \mathbf{k}[Y]\langle X|S^{(-)}, y_i^2x_{kl} = 0\ (\forall i, k, l)\rangle.$$

In $U_K(\mathcal{L})$, we have

$$0 = y_3^2x_{33}^2 = (y_2x_{22} + y_1x_{11})^2 = y_2^2x_{22}^2 + y_1^2x_{11}^2 + y_2y_1[x_{22}, x_{11}] = y_2y_1x_{12}.$$

On the other hand, since $y_2y_1x_{12} \in Irr(S')$, $y_2y_1x_{12} \neq 0$ in $\mathcal{L}$ by Theorem 2.1.4. Thus, $\mathcal{L}$ is not special as a K-algebra.

Conjecture (Cohn [44]) Let $K = \mathbf{k}[y_1, y_2, y_3|y_i^p = 0, i = 1, 2, 3]$ be an algebra of truncated polynomials over a field k of characteristic $p > 0$. Let

$$\mathcal{L}_p = Lie_K(x_1, x_2, x_3 \mid y_3x_3 = y_2x_2 + y_1x_1).$$

Then $\mathcal{L}_p$ is not special. $\mathcal{L}_p$ is called Cohn's Lie algebras.

Let $Y = \{y_1, y_2, y_3\}$, $X = \{x_1, x_2, x_3\}$ and $S = \{y_3x_3 = y_2x_2 + y_1x_1,\ y_i^px_j = 0,\ 1 \leq i, j \leq 3\}$. Then $\mathcal{L}_p \cong Lie_{\mathbf{k}[Y]}(X|S)$ and $U_K(\mathcal{L}_p) \cong \mathbf{k}[Y]\langle X|S^{(-)}\rangle$. Suppose that S^C is the Gröbner-Shirshov complement of S in $Lie_{\mathbf{k}[Y]}(X)$. Let $S_{X^p} \subset \mathcal{L}_p$ be the set of all the elements of S^C whose X-degrees do not exceed p. It is clear that S_{X^p} is a finite set for any p. Although it is difficult to find the Gröbner-Shirshov complement of S in $Lie_{\mathbf{k}[Y]}(X)$, it is possible to find the set S_{X^p}. By using this idea and the Theorem 2.1.4, we have the following theorem:

Theorem 2.1.7 *([14]) Cohn's Lie algebras $\mathcal{L}_2$, $\mathcal{L}_3$ and $\mathcal{L}_5$ are non-special.*

Theorem 2.1.8 *([14]) For an arbitrary commutative k-algebra $K = k[Y|R]$, if S is a Gröbner-Shirshov basis in $Lie_{k[Y]}(X)$ such that for any $s \in S$, s is $k[Y]$-monic, then $\mathcal{L} = Lie_K(X|S)$ is special.*

Corollary 2.1.9 *([14]) Any Lie K-algebra $L_K = Lie_K(X|f)$ with one monic defining relation $f = 0$ is special.*

Theorem 2.1.10 *([14]) Suppose that S is a finite homogeneous subset of $Lie_k(X)$. Then the word problem of $Lie_K(X|S)$ is solvable for any finitely generated commutative k-algebra K.*

Theorem 2.1.11 *([14]) Every finitely or countably generated Lie K-algebra can be embedded into a two-generated Lie K-algebra, where K is an arbitrary commutative* **k**-*algebra.*

2.2 Associative algebras with multiple operations

A Ω-algebra A is a k-space with the linear operator set Ω on A.

V. Drensky and R. Holtkamp [47] constructed Gröbner bases theory for Ω-algebras, where Ω consists of n-ary operations, $n \geq 2$.

In this subsection, we consider associative Ω-algebras, where Ω consists of n-ary operations, $n \geq 1$.

An associative algebra with multiple linear operators is an associative K-algebra R with a set Ω of multilinear operators (operations).

Let X be a set and

$$\Omega = \bigcup_{n=1}^{\infty} \Omega_n$$

where Ω_n is the set of n-ary operations, for example, ary $(\delta) = n$ if $\delta \in \Omega_n$.

Define

$$\begin{aligned}
&\mathfrak{S}_0 = S(X_0),\ X_0 = X,\\
&\mathfrak{S}_1 = S(X_1),\ X_1 = X \cup \Omega(\mathfrak{S}_0),\\
&\cdots\cdots\\
&\mathfrak{S}_n = S(X_n),\ X_n = X \cup \Omega(\mathfrak{S}_{n-1}),
\end{aligned}$$

where $\Omega(\mathfrak{S}_j) = \bigcup_{t=1}^{\infty} \{\delta(u_1, u_2, \ldots, u_t) | \delta \in \Omega_t, u_i \in \mathfrak{S}_j,\ i = 1, 2, \ldots, t\}$ and $S(X_j)$ is the free semigroup generated by $X_j,\ j = 0, 1, \ldots$.

Let

$$\mathfrak{S}(X) = \bigcup_{n \geq 0} \mathfrak{S}_n.$$

Then, it is easy to see that $\mathfrak{S}(X)$ is a semigroup such that $\Omega(\mathfrak{S}(X)) \subseteq \mathfrak{S}(X)$.

For any $u \in \mathfrak{S}(X)$, $dep(u) = \min\{n | u \in \mathfrak{S}_n\}$ is called the depth of u.

Let $K\langle X;\Omega\rangle$ be the K-algebra spanned by $\mathfrak{S}(X)$. Then, the element in $\mathfrak{S}(X)$ (resp. $K\langle X;\Omega\rangle$) is called a Ω-word (resp. Ω-polynomial). If $u \in X\cup\Omega(\mathfrak{S}(X))$, we call u a prime Ω-word and define $bre(u) = 1$ (the breadth of u). If $u = u_1u_2\cdots u_n \in \mathfrak{S}(X)$, where u_i is prime Ω-word for all i, then we define $bre(u) = n$.

Extend linearly each $\delta \in \Omega_n$,

$$\delta : \mathfrak{S}(X)^n \to \mathfrak{S}(X),\ (x_1, x_2, \cdots, x_n) \mapsto \delta(x_1, x_2, \cdots, x_n)$$

to $K\langle X;\Omega\rangle$. Then, $K\langle X;\Omega\rangle$ is a free associative algebra with multiple linear operators Ω on set X.

Assume that $\mathfrak{S}(X)$ is equipped with a monomial order $>$.

Let f, g be two monic Ω-polynomials. Then, there are two kinds of *compositions*.

(i) If there exists a Ω-word $w = \bar{f}a = b\bar{g}$ for some $a, b \in \mathfrak{S}(X)$ such that $bre(w) < bre(\bar{f}) + bre(\bar{g})$, then we call $(f, g)_w = fa - bg$ the *intersection composition* of f and g with respect to w.

(ii) If there exists a Ω-word $w = \bar{f} = u|_{\bar{g}}$ for some $u \in \mathfrak{S}(X)$, then we call $(f, g)_w = f - u|_g$ the *including composition* of f and g with respect to w.

It is noted that each S-word is normal.

Theorem 2.2.1 ([20], CD-lemma for associative Ω-algebras) *Let S be a set of monic Ω-polynomials in $K\langle X;\Omega\rangle$ and $>$ a monomial order on $\mathfrak{S}(X)$. Then the following statements are equivalent.*

(i) S is a Gröbner-Shirshov basis in $K\langle X;\Omega\rangle$.

(ii) $f \in Id(S) \Rightarrow \bar{f} = u|_{\bar{s}}$ for some $u \in \mathfrak{S}(X)$ and $s \in S$.

(iii) $Irr(S) = \{w \in \mathfrak{S}(X) | w \neq u|_{\bar{s}}$ for any $u \in \mathfrak{S}(X)$ and $s \in S\}$ is a K-basis of $K\langle X;\Omega|S\rangle = K\langle X;\Omega\rangle/Id(S)$.

Now, we give some applications of Theorem 2.3.1.

First of all, we define an order on $\mathfrak{S}(X)$. Let X and Ω be well-ordered sets. We order X^* by the deg-lex order. For any $u \in \mathfrak{S}(X)$, u can be uniquely expressed without brackets as

$$u = u_0\delta_{i_1}\overrightarrow{x_{i_1}}u_1\cdots\delta_{i_t}\overrightarrow{x_{i_t}}u_t,$$

where each $u_i \in X^*, \delta_{i_k} \in \Omega_{i_k}$, $\overrightarrow{x_{i_k}} = (x_{k_1}, x_{k_2}, \cdots, x_{k_{i_k}}) \in \mathfrak{S}(X)^{i_k}$. It is reminded that for each i_k, $dep(u) > dep(\overrightarrow{x_{i_k}})$.

Denote by

$$wt(u) = (t, \delta_{i_1}, \overrightarrow{x_{i_1}}, \cdots, \delta_{i_t}, \overrightarrow{x_{i_t}}, u_0, u_1, \cdots, u_t).$$

Then, we order $\mathfrak{S}(X)$ as follows: for any $u, v \in \mathfrak{S}(X)$,

$$u > v \Longleftrightarrow wt(u) > wt(v) \text{ lexicographically} \tag{1}$$

by induction on $dep(u) + dep(v)$.

It is clear that the order (1) is a monomial order on $\mathfrak{S}(X)$.

A Rota-Baxter K-algebra of weight λ ([3, 77]) is an associative algebra R with a K-linear operation $P: R \to R$ satisfying the Rota-Baxter relation:

$$P(x)P(y) = P(P(x)y + xP(y) + \lambda xy), \ \forall x, y \in R.$$

Thus, any Rota-Baxter algebra is a special case of associative algebra with multiple operators when $\Omega = \{P\}$.

Now, let $\Omega = \{P\}$ and $\mathfrak{S}(X)$ be as before. Let $K\langle X; P\rangle$ be the free associative algebra with one operator $\Omega = \{P\}$ on a set X.

Theorem 2.2.2 *([20]) With the order (1) on $\mathfrak{S}(X)$,*

$$S = \{P(x)P(y) - P(P(x)y) - P(xP(y)) - \lambda P(xy)| \ x, y \in \mathfrak{S}(X)\}$$

is a Gröbner-Shirshov basis in $K\langle X; P\rangle$.

By Theorems 2.3.1 and 2.2.2, we obtain a normal form $Irr(S)$ of the free Rota-Baxter algebra $K\langle X; P|S\rangle$ which is the same as in [48].

A λ-differential algebra over K ([54]) is an associative K-algebra R together with a K-linear operator $D: R \to R$ such that

$$D(xy) = D(x)y + xD(y) + \lambda D(x)D(y), \ \forall x, y \in R.$$

Any λ-differential algebra is also an associative algebra with one operator $\Omega = \{D\}$.

Let X be well-ordered and $K\langle X; D\rangle$ the free associative algebra with one operator $\Omega = \{D\}$ defined as before.

For any $u \in \mathfrak{S}(X)$, u has a unique expression

$$u = u_1 u_2 \cdots u_n,$$

where each $u_i \in X \cup D(\mathfrak{S}(X))$. Denote by $deg_X(u)$ the number of $x \in X$ in u. Let

$$wt(u) = (deg_X(u), u_1, u_2, \cdots, u_n).$$

For any $u, v \in \mathfrak{S}(X)$, define

$$u > v \Longleftrightarrow wt(u) > wt(v) \ \text{ lexicographically}, \tag{2}$$

where for each t, $u_t > v_t$ if one of the following holds:

(a) $u_t, v_t \in X$ and $u_t > v_t$;

(b) $u_t = D(u_t^{'}), v_t \in X$;

(c) $u_t = D(u_t^{'}), v_t = D(v_t^{'})$ and $u_t^{'} > v_t^{'}$.

It is easy to see the order (2) is a monomial order on $\mathfrak{S}(X)$.

Theorem 2.2.3 *([20]) With the order (2) on $\mathfrak{S}(X)$,*

$$S = \{D(xy) - D(x)y - xD(y) - \lambda D(x)D(y) |\ x, y \in \mathfrak{S}(X)\}$$

is a Gröbner-Shirshov basis in $K\langle X; D\rangle$.

By Theorems 2.3.1 and 2.2.3, we obtain a normal form $Irr(S)$ of the free λ-differential algebra $K\langle X; D|S\rangle$ which is the same as in [54].

A differential Rota-Baxter algebra of weight λ ([54]), called also λ-differential Rota-Baxter algebra, is an associative K-algebra R with two K-linear operators $P, D : R \to R$ such that for any $x, y \in R$,

(I) (Rota-Baxter relation) $P(x)P(y) = P(xP(y)) + P(P(x)y) + \lambda P(xy)$;

(II) (λ-differential relation) $D(xy) = D(x)y + xD(y) + \lambda D(x)D(y)$;

(III) $D(P(x)) = x$.

Hence, any λ-differential Rota-Baxter algebra is an associative algebra with two linear operators $\Omega = \{P, D\}$.

Let $K\langle X; \Omega\rangle$ be the free associative algebra with multiple linear operators Ω on X, where $\Omega = \{P, D\}$. For any $u \in \mathfrak{S}(X)$, u has a unique expression

$$u = u_0 P(b_1) u_1 P(b_2) u_2 \cdots P(b_n) u_n,$$

where each $u_i \in (X \cup D(\mathfrak{S}(X)))^*$ and $b_i \in \mathfrak{S}(X)$. Denote by

$$wt(u) = (deg_X(u), deg_P(u), n, b_1, \cdots, b_n, u_0, \cdots, u_n),$$

where $deg_P(u)$ is the number of P in u. Also, for any $u_t \in (X \cup D(\mathfrak{S}(X)))^*$, u_t has a unique expression

$$u_t = u_{t_1} \cdots u_{t_k}$$

where each $u_{t_j} \in X \cup D(\mathfrak{S}(X))$.

Let X be well-ordered and $u, v \in \mathfrak{S}(X)$. Order $\mathfrak{S}(X)$ as follows:

$$u > v \Longleftrightarrow wt(u) > wt(v) \text{ lexicographically} \tag{3}$$

where for each t, $u_t > v_t$ if

$$(deg_X(u_t), deg_P(u_t), u_{t_1}, \cdots, u_{t_k}) > (deg_X(v_t), deg_P(v_t), v_{t_1}, \cdots, v_{t_l}) \text{ lexicographically}$$

where for each j, $u_{t_j} > v_{t_j}$ if one of the following holds:

(a) $u_{t_j}, v_{t_j} \in X$ and $u_{t_j} > v_{t_j}$;

(b) $u_{t_j} = D(u'_{t_j}), v_{t_j} \in X$;

(c) $u_{t_j} = D(u'_{t_j}), v_{t_j} = D(v'_{t_j})$ and $u'_{t_j} > v'_{t_j}$.

Clearly, the order (3) is a monomial order on $\mathfrak{S}(X)$.

Let S be the set consisting of the following Ω-polynomials:

1. $P(x)P(y) - P(xP(y)) - P(P(x)y) - \lambda P(xy)$,
2. $D(xy) - D(x)y - xD(y) - \lambda D(x)D(y)$,
3. $D(P(x)) - x$,

where $x, y \in \mathfrak{S}(X)$.

Theorem 2.2.4 *([20]) With the order (3) on $\mathfrak{S}(X)$, S is a Gröbner-Shirshov basis in $K\langle X; \Omega\rangle$.*

By Theorems 2.3.1 and 2.2.4, we obtain a normal form $Irr(S)$ of the free λ-differential Rota-Baxter algebra $K\langle X; P, D|S\rangle$ which is a similar one in [54].

2.3 Rota-Baxter algebras

In the subsection, we consider Rota-Baxter algebras over a field of characteristic 0.

The free non-commutative Rota-Baxter algebra is given by K. Ebrahimi-Fard and L. Guo [48]. The free commutative Rota-Baxter algebra is given by G. Rota [77] and P. Cartier [28].

Let X be a nonempty set, $S(X)$ the free semigroup generated by X without identity and P a symbol of a unary operation. For any two nonempty sets Y and Z, denote by

$$\Lambda_P(Y, Z) = (\cup_{r\geq 0}(YP(Z))^r Y)\cup(\cup_{r\geq 1}(YP(Z))^r)\cup(\cup_{r\geq 0}(P(Z)Y)^r P(Z))\cup(\cup_{r\geq 1}(P(Z)Y)^r),$$

where for a set T, T^0 means the empty set.

Define

$$\begin{aligned} \Phi_0 &= S(X) \\ &\vdots \\ \Phi_n &= \Lambda_P(\Phi_0, \Phi_{n-1}) \\ &\vdots \end{aligned}$$

Let

$$\Phi(X) = \cup_{n\geq 0}\Phi_n.$$

Clearly, $P(\Phi(X)) \subset \Phi(X)$. If $u \in X \cup P(\Phi(X))$, then u is called prime. For any $u \in \Phi(X)$, u has a unique form $u = u_1u_2\cdots u_n$ where u_i is prime, $i = 1, 2, \ldots, n$, and u_i, u_{i+1} can not both have forms as $p(u_i')$ and $p(u_{i+1}')$. If this is the case, then we define the breath of u to be n, denoted by $bre(u) = n$.

For any $u \in \Phi(X)$ and for a set $T \subseteq X \cup \{P\}$, denote by $deg_T(u)$ the number of occurrences of $t \in T$ in u. Let

$$Deg(u) = (deg_{\{P\}\cup X}(u), deg_{\{P\}}(u)).$$

We order $Deg(u)$ lexicographically.

Let $k\Phi(X)$ be a free k-module with k-basis $\Phi(X)$ and $\lambda \in k$ a fixed element. Extend linearly $P: \ k\Phi(X) \to k\Phi(X), \ u \mapsto P(u)$ where $u \in \Phi(X)$.

Now we define the multiplication in $k\Phi(X)$.

Firstly, for $u, v \in X \cup P(\Phi(X))$, define

$$u \cdot v = \begin{cases} P(P(u') \cdot v') + P(u' \cdot P(v')) + \lambda P(u' \cdot v'), & \text{if } u = P(u'), v = P(v'); \\ uv, & \text{otherwise.} \end{cases}$$

Secondly, for any $u = u_1 u_2 \cdots u_s, v = v_1 v_2 \cdots v_l \in \Phi(X)$ where u_i, v_j are prime, $i = 1, 2, \ldots, s, j = 1, 2, \ldots, l$, define $u \cdot v = u_1 u_2 \cdots u_{s-1}(u_s \cdot v_1) v_2 \cdots v_l$.

Equipping with the above concepts, $k\Phi(X)$ is the free Rota-Baxter algebra with weight λ generated by X (see [48]), denoted by $RB(X)$.

We have to order $\Phi(X)$. Let X be a well-ordered set. Let us define an order $>$ on $\Phi(X)$ by induction on the Deg-function.

For any $u, v \in \Phi(X)$, if $Deg(u) > Deg(v)$, then $u > v$.

If $Deg(u) = Deg(v) = (n, m)$, then we define $u > v$ by induction on (n, m).

If $(n, m) = (1, 0)$, then $u, v \in X$ and we use the order on X. Suppose that for (n, m) the order is defined where $(n, m) \geq (1, 0)$. Let $(n, m) < (n', m') = Deg(u) = Deg(v)$. If $u, v \in P(\Phi(X))$, say $u = P(u')$ and $v = P(v')$, then $u > v$ if and only if $u' > v'$ by induction. Otherwise $u = u_1 u_2 \cdots u_l$ and $v = v_1 v_2 \cdots v_s$ where $l > 1$ or $s > 1$, then $u > v$ if and only if $(u_1, u_2, \ldots, u_l) > (v_1, v_2, \ldots, v_s)$ lexicographically by induction.

It is clear that $>$ is a monomial order on $\Phi(X)$. Throughout this subsection, we will use this order.

It is noted that not each S-word is normal.

Let $f, g \in RB(X)$ be monic with $\overline{f} = u_1 u_2 \cdots u_n$ where each u_i is prime. Then, there are four kinds of compositions.

(i) If $u_n \in P(\Phi(X))$, then we define composition of right multiplication as $f \cdot u$ where $u \in P(\Phi(X))$.

(ii) If $u_1 \in P(\Phi(X))$, then we define composition of left multiplication as $u \cdot f$ where $u \in P(\Phi(X))$.

(iii) If there exits a $w = \overline{f}a = b\overline{g}$ where fa is normal f-word and bg is normal g-word, $a, b \in \Phi(X)$ and $deg_{\{P\} \cup X}(w) < deg_{\{P\} \cup X}(\overline{f}) + deg_{\{P\} \cup X}(\overline{g})$, then we define the intersection composition of f and g with respect to w as $(f, g)_w = f \cdot a - b \cdot g$.

(iv) If there exists a $w = \overline{f} = u|_{\overline{g}}$ where $u \in \Phi^\star(X)$, then we define the inclusion composition of f and g with respect to w as $(f, g)_w = f - u|_g$.

Theorem 2.3.1 ([15], CD-lemma for Rota-Baxter algebras) *Let $RB(X)$ be a free Rota-Baxter algebra over a field of characteristic 0 and S a set of monic polynomials in $RB(X)$ and $>$ the monomial order on $\Phi(X)$ defined as before. Then the following statements are equivalent.*

(i) S is a Gröbner-Shirshov basis in $RB(X)$.

(ii) $f \in Id(S) \Rightarrow \bar{f} = u|_{\bar{s}}$ for some $u \in \Phi(X)$, $s \in S$.

(iii) $Irr(S) = \{u \in \Phi(X) | u \neq v|_{\bar{s}}, s \in S, v|_s$ *is normal s-word*$\}$ *is a k-basis of* $RB(X|S) = RB(X)/Id(S)$.

The following are some applications of the Theorem 2.3.1.

Theorem 2.3.2 *([15]) Let I be a well-ordered set,* $X = \{x_i | i \in I\}$ *and the order on* $\Phi(X)$ *defined as before. Let*

$$f = x_i x_j - x_j x_i, \quad i > j, \ i, j \in I \tag{4}$$

$$g = P(u)x_i - x_i P(u), \quad u \in \Phi(X), \ i \in I \tag{5}$$

Let S consist of (4) and (5). Then S is a Gröbner-Shirshov basis in $RB(X)$.

By using Theorems 2.3.1 and 2.3.2, we get a normal form $Irr(S)$ of the free commutative Rota-Baxter algebra $RB(X|S)$ which is the same as one in [53].

Theorem 2.3.3 *([15]) Every countably generated Rota-Baxter algebra with weight 0 can be embedded into a two-generated Rota-Baxter algebra.*

An important application of Theorem 2.3.1 is PBW theorem for dendriform algebra which is a conjecture of L. Guo [52].

A dendriform algebra (see [67]) is a k-space D with two binary operations $\prec$ and $\succ$ such that for any $x, y, z \in D$,

$$(x \prec y) \prec z = x \prec (y \prec z + y \succ z)$$
$$(x \succ y) \prec z = x \succ (y \prec z)$$
$$(x \prec y + x \succ y) \succ z = x \succ (y \succ z)$$

Suppose that $(D, \prec, \succ)$ is a dendriform algebra over k with a linear basis $X = \{x_i | i \in I\}$. Let $x_i \prec x_j = \{x_i \prec x_j\}, x_i \succ x_j = \{x_i \succ x_j\}$, where $\{x_i \prec x_j\}$ and $\{x_i \succ x_j\}$ are linear combinations of $x \in X$. Then D has an expression by generators and defining relations

$$D = D(X | x_i \prec x_j = \{x_i \prec x_j\}, x_i \succ x_j = \{x_i \succ x_j\}, x_i, x_j \in X).$$

Denote by

$$U(D) = RB(X | x_i P(x_j) = \{x_i \prec x_j\}, P(x_i)x_j = \{x_i \succ x_j\}, x_i, x_j \in X).$$

Then $U(D)$ is the universal enveloping Rota-Baxter algebra of D, see [48].

The following is the PBW theorem for dendriform algebras which is proved by Yuqun Chen, Qiuhui Mo [38].

Theorem 2.3.4 *([38]) Every dendriform algebra over a field of characteristic 0 can be embedded into its universal enveloping Rota-Baxter algebra.*

2.4 Right-symmetric (or pre-Lie) algebras

A non-associative A is called a right-symmetric (or pre-Lie) algebra if A satisfies the following identity $(x, y, z) = (x, z, y)$ for the associator $(x, y, z) = (xy)z - x(yz)$. It is a Lie admissible algebra in a sense that $A^{(-)} = (A, [xy] = xy - yx)$ is a Lie algebra.

Let $X = \{x_i | i \in I\}$ be a set and for any $(u) \in X^{**}$, $|(u)|$ the length of the word (u).

Let I be a well-ordered set. We order X^{**} by the induction on the lengths of the words (u) and (v):

(i) If $|((u)(v))| = 2$, then $(u) = x_i > (v) = x_j$ if and only if $i > j$.

(ii) If $|((u)(v))| > 2$, then $(u) > (v)$ if and only if one of the following cases holds:

  (a) $|(u)| > |(v)|$.

  (b) If $|(u)| = |(v)|$, $(u) = ((u_1)(u_2))$ and $(v) = ((v_1)(v_2))$, then $(u_1) > (v_1)$ or $((u_1) = (v_1)$ and $(u_2) > (v_2))$.

It is clear that the order $<$ on X^{**} is well-ordered which is called deg-lex order on non-associative words. We use this order throughout this subsection.

We now cite the definition of good words (see [78]) by induction on length:

1) x is a good word for any $x \in X$.

   Suppose that we define good words of length $< n$.

2) non-associative word $((v)(w))$ is called a good word if

  (a) both (v) and (w) are good words,

  (b) if $(v) = ((v_1)(v_2))$, then $(v_2) \leq (w)$.

We denote (u) by $[u]$, if (u) is a good word.

Let W be the set of all good words in the alphabet X and $RS\langle X\rangle$ the free right-symmetric algebra over a field k generated by X. Then W forms a linear basis of the free right-symmetric algebra $RS\langle X\rangle$, see [78]. Daniyar Kozybaev, Leonid Makar-Limanov, Ualbai Umirbaev [63] has proved that the deg-lex order on W is monomial.

Let $S \subset RS\langle X\rangle$ be a set of monic polynomials and $s \in S$. An S-word $(u)_s$ is called a normal S-word if $(u)_{\bar{s}} = (a\bar{s}b)$ is a good word.

Let $f, g \in S$, $[w] \in W$ and $a, b \in X^*$. Then there are two kinds of compositions.

(i) If $\bar{f} = [a\bar{g}b]$, then $(f, g)_{\bar{f}} = f - [agb]$ is called composition of inclusion.

(ii) If $(\bar{f}[w])$ is not good, then $f \cdot [w]$ is called composition of right multiplication.

Theorem 2.4.1 ([16], CD-lemma for right-symmetry algebras) *Let $S \subset RS\langle X\rangle$ be a nonempty set of monic polynomials and the order $<$ be defined as before. Then the following statements are equivalent.*

(i) S is a Gröbner-Shirshov basis in $RS\langle X\rangle$.

(ii) $f \in Id(S) \Rightarrow \bar{f} = [a\bar{s}b]$ *for some* $s \in S$ *and* $a, b \in X^*$, *where* $[asb]$ *is a normal* S*-word.*

(iii) $Irr(S) = \{[u] \in W | [u] \neq [a\bar{s}b]\ a, b \in X^*,\ s \in S$ *and* $[asb]$ *is a normal* S*-word*$\}$ *is a linear basis of the algebra* $RS\langle X|S\rangle = RS\langle X\rangle/Id(S)$.

As an application, we have a GSB for universal enveloping right-symmetric algebra of a Lie algebra.

Theorem 2.4.2 *([16]) Let* $(\mathcal{L}, [,])$ *be a Lie algebra with a well-ordered basis* $\{e_i |\ i \in I\}$. *Let* $[e_i, e_j] = \sum_m \alpha_{ij}^m e_m$, *where* $\alpha_{ij}^m \in k$. *We denote* $\sum_m \alpha_{ij}^m e_m$ *by* $\{e_i e_j\}$. *Let*

$$U(\mathcal{L}) = RS\langle \{e_i\}_I |\ e_i e_j - e_j e_i = \{e_i e_j\},\ i, j \in I\rangle$$

be the universal enveloping right-symmetric algebra of $\mathcal{L}$. *Let*

$$S = \{f_{ij} = e_i e_j - e_j e_i - \{e_i e_j\},\ i, j \in I \text{ and } i > j\}.$$

Then the set S *is a Gröbner-Shirshov basis in* $RS\langle X\rangle$ *where* $X = \{e_i\}_I$.

By Theorems 2.4.1 and 2.4.2, we immediately have the following PBW theorem for Lie algebra and right-symmetric algebra.

Corollary 2.4.3 *(D. Segal [78]) A Lie algebra* $\mathcal{L}$ *can be embedded into its universal enveloping right-symmetric algebra* $U(\mathcal{L})$ *as a subalgebra of* $U(\mathcal{L})^{(-)}$.

2.5 *L*-algebras

An L-algebra (see [66]) is a k-space L equipped with two binary k-linear operations $\prec, \succ: L \otimes L \to L$ verifying the so-called entanglement relation:

$$(x \succ y) \prec z = x \succ (y \prec z), \quad \forall x, y, z \in L.$$

Let $\Omega = \{\succ, \prec\} = \Omega_2$ and $\mathfrak{S}(X) = \bigcup_{n\geq 0} \mathfrak{S}_n$ (see §2.2). For any $\delta \in \Omega,\ x, y \in X$, we write $\delta(x, y) = x\delta y$.

An Ω-word u is an L-word if u is one of the following:

i) $u = x$, where $x \in X$.

ii) $u = v \succ w$, where v and w are L-words.

iii) $u = v \prec w$ with $v \neq v_1 \succ v_2$, where $v_1,\ v_2,\ v,\ w$ are L-words.

Let $L(X)$ be the free L-algebra generated by X. Then the set N of all L-words forms a normal form of $L(X)$, see [34], Corollary 3.3.

Let X be a well-ordered set. Denote $\succ$ by δ_1 and $\prec$ by δ_2. Let $\delta_1 < \delta_2$. For any $u \in \mathfrak{S}(X)$, if $u = x \in X$, denote by $wt(u) = (1, x)$; if $u = \delta_i(u_1, u_2)$ for some $u_1,\ u_2 \in \mathfrak{S}(X)$, denote by $wt(u) = (deg_X(u), \delta_i, u_1, u_2)$. For any $u, v \in \mathfrak{S}(X)$, define

$$u > v \Longleftrightarrow wt(u) > wt(v) \qquad \textit{lexicographically}$$

by induction on $|u| + |v|$.

It is clear that $>$ is a monomial order on N.

It is noted that not each S-word is normal.

In the paper [34], CD-lemma for Ω-algebras is established, where Ω consists of n-ary operations, $n \geq 1$. This generalizes the result in V. Drensky and R. Holtkamp [47]. As a result, the linear basis N (the set of all L-words) for the free L-algebra is obtained by using CD-lemma for Ω-algebras. Then the CD-lemma for L-algebras is given. There are two kinds compositions: inclusion composition and right multiplication composition. As applications, the following embedding theorems for L-algebras are obtained:

Theorem 2.5.1 *([34]) 1) Every countably generated L-algebra over a field k can be embedded into a two-generated L-algebra.*

2) Every L-algebra over a field k can be embedded into a simple L-algebra.

3) Every countably generated L-algebra over a countable field k can be embedded into a simple two-generated L-algebra.

4) Three arbitrary L-algebras A, B, C over a field k can be embedded into a simple L-algebra generated by B and C if $|k| \leq \dim(B * C)$ *and* $|A| \leq |B * C|$, *where* $B * C$ *is the free product of B and C.*

GSB of a free dialgebra and the free product of two L-algebras, respectively are also given, and then the normal forms of such algebras are obtained in [34].

2.6 Differential algebras

Let $\mathcal{D}$ be a set of symbols, A an associative algebra over K. Then A is a differential algebra with differential operators $\mathcal{D}$ or $\mathcal{D}$-algebra for short if for any $\delta \in \mathcal{D}$, $a, b \in A$,

$$\delta(ab) = \delta(a) \cdot b + a \cdot \delta(b).$$

Let $\mathcal{D} = \{D_j | j \in J\}$ and $\mathbb{N}$ the set of non-negative integers. For any $m \in \mathbb{N}$ and $\bar{j} = (j_1, \cdots, j_m) \in J^m$, denote by $D^{\bar{j}} = D_{j_1} D_{j_2} \cdots D_{j_m}$ and $D^{\omega}(X) = \{D^{\bar{j}}(x) | x \in X, \bar{j} \in J^m, m \in \mathbb{N}\}$, where $D^{\bar{j}}(x) = x$ if $\bar{j} \in J^0$. Let $T = (D^{\omega}(X))^*$ be the free monoid generated by $D^{\omega}(X)$.

Let $\mathcal{D}(X)$ be the free differential algebra with differential operators $\mathcal{D}$ generated by X. Then T is a linear basis of $\mathcal{D}(X)$, see [31].

Let X, J be well-ordered sets, $D^{\bar{i}}(x) = D_{i_1} D_{i_2} \cdots D_{i_m}(x) \in D^{\omega}(X)$ and

$$wt(D^{\bar{i}}(x)) = (x; m, i_1, i_2, \cdots, i_m).$$

Then, we order $D^{\omega}(X)$ as follows:

$$D^{\bar{i}}(x) > D^{\bar{j}}(y) \Longleftrightarrow wt(D^{\bar{i}}(x)) > wt(D^{\bar{j}}(y)) \text{ lexicographically.}$$

It is easy to check this order is a well ordering on $D^{\omega}(X)$.

Then deg-lex order on $T = (D^{\omega}(X))^*$ is monomial.

It is noted that each S-word is normal.

Let $f, g \in \mathcal{D}(X)$ be monic polynomials and $w, a, b \in T$. Then there are two kinds of compositions.

(i) There are two sorts of composition of inclusion:

If $w = \bar{f} = a \cdot d^{\bar{j}}(\bar{g}) \cdot b$, then the composition is $(f,g)_w = f - a \cdot D^{\bar{j}}(g) \cdot b$. If $w = d^{\bar{i}}(\bar{f}) = \bar{g} \cdot b$, then the composition is $(f,g)_w = D^{\bar{i}}(f) - g \cdot b$.

(ii) Composition of intersection:

If $w = \bar{f} \cdot b - a \cdot d^{\bar{j}}(\bar{g})$ such that $|\bar{f}| + |\bar{g}| > |w|$, then the composition is $(f,g)_w = f \cdot b - a \cdot D^{\bar{j}}(g)$. In this case, we assume that $a, b \neq 1$.

Theorem 2.6.1 ***([31], CD-lemma for differential algebras)*** *Let $\mathcal{D}(X)$ be the free differential algebra with differential operators $\mathcal{D} = \{D_j | j \in J\}$, $S \subset \mathcal{D}(X)$ a monic subset and $<$ the order on $T = (D^\omega(X))^*$ as before. Then the following statements are equivalent.*

(i) S is a Gröbner-Shirshov basis in $\mathcal{D}(X)$.

(ii) $f \in Id(S) \Rightarrow \bar{f} = a \cdot d^{\bar{i}}(\bar{s}) \cdot b$ for some $s \in S$, $\bar{i} \in J^m$, $m \in \mathbb{N}$ and $a, b \in T$.

(iii) $Irr(S) = \{u \in T \mid u \neq a \cdot d^{\bar{i}}(\bar{s}) \cdot b$ for all $s \in S$, $a, b \in T$, $\bar{i} \in J^m$, $m \in \mathbb{N}\}$ is a k-linear basis of $\mathcal{D}(X|S) = \mathcal{D}(X)/Id(S)$.

In the paper [31], as applications of Theorem 2.6.1, there are given GSB for free Lie-differential algebras and free commutative-differential algebras, respectively.

2.7 Non-associative algebras over a commutative algebra

Let K be a commutative associative $k-$algebra with unit, X a set and $K(X)$ the free non-associative algebra over K generated by X.

Let $[Y]$ denote the free abelian monoid generated by Y. Denote by

$$N = [Y]X^{**} = \{u = u^Y u^X | u^Y \in [Y], u^X \in X^{**}\}.$$

Then N is a k-linear basis of the "double" free non-associative algebra $k[Y](X) = k[Y] \otimes k(X)$ over $k[Y]$ generated by X.

Suppose that both $>_X$ and $>_Y$ are monomial orders on X^{**} and $[Y]$, respectively. For any $u = u^Y u^X, v = v^Y v^X \in N$, define

$$u > v \Leftrightarrow u^X >_X v^X \text{ or } (u^X = v^X \text{ and } u^Y >_Y v^Y).$$

It is obvious that $>$ is a monomial order on N.

It is noted that each S-word is normal.

Let f and g be monic polynomials of $k[Y](X)$, $w = w^Y w^X \in [Y]X^{**}$ and $a, b, c \in X^*$, where $w^Y = L(\bar{f}^Y, \bar{g}^Y) = L$ and $L(\bar{f}^Y, \bar{g}^Y)$ is the least common multiple of $\bar{f}^Y$ and $\bar{g}^Y$ in $k[Y]$. Then we have the following two kinds compositions.

(i) X-inclusion

If $w^X = \bar{f}^X = (a(\bar{g}^X)b)$, then $(f,g)_w = \frac{L}{\bar{f}^Y} f - \frac{L}{\bar{g}^Y}(a(g)b)$ is called the composition of X-inclusion.

(ii) Y-intersection only

If $|\bar{f}^Y| + |\bar{g}^Y| > |w^Y|$ and $w^X = (a(\bar{f}^X)b(\bar{g}^X)c)$, then $(f,g)_w = \frac{L}{\bar{f}^Y}(a(f)b(\bar{g}^X)c) - \frac{L}{\bar{g}^Y}(a(\bar{f}^X)b(g)c)$ is called the composition of Y-intersection only, where for $u \in [Y]$, $|u|$ means the degree of u.

A CD-lemma for non-associative algebras over commutative algebras is given ([36], Theorem 2.6). As an application, the following embedding theorem is obtained.

Theorem 2.7.1 *([36]) Each countably generated non-associative algebra over an arbitrary commutative algebra K can be embedded into a two-generated non-associative algebra over K.*

2.8 n-conformal algebras

Definition 2.8.1 *Let k be a field with characteristic 0 and C a vector space over k. Let Z_+ be the set of non-negative integers, Z the integer ring and n a positive integer number. We associate to each $\vec{m} = (m_1, \cdots, m_n) \in Z_+^n$ a bilinear product $\langle \vec{m} \rangle$ on C. Let $D_i : C \to C$ be linear mappings such that $D_iD_j = D_jD_i$, $1 \leqslant i, j \leqslant n$. Then C is an n-conformal algebra with derivations $D = \{D_1, \ldots, D_n\}$ if the following axioms are satisfied:*

(i) If $a, b \in C$, then there is an $\vec{N}(a,b) \in Z_+^n$ such that $a\langle \vec{m} \rangle b = 0$ if $\vec{m} \nprec \vec{N}(a,b)$, where for any $\vec{m} = (m_1, \cdots, m_n)$, $\vec{l} = (l_1, \cdots, l_n) \in Z_+^n$,

$$\vec{m} = (m_1, \cdots, m_n) \prec \vec{l} = (l_1, \cdots, l_n) \Leftrightarrow m_i \le l_i, \ i = 1, \ldots, n$$
$$\text{and there exists} \quad i_0, \ 1 \le i_0 \le n \quad \text{such that} \quad m_{i_0} < l_{i_0}.$$

$\vec{N} : C \times C \to Z_+^n$ is called the locality function.

(ii) For any $a, b \in C$, $\vec{m} \in Z_+^n$, $D_i(a\langle \vec{m} \rangle b) = D_i a\langle \vec{m} \rangle b + a\langle \vec{m} \rangle D_i b$, $i = 1, \ldots, n$.

(iii) For any $a, b \in C$, $\vec{m} \in Z_+^n$, $D_i a\langle \vec{m} \rangle b = -m_i a\langle \vec{m} - \vec{e_i} \rangle b$, $i = 1, \ldots, n$, where $\vec{e_i} = (\underbrace{0, \cdots, 0}_{i-1}, 1, 0, \cdots, 0)$.

For $\vec{m}, \vec{s} \in Z_+^n$, put $(-1)^{\vec{s}} = (-1)^{s_1 + \cdots + s_n}$ and $\binom{\vec{m}}{\vec{s}} = \binom{m_1}{s_1} \cdots \binom{m_n}{s_n}$.

An n-conformal algebra C is associative if in addition the associativity condition holds.
Associativity Condition: For any $a, b, c \in C$ and $\vec{m}, \vec{m}' \in Z_+^n$,

$$(a\langle \vec{m} \rangle b)\langle \vec{m}' \rangle c = \sum_{\vec{s} \in Z_+^n} (-1)^{\vec{s}} \binom{\vec{m}}{\vec{s}} a\langle \vec{m} - \vec{s} \rangle (b\langle \vec{m}' + \vec{s} \rangle c),$$

or equivalently a *right* analogy

$$a\langle \vec{m} \rangle (b\langle \vec{m}' \rangle c) = \sum_{\vec{s} \in Z_+^n} (-1)^{\vec{s}} \binom{\vec{m}}{\vec{s}} (a\langle \vec{m} - \vec{s} \rangle b)\langle \vec{m}' + \vec{s} \rangle c.$$

We fix a locality function $\overrightarrow{N}: B \times B \to Z_+^n$.

In the paper [21], by a close analogy with the paper by L. A. Bokut, Y. Fong, W.-F. Ke [24], the free associative n-conformal algebra is constructed and CD-lemma for associative n-conformal algebras is established. There are five kinds compositions.

Let $C(B, \overrightarrow{N}, D_1, \cdots, D_n)$ be the free associative n-conformal algebra generated by B with a fixed locality function $\overrightarrow{N}$. Then the set of elements of the form

$$[u] = a_1 \langle \overrightarrow{m}^{(1)} \rangle (a_2 \langle \overrightarrow{m}^{(2)} \rangle (\cdots (a_k \langle \overrightarrow{m}^{(k)} \rangle D^{\overrightarrow{i}} a_{k+1}) \cdots))$$

forms a linear basis of $C(B, \overrightarrow{N}, D_1, \cdots, D_n)$, where $a_l \in B$, $\overrightarrow{m}^{(j)} \in Z_+^n$, $\overrightarrow{m}^{(j)} \prec \overrightarrow{N}$, $1 \leqslant l \leqslant k+1, 1 \leqslant j \leqslant k$, ind(u)$= \overrightarrow{i} \in Z_+^n$, $|u| = k+1, k \geqslant 0$. We shall refer to $[u]$ as D-free if ind(u)$=(0, \cdots, 0)$, and we shall say that f is D-free if every normal word in f is D-free.

Theorem 2.8.2 ([21], CD-lemma for associative n-conformal algebras) *Let S be a Gröbner-Shirshov basis in $C(B, \overrightarrow{N}, D_1, \cdots, D_n)$. Then the set $Irr(S)$ forms a linear basis of the n-conformal algebra $C(B, \overrightarrow{N}, D_1, \cdots, D_n|S)$ with defining relations S.*

If S is D-free, then the converse is true as well.

Remark that the condition "the set S is a Gröbner-Shirshov basis" is not equivalent to "the set $Irr(S)$ forms a linear basis of the n-conformal algebra $C(B, \overrightarrow{N}, D_1, \cdots, D_n|S)$".

As an application of Theorem 2.8.2, there are constructed GSB for loop Lie n-conformal algebras presented by generators and defining relations in [21].

2.9 λ-differential associative algebras with multiple operators

A λ-differential associative algebra with multiple operators is a λ-differential algebra R with a set Ω of multi-linear operators.

Free λ-differential associative algebra on X with multiple operators Ω is constructed in [73].

There are two kinds of compositions: intersection and inclusion.

A CD-lemma for λ-differential associative algebras with multiple operators is established in [73]. As an application, a Gröbner-Shirshov basis for the free λ-differential Rota-Baxter algebra is given and then normal forms is obtained for such an algebra which is the same as one in [20].

2.10 Categories

Let $\mathcal{C}$ be a category. Let

$$\begin{aligned} k\mathcal{C} &= \{f = \sum_{i=1}^{n} \alpha_i \mu_i | \alpha_i \in k, \ \mu_i \in mor(\mathcal{C}), \ n \geq 0, \\ &\qquad \mu_i \ (0 \leq i \leq n) \text{ have the same domains and the same codomains}\}. \end{aligned}$$

Note that in kC, for $f,\ g \in kC,\ f+g$ is defined only if $f,\ g$ have the same domain and the same codomain.

A multiplication $\cdot$ in $k\mathcal{C}$ is defined by linearly extending the usual compositions of morphisms of the category $\mathcal{C}$. Then $(k\mathcal{C},\ \cdot)$ is called the category partial algebra over k relative to $\mathcal{C}$ and $kC(X)$ the free category partial algebra generated by the graph X.

Just like the case of associative algebras, a CD-lemma for categories is established, see [17], Theorem 4.4. As applications, they give GSB for the simplicial category and the cyclic category respectively, cf. [17], Theorems 5.1 and 5.4, see S. I. Gelfand, Y. I. Manin, [51], Chapter 3.1.

2.11 Metabelian Lie algebras

Suppose that L is a Lie algebra over a field. As usual we set $L^{(0)} = L,\ L^{(n+1)} = [L^{(n)}, L^{(n)}]$. Then $L/L^{(2)}$ is called a metabelian Lie algebra. More precisely, the variety of metabelian Lie algebras is given by the identity

$$(x_1x_2)(x_3x_4) = 0.$$

Let us begin with the construction of a free metabelian Lie algebra. Let X be a linear ordered set, $Lie(X)$ the free Lie algebra generated by X. Then $L_{(2)}(X) = Lie(X)/Lie(X)^{(2)}$ is the free metabelian Lie algebra generated by $X = \{a_i\}$. We call a monomial is left-normed if it is of the form $(\cdots((ab)c)\cdots)d$. In the sequel, the left-normed brackets in the expression is omitted without confusion. For an arbitrary set of indices $j_1, j_2, \cdots, j_m$, define

$$\langle a_{j_1} \cdots a_{j_m} \rangle = a_{i_1} \cdots a_{i_m},$$

where $a_{i_1} \leq \cdots \leq a_{i_m}$ and $i_1, i_2, \cdots, i_m$ is a permutation of the indices $j_1, j_2, \cdots, j_m$.

Let

$$R = \{u = a_0a_1a_2\cdots a_n \mid u \text{ is left-normed}, a_i \in X,\ a_0 > a_1 \leq a_2 \leq \cdots \leq a_n, n \geq 1\}$$

and $N = X \cup R$. Then N forms a linear basis of the free metabelian Lie algebra $L_{(2)}(X)$. Therefore, for any $f \in L_{(2)}(X)$, f has a presentation $f = f^{(1)} + f^{(0)}$, where $f^{(1)} \in kR$ and $f^{(0)} \in kX$.

Moreover, the multiplication of the elements of N is the following, $u \cdot v = 0$ if both $u, v \in R$, and

$$a_0a_1a_2\cdots a_n \cdot b = \begin{cases} a_0\langle a_1a_2\cdots a_nb\rangle & \text{if } a_1 \leq b, \\ a_0ba_1a_2\cdots a_n - a_1b\langle a_0a_2\cdots a_n\rangle & \text{if } a_1 > b. \end{cases}$$

If $u = a_0a_1\cdots a_n \in R$, then the words $a_i,\ a_0\langle a_{i_1}\cdots a_{i_s}\rangle$ $(s \leq n,\ a_{i_1}\cdots a_{i_s}$ is a subsequence of the sequence $a_1\cdots a_n)$ are called subwords of the word u.

Define the length of monomials in N:

$$|a| = 1,\ |a_0a_1a_2\cdots a_n| = n+1.$$

Now we order the set N degree-lexicographically.

Then we have $\overline{a_0a_1a_2\cdots a_n\cdot b} = a_0\langle a_1a_2\cdots a_nb\rangle$ and $|u{\cdot}b| = |u|+1$. For any $f \in L_{(2)}(X)$, we call f to be monic, (1)-monic and (0)-monic if the coefficients of $\bar{f}$, $\overline{f^{(1)}}$ and $\overline{f^{(0)}}$ is 1 respectively.

Let $S \subset L_{(2)}(X)$. Then the following two kinds of polynomial are called normal S-words:

(i) $sa_1a_2\cdots a_n$, where $a_1 \leq a_2 \leq \cdots \leq a_n$, $s \in S$, $\bar{s} \neq a_1$ and $n \geq 0$;

(ii) us, where $u \in R$, $s \in S$ and $\bar{s} \neq u$.

Let f and g be momic polynomials of $L_{(2)}(X)$. We define seven different types of compositions as follows:

1. If $\bar{f} = a_0a_1\cdots a_n$, $\bar{g} = a_0b_1\cdots b_m$, $(n,m \geq 0)$, then let $w = a_0\langle lcm(ab)\rangle$, where $lcm(ab)$ denotes the least common multiple of associative words $a_1\cdots a_n$ and $b_1\cdots b_m$. The composition of type I of f and g relative to w is defined by

$$C_I(f,g)_w = f\langle\frac{lcm(ab)}{a_1\cdots a_n}\rangle - g\langle\frac{lcm(ab)}{b_1\cdots b_m}\rangle.$$

2. If $\bar{f} = \overline{f^{(1)}} = a_0a_1\cdots a_n$, $\overline{g^{(0)}} = a_i$ for some $i \geq 2$ or $\overline{g^{(0)}} = a_1$ and $a_0 > a_2$, then let $w = \bar{f}$ and the composition of type II of f and g relative to w is defined by

$$C_{II}(f,g)_w = f - \alpha^{-1}a_0a_1\cdots\hat{a}_i\cdots a_n\cdot g,$$

where α is the coefficient of $\overline{g^{(0)}}$.

3. If $\bar{f} = \overline{f^{(1)}} = a_0a_1\cdots a_n$, $\bar{g} = \overline{g^{(0)}} = a_1$ and $a_0 \leq a_2$ or $n = 1$, then let $w = \bar{f}$ and the composition of type III of f and g relative to w is defined by

$$C_{III}(f,g)_{\bar{f}} = f + ga_0a_2\cdots a_n.$$

4. If $\bar{f} = \overline{f^{(1)}} = a_0a_1\cdots a_n$, $g^{(1)} \neq 0$, $\overline{g^{(0)}} = a_1$ and $a_0 \leq a_2$ or $n = 1$, then for any $a < a_0$ and $w = a_0\langle a_1\cdots a_na\rangle$, the composition of type IV of f and g relative to w is defined by

$$C_{IV}(f,g)_w = fa - \alpha^{-1}a_0aa_2\cdots a_n\cdot g,$$

where α is the coefficient of $\overline{g^{(0)}}$.

5. If $\bar{f} = \overline{f^{(1)}} = a_0a_1\cdots a_n$, $g^{(1)} \neq 0$ and $\overline{g^{(0)}} = b \notin \{a_i\}_{i=1}^n$, then let $w = a_o\langle a_1\cdots a_nb\rangle$ and the composition of type V of f and g relative to w is defined by

$$C_V(f,g)_w = fb - a_0a_1\cdots a_n\cdot g,$$

where α is the coefficient of $\overline{g^{(0)}}$.

6. If $\overline{f^{(0)}} = \overline{g^{(0)}} = a$ and $f^{(1)} \neq 0$, then for any $a_0a_1 \in R$ and $w = a_0\langle a_1a\rangle$, the composition of type VI of f and g relative to w is defined by

$$C_{VI}(f,g)_w = (a_0a_1)(\alpha^{-1}f - \beta^{-1}g),$$

where α and β are the coefficients of $\overline{f^{(0)}}$ and $\overline{g^{(0)}}$ respectively.

7. If $f^{(1)} \neq 0$, $g^{(1)} \neq 0$ and $\overline{f^{(0)}} = a > \overline{g^{(0)}} = b$, then for any $a_0 > a$ and $w = a_0ba$, the composition of type VII of f and g relative to w is defined by

$$C_{VII}(f,g)_w = \alpha^{-1}(a_0b)f - \beta^{-1}(a_0a)g,$$

where α and β are the coefficients of $\overline{f^{(0)}}$ and $\overline{g^{(0)}}$ respectively.

Then, in [30], CD-lemma for metabelian Lie algebras is given.

Remark: In V. V. Talapov [84], there is a mistake for compositions that is now modified.

2.12 S-act algebras

Let A be an associative algebra over k and S a monoid of linear operators on A. Then A is called an S-act algebra if A is an S-act with the action $s(a)$ satisfying

$$s(ab) = s(a)s(b), \quad s \in S, \ a, b \in A.$$

In the paper [87], the "double free" S-act algebra (i.e., a free S-act algebra, where S is a free semigroup) is constructed. Then Gröbner-Shirshov bases theory for S-act algebras is established, where S is an arbitrary semigroup. As an application, a Gröbner-Shirshov basis of free Chinese monoid-act algebra is given and hence a linear basis of free Chinese monoid-act algebra is obtained.

2.13 Operads

We present elements of the free operad by trees.

Let $\mathscr{V} = \bigcup_{n=1}^{\infty} \mathscr{V}_n$, where $\mathscr{V}_n = \{\delta_i^{(n)} | i \in I_n\}$ is the set of n-ary operations.

A tree with n leaves is called decorated if we label the leaves by $[n] = \{1, 2, 3, \ldots, n\}, n \in \mathbb{N}$ and each vertex by an element in $\mathscr{V}$.

A decorated tree is called tree monomial if for each vertex, the minimal value on the leaves of the left subtree is always less than that of the right subtree.

For example, $\delta_1^{(2)}(\delta_2^{(2)}(1,3),2)$, $\delta_1^{(2)}(1,\delta_2^{(2)}(2,3))$ are tree monomials, but $\delta_1^{(2)}(1,\delta_2^{(2)}(3,2))$ is not a tree monomial.

Let $\mathscr{F}_{\mathscr{V}}(n)$ be the set of all tree monomials with n leaves and $T = \cup_{n\geq 1}\mathscr{F}_{\mathscr{V}}(n)$. For any $\alpha = \alpha(x_1, \ldots, x_n) \in \mathscr{F}_{\mathscr{V}}(n)$, $\beta \in \mathscr{F}_{\mathscr{V}}(m)$, define the shuffle composition $\alpha \circ_{i,\sigma} \beta$ as follows

$$\alpha(x_1, \ldots, x_{i-1}, \beta(x_i, x_{\sigma(i+1)}, \ldots, x_{\sigma(i+m-1)}), x_{\sigma(i+m)}, \ldots, x_{\sigma(m+n-1)})$$

which is in $\mathscr{F}_{\mathscr{V}}(n+m-1)$, where $1 \leq i \leq n$ and the bijection $\sigma : \{i+1, \ldots, m+n-1\} \to \{i+1, \ldots, m+n-1\}$ is an $(m-1, n-i)$-shuffle; that is

$$\sigma(i+1) < \sigma(i+2) < \cdots < \sigma(i+m-1),$$
$$\sigma(i+m) < \sigma(i+m+1) < \cdots < \sigma(n+m-1).$$

For example,

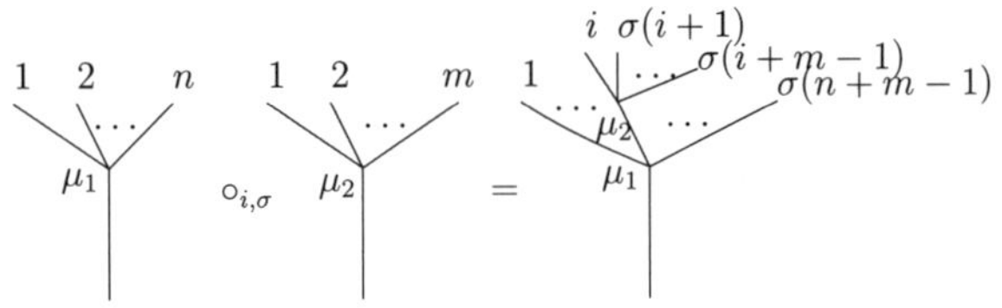

Then with the shuffle composition, the set T is freely generated by $\mathscr{V}$.

Let $\mathscr{F}_{\mathscr{V}} = kT$ be a k-space with k-basis T. Then with the shuffle compositions $\circ_{i,\sigma}$, $\mathscr{F}_{\mathscr{V}}$ is called the free shuffle operad.

Let S be a homogeneous subset of $\mathscr{F}_{\mathscr{V}}$, $s \in S$. The S-word $u|_s$ is defined as in §1 (in this case, the $\star$ depends on s). That means

(i) s is an S-word.

(ii) If $u|_s$ is an S-word, then for any $v \in T$, $u|_s \circ_{i,\sigma} v$ and $v \circ_{j,\tau} u|_s$ are both S-words if the shuffle compositions are defined.

It is clear that the ideal $Id(S)$ of $\mathscr{F}_{\mathscr{V}}$ is the set of linear combination of S-words.

A well ordering $>$ on T is called monomial (admissible) in the sense that

$$\alpha > \beta \Rightarrow u|_\alpha > u|_\beta \quad \text{for any } u \in T.$$

Suppose that T equips with a monomial order. Then each S-word is a normal S-word.

For example, the following order $>$ on T is monomial, see [45], Proposition 5.

For any $\alpha = \alpha(x_1, \ldots, x_n) \in \mathscr{F}_{\mathscr{V}}(n)$, α has an unique expression

$$\alpha = (path(1), \ldots, path(n), x_1, \ldots, x_n),$$

where each $path(i) \in \mathscr{V}^*$ is the unique path from the root to the leaf i. If this is the case, we denote by

$$wt(\alpha) = (n, path(1), \ldots, path(n), x_1, \ldots, x_n).$$

Let $\mathscr{V}$ be a well-ordered set. We order $\mathscr{V}^*$ by deg-lex order. We order numbers by inverse order, i.e., $x > y$ if and only if as natural number, x is less than y.

Now, for any $\alpha, \beta \in T$, we define

$$\alpha > \beta \Leftrightarrow wt(\alpha) > wt(\beta) \quad \textit{lexicographically}.$$

For $a, b \in T$, we say a is divisible by b, if there exists a subtree of the underlying tree of a for which the corresponding tree monomial a' is equal to b.

An element of $\mathscr{F}_{\mathscr{V}}$ is said to be homogeneous if all tree monomials that occur in this element with nonzero coefficients have the same arity degree (but not necessarily the same operation degree).

Definition 2.13.1 *Let f, g be two monic homogeneous elements of $\mathscr{F}_{\mathscr{V}}$. We define two types of compositions as follows:*

1. *If* $w=\bar{f}=u|_{\bar{g}}$ *for some g-word* $u|_g$, *then* $(f,g)_w=f-u|_g$ *is called the inclusion composition of* f *and* g.

2. *If* $w=a\circ_{i,\sigma}\bar{f}=\bar{g}\circ_{j,\tau}b$ *(or* $w=\bar{f}\circ_{i,\sigma}a=\bar{g}\circ_{j,\tau}b$*) for some* $a,b\in T$ *and the number of vertices of the underlying tree of* w *is less than the total number of vertices for* f *and* g, *then* $(f,g)_w=a\circ_{i,\sigma}f-g\circ_{j,\tau}b$ *(*$(f,g)_w=f\circ_{i,\sigma}a-g\circ_{j,\tau}b$*) is called the intersection composition of* f *and* g.

Theorem 2.13.2 ([45], Theorem 1, CD-lemma for shuffle operads) *Let the notations be as above.* $S\subset\mathscr{F}_{\mathscr{V}}$ *the nonempty set of monic homogeneous elements and* $<$ *a monomial order on* T. *Then the following statements are equivalent.*

(i) S *is a Gröbner-Shirshov basis in* $\mathscr{F}_{\mathscr{V}}$.

(ii) $f\in Id(S)\Rightarrow\bar{f}=u|_{\bar{s}}$ *for some S-word* $u|_s$.

(iii) $Irr(S)=\{u\in T|u\neq v|_{\bar{s}}$ *for any S-word* $v|_s\}$ *is a k-basis of* $\mathscr{F}_{\mathscr{V}}/Id(S)$.

As applications, in [45], the authors compute the Gröbner-Shirshov bases for some well-known operads, such as the operad Lie of Lie algebras, the operad As of associative algebras, the operad PreLie of pre-Lie algebras and so on.

In the paper of V. Dotsenko, M. V. Johansson [46], an implementation of the algorithm for computing Gröbner-Shirshov bases for operads is given.

3 Applications of known CD-lemmas

3.1 Artin-Markov normal form for braid groups in Artin-Burau generators

Let B_n denote the braid group of type $\mathbf{A_n}$. Then

$$B_n=gp\langle\sigma_1,\dots,\sigma_n\mid\sigma_j\sigma_i=\sigma_i\sigma_j\ (j-1>i),\ \sigma_{i+1}\sigma_i\sigma_{i+1}=\sigma_i\sigma_{i+1}\sigma_i,\ 1\leq i\leq n-1\rangle.$$

Let $X=Y\dot{\cup}Z$, Y^* and Z be well-ordered. Suppose that the order on Y^* is monomial. Then, any word in X has the form $u=u_0z_1u_1\cdots z_ku_k$, where $k\geq 0,\ u_i\in Y^*,\ z_i\in Z$. Define the inverse weight of the word $u\in X^*$ by

$$inwt(u)=(k,u_k,z_k,\cdots,u_1,z_1,u_0).$$

Now we order the inverse weights lexicographically as follows

$$u>v\Leftrightarrow inwt(u)>inwt(v).$$

Then we call the above order the inverse tower order. Clearly, this order is a monomial order on X^*.

In case $Y=T\dot{\cup}U$ and Y^* is endowed with the inverse tower order, we call the order of words in X the inverse tower order of words relative to the presentation

$$X=(T\dot{\cup}U)\dot{\cup}Z.$$

In general, we can define the inverse tower order of X-words relative to the presentation

$$X = (\cdots (X^{(n)} \dot{\cup} X^{(n-1)}) \dot{\cup} \cdots) \dot{\cup} X^{(0)},$$

where $X^{(n)}$-words are endowed by a monomial order.

In the braid group B_n, we now introduce a new set of generators which are called the Artin Burau generators. We set

$$s_{i,i+1} = \sigma_i^2, \quad s_{i,j+1} = \sigma_j \cdots \sigma_{i+1} \sigma_i^2 \sigma_{i+1}^{-1} \cdots \sigma_j^{-1}, \quad 1 \le i < j \le n-1;$$
$$\sigma_{i,j+1} = \sigma_i^{-1} \cdots \sigma_j^{-1}, \quad 1 \le i \le j \le n-1; \quad \sigma_{ii} = 1, \quad \{a, b\} = b^{-1}ab.$$

Form the set

$$S_j = \{s_{i,j}, s_{i,j}^{-1}, \ 1 \le i, j < n\} \ \text{and} \ \Sigma^{-1} = \{\sigma_1^{-1}, \cdots \sigma_{n-1}^{-1}\}.$$

Then the set

$$S = S_n \cup S_{n-1} \cup \cdots \cup S_2 \cup \Sigma^{-1}$$

generates B_n as a semigroup.

Now we order the set S in the following way:

$$S_n < S_{n-1} < \cdots < S_2 < \Sigma^{-1},$$

and

$$s_{1,j}^{-1} < s_{1,j} < s_{2,j}^{-1} < \cdots < s_{j-1,j} \ , \quad \sigma_1^{-1} < \sigma_2^{-1} < \cdots \sigma_{n-1}^{-1}.$$

With the above notation, we now order the S-words by using the inverse tower order, according to the fixed presentation of S as the union of S_j and Σ^{-1}. We order the S_n-words by the *deg − inlex* order, i.e., we first compare the words by length and then by inverse lexicographical order, starting from their last letters.

Lemma 3.1.1 *The following Artin-Markov relations hold in the braid group B_n. For $\delta = \pm 1$,*

$$\sigma_k^{-1} s_{i,j}^{\delta} = s_{i,j}^{\delta} \sigma_k^{-1}, k \neq i-1, i, j-1, j \tag{6}$$
$$\sigma_i^{-1} s_{i,i+1}^{\delta} = s_{i,i+1}^{\delta} \sigma_1^{-1} \tag{7}$$
$$\sigma_{i-1}^{-1} s_{i,j}^{\delta} = s_{i-1,j}^{\delta} \sigma_{i-1}^{-1} \tag{8}$$
$$\sigma_i^{-1} s_{i,j}^{\delta} = \{s_{i+1,j}^{\delta}, s_{i,i+1}\} \sigma_i^{-1} \tag{9}$$
$$\sigma_{j-1}^{-1} s_{i,j}^{\delta} = s_{i,j-1}^{\delta} \sigma_{j-1}^{-1} \tag{10}$$
$$\sigma_j^{-1} s_{i,j}^{\delta} = \{s_{i,j+1}^{\delta}, s_{j,j+1}\} \sigma_j^{-1} \tag{11}$$

for $i < j < k < l, \ \varepsilon = \pm 1,$

$$s_{j,k}^{-1} s_{k,l}^{\varepsilon} = \{s_{k,l}^{\varepsilon}, s_{j,l}^{-1}\} s_{j,k}^{-1} \tag{12}$$
$$s_{j,k} s_{k,l}^{\varepsilon} = \{s_{k,l}^{\varepsilon}, s_{j,l} s_{k,l}\} s_{j,k} \tag{13}$$
$$s_{j,k}^{-1} s_{j,l}^{\varepsilon} = \{s_{j,l}^{\varepsilon}, s_{k,l}^{-1} s_{j,l}^{-1}\} s_{j,k}^{-1} \tag{14}$$
$$s_{j,k} s_{j,l}^{\varepsilon} = \{s_{j,l}^{\varepsilon}, s_{k,l}\} s_{j,k} \tag{15}$$
$$s_{i,k}^{-1} s_{j,l}^{\varepsilon} = \{s_{j,l}^{\varepsilon}, s_{k,l} s_{i,l} s_{k,l}^{-1} s_{i,l}^{-1}\} s_{i,k}^{-1} \tag{16}$$
$$s_{i,k} s_{j,l}^{\varepsilon} = \{s_{j,l}^{\varepsilon}, s_{i,l}^{-1} s_{k,l}^{-1} s_{i,l} s_{k,l}\} s_{i,k} \tag{17}$$

for $j < i < k < l$ *or* $i < k < j < l$, *and* $\varepsilon,\ \delta = \pm 1$,

$$s_{i,k}^{\delta} s_{j,l}^{\varepsilon} = s_{j,l}^{\varepsilon} s_{i,k}^{\delta} \tag{18}$$

and

$$\sigma_j^{-1}\sigma_k^{-1} = \sigma_k^{-1}\sigma_j^{-1}, \quad j < k-1 \tag{19}$$

$$\sigma_{j,j+1}\sigma_{k,j+1} = \sigma_{k,j+1}\sigma_{j-1,j}\ , \quad j < k \tag{20}$$

$$\sigma_i^{-2} = s_{i,i+l}^{-1} \tag{21}$$

$$s_{i,j}^{\pm 1} s_{i,j}^{\mp 1} = 1 \tag{22}$$

Theorem 3.1.2 *([10]) The Artin-Markov relations (6)-(22) form a Gröbner-Shirshov basis of the braid group B_n in terms of the Artin-Burau generators relative to the inverse tower order of words.*

In the paper [10], it was claimed that some compositions are trivial. The paper [37] supported the claim and all compositions worked out explicitly.

3.2 Braid groups in Artin-Garside generators

The Artin-Garside generators of the braid group B_{n+1} are $\sigma_i,\ 1 \le i \le n,\ \triangle,\ \triangle^{-1}$ (Garside 1969), where $\triangle = \Lambda_1 \cdots \Lambda_n$, with $\Lambda_i = \sigma_1 \cdots \sigma_i$.

Let us order $\triangle^{-1} < \triangle < \sigma_1 < \cdots < \sigma_n$. We order $\{\triangle^{-1}, \triangle, \sigma_1, \dots, \sigma_n\}^*$ by deg-lex order.

By $V(j,i), W(j,i), \dots,$ where $j \le i$, we understand positive words in the letters $\sigma_j,\ \sigma_{j+1}, \dots, \sigma_i$. Also $V(i+1,i) = 1,\ W(i+1,i) = 1, \dots$.

Given $V = V(1,i)$, let $V^{(k)},\ 1 \le k \le n-i$ be the result of shifting in V all indices of all letters by k, $\sigma_1 \mapsto \sigma_{k+1}, \dots, \sigma_i \mapsto \sigma_{k+i}$, and we also use the notation $V^{(1)} = V'$. We write $\sigma_{ij} = \sigma_i\sigma_{i-1}\dots\sigma_j,\ j \le i-1,\ \sigma_{ii} = \sigma_i,\ \sigma_{ii+1} = 1$.

Theorem 3.2.1 *([8]) A Gröbner-Shirshov basis of B_{n+1} in the Artin-Garside generators consists of the following relations:*

$$\sigma_{i+1}\sigma_i V(1,i-1) W(j,i) \sigma_{i+1j} = \sigma_i\sigma_{i+1}\sigma_i V(1,i-1)\sigma_{ij} W(j,i)',$$
$$\sigma_s\sigma_k = \sigma_k\sigma_s,\ s-k \ge 2,$$
$$\sigma_1 V_1 \sigma_2\sigma_1 V_2 \cdots V_{n-1}\sigma_n \cdots \sigma_1 = \triangle V_1^{(n-1)} V_2^{(n-2)} \cdots V'_{(n-1)},$$
$$\sigma_l \triangle^{-1} = \triangle^{-1}\sigma_{n-l+1},\ 1 \le l \le n,$$
$$\triangle\triangle^{-1} = 1,\ \triangle^{-1}\triangle = 1,$$

where $1 \le i \le n-1$, $1 \le j \le i+1$, W begins with σ_i if it is not empty, and $V_i = V_i(1,i)$.

As results, we have the following corollaries.

Corollary 3.2.2 *The S-irreducible normal form of each word of B_{n+1} coincides with the Garside normal form of the word.*

Corollary 3.2.3 *(Garside (1969)) The semigroup of positive braids B_{n+1}^+ can be embedded into a group.*

3.3 Braid groups in Birman-Ko-Lee generators

Recall that the Birman-Ko-Lee generators σ_{ts} of the braid group B_n are the elements

$$\sigma_{ts} = (\sigma_{t-1}\sigma_{t-2}\dots\sigma_{s+1})\sigma_s(\sigma_{s+1}^{-1}\cdots\sigma_{t-2}^{-1})\sigma_{t-1}^{-1}.$$

Then B_n has an expression.

$$\begin{aligned} B_n &= gp\langle \sigma_{ts},\ n \geq t > s \geq 1 | \sigma_{ts}\sigma_{rq} = \sigma_{rq}\sigma_{ts},\ (t-r)(t-q)(s-r)(s-q) > 0, \\ &\qquad \sigma_{ts}\sigma_{sr} = \sigma_{tr}\sigma_{ts} = \sigma_{sr}\sigma_{tr}, n \geq t > s > r \geq 1 \rangle. \end{aligned}$$

Denote by $\delta = \sigma_{nn-1}\sigma_{n-1n-2}\cdots\sigma_{21}$.

Let us order $\delta^{-1} < \delta < \sigma_{ts} < \sigma_{rq}$ iff $(t,s) < (r,q)$ lexicographically. We order $\{\delta^{-1}, \delta, \sigma_{ts},\ n \geq t > s \geq 1\}^*$ by deg-lex order.

Instead of σ_{ij}, we write simply (i,j) or (j,i). We also set $(t_m, t_{m-1}, \dots, t_1) = (t_m, t_{m-1})(t_{m-1}, t_{m-2})\dots(t_2, t_1)$, where $t_j \neq t_{j+1},\ 1 \leq j \leq m-1$. In this notation, the defining relations of B_n can be written as

$$\begin{aligned} &(t_3, t_2, t_1) = (t_2, t_1, t_3) = (t_1, t_3, t_2),\ t_3 > t_2 > t_1, \\ &(k,l)(i,j) = (i,j)(k,l),\ k > l,\ i > j,\ k > i, \end{aligned}$$

where either $k > i > j > l$, or $k > l > i > j$.

Let us assume the following notation: $V_{[t_2,t_1]}$, where $n \geq t_2 > t_1 \geq 1$, is a positive word in (k,l) such that $t_2 \geq k > l \geq t_1$. We can use any capital Latin letter with indices instead of V, and any appropriate numbers (for example, t_3, t_0 such that $t_3 > t_0$) instead of t_2, t_1. We will also use the following notations: $V_{[t_2-1,t_1]}(t_2,t_1) = (t_2,t_1)V'_{[t_2-1,t_1]},\ t_2 > t_1$, where $V'_{[t_2-1,t_1]} = (V_{[t_2-1,t_1]})|_{(k,l)\mapsto(k,l),\ if\ l\neq t_1;\ (k,t_1)\mapsto(t_2,k)}$; $W_{[t_2-1,t_1]}(t_1,t_0) = (t_1,t_0)W^\star_{[t_2-1,t_1]},\ t_2 > t_1 > t_0$, where $W^\star_{[t_2-1,t_1]} = (W_{[t_2-1,t_1]})|_{(k,l)\mapsto(k,l),\ if\ l\neq t_1;\ (k,t_1)\mapsto(k,t_0)}$.

Theorem 3.3.1 *([9]) A Gröbner-Shirshov basis of the braid group B_{n+1} in the Birman-Ko-Lee generators consists of the following relations:*

$$\begin{aligned} &(k,l)(i,j) = (i,j)(k,l),\ k > l > i > j, \\ &(k,l)V_{[j-1,1]}(i,j) = (i,j)(k,l)V_{[j-1,1]},\ k > i > j > l, \\ &(t_3,t_2)(t_2,t_1) = (t_2,t_1)(t_3,t_1), \\ &(t_3,t_1)V_{[t_2-1,1]}(t_3,t_2) = (t_2,t_1)(t_3,t_1)V_{[t_2-1,1]}, \\ &(t,s)V_{[t_2-1,1]}(t_2,t_1)W_{[t_3-1,t_1]}(t_3,t_1) = (t_3,t_2)(t,s)V_{[t_2-1,1]}(t_2,t_1)W'_{[t_3-1,t_1]}, \\ &(t_3,s)V_{[t_2-1,1]}(t_2,t_1)W_{[t_3-1,t_1]}(t_3,t_1) = (t_2,s)(t_3,s)V_{[t_2-1,1]}(t_2,t_1)W'_{[t_3-1,t_1]}, \\ &(2,1)V_{2[2,1]}(3,1)\dots V_{n-1[n-1,1]}(n,1) = \delta V'_{2[2,1]}\cdots V'_{n-1[n-1,1]}, \\ &(t,s)\delta = \delta(t+1,s+1),\quad (t,s)\delta^{-1} = \delta^{-1}(t-1,s-1),\ t\pm 1,\ s\pm 1\ (modn), \\ &\delta\delta^{-1} = 1,\ \delta^{-1}\delta = 1, \end{aligned}$$

where $V_{[k,l]}$ means as before any word in (i,j) such that $k \geq i > j \geq l,\ t > t_3,\ t_2 > s$.

As results, we have the following corollaries.

Corollary 3.3.2 *The semigroup of positive braids BB_n^+ in Birman-Ko-Lee generators can be embedded into a group.*

Corollary 3.3.3 *The S-irreducible normal form of a word of B_n in Birman-Ko-Lee generators coincides with the Birman-Ko-Lee-Garside normal form $\delta^k A,\ A \in BB_n^+$ of the word.*

3.4 Braid groups in Adyan-Thurston generators

The symmetry group is as follow:

$$S_{n+1} = gp\langle s_1, \ldots, s_n \mid s_i^2 = 1, s_j s_i = s_i s_j \ (j-1 > i), s_{i+1}s_i s_{i+1} = s_i s_{i+1} s_i \rangle.$$

L. A. Bokut, L.-S. Shiao [27] found the normal form for S_{n+1} in the following theorem.

Theorem 3.4.1 *([27]) $N = \{s_{1i_1}s_{2i_2}\cdots s_{ni_n} | \ i_j \leq j+1\}$ is the Gröbner-Shirshov normal form for S_{n+1} in generators $s_i = (i, i+1)$ relative to the deg-lex order, where $s_{ji} = s_j s_{j-1}\cdots s_i \ (j \geq i), \ s_{jj+1} = 1$.*

Let $\alpha \in S_{n+1}$ and $\overline{\alpha} = s_{1i_1}s_{2i_2}\cdots s_{ni_n} \in N$ be the normal form of α. Define the length of α as $|\overline{\alpha}| = l(s_{1i_1}s_{2i_2}\cdots s_{ni_n})$ and $\alpha \perp \beta$ if $|\overline{\alpha\beta}| = |\overline{\alpha}| + |\overline{\beta}|$. Moreover, each $\overline{\alpha} \in N$ has a unique expression $\overline{\alpha} = s_{l_1 i_{l_1}} s_{l_2 i_{l_2}} \cdots s_{l_t i_{l_t}}$, where each $s_{l_j i_{l_j}} \neq 1$. Such a t is called the breath of α.

Now, we let

$$B'_{n+1} = gp\langle r(\overline{\alpha}), \ \alpha \in S_{n+1} \setminus \{1\} \mid r(\overline{\alpha})r(\overline{\beta}) = r(\overline{\alpha\beta}), \ \alpha \perp \beta \rangle,$$

where $r(\overline{\alpha})$ means a letter with the index $\overline{\alpha}$.

Then for the braid group with $n+1$ generators (see §3.1), $B_{n+1} \cong B'_{n+1}$. Indeed, define $\theta : B_{n+1} \to B'_{n+1}, \ \sigma_i \mapsto r(s_i)$ and $\theta' : B'_{n+1} \to B_{n+1}, \ r(\overline{\alpha}) \mapsto \overline{\alpha}|_{s_i \mapsto \sigma_i}$. Then two mappings are homomorphisms and $\theta\theta' = l_{B'_{n+1}}, \theta'\theta = l_{B_{n+1}}$. Hence,

$$B_{n+1} = gp\langle r(\overline{\alpha}), \ \alpha \in S_{n+1} \setminus \{1\} \mid r(\overline{\alpha})r(\overline{\beta}) = r(\overline{\alpha\beta}), \ \alpha \perp \beta \rangle.$$

Let $X = \{r(\overline{\alpha}), \ \alpha \in S_{n+1} \setminus \{1\}\}$. The generator X of B_{n+1} is called Adyan-Thurston generator.

Then the positive braid semigroup in generator X is

$$B^+_{n+1} = sgp\langle X \mid r(\overline{\alpha})r(\overline{\beta}) = r(\overline{\alpha\beta}), \ \alpha \perp \beta \rangle.$$

Let $s_1 < s_2 < \cdots < s_n$. Define $r(\overline{\alpha}) < r(\overline{\beta})$ if and only if $|\overline{\alpha}| > |\overline{\beta}|$ or $|\overline{\alpha}| = |\overline{\beta}|, \ \overline{\alpha} <_{lex} \overline{\beta}$. It is clear that such an order on X is well-ordered. We will use the deg-lex order on X^* in this subsection.

Theorem 3.4.2 *([42]) A Gröbner-Shirshov basis of B^+_{n+1} in Adyan-Thurston generator X relative to the deg-lex order on X^* is:*

$$\begin{aligned} r(\overline{\alpha})r(\overline{\beta}) &= r(\overline{\alpha\beta}), \quad \alpha \perp \beta, \\ r(\overline{\alpha})r(\overline{\beta\gamma}) &= r(\overline{\alpha\beta})r(\overline{\gamma}), \quad \alpha \perp \beta \perp \gamma. \end{aligned}$$

Theorem 3.4.3 *([42]) A Gröbner-Shirshov basis of B_{n+1} in Adyan-Thurston generator X relative to the deg-lex order on X^* is:*

1) $r(\overline{\alpha})r(\overline{\beta}) = r(\overline{\alpha\beta}), \quad \alpha \perp \beta,$
2) $r(\overline{\alpha})r(\overline{\beta\gamma}) = r(\overline{\alpha\beta})r(\overline{\gamma}), \quad \alpha \perp \beta \perp \gamma,$
3) $r(\overline{\alpha})\Delta^{\varepsilon} = \Delta^{\varepsilon} r(\overline{\alpha}'), \quad \overline{\alpha}' = \overline{\alpha}|_{s_i \mapsto s_{n+1-i}},$
4) $r(\overline{\alpha\beta})r(\overline{\gamma\mu}) = \Delta r(\overline{\alpha}')r(\overline{\mu}), \quad \alpha \perp \beta \perp \gamma \perp \mu, \ r(\overline{\beta\gamma}) = \Delta,$
5) $\Delta^{\varepsilon}\Delta^{-\varepsilon} = 1.$

3.5 GSB and normal forms for free inverse semigroups

We recall that an inverse semigroup is a semigroup in which every element a has a uniquely determined a^{-1} such that $aa^{-1}a = a$ and $a^{-1}aa^{-1} = a^{-1}$. Let $\mathcal{FI}(X)$ be a free inverse semigroup generated by a set X, $X^{-1} = \{x^{-1}|x \in X\}$ with $X \cap X^{-1} = \varnothing$. Denote $X \cup X^{-1}$ by Y. Then $\mathcal{FI}(X)$ has the following semigroup presentation

$$\mathcal{FI}(X) = sgp\langle Y|\ aa^{-1}a = a,\ aa^{-1}bb^{-1} = bb^{-1}aa^{-1},\ a, b \in Y^*\rangle$$

where $1^{-1} = 1$, $(x^{-1})^{-1} = x$ $(x \in X)$ and $(y_1y_2\cdots y_n)^{-1} = y_n^{-1}\cdots y_2^{-1}y_1^{-1}$ $(y_1, y_2, \cdots, y_n \in Y)$.

Let us assume that the set Y is well-ordered by an order $<$. Let $<$ be also the corresponding deg-lex order of Y^*. For any $u = y_1y_2\cdots y_n$ $(y_1, y_2, \cdots, y_n \in Y)$, let $fir(u) = y_1$.

We give inductively definitions in Y^* of an idempotent, canonical idempotent, prime canonical idempotent, ordered (prime) canonical idempotent and factors of a canonical idempotent, all of which but (prime) idempotent and ordered (prime) canonical idempotent are defined in O. Poliakova, B. M. Schein [71].

(i) The empty word 1 is an idempotent, a canonical idempotent, and an ordered canonical idempotent. This canonical idempotent has no factors.

(ii) If h is an idempotent and $x \in Y$, then $x^{-1}hx$ is both an idempotent and a prime idempotent. If h is a canonical idempotent, $x \in Y$ and the first letters of factors of h are different from x, then $x^{-1}hx$ is both a canonical idempotent and a prime canonical idempotent. This canonical idempotent is its own factor. Moreover, if the subword h in this canonical idempotent is an ordered canonical idempotent, then $x^{-1}hx$ is both an ordered canonical idempotent and an ordered prime canonical idempotent.

(iii) If $e_1, e_2, \cdots, e_m$ $(m > 1)$ are prime idempotents, then $e = e_1e_2\cdots e_m$ is an idempotent. Moreover, if $e_1, e_2, \cdots, e_m$ are prime canonical idempotents and their first letters are pairwise distinct, then $e = e_1e_2\cdots e_m$ is a canonical idempotent and $e_1, e_2, \cdots, e_m$ are factors of e. For this canonical idempotent, if $e_1, e_2, \cdots, e_m$ are ordered canonical idempotents and $e \leq e_{i_1}e_{i_2}\cdots e_{i_m}$ for any permutation $(i_1, i_2, \cdots, i_m)$ of $(1, 2, \cdots, m)$, then e is an ordered canonical idempotent.

Theorem 3.5.1 *([22]) Let X be a set, $X^{-1} = \{x^{-1}|x \in X\}$ with $X \cap X^{-1} = \varnothing$, $<$ a well ordering on $Y = X \cup X^{-1}$ and also the deg-lex order of Y^*, $\mathcal{FI}(X)$ the free inverse semigroup generated by X. Let $S \subset k\langle Y\rangle$ be the set of the following defining relations (a) and (b):*

(a) $ef = fe$, where both e and f are ordered prime canonical idempotents, ef is a canonical idempotent and $fe < ef$;

(b) $x^{-1}e'xf'x^{-1} = f'x^{-1}e'$, where $x \in Y$, both $x^{-1}e'x$ and $xf'x^{-1}$ are ordered prime canonical idempotents.

Then $\mathcal{FI}(X) = sgp\langle Y|S\rangle$ and S is a Gröbner-Shirshov basis in $k\langle Y\rangle$.

By Theorem 3.5.1 and CD-lemma for associative algebras, $Irr(S)$ is normal forms of the free inverse semigroup $\mathcal{FI}(X)$. It is easy to see that $Irr(S) = \{u \in (X \cup X^{-1})^*|u \neq a\bar{s}b, s \in S, a, b \in (X \cup X^{-1})^*\}$ consists of the word $u_0e_1u_1\cdots e_mu_m \in (X \cup X^{-1})^*$, where

$m \geq 0$, $u_1, \cdots, u_{m-1} \neq 1$, $u_0u_1 \cdots u_m$ has no subword of form yy^{-1} for $y \in X \cup X^{-1}$, $e_1, \cdots, e_m$ are ordered canonical idempotents, and the first (last, respectively) letters of the factors of e_i $(1 \leq i \leq m)$ are not equal to the first (last, respectively) letter of u_i (u_{i-1}, respectively). Thus $Irr(S)$ is a set of canonical words in the sense of [71], and different words in $Irr(S)$ represent different elements in $\mathcal{FI}(X)$.

3.6 Embeddings of algebras

By using CD-lemma for associative algebras, in [19], they give another proofs for the following theorems (G. Higman, B. H. Neumann, H. Neumann [55], A. I. Malcev [69], T. Evans [50]): every countably generated group (resp. associative algebra, semigroup) can be embedded into a two-generated group (resp. associative algebra, semigroup). Also some new results are proved.

Theorem 3.6.1 *([19]) (i) Every countably generated associative algebra over a countable field k can be embedded into a simple two-generated associative algebra.*

(ii) Every countably generated semigroup can be embedded into a (0-)simple two-generated semigroup.

By using CD-lemma for Lie algebras over a field (see §2.1), in [19], they give another proof of the Shirshov's theorem (A. I. Shirshov [80]): every countably generated Lie algebra can be embedded into a two-generated Lie algebra. Namely, let $L = Lie(X|S)$ is a Lie algebra generated by X with relations S, where $X = \{x_i, i = 1, 2, \dots\}$ and S is a GSB in the free Lie algebra $Lie(X)$ on deg-lex order. Let $H = Lie(X, a, b|S, [aab^iab] = x_i,\ i = 1, 2, \dots)$. Then $\{S, [aab^iab] = x_i,\ i = 1, 2, \dots\}$ is a GSB in $Lie(X, a, b)$ on deg-lex ordering with $a > b > x_i$. By CD-lemma for Lie algebras, L is a subalgebra of H which is generated by $\{a, b\}$.

Theorem 3.6.2 *([19]) Every every countably generated Lie algebra over a countable field k can be embedded into a simple two-generated Lie algebra.*

CD-lemma for associative differential algebras with unit is established in [31], see §2.6. By applying this lemma, they prove the following theorem.

Theorem 3.6.3 *([19]) (i) Every countably generated associative differential algebra can be embedded into a two-generated associative differential algebra.*

(ii) Any associative differential algebra can be embedded into a simple associative differential algebra.

(iii) Every countably generated associative differential algebra with countable set $\mathcal{D}$ of differential operations over a countable field k can be embedded into a simple two-generated associative differential algebra.

CD-lemma for associative algebra with multiple operations is established in [20], see §2.2. By applying this lemma, they prove the following theorems.

Theorem 3.6.4 *([19]) (i) Every countably generated associative Ω-algebra can be embedded into a two-generated associative Ω-algebra.*

(ii) Any associative Ω-algebra can be embedded into a simple associative Ω-algebras.

(iii) Each countably generated associative Ω-algebra with countable multiple operations Ω over a countable field k can be embedded into a simple two-generated associative Ω-algebra.

The definition of associative λ-differential algebra is mentioned in §2.2.

Theorem 3.6.5 *([19]) (i) Each countably generated associative λ-differential algebra can be embedded into a two-generated associative λ-differential algebra.*

(ii) Each associative λ-differential algebra can be embedded into a simple associative λ-differential algebra.

(iii) Each countably generated associative λ-differential algebra over a countable field k can be embedded into a simple two-generated associative λ-differential algebra.

3.7 Word problem for Novikov's and Boone's group

Let us call the set of letters $a_1, \cdots, a_n$ the principal alphabet and refer to the letters

$$q_1, \cdots, q_\lambda, r_1, \cdots, r_\lambda, l_1, \cdots, l_\lambda$$

as signal $(n, \lambda > 0)$. Append a copy to the previous alphabet, namely, the letters

$$q_1^+, \cdots, q_\lambda^+, r_1^+, \cdots, r_\lambda^+, l_1^+, \cdots, l_\lambda^+.$$

Consider the set $\{(A_i, B_i) | 1 \le i \le \lambda\}$ which is constituted by pairs of nonempty words in the principal alphabet. Consider the chain of the following groups: G_0, G_1, G_2 and $\mathcal{A}_{p_1p_2}$ given as follows.

$$\begin{aligned}
G_0 &= gp\langle\, q_i,\ r_i,\ q_i^+,\ r_i^+\ (1 \le i \le \lambda)\ |\emptyset\rangle \\
G_1 &= gp\langle\, G_0, a_j, a_j^+\ (\ 1 \le j \le n)\ |\ q_i a_j = a_j q_i q_i, \\
&\qquad r_i r_i a_j = a_j r_i,\ q_i^+ q_i^+ a_j^+ = a_j^+ q_i^+,\ r_i^+ a_j^+ = a_j^+ r_i^+ r_i^+\ \rangle \\
G_2 &= gp\langle\, G_1,\ l_i, l_i^+\ (\ 1 \le i \le \lambda\ )\ |\ a_j l_i = l_i a_j,\ a_j^+ l_i^+ = l_i^+ a_j^+\ \rangle \\
\mathcal{A}_{p_1p_2} &= gp\langle\, G_2,\ p_1,\ p_2\ |\ q_i^+ l_i^+ p_1 l_i q_i = A_i^+ p_1 A_i,\ r_i^+ p_1 r_i = p_1, \\
&\qquad r_i l_i p_2 l_i^+ r_i^+ = B_i p_2 B_i^+,\ q_i p_2 q_i^+ = p_2\ \rangle
\end{aligned}$$

Then $\mathcal{A}_{p_1p_2}$ is called the Novikov group.

Now, the principal alphabet is constituted by letters s_j's and q_j's, $1 \le j \le m$. The letters x, y, l_i's, and r_i's, $1 \le i \le n$, form an auxiliary alphabet. Denote by $\{(\Sigma_i, \Gamma_i),\ 1 \le i \le n\}$ the set of pairs of special words, i.e., the words of the form sq_js', where s and s' are words in the alphabet $\{s_j\}$. Designate $q_1 = q$. The Boone group $G(T, q)$ is given by the above generators and the following defining relations:

(i) $y^2 s_j = s_j y,\ \ x s_j = s_j x^2$;

(ii) $s_j l_i = y l_i y s_j, \quad s_j x r_i x = r_i s_j$;

(iii) $l_i \Gamma_i r_i = \Sigma_i$;

(iv) $l_i t = t l_i, \quad yt = ty$;

(v) $r_i k = k r_i, \quad xk = kx$;

(vi) $q^{-1} t q k = k q^{-1} t q$.

In the paper [32], direct proofs of GSB for the Novikov group and Boone group are given, respectively. Then the corresponding normal forms are obtained. These proofs are different from those in L. A. Bokut [5, 6].

3.8 PBW basis of $U_q(A_N)$

Let $A = (a_{ij})$ be an integral symmetrizable $N \times N$ Cartan matrix so that $a_{ii} = 2$, $a_{ij} \leq 0$ $(i \neq j)$ and there exists a diagonal matrix D with diagonal entries d_i which are nonzero integers such that the product DA is symmetric. Let q be a nonzero element of k such that $q^{4d_i} \neq 1$ for each i. Then the quantum enveloping algebra is

$$U_q(A) = k\langle X \cup H \cup Y | S^+ \cup K \cup T \cup S^- \rangle,$$

where

$$\begin{aligned}
X &= \{x_i\}, \\
H &= \{h_i^{\pm 1}\}, \\
Y &= \{y_i\}, \\
S^+ &= \{\sum_{\nu=0}^{1-a_{ij}} (-1)^\nu \begin{pmatrix} 1-a_{ij} \\ \nu \end{pmatrix}_t x_i^{1-a_{ij}-\nu} x_j x_i^\nu, \ \mathit{where}\ i \neq j,\ t = q^{2d_i}\}, \\
S^- &= \{\sum_{\nu=0}^{1-a_{ij}} (-1)^\nu \begin{pmatrix} 1-a_{ij} \\ \nu \end{pmatrix}_t y_i^{1-a_{ij}-\nu} y_j y_i^\nu, \ \mathit{where}\ i \neq j,\ t = q^{2d_i}\}, \\
K &= \{h_i h_j - h_j h_i, h_i h_i^{-1} - 1, h_i^{-1} h_i - 1, x_j h_i^{\pm 1} - q^{\mp 1} d_i a_{ij} h^{\pm 1} x_j, h_i^{\pm 1} y_j - q^{\mp 1} y_j h^{\pm 1}\}, \\
T &= \{x_i y_j - y_j x_i - \delta_{ij} \frac{h_i^2 - h_i^{-2}}{q^{2d_i} - q^{-2d_i}}\} \quad \text{and}
\end{aligned}$$

$$\begin{pmatrix} m \\ n \end{pmatrix}_t = \begin{cases} \prod_{i=1}^{n} \frac{t^{m-i+1} - t^{i-m-1}}{t^i - t^{-i}} & (\mathit{for}\ m > n > 0), \\ 1 & (\mathit{for}\ n = 0\ \mathit{or}\ m = n). \end{cases}$$

Let

$$A = A_N = \begin{pmatrix} 2 & -1 & 0 & \cdots & 0 \\ -1 & 2 & -1 & \cdots & 0 \\ 0 & -1 & 2 & \cdots & 0 \\ \cdot & \cdot & \cdot & \cdot & \cdot \\ 0 & 0 & 0 & \cdots & 2 \end{pmatrix} \quad \text{and} \quad q^8 \neq 1.$$

We introduce some new variables defined by Jimbo (see [85]) which generate $U_q(A_N)$:

$$\widetilde{X} = \{x_{ij}, 1 \leq i < j \leq N+1\},$$

where

$$x_{ij} = \begin{cases} x_i & j = i+1, \\ qx_{i,j-1}x_{j-1,j} - q^{-1}x_{j-1,j}x_{i,j-1} & j > i+1. \end{cases}$$

We now order the set $\widetilde{X}$ in the following way.

$$x_{mn} > x_{ij} \Longleftrightarrow (m,n) >_{lex} (i,j).$$

Let us recall from Yamane [85] the following notation:

$$\begin{aligned}
C_1 &= \{((i,j),(m,n)) | i = m < j < n\}, \\
C_2 &= \{((i,j),(m,n)) | i < m < n < j\}, \\
C_3 &= \{((i,j),(m,n)) | i < m < j = n\}, \\
C_4 &= \{((i,j),(m,n)) | i < m < j < n\}, \\
C_5 &= \{((i,j),(m,n)) | i < j = m < n\}, \\
C_6 &= \{((i,j),(m,n)) | i < j < m < n\}.
\end{aligned}$$

Let the set $\widetilde{S}^+$ consist of Jimbo relations:

$$\begin{array}{lll}
x_{mn}x_{ij} & - & q^{-2}x_{ij}x_{mn} \qquad ((i,j),(m,n)) \in C_1 \cup C_3, \\
x_{mn}x_{ij} & - & x_{ij}x_{mn} \qquad ((i,j),(m,n)) \in C_2 \cup C_6, \\
x_{mn}x_{ij} & - & x_{ij}x_{mn} + (q^2 - q^{-2})x_{in}x_{mj} \qquad ((i,j),(m,n)) \in C_4, \\
x_{mn}x_{ij} & - & q^2x_{ij}x_{mn} + qx_{in} \qquad ((i,j),(m,n)) \in C_5.
\end{array}$$

It is easily seen that $U_q^+(A_N) = k\langle \widetilde{X} | \widetilde{S^+} \rangle$.

In the paper [40], a direct proof is given that $\widetilde{S}^+$ is a Gröbner-Shirshov basis for $k\langle \widetilde{X} | \widetilde{S^+} \rangle = U_q^+(A_N)$ ([26]). The proof is different from one in L. A. Bokut, P. Malcolmson [26].

3.9 GSB for free dendriform algebras

The L-algebra and the dendriform algebra are mentioned in §2.5 and §2.3, respectively.

Let $L(X)$ be the free L-algebra generated by a set X and N the set of all L-words, see §2.5. Let the order on N be as in §2.5.

It is clear that the free dendriform algebra generated by X, denoted by $DD(X)$, has an expression

$$\begin{aligned}
L(X \mid\ & (x \prec y) \prec z = x \prec (y \prec z) + x \prec (y \succ z), \\
& x \succ (y \succ z) = (x \succ y) \succ z + (x \prec y) \succ z,\ x, y, z \in N).
\end{aligned}$$

The following theorem gives a GSB for $DD(X)$.

Theorem 3.9.1 *([41]) Let the order on N be as before. Let*

$$\begin{aligned} f_1(x,y,z) &= (x \prec y) \prec z - x \prec (y \prec z) - x \prec (y \succ z), \\ f_2(x,y,z) &= (x \prec y) \succ z + (x \succ y) \succ z - x \succ (y \succ z), \\ f_3(x,y,z,v) &= ((x \succ y) \succ z) \succ v - (x \succ y) \succ (z \succ v) + (x \succ (y \prec z)) \succ v. \end{aligned}$$

Then, $S = \{f_1(x,y,z),\ f_2(x,y,z),\ f_3(x,y,z,v) |\ x,y,z,v \in N\}$ is a Gröbner-Shirshov basis in $L(X)$.

Then, by using the CD-lemma for L-algebras, see [34], $Irr(S)$ is normal words of free dendriform algebra $DD(X)$.

3.10 Anti-commutative GSB of a free Lie algebra relative to Lyndon-Shirshov words

Let X be a well-ordered set, $>_{lex}$, $>_{dex-lex}$ lexicographical order and degree-lexicographical order on X^* respectively.

Let $>_{n-lex}$ be the nonassociative lexicographical order on X^{**}. For example, let $X = \{x_i | 1 \leq i \leq n \text{ and } x_1 < x_2 < \ldots < x_n\}$. Then $x_1(x_2x_3) >_{n-lex} (x_1x_2)x_3$ since $x_1 >_{n-lex} (x_1x_2)$.

Let $>_{n-dex-lex}$ be the nonassociative degree-lexicographical order on X^{**}. For example, $x_1(x_2x_3) <_{n-deg-lex} (x_1x_2)x_3$ since $x_1 <_{n-deg-lex} (x_1x_2)$ (the degree of x_1 is one and the degree of (x_2x_3) is two).

An order $\succ_1$ on X^{**}:

$$(u) \succ_1 (v) \Leftrightarrow \textit{either} \quad u >_{lex} v \quad \textit{or} \quad (u = v \quad \textit{and} \quad (u) >_{n-dex-lex} (v)).$$

An order $\succ_2$ on X^{**}:

$$(u) \succ_2 (v) \Leftrightarrow \textit{either} \quad u >_{deg-lex} v \quad \textit{or} \quad (u = v \quad \textit{and} \quad (u) >_{n-dex-lex} (v)).$$

We define normal words $(u) \in X^{**}$ by induction:

(i) $x_i \in X$ is a normal word.

(ii) $(u) = (u_1)(u_2)$ is normal if and only if both (u_1) and (u_2) are normal and $(u_1) \succ_1 (u_2)$.

Denote N the set of all normal words $[u]$ on X.

Let $AC(X)$ be a k-space spanned by N. Now define the product of normal words by the following way: for any $[u],[v] \in N$,

$$[u][v] = \begin{cases} [[u][v]], & \text{if } [u] \succ_1 [v] \\ -[[v][u]], & \text{if } [u] \prec_1 [v] \\ 0, & \text{if } [u] = [v] \end{cases}$$

Remark By definition, for any $(u) \in X^{**}$, there exists a unique $[v] \in N$ such that, in $AC(X)$, $(u) = \pm[v]$ or 0.

Then $AC(X)$ is a free anti-commutative algebra generated by X.

The order $\succ_2$ is a monomial order on N.

Shirshov [81] proved a CD-lemma for free anti-commutative algebra $AC(X)$, where there is only one composition: inclusion. By using the above order $\succ_2$ on N, Shirshov's ([81]) CD-lemma holds.

Theorem 3.10.1 *([18]) Let the order $\succ_2$ on N be defined as before and*

$$S = \{(([[u]][[v]])[[w]] - ([[u]][[w]])[[v]] - [[u]]([[v]][[w]]) \mid u >_{lex} v >_{lex} w \text{ and } [[u]], [[v]], [[w]] \text{ are } NLSWs\}.$$

Then S is a Gröbner-Shirshov basis in $AC(X)$.

By CD-lemma for anti-commutative algebras, $Irr(S) = NLSW(X)$ is linear basis of the free Lie algebra $Lie(X)$ which is due to A. I. Shirshov [80] and K.-T. Chen, R. Fox, R. Lyndon [33].

3.11 Free partially commutative groups, associative algebras and Lie algebras

Let X be a set, $Free(X)$ the free algebra over a field k generated by X, for example, free monoid, group, associative algebra over a field, Lie algebra and so on. Let $\vartheta \subseteq (X \times X) \backslash \{(x, x) \mid x \in X\}$. Then $Free(X|R) = Free(X)/Id(R)$ with generators X and defining relations $R = \{ab = ba,\ (a, b) \in \vartheta\}$ (in the Lie case $R = \{(a, b) = 0,\ (a, b) \in \vartheta\}$) is the free partially commutative algebra.

Denote by

— $gp\langle X|\vartheta\rangle$ the free partially commutative group.

— $sgp\langle X|\vartheta\rangle$ the free partially commutative monid.

— $k\langle X|\vartheta\rangle$ the free partially commutative associative algebra.

— $Lie(X|\vartheta)$ the free partially commutative Lie algebra.

Let $<$ be a well ordering on X. Throughout this subsection, if $a > b$ and $(a, b) \in \vartheta$ or $(b, a) \in \vartheta$, we denote $a \rhd b$. Generally, for any set Y, $a \rhd Y$ means $a \rhd y$ for any $y \in Y$. For any $u = x_{i_1} \cdots x_{i_n} \in X^*$ where $x_{i_j} \in X$, we denote the set $\{x_{i_j}, j = 1, \ldots, n\}$ by $supp(u)$.

For free partially commutative group $gp\langle X|\vartheta\rangle$, let $\prec$ be a well ordering on X. We extend this order to $X \cup X^{-1}$ as follows: for any $x, y \in X$, $\varepsilon, \eta = \pm 1$

(i) $x > x^{-1}$, if $x \in X$;

(ii) $x^{\varepsilon} > y^{\eta} \Longleftrightarrow x \succ y$.

For any $a, b \in X$, if $a^{\varepsilon} > b^{\eta}$ $(\varepsilon, \eta = \pm 1)$ and $(a, b) \in \vartheta$ or $(b, a) \in \vartheta$, we denote $a^{\varepsilon} \rhd b^{\eta}$. Moreover, for any set Y, $a \rhd Y$ means $a \rhd y$ for any $y \in Y$.

It is obvious that

$$gp\langle X|\vartheta\rangle = sgp\langle X \cup X^{-1}|x^{\varepsilon}x^{-\varepsilon} = 1,\ a^{\varepsilon}b^{\eta} = b^{\eta}a^{\varepsilon},\ x \in X,\ (a, b) \in \vartheta,\ \varepsilon, \eta = \pm 1\rangle.$$

By using CD-lemma for associative algebras, the following results are proved in [39].

Theorem 3.11.1 *([27]) With the deg-lex order on X^*, the set $S = \{xuy - yxu \mid x, y \in X, u \in X^*, x \rhd y \rhd supp(u)\}$ forms a Gröbner-Shirshov basis of the free partially commutative associative algebra $k\langle X|\vartheta\rangle$. As a result, $Irr(S) = \{u \in X^* | u \neq a\bar{s}b, s \in S, a, b \in X^*\}$ is a k-basis of $k\langle X|\vartheta\rangle$. Then, $Irr(S)$ is also a normal form of the free partially commutative monoid $sgp\langle X|\vartheta\rangle$.*

Theorem 3.11.2 *([49]) Let the notations be as the above. Then with the deg-lex order on $(X \cup X^{-1})^*$, the set $S = \{x^{\varepsilon}uy^{\eta} = y^{\eta}x^{\varepsilon}u,\ z^{\gamma}z^{-\gamma} = 1 \mid \varepsilon, \eta, \gamma = \pm 1,\ x, y, z \in X \cup X^{-1},\ u \in (X \cup X^{-1})^*,\ x^{\varepsilon} \rhd y^{\eta} \rhd supp(u)\}$ forms a Gröbner-Shirshov basis of the free partially commutative group $gp\langle X|\vartheta\rangle$. As a result, $Irr(S) = \{u \in (X \cup X^{-1})^* | u \neq a\bar{s}b, s \in S, a, b \in (X \cup X^{-1})^*\}$ is a normal form of $gp\langle X|\vartheta\rangle$.*

By using CD-lemma for Lie algebras, we have

Theorem 3.11.3 *([39]) With deg-lex order on X^*, the set $S = \{[xuy] \mid x, y, z \in X,\ u \in X^*, x \rhd y \rhd supp(u)\}$ forms a Gröbner-Shirshov basis of the partially commutative Lie algebra $Lie(X|\vartheta)$. As a result, $Irr(S) = \{[u] \in NLSW(X) \mid u \neq a\bar{s}b,\ s \in S,\ a, b \in X^*\}$ is a k-basis of $Lie(X|\vartheta)$.*

Theorem 3.11.3 is also proved by E. Poroshenko [72].

3.12 Plactic monoids in row generators

Let $X = \{x_1, \ldots, x_n\}$ be a set of n elements with the order $x_1 < \cdots < x_n$. The plactic monoid (see M. Lothaire [68], Chapter 5) on the alphabet X is $P_n = sgp\langle X|T\rangle$, where T consists of the Knuth relations

$$x_i x_k x_j = x_k x_i x_j \quad (x_i \leq x_j < x_k),$$
$$x_j x_i x_k = x_j x_k x_i \quad (x_i < x_j \leq x_k).$$

A nondecreasing word $R \in X^*$ is called a row, for example, $x_1x_1x_3x_5x_5x_5x_6$ is a row when $n \geq 6$. Let Y be the set of all rows in X^*. For any $R = r_1 \ldots r_u$, $S = s_1 \ldots s_v \in Y$, $r_i, s_j \in X$, we say R dominates S if $u \leq v$ and for $i = 1, \ldots, u$, $r_i > s_i$.

The multiplication of two rows is defined by Schensted's algorithm: for a row R and $x \in X$,

$$R \cdot x = \begin{cases} Rx, & if\ Rx\ is\ a\ row \\ y \cdot R', & otherwise \end{cases}$$

where y is the leftmost letter in R which is strictly larger than x, and $R' = R|_{y \mapsto x}$.

Then, for any $R, S \in Y$, it is clear that there exist uniquely $R', S' \in Y$ such that $R \cdot S = R' \cdot S'$ in P_n, where R' dominates S', i.e., $|R'| \leq |S'|$ and each letter of R' is larger than the corresponding letter of S'.

We express the P_n as follows:

$$P_n = sgp\langle Y | R \cdot S = R' \cdot S', \ R, S \in Y \rangle.$$

For any $R, S \in Y$, we define $R > S$ by deg-lex order on X^*. Then we have a well ordering on Y.

It follows from CD-lemma for associative algebras that with the deg-lex order on Y^*, the set $\{R \cdot S = R' \cdot S' | R, S \in Y\}$ is a GSB in $k\langle Y \rangle$, see M. Lothaire [68], Chapter 5. For $n = 4$, we give a direct proof. Namely, any composition $(f, g)_w = (RS)T - R(ST)$ is trivial, where $f = RS - R'S'$, $g = ST - S''T'$, $w = RST$.

Let $R = 1^{r_1}2^{r_2}3^{r_3}4^{r_4}$ be a row in P_4, where each r_i is non-negative integer. For convenience, denote by $R = (r_1, r_2, r_3, r_4)$. Then for three rows $R = (r_1, r_2, r_3, r_4)$, $S = (s_1, s_2, s_3, s_4)$, $T = (t_1, t_2, t_3, t_4)$, we have

$$\begin{aligned}
R(ST) &= \left(\begin{array}{cccc} r_1 & r_2 & r_3 & r_4 \\ \multicolumn{4}{c}{\left(\begin{array}{cccc} s_1 & s_2 & s_3 & s_4 \\ t_1 & t_2 & t_3 & t_4 \end{array} \right)} \end{array} \right) = \left(\begin{array}{c} \left(\begin{array}{cccc} r_1 & r_2 & r_3 & r_4 \\ 0 & a_2 & a_3 & a_4 \end{array} \right) \\ \begin{array}{cccc} b_1 & b_2 & b_3 & b_4 \end{array} \end{array} \right) \\
&= \left(\begin{array}{c} \begin{array}{cccc} 0 & 0 & c_3 & c_4 \end{array} \\ \left(\begin{array}{cccc} d_1 & d_2 & d_3 & d_4 \\ b_1 & b_2 & b_3 & b_4 \end{array} \right) \end{array} \right) = \left(\begin{array}{cccc} 0 & 0 & c_3 & c_4 \\ 0 & e_2 & e_3 & e_4 \\ f_1 & f_2 & f_3 & f_4 \end{array} \right) \\
(RS)T &= \left(\begin{array}{c} \left(\begin{array}{cccc} r_1 & r_2 & r_3 & r_4 \\ s_1 & s_2 & s_3 & s_4 \end{array} \right) \\ \begin{array}{cccc} t_1 & t_2 & t_3 & t_4 \end{array} \end{array} \right) = \left(\begin{array}{c} \begin{array}{cccc} 0 & g_2 & g_3 & g_4 \end{array} \\ \left(\begin{array}{cccc} h_1 & h_2 & h_3 & h_4 \\ t_1 & t_2 & t_3 & t_4 \end{array} \right) \end{array} \right) \\
&= \left(\begin{array}{c} \left(\begin{array}{cccc} 0 & g_2 & g_3 & g_4 \\ 0 & i_2 & i_3 & i_4 \end{array} \right) \\ \begin{array}{cccc} j_1 & j_2 & j_3 & j_4 \end{array} \end{array} \right) = \left(\begin{array}{cccc} 0 & 0 & k_3 & k_4 \\ 0 & l_2 & l_3 & l_4 \\ j_1 & j_2 & j_3 & j_4 \end{array} \right)
\end{aligned}$$

Now, one needs only to prove $l_2 = e_2$, $l_3 = e_3$, $l_4 = e_4$, $c_3 = k_3$ and $c_4 = k_4$. Let us mention that it takes almost 20 pages to calculate.

As the result, in [35], a new direct elementary proof (but not simple) of the following theorem is given.

Theorem 3.12.1 *Let the notations be as above. Then with deg-lex order on* Y^*, $\{R \cdot S = R' \cdot S' | R, S \in Y\}$ *is a Gröbner-Shirshov basis in* $k\langle Y \rangle$ *if* $n = 4$.

3.13 Plactic algebras in standard generators

Let $P_n = sgp\langle X | T \rangle$ be the plactic monoid with n-generators, see §3.12. Then the semigroup algebra $k\langle X | T \rangle$ is called the plactic algebra. For any $x_{i_1} \cdots x_{i_t} \in X^*$, we denote $x_{i_1} \cdots x_{i_t}$ by $i_1 \cdots i_t$.

In the paper [64], the following theorem is given.

Theorem 3.13.1 *([64], Theorem 1, Theorem 3) Let the notation be as above. Then with the deg-lex order on X^*, the following statements hold.*

(i) If $|X| = 3$, then the set $S = \{332 - 323,\ 322 - 232,\ 331 - 313,\ 311 - 131,\ 221 - 212,\ 211 - 121,\ 231 - 213,\ 312 - 132,\ 3212 - 2321,\ 32131 - 31321,\ 32321 - 32132\}$ is a Gröbner-Shirshov basis in $k\langle x_1, x_2, x_3\rangle$.

(ii) If $|X| > 3$, then the Gröbner-Shirshov complement T^c of T is infinite.

3.14 Filtrations and distortion in infinite-dimensional algebras

Let A be a linear algebra over a field k. An ascending filtration $\alpha = \{A_n\}$ on A is a sequence of subspaces $A_0 \subset A_1 \subset \ldots \subset A_n \subset \ldots$ such that $A = \sum_{n=0}^{\infty} A_n$ and $A_k A_l \subset A_{k+l}$ for all $k, l = 0, 1, 2, \ldots$. Given $a \in A$, the α-degree of a, denoted by $deg_\alpha a$, is defined as the least n such that $a \in A_n$. If B is a subalgebra of A with a filtration α, as above, then we call the filtration $\beta = \{B \cap A_n\}$ the restriction of α to B and we write $\beta = \alpha \cap B$. In the case of monoids the terms of filtrations are simply subsets, with all other conditions being the same. In the case of groups the terms of filtrations must be closed under inverses.

A generic example is as follows. Let A be a unital associative algebra generated by a finite set X. Then a filtration $\alpha = \{A_n\}$ arises on A, if one sets $A_0 = Span\{1\}$ and $A_n = A_{n-1} + Span\{X^n\}$, for any $n > 1$. The α-degree of $a \in A$ in this case is an ordinary degree with respect to the generating set X, that is the least degree of a polynomial in X equal a. We write $deg_\alpha a = deg_X a$. Such filtration α is called the degree filtration defined by the generating set X.

Definition 3.14.1 *([2]) Given two filtrations $\beta = \{B_n\}$ and $\beta' = \{B'_n\}$ on the same algebra B, we say that β is majorated by β' if there is an integer $t > 0$ such that $B_n \subseteq B'_{tn}$, for all $n \geq 0$. We then write $\beta \preceq \beta'$. If $\beta \preceq \beta'$ and $\beta' \preceq \beta$ then we say that β and β' are equivalent and write $\beta \sim \beta'$.*

If B is a finitely generated subalgebra of a finitely generated algebra A then we say then B is embedded in A without distortion (or that B is an undistorted subalgebra of A) if a degree filtration of B is equivalent to the restriction to B of a degree filtration of A

Definition 3.14.2 *([2]) A filtration $\alpha = \{A_n\}$ on an algebra A is called a tame filtration if it satisfies: there is $c > 0$ such that $dim(A_n) < c^n$, for all $n = 1, 2, \ldots$.*

In the paper [2], the following questions are discussed.

(1) Is it true that every tame filtration of an algebra B is equivalent (or equal) to a filtration restricted from the degree filtration of a finitely generated algebra A where B is embedded as a subalgebra?

(2) If the answer to the previous question is "yes", can one choose A finitely presented? If not, indicate conditions ensuring that the answer is still "yes".

By using CD-lemmas for associative and Lie algebras, Y. Bahturin, A. Olshanskii [2] proved the following theorems.

Theorem 3.14.3 *([2], Theorem 7) Let B be a unital associative algebra over a field F.*

(1) A filtration β on B is tame if and only if $\beta \sim \alpha \cap B$ where α is a degree filtration on a unital 2-generator associative algebra A where B is embedded as a subalgebra.

(2) A filtration β on B is tame if and only if $\beta = \alpha \cap B$ where α is a degree filtration on a unital finitely generated associative algebra A where B is embedded as a subalgebra.

Theorem 3.14.4 *([2], Theorem 8) Let N be a countable monoid.*

(i) There exist a 3-generator monoid M where N is embedded as a submonoid.

(ii) If N is finitely generated then the embedding of N in a 3-generator monoid M can be done without distortion.

(iii) A filtration β on N is a tame filtration if and only if there is a finitely generated monoid M with a degree filtration α such that $\beta \sim \alpha \cap N$.

Theorem 3.14.5 *([2], Theorem 9) (1) A filtration χ on a Lie algebra H is tame if and only if $\chi \sim \gamma \cap H$ where χ is the degree filtration on a 2-generator Lie algebra G where H is embedded as a subalgebra, if and only if $\chi \sim \rho \cap H$ where ρ is the degree filtration on a 2-generator associative algebra R where H is embedded as a Lie subalgebra.*

(2) A filtration χ on a Lie algebra H is tame if and only if $\chi = \gamma \cap H$ where γ is the degree filtration on a finitely generated Lie algebra G where H is embedded as a subalgebra, if and only if $\chi = \rho \cap H$ where ρ is the degree filtration on a finitely generated associative algebra R where H is embedded as a Lie subalgebra.

Theorem 3.14.6 *([2], Theorem 10) Any finitely generated associative, respectively, Lie algebra can be embedded without distortion in a simple 2-generator associative, respectively, Lie algebra.*

Theorem 3.14.7 *([2], Theorem 11) Any finitely generated monoid M with a recursively enumerable set of defining relations can be embedded in a finitely presented monoid M' as an undistorted submonoid.*

Theorem 3.14.8 *([2], Theorem 13) Let B be an arbitrary finitely generated unital associative algebra with a recursively enumerable set of defining relations, over a field k which is finitely generated over prime subfield. Then there exists a finitely presented unital associative k-algebra A in which B is contained as an undistorted unital subalgebra.*

3.15 Sufficiency conditions for Bokut' normal forms

L. A. Bokut [5, 6] (see also L. A. Bokut, D. J. Collins [23]) introduced a method for producing normal forms, called Bokut's normal forms, for groups obtained as a sequence of HNN extensions starting with a free group. K. Kalorkoti [57, 58, 59] has used these in several applications. In any application we need to prove first of all that normal forms exists, i.e., a certain rewriting process terminates. Uniqueness is then established separately, it is not guaranteed. Termination is usually ensured for both requirements that lead to a fairly uniform approach.

In the paper [60], the author provides sufficient conditions for the existence and uniqueness of normal forms of sequences of HNN extentions defined by Bokut'. Furthermore, the author shows that under an assumption, which holds for various applications, such normal forms always exist (but might not be unique). The conditions are amenable to be used in automatic theorem provers. It is discussed also how to obtain a Gröbner-Shirshov basis from the rewrite rules of Bokut's normal forms under certain assumptions. An application drawn from a paper of S. Aaanderaa, D. E. Cohen [1] to illustrate to sufficiency conditions is also given.

3.16 Quantum groups of type G_2 and D_4

First, we recall some basic notions about Ringel-Hall algebra from [75].

Let k be a finite field with q elements, Λ an $k-$algebra. By $\Lambda-$mod we denote the category of finite dimensional right $\Lambda-$modules. For $M, N_1, \cdots, N_t \in \Lambda-$mod, let $F^M_{N_1,\dots,N_t}$ be the number of filtrations

$$M = M_0 \supseteq M_1 \supseteq \cdots \supseteq M_{t-1} \supseteq M_t = 0,$$

such that $M_{i-1}/M_i \cong N_i$ for all $1 \leq i \leq t$.

Now, let Λ be a finitary $k-$algebra, i.e., for any $M, N \in \Lambda-$mod, $\mathrm{Ext}^1_\Lambda(M,N)$ is finite dimensional $k-$vector space. For each $M \in \Lambda-$mod, we denote by $[M]$ the isomorphism class of M and by $\mathbf{dim}M$ the dimension vector of the Λ-module M. We have the well-known Euler form $\langle -,- \rangle$ defined by

$$\langle \mathbf{dim}M, \mathbf{dim}N \rangle = \sum_{i=0}^{\infty} \mathbf{dim}\mathrm{Ext}^i_\Lambda(M,N).$$

Note that $(-,-)$ is the symmetrization of $\langle -,- \rangle$. It is well-known that if Λ is Dynkin type, that is one of the types $A_n, B_n, C_n, D_n, E_6, E_7, E_8, F_4$ and G_2, then $\langle \mathbf{dim}M, \mathbf{dim}N \rangle = \mathbf{dim}\mathrm{Hom}_\Lambda(M,N) - \mathbf{dim}\mathrm{Ext}^1_\Lambda(M,N)$.

The (twisted) Ringel-Hall algebra $\mathcal{H}(\Lambda)$ is the free $\mathbb{Q}(v)$-algebra with bases $\{u_{[M]} \mid M \in \Lambda - \mathrm{mod}\}$ indexed by the set of isomorphism classes of Λ-modules, with multiplication given by

$$u_{[M]}u_{[N]} = v^{\langle \mathbf{dim}M, \mathbf{N} \rangle} \sum_{[L]} F^L_{M,N} u_{[L]} \qquad \text{for all } M, N \in \Lambda - \mathrm{mod},$$

where $\mathbb{Q}$ is the field of quotient numbers, v an indeterminate and $\mathbb{Q}(v)$ the factor field of the polynomial algebra $\mathbb{Q}[v]$.

Then $\mathcal{H}(\Lambda)$ is an associative algebra with identity $1 = u_0$, where 0 denotes the isomorphism class of the trivial $\Lambda-$module 0.

Let Δ be the graph defined by the symmetrizable Cartan matrix A and $\overrightarrow{\Delta}$ be obtained by choosing some orientation of the edges of the graph Δ. Choose a $k-$species $\mathfrak{S} = (F_i, {}_iM_j)$ of type $\overrightarrow{\Delta}$ (see [76]). Then the main result in [75] is

Theorem 3.16.1 *([75]) The map $\eta : U_q^+(A) \longrightarrow \mathcal{H}(\overrightarrow{\Delta})$ given by*

$$\eta(E_i) = [S_i]$$

is an $\mathbb{Q}(v)-$algebra isomorphism, where $[S_i]$ is the isomorphism class of i'th simple representations of $\mathfrak{S}$-module.

We take the set $B = \{u_{[M]} \mid M \text{ is indecomposable}\}$ as a generating set for the Ringel-Hall algebra $\mathcal{H}(\overrightarrow{\Delta})$ and denote by T^+ the set of all skew-commutator relations between the elements of B. Then by the isomorphism η in Theorem 3.16.1, we get a corresponding set S^{+c} of relations in the positive part of quantum group $U_q^+(A)$.

Let $\Delta = G_2$ (resp. D_4) and $\overrightarrow{\Delta}$ be obtained by choosing the following orientations: for G_2 :

$$\underset{1}{\bullet} \xrightarrow{(1\ ,\ 3)} \underset{2}{\bullet}$$

and for D_4 :

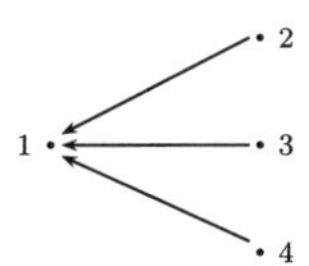

Then under the deg-lex order defined in [74] and [86], we get

Theorem 3.16.2 *([74, 86]) The set S^{+c} is a Gröbner-Shirshov basis of the algebra $U_q^+(G_2)$ (resp. $U^+(D_4)$).*

If we replace all x's in $U_q^+(G_2)$ (resp. $U^+(D_4)$) with y's, then we get a similar result for the negative part of the quantum group. Then, by Theorem 2.7 in [26], we have

Theorem 3.16.3 *([74, 86]) The set $S^{+c} \cup K \cup T \cup S^{-c}$ is a Gröbner-Shirshov basis of the quantum group $U_q(G_2)$ (resp. $U^+(D_4)$).*

3.17 An embedding of recursively presented Lie algebras

In 1961 G. Higman [56] proved an important Embedding Theorem which states that every recursively presented group can be embedded in a finitely presented group. Recall that a group (or an algebra) is called recursively presented if it can be given by a finite set of generators and a recursively enumerable set of defining relations. If a group (or an algebra) can be given by finite sets of generators and defining relations it is called finitely presented. As a corollary to this theorem G. Higman proved the existence of a universal finitely presented group containing every finitely presented group as a subgroup. In fact, its finitely generated subgroups are exactly the finitely generated recursively presented groups.

In [4] V.Ya.Belyaev proved an analog of Higman's theorem for associative algebras over a field which is a finite extension of its simple subfield. The proof was based on his theorem stating that every recursively presented associative algebra over a field as above can be embedded in a recursively presented associative algebra with defining relations which are equalities of words of generators and $\alpha + \beta = \gamma$, where α, β, γ are generators.

In recent paper [2] Y. Bahturin and A. Olshanskii, see §3.14, showed that such embedding can be performed distortion-free. The idea of transition from algebras to semigroups was also used by G. P. Kukin in [65] (see also [25]).

The paper [43] appears as a byproduct of the author's joint attempts with Y. Bahturin to prove Lie algebra analog of Higman's Theorem. In particular, Y. Bahturin suggested to prove that any recursively presented Lie algebra can be embedded in a Lie algebra given by Lie relations of the type mentioned above. In the paper [43], the author shows that this is indeed true. Namely, every recursively presented Lie algebra over a field which is a finite extension of its simple subfield can be embedded in a recursively presented Lie algebra defined by relations which are equalities of (nonassociative) words of generators and $\alpha+\beta=\gamma$ (α,β,γ are generators). Note that an existence of Higman's embedding for Lie algebras is still an open problem (see [61]). It is worth to mention a result by L. A. Bokut [7] that for every recursively enumerable set M of positive integers, the Lie algebra

$$L_M = \mathrm{Lie}(a,b,c \mid [ab^nc]=0, n\in M),$$

where $[ab^nc]=[[\ldots[[ab]b]\ldots b]c]$, can be embedded into a finitely presented Lie algebra.

By using Grobner-Shirshov basis theory for Lie algebras, in [43], the following theorem is given.

Theorem 3.17.1 *([43]) A recursively presented Lie algebra over a field which is a finite extension of its simple subfield can be embedded into a recursively presented Lie algebra defined by relations which are equalities of (nonassociative) words of generators and $\alpha+\beta=\gamma$, where α,β,γ are generators.*

3.18 GSB for some monoids

In the paper [62], Gröbner-Shirshov bases for the graph product, the Schützenberger product and the Bruck-Reilly extention of monoids are given, respectively, by using Shirshov algorithm for associative algebras.

References

[1] S. Aaanderaa, D. E. Cohen, Modular machines and Higman-Clapham-Valiev embedding theorem, in Word Problems II, A. I. Adian, W. Boone, G. Higman (eds), Studies in Logic and the Foundations of Mathematic, Vol. 95, North-Holland.

[2] Y. Bahturin, A. Olshanskii, Filtrations and distortion in infinite-dimensional algebras, J. Algebra, 327(1)(2011), 251-291.

[3] G. Baxter, An analytic problem whose solution follows from a simple algebraic identity, Pacific J. Math., 10(1960), 731-742.

[4] V. Ya. Belyaev, Subrings of finitely presented associative rings, Algera i Logica, 17(1978), 627-638.

[5] L. A. Bokut, On one property of the Boone group, Algebra i Logika, 5(5)(1966), 5-23.

[6] L. A. Bokut, On the Novikov groups, Algebra i Logika, 6(1)(1967), 25-38.

[7] L. A. Bokut, Unsolvability of the equality problem and subalgebras of finitely presented Lie algebras, Math. USSR Izvestia, 6(1972), 1153-1199.

[8] L. A. Bokut, Gröbner-Shirshov bases for braid groups in Artin-Garside generators, J. Symbolic Computation, 43(2008), 397-405.

[9] L. A. Bokut, Gröbner-Shirshov bases for the braid group in the Birman-Ko-Lee generators, J. Algebra, 321(2009), 361-379.

[10] L. A. Bokut, V. V. Chainikov, K. P. Shum, Markov and Artin normal form theorem for braid groups, Comm. Algebra, 35(2007), 2105-2115.

[11] L. A. Bokut, Yuqun Chen, Gröbner-Shirshov bases for Lie algebras: after A. I. Shirshov, Southeast Asian Bull. Math., 31(2007), 1057-1076.

[12] L. A. Bokut, Yuqun Chen, Gröbner-Shirshov bases: some new results, Advance in Algebra and Combinatorics, Proceedings of the Second International Congress in Algebra and Combinatorics, Eds. K. P. Shum, E. Zelmanov, Jiping Zhang, Li Shangzhi, World Scientific, 2008, 35-56.

[13] L. A. Bokut, Yuqun Chen, Yongshan Chen, Composition-Diamond lemma for tensor product of free algebras, J. Algebra, 323(2010), 2520-2537.

[14] L. A. Bokut, Yuqun Chen, Yongshan Chen,Gröbner-Shirshov bases for Lie algebras over a commutative algebra, arXiv:1006.3217

[15] L. A. Bokut, Yuqun Chen, Xueming Deng, Gröbner-Shirshov bases for Rota-Baxter algebras, Siberian Mathematical Journal, 51(6)(2010), 978-988. Sibirskii Matematicheskii Zhurnal (in Russian), 51(6)(2010), 1237-1250.

[16] L. A. Bokut, Yuqun Chen, Yu Li, Gröbner-Shirshov bases for Vinberg-Koszul-Gerstenhaber right-symmetric algebras, Journal of Mathematical Sciences, 166(2010), 603-612. Fundamental and Applied Mathematics (in Russian), 14(8)(2008), 55-67.

[17] L. A. Bokut, Yuqun Chen, Yu Li, Gröbner-Shirshov bases for categories, Proceeding of Operads and Universal Algebra, Nankai University, 2010, to appear.

[18] L. A. Bokut, Yuqun Chen, Yu Li, Anti-commutative Gröbner-Shirshov basis of a free Lie algebra relative to Lyndon-Shirshov words, preprint.

[19] L. A. Bokut, Yuqun Chen, Qiuhui Mo, Gröbner-Shirshov bases and embeddings of algebras, International Journal of Algebra and Computation, 20(2010), 875-900.

[20] L. A. Bokut, Yuqun Chen, Jianjun Qiu, Gröbner-Shirshov bases for associative algebras with multiple operations and free Rota-Baxter algebras, Journal of Pure and Applied Algebra, 214(2010), 89-100.

[21] L.A. Bokut, Yuqun Chen, Guangliang Zhang, Composition-Diamond lemma for associative n-conformal algebras, arXiv:0903.0892

[22] L. A. Bokut, Yuqun Chen, Xiangui Zhao, Gröbner-Shirshov beses for free inverse semigroups, International Journal of Algebra and Computation, 19(2)(2009), 129-143.

[23] L. A. Bokut, D. J. Collins, Malcev's problem and groups with a normal form, in Word Problems II, A. I. Adian, W. Boone, G. Higman (eds), Studies in Logic and the Foundations of Mathematic, Vol. 95, North-Holland.

[24] L. A. Bokut, Y. Fong, W.-F. Ke, Composition Diamond lemma for associative conformal algebras, J. Algebra, 272(2004), 739-774.

[25] L. A. Bokut, G. P. Kukin, Algorithmic and Combinatorial Algebra, Mathematics and its Applications, Kluwer Academic Publishers Group, Dordrecht, 1994.

[26] L. A. Bokut, P. Malcolmson, Gröbner-Shirshov bases for quantum enveloping algebras, Israel Journal of Mathematics, 96(1996), 97-113.

[27] L. A. Bokut, L.-S. Shiao, Gröbner-Shirshov bases for Coxeter groups, Comm. Algebra, 29(9)(2001), 4305-4319.

[28] P. Cartier, On the structure of free Baxter algebras, Adv. Math., 9(1972), 253-265.

[29] P. Cartier, Remarques sur le théorème de Birkhoff-Witt, Annali della Scuola Norm. Sup. di Pisa série III vol XII(1958), 1-4.

[30] Yuqun Chen, Yongshan Chen, Gröbner-Shirshov bases for metabelian Lie algebras, preprint.

[31] Yuqun Chen, Yongshan Chen, Yu Li, Composition-Diamond lemma for differential algebras, The Arabian Journal for Science and Engineering, 34(2A)(2009), 135-145.

[32] Yuqun Chen, Wenshu Chen, Runai Luo, Word problem for Novikov's and Boone's group via Gröbner-Shirshov bases, Southeast Asian Bull. Math., 32(5)(2008), 863-877.

[33] K.-T. Chen, R. Fox, R. Lyndon, Free differential calculus IV: The quotient group of the lower central series, Ann. Math., 68(1958) 81-95.

[34] Yuqun Chen, Jiapeng Huang, Gröbner-Shirshov bases for L-algebras, International Journal of Algebra and Computation, to appear. arXiv:1005.0118

[35] Yuqun Chen, Jing Li, Gröbner-Shirshov bases for plactic monoid with four genrators, preprint.

[36] Yuqun Chen, Jing Li, Mingjun Zeng, Composition-Diamond lemma for non-associative algebras over a polynomial algebra, Southeast Asian Bull. Math., 34(2010), 629-638.

[37] Yuqun Chen, Qiuhui Mo, Artin-Markov normal form for braid group, Southeast Asian Bull. Math, 33(2009), 403-419.

[38] Yuqun Chen, Qiuhui Mo, Embedding dendriform algebra into its universal enveloping Rota-Baxter algebra, Proc. Amer. Math. Soc., to appear. arXiv:1005.2717

[39] Yuqun Chen, Qiuhui Mo, Gröbner-Shirshov bases for free partially commutative groups, associative algebras and Lie algebras, preprint.

[40] Yuqun Chen, Hongshan Shao, K. P. Shum, On Rosso-Yamane theorem on PBW basis of $U_q(A_N)$, CUBO A Mathematical Journal, 10(3)(2008), 171-194.

[41] Yuqun Chcn, Bin Wang, Gröbner-Shirshov bases and Hilbert series of free dendriform algebras, Southeast Asian Bull. Math., 34(2010), 639-650.

[42] Yuqun Chen, Chanyan Zhong, Gröbner-Shirshov bases for braid groups in Adyan-Thurston generators, Algebra Colloq., to appear. arXiv:0909.3639

[43] E. Chibrikov, On some embedding of Lie algebras, preprint.

[44] P. M. Cohn, A remark on the Birkhoff-Witt theorem, J. London Math. Soc., 38(1963), 197-203

[45] V. Dotsenko, A. Khoroshkin, Gröbner bases for operads, Duke Mathematical Journal, 153(2)(2010), 363-396.

[46] V. Dotsenko, M. V. Johansson, Implementing Gröbner bases for operads, arXiv:0909.4950v2

[47] V. Drensky, R. Holtkamp, Planar trees, free nonassociative algebras, invariants, and elliptic integrals, Algebra and Discrete Mathmatics, 2(2008), 1-41.

[48] K. Ebrahimi-Fard, L. Guo, Rota-Baxter algebras and dendriform dialgebras, Journal of Pure and Applied Algebra, 212(2)(2008), 320-339.

[49] E. S. Esyp, I. V. Kazachkov, V. N. Remeslennikov, Divisibility Theory and Complexity of Algorithms for Free Partially Commutative Groups, Cont. Math. 378, AMS, 2004, 319-348.

[50] T. Evans, Embedding theorems for multiplicative systems and projective geometries, Proc. Amer. Math. Soc., 3(1952), 614-620.

[51] S. I. Gelfand, Y. I. Manin, Homological Algebra, Springer-Verlag, 1999.

[52] Li Guo, private communication, 2009.

[53] L. Guo, W. Keigher, On free Baxter algebras: completions and the internal construction, Adv. Math., 151(2000), 101-127.

[54] L. Guo, W. Keigher, On differential Rota-Baxter algebras, Journal of Pure and Applied Algebra, 212(2008), 522-540.

[55] G. Higman, B. H. Neumann, H. Neumann, Embedding theorems for groups, J. London Math. Soc., 24(1949), 247-254.

[56] G. Higman, Subgroups of finitely presented groups, Proc. Royal Soc. London (Series A), 262(1961), 455-475.

[57] K. Kalorkoti, Decision problems in group theory, Proceedings of the London Mathematical Society, 44(3)(1982), 312-332.

[58] K. Kalorkoti, Turing degree and the word and conjugacy problems for finitely presented groups, Southeast Asian Bull. Math., 30(2006), 855-887.

[59] K. Kalorkoti, A finite presented group with almost solvable conjugacy problem, Asian-European Journal of Mathematics, 2(4)92009), 603-627.

[60] K. Kalorkoti, Sufficiency conditions for Bokut' normal forms, Comm. Algebra, to appear.

[61] O. G. Kharlampovich, M. V. Sapir, Algorithmic problems in varieties, Inter. J. Algebra Compt., 5 (1995), 379-602.

[62] Canan Kocapinar, Firat Ates, A. Sinan Cevik, Gröbner-Shirshov bases of some monoids, preprint.

[63] Daniyar Kozybaev, Leonid Makar-Limanov, Ualbai Umirbaev, The Freiheitssatz and autoumorphisms of free right-symmetric algebras, Asian-European Journal of Mathematics, 1(2)(2008), 243-254.

[64] Lukasz Kubat, Jan Okniński, Gröbner-Shirshov bases for plactic algebras, arXiv:1010.3338v1

[65] G. P. Kukin, On the equality problem for Lie algebras, Sib. Mat. Zh., 18(5)(1977), 1194-1197.

[66] P. Leroux, L-algebras, triplicial-algebras, within an equivalence of categories motivated by graphs, arXiv:0709.3453v2

[67] J.-L. Loday, Dialgebras, in dialgebras and related operads, Lecture Notes in Math., 1763(2001), 7-66.

[68] M. Lothaire, Algebraic Combinatorics on Words, Cambridge University Press, 2002.

[69] A. I. Malcev, On a representation of nonassociative rings [in Russian], Uspekhi Mat. Nauk N.S., 7(1952), 181-185.

[70] A. A. Mikhalev, A. A. Zolotykh, Standard Gröbner-Shirshov bases of free algebras over rings, I. Free associative algebras, International Journal of Algebra and Computation, 8(6)(1998), 689-726.

[71] O. Poliakova, B. M. Schein, A new construction for free inverse semigroups, J. Algebra, 288(2005) 20-58.

[72] E. N. Poroshenko, Bases for partially commutative Lie algebras, arXiv:1012.1089v1

[73] Jianjun Qiu, Yuqun Chen, Composition-Diamond lemma for λ-differential associative algebras with multiple operators, Journal of Algebra and its Applications, 9(2010), 223-239.

[74] Yanhua Ren, Abdukadir Obul, Gröbner-Shirshov basis of quantum group of type G_2, Comm. Algebra, to appear.

[75] C. M. Ringel, Hall algebras and quantum groups, Invent. Math., 101(1990), 583-592.

[76] C. M. Ringel, PBW-bases of quantum groups, J.reine angew. Math., 470(1996), 51-88.

[77] G.-C. Rota, Baxter algebras and combinatorial identities I, Bull. Amer. Math. Soc., 5(1969), 325-329.

[78] Dan Segal, Free Left-Symmetric algebras and an analogue of the Poincaré-Birkhoff-Witt Theorem, J. Algebra, 164(1994), 750-772.

[79] A. I. Shirshov, On the representation of Lie rings in associative rings, Uspekhi Mat. Nauk N.S., 8(1953), No.5(57), 173-175.

[80] A. I. Shirshov, On free Lie rings, Mat. Sb., 45(1958), 2, 113–122. (in Russian)

[81] A. I. Shirshov, Some algorithmic problem for ε-algebras, Sibirsk. Mat. Z., 3(1962), 132-137.

[82] A. I. Shirshov, Some algorithmic problem for Lie algebras, Sibirsk. Mat. Z., 3(2)(1962), 292-296 (in Russian); English translation in SIGSAM Bull., 33(2)(1999), 3-6.

[83] Selected works of A. I. Shirshov, Eds. L. A. Bokut, V. Latyshev, I. Shestakov, E. Zelmanov, Trs. M. Bremner, M. Kochetov, Birkhäuser, Basel, Boston, Berlin, 2009.

[84] V. V. Talapov, Algebraically closed metabelian Lie algebras, Algebra i Logika, 21(3)(1982), 357-367.

[85] I. Yamane, A Poincare-Birkhoff-Witt theorem for quantized universal enveloping algebras of type A_N, Publ., RIMS. Kyoto Univ., 25(3)(1989), 503-520.

[86] Gulshadam Yunus, Abdukadir Obul, Gröbner-Shirshov basis of quantum group of type D_4, preprint.

[87] Xia Zhang, Gröbner-Shirshov bases for S-act algebras, Southeast Asian Bull. Math., 34(2010), 791-800.

A Note on Some Properties of the Least Common Multiple of Conjugacy Class Sizes

Alan Camina [1] and Rachel D Camina [2]

[1]School of Mathematics, University of East Anglia, Norwich, NR4 7TJ, UK; a.camina@uea.ac.uk

[2]Fitzwilliam College, Cambridge, CB3 0DG, UK; rdc26@dpmms.cam.ac.uk

Abstract

We present some comments about the least common multiple of the sizes of conjugacy classes. It is noted how the problems related to this least common multiple connect to questions concerning the existence of regular orbits of linear groups.

Dedicated to Prof. K. P. Shum on his 70th birthday

1 Introduction

Let $\mathrm{cs}(G)$ be the set of conjugacy class sizes of a finite group G and $\mathrm{lc}(G)$ be the least common multiple of $\mathrm{cs}(G)$. In this note we wish to consider when $\mathrm{lc}(G)$ is equal to the order of G and what happens when it is not. We will use the language of referring to the size of a conjugacy class containing an element $x \in G$ as the *index* of x, and the notation x^G for the conjugacy class containing x. Clearly, if the centre of G, denoted $Z(G)$, is not trivial, then $\mathrm{lc}(G)$ cannot be equal to the order of G, so we concentrate on groups with trivial centre. Using these ideas we introduce the following:-

Definition 1.1 *We say a group G is a full index group if its order $|G|$ is given by the least common multiple of the indices of the elements of G i.e.*

$$|G| = \mathrm{lcm}\{|x^G| : x \in G\}.$$

2000 Mathematics Subject Classification: 20D10
Finite groups, Conjugacy classes, Simple groups

We denote the centraliser of x in G by $C_G(x)$ and note that $|x^G|$ is given by the index of $C_G(x)$ in G. Clearly if G is not a full index group there must be at least one prime which divides $|C_G(x)|$ for all $x \in G$. We also give this property a name:-

Definition 1.2 *Let G be a group without centre. We say that a prime p is quasi-central in a group G if $|C_G(x)|$ is divisible by p for all $x \in G$.*

One might hope that there are no such primes, however in the next section we give some examples to show that this is not true. We are left with the following question:-

Question 1.3 *Can we describe those groups without centre that do not have full index? (Equivalently, those groups with a quasi-central prime.)*

It would seem from the examples given in the next section that the prime 2 and Fermat and Mersenne primes might play a role in the solution to this problem.

We observe that for a p-group G the breadth, $b(G)$ is defined to be such that $p^{b(G)}$ is the size of the largest conjugacy class in G. Note that $p^{b(G)} = \mathrm{lcm}\{|x^G| : x \in G\}$. The breadth has been the subject of many papers, we mention the most recent [4].

2 Some groups with quasi-central primes

Let G be a group with a faithful irreducible representation over $\mathrm{GF}(q)$ for some prime q. Let the module which gives rise to this representation be V. There has been considerable research into when V contains a regular orbit for G. That is there is an element $v \in V$ so that $|v^G| = |G|$.

Consider the situation when G is a p-group, $p \neq q$. We have the following simple lemma:

Lemma 2.1 *Let G be a p-group with an irreducible faithful module V. Then the extension $H = VG$ will be a full index group if and only if there is a regular orbit.*

Proof. We begin by showing that there is always a p-element whose centraliser is a p-group. Since G is a p-group it has a non-trivial centre. Since G has a faithful irreducible representation any non-trivial element, say z, of the centre acts fixed-point freely on V, in particular $C_V(z) = 1$. Thus we only need to consider whether p is quasi-central.

Assume first that there is no regular orbit. Let $v \in V$. Then $|v^G| < |G|$. So $C_G(v)$ is not trivial. Thus p divides the order of the centraliser of every element of H as all other elements have order divisible by p. The converse is straightforward since if $|v^G| = |G|$ then p does not divide $|C_G(v)|$. □

The existence of regular orbits has been studied by many authors. In particular Huppert and Manz [5] constructed some example of groups which do not have regular orbits. In Example 5 (a), (b) and (c) they find groups $H = VG$ where G is a p-group and V is an irreducible GF(q) module for G and V has no regular orbit. We use the notation C_n to denote the cyclic group of order n.

Example 2.2 *Using the examples above we obtain three classes of groups without centre which are not of full index.*

a) In this example p is a Mersenne prime $p = 2^f - 1$. We have $G = C_p \wr C_p$ with ???$V = 2^{fp}$ and 2 is a quasi-central prime.

b) In this example $q = 2^f + 1$ is a Fermat prime. We have $G = C_{2^f} \wr C_2$ with ???$V = q^{2f+1}$ and q is a quasi-central prime.

c) In this example q is a Mersenne prime $q = 2^f - 1$. We have G is a group of order $2^f + 1$ with ???$V = q^2$ and q is a quasi-central prime.

We can vary Lemma 2.1 a little.

Lemma 2.3 *Let G be a group with an irreducible faithful module V over GF(q) for some prime q. Let $H = VG$. If there is a regular orbit and H does not have full index then q is quasi-central in H.*

Proof. Since there is a regular orbit, there is an element $v \in V$ so that ???v^G = ???G. Then $C_H(v)$ is q-group. So the only possible quasi-prime is q. □

We end this section with a slightly more complex example of a group where 2 is quasi-central.

Example 2.4 *Let G be a group isomorphic to a direct product of the symmetric group of degree 3 and a group of order 2. G can be realised as a group of 2×2 matrices over GF(5). Specifically it can be generated by the following matrices:*

$$\begin{pmatrix} 0 & 1 \\ -1 & -1 \end{pmatrix}, \begin{pmatrix} 0 & 1 \\ 1 & 0 \end{pmatrix}, \begin{pmatrix} -1 & 0 \\ 0 & -1 \end{pmatrix}$$

Let V be a the vector space of dimension 2 over GF(5). Then V becomes a G-module with the action defined by the matrices. Let $H = VG$ then a straightforward calculation shows that 2 is a quasi-central prime for H.

Which suggests the following question:-

Question 2.5 *Does there exist a group G of odd order with trivial centre which does not have full index?*

3 Some positive results

First we comment that if $\gcd(|C_G(x)|, |C_G(y)|) = 1$ for two elements x, y then G has full index. Now we give some examples of groups with full index.

Example 3.1 *The full symmetric group S_n for $n \geq 3$. Consideration of the conjugacy class sizes of an n-cycle and an $(n-1)$-cycle yield the result.*

Example 3.2 *The Alternating groups A_n for $n \geq 4$. For n odd consider the n-cycle and $(n-2)$-cycle. For n even consider the $(n-1)$-cycle and the $(n-3)$-cycle, this is sufficient if $n-1$ is not divisible by 3. If 3 does divide $n-1$ also consider the element consisting of an $(n-5)$-cycle and two transpositions.*

Example 3.3 *Frobenius groups, of the form $G = K \rtimes H$, where H and K have coprime orders.*

In the next proposition we show ways to construct more full index groups.

Proposition 3.4 *(i) Let $G = H \times K$. Then G is a full index group if and only if both H and K are full index groups.*

(ii) Suppose $N \trianglelefteq G$, both N and G/N are full index groups and the orders of N and G/N are coprime. Then G is a full index group.

Proof. (i) Assume that H and K are full index groups. Let p be a prime. There exist $h \in H$ and $k \in K$ so that $p \nmid ???C_H(h)$ and $p \nmid |C_K(k)|$. Let $g = (h,k) \in G$, then $C_G(g) = C_H(h) \times C_K(k)$ and the result follows.

Suppose that H is not a full index group. So there is a prime, say p, so that p divides $???C_H(h)$ for all $h \in H$. But for $g = (h,k) \in H \times K$, we have $C_G(g) = C_H(h) \times C_K(k)$ and thus p divides $???C_G(g)$ for all $g \in G$. Hence G is not a full index group.

(ii) From the hypotheses we have

$$|N| = \text{lcm}\{|x^N| : x \in N\}$$

and

$$|G/N| = \text{lcm}\{|xN^{G/N}| : xN \in G/N\}.$$

Recalling that $|x^N|$ divides $|x^G|$ and $|x^{G/N}|$ divides $|x^G|$ we get the result $|G| = \text{lcm}\{|x^G| : x \in G\}$. □

We would like to thank Jan Saxl for showing us how to prove the next theorem.

Theorem 3.5 *Simple groups are full index groups.*

PROOF. The proof is a case-by-case study using the classification. For Alternating groups see Example 3.2 above. For G a simple group of Lie type consider a regular unipotent element, such an element is a p-element, where p is the characteristic and has centraliser a p-group, thus class size divisible by the full p'-part of $|G|$ see [2, §5.1]. Furthermore, the centraliser of a regular semi-simple element is a torus, so its class size is divisible by the full power of p, [2, §1.14].

A check through the Atlas [3], yields the result for the 26 sporadic groups. □

4 Concluding remarks

It would be very satisfactory if we could classify groups with a quasi-central prime or, equivalently, full index groups. As a small contribution to this we prove the following:-

Proposition 4.1 *Suppose G is a group with trivial centre and with a cyclic Sylow p-subgroup. Then p is not quasi-central in G.*

PROOF. Suppose p divides $C_G(x)$ for all elements x of G. We show that the centre of G is not trivial. Let $x \in G$ and denote by y_x an element of order p in $C_G(x)$. Note that, as the Sylow p-subgroup of G is cyclic and hence has a unique cyclic subgroup of order p, it follows that subgroups of order p in G are conjugate. Thus, if y is an element of order p in G the centraliser of y contains a conjugate of any element of G and so, by Burnside, [1, §26], is central. □

Note from this that the class of full index groups contains the simple groups and groups with trivial centre all of whose Sylow subgroups are cyclic. This suggests the next question.

Question 4.2 *Let G be a supersoluble group with trivial centre. Does G have full index?*

Note that $\mathrm{lc}(G)$ always divides the order of G. Consider $???G/\mathrm{lc}(G)$ and observe that G has full index if and only if $???G/\mathrm{lc}(G) = 1$. We ask

Question 4.3 *If G is a group with trivial centre, how big can $\frac{???G}{\mathrm{lc}(G)}$ be?*

References

[1] W. Burnside. *Theory of groups of finite order.* Dover Publications Inc., New York, 1955.

[2] Roger W Carter. *Finite groups of Lie type:Conjugacy classes and complex characters* A Wiley-Interscience Publication, John Wiley & Sons Ltd, Chichester,1993

[3] J. H. Conway, R. T. Curtis, S. P. Norton, R. A. Parker, and R. A. Wilson. *Atlas of finite groups.* Oxford University Press, Eynsham, 1985. Maximal subgroups and ordinary characters for simple groups, With computational assistance from J. G. Thackray.

[4] Giovanni Cutolo, Howard Smith, and James Wiegold. On finite p-groups with subgroups of breadth 1. *Bulletin of the Australian Mathematical Society*, 82:84–98, 2010.

[5] Bertram Huppert. and Olaf Manz. Orbit sizes of p-groups. *Arch. Math (Basel)*, 54:105–110, 1987.

Quotients and Homomorphisms of Relational Systems*

Ivan Chajda and Helmut Länger

Abstract

Relational systems containing one binary relation are investigated. Quotient relational systems are introduced and some of their properties are characterized. Moreover, homomorphisms, strong mappings and cone preserving mappings are introduced and the interplay between these notions is considered. Finally, the connection between directed relational systems and corresponding groupoids is investigated.

Keywords: relational system, quotient relational system, cone, homomorphism, strong mapping, cone preserving mapping, groupoid, g-homomorphism, quotient groupoid

1 Introduction

The theory of relational systems and of binary relations was settled in the pioneering work by J. Riguet ([6]) and the classical algebraic approach by A. I. Maltsev ([5]). Several properties of equivalences and homomorphisms were treated by the first author in [2] for equivalence systems, in [4] for quasiordered sets and in [3] for posets. It turns out that these results can be extended to arbitrary relational systems. This is the aim of the present paper. In particular, we set up several properties of quotient relational systems and relate the concept of strong homomorphism to that of cone preserving mapping.

We start with the definition of a relational system.

Definition 1.1 *By a* relational system *we mean an ordered pair $\mathcal{A} = (A, R)$ consisting of a set A and a binary relation R on A. $\mathcal{A}$ is called* connex *if for all $a, b \in A$ either $(a, b) \in R$ or $(b, a) \in R$ (or both). $\mathcal{A}$ is called* directed *if $U_R(a, b) \neq \emptyset$ for all $a, b \in A$ where*

$$U_R(a, b) := \{c \in A \mid (a, c), (b, c) \in R\}$$

is called the (upper) cone *of $\{a, b\}$ with respect to R. If $a = b$ then $U_R(a, b)$ is simply denoted by $U_R(a)$.*

*Support of the research of both authors by ÖAD, Cooperation between Austria and Czech Republic in Science and Technology, grant No. 2009/12 and of the first author by the Project MSM 6198959214 of the Research and Development Council of the Czech Government is gratefully acknowledged.

2 Quotient relational systems

In this section we investigate quotients of relational systems by equivalence relations. First we define a quotient relational system as follows:

Definition 2.1 *Let $\mathcal{A} = (A, R)$ be a relational system and Θ an equivalence relation on A. Then $\mathcal{A}/\Theta := (A/\Theta, R/\Theta)$ where*

$$R/\Theta := \{([a]\Theta, [b]\Theta) \mid (a, b) \in R\}$$

is called the quotient relational system *of $\mathcal{A}$ by Θ.*

The following lemma will be useful in the proof of the next theorem.

Lemma 2.2 *Let $\mathcal{A} = (A, R)$ be a relational system, $a, b \in A$ and Θ an equivalence relation on A. Then the following are equivalent:*

(i) $([a]\Theta, [b]\Theta) \in R/\Theta$

(ii) $(a, b) \in \Theta \circ R \circ \Theta$

Proof Both are equivalent to the fact that there exists $(c, d) \in R$ with $([c]\Theta, [d]\Theta) = ([a]\Theta, [b]\Theta)$. □

Now we characterize some properties of the relation R/Θ.

Theorem 2.3 *Let $\mathcal{A} = (A, R)$ be a relational system and Θ an equivalence relation on A. Then* (i) – (v) *hold:*

(i) *R/Θ is reflexive if and only if $([a]\Theta)^2 \cap R \neq \emptyset$ for all $a \in A$.*

(ii) *R/Θ is symmetric if and only if $R \subseteq \Theta \circ R^{-1} \circ \Theta$.*

(iii) *R/Θ is antisymmetric if and only if $(\Theta \circ R \circ \Theta) \cap (\Theta \circ R^{-1} \circ \Theta) \subseteq \Theta$.*

(iv) *R/Θ is transitive if and only if $R \circ \Theta \circ R \subseteq \Theta \circ R \circ \Theta$.*

(v) *R/Θ is connex if and only if so is $\Theta \circ R \circ \Theta$.*

Proof Let $a, b \in A$. We use Lemma 2.2.
(i) R/Θ is reflexive if and only if $(a, a) \in \Theta \circ R \circ \Theta$ for all $a \in A$.
(ii) R/Θ is symmetric if and only if $\Theta \circ R \circ \Theta$ has this property.
(iii) R/Θ is antisymmetric if and only if $(a, b) \in \Theta \circ R \circ \Theta$ and $(b, a) \in \Theta \circ R \circ \Theta$ together imply $(a, b) \in \Theta$.
(iv) The following are equivalent:
R/Θ is transitive.
$\Theta \circ R \circ \Theta$ is transitive.
$\Theta \circ R \circ \Theta \circ \Theta \circ R \circ \Theta \subseteq \Theta \circ R \circ \Theta$
$\Theta \circ R \circ \Theta \circ R \circ \Theta \subseteq \Theta \circ R \circ \Theta$
$R \circ \Theta \circ R \subseteq \Theta \circ R \circ \Theta$
(v) This is clear. □

Corollary 2.4 *Let $\mathcal{A} = (A, R)$ be a relational system and Θ an equivalence relation on A. Then* (i) – (iii) *hold:*

(i) *If R is a reflexive then so is R/Θ.*

(ii) *If R is a symmetric then so is R/Θ.*

(iii) *If R is transitive and either $\Theta \circ R \subseteq R \circ \Theta$ or $R \circ \Theta \subseteq \Theta \circ R$ (or both) then R/Θ is transitive, too.*

Remark 2.5 *It was proved in Theorem 2 of* [1] *that if R is transitive then R/Θ need not have this property.*

3 Homomorphisms of relational systems

In this section we investigate the interplay between different properties of mappings between relational systems.

In the following, if $\mathcal{A} = (A, F)$ and $\mathcal{B} = (B, S)$ are relational systems and $f : A \to B$ then

$$f^{-1}(S) := \{(a, b) \in A^2 \mid (f(a), f(b)) \in S\}.$$

Moreover, if Θ is an equivalence relation on A then f_Θ denotes the canonical mapping $x \mapsto [x]\Theta$ from A to A/Θ.

Definition 3.1 (cf. [4] and [5]) Let $\mathcal{A} = (A, R)$ and $\mathcal{B} = (B, S)$ be relational systems and $f : A \to B$.

(i) f is called a *homomorphism* from $\mathcal{A}$ to $\mathcal{B}$ if $(f(a), f(b)) \in S$ for all $(a, b) \in R$.

(ii) f is called *strong* if for all $(c, d) \in S$ there exists $(a, b) \in R$ with $f(a) = c$ and $f(b) = d$.

(iii) f is called *cone preserving* if $f(U_R(a)) = U_S(f(a))$ for all $a \in A$.

Remark 3.2 (iii) *extends the concept defined in* [4] *for quasiordered sets. Obviously, f is a homomorphism from $\mathcal{A}$ to $\mathcal{B}$ if and only if $f(U_R(a)) \subseteq U_S(f(a))$ for all $a \in A$. Hence, if f is cone preserving then it is a homomorphism from $\mathcal{A}$ to $\mathcal{B}$.*

Lemma 3.3 *Let $\mathcal{A} = (A, F)$ be a relational system and Θ an equivalence relation on A. Then f_Θ is a strong homomorphism from $\mathcal{A}$ onto $\mathcal{A}/\Theta$ and R/Θ is the smallest binary relation T on A/Θ such that f_Θ is a homomorphism from $\mathcal{A}$ to $(A/\Theta, T)$.*

Proof The proof is straightforward. □

For all sets A, B and every $f : A \to B$ let $\ker f$ denote the equivalence relation on A induced by f.

Now we state some characterizations of the notion of homomorphism which will be used later.

Lemma 3.4 *Let $\mathcal{A} = (A, R)$ and $\mathcal{B} = (B, S)$ be relational systems and $f : A \to B$. Then the following are equivalent:*

(i) *f is a homomorphism from $\mathcal{A}$ to $\mathcal{B}$.*

(ii) $f^{-1}(S) \supseteq R$

(iii) $f^{-1}(S) \supseteq (\ker f) \circ R$

(iv) $f^{-1}(S) \supseteq R \circ (\ker f)$

(v) $f^{-1}(S) \supseteq (\ker f) \circ R \circ (\ker f)$

Proof (i) $\Rightarrow$ (v) $\Rightarrow$ (iv) $\Rightarrow$ (ii) $\Rightarrow$ (i) $\Rightarrow$ (iii) $\Rightarrow$ (i) □

Lemma 3.5 *Let $\mathcal{A} = (A, R)$ and $\mathcal{B} = (B, S)$ be relational systems and f a mapping from A onto B. Then the following are equivalent:*

(i) *f is strong.*

(ii) $f^{-1}(S) \subseteq (\ker f) \circ R \circ (\ker f)$

Proof Both are equivalent to the fact that for all $a, b \in A$ with $(f(a), f(b)) \in S$ there exists a $(c, d) \in R$ with $f(c) = f(a)$ and $f(d) = f(b)$. □

Theorem 3.6 *Let $\mathcal{A} = (A, R)$ and $\mathcal{B} = (B, S)$ be relational systems and f a homomorphism from $\mathcal{A}$ onto $\mathcal{B}$. Then the following are equivalent:*

(i) *f is strong.*

(ii) $f^{-1}(S) = (\ker f) \circ R \circ (\ker f)$

Proof This follows from Lemmata 3.4 and 3.5. □

Lemma 3.7 *Let $\mathcal{A} = (A, R)$ and $\mathcal{B} = (B, S)$ be relational systems and f a mapping from A onto B. Then the following are equivalent:*

(i) *f is cone preserving.*

(ii) $f^{-1}(S) = R \circ (\ker f)$

Proof The following are equivalent:
(i)
$f(U_R(a)) = U_S(f(a))$ for all $a \in A$
For all $a, b \in A$, $f(b) \in U_S(f(a))$ if and only if $f(b) \in f(U_R(a))$
For all $a, b \in A$, $f(b) \in U_S(f(a))$ if and only if there exists a $c \in U_R(a)$ with $f(c) = f(b)$
(ii) □

Theorem 3.8 *Let $\mathcal{A} = (A, R)$ and $\mathcal{B} = (B, S)$ be relational systems and f a homomorphism from $\mathcal{A}$ onto $\mathcal{B}$. Then the following are equivalent:*

(i) *f is cone preserving.*

(ii) $f^{-1}(S) \subseteq R \circ (\ker f)$

Proof This follows from Lemmata 3.4 and 3.7. □

Theorem 3.9 *Let $\mathcal{A} = (A, R)$ and $\mathcal{B} = (B, S)$ be relational systems and f a strong homomorphism from $\mathcal{A}$ onto $\mathcal{B}$. Then the following are equivalent:*

(i) *f is cone preserving.*

(ii) $(\ker f) \circ R \subseteq R \circ (\ker f)$

Proof (i) $\Rightarrow$ (ii):
This follows from Lemmata 3.7 and 3.4.
(ii) $\Rightarrow$ (i):
$f^{-1}(S) \subseteq (\ker f) \circ R \circ (\ker f) \subseteq R \circ (\ker f) \circ (\ker f) = R \circ (\ker f)$ and hence f is cone preserving according to Theorem 3.8. □

Corollary 3.10 *Let $\mathcal{A} = (A, R)$ and $\mathcal{B} = (B, S)$ be relational systems, assume R to be symmetric and let f be a strong homomorphism from $\mathcal{A}$ onto $\mathcal{B}$. Then the following are equivalent:*

(i) *f is cone preserving.*

(ii) $(\ker f) \circ R = R \circ (\ker f)$

Proof (i) $\Rightarrow$ (ii):
According to Theorem 3.9, $(\ker f) \circ R \subseteq R \circ (\ker f)$. Since R is symmetric, we have $R \circ (\ker f) = R^{-1} \circ (\ker f)^{-1} = ((\ker f) \circ R)^{-1} \subseteq (R \circ (\ker f))^{-1} = (\ker f)^{-1} \circ R^{-1} = (\ker f) \circ R$
(ii) $\Rightarrow$ (i):
This follows from Theorem 3.9. □

Corollary 3.11 *Let $\mathcal{A} = (A, R)$ be a relational system and Θ an equivalence relation on A. Then the following are equivalent:*

(i) *f_Θ is cone preserving.*

(ii) $\Theta \circ R \subseteq R \circ \Theta$

Proof This follows from Lemma 3.3 and Theorem 3.9. □

Corollary 3.12 *Let $\mathcal{A} = (A, R)$ and $\mathcal{B} = (B, S)$ be relational systems and f a cone preserving mapping from A onto B. Then f is strong.*

Proof This follows from Lemmata 3.7 and 3.5. □

4 Groupoids corresponding to relational systems

In this section we associate to certain relational systems certain groupoids and investigate relationships between homomorhisms and congruence relations of certain relational systems on the one hand and of the corresponding groupoids on the other hand.

Definition 4.1 *Let $\mathcal{A} = (A, R)$ be a relational system. We say that a* groupoid $\mathcal{G}(\mathcal{A}) = (A, \cdot)$ corresponds *to $\mathcal{A}$ if $ab = b$ in case $(a, b) \in R$ and $ab \in U_R(a, b)$ otherwise.*

Remark 4.2 *Every relational system $\mathcal{A} = (A, R)$ can be extended to a directed one by adjoining one element: If $1 \notin A$ then put $A_d := A \cup \{1\}$ and $R_d := R \cup (A_d \times \{1\})$. Then $\mathcal{A}_d := (A_d, R_d)$ is directed and $\mathcal{A}$ is the restriction of $\mathcal{A}_d$ to A.*

Remark 4.3 *Every relational system $\mathcal{A} = (A, R)$ can be considered as a directed graph with vertex-set A and edge-set R. For $1 \notin A$ the groupoid $(A, \cdot)$ defined by $ab := b$ if $(a, b) \in R$ and $ab := 1$ otherwise is called the corresponding* graph algebra. *It is in fact a groupoid corresponding to $\mathcal{A}_d$. Hence the groupoids corresponding to relational systems are generalizations of graph algebras.*

Remark 4.4 *Although a groupoid $\mathcal{G}(\mathcal{A}) = (A, \cdot)$ corresponding to a relational system $\mathcal{A} = (A, R)$ need not be uniquely determined by $\mathcal{A}$, for $a, b \in A$, $ab = b$ if and only if $(a, b) \in R$.*
In what follows, if we consider a groupoid $\mathcal{G}$ corresponding to a relational system $\mathcal{A} = (A, R)$ we assume that $\mathcal{G}$ exists, it means that $\mathcal{A}$ is (partly) directed, i. e. if $a, b \in A$ and $(a, b) \notin R$ then the upper cone $U_R(a, b)$ of $\{a, b\}$ with respect to R is not empty.

Theorem 4.5 *Let $\mathcal{A} = (A, R)$ be a relational system and $\mathcal{G}(\mathcal{A}) = (A, \cdot)$ a groupoid corresponding to $\mathcal{A}$. Then* (i) – (vi) *hold:*

(i) *R is reflexive if and only if $\mathcal{G}$ is idempotent.*

(ii) *R is symmetric if and only if $\mathcal{G}(\mathcal{A})$ satisfies the identity $(xy)x = x$.*

(iii) *If $\mathcal{G}(\mathcal{A})$ is commutative then R is antisymmetric.*

(iv) *If $\mathcal{G}(\mathcal{A})$ satisfies the identity $(xy)x = xy$ then R is antisymmetric.*

(v) *If $\mathcal{G}(\mathcal{A})$ is a semigroup then R is transitive.*

(vi) *R is transitive if and only if $\mathcal{G}(\mathcal{A})$ satisfies the identity $x((xy)z) = (xy)z$.*

Proof (i), (iii) and (v) follow immediately from Remark 4.4.
Let $a, b, c \in A$.
(ii): First assume R to be symmetric. We have $(a, ab) \in R$ and hence, by symmetry of R, $(ab, a) \in R$ whence $(ab)a = a$. If, conversely, $\mathcal{G}(\mathcal{A})$ satisfies the identity $(xy)x = x$ and if $(a, b) \in R$ then $ab = b$ and hence $ba = (ab)a = a$ which shows $(b, a) \in R$ proving symmetry of R.
(iv): If $(a, b), (b, a) \in R$ then $ab = b$ and $ba = a$ and hence $a = ba = (ab)a = ab = b$.
(vi): First assume R to be transitive. Since $(a, ab), (ab, (ab)c) \in R$ one has, by transitivity

of R, $(a,(ab)c) \in R$ whence $a((ab)c) = (ab)c$. Conversely, assume $\mathcal{G}(\mathcal{A})$ to satisfy the identity $x((xy)z) = (xy)z$ and assume $(a,b),(b,c) \in R$. Then $ab = b$ and $bc = c$ and hence $ac = a(bc) = a((ab)c) = (ab)c = bc = c$ whence $(a,c) \in R$. □

Now we define the notion of a g-homomorphism.

Definition 4.6 *Let $\mathcal{A} = (A,R)$ and $\mathcal{B} = (B,S)$ be relational systems and f a homomorphism from $\mathcal{A}$ to $\mathcal{B}$. Then f is called a* g-homomorphism *from $\mathcal{A}$ to $\mathcal{B}$ if there exists a groupoid $\mathcal{G}(\mathcal{A}) = (A,\cdot)$ corresponding to $\mathcal{A}$ such that $a,b,c,d \in A$, $f(a) = f(c)$ and $f(b) = f(d)$ together imply $f(ab) = f(cd)$.*

The following theorem states that homomorphisms between groupoids corresponding to certain relational systems are g-homomorphisms between these relational systems.

Theorem 4.7 *Let $\mathcal{A} = (A,R)$ and $\mathcal{B} = (B,S)$ be relational systems, $\mathcal{G}(\mathcal{A}) = (A,\cdot)$ respectively $\mathcal{G}(\mathcal{B}) = (B,\circ)$ corresponding groupoids and f a homomorphism from $\mathcal{G}(\mathcal{A})$ to $\mathcal{G}(\mathcal{B})$. Then f is a g-homomorphism from $\mathcal{A}$ to $\mathcal{B}$.*

Proof Let $a,b,c,d \in A$. If $(a,b) \in R$ then $ab = b$ and hence $f(a) \circ f(b) = f(ab) = f(b)$ showing $(f(a),f(b)) \in S$. Moreover, if $a,b,c,d \in A$, $f(a) = f(c)$ and $f(b) = f(d)$ then $f(ab) = f(a) \circ f(b) = f(c) \circ f(d) = f(cd)$. □

The next theorem states that in some sense homomorphisms between relational systems are homomorphisms between corresponding groupoids.

Theorem 4.8 *Let $\mathcal{A} = (A,R)$ and $\mathcal{B} = (B,S)$ be relational systems and f a strong g-homomorphism from $\mathcal{A}$ onto $\mathcal{B}$ with the groupoid $\mathcal{G}(\mathcal{A}) = (A,\cdot)$ corresponding to $\mathcal{A}$. Then there exists a groupoid $\mathcal{G}(\mathcal{B}) = (B,\circ)$ corresponding to $\mathcal{B}$ such that f is a homomorphism from $\mathcal{G}(\mathcal{A})$ to $\mathcal{G}(\mathcal{B})$.*

Proof According to Definition 4.6 there exists a groupoid $\mathcal{G}(\mathcal{A}) = (A,\cdot)$ corresponding to $\mathcal{A}$ such that for each $a,b,c,d \in A$, if $f(a) = f(c)$ and $f(b) = f(d)$ then $f(ab) = f(cd)$. Define $f(x) \circ f(y) := f(xy)$ for all $x,y \in A$. According to Definition 4.6, $\circ$ is well-defined. Let $a,b \in A$. If $(f(a),f(b)) \in S$ then, since f is strong, there exists $(c,d) \in R$ with $f(c) = f(a)$ and $f(d) = f(b)$. Now $f(a) \circ f(b) = f(ab) = f(cd) = f(d) = f(b)$ according to Definition 4.6. If $(f(a),f(b)) \notin S$ then $(a,b) \notin R$ according to Definition 4.6 and hence $ab \in U_R(a,b)$, i. e. $(a,ab),(b,ab) \in R$ whence $(f(a), f(a) \circ f(b)) = (f(a), f(ab)) \in S$ and $(f(b), f(a) \circ f(b)) = (f(b), f(ab)) \in S$, i. e. $f(a) \circ f(b) \in U_S(f(a),f(b))$. This shows that $\mathcal{G}(\mathcal{B})$ corresponds to $\mathcal{B}$. Finally, f is a homomorphism from $\mathcal{G}(\mathcal{A})$ to $\mathcal{G}(\mathcal{B})$ since $f(xy) = f(x) \circ f(y)$ for all $x,y \in A$. □

Our next theorem contains some assertions concerning factor groupoids.

Theorem 4.9 *Let $\mathcal{A} = (A,R)$ and $\mathcal{B} = (B,S)$ be relational systems. Then* (i) *and* (ii) *hold:*

(i) *If f is a g-homomorphism from $\mathcal{A}$ to $\mathcal{B}$ then there exists a groupoid $\mathcal{G}(\mathcal{A}) = (A,\cdot)$ corresponding to $\mathcal{A}$ such that* $\ker f \in \mathrm{Con}\mathcal{G}(\mathcal{A})$.

(ii) *If $\mathcal{G}(\mathcal{A}) = (A, \cdot)$ is a groupoid corresponding to $\mathcal{A}$ and $\Theta \in \mathrm{Con}\mathcal{G}(\mathcal{A})$ then f_Θ is a g-homomorphism from $\mathcal{A}$ to $\mathcal{A}/\Theta$.*

Proof Let $a, b, c, d \in A$.
(i) Obviously, $\ker f \in \mathrm{Equ}A$. According to Definition 4.6 there exists a groupoid $\mathcal{G}(\mathcal{A}) = (A, \cdot)$ corresponding to $\mathcal{A}$ such that for each $a, b, c, d \in A$, if $f(a) = f(c)$ and $f(b) = f(d)$ then $f(ab) = f(cd)$, i. e. $(a, c), (b, d) \in \ker f$ implies $(ab, cd) \in \ker f$.
(ii) If $(a, b) \in R$ then $(f_\Theta(a), f_\Theta(b)) = ([a]\Theta, [b]\Theta) \in R/\Theta$. Moreover, if $f_\Theta(a) = f_\Theta(c)$ and $f_\Theta(b) = f_\Theta(d)$ then $[a]\Theta = f_\Theta(a) = f_\Theta(c) = [c]\Theta$ and $[b]\Theta = f_\Theta(b) = f_\Theta(d) = [d]\Theta$ and hence $f_\Theta(ab) = [ab]\Theta = [a]\Theta \cdot [b]\Theta = [c]\Theta \cdot [d]\Theta = [cd]\Theta = f_\Theta(cd)$. □

Finally, we assign to each groupoid a relational system.

Definition 4.10 *For every groupoid $\mathcal{G} = (G, \cdot)$ we define two* corresponding relational systems *$\mathcal{A}(\mathcal{G})$ and $\mathcal{A}^*(\mathcal{G})$ by $\mathcal{A}(\mathcal{G}) := (G, R(\mathcal{G}))$ respectively $\mathcal{A}^*(\mathcal{G}) := (G, R^*(\mathcal{G}))$ where $R(\mathcal{G}) := \{(x, y) \in G^2 \mid xy = y\}$ respectively $R^*(\mathcal{G}) := \bigcup_{x,y \in G} \{(x, xy), (y, xy)\}$.*

Lemma 4.11 *Let $\mathcal{G} = (G, \cdot)$ be a groupoid. Then $\mathcal{A}^*(\mathcal{G})$ is directed.*

Proof $U_{R^*(\mathcal{G})}(x, y) \supseteq \{xy\} \neq \emptyset$ for every $x, y \in G$. □

Lemma 4.12 *Let $\mathcal{A} = (A, R)$ be a relational system and $\mathcal{G}(\mathcal{A}) = (A, \cdot)$ a corresponding groupoid. Then $\mathcal{A}(\mathcal{G}(\mathcal{A})) = \mathcal{A}$ and $R \subseteq R^*(\mathcal{G}(\mathcal{A}))$.*

Proof Let $a, b \in A$. If $(a, b) \in R(\mathcal{G}(\mathcal{A}))$ then $ab = b$. Since $\mathcal{G}(\mathcal{A})$ corresponds to $\mathcal{A}$ we have $(a, b) \in R$. If, conversely, $(a, b) \in R$ then $ab = b$ in $\mathcal{G}(\mathcal{A})$ and hence $(a, b) \in R(\mathcal{G}(\mathcal{A}))$ and $(a, b) = (a, ab) \in R^*(\mathcal{G}(\mathcal{A}))$. □

Remark 4.13 *As mentioned above, a groupoid $\mathcal{G}(\mathcal{A})$ corresponding to the relational system $\mathcal{A} = (A, R)$ need not be uniquely determined by $\mathcal{A}$. However, Lemma 4.12 shows this not to be essential since $\mathcal{A}$ can be reconstructed from every groupoid $\mathcal{G}(\mathcal{A})$ corresponding to $\mathcal{A}$ due to the fact $\mathcal{A}(\mathcal{G}(\mathcal{A})) = \mathcal{A}$. Hence $\mathcal{G}(\mathcal{A})$ contains all the information on $\mathcal{A}$. Therefore one can switch between $\mathcal{A}$ and $\mathcal{G}(\mathcal{A})$ whenever it is suitable. Of course, it is more convenient to work with groupoids instead of relational systems since the theory of groupoids is more advanced.*

Theorem 4.14 *Let $\mathcal{G} = (G, \cdot)$ and $\mathcal{H} = (H, \circ)$ be groupoids and f a homomorphism from $\mathcal{G}$ to $\mathcal{H}$. Then f is a homomorphism from $\mathcal{A}(\mathcal{G})$ to $\mathcal{A}(\mathcal{H})$ and from $\mathcal{A}^*(\mathcal{G})$ to $\mathcal{A}^*(\mathcal{H})$.*

Proof Let $a, b \in G$. If $(a, b) \in R(\mathcal{G})$ then $ab = b$ and hence $f(a) \circ f(b) = f(ab) = f(b)$, i. e. $(f(a), f(b)) \in R(\mathcal{H})$. This shows that f is a homomorphism from $\mathcal{A}(\mathcal{G})$ to $\mathcal{A}(\mathcal{H})$. If, on the other hand, $(a, b) \in R^*(\mathcal{G})$ then there exist $e, h \in G$ with $(a, b) \in \{(e, eh), (h, eh)\}$. Now $f(e), f(h) \in H$ and

$$(f(a), f(b)) \in \{(f(e), f(eh)), (f(h), f(eh))\} = \{(f(e), f(e) \circ f(h)), (f(h), f(e) \circ f(h))\}$$

and hence $(f(a), f(b)) \in R^*(\mathcal{H})$ showing that f is a homomorphism from $\mathcal{A}^*(\mathcal{G})$ to $\mathcal{A}^*(\mathcal{H})$. □

Lemma 4.15 *Let $\mathcal{G} = (G, \cdot)$ be a groupoid satisfying the identities $x(xy) = y(xy) = xy$. Then $\mathcal{G}$ corresponds to $\mathcal{A}(\mathcal{G})$.*

Proof Let $a, b \in G$. If $(a, b) \in R(\mathcal{G})$ then $ab = b$. If $(a, b) \notin R(\mathcal{G})$ then $a(ab) = b(ab) = ab$ and hence $(a, ab), (b, ab) \in R(\mathcal{G})$. □

Theorem 4.16 *Let $\mathcal{G} = (G, \cdot)$ be a groupoid. Then the following are equivalent:*

(i) *There exists a relational system $\mathcal{A} = (G, R)$ with reflexive R such that $\mathcal{G}$ corresponds to $\mathcal{A}$.*

(ii) *$\mathcal{G}$ is idempotent and satisfies the identities $x(xy) = y(xy) = xy$.*

Proof Let $a, b \in G$.
(i) $\Rightarrow$ (ii):
Since $(b, b) \in R$, $bb = b$. If $(a, b) \in R$ then $ab = b$ and hence $a(ab) = ab$ and $b(ab) = bb = b = ab$. If $(a, b) \notin R$ then $(a, ab), (b, ab) \in R$ and hence $a(ab) = b(ab) = ab$.
(ii) $\Rightarrow$ (i):
Put $R := \{(x, y) \in G^2 \,|\, xy = y\} \cup \{(x, x) \,|\, x \in G\}$. Then R is reflexive. Moreover, if $(a, b) \in R$ then $ab = b$ or $a = b$. In the second case $ab = bb = b$. If $(a, b) \notin R$ then $a(ab) = b(ab) = ab$ and hence $(a, ab), (b, ab) \in R$. □

Remark 4.17 *Let $\mathcal{G} = (G, \cdot)$ and $\mathcal{H} = (H, \circ)$ be groupoids and f a homomorphism from $\mathcal{G}$ to $\mathcal{H}$. Then f need not be a g-homomorphism from $\mathcal{A}(\mathcal{G})$ to $\mathcal{A}(\mathcal{H})$ as can be seen from the following example.*

Example 4.18 *Put $\mathcal{G} := (\{-1, 0, 1\}, \cdot)$ and $\mathcal{H} := (\{0, 1\}, \cdot)$ where $\cdot$ denotes the multiplication of integers and let f denote the mapping $x \mapsto |x|$ from $\{-1, 0, 1\}$ to $\{0, 1\}$. Then f is a homomorphism from $\mathcal{G}$ to $\mathcal{H}$ and $R(\mathcal{G}) = \{(-1, 0), (0, 0), (1, -1), (1, 0), (1, 1)\}$. Let $(\{-1, 0, 1\}, *)$ be a groupod corresponding to $\mathcal{A}(\mathcal{G})$. Then for $x, y \in \{-1, 0, 1\}$, $1 * x = x$ and $x * y = 0$ otherwise. Now $f(-1) = f(1)$ but $f((-1) * (-1)) = f(0) = 0 \neq 1 = f(1) = f(1 * 1)$ and hence f is not a g-homomorphism from $\mathcal{A}(\mathcal{G})$ to $\mathcal{A}(\mathcal{H})$. Observe e. g. $0 \cdot (1 \cdot 1) = 0 \neq 1 = 1 \cdot 1$.*

Theorem 4.19 *Let $\mathcal{G} = (G, \cdot)$ and $\mathcal{H} = (H, \circ)$ be groupoids, assume $\mathcal{G}$ to satisfy the identities $x(xy) = y(xy) = xy$ and let f be a homomorphism from $\mathcal{G}$ to $\mathcal{H}$. Then f is a g-homomorphism from $\mathcal{A}(\mathcal{G})$ to $\mathcal{A}(\mathcal{H})$.*

Proof According to Lemma 4.15, $\mathcal{G}$ corresponds to $\mathcal{A}(\mathcal{G})$. If $a, b, c, d \in G$, $f(a) = f(c)$ and $f(b) = f(d)$ then $f(ab) = f(a) \circ f(b) = f(c) \circ f(d) = f(cd)$ and hence f is a g-homomorphism from $\mathcal{A}(\mathcal{G})$ to $\mathcal{A}(\mathcal{H})$. □

References

[1] I. Chajda, Congruences in transitive relational systems. Miskolc Math. Notes **5** (2004), 19 – 23.

[2] I. Chajda, Class preserving mappings of equivalence systems. Acta Univ. Palacki. Olomuc., Fac. rer. nat., Math. **43** (2004), 61 – 64.

[3] I. Chajda, Homomorphisms of directed posets. Asian-European J. Math. **1** (2008), 45 – 51.

[4] I. Chajda and Š. Hošková, A characterization of cone preserving mappings of quasiordered sets. Miskolc Math. Notes **6** (2005), 147 – 152.

[5] A. I. Mal'cev, Algebraic systems. Springer, New York 1973.

[6] J. Riguet, Relations binaires, fermetures, correspondances de Galois. Bull. Soc. Math. France **76** (1948), 114 – 155.

Authors' addresses:

Ivan Chajda
Palacký University Olomouc
Department of Algebra and Geometry
Trida 17.listopadu 12
77146 Olomouc
Czech Republic
chajda@inf.upol.cz

Helmut Länger
Vienna University of Technology
Institute of Discrete Mathematics and Geometry
Wiedner Hauptstraße 8-10
1040 Vienna
Austria
h.laenger@tuwien.ac.at

Automorphism Groups of Some Stable Lie Algebras with Exponential Functions I

Seul Hee Choi

Dept. of Mathematics, Univ. of Jeonju,
Jeonju 560-759, Korea
E-mail: chois@jj.ac.kr

Xueqing Chen* and Ki-Bong Nam†

Dept. of Mathematical and Computer Sciences, Univ. of Wisconsin-Whitewater,
Whitewater, WI 53190, USA
** Email: chenx@uww.edu*
† E-mail: namk@uww.edu

A degree stable Lie algebra is defined in the paper.[13] The Lie algebra automorphism group of the Lie algebra $W(1,0,2)$ is also found in this paper.[2] We find the algebra automorphism groups of the Lie algebras $W(1^3,1,1)$ and $W(1^3,2,0)$ in this work.

Keywords: Simple, Witt algebra, degreeing Lie algebra, Cartan subalgebra

1. Preliminaries

Let $\mathbb{F}$ be the field of characteristic zero (not necessarily algebraically closed). Throughout the paper, $\mathbb{N}$ and $\mathbb{Z}$ denote the non-negative integers and the integers, respectively. Let $\mathbb{F}^\bullet$ be the multiplicative group of non-zero elements of $\mathbb{F}$. Let L be a Lie algebra over $\mathbb{F}$ with a basis $S = \{s_u | u \in I\}$ where I is an index set. The Lie algebra L is degreeing if for any $s \in S$ we define the Lie degree $deg_{Lie}(s) \in \mathbb{Z}$ of s. Thus for any l of L, we may define $deg_{Lie}(l)$ as the highest Lie degree of non-zero basis terms of l. An element l of L is degree stable if for any $l_1 \in L$ $deg_{Lie}([l, l_1]) \leq deg_{Lie}(l_1)$ holds. For a degreeing Lie algebra L, the degree stabilizer $St_{Lie}(L)$ of the Lie algebra L is the vector subspace of L spanned by all the elements which are degree stable. For any $\theta \in Aut_{Lie}(L)$ we have the following diagram:

$$\begin{array}{ccc} St_{Lie}(L) & \longrightarrow & St_{Lie}(L) \\ \downarrow \iota & & \downarrow \iota \\ L & \longrightarrow & L \end{array}$$

Figure 1

where $Aut_{Lie}(L)$ is the automorphism group of the Lie algebra L and ι is an embedding from $St_{Lie}(L)$ to L as vector spaces. It is an interesting to note that the equality

$$St_{Lie}(L) = \theta(St_{Lie}(L)) \tag{1}$$

sometimes holds and sometimes does not hold for any $\theta \in Aut_{Lie}(L)$. A Lie algebra L is degree-stabilizing if $St_{Lie}(L)$ is auto-invariant, i.e., the equality (1) holds. Kaplansky generalizes the Witt algebra as follows:
Let $\mathbf{V}$ be a vector space over $\mathbb{F}$ and G a total additive group of functionals on $\mathbf{V}$. Let A be the vector space direct sum of copies of $\mathbf{V}$, one for each element of A. An element of A is $\sum_{x \in \mathbf{V}, \alpha \in G} c_{x,\alpha}(x, \alpha)$ where $c_{x,\alpha} \in \mathbb{F}$. If we define the multiplication as $[(x, \alpha), (y, \beta)] = \alpha(y)(x, \alpha+\beta) - \beta(x)(y, \alpha+\beta)$, then we have a Lie algebra (see[7]). Kaplansky shows that if $dim(\mathbf{V}) \neq 1$, then the Lie algebra is simple in the paper.[7] Kawamoto defines an infinite dimensional generalized Witt Lie algebra which is simple in his paper.[8] Đoković and K. Zhao also define a class of infinite dimensional generalized Witt Lie algebras which are simple in the papers.[4,5,9,14–17] The other generalized Witt algebra are defined on a stable algebra in the formal power series ring $\mathbb{F}[[x_1, \cdots, x_n]]$ or on the localization of the stable algebra (see[3,6,12]). One of those kinds of algebras is defined as follows:
For fixed positive integers $t_{11} > \cdots > t_{1p}, \cdots, t_{n1} > \cdots > t_{np}$, we define the $\mathbb{F}$-algebra $\mathbb{F}[n^{p+\cdots+q}, m, s]$ is spanned by

$$\begin{aligned} &\{e^{a_{11}x_1^{t_{11}}} \cdots e^{a_{1p}x_1^{t_{1p}}} \cdots e^{a_{n1}x_n^{t_{n1}}} \cdots e^{a_{nq}x_n^{t_{nq}}} x_1^{i_1} \cdots x_m^{i_m} x_{m+1}^{i_{m+1}} \cdots x_{m+s}^{i_{m+s}} \mid \\ &a_{11}, \cdots, a_{np}, i_1, \cdots, i_m \in \mathbb{Z}, i_{m+1}, \cdots, i_{m+s} \in \mathbb{N}\} \end{aligned} \tag{2}$$

such that $\mathbb{F}[n^{p+\cdots+q}, m, s] := \mathbb{F}[n^*, m, s]$ contains the polynomial ring $\mathbb{F}[x_1, \cdots, x_{m+s}]$ where e^{x_r} is the exponential function for $r \in \{1, \cdots, n\}$ etc. (see[1,6,10,11]). For $n, m, s \in \mathbb{N}$, the Lie admissible algebra

$NW(n^{p+\cdots+q}, m, s) := NW(n^*, m, s)$ has the standard basis

$$\begin{aligned}
&B_{W(n,m,s)} = \\
&\{e^{a_{11}x_1^{t_{11}}} \cdots e^{a_{1p}x_1^{t_{1p}}} \cdots e^{a_{n1}x_n^{t_{n1}}} \cdots e^{a_{nq}x_n^{t_{nq}}} x_1^{i_1} \cdots x_m^{i_m} x_{m+1}^{i_{m+1}} \cdots x_{m+s}^{i_{m+s}} \partial_u | \\
&a_{11}, \cdots, a_{np}, i_1, \cdots, i_m \in \mathbb{Z}, i_{m+1}, \cdots, i_{m+s} \in \mathbb{N}, \\
&1 \leq u \leq m+s, n \leq max\{m, s\}\} \qquad (3)
\end{aligned}$$

with the obvious addition such that the multiplication $*$ is defined as follows:

$$f\partial_u * g\partial_v = f\partial_u(g)\partial_v$$

for $f, g \in NW(n^*, m, s)$ where ∂_u is the partial derivative on $\mathbb{F}[n^*, m, s]$ with respect to x_u, $1 \leq u \leq m+s$. The antisymmetrized algebra of $NW(n^*, m, s)$ is the Witt type Lie algebra $W(n^*, m, s)$. The Lie algebra $W(n^*, m, s)$ is $\mathbb{Z}^{p+\cdots+q}$-graded as follows:

$$W(n^*, m, s) = \bigoplus_{a_{11}, \cdots, a_{nq}} W_{a_{11}, \cdots, a_{nq}} \qquad (4)$$

where $W_{a_{11}, \cdots, a_{nq}}$ is the vector subspace of $W(n^*, m, s)$ spanned by

$$\begin{aligned}
&\{e^{a_{11}x_1^{t_{11}}} \cdots e^{a_{1p}x_1^{t_{1p}}} \cdots e^{a_{n1}x_n^{t_{n1}}} \cdots e^{a_{nq}x_n^{t_{nq}}} x_{i_1}^{i_1} \cdots x_{m+s}^{i_{m+s}} \partial_u | \\
&i_1, \cdots, i_m \in \mathbb{Z}, i_{m+1}, \cdots, i_{m+s} \in \mathbb{N}, \\
&1 \leq u \leq m+s, n \leq max\{m, s\}\}
\end{aligned}$$

(see[16]). For each basis

$$e^{a_{11}x_1^{t_{11}}} \cdots e^{a_{1p}x_1^{t_{1p}}} \cdots e^{a_{n1}x_n^{t_{n1}}} \cdots e^{a_{nq}x_n^{t_{nq}}} x_1^{i_1} \cdots x_m^{i_m} x_{m+1}^{i_{m+1}} \cdots x_{m+s}^{i_{m+s}} \partial_u$$

of $W(n^*, m, s)$, we define the Lie degree of the basis as follow:

$$\begin{aligned}
°_{Lie}(e^{a_{11}x_1^{t_{11}}} \cdots e^{a_{1p}x_1^{t_{1p}}} \cdots e^{a_{n1}x_n^{t_{n1}}} \\
&\cdots e^{a_{nq}x_n^{t_{nq}}} x_1^{i_1} \cdots x_m^{i_m} x_{m+1}^{i_{m+1}} \cdots x_{m+s}^{i_{m+s}} \partial_u) \\
&= |i_1| + \cdots + |i_m| + i_{m+1} + \cdots + i_{m+s}
\end{aligned}$$

(see[16]). For any l of $W(n^*, m, s)$, we can define the Lie degree $deg_{Lie}(l)$ as the highest degree of non-zero terms of l. The Witt algebra $W(0, 0, 1)$ and the centerless Virasoro algebra $W(0, 1, 0)$ are self-centralizing (see[15]). Furthermore they are degree-stabilizing (see[7]). Let A be a subset of a Lie algebra L. The centralizer $Cl_L(A)$ is the set $\{l \in L | [l, l_1] = 0$ for any $l_1 \in A\}$. For any l in the Lie algebra L, l_1 is ad-diagonal with respect to l, if $[l, l_1] = cl$ holds where $c \in \mathbb{F}$. For a Lie algebra L, an element l in L is ad-diagonal of the set A in L, if for any $x \in A$, $[l, x] = c_x x$ holds where

$c_x \in \mathbb{F}$. For a given basis B of a Lie algebra L, the toral $tor_L(B) = tor(B)$ of B is n, if there are n ad-diagonal elements $\{l_1, \cdots l_n\}$ with respect to B such that the set $\{l_1, \cdots, l_n\}$ is the linearly independent maximal subset of L.

2. Automorphism group of $W(1^3, 1, 1)$

Note 2.1. It is well know that the non-associative algebra $NW(n^*, m, s)$ and the Lie (or its antisymmetrized) algebra $W(n^*, m, s)$ are simple (see[3,11,12]). Thus every non-zero endomorphism of $NW(n^*, m, s)$ or $W(n^*, m, s)$ is injective. □

Note that the standard basis of $W(1^3, 1, 1)$ is $\{e^{ax^{t_1}} e^{bx^{t_2}} e^{kx^{t_3}} x^i y^j \partial_u | a, b, k, i \in \mathbb{Z}, j \in \mathbb{N}, 1 \leq u \leq 2\}$. Generally, it is not easy to prove that $St_{Lie}(L)$ is a Lie subalgebra of L or not, i.e., it depends on the Lie algebra. For any basis elements $e^{a_1 x^{t_1}} e^{b_1 x^{t_2}} e^{k_1 x^{t_3}} x^{i_1} y^{j_1} \partial_u$ and $e^{a_2 x} e^{b_2 x^{t_2}} e^{k_2 x^{t_3}} x^{i_2} y^{j_2} \partial_v$ of $W(1^3, 1, 1)$, let us define the natural order $>_{Lie}$ as follows:

$$c_1 e^{a_1 x^{t_1}} e^{b_1 x^{t_2}} e^{k_1 x^{t_3}} x^{i_1} y^{j_1} \partial_u >_{Lie} c_2 e^{a_2 x^{t_1}} e^{b_2 x^{t_2}} e^{k_2 x^{t_3}} x^{i_2} y^{j_2} \partial_v, \quad (5)$$

if and only if $a_1 > a_2$, or $a_1 = a_2$ and $b_1 > b_2$, or $a_1 = a_2$, $b_1 = b_2$, and $k_1 > k_2$, or $\cdots$, and $a_1 = a_2$, $b_1 = b_2$, $k_1 = k_2$, $i_1 = i_2$, $j_1 = j_2$, and $u < v$ for any non-zero scalars c_1 and c_2. Thus we can define the natural order on $W(1^3, 1, 1)$. In (5), note that a coefficient of a basis element does not affect the order $>_{Lie}$ of $W(1^3, 1, 1)$. Thus we may define $deg_{Lie}(l)$ of any element $l \in W(1^3, 1, 1)$ as the highest Lie degree of non-zero basis terms of l. Note that $W(1^3, 1, 1)$ is simple (see[12]). From now on, let us assume that $t_1 > t_2 > t_3$.

Lemma 2.1. *$St_{Lie}(W(1^3, 1, 1))$ is a Lie subalgebra of the Lie algebra $W(1^3, 1, 1)$ spanned by $\{x\partial_2, y\partial_2, \partial_2\}$.*

Proof. It is obvious that the Lie subalgebra $< \{x\partial_2, y\partial_2, , \partial_2\} >$ of $W(1^3, 1, 1)$ spanned by $\{x\partial_2, y\partial_2, \partial_2\}$ is in $St_{Lie}(W(1^3, 1, 1))$. It is trivial to prove that every element which is not in $< \{x\partial_2, y\partial_2, \partial_1, \partial_2\} >$ cannot be degree stable. This implies that $St_{Lie}(W(1^3, 1, 1)) =< \{x\partial_2, y\partial_2, \partial_2\} >$. Therefore we have proven the lemma. □

To find the automorphism group $Aut_{Lie}(W(1^3, 1, 1))$ of the Lie algebra $W(1^3, 1, 1)$, we will find the stable Lie subalgebra of $W(1^3, 1, 1)$ and an auto-invariant set of $W(1^3, 1, 1)$.

Lemma 2.2. *For any $\theta \in Aut_{Lie}(W(1^3,1,1))$, the element $\theta(y\partial_2)$ is in the stabilizer $St_{Lie}(W(1^3,1,1))$ of the Lie algebra $W(1^3,1,1)$.*

Proof. For any $\theta \in Aut_{Lie}(W(1^3,1,1))$ and a basis element $e^{ax^{t_1}}e^{bx^{t_2}}e^{kx^{t_3}}x^iy^j\partial_u$ of the algebra $W(1^3,1,1)$, we have that

$$\theta([y\partial_2, e^{ax^{t_1}}e^{bx^{t_2}}e^{kx^{t_3}}x^iy^j\partial_u] = (j-\delta_{2,u})\theta(e^{ax^{t_1}}e^{bx^{t_2}}e^{kx^{t_3}}x^iy^j\partial_u) \quad (6)$$

where $\delta_{2,u}$ is the Kronecker delta. By (6) and the fact that $W(1^3,1,1)$ is simple, for any $l \in W(1^3,1,1)$, we have that

$$deg_{Lie}([\theta(y\partial_2), \theta(l)]) \leq deg_{Lie}(\theta(l)).$$

This implies that $\theta(y\partial_2) \in St_{Lie}(W(1^3,1,1))$ and so $\theta(y\partial_2)$ can be written as follows:

$$\theta(y\partial_2) = d_1x\partial_2 + d_2y\partial_2 + d_3\partial_2 \quad (7)$$

where $d_1, d_2, d_3 \in \mathbb{F}$. □

Lemma 2.3. *There is no automorphism θ of $W(1^3,1,1)$ such that*

$$\theta(y\partial_2) = d_1x\partial_2 + d_2y\partial_2 + d_3\partial_2 \quad (8)$$

where $d_1, d_2 \in \mathbb{F}^\bullet$ and $d_3 \in \mathbb{F}$.

Proof. Let θ be the automorphism of $W(1^3,1,1)$ such that it holds the conditions of the lemma for the element in the lemma. $\theta(x^u\partial_1)$ can be written as follow:

$$\theta(x^u\partial_1) = c(a_{u1}, b_{u1}, k_{u1}, i_{u1}, j_{u1}, 1)e^{a_{u1}x^{t_1}}e^{b_{u1}x^{t_2}}e^{k_{u1}x^{t_3}}x^{i_{u1}}y^{j_{u1}}\partial_1 + c(a_{u1}, b_{u1}, k_{u1}, i_{u1}, j_{u1}, 1)e^{a_{u1}x^{t_1}}e^{b_{u1}x^{t_2}}e^{k_{u1}x^{t_3}}x^{i_{u1}}y^{j_{u1}}\partial_2 + \#_1 \quad (9)$$

where either $e^{a_1x^{t_1}}e^{b_1x^{t_2}}e^{k_{u1}x^{t_3}}x^{i_{u1}}y^{j_{u1}}\partial_1$ or $e^{a_1x^{t_1}}e^{b_1x^{t_2}}e^{k_{u1}x^{t_3}}x^{i_{u1}}y^{j_{u1}}\partial_2$ is the maximal term of the element $\theta(x^u\partial_1)$ depending on their coefficients and $\#_1$ is the sum of the remaining terms of $\theta(\partial_1)$ with appropriate coefficients using the order $>_{Lie}$ and $u \in \mathbb{N}$. Furthermore, by Lemma 3 of,[2] we can assume that at least one of a_{u1}, b_{u1} or k_{u1} is not zero. If $j_{u1} \neq 0$, then $x^u\partial_1$ cannot centralize $y\partial_2$. We have that

$$\theta(x^u\partial_1) = c(a_{u1}, b_{u1}, k_{u1}, i_{u1}, 0, 1)e^{a_{u1}x^{t_1}}e^{b_{u1}x^{t_2}}e^{k_{u1}x^{t_3}}x^{i_{u1}}\partial_1 + c(a_{u1}, b_{u1}, k_{u1}, i_{u1}, 0, 1)e^{a_{u1}x^{t_1}}e^{b_{u1}x^{t_2}}e^{k_{u1}x^{t_3}}x^{i_{u1}}\partial_2 + \#_1. \quad (10)$$

Similarlily we can prove that $\theta(x^u\partial_1)$ does not have a non-zero term such that y is a factor of the term. Since $\theta(x\partial_1)$ is an ad-diagonal element with respect to $\{\theta(x^v\partial_1)|v \in \mathbb{N}\}$, every maximal term of $\theta(x^v\partial_1)$ is in the (a_{u1}, b_{u1}, k_{u1})-homogeneous component $W_{a_{u1},b_{u1},k_{u1}}$. Since $\theta(y\partial_2)$ centralizes $\theta(x^u\partial_1)$ and $d_1, d_2 \neq 0$, if $c(a_{u1}, b_{u1}, k_{u1}, i_{u1}, j_{u1}, 1) \neq 0$, then $c(a_{u1}, b_{u1}, k_{u1}, i_{u1}, j_{u1}, 2) \neq 0$ and vice versa. Since $\theta(x\partial_1)$ is an ad-diagonal element with respect to $\{\theta(x^v\partial_1)|v \in \mathbb{N}\}$, $\theta(x^v\partial_1)$ and $\theta(\partial_1)$ have terms in the same homogeneous components. This implies that all terms of the elements $\theta(x^u\partial_1)$, $u \in \mathbb{N}$, have the same maximal terms with appropriate coefficients. Let us prove the lemma by induction on the number $H(\theta(x\partial_1))$ of homogeneous components of $\theta(x\partial_1)$ such that the homogeneous components have a non-zero term of $\theta(x\partial_1)$. Let us assume that $H(\theta(x\partial_1))$ is one. Since $\theta(x\partial_1)$ is an ad-diagonal element with respect to $\{\theta(x^v\partial_1)|v \in \mathbb{N}\}$, it has a term in the $(0,0,0)$-homogeneous component $W_{0,0,0}$. By assumption, there is no room of $\theta(x\partial_1)$ to have a term of $W_{0,0,0}$. This contradiction shows that we can assume that $H(\theta(x\partial_1)) \geq 2$. This implies that $\theta(x\partial_1)$ has a non-zero term of $W_{0,0,0}$. There is an element $\theta(x^u\partial_1)$ which also has a non-zero term of $W_{0,0,0}$ such that the degree of the maximal term of $\theta(x^u\partial_1)$ is greater than zero where $u \neq 1$. $\theta(x\partial_1)$ and $\theta(x^u\partial_1)$ have the same maximal terms of $W_{0,0,0}$ with appropriate scalars. Thus every non-zero term of $\theta(x\partial_1)$ which is not in $W_{0,0,0}$ is a non-zero term of $\theta(x^u\partial_1)$ with appropriate coefficients and vice versa. Since $H(\theta(x\partial_1)) = H(\theta(x\partial_1)) \geq 2$, there are $c \in \mathbb{F}$ and $u \in \mathbb{N}$ such that

$$[\theta(x\partial_1) - c\theta(x^u\partial_1), \theta(x^u\partial_1)] \neq (u-1)\theta(x^u\partial_1). \tag{11}$$

This contradiction shows that we can assume that $x^r\partial_1$ is the maximal term of $\theta(x\partial_1)$ for an integer $r > 1$. This gives a similar contradiction as (11). This implies that $\theta(x\partial_1) \in W_{0,0,0}$. This implies that $\theta(y\partial_2)$ cannot centralize $\theta(x^u\partial_1)$. This contradiction shows that there is no automorphism θ of $W(1^3, 0, 2)$ which holds (2.7). Therefore we have proven the lemma. □

Lemma 2.4. *There is no automorphism θ of $W(1^3, 1, 1)$ such that*

$$\theta(y\partial_2) = d_1 x\partial_2 + d_2\partial_2 = (d_1 x + d_2)\partial_2 \tag{12}$$

holds where $d_1 \in \mathbb{F}^\bullet$ and $d_2 \in \mathbb{F}$.

Proof. Let θ be the automorphism of $W(1^3, 0, 2)$ such that it holds (12). By Lemma 2.3, we are able to prove that $\theta(y\partial_2)$ cannot centralize an element $\theta(x^u\partial_1)$, $u > 1$. This contradiction shows that there is no automorphism θ of $W(1^3, 0, 2)$ which holds (12). Therefore we have proven the lemma. □

Lemma 2.5. *For any automorphism θ of $W(1^3,1,1)$ and any basis element $y^k\partial_2$ of $W(1^3,1,1)$,*

$$\theta(y^k\partial_2) = d^{1-k}(y+d_1)^k\partial_2 \tag{13}$$

holds where $d_1 \in \mathbb{F}$ and $d \in \mathbb{F}^\bullet$.

Proof. Let θ be the automorphism of $W(1^3,1,1)$. By Lemmas 1-4, we have that $\theta(y\partial_2) = (y+d_1)\partial_2$ holds for $d_1 \in \mathbb{F}$. This implies that $\theta(\partial_2) = d\partial_2$ holds for $d \in \mathbb{F}^\bullet$. By induction on $k \in \mathbb{N}$ of $y^k\partial_2$, we are able to prove that $\theta(y^k\partial_2) = d^{1-k}(y+d_1)^k\partial_2$ holds. Therefore we have proven the lemma. □

Lemma 2.6. *For any automorphism θ of $W(1^3,1,1)$ and any basis element $x^i\partial_1$ of $W(1^3,1,1)$,*

$$\theta(x^i\partial_1) = c^{1-i}x^i\partial_1 \tag{14}$$

holds where $c \in \mathbb{F}^\bullet$.

Proof. Let θ be the automorphism of $W(1^3,1,1)$. By Lemma 5, we have that $\theta(y^k\partial_2) = d^{1-k}(y+d_1)^k\partial_2$ holds for $d_1 \in \mathbb{F}$ and $d \in \mathbb{F}^\bullet$. So we are able to prove that $\theta(\partial_1) = c\partial_1$ holds for $c \in \mathbb{F}^\bullet$. Since the Lie subalgebra $W(0,1,0)$ of $W(1^3,1,1)$ spanned by $\{x^u\partial_1 | u \in \mathbb{Z}\}$ is a self-centralizing Lie algebra, we have two cases, Case I: $\theta(x\partial_1) = -(x+c_1)\partial_1$ and Case II: $\theta(x\partial_1) = (x+c_1)\partial_1$ for $c_1 \in \mathbb{F}$.

Case I. Let us assume that $\theta(x\partial_1) = -(x+c_1)\partial_1$ holds. By $\theta([\partial_1, x\partial_1]) = \theta(\partial_1)$, we have that $-[\theta(\partial), (x+c_1)\partial_1] = \theta(\partial_1)$, we have that $\theta(\partial_1) = \alpha_0(x+c_1)^2\partial_1$ for $\alpha_0 \in \mathbb{F}^\bullet$. This implies that $\theta(x^2\partial_1) = \alpha_2\partial_1$ for $\alpha_2 \in \mathbb{F}^\bullet$. By $\theta([x^{-1}\partial_1, x^2\partial_1]) = 3\theta(\partial_1)$, we have that $c_1 = 0$ and $\theta(x^{-1}\partial_1) = \alpha_1 x^3\partial_1$ $\alpha_1 \in \mathbb{F}^\bullet$. By induction on i of $x^{-i}\partial_1$, we have that $\theta(x^{-i}\partial_1) = \alpha_{-i}x^{i+2}\partial_1$ for $\alpha_{-i} \in \mathbb{F}^\bullet$. By $\theta([x^{-t_1+1}\partial_1, e^{x^{t_1}}\partial_1]) = \theta(e^{x^{t_1}}\partial_1)$, we are able to prove that

$$[\alpha_{-t_1+1}x^{t_1+1}\partial_1, \theta(e^{x^{t_1}}\partial_1)] = \theta(e^{x^{t_1}}\partial_1) \tag{15}$$

holds. Since $\theta(e^{x^{t_1}}\partial_1) \notin W_{0,0,0}$ and t_1+1 is positive, there is no element $\theta(e^{x^{t_1}}\partial_1)$ of the algebra which holds the equality (15). This contradiction shows that there is no automorphism which holds $\theta(x\partial_1) = -(x+c_1)\partial_1$.

Case II. Let us assume that $\theta(x\partial_1) = (x+c_1)\partial_1$ holds. By induction on $i \in \mathbb{N}$, we are able to prove that $\theta(x^i\partial_1) = c^{1-i}(x+c_1)^i\partial_1$ holds. By

$\theta([x^{-2}\partial_1, x^3\partial_1]) = 5\theta(\partial_1)$, we have that $[\theta(x^{-2}\partial_1), c^{-2}(x+c_1)^3\partial_1] = 5c\partial_1$. This implies that $c_1 = 0$ and $\theta(x^{-2}\partial_1) = c^3(x^{-2}\partial_1$. By induction on i of $x^i\partial_1$, we can prove that $\theta(x^i\partial_1) = c^{1-i}x^i\partial_1$ easily. Therefore we have proven the lemma. □

Note 2.2.
For any basis elements $e^{ax^{t_1}}e^{bx^{t_2}}e^{kx^{t_3}}x^iy^j\partial_1$ and $e^{ax^{t_1}}e^{bx^{t_2}}e^{kx^{t_3}}x^iy^j\partial_1$ of $W(1^3,1,1)$, c_{11}, c_{12}, $d_{11}, d_{12}, d_{13} \in \mathbb{F}^\bullet$, and $c_{13} \in \mathbb{F}$, if we define a linear map $\theta_{c_{11},c_{12},c_{13},d_{11},d_{12},d_{13},1}$ from $W(1^3,1,1)$ to itself as follows:

$$\begin{aligned}
&\theta_{c_{11},c_{12},c_{13},d_{11},d_{12},d_{13},1}(e^{ax^{t_1}}e^{bx^{t_2}}e^{kx^{t_3}}x^iy^j\partial_1)\\
&= c_{11}^{1-i}c_{12}^{-j}d_{11}^a d_{12}^b d_{13}^k e^{ax^{t_1}}e^{bx^{t_2}}e^{kx^{t_3}}x^i(y+c_{13})^j\partial_1,\\
&\theta_{c_{11},c_{12},c_{13},d_{11},d_{12},d_{13},1}(e^{ax^{t_1}}e^{bx^{t_2}}e^{kx^{t_3}}x^iy^j\partial_2)\\
&= c_{11}^{-i}c_{12}^{1-j}d_{11}^a d_{12}^b d_{13}^k e^{ax^{t_1}}e^{bx^{t_2}}e^{kx^{t_3}}x^i(y+c_{13})^j\partial_2,
\end{aligned}$$

then $\theta_{c_{11},c_{12},c_{13},d_{11},d_{12},d_{13},1}$ can be linearly extended to a Lie automorphism of $W(1^3,1,1)$ such that $c_{11}^{t_1} = c_{11}^{t_2} = c_{11}^{t_3} = 1$. □

Note 2.3.
For any basis elements $e^{ax^{t_1}}e^{bx^{t_2}}e^{kx^{t_3}}x^iy^j\partial_1$ and $e^{ax^{t_1}}e^{bx^{t_2}}e^{kx^{t_3}}x^iy^j\partial_1$ of $W(1^3,1,1)$, c_{21}, c_{22}, $d_{21}, d_{22}, d_{23} \in \mathbb{F}^\bullet$, and $c_{23} \in \mathbb{F}$, if we define a linear map $\theta_{c_{21},c_{22},c_{23},d_{21},d_{22},d_{23},2}$ from $W(1^3,1,1)$ to itself as follows:

$$\begin{aligned}
&\theta_{c_{21},c_{22},c_{23},d_{21},d_{22},d_{23},2}(e^{ax^{t_1}}e^{bx^{t_2}}e^{kx^{t_3}}x^iy^j\partial_1)\\
&= c_{21}^{1-i}c_{22}^{-j}d_{21}^a d_{22}^b d_{23}^k e^{-ax^{t_1}}e^{bx^{t_2}}e^{kx^{t_3}}x^i(y+c_{23})^j\partial_1,\\
&\theta_{c_{21},c_{22},c_{23},d_{21},d_{22},d_{23},2}(e^{ax^{t_1}}e^{bx^{t_2}}e^{kx^{t_3}}x^iy^j\partial_2)\\
&= c_{21}^{-i}c_{22}^{1-j}d_{21}^a d_{22}^b d_{23}^k e^{-ax^{t_1}}e^{bx^{t_2}}e^{kx^{t_3}}x^i(y+c_{23})^j\partial_2,
\end{aligned}$$

then $\theta_{c_{21},c_{22},c_{23},d_{21},d_{22},d_{23},2}$ can be linearly extended to a Lie automorphism of $W(1^3,1,1)$ such that $c_{21}^{t_1} = -1$ and $c_{21}^{t_2} = 1 = c_{21}^{t_3}$. □

Note 2.4.
For any basis elements $e^{ax^{t_1}}e^{bx^{t_2}}e^{kx^{t_3}}x^iy^j\partial_1$ and $e^{ax^{t_1}}e^{bx^{t_2}}e^{kx^{t_3}}x^iy^j\partial_1$ of $W(1^2,1,1)$, c_{31}, c_{32}, $d_{31}, d_{32} \in \mathbb{F}^\bullet$, and $c_{33} \in \mathbb{F}$, if we define a linear map $\theta_{c_{31},c_{32},c_{33},d_{31},d_{32},d_{33},3}$ from $W(1^3,1,1)$ to itself as follows:

$$\begin{aligned}
&\theta_{c_{31},c_{32},c_{33},d_{31},d_{32},d_{33},3}(e^{ax^{t_1}}e^{bx^{t_2}}e^{kx^{t_3}}x^iy^j\partial_1)\\
&= c_{31}^{1-i}c_{32}^{-j}d_{31}^a d_{32}^b d_{33}^k e^{ax^{t_1}}e^{-bx^{t_2}}e^{kx^{t_3}}x^i(y+c_{33})^j\partial_1,\\
&\theta_{c_{31},c_{32},c_{33},d_{31},d_{32},d_{33},3}(e^{ax^{t_1}}e^{bx^{t_2}}e^{kx^{t_3}}x^iy^j\partial_2)\\
&= c_{31}^{-i}c_{32}^{1-j}d_{31}^a d_{32}^b d_{33}^k e^{ax^{t_1}}e^{-bx^{t_2}}e^{kx^{t_3}}x^i(y+c_{33})^j\partial_2,
\end{aligned}$$

then $\theta_{c_{31},c_{32},c_{33},d_{31},d_{32},d_{33},3}$ can be linearly extended to a Lie automorphism of $W(1^3,1,1)$ such that $c_{31}^{t_1} = 1 = c_{31}^{t_3}$ and $c_{31}^{t_2} = -1$. $\square$

Note 2.5.
For any basis elements $e^{ax^{t_1}}e^{bx^{t_2}}e^{kx^{t_3}}x^iy^j\partial_1$ and $e^{ax^{t_1}}e^{bx^{t_2}}e^{kx^{t_3}}x^iy^j\partial_1$ of $W(1^3,1,1)$, c_{41}, c_{42}, $d_{41}, d_{42}, d_{43} \in \mathbb{F}^\bullet$, and $c_{43} \in \mathbb{F}$, if we define a linear map $\theta_{c_{41},c_{42},c_{43},d_{41},d_{42},d_{43},4}$ from $W(1^3,1,1)$ to itself as follows:

$$\begin{aligned}
&\theta_{c_{41},c_{42},c_{43},d_{41},d_{42},4}(e^{ax^{t_1}}e^{bx^{t_2}}e^{kx^{t_3}}x^iy^j\partial_1)\\
&= c_1^{1-i}c_2^{-j}d_{41}^a d_{42}^b d_{43}^k e^{ax^{t_1}}e^{bx^{t_2}}e^{-kx^{t_3}}x^i(y+c_{43})^j\partial_1,\\
&\theta_{c_{41},c_{42},c_{43},d_{41},d_{42},d_{43},4}(e^{ax^{t_1}}e^{bx^{t_2}}e^{kx^{t_3}}x^iy^j\partial_2)\\
&= c_1^{-i}c_2^{1-j}d_{41}^a d_{42}^b d_{43}^k e^{ax^{t_1}}e^{bx^{t_2}}e^{-kx^{t_3}}x^i(y+c_{43})^j\partial_2,
\end{aligned}$$

then $\theta_{c_{41},c_{42},c_{43},d_{41},d_{42},d_{43},4}$ can be linearly extended to a Lie automorphism of $W(1^3,1,1)$ such that $c_{41}^{t_1} = 1 = c_{41}^{t_2}$ and $c_{41}^{t_3} = -1$. $\square$

Note 2.6.
For any basis elements $e^{ax^{t_1}}e^{bx^{t_2}}e^{kx^{t_3}}x^iy^j\partial_1$ and $e^{ax^{t_1}}e^{bx^{t_2}}e^{kx^{t_3}}x^iy^j\partial_1$ of $W(1^3,1,1)$, c_{51}, c_{52}, $d_{51}, d_{52} \in \mathbb{F}^\bullet$, and $c_{53} \in \mathbb{F}$, if we define a linear map $\theta_{c_{51},c_{52},c_{53},d_{51},d_{52},d_{53},5}$ from $W(1^3,1,1)$ to itself as follows:

$$\begin{aligned}
&\theta_{c_{51},c_{52},c_{53},d_{51},d_{52},d_{53},5}(e^{ax^{t_1}}e^{bx^{t_2}}e^{kx^{t_3}}x^iy^j\partial_1)\\
&= c_{51}^{1-i}c_{52}^{-j}d_{51}^a d_{52}^b d_{53}^k e^{-ax^{t_1}}e^{-bx^{t_2}}e^{kx^{t_3}}x^i(y+c_{53})^j\partial_1,\\
&\theta_{c_{51},c_{42},c_{53},d_{51},d_{52},d_{53},5}(e^{ax^{t_1}}e^{bx^{t_2}}e^{kx^{t_3}}x^iy^j\partial_2)\\
&= c_1^{-i}c_2^{1-j}d_{51}^a d_{52}^b d_{53}^k e^{-ax^{t_1}}e^{-bx^{t_2}}e^{kx^{t_3}}x^i(y+c_{53})^j\partial_2,
\end{aligned}$$

then $\theta_{c_{51},c_{52},c_{53},d_{51},d_{52},d_{53},5}$ can be linearly extended to a Lie automorphism of $W(1^3,1,1)$ such that $c_{51}^{t_1} = c_{51}^{t_2} = -1$ and $c_{51}^{t_3} = 1$. $\square$

Note 2.7.
For any basis elements $e^{ax^{t_1}}e^{bx^{t_2}}e^{kx^{t_3}}x^iy^j\partial_1$ and $e^{ax^{t_1}}e^{bx^{t_2}}e^{kx^{t_3}}x^iy^j\partial_1$ of $W(1^3,1,1)$, c_{61}, c_{62}, $d_{61}, d_{62} \in \mathbb{F}^\bullet$, and $c_{63} \in \mathbb{F}$, if we define a linear map $\theta_{c_{61},c_{62},c_{63},d_{61},d_{62},d_{63},6}$ from $W(1^3,1,1)$ to itself as follows:

$$\begin{aligned}
&\theta_{c_{61},c_{62},c_{63},d_{61},d_{62},d_{63},6}(e^{ax^{t_1}}e^{bx^{t_2}}e^{kx^{t_3}}x^iy^j\partial_1)\\
&= c_{61}^{1-i}c_{62}^{-j}d_{61}^a d_{62}^b d_{63}^k e^{ax^{t_1}}e^{-bx^{t_2}}e^{-kx^{t_3}}x^i(y+c_{63})^j\partial_1,\\
&\theta_{c_{61},c_{62},c_{63},d_{61},d_{62},d_{63},6}(e^{ax^{t_1}}e^{bx^{t_2}}e^{kx^{t_3}}x^iy^j\partial_2)\\
&= c_1^{-i}c_2^{1-j}d_{61}^a d_{62}^b d_{63}^k e^{ax^{t_1}}e^{-bx^{t_2}}e^{-kx^{t_3}}x^i(y+c_{63})^j\partial_2,
\end{aligned}$$

then $\theta_{c_{61},c_{62},c_{63},d_{61},d_{62},d_{63},6}$ can be linearly extended to a Lie automorphism of $W(1^3,1,1)$ such that $c_{61}^{t_1} = 1$ and $c_{61}^{t_2} = -1 = c_{61}^{t_3}$. $\square$

Note 2.8.
For any basis elements $e^{ax^{t_1}}e^{bx^{t_2}}e^{kx^{t_3}}x^iy^j\partial_1$ and $e^{ax^{t_1}}e^{bx^{t_2}}e^{kx^{t_3}}x^iy^j\partial_1$ of $W(1^3,1,1)$, c_{71}, c_{72}, $d_{71}, d_{72} \in \mathbb{F}^\bullet$, and $c_{73} \in \mathbb{F}$, if we define a linear map $\theta_{c_{71},c_{72},c_{73},d_{71},d_{72},d_{73},7}$ from $W(1^3,1,1)$ to itself as follows:

$$\begin{aligned}
&\theta_{c_{71},c_{72},c_{73},d_{71},d_{72},d_{73},7}(e^{ax^{t_1}}e^{bx^{t_2}}e^{kx^{t_3}}x^iy^j\partial_1)\\
&= c_{71}^{1-i}c_{72}^{-j}d_{71}^a d_{72}^b d_{73}^k e^{-ax^{t_1}}e^{bx^{t_2}}e^{-kx^{t_3}}x^i(y+c_{63})^j\partial_1,\\
&\theta_{c_{71},c_{72},c_{73},d_{71},d_{72},d_{73},7}(e^{ax^{t_1}}e^{bx^{t_2}}e^{kx^{t_3}}x^iy^j\partial_2)\\
&= c_{71}^{-i}c_{72}^{1-j}d_{71}^a d_{72}^b d_{73}^k e^{-ax^{t_1}}e^{bx^{t_2}}e^{-kx^{t_3}}x^i(y+c_{73})^j\partial_2,
\end{aligned}$$

then $\theta_{c_{71},c_{72},c_{73},d_{71},d_{72},d_{73},7}$ can be linearly extended to a Lie automorphism of $W(1^3,1,1)$ such that $c_{71}^{t_1} = 1 = c_{71}^{t_3}$ and $c_{71}^{t_2} = -1$. □

Note 2.9.
For any basis elements $e^{ax^{t_1}}e^{bx^{t_2}}e^{kx^{t_3}}x^iy^j\partial_1$ and $e^{ax^{t_1}}e^{bx^{t_2}}e^{kx^{t_3}}x^iy^j\partial_1$ of $W(1^3,1,1)$, c_{81}, c_{82}, $d_{81}, d_{82} \in \mathbb{F}^\bullet$, and $c_{83} \in \mathbb{F}$, if we define a linear map $\theta_{c_{81},c_{82},c_{83},d_{81},d_{82},d_{83},8}$ from $W(1^3,1,1)$ to itself as follows:

$$\begin{aligned}
&\theta_{c_{81},c_{82},c_{83},d_{81},d_{82},d_{83},8}(e^{ax^{t_1}}e^{bx^{t_2}}e^{kx^{t_3}}x^iy^j\partial_1)\\
&= c_{81}^{1-i}c_{82}^{-j}d_{81}^a d_{82}^b d_{83}^k e^{-ax^{t_1}}e^{-bx^{t_2}}e^{-kx^{t_3}}x^i(y+c_{83})^j\partial_1,\\
&\theta_{c_{81},c_{82},c_{83},d_{81},d_{82},d_{83},8}(e^{ax^{t_1}}e^{bx^{t_2}}e^{kx^{t_3}}x^iy^j\partial_2)\\
&= c_{81}^{-i}c_{82}^{1-j}d_{81}^a d_{82}^b d_{83}^k e^{-ax^{t_1}}e^{-bx^{t_2}}e^{-kx^{t_3}}x^i(y+c_{73})^j\partial_2,
\end{aligned}$$

then $\theta_{c_{81},c_{82},c_{83},d_{81},d_{82},d_{83},8}$ can be linearly extended to a Lie automorphism of $W(1^3,1,1)$ such that $c_{81}^{t_1} = c_{81}^{t_3} = c_{81}^{t_2} = -1$. □

Lemma 2.7. *For any automorphism θ of $W(1^3,1,1)$, θ is one of the automorphisms*
$\theta_{c_{11},c_{12},c_{13},d_{11},d_{12},d_{13},1}$, $\theta_{c_{21},c_{22},c_{23},d_{21},d_{22},d_{23},2}$, $\theta_{c_{31},c_{32},c_{33},d_{31},d_{32},d_{33},3}$, $\theta_{c_{41},c_{42},c_{43},d_{41},d_{42},d_{43},4}$, $\theta_{c_{51},c_{52},c_{53},d_{51},d_{52},d_{53},5}$, $\theta_{c_{61},c_{62},c_{63},d_{61},d_{62},d_{63},6}$, $\theta_{c_{71},c_{72},c_{73},d_{71},d_{72},d_{73},7}$, *and* $\theta_{c_{81},c_{82},c_{83},d_{81},d_{82},d_{83},8}$ *as shown in Notes 2.2– 2.5 with appropriate constant conditions.*

Proof. Let θ be the automorphism of $W(1^3,1,1)$ in the theorem. By Lemmas 5-6, we can assume that (13) and (14) hold with the same constants. Thus by induction on i,j of $x^iy^j\partial_u$, $1 \leq u \leq 2$, we are able to prove that $\theta(W(0,0,2)) = W(0,0,2)$ holds, i.e., $W(0,0,2)$ is θ-invariant or auto-invariant. Since $y^u\partial_2$ centralizes $e^{x^{t_1}}\partial_1$ and $y\partial_2 \in St_{Lie}(W(1^3,1,1))$, we have that

$$\theta(e^{x^{t_1}}\partial_1) = c_{a,b,i,0,1}e^{ax^{t_1}}e^{bx^{t_2}}x^i\partial_1 + \#_1 \tag{16}$$

holds where $e^{ax^{t_1}}e^{bx^{t_2}}x^i\partial_1$ is the maximal term of $\theta(e^{x^{t_1}}\partial_1)$ and $\#_1$ does not have a term with ∂_2. We have three cases, Case I: $a, b \neq 0$, Case II: $a = 0$ and $b \neq 0$, and Case III: $a \neq 0$ and $b = 0$.

Case I. Let us assume that $a, b \neq 0$. We have that $\theta(e^{-x^{t_1}}\partial_1)$ has a similar form as (16). By

$$\theta([e^{x^{t_1}}\partial_1, e^{x^{t_1}}\partial_1]) \in W(0,0,2), \tag{17}$$

we have that the maximal term of $\theta(e^{-x^{t_1}}\partial_1)$ is in W_{a_1,b_1,k_1} or in $W_{-a_1,-b_1,-k_1}$. Let us assume that the maximal term of $\theta(e^{-x^{t_1}}\partial_1)$ is in W_{a_1,b_1,k_1}. Thus by (18), $\theta(e^{x^{t_1}}\partial_1)$ and $\theta(e^{-x^{t_1}}\partial_1)$ have terms in the same homogeneous components. Furthermore we can assume that $H(\theta(e^{x^{t_1}}\partial_1) = \theta(e^{-x^{t_1}}\partial_1) \geq 2$. This implies that there is non-zero constant c such that

$$[\theta(e^{x^{t_1}}\partial_1), \theta(e^{x^{t_1}}\partial_1 - ce^{-x^{t_1}}\partial_1)] \neq -2ct_1\theta(x^{t_1-1}\partial_1). \tag{18}$$

Thus we can assume that the maximal term of $\theta(e^{-x^{t_1}}\partial_1)$ is in $W_{-a_1,-b_1,-k_1}$. If $H(\theta(e^{x^{t_1}}\partial_1) \neq 1$, then we can derive a contradiction because of the minimal term of $\theta([e^{t_1}\partial_1, e^{-x^{t_1}}])$. If $H(\theta(e^{x^{t_1}}\partial_1) = 1$, then we have that $\theta([e^{t_1}\partial_1, e^{-x^{t_1}}]) \neq -2t_1\theta(x^{t_1-1}\partial_1)$. This gives a contradiction. Thus $a, b \neq 0$ does not hold.

Case II. Let us assume that $a = 0$ and $b \neq 0$. This implies that $\theta(e^{x^{t_1}}\partial_1) = c'e^{x^{bt_2}}\partial_1 + \#_2$ holds. By $\theta([\partial_1, e^{x^{t_1}}\partial_1]) = t_1\theta(e^{x^{t_1}}x^{t_1-1}\partial_1)$, we have that $[c\partial_1, c'e^{x^{bt_2}}\partial_1 + \#_2] = cc'bt_2e^{x^{t_1}}x^{i+t_1-1}\partial_1 + \#_3$ holds. This implies that

$$\theta(e^{x^{t_1}}x^{t_1-1}\partial_1) = cc'b\frac{t_2}{t_1}e^{x^{t_1}}x^{i+t_1-1}\partial_1 + \frac{\#_3}{t_1}.$$

This implies that

$$\begin{aligned}&\theta([x\partial_1, e^{x^{t_1}}x^{t_1-1}\partial_1]) = t_1\theta(e^{x^{t_1}}x^{2t_1-1}\partial_1) + (t_1-2)\theta(e^{x^{t_1}}x^{t_1-1}\partial_1)\\ &= [(x+c_1)\partial_1, cc'b\frac{t_2}{t_1}e^{x^{t_1}}x^{i+t_1-1}\partial_1 + \frac{\#_3}{t_1}]\end{aligned}$$

holds. This implies that $\theta(e^{x^{t_1}}x^{2t_1-1}\partial_1) = cc'b^2\frac{t_2^2}{t_1^2}e^{bx^{t_1}}x^{i+2t_1-1}\partial_1 + \#_4$. By $\theta([x^{t_1}\partial_1, e^{x^{t_1}}\partial_1]) = t_1\theta(e^{x^{t_1}}x^{2t_1-1}\partial_1) - t_1 + \theta(e^{x^{t_1}}x^{t_1-1}\partial_1)$. This implies that

$$[c^{1-t_1}(x+c_1)^{t_1}\partial_1, c_1e^{bx^{t_2}}x^i\partial_1] = cc'b^2\frac{t_2^2}{t_1^2}e^{bx^{t_1}}x^{i+2t_2-1}\partial_1 + \#_5 \tag{19}$$

holds. Since $i + t_1 + t_2 - 1 \neq i + 2t_2 - 1$, the equality (19) does not hold. So we have a contradiction. Thus there is no automorphism of $W(1^3, 1, 1)$

which holds $a = 0$ and $b \neq 0$.

Case III. Let us assume that $a \neq 0$ and $b = 0$. This implies that $\theta(e^{x^{t_1}}\partial_1) = c'e^{x^{at_1}}\partial_1 + \#_6$. Let us assume that $\#_6 \neq 0$. Let us assume that $H(e^{x^{t_1}}\partial_1) \geq 1$. This implies that $H(e^{x^{\ \ t_1}}\partial_1) = H(e^{x^{t_1}}\partial_1)$ holds. This implies that there is a non-zero scalar c such that

$$\theta([e^{-x^{-t_1}}\partial_1, e^{x^{-t_1}}\partial_1 - ce^{x^{-t_1}}\partial_1]) \neq 2t_1\theta(e^{x^{-t_1}}\partial_1). \tag{20}$$

This contradiction shows that $H(e^{x^{-t_1}}\partial_1) = 1$. Similarly we can prove that $\theta(e^{x^{t_1}}\partial_1) = de^{ax^{t_1}}\partial_1$ and $\theta(e^{-x^{t_1}}\partial_1) = d_1e^{-ax^{t_1}}\partial_1$. This implies that $\theta(x^i\partial_1) = c^{1-i}x^i\partial_1$. By induction on i, k of $x^iy^k\partial_u$, $1 \leq u \leq 2$, we are able to prove that

$$\theta(x^iy^k\partial_u) = c^{\delta_{1u}-i}d^{\delta_{2u}-k}x^i(y+c)^k\partial_u \tag{21}$$

where δ_{1u} and δ_{2u} are Kronecker deltas. Since

$$\{x^py^j\partial_u, e^{x^{t_1}}\partial_1, e^{-x^{t_1}}\partial_1 | a, i \in \mathbb{Z}, u, i \in \mathbb{N}, 1 \leq u \leq 2\}$$

is a generator of the Lie subalgebra $W(1^1, 1, 1)$ of $W(1^3, 1, 1)$, we have that a is either 1 or -1. So we have two subcases, Subcase I: $a = 1$ and Subcase II: $a = -1$.

Subcase I. Let us assume that $\theta(e^{x^{t_1}}\partial_1) = d_1e^{x^{t_1}}\partial_1$ holds for $d_1 \in \mathbb{F}^\bullet$. By $\theta([e^{-x^{t_1}}\partial_1, e^{x^{t_1}}\partial_1]) = 2t_1\theta(x^{t_1-1}\partial_1)$, we have that $[\theta(e^{-x^{t_1}}\partial_1), d_1e^{x^{t_1}}\partial_1] = 2c^{2-t_1}t_1(x+c_1)^{t_1-1}\partial_1$ holds. This implies that $c_1 = 0$ and $\theta(e^{-x^{t_1}}\partial_1) = \frac{c^{2-t_1}}{d_1}e^{-x^{t_1}}\partial_1$ hold. By $\theta([x\partial_1, e^{x^{t_1}}\partial_1]) = t_1\theta(e^{x^{t_1}}x^{t_1}\partial_1) - \theta(e^{x^{t_1}}\partial_1)$, we have that $\theta(e^{x^{t_1}}x^{t_1}\partial_1) = de^{x^{t_1}}x^{t_1}\partial_1$ holds. By Lemma 6 and $\theta([x^{-t_1+1}\partial_1, e^{x^{-t_1}}\partial_1]) = t_1\theta(e^{x^{t_1}}\partial_1) - (-t_1+1)\theta(e^{x^{t_1}}x^{-t_1}\partial_1)$, we have that $[c^{t_1}x^{-t_1+1}\partial_1, d_1e^{x^{-t_1}}\partial_1] = t_1d_1e^{x^{t_1}}\partial_1 + (t_1-1)\theta(e^{x^{t_1}}x^{-t_1}\partial_1)$, This implies that $c = 1$. Similarly we can prove that A: $\theta(e^{x^{t_2}}\partial_1) = d_2e^{x^{t_2}}\partial_1$ and B: $\theta(e^{x^{t_2}}\partial_1) = d_2e^{-x^{t_2}}\partial_1$ where $d_2 \in \mathbb{F}^\bullet$.

A. Let us assume that $\theta(e^{x^{t_2}}\partial_1) = d_2e^{x^{t_2}}\partial_1$ holds. Since $x^{-t_2+1}\partial_1$ is an ad-diagonal element with respect to $e^{x^{t_2}}\partial_1$, we can prove that $c^{t_2} = 1$ holds. By induction on b of $e^{x^{bt_2}}\partial_1$, we can prove that $\theta(e^{x^{bt_2}}\partial_1) = d_2^be^{x^{bt_2}}\partial_1$. We have that either $\theta(e^{x^{t_3}}\partial_1) = d_3e^{x^{t_3}}\partial_1$ or $\theta(e^{x^{t_3}}\partial_1) = d_3e^{x^{-t_3}}\partial_1$. Let us we assume that $\theta(e^{x^{t_3}}\partial_1) = d_3e^{x^{t_3}}\partial_1$. So we have that

$$\theta(e^{x^{at_1}}e^{x^{bt_2}}e^{x^{kt_3}}x^i\partial_1) = c^{1-i}d_1^ad_2^bd_3^be^{x^{at_1}}e^{x^{bt_2}}e^{kx^{t_3}}x^i\partial_1 \tag{22}$$

holds. So by (21) and (22), we can prove that θ can be linearly extended to the automorphism $\theta_{c_{11},c_{12},c_{13},d_{11},d_{12},d_{13},1}$ as shown Note 2.2 with appropriate constants. If we assume that $\theta(e^{x^{t_3}}\partial_1) = d_3e^{-x^{t_3}}\partial_1$, then we can prove that θ can be linearly extended to the automorphism $\theta_{c_{41},c_{42},c_{43},d_{41},d_{42},d_{43},4}$ as shown Note 2.5 with appropriate constants.

B. Let us assume that $\theta(e^{x^{t_2}}\partial_1) = d_2e^{-x^{t_2}}\partial_1$. Similarly we can prove that $c^{t_2} = -1$. We have that either $\theta(e^{x^{t_3}}\partial_1) = d_3e^{x^{t_3}}\partial_1$ or $\theta(e^{x^{t_3}}\partial_1) = d_3e^{x^{-t_3}}\partial_1$. Similarly to A, we are able to prove that θ can be linearly extended to the automorphism $\theta_{c_{31},c_{32},c_{33},d_{31},d_{32},3}$ as shown Note 2.4 with appropriate constants or $\theta_{c_{61},c_{62},c_{63},d_{61},d_{62},d_{63},6}$ as shown Note 2.7 with appropriate constants respectively.

Subcase II. Let us assume that $\theta(e^{x^{t_1}}\partial_1) = d_1e^{-x^{t_1}}\partial_1$ holds for $d_1 \in \mathbb{F}^\bullet$. Similarly to Subcase I, we have that C: $\theta(e^{x^{t_2}}\partial_1) = d_2e^{x^{t_2}}\partial_1$ and D: $\theta(e^{x^{t_2}}\partial_1) = d_2e^{-x^{t_2}}\partial_1$ where $d_2 \in \mathbb{F}^\bullet$.

C. If we assume that $\theta(e^{x^{t_2}}\partial_1) = d_2e^{x^{t_2}}\partial_1$, then we have that either $\theta(e^{x^{t_3}}\partial_1) = d_3e^{x^{t_3}}\partial_1$ or $\theta(e^{x^{t_3}}\partial_1) = d_3e^{x^{-t_3}}\partial_1$. Similarly to A, θ can be linearly extended to the automorphism $\theta_{c_{21},c_{22},c_{23},d_{21},d_{22},d_{23},2}$ as shown Note 2.3 with appropriate constants or $\theta_{c_{71},c_{72},c_{73},d_{71},d_{72},d_{73},7}$ as shown Note 2.8 with appropriate constants respectively.

D. If we assume that $\theta(e^{x^{t_2}}\partial_1) = d_2e^{-x^{t_2}}\partial_1$, then we have that either $\theta(e^{x^{t_3}}\partial_1) = d_3e^{x^{t_3}}\partial_1$ or $\theta(e^{x^{t_3}}\partial_1) = d_3e^{x^{-t_3}}\partial_1$. Similarly to A, θ can be linearly extended to the automorphism $\theta_{c_{51},c_{52},c_{53},d_{51},d_{52},d_{53},5}$ as shown Note 2.6 with appropriate constants or $\theta_{c_{81},c_{82},c_{83},d_{81},d_{82},d_{83},8}$ as shown Note 2.9 with appropriate constants respectively.

This implies that θ can be linearly extended to one of the automorphisms $\theta_{c_{11},c_{12},c_{13},d_{11},d_{12},d_{13},1}$, $\theta_{c_{21},c_{22},c_{23},d_{21},d_{22},d_{23},2}$, $\theta_{c_{31},c_{32},c_{33},d_{31},d_{32},d_{33},3}$, $\theta_{c_{41},c_{42},c_{43},d_{41},d_{42},d_{43},4}$, $\theta_{c_{51},c_{52},c_{53},d_{51},d_{52},d_{53},5}$, $\theta_{c_{61},c_{62},c_{63},d_{61},d_{62},d_{63},6}$, $\theta_{c_{71},c_{72},c_{73},d_{71},d_{72},d_{73},7}$, and $\theta_{c_{81},c_{82},c_{83},d_{81},d_{82},d_{83},8}$ as shown in Notes 2.2– 2.9. Therefore we have proven the lemma. $\square$

Theorem 2.10. *The automorphism group $Aut(W(1^3,1,1))$ of the algebra $W(1^3,1,1)$ is generated by the automorphisms $\theta_{c_{11},c_{12},c_{13},d_{11},d_{12},d_{13},1}$, $\theta_{c_{21},c_{22},c_{23},d_{21},d_{22},d_{23},2}$, $\theta_{c_{31},c_{32},c_{33},d_{31},d_{32},d_{33},3}$, $\theta_{c_{41},c_{42},c_{43},d_{41},d_{42},d_{43},4}$, $\theta_{c_{51},c_{52},c_{53},d_{51},d_{52},d_{53},5}$, $\theta_{c_{61},c_{62},c_{63},d_{61},d_{62},d_{63},6}$, $\theta_{c_{71},c_{72},c_{73},d_{71},d_{72},d_{73},7}$, and $\theta_{c_{81},c_{82},c_{83},d_{81},d_{82},d_{83},8}$ as shown in Notes 2.2–2.9 with appropriate constant conditions.*

Proof. Let θ be an automorphism of $W(1^3,1,1)$. By Lemma 7, θ is one of the automorphisms $\theta_{c_{11},c_{12},c_{13},d_{11},d_{12},d_{13},1}$, $\theta_{c_{21},c_{22},c_{23},d_{21},d_{22},d_{23},2}$, $\theta_{c_{31},c_{32},c_{33},d_{31},d_{32},d_{33},3}$, $\theta_{c_{41},c_{42},c_{43},d_{41},d_{42},d_{43},4}$, $\theta_{c_{51},c_{52},c_{53},d_{51},d_{52},d_{53},5}$, $\theta_{c_{61},c_{62},c_{63},d_{61},d_{62},d_{63},6}$, $\theta_{c_{71},c_{72},c_{73},d_{71},d_{72},d_{73},7}$, and $\theta_{c_{81},c_{82},c_{83},d_{81},d_{82},d_{83},8}$ as shown in Notes 2.2–2.9 with appropriate constant conditions. Thus the automorphism group $Aut(W(1^3,1,1))$ of the algebra $W(1^3,1,1)$ is generated by the automorphisms $\theta_{c_{11},c_{12},c_{13},d_{11},d_{12},d_{13},1}$, $\theta_{c_{21},c_{22},c_{23},d_{21},d_{22},d_{23},2}$, $\theta_{c_{31},c_{32},c_{33},d_{31},d_{32},d_{33},3}$, $\theta_{c_{41},c_{42},c_{43},d_{41},d_{42},d_{43},4}$, $\theta_{c_{51},c_{52},c_{53},d_{51},d_{52},d_{53},5}$, $\theta_{c_{61},c_{62},c_{63},d_{61},d_{62},d_{63},6}$, $\theta_{c_{71},c_{72},c_{73},d_{71},d_{72},d_{73},7}$, and $\theta_{c_{81},c_{82},c_{83},d_{81},d_{82},d_{83},8}$ as shown in Notes 2.2–2.9 with appropriate constant conditions. Therefore we have proven the theorem. □

Remark I. Thanks to Theorem 1, we have that the automorphism group of $W(1^3,2,0)$ is generated by the automorphisms $\theta_{c_{11},c_{12},0,d_{11},d_{12},d_{13},1}$, $\theta_{c_{21},c_{22},0,d_{21},d_{22},d_{23},2}$, $\theta_{c_{31},c_{32},0,d_{31},d_{32},d_{33},3}$, $\theta_{c_{41},c_{42},0,d_{41},d_{42},d_{43},4}$, $\theta_{c_{51},c_{52},0,d_{51},d_{52},d_{53},5}$, $\theta_{c_{61},c_{62},0,d_{61},d_{62},d_{63},6}$, $\theta_{c_{71},c_{72},0,d_{71},d_{72},d_{73},7}$, and $\theta_{c_{81},c_{82},0,d_{81},d_{82},d_{83},8}$ which are defined on the algebra $W(1^3,2,0)$ as similar to Notes 2.2–2.9, of $W(1^3,2,0)$. □

References

1. H. J. Chang, *Über Wittsche Lie-Ringe* (German), Abh. Math. Sem. Hansischen Univ. **14**, (1941). 151–184.
2. Seul Hee Choi and Ki-Bong Nam, *Grade Stable Lie Algebras I*, Rocky Mountain Journal of Mathematics, Volume 40, Number 3, 2010, 813-824.
3. Seul Hee Choi, Jongwoo Lee and Ki-Bong Nam, *Derivations of a restricted Weyl-type algebra containing the polynomial ring*, Comm. Algebra 36 (2008), no. 9, 3435–3446.
4. D. Ž. Đoković and K. Zhao, *Derivations, isomorphisms, and second cohomology of generalized Witt algebras* Trans. Amer. Math. Soc. **350** (1998), no. 2, 643–664.
5. D. Ž. Đoković and Zhao, Kaiming, *Generalized Cartan type W Lie algebras in characteristic zero*, J. Algebra 195 (1997), no. 1, 170–210.

6. V. G. Kac, *Description of Filtered Lie Algebra with which Graded Lie algebras of Cartan type are Associated*, Izv. Akad. Nauk SSSR, Ser. Mat. Tom, **38**, 1974, 832–834.
7. I. Kaplansky, *Seminar on simple Lie algebras*, Bull. Amer. Math. Soc. 60 (1954)
8. N. Kawamoto, *Generalizations of Witt algebras over a field of characteristic zero*, Hiroshima Math. J. **16** (1986), no. 2, 417–426.
9. N. Kawamoto, *On G-graded automorphisms of generalized Witt algebras*, Contem. Math. A.M.S., **184**(1995), 225-230.
10. Naoki Kawamoto, Atsushi Mitsukawa, Ki-Bong Nam and Moon-Ok Wang, *The automorphisms of generalized Witt type Lie algebras*, Journal of Lie Theory, Vol (**13**) 2003, 571–576.
11. Jongwoo Lee and Ki-Bong Nam, *Non-associative algebras containing the matrix ring*, Linear Algebra Appl. 429 (2008), no. 1, 72–78.
12. Ki-Bong Nam, *Generalized W and H Type Lie Algebras*, Algebra Colloquium **6**:3, (1999), 329-340.
13. Ki-Bong Nam and Seul Hee Choi, *Degree Stable Lie Algebras I*, Algebra Colloquium 13 (2006), no. 3, 487–494.
14. Ki-Bong Nam, *Generalized S-type Lie Algebras*, Vol. 37, No. 4, Rocky Mountain Journal of Mathematics, 2007, 1291–1230.
15. Ki-Bong Nam and Jonathan Pakianathan, *On generalized Witt algebras in one variable*, Appear, Turkish Journal of Mathematics, 2010.
16. A. N. Rudakov, *Groups of automorphisms of infinite-dimensional simple Lie algebras*, Math. USSR-Izvestija, **3**(1969), 707–722.
17. Kaiming Zhao, *Isomorphisms between generalized Cartan type W Lie algebras in characteristic* 0, Canad. J. Math. 50 (1998), no. 1, 210–224.

Join Spaces, Multivalued Functions and Soft Sets

Piergiulio Corsini
Department of Civil Engineering
Udine University
Via delle Scienze 208, 33100 Udine, Italy
http://ijpam.uniud.it/journal/curriculum_corsini.htm
e-mail: corsini2002@yahoo.com

Abstract

Soft Sets have been introduced by Molodtsov in 1999 as a mathematical instrument to deal Uncertainty Theory.

Some scientists have utilized this notion to tackle problems of Decision Making and others, of BCK/BCI Algebras.

Connections between Soft Sets, (and in particular, Soft Groups), Algebraic Hyperstructures and Fuzzy Sets, have been studied for the first time in [24]. In this paper one goes on in this theme, calculating the fuzzy grade of the hypergroupoids (or partial hypergroupoids) associated with Soft Sets in seven different cases.

Keywords: fuzzy grade; hypergroupoid

1 Introduction

Multivalued functions were considered for the first time by Berge in 1963 (Espaces Topologique). They were utilized in the study of Topological Spaces, Topological methods for set-valued nonlinear analysis, in Binary Relation Theory, in Analysis (optimal control theory, especially differential inclusions and related subjects as game theory).

In physics, multivalued functions play an important role, in the study of the theory of defects in crystal and the resulting plasticity of materials, in the study of superfluids and superconductors, and quark confinement.

Finally a particular type of multivalued functions, the multivalued operations are fundamental in Algebra. They are used in Hyperstructure Theory. Born in

1934 [25], they are actually studied in many countries of Europe, Asia, America, Australia. After the foundation of Fuzzy Set Theory [32], it appeared natural to look for interconnections between Fuzzy Sets and Hyperstructures as they gauge both, in some measure, the vagueness of a situation.

They began to do it in the 70's, considering groups endowed with a fuzzy structure [28], and afterwards, many continued in this direction [18]-[22], studying Fuzzy Rings, Fuzzy Algebras and so on. In 1994 Corsini associated a join spaces with a fuzzy set [3], and in 2003, a fuzzy set with a hypergroupoid [4]. He and several others, in Italy [6]-[12], [13], in Romania [30], [9]-[12], in Greece [23], [29], [31] in Iran [1], [12] and in Canada [27] went on this way.

Another subject, connected with multivalued functions, is that one of Soft Sets and Soft Groups. This also has been introduced to represent the vagueness of several situations, see [24].

2 Preliminaries

A remarkable class of soft sets has been considered and studied in [24], that one denoted by $\mathcal{F}(H)$ of the functions $f : H \to \mathcal{P}(H)$ such that

$$(\delta) \qquad \forall x \in H,\ f(x) \neq \emptyset \text{ and } \cup_{x\in H} f(x) = H.$$

Here one supposes $H = \{1,2\}$ and one endowes $\mathcal{F}(H) = \{f_i \mid 1 \leq i \leq 7\}$ with four different hyperoperations and one calculates the Fuzzy Grade of all the hypergroupoids (in one case, the partial hypergroupoid) defined by these hyperoperations.

Let us remember how one associates a fuzzy set with a hypergroupoid $(H, \circ)$. Set

$$(I) \qquad \begin{cases} \mu_{x,y}(u) = \dfrac{1}{|x \circ y|}, \quad \text{if } u \in x \circ y \text{ and } \mu_{x,y}(u) = 0, \quad \text{if } u \notin x \circ y. \\ \text{Then } Q(u) = \{(a,b) \in H^2 \mid u \in a \circ b\},\ q(u) = |Q(u)|, \\ A(u) = \displaystyle\sum_{(x,y)\in H^2} \mu_{x,y}(u) \text{ and } \bar{\mu}(u) = \dfrac{A(u)}{q(u)}. \end{cases}$$

Now, let μ be a fuzzy set on H. One may associate with H a join space, by defining a hyperproduct "$\circ_\mu$", introduced by Corsini [3], as follows

$$(II) \qquad x \circ_\mu y = \{u \in H \mid \mu(x) \wedge \mu(y) \leq \mu(u) \leq \mu(x) \vee \mu(y)\}.$$

Using repeatedly the formulas (I) and (II), one obtains two sequences $H^i = \langle H^i, \circ_{\bar{\mu}_i} \rangle$ and $(\bar{\mu}_i)_{i \geq 1}$ of join spaces and fuzzy sets associated with H.

We shall denote this pair of sequences by F.J.s.

Set $(H^0, H^1, ..., H^m)$ the sequence of minimum length of hypergroupoids constructed as we showed above. Then we have that

$$\forall (i,j) : i \neq j,\ m \geq i \geq 0 \leq j \leq m,\ H^i \text{ is not isomorphic to } H^j$$

$$\text{and } H^{m+1} \text{ is isomorphic to } H^m.$$

We say that m is the *fuzzy grade* of H $(f.g.(H) = m)$. If $H^{m+1} = H^m$, then we say that m is the *strong fuzzy grade* of H $(s.f.g.(H) = m)$.

Remark. It is clear that if $s.f.g.(H) = m$, then we have also $f.g.(H) = m$. The converse, generally is not true as it can be seen at the end of paragraph 8.

Now, we set
$f_1(1) = 1,\ f_1(2) = 2,\ f_2(1) = 2,\ f_2(2) = 1,\ f_3(1) = 1,$
$f_3(2) = H,\ f_4(1) = 2,\ f_4(2) = H,\ f_5(1) = H,\ f_5(2) = 1,\ f_6(1) = H,$
$f_6(2) = 2,\ f_7(1) = H = f_7(2).$
Let us denote $\mathcal{F} = \{f_i \mid 1 \leq i \leq 7\}$.

Set now $H = \{1, 2\}$. The soft sets (f_i, H) satisfy the condition (δ). In what follows, we shall simply write f_i instead of (f_i, H) .

One can check (see [24], pag. 86, 87) that $(\mathcal{F}_1^0, \circ_1)$; $(\mathcal{F}_2^0, \circ_2)$ are hypergroups.

3 The "Union" hypergroupoid

Let $(\circ_1)$ be the hyperoperation defined in $\mathcal{F}$ as follows:

$$f \circ_1 g = \{h \in \mathcal{F} \mid \forall x,\ h(x) \subseteq f(x) \cup g(x)\}.$$

Calculating the hyperproducts among the f_i's, we get the following comutative hypergroupoid

$\mathcal{F}_1^0$	f_1	f_2	f_3	f_4	f_5	f_6	f_7
f_1	f_1	$\mathcal{F}$	f_1, f_3	$\mathcal{F}$	$\mathcal{F}$	f_1, f_6	$\mathcal{F}$
f_2		f_2	$\mathcal{F}$	f_2, f_4	f_2, f_5	$\mathcal{F}$	$\mathcal{F}$
f_3			f_1, f_3	$\mathcal{F}$	$\mathcal{F}$	$\mathcal{F}$	$\mathcal{F}$
f_4				f_4, f_2	$\mathcal{F}$	$\mathcal{F}$	$\mathcal{F}$
f_5					f_5, f_2	$\mathcal{F}$	$\mathcal{F}$
f_6						f_6, f_1	$\mathcal{F}$
f_7							$\mathcal{F}$

So, by (I) we obtain
$\mu_1(f_1) = 0.21428 = \mu_1(f_2),\ \mu_1(f_3) = 0.17105 = \mu_1(f_4) = \mu_1(f_5) = \mu_1(f_6),$
$\mu_1(f_7) = 0.142857.$

Setting $\{f_1, f_2\} = F_{12},\ D = \{f_3, f_4, f_5, f_6\}$ and $\mathcal{F}' = D \cup \{f_7\}$, from μ_1 we obtain

$\mathcal{F}_1^1$	f_1	f_2	f_3	f_4	f_5	f_6	f_7
f_1	F_{12}	F_{12}	F_{12}, D	F_{12}, D	F_{12}, D	F_{12}, D	$\mathcal{F}$
f_2		F_{12}	F_{12}, D	F_{12}, D	F_{12}, D	F_{12}, D	$\mathcal{F}$
f_3			D	D	D	D	$\mathcal{F}'$
f_4				D	D	D	$\mathcal{F}'$
f_5					D	D	$\mathcal{F}'$
f_6						D	$\mathcal{F}'$
f_7							f_7

From $\mathcal{F}_1^1$ we get μ_2 : $\mu_2(f_1) = \mu_2(f_2) = 0.218,$
$\mu_2(f_3) = \mu_2(f_4) = \mu_2(f_5) = \mu_2(f_6) = 0.2008658,$
$\mu_2(f_7) = 0.243956.$

Setting $\mathcal{F}^* = \mathcal{F} - \{f_7\}$, from μ_2 we obtain

$\mathcal{F}_1^2$	f_7	f_1	f_2	f_3	f_4	f_5	f_6
f_7	f_7	f_7, F_{12}	f_7, F_{12}	$\mathcal{F}$	$\mathcal{F}$	$\mathcal{F}$	$\mathcal{F}$
f_1		F_{12}	F_{12}	$\mathcal{F}^*$	$\mathcal{F}^*$	$\mathcal{F}^*$	$\mathcal{F}^*$
f_2			F_{12}	$\mathcal{F}^*$	$\mathcal{F}^*$	$\mathcal{F}^*$	$\mathcal{F}^*$
f_3				D	D	D	D
f_4					D	D	D
f_5						D	D
f_6							D

Then $\mu_3(f_7) = 0.2674 > \mu_3(f_1) = \mu_3(f_2) = 0.2232 > \mu_3(f_3) = \mu_3(f_4) = \mu_3(f_5) = \mu_3(f_6) = 0.195$, whence $s.f.g.(\mathcal{F}_1) = 2$.

4 The "Function product" hypergroupoid

We consider now the following hyperoperation defined in $\mathcal{F}$:

$$\forall (f, g) \in \mathcal{F} \times \mathcal{F}, \quad f \circ_2 g = \{h \mid \forall x \in H, f(g(x)) \supseteq h(x)\}.$$

We find the hypergroupoid $\mathcal{F}_2^0$:

$\mathcal{F}_2^0$	f_1	f_2	f_3	f_4	f_5	f_6	f_7
f_1	f_1	f_2	f_1, f_3	f_2, f_4	f_2, f_5	f_1, f_6	$\mathcal{F}$
f_2	f_2	f_1	f_2, f_4	f_1, f_3	f_1, f_6	f_2, f_5	$\mathcal{F}$
f_3	f_1, f_3	f_2, f_5	f_1, f_3	$\mathcal{F}$	f_2, f_5	$\mathcal{F}$	$\mathcal{F}$
f_4	f_2, f_4	f_1, f_6	f_2, f_4	$\mathcal{F}$	f_1, f_6	$\mathcal{F}$	$\mathcal{F}$
f_5	f_2, f_5	f_1, f_3	$\mathcal{F}$	f_1, f_3	$\mathcal{F}$	f_2, f_5	$\mathcal{F}$
f_6	f_1, f_6	f_2, f_4	$\mathcal{F}$	f_2, f_4	$\mathcal{F}$	f_1, f_6	$\mathcal{F}$
f_7	$\mathcal{F}$	$\mathcal{F}$	$\mathcal{F}$	$\mathcal{F}$	$\mathcal{F}$	$\mathcal{F}$	$\mathcal{F}$

We find $\mu_1(f_1) = \mu_1(f_2) = 0.3143$,
$\mu_1(f_3) = \mu_1(f_4) = \mu_1(f_5) = \mu_1(f_6) = 0.222$,
$\mu_1(f_7) = 0.143$.

Setting $B = \{f_3, f_4, f_5, f_6\}$, $C = B \cup \{f_1, f_2\}$, $D = B \cup \{f_7\}$, from μ_1 we find $\mathcal{F}_2^1$:

$\mathcal{F}_2^1$	f_1	f_2	f_3	f_4	f_5	f_6	f_7
f_1	f_1, f_2	f_1, f_2	C	C	C	C	$\mathcal{F}$
f_2		f_1, f_2	C	C	C	C	$\mathcal{F}$
f_3			B	B	B	B	D
f_4				B	B	B	D
f_5					B	B	D
f_6						B	D
f_7							f_7

We have $\mu_2(f_1) = \mu_2(f_2) = 0.21825$,
$\mu_2(f_3) = \mu_2(f_4) = \mu_2(f_5) = \mu_2(f_6) = 0.2008658$, $\mu_2(f_7) = 0.24396$.

So we obtain $\mathcal{F}_2^2$:

$\mathcal{F}_2^2$	f_7	f_1	f_2	f_3	f_4	f_5	f_6
f_7	f_7	f_7, f_1, f_2	f_7, f_1, f_2	$\mathcal{F}$	$\mathcal{F}$	$\mathcal{F}$	$\mathcal{F}$
f_1		f_1, f_2	f_1, f_2	C	C	C	C
f_2			f_1, f_2	C	C	C	C
f_3				B	B	B	B
f_4					B	B	B
f_5						B	B
f_6							B

We find $\mu_3(f_7) = 0.2674$, $\mu_3(f_1) = \mu_3(f_2) = 0.2232$,

$\mu_3(f_3) = \mu_3(f_4) = \mu_3(f_5) = \mu_3(f_6) = 0.19524.$

So, $\mathcal{F}_2^3 = \mathcal{F}_2^2$, whence $s.f.g.(\mathcal{F}_2) = 2$.

In what follows, we shall construct F.J.s.'s of $\mathcal{F}(H)$, starting from a hypergroupoid H which in some cases is a hypergroup - as in [24], page 88, in another a quasi-hypergroup, in another one a quasigroup, and finally in another one it is a hypergroupoid (neither associative, nor a quasi-hypergroup).

5 The hypergroupoid $\mathcal{F}(H)$ constructed from H_1

Let $H_1 = (H, \circ_1)$ be the following hypergroup

H_1	1	2
1	1	1,2
2	1,2	2

Let us establish in $\mathcal{F}(H)$ the following hyperoperation:

$$f \circ_3 g = \{h \mid \forall x,\ h(x) \subseteq f(x) \circ_1 g(x)\}.$$

One obtains the following hypergroupoid:

$\mathcal{F}^0(H_1)$	f_1	f_2	f_3	f_4	f_5	f_6	f_7
f_1	f_1	$\mathcal{F}$	f_1, f_3	$\mathcal{F}$	$\mathcal{F}$	f_1, f_6	$\mathcal{F}$
f_2		f_2	$\mathcal{F}$	f_2, f_4	f_2, f_5	$\mathcal{F}$	$\mathcal{F}$
f_3			f_1, f_3	$\mathcal{F}$	$\mathcal{F}$	$\mathcal{F}$	$\mathcal{F}$
f_4				f_2, f_4	$\mathcal{F}$	$\mathcal{F}$	$\mathcal{F}$
f_5					f_2, f_5	$\mathcal{F}$	$\mathcal{F}$
f_6						f_1, f_6	$\mathcal{F}$
f_7							$\mathcal{F}$

Calculating $A(f_i)$, $q(f_i)$ one obtains
$\mu_1(f_1) = \mu_1(f_2) = 0.21428$
$\mu_1(f_3) = 0.17105 = \mu_1(f_4) = \mu_1(f_5) = \mu_1(f_6)$
$\mu_1(f_7) = 0.142857.$

Setting $\{f_3, f_4, f_5, f_6\} = F^*$, $F^* \cup \{f_7\} = F^{**}$, $\{f_1, f_2\} \cup F^* = F'$ one obtains the following join space:

$\mathcal{F}^1(H_1)$	f_1	f_2	f_3	f_4	f_5	f_6	f_7
f_1	f_1, f_2	f_1, f_2	F'	F'	F'	F'	$\mathcal{F}$
f_2		f_1, f_2	F'	F'	F'	F'	$\mathcal{F}$
f_3			F^*	F^*	F^*	F^*	F^{**}
f_4				F^*	F^*	F^*	F^{**}
f_5					F^*	F^*	F^{**}
f_6						F^*	F^{**}
f_7							f_7

From $\mathcal{F}^1(H_1)$ we have $\mu_2(f_1) = \mu_2(f_2) = 0.21825$,
$\mu_2(f_3) = \mu_2(f_4) = \mu_2(f_5) = \mu_2(f_6) = 0.2008658$, $\mu_2(f_7) = 0.243956$.

So, we find the following join space:

$\mathcal{F}^2(H_1)$	f_7	f_1	f_2	f_3	f_4	f_5	f_6
f_7	f_7	f_7, f_1, f_2	f_7, f_1, f_2	$\mathcal{F}$	$\mathcal{F}$	$\mathcal{F}$	$\mathcal{F}$
f_1		f_1, f_2	f_1, f_2	F'	F'	F'	F'
f_2			f_1, f_2	F'	F'	F'	F'
f_3				F^*	F^*	F^*	F^*
f_4					F^*	F^*	F^*
f_5						F^*	F^*
f_6							F^*

We obtain $\mu_3(f_7) = 0.2674$, $\mu_3(f_1) = \mu_3(f_2) = 0.2232$,
$\mu_3(f_3) = \mu_3(f_4) = \mu_3(f_5) = \mu_3(f_6) = 0.19524$.

So we have $\mathcal{F}^3(H_1) = \mathcal{F}^2(H_1)$. Therefore $s.f.g.(\mathcal{F}^0(H_1)) = 2$.

6 The first partial hypergroupoid

Let H_2 be the hypergroup

H_2	1	2
1	1	2
2	2	1,2

Let $(\mathcal{F}, \circ_4)$ be the partial hypergroupoid defined as follows:

$$\forall (i,j), \quad f_i \circ_4 f_j = \{h \in \mathcal{F} \mid h(x) \subseteq f_i(x) \circ_2 f_j(x)\}.$$

We have $f_1 \circ_4 f_2 = f_2 \circ_4 f_1 = \emptyset$, because

$$\forall x \in H_2,\ f_1(x) \circ_2 f_2(x) = f_2(x) \circ_2 f_1(x) = 2,$$

so $f_1 \circ_4 f_2 \cap \mathcal{F} = \emptyset$, whence the following structure is obtained:

$\mathcal{F}^0(H_2)$	f_1	f_2	f_3	f_4	f_5	f_6	f_7
f_1	f_1, f_3	$\emptyset$	f_1, f_3	f_2, f_4	f_1, f_6	$\mathcal{F}$	$\mathcal{F}$
f_2	$\emptyset$	f_2, f_5	f_2, f_4	$\mathcal{F}$	f_2, f_5	f_1, f_6	$\mathcal{F}$
f_3	f_1, f_3	f_2, f_4	f_1, f_3	f_2, f_4	$\mathcal{F}$	$\mathcal{F}$	$\mathcal{F}$
f_4	f_2, f_4	$\mathcal{F}$	f_2, f_4	$\mathcal{F}$	$\mathcal{F}$	$\mathcal{F}$	$\mathcal{F}$
f_5	f_1, f_6	f_2, f_5	$\mathcal{F}$	$\mathcal{F}$	f_2, f_5	f_1, f_6	$\mathcal{F}$
f_6	$\mathcal{F}$	f_1, f_6	$\mathcal{F}$	$\mathcal{F}$	f_1, f_6	$\mathcal{F}$	$\mathcal{F}$
f_7	$\mathcal{F}$	$\mathcal{F}$	$\mathcal{F}$	$\mathcal{F}$	$\mathcal{F}$	$\mathcal{F}$	$\mathcal{F}$

One obtains $\mu_1(f_1) = \mu_1(f_2) = 0.23938$,
$\mu_1(f_3) = \mu_1(f_5) = 0.18894,\ \mu_1(f_4) = \mu_1(f_6) = 0.20779$
$\mu_1(f_7) = 0.142857$.

$\mathcal{F}^1(H_2)$	f_1	f_2	f_4	f_6	f_3	f_5	f_7
f_1	f_1, f_2	f_1, f_2	f_1, f_2, f_4, f_6	f_1, f_2, f_4, f_6	$\mathcal{F}^*$	$\mathcal{F}^*$	$\mathcal{F}$
f_2		f_1, f_2	f_1, f_2, f_4, f_6	f_1, f_2, f_4, f_6	$\mathcal{F}^*$	$\mathcal{F}^*$	$\mathcal{F}$
f_4			f_4, f_6	f_4, f_6	f_4, f_6, f_3, f_5	f_4, f_6, f_3, f_5	$\mathcal{F}^{**}$
f_6				f_4, f_6	f_4, f_6, f_3, f_5	f_4, f_6, f_3, f_5	$\mathcal{F}^{**}$
f_3					f_3, f_5	f_3, f_5	f_3, f_5, f_7
f_5						f_3, f_5	f_3, f_5, f_7
f_7							f_7

where $\mathcal{F}^* = \mathcal{F} - \{f_7\},\ \mathcal{F}^{**} = \mathcal{F} - \{f_1, f_2\}$.

We have clearly $\mathcal{F}^1(H_2) = \mathcal{F}^1(H_4)$, (see paragraph 8), whence

$$\forall k \geq 2,\ \ \mathcal{F}^{2k}(H_2) = \mathcal{F}^2(H_2),\ \ \mathcal{F}^{2k+1}(H_2) = \mathcal{F}^3(H_2)$$

and $\mathcal{F}^3(H_2) \simeq \mathcal{F}^2(H_2)$, so $f.g.(\mathcal{F}^0(H_2)) = 2$, while $s.f.g.(\mathcal{F}^0(H_2))$ is not definable.

7 The second partial hypergroupoid

Let H_3 be the following hypergroupoid:

H_3	1	2
1	1,2	2
2	2	1

It is clearly a quasihypergroup, but it is not associative. Endowing $\mathcal{F}(H_3)$ with the same hyperoperation as in the above paragraph, we obtain

$\mathcal{F}^0(H_3)$	f_1	f_2	f_3	f_4	f_5	f_6	f_7
f_1	f_2, f_5	$\emptyset$	$\mathcal{F}$	f_2, f_4	f_1, f_6	f_2, f_5	$\mathcal{F}$
f_2		f_1, f_3	f_2, f_4	f_1, f_3	$\mathcal{F}$	f_1, f_6	$\mathcal{F}$
f_3			$\mathcal{F}$	f_2, f_4	$\mathcal{F}$	$\mathcal{F}$	$\mathcal{F}$
f_4				f_1, f_3	$\mathcal{F}$	$\mathcal{F}$	$\mathcal{F}$
f_5					$\mathcal{F}$	f_1, f_6	$\mathcal{F}$
f_6						f_2, f_5	$\mathcal{F}$
f_7							$\mathcal{F}$

The values of membership functions are
$\mu_1(f_1) = \mu_1(f_2) = 0.23938, \quad \mu_1(f_6) = 0.207792 = \mu_1(f_4),$
$\mu_1(f_7) = 0.142857, \quad \mu_1(f_3) = \mu_1(f_5) = 0.18894.$

So, we obtain $\mathcal{F}^1(H_3) = \mathcal{F}^1(H_2) = \mathcal{F}^1(H_4)$, by consequence $f.g.(\mathcal{F}^0(H_3)) = 2$.

8 The hypergroupoid $\mathcal{F}(H)$ constructed from H_4

Let us consider now the hypergroupoid H_4:

H_4	1	2
1	1,2	2
2	1	2

H_4 is not clearly either associative or a quasi-hypergroup.

Endowing $\mathcal{F}^0(H_4)$ with the same hyperproduct defined in paragraph 6 and in in paragraph 7, we obtain the following structure:

$\mathcal{F}^0(H_4)$	f_1	f_2	f_3	f_4	f_5	f_6	f_7
f_1	f_1, f_6	f_2	$\mathcal{F}$	f_2, f_4	f_2, f_5	f_1, f_6	$\mathcal{F}$
f_2	f_1	f_2, f_4	f_1, f_3	f_2, f_4	$\mathcal{F}$	f_1, f_6	$\mathcal{F}$
f_3	f_1, f_6	f_2, f_4	$\mathcal{F}$	f_2, f_4	$\mathcal{F}$	f_1, f_6	$\mathcal{F}$
f_4	f_1	f_2, f_4	f_1, f_3	f_2, f_4	$\mathcal{F}$	f_1, f_6	$\mathcal{F}$
f_5	f_1, f_6	f_2, f_4	$\mathcal{F}$	f_2, f_4	$\mathcal{F}$	f_1, f_6	$\mathcal{F}$
f_6	f_1, f_6	f_2	$\mathcal{F}$	f_2, f_4	f_2, f_5	f_1, f_6	$\mathcal{F}$
f_7	f_1, f_6	f_2, f_4	$\mathcal{F}$	f_2, f_4	$\mathcal{F}$	f_1, f_6	$\mathcal{F}$

Calculating the membership function we obtain

$\mu_1(f_1) = \mu_1(f_2) = 0.346320,$

$\mu_1(f_3) = \mu_1(f_5) = 0.18045,$

$\mu_1(f_4) = \mu_1(f_6) = 0.29064, \ \mu_1(f_7) = 0.142857,$

whence setting $\mathcal{F}^* = \mathcal{F} - \{f_7\}$, we obtain the join space $\mathcal{F}^1(H_4)$:

$\mathcal{F}^1(H_4)$	f_1	f_2	f_4	f_6	f_3	f_5	f_7
f_1	f_1, f_2	f_1, f_2	f_1, f_2, f_4, f_6	f_1, f_2, f_4, f_6	$\mathcal{F}^*$	$\mathcal{F}^*$	$\mathcal{F}$
f_2		f_1, f_2	f_1, f_2, f_4, f_6	f_1, f_2, f_4, f_6	$\mathcal{F}^*$	$\mathcal{F}^*$	$\mathcal{F}$
f_4			f_4, f_6	f_4, f_6	f_4, f_6, f_3, f_5	f_4, f_6, f_3, f_5	$\mathcal{F} - \{f_1, f_2\}$
f_6				f_4, f_6	f_4, f_6, f_3, f_5	f_4, f_6, f_3, f_5	$\mathcal{F} - \{f_1, f_2\}$
f_3					f_3, f_5	f_3, f_5	f_3, f_5, f_7
f_5						f_3, f_5	f_3, f_5, f_7
f_7							f_7

Calculating the membership function we have

$\mu_2(f_1) = \mu_2(f_2) = 0.24603,$

$\mu_2(f_4) = \mu_2(f_6) = 0.2418,$

$\mu_2(f_3) = \mu_2(f_5) = 0.25119, \ \mu_2(f_7) = 0.28498.$

So we obtain

$\mathcal{F}^2(H_4)$	f_7	f_3	f_5	f_1	f_2	f_4	f_6
f_7	f_7	f_7, f_3, f_5	f_7, f_3, f_5	$\mathcal{F} - \{f_4, f_6\}$	$\mathcal{F} - \{f_4, f_6\}$	$\mathcal{F}$	$\mathcal{F}$
f_3		f_3, f_5	f_3, f_5	f_3, f_5, f_1, f_2	f_3, f_5, f_1, f_2	$\mathcal{F}^*$	$\mathcal{F}^*$
f_5			f_3, f_5	f_3, f_5, f_1, f_2	f_3, f_5, f_1, f_2	$\mathcal{F}^*$	$\mathcal{F}^*$
f_1				f_1, f_2	f_1, f_2	f_1, f_2, f_4, f_6	f_1, f_2, f_4, f_6
f_2					f_1, f_2	f_1, f_2, f_4, f_6	f_1, f_2, f_4, f_6
f_4						f_4, f_6	f_4, f_6
f_6							f_4, f_6

The membership functions corresponding to $\mathcal{F}^2(H_4)$ are the following:

$\mu_3(f_7) = 0.28498,\ \mu_3(f_3) = \mu_3(f_5) = 0.25119,$
$\mu_3(f_4) = \mu_3(f_6) = 0.24603,$
$\mu_3(f_1) = \mu_3(f_2) = 0.2418.$

So we obtain $\mathcal{F}^3(H_4)$ as it follows

$\mathcal{F}^3(H_4)$	f_7	f_3	f_5	f_4	f_6	f_1	f_2
f_7	f_7	f_7, f_3, f_5	f_7, f_3, f_5	$\mathcal{F} - \{f_1, f_2\}$	$\mathcal{F} - \{f_1, f_2\}$	$\mathcal{F}$	$\mathcal{F}$
f_3		f_3, f_5	f_3, f_5	f_3, f_5, f_4, f_6	f_3, f_5, f_4, f_6	$\mathcal{F}^*$	$\mathcal{F}^*$
f_5			f_3, f_5	f_3, f_5, f_4, f_6	f_3, f_5, f_4, f_6	$\mathcal{F}^*$	$\mathcal{F}^*$
f_4				f_4, f_6	f_4, f_6	f_4, f_6, f_1, f_2	f_4, f_6, f_1, f_2
f_6					f_4, f_6	f_4, f_6, f_1, f_2	f_4, f_6, f_1, f_2
f_1						f_1, f_2	f_1, f_2
f_2							f_1, f_2

Therefore we have $\forall k \geq 2,\ \mathcal{F}^{2k}(H_4) = \mathcal{F}^2(H_4),\ \mathcal{F}^{2k+1}(H_4) = \mathcal{F}^3(H_4).$

Since $\mathcal{F}^3(H_4)$ is isomorphic to $\mathcal{F}^2(H_4)$, we have $f.g.(\mathcal{F}^0(H_4)) = 2$, whereas $s.f.g.(\mathcal{F}^0(H_4))$ is not definable in this case.

9 The third partial hypergroupoid

Let us consider now the quasigroup H_5:

H_5	1	2
1	2	1
2	1	2

We obtain the following $\mathcal{F}^0(H_5)$:

$\mathcal{F}^0(H_5)$	f_1	f_2	f_3	f_4	f_5	f_6	f_7
f_1	$\emptyset$	$\emptyset$	f_2, f_4	f_1, f_3	f_2, f_5	f_1, f_6^2	$\mathcal{F}$
f_2		$\emptyset$	f_1, f_3	f_2, f_4	f_1, f_6	f_2, f_5	$\mathcal{F}$
f_3			f_2, f_4	f_1, f_3	$\mathcal{F}$	$\mathcal{F}$	$\mathcal{F}$
f_4				f_2, f_4	$\mathcal{F}$	$\mathcal{F}$	$\mathcal{F}$
f_5					f_1, f_6	f_2, f_5	$\mathcal{F}$
f_6						f_1, f_6	$\mathcal{F}$
f_7	$\mathcal{F}$	$\mathcal{F}$	$\mathcal{F}$	$\mathcal{F}$	$\mathcal{F}$	$\mathcal{F}$	$\mathcal{F}$

We find

$\mu_1(f_1) = \mu_1(f_2) = 0.2727,$

$\mu_1(f_3) = \mu_1(f_4) = \mu_1(f_5) = \mu_1(f_6) = 0.2222,\ \mu_1(f_7) = 0.142857.$

So, we obtain:

$\mathcal{F}^1(H_5)$	f_1	f_2	f_3	f_4	f_5	f_6	f_7
f_1	f_1, f_2	f_1, f_2	$\mathcal{F}^*$	$\mathcal{F}^*$	$\mathcal{F}^*$	$\mathcal{F}^*$	$\mathcal{F}$
f_2		f_1, f_2	$\mathcal{F}^*$	$\mathcal{F}^*$	$\mathcal{F}^*$	$\mathcal{F}^*$	$\mathcal{F}$
f_3			$\mathcal{F}''$	$\mathcal{F}''$	$\mathcal{F}''$	$\mathcal{F}''$	$\mathcal{F}'', f_7$
f_4				$\mathcal{F}''$	$\mathcal{F}''$	$\mathcal{F}''$	$\mathcal{F}'', f_7$
f_5					$\mathcal{F}''$	$\mathcal{F}''$	$\mathcal{F}'', f_7$
f_6						$\mathcal{F}''$	$\mathcal{F}'', f_7$
f_7							f_7

where $\mathcal{F}^* = \mathcal{F} - \{f_7\}$ and $\mathcal{F}'' = \{f_3, f_4, f_5, f_6\}$.

We find
$\mu_2(f_1) = \mu_2(f_2) = 0.21825,$
$\mu_2(f_3) = \mu_2(f_4) = \mu_2(f_5) = \mu_2(f_6) = 0.208658,\ \mu_2(f_7) = 0.243956.$

Finally,

$\mathcal{F}^2(H_5)$	f_7	f_1	f_2	f_3	f_4	f_5	f_6
f_7	f_7	f_7, f_1, f_2	f_7, f_1, f_2	$\mathcal{F}$	$\mathcal{F}$	$\mathcal{F}$	$\mathcal{F}$
f_1		f_1, f_2	f_1, f_2	$\mathcal{F}^*$	$\mathcal{F}^*$	$\mathcal{F}^*$	$\mathcal{F}^*$
f_2			f_1, f_2	$\mathcal{F}^*$	$\mathcal{F}^*$	$\mathcal{F}^*$	$\mathcal{F}^*$
f_3				$\mathcal{F}''$	$\mathcal{F}''$	$\mathcal{F}''$	$\mathcal{F}''$
f_4					$\mathcal{F}''$	$\mathcal{F}''$	$\mathcal{F}''$
f_5						$\mathcal{F}''$	$\mathcal{F}''$
f_6							$\mathcal{F}''$

So
$\mu_3(f_7) = 0.2674,\quad \mu_3(f_1) = \mu_3(f_2) = 0.22321,$
$\mu_3(f_3) = \mu_3(f_4) = \mu_3(f_5) = \mu_3(f_6) = 0.195238$, whence $\mathcal{F}^3(H_5) = \mathcal{F}^2(H_5)$, therefore $s.f.g.(\mathcal{F}^0(H_5)) = 2$.

References

[1] Ameri R. Zahedi M. M. - Hypergroup and join space induced by a fuzzy subset, PU.M.A. , 8, (2-3-4) (1997), pp. 155-168

[2] P. Corsini, Prolegomena of hypergroup theory, Aviani Editore, 1993.

[3] P. Corsini, Join spaces, power sets, fuzzy sets, in: Proc. Fifth Int. Cong. A.H.A. 1993, Iasi, Romania, Hadronic Press (1994), 45-52.

[4] P. Corsini, A new connection between hypergroups and fuzzy sets, Southeast Asian Bull. Math., 27(2003), 221-229.

[5] P. Corsini, Hypergraphs and hypergroups, Algebra Universalis 36(1996) 548-555.

[6] P. Corsini, I. Cristea, Fuzzy grade of i.p.s hypergroups of order less than or equal to 6, PU. M. A., 14(2003), no.4, 275-288.

[7] P. Corsini, I. Cristea, Fuzzy grade of i.p.s hypergroups of order 7, Iran. J. Fuzzy Syst., 1(2004), 15-32.

[8] P. Corsini, I. Cristea, Fuzzy sets and non complete 1-hypergroups, An. St. Univ. Ovidius Constanta Ser. Mat., 13(2005), no.1, 27-53.

[9] P. Corsini, V. Leoreanu, Join spaces associated with fuzzy sets, J. Combin. Inform. Syst. Sci., 20(1995), no. 1-4, 293-303.

[10] P. Corsini, V. Leoreanu, Applications of hyperstructure theory, Advances in Mathematics, Kluwer Academic Publishers, Dordercht, 2003.

[11] P. Corsini, V. Leoreanu-Fotea, On the grade of a sequence of fuzzy sets and join spaces determined by a hypergraph, Southeast Asian Bull. Math., 34(2010), no.2, 231-242.

[12] P. Corsini, V. Leoreanu-Fotea, A. Iranmanesh, On the sequence of hypergroups and membership functions determined by a hypergraph, J. Mult.-Valued Logic Soft Comput. 14 (2008), no.6, 565-577

[13] P. Corsini, R. Mahjoob - Multivalued Functions, fuzzy subsets and join spaces, Ratio Mathematica, 20 (2010), pp. 1-41

[14] I. Cristea, A property of the connection between fuzzy sets and hypergroupoids, Italian J. of Pure and Appl. Math. , 21, (2007) pp. 73-82

[15] I. Cristea, Complete hypergroups, 1-hypergroups and fuzzy sets, An. St. Univ. Ovidius Constanta Ser. Mat., 10(2002), no. 2, 25-37.

[16] I. Cristea, About the fuzzy grade of the direct product of two hypergroupoids, Iran. J. Fuzzy Syst, 7(2010), no.2, 95-108.

[17] I. Cristea, B. Davvaz, Atanassov's intuitionistic fuzzy grade of hypergroups, Inform. Sci., 180(2010), 1506-1517.

[18] B. Davvaz, Fuzzy H_v-groups, Fuzzy Sets and Systems, 101(1999), 191-195.

[19] B. Davvaz, Fuzzy H_v-submodules, Fuzzy Sets and Systems, 117(2001), 477-484.

[20] B. Davvaz, P. Corsini, V. Leoreanu-Fotea, Atanassov's intuitionistic (S,T)-fuzzy n-ary sub-hypergroups and their properties, Inform. Sci., 179(2009), 654-666.

[21] B. Davvaz, W.A. Dudek, Intuitionistic H_v-ideals, Int. J. Math. Sci., (2006) Article ID 65921, 11 pp.

[22] B. Davvaz, V. Leoreanu-Fotea, Hyperring theory and Applications, Hadronic Press, Inc, 115, Palm Harbor, USA, 2007.

[23] Ath. Kehagias, An example of L-fuzzy join space, Circolo Matematico di Palermo 52, (2003), pp. 322-350.

[24] V. Leoreanu-Fotea, P. Corsini, Soft hypergroups, Critical Review. A publication of Society for Mathematics of Uncertainty, Creighton University - USA, July 2010, vol. IV, 81-97;

[25] F. Marty, Sur une generalization de la notion de groupe, Eight Congress Math. Scandenaves, Stockholm, (1934), 45-49.

[26] W. Prenowits, J. Jantosciak, Geometries and join spaces, J. Reind und Angew Math., 257(1972), 100-128.

[27] I.G. Rosenberg, Hypergroups and join spaces determined by relations, Italian J. Pure and Applied Mathematics (4) (1998), 93101.

[28] A. Rosenfeld, Fuzzy groups, J. Math. Anal. Appl., 35(1971), 512-517.

[29] K. Serafimidis, Ath. Kehagias, M. Konstantinidou, The L-fuzzy Corsini Join Hyperoperation, Italian J. of Pure and Appl. Math.,12, (2003), pp.83-90.

[30] M. Stefanescu, I. Cristea, On the fuzzy grade of hypergroups, Fuzzy Sets and Systems, 159(2008), 1097-1106.

[31] T. Vougiouklis, Hyperstructures and their representations, Hadronic Press, Inc, 115 Palm Harbor, USA. 1994.

[32] L.A. Zadeh, Fuzzy sets, Inform. Control, 8(1965), 338-353.

A Survey on Polygroups and Their Properties

B. Davvaz

Department of Mathematics, Yazd University,
Yazd, Iran
E-mail: davvaz@yazduni.ac.ir
bdavvaz@yahoo.com

The overall aim of this paper is to present an introduction to some of the results, methods and ideas about polygroups. We review: (1) Definition and examples of polygroups; (2) Isomorphism theorems of polygroups; (3) Fuzzy polygroups; (4) n-ary polygroups and fuzzy n-ary subpolygroups.

Keywords: Polygroup; Subpolygroup.

1. Definition and examples of polygroups

A *polygroup*[1] is a system $\wp =< P, \cdot, e, ^{-1} >$, where $e \in P$, $^{-1}$ is a unitary operation on P, $\cdot$ maps $P \times P$ into the non-empty subsets of P, and the following axioms hold for all x, y, z in P: $(x \cdot y) \cdot z = x \cdot (y \cdot z)$, $e \cdot x = x \cdot e = x$, $x \in y \cdot z$ implies $y \in x \cdot z^{-1}$ and $z \in y^{-1} \cdot x$. The following elementary facts about polygroups follow easily from the axioms: $e \in x \cdot x^{-1} \cap x^{-1} \cdot x$, $e^{-1} = e$, $(x^{-1})^{-1} = x$, and $(x \cdot y)^{-1} = y^{-1} \cdot x^{-1}$, where $A^{-1} = \{a^{-1} | \ a \in A\}$.

Suppose that H is a subgroup of a group G. Define a system $G//H =< \{HgH \mid g \in G\}, *, H, ^{-I} >$, where $(HgH)^{-I} = Hg^{-1}H$ and $(Hg_1H) * (Hg_2H) = \{Hg_1hg_2H \mid h \in H\}$. The algebra of double cosets $G//H$ is a polygroup introduced in (Dresher and Ore[2]).

Suppose G is a projective geometry with a set P of points and suppose, for $p \neq q$, $\overline{pq}$ denoted the set of all points on the unique line through p and q. Choose an object $I \notin P$ and form the system $P_G =< P \cup \{I\}, \cdot, I, ^{-1} >$, where $x^{-1} = x$ and $I \cdot x = x \cdot I = x$ for all $x \in P \cup \{I\}$ and for $p, q \in P$,

$$p \cdot q = \begin{cases} \overline{pq} - \{p, q\} & \text{if } p \neq q \\ \{p, I\} & \text{if } p = q. \end{cases}$$

P_G is a polygroup (Prenowitz[3]).

Let $< P_1, \cdot, e_1, ^{-1} >$ and $< P_2, *, e_2, ^{-I} >$ be two polygroups, then on

$P_1 \times P_2$ we can define a hyperproduct as follows: $(x_1, y_1) \circ (x_2, y_2) = \{(x, y) \mid x \in x_1x_2,\ y \in y_1 * y_2\}$. We recall this the *direct hyperproduct* of P_1 and P_2. Clearly, $P_1 \times P_2$ equipped with the usual direct hyperproduct becomes a polygroup. In Comer,[4] extensions of polygroups by polygroups were introduced in the following way. Suppose that $\mathcal{A} =< A, \cdot, e, ^{-1} >$ and $\mathcal{B} =< B, \cdot, e, ^{-1} >$ are two polygroups whose elements have been renamed so that $A \cap B = \{e\}$. A new system $\mathcal{A}[\mathcal{B}] =< M, *, e, ^I >$ called the *extension* of $\mathcal{A}$ by $\mathcal{B}$ is formed in the following way: Set $M = A \cup B$ and let $e^I = e$, $x^I = x^{-1}$, $e * x = x * e = x$ for all $x \in M$, and for all $x, y \in M - \{e\}$

$$x * y = \begin{cases} x \cdot y & \text{if } x, y \in A \\ x & \text{if } x \in B,\ y \in A \\ y & \text{if } x \in A,\ y \in B \\ x \cdot y & \text{if } x, y \in B,\ y \neq x^{-1} \\ x \cdot y \cup A & \text{if } x, y \in B,\ y = x^{-1}. \end{cases}$$

$\mathcal{A}[\mathcal{B}]$ always yields a polygroup.[4]

2. Isomorphism theorems of polygroups

A nonempty subset K of a polygroup P is a *subpolygroup* of P if $a, b \in K$ implies $ab \subseteq K$ and $a \in K$ implies $a^{-1} \in K$. A subpolygroup N of a polygroup P is *normal* in P if $a^{-1}Na \subseteq N$, for all $a \in P$. If N is a normal subpolygroup of a polygroup P, then $Na = aN$ and $(Na)(Nb) = Nab$, for all $a, b \in P$. Moreover, we have $Na = Nb$, for all $b \in Na$. If N is a normal subpolygroup of P, then we define the relation $x \equiv y(modN)$ if and only if $xy^{-1} \cap N \neq \emptyset$. This relation is denoted by N_P and it is an equivalence relation. Let $[P : N]$ be the set of all equivalence classes. Then $< [P : N], \odot, N_P(e), ^{-I} >$ is a polygroup, where $N_P(a)^{-I} = N_P(a^{-1})$. Note that $[P : N] = P/N$.

We recall that a strong homomorphism $\varphi : P_1 \longrightarrow P_2$ is an *isomorphism* if φ is one to one and onto. We write $P_1 \cong P_2$ if P_1 is isomorphic to P_2. Because P_1 is a polygroup, $e \in aa^{-1}$ for all $a \in P_1$, then we have $\varphi(e_1) \in \varphi(a) * \varphi(a^{-1})$ or $e_2 \in \varphi(a) * \varphi(a^{-1})$ which implies $\varphi(a^{-1}) \in \varphi(a)^{-1} * e_2$, therefore $\varphi(a^{-1}) = \varphi(a)^{-1}$ for all $a \in P_1$. Moreover, if φ is a strong homomorhism from P_1 into P_2, then the kernel of φ is the set $ker\varphi = \{x \in P_1 \mid \varphi(x) = e_2\}$. It is trivial that $ker\varphi$ is a subpolygroup of P_1 but in general is not normal in P_1. The strong homomorphism φ is injective if and only if $ker\varphi = \{e_1\}$. Let φ be a strong homomorphism from P_1 into P_2 with kernel K such that K is a normal subpolygroup of P_1, then $P_1/K \cong Im\varphi$.

Let P be a polygroup. We define the relation β^* as the smallest equivalence relation on P such that the quotient P/β^*, the set of all equivalence classes, is a group. In this case β^* is called the *fundamental equivalence relation* on P and P/β^* is called the *fundamental group*. This relation was introduced by Koskas[5] and studied by others, for example see.[6–8] The product $\otimes$ in P/β^* is defined as follows: $\beta^*(x) \otimes \beta^*(y) = \beta^*(z)$ for all $z \in \beta^*(x) \cdot \beta^*(y)$. Let $\mathcal{U}_P$ be the set of all finite products of elements of P. We define the relation β as follows: $x\beta y$ if and only if $\{x, y\} \subseteq u$ for some $u \in \mathcal{U}_P$. We have $\beta^* = \beta$ for hypergroups. Since polygroups are certain subclasses of hypergroups, we have $\beta^* = \beta$. The kernel of the canonical map $\varphi : P \longrightarrow P/\beta^*$ is called the *core* of P and is denoted by ω_P. Here we also denote by ω_P the unit of P/β^*. It is easy to prove that the following statements: $\omega_P = \beta^*(e)$ and $\beta^*(x)^{-1} = \beta^*(x^{-1})$ for all $x \in P$.

Theorem 2.1.[9] *Let β_1^*, β_2^* and β^* be fundamental equivalence relations on polygroups P_1, P_2 and $P_1 \times P_2$ respectively. Then $(P_1 \times P_2)/\beta^* \cong P_1/\beta_1^* \times P_2/\beta_2^*$.*

Corollary 2.1.[9] *If N_1, N_2 are normal subpolygroups of P_1, P_2 respectively, and β_1^*, β_2^* and β^* fundamental equivalence relations on P_1/N_1, P_2/N_2 and $(P_1 \times P_2)/(N_1 \times N_2)$ respectively, then $((P_1 \times P_2)/(N_1 \times N_2))/\beta^* \cong (P_1/N_1)/\beta_1^* \times (P_2/N_2)/\beta_2^*$.*

Let f be a strong homomorphism from P_1 into P_2 and let β_1^*, β_2^* be fundamental relations on P_1, P_2 respectively. Then we define $\overline{kerf} = \{\beta_1^*(x) \mid x \in P_1,\ \beta_2^*(f(x)) = \omega_{P_2}\}$.

Lemma 2.1.[9] *$\overline{kerf}$ is a normal subgroup of the fundamental group P_1/β_1^*.*

Theorem 2.2.[9] *Let P be a polygroup, M, N two normal subpolygroups of P with $N \subseteq M$ and $\phi : P/N \longrightarrow P/M$ canonical map. Suppose that β_M^*, β_N^* are the fundamental equivalence relations on P/M, P/N, respectively. Then $((P/N)/\beta_N^*)/\overline{ker\phi} \cong (P/M)/\beta_M^*$.*

3. Fuzzy polygroups

Let P be a polygroup. A fuzzy subset μ of P is called a *fuzzy subpolygeoup*[10] if for all $x, y \in P$, $\min\{\mu(x), \mu(y)\} \leq \mu(z)$ for all $z \in x \cdot y$ and $\mu(x) \leq \mu(x^{-1})$. The following elementary facts about fuzzy subpolygroups follow easily from the axioms: $\mu(x) = \mu(x^{-1})$ and $\mu(x) \leq \mu(e)$, for all $x \in P$. A *strong level set* $\mu_t^>$ of a fuzzy subset μ in P is defined by $\mu_t^> = \{x \in P \mid \mu(x) > t\}$. Let P be a polygroup and μ be a fuzzy subset of P. Then μ is a fuzzy subpolygroup of P if and only if each non-empty strong level set of μ is a subpolygroup of P.[10] Let μ be a fuzzy subpolygroup of P. Then μ is said to be *normal* if for all $x, y \in P$, $\mu(z) = \mu(z')$, for all $z \in x \cdot y$ and $z' \in y \cdot x$. It is obvious that if μ is a fuzzy normal subpolygroup of P, then for all $x, y \in P$, $\mu(z) = \mu(z')$, for all $z, z' \in x \cdot y$.

Let P be a polygroup, μ be a fuzzy subpolygroup of P and $t \in [0, \mu(e)]$. Then the fuzzy subset $\mu^t a$ of P which is defined by $\mu^t a(x) = \min\{\inf_{z \in x \cdot a} \mu(z), t\}$ is called the *fuzzy right t-coset* of μ. The *fuzzy left t-coset* $a\mu^t$ of μ is defined similarly. If μ is a fuzzy normal subpolygroup of P, then it is easy to see that $a\mu^t = \mu^t a$. We denote the set of all fuzzy right t-cosets of μ by P/μ^t, i.e., $P/\mu^t = \{\mu^t a \mid a \in P\}$.

Theorem 3.1.[10] *Let P be a polygroup, μ be a fuzzy subpolygroup of P and $t \in [0, \mu(e)]$. Then, there is a bijection between P/μ^t and P/μ_t.*

Theorem 3.2.[10] *Let P be a polygroup, μ be a fuzzy normal subpolygroup of P and $t \in [0, \mu(e)]$. Then $(P/\mu^t, *)$ is a polygroup (called the polygroup of fuzzy t-cosets induced by μ and t), where the hyperoperation $*$ is defined as follows: $* : P/\mu^t \times P/\mu^t \longrightarrow \wp^*(P/\mu^t)$ by $(\mu^t a, \mu^t b) \mapsto \{\mu^t z \mid z \in \mu_t \cdot a \cdot b\}$.*

Let $\{P_\alpha | \ \alpha \in I\}$ be a collection of polygroups and let μ_α be a fuzzy subset of P_α, for all $\alpha \in I$. We define *direct product* μ_α by $(\prod_{\alpha \in I} \mu_\alpha)(x) = \inf\{\mu_\alpha(x_\alpha) | \ \alpha \in I\}$, where $x =< x_\alpha >$ and $< x_\alpha >$ denotes an element of the direct product $\prod_{\alpha \in I} P_\alpha$. Let $\{P_\alpha \mid \alpha \in I\}$ be a collection of polygroups and let μ_α be a fuzzy subpolygroup of P_α. Then $\prod_{\alpha \in I} \mu_\alpha$ is a fuzzy subpolygroup of $\prod_{\alpha \in I} P_\alpha$. Let P_1, P_2 be two polygroups and μ, λ be fuzzy subpolygroups of P_1, P_2 respectively. Then $(\mu \times \lambda)_t = \mu_t \times \lambda_t$.

Proposition 3.1.[11] *Let P_1, P_2 be two polygroups and μ be a fuzzy subpolygroup of $P_1 \times P_2$. Then μ_1, μ_2 are fuzzy subpolygroups of P_1, P_2 respectively, where $\mu_1(x) = \mu(x, e_2)$, $x \in P_1$ and $\mu_2(x) = \mu(e_1, y)$, $x \in P_2$. Moreover, $\mu_1 \times \mu_2 \subseteq \mu$.*

Proposition 3.2.[11] *Let P_1, P_2 be two polygroups and μ, λ be fuzzy subsets of P_1, P_2, respectively. If $\mu(e_1) = \lambda(e_2) = 1$ and $\mu \times \lambda$ is a fuzzy subpolygroup of $P_1 \times P_2$, then μ and λ are fuzzy subpolygroups of P_1, P_2, respectively.*

Let P be a polygroup and μ be a fuzzy subset of P. The fuzzy subset μ_{β^*} on P/β^* is defined as follows: $\mu_{\beta^*} : P/\beta^* \longrightarrow [0, 1]$ by $\mu_{\beta^*}(\beta^*(x)) = \sup\limits_{a \in \beta^*(x)} \{\mu(a)\}$.

Theorem 3.3.[11] *Let P be a polygroup and μ be a fuzzy subpolygroup of P. Then μ_{β^*} is a fuzzy subgroup of the group P/β^*.*

Theorem 3.4.[11] *Let P_1, P_2 be plygroups and β_1^*, β_2^* and β^* be fundamental equivalence relations on P_1, P_2 and $P_1 \times P_2$ respectively. If μ, λ are fuzzy subpolygroups of P_1, P_2, respectively. Then $(\mu \times \lambda)_{\beta^*} = \mu_{\beta_1^*} \times \lambda_{\beta_2^*}$.*

A *t-norm* is a mapping $T : [0,1] \times [0,1] \to [0,1]$ satisfying, for all $x, y, z \in [0,1]$, $T(x,1) = x$, $T(x,y) = T(y,x)$, $T(x, T(y,z)) = T(T(x,y),z))$, $T(x,y) \leq T(x,z)$ whenever $y \leq z$. Let T be a t-norm on $[0,1]$. We define $T_{i=1}^n x_i$, and D_T as follows: $T_{i=1}^n x_i = T(T_{i=1}^{n-1} x_i, x_n) = T(x_1, \ldots, x_n)$ and $D_T = \{x \in [0,1] \mid T(x,x) = x\}$. It is clear that when $T = \wedge$, D_T coincides with $[0,1]$. In particular, for every t-norm T, we have $T(x_1, \ldots, x_n) \leq x_1 \wedge \ldots \wedge x_n$ for all $x_1, \ldots, x_n \in [0,1]$. There exist uncountably t-norms. The following are the three basic t-norms T_M, T_P, and T_L given by, respectively: $T_M(x,y) = \min(x,y)$, $T_P(x,y) = x \cdot y$, $T_L(x,y) = \max(x+y-1, 0)$. Let T_1 and T_2 be two *t*-norms. T_2 is said to *dominate* T_1 and write $T_1 \ll T_2$ if for all $a,b,c,d \in [0,1]$, $T_1(T_2(a,c), T_2(b,d)) \leq T_2(T_1(a,b), T_1(c,d))$ and T_1 is said weaker then T_2 or T_2 is stronger than T_1 and write $T_1 \leq T_2$ if for all $x, y \in [0,1]$, $T_1(x,y) \leq T_2(x,y)$. Since a triangular norm T is a generalization of the minimum function, We can replace the axiom $\min\{\mu(x), \mu(y)\} \leq \mu(z)$, for all $z \in x \cdot y$, occurring in the definition of a fuzzy subpolygroup by the inequality $\mu(x) T \mu(y) \leq \mu(z)$, for all $z \in x \cdot y$. Also, we can generalize the fuzzy subsets of X, to L-subsets, as a function from X to a lattice L. From now on this section, L is a complete lattice, i.e., there is a partial order $\leq$ on L such that, for any $S \subseteq L$, infimum of S and supremum of S exist and these are denoted by $\bigwedge_{s \in S}\{s\}$ and $\bigvee_{s \in S}\{s\}$, respectively. In particular, for any elements $a, b \in L$, $\inf\{a,b\}$ and $\sup\{a,b\}$ is denoted by $a \wedge b$ and $a \vee b$, respectively. Also L is a distributive lattice with a least element 0 and a greatest element 1.

Let T be a triangular norm on the complete lattice $(L, \leq, \vee, \wedge)$. An L-subset $\mu \in F^L(P)$ of the polygroup P is a *TL-subpolygroup* of P if the following axioms hold: $Im(\mu) \subseteq I_T$, $T(\mu(x), \mu(y)) \leq \bigwedge_{\alpha \in x \cdot y} \mu(\alpha)$, for all $x, y \in P$, $\mu(x) \leq \mu(x^{-1})$, for all $x \in P$. If μ is a TL-subpolygroup of P, then $\mu(e) \geq \mu(x)$ for all $x \in P$.

Theorem 3.5.[12] *Let T be a t-norm on the complete lattice $(L, \leq, \vee, \wedge)$, μ be an L-subset of P such that $Im(\mu) \subseteq I_T$ and $b = \bigvee Im(\mu)$. Then μ is a TL-subpolygroup of P if and only if $\mu^{-1}[a,b]$ is a subpolygroup of P, whenever $a \in I_T$ and $0 < a \leq b$.*

Let P_1, P_2 be polygroups and μ, λ be TL-subpolygroups of P_1, P_2 respectively. The *product* of μ, λ is defined to be the TL-subset $\mu \times \lambda$ of $P_1 \times P_2$ with $(\mu \times \lambda)(x, y) = T(\mu(x), \lambda(x))$, for all $(x, y) \in P_1 \times P_2$. By this definition, $\mu \times \lambda$ is a TL-subpolygroup of $P_1 \times P_2$.

Let μ be a TL-normal subpolygroup of a polygroup P. We define $x \sim y(mod\mu)$ if there exists $a \in x \cdot y^{-1}$ such that $\mu(a) = \mu(e)$. The relation $\sim$ is an equivalence relation. If $x \sim y(mod\mu)$ then $\mu(x) = \mu(y)$. Suppose that $\mu[x]$ is the equivalence class containing x, we denote $P/\sim$ the set of all equivalence classes, i.e., $P/\sim = \{\mu[x] \mid x \in P\}$. We define $\mu[x] \odot \mu[y] = \{\mu[z] \mid z \in \mu[x] \cdot \mu[y]\}$, for all $\mu[x],\ \mu[y] \in P/\sim$.

Corollary 3.1.[12] *Let P be an abelian polygroup. Then $< P/\sim, \odot, \mu[e], ^I >$ is a polygroup, where $\mu[x]^I = \mu[x^{-1}]$.*

Lemma 3.1.[12] *Let T be a t-norm on the complete lattice $(L, \leq, \vee, \wedge)$ and let $< P_1, \cdot, e_1, ^{-1} >$ and $< P_2, *, e_2, ^{-1} >$ be polygroups and μ, λ be TL-subpolygroups of P_1, P_2 respectively. Then $(x_1, y_1) \sim (x_2, y_2)(mod(\mu \times \lambda))$ if and only if $x_1 \sim x_2(mod\mu)$, $y_1 \sim y_2(mod\lambda)$, for all (x_1, y_1), $(x_2, y_2) \in P_1 \times P_2$.*

Theorem 3.6.[12] *Let T be a t-norm on the complete lattice $(L, \leq, \vee, \wedge)$ and let $< P_1, \cdot, e_1, ^{-1} >$ and $< P_2, *, e_2, ^{-1} >$ be polygroups and μ, λ be TL-subpolygroups of P_1, P_2 respectively. Then $P_1/\sim \times P_2/\sim \cong (P_1 \times P_2)/\sim$.*

Theorem 3.7.[12] *Let P be a polygroup and μ be a TL-subpolygroup of P. Then, μ_{β^*} is a TL-subgroup of P/β^*.*

4. n-ary polygroups and fuzzy n-ary subpolygroups

The concept of n-ary hypergroup is defined by Davvaz and Vougiouklis[13] in 2006, which is a generalization of the concept of hypergroup in the sense of Marty and a generalization of n-ary group, too. Then this concept studied by many authors, for example see.[14–18] We denote by H^n the cartesian product $H \times \ldots \times H$ where H appears n times. An element of H^n is denoted by $(x_1, \ldots, x_n)$ where $x_i \in H$ for any i with $1 \leq i \leq n$. In general, a mapping $f : H^n \longrightarrow \wp^*(H)$ is called an *n-ary hyperoperation*. Let f be an n-ary hyperoperation on H and $A_1, \ldots, A_n$ be non-empty subsets of H. We define $f(A_1, \ldots, A_n) = \cup\{f(x_1, \ldots, x_n) \mid x_i \in A_i, i = 1, \ldots, n\}$. We use the following abbreviated notation: The sequence $x_i, x_{i+1}, \ldots, x_j$ is denoted by x_i^j. For $j < i$, x_i^j is the empty set. A non-empty set H with an n-ary hyperoperation $f : H^n \longrightarrow \wp^*(H)$ will be called an *n-ary hypergroupoid* and will be denoted by (H, f). An n-ary hypergroupoid (H, f) will be called an *n-ary semihypergroup* if and only if the following associative axiom holds: $f(x_1^{i-1}, f(x_i^{n+i-1}), x_{n+i}^{2n-1}) = f(x_1^{j-1}, f(x_j^{n+j-1}), x_{n+j}^{2n-1})$ for every $i, j \in \{1, 2, \ldots, n\}$ and $x_1, x_2, \ldots x_{2n-1} \in H$. If for all $(a_1, a_2, \ldots, a_n) \in H^n$, the set $f(a_1, a_2, \ldots, a_n)$ is singleton, then f is called an *n-ary operation* and (H, f) is called an *n-ary groupoid* (respectively, *n-ary semigroup*). An n-ary semihypergroup (H, f) in which equation $b \in f(a_1^{i-1}, x_i, a_{i+1}^n)$ $(*)$ has a solution $x_i \in H$ for every $a_1, \ldots, a_{i-1}, a_{i+1}, \ldots, a_n, b \in H$ and $1 \leq i \leq n$, is called an *n-ary hypergroup.* If f is n-ary operation then the equation $(*)$ becomes: $b = f(a_1^{i-1}, x_i, a_{i+1}^n)$ $(**)$. In this case (H, f) is an *n-ary group.*

An n-ary polygroup[19] is a multivalued system $< P, f, e, ^{-1} >$, where $e \in P$, $^{-1}$ is a unitary operation on P, (P, f) is an n-ary semihypergroup, and the following axioms hold: e is a unique element such that $f(\underbrace{e, \ldots, e}_{i-1}, x, \underbrace{e, \ldots, e}_{n-i}) = x$ and $x \in f(x_1^n)$ implies $x_i \in f(x_{i-1}^{-1}, \ldots, x_1^{-1}, x, x_n^{-1}, \ldots, x_{i+1}^{-1})$. It is clear that every 2-ary polygroup is a polygroup. An n-ary subpolygroup N of P is said to be *normal* in P if for every $a \in P$, $f(a^{-1}, N, a, e*) \subseteq N$. Suppose that K is an n-ary subpolygroup of P. Ghadiri and Waphare[19] defined the relation $\equiv_K$ on $P^{(n-1)}$ as follows: $(x_2^n) \equiv_k (y_2^n)$ if and only if $f(K, x_2^n) = f(K, y_2^n)$, for $(x_2^n), (y_2^n) \in P^{(n-1)}$. It is clear that the relation $\equiv_k$ is an equivalence relation on $P^{(n-1)}$. The class of $(x_2^n) \in P^{(n-1)}$ is denoted by $K[x_2^n]$, and the set of all equivalence classes is denoted by $P^{(n-1)}/K$.

Lemma 4.1.[19] *For sequences a_2^n, b_2^n in P, (1) $f(K, a_2^n) \cap f(K, b_2^n) \neq \emptyset$ implies $f(K, a_2^n) = f(K, b_2^n)$ and $K[a_2^n] = K[b_2^n]$, (2) if $x \in f(K, a_2^n)$, then $f(K, x, e*) = f(K, a_2^n)$ and $K[x, \underbrace{e, \ldots, e}_{n-2}] = K[a_2^n]$.*

We denote $K[x, e, \ldots, e]$ by $K[x, e*]$.

Theorem 4.1.[19] *Let $< P, f, e, ^{-1} >$ be an n-ary polygroup and N be an n-ary normal subpolygroup of P. For $N[a, e*]$ and $N[a_1, e*], \ldots, N[a_n, e*]$ in $P^{(n-1)}/N$, we define $F : P^{(n-1)}/N \times \ldots P^{(n-1)}/N \longrightarrow \mathcal{P}^*(P^{(n-1)}/N)$ by $F(N[a_1, e*], \ldots, N[a_n, e*]) = \{N[t, e*] \mid t \in f(a_1^n)\}$ and $^{-I} : P^{(n-1)}/N \longrightarrow P^{(n-1)}/N$ by $N[a, e*]^{-I} = N[a^{-1}, e*]$. Then $< P^{(n-1)}/N, F, N, ^{-I} >$ is an n-ary polygroup.*

Let P be an n-ary polygroup. A fuzzy subset of P is called a *fuzzy n-ary subpolygroup*[20] of P if the following axioms hold: (1) $\min\{\mu(x_1), \ldots, \mu(x_n)\} \leq \inf_{z \in f(x_1^n)}\{\mu(z)\}$ and (2) $\mu(x) \leq \mu(x^{-1})$ for all $x \in P$.

Theorem 4.2.[20] *A fuzzy subset μ on an n-ary polygroup P is a fuzzy n-ary subpolygroup if and only if each its non-empty level subset is an n-ary subpolygroup of P.*

Let μ be a fuzzy n-ary subpolygroup of P. Then μ is said to be *normal* if for all $x, y \in P$, $\mu(z) = \mu(z')$, for all $z \in f(x, y, e*)$ and for all $z' \in f(y, x, e*)$.

Theorem 4.3.[20] *Let μ be a fuzzy n-ary subpolygroup of P. Then the following conditions are equivalent: (1) μ is normal, (2) for all $x, y \in P$ and for all $z \in f(x, y, x^{-1}, e*)$, $\mu(z) = \mu(y)$,(3) fFor all $x, y \in P$ and for all $z \in f(x, y, x^{-1}, e*)$, $\mu(z) \geq \mu(y)$, (4) for all $x, y \in P$ and for all $z \in f(x^{-1}, y^{-1}, x, y, e*)$, $\mu(z) \geq \mu(y)$.*

Theorem 4.4.[20] *Let μ be a fuzzy n-ary subpolygroup of P. Then μ is normal if and only if for all $x, y \in P$, $\mu(x) = \mu(y)$ if and only if $\mu(z) = \mu(z')$, for all $z \in f(x, y, e*)$ and for all $z' \in f(y, x, e*)$.*

Let $\varphi : P_1 \longrightarrow P_2$ be a function and μ be a fuzzy set subset of P_1. Then φ induces a fuzzy subset $\varphi(\mu)$ in $\varphi(\mu)$ in P_2 defined by:

$$\varphi(\mu)(y) = \begin{cases} \sup\limits_{x \in \varphi^{-1}(y)} \{\mu(x)\} & \text{if } y \in \varphi(P_1) \\ 0 & \text{otherwise.} \end{cases}$$

Here $\varphi(\mu)$ is called the *image* of μ under φ. Let λ be a fuzzy subset in P_2. Then φ induces a fuzzy subset $\varphi^{-1}(\lambda)$ in P_1 defined by $\varphi^{-1}(\lambda)(x) = \lambda(\varphi(x))$ for all $x \in P_1$. Here $\varphi^{-1}(\lambda)$ is called the *inverse image* of λ under μ.

Theorem 4.5.[20] *Let $\varphi : P_1 \longrightarrow P_2$ be a strong homomorphism. (1) If μ is a fuzzy n-ary subpolygroup of P_1, then $\varphi(\mu)$ is a fuzzy n-ary subpolygroup of P_2. (2) If λ is a fuzzy n-ary subpolygroup of P_2, then $\varphi^{-1}(\lambda)$ is a fuzzy n-ary subpolygroup of P_1.*

References

1. S.D. Comer, Polygroups derived from cogroups, J. Algebra **89** (1984), 397-405.
2. M. Dresher and O. Ore, Theory of multigroups, *Amer. J. Math.* **60** (1938), 705-733.
3. W. Prenowitz, Projective geometries as multigroups, *Amer. J. Math.* **65** (1943), 235-256.
4. S.D. Comer, Extension of polygroups by polygroups and their representations using colour schemes, in: *Lecture notes in Meth.*, No 1004, Universal Algebra and Lattice Theory, (1982), 91-103.

5. M. Koskas, Groupoides, Demi-hypergroupes et hypergroupes, *J. Math. Pures et Appl.* **49** (1970), 155-192.
6. P. Corsini, *Prolegomena of hypergroup theory*, Second edition, Aviani editor, 1993.
7. T. Vougiouklis, *Hyperstructures and their representations*, Hadronic Press Inc, USA, 1994.
8. B. Davvaz and V. Leoreanu-Fotea, *Hyperring theory and applications*, International Academic Press, USA, 2007.
9. B. Davvaz, Isomorphism theorems of polygroups, *Bulletin of the Malaysian Mathematical Sciences Society (2)* **33** (2010), 385-392.
10. M.M. Zahedi, M. Bolurian and A. Hasankhani, On polygroups and fuzzy subpolygroups, *J. Fuzzy Math.* **3** (1995), 1-15.
11. B. Davvaz, Fuzzy weak polygroups, in: *Algebraic hyperstructures and applications (Alexandroupoli-Orestiada, 2002)*, 127–135, Spanidis, Xanthi, 2003.
12. B. Davvaz, TL-subpolygroups of a polygroup, *Pure Math. Appl.* **12** (2001), 137-145.
13. B. Davvaz and T. Vougiouklis, n-Ary hypergroups, *Iran. J. Sci. Technol. Trans. A Sci.*, **30** (2006), 165-174.
14. B. Davvaz, W. A. Dudek and T. Vougiouklis, A Generalization of n-ary algebraic systems, *Comm. Algebra* **37** (2009), 1248-1263.
15. B. Davvaz, W. A. Dudek and S. Mirvakili, Neutral elements, fundamental relations and n-ary hypersemigroups, *Internat. J. Algebra Comput.* **19** (2009), 567-583.
16. V. Leoreanu-Fotea and B. Davvaz, Join n-spaces and lattices, *J. Mult.-Valued Logic Soft Comput.* **15** (2009), 421-432.
17. V. Leoreanu-Fotea and B. Davvaz, n-hypergroups and binary relations, *European J. Combin.* **29** (2008), 1207-1218.
18. J. Zhan, B. Davvaz and K.P. Shum, Probabilistic n-ary hypergroups, *Information Sciences* **180** (2010), 1159-1166.
19. M. Ghadiri and B.N. Waphare, n-Ary polygroups, *Iran. J. Sci. Technol. Trans. A Sci.*, to appear.
20. B. Davvaz, P. Corsini and V. Leoreanu-Fotea, Fuzzy n-ary subpolygroups, *Comput. Math. Appl.* **57** (2009), 141-152.

Semigroups of n-ary Operations on Finite Sets

K. Denecke and Y. Susanti

Dedicated to Prof. K. P. Shum's 70th birthday

Abstract

On the set $O^n(A)$ of all n-ary operations defined on an arbitrary set A, a binary associative operation $+$ can be defined by $f + g := f(g, \ldots, g)$ giving a semigroup $(O^n(A); +)$. We characterize all idempotent, regular and completely regular elements in $(O^n(A); +)$. We also describe idempotent, regular, completely regular, right-zero, left-zero and constant subsemigroups of the semigroup $(O^n(A); +)$. Moreover we determine all normal bands, regular bands, rectangular bands, semilattices, groups and consider Green's relations in $(O^n(A); +)$.

1 Introduction

Transformation semigroups belong to the most important classes of semigroups since by Cayley's theorem any semigroup is isomorphic to a transformation semigroup. Let $O^1(A)$ be the set of all transformations on A, i.e. operations with domain A and values in A. Together with the composition $\circ$ of operations defined on A one obtains the full transformation semigroup $\mathcal{O}^1(A) := (O^1(A); \circ)$. Any subsemigroup of $\mathcal{O}^1(A)$ is called a transformation semigroup. Let $O^{1-1}(A)$ be the set of all bijective transformations on A. The full transformation semigroup is regular, i.e. every element $f \in O^1(A)$ is regular in the sense that there is an operation $g \in O^1(A)$ such that $f \circ g \circ f = f$. Idempotent elements f of $\mathcal{O}^1(A)$, i. e. with $f \circ f = f$, are particular regular elements. An element $f \in O^1(A)$ is idempotent if and only the restriction of f onto the image of f is the identity operation on the image, i.e. $f|Imf = id_{Imf}$.

Let $O^n(A)$ for a fixed natural number $n \geq 1$ be the set of all n-ary operations defined on A, i.e. operations with domain A^n and values in A. The composition of $f \in O^n(A)$ with $g_1, \ldots, g_n \in O^n(A)$ is defined by

$$f(g_1, \ldots, g_n)(x_1, \ldots, x_n) := f(g_1(x_1, \ldots, x_n), \ldots, g_n(x_1, \ldots x_n)) \text{ for all } x_1, \ldots, x_n \in A.$$

Sets of n-ary operations defined on A which are closed under this $(n+1)$-ary composition operation and contain all n-ary projections $e_i^{n,A}, i = 1, \ldots, n$ (with $e_i^{n,A}(a_1, \ldots, a_n) := a_i$) are called n-ary clones. From the $(n+1)$-ary composition operation one can derive a binary operation $+ : O^n(A) \times O^n(A) \to O^n(A)$ by

$$f + g := f(g, \ldots, g) \text{ for any } f, g \in O^n(A).$$

Since the operation $+$ is associative, as generalization of $\mathcal{O}^1(A)$ one obtains a semigroup $\mathcal{O}^n(A) := (O^n(A); +)$. Clearly, all n-ary clones are subsemigroups of this semigroup. Therefore, studying the properties of the semigroup $\mathcal{O}^n(A)$ connects clone theory with the theory of semigroups.

In some cases we assume that A is a finite set with $|A| \geq 2$. Without restriction of the generality we use $A = \{0, 1, \ldots, m-1\}$. Many results are also valid if A is infinite. For simplification we use $\bar{x}$ instead of $(x_1, x_2, \ldots, x_n)$. We also use $\hat{x}$ for the n-tuple consisting of the same element $x \in A$ i.e. $\hat{x} = (x, \ldots, x)$. Let $\Delta_{A^n} := \{\hat{x} | x \in A\}$. With this notation, using the composition of operations, the binary operation $+$ on $O^n(A)$ can be written as

$$(f + g)(\bar{x}) := f(\widehat{g(\bar{x})}).$$

Notice that some properties of $\mathcal{O}^n(A)$ were already considered in [5] and [6], especially for $A = \{0, 1\}$. For $f \in O^n(A)$ we consider its restriction to $\Delta_{A^n} : f|_{\Delta_{A^n}}$. This restriction can be mapped to a unary operation $\pi_f \in O^1(A)$ which is defined by $\pi_f(a) = b$ if and only if $f(\hat{a}) = b$. Clearly, the mapping $\Psi : \{f|_{\Delta_{A^n}} | f \in O^n(A)\} \to O^1(A)$ defined by $f|_{\Delta_{A^n}} \mapsto \pi_f$ is one-to-one. π_f is a permutation i.e., $\pi_f \in O_A^{1-1}$ iff $f(\Delta_{A^n}) = A$. Consider the mapping $\Phi : O^n(A) \to O^1(A)$ defined by $f \mapsto \pi_f$. Then we get:

Proposition 1.1 *The mapping $\Phi : O^n(A) \to O^1(A)$ is a homomorphism of $(O^n(A); +)$ onto $(O^1(A); \circ)$.*

Proof: Φ is a function and $\Phi(f + g) = \pi_{f+g}$. Moreover, for every $x \in A$ we have $\pi_{f+g}(x) = (f + g)(\hat{x}) = f(\widehat{g(\hat{x})}) = \pi_f(\pi_g(x)) = (\pi_f \circ \pi_g)(x)$ and hence $\pi_{f+g} = \pi_f \circ \pi_g = \Phi(f) \circ \Phi(g)$ and therefore Φ is a homomorphism. Clearly, Φ is surjective. ■

Now consider $Ker\Phi = \{(f, g) | \Phi(f) = \Phi(g)$, i.e. $\pi_f = \pi_g\}$. By the homomorphism theorem $(O^1(A); \circ) \cong (O^n(A); +)/_{Ker\Phi}$. Let C_{π_f} be the set $\{g \in O^n(A) | g \in \Phi^{-1}(\pi_f)\} = \{g \in O^n(A) | (g, f) \in Ker\Phi\} = [f]_{Ker\Phi}$. Thus $g \in C_{\pi_f}$ iff $f(\hat{x}) = g(\hat{x})$ for every $x \in A$. For $S \subseteq O^1(A)$ let C_S be the set $C_S := \bigcup_{\pi \in S} C_\pi$. Moreover, $|[f]_{Ker\Phi}| = m^{m^n - m}$ if $|A| = m$. We use id for the identity mapping on A and π_i for the constant mapping which maps every $x \in A$ to a fixed element $i \in A$. Let ω be the permutation on A which maps every $k \in \{0, 1, \dots, m-1\}$ to $m-1-k$. For arbitrary $f \in O^n(A)$ by $\lhd f$ we mean an operation that maps $\bar{x}$ to $\omega(f(\bar{x}))$. It is easy to see that $\lhd \lhd f = f$, $\omega\pi_i = \pi_{m-1-i}$, $\pi_{\lhd f} = \omega \circ \pi_f$ and moreover, $f \in C_{id}$ iff $\lhd f \in C_\omega$. For $B \subseteq O^n(A)$ we define $\lhd B = \{\lhd f | f \in B\}$ and obtain $C_{\omega\circ\pi} = \lhd C_\pi$. We use c_i^n for the constant n-ary operation with value i i.e. $c_i^n(\bar{x}) = i$ for all $\bar{x} \in A^n$ and for all $i \in A$. Obviously, c_i^n is in C_{π_i}.

2 Idempotent and Regular Elements in $O^n(A)$

The surjective homomorphism Φ allows us to derive several elementary properties of the semigroup $\mathcal{O}^n(A)$. By Proposition 1.1 for every subsemigroup $\mathcal{C}$ in $\mathcal{O}^n(A)$ its image $\Phi(\mathcal{C})$ is a subsemigroup of $\mathcal{O}^1(A)$ and for any subsemigroup $\mathcal{S}$ of $\mathcal{O}^1(A)$ its preimage $\Phi^{-1}(\mathcal{S})$ is a subsemigroup of $\mathcal{O}^n(A)$.

Lemma 2.1 *Let C be a subset of $O^n(A)$. Then $C \subseteq C_{\Phi(C)}$.*

Proof: Let f be an arbitrary element in C. Since $f \in C_{\pi_f}$ and $\pi_f \in \Phi(C)$ then $C \subseteq \bigcup\limits_{\pi \in \Phi(C)} C_\pi = C_{\Phi(C)}$. ■

Proposition 2.2 *Let f, g be two elements of $O^n(A)$ and π be an element in $O^1(A)$. Then the following propositions are true:*

(i) *If $f \in C_{id}$ and $g \in C_\pi$, then $f + g, g + f \in C_\pi$.*

(ii) *If $f \in C_\omega$ and $g \in C_\pi$, then $f + g \in \lhd C_\pi$.*

(iii) *If $f \in C_{\pi_i}$ and $g \in O^n(A)$, then $f + g \in C_{\pi_i}$.*

(iv) *If $f \in C_{id}$, then $f + g = g$.*

(v) *If $f \in C_\omega$, then $f + g = \lhd g$.*

(vi) *If $f \in C_{\pi_i}$, then $f + g = c_i^n$.*

(vii) $\lhd(f + g) = \lhd f + g$.

The proofs are straightforward.

The question of whether a block C_π forms a subsemigroup of $\mathcal{O}^n(A)$ can be answered as follows: If π is an idempotent element of $\mathcal{O}^1(A)$, then the set $\{\pi\}$ forms a one-element subalgebra of $\mathcal{O}^1(A)$ and its pre-image C_π forms a subalgebra of $\mathcal{O}^n(A)$. Conversely, if the block C_π forms a subalgebra of $\mathcal{O}^n(A)$, then its image $\{\pi\}$ forms a subalgebra of $\mathcal{O}^1(A)$ and thus π is idempotent. Therefore, we have:

Proposition 2.3 *Let π be an element in $O^1(A)$. Then C_π forms a subsemigroup in $(O^n(A);+)$ if and only if π is an idempotent element in $(O^1(A);\circ)$.*

Proof: Assume that C_π forms a subsemigroup of $(O^n(A);+)$. Then with $f,g \in C_\pi$ we have $f+g \in C_\pi$, thus $\pi = \Phi(f+g) = \Phi(f)\circ\Phi(g) = \pi\circ\pi$ and π is idempotent. If conversely π is idempotent and $f,g \in C_\pi$, then $\Phi(f+g) = \Phi(f)\circ\Phi(g) = \pi\circ\pi = \pi$ and $f+g \in C_\pi$. ∎

It is well-known that $f \in O^n(A)$ is an idempotent element of $\mathcal{O}^n(A) = (O^n(A);+)$ iff for all $a \in Imf$ there holds $f(\hat{a}) = a$ (see e.g. [5]). For another characterization of idempotent elements and idempotent subsemigroups of $(O^n(A);+)$ we consider the following subset of $O^n(A)$:

$$\Sigma := \{f \in O^n(A) | f(A^n) = f(\Delta_{A^n})\}.$$

The set Σ consists of all n-ary operations defined on A with $Imf = Im(f|\Delta_{A^n})$ or, since $Im(f|\Delta_{A^n}) = Im\pi_f$, with $Imf = Im\pi_f$.

The set Σ has the following properties:

Proposition 2.4 (i) *Σ is the universe of a subsemigroup of $(O^n(A);+)$ which contains all idempotent elements of $(O^n(A);+)$.*

(ii) *f is idempotent in $(O^n(A);+)$ iff π_f is idempotent in $(O^1(A);\circ)$ and $f \in \Sigma$.*

(iii) *If S is the universe of an idempotent semigroup in $(O^1(A);\circ)$, then $C_S \cap \Sigma$ forms an idempotent semigroup in $(O^n(A);+)$.*

(iv) *If $\mathcal{C}$ is an idempotent subsemigroup of $(O^n(A);+)$, then there exists an idempotent subsemigroup $\mathcal{S}$ such that $C \subseteq C_S \cap \Sigma$.*

Proof: (i) Let $f,g \in \Sigma$. Then for any $\bar{x} \in A^n$ there exists an n-tuple $\hat{y} \in \Delta_{A^n}$ such that $g(\bar{x}) = g(\hat{y})$. This implies $(f+g)(\bar{x}) = f(\widehat{g(\bar{x})}) = f(\widehat{g(\hat{y})}) = (f+g)(\hat{y})$ and then $f+g \in \Sigma$.

If f is idempotent then $f(\bar{x}) = f(\widehat{f(\bar{x})})$. This means that $f \in \Sigma$.
(ii) If f is idempotent, then by Proposition 1.1 the unary operation π_f is idempotent and $f \in \Sigma$ by (i). Conversely, if π_f is idempotent, then the restriction $f|_{\Delta_{A^n}}$ is also idempotent, i.e. for any $\hat{y} \in \Delta_{A^n}$ we get $f(\hat{y}) = (f+f)(\hat{y})$. Since $f \in \Sigma$, for any $\bar{x} \in A^n$ there exists an $\hat{y} \in \Delta_{A^n}$ such that $f(\bar{x}) = f(\hat{y})$. By substitution we get $f(\bar{x}) = f(\hat{y}) = f(\widehat{f(\hat{y})}) = f(\widehat{f(\bar{x})}) = (f+f)(\bar{x})$ and therefore f is idempotent.
(iii) Since S is the universe of an idempotent subsemigroup of $(O^1(A); \circ)$, by Proposition 1.1 the set $C_S = \phi^{-1}(S)$ forms a semigroup and thus $C_S \cap \Sigma$ forms a semigroup. Let $f \in C_S \cap \Sigma$, then $\pi_f \in S$ is idempotent and by (ii) f is idempotent.
(iv) By (ii), $C \subseteq \Sigma$. Now, if $\mathcal{C}$ is an idempotent subsemigroup of $(O^n(A); +)$ then $S = \Phi(C)$ forms an idempotent subsemigroup of $\mathcal{O}^1(A)$ and thus by Lemma 2.1 we get $C \subseteq C_{\Phi(C)} = C_S$ and then $C \subseteq C_S \cap \Sigma$. ∎

Then we get the following characterization of idempotent subsemigroups of $(O^n(A); +)$.

Theorem 2.5 *Let $\mathcal{C}$ be a subsemigroup of $\mathcal{O}^n(A)$. Then $\mathcal{C}$ is an idempotent semigroup if and only if there exists an idempotent semigroup $\mathcal{S}$ in $\mathcal{O}^1(A)$ such that $C \subseteq C_S \cap \Sigma$.*

Proof: One direction follows from Proposition 2.4 (iv). Conversely, by Proposition 2.4 (iv) we know that $C_S \cap \Sigma$ forms an idempotent semigroup. Since $C \subseteq C_S \cap \Sigma$, C forms an idempotent semigroup. ∎

Now we come to regular elements and regular subsemigroups of $(O^n(A); +)$. We mentioned already that $(O^1(A); \circ)$ is a regular semigroup. Let $Reg(\mathcal{O}^n(A))$ be the set of all regular elements of $\mathcal{O}^n(A)$ and let $Idem(\mathcal{O}^n(A))$ be the set of all idempotent elements of $\mathcal{O}^n(A)$.

The following proposition has already been proved in [5]. Now we prove it in another way.

Lemma 2.6 $Reg(\mathcal{O}^n(A)) = \Sigma$.

Proof: Let $f \in Reg(\mathcal{O}^n(A))$, then there is an element $g \in O^n(A)$ such that $f + g + f = f$, thus for every $\bar{x} \in A^n$ we have $f(\bar{x}) = (f + g + f)(\bar{x}) = f(\widehat{g(\widehat{f(\bar{x})})})$. Thus $f(A^n) \subseteq f(\Delta_{A^n})$ and hence $f(A^n) = f(\Delta_{A^n})$. Therefore $f \in \Sigma$. Conversely, let f be in Σ. Since π_f is regular then there exists $\pi \in O^1(A)$ such that $\pi_f \circ \pi \circ \pi_f = \pi_f$. Now, let g be an arbitrary element in C_π. Since $f \in \Sigma$, for every $\bar{x}$ there exists $\hat{y} \in \Delta_{A^n}$ such that $f(\bar{x}) = f(\hat{y})$. Thus for every $\bar{x}$ we have $(f+g+f)(\bar{x}) = f(\widehat{g(\widehat{f(\bar{x})})}) = f(\widehat{g(\widehat{f(\hat{y})})}) = (\pi_f \circ \pi \circ \pi_f)(y) = \pi_f(y) = f(\hat{y}) = f(\bar{x})$. Therefore f is regular. ∎

By Proposition 2.4 (i) $Reg(\mathcal{O}^n(A); +)$ forms a subsemigroup of $(O^n(A); +)$.

Recall that an element a in semigroup $\mathcal{S} = (S; *)$ is said to be completely regular if there exists b in S such that $a * b * a = a$ and $a * b = b * a$. The following theorem characterizes completely regular elements in $(O^n(A); +)$.

Theorem 2.7 *Let f be an element in $O^n(A)$. Then f is a completely regular element in $(O^n(A); +)$ if and only if there is an element $g \in O^n(A)$ such that $f + g + f = f$ satisfying*

(i) $\pi_f \circ \pi_g = \pi_g \circ \pi_f$ *and*

(ii) *for all $\bar{x} \in A^n$ there exists $\hat{y} \in \Delta_{A^n}$ satisfying $f(\bar{x}) = f(\hat{y})$ and $g(\bar{x}) = g(\hat{y})$.*

Proof: $\Leftarrow$ It remains to show that $f + g = g + f$. Let $\bar{x} \in A^n$ and $\hat{y}$ be in Δ_{A^n} such that $f(\bar{x}) = f(\hat{y})$ and $g(\bar{x}) = g(\hat{y})$. Then we have $(f + g)(\bar{x}) = f(\widehat{g(\bar{x})}) = f(\widehat{g(\hat{y})}) = \pi_f(g(\hat{y})) = \pi_f(\pi_g(y)) = (\pi_f \circ \pi_g)(y) = (\pi_g \circ \pi_f)(y) = \pi_g(\pi_f(y)) = \pi_g(f(\hat{y})) = g(\widehat{f(\hat{y})}) = g(\widehat{f(\bar{x})}) = (g + f)(\bar{x})$. Therefore f is a completely regular element.
$\Rightarrow$ Let f be a completely regular element in $O^n(A)$ and g be an element in $\in O^n(A)$ which satisfies $f + g + f = f$ and $f + g = g + f$. Therefore by Proposition 1.1 $\pi_f \circ \pi_g = \pi_g \circ \pi_f$. Now assume that there exists $\bar{x}$ such that for every $\hat{y} \in \Delta_{A^n}$, $f(\bar{x}) \neq f(\hat{y})$ or $g(\bar{x}) \neq g(\hat{y})$. If $f(\bar{x}) \neq f(\hat{y})$ then $f(A^n) \neq f(\Delta_{A^n})$. By Lemma 2.6 this implies that f is not regular, a contradiction. Now, assume that there exists $\bar{x}$ such that for every $\hat{y} \in \Delta_{A^n}$, $f(\bar{x}) = f(\hat{y})$ and $g(\bar{x}) \neq g(\hat{y})$. Since for every $\hat{y} \in \Delta_{A^n}$, $f(\bar{x}) = f(\hat{y})$ we get $f \in C_{\pi_{f(\bar{x})}}$. Because of $f + g + f = f$ and $f \in C_{\pi_{f(\bar{x})}}$ by Proposition 2.2 (iii) we get $f = f + g + f = c^n_{f(\bar{x})}$. Moreover, since $f + g = g + f$ and $f + g = c^n_{f(\bar{x})} + g = c^n_{f(\bar{x})}$ we have $g + f = c_{f(\bar{x})} = f$, i.e., $g(\widehat{f(\bar{x})}) = f(\bar{x})$. Since $\widehat{f(\bar{x})} \in \Delta_{A^n}$ then $g(\bar{x}) \neq g(\widehat{f(\bar{x})}) = f(\bar{x})$ and hence $f \neq g$ for any completely regular element f. But $f = c^n_{f(\bar{x})}$ and $g = c^n_{f(\bar{x})}$ satisfy $f + g + f = f$ and $f + g = g + f$ and are equal. ∎

Proposition 2.8 *If $\mathcal{C} \subseteq \mathcal{O}^n(A)$ is a completely regular semigroup then there exists a completely regular semigroup $\mathcal{S}$ in $(O^1(A); +)$ such that $C \subseteq C_S$.*

Proof: By Proposition 1.1, $\Phi(C)$ forms a completely regular semigroup in $(O^1(A); +)$. Moreover, by Proposition 2.3, $C \subseteq C_{\Phi(C)}$ and we complete the proof by choosing $S = \Phi(C)$. ∎

3 Constant Semigroups and Generalizations

The semigroups considered in this section occur as subsemigroups of $\mathcal{O}^n(\{0, 1\})$ ([5], [6]).

A semigroup $\mathcal{S} = (S; *)$ is called a right-zero semigroup if $a * b = b$ for every $a, b \in S$. In [5] all right-zero semigroups of $\mathcal{O}^n(A)$ are characterized as follows:

Theorem 3.1 ([5]) *$C \subseteq (O^n(A); +)$ forms a right-zero semigroup iff $C \subseteq Idem(O^n(A))$ and for all $f, g \in C$ we have $Imf = Img$.*

For $C \subseteq O^n(A)$ by ImC we mean $ImC := \bigcup_{f \in C} Imf$. We have the following theorem:

Theorem 3.2 *Let $\mathcal{C}$ be a subsemigroup of $(O^n(A); +)$. Then the following propositions are equivalent:*

(i) *$\mathcal{C}$ is a right-zero subsemigroup of $\mathcal{O}^n(A)$*

(ii) *$f(\hat{x}) = x$ for every $f \in C$ and for every $x \in ImC$*

(iii) *$\Phi(C)$ forms a right-zero semigroup in $(O^1(A), \circ)$ and $C \subseteq \Sigma$.*

Proof: (i)⇔(ii) Let f be an element in C and x be arbitrary in ImC. Thus we can find $g \in C$ and $\bar{y} \in A^n$ such that $x = g(\bar{y})$. Moreover, since C is right-zero, we get $(f+g)(\bar{y}) = g(\bar{y}) = x$. On the other hand side, $(f+g)(\bar{y}) = f(\widehat{g(\bar{y})}) = f(\hat{x})$. Therefore we have $f(\hat{x}) = x$. Conversely, let f and g be arbitrary elements in C and $\bar{x}$ be an arbitrary element in A^n. Thus we get $g(\bar{x}) \in ImC$ and hence by assumption $(f+g)(\bar{x}) = f(\widehat{g(\bar{x})}) = g(\bar{x})$. Thus C is a right zero semigroup.
(i)⇔(iii) If $\mathcal{C}$ is right-zero, then by Proposition 1.1, $\Phi(C)$ forms a right-zero semigroup in $(O^1(A); \circ)$. Since every right-zero element is regular by Lemma 2.4, $C \subseteq \Sigma$. Conversely, let $f, g \in C$ and $\bar{x} \in A^n$. Since $C \subseteq \Sigma$ there exist $\hat{y} \in \Delta_{A^n}$ such that $g(\bar{x}) = g(\hat{y})$. Thus by assumption we have $(f+g)(\bar{x}) = f(\widehat{g(\bar{x})}) = f(\widehat{g(\hat{y})}) = \pi_f(\pi_g(y)) = (\pi_f \circ \pi_g)(y) = \pi_g(y) = g(\hat{y}) = g(\bar{x})$. Therefore $\mathcal{C}$ is a right-zero semigroup. ■

A semigroup $\mathcal{S} = (S; *)$ is called a constant semigroup if $a * a' = b * b'$ for every a, a', b and b' in S. From now on we say $\mathcal{S} = (S; *)$ is a constant semigroup with the constant element c if $a * b = c$ for all $a, b \in S$. For $C \subseteq O^n(A)$ and $\bar{x} \in A^n$ let $C(\bar{x})$ be defined by $C(\bar{x}) := \{f(\bar{x}) | f \in C\}$. Now we characterize all constant semigroups as follows:

Proposition 3.3 *Let $\mathcal{C}$ be a subsemigroup of $\mathcal{O}^n(A)$. Then $\mathcal{C}$ is a constant semigroup if and only if $f(\hat{y}) = g(\hat{z})$ for every $f, g \in C$ and $y, z \in C(\bar{x})$ for every $\bar{x} \in A^n$.*

Proof: Let $f, g \in C$ and $y, z \in C(\bar{x})$ for an arbitrary $\bar{x} \in A^n$. Thus there exist f' and g' in C such that $y = f'(\bar{x})$ and $z = g'(\bar{x})$. By assumption we have $f(\hat{y}) = f(\widehat{f'(\bar{x})}) = (f + f')(\bar{x}) = (g + g')(\bar{x}) = g(\widehat{g'(\bar{x})}) = g(\hat{z})$. Conversely, let f, f', g, g' be arbitrary in C and $\bar{x}$ be arbitrary in A^n. Thus $y = f'(\bar{x}), z = g'(\bar{x}) \in C(\bar{x})$. By the assumption we have $f(\hat{y}) = g(\hat{z})$. But $f(\hat{y}) = f(\widehat{f'(\bar{x})}) = (f + f')(\bar{x})$ and $g(\hat{z}) = g(\widehat{g'(\bar{x})}) = (g + g')(\bar{x})$. Therefore $f + f' = g + g'$. Thus $\mathcal{C}$ is a constant semigroup. ■

A semigroup $\mathcal{S} = (S; *)$ is called a two-constant semigroup if there are subsets S_1, S_2 of S such that $S = S_1 \cup S_2$, $S_1 \cap S_2 = \emptyset, S_i \neq \emptyset, i = 1, 2$ and if there are two fixed elements $b^* \in S_1, b^{**} \in S_2$ such that

$$a * b = \begin{cases} b^* \in S_1 & \text{if } a \in S_1 \\ b^{**} \in S_2 & \text{if } a \in S_2. \end{cases}$$

(see [5], [2]).

We generalize the concept of a two-constant semigroup to a k-constant semigroup for an arbitrary natural number $k \geq 2$ as follows:

The following kinds of semigroups became important when subsemigroups of $\mathcal{O}^n(\{0, 1\})$ were investigated (see [5]).

Definition 3.4 A semigroup $\mathcal{S} = (S; *)$ is called a k-constant semigroup if there exist subsets S_i of S such that $S_i \neq \emptyset$, $i \in I = \{1, 2, \ldots, k\}$, $\bigcup_{i \in I} S_i = S$ and $S_i \cap S_j = \emptyset$ for $i, j \in I$ and $i \neq j$ and if there exist k fixed elements $c_i \in S_i$ for all $i \in I$ such that $a * b = c_i$ for all $a \in S_i$ and $b \in S$.

Proposition 3.5 *For every natural number $k \geq 2$ and every finite k-constant semigroup $\mathcal{S}$ there exist natural numbers m and n such that $\mathcal{S}$ is isomorphic to some semigroups in $O^n(A)$ for an arbitrary finite set A of cardinality m.*

Proof: Let $S_i, i \in I = \{1, 2, \ldots, k\}$ be nonempty disjoint sets such that $S = \bigcup_{i \in I} S_i$ and $c_i \in S_i$ be constant elements and $a * b = c_i$ for all $a \in S_i$ and $b \in S$. We take natural numbers $m \geq k$ and n such that $max\{|S_i|\} \leq m^{m^n - m}$, choose k different indices $j_1, j_2, \ldots, j_k \in \{0, 1, \ldots, m-1\}$ and define arbitrary one-to-one mappings $\phi_i : S_i \to C_{\pi_{j_i}} \subseteq O^n(A)$ with $|A| = m$ such that $\phi_i(c_i) = c^n_{j_i}$. By choosing $C = \bigcup_{i \in I} \phi_i(S_i)$ and defining a mapping $\phi : S \to C$ by $\phi(x) = \phi_i(x)$ if $x \in S_i$ then we get $\mathcal{S} \cong \mathcal{C}$. ■

Recall that a semigroup $\mathcal{S} = (S; *)$ is called a right-zero constant semigroup if there are subsets S_1, S_2 of S such that $S = S_1 \cup S_2$, $S_1 \cap S_2 = \emptyset, S_i \neq \emptyset, i = 1, 2$ and if there is a fixed element b^* in S_2 such that

$$a * b = \begin{cases} b & \text{if } a \in S_1 \\ b^* \in S_2 & \text{if } a \in S_2. \end{cases}$$

A semigroup $\mathcal{S} = (S; *)$ is called a right-zero two-constant semigroup if there are subsets S_1, S_2 of S such that $S = \bigcup_{n=1}^{3} S_n$, $S_i \neq \emptyset, S_i \cap S_j = \emptyset, i \neq j \in \{1,2,3\}$ and if there are distinguished elements $b^* \in S_2, b^{**} \in S_3$ such that

$$a * b = \begin{cases} b & \text{if } a \in S_1 \\ b^* \in S_2 & \text{if } a \in S_2 \\ b^{**} \in S_3 & \text{if } a \in S_3. \end{cases}$$

In the following definition we generalize the concept of right-zero constant and right-zero two-constant semigroups (see [5], [2]) to right-zero k-constant semigroups.

Definition 3.6 A semigroup $\mathcal{S} = (S; *)$ is called a right-zero k-constant semigroup if there are nonempty subsets S_i of S, $i \in I = \{1, 2, \ldots, k+1\}$ such that $S = \bigcup_{i \in I} S_i$, $S_i \cap S_j = \emptyset$ for $i \neq j$ and if there are distinguished elements $b_i^* \in S_{i+1}$ for $i \in I - \{k+1\}$ such that

$$a * b = \begin{cases} b & \text{if } a \in S_1 \\ b_i^* \in S_{i+1} & \text{if } a \in S_{i+1}, i \in I - \{k+1\}. \end{cases}$$

The union of a constant semigroup and a right-zero semigroup forms a right-zero semigroup under some conditions. The following proposition gives these conditions.

Proposition 3.7 *Let $\mathcal{C}_1$ and $\mathcal{C}_2$ be a right-zero and a constant subsemigroup, respectively, of $\mathcal{O}^n(A)$ such that $C_1 \cap C_2 = \emptyset$ and let c be the constant element of $\mathcal{C}_2$. Then $C = C_1 \cup C_2$ forms a right-zero constant semigroup if and only if $ImC_2 \subseteq ImC_1$ and $f + g = c$ for every $f \in C_2$ and $g \in C_1$.*

Proof: Let $f, g \in C$. If they belong to the same set C_1 or C_2 then it is clear that $f + g = g$ or $f + g = c$. Now, let f be in C_2 and g be in C_1. Then by the assumption we have $f + g = c$. If $f \in C_1$ and $g \in C_2$ then by $ImC_2 \subseteq ImC_1$ we get $f + g = g$. Therefore C is a right-zero constant semigroup.
Conversely, if $\mathcal{C}$ is a right-zero constant semigroup, then $f + g = g$ if $f \in C_1$ and $g \in C_2$, i.e. $ImC_2 \subseteq ImC_1$ and $f + g = c$ if $f \in C_1$. This completes the proof. ■

Generally we have the following:

Proposition 3.8 *Let $\mathcal{A}$ be a right-zero semigroup, let the $\mathcal{C}_i$'s, $i \in I = \{i, \ldots, k\}$, be k pairwise disjoint constant subsemigroups of $\mathcal{O}^n(A)$ with constant elements c_i such that $C = \bigcup_{i \in I} C_i$ is the universe of a k-constant semigroup and $A \cap C = \emptyset$. Then $A \cup C$ is the universe of a right-zero k-constant semigroup if and only if $ImC \subseteq ImA$ and $g_i + f = c_i$ for every $f \in A$ and arbitrary $g_i \in C_i$ for every $i \in I$.*

Proposition 3.9 *For every natural number $k \geq 2$ and every finite right-zero k-constant semigroup $\mathcal{S}$ there exist natural numbers m and n such that $\mathcal{S}$ is isomorphic to some semigroups in $\mathcal{O}^n(A)$ for an arbitrary finite set A of cardinality m.*

Proof: Let $S_i, i \in I = \{1, 2, \ldots, k+1\}$ be nonempty pairwise disjoint sets such that $S = \bigcup_{i \in I} S_i$ and $c_i \in S_{i+1}$ are constant elements, $f + g = g$ if $f \in S_1$ and $f + g = c_i$ if $f \in S_i$ for $i = 2, \ldots, k+1$. We take natural numbers $m \geq k$ and n such that $max\{|S_i|\} \leq m^{m^n - m}$ and choose k different indices $j_2, j_3, \ldots, j_{k+1} \in \{0, 1, \ldots, m-1\}$ and define arbitrary one-to-one mappings $\phi_1 : S_1 \to C_{id} \subseteq O^n(A)$ and $\phi_i : S_i \to C_{\pi_{j_i}} \subseteq O^n(A)$ with $|A| = m$ for $i = 2, 3, \ldots, k+1$ such that $\phi_i(c_i) = c^n_{j_i}$. By choosing $C = \bigcup_{i \in I} \phi_i(S_i)$ and defining a mapping $\phi : S \to C$ by $\phi(x) = \phi_i(x)$ if $x \in S_i$ we get $\mathcal{S} \cong \mathcal{C}$. ■

Recall that a semigroup $\mathcal{S} = (S; *)$ is said to be left-zero if $a * b = a$ for every a and b in S. We introduce the concept of right-left-zero semigroup and show that there exist such semigroups in $(O^n(A); +)$.

Definition 3.10 A semigroup $\mathcal{S} = (S; *)$ is called a right-left-zero semigroup if there exist two nonempty disjoint sets S' and S'' such that $S' \cup S'' = S$ and $a * b = b$ for every $a \in S'$ and $b \in S$ and $a * b = a$ for every $a \in S''$ and $b \in S$.

We denote S' as the right-zero part and S'' as the left-zero part of S.

Proposition 3.11 *Let $C \subseteq O^n(A)$ form a right-left-zero semigroup. Then $\Phi(C)$ forms a right-left-zero semigroup in $(O^1(A); \circ)$ and $C \subseteq \Sigma$.*

Proof: Let $C \subseteq O^n(A)$ form a right-left-zero semigroup. By Proposition 1.1, $\Phi(C)$ forms a right-left-zero semigroup in $(O^1(A); \circ)$. Moreover, for every $f \in C$ we have $f + f + f = f$ and hence f is regular. Therefore by Lemma 2.6, $C \subseteq \Sigma$. ■

Proposition 3.12 *Let $C_1, C_2 \subseteq O^n(A)$ form a right-zero and a left-zero semigroup respectively. Then $C = C_1 \cup C_2$ forms a right-left-zero semigroup if and only if $ImC_2 \subseteq ImC_1$, $\pi_g \circ \pi_f = \pi_g$ for all $\pi_f \in \Phi(C_1), \pi_g \in \Phi(C_2)$ and for all $f \in C_1 \cup C_2$ and $\bar{x} \in A^n$ there exists $\hat{y} \in \Delta_{A^n}$ such that $f(\bar{x}) = f(\hat{y})$.*

Proof: $\Rightarrow$ Let $C_1, C_2 \subseteq O^n(A)$ form a right-zero and a left-zero semigroup, respectively such that $C_1 \cup C_2$ forms a right-left-zero semigroup. Let $y \in ImC_2$, then there exist $g \in C_2$ and $\bar{x} \in A^n$ such that $y = g(\bar{x})$ and therefore $y = g(\bar{x}) = (f+g)(\bar{x}) = f(\widehat{g(\bar{x})})$ for some $f \in C_1$, i.e. $y \in ImC_1$ and thus $ImC_2 \subseteq ImC_1$. By Proposition 1.1, $\pi_g \circ \pi_f = \pi_g$ for all $\pi_f \in \Phi(C_1), \pi_g \in \Phi(C_2)$. By Proposition 3.11 for every $f \in C_1$ there exists $\hat{y} \in \Delta_{A^n}$ such that $f(\bar{x}) = f(\hat{y})$ and by assumption, $(g+f)(\bar{x}) = g(\bar{x})$ for every $g \in C_2$. On the other hand side, $(g+f)(\bar{x}) = g(\widehat{f(\bar{x})}) = g(\widehat{f(\hat{y})}) = \pi_g(\pi_f(y)) = \pi_g(y) = g(\hat{y})$ and therefore $g(\bar{x}) = g(\hat{y})$. $\Leftarrow$ Conversely, if $f, g \in C_1$ or $f, g \in C_2$ then it is clear that $f+g = g$ or $f+g = f$. Now, let $f \in C_1$ and let $g \in C_2$. Then for every $\bar{x} \in A^n$ we have $g(\bar{x}) \in ImC_2 \subseteq ImC_1$ and hence $(f+g)(\bar{x}) = f(\widehat{g(\bar{x})}) = g(\bar{x})$. Moreover, by assumption we obtain $(g+f)(\bar{x}) = g(\widehat{f(\bar{x})}) = g(\widehat{f(\hat{y})}) = \pi_g(\pi_f(y)) = \pi_g(y) = g(\hat{y}) = g(\bar{x})$ for some $\hat{y} \in \Delta_{A^n}$. This completes the proof. ■

Proposition 3.13 *For an arbitrary right-left-zero semigroup $\mathcal{S}$ there exist natural numbers m and n such that $\mathcal{S}$ is isomorphic to some subsemigroups of $(O^n(A); +)$.*

Proof: Let $\mathcal{S}$ be a right-left-zero semigroup with S_1 and S_2 be the right-zero and the left-zero parts respectively. Take natural numbers m and n such that $m \geq |S_2|$ and $|S_1| \leq m^{m^n - m}$ and let $A = \{0, 1, \cdots, m-1\}$. We choose $|S_2|$ different indices $j_1, \cdots, j_{|S_2|} \in A$ and define arbitrary one-to-one mappings $\phi_1 : S_1 \to C_{id} \subseteq O^n(A)$ and $\phi_2 : S_2 \to \{c_i^n | i = j_1, \cdots, j_{|S_2|}\} \subseteq O^n(A)$. By choosing $C = \phi_1(S_1) \cup \{c_i^n | i = j_1, \cdots, j_{|S_2|}\}$ and defining a mapping $\phi : S \to C$ such that $\phi(x) = \phi_i(x)$ if $x \in S_i$, $i = 1, 2$ then we have $\mathcal{C} \cong \mathcal{S}$. ■

Recall that a four-part semigroup is defined in the following way. We use four non-empty, finite and pairwise disjoint sets with elements denoted as follows:

$$\begin{aligned} S_1 &= \{a_{11}, a_{12}, \ldots, a_{1n_r}\}, \\ S_2 &= \{a_{21}, a_{22}, \ldots, a_{2n_r}\}, \\ S_3 &= \{a_{31}, a_{32}, \ldots, a_{3n_s}\}, \\ S_4 &= \{a_{41}, a_{42}, \ldots, a_{4n_s}\}. \end{aligned}$$

We define a binary operation $*$ on $S = S_1 \cup S_2 \cup S_3 \cup S_4$ by

$$a_{ij} * a_{lk} = \begin{cases} a_{lk} & \text{if } a_{ij} \in S_1 \\ a_{tk} & \text{if } a_{ij} \in S_2 \text{ where } t = \begin{cases} 1 & \text{if } l = 2 \\ 2 & \text{if } l = 1 \\ 3 & \text{if } l = 4 \\ 4 & \text{if } l = 3 \end{cases} \\ a^* \in S_3 & \text{if } a_{ij} \in S_3 \\ a^{**} \in S_4 & \text{if } a_{ij} \in S_4. \end{cases}$$

The binary operation $*$ is well-defined, and it can be checked that it is associative, giving us a semigroup $(S;*)$ called a four-part semigroup.

Theorem 3.14 *Let $\mathcal{C}$ be a right-zero constant subsemigroup of $\mathcal{O}^n(A)$, let C_1 and C_2 be the right-zero and the constant parts of $\mathcal{C}$, respectively, and let c be the constant element of $\mathcal{C}_2$. Then $C \cup \lhd C$ forms a four part subsemigroup of $(O^n(A);+)$ iff $ImC_1 = Im \lhd C_1$.*

Proof: Let $C \cup \lhd C$ be the universe of a four-part semigroup. Then for every $f \in C_1$ and every $\bar{x} \in A^n$ we get $f(\bar{x}) = (\lhd f + \lhd f)(\bar{x}) = \lhd f(\widehat{\lhd f(\bar{x})}) \in Im \lhd C_1$ and $\lhd f(\bar{x}) = (f + \lhd f)(\bar{x}) = f(\widehat{\lhd f(\bar{x})}) \in ImC_1$. Thus $ImC_1 = Im \lhd C_1$. Conversely, let f be in C_1. Then $f + g = g$ for every $g \in C$ and thus by Proposition 2.2 (vii) we get $\lhd f + g = \lhd g$ for every $g \in C$. Now, if $f \in C_2$ then $f + g = c$ for every $g \in C$ and by Proposition 2.2 (vii) we get $\lhd f + g = \lhd c$. Since $\mathcal{C}$ is a right-zero constant semigroup, by Proposition 3.7 we have $ImC_2 \subseteq ImC_1$ and thus $ImC = ImC_1$. This implies $Im \lhd C = Im \lhd C_1 = ImC_1$. Now, let $f \in C_1$ and $\lhd g \in \lhd C$. Since C_1 is the universe of a right-zero semigroup then by Theorem 3.2 we have $(f + \lhd g)(\bar{x}) = f(\widehat{\lhd g(\bar{x})}) = \lhd g(\bar{x})$. Therefore by Proposition 2.2 (vii) we obtain $\lhd f + \lhd g = \lhd \lhd g = g$. Now let $f \in C_2$ and $\lhd g \in \lhd C$, then $\lhd g(\bar{x}) \in ImC_1$ for every $\bar{x}$. Therefore by Proposition 3.7 we get $(f + \lhd g)(\bar{x}) = f(\widehat{\lhd g(\bar{x})}) = c(\bar{x})$. By Proposition 2.2 (vii) this implies $\lhd f + \lhd g = \lhd c$. This completes the proof. ■

4 Particular Semigroups in $(O^n(A);+)$

Recall that a semigroup $\mathcal{S} = (S;*)$ is called a band if it is idempotent, a rectangular band, if $a * b * c = a * c$, a regular band, if $a * b * c * a = a * b * a * c * a$ and a normal band, if $a * b * c * d = a * c * b * d$ for all $a, b, c, d \in S$. $\mathcal{S}$ is called a semilattice if it is commutative and idempotent. The following results characterize all rectangular, regular and normal bands in $\mathcal{O}^n(A)$. If $\mathcal{C} \subseteq \mathcal{O}^n(A)$ is one of these semigroups then $C \subseteq \Sigma$ since all these semigroups are idempotent.

Theorem 4.1 *Let $\mathcal{C}$ be a subsemigroup of $\mathcal{O}^n(A)$. Then*

(i) *C forms a normal band if and only if $C \subseteq \Sigma$ and $\Phi(C)$ forms a normal band in $\mathcal{O}^1(A)$.*

(ii) *C forms a rectangular band if and only if $C \subseteq \Sigma$ and $\Phi(C)$ forms a rectangular band in $\mathcal{O}^1(A)$.*

(iii) *C forms a regular band if and only if $C \subseteq \Sigma$ and $\Phi(C)$ forms a regular band in $\mathcal{O}^1(A)$.*

Proof: (i) Let $\mathcal{C} \subseteq \mathcal{O}^n(A)$ be a normal band. Then by Proposition 1.1 $\Phi(\mathcal{C})$ is also a normal band. Moreover, $C \subseteq \Sigma$.
Conversely, let the two conditions be satisfied. For every $f, g, h, k \in C$ we have $\pi_f, \pi_g, \pi_h, \pi_k \in \Phi(C)$ and hence π_f, π_g, π_h and π_k are idempotent and $\pi_f \circ \pi_g \circ \pi_h \circ \pi_k = \pi_f \circ \pi_h \circ \pi_g \circ \pi_k$. By the assumption, for all $\bar{x} \in A^n$ and arbitrary $k \in C$ we can find $\hat{y} \in \Delta_{A^n}$ such that $k(\bar{x}) = k(\hat{y})$ and hence $(k+k)(\bar{x}) = k(\widehat{k(\bar{x})}) = k(\widehat{k(\hat{y})}) = \widehat{(\pi_k \circ \pi_k)(y)} = \pi_k(y) = \widehat{k(\hat{y})} = k(\bar{x})$ and then $k + k = k$. Moreover, $(f+g+h+k)(\bar{x}) = f(\widehat{g(\widehat{h(\widehat{k(\bar{x})})})}) = f(\widehat{g(\widehat{h(\widehat{k(\hat{y})})})}) = \pi_f(\pi_g(\pi_h(\pi_k(y)))) = (\pi_f \circ \pi_g \circ \pi_h \circ \pi_k)(y) = (\pi_f \circ \pi_h \circ \pi_g \circ \pi_k)(y) = \pi_f(\pi_h(\pi_g(\pi_k(y)))) = f(\widehat{h(\widehat{g(\widehat{k(\hat{y})})})}) = f(\widehat{h(\widehat{g(\widehat{k(\bar{x})})})}) = (f+h+g+k)(\bar{x})$. The proofs of the other propositions are similar. ∎

Proposition 4.2 *If for all $\bar{x} \in A^n$ there exists $\hat{y} \in \Delta_{A^n}$ such that $f(\bar{x}) = f(\hat{y})$ for all $f \in C$ then $[h]_{Ker\Phi} \cap C$ is a singleton or the empty set for every $h \in A^n$.*

Proof: Let $[h]_{Ker\Phi} \cap C \neq \emptyset$. Let f and g be in $[h]_{Ker\Phi} \cap C$ and $\bar{x}$ be arbitrary in A^n. Then $\pi_f = \Phi(f) = \Phi(g) = \pi_g$ and there exists $\hat{y} \in \Delta_{A^n}$ such that $f(\bar{x}) = f(\hat{y})$ and $g(\bar{x}) = g(\hat{y})$. Therefore we get $f(\bar{x}) = f(\hat{y}) = \pi_f(y) = \pi_g(y) = g(\hat{y}) = g(\bar{x})$. ∎

The following results characterize all left-zero semigroups and semilattices in $(O^n(A); +)$.

Theorem 4.3 *Let C be a subset of $O^n(A)$. Then C forms a left-zero semigroup in $(O^n(A); +)$ if and only if these conditions are satisfied:*

(i) *$\Phi(C)$ forms a left-zero semigroup in $(O^1(A); \circ)$*

(ii) *For all $\bar{x} \in A^n$ there exists $\hat{y} \in \Delta_{A^n}$ such that $f(\bar{x}) = f(\hat{y})$ for all $f \in C$.*

Proof: Since C forms a left-zero semigroup, by Proposition 1.1 $\Phi(C)$ forms a left-zero semigroup. Moreover, for all $f, g \in C$ we have $f + g = f$ and $g + f = g$ and hence $f + g + f = f$. Thus every element in C is regular and therefore by Lemma 2.6 we have $C \subseteq \Sigma$. Hence, for an arbitrary $\bar{x} \in A^n$ there exists $\hat{y} \in \Delta_{A^n}$ such that $g(\bar{x}) = g(\hat{y})$ and thus $f(\bar{x}) = (f+g)(\bar{x}) = f(\widehat{g(\bar{x})}) = f(\widehat{g(\hat{y})}) = (f+g)(\hat{y}) = f(\hat{y})$. Therefore $f(\bar{x}) = f(\hat{y})$ and $g(\bar{x}) = g(\hat{y})$.
Let conversely the two conditions be satisfied. Then by Proposition 1.1. and Proposition 4.2 we get $C \cong \Phi(C)$ and thus $\mathcal{C}$ is a left-zero semigroup. ∎

Proposition 4.4 *For every finite left-zero constant semigroup $\mathcal{S}$ there exist natural numbers m and n such that $\mathcal{S}$ is isomorphic to some semigroup in $O^n(A)$ for an arbitrary finite set A of cardinality m.*

Proof: We choose $n \geq 2$, $m \geq |S|$ and $|S|$ different indices $j_1, j_2, \ldots, j_{|S|} \in \{0, 1, \ldots, m-1\}$ and take $C = \{c_i^n | i = j_1, j_2, \ldots, j_{|S|}\} \subseteq O^n(A)$ with $|A| = m$. By defining an arbitrary one-to-one mapping $\phi : S \to C$ we get an isomorphism between $\mathcal{S}$ and $\mathcal{C}$. ■

Theorem 4.5 *Let C form a subsemigroup of $(O^n(A); +)$. Then C forms a semilattice if and only if the following conditions are satisfied:*

(i) *$\Phi(C)$ forms a semilattice in $(O^1(A); \circ)$.*

(ii) *For all $\bar{x} \in A^n$ there exists $\hat{y} \in \Delta_{A^n}$ such that $f(\bar{x}) = f(\hat{y})$ for all $f \in C$.*

Proof: Let C form a semilattice. Then by Proposition 1.1, $\Phi(C)$ forms a semilattice. Now, assume that the second condition is not satisfied i.e. there exists $\bar{x} \in A^n$ such that for every $\hat{y} \in \Delta_{A^n}$ there is an operation $f \in C$ such that $f(\bar{x}) \neq f(\hat{y})$. Then $f \notin \Sigma$. But $\mathcal{C}$ is idempotent and thus by Proposition 2.4, $C \subseteq \Sigma$, a contradiction.
Conversely, let the two conditions be satisfied. Then by Proposition 1.1 and Proposition 4.2, $\mathcal{C} \cong \Phi(\mathcal{C})$. Thus $\mathcal{C}$ is a semilattice. ■

Theorem 4.6 *Let C be a subset of $O^n(A)$. Then C forms a (commutative) group in $(O^n(A); +)$ if and only if*

(i) *$\Phi(C)$ forms a (commutative) group in $(O^1(A); \circ)$ and*

(ii) *for all $\bar{x} \in A^n$ there exists $\hat{y} \in \Delta_{A^n}$ such that $f(\bar{x}) = f(\hat{y})$ for all $f \in C$.*

Proof: Let $\mathcal{C}$ be a (commutative) subgroup of $\mathcal{O}^n(A)$. Then by Proposition 1.1 $\Phi(C)$ forms a (commutative) group in $(O^1(A); \circ)$. Now, assume that the second condition is not satisfied i.e. there exist $\bar{x} \in A^n$ such that for every $\hat{y} \in \Delta_{A^n}$ there exists $f \in C$ such that $f(\bar{x}) \neq f(\hat{y})$. Then $f \notin \Sigma$. But $\mathcal{C}$ is regular and thus by Proposition 2.4 $C \subseteq \Sigma$, a contradiction.
Let the two conditions be satisfied. Then By Proposition 1.1 and Proposition 4.2 we get $\mathcal{C} \cong \Phi(\mathcal{C})$ and thus $\mathcal{C}$ is a (commutative) group. ■

The following proposition gives necessary and sufficient conditions for the second condition of Theorem 4.3, Theorem 4.5 and Theorem 4.6.

Proposition 4.7 *Let $f, g \in O^n(A)$. For all $\bar{x} \in A^n$ there exists $\hat{y} \in \Delta_{A^n}$ such that $f(\bar{x}) = f(\hat{y})$ and $g(\bar{x}) = g(\hat{y})$ if and only if these two conditions are satisfied:*

(i) $Kerf = Kerg$.

(ii) $[\bar{x}]_{Kerf} \cap \Delta_{A^n} \neq \emptyset$ *for all* $[\bar{x}]_{Kerf} \in A^n/_{Kerf}$.

Proof: Let $f, g \in O^n(A)$ and $(\bar{x}_1, \bar{x}_2) \in Kerf$ i.e. $f(\bar{x}_1) = f(\bar{x}_2)$. Then we can find $\hat{y} \in \Delta_{A^n}$ such that $f(\bar{x}_1) = f(\hat{y}) = f(\bar{x}_2)$. By assumption we have $g(\bar{x}_1) = g(\hat{y}) = g(\bar{x}_2)$ and hence $Kerf \subseteq Kerg$. Similarly we get $Kerg \subseteq Kerf$ and thus $Kerf = Kerg$. Now, let $\bar{x} \in A^n$. By the assumption we can find $\hat{y} \in \Delta_{A^n}$ such that $f(\bar{x}) = f(\hat{y})$. Therefore $[\bar{x}]_{Kerf} \cap \Delta_{A^n} \neq \emptyset$. Let conversely $\bar{x} \in A^n$ and $f, g \in O^n(A)$ and assume that (i) and (ii) are satisfied. Then $[\bar{x}]_{Kerf} \cap \Delta_{A^n} \neq \emptyset$ and hence there exists $\hat{y} \in [\bar{x}]_{Kerf}$. Therefore, $f(\bar{x}) = f(\hat{y})$ i.e. $(\bar{x}, \hat{y}) \in Kerf$. Since $Kerf = Kerg$, then $g(\bar{x}) = g(\hat{y})$. ■

5 Green's Relations in $\mathcal{O}^n(A)$

Let a and b be two elements in semigroup $\mathcal{S} = (S; *)$. Green's relations are defined in the following way: $a\mathcal{L}b$ iff $a = b$ or there exist $c, d \in S$ such that $c * a = b$ and $d * b = a$, $a\mathcal{R}b$ iff $a = b$ or there exist $c, d \in S$ such that $a * c = b$ and $b * d = a$, $\mathcal{H} = \mathcal{L} \cap \mathcal{R}$, $\mathcal{D} = \mathcal{L} \circ \mathcal{R}$ and $a\mathcal{J}b$ iff $a = b$ or there exist c, d, e, f in S such that $a = c * b * d$ and $b = e * a * f$. The following theorem has been already proved in [5]. Now we want to prove it in another way. We recall that Green's relations $\mathcal{L}$ and $\mathcal{R}$ on $\mathcal{O}^1(A)$ are defined by equal kernels and equal images, respectively, of the transformations f and g. For $\mathcal{O}^n(A)$ we have similar characterizations.

Theorem 5.1 ([5]) *Let f, g be two elements in $O^n(A)$. Then $f\mathcal{L}g$ if and only if $Kerf = Kerg$.*

Proof: Let $f\mathcal{L}g$. Then $f = g$ or there exist $h, k \in O^n(A)$ such that $h + f = g$ and $k + g = f$. If $f = g$, then everything is clear. Now assume that there exist $h, k \in O^n(A)$ such that $h + f = g$ and $k + g = f$ that is $h(\widehat{f(\bar{x})}) = g(\bar{x})$ and $k(\widehat{g(\bar{x})}) = f(\bar{x})$ for every $\bar{x} \in A^n$. Thus there exist π_h and π_k in $O^1(A)$ such that $\pi_h(f(\bar{x})) = g(\bar{x})$ and $\pi_k(g(\bar{x})) = f(\bar{x})$. Therefore for all $\bar{x}_1, \bar{x}_2$ such that $f(\bar{x}_1) = f(\bar{x}_2)$ we get $g(\bar{x}_1) = g(\bar{x}_2)$ and for all $\bar{x}_1, \bar{x}_2$ such that $g(\bar{x}_1) = g(\bar{x}_2)$ we get $f(\bar{x}_1) = f(\bar{x}_2)$. Thus $Kerf = Kerg$.
Conversely, if $Kerf = Kerg$ then $f(\bar{x}_1) = f(\bar{x}_2)$ iff $g(\bar{x}_1) = g(\bar{x}_2)$. Thus we can get well-defined n-ary operations h and k such that $h(\widehat{f(\bar{x})}) = g(\bar{x})$ and $k(\widehat{g(\bar{x})}) = f(\bar{x})$ and hence $f\mathcal{L}g$. ■

Theorem 5.2 *Let f and g be two elements in $O^n(A)$ such that $f \neq g$. Then these three propositions are equivalent:*

(i) $f\mathcal{R}g$.

(ii) $f^{-1}(g(\bar{x})) \cap \Delta_{A^n} \neq \emptyset$ *and* $g^{-1}(f(\bar{x})) \cap \Delta_{A^n} \neq \emptyset$ *for all* $\bar{x} \in A^n$.

(iii) $Imf = Img$ *and* $f^{-1}(f(\bar{x})) \cap \Delta_{A^n} \neq \emptyset$ *and* $g^{-1}(g(\bar{x})) \cap \Delta_{A^n} \neq \emptyset$ *for all* $\bar{x} \in A^n$.

Proof: (i) $\Longleftrightarrow$ (ii) Let $f\mathcal{R}g$. Then there exist $h, k \in O^n(A)$ such that $f+h = g$ and $g+k = f$, that is $f(\widehat{h(\bar{x})}) = g(\bar{x})$ and $g(\widehat{k(\bar{x})}) = f(\bar{x})$ for every $\bar{x} \in A^n$. These mean that there exist h and k such that $\widehat{h(\bar{x})} \in f^{-1}(g(\bar{x}))$ and $\widehat{k(\bar{x})} \in g^{-1}(f(\bar{x}))$ for all $\bar{x} \in A^n$. Since $\widehat{h(\bar{x})}, \widehat{k(\bar{x})} \in \Delta_{A^n}$ thus $f^{-1}(g(\bar{x})) \cap \Delta_{A^n} \neq \emptyset$ and $g^{-1}(f(\bar{x})) \cap \Delta_{A^n} \neq \emptyset$. Conversely, let $\hat{y}$ be in $f^{-1}(g(\bar{x})) \cap \Delta_{A^n}$ and $\hat{z}$ be in $g^{-1}(f(\bar{x})) \cap \Delta_{A^n}$. Define n-ary operations h and k on A by $h(\bar{x}) = y$ and $k(\bar{x}) = z$. Since $\hat{y} \in f^{-1}(g(\bar{x}))$ and $\hat{z} \in g^{-1}(f(\bar{x}))$ we get $(f+h)(\bar{x}) = f(\widehat{h(\bar{x})}) = f(\hat{y}) = g(\bar{x})$ and $(g+k)(\bar{x}) = g(\widehat{k(\bar{x})}) = g(\hat{z}) = f(\bar{x})$. Therefore $f\mathcal{R}g$.
(ii) $\Longleftrightarrow$ (iii) Let $\bar{x}$ be arbitrary in A^n, $\hat{y}$ be in $f^{-1}(g(\bar{x})) \cap \Delta_{A^n}$ and $\hat{z}$ be in $g^{-1}(f(\bar{x})) \cap \Delta_{A^n}$. Thus we get $f(\hat{y}) = g(\bar{x})$ and $g(\hat{z}) = f(\bar{x})$ and hence $f(A^n) \subseteq g(A^n)$ and $g(A^n) \subseteq f(A^n)$ and therefore $Imf = Img$. By assumption we also get $f^{-1}(f(\bar{x})) \cap \Delta_{A^n} \neq \emptyset$ and $g^{-1}(g(\bar{x})) \cap \Delta_{A^n} \neq \emptyset$ for all $\bar{x} \in A^n$. Conversely, if $Imf = Img$ and $f^{-1}(f(\bar{x})) \cap \Delta_{A^n} \neq \emptyset$ and $g^{-1}(g(\bar{x})) \cap \Delta_{A^n} \neq \emptyset$ for all $\bar{x} \in A^n$ then $f^{-1}(g(\bar{x})) \cap \Delta_{A^n} \neq \emptyset$ and $g^{-1}(f(\bar{x})) \cap \Delta_{A^n} \neq \emptyset$ for all $\bar{x} \in A^n$. ■

Theorem 5.3 *Let f and g be two elements in $O^n(A)$ such that $f \neq g$. Then $f\mathcal{H}g$ if and only if the following conditions are satisfied:*

(i) $Imf = Img$.

(ii) $f^{-1}(f(\bar{x})) \cap \Delta_{A^n} \neq \emptyset$ *and* $g^{-1}(g(\bar{x})) \cap \Delta_{A^n} \neq \emptyset$ *for all* $\bar{x} \in A^n$.

(iii) $Kerf = Kerg$.

Proof: This is clear by Theorem 5.1 and Theorem 5.2. ■

It is well known (see [1]) that $\mathcal{D} = \mathcal{J}$ for an arbitrary finite subsemigroup S. We have the following theorem.

Theorem 5.4 *Let f and g be two elements in $O^n(A)$ such that $f \neq g$. Then these three propositions are equivalent:*

(i) $f\mathcal{J}g$ $(f\mathcal{D}g)$.

(ii) $|Imf| = |Img| = |Imf|_{\Delta_{A^n}}| = |Img|_{\Delta_{A^n}}|$.

(iii) $f^{-1}(f(\bar{x})) \cap \Delta_{A^n} \neq \emptyset$ *and* $g^{-1}(g(\bar{x})) \cap \Delta_{A^n} \neq \emptyset$ *for all* $\bar{x} \in A^n$ *and* $|Imf| = |Img|$.

Proof: (i)⇔(ii) Let $f\mathcal{J}g$. Then there exist $p, q, r, s \in O^n(A)$ such that $f = \widehat{p+g+q}$ and $g = r + f + s$ and for arbitrary $\bar{x} \in A^n$ we have $f(\bar{x}) = (p+g+q)(\bar{x}) = p(\widehat{g(q(\bar{x}))})$. Therefore there exists $\pi_p \in O^1(A)$ such that $\pi_p((g+q)(\bar{x})) = f(\bar{x})$. Thus for all $\bar{x}_1, \bar{x}_2 \in A^n$ if $(g+q)(\bar{x}_1) = (g+q)(\bar{x}_2)$ then $f(\bar{x}_1) = f(\bar{x}_2)$. This means $Ker(g+q) \subseteq Kerf$ and hence $|Imf| \leq |Im(g+q)|$ and we obtain $|Imf| \leq |Im(g+q)| \leq |Img|_{\Delta_{A^n}}| \leq |Img|$. Similarly we have $|Img| \leq |Im(f+s)| \leq |Imf|_{\Delta_{A^n}}| \leq |Imf|$ and therefore $|Imf| = |Img| = |Imf|_{\Delta_{A^n}}| = |Img|_{\Delta_{A^n}}|$.
Conversely, if $|Imf| = |Img| = |Imf|_{\Delta_{A^n}}| = |Img|_{\Delta_{A^n}}|$, then $|\Delta_{A^n}/_{Kerf|_{\Delta_{A^n}}}| = |\Delta_{A^n}/_{Kerg|_{\Delta_{A^n}}}| = |Imf| = |Img|$. Now, define two arbitrary one-to-one mappings $\phi_1 : \Delta_{A^n}/_{Kerg|_{\Delta_{A^n}}} \rightarrow Imf$ and $\phi_2 : \Delta_{A^n}/_{Kerf|_{\Delta_{A^n}}} \rightarrow Img$. Then define two n-ary operations on A^n namely q and s as follows: $q(\bar{x}) = y$ for every $\bar{x} \in A^n$ such that $f(\bar{x}) = \phi_1([\hat{y}]_{Kerg|_{\Delta_{A^n}}})$ for $[\hat{y}]_{Kerg|_{\Delta_{A^n}}} \in \Delta_{A^n}/_{Kerg|_{\Delta_{A^n}}}$ and $s(\bar{x}) = y$ for every $\bar{x} \in A^n$ such that $g(\bar{x}) = \phi_2([\hat{y}]_{Kerf|_{\Delta_{A^n}}})$ for $[\hat{y}]_{Kerf|_{\Delta_{A^n}}} \in \Delta_{A^n}/_{Kerf|_{\Delta_{A^n}}}$. By these constructions we will show that $Ker(g+q) = Kerf$ and $Ker(f+s) = Kerg$. Let $(\bar{x}_1, \bar{x}_2)$ be in $Kerf$ i.e. $f(\bar{x}_1) = f(\bar{x}_2)$. Since ϕ_1 is one-to-one then there exist $[\hat{z}]_{Kerg|_{\Delta_{A^n}}} \in \Delta_{A^n}/_{Kerg|_{\Delta_{A^n}}}$ such that $\phi_1([\hat{z}]_{Kerg|_{\Delta_{A^n}}}) = f(\bar{x}_1) = f(\bar{x}_2)$. Thus $(g+q)(\bar{x}_1) = g(\widehat{q(\bar{x}_1)}) = g(\widehat{q(\bar{x}_2)}) = (g+q)(\bar{x}_2)$ and we obtain $Kerf \subseteq Ker(g+q)$. Now, we will show that $Ker(g+q) \subseteq Kerf$. Assume there exists $(\bar{x}_1, \bar{x}_2)$ such that $(\bar{x}_1, \bar{x}_2) \in Ker(g+q)$ and $(\bar{x}_1, \bar{x}_2) \notin Kerf$. Thus $(g+q)(\bar{x}_1) = (g+q)(\bar{x}_2)$ i.e. $g(\widehat{q(\bar{x}_1)}) = g(\widehat{q(\bar{x}_2)})$ and hence $(\widehat{q(\bar{x}_1)}, \widehat{q(\bar{x}_2)}) \in Kerg$. Therefore $[\widehat{q(\bar{x}_1)}]_{Kerg|_{\Delta_{A^n}}} = [\widehat{g(\bar{x}_2)}]_{Kerg|_{\Delta_{A^n}}}$. But $(\bar{x}_1, \bar{x}_2) \notin Kerf$ i.e $f(\bar{x}_1) \neq f(\bar{x}_2)$ implies that there exist $[\hat{z}_1]_{Kerg|_{\Delta_{A^n}}}, [\hat{z}_2]_{Kerg|_{\Delta_{A^n}}} \in \Delta_{A^n}/_{Kerg|_{\Delta_{A^n}}}$, $[\hat{z}_1]_{Kerg|_{\Delta_{A^n}}} \neq [\hat{z}_2]_{Kerg|_{\Delta_{A^n}}}$ such that $\phi_1([\hat{z}_1]_{Kerg|_{\Delta_{A^n}}}) = f(\bar{x}_1)$ and $\phi_1([\hat{z}_2]_{Kerg|_{\Delta_{A^n}}}) = f(\bar{x}_2)$. Therefore $q(\bar{x}_1) = z_1$ and $q(\bar{x}_2) = z_2$. Thus $[\widehat{q(\bar{x}_1)}]_{Kerg|_{\Delta_{A^n}}} = [\hat{z}_1]_{Kerg|_{\Delta_{A^n}}} \neq [\hat{z}_2]_{Kerg|_{\Delta_{A^n}}} = [\widehat{q(\bar{x}_2)}]_{Kerg|_{\Delta_{A^n}}}$, a contradiction. Therefore $Ker(g+q) = Kerf$. Similarly we get $Ker(f+s) = Kerg$. By Theorem 5.1 these imply $(g+q)\mathcal{L}f$ and $(f+s)\mathcal{L}g$. Thus we can find $p, r \in O^n(A)$ such that $p+g+q = f$ and $r+f+s = g$ and hence $f\mathcal{J}g$.
(ii)⇔(iii) By assumption we get $Imf = Imf|_{\Delta_{A^n}}$ and $Img = Img|_{\Delta_{A^n}}$. Thus for every $\bar{x} \in A^n$ there exist $\hat{y}_1, \hat{y}_2 \in \Delta_{A^n}$ such that $f(\hat{y}_1) = f(\bar{x})$ and $g(\hat{y})_2 = g(\bar{x})$. Therefore $f^{-1}(f(\bar{x})) \cap \Delta_{A^n} \neq \emptyset$ and $g^{-1}(g(\bar{x})) \cap \Delta_{A^n} \neq \emptyset$.
For the converse, let $\bar{x} \in A^n$. Then $f(\bar{x}) \in Imf$ and $g(\bar{x}) \in Img$. By assumption we can find $\hat{y}_1, \hat{y}_2 \in \Delta_{A^n}$ such that $\hat{y}_1 \in f^{-1}(f(\bar{x}))$ and $\hat{y}_2 \in g^{-1}(g(\bar{x}))$ and thus $f(\hat{y}_1) = f(\bar{x})$ and $g(\hat{y})_2 = g(\bar{x})$. Therefore $Imf \subseteq Imf|_{\Delta_{A^n}}$ and $Img \subseteq Img|_{\Delta_{A^n}}$ and hence $|Imf| = |Img| = |Imf|_{\Delta_{A^n}}| = |Img|_{\Delta_{A^n}}|$. ∎

Corollary 5.5 *Let f and g be in $O^n(A)$. Then $f\mathcal{R}g$ iff $f\mathcal{J}g$ $(f\mathcal{D}g)$ and $Imf \subseteq Img$.*

Proof: If $f\mathcal{R}g$, then by Theorem 5.2 and Theorem 5.4 $f\mathcal{D}g$ and $Imf \subseteq Img$ because of $g(\widehat{k(\bar{x})}) = f(\bar{x})$ for some k and all $\bar{x} \in A^n$. Now, if $f\mathcal{J}g$ then by Theorem 5.4 $f^{-1}(f(\bar{x})) \cap \Delta_{A^n}$, $g^{-1}(g(\bar{x})) \cap \Delta_{A^n}$ are not empty for all $\bar{x} \in A^n$ and $|Imf| = |Img|$. Since $|Imf| = |Img|$ and $Im(f) \subseteq Im(g)$ then $Im(f) = Im(g)$ and hence $f\mathcal{R}g$ by Theorem 5.2. ∎

Corollary 5.6 *Let $f, g \in O^n(A)$. Then $f\mathcal{J}g$ $(f\mathcal{D}g)$ iff f and g are regular elements and $|Imf| = |Img|$.*

Proof: By Theorem 5.6 we have $f\mathcal{J}g$ iff $|Imf| = |Img| = |Imf|_{\Delta_{A^n}}| = |Img|_{\Delta_{A^n}}|$. But $Imf|_{\Delta_{A^n}} \subseteq Imf$ and $Imf|_{\Delta_{A^n}} \subseteq Imf$ and thus $Imf = Imf|_{\Delta_{A^n}}$ and $Imf = Imf|_{\Delta_{A^n}}$ i.e. $f \in \Sigma$. But by Lemma 2.6 $\Sigma = Reg(\mathcal{O}^n(A))$. Therefore $f\mathcal{J}g$ iff f and g are regular and $|Imf| = |Img|$. ∎

6 Examples

To illustrate our results we consider n-ary operations on the three-element set $A = \{0, 1, 2\}$. Let $f_{(a,b,c)} \in O^1(\{0, 1, 2\})$ be defined by $f_{(a,b,c)}(0) = a$, $f_{(a,b,c)}(1) = b$, $f_{(a,b,c)}(2) = c$. Then

$$Idem(\mathcal{O}^1(\{0,1,2\})) = \{f_{(0,0,0)}, f_{(1,1,1)}, f_{(0,1,0)}, f_{(0,1,0)}, f_{(0,1,1)}, f_{(0,0,2)}, f_{(0,2,2)}, f_{(1,1,2)}, f_{(2,1,2)}, f_{(0,1,2)}\}.$$

We consider the following sets of binary operations on $\{0, 1, 2\}$

$$\begin{array}{lcl}
C_{(0,0,0)} & := & \{f \in O^2(\{0,1,2\}) | f(\Delta_{A^2}) = \{0\}\} \\
C_{(1,1,1)} & := & \{f \in O^2(\{0,1,2\}) | f(\Delta_{A^2}) = \{1\}\} \\
C_{(2,2,2)} & := & \{f \in O^2(\{0,1,2\}) | f(\Delta_{A^2}) = \{2\}\} \\
C_{(0,1,0)} & := & \{f \in O^2(\{0,1,2\}) | f(\hat{0}) = 0, f(\hat{1}) = 1, f(\hat{2}) = 0\} \\
C_{(0,1,1)} & := & \{f \in O^2(\{0,1,2\}) | f(\hat{0}) = 0, f(\hat{1}) = 1, f(\hat{2}) = 1\} \\
C_{(0,0,2)} & := & \{f \in O^2(\{0,1,2\}) | f(\hat{0}) = 0, f(\hat{1}) = 0, f(\hat{2}) = 2\} \\
C_{(0,2,2)} & := & \{f \in O^2(\{0,1,2\}) | f(\hat{0}) = 0, f(\hat{1}) = 2, f(\hat{2}) = 2\} \\
C_{(1,1,2)} & := & \{f \in O^2(\{0,1,2\}) | f(\hat{0}) = 1, f(\hat{1}) = 1, f(\hat{2}) = 2\} \\
C_{(2,1,2)} & := & \{f \in O^2(\{0,1,2\}) | f(\hat{0}) = 2, f(\hat{1}) = 1, f(\hat{2}) = 2\} \\
C_{(0,1,2)} & := & \{f \in O^2(\{0,1,2\}) | f(\hat{0}) = 0, f(\hat{1}) = 1, f(\hat{2}) = 2\}.
\end{array}$$

Thus we have the following characterization:

Proposition 6.1 *Let f be in $O^2(\{0,1,2\})$. Then f is idempotent if and only if $f \in (C_{(0,0,0)} \cup C_{(1,1,1)} \cup C_{(2,2,2)} \cup C_{(0,1,0)} \cup C_{(0,1,1)} \cup C_{(0,0,2)} \cup C_{(0,2,2)} \cup C_{(1,1,2)} \cup C_{(2,1,2)} \cup C_{(0,1,2)}) \cap \Sigma$.*

Proof: By Proposition 2.4 (ii) $f \in O^2(\{0,1\})$ is idempotent if and only if π_f is idempotent and $f \in \Sigma$. ■

Counting all operations from Proposition 6.1 we obtain precisely $3 + 3 \cdot 2^7 + 3^6$ idempotent elements in $\mathcal{O}^2(\{0,1,2\})$.

In general we have the following proposition:

Proposition 6.2 *Let A be $\{0,1,\cdots,m-1\}$. Then there are precisely*

$$\sum_{i=1}^{m} \binom{m}{i} \cdot i^{m^n - i}$$

idempotent elements and precisely

$$\sum_{i=1}^{m} \binom{m}{i} \cdot i! \cdot i^{m^n - i}$$

regular elements in $(O^n(A);+)$.

The proof is by simple counting arguments.

References

[1] J. M. Howie, *Fundamentals of Semigroup Theory*, Oxford Science Publication, 1995.

[2] K. Deneck and S. L. Wismath, *Universal Algebra and Coalgebra*, World Scientific, 2009.

[3] R. Butkote, K. Denecke and Ch. Ratanaprasert, Semigroup Properties of n-ary Operations on Finite Sets, *Asian-European J. Math.* Vol. **1**, 1 (2008) 27–44.

[4] R. Butkote and K. Denecke, Semigroup Properties of Boolean Operations, *Asian-European J. Math.* Vol. **1**, 2 (2008) 157–175.

[5] R. Butkote, *Universal-algebraic and Semigroup-theoretical Properties of Boolean Operations*, Dissertation, Universität Potsdam 2009.

[6] K. Denecke and Y. Susanti, Semigroup-theoretical Properties of Boolean Operations, preprint 2009 (submitted).

K. Denecke, Universität Potsdam,
14469 Potsdam, Am Neuen Palais 10
kdenecke@rz.uni-potsdam.de
Y. Susanti, Gadjah Mada University
Yogyakarta, Indonesia and
Universität Potsdam,
14469 Potsdam, Am Neuen Palais 10
inielsusan@yahoo.com

Cones of Weighted and Partial Metrics

Michel Deza* Elena Deza† and Janoš Vidali‡

Abstract

A *partial semimetric* on $V_n = \{1, \ldots, n\}$ is a function $f = ((f_{ij})) : V_n^2 \longrightarrow \mathbb{R}_{\geq 0}$ satisfying $f_{ij} = f_{ji} \geq f_{ii}$ and $f_{ij} + f_{ik} - f_{jk} - f_{ii} \geq 0$ for all $i, j, k \in V_n$. The function f is a *weak partial semimetric* if $f_{ij} \geq f_{ii}$ is dropped, and it is a *strong partial semimetric* if $f_{ij} \geq f_{ii}$ is complemented by $f_{ij} \leq f_{ii} + f_{jj}$.

We describe the cones of weak and strong partial semimetrics via corresponding weighted semimetrics and list their 0, 1-valued elements, identifying when they belong to extreme rays. We consider also related cones, including those of *partial hypermetrics*, *weighted hypermetrics*, ℓ_1-*quasi semimetrics* and weighted/partial cuts.

Key Words and Phrases: weighted metrics; partial metrics; hypermetrics; cuts; convex cones; computational experiments.

1 Convex cones under consideration

There are following two main motivations for this study. One is to extend the rich theory of metric, cut and hypermetric cones on weighted, partial and non-symmetric generalizations of metrics. Another is a new appoach to partial semimetrics (having important applications in Computer Science) via cones formed by them.

*michel.deza@ens.fr, École Normale Supérieure, Paris

†elena.deza@gmail.com, Moscow State Pedagogical University, Moscow

‡janos.vidali@fri.uni-lj.si, University of Ljubljana, Slovenia

A *convex cone in* $\mathbb{R}^m$ (see, for example, [Sc86]) is defined either by *generators* $v_1, \dots, v_N$, as $\{\sum \lambda_i v_i : \lambda_i \geq 0\}$, or by *linear inequalities* $f^1, \dots, f^M$, as $\cap_{j=1}^M \{x \in \mathbb{R}^m : f^j(x) = \sum_{i=1}^m f_i^j x_i \geq 0\}$.

Let C be an m'-dimensional convex cone in $\mathbb{R}^m$. Given $f \in \mathbb{R}^m$, the linear inequality $f(x) = \sum_{i=1}^m f_i x_i = \langle f, x \rangle \geq 0$ is said to be *valid* for C if it holds for all $x \in C$. Then the set $\{x \in C : \langle f, x \rangle = 0\}$ is called the *face* of C, *induced by* F. A face of dimension $m' - 1$, $m' - 2$, 1 is called a *facet, ridge, extreme ray* of C, respectively (a *ray* is a set $\mathbb{R}_{\geq 0}x$ with $x \in C$). Denote by $F(C)$ the set of facets of C and by $R(C)$ the set of its extreme rays. We consider only *polyhedral* (i.e., $R(C)$ and, alternatively, $F(C)$ is finite) *pointed* (i.e., $(0) \in C$) convex cones. Each ray $r \subset C$ below contains a unique *good representative*, i.e., an integer-valued vector $v(r)$ with g.c.d. 1 of its entries; so, by abuse of language, we will identify r with $v(r)$.

For a ray $r \subset C$ denote by $F(r)$ the set $\{f \in F(C) : r \subset f\}$. For a face $f \subset C$ denote by $R(f)$ the set $\{r \in R(C) : r \subset f\}$. The *incidence number* $\mathrm{Inc}(f)$ of a face f (or $\mathrm{Inc}(r)$ of a ray r) is the number $|\{r \in R(C) : r \subset f\}|$ (or, respectively, $|\{f \in F(C) : r \subset f\}|$). The $\mathrm{rank}(f)$ of a face f (or $\mathrm{rank}(r)$ of a ray r) is the dimension of $\{r \in R(C) : r \subset f\}$ (or of $\{f \in F(C) : r \subset f\}$).

Two extreme rays (or facets) of C are *adjacent on* C if they span a 2-dimensional face (or, respectively, their intersection has dimension $m' - 2$). The *skeleton* $\mathrm{Sk}(C)$ is the graph whose vertices are the extreme rays of C and with an edge between two vertices if the corresponding rays are adjacent on C. The *ridge graph* $\mathrm{Ri}(C)$ is the graph whose vertices are facets of C and with an edge between two vertices if the corresponding facets are adjacent on C. Let $D(G)$ denote the diameter of the graph G

Given a cone C_n of some functions, say, $d = ((d_{ij})) : V_n^2 \longrightarrow \mathbb{R}_{\geq 0}$ the 0-*extension* of the inequality $\sum_{1 \leq i \neq j \leq n-1} F_{ij} d_{ij} \geq 0$ is the inequality

$$\sum_{1 \leq i \neq j \leq n} F'_{ij} d_{ij} \geq 0 \text{ with } F'_{ni} = F'_{in} = 0 \text{ and } F'_{ij} = F_{ij}, \text{ otherwise.}$$

Clearly, the 0-extension of any facet-defining inequality of a cone C_n is a valid inequality (usually, facet-defining) of C_{n+1}. The 0-extension of an extreme ray is defined similarly. For any cone C denote by $0,1$-C the cone generated by all extreme rays of C containing a non-zero $0,1$-valued point.

The cones C considered here will be symmetric under permutations and usually $\mathrm{Aut}(C) = \mathrm{Sym}(n)$. All orbits below are under $\mathrm{Sym}(n)$.

Set $V_n = \{1, \ldots, n\}$. The function $f = ((f_{ij})) : V_n^2 \longrightarrow \mathbb{R}$ is called *weak partial semimetric* if the following holds:

(1) $f_{ij} = f_{ji}$ (*symmetry*) for all $i, j \in V_n$,

(2) $L_{ij} : f_{ij} \geq 0$ (*non-negativity*) for all $i, j \in V_n$, and

(3) $Tr_{ij,k} : f_{ik} + f_{kj} - f_{ij} - f_{kk} \geq 0$ (*triangle inequality*) for all $i, j, k \in V_n$.

Weak partial semimetrics were introduced in [He99] as a generalization of *partial semimetrics* introduced in [Ma92]. Clearly, all $Tr_{ij,i} = 0$ and $Tr_{ii,k} = 2f_{ik} - f_{ii} - f_{kk} = Tr_{ij,k} + Tr_{kj,i}$. So, it is sufficient to require (2) only for $i = j$ and (3) only for different i, j, k. The weak partial semimetrics on V_n form a $\binom{n+1}{2}$-dimensional convex cone with n facets L_{ii} and $3\binom{n}{3}$ facets $Tr_{ij,k}$. Denote this cone by $wPMET_n$.

A weak partial semimetric f is called *partial semimetric* if it holds that

(4) $M_{ij} : f_{ij} - f_{ii} \geq 0$ (*small self-distances*) for all different $i, j \in V_n$.

The partial semimetrics on V_n form a $\binom{n+1}{2}$-dimensional subcone, denote it by $PMET_n$, of $wPMET_n$. This cone has n facets L_{ii}, $2\binom{n}{2}$ facets $M_{ij,i}$ and $3\binom{n}{3}$ facets $Tr_{ij,k}$. Partial metrics were introduced by Matthews in [Ma92] for treatment of partially defined objects in Computer Science; see also [Ma08, Hi01, Se97]. The cone $PMET_n$ was considered in [DeDe10].

A partial semimetric f is called *strong partial semimetric* if it holds that

(5) $N_{ij} : f_{ii} + f_{jj} - f_{ij} \geq 0$ (*large self-distances*) for all $i, j \in V_n$.

So, $f_{ii} = N_{ij} + M_{ji} \geq 0$, i.e., (5) and (4) imply L_{ii} for all. i. The strong partial semimetrics on V_n form a $\binom{n+1}{2}$-dimensional subcone, denote it by $sPMET_n$, of $PMET_n$. This cone has $3\binom{n+1}{3}$ facets: $2\binom{n}{2}$ facets M_{ij}, $\binom{n}{2}$ facets N_{ij} and $3\binom{n}{3}$ facets $Tr_{ij,k}$.

A partial semimetric f is called *semimetric* if it holds that

(6) $f_{ii} = 0$ (*reflexivity*) for all $i \in V_n$.

The semimetrics on V_n form a $\binom{n}{2}$-dimensional convex cone, denoted by MET_n, which has $3\binom{n}{3}$ facets $Tr_{ij,k}$ (clearly, $f_{ij} = \frac{Tr_{ij,k}+Tr_{jk,i}}{2} \geq 0$). This cone is well-known; see, for example, [DeLa97] and references there.

The function f is *quasi-semimetric* if only (2), (3), (6) are required. The quasi-semimetrics on V_n form a $n(n-1)$-dimensional convex cone, denoted by $QMET_n$, which has $2\binom{n}{2}$ facets L_{ij} and $6\binom{n}{3}$ facets $OTr_{ij,k} : f_{ik} + f_{kj} - f_{ij} \geq 0$ (*oriented triangle inequality*). But other oriented versions of $Tr_{ij,k}$ (for example, $f_{ik} + f_{kj} - f_{ji}$) are not valid on $QMET_n$.

A quasi-semimetric f is *weightable* if there exist a (weight) function $w = (w_i) : V_n \longrightarrow \mathbb{R}_{\geq 0}$ such that $f_{ij} + w_i = f_{ji} + w_j$ for all $i, j \in V_n$. Such quasi-semimetrics f (or, equivalently, pairs (f, w)) on V_n form a $\binom{n+1}{2}$-dimensional cone, denote it by $WQMET_n$, with $2\binom{n}{2}$ facets L_{ij} and $3\binom{n}{3}$ facets $OTr_{ij,k}$ since, for a quasi-semimetric, $OTr_{ij,k} = OTr_{ji,k}$ if it is weightable.

A weightable quasi-semimetric (f, w) with all $f_{ij} \leq w_j$ is a *weightable strong quasi-semimetric.* But if, on the contrary, (2) is weakened to $f_{ij} + f_{ji} \geq 0$ (so, $f_{ij} < 0$ is allowed), (f, w) is a *weightable weak quasi-semimetric.* Denote by $sWQMET_n$ and $wWQMET_n$ the corresponding cones.

Let us denote the function f by p, d, or q if it is a weak partial semimetric, semimetric, or weightable weak quasi-semimetric, respectively.

A *weighted semimetric* $(d; w)$ is a semimetric d with a weight function $w : V_n \to \mathbb{R}_{\geq 0}$ on its points. Denote by $(d; w)$ the matrix $((d'_{ij}))$, $0 \leq i, j \leq n$, with $d'_{00} = 0$, $d'_{0i} = d'_{i0} = w_i$ for $i \in V_n$ and $d'_{ij} = d_{ij}$ for $i, j \in V_n$. The weighted semimetrics $(d; w)$ on V_n form a $\binom{n+1}{2}$-dimensional convex cone with n facets $w_i \geq 0$ and $3\binom{n}{3}$ facets $Tr_{ij,k}$. Denote this cone by $WMET_n$. So, $MET_n \simeq \{(d; (k, \ldots, k)) : d \in WMET_n\}$. Also, $MET_n = QMET_n \cap PMET_n$.

Call a weighted semimetric $(d; w)$ *down-* or *up-weighted* if

(4') $d_{ij} \geq w_i - w_j$, or

(5') $d_{ij} \leq w_i + w_j$

holds (for all distinct $i, j \in V_n$), respectively. Denote by $dWMET_n$ the cone of down-weighted semimetrics on V_n and by $sWMET_n$ the cone of *strongly*, i.e., both, down- and up-, weighted semimetrics. So, $sWMET_n = MET_{n+1}$.

2 Maps P, Q and semimetrics

Given a weighted semimetric $(d; w)$, define the map P by the function $p = P(d; w)$ with $p_{ij} = \frac{d_{ij}+w_i+w_j}{2}$. Clearly, P is an *automorphism* (invertible linear operator) of the vector space $\mathbb{R}^{\binom{n+1}{2}}$, and $(d; w) = P^{-1}(p)$, where the inverse map P^{-1} is defined by $d_{ij} = 2p_{ij} - p_{ii} - p_{jj}$, $w_i = p_{ii}$.

Define the map Q by the function $(q, w) = Q(d; w)$ with $q_{ij} = \frac{d_{ij}-w_i+w_j}{2}$. So, $Q(d; w) = P(d; w) - ((1))w$ (i.e., $q_{ij} = p_{ij} - p_{ii}$) and $d_{ij} = q_{ij} + q_{ji}$, is the *symmetrization semimetric* of q.

Example. Below are given: the semimetric $d = 2\delta(\{56\}, \{1\}, \{23\}, \{4\})$ $-\delta(\{56\}) \in MET_6$, and, taking weight $w = (1_{i \in \{56\}}) = (0, 0, 0, 0, 1, 1)$, the partial semimetric $P(d; w) = J(\{56\}) + \delta(\{56\}, \{1\}, \{23\}, \{4\})$ (its ray is extreme in $PMET_6$) and the weightable quasi-semimetric $Q(d; w) = \delta'(\{1\}) + \delta'(\{23\}) + \delta'(\{4\})$ (its ray is not extreme in $WQMET_6$).

$$
\begin{array}{cccccc}
\mathbf{0} & 2 & 2 & 2 & 1 & 1 \\
2 & \mathbf{0} & 0 & 2 & 1 & 1 \\
2 & 0 & \mathbf{0} & 2 & 1 & 1 \\
2 & 2 & 2 & \mathbf{0} & 1 & 1 \\
1 & 1 & 1 & 1 & \mathbf{0} & 0 \\
1 & 1 & 1 & 1 & 0 & \mathbf{0}
\end{array}
\qquad
\begin{array}{cccccc}
\mathbf{0} & 1 & 1 & 1 & 1 & 1 \\
1 & \mathbf{0} & 0 & 1 & 1 & 1 \\
1 & 0 & \mathbf{0} & 1 & 1 & 1 \\
1 & 1 & 1 & \mathbf{0} & 1 & 1 \\
1 & 1 & 1 & 1 & \mathbf{1} & 1 \\
1 & 1 & 1 & 1 & 1 & \mathbf{1}
\end{array}
\qquad
\begin{array}{cccccc}
\mathbf{0} & 1 & 1 & 1 & 1 & 1 \\
1 & \mathbf{0} & 0 & 1 & 1 & 1 \\
1 & 0 & \mathbf{0} & 1 & 1 & 1 \\
1 & 1 & 1 & \mathbf{0} & 1 & 1 \\
0 & 0 & 0 & 0 & \mathbf{0} & 0 \\
0 & 0 & 0 & 0 & 0 & \mathbf{0}
\end{array}
$$

Clearly, $d_{ij} + d_{ik} - d_{jk} = p_{ij} + p_{ik} - p_{jk} - p_{ii} = q_{ji} + q_{ik} - q_{jk}$, i.e., the triangle inequalities are equivalent on all three levels: d - of semimetrics, p - of would-be partial semimetrics and q - of would-be quasi-semimetrics.

Now, $p_{ij} \geq p_{ii}$ iff $d_{ij} \geq w_i - w_j$ iff $q_{ij} \geq 0$; so, (4) is equivalent to (4'), $p_{ij} \leq p_{ii} + p_{jj}$ iff $d_{ij} \leq w_i + w_j$ iff $q_{ij} \leq w_j$; so, (5) is equivalent to (5'), and $2p_{ij} \geq p_{ii} + p_{jj}$ iff $d_{ij} \geq 0$ iff $q_{ij} + q_{ji} \geq 0$. This implies

Lemma 1 *The following statements hold.*
(i) $wPMET_n = P(WMET_n)$, $PMET_n = P(dWMET_n)$ *and* $sPMET_n = P(sWMET_n)$,
(ii) $wWQMET_n = Q(WMET_n)$, $WQMET_n = Q(dWMET_n)$ *and* $sWQMET_n = Q(sWMET_n)$.

The metric cone $MET_n \in \mathbb{R}^{\binom{n}{2}}$ has a unique orbit of $3\binom{n}{3}$ facets $Tr_{ij,k}$. Its symmetry group $\text{Aut}(MET_n)$ is $\text{Sym}(n)$ $n \neq 4$. The number of extreme rays (orbits) of MET_n is 3 (1), 7 (2), 25 (3), 296 (7), 55226 (46) for $3 \leq n \leq 7$. $D(\text{Ri}(MET_n)) = 2$ for $n > 3$, while $\text{Ri}(MET_3) = \text{Sk}(MET_3) = K_3$. $D(\text{Sk}(MET_n))$ is 1 for $n = 4$, 2 for $5 \leq n \leq 6$ and 3 for $n = 7$.

For a partition $\mathcal{S} = \{S_1, \dots, S_t\}$ of V_n, the *multicut* $\delta(\mathcal{S}) \in MET_n$ has $\delta_{ij}(\mathcal{S}) = 1$ if $|\{i,j\}| = 2 > |\{i,j\} \cap S_h|, 1 \leq h \leq t$ and $\delta_{ij}(\mathcal{S}) = 0$, otherwise. Call $\delta(\mathcal{S})$ a *t-cut* if $S_h \neq \emptyset$ for $1 \leq h \leq t$. Clearly,

$$\delta(S_1, \dots, S_t) = \frac{1}{2} \sum_{h=1}^{t} \delta(S_h, \overline{S_h}).$$

Denote by CUT_n the cone generated by all $2^{n-1} - 1$ 2-cuts $\delta(S, \overline{S}) = \delta(S)$. $CUT_n = MET_n$ holds for $n \leq 4$ and $\text{Aut}(CUT_n) = \text{Aut}(MET_n)$. The number of facets (orbits) of CUT_n is 3 (1), 12 (1), 40 (2), 210 (4), 38780 (36) for $3 \leq n \leq 7$. $D(\text{Sk}(CUT_n)) = 1$ and $D(\text{Ri}(CUT_n)) = 2, 3, 3$ for $n = 5, 6, 7$. See, for example, [DeLa97, DDF96, Du08] for details on MET_n and CUT_n.

The number of t-cuts of V_n is the number of ways to partition a set of n objects into t groups, i.e., the *Stirling number of the second kind* $S(n,t) = \frac{1}{t!}\sum_{j=0}^{t}(-1)^j\binom{t}{j}(t-j)^n$. So, $S(n,2) = 2^{n-1}-1$ and $S(n,n-1) = \binom{n}{2}$. The number of multicuts of V_n is the *Bell number* $B(n) = \sum_{t=0}^{n} S(n,t) = \sum_{t=0}^{n-1}(t+1)S(n,t) = \sum_{t=0}^{n-1}\binom{n-1}{t}B(t)$ (the sequence A000110 $= 1,1,2,5,$ $15,52,203,877,\ldots$ in [Sl10]). The number of ways to write i as a sum of positive integers is i-th *partition number* Q_i (the sequence $A000041$ in [Sl10]).

The $0,1$-valued elements $d \in MET_n$ are all $B(n)$ multicuts $\delta(\{S_1,\ldots,S_t\})$ of V_n. It follows by induction using that $d_{1i} = d_{1j} = 0$ implies $d_{ij} = 0$ and $d_{1i} \neq d_{1j}$ implies $d_{ij} = 1$. In fact, $S_1,\ldots,S_t$ are the equivalence classes of the equivalence $\sim$ on V_n, defined by $i \sim j$ if $d_{ij} = 0$.

$R(0,1\text{-}MET_n)$ consists of all $S(n,2)$ 2-cuts; so, $0,1\text{-}MET_n = CUT_n$.

3 Description of $wPMET_n$ and $sPMET_n$

Denote $MET_{n;0} = \{(d;(0)) : d \in MET_n\}$ and $CUT_{n;0} = \{(d;(0)) : d \in CUT_n\}$. So, $MET_n \simeq MET_{n;0} \simeq P(MET_{n;0}) \simeq Q(MET_{n;0})$ and $CUT_n \simeq CUT_{n;0} \simeq P(CUT_{n;0}) \simeq Q(CUT_{n;0})$. Denote by $WCUT_n$ the cone $\{d;w) \in WMET_n : d \in CUT_n\}$ of *weighted ℓ_1-semimetrics* on V_n.

Denote $e_j = (((0));w = (w_i = 1_{i=j})) \in WMET_n$. So, $2P(e_j) = 2$ on the position (jj), 1 on $(ij),(ji)$ with $i \neq j$ and 0, else; $2Q(e_j) = -1$ on the positions (ji), 1 on (ij) (with $i \neq j$ again) and 0, else.

Theorem 1 *The following statements hold.*

(i) $R(WMET_n) = \{e_j : j \in V_n\} \cup R(MET_{n;0})$,
$R(\mathrm{w}PMET_n) = \{2P(e_j) : j \in V_n\} \cup P(R(MET_{n;0}))$,
$R(\mathrm{w}WQMET_n) = \{2Q(e_j) : j \in V_n\} \cup Q(R(MET_{n;0}))$.

(ii) $F(WMET_n) = \{w_j \geq 0 : j \in V_n\} \cup F(MET_{n;0})$,
$F(\mathrm{w}PMET_n) = \{L_{jj} = p_{jj} \geq 0 : j \in V_n\} \cup F(P(MET_{n;0}))$,
$F(\mathrm{w}WQMET_n) = \{w_j \geq 0 : j \in V_n\} \cup F(Q(MET_{n;0}))$.

(iii) $\mathrm{Inc}(2P(e_j)) = |F(\mathrm{w}PMET_n)| - 1$ *and* $\mathrm{Inc}(L_{jj}) = |R(\mathrm{w}PMET_n)| - 1$.

(iv) $\mathrm{Ri}(WMET_n) = \mathrm{Ri}(\mathrm{w}PMET_n) = \mathrm{Ri}(\mathrm{w}WQMET_n) = K_n \times \mathrm{Ri}(MET_n)$,
$\mathrm{Sk}(WMET_n) = \mathrm{Sk}(wPMET_n) = \mathrm{Sk}(\mathrm{w}WQMET_n) = K_n \times \mathrm{Sk}(MET_n)$.

(v) $\mathrm{w}PMET_n$ *has* Aut, $D(\mathrm{Sk})$, $D(\mathrm{Ri})$ *and edge-connectivity of* MET_n.

(vi) The $0,1$*-valued elements of* $\mathrm{w}PMET_n$ *are the* $B(n+1)$ $0,1$*-valued elements of* $PMET_n$ *and* $0,1\text{-}\mathrm{w}PMET_n = CUT_{n;0}$.

(vii) The $0,1$*-valued elements of* $WMET_n$ *are* $2^n B(n)$ $0,1$*-weighted multicuts of* V_n *and* $0,1$-$WMET_n = WCUT_n$.

$R(WCUT_n) = \{e_j : j \in V_n\} \cup R(CUT_{n;0})$ *and* $\mathrm{Sk}(WCUT_n) = K_{n+S(n,2)}$.

$F(WCUT_n) = \{w_j \geq 0 : j \in V_n\} \cup F(CUT_{n;0})$ *and* $\mathrm{Ri}(WCUT_n) = K_n \times \mathrm{Ri}(CUT_n)$ *has diameter* 2.

Proof.

(i). Let $p \in wPMET_n$. We will show that $p' = p - \frac{1}{2}\sum_{t=1}^{n} p_{tt} 2P(e_t) \in MET_{n;0}$. (For example, a well-known weak partial semimetric $i+j$ is the sum of $\sum_t t2P(e_t)$ and the all-zero semimetric $((0))$.)

In fact, $p'_{ii} = p_{ii} - \frac{1}{2}p_{ii}2P(e_i)_{ii} = 0$. Also, p' satisfies to all triangle inequalities (3), since for different $i,j,k \in V_n$, we have $Tr_{ij,k} = p'_{ij} + p'_{ik} - p'_{jk} =$

$$= \left(p_{ij} - \frac{p_{ii}+p_{jj}}{2}\right) + \left(p_{ik} - \frac{p_{ii}+p_{kk}}{2}\right) - \left(p_{jk} - \frac{p_{jj}+p_{kk}}{2}\right) =$$

$$= p_{ij} + p_{ik} - p_{jk} - p_{ii} \geq 0.$$

So, $2P(e_i)$, $1 \leq i \leq n$, and the generators of $P(MET_{n;0}) \simeq MET_{n;0}$ (i.e., the 0-extensions of the generators of MET_n) generate $wPMET_n$. They are, moreover, the generators of $wPMET_n$ since they belongs to all n (linearly independent) facets L_{ii}; so, their rank in $\mathbb{R}^{\binom{n+1}{2}}$is $(\binom{n}{2} - 1) + n = \binom{n+1}{2} - 1$.

Clearly, any $2P(e_i)$ belongs to all facets of $wPMET_n$ except L_{ii}, i.e., its incidence is $(n-1) + 3\binom{n}{3}$. So, its rank in $\mathbb{R}^{\binom{n+1}{2}}$ is $\binom{n+1}{2} - 1$. For $WMET_n$ and $wWQMET_n$, (i) follows similarly, as well as (ii).

(iii), (iv). The ray of $2P(e_i)$ is adjacent to any other extreme ray r, as the set of facets that contain r (with rank $\binom{n+1}{2} - 1$) only loses one element if we intersect it with the set of facets that contain $2P(e_i)$.

(v). The diameters of $\mathrm{Ri}(wPMET_n)$ and $\mathrm{Sk}(wPMET_n)$ being 2, their edge-connectivity is equal to their minimal degrees [Pl75]. But this degree is the same as of $\mathrm{Ri}(MET_n)$ (which is regular of degree $\frac{(n-3)(n^2-6)}{2}$ if $n > 3$) and of $\mathrm{Sk}(MET_n)$, respectively. $\mathrm{Aut}(wPMET_n)$ for $n \geq 5$ is $\mathrm{Sym}(n)$, because it contains $\mathrm{Sym}(n)$ but cannot be larger than $\mathrm{Aut}(MET_n) = \mathrm{Sym}(n)$.

(vi). If $p \in wPMET_n$ is $0,1$-valued, then $p_{ij} = 0 < p_{ii} = 1$ is impossible because $2p_{ij} \geq p_{ii} + p_{jj}$; so, $p \in PMET_n$. (vii) is implied by (i), (ii). □

Any partial semimetric $p \in PMET_n$ induces the partial order on V_n by defining $i \preceq j$ if $p_{ii} = p_{ij}$. This *specialization order* is important in Computer Science applications, where the partial metrics act on certain posets called

Scott domains. In particular, $i_0 \in V_n$ is a *p-maximal* element in V_n if $p_{ii} = p_{ii_0}$ for all $i \neq i_0$. It is a *p-minimal* element in V_n if $p_{i_0i_0} = p_{ii_0}$ for all $i \neq i_0$. The *lifting* of $p \in PMET_n$ is the function $p^+ = ((p^+_{ij}))$, $i,j \in V_n$, with $p^+_{00} = 0$, $p^+_{0i} = p^+_{i0} = p^+_{ii}$ for $i \in V_n$ and $p^+_{ij} = p_{ij}$ for $i,j \in V_n$. Clearly, 0 is a p^+-maximal element in the specialization order, induced on $\{0\} \cup V_n = \{0, 1, \ldots, n\}$ by p^+, since $p^+_{ii} = p_{ii}$ as well as $p^+_{i0} = p_{ii}$ for all $i \in V_n$.

Theorem 2 *The following statements hold.*

(i) $sPMET_n) = \{p \in PMET_n : p^+ \in PMET_n\}$.

(ii) $sWMET_n = MET_{n+1} \simeq P(MET_{n+1}) = sPMET_n$.

(iii) The $0,1$*-valued elements of* $sPMET_n$ *are* $((0))$ *and* $2^n - 1$ *partial 2-cuts* $\gamma(S \neq \emptyset; \overline{S})$ *generating* $0,1\text{-}sPMET_n = CUT_{n+1}$,

$CUT_{n+1} \simeq P(CUT_{n+1}) = 0,1\text{-}sPMET_n$ *and* $Q(CUT_{n+1}) = OCUT_n$.

Proof. We should check for p^+ only inequalities (2), (3), (4) involving the new point 0. $2n+1$ of the required inequalities hold as equalities: $p^+_{00} = 0$ and $p^+_{0i} = p^+_{i0} = p^+_{ii} = p_{ii}$ for $i \in V_n$. All $Tr_{0j,i} = p_{ij} - p_{ij} \geq 0$ hold since (4) is satisfied. All $Tr_{ij,0} = p_{ii} + p_{jj} - p_{ij} \geq 0$ hold whenever p satisfies (5), i.e, $p \in PMET_n$. □

Given $p \in sPMET_n$, the semimetric $P^{-1}(p^+) \in MET_{n+2}$ is $P^{-1}(p) \in MET_{n+1}$ with the first point split in two coinciding points. The cone $sPMET_n$ is nothing but the linear image $P(sWPMET_n = MET_{n+1})$. So, for $n \geq 4$, $\text{Aut}(sWPMET_n) = \text{Sym}(n+1)$ on $\{0, 1, \ldots, n\}$ acting as $p' = P(\tau(P^{-1}(p)))$ on $sPMET_n$ for any $\tau \in \text{Sym}(n+1)$. If τ fixes 0, then $p' = \tau(p)$.

4 $0,1$-valued elements of $PMET_n$ and $dWMET_n$

For a partition $\mathcal{S} = \{S_1, \ldots, S_t\}$ of V_n and $A \subseteq \{1, \ldots, t\}$, let us denote $\hat{A} = \bigcup_{h \in A} S_h$ and $w(\hat{A}) = (w_i = 1_{i \in \hat{A}})$. So, weight is constant on each S_h.

For any $S \subset V_n$, denote by $J(S)$ the $0,1$-valued function with $J(S)_{ij} = 1$ exactly when $i,j \in S_0$. So, $J(V_n)$ and $J(\emptyset)$ are all-ones and all-zeros partial semimetrics, respectively.

For any $S_0 \subset V_n$ and partition $\mathcal{S} = \{S_1, \ldots, S_t\}$ of $\overline{S_0}$, denote $J(S_0) + \delta(S_0, S_1, \ldots, S_t)$ by $\gamma(S_0; S_1, \ldots, S_t)$ and call it a *partial multicut* or, specifically, a *partial t-cut*. Clearly, $\gamma(S_0; S_1, \ldots, S_t) \in PMET_n$ and it is $P(d; w)$, where $d = 2\delta(S_0, S_1, \ldots, S_t) - \delta(S_0) = \sum_{i=1} \delta(S_i)$ and $w = (w_i = 1_{i \in S_0})$.

Theorem 3 *The following statements hold.*

(i) The $0,1$*-valued elements of* $dWMET_n$ *are* $\sum_{t=1}^{n} 2^t S(n,t)$ $(\delta(\mathcal{S}); w(\hat{A}))$.

$R(0,1\text{-}dWMET_n)$ *consists of all such elements with* $(|A|, t-|A|) = (1,0)$, $(0,2)$ *or* $(1,1)$, *i.e.,* $(((0)); (1))$ *and 2-cuts* $(\delta(S); w)$ *with weight* $(0), w' = (1_{i \in S})$ *or* $w'' = (1_{i \notin S})$. *There are* $1 + 3(2^{n-1} - 1)$ *of them, in* $\lfloor \frac{3n}{2} \rfloor$ *orbits.*

(ii) The $0,1$*-valued elements of* $WQMET_n$ *are* $Q(2\delta(\mathcal{S}) - \delta(\hat{A}; w(A)) = \delta(\mathcal{S}) - \delta'(\hat{A})$.

$R(0,1\text{-}dWMET_n)$ *consists of all such elements with either* $|A| = t - |A| = 1$ *(o-2-cuts), or* $2 \leq |A|, t - |A| \leq n-2$.

(iii) The $0,1$*-valued elements of* $PMET_n$ *are the partial multicuts* $P((2\delta(\mathcal{S}) - \delta(\hat{A}); w(A)) = \delta(\mathcal{S}) + J(\hat{A})$ *with* $|A| \leq 1$.

There are $B(n+1) = \sum_{i=0}^{n} \binom{n}{i} B(i)$ *of them, in* $\sum_{i=0}^{n} Q(i)$ *orbits.*

$R(0,1\text{-}PMET_n)$ *consists of all such elements except* $B(n) - (2^{n-1} - 1)$, *in* $Q(n) - \lfloor \frac{n}{2} \rfloor$ *orbits, those (*partial t-cuts*) with* $|A| = 0, t \neq 2$.

Proof.

(i). $(((0)); (1))$ belongs to $R(0,1\text{-}dWMET_n)$ since its rank is $\binom{n}{2}$ plus $n-1$, the rank of the set of equalities $d_{ij} = w_i - w_j$. The same holds for $(\delta(S); w)$ with weight $(0), w' = (1_{i \in S})$ or $w'' = (1_{i \notin S})$, since their rank is $\binom{n}{2} - 1$ plus $k \geq 1$ equalities $w_i = 0$ plus, if $k < n$, $n-k$ equalities $d_{ij} = w_j - w_{i'} = 1$, where $w_{i'} = 0$ and $w_j = 1$. But the all-ones-weighted 2-cut δ is equal to $\frac{1}{2}((\delta; w') + (\delta; w'') + (((0))); (1))$. No other $(\delta(S_1, \dots, S_t); w)$ belongs to $R(0,1\text{-}dWMET_n)$ since t should be 2 (otherwise, the rank will be $< \binom{n}{2} - 1 + n$) and the weight should be constant on each S_h, $1 \leq h \leq t$.

(ii). The $0,1$-valued elements of $WMET_n$ should be $0,1$-weighted multicuts $\delta(\mathcal{S})$. Now, the inequality (4') $d_{ij} \geq w_i - w_j$, valid on $dWMET_n$, implies (i).

Let $q \in WQMET$ be $0,1$-valued. Without loss of generality, $\min_{i=1}^{n}(w_i) = w_1 = 0$. But $q_{1i} + w_1 = q_{i1} + w_i$ for any $i > 1$. So, $w_i = 1$ if and only if $q_{1i} \neq q_{i1}$. The quasi-semimetrics q, restricted on the sets $\{i : w_i = 0\}$ and $\{i : w_i = 1\}$, should be $0,1$-valued semimetrics, i.e., multicuts.

(iii) is proven in [DeDe10]. For example, there are $52 = 1 \times 1 + 4 \times 1 + 6 \times 2 + 4 \times 5 + 1 \times 15$ (1+1+2+3+5 orbits) $0,1$-valued elements of $PMET_4$. Among them, only $\delta(S_1, \dots, S_t)$ with $t = 1, 3, 4$, i.e., $((0))$, $\delta(\{1\}, \{2\}, \{3\}, \{4\})$ and 6 elements of the orbit with $t = 3$ are not representatives of extreme rays. □

5 Two generalized hypermetric cones

For a sequence $b = (b_1, \dots, b_n)$ of integers, where Σ_b denotes $\sum_{i=1}^n b_i$, and a symmetric $n \times n$ matrix $((a_{ij}))$, denote by $H_b(a)$ the sum $-\sum_{1\le i,j\le n} b_i b_j a_{ij}$ of the entries of the matrix $-b^T ab$.

The cone HYP_n of all *hypermetrics*, i.e., semimetrics $d \in MET_n$ with $H_b(d) \ge 0$, whenever $\Sigma_b = 1$, was introduced in [De60].

This cone is polyhedral [DGL93]; $HYP_n \subseteq MET_n$ with equality for $n \le 4$ and $CUT_n \subseteq HYP_n$ with equality for $n \le 6$. HYP_7 was described in [DeDu04].

The hypermetrics have deep connections with Geometry of Numbers and Analysis; see, for example, [DeTe87, DeGr93, DGL95] and Chapters 13-17, 28 in [DeLa97]. So, generalizations of HYP_n can put those connections in a more general setting.

For a weighted semimetric $(d; w) \in WMET_n$, we will use the notation:
$\mathrm{Hyp}_b(d; w) = \frac{1}{2}H_b(d) + (1 - \Sigma_b)\langle b, w\rangle \ge 0$ and
$\mathrm{Hyp}'_b(d; w) = \frac{1}{2}H_b(d) + (1 + \Sigma_b)\langle b, w\rangle \ge 0$.

Denote by $WHYP_n$ the cone of all *weighted hypermetrics*, i.e., $(d; w) \in WMET_n$ with $\mathrm{Hyp}_b(d; w) \ge 0$ and $\mathrm{Hyp}'_b(d; w) \ge 0$ for all b with $\Sigma_b = 1$ or 0. Denote by $PHYP_n$ the cone of all *partial hypermetrics*, i.e., $p \in wPMET_n$ with $\mathrm{Hyp}_b(P^{-1}(p)) \ge 0$ for all b with $\Sigma_b = 1$ or 0. For $p = P(d; w)$ and $(q, w) = Q(d; w)$, we have

$$\mathrm{Hyp}_b(d; w) = H_b(p) + \sum_{i=1}^n b_i p_{ii} = H_b(q) + (1 - \Sigma_b)\langle b, w\rangle.$$

$WHYP_n \subset dWMET_n$ and $PMET_n \supset PHYP_n$ hold since the needed inequalities $w_i \ge 0$, (4') (and (4)) are provided by permutations of $\mathrm{Hyp}'_{(1,0,\dots,0)}(d; w) \ge 0$ and $\mathrm{Hyp}_{(1,-1,0,\dots,0)}(d; w) \ge 0$.

Lemma 2 *Besides the cases* $PMET_3 = PHYP_3 = 0,1\text{-}PMET_3$ *and* $0,1\text{-}dWMET_n = WHYP_n$ *for* $n = 3, 4$, $0,1\text{-}dWMET_n \subset WHYP_n \subset dWMET_n \simeq PMET_n \supset PHYP_n \supset 0,1\text{-}PMET_n$ *holds.*

Proof.
Denoting $\langle b, (1_{i\in S_h})\rangle$ by r_h, we have $r_0 = \Sigma_b - \sum_{h=1}^t r_h$ and

$$H_b(\delta(S_0, S_1, \dots, S_t)) = \frac{1}{2}\sum_{h=0}^t H_b(\delta(S_h, \overline{S_h})) = \sum_{h=0}^t r_h(r_h - \Sigma_b).$$

Let $(d = 2\delta(S_0, S_1, \ldots, S_t) - \delta(S_0); w = (1_{i \in S_0}))$ be a generic $P^{-1}(p)$, where p is $0,1$-valued element of *PMET* belonging to its extreme ray. Then $\frac{1}{2}H_b(d) = \frac{1}{2}(2\sum_{h=0}^{t} r_h(r_h - \Sigma_b) - 2r_0(r_0 - \Sigma_b) = \sum_{h=1}^{t} r_h(r_h - \Sigma_b)$ implies

$\mathrm{Hyp}_b(d; w) = \sum_{h=1}^{t} r_h(r_h - 1) - \Sigma_b(\Sigma_b - 1) \geq 0$ for our $\Sigma = 0, 1$.

All $0,1$-valued elements $(d; w)$ of $dWMET_n$ belonging to its extreme rays are $(((0)); (1))$, $(\delta(S); (0))$, $(\delta(S,); w' = (1_{i \in S}))$ and $(\delta(S); w'' = (1_{i \notin S}))$. For them, $\mathrm{Hyp}'_b(d; w) = (\Sigma_b + 1)\Sigma_b$, $r_S(r_S - \Sigma)$, $r_S(r_S + 1)$ and $(\Sigma_b - r_S)(\Sigma_b - r_S + 1)$ hold, respectively, and so, $\mathrm{Hyp}'_b(d; w) \geq 0$ for our $\Sigma = 0, 1$.

The cases $n = 3, 4$ were checked by computation; see Lemma below. □

Lemma 3 *The following statements hold.*

(i) All facets of WHYP$_n$, $n \leq 4$, up to $\mathrm{Sym}(n)$ *and 0-extensions, are* Hyp_b *with* $b = (1, -1), (1, 1, -1), (1, 1, -1, -1)$ *and* Hyp'_b *with* $b = (1)$, $(1, 1, -1)$, $(1, 1, 1, -2)$, $(2, 1, -1, -1)$.

(ii) Besides $w_i \geq 0$, *among the facets of* $P^{-1}(PHYP_n)$, $n \leq 5$, *up to* $\mathrm{Sym}(n)$ *and 0-extensions, are:* Hyp_b *with* $b = (1, -1)$, $(1, 1, -1)$, $(1, 1, -1, -1)$, $(1, 1, 1, -1, -1)$, $(1, 1, 1, -1, -2)$, $(2, 1, 1, -1, -1)$.

Proof.

It was obtained by direct computation. The equality $WHYP_n = 0,1$-$WMET_n$ for $n = 3, 4$ holds, because only inequalities which are requested in $WHYP_n$ appeared among those of $0,1$-$WMET_n$.

The facets of $PHYP_4$ were deduced by computation using the tightness of the inclusions $0,1$-$PMET_4 \subset PHYP_4 \subset PMET_4$ (see Table 1): $0,1$-$PMET_4$ contained exactly one facet (orbit F_5) different from Hyp_b and $p_{ii} \geq 0$, and $PMET_4$ contained exactly two (orbits R_{10} and R_{11}) non-$0,1$-valued extreme ray representatives. The 6 rays from R_{10} are removed by 6 respective Hyp_b with $b = (1, 1, -1, -1)$, while the 12 rays from R_{11} are removed by 12 F_5. □

6 Oriented multicuts and quasi-semimetrics

For an ordered partition $(S_1, \ldots, S_t)$ of V_n into non-empty subsets, the *oriented multicut* (or *o-multicut, o-t-cut*) $\delta'(S_1, \ldots, S_t)$ on V_n is defined by:

$$\delta'_{ij}(S_1, \ldots, S_t) = \begin{cases} 1, & \text{if } i \in S_h, j \in S_m, m > h, \\ 0, & \text{otherwise.} \end{cases}$$

R_i	Representative p	11	21	22	31	32	33	41	42	43	44	Inc.	Adj.	$\lvert O_i \rvert$
R_1	$\gamma(\{1,2,3,4\};)$	1	1	1	1	1	1	1	1	1	1	24	20	1
R_2	$\gamma(\{2\};\overline{\{2\}})$	0	1	1	0	1	0	0	1	0	0	21	38	4
R_3	$\gamma(\overline{\{3\}};\{3\})$	1	1	1	1	1	0	1	1	1	1	19	17	4
R_4	$\gamma(\emptyset;\overline{\{3\}},\{3\})$	0	0	0	1	1	0	0	0	1	0	19	32	4
R_5	$\gamma(\{1,2\};\overline{\{1,2\}})$	1	1	1	1	1	0	1	1	0	0	18	31	6
R_6	$\gamma(\emptyset;\{1,2\},\overline{\{1,2\}})$	0	0	0	1	1	0	1	1	0	0	16	32	3
R_7	$\gamma(\{1,4\};\{2\},\{3\})$	1	1	0	1	1	0	1	1	1	1	14	14	6
R_8	$\gamma(\{1\};\{2\},\{3,4\})$	1	1	0	1	1	0	1	1	0	0	14	20	12
R_9	$\gamma(\{4\};\{1\},\{2\},\{3\})$	0	1	0	1	1	0	1	1	1	1	9	9	4
R_{10}		1	1	0	1	1	0	2	1	1	1	10	18	6
R_{11}		0	2	0	1	1	0	2	2	3	2	9	9	12
F_1	$L_{11}: p_{11} \geq 0$	1	0	0	0	0	0	0	0	0	0	29	36	4
F_2	$\text{Hyp}_{(-1,1,1,0)} \geq 0$	-1	1	0	1	-1	0	0	0	0	0	26	24	12
F_3	$M_{12} = \text{Hyp}_{(-1,1,0,0)} \geq 0$	-1	1	0	0	0	0	0	0	0	0	23	23	12
F_4	$\text{Hyp}_{(1,1,-1,-1)} \geq 0$	0	-1	0	1	1	-1	1	1	-1	-1	16	12	6
F_5	$H_{(2,1,-1,-1)} - 2p_{11} \geq 0$	-2	-2	0	2	1	0	2	1	-1	0	9	9	12

Table 1: The orbits of extreme rays in $PMET_4$ and facets in $0,1$-$PMET_4$

The o-2-multicuts $\delta'(S,\overline{S})$ are called *o-cuts* and denoted by $\delta'(S)$. Clearly,

$$\delta(S_1,\dots,S_t) = \sum_{i=1}^{t} \delta'(S_i) = \sum_{i=1}^{t} \delta'(\overline{S_i}) = \frac{1}{2}\sum_{i=1}^{t} \delta(S_i).$$

Denote by $OCUT_n$ and $OMCUT_n$ the cones generated by $2^n - 2$ non-zero o-cuts and $Bo(n) - 1$ non-zero o-multicuts, respectively. (Here, $Bo(n)$ are the *ordered Bell numbers* given by the sequence *A000670* in [Sl10].) So, $CUT_n = \{q + q^T : q \in OMCUT_n\}$. In general, $Z_2 \times \text{Sym}(n)$ is a symmetry group of $QMET_n$, $OMCUT_n$, $WQMET_n$, $OCUT_n$; Dutour, 2002, proved that it is the full group of those cones. The cones $QMET_n$ and $OMCUT_n$ were studied in [DDD03]. Clearly, $\delta'_{ij}(S_1,\dots,S_t) \in WQMET_n$ if and only if $t = 2$ and then $w = (1_{i \notin S_1})$. So, $OCUT_n = OMCUT_n \cap WQMET_n$.

Theorem 4 *The following statement holds.*
$OCUT_n = Q(CUT_{n+1} = 0,1\text{-}sWMET_n) = Q(0,1\text{-}dWMET_n) = 0,1\text{-}Q(dWMET_n)$.

Proof. Given a representative $(d;w) = (\delta(S);w')$, $(\delta(S);w'')$, $(\delta(S);(0))$, $(\delta(\emptyset);(1))$ of an extreme ray of $0,1\text{-}dWMET_n$, we have $Q(q;w) = (\delta'(S),w')$, $(\delta'(\overline{S}),w'')$, $(\delta(S),(0))$, $(\delta(\emptyset),(1))$, respectively. But $\delta(S) = \delta'(S) + \delta'(\overline{S})$ and $(((0)),t(1))$ are not extreme rays. $\square$

The above equality $OCUT_n = Q(0,1\text{-}sWMET_n)$ means that $q \in OCUT_n$ are $Q(d;w)$, where (d,w) is a semimetric $d' \in CUT_{n+1}$ on $V_n \cup \{0\}$. So,

$q_{ij} = \frac{1}{2}(d'_{ij} - d'_{0i} + d'_{0j})$. But CUT_n is the set of ℓ_1-*semimetrics* on V_n, see [DeLa97]. So, $q \in OCUT_n$ can be seen as ℓ_1-*quasi-semimetrics*; it was realized in [DDD03, CMM06]. In fact, $OCUT_n$ is the set of quasi-semimetrics q on V_n, for which there is some $x_1, \ldots, x_m \in \mathbb{R}^m$ with all $q_{ij} = ||x_i - x_j||_{1.or}$, where the *oriented* ℓ_1-*norm* is defined as $||x-y||_{1.or} = \sum_{k=1}^{m} \max(x_k - y_k, 0)$; the proof is the same as in Proposition 4.2.2 of [DeLa97].

Let C be any cone closed under *reversal*, i.e., $q \in C$ implies $q^T \in C$. If the linear inequality $\sum_{1 \le i,j \le n} f_{ij} q_{ij} = \langle F, q \rangle \ge 0$ is valid on C, then F also defines a face of $\{q+q^T : q \in C\}$. Given a valid inequality $G : \sum_{1 \le i < j \le n} g_{ij} d_{ij}$ of $\{q + q^T : q \in C\}$ and an *oriented* K_n (i.e., exactly one arc connects any i and j) O, let $G^O = ((g^O_{ij}))$ where $g^O_{ij} = g_{ij}$ if the arc (ij) belongs to O and $= 0$, otherwise. Call G^O *standard* if there exists $\tau \in \mathrm{Sym}(n)$ with $(ij) \in O$ if and only if $\tau(i) < \tau(j)$, and *reversal-stable* (*rs* for short) if $\langle G^O, q \rangle = \langle G^O, q^T \rangle$. In general, G^O is not valid on C and does not preserve the rank of G.

For example, the standard $Tr_{12,3} : q_{13} + q_{23} - q_{12} \ge 0$ is not valid on $OCUT_n$, and the standard $L_{ij} : q_{ij} \ge 0$ defines a facet in $OCUT_n$, while $G : d_{ij} \ge 0$ only defines a face in MET_n. If $F = G^O$ is rs, then $\langle F, q \rangle = \frac{1}{2}\langle G, q + q^T \rangle$, i.e., F is valid on C if G is valid on $\{q + q^T : q \in C\}$.

Let E be an equality that holds on C, i.e., $\sum_{1 \le i,j \le n} e_{ij} q_{ij} = \langle E, q \rangle = 0$ holds for any $q \in C$. If the dimension of the subspace $\mathcal{E}$, spanned by all such equalities, is greater than zero, and $F \ge 0$ is a facet-defining inequality, then for any $E \in \mathcal{E}$, $F + E \ge 0$ defines the same facet. We call a facet standard or rs if one of its defining inequalities is standard or rs. Of all the defining inequalities we can choose one of them (up to a positive factor) to be the *canonical representative* – let it be such a $G = F + E$, $E \in \mathcal{E}$ that $\langle G, E \rangle = 0$ holds for all $E \in \mathcal{E}$, i.e., G is orthogonal to $\mathcal{E}$.

Lemma 4 *Let $C \subseteq WQMET_n$ be a cone of the same dimension as $WQMET_n$ and $\mathcal{E}$ the subspace of its equalities. Then, the following statements hold.*

(i) $\mathcal{E}$ is spanned by the equalities $q_{ij} + q_{jk} + q_{ki} = q_{ji} + q_{kj} + q_{ik}$ for $i, j, k \in V_n$ and its dimension is $\binom{n-1}{2}$.

(ii) For each $E \in \mathcal{E}$, $E^T = -E$ holds and E is rs.

(iii) If a facet of C is rs, then all of its defining inequalities are rs.

Proof.

(i). The equalities $E_{ijk} = q_{ij} + q_{jk} + q_{ki} - q_{ji} - q_{kj} - q_{ik} = 0$ for $i, j, k \in V_n$ follow directly from the weightability condition $q_{ij} + w_i = q_{ji} + w_j$. Since $E_{ijk} = E_{jki} = -E_{kji}$ and $E_{jk\ell} = E_{ijk} - E_{ij\ell} + E_{ik\ell}$ hold, we can choose a

basis of $\mathcal{E}$ such that the indices (ijk) of the basis elements E_{ijk} are ordered triples that all contain a fixed element of V_n (say, n). There are $\binom{n-1}{2}$ such triples, and since all such E_{ijk} are linearly independent, the subspace $\mathcal{E}$ has dimension $\binom{n-1}{2}$.

(ii). As $E_{ijk}^T = E_{kji} = -E_{ijk}$ for all $i, j, k \in V_n$, $E^T = -E$ holds for all $E \in \mathcal{E}$. Then $E^T \in \mathcal{E}$, and as $0 = \langle E, q\rangle = \langle E^T, q\rangle = \langle E, q^T\rangle$, E is also rs.

(iii). If F is a defining inequality of a facet and is rs, then for each $E \in \mathcal{E}$ and $q \in C$, $\langle (F+E)^T, q\rangle = \langle F^T + E^T, q\rangle = \langle F + E, q\rangle$, so $F+E$, and by extension any defining inequality, is also rs. □

The facets $OTr_{ij,k}$ and L_{ij} (only 1st is rs) of $WQMET_n$ are standard and of the form Hyp_b where b is a permutation of $(1,1,-1,0,\ldots,0)$ or $(1,-1,0,\ldots,0)$. $OCUT_4$ has one more orbit: six standard, non-rs facets of the form Hyp_b where b is a permutation of $(1,1,-1,-1)$, or $q_{13}+q_{14}+q_{23}+q_{24}-(q_{12}+q_{34}) \geq 0$.

$OCUT_5$ has, up to $\text{Sym}(n)$, 3 *new* (i.e., in addition to 0-extensions of the facets of $OCUT_4$) such orbits: one standard rs $(1,1,1,-1,-1)$ and two non-standard, non-rs orbits. $OCUT_6$ has, among its 56 new orbits, two non-standard rs orbits for $b = (2,1,1,-1,-1,-1)$ and $(1,1,1,1,-1,-2)$.

The adjacencies of cuts in CUT_n are defined only by the facets $Tr_{ij,k}$, and adjacencies of those facets are defined only by cuts. It gives at once $\binom{n}{2}-1$ linearly independent facets $OTr_{ij,k}$ containing any given pair $(\delta'(S_1), \delta'(S_2))$, using that $OTr_{ij,k}$ are rs facets. So, only n more facets are needed to get the adjacencies of o-cuts. It is a way to prove Conjecture 1 (i) below.

Call a *tournament* (K_n with unique arc between any i, j) *admissible* if its arcs can be partitioned into arc-disjoint directed cycles. It does not exists for even n, because then the number of arcs involving each vertex is odd, while each cycle provides 0 or 2 such arcs. But for odd n, there are at least $2^{\frac{n-3}{2}}$ admissible tournaments: take the decomposition of K_n into $\frac{n-1}{2}$ disjoint Hamiltonian cycles and, fixing the order on one them, all possible orders on remaining cycles. For odd n, denote by Oc the *canonic admissible tournament* consiting of all $(i, i+k)$ with $1 \leq i \leq n-1, 1 \leq k \leq \lceil\frac{n}{2}\rceil + 1 - i$ and $(i+k, i)$ with $1 \leq i \leq \lfloor\frac{n}{2}\rfloor, \lceil\frac{n}{2}\rceil \leq k \leq n-i$, i.e., $0 = C_{1,2,3,4,5,6,7,\ldots} + C_{1,3,5,7,\ldots} + C_{1,4,7,\ldots} + \ldots$. The *Kelly conjecture* state that the arcs of every *regular* (i.e., the vertices have the same outdegree) tournament can be partitioned into arc-disjoint directed Hamiltonian cycles.

0-extensions of $q_{ij} \geq 0$ and $q_{13}+q_{14}+q_{23}+q_{24}-(q_{12}+q_{34}) \geq 0$ can be seen, as the first instances (for $b = (1,-1,0,\ldots,0)$, $(1,1,-1,-1,0,\ldots,0)$) of the *oriented negative type inequality* $\mathrm{ONeg}_{b,O}(q) = -\sum_{1\le i<j\le n} b_i b_j q_{a(ij)} \ge 0$, where for a given $b = (b_1, \dots, b_n) \in \mathbb{Z}^n$, $\Sigma_b = 0$, and the arcs $a(ij)$ on the edges (ij) by some rule defined by a given tournament O.

Denote by $OWHYP_n$ the cone consisting of all $q \in WQMET_n$, satisfying the two above orbits and all *oriented hypermetric inequalities*

$$\mathrm{OHyp}_{b,O}(q) = -\sum_{1\le i<j\le n} b_i b_j q_{a(ij)} \ge 0,$$

where $b = (b_1, \dots, b_n) \in \mathbb{Z}^n$, $\Sigma_b = 1$, O is an admissible tournament, and the arc $a(ij)$ on the edge (ij) is the same as in O if $b_i b_j \ge 0$, or the opposite one otherwise. So, $OWHYP_n = OCUT_n$ for $n = 3, 4$.

Theorem 5 *$OCUT_n \subset OWHYP_n \subset WQMET_n$ holds for $n \ge 5$.*

Proof.
Without loss of generality, let $b_i = 1$ for $1 \le i \le \lfloor \frac{n}{2} \rfloor$ and $b_i = -1$, otherwise. The general case means only that we have sets of $|b_i|$ coinciding points. $\mathrm{OHyp}_{b,O}$ is rs, because Lemma 4 in [DeDe10] implies that any inequality on a $q \in WQMET_n$ is preserved by the reversal of q. So, $\mathrm{OHyp}_{b,O}(q) = \frac{1}{2}\mathrm{Hyp}_b(q + q^T)$. On an o-cut $\delta'(S)$ it gives, putting $r = \langle b, (1_{i\in S})\rangle$,

$$\frac{1}{2}\mathrm{Hyp}_b(\delta'(S) + \delta'(\overline{S})) = \mathrm{Hyp}_b(\delta(S)) = r(r - \Sigma_b) \ge 0.$$

$OWHYP_5$ has, besides o-cuts, 40 extreme rays in two orbits: F_{ab}, F'_{ab}, having 2 on the position (ab), 1 on ba, 0 on three other (ka) for $k \ne b$ in F_{ab}, or on three other (bk) for $k \ne a$ in F'_{ab}, and ones on other non-diagonal positions. Also, $D(\mathrm{Sk}(OWHYP_5)) = D(\mathrm{Ri}(OWHYP_5)) = 2$.

The cone $QHYP_n = \{q \in QMET_n : ((q_{ij} + q_{ji})) \in HYP_n\}$ was considered in [DDD03]. Clearly, it is polyhedral and coincides with $QMET_n$ for $n = 3, 4$; $QHYP_5$ has 90 facets ($20 + 60$ from $QMET_5$ and those with $b = (1, 1, 1, -1, -1)$) and 78810 extreme rays; $D(\mathrm{Ri}(QHYP_5)) = 2$.

Besides $OCUT_3 = 0,1\text{-}WQMET_3 = WQMET_3$ and $0,1\text{-}WQMET_4 = WQMET_4$, $OCUT_n \subset 0,1\text{-}WQMET_n \subset WQMET_n$ holds. We conjecture $\mathrm{Sk}(OCUT_n) \subset \mathrm{Sk}(0,1\text{-}WQMET_n) \subset \mathrm{Sk}(WQMET_n)$ and $\mathrm{Ri}(0,1\text{-}WQMET_n) \supset \mathrm{Ri}(WQMET_n) \supset \mathrm{Ri}(MET_n)$. $0,1\text{-}WQMET_5$ has $OTr_{ij,k}$, L_{ij} and 3 other, all standard, orbits. Those facets give, for permutations of $b = (1, -1, 1, -1, 1)$, the non-negativity of $-\sum_{1\le i<j\le 5} b_i b_j q_{ij}$ plus q_{24}, q_{23} or $q_{12} + q_{45}$.

The cone $\{q+q^T : q \in 0,1\text{-}WQMET_n\}$ coincides with MET_n for $n \le 5$, but for $n=6$ it has 7 orbits of extreme rays (all those of MET_6 except the one, good representatives of which are not $0,1,2$-valued as required); its skeleton, excluding another orbit of 90 rays, is an induced subgraph of Sk(MET_6). It has 3 orbits of facets including $Tr_{ij,k}$ (forming Ri(MET_6) in its ridge graph) and the orbit of $\sum_{(ij)\in C_{123456}} d_{ij}+d_{14}+d_{35}-d_{13}-d_{46}-2d_{25} \ge 0$.

If $q \in QMET_n$ is $0,1$-valued with $S=\{i : q_{i1}=1\}$, $S'=\{i : q_{1i}=1\}$, then $q_{ij}=0$ for $i,j \in \overline{S}\cap\overline{S'}$ (since $q_{i1}+q_{1j} \ge q_{ij}$) and $q_{ij}=q_{ji}=1$ for $i \in S, j \in \overline{S'}$ (since $q_{ij}+q_{j1} \ge q_{i1}$); so, $|\overline{S}\cap\overline{S'}|(|\overline{S}\cap\overline{S'}|-1)+|S|(|\overline{S}|-1)+|S'|(|\overline{S'}|-1)-|S\cap\overline{S'}||\overline{S}\cap S'|$ elements q_{ij} with $2 \le i \ne j \le n$ are defined.

7 The cases of $3,4,5,6$ points

In Table 2 we summarize the most important numeric information on cones under consideration for $n \le 6$. The column 2 indicates the dimension of the cone, the columns 3 and 4 give the number of extreme rays and facets, respectively; in parentheses are given the numbers of their orbits. The columns 5 and 6 give the diameters of the skeleton and the ridge graph. The expanded version of the data can be found on the third author's homepage [Vi10].

In the simplest case $n=3$ the numbers of extreme rays and facets are:

$0,1\text{-}WMET_3 = WHYP_3 = WMET_3 \simeq wPMET_3$: $(6,6$, simplicial) and $0,1\text{-}wPMET_3$: $(3,3$, simplicial);

$0,1\text{-}sWMET_3 = sWMET_3 = CUT_4 = HYP_4 = MET_4 \simeq 0,1\text{-}sPMET_3 = sPMET_3$: $(7,12)$;

$0,1\text{-}PMET_3 = PHYP_3 = PMET_3$ $(13,12)$ and $0,1\text{-}dWMET_3$: $(10,15)$;

$0,1\text{-}QMET_3 = QHYP_3 = QMET_3$: $(12,12$, simplicial) and $OCUT_3 = 0,1\text{-}WQMET_3 = WQMET_3$: $(6,9)$.

$R(dWMET_3) \setminus R(0,1\text{-}dWMET_3)$ and $F(0,1\text{-}dWMET_3) \setminus F(dWMET_3)$ consist of 3 simplicial elements forming $\overline{K_3}$ in the graph. But only Ri$(0,1\text{-}dWMET_3)$ is an induced subgraph of Ri$(dWMET_3)$.

Recall that $2^{n-1}-1$ is the Stirling number $S(n,2)$, Sk$(CUT_n)=K_{S(n,2)}$, and [DeDe94] Ri(MET_n), $n \ge 4$, has diameter 2 with $Tr_{ij,k} \not\sim Tr_{i'j',k'}$ whenever they are *conflicting*, i.e., have values of different sign on a position (p,q), $p,q \in \{i,j,k\}\cap\{i',j',k'\}$. Clearly, $|\{i,j,k\}\cap\{i',j',k'\}|$ should be 3 or 2, and $Tr_{ij,k}$ conflicts with 2 and $4(n-3)$ $Tr_{i'j',k'}$'s, respectively. The proofs of the conjectures below should be tedious but easy.

cone	dim.	Nr. ext. rays (orbits)	Nr. facets (orbits)	diam.	diam. dual
$wPMET_3$	6	6 (2)	6 (2)	1	1
$wPMET_4$	10	11 (3)	16 (2)	1	2
$wPMET_5$	15	30 (4)	35 (2)	2	2
$wPMET_6$	21	302 (8)	66 (2)	2	2
$sPMET_3 = 0,1\text{-}sPMET_3$	6	7 (2)	12 (1)	1	2
$sPMET_4$	10	25 (3)	30 (1)	2	2
$sPMET_5$	15	296 (7)	60 (1)	2	2
$sPMET_6$	21	55226 (46)	105 (1)	3	2
$0,1\text{-}sPMET_4$	10	15 (2)	40 (2)	1	2
$0,1\text{-}sPMET_5$	15	31 (3)	210 (4)	1	3
$0,1\text{-}sPMET_6$	21	63 (3)	38780 (36)	1	3
$PMET_3 = 0,1\text{-}PMET_3$	6	13 (5)	12 (3)	3	2
$PMET_4$	10	62 (11)	28 (3)	3	2
$PMET_5$	15	1696 (44)	55 (3)	3	2
$PMET_6$	21	337092 (734)	96 (3)	3	2
$PHYP_4$	10	56 (10)	34 (4)	3	2
$0,1\text{-}PMET_4$	10	44 (9)	46 (5)	3	2
$0,1\text{-}PMET_5$	15	166 (14)	585 (15)	3	3
$0,1\text{-}PMET_6$	21	705 (23)		3	
$0,1\text{-}dWMET_3$	6	10 (4)	15 (4)	2	2
$0,1\text{-}dWMET_4$	10	22 (6)	62 (7)	2	3
$0,1\text{-}dWMET_5$	15	46 (7)	1165 (27)	2	3
$0,1\text{-}dWMET_6$	21	94 (9)	369401 (806)	2	
$WQMET_3 = OCUT_3$	5	6 (2)	9 (2)	1	2
$WQMET_4 = 0,1\text{-}WQMET_4$	9	20 (4)	24 (2)	2	2
$WQMET_5$	14	190 (11)	50 (2)	2	2
$WQMET_6$	20	18502 (77)	90 (2)		2
$0,1\text{-}WQMET_5$	14	110 (8)	250 (5)	2	2
$0,1\text{-}WQMET_6$	20	802 (17)			
$\{q+q^T : q \in 0,1\text{-}WQMET_6\}$	15	206 (7)	510 (3)	2	3
$OWHYP_5$	14	70 (6)	90 (4)	2	2
$OCUT_4$	9	14 (3)	30 (3)	1	2
$OCUT_5$	14	30 (4)	130 (6)	1	3
$OCUT_6$	20	62 (5)	16460 (62)	1	

Table 2: Main parameters of cones with $n \le 6$

Conjecture 1 *(i)* $\text{Sk}(OCUT_n) = K_{2S(n,2)}$ *and belongs to* $\text{Sk}(WQMET_n)$.
(ii) $\overline{\text{Sk}(0,1\text{-}dWMET_n)} = K_{1,S(n,2)} + S(n,2)K_2$;
$\text{Sk}(0,1\text{-}dWMET_n)$ *has diameter* 2, *all non-adjacencies are of the form:* $(((0));(1)) \not\sim (\delta'(S);(0))$ *and* $(\delta'(S);w') \not\sim (\delta'(S);w')$.

Conjecture 2 *(i)* $\text{Ri}(PMET_n)$ *has diameter* 2, *all non-adjacencies are:* $L_{ii} \not\sim M_{ik}$; $M_{ij} \not\sim M_{ji}, M_{ki}, M_{jk}, Tr_{ij,k}$; $Tr_{ij,k} \not\sim Tr_{i'j',k'}$ *if they conflict.*
(ii) $\text{Ri}(WQMET_n)$ *has diameter* 2; *it is* $\text{Ri}(PMET_n)$ *without vertices* L_{ii}.

R_i	Representative	11	21	22	31	32	33	Inc.	Adj.	$\|R_i\|$
R_1 ▲	$\gamma(\{1,2,3\};)$	1	1	1	1	1	1	9	6	1
R_2 ○	$\gamma(\{1\};\{2,3\})$	1	1	0	1	0	0	8	9	3
R_3 ●	$\gamma(\{2,3\};\{1\})$	0	1	1	1	1	1	7	6	3
R_4 □	$\gamma(\emptyset;\{1\},\{2,3\})$	0	1	0	1	0	0	7	8	3
R_5 ■	$\gamma(\{3\};\{1\},\{2\})$	0	1	0	1	1	1	5	5	3
F_1 ○	$L_{11}: p_{11} \geq 0$	1	0	0	0	0	0	8	9	3
F_2 ▲	$Tr_{12,3}: p_{13} + p_{23} - p_{12} - p_{33} \geq 0$	0	-1	0	1	1	-1	8	7	3
F_3 ●	$M_{12}: p_{12} - p_{11} \geq 0$	-1	1	0	0	0	0	7	6	6

Table 3: The orbits of extreme rays and facets in $PMET_3 = 0,1\text{-}PMET_3$

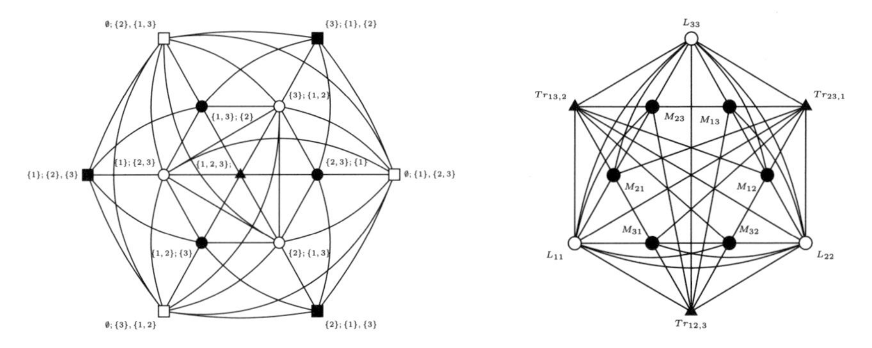

Figure 1: The skeleton and ridge graph of $PMET_3 = 0,1\text{-}PMET_3$

R_i	Representative	1	21	2	31	32	3	Inc.	Adj.	$\|R_i\|$
R_1 ▲	$(\delta(\emptyset);(1))$	1	0	1	0	0	1	9	6	1
R_2 ○	$(\delta(\{1\};w'')$	0	1	1	1	0	1	9	8	3
R_3 ●	$(\delta(\{1\};w')$	1	1	0	1	0	0	9	8	3
R_4 □	$(\delta(\{1\};(0))$	0	1	0	1	0	0	9	8	3
F_1 ○	$L_1: w_1 \geq 0$	1	0	0	0	0	0	6	9	3
F_2 ▲	$Tr_{12,3}: d_{13}+d_{23}-d_{12} \geq 0$	0	-1	0	1	1	0	7	8	3
F_3 ●	$M'_{12}: d_{12}+(w_2-w_1) \geq 0$	-1	1	1	0	0	0	6	6	6
F_4 △	$Tr'_{12,3}: (d_{13}+d_{23}-d_{12})+2(w_1+w_2-w_3) \geq 0$	2	-1	2	1	1	-2	5	5	3

Table 4: The orbits of extreme rays and facets in $0,1\text{-}dWMET_3$

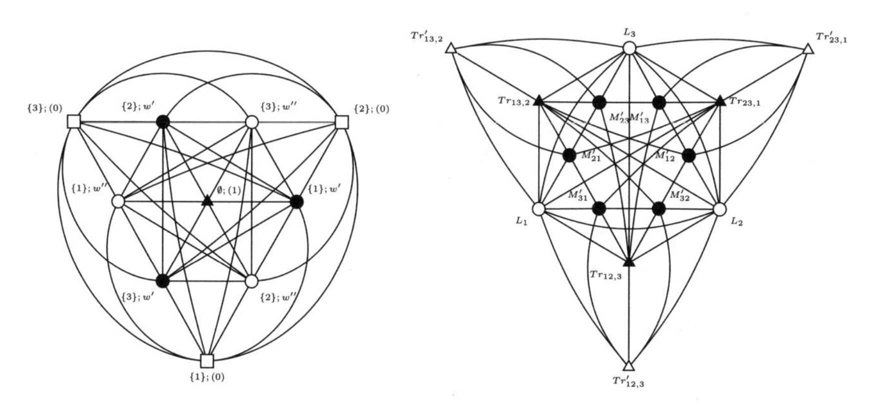

Figure 2: The skeleton and ridge graph of $0,1\text{-}dWMET_3$

References

[CMM06] M. Charikar, K. Makarychev and Y. Makarychev, *Directed Metrics and Directed Graph Partitioning Problems*, Proc. of 17th ACM-SIAM Symposium on Discrete Algorithms (2006) 51–60.

[DeDe94] A. Deza and M. Deza, *The ridge graph of the metric polytope and some relatives*, in T. Bisztriczky, P. McMullen, R. Schneider and A. Ivic Weiss eds. Polytopes: Abstract, Convex and Computational (1994) 359–372.

[DDF96] A. Deza, M. Deza and K. Fukuda, *On Skeletons, Diameters and Volumes of Metric Polyhedra*, in Combinatorics and Computer Science, Lecture Notes in Computer Science **1120**, Springer (1996) 112–127.

[De60] M. Tylkin (=M. Deza), *Hamming geometry of unitary cubes*, Doklady Akademii Nauk SSSR **134-5** (1960) 1037–1040. (English translation in *Cybernetics and Control Theory* **134-5** (1961) 940–943.

[DeDe10] M. Deza and E. Deza, *Cones of Partial Metrics*, Contributions in Discrete Mathematics, 2010.

[DDD03] M. Deza, M. Dutour and E. Deza, *Small cones of oriented semi-metrics*, American Journal of Mathematics and Management Science **22-3,4** (2003) 199–225.

[DeDu04] M. Deza and M. Dutour, *The hypermetric cone on seven vertices*, Experimental Mathematics **12** (2004) 433–440.

[DGL93] M. Deza, V. P. Grishukhin and M. Laurent, *The hypermetric cone is polyhedral*, Combinatorica 13 (1993) 397–411.

[DeGr93] M. Deza and V. P. Grishukhin, *Hypermetric graphs*, The Quarterly Journal of Mathematics Oxford, 2 (1993) 399–433.

[DGL95] M. Deza, V. P. Grishukhin and M. Laurent, *Hypermetrics in geometry of numbers*. In W. Cook, L. Lovász and P. Seymour, editors, *Combinatorial Optimization*, DIMACS Series in Discrete Mathematics and Theoretical Computer Science **20** AMS (1995) 1–109.

[DeLa97] M. Deza and M. Laurent, *Geometry of cuts and metrics*, Springer-Verlag, Berlin, 1997.

[DeTe87] M. Deza and P. Terwilliger, *The classification of finite connected hypermetric spaces*, Graphs and Combinatorics **3** (1987) 293–298.

[Du08] M. Dutour Sikirić, *Cut and Metric Cones*,
`http://www.liga.ens.fr/~dutour/Metric/CUT_MET/index.html`.

[Du10] M. Dutour Sikirić, *Polyhedral*,
`http://www.liga.ens.fr/~dutour/polyhedral`.

[Fu95] K. Fukuda, *The cdd program*,
`http://www.ifor.math.ethz.ch/~fukuda/cdd_home/cdd.html`.

[Gr92] V. P. Grishukhin, *Computing extreme rays of the metric cone for seven points*, European Journal of Combinatorics **13** (1992) 153–165.

[He99] R. Heckmann, *Approximation of Metric Spaces by Partial Metric Spaces*, Applied Categorical Structures **7** (1999) 7–83.

[Hi01] P. Hitzler, *Generalized Metrics and Topology in Logic Programming Semantics*, PhD Thesis, Dept. Mathematics, National University of Ireland, University College Cork, 2001.

[Ma92] S. G. Matthews, *Partial metric topology*, Research Report 212, Dept. of Computer Science, University of Warwick, 1992.

[Ma08] S. G. Matthews, A collection of resources on partial metric spaces, available at `http://partialmetric.org`, 2008.

[Pl75] J. Plesník, *Critical graphs of given diameter*, Acta Math. Univ. Comenian **30** (1975) 71–93.

[Sc86] A. Schrijver, *Theory of Linear and Integer Programming*, Wiley, 1986.

[Se97] A. K. Seda, *Quasi-metrics and the semantic of logic programs*, Fundamenta Informaticae **29** (1997) 97–117.

[Sl10] N. Sloane, *The On-Line Encyclopedia of Integer Sequences*, published electronically at `http://oeis.org`, 2010.

[Vi10] J. Vidali, *Cones of Weighted and Partial Metrics*,
`http://lkrv.fri.uni-lj.si/~janos/cones/`.

Menger Algebras of n-Place Functions

WIESŁAW A. DUDEK AND VALENTIN S. TROKHIMENKO

ABSTRACT. It is a survey of the main results on abstract characterizations of algebras of n-place functions obtained in the last 40 years. A special attention is paid to those algebras of n-place functions which are strongly connected with groups and semigroups, and to algebras of functions closed with respect natural relations defined on their domains.

Dedicated to Professor K.P. Shum's 70th birthday

1. Introduction

Every group is known to be isomorphic to some group of set substitutions, and every semigroup is isomorphic to some semigroup of set transformations. It accounts for the fact that the group theory (consequently, the semigroup theory) can be considered as an abstract study about the groups of substitutions (consequently, the semigroups substitutions). Such approach to these theories is explained, first of all, by their applications in geometry, functional analysis, quantum mechanics, etc. Although the group theory and the semigroup theory deal in particular only with the functions of one argument, the wider, but not less important class of functions remains without their attention – the functions of many arguments (in other words – the class of multiplace functions). Multiplace functions are known to have various applications not only in mathematics itself (for example, in mathematical analysis), but are also widely used in the theory of many-valued logics, cybernetics and general systems theory. Various natural operations are considered on the sets of multiplace functions. The main operation is the superposition, i.e., the operation which as a result of substitution of some functions into other instead of their arguments gives a new function. Although the algebraic theory of superpositions of multiplace functions has not developed during a long period of time, K. Menger paid attention to abnormality of this state in the middle of 40's in the previous century. He stated, in particular, that the superposition of n-place functions, where n is a fixed natural number, has the property resembling the associativity, which he called *superassociativity* [34, 35, 36]. As it has been found later, this property appeared to be fundamental in this sense that every set with superassociative $(n+1)$-ary operation can be represented as a set of n-place functions with the operation of superposition. This fact was first proved by R. M. Dicker [3] in 1963, and the particular case of the given theorem was received by H. Whitlock [58] in 1964.

2010 Mathematics Subject Classification. 20N15

Key words and phrases. Menger algebra, algebra of multiplace functions, representation, group-like Menger algebra, diagonal semigroup

The theory of algebras of multiplace functions, which now are called *Menger algebras* (if the number of variables of functions is fixed) or *Menger systems* (if the number of variables is arbitrary), has been studied by (in alphabetic order) M. I. Burtman [1, 2], W. A. Dudek [6] – [10], F. A. Gadzhiev [16, 17, 18], L. M. Gluskin [19], Ja. Henno [21] – [25], F. A. Ismailov [27, 28], H. Länger [30, 31, 32], F. Kh. Muradov [37, 38], B. M. Schein [42, 43], V. S. Trokhimenko (Trohimenko) [47] – [57] and many others.

The first survey on algebras of multiplace functions was prepared by B. M. Schein and V. S. Trokhimenko [43] in 1979, the first monograph (in Russian) by W. A. Dudek and V. S. Trokhimenko [8]. Extended English version of this monograph was edited in 2010 [10]. This survey is a continuation of the previous survey [43] prepared in 1979 by B. M. Schein and V. S. Trokhimenko.

2. Basic definitions and notations

An *n-ary relation* (or an *n-relation*) between elements of the sets $A_1, A_2, \ldots, A_n$ is a subset ρ of the Cartesian product $A_1 \times A_2 \times \ldots \times A_n$. If $A_1 = A_2 = \cdots = A_n$, then the n-relation ρ is called *homogeneous*. Later on we shall deal with $(n+1)$-ary relations, i.e., the relations of the form $\rho \subset A_1 \times A_2 \times \ldots \times A_n \times B$. For convenience we shall consider such relations as binary relations of the form $\rho \subset (A_1 \times A_2 \times \ldots \times A_n) \times B$ or $\rho \subset \Big(\mathop{\times}\limits_{i=1}^{n} A_i\Big) \times B$. In the case of a homogeneous $(n+1)$-ary relation we shall write $\rho \subset A^n \times A$ or $\rho \subset A^{n+1}$.

Let $\rho \subset \Big(\mathop{\times}\limits_{i=1}^{n} A_i\Big) \times B$ be an $(n+1)$-ary relation, $\bar{a} = (a_1, \ldots, a_n)$ an element of $A_1 \times A_2 \times \ldots \times A_n$, $H_i \subset A_i$, $i \in \{1, \ldots, n\} = \overline{1,n}$, then $\rho\langle\bar{a}\rangle = \{b \in B \mid (\bar{a}, b) \in \rho\}$ and

$$\rho(H_1, \ldots, H_n) = \bigcup \{\, \rho\langle\bar{a}\rangle \mid \bar{a} \in H_1 \times H_2 \times \ldots \times H_n\}.$$

Moreover, let

$$\mathrm{pr}_1\rho = \{\bar{a} \in \mathop{\times}\limits_{i=1}^{n} A_i \mid (\exists b \in B)\, (\bar{a}, b) \in \rho\},$$

$$\mathrm{pr}_2\rho = \{b \in B \mid (\exists \bar{a} \in \mathop{\times}\limits_{i=1}^{n} A_i)\, (\bar{a}, b) \in \rho\}.$$

To every sequence of $(n+1)$-relations $\sigma_1, \ldots, \sigma_n, \rho$ such that $\sigma_i \subset A_1 \times \ldots \times A_n \times B_i$, $i \in \overline{1,n}$, and $\rho \subset B_1 \times \ldots \times B_n \times C$, we assign an $(n+1)$-ary relation

$$\rho[\sigma_1 \ldots \sigma_n] \subset A_1 \times \ldots \times A_n \times C,$$

which is defined as follows:

$$\rho[\sigma_1 \ldots \sigma_n] = \{(\bar{a}, c) \mid (\exists \bar{b})\, (\bar{a}, b_1) \in \sigma_1 \;\&\; \ldots \;\&\; (\bar{a}, b_n) \in \sigma_n \;\&\; (\bar{b}, c) \in \rho\},$$

where $\bar{b} = (b_1, \ldots, b_n) \in B_1 \times \ldots \times B_n$. Obviously:

$$\rho[\sigma_1 \ldots \sigma_n](H_1, \ldots, H_n) \subset \rho(\sigma_1(H_1, \ldots, H_n), \ldots, \sigma_n(H_1, \ldots, H_n)),$$

$$\rho[\sigma_1 \ldots \sigma_n][\chi_1 \ldots \chi_n] \subset \rho[\sigma_1[\chi_1 \ldots \chi_n] \ldots \sigma_n[\chi_1 \ldots \chi_n]],$$

where[1] $\chi_i \subset A_1 \times \ldots \times A_n \times B_i$, $\sigma_i \subset B_1 \times \ldots \times B_n \times C_i$, $i = 1, \ldots, n$ and $\rho \subset C_1 \times \ldots \times C_n \times D$.

[1] It is clear that the symbol $\rho[\sigma_1 \ldots \sigma_n][\chi_1 \ldots \chi_n]$ must be read as $\mu[\chi_1 \ldots \chi_n]$, where $\mu = \rho[\sigma_1 \ldots \sigma_n]$.

The $(n+1)$-operation $O : (\rho, \sigma_1, \ldots, \sigma_n) \mapsto \rho[\sigma_1 \ldots \sigma_n]$ defined as above is called a *Menger superposition* or a *Menger composition* of relations.

Let

$$\overset{n}{\triangle}_A = \{(\underbrace{a, \ldots, a}_{n}) \mid a \in A\},$$

then the homogeneous $(n+1)$-relation $\rho \subset A^{n+1}$ is called

- *reflexive* if $\overset{n+1}{\triangle}_A \subset \rho$,
- *transitive* if $\rho[\rho \ldots \rho] \subset \rho$,
- an *n-quasi-order* (*n-preorder*) if it is reflexive and transitive. For $n = 2$ it is a *quasi-order*.

An $(n+1)$-ary relation $\rho \subset A^n \times B$ is an *n-place function* if it is one-valued, i.e.,

$$(\forall \bar{a} \in A^n)(\forall b_1, b_2 \in B)\,(\,(\bar{a}, b_1) \in \rho \ \&\ (\bar{a}, b_2) \in \rho \longrightarrow b_1 = b_2).$$

Any mapping of a subset of A^n into B is a *partial n-place function.* The set of all such functions is denoted by $\mathcal{F}(A^n, B)$. The set of all *full n-place functions* on A, i.e., mappings defined for every $(a_1, \ldots, a_n) \in A^n$, is denoted by $\mathcal{T}(A^n, B)$. Elements of the set $\mathcal{T}(A^n, A)$ are also called *n-ary transformations of A*. Obviously $\mathcal{T}(A^n, B) \subset \mathcal{F}(A^n, B)$. Many authors instead of a full n-place function use the term an *n-ary operation.*

The superposition of n-place functions is defined by

$$f[g_1 \ldots g_n](a_1, \ldots, a_n) = f(g_1(a_1, \ldots, a_n), \ldots, g_n(a_1, \ldots, a_n)), \tag{2.1}$$

where $a_1, \ldots, a_n \in A$, $f, g_1, \ldots, g_n \in \mathcal{F}(A^n, A)$. This superposition is an $(n+1)$–ary operation $\mathcal{O}$ on the set $\mathcal{F}(A^n, A)$ determined by the formula $\mathcal{O}(f, g_1, \ldots, g_n) = f[g_1 \ldots g_n]$. Sets of n-place functions closed with respect to such superposition are called *Menger algebras of n-place functions* or *n-ary Menger algebras of functions.*

According to the general convention used in the theory of n-ary systems, the sequence $a_i, a_{i+1}, \ldots, a_j$, where $i \leqslant j$, can be written in the abbreviated form as a_i^j (for $i > j$ it is the empty symbol). In this convention (2.1) can be written as

$$f[g_1^n](a_1^n) = f(g_1(a_1^n), \ldots, g_n(a_1^n)).$$

For $g_1 = g_2 = \ldots = g_n = g$ instead of $f[g_1^n]$ we will write $f[g^n]$.

An $(n+1)$-ary groupoid $(G; o)$, i.e., a non-empty set G with one $(n+1)$-ary operation $o : G^{n+1} \to G$, is called a *Menger algebra of rank n*, if it satisfies the following identity (called the *superassociativity*):

$$o(\,o(x, y_1^n), z_1^n) = o(x, o(y_1, z_1^n), o(y_2, z_1^n), \ldots, o(y_n, z_1^n)).$$

A Menger algebra of rank 1 is a semigroup.

Since a Menger algebra (as we see in the sequel) can be interpreted as an algebra of n-place functions with a Menger composition of such functions, we replace the symbol $o(x, y_1^n)$ by $x[y_1^n]$ or by $x[\bar{y}]$, i.e., these two symbols will be interpreted as the result of the operation o applied to the elements $x, y_1, \ldots, y_n \in G$.

In this convention the above superassociativity has the form

$$x[y_1^n][z_1^n] = x[y_1[z_1^n] \ldots y_n[z_1^n]]$$

or shortly

$$x[\bar{y}][\bar{z}] = x[y_1[\bar{z}] \ldots y_n[\bar{z}]],$$

where the left side can be read as in the case of functions, i.e., $x[y_1^n][z_1^n] = \big(x[y_1^n]\big)\,[z_1^n]$.

Theorem 2.1. (R. M. Dicker, [3])
Every Menger algebra of rank n is isomorphic to some Menger algebra of full n-place functions.

Indeed, for every element g of a Menger algebra $(G; o)$ of rank n we can put into accordance the full n-place function λ_g defined on the set $G' = G \cup \{a, b\}$, where a, b are two different elements not belonging to G, such that

$$\lambda_g(x_1^n) = \begin{cases} g[x_1^n] & \text{if } x_1, \dots, x_n \in G, \\ g & \text{if } x_1 = \dots = x_n = a, \\ b & \text{in all other cases.} \end{cases}$$

Using the set $G'' = G \cup \{a\}$, where $a \notin G$, and the partial n-place functions

$$\lambda'_g(x_1^n) = \begin{cases} g[x_1^n] & \text{if } x_1, \dots, x_n \in G, \\ g & \text{if } x_1 = \dots = x_n = a, \end{cases}$$

we can see that every Menger algebra of rank n is isomorphic to a Menger algebra of partial n-place functions too [42].

Theorem 2.2. (J. Henno, [22])
Every finite or countable Menger algebra of rank $n > 1$ can be isomorphically embedded into a Menger algebra of the same rank generated by a single element.

A Menger algebra $(G; o)$ containing *selectors*, i.e., elements $e_1, \dots, e_n \in G$ such that

$$x[e_1^n] = x \quad \text{and} \quad e_i[x_1^n] = x_i$$

for all $x, x_i \in G$, $i = 1, \dots, n$, is called *unitary.*

Theorem 2.3. (V.S. Trokhimenko, [47])
Every Menger algebra $(G; o)$ of rank n can be isomorphically embedded into a unitary Menger algebra $(G^*; o^*)$ of the same rank with selectors $e_1, \dots, e_n$ and a generating set $G \cup \{e_1, \dots, e_n\}$, where $e_i \notin G$ for all $i \in \overline{1, n}$.

J. Hion [26] and J. Henno [23] have proven that Menger algebras with selectors can be identified with some set of multiplace endomorphisms of a universal algebra. Moreover E. Redi proved in [41] that a Menger algebra is isomorphic to the set of all multiplace endomorphisms of some universal algebra if and only if it contains all selectors.

W. Nöbauer and W. Philipp consider [39] the set of all one-place mappings of a fixed universal algebra into itself with the Menger composition and proved that for $n > 1$ this algebra is simple in the sense that it possesses no congruences other than the equality and the universal relation [40].

3. Semigroups

The close connection between semigroups and Menger algebra was stated already in 1966 by B. M. Schein in his work [42]. He found other type of semigroups, which simply define Menger algebras. Thus, there is the possibility to study semigroups of such type instead of Menger algebras. But the study of these semigroups is quite difficult, that is why in many questions it is more advisable simply to study Menger algebras, than to substitute them by the study of similar semigroups.

Definition 3.1. Let $(G;o)$ be a Menger algebra of rank n. The set G^n together with the binary operation $*$ defined by the formula:

$$(x_1,\ldots,x_n)*(y_1,\ldots,y_n)=(x_1[y_1\ldots y_n],\ldots,x_n[y_1\ldots y_n])$$

is called the *binary comitant* of a Menger algebra $(G;o)$.

An $(n+1)$-ary operation o is superassociative if and only if the operation $*$ is associative [42].

Theorem 3.2. (B. M. Schein, [43])
The binary comitant of a Menger algebra is a group if and only if the algebra is of rank 1 *and is a group or the algebra is a singleton.*

It is evident that binary comitants of isomorphic Menger algebras are isomorphic. However, as it was mentioned in [42], from the isomorphism of binary comitants in the general case the isomorphism of the corresponding Menger algebras does not follow. This fact, due to necessity, leads to the consideration of binary comitants with some additional properties such that the isomorphism of these structures implies the isomorphism of initial Menger algebras.

L. M. Gluskin observed (see [19] and [20]) that the sets

$$M_1[G]=\{\bar{c}\in G^n\,|\,x[\bar{y}][\bar{c}]=x[\bar{y}*\bar{c}]\ \text{ for all } x\in G,\ \bar{y}\in G^n\}$$

and

$$M_2[G]=\{\bar{c}\in G^n\,|\,x[\bar{c}][\bar{y}]=x[\bar{c}*\bar{y}]\ \text{ for all } x\in G,\ \bar{y}\in G^n\}$$

are either empty or subsemigroups of the binary comitant $(G^n;*)$. The set

$$M_3[G]=\{a\in G\,|\,a[\bar{x}][\bar{y}]=a[\bar{x}*\bar{y}]\ \text{ for all } \bar{x},\bar{y}\in G^n\}$$

is either empty or a Menger subalgebra of $(G;o)$.

Let us define on the binary comitant $(G^n;*)$ the equivalence relations $\pi_1,\ldots,\pi_n$ putting

$$(x_1,\ldots,x_n)\equiv(y_1,\ldots,y_n)(\pi_i)\longleftrightarrow x_i=y_i$$

for all $x_i,y_i\in G$, $i\in\overline{1,n}$. It is easy to check, that these relations have the following properties:

(a) for any elements $\bar{x}_1,\ldots,\bar{x}_n\in G^n$ there is an element $\bar{y}\in G^n$ such that $\bar{x}_i\equiv\bar{y}(\pi_i)$ for all $i\in\overline{1,n}$,

(b) if $\bar{x}\equiv\bar{y}(\pi_i)$ for some $i\in\overline{1,n}$, then $\bar{x}=\bar{y}$, where $\bar{x},\bar{y}\in G^n$,

(c) relations π_i are *right regular*, i.e.,

$$\bar{x}_1\equiv\bar{x}_2\longrightarrow\bar{x}_1*\bar{y}\equiv\bar{x}_2*\bar{y}(\pi_i)$$

for all $\bar{x}_1,\bar{x}_2,\bar{y}\in G^n$, $i\in\overline{1,n}$,

(d) $\overset{n}{\triangle}_G$ is a *right ideal* of $(G^n;*)$, i.e.,

$$\bar{x}\in\overset{n}{\triangle}_G\wedge\bar{y}\in G^n\longrightarrow\bar{x}*\bar{y}\in\overset{n}{\triangle}_G$$

(e) for any $i\in\overline{1,n}$ every π_i-class contains precisely one element from $\overset{n}{\triangle}_G$.

All systems of the type $(G^n;*,\pi_1,\ldots,\pi_n,\overset{n}{\triangle}_G)$ will be called the *rigged binary comitant of a Menger algebra* $(G;o)$.

Theorem 3.3. *Two Menger algebras of the same rank are isomorphic if and only if their rigged binary comitants are isomorphic.*

This fact is a consequence of the following more general theorem proved in [42].

Theorem 3.4. (B. M. Schein, [42])
The system $(G;\cdot,\varepsilon_1,\dots,\varepsilon_n,H)$, where $(G;\cdot)$ is a semigroup, $\varepsilon_1,\dots,\varepsilon_n$ are binary relations on G and $H \subset G$ is isomorphic to the rigged binary comitant of some Menger algebra of rank n if and only if

1) *for all $i = 1,\dots,n$ the relations ε_i are right regular equivalence relations and for any $(g_1,\dots,g_n) \in G^n$ there is exactly one $g \in G$ such that $g_i \equiv g(\varepsilon_i)$ for all $i \in \overline{1,n}$,*
2) *H is a right ideal of a semigroup $(G;\cdot)$ and for any $i \in \overline{1,n}$ every ε_i-class contains exactly one element of H.*

In the literature the system of the type $(G;\cdot,\varepsilon_1,\dots,\varepsilon_n,H)$, satisfying all the conditions of the above theorem, is called a *Menger semigroup of rank n*. Of course, a Menger semigroup of rank 1 coincides with the semigroup $(G;\cdot)$.

So, the theory of Menger algebras can be completely restricted to the theory of Menger semigroups. But we cannot use this fact. It is possible that in some cases it would be advisable to consider Menger algebras and in other – Menger semigroups. Nevertheless, in our opinion, the study of Menger semigroups is more complicated than study of Menger algebras with one $(n+1)$-ary operation because a Menger semigroup besides one binary operation contains $n+1$ relations, which naturally leads to additional difficulties.

Let $(G;o)$ be a Menger algebra of rank n. Let us define on its binary comitant $(G^n;*)$ the unary operations $\rho_1,\dots,\rho_n$ such that

$$\rho_i(x_1,\dots,x_n) = (x_i,\dots,x_i)$$

for any $x_1,\dots,x_n \in G$, $i \in \overline{1,n}$. Such obtained system $(G^n;*,\rho_1,\dots,\rho_n)$ is called the *selective binary comitant* of Menger algebra $(G;o)$.

Theorem 3.5. (B. M. Schein, [42])
For the system $(G;\cdot,p_1,\dots,p_n)$, where $(G;\cdot)$ is a semigroup and $p_1,\dots,p_n$ are unary operations on it, a necessary and sufficient condition that this system be isomorphic to the selective binary comitant of some Menger algebra of rank n is that the following conditions hold:

1) *$p_i(x)y = p_i(xy)$ for all $i \in \overline{1,n}$ and $x,y \in G$,*
2) *$p_i \circ p_j = p_j$ for all $i,j \in \overline{1,n}$,*
3) *for every vector $(x_1,\dots,x_n) \in G^n$ there is exactly only one $g \in G$ such that $p_i(x_i) = p_i(g)$ for all $i \in \overline{1,n}$.*

The system $(G;\cdot,p_1,\dots,p_n)$ satisfying the conditions of this theorem is called *selective semigroups of rank n*.

Theorem 3.6. (B. M. Schein, [42])
For every selective semigroup of rank n there exists a Menger algebra of the same rank with which the selective semigroup is associated. This Menger algebra is unique up to isomorphism.

These two theorems give the possibility to reduce the theory of Menger algebras to the theory of selective semigroups. In this way, we received three independent methods to the study of superposition of multiplace functions: Menger algebras, Menger semigroups and selective semigroups. A great number of papers dedicated to the study of Menger algebras have been released lately, but unfortunately the same cannot be said about Menger semigroups and selective semigroups.

Defining on a Menger algebra $(G;o)$ the new binary operation

$$x \cdot y = x[y \dots y],$$

we obtain the so-called *diagonal semigroup* $(G;\cdot)$. An element $g \in G$ is called *idempotent*, if it is idempotent in the diagonal semigroup of $(G;o)$, i.e., if $g[g^n] = g$. An element $e \in G$ is called a *left (right) diagonal unit* of a Menger algebra $(G;o)$, if it is a left (right) unit of the diagonal semigroup of $(G;o)$, i.e., if the identity $e[g^n] = g$ (respectively, $g[e^n] = g$) holds for all $g \in G$. If e is both a left and a right unit, then it is called a *diagonal unit*. It is clear that a Menger algebra has at most only one diagonal unit. Moreover, if a Menger algebra has an element which is a left diagonal unit and an element which is a right diagonal unit, then these elements are equal and no other elements which are left or right diagonal units.

An $(n+1)$-ary groupoid $(G;o)$ with the operation $o(x_0^n) = x_0$ is a simple example of a Menger algebra of rank n in which all elements are idempotent and right diagonal units. Of course, this algebra has no left units. In the Menger algebra $(G;o_n)$, where $o_n(x_0^n) = x_n$, all elements are left diagonal units, but this algebra has no right diagonal units.

If a Menger algebra $(G;o)$ has a right diagonal unit e, then its every element $c \in G$ satisfying the identity $e = e[c^n]$ is also a right diagonal unit.

Non-isomorphic Menger algebras may have the same diagonal semigroup. Examples are given in [6].

Theorem 3.7. *A semigroup $(G;\cdot)$ is a diagonal semigroup of some Menger algebra of rank n only in the case when on G can be defined an idempotent n-ary operation f satisfying the identity*

$$f(g_1, g_2, \dots, g_n) \cdot g = f(g_1 \cdot g, g_2 \cdot g, \dots, g_n \cdot g).$$

The operation of diagonal semigroup is in some sense distributive with respect to the Menger composition. Namely, for all $x, y, z_1, \dots, z_n \in G$ we have

$$(x \cdot y)[\bar{z}] = x \cdot y[\bar{z}] \quad \text{and} \quad x[\bar{z}] \cdot y = x[(z_1 \cdot y) \dots (z_n \cdot y)].$$

In some cases Menger algebras can be completely described by their diagonal semigroups. Such situation takes place, for example, in the case of algebras of closure operations.

Theorem 3.8. (V. S. Trokhimenko, [54])
A Menger algebra $(G;o)$ of rank n is isomorphic to some algebra of n-place closure operations on some ordered set if and only if its diagonal semigroup $(G;\cdot)$ is a semilattice and

$$x[y_1 \dots y_n] = x \cdot y_1 \cdot \ldots \cdot y_n$$

for any $x, y_1, \dots, y_n \in G$.

A non-empty subset H of a Menger algebra $(G;o)$ of rank n is called

- an *s-ideal* if $h[x_1^n] \in H$,
- a *v-ideal* if $x[h_1^n] \in H$,
- an *l-deal* if $x[x_1^{i-1}, h_i, x_{i+1}^n] \in H$

for all $x, x_1, \ldots, x_n \in G$, $h, h_1, \ldots, h_n \in H$ and $i \in \overline{1,n}$.

An s-ideal which is a v-ideal is called an *sv-ideal.* A Menger algebra is *s-v-simple* if it possesses no proper s-ideals and v-ideals, and *completely simple* if it posses minimal s-ideals and v-ideals but has no proper sv-ideals.

Theorem 3.9. (Ja. N. Yaroker, [60])
A Menger algebra is completely simple (s-v-simple) if and only if its diagonal semigroup is completely simple (a group).

Theorem 3.10. (Ja. N. Yaroker, [60])
A completely simple Menger algebra of rank n can be decomposed into disjoint union of s-v-simple Menger subalgebras with isomorphic diagonal groups.

Subalgebras obtained in this decomposition are classes modulo some equivalence relation. This decomposition gives the possibility to study connections between isomorphisms of completely simple Menger algebras and isomorphisms of their diagonal semigroups (for details see [60]).

If for $\bar{g} \in G^n$ there exists $x \in G$ such that $g_i[x^n][\bar{g}] = g_i$ holds for each $i \in \overline{1,n}$, then we say that $\bar{g}$ is a *v-regular* vector. The diagonal semigroup of a Menger algebra in which each vector is v-regular is a regular semigroup [55].

An element $x \in G$ is an *inverse element* for $\bar{g} \in G^n$, if $x[\bar{g}][x^n] = x$ and $g_i[x^n][\bar{g}] = g_i$ for all $i \in \overline{1,n}$. Every v-regular vector has an inverse element [55]. Moreover, if each vector of a Menger algebra $(G; o)$ is v-regular and any two elements of $(G; o)$ are commutative in the diagonal semigroup of $(G; o)$, then this semigroup is inverse.

Theorem 3.11. (V. S. Trokhimenko, [55])
If in a Menger algebra $(G; o)$ each vector is v-regular, then the diagonal semigroup of $(G; o)$ is a group if and only if $(G; o)$ has only one idempotent.

4. Group-like Menger algebras

A Menger algebra $(G; o)$ of rank n in which the following two equations

$$x[a_1 \ldots a_n] = b, \tag{4.1}$$

$$a_0[a_1 \ldots a_{i-1} x_i a_{i+1} \ldots a_n] = b \tag{4.2}$$

have unique solutions $x, x_i \in G$ for all $a_0, a_1^n, b \in G$ and some fixed $i \in \overline{1,n}$, is called *i-solvable.* A Menger algebra which is i-solvable for every $i \in \overline{1,n}$ is called *group-like.* Such algebra is an $(n+1)$-ary quasigroup. It is associative (in the sense of n-ary operations) only in some trivial cases (see [4, 5, 7]).

A simple example of a group-like Menger algebra is the set of real functions of the form

$$f_\alpha(x_1, \ldots, x_n) = \frac{x_1 + \cdots + x_n}{n} + \alpha.$$

Every group-like Menger algebra of rank n is isomorphic to some group-like Menger algebra of *reversive* n-place functions, i.e., a Menger algebra of functions $f \in \mathcal{F}(G^n, G)$ with the property:

$$f(x_1^{i-1}, y, x_{i+1}^n) = f(x_1^{i-1}, z, x_{i+1}^n) \longrightarrow y = z$$

for all $x_1^n, y, z \in G$ and $i = 1, \ldots, n$.

The investigation of group-like Menger algebras was initiated by H. Skala [45], the study of i-solvable Menger algebras by W. A. Dudek (see [4], [5] and [6]). A simple example of i-solvable Menger algebras is a Menger algebra $(G;o_i)$ with an $(n+1)$-ary operation $o_i(x_0,x_1,\ldots,x_n) = x_0 + x_i$ defined on a non-trivial commutative group $(G;+)$. It is obvious that the diagonal semigroup of this Menger algebra coincides with the group $(G;+)$. This algebra is j-solvable only for $j = i$, but the algebra $(G;o)$, where $o(x_0,x_1,\ldots,x_n) = x_0 + x_1 + \ldots + x_{k+1}$ and $(G;+)$ is a commutative group of the exponent $k \leqslant n-1$, is i-solvable for every $i = 1,\ldots,k+1$. Its diagonal semigroup also coincides with $(G;+)$. Note that in the definition of i-solvable Menger algebras one can postulate the existence of solutions of (4.1) and (4.2) for all $a_0,\ldots,a_n \in G$ and some fixed $b \in G$ (see [6]). The uniqueness of solutions cannot be dropped, but it can be omitted in the case of finite algebras.

Theorem 4.1. (W. A. Dudek, [6])
A finite Menger algebra $(G;o)$ of rank n is i-solvable if and only if it has a left diagonal unit e and for all $a_1,\ldots,a_n \in G$ there exist $x,y \in G$ such that

$$x[a_1\ldots a_n] = a_0[a_1\ldots a_{i-1}y\,a_{i+1}\ldots a_n] = e.$$

A diagonal semigroup of an i-solvable Menger algebra is a group [4]. The question when a given group is a diagonal group of some Menger algebra is solved by the following two theorems proved in [6].

Theorem 4.2. *A group $(G;\cdot)$ is a diagonal group of some i-solvable Menger algebra if and only if on G can be defined an n-ary idempotent operation f such that the equation $f(a_1^{i-1},x,a_{i+1}^n) = b$ has a unique solution for all $a_1^n, b \in G$ and the identity*

$$f(g_1,g_2,\ldots,g_n)\cdot g = f(g_1\cdot g, g_2\cdot g,\ldots,g_n\cdot g)$$

is satisfied.

Theorem 4.3. *A group $(G;\cdot)$ is the diagonal group of some n-solvable Menger algebra of rank n if and only if on G can be defined an $(n-1)$-ary operation f such that $f(e,\ldots,e) = e$ for the unit of the group $(G;\cdot)$ and the equation*

$$f(a_1\cdot x,\ldots,a_{n-1}\cdot x) = a_n\cdot x \tag{4.3}$$

has a unique solution $x \in G$ for all $a_1,\ldots,a_n \in G$.

On the diagonal semigroup $(G;\cdot)$ of a Menger algebra $(G;o)$ of rank n with the diagonal unit e we can define a new $(n-1)$-ary operation f putting

$$f(a_1,\ldots,a_{n-1}) = e[a_1\ldots a_{n-1}e]$$

for all $a_1,\ldots,a_{n-1} \in G$. The diagonal semigroup with such defined operation f is called a *rigged diagonal semigroup* of $(G;o)$. In the case when $(G;\cdot)$ is a group, the operation f satisfies the condition

$$a[a_1\ldots a_n] = a\cdot f(a_1\cdot a_n^{-1},\ldots,a_{n-1}\cdot a_n^{-1})\cdot a_n, \tag{4.4}$$

where a_n^{-1} is the inverse of a_n in the group $(G;\cdot)$.

Theorem 4.4. (W. A. Dudek, [6])
A Menger algebra $(G;o)$ of rank n is i-solvable for some $1 \leqslant i < n$ if and only if in its rigged diagonal group $(G;\cdot,f)$ the equation

$$f(a_1,\ldots,a_{i-1},x,a_{i+1},\ldots,a_{n-1}) = a_n$$

has a unique solution $x \in G$ for all $a_1,\ldots,a_n \in G$.

Theorem 4.5. (W. A. Dudek, [6])
A Menger algebra $(G; o)$ of rank n is n-solvable if and only if in its rigged diagonal group $(G; \cdot, f)$ for all $a_1, \ldots, a_{n-1} \in G$ there exists exactly one element $x \in G$ satisfying (4.3).

Theorem 4.6. (B. M. Schein, [43])
A Menger algebra $(G; o)$ of rank n is group-like if and only if on its diagonal semigroup $(G; \cdot)$ is a group with the unit e, the operation $f(a_1^{n-1}) = e[a_1^{n-1}e]$ is a quasigroup operation and for all $a_1, \ldots, a_{n-1} \in G$ there exists exactly one $x \in G$ satisfying the equation (4.3).

Non-isomorphic group-like Menger algebras may have the same diagonal group [45], but group-like Menger algebras of the same rank are isomorphic only in the case when their rigged diagonal groups are isomorphic.

In [45] conditions are considered under which a given group is a diagonal group of a group-like Menger algebras of rank n. This is always for an odd n and a finite group. However, if both n and the order of a finite group are even, then a group-like Menger algebra of rank n whose diagonal group is isomorphic to the given group need exists. If n is even, then such algebra exists only for finite orders not of the form $2p$, where p is an odd prime. There are no group-like Menger algebras of rank 2 and finite order $2p$. The existence of group-like Menger algebras of order $2p$ and even rank n greater than 2 is undecided as yet.

5. Representations by n-place functions

Any homomorphism P of a Menger algebra $(G; o)$ of rank n into a Menger algebra $(\mathcal{F}(A^n, A); O)$ of n-place functions (respectively, into a Menger algebra $(\mathfrak{R}(A^{n+1}); O)$ of $(n+1)$-ary relations), where A is an arbitrary set, is called a *representation of* $(G; o)$ *by n-place functions* (respectively, *by $(n+1)$-ary relations*). In the case when P is an isomorphism we say that this representation is *faithful.* If P and P_i, $i \in I$, are representations of $(G; o)$ by functions from $\mathcal{F}(A^n, A)$ (relations from $\mathfrak{R}(A^{n+1})$) and $P(g) = \bigcup_{i \in I} P_i(g)$ for every $g \in G$ P, then we say that P is the *union* of the family $(P_i)_{i \in I}$. If $A = \bigcup_{i \in I} A_i$, where A_i are pairwise disjoint, then the union $\bigcup_{i \in I} P_i(g)$ is called the *sum* of $(P_i)_{i \in I}$.

Definition 5.1. A *determining pair* of a Menger algebra $(G; o)$ of rank n is any pair (ε, W), where ε is a partial equivalence on $(G^*; o^*)$, W is a subset of G^* and the following conditions hold:

1) $G \cup \{e_1, \ldots, e_n\} \subset \mathrm{pr}_1 \varepsilon$, where $e_1, \ldots, e_n$ are the selectors of $(G^*; o^*)$,
2) $e_i \notin W$ for all $i = 1, \ldots, n$,
3) $g[\varepsilon\langle e_1 \rangle \ldots \varepsilon\langle e_n \rangle] \subset \varepsilon\langle g \rangle$ for all $g \in G$,
4) $g[\varepsilon\langle g_1 \rangle \ldots \varepsilon\langle g_n \rangle] \subset \varepsilon\langle g[g_1 \ldots g_n] \rangle$ for all $g, g_1, \ldots, g_n \in G$,
5) if $W \neq \varnothing$, then W is an ε-class and $W \cap G$ is an l-ideal of $(G; o)$.

Let $(H_a)_{a \in A}$ be the family of all ε-classes indexed by elements of A and distinct from W. Let $e_i \in H_{b_i}$ for every $i = 1, \ldots, n$, $A_0 = \{a \in A \,|\, H_a \cap G \neq \varnothing\}$, $\mathfrak{A} = A_0^n \cup \{(b_1, \ldots, b_n)\}$, $B = G^n \cup \{(e_1, \ldots, e_n)\}$. Every $g \in G$ is associated with an n-place function $P_{(\varepsilon, W)}(g)$ on A, which is defined by

$$(\bar{a}, b) \in P_{(\varepsilon, W)}(g) \longleftrightarrow \bar{a} \in \mathfrak{A} \wedge g[H_{a_1} \ldots H_{a_n}] \subset H_b,$$

where $\bar{a} \in A^n$, $b \in A$. The representation $P_{(\varepsilon,W)} : g \to P_{(\varepsilon,W)}(g)$ is called the *simplest representation* of $(G;o)$.

Theorem 5.2. *Every representation of a Menger algebra of rank n by n-place functions is the union of some family of its simplest representations.*

The proof of this theorem can be found in [10] and [14].

With every representation P of $(G;o)$ by n-place functions ($(n+1)$-ary relations) we associate the following binary relations on G:

$$\begin{aligned}
\zeta_P &= \{(g_1,g_2) \mid P(g_1) \subset P(g_2)\},\\
\chi_P &= \{(g_1,g_2) \mid \mathrm{pr}_1 P(g_1) \subset \mathrm{pr}_1 P(g_2)\},\\
\pi_P &= \{(g_1,g_2) \mid \mathrm{pr}_1 P(g_1) = \mathrm{pr}_1 P(g_2)\},\\
\gamma_P &= \{(g_1,g_2) \mid \mathrm{pr}_1 P(g_1) \cap \mathrm{pr}_1 P(g_2) \neq \varnothing\},\\
\kappa_P &= \{(g_1,g_2) \mid P(g_1) \cap P(g_2) \neq \varnothing\},\\
\xi_P &= \{(g_1,g_2) \mid P(g_1) \circ \triangle_{\mathrm{pr}_1 P(g_2)} = P(g_2) \circ \triangle_{\mathrm{pr}_1 P(g_1)}\}.
\end{aligned}$$

A binary relation ρ defined on $(G;o)$is *projection representable* if there exists a representation P of $(G;o)$ by n-place functions for which $\rho = \rho_P$.

It is easy to see that if P is a sum of the family of representations $(P_i)_{i\in I}$ then $\sigma_P = \bigcap_{i\in I} \sigma_{P_i}$ for $\sigma \in \{\zeta,\chi,\pi,\xi\}$ and $\sigma_P = \bigcup_{i\in I} \sigma_{P_i}$ for $\sigma \in \{\kappa,\gamma\}$.

Algebraic systems of the form $(\Phi;O,\zeta_\Phi,\chi_\Phi)$, $(\Phi;O,\zeta_\Phi)$, $(\Phi;O,\chi_\Phi)$ are called: *fundamentally ordered projection (f.o.p.) Menger algebras, fundamentally ordered (f.o.) Menger algebras and projection quasi-ordered (p.q-o.) Menger algebras.*

A binary relation ρ defined on a Menger algebra $(G;o)$ of rank $n \geqslant 2$ is

- *stable*, if $(x,y),(x_1,y_1),\ldots,(x_n,y_n) \in \rho \longrightarrow (x[x_1^n],\, y[y_1^n]) \in \rho$,
- *l-regular*, if $(x,y) \in \rho \longrightarrow (x[x_1^n], y[x_1^n]) \in \rho$,
- *v-negative*, if $(x,\, u[x_1^{i-1}, y, x_{i+1}^n]) \in \rho \longrightarrow (x,y) \in \rho$

for all $u,x,y,x_1,\ldots,x_n,y_1,\ldots,x_n \in G$ and $i \in \overline{1,n}$.

Theorem 5.3. (V. S. Trokhimenko, [51]**)**
An algebraic system $(G;o,\zeta,\chi)$, where $(G;o)$ is a Menger algebra of rank n and ζ, χ are relations defined on G, is isomorphic to a f.o.p. Menger algebra of $(n+1)$-relations if and only if ζ is a stable order and χ is an l-regular and v-negative quasi-order containing ζ.

From this theorem it follows that each stable ordered Menger algebra of rank n is isomorphic to some f.o. Menger algebra of $(n+1)$–relations; each p.q-o. Menger algebra of $(n+1)$-relations is a Menger algebra of rank n with the l-regular and v-negative quasi-order.

Let $T_n(G)$ be the set of all polynomials defined on a Menger algebra $(G;o)$ of rank n. For any binary relation ρ on $(G;o)$ we define the relation $\zeta(\rho) \subset G \times G$ putting $(g_1,g_2) \in \zeta(\rho)$ if and only if there exist polynomials $t_i \in T_n(G)$, vectors $\bar{z}_i \in B = G^n \cup \{\bar{e}\}$ and pairs $(x_i,y_i) \in \rho \cup \{(e_1,e_1),\ldots,(e_n,e_n)\}$, where $e_1,\ldots,e_n$ are the selectors from $(G^*;o^*)$, such that

$$g_1 = t_1(x_1[\bar{z}_1]) \wedge \bigwedge_{i=1}^{m} (t_i(y_i[\bar{z}_i]) = t_{i+1}(x_{i+1}[\bar{z}_{i+1}])) \wedge t_{m+1}(y_{m+1}[\bar{z}_{m+1}]) = g_2$$

for some natural number m. Such defined relation $\zeta(\rho)$ is the least stable quasi-order on $(G; o)$ containing ρ.

Theorem 5.4. (V. S. Trokhimenko, [51]**)**
An algebraic system $(G; o, \chi)$, where $(G; o)$ is a Menger algebra of rank $n \geqslant 2$, $\chi \subset G \times G$, $\sigma = \{(x, g[x^n]) \mid x, g \in G\}$ is isomorphic to some p.q-o. Menger algebra of reflexive $(n+1)$-relations if and only if $\zeta(\sigma)$ is an antisymmetric relation, χ is an l-regular v-negative quasi-order and $(t(x[\bar{y}]), t(g[x^n][\bar{y}])) \in \chi$ for all $t \in T_n(G)$, $x, g \in G$, $\bar{y} \in G^n \cup \{\bar{e}\}$.

The system $(G; o, \zeta, \chi)$, where $(G; o)$ is a Menger algebra of rank n and $\zeta, \chi \subset G \times G$ satisfy all the conditions of Theorem 5.3 can be isomorphically represented by transitive $(n+1)$-relations if and only if for each $g \in G$ we have $(g[g^n], g) \in \zeta$.

Analogously we can prove that each stable ordered Menger algebra of rank n satisfying the condition $(g[g^n], g) \in \zeta$, is isomorphic to some f.o. Menger algebra of transitive $(n+1)$-relations [51]. Every idempotent stable ordered Menger algebra of rank $n \geqslant 2$ in which $(x, y[x^n]) \in \zeta$ is isomorphic to some f.o. Menger algebra of n-quasi-orders.

Theorem 5.5. (V. S. Trokhimenko, [47]**)**
An algebraic system $(G; o, \zeta, \chi)$, where o is an $(n+1)$-operation on G and $\zeta, \chi \subset G \times G$, is isomorphic to a fundamentally ordered projection Menger algebra of n-place functions if and only if o is a superassociative operation, ζ is a stable order, χ – an l-regular v-negative quasi-order containing ζ, and for all $i \in \overline{1,n}$ $u, g, g_1, g_2 \in G$, $\bar{w} \in G^n$ the following two implications hold

$$(g_1, g), (g_2, g) \in \zeta \ \&\ (g_1, g_2) \in \chi \longrightarrow (g_1, g_2) \in \zeta,$$

$$(g_1, g_2) \in \zeta \ \&\ (g, g_1), (g, u[\bar{w}|_i g_2]) \in \chi \longrightarrow (g, u[\bar{w}|_i g_1]) \in \chi,$$

where $u[\bar{w}|_i g] = u[w_1^{i-1}, g, w_{i+1}^n]$.

Theorem 5.6. (V. S. Trokhimenko, [47]**)**
The necessary and sufficient condition for an algebraic system $(G; o, \zeta)$ to be isomorphic to some f.o. Menger algebra of n-place functions is that o is a superassociative $(n+1)$-ary operation and ζ is a stable order on G such that

$$(x, y), (z, t_1(x)), (z, t_2(y)) \in \zeta \longrightarrow (z, t_2(x)) \in \zeta$$

for all $x, y, z \in G$ and $t_1, t_2 \in T_n(G)$.

Replacing the last implication by the implication

$$(z, t_1(x), (z, t_1(y)), (z, t_2(y)) \in \zeta \longrightarrow (z, t_2(x)) \in \zeta$$

we obtain a similar characterization of algebraic systems $(G; o, \zeta)$ isomorphic to f.o. Menger algebras of reversive n-place functions (for details see [43] or [10, 14]).

6. Menger algebras with additional operations

Many authors investigate sets of multiplace functions closed with respect to the Menger composition of functions and other naturally defined operations such as, for example, the set-theoretic intersection of functions considered as subsets of the corresponding Cartesian product. Such obtained algebras are denoted by $(\Phi; O, \cap)$

and are called $\cap$-*Menger algebras*. Their abstract analog is called a $\curlywedge$-*Menger algebra of rank* n or a *Menger* $\mathcal{P}$-*algebra*.

An abstract characterization of $\cap$-Menger algebras is given in [50] (see also [15]).

Theorem 6.1. (V. S. Trokhimenko, [50])
For an algebra $(G; o, \curlywedge)$ *of type* $(n+1, 2)$ *the following statements are true:*

(*a*) $(G; o, \curlywedge)$ *is a* $\curlywedge$-*Menger algebra of rank* n, *if and only if*

(*i*) o *is a superassociative operation,*

(*ii*) $(G; \curlywedge)$ *is a semilattice, and the following two conditions*

$$(x \curlywedge y)[\bar{z}] = x[\bar{z}] \curlywedge y[\bar{z}], \tag{6.1}$$

$$t_1(x \curlywedge y \curlywedge z) \curlywedge t_2(y) = t_1(x \curlywedge y) \curlywedge t_2(y \curlywedge z) \tag{6.2}$$

hold for all $x, y, z \in G$, $\bar{z} \in G^n$ *and* $t_1, t_2 \in T_n(G)$,

(*b*) *if* $n > 1$, *then* $(G; o, \curlywedge)$ *is isomorphic to a* $\cap$-*Menger algebra of reversive* n-*place functions if the operation* o *is superassociative,* $(G; \curlywedge)$ *is a semilattice,* (6.1) *and*

$$u[\bar{z}|_i(x \curlywedge y)] = u[\bar{z}|_i x] \curlywedge u[\bar{z}|_i y], \tag{6.3}$$

$$t_1(x \curlywedge y) \curlywedge t_2(y) = t_1(x \curlywedge y) \curlywedge t_2(x) \tag{6.4}$$

are valid for all $i \in \overline{1,n}$, $u, x, y \in G$, $\bar{z} \in G^n$, $t_1, t_2 \in T_n(G)$.

Theorem 6.2. (B. M. Schein, V. S. Trokhimenko, [57])
An algebra $(G; o, \curlyvee)$ *of type* $(n+1, 2)$ *is isomorphic to some Menger algebra* $(\Phi; O)$ *of* n-*place functions closed under the set-theoretic union of functions if and only if the operation* o *is superassociative,* $(G; \curlyvee)$ *is a semilattice with the semillatice order* $\leqslant$ *such that* $x[\bar{z}] \leqslant z_i$ *and*

$$(x \curlyvee y)[\bar{z}] = x[\bar{z}] \curlyvee y[\bar{z}], \tag{6.5}$$

$$u[\bar{z}|_i(x \curlyvee y)] = u[\bar{z}|_i x] \curlyvee u[\bar{z}|_i y], \tag{6.6}$$

$$x \leqslant y \curlyvee u[\bar{z}|_i z] \longrightarrow x \leqslant y \curlyvee u[\bar{z}|_i x], \tag{6.7}$$

for all $x, y, z \in G$, $\bar{z} \in G^n$, $u \in G \cup \{e_1, \ldots, e_n\}$, $i \in \overline{1,n}$.

Theorem 6.3. (B. M. Schein, V. S. Trokhimenko, [57])
An abstract algebra $(G; o, \curlywedge, \curlyvee)$ *of type* $(n+1, 2, 2)$ *is isomorphic to some Menger algebra of* n-*place functions closed with respect to the set-theoretic intersection and union of functions if and only if the operation* o *is superassociative,* $(G, \curlywedge, \curlyvee)$ *is a distributive latticc, the identities* (6.1), (6.5), (6.6) *and*

$$x[(y_1 \curlywedge z_1) \ldots (y_n \curlywedge z_n)] = x[\bar{y}] \curlywedge z_1 \curlywedge \ldots \curlywedge z_n \tag{6.8}$$

are satisfied for all $x, y \in G$, $\bar{y}, \bar{z} \in G^n$, $u \in G \cup \{e_1, \ldots, e_n\}$, $i \in \overline{1,n}$.

The restriction of an n-place function $f \in \mathcal{F}(A^n, B)$ to the subset $H \subset A^n$ can be defined as a composition of f and $\triangle_H = \{(\bar{a}, \bar{a}) | \bar{a} \in H\}$, i.e., $f|_H = f \circ \triangle_H$. The *restrictive product* $\rhd$ of two functions $f, g \in \mathcal{F}(A^n, B)$ is defined as

$$f \rhd g = g \circ \triangle_{\mathrm{pr}_1} f.$$

Theorem 6.4. (V. S. Trokhimenko, [48])
An algebra $(G; o, \blacktriangleright)$ of type $(n+1, 2)$ is isomorphic to a Menger algebra of n-place functions closed with respect to the restrictive product of functions if and only if $(G; o)$ is a Menger algebra of rank n, $(G; \blacktriangleright)$ is an idempotent semigroup and the following three identities hold:

$$x[(y_1 \blacktriangleright z_1) \ldots (y_n \blacktriangleright z_n)] = y_1 \blacktriangleright \ldots \blacktriangleright y_n \blacktriangleright x[z_1 \ldots z_n], \tag{6.9}$$

$$(x \blacktriangleright y)[\bar{z}] = x[\bar{z}] \blacktriangleright y[\bar{z}], \tag{6.10}$$

$$x \blacktriangleright y \blacktriangleright z = y \blacktriangleright x \blacktriangleright z. \tag{6.11}$$

An algebra $(G; o, \curlywedge, \blacktriangleright)$ of type $(n+1, 2, 2)$ is isomorphic to a Menger algebra of n-place functions closed with respect to the set-theoretic intersection and the restrictive product of functions, i.e., to $(\Phi; O, \cap, \rhd)$, if and only if $(G; o, \blacktriangleright)$ satisfies the conditions of Theorem 6.4, $(G; \curlywedge)$ is a semilattice and the identities $(x \curlywedge y)[\bar{z}] = x[\bar{z}] \curlywedge y[\bar{z}]$, $x \curlywedge (y \blacktriangleright z) = y \blacktriangleright (x \curlywedge z)$, $(x \curlywedge y) \blacktriangleright y = x \curlywedge y$ are satisfied [50]. Moreover, if $(G; o, \curlywedge, \blacktriangleright)$ also satisfies (6.3), then it is isomorphic to an algebra $(\Phi; O, \cap, \rhd)$ of reversive n-place functions.

More results on such algebras can be found in [10] and [14].

7. $(2, n)$-SEMIGROUPS

On $\mathcal{F}(A^n, A)$ we can define n binary compositions $\underset{1}{\oplus}, \ldots, \underset{n}{\oplus}$ of two functions by putting

$$(f \underset{i}{\oplus} g)(a_1^n) = f(a_1^{i-1}, g(a_1^n), a_{i+1}^n).$$

for all $f, g \in \mathcal{F}(A^n, A)$ and $a_1, \ldots, a_n \in A$. Since all such defined compositions are associative, the algebra $(\Phi; \underset{1}{\oplus}, \ldots, \underset{n}{\oplus})$, where $\Phi \subset \mathcal{F}(A^n, A)$, is called a $(2, n)$-*semigroup of n-place functions.*

The study of such compositions of functions for binary operations was initiated by Mann [33] . Nowadays such compositions are called *Mann's compositions* or *Mann's superpositions.* Mann's compositions of n-ary operations, i.e., full n-place functions, were described by T. Yakubov in [59]. Abstract algebras isomorphic to some sets of operations closed with respect to these compositions are described in [46].

Menger algebras on n-place functions closed with respect to Mann's superpositions are called *Menger $(2, n)$-semigroups.* Their abstract characterizations are given in [9].

Any non-empty set G with n binary operations defined on G is also called an abstract $(2, n)$-semigroup. For simplicity, for these operations the same symbols as for Mann's compositions of functions will be used. An abstract $(2, n)$-semigroup having a representation by n-place function is called *representable.*

Further, for simplicity, all expressions of the form $(\cdots((x \underset{i_1}{\oplus} y_1) \underset{i_2}{\oplus} y_2) \cdots) \underset{i_k}{\oplus} y_k$ are denoted by $x \underset{i_1}{\oplus} y_1 \underset{i_2}{\oplus} \cdots \underset{i_k}{\oplus} y_k$ or, in the abbreviated form, by $x \overset{i_k}{\underset{i_1}{\oplus}} y_1^k$. The symbol $\mu_i(\overset{i_s}{\underset{i_1}{\oplus}} x_1^s)$ will be reserved for the expression $x_{i_k} \overset{i_s}{\underset{i_{k+1}}{\oplus}} x_{k+1}^s$, if $i \neq i_1, \ldots, i \neq i_{k-1}$, $i = i_k$ for some $k \in \{1, \ldots, s\}$. In any other case this symbol is empty. For example, $\mu_1(\underset{2}{\oplus} x \underset{1}{\oplus} y \underset{3}{\oplus} z) = y \underset{3}{\oplus} z$, $\mu_2(\underset{2}{\oplus} x \underset{1}{\oplus} y \underset{3}{\oplus} z) = x \underset{1}{\oplus} y \underset{3}{\oplus} z$, $\mu_3(\underset{2}{\oplus} x \underset{1}{\oplus} y \underset{3}{\oplus} z) = z$. The symbol $\mu_4(\underset{2}{\oplus} x \underset{1}{\oplus} y \underset{3}{\oplus} z)$ is empty.

Theorem 7.1. (W. A. Dudek, V. S. Trokhimenko, [11])
A $(2,n)$-semigroup $(G; \underset{1}{\oplus}, \ldots, \underset{n}{\oplus})$ has a faithful representation by partial n-place functions if and only if for all $g, x_1, \ldots, x_s, y_1, \ldots, y_k \in G$ and $i_1, \ldots, i_s, j_1, \ldots, j_k \in \overline{1,n}$ the following implication

$$\bigwedge_{i=1}^{n} \left(\mu_i(\overset{i_s}{\underset{i_1}{\oplus}} x_1^s) = \mu_i(\overset{j_k}{\underset{j_1}{\oplus}} y_1^k)\right) \longrightarrow g \overset{i_s}{\underset{i_1}{\oplus}} x_1^s = g \overset{j_k}{\underset{j_1}{\oplus}} y_1^k \tag{7.1}$$

is satisfied.

A $(2,n)$-semigroup $(G; \underset{1}{\oplus}, \ldots, \underset{n}{\oplus})$ satisfying the above implication has also a faithful representation by full n-place functions.

Moreover, for any $(2,n)$-semigroup any its representation by n-place functions is a union of some family of its simplest representations [11].

Theorem 7.2. (W. A. Dudek, V. S. Trokhimenko, [11])
An algebraic system $(G; \underset{1}{\oplus}, \ldots, \underset{n}{\oplus}, \chi)$, where $(G; \underset{1}{\oplus}, \ldots, \underset{n}{\oplus})$ is a $(2,n)$-semigroup and χ is a binary relation on G, is isomorphic to a p.q-o. $(2,n)$-semigroup of partial n-place functions if and only if the implication (7.1) is satisfied and χ is a quasi-order such that $(x \overset{i_s}{\underset{i_1}{\oplus}} z_1^s,\ \mu_j(\overset{i_s}{\underset{i_1}{\oplus}} z_1^s)) \in \chi$ and $(x \underset{i}{\oplus} z,\ y \underset{i}{\oplus} z) \in \chi$ for all $(x,y) \in \chi$.

In the case of Menger $(2,n)$-semigroups the situation is more complicated since conditions under which a Menger $(2,n)$-semigroup is isomorphic to a Menger $(2,n)$-semigroup of n-place functions are not simple.

Theorem 7.3. (W. A. Dudek, V. S. Trokhimenko, [9])
A Menger $(2,n)$-semigroup $(G; o, \underset{1}{\oplus}, \ldots, \underset{n}{\oplus})$ is isomorphic to a Menger $(2,n)$-semigroup of partial n-place functions if and only if it satisfies the implication (7.1) and the following three identities

$$x \underset{i}{\oplus} y[z_1^n] = x[z_1^{i-1}, y[z_1^n], z_{i+1}^n]\,,$$

$$x[y_1^n] \underset{i}{\oplus} z = x[y_1 \underset{i}{\oplus} z \ldots y_n \underset{i}{\oplus} z]\,,$$

$$x \overset{i_s}{\underset{i_1}{\oplus}} y_1^s = x[\mu_1(\overset{i_s}{\underset{i_1}{\oplus}} y_1^s) \ldots \mu_n(\overset{i_s}{\underset{i_1}{\oplus}} y_1^s)]\,,$$

where $\{i_1, \ldots, i_s\} = \{1, \ldots, n\}$.

Using this theorem one can prove that any Menger $(2,n)$-semigroup of n-place functions is isomorphic to some Menger $(2,n)$-semigroup of full n-place functions (for details see [9]). Moreover, for any Menger $(2,n)$-semigroup satisfying all the assumptions of Theorem 7.3 one can find the necessary and sufficient conditions under which the triplet (χ, γ, π) of binary relations defined on this Menger $(2,n)$-semigroup be a projection representable by the triplet of relations $(\chi_P, \gamma_P, \pi_P)$ defined on the corresponding Menger $(2,n)$-semigroup of n-place functions [12]. Similar conditions were found for pairs (χ, γ), (χ, π) and (γ, π). But conditions under which the triplet (χ, γ, π) will be a faithful projection representable have not yet been found.

8. Functional Menger systems

On the set $\mathcal{F}(A^n, A)$ we also can consider n unary operations $\mathcal{R}_1, \ldots, \mathcal{R}_n$ such that for every function $f \in \mathcal{F}(A^n, A)$ $\mathcal{R}_i f$ is the restriction of n-place projectors defined on A to the domain of f, i.e., $\mathcal{R}_i f = f \rhd I_i^n$ for every $i = 1, \ldots, n$, where $I_i^n(a_1^n) = a_i$ is the ith n-place projector on A. In other words, $\mathcal{R}_i f$ is such n-place function from $\mathcal{F}(A^n, A)$, which satisfies the conditions

$$\mathrm{pr}_1 \mathcal{R}_i f = \mathrm{pr}_1 f, \qquad \bar{a} \in \mathrm{pr}_1 f \longrightarrow \mathcal{R}_i f(\bar{a}) = a_i$$

for any $\bar{a} \in A^n$. Algebras of the form $(\Phi; O, \mathcal{R}_1, \ldots, \mathcal{R}_n)$, where $\Phi \subset \mathcal{F}(A^n, A)$, are called *functional Menger system of rank n*. Such algebras (with some additional operations) were firstly studied in [29] and [44]. For example, V. Kafka considered in [29] the algebraic system of the form $(\Phi; O, \mathcal{R}_1, \ldots, \mathcal{R}_n, \mathcal{L}, \subset)$, where $\Phi \subset \mathcal{F}(A^n, A)$ and $\mathcal{L} f = \overset{n+1}{\triangle}_{\mathrm{pr}_2 f}$ for every $f \in \Phi$. Such algebraic system satisfies the conditions:

(A_1) $(\Phi; O)$ is a Menger algebra of functions,
(A_2) $\subset$ is an order on Φ,
(A_3) the following identities are satisfied

$$\begin{cases} f[\mathcal{R}_1 f \ldots \mathcal{R}_n f] = f, \\ \mathcal{R}_i(f[g_1 \ldots g_n]) = \mathcal{R}_i((\mathcal{R}_j f)[g_1 \ldots g_n]), \\ (\mathcal{L} f)[f \ldots f] = f, \end{cases}$$

(A_4) $\mathcal{L}(f[g_1 \ldots g_n]) \subset \mathcal{L} f$ for all $f, g_1 \ldots g_n \in \Phi$,
(A_5) Φ has elements $I_1, \ldots, I_n$ such that $f[I_1 \ldots I_n] = f$ and

$$\begin{cases} g_i \subset I_i \longrightarrow \mathcal{R}_i g_i \subset g_i, \\ f \subset \bigcap\limits_{j=1}^{n} I_j \longrightarrow \mathcal{L} f \subset f, \\ g_k \subset I_k \longrightarrow g_k[f_1 \ldots f_n] \subset f_k, \\ \bigwedge\limits_{k=1}^{n} (g_k \subset I_k) \longrightarrow f[g_1 \ldots g_n] \subset f, \\ \mathcal{R}_i g_j \subset \bigcap\limits_{k=1}^{n} \mathcal{R}_i g_k \longrightarrow I_j[g_1 \ldots g_n] = g_j, \\ f \subset h \longrightarrow f = h[p_1 \ldots p_n] \ \text{ for some } p_1 \subset I_1, \ldots, p_n \subset I_n \end{cases}$$

for all $f, f_1, \ldots, f_n, g_1, \ldots, g_n \in \Phi$ and $i, j, k \in \{1, \ldots, n\}$.

It is proved in [29] that for any algebraic system $(G; o, R_1, \ldots, R_n, L, \leqslant)$ satisfying the above conditions there exists $\Phi \subset \mathcal{F}(A^n, A)$ and an isomorphism $(G; o)$ onto $(\Phi; O)$ which transforms the order $\leqslant$ into the set-theoretical inclusion of functions. However, $A_1 - A_5$ do not give a complete characterization of systems of the form $(\Phi; O, \mathcal{R}_1, \ldots, \mathcal{R}_n, \mathcal{L}, \subset)$. Such characterization is known only for systems of the form $(\Phi; O, \mathcal{R}_1, \ldots, \mathcal{R}_n)$.

Theorem 8.1. (V. S. Trokhimenko, [49]**)**
A functional Menger system $(\Phi; O, \mathcal{R}_1, \ldots, \mathcal{R}_n)$ of rank n is isomorphic to an algebra $(G; o, R_1, \ldots, R_n)$ of type $(n+1, 1, \ldots, 1)$, where $(G; o)$ is a Menger algebra,

if and only if for all $i, k \in \overline{1,n}$ *it satisfies the identities*

(8.1) $x[R_1x \ldots R_nx] = x,$

(8.2) $x[\bar{u}|_i z][R_1y \ldots R_ny] = x[\bar{u}|_i\, z[R_1y \ldots R_ny]],$

(8.3) $R_i(x[R_1y \ldots R_ny]) = (R_ix)[R_1y \ldots R_ny],$

(8.4) $x[R_1y \ldots R_ny][R_1z \ldots R_nz] = x[R_1z \ldots R_nz][R_1y \ldots R_ny],$

(8.5) $R_i(x[\bar{y}]) = R_i((R_kx)[\bar{y}]),$

(8.6) $(R_ix)[\bar{y}] = y_i[R_1(x[\bar{y}]) \ldots R_n(x[\bar{y}])].$

It is interesting to note that defining on $(G; o, R_1, \ldots, R_n)$ a new operation $\blacktriangleright$ by putting $x \blacktriangleright y = y[R_1x \ldots R_nx]$ we obtain an algebra $(G; o, \blacktriangleright)$ isomorphic to a Menger algebra of n-place functions closed with respect to the restrictive product of functions [14].

Another interesting fact is

Theorem 8.2. (W. A. Dudek, V. S. Trokhimenko, [8]**)**
An algebra $(G; o, \curlywedge, R_1, \ldots, R_n)$ *of type* $(n+1, 2, 1, \ldots, 1)$ *is isomorphic to some functional Menger* $\cap$*-algebra of* n*-place functions if and only if* $(G; o, R_1, \ldots, R_n)$ *is a functional Menger system of rank* n, $(G; \curlywedge)$ *is a semilattice, and the identities*

$$(x \curlywedge y)[z_1 \ldots z_n] = x[z_1 \ldots z_n] \curlywedge y[z_1 \ldots z_n],$$

$$(x \curlywedge y)[R_1z \ldots R_nz] = x \curlywedge y[R_1z \ldots R_nz],$$

$$x[R_1(x \curlywedge y) \ldots R_n(x \curlywedge y)] = x \curlywedge y$$

are satisfied.

Now we present abstract characterizations of two important sets used in the theory of functions. We start with the set containing functions with the same fixed point.

Definition 8.3. *A non-empty subset* H *of* G *is called a stabilizer of a functional Menger system* $(G; o, R_1, \ldots, R_n)$ *of rank* n *if there exists a representation* P *of* $(G; o, R_1, \ldots, R_n)$ *by* n*-place functions* $f \in \mathcal{F}(A^n, A)$ *such that*

$$H = H_P^a = \{g \in G \,|\, P(g)(a, \ldots, a) = a\}$$

for some point $a \in A$ *common for all* $g \in H$.

Theorem 8.4. (W. A. Dudek, V. S. Trokhimenko, [13]**)**
A non-empty $H \subset G$ *is a stabilizer of a functional Menger system* $(G; o, R_1, \ldots, R_n)$ *of rank* n *if and only if there exists a subset* U *of* G *such that*

(8.7) $H \subset U, \;\; R_iU \subset H, \;\; R_i(G \backslash U) \subset G \backslash U,$

(8.8) $x, y \in H \;\&\; t(x) \in U \longrightarrow t(y) \in U,$

(8.9) $x = y[R_1x \ldots R_nx] \in U \;\&\; u[\bar{w}\,|_iy] \in H \longrightarrow u[\bar{w}\,|_ix] \in H,$

(8.10) $x = y[R_1x \ldots R_nx] \in U \;\&\; u[\bar{w}\,|_iy] \in U \longrightarrow u[\bar{w}\,|_ix] \in U,$

(8.11) $y \in H \;\&\; x[y \ldots y] \in H \longrightarrow x \in H,$

(8.12) $x, y \in H \;\&\; t(x) \in H \longrightarrow t(y),$

(8.13) $x \in H \longrightarrow x[x \ldots x] \in H$

for all $x, y \in G$, $\bar{w} \in G^n$, $t \in T_n(G)$ *and* $i \in \overline{1,n}$, *where the symbol* $u[\bar{w}\,|_i\,]$ *may be empty.*[2]

In the case of functional Menger algebras isomorphic to functional Menger $\cap$-algebras of n-place functions stabilizers have simplest characterization. Namely, as it is proved in [13], the following theorem is valid.

Theorem 8.5. *A non-empty subset* H *of* G *is a stabilizer of a functional Menger algebra* $(G; o, \curlywedge, R_1, \ldots, R_n)$ *of rank* n *if and only if* H *is a subalgebra of* $(G; o, \curlywedge)$, $R_iH \subset H$ *for every* $i \in \overline{1,n}$, *and*

$$x[y_1 \ldots y_n] \in H \longrightarrow x \in H$$

for all $x \in G$, $y_1, \ldots, y_n \in H$.

Another important question is of abstract characterizations of stationary subsets of Menger algebras of n-place functions. Such characterizations are known only for some types of such algebras.

As it is well known, by the *stationary subset* of $\Phi \subset \mathcal{F}(A^n, A)$ we mean the set

$$\mathbf{St}(\Phi) = \{f \in \Phi \,|\, (\exists a \in A)\, f(a, \ldots, a) = a\}.$$

Note that in a functional Menger $\cap$-algebra $(\Phi; \mathcal{O}, \cap, \mathcal{R}_1, \ldots, \mathcal{R}_n)$ of n-place functions without a zero $\mathbf{0}$ the subset $\mathbf{St}(\Phi)$ coincides with Φ. Indeed, in this algebra $f \neq \emptyset$ for any $f \in \Phi$. Consequently, $f \cap g[f \ldots f] \neq \emptyset$ for all $f, g \in \Phi$. Hence, $g[f \ldots f](\bar{a}) = f(\bar{a})$, i.e., $g(f(\bar{a}), \ldots, f(\bar{a})) = f(\bar{a})$ for some $\bar{a} \in A^n$. This means that $g \in \mathbf{St}(\Phi)$. Thus, $\Phi \subseteq \mathbf{St}(\Phi) \subseteq \Phi$, i.e., $\mathbf{St}(\Phi) = \Phi$. Therefore, below we will consider only functional Menger $\cap$-algebras with a zero.

Definition 8.6. A non-empty subset H of G is called a *stationary subset* of a functional Menger algebra $(G; o, \curlywedge, R_1, \ldots, R_n)$ of rank n if there exists its faithful representation P by n-place functions such that

$$g \in H \longleftrightarrow P(g) \in \mathbf{St}(P(G))$$

for every $g \in G$, where $P(G) = \{P(g) \,|\, g \in G\}$.

Theorem 8.7. (W. A. Dudek, V. S. Trokhimenko, [15])
A non-empty subset H *of* G *is a stationary subset of a functional Menger algebra* $(G; o, \curlywedge, R_1, \ldots, R_n)$ *with a zero* $\mathbf{0}$ *if and only if*

$$x[x \ldots x] \curlywedge x \in H,$$

$$z[y \ldots y] = y \in H \longrightarrow z \in H,$$

$$z[y \ldots y] \curlywedge y \neq \mathbf{0} \longrightarrow z \in H,$$

$$\mathbf{0} \notin H \longrightarrow R_i\mathbf{0} = \mathbf{0},$$

for all $x \in H$, $y, z \in G$ *and* $i \in \overline{1,n}$.

Conditions formulated in the above theorem are not identical with the conditions used for the characterization of stationary subsets of restrictive Menger $\mathcal{P}$-algebras (see Theorem 8 in [56]). For example, the implication

$$y[R_1x \ldots R_nx] = x \in H \longrightarrow y \in H$$

[2] If $u[\bar{w}\,|_i\,]$ is the empty symbol, then $u[\bar{w}\,|_i x]$ is equal to x.

is omitted. Nevertheless, stationary subsets of functional Menger $\curlywedge$-algebras with a zero have the same properties as stationary subsets of restrictive Menger $\mathcal{P}$-algebras.

More results on stationary subsets and stabilizers in various types of Menger algebras can be found in [10] and [14]. In [14] one can find abstract characterizations of algebras of vector-valued functions, positional algebras of functions, Mal'cev–Post iterative algebras and algebras of multiplace functions partially ordered by various types of natural relations connected with domains of functions.

References

[1] M. I. Burtman, Congruences of the Menger algebra of linear mappings, (Russian), *Izv. Akad. Nauk Azerb. SSR, ser. Fiz.–Tekh. Mat. Nauk* **3** (1984), 8 – 14.

[2] M. I. Burtman, Finitely generated subalgebras of the Menger algebra of linear mappings, (Russian), *Izv. Akad. Nauk Azerb. SSR, ser. Fiz.–Tekh. Mat. Nauk* **2** (1984), 3 – 9.

[3] R. M. Dicker, The substitutive law, *Proc. London Math. Soc.* **13** (1963), 493 – 510.

[4] W. A. Dudek, On some group-like Menger n-groupoids, *Proc. II International Symp. "n-ary Structures"*, Varna 1983, 83 – 94, Center Appl. Math. Sofia 1985.

[5] W. A. Dudek, Remarks on alternating symmetric n-quasigroups, *Zbornik Radova Prirod.-Mat. Fak. Univ. u Novom Sadu* **15.2** (1985), 67 – 78.

[6] W. A. Dudek, On group-like Menger n-groupoids, *Radovi Matematički* (Sarajevo) **2** (1986), 81 – 98.

[7] W. A. Dudek, *On some Menger groupoids*, Mathematica (Cluj-Napoca) **43(66)** (2001), 195 – 201.

[8] W. A. Dudek, V. S. Trokhimenko, *Functional Menger $\mathcal{P}$-algebras*, Commun. Algebra **30** (2002), 5921 – 5931.

[9] W. A. Dudek, V. S. Trokhimenko, *Representations of Menger $(2,n)$-semigroups by multiplace functions*, Commun. Algebra **34** (2006), 259 – 274.

[10] W. A. Dudek, V. S. Trokhimenko, Menger algebras of multiplace functions, (Russian), *Centrul Ed. USM, Chişinău*, 2006.

[11] W. A. Dudek, V. S. Trokhimenko, *Representations of $(2,n)$-semigroups by multiplace functions*, Studia Sci. Math. Hungar. **44** (2007), 131 – 146.

[12] W. A. Dudek, V. S. Trokhimenko, *Projection representable relations on Menger $(2,n)$-semigroups*, Czechoslovak Math. J. **53(133)** (2008), 1015 – 1037.

[13] W. A. Dudek, V. S. Trokhimenko, *Stabilizers of functional Menger systems*, Commun. Algebra **37** (2009), 985 – 1000.

[14] W. A. Dudek, V. S. Trokhimenko, Algebras of multiplace functions, *Christian Dawn, Kremenchuk*, 2010.

[15] W. A. Dudek, V. S. Trokhimenko, *Stationary subsets of functional Menger $\cap$-algebras of multiplace functions*, Algebra Colloq. **17** (2010), 65 – 73.

[16] F. A. Gadzhiev, On Menger algebras, *Proc. Steklov Inst. Math.* **163** (1984), 93 – 96, (translation from *Trudy Mat. Inst. Steklova* **163** (1984), 78 – 80).

[17] F. A. Gadzhiev, Functions of four variables on the three-dimensional sphere, which are not representable as a superposition of functions of a smaller number variables, (Russian), *Doklady AN SSSR* **289** (1986), 785 – 788.

[18] F. A. Gadzhiev, A. A. Mal'tsev, On dense subalgebras of Post algebras and Menger algebras of continuous functions, *Proc. Int. Conf., Leningrad 1982, Lecture Notes in Math.* **1060**, 258 – 266, Springer 1984.

[19] L. M. Gluskin, Composition of multidimensional functions, (Russian), *Publ. Math.* (Debrecen) **17** (1970), 349 – 378.

[20] L. M. Gluskin, Function algebras, (Ukrainian), *Dopovidi Akad. Nauk Ukrain. RSR, ser.* **A** (1971), 394 – 396.

[21] J. Henno, Green's equivalences in Menger systems, (Russian), *Tartu Riikl. Ul. Toimetised* (*Uchen. Zapiski Tartu Gosudarstv. Univ. Trudy Mat. Meh.*) **277** (1971), 37 – 46.

[22] J. Henno, Free Ω–systems, (Russian), *Tallin. Politehn. Inst. Toimestised* (*Trudy Tallin. Politeh. Inst.*) **345** (1973), 29 – 40.

[23] J. Henno, Embeddings of Ω–systems in Ω–systems with a minimal number of generators, (Russian), *Eesti NSV Tead. Akad. Toimetised Fuus.-Mat.* (*Izv. Akad. Nauk Eston. SSR*) **23** (1974), 326 – 335.
[24] J. Henno, On idempotents and Green relations in the algebras of many-placed functions, *Ann. Acad. Sci. Fennicae, ser. Math.* **3** (1977), 169 – 173.
[25] J. Henno, On the completeness of associative idempotent functions, *Z. Math. Logik Grundlag. Math.* **25** (1979), 37 – 43.
[26] J. V. Hion, m-ary Ω-ringoids, *Siberian Math. J.* **8** (1967), 131–146 (translation from *Sibirsk. Mat. Zh.* **8** (1967), 174 – 194).
[27] F. A. Ismailov, Superassociative algebras of partial homomorphisms of graphs, (Russian), *Doklady Akad. Nauk Azerb. SSR* **42** (1986), no. 8, 3 – 6.
[28] F. A. Ismailov, L. G. Mustafaev, Abstract characteristics of superassociative algebras of homomorphisms, (Russian), *Izv. Akad. Nauk Azerb. SSR, ser. Fiz.–Tekh. Mat. Nauk* **7** (1986), 3 – 7.
[29] V. Kafka, Axiomatics for systems of multiplace functions, *Master of Science Thesis*, Ill. Inst. of Technol., Chicago, 1965.
[30] H. Länger, A generalization of a theorem of Skala, *Fund. Math.* **112** (1981), 121 – 123.
[31] H. Länger, The independence of certain associative laws, *Proc. Roy. Soc. Edinburg* **91** (1981/82), 179 – 180.
[32] H. Länger, A characterization of full function algebras, *J. Algebra* **119** (1988), 261 – 264.
[33] H. Mann, On orthogonal Latin squares, *Bull. Amer. Math. Soc.* **50** (1944), 240 – 257.
[34] K. Menger, The algebra of functions: past, present, future, *Rend. Math.* **20** (1961), 409–430.
[35] K. Menger, Superassociative systems and logical functors, *Math. Ann.* **157** (1964), 278–295.
[36] K. Menger, H. Whitlock, Two theorems on the generations of systems of functions, *Fund. Math.* **58** (1966), 229 – 240.
[37] F. Kh. Muradov, Superassociative algebras of continuous mappings, (Russian), *Izv. Akad. Nauk Azerb. SSR, ser. Fiz.–Tekh. Mat. Nauk* **8** (1985), no.6, 8 – 13.
[38] F. Kh. Muradov, Superassociative algebras of continuous multiplace functions, (Russian), *Izv. Akad. Nauk Azerb. SSR, ser. Fiz.–Tekh. Mat. Nauk* **10** (1989), no.2 – 3, 3 – 8.
[39] W. Nöbauer, W. Philipp, Über die Einfachheit von Funktionenalgebren, *Monatsh. Math.* **66** (1962), 441 – 452.
[40] W. Nöbauer, W. Philipp, Die Einfachheit der mehrdimensionalen Funktionenalgebren, *Arch. Math.* **15** (1964), 1 – 5.
[41] E. Redi, Representation of Menger systems by multiplace endomorphisms, (Russian), *Tartu Riikl. Ul. Toimetised* (*Uchen. Zapiski Tartu Gos. Univ.*) **277** (1971), 47 – 51.
[42] B. M. Schein, Theory of semigroups as a theory of superpositions of multilace functions, (Russian), *Interuniv. Sci. Sympos. General Algebra, Tartu* 1966, 169 – 192, (Izdat. Tartu Gos. Univ.)
[43] B. M. Schein, V. S. Trokhimenko, Algebras of multiplace functions, *Semigroup Forum* **17** (1979), 1 – 64.
[44] B. Schweizer, A. Sklar, A Grammar of functions, I, *Aequat. Math.* **2** (1968), 62 – 85.
[45] H. L. Skala, Grouplike Menger algebras, *Fund. Math.* **79** (1973), 199 – 207.
[46] F. N. Sokhatsky, An abstract characterization of $(2, n)$-semigroups of n-ary operations, (Russian), *Mat. Issled.* **65** (1082), 132 – 139.
[47] V. S. Trokhimenko, Ordered algebras of multiplace functions, (Russian), Izv. Vyssh. Uchebn. Zaved. Matematika **104** (1971), 90 – 98.
[48] V. S. Trokhimenko, Restrictive algebras of multiplace functions, (Ukrainian), Dopovidi Akad. Nauk Ukrain. RSR, ser. A (1972), 340 – 342.
[49] V. S. Trokhimenko, Menger's function systems, (Russian), Izv. Vyssh. Uchebn. Zaved. Matematika **11** (1973), 71 – 78.
[50] V. S. Trokhimenko, A characterization of $\mathcal{P}$-algebras of multiplace functions, Sib. Math. J. **16** (1975), 461 – 470 (1976) (translation from Sibir. Matem. Zhurn. **16** (1975), 599 – 611).
[51] V. S. Trokhimenko, Some algebras of Menger relations, (Russian), Izv. Vyssh. Uchebn. Zaved. Matematika **2(189)** (1978), 87 – 95.
[52] V. S. Trokhimenko, Stationary subsets of ordered Menger algebras, (Russian), Izv. Vyssh. Uchebn. Zaved. Matematika **8** (1983), 82 – 84.
[53] V. S. Trokhimenko, Stabilizers of ordered Menger algebras, (Russian), Izv. Vyssh. Uchebn. Zaved. Matematika **6** (1986), 74 – 76.

[54] V. S. Trokhimenko, On some Menger algebras of multiplace transformations of ordered sets, *Algebra Universalis* **33** (1995), 375 – 386.
[55] V. S. Trokhimenko, v-regular Menger algebras, *Algebra Universalis* **38** (1997), 150 – 164.
[56] V. S. Trokhimenko, Stationary subsets and stabilizers of restrictive Menger P-algebras of multiplace functions, Algebra Universalis **44** (2000), 129 – 142.
[57] V. S. Trohimenko, B. M. Schein, Compatible algebras of multiplace functions, (Russian), Teor. Polugrupp Prilozh. **4** (1978), 89 – 98, (Izdat. Saratov. Gos. Univ.)
[58] H. Whitlock, A composition algebra for multiplace functions, *Math. Ann.* **157** (1964), 167 – 178.
[59] T. Yakubov, On $(2, n)$-semigroups of n-ary operations, (Russian), *Bull. Akad. Ştiinţe SSR Moldov.* **1** (1974), 29 – 46.
[60] Ja. N. Yaroker, Complete simple Menger operatives, (Ukrainian), *Dopovidi Akad. Nauk Ukrain. RSR, ser.* **A** (1972), 34 – 36.

Institute of Mathematics and Computer Science, Wrocław University of Technology, Wybrzeże Wyspiańskiego 27, 50-370 Wrocław, Poland
E-mail address: wieslaw.dudek@pwr.wroc.pl

Department of Mathematics, Pedagogical University, 21100 Vinnitsa, Ukraine
E-mail address: vtrokhim@gmail.com

A General Type of Regularity and Semiprime Ideals in Ternary Semiring

T. K. Dutta* **and S. Mandal**[†]

Department of Pure Mathematics, University of Calcutta.

35, Ballygunge Circular Road, Kolkata - 700019, India.

Abstract

We extend the notion of a general type of regularity of rings to the general setting of ternary semirings. Also, we use this concept of general type of regularity in semirings to introduce a general semiprime ideals of ternary semirings.

AMS Mathematics Subject Classification (2000): 16Y60.

Keywords: Von Neumann regularity; λ-regularity; D-regularity; f-regularity; λ-regular semiring; D-regular semiring; f-regular semiring; D-semiprime ideal; λ-semiprime ideal; f-semiprime ideal; λ_p-system; D_p-system; f_p-system.

1 Introduction

In ring theory many so-called regularities appear, namely the Von Nuemann regularity, D-regularity, λ-regularity, f-regularity etc. The oldest one seems to be the von Neumann regularity. Von Neumann (1936) defined a ring R (with identity) to be regular if for each element r of R there exists an element s of R such that $r = rsr$. Brown and McCoy (1950) generalized this notion to ring without identity, and they succeeded in proving that every ring contains a greatest regular ideal. There had always been an attempt to unify all these regularities. Roos [11] introduced the notion of a general type of regularity for rings which

E-mail :** ***duttatapankumar@yahoo.co.in

[†]**E-mail :** ***shobhan139@gmail.com*****(thanks to CSIR, India for financial support)**

includes each of the above mentioned regularities as a particular case. It is known that I is a semiprime ideal of a ring R if and only if $aRa \subseteq I$ implies $a \in I$. Now if we define $F_R(a) = aRa$ for all $a \in R$. Then the F_R, as defined above, gives rise to a type of regularity called Von Neumann regularity or simply regularity. Taking this as a motivation, Buys and De La Rosa [10] generalized the concept of semiprime ideals in rings by using different types of regularities in the following manner. They defined an ideal I of a ring R to be F-semiprime if $F_R(a) \subseteq I$ implies $a \in I$ where $\{F_R$: R is a ring$\}$ is a regularity(definition follows) for rings.

After these introductory historical remarks we now come to a brief description of the substance of this paper. The literature of the theory of ternary operations is vast and scatter over diverse areas of mathematics. The notion of ternary semiring is introduced by T. K. Dutta and S. Kar [1] in the year 2003 as a natural generalization of ternary ring introduced by W. G. Lister [7] in 1971. In 2009, S. K. Sardar and B. C. Saha [12] introduced and studied general type of regularity and semiprime ideals in semirings. We introduce here a general type of regularity for ternary semirings which is analogous to that for rings. We obtain characterization of some types of regular ternary semirings viz., λ-regular, D-regular and f-regular ternary semirings. We then use this concept of general type of regularity for ternary semirings to introduce the notion of generalized semiprime ideals in ternary semirings and we obtain some characterizations of general type of semiprime ideals.

2 Preliminaries

Definition 2.1 A non-empty set S together with a binary operation, called addition and a ternary multiplication, denoted by juxtaposition, is said to be a ternary semiring if S is an additive commutative semigroup satisfying the following conditions :

(*i*) $(abc)de = a(bcd)e = ab(cde)$;

(*ii*) $(a+b)cd = acd + bcd$,

(*iii*) $a(b+c)d = abd + acd$,

(*iv*) $ab(c+d) = abc + abd$ for all $a, b, c, d, e \in S$.

Definition 2.2 [1] Let S be a ternary semiring. If there exists an element $0 \in S$ such that $0 + x = x$ and $0xy = x0y = xy0 = 0$ for all $x, y \in S$ then '0' is called the zero element or simply the zero of the ternary semiring S. In this case we say that S is a ternary semiring with zero.

Throughout the paper S will always denote a ternary semiring with zero and $S^* = S\backslash\{0\}$.

We note that a ternary semiring may not contain an identity but there are certain ternary semirings which generate identities in the sense defined below :

Definition 2.3 [5] A ternary semiring S admits an identity provided that there exist elements $\{(e_i, f_i) \in S \times S \ (i = 1, 2, ..., n)\}$ such that $\sum_{i=1}^{n} e_i f_i x = \sum_{i=1}^{n} e_i x f_i = \sum_{i=1}^{n} x e_i f_i = x$ for all $x \in S$. In this case, the ternary semiring S is said to be a ternary semiring with identity $\{(e_i, f_i) : i = 1, 2, ..., n\}$.

In particular, if there exists an element $e \in S$ such that $eex = exe = xee = x$ for all $x \in S$, then 'e' is called a unital element of the ternary semiring S.

It is easy to see that $xye = (exe)ye = ex(eye) = exy$ and $xye = x(eye)e = xe(yee) = xey$ for all $x, y \in S$. So we have the following result :

Proposition 2.4 *If e is a unital element of a ternary semiring S, then $exy = xey = xye$ for all $x, y \in S$.*

Definition 2.5 [1] An additive subsemigroup T of a ternary semiring S is called a ternary subsemiring if $t_1t_2t_3 \in T$ for all $t_1, t_2, t_3 \in T$.

Definition 2.6 [1] An additive subsemigroup I of a ternary semiring S is called a left (right, lateral) ideal of S if s_1s_2i (respectively is_1s_2, s_1is_2) $\in I$ for all $s_1, s_2 \in S$ and $i \in I$. If I is a left, a right, a lateral ideal of S, then I is called an ideal of S.

Proposition 2.7 [1] Let S be a ternary semiring and $a \in S$. Then the principal

(i) left ideal generated by 'a' is given by $< a >_l = SSa + na$

(ii) right ideal generated by 'a' is given by $< a >_r = aSS + na$

(iii) two-sided ideal generated by 'a' is given by $< a >_t = SSa + aSS + SSaSS + na$

(iv) lateral ideal generated by 'a' is given by $< a >_m = SaS + SSaSS + na$

(v) ideal generated by 'a' is given by $< a > = SSa + aSS + SaS + SSaSS + na$,

where $n \in \mathbb{Z}_0^+$ (set of all positive integers with zero).

Definition 2.8 An ideal I of a ternary semiring S is called a k-ideal if $x + y \in I$; $x \in S$, $y \in I$ imply that $x \in I$.

Definition 2.9 [3] A proper ideal P of a ternary semiring S is called a prime ideal of S if for any three ideals A, B, C of S; $ABC \subseteq P$ implies $A \subseteq P$ or $B \subseteq P$ or $C \subseteq P$.

Definition 2.10 [3] A ternary semiring S is called a prime ternary semiring if the zero ideal (0) is a prime ideal of S.

Definition 2.11 [4] A proper ideal Q of a ternary semiring S is called a semiprime ideal of S if for any three ideals I of S; $I^3 \subseteq Q$ implies $I \subseteq Q$.

3 F-regularity in ternary semirings

Definition 3.1 For every ternary semiring S, let us associate a mapping $F_S : S \to G(S, +)$, where $G(S, +)$ denotes the set of all submonoids of $(S, +)$. Then
$\{F_S : S$ is a ternary semiring $\}$ is called a regularity for ternary semirings if the following conditions are satisfied :

(a) if $\Phi : S \to T$ is a ternary semiring epimorphism then $F_T(\Phi(a)) = \Phi(F_S(a))$ for all $a \in S$,

(b) if A is an ideal of S and $a \in A$, then $F_A(a) \subseteq F_S(a)$,

and (c) if $r, s \in S$ and $s \in F_S(r)$, then $F_S(r + s) \subseteq F_S(r)$

Theorem 3.2 *Let S be a ternary semiring and $F_S : S \to G(S, +)$ be defined by*
(i) $F_S(a) = aSa$
(ii) $F_S(a) = SSaSS$
(iii) $F_S(a) = aSS$
(iv) $F_S(a) = SSa$
(v) $F_S(a) = SSaSaSS$
(vi) $F_S(a) = SSaSaSaSS$
Then in each case F_S gives rise to a regularity for ternary semirings.

Proof : (i) (a) Let $\phi : S \to T$ be a ternary semiring epimorphism. Then $F_T(\phi(a)) = \phi(a)T\phi(a) = \phi(a)\phi(S)\phi(a) = \phi(aSa) = \phi(F_S(a))$.

(b) Let A be an ideal of S and $a \in A$. Then $F_A(a) = aAa \subseteq aSa = F_S(a)$

(c) Let $r, s \in S$ and $s \in F_S(r)$. Then there exists $t \in S$ such that $s = rtr$. Suppose $x \in F_S(r+s)$. Then there exists $w \in S$ such that

$x = (r+s)w(r+s)$

$= (r+rtr)w(r+rtr)$

$= rwr + rtrwr + rwrtr + rtrwrtr$

$= rwr + r(trw)r + r(wrt)r + r(trwrt)r \in rSr$. Consequently, $F_S(r+s) \subseteq F_S(r)$. Thus F_S gives rise to a regularity for ternary semirings.

Similarly we can prove (ii) to (iv).

Definition 3.3 If $\{F_S : S$ is a ternary semiring $\}$ is a regularity for ternary semirings, then an element a of a ternary semiring S is called F_S-regular if $a \in F_S(a)$. A ternary semiring S is called F-regular if each element of S is F_S-regular. An ideal A of S is called F_S-regular if each a of A is F_S-regular and F-regular if each a of A is F_A-regular.

Note 3.4 *For any ideal A of a ternary semiring S, $F_A(a) \subseteq F_S(a)$ for all $a \in A$. Thus if A is F-regular then it is F_S-regular.*

Drawing analogy with the terminology of ring theory we give the following definitions.

Definition 3.5 The regularities in theorem (i), (ii), (iii), (iv), (v) and (vi) are respectively called Von Neumann regularity or simply regularity, λ-regularity, right D-regularity, left D-regularity, weak f-regularity, f-regularity in ternary semirings.

Theorem 3.6 *Let S be a ternary semiring which is right D-regular and left D-regular. Then S is λ-regular.*

Proof : Let $a \in S$. As S is right and left D-regular, there exist $s_1, s_2, t_1, t_2 \in S$ such that $a = as_1s_2$ and $a = t_1t_2a$. So, $a = as_1s_2 = t_1t_2as_1s_2 \in SSaSS$. Thus S is λ-regular. ♯

Proposition 3.7 *If an element of a ternary semiring S is regular then it is (i) λ-regular, (ii) right(left) D-regular, (iii) weak f-regular (iv) f-regular.*

Proof : Let a be a regular element of S. Then there exist $s \in S$ such that $a = asa$. (1)
Hence $a = asa = (asa)sa \in SSaSS$. Consequently, a is a λ-regular element of S.

Again from (1), we have $a = asa \in aSS$. Consequently, a is a right D-regular element of S. Also $a = asa \in SSa$. Consequently, a is a left D-regular element of S.

Also from (1), we have $a = asa = (asa)sa = asas(asa) \in SSaSaSS$. Consequently, a is a weak f-regular element of S.

Lastly from (1), we have $a = asa = (asa)sa = asas(asa) = (asa)sasasa \in SSaSaSaSS$. Consequently, a is a f-regular element of S. ♯

Proposition 3.8 *If an element of a ternary semiring S is f-regular, then it is weak f-regular.*

Proof : Follows from definition. ♯

Theorem 3.9 *Let S be a ternary semiring. Then the following statements are equivalent:*
(i) S is λ-regular.
(ii) $I = SSISS$ for any two sided ideal I of S.
(iii) $K \subseteq KSS$ and $J \subseteq SSJ$ for any left ideal K and right ideal J of S.
(iv) $\langle a \rangle_t = SSaSS$ for any element a of S, where $\langle a \rangle_t$ denotes the two sided ideal generated by a.

Proof : (i) $\Rightarrow$ (ii) Suppose (i) holds. Let I be a two sided ideal of S and $a \in I$. Since S is λ-regular, $a \in SSaSS \subseteq SSISS$. So $I \subseteq SSISS$. Again since I is an ideal, $SSISS \subseteq I$. Hence $I = SSISS$.
(ii) $\Rightarrow$ (iii) Let (ii) hold. Suppose K is a left ideal of S. Then $K + KSS$ is a two sided ideal of S. Then by (ii), $K + KSS = SS(K + KSS)SS \subseteq KSS$ whence $K \subseteq KSS$. Similarly, $J \subseteq SSJ$ for any right J of S.
(iii) $\Rightarrow$ (iv) Suppose (iii) holds. Let $a \in S$. Since $\langle a \rangle_t$ is a two sided ideal of S, by repeated use of (iii), we obtain $\langle a \rangle_t \subseteq SS\langle a \rangle_t SS = SS(na + SSa + aSS + SSaSS)SS \subseteq SSaSS$. Now by definition of $\langle a \rangle_t$, $SSaSS \subseteq \langle a \rangle_t$. Hence $\langle a \rangle_t = SSaSS$.
(iv) $\Rightarrow$ (i) Suppose (iv) holds. Then by definition of $\langle a \rangle_t$, $a \in \langle a \rangle_t$. Hence by (iv), $a \in SSaSS$. Consequently, S is λ-regular. ♯

Theorem 3.10 *Let S be a ternary semiring. Then the following statements are equivalent:*
(i) S is right D-regular.

(ii) $I = ISS$ *for any right ideal* I *of* S.

(iii) $\langle a \rangle_r = aSS$ *for any element* a *of* S, *where* $\langle a \rangle_r$ *denotes the right ideal generated by* a.

Proof : Proof is similar to above theorem and so we omit it. ♯

Theorem 3.11 *Let* S *be a ternary semiring with a unital element* e. *Then the following statements are equivalent:*

(i) S *is* f*-regular;*

(ii) Every ideal of S *is semiprime;*

(iii) $I = I^3$ *for any ideal* I *of* S*;*

(iv) $IJK = I \cap J \cap K$ *for any three ideal* I, J, K *of* S*;*

(v) $\langle a \rangle = \langle a \rangle^3$ *for any element* a *of* S*;.*

(vi) $\langle a \rangle = SSaSaSaSS$ *for any element* a *of* S, *where* $\langle a \rangle$ *denotes the ideal generated by* a.

Proof :$(i) \Rightarrow (ii)$ Let S be an f-regular ternary semiring and I be an ideal of S. Suppose for any ideal A of S, $A^3 \subseteq I$. Let $a \in A$. Now as S is an f-regular ternary semiring, $a \in SSaSaSaSS \subseteq A^3 \subseteq I$. Thus $A \subseteq I$. So I is a semiprime ideal of S.

$(ii) \Rightarrow (iii)$ Let I be an ideal of S. Then I^3 is an ideal of S and so by (ii), it is a semiprime ideal of S. Hence from $I^3 \subseteq I^3$, we have $I \subseteq I^3$. Also as I is an ideal, $I^3 \subseteq I$. Consequently, $I = I^3$.

$(iii) \Rightarrow (iv)$ Let I, J, K be three ideals of S. Then by (iii), $I \cap J \cap K = (I \cap J \cap K)^3 \subseteq IJK$. Also $IJK \subseteq I \cap J \cap K$. Hence $IJK = I \cap J \cap K$

$(iv) \Rightarrow (v)$ Taking $I = \langle a \rangle$, $J = \langle a \rangle$, $K = \langle a \rangle$, we obtain from (iv), $\langle a \rangle = \langle a \rangle^3$

$(v) \Rightarrow (vi)$ Let $a \in S$. From (v), we have $\langle a \rangle = \langle a \rangle^3 = \langle a \rangle^5 \subseteq SS\langle a \rangle SS$. Now as S contains a unital element e, by definition of $\langle a \rangle$, we obtain $SS\langle a \rangle SS = SS(na + aSS + SSa + SSaSS + SaS)SS = SS(na + aSS + SSa + SSaSS + SeaeS)SS \subseteq SS(na + aSS + SSa + SSaSS + SSaSS)SS \subseteq SSaSS$. Again by definition of $\langle a \rangle$, we have $SSaSS \subseteq \langle a \rangle$. Hence

$$\langle a \rangle = SSaSS. \tag{1}$$

Now using (v) and (1), we obtain $\langle a \rangle = \langle a \rangle^3 = \langle a \rangle^5 \subseteq \langle a \rangle S \langle a \rangle S \langle a \rangle \subseteq (SSaSS)S(SSaSS) S(SSaSS) \subseteq SSaSaSaSS$. Further by definition of $\langle a \rangle$, we have $SSaSaSaSS \subseteq \langle a \rangle$. Hence $\langle a \rangle = SSaSaSaSS$.

$(iv) \Rightarrow (i)$ Suppose (vi) holds. Then by definition of $\langle a \rangle$, $a \in \langle a \rangle$. Hence by (vi), $a \in SSaSaSaSS$. Consequently, by definition of f-regularity, S is f-regular. ♯

Definition 3.12 [8] A proper ideal P of a ternary semiring S called irreducible if $I \cap J \cap K = P \Rightarrow I = P$ or $J = P$ or $K = P$ for any three ideals $I,\ J,\ K$ of S.

Definition 3.13 [8] A proper ideal P of a ternary semiring S called strongly irreducible if $I \cap J \cap K \subseteq P \Rightarrow I \subseteq P$ or $J \subseteq P$ or $K \subseteq P$ for any three ideals $I,\ J,\ K$ of S.

Definition 3.14 [8] A ternary semiring S is called fully idempotent if each ideal I of S is idempotent i.e. $I^3 = I$.

Definition 3.15 A lattice $\mathcal{L}$ is called Brouwerian if, for any $a, b \in \mathcal{L}$, the set of all $x \in \mathcal{L}$ satisfying $a \wedge x \leq b$ contains a greatest element c, which is called the pseudo-complement of a relative to b.

Proposition 3.16 *The following assertions for a ternary semiring S are equivalent :*
(1) S is fully idempotent;
(2) For any three ideals $I,\ J,\ K$ of S, $I \cap J \cap K = IJK$.

Proof : (1) $\Rightarrow$ (2) Let S be a fully idempotent ternary semiring. Let $I,\ J,\ K$ are ideals of S, then $I \cap J \cap K$ is an ideal of S. As S is fully idempotent $I \cap J \cap K = (I \cap J \cap K)^3 \subseteq IJK$. Also we have, $IJK \subseteq I \cap J \cap K$. Hence $I \cap J \cap K = IJK$.
(2) $\Rightarrow$ (1) Let (2) hold. Then for each ideals I, we have $I \cap I \cap I = III$ i.e. $I = I^3$. Thus S is a fully idempotent ternary semiring.

Proposition 3.17 *If S is a fully idempotent ternary semiring, then the ideal lattice $\mathcal{L}_S$ of S is a complete Brouwerian lattice.*

Proof : Clearly, $\mathcal{L}_S$ is a complete lattice under the sum and intersection of ideals. Let B and C be two ideals of S. By Zorn's lemma, there is an ideal M of S which is maximal in the family of ideals I satisfying $B \cap I \subseteq C$. Thus if I is any such ideal then $BBI = B \cap B \cap I = B \cap I \subseteq C$, and moreover, $BB(I + M) \subseteq BBI + BBM \subseteq C$. Hence $B \cap (I + M) \subseteq C$, by the above proposition. By the maximality of M we get $I + M = M$, and therefore $I \subseteq M$. Thus $\mathcal{L}_S$ is a complete Brouwerian lattice. ♯

Proposition 3.18 *If S is a fully idempotent ternary semiring, then the ideal lattice $\mathcal{L}_S$ of S is distributive i.e. if $A,\ B,\ C,\ D$ are four ideals of S then $A + (B \cap C) = (A + B) \cap (A + C)$.*

Proof : Let S be a fully idempotent ternary semiring and A, B, C are three ideals of S. Now, we have $A + (B \cap C) = [A + (B \cap C)]^3 \subseteq (A + B)(A + B)(A + C) = (A + B) \cap (A + B) \cap (A + C) = (A + B) \cap (A + C)$, by Proposition 3.16. Again by Proposition 3.16, $(A+B)\cap(A+C) = (A+B)\cap(A+B)\cap(A+C) = (A+B)(A+B)(A+C) \subseteq AAA+AAC+ABA+ABC+BAA+BAC+BBA+BBC \subseteq A+BBC \subseteq A+(B\cap B\cap C) = A+(B\cap C)$. Thus the ideal lattice $\mathcal{L}_S$ of S is distributive. ♯

Proposition 3.19 *Let S be a fully idempotent ternary semiring. Then the following assertions for an ideal P of S are equivalent:*

(1) *P is irreducible;*

(2) *P is prime.*

Proof : As (2) $\Rightarrow$ (1) is clear from definition, it suffices to show that (1) $\Rightarrow$ (2). Suppose $IJK \subseteq P$ for ideals I, J, K of S. Hence $I \cap J \cap K \subseteq P$, by Proposition 3.16. Thus it follows that $(I \cap J \cap K) + P = P$. Since the ideal lattice of S is distributive by Corollary 2 of the above proposition, we have $P = (I \cap J \cap K) + P = (I + P) \cap (J + P) \cap (K + P)$. Since P is irreducible, so $I + P = P$ or $J + P = P$ or $K + P = P$. This implies that $I \subseteq P$ or $J \subseteq P$ or $K \subseteq P$. Hence P is a prime ideal. ♯

Theorem 3.20 *Let S be a ternary semiring. Then the following assertions are equivalent:*

(1) *S is fully idempotent;*

(2) *each proper ideal of S is the intersection of prime ideals which contain it.*

Proof : (1) $\Rightarrow$ (2). Let I be a proper ideal of S and let $\{P_\alpha : \alpha \in \Lambda\}$ (Λ is an index set) be a family of prime ideals of S which contain I. Clearly $I \subseteq \bigcap_{\alpha\in\Lambda} P_\alpha$. To prove the converse, suppose $a \notin I$. By Zorn's lemma, there exists an ideal P such that P is proper, $I \subseteq P$, $a \notin P$, and P is maximal with these properties. Then P is irreducible. For, suppose on the contrary, $P = J \cap K \cap L$ and J, K and L properly contain P. Then each of the ideals J, K and L contains a. Hence $a \in J \cap K \cap L = P$. This contradicts the assumption that $a \notin P$. Hence P is irreducible, and so it is prime by Proposition 3.19. This establishes the existence of a prime ideal P such that $a \notin P$ and $I \subseteq P$. Hence $a \notin \bigcap_{\alpha\in\Lambda} P_\alpha$. This implies that $\bigcap_{\alpha\in\Lambda} P_\alpha \subseteq I$. Hence $\bigcap_{\alpha\in\Lambda} P_\alpha = I$.

(2) $\Rightarrow$ (1). Let I be any ideal of S. If $I^3 = S$ then I is certainly idempotent, since $S = I^3 \subseteq I \subseteq S \Rightarrow I = S \Rightarrow I^3 = S = I$. If $I^3 \neq S$ then I^3 is a proper ideal of S, and so

it is the intersection of prime ideals P_α of S, which contains I^3, by our assumption. Hence $I^3 = \bigcap_{\alpha \in \Lambda} P_\alpha \subseteq P_\alpha$, for each α. This implies that $I \subseteq P_\alpha$ for every α, since P_α is a prime ideal. Thus we have $I \subseteq \bigcap_{\alpha \in \Lambda} P_\alpha = I^3$. Hence $I = I^3$, and so S is fully idempotent. ♯

Corollary 3.21 *Let S be a f-regular ternary semiring. Then the ideal lattice $\mathcal{L}_S$ of S is a complete Brouwerian lattice.*

Proof : By Theorem 3.11 (*iii*), S is fully idempotent. Hence by Proposition 3.17, $\mathcal{L}_S$ is a complete Brouwerian lattice. ♯

Corollary 3.22 *Let S be a f-regular ternary semiring. Then an ideal I of S is irreducible if and only if I is prime.*

Proof : By Theorem 3.11 (*iii*), S is fully idempotent. Hence by Proposition 3.19 the result follows. ♯

Corollary 3.23 *Let S be a ternary semiring. Then S is f-regular if and only if every proper ideal of S is the intersection of prime ideals which contains it.*

Proof: By Theorem 3.11(*iii*), S is fully idempotent. Hence by Proposition 3.20 the result follows. ♯

4 F-semiprime ideals

Definition 4.1 Let $\{F_S : S \text{ is a ternary semiring }\}$ be a regularity for ternary semirings. An ideal I of a ternary semiring S is called F-semiprime if for $r \in S$ and $F_S(r) \subseteq I$ implies $r \in I$.

Theorem 4.2 *Let $\{F_S : S$ is a ternary semiring $\}$ be a regularity for ternary semirings such that $F_S(a)$ is an ideal of S. Then S is F-regular if and only if every ideal of S is F-semiprime.*

Proof : Let every ideal of S be F-semiprime. Suppose $a \in S$. By hypothesis $F_S(a)$ is an ideal of S, so it is F-semiprime. Also $F_S(a) \subseteq F_S(a)$. Hence by definition of F-semiprime ideal, $a \in F_S(a)$. Consequently, S is F-regular.

Conversely, let S be an F-regular ternary semiring. Then for any $a \in S$, $a \in F_S(a)$. Hence for any ideal I of S and for any $a \in S$ with $F_S(a) \subseteq I$, we see that $a \in I$. Hence every ideal of S is F-semiprime. ♯

Notation 4.3 *Let $\{F_S : S$ is a ternary semiring $\}$ be a regularity for ternary semirings. Then for any subset A of S, let us denote ASA by $F_S(A)$ where F_S is defined by (i) of Theorem 3.2. Similar is the meaning for F_S's defined in (ii), (iii), (iv), (v) and (vi).*

Theorem 4.4 *Let S be a ternary semiring with a unitary element e and I be an ideal of S. Then the following statements are equivalent:*
(a) I is an F-semiprime ideal of S.
(b) For an ideal J of S, $F_S(J) \subseteq I \Rightarrow J \subseteq I$.
(c) For $a \in S$, $F_S(\langle a \rangle) \subseteq I \Rightarrow a \in I$ ($\langle a \rangle$ is the ideal generated by a) where is a regularity for ternary semirings as defined in the Theorem 3.2 and meaning of is as explain in the Notation 4.3.

Proof : (a) $\Rightarrow$ (b). Let I be an F-semiprime ideal of the ternary semiring S and J be an ideal of S such that $F_S(J) \subseteq I$. Let $a \in J$. Then $F_S(a) \subseteq I$. Since I is F-semiprime, $a \in I$. Hence $J \subseteq I$.
(b) $\Rightarrow$ (c). Let (b) hold. Let $J = \langle a \rangle$. Then (c) follows immediately.
(c) $\Rightarrow$ (a). Suppose (c) hold. Let $F_S(a) \subseteq I$ where $a \in S$, Then in view of definition of $\langle a \rangle$ and definition of F_S's (cf. Theorem 3.2), $F_S(\langle a \rangle) \subseteq I$. Hence by (c), $a \in I$. Consequently, I is an F-semiprime ideal of S. ♯

Definition 4.5 A nonempty subset A of a ternary semiring S is called a λ_p-system if $a \in A$ implies that there exist $s_1, s_2, t_1, t_2 \in S$ such that $s_1 s_2 a t_1 t_2 \in A$.

Theorem 4.6 *An ideal I of a ternary semiring S is λ-semiprime if and only if I^c, the complement of I in S, is a λ_p-system.*

Proof : Suppose S is ternary semiring and I be a λ-semiprime ideal of S. Let $a \in I^c$. Then $a \notin I$ and hence $SSaSS \not\subseteq I$. Thus there exist $s_1, s_2, t_1, t_2 \in S$ such that $s_1 s_2 a t_1 t_2 \notin I$ i.e. $s_1 s_2 a t_1 t_2 \in I^c$. Hence I^c is a λ_p-system.

Conversely, let I^c be a λ_p-system in S. Suppose $a \in S$ such that $SSaSS \subseteq I$. (1)
If possible, let $a \notin I$ i.e. $a \in I^c$. Hence there exist $s_1, s_2, t_1, t_2 \in S$ such that $s_1 s_2 a t_1 t_2 \in I^c$ whence $s_1 s_2 a t_1 t_2 \notin I$. This is a contradiction to (1). Thus I is a λ-semiprime ideal of S. ♯

Theorem 4.7 *[6] Let S be a ternary semiring and S_n be the ternary semiring of matrices of order $n \times n$ over S. Then I is an ideal of S_n if and only if $I = J_n$ for some ideal J of S.*

We use this theorem to obtain the following result wherein it has been shown that λ-semiprimeness is a Morita invariant i.e. it remains invariant while passing to the matrix semiring.

Theorem 4.8 *Let S be a ternary semiring. Then any ideal I of S is λ-semiprime if and only if I_n is λ-semiprime ideal of S_n, where S_n is the ternary semiring of matrices of order n over S and I_n has similar meaning.*

Proof : Let I be a λ-semiprime ideal of S and $A = \sum\limits_{i,j=1}^{n} a_{ij}E_{ij}($ where $a_{ij} \in S$ and $a_{ij}E_{ij}$ is the $n \times n$ matrix of which $(i,j)^{th}$ element is a_{ij} and the others are zero) such that $S_nS_nAS_nS_n \subseteq I_n$. If possible, let $A \notin I_n$. Then there exists an $a_{kl} \in A$ such that $a_{kl} \notin I$. Since I is a λ-semiprime ideal of S, $SSa_{kl}SS \nsubseteq I$. Hence there exist $s_1, s_2, t_1, t_2 \in S$ such that $s_1s_2at_1t_2 \notin I$. Then $(s_1s_2at_1t_2)E_{lk} \notin I_n$ i.e. $(s_1E_{ll})(s_2E_{lk})A(t_1E_{lk})(t_2E_{kk}) \notin I_n$ This is a contradiction to $S_nS_nAS_nS_n \subseteq I_n$. Hence $A \in I_n$. This prove that I_n is a λ-semiprime.

Conversely, let I be an ideal of S such that is a λ-semiprime ideal of S_n. Let $a \in S$ such that $SSaSS \subseteq I$ and $A = \sum\limits_{i,j=1}^{n} aE_{ij} \in S_n$. If $\Delta \in S_nS_nAS_nS_n$ then Δ is of the form $(\sum\limits_{i,j=1}^{n} x_{ij}E_{ij})(\sum\limits_{i,j=1}^{n} y_{ij}E_{ij})(\sum\limits_{i,j=1}^{n} aE_{ij})(\sum\limits_{i,j=1}^{n} s_{ij}E_{ij})(\sum\limits_{i,j=1}^{n} t_{ij}E_{ij})$ which is the sum of matrices of the form $(\sum\limits_{i,j,k,l=1}^{n} x_{pi}y_{ij}as_{kl}t_{lq})E_{pq}$. Since $SSaSS \subseteq I$, $x_{pi}y_{ij}as_{kl}t_{lq} \in I$ for all $i,j,k,l = 1,2,\ldots,n$. Hence $\sum\limits_{i,j,k,l=1}^{n} x_{pi}y_{ij}as_{kl}t_{lq} \in I$. This implies that $S_nS_nAS_nS_n \subseteq I_n$. Hence $A \in I_n$ and so $a \in I$. This proves that I is a λ-semiprime ideal of S. ♯

Theorem 4.9 *Let S be a ternary semiring. Then any ideal I of S is right (left) D-semiprime if and only if I_n is right (left) D-semiprime ideal in the ternary semiring S_n.*

Proof : The proof is analogous to that of the above theorem. ♯

Theorem 4.10 *Let S be a ternary semiring. Then any ideal I of S is (weak) f-semiprime if and only if I_n is (weak) f-semiprime ideal in the ternary semiring S_n.*

Proof : The proof is analogous to that of the above theorem. ♯

Definition 4.11 A nonempty subset A of a ternary semiring S is called a right D_p-system if $a \in A$ implies that there exist $s_1, s_2 \in S$ such that $as_1s_2 \in A$.

Similar is the definition of left D_p-system.

Theorem 4.12 *An ideal I of a semiring S is right(left) D-semiprime if and only if I^c, the complement of I in S, is a right(left) D_p-system .*

Proof : Follows from definitions of right(left) D-semiprime and right(left) D_p-system. ♯

Definition 4.13 A nonempty subset A of a ternary semiring S is called a weak f_p-system if $a \in A$ implies that there exist $s_1, s_2, t, t_1, t_2 \in S$ such that $s_1s_2atat_1t_2 \in A$.

Theorem 4.14 *An ideal I of a ternary semiring S is weak f-semiprime if and only if I^c is a weak f_p-system.*

Proof : Follows from definitions of weak f-semiprime and weak f_p-system. ♯

Definition 4.15 A nonempty subset A of a ternary semiring S is called a f_p-system if $a \in A$ implies that there exist $s_1, s_2, r, t, t_1, t_2 \in S$ such that $s_1s_2aratat_1t_2 \in A$.

Theorem 4.16 *An ideal I of a ternary semiring S is f-semiprime if and only if I^c is a f_p-system.*

Proof : Follows from definitions of f-semiprime and f_p-system. ♯

References

[1] Dutta, T. K. and Kar, S. : *On Regular Ternary Semirings*; Advances in Algebra, Proceedings of the ICM Satellite Conference in Algebra and Related Topics, World Scientific (2003), 343 - 355.

[2] Dutta, T. K. and Kar, S. : *A Note On Regular Ternary Semirings*; Kyungpook Mathematical Journal, Vo. 46, No. 3 (2006), 357 - 365.

[3] Dutta, T. K. and Kar, S. : *On Prime Ideals And Prime Radical Of Ternary Semirings*; Bull. Cal. Math. Soc., Vol. 97, No. 5 (2005), 445 - 454.

[4] Dutta, T. K. and Kar, S. : *On Semiprime Ideals And Irreducible Ideals Of Ternary Semirings*; Bull. Cal. Math. Soc., Vol. 97, No. 5 (2005), 467 - 476.

[5] Dutta, T. K. and Kar, S. : *On Ternary Semifields*; Discussiones Mathematicae - General Algebra and Applications, Vol. 24, No. 2 (2004), 185 - 198.

[6] Dutta, T. K. and Kar, S. : *On Matrix Ternary Semirings*; International Journal of Mathematics and Analysis, Vol. 1, No. 1 (2006), 81 - 95.

[7] Lister, W. G. : *Ternary Rings*; Trans. Amer. Math. Soc. 154 (1971), 37-55.

[8] Javed, Ahasan: *Fully Idempotent Semirings*; Proc. Japan Acad. 69. Ser. A 1993.

[9] C. Roos : *A class of regularities for rings*; J. Austral. Math.Soc.(Series A) 27 (1979), 437-453.

[10] De la Rosa, Buys. A.: *A radical class which is fully determined by a lattice isomorphism*; Acta Sci. Math(Syeged), 33(1972), 337-341.

[11] C. Roos : *A class of regularities for rings*; J. Austral. Math.Soc.(Series A) 27 (1979), 437-453.

[12] Sardar S. K. and Saha B. C. : *A General Type of Regularity and Semiprime Ideals in Semirings*; Southeast Asian Bulletin of Mathematics, Vol. 33(2009), 769-780.

Projective Cover in $\sigma[M]$

Fitriani

Mathematics Department, Universitas Lampung,
Bandar Lampung, Indonesia
E-mail:fitriani_mathunila@yahoo.co.id
www.unila.ac.id

Indah Emilia Wijayanti

Mathematics Department, Universitas Gadjah Mada,
Sekip Utara Yogyakarta, Indonesia
E-mail: ind_wijayanti@ugm.ac.id, ind_wijayanti@yahoo.com

Let M be an R-module and $N \in \sigma[M]$. A projective module P with a superfluous epimorphism $\pi : P \to N$ is called projective cover of N in $\sigma[M]$. Even if there are enough projective in $\sigma[M]$, a module need not have a projective cover. In this paper, we will investigate the properties of projective cover in $\sigma[M]$.

Keywords: M-projective module; superfluous epimorphism; projective cover.

1. Intoduction

Let M and P be R-modules. P is called M-projective if P relative projective to exact sequence $M \to N \to 0$, for every R-module N. Related to M-projective module, there are superfluous submodule, superfluous epimorphism and projective cover. Even if there are enough projectives in $\sigma[M]$, which is a full subkategory of the category of R-modules *R-MOD*, a module need not has a projective cover.

Projective cover have been discussed by several authors (see [1], [3], [4] and [6]) in the literature. Katayama [5] was observed a necessary and sufficient condition a module have a projective cover in category R-MOD which is related to the existence idempotent element in endomorphism rings.

In this paper, we give some properties of projective cover in $\sigma[M]$. Futhermore, we show that the necessary and sufficient condition a module have projective cover in *R-MOD* is valid in category $\sigma[M]$.

2. M-projective module

Let M and N be R-modules. The module N is called M-generated, or M is a generator of N, if there is an epimorphism from direct sum of M to N. The module N is called M-subgenerated, or M is a subgenerator of N, if N is isomorph with a submodule of an M-generated module. The category of M-subgenerated modules is denoted by $\sigma[M]$, which is a full subcategory of the category of R-modules R-MOD.

Let M be an R-module. M-projective module is defined by the following definition.

Definition 2.1. Let M and P be R-modules. P is called M-projective if every diagram in R-MOD with exact row

$$\begin{array}{c} P \\ \downarrow \\ M \to N \to 0 \end{array}$$

can be extended commutatively by a morphism $P \to M$. This condition is equivalent to the surjectivity of the map

$$Hom_R(P, g) : Hom_R(P, M) \to Hom_R(P, N), \forall N \in R - MOD.$$

for every epimorphism $g : M \to N$. If P is P-projective, then P is also called self- (or quasi-) projective

Definition 2.2. A submodule K of an R-module M is called superfluous or small in M, written $K \ll M$, if, for every submodule $L \subset M$, the equality $K + L = M$ implies $L = M$. An epimorphism $f : M \to N$ is called superfluous if $Ker f \ll M$.

Obviously $K \ll M$ if and only if the canonical projection $M \to M/K$ is a superfluous epimorphism.

Example 2.1. Any nil (left) ideal I in a ring R is superfluous as a left module. Assume $R = I + L$ for some left ideal $L \subset R$. Then $1 = i + l$, for suitable $i \in I, l \in L$, and hence, for some $k \in N$ we get $0 = i^k = (1 - l)^k = 1 + l$, for some $l \in L$, so that $1 \in L = R$.

The properties of superfluous submodules and superfluous epimorphisms are given by the following proposition.

Proposition 2.1. *Let K, L, N and M be R-modules.*

(1) If $f : M \to N$ and $g : N \to L$ are two epimorphisms, then $g \circ f$ is superfluous if and only if f and g are superfluous.

(2) If $K \ll L \ll M$, then $L \ll M$ if and only if $K \ll M$ and $L/K \ll M/K$.
(3) If $K_1, \cdots, K_n$ are superfluous submodules of M, then $K_1+K_2+\cdots+K_n$ is also superfluous in M.
(4) For $K \ll M$ and $f : M \to N$ we get $f(K) \ll N$.
(5) If $K \subset L \subset M$ and L is a direct summand in M, then $K \ll M$ if and only if $K \ll L$.

Proof.

(1) If f and g are superfluous epimorphisms and $h : M \to M$ is a morphism with $(g \circ f) \circ h$ epimorhism, then $f \circ h$ and also h is epimorhism, i.e $g \circ f$ is superfluous epimorphism. Let $g \circ f$ is superfluous epimorphism. If $h : M \to M$ is a morphism with $f \circ h$ is epimorphism, then $g \circ f \circ h$ is epimorphism and hence h is epimorpism. Therefore f is superfluous epimorphism. For any $k : N^{'} \to N$ with $g \circ k$ is epimorphism, we form the *pullback* diagram

$$\begin{array}{ccccc} Q & \xrightarrow{\pi_2} & N^{'} & & \\ \downarrow^{\pi_1} & & \downarrow^{k} & & \\ M & \xrightarrow{f} & N & \xrightarrow{g} & L \end{array}$$

With $g \circ k$ and $g \circ f \circ \pi_1 = g \circ k \circ \pi_2$ are epimorphisms. Therefore, π_1 is epimorphism ($g \circ f$ is superfluous epimorphism) and $k \circ \pi_2 = f \circ \pi_1$ implies k is epimorphism. Hence, g is superfluous epimorphism.

(2) Let $K \ll L \ll M$ and $L \ll M$. With the canonical mapping

$$M \xrightarrow{f} M/K \xrightarrow{g} M/L,$$

$g \circ f$ is superfluous epimorphism ($Ker(g \circ f) = L \ll M$) and hence f and g are superfluous epimorphisms. Therefore, $Ker\, f = K \ll M$ and $Ker\, g = L/K \ll M/K$. If $K \subset L \subset M$, $K \ll M$ and $L/K \ll M/K$, then with the canonical mapping

$$M \xrightarrow{f} M/K \xrightarrow{g} M/L$$

f and g are superfluous epimorphisms. Therefore, $g \circ f$ is superfluous epimorphism, i. e $Ker(g \circ f) = L \ll M$.

(3) This is shown by induction on the number of superfluous submodules of M. For $k = 1$, it is obvious. Assume this assertion is true for $n \in \mathbb{N}$. Let $K_1, K_2, , K_{n+1}$ are superfluous epimorphisms and $K_1 + K_2 + \cdots + K_n + K_{n+1} + L = M$, for some submodule $L \subset M$. By hypothesis, $K_1+K_2+\cdots+K_n$ are superfluous epimorphisms and hence $K_{n+1}+L =$

M. Since K_{n+1} is superfluous epimorphism, $K_{n+1} + L = M$ implies $L = M$ and the assertion is verified.

(4) Assume $X \subset N$ with $f(K) + X = N$. Then $M = K + f^{-1}(X) = f^{-1}(X)$, since $K \ll M$. Hence $K \subset M = f^{-1}(X)$. Therefore $f(K) \subset X$ and $X = N$.

(5) Assume $K \subset L \subset M$, L is a direct summand in M and $K \ll M$. Since L is a direct summand in M, then, there exist submodule $N \subset M$ such that $L + N = M$ and $L \cap N = \{0\}$. With the canonical mapping $f : M \to L, m = l + n \mapsto l$, for any $m \in M$, we have $f(K) = K$ (since $K \subset L$). Hence from (4), $K \ll L$. Let $K \subset L \subset M$, L is a direct summand in M and $K \ll L$. With the canonical mapping $g : L \to M$, $l \mapsto l + 0$, for any $l \in L$. Then $g(K) = K$ (since $K \subset L$) and hence $K \ll M$. □

3. Projective cover in $\sigma[M]$

Related to M-projective module and superfluous epimorphism, projective cover in $\sigma[M]$ is defined by the following definition.

Definition 3.1. Let M be an R-module and $N \in \sigma[M]$. A projective module P in $\sigma[M]$ together with a superfluous epimorphism $\pi : P \to N$ is called a projective cover of N in $\sigma[M]$ or a $\sigma[M]$-projective cover of N.

The next result shows that if a module M has projective cover in $\sigma[M]$, then the projective cover of M in $\sigma[M]$ is unique up to isomorphism.

Proposition 3.1. *Let M be an R-module and $\pi : P \to N$ a projective cover of N in $\sigma[M]$.*

(1) If $f : Q \to N$ is epic with Q projective in $\sigma[M]$, then there is a decompositition $Q = Q_1 \oplus Q_2$, with $Q_1 \simeq P$, $Q_2 \subset Ker\, f$ and $f|_{Q_1} : Q_1 \to N$ is a $\sigma[M]$-projective cover of N.

(2) If (Q, f) is another projective cover of N in $\sigma[M]$, then there is an isomorphism $h : Q \to P$ with $\pi \circ h = f$.

Proof.

(1) Because of the projectivity of Q, there exists $h : Q \to P$ with $\pi \circ h = f$. Since π is superfluous, h is epic and hence h splits (P is projective). Therefore there exists some $g : P \to Q$ with $h \circ g = id_P$ and hence $Q = Img \oplus Ker\, h$. Putting $Q_1 = Im\, g$ and $Q_2 = Ker\, h$, we get the desired decomposition. Q_1 is projective in $\sigma[M]$ and, since $\pi = f \circ g|_{Q_1}$, the epimorphism $f|_{Q_1}$ is superfluous.

(2) Let $\pi : P \to N$ a projective cover of N in $\sigma[M]$ and (Q, f) is another projective cover of N in $\sigma[M]$, we will show there is an isomorphism $h : Q \to P$ with $\pi \circ h = f$. From (1), we get $Q = Q_1 \oplus Q_2$, with $Q_1 \simeq P$, $Q_2 \subset Ker\, f$ and the following diagram is commutative:

$$\begin{array}{ccc} & & Q \\ & \swarrow_h & \downarrow^f \\ P & \stackrel{\pi}{\to} & N \to 0 \end{array}$$

i.e $\pi \circ h = f$. Since f is superfluous epimorphism, then $Ker\, f$ cannot contain a non-zero direct summand and hence $Q_2 \subset Ker\, h \subset Ker\, f = 0$ in (1). Therefore, h is monomorphism and h is epimorphism (from (1)), i.e h is isomorphism such that $\pi \circ h = f$. □

If (P, π) is a projective cover of N in $\sigma[M]$, then the properties of N influence the properties of P.

Proposition 3.2. *Let M be an R-module and $\pi : P \to N$ a projective cover of N in $\sigma[M]$.*

(1) If N is finitely generated, then P is also finitely generated.
(2) If M is projective in $\sigma[M]$ and N (finitely) M-generated, then P is also (finitely) M-generated.

Proof.

(1) By assumption, there exist epimorphism $\alpha : R^n \to N$, with $n \in \mathbb{N}$. Since P is projective in $\sigma[M]$, then the following diagram is commutative and β is epimorphism.

$$\begin{array}{ccc} & & R^n \\ & \swarrow_\beta & \downarrow^\alpha \\ P & \stackrel{f}{\to} & N \to 0 \end{array}$$

Therefore, there exist epimorphism $\beta : R^n \to P$, with $n \in \mathbb{N}$, i.e P is finitely generated.

(2) By assumption, there exist epimorphism $f : M^{(\Lambda)} \to N$, with Λ is finite index set. Since M is projective in $\sigma[M]$, then $M^{(\Lambda)}$ is also projective in $\sigma[M]$ and hence there exist a decompositition $M^{(\Lambda)} = Q_1 \oplus Q_2$, with $Q_1 \simeq P$, $Q_2 \subset Ker\, f$. Therefore, there exist epimorphism $g : M^{(\Lambda)} \to P$, for finite index set Λ, i.e P is finitely M-generated. □

The following proposition shows that if N_1 and N_2 have projective cover in $\sigma[M]$, then $N_1 \oplus N_2$ has a projective cover in $\sigma[M]$.

Proposition 3.3. *If $\pi_1 : P_1 \to N_1$, $\pi_2 : P_2 \to N_2$ are projective covers of N_1, N_2 in $\sigma[M]$, then $\pi_1 \oplus \pi_2 : P_1 \oplus P_2 \to N_1 \oplus N_2$ is a projective cover of $N_1 \oplus N_2$ in $\sigma[M]$.*

Proof. Observe that $Ker\,\pi_1 \subset Q_1 \subset Q_1 \oplus Q_2$. By assumption, $\pi_1 : P_1 \to N_1$ is superfluous epimorphism, then $Ker\,\pi_1 \ll Q_1$ and hence from Proposition 2.1 (5), we have $Ker\,\pi_1 \ll Q_1 \oplus Q_2$. In a similar way, we have $Ker\,\pi_2 \ll Q_1 \oplus Q_2$. Hence, from Proposition 2.1 (3),

$$Ker(\pi_1 \oplus \pi_2) \simeq Ker\,\pi_1 \oplus Ker\,\pi_2 \ll Q_1 \oplus Q_2$$

Since P_1 and P_2 are projective modules in $\sigma[M]$, then $P_1 \oplus P_2$ is also projective module in $\sigma[M]$. Therefore, $\pi_1 \oplus \pi_2 : P_1 \oplus P_2 \to N_1 \oplus N_2$ is a projective cover of $N_1 \oplus N_2$ in $\sigma[M]$. □

The following corrolary is an immediate consequence of Proposition 3.3.

Corollary 3.1. *If $\pi_1 : P_1 \to N_1, \pi_2 : P_2 \to N_2, \cdots, \pi_k : P_k \to N_k$ are projective covers of $N_1, N_2, \cdots, N_k$ in $\sigma[M]$, then*

$$\pi_1 \oplus \pi_2 \oplus \cdots \pi_k : P_1 \oplus P_2 \oplus \cdots \oplus P_k \to N_1 \oplus N_2 \oplus \cdots \oplus N_k$$

is a projective cover of $N_1 \oplus N_2 \oplus \cdots \oplus N_k$ in $\sigma[M]$.

Katayama was observed a necessary and sufficient condition a module has projective cover in category *R-MOD*. In the next theorem shows that the necessary and sufficient condition a module has projective cover is valid in category $\sigma[M]$.

Theorem 3.1. *Let N be projective module in $\sigma[M]$ and I a maximal submodule of M. Then N/I has a projective cover if and only if there exist a nonzero idempotent $e \in S$ such that $e(I)$ is small in N.*

Proof. Let (P, π) is a projective cover of N/I in $\sigma[M]$ and $\upsilon_I : N \to N/I$ is canonical projection mapping. Since N is projective in $\sigma[M]$, there exists a homomorphism $\alpha : N \to P$ in $\sigma[M]$ such that $\pi \circ \alpha = \upsilon_I$;

$$\begin{array}{ccc} & & N \\ & \swarrow_{\alpha} & \downarrow^{\upsilon_I} \\ P & \xrightarrow{\pi} & N/I \to 0 \end{array}$$

Then $P = Im\,\alpha + Ker\,\pi$. Since $Ker\,\pi$ is small in P, then $Im\,\alpha = P$. So, α is epimorphism. Since P is projective, the exact sequence

$$0 \to Ker\,\alpha \to N \xrightarrow{\alpha} P \to 0$$

splits, so there exists a homomorphism $\beta : P \to N$ such that $M = Ker\,\alpha \oplus Im\,\beta$. If $\alpha(I) = 0$, then $I = Ker\,\alpha$. Let $f : N \to I$ be the projection and let $e = 1-f \in S$, then we have $e(I) = (1-f)\circ f(N) = 0$, and hence $e(I)$ is small in M. If $\alpha(I) \neq 0$, then $\alpha(I) \subset Ker\,\pi$, since $(\pi \circ \alpha)(I) = v_I(I) = 0$. Hence $\alpha(I)$ is small in P. Therefore, $(\beta \circ \alpha)(I)$ is small in M. Put $e = \beta \circ \alpha \in S$. Then $e^2 = \beta \circ \alpha \circ \beta \circ \alpha = \beta \circ \alpha = e$. Hence e is a idempotent.

Conversely, suppose that there is a nonzero idempotent $e \in S$ such that $e(I)$ is small in M. Put $(I : e) = \{x \in M | e(x) \in I\}$. Since $e(I) \subseteq J(M)$, the maximality of I implies $(I : e) = I$. Now define a mapping $g : e(M) \to M/I$, by $g(e(x)) = x + I$. Then, the mapping g is well defined and is an epimorphism with kernel $e(I)$. Since $e(M)$ is a direct summand of M, $e(M)$ is projective. Thus g is a projective cover of M/I. □

References

1. Anderson, W. and Fuller, K., 1992, *Rings and Categories of Modules*, Springer-Verlag Berlin Heidelberg, New York.
2. Bilhan, G., 2005, Amply Fws Modules, *Journal of Arts and Sciences Sayi*, Vol.4, 2005.
3. Buyukasik, E., 2008, On Cofinitely Weak Supplemented Modules, *The Arabian Journal for Science and Engineering*, Vol.34, No.1A January 2009, page.159-164.
4. Clark, J., Lomp, C., Vanaja, N., and Wisbauer, R., 2006, *Lifting Modules (Supplement and Projectivity in Modul Theory)*, Birkhauser Verlag, Basel Switzerland.
5. Katayama, H., 1969, On The Projective Cover of a Factor Module Modulo a Maximal Submodule, *Proc. Japan Acad*, Vol.45, 12 September 1969, page.513-516.
6. Wisbauer, R., 1991, *Foundation of Modul and Ring Theory*, Gordon and Breach.

Bialgebras, Defined on Simple Alternative and Malcev Algebras

M. E. GONCHAROV

Sobolev Institute of Mathematics,
Novosibirsk, 630090, Russia
E-mail: gme@math.nsc.ru
http://math.nsc.ru/

This survey covers resent results about alternative, Jordan and Malcev D-bialgebras, with special emphasis to the one whose comultiplication is given by a skew-symmetric solution of the classical Yang–Baxter equation. Also, we show a connection between simple alternative D-bialgebras and Lie bialgebras.

Keywords: Lie bialgebra, alternative bialgebra, Jordan bialgebra, Malcev bialgebra, classical Yang-Baxter equation, nonassociative coalgebra, Cayley-Dickson algebra, classical Yang–Baxter equation, nonassociative coalgebra, simple non-Lie Malcev algebra.

1. Definitions and preliminaries

Lie bialgebras are Lie algebras and Lie coalgebras at the same time, such that comultiplication is a 1-cocycle. These bialgebras were introduced by Drinfeld 1 in studying the solutions to the classical Yang-Baxter equation.

In [2,3], the definition of a bialgebra in the sense of Drinfeld (D-bialgebra) related with any variety of algebras was stated. In particular, the associative and Jordan D-bialgebras were introduced, and an associative analogue of the Yang-Baxter equation was considered as well as the associative D-bialgebras related with the solutions to this equation. In the same papers, the associative algebras that admit a nontrivial structure of a D-bialgebra with cocommutative comultiplication on the center were also described. The comultiplication in an associative D-bialgebra is a derivation of an initial algebra into its tensor square considered as a bimodule over the initial algebra. These bialgebras were introduced in [4] and studied in [5]. The paper [5] is devoted to some properties of solutions to an associative analogue of the Yang-Baxter equation and the balanced bialgebras (i.e., D-bialgebras). The associative classical Yang-Baxter equations with parameters were considered in [6]. A class of Jordan D-bialgebras related to the "Jordan analogue" of the Yang-Baxter equation was introduced in [7], where it was proved that every finite-dimensional Jordan D-bialgebra, semisimple as an algebra, belongs to this class.

Given vector spaces V and U over a field F, denote by $V \otimes U$ its tensor product over F. Define the linear mapping τ on V by $\tau(\sum_i a_i \otimes b_i) = \sum_i b_i \otimes a_i$. Define the linear mapping ξ on $V \otimes V \otimes V$ by $\xi(\sum_i a_i \otimes b_i \otimes c_i) = \sum_i b_i \otimes c_i \otimes a_i$. Denote by V^* the dual space of V. Given $f \in V$ and $v \in V$, the symbol $\langle f, v\rangle$ denotes the linear functional f evaluated at v (i.e., $\langle f, a\rangle = f(a)$).

Definition 1.1. A pair (A, Δ), where A is a vector space over F and $\Delta : A \to A \otimes A$ is a linear mapping, is called a *coalgebra*, while Δ is a *comultiplication*.

Given $a \in A$, we will you the following notation: $\Delta(a) = \sum a_{(1)} \otimes a_{(2)}$. Define some multiplication on A^* by

$$\langle fg, a\rangle = \sum \langle f, a_{(1)}\rangle \langle g, a_{(2)}\rangle,$$

where $f, g \in A^*$, $a \in A$ and $\Delta(a) = \sum a_{(1)} \otimes a_{(2)}$. The algebra obtained is called the dual algebra of the coalgebra (A, Δ).

The dual algebra A^* of (A, Δ) gives rise to the following bimodule actions on A:

$$f \rightharpoonup a = \sum a_{(1)} \langle f, a_{(2)}\rangle \text{ and } a \leftharpoonup f = \sum \langle f, a_{(1)}\rangle a_{(2)},$$

where $a \in A,\ f \in A^*$ and $\Delta(a) = \sum a_{(1)} \otimes a_{(2)}$.

The following definition of a coalgebra related to some variety of algebras was given in [8].

Definition 1.2. Let $\mathcal{M}$ be an arbitrary variety of algebras. The pair (A, Δ) is called a $\mathcal{M}$-coalgebra if A^* belongs to M.

Let A be an arbitrary algebra with a comultiplication Δ, and let A^* be the dual algebra for (A, Δ). Then A induces the bimodule action on A^* by the formulas

$$\langle f \leftharpoondown a, b\rangle = \langle f, ab\rangle \text{ and } \langle b \rightharpoondown f, a\rangle = \langle f, ab\rangle,$$

where $a, b \in A,\ f \in A^*$.

Consider the space $D(A) = A \oplus A$ and equip it with the multiplication by putting

$$(a+f)(b+g) = (ab + f \rightharpoonup b + a \leftharpoonup g) + (fg + f \leftharpoondown b + a \rightharpoondown g).$$

Then $D(A)$ is an ordinary algebra over F, A and A^* are some subalgebras in $D(A)$. It is called the Drinfeld double.

Let Q be a bilinear form on $D(A)$ defined by

$$Q(a+f, b+g) = \langle g, a\rangle + \langle f, b\rangle$$

for all $a, b \in A$ and $f, g \in A^*$. It is easy to check that Q is a nondegenerate symmetric associative form, that is $Q(xy, z) = Q(x, yz)$.

Let us recall the definition of a Lie bialgebra. Let L be a Lie algebra with a comultiplication Δ. The pair (A, Δ) is called a Lie bialgebra if and only if (L, Δ) is a Lie coalgebra and Δ is a 1-cocycle, i.e.,

$$\Delta([a,b]) = \sum([a_{(1)}, b]\otimes a_{(2)} + a_{(1)}\otimes[a_{(2)}, b]) + \sum([a, b_{(1)}]\otimes b_{(2)} + b_{(1)}\otimes[a, b_{(2)}])$$

for all $a, b \in L$.

In [1], it was proved that the pair (L, Δ) is a Lie bialgebra if and only if its Drinfeld double $D(L)$ is a Lie algebra. This observation inspired the following definition.

Definition 1.3 ([2]). *Let $\mathcal{M}$ be an arbitrary variety of algebras and let A be an algebra from $\mathcal{M}$ with a comultiplication Δ. The pair (A, Δ) is called an $\mathcal{M}$-bialgebra in the sense of Drinfeld if its Drinfeld double $D(A)$ belongs to $\mathcal{M}$.*

Note that this definition corresponds with the definition of coalgebra given in [8].

There is an important type of Lie bialgebras called *coboundary* bialgebras. Namely, let L be a Lie algebra and $R = \sum_i a_i \otimes b_i$ from $(id-\tau)(L\otimes L)$, that is, $\tau(R) = -R$. Define a comultiplication Δ_R on L by

$$\Delta_R(a) = \sum_i [a_i, a] \otimes b_i - a_i \otimes [a, b_i]$$

for all $a \in L$. It is easy to see that Δ_R is a 1-cocycle. In [9] it was proved that (L, Δ_R) is a Lie coalgebra if and only if the element

$$C_L(R) = [R_{12}, R_{13}] + [R_{12}, R_{23}] + [R_{13}, R_{23}]$$

is L-invariant. Here $[R_{12}, R_{13}] = \sum_{ij}[a_i, a_j] \otimes b_i \otimes b_j$, $[R_{12}, R_{23}] = \sum_{ij} b_i \otimes [a_i, a_j] \otimes b_j$, and $[R_{13}, R_{23}] = \sum_{ij} a_i \otimes a_j \otimes [b_i, b_j]$. In particular, if

$$[R_{12}, R_{13}] + [R_{12}, R_{23}] + [R_{13}, R_{23}] = 0, \tag{1}$$

then the pair (L, Δ_R) is a Lie bialgebra. In this case, we say that (L, Δ_R) is a *triangular* Lie bialgebra. The equation (1) is called *the classical Yang-Baxter equation.*

Let B be an arbitrary algebra and $r = \sum_i a_i \otimes b_i \in B \otimes B$. Then the equation

$$C_B(r) = r_{12}r_{13} + r_{13}r_{23} - r_{23}r_{12} = 0 \tag{2}$$

is called the classical Yang-Baxter equation on B. Here the subscripts specify the way of embedding $B \otimes B$ into $B \otimes B \otimes B$, that is, $r_{12} = \sum_i a_i \otimes b_i \otimes 1$, $r_{13} = \sum_i a_i \otimes 1 \otimes b_i$, $r_{23} = \sum_i 1 \otimes a_i \otimes b_i$. Note that $C_B(r)$ is well defined even if B is non-unital. This equation for different varieties of algebras were considered in [3,5–7,10]. Usually, antisymmetric solutions $(\tau(r) = -r)$ to the equation (2) are considered.

Definition 1.4. Let B be an arbitrary algebra and ω be a non-degenerate skew-symmetric bilinear form on algebra B. Then ω is called *symplectic* if for all $x, y, z \in B$

$$\omega(xy, z) + \omega(yz, x) + \omega(zx, y) = 0.$$

In this case the pair (B, ω) is called a symplectic algebra.

Proposition 1.1 (10–12). *Let B be an algebra. Then antisymmetric solutions of the classical Yang–Baxter equation $C_B(r) = 0$ are in one to one correspondence with symplectic subalgebras in B.*

Proof. We give the sketch of the proof.

Let r be an antisymmetric solution of $C_B(r) = 0$. Assume that $\phi_r : B^* \mapsto B$ is given by the formula $\phi_r(f) = \sum_i \langle f, a_i \rangle b_i = -\sum_i a_i \langle f, b_i \rangle$. Then the map ϕ_r is a homomorphism of algebras. Let A be the image of B^* under the map ϕ_r([12]). We may assume that the elements a_i(respectively b_i), appearing in the expression for r, are a base of the space A. Assume that dim$A$=n. Expressing each b_i through $a_1, ..., a_n$ we get $r = \sum_{ij} \alpha_{ij}(a_i \otimes a_j)$, where $\alpha_{ij} \in F$. Let (β_{ij}) be the inverse matrix to (α_{ij}).

Let $a_1^*, \ldots, a_n^* \in A^*$ be a dual base to $a_1, ..., a_n$. Define some form ω_r on A, by putting

$$\omega_r(a, b) = \sum_{ij} \beta_{ij} \langle a_i^* \otimes a_j^*, a \otimes b \rangle.$$

Then ([12]) ω_r is a symplectic form on A.

Now let (A, ω) be a symplectic subalgebra in A. Take $a_1, \ldots a_n$ to be a basis of B. Let $(\beta)_{ij}$ be the inverse matrix to $(\omega(a_i, a_j))$. Consider an element $r = \sum_{ij} \beta_{ij} a_i \otimes a_j$. Then $\tau(r) = -r$, r is an antisymmetric solution of the classical Yang–Baxter equation on B and $\omega = \omega_r([12])$. □

An element $r = \sum_i a_i \otimes b_i \in B \otimes B$ induces a comultiplication Δ_r on B:

$$\Delta_r(a) = \sum a_i a \otimes b_i - a_i \otimes ab_i$$

for all $a \in B$.

2. Alternative D-bialgebras

2.1. *Some basic results about alternative D-bialgebras*

Definition 2.1. An algebra A is called *alternative* if for all $a, b, c \in A$ the following equations hold:

$$(x, x, y) = 0 \text{ and } (y, x, x) = 0,$$

where $(x, y, z) = (xy)z - x(yz)$ — the associator of elements x, y, z.

It is easy to see that every associative algebra is alternative. On the other hand, every two-generated subalgebra in an alternative algebra is associative due to Artin's theorem.

Let A be an alternative algebra. The following theorem gives us necessary and sufficient condition for a coalgebra (A, Δ) to be an alternative D-bialgebra.

Theorem 2.1. *Let A be an alternative algebra ofer a field of characteristic not equal 2 with a comultiplication Δ. Then the pair (A, Δ) is an alternative D-bialgebra if and only if the pair (A, Δ) is an alternative coalgebra and Δ satisfies the following conditions:*

(1) $\Delta(ab) = \sum_a (a_{(1)} b \otimes a_{(2)} + a_{(2)} b \otimes a_{(1)} - a_{(2)} \otimes b a_{(1)})$
$+ \sum_b (b_{(1)} \otimes a b_{(2)} + b_{(1)} \otimes b_{(2)} a - a b_{(1)} \otimes b_{(2)});$

(2) $\sum_{ab}((ab)_{(1)} \otimes (ab)_{(2)} + (ab)_{(2)} \otimes (ab)_{(1)})$
$= \sum_a (a_{(1)}b \otimes a_{(2)} + a_{(2)} \otimes a_{(1)}b) + \sum_b (b_{(1)} \otimes ab_{(2)} + ab_{(2)} \otimes b_{(1)})$.

Note, that theorem 2.1 is an alternative analogue of theorem 1 from [3].

In the class of alternative bialgebras one can consider those related to the solutions of the classical Yang–Baxter equation (2).

Let us fix an element $r = \sum_i a_i \otimes b_i$ from $(id - \tau)(A \otimes A)$, that is $\tau(\sum_i a_i \otimes b_i) = -\sum_i a_i \otimes b_i$, and define a comultiplication Δ_r on A by putting

$$\Delta_r(a) = \sum_i a_i a \otimes b_i - \sum_i a_i \otimes ab_i. \tag{3}$$

Theorem 2.2. *Let A be an alternative algebra over a field of characteristic not equal 2, $r \in (id - \tau)(A \otimes A)$. If for all $a \in A$*

$$C_A(r)(a \otimes 1 \otimes 1) - (1 \otimes 1 \otimes a)C_A(r) = 0,$$

then the pair (A, Δ_r) is an alternative D-bialgebra.

As a corollary of theorem 2.2 we obtain that antisymmetric solutions of the classical Yang-Baxter equation on an alternative algebra A induces a structure of an alternative D-bialgebra on A.

2.2. *Structures of an alternative D-bialgebra on a Cayley–Dickson algebra*

Let F_2 be the algebra of 2×2 - matrices over an algebraically closed field F. Denote by e_{ij} the matrix units of F_2.

Let $\mathcal{C} = F_2 \oplus vF_2$ be the Cayley–Dickson matrix algebra. Recall that the product in F_2 may be extended to a product in $\mathcal{C}$ by the following relations:

$$a(vb) = v\bar{a}b, \ (vb)a = v(ab),$$

$$(va)(vb) = b\bar{a},$$

where $a, b \in F_2$, and the bar over the letter denotes the standard involution in F_2. We will describe all alternative D-bialgebra structures on $\mathcal{C}$. In what follows, we assume that the characteristic of F is different from 2.

Proposition 2.1. *Let $\mathcal{C}$ be a Cayley-Dickson algebra with a comultiplication Δ. If $(\mathcal{C}, \Delta)$ is an alternative D-bialgebra, then there exists an element r in $(id - \tau)(\mathcal{C} \otimes \mathcal{C})$, such that $C_{\mathcal{C}}(r) = 0$ and $\Delta = \Delta_r$.*

Thus, our problem is reduced to the determination of the elements $r = \sum_i a_i \otimes b_i \in (id-\tau)(\mathcal{C}\otimes\mathcal{C})$ satisfying the equation $C_{\mathcal{C}}(r) = 0$. By proposition 1.1 it is to find all symplectic subalgebras in $\mathcal{C}$.

Theorem 2.3 ([10]). *Let A be a symplectic subalgebra in $\mathcal{C}$. Then A is isomorphic to one of the following subalgebras:*

1. Subalgebra $A(2,1,0)$ with base e_{11}, e_{12}.

2. Subalgebra $A(2,0,1)$ with base e_{11}, e_{21}.

3. Subalgebra $N(2)$ with base e_{12}, ve_{11}.

In this cases each nondegenerate skew-symmetric bilinear form is a symplectic form.

4. Subalgebra $A(4,1,2)$ with base e_{11}, e_{12}, ve_{11}, ve_{12}. A nondegenerate skew-symmetric bilinear form ω on $A(4,1,2)$ is a symplectic form if and only if it satisfies

$$\omega(e_{11}, e_{12}) = -\omega(ve_{11}, ve_{12}).$$

5. Subalgebra $A(4,2,1)$ with base e_{11}, ve_{21}, ve_{22}, e_{21}. A nondegenerate skew-symmetric bilinear form ω on $A(4,2,1)$ is a symplectic form if and only if it satisfies

$$\omega(e_{11}, e_{21}) = \omega(ve_{21}, ve_{22}).$$

3. On a Lie bialgebras that arise from alternative and Jordan bialgebras

In this section we consider a connection between simple alternative D-bialgebras and Lie bialgebras.

Definition 3.1. A commutative algebra J is called a *Jordan* algebra if for all $a, b \in J$ the following equation holds

$$(x^2y)x = x^2(yx). \tag{4}$$

Let J be a Jordan algebra with 1. To any pair of elements $x, y \in J$ one can assign the linear mapping $D_{x,y}$: $z \to (x,y,z)$, where $(x,y,z) = (xy)z - x(yz)$ is the associator of elements x, y, z. It is well known that $D_{x,y}$ is a derivation of the algebra J([13]), that is

$$D_{x,y}(ab) = D_{x,y}(a)b + aD_{x,y}(b).$$

The space generated by all these derivations is a Lie subalgebra of the Lie algebra of all derivations on J. Elements of this subalgebra are called *inner* derivations.

Let D be an arbitrary Lie algebra of derivations of a Jordan algebra J and D contains inner derivations. Consider $L(J)$ to be the Kantor-Koecher-Tits construction of a Jordan algebra J, that is

$$L(J) = J \oplus (J^{'} \oplus D) \oplus \overline{J},$$

where $J' = \{a' | a'(b) = ba\}$ is the space of operators of right multiplications on J and $\overline{J}$ is an isomorphic copy of the space J. Let $\epsilon : L(J) \to L(J)$ be a liner mapping defined by

$$\epsilon(a + b' + d + \overline{c}) = c - b' + d + \overline{a},$$

where $a, b, c \in J$ and $d \in D$. Define a multiplication $[,]_L$ on $L(J)$ by putting

$$[a_1 + V_1 + \overline{b_1}, a_2 + V_2 + \overline{b_2}]_L =$$

$$= a_1 V_2 - a_2 V_1 + \overline{b_1 \epsilon(V_2)} - \overline{b_2 \epsilon(V_1)} + a_1 \bigtriangledown b_2 - a_2 \bigtriangledown b_1 + V_1 V_2 - V_2 V_1,$$

where $a_i, b_i \in J,\ V_i \in J' \oplus D,\ i = 1, 2$. In this case $L(J)$ is a -1, 1 graded algebra($L_{-1} = J$, $L_0 = (J^{'} \oplus D)$, $L_1 = \overline{J}$).

Let U be a subalgebra in $L(J)$, generated by $1,\ 1^{'},\ \overline{1}$. Then U is isomorphic to sl_2.

A comultiplication Δ_L on $L(J)$ *agrees* with the grading of $L(J)$ if

$$\Delta_L(L_i) \subset \sum_{i+k=i} L_j \otimes L_k.$$

Let Δ be a comultiplication on J. We say that that Δ_L is *associated* with Δ if

$$\Delta_L(a) \in \sum a_{(1)} \otimes a'_{(2)} - a'_{(2)} \otimes a_{(1)} + L_1 \otimes D + D \otimes L_1,$$

where $\Delta(a) = \sum a_{(1)} \otimes a_{(2)}$.

Theorem 3.1 ([2, Theorem 4]). *Let* $charF \neq 2, 3$. *If a pair* $(L(J), \Delta_L)$ *is a Lie bialgebra, comultiplication* Δ_L *agrees with the grading and* $\Delta(U) = 0$, *then* J *admits the structure of a Jordan D-bialgebra* (J, Δ) *with a comultiplication* Δ *such, that* Δ_L *is associated with* Δ.

In general, the converse to theorem 3.1 is not true([14, example 1]). But in some particular cases of Jordan D-bialgebras (J, Δ), the algebra $L(J)$ admits a structure of a Lie bialgebra $(L(J), \Delta_L)$ with Δ_L agrees with the grading of $L(J)$ and Δ_L is associated with Δ.

Let A be an associative(alternative) algebra with 1, and (A, Δ) is a associative(alternative) D-bialgebra. Let $A^{(+)}$ be the adjoint Jordan algebra, that is the algebra with new multiplication

$$a \circ b = \frac{1}{2}(ab + ba).$$

Let $\Delta^{(+)}$ be the adjoint comultiplication:

$$\Delta^{(+)}(a) = \frac{1}{2}(\Delta(a) + \tau(\Delta(a))).$$

Then $(A^{(+)}, \Delta^{(+)})$ is a Jordan D-bialgebra.

Definition 3.2. Let B be an arbitrary coalgebra with a comultiplication Δ. Then Δ is sad to be cocommutative on the center Z of B if for all $z \in Z$ $\Delta(z) = \tau(\Delta(z)$.

It is easy to see that if A — associative(alternative) algebra and $r \in (id - \tau)(A \otimes A)$ then Δ_r is cocommutative on the center of A.

Theorem 3.2 ([2, Theorem 3]). *Let (A, Δ) be an associative D-bialgebra and let $(A^{(+)}, A^{(+)})$ be the adjoint Jordan D-bialgebra. Assume that comultiplication Δ is cocommutative on the center of A. Then the algebra $L(A^{(+)})$ can be endowed with the structure of a Lie bialgebra with comultiplication Δ_L, such that Δ_L agrees with the grading of $L(A^{(+)})$, Δ_L is associated with $\Delta^{(+)}$ and $\Delta_L(U) = 0$.*

The situation changes if in theorem 3.2 one take A to be an alternative algebra. In this case we can only proof the following

Theorem 3.3 ([14]). *Let A be a finite-dimensional simple alternative algebra over an algebraically closed field. Assume that (A, Δ) is an alternative D-bialgebra and $(A^{(+)}, \Delta^{(+)})$ is the adjoint Jordan D-bialgebra. Then $L(A^{(+)})$ may be equipped with the structure of a Lie bialgebra with comultiplication Δ_L such that Δ_L agrees with the grading of $L(A^{(+)})$, Δ_L is associated with $\Delta^{(+)}$, and $\Delta_L(U) = 0$.*

At the same time there is an example of an alternative D-bialgebra (A, Δ), where A is a semisimple alternative algebra, for which the statement of theorem 3.3 is not true ([14, example 1]).

4. Malcev bialgebras

Definition 4.1. An anticommutative algebra is called a Malcev algebra if for all $x, y, z \in M$ the following equation holds:

$$J(x, y, xz) = J(x, y, z)x, \tag{5}$$

where $J(x, y, z) = (xy)z + (yz)x + (zx)y$ is the jacobian of elements x, y, z.

Malcev algebras were introduced by A.I. Malcev [15] as tangent algebras for local analytic Moufang loops. The class of Malcev algebras generalizes the class of Lie algebras and has a well developed theory [13].

An important example of a non-Lie Malcev algebra is the vector space of zero trace elements of a Caley-Dickson algebra with the commutator bracket multiplication [16,17]. It turns out that it is the only simple non-Lie Malcev algebra.

4.1. *Coboundary and triangular Malcev bialgebras*

In [18], the following statement was proved.

Theorem 4.1. *A Malcev algebra M with a multiplication μ and a comultiplication Δ is a Malcev bialgebra if and only if the dual algebra M^* is a Malcev algebra and the comultiplication satisfies*

(1) $\Delta((ab)c) + \Delta(bc)(a \otimes 1) + (b \otimes 1)\Delta(ac) = \sum a_{(1)}(bc) \otimes a_{(2)} + \sum a_{(1)}c \otimes a_{(2)}b + \sum a_{(1)} \otimes (a_{(2)}b)c + \sum ab_{(1)} \otimes b_{(2)}c - \sum b_{(1)} \otimes b_{(2)}(ac) + \sum(ab_{(1)})c \otimes b_{(2)} + \sum a(bc_{(1)}) \otimes c_{(2)} - \sum c_{(1)} \otimes (c_{(2)}a)b,$

(2) $(1 \otimes \Delta)\Delta(ab) = (1 \otimes 1 \otimes a)((1 \otimes \Delta)\Delta(b)) + (\Delta \otimes 1)((a \otimes 1)\Delta(b)) - (1 \otimes \tau)((1 \otimes 1 \otimes a)(\Delta \otimes 1)\Delta(b)) + (1 \otimes \tau)(((\Delta \otimes 1)\Delta(b))(a \otimes 1 \otimes 1)) + ((\Delta \otimes 1)\Delta(a))(b \otimes 1 \otimes 1) + ((\Delta \otimes 1)\Delta(a))(1 \otimes 1 \otimes b) - (1 \otimes \tau)((\Delta \otimes 1)(\Delta(a)(b \otimes 1))) - (1 \otimes b \otimes 1)((1 \otimes \Delta)\Delta(a)) + (1 \otimes \tau)(1 \otimes 1 \otimes \mu)((1 \otimes \tau \otimes 1)(\Delta(b) \otimes \Delta(a))) + (1 \otimes 1 \otimes \mu)((1 \otimes \tau \otimes 1)(\Delta(a) \otimes \Delta(b))).$

For Malcev bialgebras, one can also consider the class of coboundary bialgebras.

Let M be a Malcev algebra, $r \in (id - \tau)(M \otimes M)$. Then r induces a comultiplication Δ_r on M defined by

$$\Delta_r(a) = [r, a] = \sum_i a_i a \otimes b_i - a_i \otimes ab_i$$

for all $a \in M$. The following theorem gives the necessary and sufficient conditions for the pair (M, Δ_r) to be a Malcev bialgebra.

Theorem 4.2 ([19]). *Let M be a Malcev algebra over a field characteristic not equal 2, $r \in (id - \tau)(M \otimes M)$. The pair (M, Δ_r) is a Malcev bialgebra if and only if for all $a, b \in M$*

$$(C_M(r)(1 \otimes b \otimes 1))(1 \otimes a \otimes 1) - C_M(r)(ab \otimes 1 \otimes 1) - (C_M(r)(1 \otimes 1 \otimes a)), (1 \otimes 1 \otimes b) =$$

$$= C_M(r)(b \otimes 1 \otimes a) - C_M(r)(a \otimes b \otimes 1) \qquad (6)$$

or

$$C_M(r)(1 \otimes J_{b,a} \otimes 1 - 1 \otimes 1 \otimes J_{a,b}) =$$

$$= [C_M(r), ab] + [C_M(r), b](1 \otimes 1 \otimes a) - [C_M(r), a](1 \otimes b \otimes 1),$$

where the operator $J_{a,b}$ is defined by $cJ_{a,b} = J(c, a, b)$, and by $[C_M(r), a]$ we denote the action M on $M \otimes M \otimes M$, that is

$$[x \otimes y \otimes z, a] = xa \otimes y \otimes z + x \otimes ya \otimes z + x \otimes y \otimes za.$$

4.2. *Structures of a Malcev bialgebra on the non-Lie simple Malcev algebra*

Let F be an algebraically closed field of characteristic different from 2 and 3. Then, up to an isomorphism, there is only one non-Lie simple Malcev algebra $\mathbb{M}$ over F [20]. The dimension of $\mathbb{M}$ is equal to seven, and it is convenient to consider a base h, x, x', y, y', z, z' of $\mathbb{M}$ with the following multiplication table:

$$hx = 2x, \; hy = 2y, \; hz = 2z,$$

$$hx' = -2x', \; hy' = -2y', \; hz' = -2z',$$

$$xx' = yy' = zz' = h,$$

$$xy = 2z', \; yz = 2x', \; zx = 2y',$$

$$x'y' = -2z, \; y'z' = -2x, \; z'x' = -2y.$$

The remaining products are zero. This basis is called a standard basis.

The algebra $\mathbb{M}$ can be constructed in the following way. Let $\mathcal{C}$ be the matrix Cayley-Dickson algebra with a multiplication $x \cdot y$. Then $\mathcal{C} = F_2 \oplus vF_2$, where F_2 is the algebra of 2×2 matrices. Define a new multiplication in $\mathcal{C}$: $xy = \frac{1}{2}(x \cdot y - y \cdot x)$. Then the vector space $\mathcal{C}$ with this multiplication turns into a Malcev algebra denoted by $\mathcal{C}^{(-)}$. The set $F \cdot 1$, where 1 is the unit of $\mathcal{C}$, is the center of $\mathcal{C}^{(-)}$, and the quotient algebra $\mathcal{C}^{(-)}/F \cdot 1$ is isomorphic to $\mathbb{M}$.

From the multiplication table one can see that the space $M(4)$ with the basis h, x, y', z is a four-dimensional subalgebra in $\mathbb{M}$. This algebra first appeared in [21].

Suppose $(\mathbb{M}, \Delta)$ is a Malcev bialgebra. Consider the Drinfeld double $D(\mathbb{M})$. Obviously, $D(\mathbb{M})$ is not a simple algebra. Therefore $D(\mathbb{M})$ is either a semisimple Malcev algebra or it possesses a non-zero radical(=solvable radical) R. We will consider each of these cases separately.

Suppose that the radical R of the Drinfeld double $D(\mathbb{M})$ is nonzero.

Theorem 4.3. *If the radical R of the Drinfeld double $D(M)$ is nonzero, then*

1. $R^2 = 0$ and $D(M) = M + R$ is a direct sum of vector spaces M and R.

2. There is an element $r = \sum_i a_i \otimes b_i \in M \otimes M$, such that $\tau(r) = -r$,

$$C_M(r) = \sum_{ij} a_i a_j \otimes b_i \otimes b_j - a_i \otimes a_j b_i \otimes b_j + a_i \otimes a_j \otimes b_i b_j = 0 \qquad (7)$$

and

$$\Delta(a) = \sum_i a_i a \otimes b - a_i \otimes ab_i.$$

Thus, in order to describe all Malcev bialgebra structures on $\mathbb{M}$ one should find all antisymmetric solutions to the classical Yang-Baxter equation (2). By proposition 1.1 it is to find all symplectic subalgebras in $\mathbb{M}$.

Lemma 4.1 ([19]). *Let B be a symplectic subalgebra of $\mathbb{M}$. Then B is isomorphic to one of the following subalgebra:*

1. The subalgebra with the base x, y'.

2. The subalgebra with the base h, x.

In these cases every non-degenerate skew-symmetric bilinear form is symplectic.

3. The subalgebra $M(4)$ with the base h, x, y', z. In this case non-degenerate skew-symmetric bilinear form is symplectic if and only if it satisfies

$$\omega(y', h) = 2\omega(x, z).$$

Consider that $D(\mathbb{M})$ is a semisimple Malcev algebra.

Theorem 4.4 ([19]). *Let $\mathbb{M}$ be a simple non-Lie Malcev algebra over an algebraically closed field of characteristic not equal 2, 3. Then in a standard basis h, x, x', y, y', z, z' of the algebra $\mathbb{M}$ the element*

$$r_0 = \alpha_{12}(h \otimes x - x \otimes h) + \alpha_{15}(h \otimes y' - y' \otimes h) + \alpha_{16}(h \otimes z - z \otimes h) +$$

$$\alpha_{25}(x \otimes y' - y' \otimes x) - 2\alpha_{15}(x \otimes z - z \otimes x) + \alpha_{56}(y' \otimes z - z \otimes y'),$$

where $\alpha_{ij} \in F$, is a skew-symmetric solution to the classical Yang-Baxter equation on $\mathbb{M}$. Moreover, the element

$$r = r_0 + \frac{1}{4} h \otimes h + x \otimes x' + y' \otimes y + z \otimes z'. \tag{8}$$

induces on $\mathbb{M}$ a structure of a Malcev bialgebra with a semisimple Drinfeld double.

Conversely, let $(\mathbb{M}, \Delta)$ be a Malcev bialgebra with a semisimple Drinfeld double. Then $\Delta = -\Delta_r$, and one can choose a standard basis h, x, x', y, y', z, z' of $\mathbb{M}$ in such a way that r has the form (8).

Note, that the elements (8) are solutions of the classical Yang-Baxter equation on $\mathbb{M}$, but they fails to be antisymmetric. Theorem 4.4 says that structures of a Malcev bialgebra on $\mathbb{M}$ with a semisimple Drinfeld double are induced by non-antisymmetric solutions of the classical Yang-Baxter equation, and this solutions have form (8) in the appropriate basis.

References

1. V.G. Drinfeld, *Sov Math Dokl* **27**, 68 (1983).
2. V.N. Zhelyabin, *Algebra and logic* **36**, 1 (1997).
3. V.N. Zhelyabin, *Siberian mathematical journal* **32**, 261 (1998).
4. S.A. Joni G.C. Rota, *Studies in Applied Mathematics* **61**, 93 (1998).
5. M. Aguiar, *Journal of Algebra* **244**, 492 (2001).
6. A. Polishchuk, *Advances in Mathematics* **168**, 56 (2002).
7. V.N. Zhelyabin, *St. Petersburg Mathematical Journal* **11**, 589 (2000).
8. J.A. Anquela, T. Cortes, F. Montaner, *Communication in Algebra.* **22**, 4693 (1994).
9. V.G. Drinfeld, *Proc. Int. Congress Math., Berkeley, 1986* , 798 (1987).
10. M.E. Goncharov, *Siberian mathematical journal* **48**, 809 (2007).
11. A.A. Belavin, V.G. Drinfeld, *Funct. Anal. Appl.* **16**, 159 (1982).
12. V.N. Zhelyabin, *Siberian Advances in Mathematics* **10**, 142 (2000).
13. E.N. Kuz'min, I.P. Shestakov, Non-associative structures, in *Algebra VI Encyclopaedia Math. Sci. 57*, (Springer Berlin, 1995) pp. 197–280.
14. M.E. Goncharov, *Siberian mathematical journal* **51**, 215 (2010).
15. A.I. Mal'cev, *Matem. Sbornik.* **36(78)**, p. 569576 (1955).
16. A.A. Sagle, *Pacific j. Math* **12**, 1047 (1962).

17. E.N. Kuz'min, *Algebra and logic* **7**, 233 (1955).
18. V.V. Vershinin, *Acta Applicandae mathematicae* **75**, 281 (2003).
19. M.E. Goncharov, Structures of Malcev Bialgebras on a simple non-Lie Malcev algebra. arXiv:1008.4214v1.
20. E.N. Kuz'min, Structure and representations of finite-dimensional simple Malcev algebras (in Russian), in *Issledovaniya po teorii kolec i algebr (trud. inst. matem. SO RAN SSSR,16)*, (Novosibirsk Nauka, 1989) pp. 75–101.
21. A.A. Sagle, *Trans. Am. Math. Soc.* **101**, 426 (1962).

A Structure Theorem for Ortho-u-monoids*

C. M. Gong, Y. Q. Guo and X. M. Ren

Abstract

Ortho-u-monoids form a generalization of orthogroups in the class of $E(S)$-semiabundant semigroups. The semilattice decomposition of ortho-u-monoids has been thoroughly described by Guo Yuqi,C M Gong and K P Shum in 2010, In this paper, a structure theorem for ortho-u-monoids is established. As a direct applications of the above result, the main structure theorem of ortho-lc-monoids given in a recent paper of Y.Q.Guo,CM.Cong and X.M. Ren will be reobtained and some other structure theorems of ortho-c-monoids and orthogroups will also be given.

Keywords and phrases: Super $E(S)$-semiabundant semigroups; Ortho-u-monoids; $E(S)$-semiabundant semigroups.

AMS Mathematics Subject Classification (2000): 20M10

Introduction

A completely regular semigroup S is an orthogroup if S is orthodox. The class of orthogroups is a very important subclass of the class of completely regular semigroups. The structures of orthogroups have been described by Petrich [1] in 1987 and also by Petrich-Reilly [2] in 1999. The papers [3], [4], [5] and [6] all related with the structuure of some kinds of orthogroups.

It is noted that the constructions of the generalized orthogroups in the class of quasi-regular semigroups were given by Ren-Shum in [7]. In the class of abundant semigroups, Ren and Shum [8] gave a structure theorem

*The research is supported by a grant of National Natural Science Foundation of China (Grant No:10871161, 10971160 and 10926031); a NSF grant of Shaanxi Province (No: SJ08A06); Natural Science Foundation Project of CQ CSTC2009BB2291.

for another kind of generalized orthogroups called cyber-groups which are superabundant semigroups in which the set $E(S)$ of idempotents forms a subsemigroup (in [9], [10], such a semigroup was called an ortho-c-monoid). Recently, Guo, Shum and Gong [9] generalized the completely regular semigroups and orthogroups to super-r-wide semigroups and ortho-lc-monoids in the class of *rpp* semigroups, respectively, by using the $(*, \sim)$-Green's relations, and established a structure theorem for ortho-lc-monoids (see [11]).

In order to extend furthermore the class of regular semigroups, Lawson [12] considered the $E(S)$-semiabundant semigroups which are natural generalizations of regular semigroups and abundant semigroups. In [14], we first introduced the concepts of super $E(S)$-semiabundant semigroups and ortho-u-monoids which generalized the completely regular semigroups and the orthogroups to the range of $E(S)$-semiabundant semigroups, respectively. For more information on generalized regular semigroups and their special subclassses ,the reader is refered to [16].

The main purpose of this paper is to establish a new construction method of ortho-u-monoids which is different from that adopted in [8] and [1]. As direct corollaries of our result, a new structure theorem for ortho-lc-monoids in [10] is reobtained and some structure theorems for ortho-c-monoids and orthogroups will also be given.

For notation and terminologies on abundant semigroups and their generalizations not given in this paper, the reader is referred to[13] and also [15].

1 Preliminaries

Let S be a semigroup and $\mathcal{E}(S)$ the lattice of all equivalences on S. For any $\sigma \in \mathcal{E}(S)$, call $A \subseteq S$ a *subset saturated by* σ if A is a union of some σ-classes of S; call S σ*-abundant* if every σ-class of S contains idempotents of S.

For any semigroup S, Lawson [12] defined two equivalences as follows:

$$(a,b) \in \widetilde{\mathcal{L}} \quad \text{if and only if} \quad (\forall e \in E(S)) \;\; ae = a \Longleftrightarrow be = b$$

and

$$(a,b) \in \widetilde{\mathcal{R}} \quad \text{if and only if} \quad (\forall e \in E(S)) \;\; ea = a \Longleftrightarrow eb = b$$

for any $a, b \in S$, where $E(S)$ is the set of idempotents of S.

A semigroup S is said to be an $E(S)$-semiabundant semigroup if S is a $\widetilde{\mathcal{L}}$-abundant semigroup and $\widetilde{\mathcal{R}}$-abundant semigroup.

In order to generalize the completely regular semigroups and orthogroups within the class of $E(S)$-semiabundant semigroups, we consider the following relations:

$$\widetilde{\mathcal{H}} \stackrel{d}{=} \widetilde{\mathcal{L}} \wedge \widetilde{\mathcal{R}},$$

$$\widetilde{\mathcal{D}} \stackrel{d}{=} \widetilde{\mathcal{L}} \vee \widetilde{\mathcal{R}},$$

$$a\widetilde{\mathcal{J}}b \stackrel{d}{\Longleftrightarrow} \widetilde{J}(a) = \widetilde{J}(b),$$

where $\widetilde{J}(a)$ is the smallest ideal containing a and saturated by $\widetilde{\mathcal{L}}$ and $\widetilde{\mathcal{R}}$.

In general, the relation $\widetilde{\mathcal{R}}[\widetilde{\mathcal{L}}]$ need not be a left[right] congruence on S. In view of [10], we call a semigroup S a *super $E(S)$-semiabundant semigroup* if it is $\widetilde{\mathcal{H}}$-abundant, $\widetilde{\mathcal{R}}$ and $\widetilde{\mathcal{L}}$ are left and right congruences on S, respectively.

In order to introduce an analogue of orthogroups within the class of $E(S)$-semiabundant semigroups, we first give the following definition.

Definition 1.1. *[[14]] A super $E(S)$-semiabundant semigroup S is called an ortho-u-monoid if $E(S) \leqslant S$.*

We now recall the following basic concepts and results.

A semigroup is called a rectangular u-monoid if it is isomorphic to a direct product of a rectangular band and a unipotent monoid.

Lemma 1.2. *[[14]] A semigroup S is an ortho-u-monoid if and only if S is a semilattice of rectangular u-monoids.*

By Lemma [14], we have the following corollaries (also see Ref. [9]).

Corollary 1.3. *A semigroup S is an ortho-lc-monoid if and only if S is rpp and is a semilattice of rectangular lc-monoids.*

Corollary 1.4. *A semigroup S is an ortho-c-monoid if and only if S is abundant and a semilattice of rectangular c-monoids.*

Corollary 1.5. *A semigroup S is an orthogroup, if and only if S is a semilattice of rectangular groups.*

Lemma 1.6. *[Ref. [14]] Let S be a semilattice Y of rectangular u-monoids $S_\alpha = I_\alpha \times M_\alpha \times \Lambda_\alpha (\alpha \in Y)$, where M_α is a unipotent monoid and $I_\alpha[\Lambda_\alpha]$ is a left [right] zero band, for each $\alpha \in Y$. Then for any $(i, a, \lambda) \in S_\alpha$, $(j, b, \mu) \in S_\beta$, the following statements hold:*

(i). $(i, a, \lambda)\widetilde{\mathcal{R}}(j, b, \mu) \Leftrightarrow \alpha = \beta, i = j.$

(ii). $(i, a, \lambda)\widetilde{\mathcal{L}}(j, b, \mu) \Leftrightarrow \alpha = \beta, \lambda = \mu.$

(iii). $(i, a, \lambda)\widetilde{\mathcal{H}}(j, b, \mu) \Leftrightarrow \alpha = \beta, i = j, \lambda = \mu.$

(iv). $(i, a, \lambda)\widetilde{\mathcal{D}}(j, b, \mu) \Leftrightarrow \alpha = \beta.$

Lemma 1.7. *[[14]] Let $S = (Y, S_\alpha)$ be an ortho-u-monoid. Then for any $\alpha, \beta \in Y, \alpha \leq \beta$, $E(S_\alpha)$ is a rectangular band and for any $a, b \in S_\alpha, e \in E(S_\beta)$, $ab = aeb$.*

Lemma 1.8. *Let S be a semilattice Y of rectangular u-monoids $S_\alpha = I_\alpha \times M_\alpha \times \Lambda_\alpha (\alpha \in Y)$. For any $\alpha \in Y$, in I_α and Λ_α choose arbitrarily an element k_α and ξ_α, respectively. And let $c_\alpha = (k_\alpha, 1_\alpha, \xi_\alpha)$. For any $\alpha, \beta \in Y, \alpha \geq \beta$, define a function $\chi_{\alpha,\beta}$ from M_α to M_β by the requirement*

$$c_\beta(k_\alpha, g, \xi_\alpha)c_\beta = (k_\beta, g\chi_{\alpha,\beta}, \xi_\beta).$$

Then one can define a strong semilattice $M = [Y, M_\alpha, \chi_{\alpha,\beta}]$ such that for any $\alpha, \beta \in Y$, $a = (i, g, \lambda) \in S_\alpha$ and $b = (j, h, \mu) \in S_\beta$,

$$(i, g, \lambda)(j, h, \mu) = (-, g\chi_{\alpha,\alpha\beta}h\chi_{\beta,\alpha\beta}, -).$$

Proof. For $\alpha \geq \beta$ and $s, t \in M_\alpha$, we have

$$\begin{aligned}
& (k_\beta, s\chi_{\alpha,\beta}t\chi_{\alpha,\beta}, \xi_\beta) \\
= \;& (k_\beta, s\chi_{\alpha,\beta}, \xi_\beta)(k_\beta, t\chi_{\alpha,\beta}, \xi_\beta) \\
= \;& c_\beta(k_\alpha, s, \xi_\alpha)c_\beta^2(k_\alpha, t, \xi_\alpha)c_\beta \\
= \;& c_\beta(k_\alpha, s, \xi_\alpha)(k_\alpha, t, \xi_\alpha)c_\beta \qquad \text{(by Lemma 1.7)} \\
= \;& c_\beta(k_\alpha, st, \xi_\alpha)c_\beta \\
= \;& (k_\beta, (st)\chi_{\alpha,\beta}, \xi_\beta),
\end{aligned}$$

and hence, $\chi_{\alpha,\beta}$ is a homomorphism.

For any $\alpha \geq \beta \geq \gamma$ and $g \in M_\alpha$, we have

$$\begin{aligned}
& (k_\gamma, g\chi_{\alpha,\beta}\chi_{\beta,\gamma}, \xi_\gamma) \\
= \;& c_\gamma(k_\beta, g\chi_{\alpha,\beta}, \xi_\beta)c_\gamma \\
= \;& c_\gamma c_\beta(k_\alpha, g, \xi_\alpha)c_\beta c_\gamma \\
= \;& c_\gamma c_\beta c_\gamma(k_\alpha, g, \xi_\alpha)c_\gamma c_\beta c_\gamma \quad \text{(by Lemma 1.7)} \\
= \;& c_\gamma(k_\alpha, g, \xi_\alpha)c_\gamma \quad \text{(by Lemma 1.7)} \\
= \;& (k_\gamma, g\chi_{\alpha,\gamma}, \xi_\gamma)
\end{aligned}$$

so that $\chi_{\alpha,\beta}\chi_{\beta,\gamma} = \chi_{\alpha,\gamma}$. It is clear that $\chi_{\alpha,\alpha} = 1_{M_\alpha}$. Therefore the conditions for a strong semilattice are fulfilled.

Let $(i, g, \lambda) \in S_\alpha$ and $(j, h, \mu) \in S_\beta$. Then

$$(i, g, \lambda)(j, h, \mu) = (k, u, \xi) \in S_{\alpha\beta}$$

for some $k \in I_{\alpha\beta}, u \in M_{\alpha\beta}, \xi \in \Lambda_{\alpha\beta}$, and

$$\begin{aligned}
(k, u, \xi) &= (k, 1_{\alpha\beta}, \xi)c_{\alpha\beta}(k, u, \xi)c_{\alpha\beta}(k, 1_{\alpha\beta}, \xi) \quad \text{(by Lemma 1.7)} \\
&= (k, 1_{\alpha\beta}, \xi)c_{\alpha\beta}(i, g, \lambda)(j, h, \mu)c_{\alpha\beta}(k, 1_{\alpha\beta}, \xi) \\
&= (k, 1_{\alpha\beta}, \xi)c_{\alpha\beta}[c_\alpha(i, g, \lambda)c_\alpha]c_{\alpha\beta}[c_\beta(j, h, \mu)c_\beta]c_{\alpha\beta}(k, 1_{\alpha\beta}, \xi) \\
&= (k, 1_{\alpha\beta}, \xi)[c_{\alpha\beta}(k_\alpha, g, \xi_\alpha)c_{\alpha\beta}][c_{\alpha\beta}(k_\beta, h, \xi_\beta)c_{\alpha\beta}](k, 1_{\alpha\beta}, \xi) \quad \text{(by Lemma 1.7)} \\
&= (k, 1_{\alpha\beta}, \xi)(k_{\alpha\beta}, g\chi_{\alpha,\alpha\beta}, \xi_{\alpha\beta})(k_{\alpha\beta}, h\chi_{\beta,\alpha\beta}, \xi_{\alpha\beta})(k, 1_{\alpha\beta}, \xi) \\
&\quad \text{(by the definition of } \chi_{\alpha,\alpha\beta}) \\
&= (k, g\chi_{\alpha,\alpha\beta}h\chi_{\beta,\alpha\beta}, \xi) \\
&= (k, gh, \xi),
\end{aligned}$$

so that $u = gh$, as required. □

2 A structure theorem for ortho-u-monoids

Let Y be a semilattice, $\alpha, \beta \in Y$; $S_\alpha = I_\alpha \times M_\alpha \times \Lambda_\alpha$ a rectangular u-monoid such that $S_\alpha \cap S_\beta \neq \emptyset$ if $\alpha \neq \beta$; $M = [Y; M_\alpha, \chi_{\alpha,\beta}]$ a strong semilattice of M_α; $S = \bigcup_{\alpha \in Y} S_\alpha$.

For any $\alpha, \beta \in Y$, we define the mappings

$$\begin{aligned}
\varphi_{\alpha,\beta}: \;& S_\alpha \times I_\beta \longrightarrow I_{\alpha\beta} \quad \text{by} \quad (a, j) \longmapsto \langle a, j \rangle, \\
\psi_{\alpha,\beta}: \;& \Lambda_\alpha \times S_\beta \longrightarrow \Lambda_{\alpha\beta} \quad \text{by} \quad (\lambda, b) \longmapsto [\lambda, a],
\end{aligned}$$

satisfying the following conditions:

for any $a = (i, g, \lambda) \in S_\alpha$, $b = (j, h, \mu) \in S_\beta$, $c = (k, l, \nu) \in S_\gamma$, $m = (\langle a, j\rangle, g\chi_{\alpha,\alpha\beta} h\chi_{\beta,\alpha\beta}, [\lambda, b]) \in S_{\alpha\beta}$ and $n = (\langle b, k\rangle, h\chi_{\beta,\beta\gamma} l\chi_{\gamma,\beta\gamma}, [\mu, c]) \in S_{\beta\gamma}$,

(P1) If $\alpha \leq \beta$, then $\langle a, j\rangle = i$;

(Q1) If $\alpha \geq \beta$, then $[\lambda, b] = \mu$;

(P2) $\langle m, k\rangle = \langle a, \langle b, k\rangle\rangle$;

(Q2) $[[\lambda, b], c] = [\lambda, n]$.

Then we define a multiplication "$\circ$" on the set $S = \bigcup_{\alpha \in Y} S_\alpha$ by

$$(i, g, \lambda)(j, h, \mu) = (\langle a, j\rangle, g\chi_{\alpha,\alpha\beta} h\chi_{\beta,\alpha\beta}, [\lambda, b]). \tag{2.1}$$

Theorem 2.1. *The above set S forms an ortho-u-monoid under the multiplication given by (2.1), denoted by $S = S(Y, S_\alpha, M, \varphi_{\alpha,\beta}, \psi_{\alpha,\beta})$, and the multiplication "$\circ$" restricted to each rectangular u-monoid S_α coincides with the given multiplication on S_α.*

Conversely, every ortho-u-monoid can be constructed in this manner.

Proof. Firstly, let S be an ortho-u-monoid. Then by Lemma 1.2, we may suppose that S is a semilattice Y of rectangular u-monoids $S_\alpha = I_\alpha \times M_\alpha \times \Lambda_\alpha$.

(1). By Lemma 1.7 and Lemma 1.8, We can define a strong semilattice of semigroups M_α, say $M = [Y; M_\alpha, \chi_{\alpha,\beta}]$ such that for any $\alpha, \beta \in Y$, $a = (i, g, \lambda) \in S_\alpha$ and $b = (j, h, \mu) \in S_\beta$,

$$(i, g, \lambda)(j, h, \mu) = (k', g\chi_{\alpha,\alpha\beta} h\chi_{\beta,\alpha\beta}, \nu') \in S_{\alpha\beta}. \tag{2.2}$$

We call the above semigroup M the u-Clifford component of S.

(2). Next, we prove that $k'[\nu']$ in (2.2) is independent of h and μ [i and g].

Let $(j', h', \mu') \in S$. If $(j, h, \mu)\widetilde{\mathcal{R}}(j', h', \mu')$, then by Lemma 1.6 (i), $j = j'$. Since $\widetilde{\mathcal{R}}$ is a left congruence on S, we have

$$(i, g, \lambda)(j, h, \mu)\widetilde{\mathcal{R}}(i, g, \lambda)(j, h', \mu').$$

By Lemma 1.6 (i), we can deduce that k' in (2.2) is independent of h and μ.

Similarly, we can prove that ν' in (2.2) is independent of i and g.

Now, we write

$$k' = \langle a, j\rangle \textit{ and } \nu' = [\lambda, b]. \tag{2.3}$$

Then by (2.3), we can define

$$\varphi_{\alpha,\beta} : S_\alpha \times I_\beta \longrightarrow I_{\alpha\beta} \text{ by } \quad (a, j) \longmapsto \langle a, j\rangle$$

and

$$\psi_{\alpha,\beta} : \Lambda_\alpha \times S_\beta \longrightarrow \Lambda_{\alpha\beta} \qquad \text{by} \quad (\lambda, b) \longmapsto [\lambda, b]$$

Thus, we can denote the semigroup operation (2.2) of S in terms of $\varphi_{\alpha,\beta}$, $\psi_{\alpha,\beta}$ and $\chi_{\alpha,\beta}$ by

$$(i, g, \lambda)(j, h, \mu) = (\langle a, j\rangle, g\chi_{\alpha,\alpha\beta}h\chi_{\beta,\alpha\beta}, [\lambda, b]). \tag{2.4}$$

If $\alpha \leq \beta$, then $(i, g, \lambda)(j, h, \mu) \in S_\alpha$, and we have

$$\begin{aligned} & (i, g, \lambda)(j, h, \mu) \\ = \; & [(i, 1_\alpha, \lambda)(i, g, \lambda)](j, h, \mu) \\ = \; & (i, 1_\alpha, \lambda)[(i, g, \lambda)(j, h, \mu)] \\ = \; & (i, -, -). \end{aligned}$$

By (2.4), $\langle a, j\rangle = i$. This shows that $\varphi_{\alpha,\beta}$ satisfies the above condition (P1).

Similarly, we can show that $\psi_{\alpha,\beta}$ satisfies the above condition (Q1).

For any $\alpha, \beta \in Y$, $a = (i, g, \lambda) \in S_\alpha$, $b = (j, h, \mu) \in S_\beta$, and $c = (k, l, \nu) \in S_\gamma$, by (2.4), we have

$$\begin{aligned} & [(i, g, \lambda)(j, h, \mu)](k, l, \nu) \\ = \; & (\langle a, j\rangle, g\chi_{\alpha,\alpha\beta}h\chi_{\beta,\alpha\beta}, [\lambda, b])(k, l, \nu) \\ = \; & (\langle m, k\rangle, g\chi_{\alpha,\alpha\beta\gamma}h\chi_{\beta,\alpha\beta\gamma}l\chi_{\gamma,\alpha\beta\gamma}, [[\lambda, b], c]), \end{aligned}$$

and

$$\begin{aligned} & (i, g, \lambda)[(j, h, \mu)(k, l, \nu)] \\ = \; & (i, g, \lambda)(\langle b, k\rangle, h\chi_{\beta,\beta\gamma}l\chi_{\gamma,\beta\gamma}, [\mu, c]) \\ = \; & (\langle a, \langle b, k\rangle\rangle, g\chi_{\alpha,\alpha\beta\gamma}h\chi_{\beta,\alpha\beta\gamma}l\chi_{\gamma,\alpha\beta\gamma}, [\lambda, n]). \end{aligned}$$

Consequently, we obtain $\langle m,k\rangle = \langle a,\langle b,k\rangle\rangle$ and $[[\lambda,b],c] = [\lambda,n]$. This shows that $\varphi_{\alpha,\beta}$, $\psi_{\alpha,\beta}$ satisfy the above conditions (P2), (Q2), respectively.

Secondly, let Y be a semilattice, $\alpha,\beta \in Y$; $S_\alpha = I_\alpha \times M_\alpha \times \Lambda_\alpha$ a rectangular u-monoid such that $S_\alpha \cap S_\beta \neq \emptyset$ if $\alpha \neq \beta$; $M = [Y; M_\alpha, \chi_{\alpha,\beta}]$ a strong semilattice of M_α; and suppose that there exist mappings

$$\varphi_{\alpha,\beta} : S_\alpha \times I_\beta \longrightarrow I_{\alpha\beta}$$

and

$$\psi_{\alpha,\beta} : \Lambda_\alpha \times S_\beta \longrightarrow \Lambda_{\alpha\beta}$$

satisfying the above conditions (P1)-(P2), (Q1)-(Q2). Then, we will claim that

$$S = \bigcup_{\alpha\in Y} (I_\alpha \times M_\alpha \times \Lambda_\alpha)$$

forms an ortho-u-monoid under the above multiplication (2.1).

We first prove that S is a semilattice of rectangular-u-monoids S_α under the operation "$\circ$".

It is clear that S is closed under the operation "$\circ$". Let $\alpha,\beta \in Y$, $a = (i,g,\lambda) \in S_\alpha$, $b = (j,h,\mu) \in S_\beta$, $c = (k,l,\nu) \in S_\gamma$, $m = (\langle a,j\rangle, g\chi_{\alpha,\alpha\beta}h\chi_{\beta,\alpha\beta}, [\lambda, l$ $S_{\alpha\beta}$ and $n = (\langle b,k\rangle, h\chi_{\beta,\beta\gamma}l\chi_{\gamma,\beta\gamma}, [\mu,c]) \in S_{\beta\gamma}$. Then we have

$$\begin{aligned}
& [(i,g,\lambda)(j,h,\mu)](k,l,\nu) & \\
= \ & (\langle a,j\rangle, g\chi_{\alpha,\alpha\beta}h\chi_{\beta,\alpha\beta}, [\lambda,b])(k,l,\nu) & \text{(by (2.1))} \\
= \ & (\langle m,k\rangle, g\chi_{\alpha,\alpha\beta\gamma}h\chi_{\beta,\alpha\beta\gamma}l\chi_{\gamma,\alpha\beta\gamma}[[\lambda,b],c]) & \text{(by (2.1))} \\
= \ & (\langle a,\langle b,k\rangle\rangle, g\chi_{\alpha,\alpha\beta\gamma}h\chi_{\beta,\alpha\beta\gamma}l\chi_{\gamma,\alpha\beta\gamma}, [\lambda,n]) & \text{(by (P2)and (Q2))} \\
= \ & (i,g,\lambda)(\langle b,k\rangle, h\chi_{\beta,\beta\gamma}l\chi_{\gamma,\beta\gamma}, [\mu,c]) & \text{(by (2.1))} \\
= \ & (i,g,\lambda)[(j,h,\mu)(k,l,\nu)] & \text{(by (2.1))}
\end{aligned}$$

Hence S is a semigroup. Also since S is the disjoint union of rectangular u-monoid $S_\alpha(\alpha \in Y)$, and satisfies $S_\alpha S_\beta \subseteq S_{\alpha\beta}$. It follows that S is a semilattice of rectangular-u-monoids S_α under the operation "$\circ$".

By Lemma 1.2, S is an ortho-u-monoid.

Let $(i,g,\lambda),(j,h,\mu) \in S_\alpha$. Then we have

$$\begin{aligned}
(i,g,\lambda)(j,h,\mu) &= (\langle a,j\rangle, g\chi_{\alpha,\alpha}h\chi_{\alpha,\alpha}, [\lambda,b]) && \text{(by (2.1))} \\
&= (i, g\chi_{\alpha,\alpha}h\chi_{\alpha,\alpha}, \mu) && \text{(by (P1) and (Q1))} \\
&= (i, gh, \mu) && \text{(by } \chi_{\alpha,\alpha} = 1_{M_\alpha})
\end{aligned}$$

Hence the multiplication "$\circ$" restricted to each rectangular u-monoid S_α coincides with the given multiplication on S_α. $\square$

Now we consider some special cases of Theorem 2.1.

In Theorem 2.1, if we take $S_\alpha = I_\alpha \times M_\alpha \times \Lambda_\alpha$ a rectangular lc-monoid, and add the following conditions:

(P3) Let $u = (p, r, \omega_1) \in S_\delta, v = (q, s, \omega_2) \in S_\varepsilon$. If

$$\langle a, p\rangle = \langle a, q\rangle, g\chi_{\alpha,\alpha\delta} r\chi_{\delta,\alpha\delta} = g\chi_{\alpha,\alpha\varepsilon} s\chi_{\varepsilon,\alpha\varepsilon}, [\lambda, u] = [\lambda, v],$$

then $\langle a^0, p\rangle = \langle a^0, q\rangle$, where a^0 is the identity of the monoid $H_a^{*,\sim}$.

By Corollary [9] and Theorem 2.1, we have the following result.

Corollary 2.2. *The set S is an ortho-lc-monoid under the multiplication given by (2.1), denoted by $S = S(Y, S_\alpha, M, \varphi_{\alpha,\beta}, \psi_{\alpha,\beta})$, and the multiplication "$\circ$" restricted to each rectangular lc-monoid S_α coincides with the given multiplication on S_α.*

Conversely, every ortho-lc-monoid can be constructed in this manner.

In Theorem 2.1, if we take $S_\alpha = I_\alpha \times M_\alpha \times \Lambda_\alpha$ a rectangular c-monoid, and add the condition (P3) and the following condition:

(Q3) Let $u = (p, r, \omega_1) \in S_\delta, v = (q, s, \omega_2) \in S_\varepsilon$. If

$$\langle u, i\rangle = \langle v, i\rangle, r\chi_{\delta,\alpha\delta} g\chi_{\alpha,\alpha\delta} = s\chi_{\varepsilon,\alpha\varepsilon} g\chi_{\alpha,\alpha\varepsilon}, [\omega_1, a] = [\omega_2, a],$$

then $[\omega_1, a^0] = [\omega_2, a^0]$.

By Corollary 1.4 and Theorem 2.1, we have the following result.

Corollary 2.3. *The set S is an ortho-c-monoid under the above multiplication (1), denoted by $S = S(Y, S_\alpha, M, \varphi_{\alpha,\beta}, \psi_{\alpha,\beta})$, and the multiplication "$\circ$" restricted to each rectangular c-monoid S_α coincides with the given multiplication on S_α.*

Conversely, every ortho-c-monoid can be constructed in this manner.

In Theorem 2.1, if we take $S_\alpha = I_\alpha \times M_\alpha \times \Lambda_\alpha$ a rectangular group, then by Corollary 1.5 and Theorem 2.1, we have the following corollary.

Corollary 2.4. *The set S is an orthogroup under the above multiplication (2.1), denoted by $S = S(Y, S_\alpha, M, \varphi_{\alpha,\beta}, \psi_{\alpha,\beta})$, and the multiplication "$\circ$" restricted to each rectangular group S_α coincides with the given multiplication on S_α.*

Conversely, every orthogroup can be constructed in this manner.

Dedication

This article is dedicated to the 70 birthday conference of Professor Kar Ping Shum who is the research collaborator of Guo Yuqi for many years standing and he is also the supervior of Xueming Ren. In this very special occasion, we wish him to have a long happy life.

References

[1] Petrich, M. A structure theorem for completely regular semigroups. Proc. Amer. Math. Soc., 1987, **99**(4): 617-622.

[2] Petrich, M. and Reilly, N. R. Completely regular semigroups. New York: Wiley-Interscience Publication, 1999.

[3] Shum, K. P., Guo, X. J. and Ren, X. M. (l)-Green's relations and perfect semigroups . Proceedings of AMC 2000, Manila, World Scientific INC 2002: 604-613

[4] Guo, Y. Q., Shum, K. P. and Zhu, P. Y. The structure of left C-*rpp* semigroups. Semigroup Forum, 1995, **50**: 9-23.

[5] Guo, Y.Q. Structure of the weakly left C-semigroups. Chinese Science Bulletin, 1996, **41**(6): 462-467.

[6] Guo, Y. Q., Shum, K. P. and Zhu, P. Y. On quasi-C-semigroups and some special subclasses. Algebra Colloquium, 1999, **6**(1): 105-120.

[7] Ren, X. M. and Shum, K. P., On generalized orthogroups. Comm. in Algebra, 2001, **29**(6): 2341-2361.

[8] Ren, X.M., Shum, K.P. On superabundant semigroups whose set of idempotents form a subsemigroup. Algebra Colloquium, 2007, **14**(2): 215-228.

[9] Guo, Y. Q., Shum, K. P. and Gong, C. M., On $(*, \sim)$-Green's Relations and Ortho-lc-monoids. To appear in Communication in Algebra.

[10] Guo, Y. Q., Gong, C. M. and Ren, X. M. A survey on the origin and developments of Green's relations on semigroups. Journal of Shandong University(Natural Science), 2010, **45**(8): 1-18.

[11] Gong, C. M., Guo, Y. Q., and Ren, X. M., The srtucture of ortho-lc-monoids. To appear in Communication in Algebra.

[12] Lawson, M. V. Rees matrix semigroups. Proc. of the Edinburgh Math. Soc., 1990, **33**: 23-39.

[13] Howie, J. M. Fundamentals of semigroup Theory. Oxford Science Publications, 1995.

[14] Gong, C. M., Guo, Y. Q., and Shum, K. P., Ortho-u-monoids. To appear in Acta Mathematica Sinica.

[15] Shum, K. P. rpp semigroups, its generalizations and special subclasses. Advances in algebra and combinatorics, 303V334, World Sci. Publ., Hackensack, NJ, 2008.

[16] Shum, K. P. and Guo, Yuqi Regular semigroups and their generalizations. Rings, groups, and algebras, 181V226, Lecture Notes in Pure and Appl. Math., 181, Dekker, New York, 1996.

Authors' address:

C. M. Gong, Department of Mathematics, Xi'an University of Architecture and Technology, Xi'an, 710055, China. Email: meigongchu@163.com

Y. Q. Guo, Department of Mathematics, Southwest University, Chongqing, 400715, China. Email: yqguo259@swu.edu.cn

X. M. Ren, Department of Mathematics, Xi'an University of Architecture and Technology, Xi'an, 710055, China. Email: xmren@xauat.edu.cn

Restriction and Ehresmann Semigroups

Victoria Gould

Department of Mathematics, University of York, Heslington, York YO10 5DD, UK,
vargl@york.ac.uk

Abstract. Inverse semigroups form a variety of unary semigroups, that is, semigroups equipped with an additional unary operation, in this case $a \mapsto a^{-1}$. The theory of inverse semigroups is perhaps the best developed within semigroup theory, and relies on two factors: an inverse semigroup S is regular, and has semilattice of idempotents. Three major approaches to the structure of inverse semigroups have emerged. Effectively, they each succeed in classifying inverse semigroups via groups (or groupoids) and semilattices (or partially ordered sets). These are (a) the Ehresmann-Schein-Nambooripad characterisation of inverse semigroups in terms of inductive groupoids, (b) Munn's use of fundamental inverse semigroups and his construction of the semigroup T_E from a semilattice E, and (c) McAlister's results showing on the one hand that every inverse semigroup has a proper (E-unitary) cover, and on the other, determining the structure of proper inverse semigroups in terms of groups, semilattices and partially ordered sets.
The aim of this article is to explain how the above techniques, which were developed to study inverse semigroups, may be adapted for certain classes of bi-unary semigroups. The classes we consider are those of restriction and Ehresmann semigroups. The common feature is that the semigroups in each class possess a semilattice of idempotents; however, there is no assumption of regularity.

For Professor K.P. Shum on the occasion of his 70th birthday

1. Introduction

A semigroup S is *(von Neumann) regular* if for every $a \in S$ there exists $b \in S$ such that $a = aba$. Much of the early push in semigroup theory was to study and characterise regular semigroups (whether by means of a structure theory, or by representations) and in particular *inverse* semigroups, that is, regular semigroups

2000 Mathematics Subject Classification: 20 M 10
Keywords: *restriction semigroup, Ehresmann semigroup, proper cover*

with commuting idempotents. This direction was entirely natural from the point of view that every semigroup S embeds into a full transformation semigroup $\mathcal{T}_S$, which is regular. Moreover, the use of Green's relations yields, in many cases, vital information about the structure and properties of regular semigroups. But, just as group theory cannot be reduced to the theory of symmetric groups, neither can semigroup theory be boiled down to the study of full transformation (and hence regular) semigroups.

This article will survey three techniques for studying each of two classes of semigroups that may be regarded as non-regular analogues of inverse semigroups. For the first class, that of restriction semigroups, the connections with the inverse case may be clearly seen. For the second, that of Ehresmann semigroups, new insights are required. In fact, we have four classes of semigroups, since, unlike the case for inverse semigroups, which are defined in a left-right dual manner, restriction and Ehresmann semigroups come in left (and, dually, right) and two-sided versions. Left restriction and left Ehresmann semigroups are varieties of unary semigroups, whereas restriction and Ehresmann semigroups are varieties of bi-unary semigroups. Note that a semigroup in each of these classes contains a distinguished semilattice, which is the image of the unary operation(s). As classes of semigroups, all of these classes contain the class of inverse semigroups. The latter is itself a variety of unary semigroups, where the unary operation is $a \mapsto a^{-1}$, but we remark that the unary operations we employ are different.

The three approaches that we consider are inspired by those in the classical theory of inverse semigroups, where they each succeed in classifying inverse semigroups via groups (or groupoids) and semilattices (or partially ordered sets). They are (a) the Ehresmann-Schein-Nambooripad characterisation of inverse semigroups in terms of inductive groupoids (hereafter referred to as the *categorical* approach), (b) Munn's use of fundamental inverse semigroups and his construction of the semigroup T_E from a semilattice E (hereafter referred to as the *fundamental* approach), and (c) McAlister's results showing on the one hand that every inverse semigroup has a proper (E-unitary) cover, and on the other, determining the structure of proper inverse semigroups in terms of groups, semilattices and partially ordered sets (hereafter referred to as the *covering* approach).

In Section 2 we introduce the varieties of (bi)-unary semigroups under consideration. We present them as both varieties, and as determined by relations that may be thought of as analogues of Green's relations. The three subsequent sections take each of the above methods - categorical, fundamental and covering - and examine how they can be used to study restriction and Ehresmann semigroups. Our aim is to give as complete a picture as possible in the two-sided case, but, in order to do so, we must make some mention of the one-sided varieties.

2. Preliminaries

The classes of semigroups we consider may be arrived at in two ways: by using relations $\widetilde{\mathcal{R}}_E$ and $\widetilde{\mathcal{L}}_E$, which may be thought of as 'generalisations' of Green's relations $\mathcal{R}$ and $\mathcal{L}$, or as varieties of (bi)-unary semigroups. We give both descriptions, beginning with the former, since this was the original route taken by the 'York school', to which the author belongs. Further details may be found in the notes [21].

Let S be a semigroup and let $E \subseteq E(S)$, where $E(S)$ will always denote the set of idempotents of S. The relation $\widetilde{\mathcal{R}}_E$ is defined on S by the rule that for any $a, b \in S$, we have $a \,\widetilde{\mathcal{R}}_E\, b$ if

$$ea = a \Leftrightarrow eb = b, \text{ for all } e \in E.$$

Where $E = E(S)$ it is usual to drop the subscript and write $\widetilde{\mathcal{R}}_{E(S)}$ more simply as $\widetilde{\mathcal{R}}$. It is easy to see that $\mathcal{R} \subseteq \widetilde{\mathcal{R}}_E$, moreover if S is regular, then $\mathcal{R} = \widetilde{\mathcal{R}}$. For, in this case, if $a \,\widetilde{\mathcal{R}}\, b$, then choosing $x, y \in S$ with $a = axa$ and $b = byb$, we have that $ax, by \in E(S)$ and so $a = bya$ and $b = axb$. In general, for any S and any $e, f \in E \subseteq E(S)$, we see that $e \,\mathcal{R}\, f$ if and only if $e \,\widetilde{\mathcal{R}}_E\, f$.

We say that the semigroup S is *weakly left E-abundant* if every $\widetilde{\mathcal{R}}_E$-class contains an element of E (the reader should be aware that some authors also insist that $\widetilde{\mathcal{R}}_E$ be a left congruence). Notice that if $a \,\widetilde{\mathcal{R}}_E\, e$ where $e \in E$, then $ea = a$. If E is a semilattice in S (by which we mean, a commutative subsemigroup of idempotents of S), then it is clear that any $\widetilde{\mathcal{R}}_E$-class contains at most one element of E. In this case we denote the unique idempotent of E in the $\widetilde{\mathcal{R}}_E$-class of a, if it exists, by a^+.

Definition 2.1. A semigroup S is *left Ehresmann*, or *left E-Ehresmann*, if E is a semilattice in S, every $\widetilde{\mathcal{R}}_E$-class contains a (unique) element of E, and $\widetilde{\mathcal{R}}_E$ is a left congruence.

Some remarks on Definition 2.1 are appropriate. El Qallali [9] introduced the relation $\widetilde{\mathcal{R}}$ and later Lawson [29] defined $\widetilde{\mathcal{R}}_E$. In addition they both used the left/right duals $\widetilde{\mathcal{L}}$ and $\widetilde{\mathcal{L}}_E$. El Qallali talks of a semigroup as being *semi-abundant* if every $\widetilde{\mathcal{R}}$- and $\widetilde{\mathcal{L}}$-class contains an idempotent, and Lawson calls a semigroup *E-semiabundant* if every $\widetilde{\mathcal{R}}_E$- and $\widetilde{\mathcal{L}}_E$-class contains an element of E. From now on we drop the hyphen that follows the prefix 'semi' in various articles. We recall that a semigroup S is *abundant* [12] if every $\mathcal{R}^*$-class and $\mathcal{L}^*$-class contains an idempotent, where $\mathcal{R}^*$ is defined by the rule that $a \,\mathcal{R}^*\, b$ if and only if for all $x, y \in S^1$,

$$xa = ya \Leftrightarrow xb = yb$$

and $\mathcal{L}^*$ is the dual. An abundant semigroup is *adequate* [10, 11] if $E(S)$ is a semilattice. It is clear that $\mathcal{R} \subseteq \mathcal{R}^* \subseteq \widetilde{\mathcal{R}}_E$, so that (E)-semiabundant semigroups were being thought of as a generalisation of abundant semigroups, which, in turn, are a generalisation of regular semigroups. Hence if E is a semilattice, then an E-semiabundant semigroup, which Lawson calls *E-semiadequate*, is a generalisation of an adequate semigroup, and again, all inverse semigroups are adequate. Note that in an inverse semigroup, $a^+ = aa^{-1}$. Whereas $\mathcal{R}$ and $\mathcal{R}^*$ are always left congruences, the same is not true of $\widetilde{\mathcal{R}}_E$, which is why the extra condition in Definition 2.1 appears.

Specialising to the case where E forms a semilattice, it was again Lawson [29] who coined the term 'Ehresmann semigroup', making the connection with the work of C. Ehresmann on small ordered categories [8], and that of Schein and Nambooripad on inductive groupoids, which we will come to later. Lawson only considered the two-sided version of Definition 2.1 and to 'keep track' of the idempotents under consideration, he talks of (S, E) as being Ehresmann; subsequent authors have said that S is 'E-Ehresmann' or just 'Ehresmann', where E is understood. Similar conventions apply in the one sided case.

From the comments above, it is clear that an inverse semigroup S is left $E(S)$-Ehresmann. Inverse semigroups form a variety, not of semigroups, but of unary semigroups, that is, of semigroups possessing a unary operation. In this case the operation is $a \mapsto a^{-1}$, where a^{-1} is the unique element such that

$$a = aa^{-1}a \text{ and } a^{-1} = a^{-1}aa^{-1}.$$

A set of three identities defining inverse semigroups is given in [1].

Let S be left E-Ehresmann. By definition, every $\widetilde{\mathcal{R}}_E$-class contains a (unique) element of E, giving a unary operation $a \mapsto a^+$ on S. We may thus regard S as a unary semigroup; the reader should bear in mind that we therefore have two different unary operations defined on an inverse semigroup.

Lemma 2.2. [21] *A semigroup S is left E-Ehresmann such that for any $a \in S$ we have that $a \,\widetilde{\mathcal{R}}_E\, a^+$ where $a^+ \in E$, if and only if it satisfies the identities*

$$x^+x = x,\ (x^+y^+)^+ = x^+y^+,\ x^+y^+ = y^+x^+,\ x^+(xy)^+ = (xy)^+,\ (xy)^+ = (xy^+)^+$$

and

$$E = \{a^+ : a \in S\}.$$

Proof. The result follows from [21], with the exception of showing that the given identities imply $(x^+)^+ = x^+$.

Suppose the identities hold. Then for any $a \in S$,

$$a^+ = (a^+a)^+ = (a^+a^+)^+ = a^+a^+,$$

using the first, fifth and second identities. Then

$$a^+ = a^+a^+ = (a^+a^+)^+ = (a^+)^+,$$

again from the second identity. □

From the above, left Ehresmann semigroups form a variety of algebras. We shall hereafter regard a left Ehresmann semigroup S as a unary semigroup and always use E (or, occasionally, E_S,) for the image of the unary operation. We refer to E as the *distinguished semilattice*. If the reader finds such semigroups a little esoteric, then we point out that the more familiar *left adequate* semigroups (where, following the standard pattern of terminology, a semigroup S is left adequate if $E(S)$ is a semilattice and every $\mathcal{R}^*$-class contains an idempotent) form a generating sub-quasivariety of the variety of left Ehresmann semigroups [18].

Right Ehresmann semigroups are defined dually, where now we use a^* to denote the idempotent in the $\widetilde{\mathcal{L}}_E$-class of a. A semigroup is *Ehresmann* if it is both left and right Ehresmann with respect to the same distinguished semilattices. Ehresmann semigroups form a variety of bi-unary semigroups (that is, of semigroups possessing two unary operations). The defining identities (other than associativity) are those of Lemma 2.2, their duals, together with

$$(x^+)^* = x^+ \text{ and } (x^*)^+ = x^*.$$

An inverse semigroup is therefore Ehresmann with $a^+ = aa^{-1}$ and $a^* = a^{-1}a$.

Definition 2.3. A semigroup S is *left restriction* (or left E-restriction), if it is left Ehresmann and satisfies the *ample condition*

$$ae = (ae)^+a$$

for all $a \in S, e \in E$.

Lemma 2.4. [21] *A semigroup S is left restriction such that for any $a \in S$ we have that $a \,\widetilde{\mathcal{R}}_E\, a^+$ where $a^+ \in E$, if and only if it satisfies the identities*

$$x^+x = x,\ x^+y^+ = y^+x^+,\ (x^+y)^+ = x^+y^+,\ xy^+ = (xy)^+x$$

and

$$E = \{a^+ : a \in S\}.$$

The presence of the last identity is therefore very strong, and corresponds to the ample condition. It tells us that in any product, we can replace an idempotent of E on the right by one on the left. We remark that any inverse semigroup S is left restriction, since for any $a, b \in S$,

$$(ab)^+a = (ab)(ab)^{-1}a = a(bb^{-1})(a^{-1}a) = a(a^{-1}a)(bb^{-1}) = abb^{-1} = ab^+.$$

A left restriction semigroup is *left ample* (formerly, *left type A*) if $\widetilde{\mathcal{R}}_E = \mathcal{R}^*$. In this case, we are forced to have $E = E(S)$, since for any $e \in E(S)$,

$$ee^+ = (ee)^+e = e^+e = e$$

so that as $e = e^+e = ee$ and $e \,\mathcal{R}^*\, e^+$,

$$e^+ = e^+e^+ = ee^+ = e.$$

Left ample semigroups have been widely studied, beginning with [10]. They form a generating sub-quasivariety of the variety of left restriction semigroups (see the description of free objects in [16]).

The following result is essentially folklore. The first time it appears in print is in the work of Trokhimenko [45]. To set the scene, we recall that the *symmetric inverse semigroup* $\mathcal{I}_X$ on a set X is the inverse semigroup of partial one-one maps of X, under left-to-right composition of partial maps. By the Vagner-Preston Representation Theorem [23], every inverse semigroup embeds in some $\mathcal{I}_X$. Indeed every left ample semigroup embeds into some $\mathcal{I}_X$ as a unary semigroup [11]. Of course, $\mathcal{I}_X$ is contained in the larger semigroup $\mathcal{PT}_X$ of partial maps of X.

Proposition 2.5. [45, 21] *A unary semigroup S is left restriction if and only if it is isomorphic to a subalgebra of $\mathcal{PT}_X$, where the unary operation on $\mathcal{PT}_X$ is given by $\alpha \mapsto \alpha^+ = I_{\mathrm{dom}\,\alpha}$.*

It is in essence the above result that determines the importance of left restriction semigroups. Owing to their intimate connections with functions, they have arisen in a number of contexts, and under a number of names, in the last half century. We cite here [5, 15, 24, 43, 31]. Further background and references are given in [21].

The term 'restriction semigroup' is taken from the work of the category theorists, Cockett and Lack [5], where in fact they are using the term for what we call *right restriction* semigroups, where such unary semigroups are defined in a dual manner to left restriction semigroups. We say that a semigroup is *restriction* if it is left and right restriction with respect to the same set of idempotents. Hence restriction semigroups form a variety of bi-unary semigroups.

Clearly inverse semigroups are restriction. We will see that the key to developing analogues of results and techniques for inverse semigroups to the class of (left) restriction semigroups is the existence of the ample condition(s). In the more general case of (left) Ehresmann semigroups, new techniques, or at least new insights, are often required. The next sections will provide ample evidence for this claim.

As mentioned earlier, our thrust here is to consider restriction and Ehresmann semigroups, that is, the two-sided versions. But, there are situations in which these semigroups are harder to deal with than their left-handed cousins. For example, an ample semigroup S (that is, S is both left, and dually, right, ample) must embed into inverse semigroups I and J in a way that preserves $^+$ and *, respectively. However, it is undecidable whether a finite ample semigroup embeds as a bi-unary semigroup into an inverse semigroup [19]. In the final section we find it useful to outline the approach in the one-sided case to inform our discussion in the two-sided.

3. The categorical approach

We begin by outlining the connection between inverse semigroups and inductive groupoids. Of all the approaches to inverse semigroups, this one makes most explicit use of the fact that every inverse semigroup possesses a natural partial order $\leq$, that is, a partial order on S that is compatible with multiplication, and restricts to the usual ordering on the semilattice $E(S)$. We recall that $\leq$ is defined on an inverse semigroup S by the rule that

$$a \leq b \text{ if and only if } a = eb \text{ for some } e \in E(S)$$

and it is easy to see that

$$a \leq b \quad \text{if and only if } a = aa^{-1}b.$$

Further, this relation may also be defined by its dual, that is,

$$a \leq b \text{ if and only if } a = bf \text{ for some } f \in E(S) \text{ if and only if } a = ba^{-1}a.$$

By Vagner-Preston, S embeds into some $\mathcal{I}_X$, and for $\alpha, \beta \in \mathcal{I}_X$, we have $\alpha \leq \beta$ if and only if α is a restriction of β.

All of the categories that we consider will be small (that is, the collection of morphisms is a *set*) unless otherwise stated. We will denote the domain and range of a morphism p in a category $\mathbf{C}$ (that is, $p \in \operatorname{Mor}\mathbf{C}$) by $\mathbf{d}(p)$ and $\mathbf{r}(p)$, respectively. Furthermore, we identify an object α of $\mathbf{C}$ (that is, $\alpha \in \operatorname{Ob}\mathbf{C}$) with the identity map I_α at that object. We think of a category $\mathbf{C}$ as a pair $\mathbf{C} = (C, \cdot)$, where C corresponds to $\operatorname{Mor}\mathbf{C}$ and $\cdot$ is the partial binary operation of 'composition' of morphisms. The existence of domains and ranges tells us when the product $p \cdot q$ exists - it exists if and only if $\mathbf{r}(p) = \mathbf{d}(q)$. For a category $\mathbf{C}$ we denote by $E_\mathbf{C}$ the set of identity maps (or objects) of $\mathbf{C}$.

Definition 3.1. A *groupoid* is a category $\mathbf{G} = (G, \cdot)$ in which for every $p \in G$ we have $p^{-1} \in G$ with

$$p \cdot p^{-1} = I_{\mathbf{d}(p)} \text{ and } p^{-1} \cdot p = I_{\mathbf{r}(p)}.$$

A one-object groupoid is precisely a group.

Let S be an inverse semigroup. It is easy to construct a groupoid $\mathcal{G}(S) = (G, \cdot)$ from S as follows:

$$G = \operatorname{Mor}\mathcal{G}(S) = S, \quad \operatorname{Ob}\mathcal{G}(S) = E(S), \ \mathbf{d}(a) = aa^{-1}, \mathbf{r}(a) = a^{-1}a$$

and when $\mathbf{r}(a) = \mathbf{d}(b)$,

$$a \cdot b = ab.$$

On the other hand, let $\mathbf{G} = (G, \cdot)$ be a small groupoid. Let S be the semigroup obtained from $\mathbf{G}$ by declaring all undefined products to be 0. Then $S = G \cup \{0\}$ is an inverse semigroup with 0 with $E(S) = E_\mathbf{G} \cup \{0\}$. The problem is, semigroups obtained in this manner are always primitive, that is, non-zero idempotents are incomparable under the natural partial order. For this reason, we need to consider *inductive groupoids*.

We begin with the more general notion of an ordered category, since it will be this we will require subsequently.

Definition 3.2. Let $\mathbf{C} = (C, \cdot)$ be a category. Suppose that $\leq$ is a partial order on C such that:

(OC1) $x \leq y$ implies that $\mathbf{d}(x) \leq \mathbf{d}(y)$ and $\mathbf{r}(x) \leq \mathbf{r}(y)$;
(OC2) $x \leq y, u \leq v, \exists x \cdot u, \exists y \cdot v$ implies that $x \cdot u \leq y \cdot v$;

(OC3) if $a \in C$ and $e \in E_{\mathbf{C}}$ with $e \leq \mathbf{d}(a)$, then there exists a unique ***restriction*** $(e|a) \in C$ with $\mathbf{d}(e|a) = e$ and $(e|a) \leq a$;

(OC4) if $a \in G$ and $e \in E_{\mathbf{C}}$ with $e \leq \mathbf{r}(a)$, then there exists a unique ***co-restriction*** $(a|e) \in C$ with $\mathbf{r}(a|e) = e$ and $(a|e) \leq a$.

Then $\mathbf{C} = (C, \cdot, \leq)$ is called an *ordered category.* If in addition

(I) $E_{\mathbf{C}}$ is a semilattice

then $\mathbf{C} = (C, \cdot, \leq)$ is called an *inductive category.*

Definition 3.3. Let $\mathbf{G} = (G, \cdot)$ be a groupoid. Suppose that $\leq$ is a partial order on G such that $\mathbf{G} = (G, \cdot, \leq)$ is an ordered (inductive) category. If in addition

(OG1) $x \leq y$ implies that $x^{-1} \leq y^{-1}$,

then $\mathbf{G} = (G, \cdot, \leq)$ is called an *ordered (inductive) groupoid.*

We note that (OG1) and (OC2) together imply (OC1), so that, if we only wished to define ordered *groupoids*, this could have been done without mention of (OC1).

Let S be an inverse semigroup. Let $\mathcal{G}(S) = (S, \cdot)$ be defined as above. Then $\mathcal{G}(S) = (S, \cdot, \leq)$ (where $\leq$ is the partial order in S) is an inductive groupoid with

$$(e|a) = ea \text{ and } (a|e) = ae.$$

Conversely, let $\mathbf{G} = (G, \cdot, \leq)$ be an inductive groupoid. The *pseudo-product* $\otimes$ is defined on G by the rule that

$$a \otimes b = (a|\mathbf{r}(a) \wedge \mathbf{d}(b)) \cdot (\mathbf{r}(a) \wedge \mathbf{d}(b)|b).$$

Then $\mathcal{S}(\mathbf{G}) = (G, \otimes)$ is an inverse semigroup (having the same partial order as G) such that the inverse of a in $\mathcal{S}(\mathbf{G})$ coincides with the inverse of a in $\mathbf{G}$. We remark that establishing associativity is the tricky part in proving this assertion.

We have outlined the object correspondence in the theorem below, which has become known as the Ehresmann-Schein-Nambooripad or ESN Theorem, due to its varied authorship. Further details may be found in [30]. The categories in the result below are, of course, general categories. An inductive functor between two inductive groupoids is a functor preserving the partial ordering and meets of identities.

Theorem 3.4. *The category* $\mathbf{I}$ *of inverse of semigroups and morphisms is isomorphic to the category* $\mathbf{G}$ *of inductive groupoids and inductive functors.*

As Lawson points out in his book [30], Theorem 3.4 is not the end of the story. If we want to work with an inverse semigroup S, we can associate to it an inductive groupoid $\mathcal{G}(S)$. We can then work in the larger category of ordered groupoids, and, having obtained a result for $\mathcal{G}(S)$, translate it back to S. Indeed, Theorem 5.3 can be proven in this manner.

A few words of background. In the 1950s Ehresmann developed what are called here inductive groupoids as a way of providing an abstract framework for pseudogroups, where a *pseudogroup* is an inverse semigroup of partial homeomorphims between open sets of a topological space. Indeed he developed a much wider categorical framework. Pseudogroups had been introduced as the suitable vehicle for characterising differentiable manifolds, in the same way that Klein's Erlanger Programme attempted to characterise geometries by their groups of symmetries. Ehresmann's work and its relation to Semigroup Theory has been championed by Lawson, as witnessed by [29, 30]. In the mid 1960s Schein [42] realised that, dropping the differential geometry framework, Ehresmann's ideas could be used to show the connection between inverse semigroups and inductive groupoids. Western semigroup theorists probably first became aware of this work when a translation appeared in 1979 [44]. Nambooripad shortly thereafter [40] extended Schein's results to regular semigroups (using ordered categories with bi-ordered sets of idempotents); he was the first to state the correspondence results at the category level. Again, further details and references may be found in [30], from which this historical sketch is taken.

We now show how this theory adapts to the wider classes of restriction and Ehresmann semigroups.

3.1. **Restriction semigroups.** For restriction semigroups and the sub-quasivariety of ample semigroups, there are smooth and natural analogues of the ESN Theorem.

We first observe that, as for inverse semigroups, a left restriction semigroup S is naturally partially ordered by $\leq$ where here

$$a \leq b \text{ if and only if } a = eb \text{ for some } e \in E \text{ if and only if } a = a^+b$$

and this partial order is 'inherited' from the embedding of S into $\mathcal{PT}_X$. Moreover, if S is restriction, then we have the equivalent characterisation that

$$a \leq b \text{ if and only if } a = bf \text{ for some } f \in E \text{ if and only if } a = ba^*.$$

Let S be a restriction semigroup. We can construct a category $\mathcal{C}(S) = (C, \cdot)$ from S as follows:

$$C = \operatorname{Mor}\mathcal{C}(S) = S, \quad \operatorname{Ob}\mathcal{C}(S) = E(S), \ \mathbf{d}(a) = a^+, \ \mathbf{r}(a) = a^*$$

and when $\mathbf{r}(a) = \mathbf{d}(b)$,

$$a \cdot b = ab$$

and then an inductive category $\mathcal{C}(S) = (C, \cdot, \leq)$ where $\leq$ is the natural partial order inherited from S, with

$$(e|a) = ea \text{ and } (a|e) = ae.$$

Conversely, let $\mathbf{C} = (C, \cdot, \leq)$ be an inductive category. The *pseudo-product* $\otimes$ is defined on C in exactly the same way as for inductive groupoids. Then $\mathcal{S}(\mathbf{C}) = (C, \otimes)$ is a restriction semigroup having the same partial order as C, with $a^+ = \mathbf{d}(a)$ and $a^* = \mathbf{r}(a)$.

Again, the 'big' categories in the result below are general categories; inductive functors are defined as between inductive groupoids.

Theorem 3.5. [29] *The category of restriction semigroups and morphisms is isomorphic to the category of inductive categories and inductive functors.*

The forerunner to Theorem 3.5 appears in the ample case in [2], albeit with different terminology. We say that a category $\mathbf{C} = (C, \cdot)$ is *cancellative* if for all $a, b, c \in C$,

$$\exists a \cdot b, \exists a \cdot c \text{ and } a \cdot b = a \cdot c \text{ implies that } b = c$$

and

$$\exists b \cdot a, \exists c \cdot a \text{ and } b \cdot a = c \cdot a \text{ implies that } b = c.$$

Corollary 3.6. [2] *The category of ample semigroups and morphisms is isomorphic to the category of inductive cancellative categories and inductive functors.*

3.2. **Ehresmann semigroups.** Let S be an Ehresmann semigroup. Again, we may define relations $\leq_r$ and $\leq_\ell$ on S by the rule that

$$a \leq_r b \text{ if and only if } a = eb \text{ for some } e \in E \text{ if and only if } a = a^+b$$

and dually

$$a \leq_\ell b \text{ if and only if } a = bf \text{ for some } f \in E \text{ if and only if } a = ba^*.$$

It is easy to see that $\leq_r$ ($\leq_\ell$) are partial orders, that restrict to the usual partial order on E and which are right (left) compatible with multiplication. However, as explained in [29], unlike the case for restriction semigroups, we need not have $\leq_r = \leq_\ell$ and hence we need not have that either relation is compatible with multiplication *on both sides*. For this reason, we cannot describe an Ehresmann semigroup by merely using an inductive category.

Instead, we consider what Lawson calls *Ehresmann categories.* Our account below is entirely taken from [29], but we use different terminology in parts, since an 'ordered category' in [29] has a weaker definition than that we have used so far in this survey, and we are composing morphisms from left to right.

Definition 3.7. A *category with order* $\mathbf{C} = (C, \cdot, \leq)$ is a category $\mathbf{C} = (C, \cdot)$ partially ordered by $\leq$ such that (OC1), (OC2) and (R) hold:

(R) if $x \leq y$, $\mathbf{d}(x) = \mathbf{d}(y)$ and $\mathbf{r}(x) = \mathbf{r}(y)$, then $x = y$.

Let S be an Ehresmann semigroup. As in the restriction case, we may define a category $\mathcal{C}(S) = (C, \cdot)$ as follows:

$$C = \operatorname{Mor} \mathcal{C}(S) = S, \quad \operatorname{Ob} \mathcal{C}(S) = E(S), \ \mathbf{d}(a) = a^+, \ \mathbf{r}(a) = a^*$$

and when $\mathbf{r}(a) = \mathbf{d}(b)$,

$$a \cdot b = ab.$$

This category is partially ordered by both $\leq_r$ and $\leq_\ell$. Moreover, we obtain categories with order $\mathcal{C}(S)_r = (C, \cdot, \leq_r)$ and $\mathcal{C}(S)_\ell = (C, \cdot, \leq_\ell)$. This is in spite of the fact that $\leq_r$ and $\leq_\ell$ may not be compatible with multiplication everywhere *on S*.

But by restricting the products under consideration in this way we find that (OC2) holds. For example, in $\mathcal{C}(S)_r$, if

$$x \leq_r y,\ u \leq_r v,\ \exists x \cdot u \text{ and } \exists y \cdot v$$

then

$$x = x^+ y,\ u = u^+ v,\ x^* = u^+ \text{ and } y^* = v^+.$$

Calculating,

$$xu = x(u^+ v) = x(x^* v) = (xx^*)v = xv = x^+ yv$$

so that

$$x \cdot u = xu \leq_r yv = y \cdot v,$$

as required.

However, $\mathcal{C}(S)_r$ and $\mathcal{C}(S)_\ell$ may not have both restrictions and co-restrictions. This leads to the following definition.

Definition 3.8. An *Ehresmann category* $\mathbf{C} = (C, \cdot, \leq_r, \leq_\ell)$ is a category $\mathbf{C} = (C, \cdot)$ equipped with two relations $\leq_r$ and $\leq_\ell$ satisfying the following axioms:

(EC1) $(C, \leq_r)$ is a category with order satisfying (OC3);
(EC21) $(C, \leq_\ell)$ is a category with order satisfying (OC4);
(EC3) if $e, f \in E_{\mathbf{C}}$, then $e \leq_r f$ if and only if $e \leq_\ell f$;
(I) $E_{\mathbf{C}}$ is a semilattice under $\leq_r$ (equivalently, $\leq_\ell$);
(EC4) $\leq_r \circ \leq_\ell = \leq_\ell \circ \leq_r$;
(EC5) if $x \leq_r y$ and $f \in E_{\mathbf{C}}$, then $(x|\mathbf{r}(x) \wedge f) \leq_r (y|\mathbf{r}(y) \wedge f)$;
(EC6) if $x \leq_\ell y$ and $f \in E_{\mathbf{C}}$, then $(\mathbf{d}(x) \wedge f|x) \leq_\ell (\mathbf{d}(y) \wedge f|y)$.

Let S be an Ehresmann semigroup and let $\mathcal{C}(S)$ be defined as above. Then $\mathcal{C}(S) = (S, \cdot, \leq_r, \leq_\ell)$ is an Ehresmann category.

Conversely, if $\mathbf{C} = (C, \cdot, \leq_r, \leq_\ell)$ is an Ehresmann category, we may define the pseudo-product in the usual way. As proven in [29], $\mathcal{S}(\mathbf{C}) = (C, \otimes)$ is an Ehresmann semigroup having the same partial orders as $\mathbf{C}$, such that $a^+ = \mathbf{d}(a)$ and $a^* = \mathbf{r}(a)$ for any $a \in C$.

We say that a functor between two Ehresmann categories is *inductive* if it preserves *both* partial orders, and meets of identities.

Theorem 3.9. [29] *The category of Ehresmann semigroups and morphisms is isomorphic to the category of Ehresmann categories and inductive functors.*

Theorem 3.9 is specialised in [29] to the case of adequate semigroups and to restriction semigroups, obtaining Theorem 3.5.

3.3. **Left restriction semigroups.** If S is a left restriction semigroup, then S possesses a natural partial order, but we cannot make a category from S in the above manner, since, although we can define a domain a^+ for $a \in S$, we cannot define a range. With this in mind, the author and her former student Hollings introduced the notion of a *left constellation.*

Definition 3.10. Let P be a set, let $\cdot$ be a partial binary operation and let $^+$ be a unary operation on P with image E, such that E consists of idempotents. We call $(P, \cdot, ^+)$ a *left constellation* if the following axioms hold:

(C1) $\exists x \cdot (y \cdot z) \Rightarrow \exists (x \cdot y) \cdot z$, in which case, $x \cdot (y \cdot z) = (x \cdot y) \cdot z$;
(C2) $\exists x \cdot (y \cdot z) \Leftrightarrow \exists x \cdot y$ and $\exists y \cdot z$;
(C3) for each $x \in P$, x^+ is the unique left identity of x in E;
(C4) $a \in P$, $g \in E$, $\exists a \cdot g \Rightarrow a \cdot g = a$.

It has been suggested by Cockett that left constellations should be called left categories, and we are inclined to agree. Hollings and the author defined the notion of an *inductive left constellation*, and proved an analogue of Theorem 3.4 for left restriction semigroups and inductive left constellations [20].

To the author's knowledge, thus far there has been no association of a left Ehresmann semigroup to a left constellation.

4. The fundamental approach

As in Section 3, we begin by outlining the techniques used originally to study inverse semigroups.

Let S be an inverse semigroup and let us denote the largest congruence contained in $\mathcal{H}$ by μ; equivalently, μ is the greatest idempotent separating congruence on S [35]. By [22], μ is given by the formula

$$a \,\mu\, b \text{ if and only if } a^{-1}ea = b^{-1}eb \text{ for all } e \in E(S).$$

We will say that a regular semigroup is *fundamental* if μ is trivial.

The seminal paper of Munn [36] shows how to construct a fundamental inverse semigroup having a given semilattice of idempotents, as follows.

Let E be a semilattice and let T_E be the subset of $\mathcal{I}_E$ consisting of all isomorphisms between principal ideals of E. It is not hard to check that T_E is a subsemigroup of $\mathcal{I}_E$, that is clearly inverse. Moreover, T_E is fundamental and has semilattice of idempotents isomorphic to E. Unsurprisingly, semigroups of the form T_E have come to be called *Munn semigroups*.

Conversely, any fundamental inverse semigroup with semilattice of idempotents E embeds in T_E, as we now explain.

Let S be an inverse semigroup with $E(S) = E$. For any $a \in S$ we define a partial map α_a of E by:

$$\operatorname{dom} \alpha_a = Eaa^{-1} \text{ and } e\alpha_a = a^{-1}ea \text{ for all } e \in Eaa^{-1}.$$

Then $\alpha_a : Eaa^{-1} \to Ea^{-1}a$ is an isomorphism between principal ideals of E, that is, it lies in T_E.

For any set X we denote by I_X the identity map of X.

Theorem 4.1. [36] *Let E be a semilattice. Then T_E is a fundamental inverse semigroup with semilattice of idempotents*

$$\overline{E} = \{I_{Ee} : e \in E\}$$

isomorphic to E.

Conversely, let S be an inverse semigroup with $E(S) = E$. Let $\theta : S \to T_E$ be given by

$$a\theta = \alpha_a.$$

Then θ is morphism with kernel μ such that

$$\theta|_E : E \to \overline{E}$$

is an isomorphism.

It follows from the above that fundamental inverse semigroups are precisely full subsemigroups of Munn semigroups, and one strategy for describing *any* inverse semigroup S with semilattice E is to solve an extension problem for a full subsemigroup of T_E by μ. Munn [37] also explains how, for example, his results can be used to obtain arbitrary 0-bisimple inverse semigroups in terms of semilattices and groups.

4.1. **Restriction semigroups.** When moving from inverse semigroups to restriction and then Ehresmann semigroups, the step to restriction semigroups is, as in Section 3, straightforward.

Let S be an Ehresmann semigroup with distinguished semilattice E. We denote by μ_E the greatest congruence contained in $\widetilde{\mathcal{H}}_E$, where $\widetilde{\mathcal{H}}_E = \widetilde{\mathcal{R}}_E \cap \widetilde{\mathcal{L}}_E$, and say that S is *fundamental* (or *E-fundamental*) if μ_E is trivial. This extends the definition given above, as in an inverse semigroup, $\mathcal{H} = \widetilde{\mathcal{H}}$. However, it is also not unusual to say that an arbitrary semigroup is fundamental if the largest congruence contained in $\mathcal{H}$ is trivial, or if the largest idempotent separating congruence is trivial - neither of which concepts is quite right for us here. We remark that for an Ehresmann semigroup, or indeed an ample semigroup, there may be no greatest idempotent separating congruence [11]. However, it is clear that any congruence contained in $\widetilde{\mathcal{H}}_E$ separates the *idempotents of E*.

Recall that in an inverse semigroup we have that $a^+ = aa^{-1}, a^* = a^{-1}a$ and $a^{-1}ea = (ea)^{-1}ea = (ea)^*$, for any idempotent e. If S is a restriction semigroup we may happily adapt the definition of α_a ($a \in S$) by putting

$$\alpha_a : Ea^+ \to Ea^*,\ e\alpha_a = (ea)^* \text{ for all } e \in Ea^+,$$

retaining the property that α_a is an isomorphism.

The following is taken from [13], where restriction semigroups are called *weakly E-ample*. It extends the corresponding result for ample semigroups, which may be found in [11].

Theorem 4.2. *Let S be a restriction semigroup with distinguished semilattice E. Then $\theta : S \to T_E$ given by $a\theta = \alpha_a$, is a morphism with kernel μ_E such that*

$$\theta|_E : E \to \overline{E}$$

is an isomorphism.

We remark that Theorem 4.2 characterises the relation μ_E on a restriction semigroup S with distinguished semilattice E by the rule that

$$a\,\mu_E\,b \Leftrightarrow a^+ = b^+ \text{ and } (ea)^* = (eb)^* \text{ for all } e \in E;$$

by duality,

$$a\,\mu_E\,b \Leftrightarrow a^* = b^* \text{ and } (ae)^+ = (be)^+ \text{ for all } e \in E.$$

4.2. **Ehresmann semigroups.** To further generalise the above to the Ehresmann case, we must analyse the method a little.

If S is an Ehresmann semigroup with distinguished semilattice E, then we may define α_a as before, and dually, β_a. For restriction semigroups, these two maps were mutually inverse, but this will not be true in general. Moreover, for a purpose that will become clear, we need to extend their domains to E^1, that is, E with an identity adjoined if necessary. To avoid confusion we call the new maps α_a^1 and β_a^1. Specifically,

$$\alpha_a^1 : E^1 \to E \text{ and } \beta_a^1 : E^1 \to E$$

are given by

$$e\alpha_a^1 = (ea)^* \text{ and } e\beta_a^1 = (ae)^+.$$

For a restriction semigroup, the maps α_a and β_a were morphisms. In general, α_a^1 and β_a^1 (and indeed α_a and β_a) need not be, but they will be order preserving. We define $\mathcal{O}_1(E^1)$ to be the semigroup of those order preserving maps of the semilattice E^1 having image contained in E, and we let $\mathcal{O}_1^*(E^1)$ be its dual (in which maps are composed right to left). It transpires that the analogue of $\mathcal{I}_E$ useful for our purposes here is $\mathcal{O}_1(E^1) \times \mathcal{O}_1^*(E^1)$. If S is Ehresmann then it is not hard to define a morphism from S to $\mathcal{O}_1(E^1) \times \mathcal{O}_1^*(E^1)$ having kernel μ_E, as we indicate below. The real difficulty is in picking out an Ehresmann subsemigroup of $\mathcal{O}_1(E^1) \times \mathcal{O}_1^*(E^1)$ that contains all the images of the relevant morphisms from Ehresmann semigroups having semilattice E, in other words, a maximal E-fundamental Ehresmann semigroup.

The semigroups $\mathcal{O}_1(E^1)$ and $\mathcal{O}_1^*(E^1)$ are partially ordered by $\leq$ where

$$\alpha \leq \beta \text{ if and only if } x\alpha \leq x\beta \text{ for all } x \in E^1.$$

It is easy to see that $\leq$ is compatible with multiplication. The subset C_E of $\mathcal{O}_1(E^1) \times \mathcal{O}_1^*(E^1)$ is then defined by

$$C_E = \{(\alpha, \beta) \in \mathcal{O}_1(E^1) \times \mathcal{O}_1^*(E^1) : \forall x \in E^1, \rho_{x\alpha} \leq \beta\rho_x\alpha \text{ and } \rho_{x\beta} \leq \alpha\rho_x\beta\}.$$

We note that for any $e \in E$, the pair $\overline{e} = (\rho_e, \rho_e) \in C_E$, where ρ_e is the order preserving map of E^1 given by multiplication with e.

Theorem 4.3. [17] *The set C_E is a fundamental Ehresmann subsemigroup of $\mathcal{O}_1(E^1) \times \mathcal{O}_1^*(E^1)$ having distinguished semilattice*

$$\overline{E} = \{\overline{e} : e \in E\}$$

isomorphic to E and such that for any $(\alpha, \beta) \in C_E$,

$$(\alpha, \beta)^* = (\rho_{1\alpha}, \rho_{1\alpha}) \text{ and } (\alpha, \beta)^+ = (\rho_{1\beta}, \rho_{1\beta}).$$

Conversely, if S is an Ehresmann semigroup with distinguished semilattice E, then $\phi : S \to C_E$ given by

$$a\phi = (\alpha_a^1, \beta_a^1)$$

is a morphism with kernel μ_E.

As in Subsection 4.1, Theorem 4.3 gives us a closed form for determining μ_E on an Ehresmann semigroup.

4.3. **Left restriction and left Ehresmann semigroups.** If S is left restriction or even left Ehresmann, then $\beta_a^1 : E^1 \to E$ is order preserving, and as $\beta_a \beta_b = \beta_{ba}$, we can certainly represent S as a subsemigroup of $\mathcal{O}_1^*(E^1)$. Moreover the kernel of the representative morphism will be μ_E^R, the largest congruence contained in $\widetilde{\mathcal{R}}_E$ (Cf [11]). Indeed, calling a left Ehresmann semigroup *left fundamental* if μ_E^R is trivial, then $\mathcal{O}_1^*(E^1)$ is itself left fundamental [25], a more straightforward situation than the two-sided case. This approach is currently being considered by Jones [25], as part of a fresh view of the varieties of semigroups related to those presented in this article.

5. **The covering approach**

Whereas Munn's theory, outlined at the beginning of Section 4, constructs an *image* $S\theta$ of an inverse semigroup S, with $E(S) \cong E(S\theta)$, McAlister's seminal papers [32, 33] construct a *preimage* $\widehat{S}$ of S such that $E(\widehat{S}) \cong E(S)$. Just as we know how to construct $T_{E(S)}$ (of which $S\theta$ is a full subsemigroup), so we can determine the structure of $\widehat{S}$ - it is 'almost' a semidirect product of $E(S)$ by the maximal group image of S.

Let S be a semigroup and let $E \subseteq E(S)$. The relation σ_E on S is the least congruence identifying all the elements of E. It is well known that if S is inverse and $\sigma = \sigma_{E(S)}$, then $a \,\sigma\, b$ if and only if $ea = eb$ for some $e \in E(S)$ [34], and further, σ is the least congruence such that S/σ is a group.

An inverse semigroup is *proper* if $\mathcal{R} \cap \sigma = \iota$, where ι denotes the trivial congruence. This definition is only apparently one sided, for it is easily seen to be equivalent to $\mathcal{L} \cap \sigma = \iota$. Moreover, an inverse semigroup is proper if and only if it is *E-unitary*, that is, $E(S)$ forms a σ-class.

The first of McAlister's results tell us that every inverse semigroup is closely related to a proper one, and is knows as *McAlister's Covering Theorem.*

Theorem 5.1. [32] *Let S be an inverse semigroup. Then there exists a proper inverse semigroup $\widehat{S}$ and an idempotent separating morphism $\theta : \widehat{S} \twoheadrightarrow S$ from $\widehat{S}$ onto S.*

The semigroup $\widehat{S}$ appearing in Theorem 5.1 is a *proper cover* of S.

The second of McAlister's results determines the structure of proper inverse semigroups, as follows. We first recall the notion of a monoid acting on a set.

Definition 5.2. Let T be a monoid and let X be a set. Then T *acts* on X (on the left) if there is a map $T \times X \to X$, $(t,x) \mapsto t \cdot x$, such that for all $x \in X$ and $s,t \in T$ we have

$$1 \cdot x = x \text{ and } st \cdot x = s \cdot (t \cdot x).$$

Let G be a group acting on the left of a partially ordered set $\mathcal{X}$ by order automorphisms, which contains $\mathcal{Y}$ as an ideal and subsemilattice, such that

$$(a)\ G \cdot \mathcal{Y} = \mathcal{X} \text{ and } (b) \text{ for all } g \in G,\ g \cdot \mathcal{Y} \cap \mathcal{Y} \neq \emptyset.$$

Then $(G, \mathcal{X}, \mathcal{Y})$ is a **McAlister triple**.

If $(G, \mathcal{X}, \mathcal{Y})$ is a McAlister triple, we put

$$\mathcal{P} = \mathcal{P}(G, \mathcal{X}, \mathcal{Y}) = \{(e,g) \in \mathcal{Y} \times G : g^{-1} \cdot e \in \mathcal{Y}\},$$

and define a binary operation on $\mathcal{P}$ by

$$(e,g)(f,h) = (e \wedge g \cdot f, gh).$$

Of course, if $\mathcal{X} = \mathcal{Y}$, then (a) and (b) would be redundant and $\mathcal{P}(G, \mathcal{X}, \mathcal{Y})$ would be a semidirect product. In general, however, we cannot dispense with the $\mathcal{X}$ and remark that although the conditions of a McAlister triple ensure that $e \wedge g \cdot f \in \mathcal{Y}$ in the definition of the binary operation, we need not have that $g \cdot f \in \mathcal{Y}$.

Theorem 5.3. [33] *An inverse semigroup is proper if and only if it is isomorphic to a semigroup of the form $\mathcal{P} = \mathcal{P}(G, \mathcal{X}, \mathcal{Y})$.*

Semigroups of the form $\mathcal{P}(G, \mathcal{X}, \mathcal{Y})$ are known as *P-semigroups* and Theorem 5.3 as the *McAlister P-Theorem.* Given a proper inverse semigroup S, the group and semilattice that appear in the corresponding McAlister triple are S/σ and $E(S)$, respectively. Finding the $\mathcal{X}$ is the major difficulty. Proper inverse semigroups form an important class in their own right - for example, the free inverse semigroup is proper [38]. Having a good structure theory for such semigroups enables the investigation of classes of proper inverse semigroups, such as 1-dimensional tiling semigroups. Further, it guides the way for the study of the corresponding classes of inverse semigroups with zero, which include tiling semigroups [30].

5.1. **Left restriction semigroups.** Unlike the case for the 'categorical' and 'fundamental' approaches of the previous two sections, for the 'covering' method it helps to deal first with the one-sided situation.

Let S be left restriction. It is clear that the least *semigroup* congruence identifying all the elements of E is also the least unary semigroup congruence to do so. Thus there is no ambiguity in talking of σ_E as the least congruence identifying all the elements of E.

We will say that a (left) restriction semigroup S is *reduced* if its distinguished semilattice is trivial. A reduced (left) restriction semigroup is simply a monoid, in augmented signature. Clearly, if S is (left) restriction, then S/σ_E is reduced.

Lemma 5.4. [21] *Let S be a left restriction semigroup. Then for any $a, b \in S$, we have that $a \,\sigma_E\, b$ if and only if $ea = eb$ for some $e \in E$.*

A left restriction semigroup S is *proper* if $\widetilde{\mathcal{R}}_E \cap \sigma_E = \iota$. A *proper cover* of S is a proper left restriction semigroup $\widehat{S}$ such that there exists an epimorphism $\theta : \widehat{S} \twoheadrightarrow S$ that separates distinguished idempotents. It follows that $E_{\widehat{S}} \cong E_S$.

Theorem 5.5. [4] *Let S be left restriction. Then there is a proper cover $\widehat{S}$ of S.*

We now give a recipe for constructing proper left restriction semigroups, inspired by the P-theorem. Naturally, the group in a McAlister triple will be replaced by a monoid.

Let T be a monoid acting on the left of a semilattice $\mathcal{X}$ via morphisms. Suppose that $\mathcal{X}$ has subsemilattice $\mathcal{Y}$ with upper bound ε such that

(a) for all $t \in T$ there exists $e \in \mathcal{Y}$ such that $e \leq t \cdot \varepsilon$;

(b) if $e \leq t \cdot \varepsilon$ then for all $f \in \mathcal{Y}$, $e \wedge t \cdot f \in \mathcal{Y}$.

Then $(T, \mathcal{X}, \mathcal{Y})$ is a **strong left M-triple**.

We remark that we have 'gained' over the formulation of McAlister triples in that we may take $\mathcal{X}$ to be a semilattice, but 'lost' in that we no longer have $T \cdot \mathcal{Y} = \mathcal{X}$.

For a strong left M-triple $(T, \mathcal{X}, \mathcal{Y})$ we put

$$\mathcal{M}(T, \mathcal{X}, \mathcal{Y}) = \{(e, t) \in \mathcal{Y} \times T : e \leq t \cdot \varepsilon\}$$

and define

$$(e, s)(f, t) = (e \wedge s \cdot f, st), \ (e, s)^+ = (e, 1).$$

Theorem 5.6. [4] *A left restriction semigroup is proper if and only if it is isomorphic to some $\mathcal{M}(T, \mathcal{X}, \mathcal{Y})$.*

Whilst properly stated for left restriction semigroups in [4], Theorems 5.5 and 5.6 may essentially be found in [15], where they are stated for 'weakly left ample' semigroups, which are left restriction semigroups S with $E = E(S)$. If S is proper left restriction, then the T and $\mathcal{Y}$ that appear in Theorem 5.6 are S/σ_E and E, respectively. In the weakly left ample case, we must insist that T be unipotent, that is, $|E(T)| = 1$.

By replacing T with a right cancellative monoid, we can specialise to the left ample case. The first result is not immediate, but easy to obtain following a similar method to that given in [14, Theorem 7.1]. It was first proven, via a different technique, in [10].

Corollary 5.7. [10] *Let S be left ample. Then there is a proper left ample cover $\widehat{S}$ of S.*

Corollary 5.8. *A left ample semigroup is proper if and only if it is isomorphic to some $\mathcal{M}(T, \mathcal{X}, \mathcal{Y})$ where T is right cancellative.*

Corollary 5.8 also appears under a different formulation in [10, 28].

5.2. **Restriction semigroups.** It follows by duality that if S is right restriction, then $a\,\sigma_E\,b$ if and only if $af = bf$ for some $f \in E$, so that if S is restriction, then either characterisation of σ_E will suffice.

A restriction semigroup S is *proper* if $\widetilde{\mathcal{R}}_E \cap \sigma_E = \iota = \widetilde{\mathcal{L}}_E \cap \sigma_E$, that is, if S is proper as a left and as a right restriction semigroup. A *proper cover* of S is a proper restriction semigroup $\widehat{S}$ such that there exists an epimorphism $\theta : \widehat{S} \twoheadrightarrow S$ that separates distinguished idempotents. The next result is stated in [14] for *monoids*, but a remark at the end of that article explains how to deduce the corresponding result for *semigroups*.

Theorem 5.9. [14] *Let S be a restriction semigroup. Then S has a proper ample cover.*

Of course the question then is, can we find a structure theorem for proper restriction semigroups? Of course, a proper ample semigroup S is proper left ample, so that there is a strong left M-triple $(T, \mathcal{X}, \mathcal{Y})$ (with T right cancellative) such that $S \cong \mathcal{M}(T, \mathcal{X}, \mathcal{Y})$. In [28] conditions are put on a strong left M-triple (which has a slightly different formulation in [28]) such that the semigroup built from it be proper ample. This is extended to the restriction case in [6]. However, there is no left-right symmetry about these descriptions.

Again, if S is proper restriction, then as it is proper left restriction, there is a strong left M-triple $(T, \mathcal{X}, \mathcal{Y})$ such that $S \cong \mathcal{M}(T, \mathcal{X}, \mathcal{Y})$, and as S is proper right restriction, there is a strong right M-triple $(T, \mathcal{X}', \mathcal{Y})$ (where a strong right M-triple is defined in a dual manner to a strong left M-triple) such that $S \cong \mathcal{M}'(T, \mathcal{X}, \mathcal{Y})$ (where $\mathcal{M}'(T, \mathcal{X}', \mathcal{Y})$ is also defined in a dual manner). In both cases, we can take $T = S/\sigma_E$ and $\mathcal{Y} = E$. For the strong left M-triple, T acts on the left of $\mathcal{X}$, and for the strong right M-triple, T acts on the right of $\mathcal{X}'$. Now, $\mathcal{X}$ and $\mathcal{X}'$ both contain $\mathcal{Y} = E$ as a subsemilattice, and clearly these actions must be linked in some way. In [14] the author and colleagues developed the notion of *double action* of a monoid acting on the left and the right of a semilattice, satisfying what were called 'compatibility conditions'. This inspired the following definitions.

Let $(T, \mathcal{X}, \mathcal{Y})$ and $(T, \mathcal{X}', \mathcal{Y})$ be strong left and right M-triples respectively. Denote the left action of T on $\mathcal{X}$ by $\cdot$ and the right action of T on $\mathcal{X}'$ by $\circ$. Suppose that for all $t \in T$ and $e \in \mathcal{Y}$, the following and the dual holds:

$$e \leq t \cdot \varepsilon \Rightarrow e \circ t \in \mathcal{Y} \text{ and } t \cdot (e \circ t) = e.$$

Then $(T, \mathcal{X}, \mathcal{X}', \mathcal{Y})$ is a *strong M-quadruple.*

Proposition 5.10. [7] *For a strong M-quadruple* $(T, \mathcal{X}, \mathcal{X}', \mathcal{Y})$,

$$\mathcal{M}(T, \mathcal{X}, \mathcal{Y}) \cong \mathcal{M}'(T, \mathcal{X}', \mathcal{Y})$$

via an isomorphism preserving the distinguished semilattices.

Consequently, if $(T, \mathcal{X}, \mathcal{X}', \mathcal{Y})$ is a *strong M-quadruple*, then putting

$$\mathcal{M}(T, \mathcal{X}, \mathcal{X}', \mathcal{Y}) = \mathcal{M}(T, \mathcal{X}, \mathcal{Y}),$$

we have that $\mathcal{M}(T, \mathcal{X}, \mathcal{X}', \mathcal{Y})$ is proper restriction.

From results of [14], every restriction monoid has a proper cover of the form $\mathcal{M}(T, \mathcal{X}, \mathcal{X}', \mathcal{Y})$, where, in fact, we can take $\mathcal{X} = \mathcal{X}' = \mathcal{Y}$ to be a semilattice with identity, and, moreover, the *free* restriction monoid has this form. However, it is shown in [6] that if $\mathcal{M}(T, \mathcal{X}, \mathcal{X}', \mathcal{Y})$ is *finite*, then it must be inverse. As there are certainly finite proper ample semigroups that are *not* inverse, it follows that not all proper ample (and hence certainly not all proper restriction) semigroups are isomorphic to some $\mathcal{M}(T, \mathcal{X}, \mathcal{X}', \mathcal{Y})$. In [7] a condition is given on a proper restriction semigroup S such that $S \cong \mathcal{M}(T, \mathcal{X}, \mathcal{X}', \mathcal{Y})$ for some strong M-quadruple $\mathcal{M}(T, \mathcal{X}, \mathcal{X}', \mathcal{Y})$.

To obtain a truly two-sided structure theorem for proper restriction semigroups, we are forced to use *partial actions.*

Definition 5.11. Let T be a monoid and let X be a set. Then T *acts partially* on X (on the left) if there is a partial map $T \times X \to X, (t, x) \mapsto t \cdot x$, such that for all $s, t \in T$ and $x \in X$,

$$\exists 1 \cdot x \text{ and } 1 \cdot x = x$$

and

$$\text{if } \exists t \cdot x \text{ and } \exists\, s \cdot (t \cdot x) \text{ then } \exists\, st \cdot x \text{ and } s \cdot (t \cdot x) = st \cdot x,$$

where we write $\exists\, u \cdot y$ to indicate that $u \cdot y$ is defined.

Of course, a partial left action of T on X with domain of the action $T \times X$ is an action. Dually, we may define the (partial) right action of T on X, using the symbol '$\circ$' to replace '$\cdot$'.

Let T be a monoid, acting partially on the left and right of a semilattice $\mathcal{Y}$, via $\cdot$ and $\circ$ respectively. Suppose that both actions preserve the partial order and the domains of each $t \in T$ are order ideals. Suppose in addition that for $e \in \mathcal{Y}$ and $t \in T$, the following and their duals hold:

(a) if $\exists e \circ t$, then $\exists t \cdot (e \circ t)$ and $t \cdot (e \circ t) = e$;

(b) for all $t \in T$, there exists $e \in \mathcal{Y}$ such that $\exists e \circ t$.

Then $(T, \mathcal{Y})$ is a **strong M-pair**.

For a strong M-pair $(T, \mathcal{Y})$ we put

$$\mathcal{M}(T, \mathcal{Y}) = \{(e, s) \in \mathcal{Y} \times T : \exists e \circ s\}$$

and define operations by

$$(e, s)(f, t) = (s \cdot (e \circ s \wedge f), st),\ (e, s)^+ = (e, 1) \text{ and } (e, s)^* = (e \circ s, 1).$$

Theorem 5.12. [7]*A semigroup is proper restriction if and only if it is isomorphic to some $\mathcal{M}(T, \mathcal{Y})$.*

Our approach to the above result is very direct, inspired by the philosophy of [39]. If S is proper restriction, then the T and $\mathcal{Y}$ that we take for Theorem 5.12 are, naturally, S/σ_E and E. The following corollaries are then almost immediate.

Corollary 5.13. [28] *A semigroup is proper ample if and only if it is isomorphic to $\mathcal{M}(C, \mathcal{Y})$ for a cancellative monoid C.*

Corollary 5.14. [41] *A semigroup is proper inverse if and only if it is isomorphic to $\mathcal{M}(G, \mathcal{Y})$ for a group G.*

Corollories 5.13 and 5.14 are presented in a slightly different form in [28] and [41], respectively. In those articles more explicit mention is made of the fact that, effectively, our compatibility conditions are insisting that the left and right actions of an element of T be mutually inverse on certain domains.

The reader might wonder why we claim that Theorem 5.12 is truly left/right dual. As we explain in [7], if $(T, \mathcal{Y})$ is a strong M-pair, then

$$\mathcal{M}'(T, \mathcal{Y}) = \{(s, e) \in T \times \mathcal{Y} : \exists s \cdot e\}$$

with operations given by

$$(s, e)(t, f) = (st, (e \wedge t \cdot f) \circ t), (s, e)^* = (1, e) \text{ and } (s, e)^+ = (1, s \cdot e)$$

is a restriction semigroup isomorphic to $\mathcal{M}(T, \mathcal{Y})$.

5.3. **Left Ehresmann semigroups.** We recall that for left Ehresmann semigroups, we do *not* have the identity $xy^+ = (xy)^+x$. This identity ensures that for a left restriction semigroup S, if we have a product

$$s = t_0 e_1 t_1 \dots t_{n-1} e_n t_n$$

where $t_0, \dots, t_n \in S$ and $e_1, \dots, e_n \in E$, then we can re-write this as

$$s = f t_0 t_1 \dots t_n$$

for some $f \in E$. Moreover, S acts by morphisms on E on the left by

$$s \cdot e = (se)^+.$$

These two facts underly the description of a proper left restriction semigroup as a semigroup of the form $\mathcal{M}(T, \mathcal{X}, \mathcal{Y})$ for some strong left M-triple $(T, \mathcal{X}, \mathcal{Y})$. Moreover, building on the first observation, if $S = \langle E \cup T \rangle$ for a *subsemigroup* T of S, then any element of S can be written as

$$t \text{ or } et$$

where $e \in E, t \in T$ and $e < t^+$. If this expression is unique, then it is easy to see that S is proper. Consider now a strong left M-triple $(T, \mathcal{Y}, \mathcal{Y})$ and put

$$T' = \{(t \cdot 1, t) : t \in T\}.$$

We know that

$$E = \{(e, 1) : e \in \mathcal{Y}\}$$

and it is easy to see that T' is a subsemigroup isomorphic to T, such that $\mathcal{M} = \mathcal{M}(T, \mathcal{Y}, \mathcal{Y}) = \langle E \cup T' \rangle$ and every element of $\mathcal{M}$ can be written uniquely as

$$(t \cdot 1, t) \text{ or as } (e, 1)(t \cdot 1, t) \text{ for some } (e, 1)^+ < (t \cdot 1, t)^+,$$

the latter condition being equivalent to $e < t \cdot 1$.

Having lost our identity, we must think again. We cannot move idempotents to the front in products, and hence we cannot find expressions of the form t/et where $E \cup T$ is a set of generators for a left Ehremann semigroup S, $t \in T$ and $e \in E$. Moreover, the action of S on E as above will be by order order preserving maps, but not usually by morphisms.

We now briefly describe work taken from recent preprints [3],[18]. These articles consider the case for monoids, which for convenience we focus on here.

Let M be a left Ehresmann monoid and let T be a submonoid of M such that $E \cup T$ generates M as a semigroup (equivalently, as a unary monoid). Then any $m \in M$ can be written as

$$m = t_0 e_1 t_1 \dots e_n t_n$$

for some $t_0, t_n \in T$, $t_i \in T \setminus \{1\}$ (for $1 \leq i \leq n-1$) and $e_j \in E \setminus \{1\}$ with $e_i < (t_i \dots e_n t_n)^+$ (for $1 \leq j \leq n$).

Such an expression is a *T-normal form.* If this expression for each m is unique, then M *has uniqueness of T-normal forms.* We remark that if If $M = \langle X \rangle$ and T is the *submonoid* generated by X, then $M = \langle E \cup T \rangle$.

From an action of T on a semilattice $\mathcal{Y}$ by order preserving maps, we can construct a semigroup $\mathcal{P}(T, \mathcal{Y})$ that has uniqueness of T-normal forms. We refer to [18] for the details.

Theorem 5.15. [18] *A left Ehresmann monoid S with set of generators $E \cup T$, for some submonoid T, has uniqueness of T-normal forms if and only if it is isomorphic to some $\mathcal{P}(T, E)$.*

The notion of *cover* for a left Ehresmann monoid is defined in the obvious way.

Theorem 5.16. [18] *Every left Ehresmann monoid has a cover of the form $\mathcal{P}(X^*, \mathcal{Y})$. Moreover, $\mathcal{P}(X^*, \mathcal{Y})$ is left adequate.*

Since left Ehresmann monoids form a variety, the free object on any set X exists, and as it is left adequate, it is also the free left adequate monoid on X.

Theorem 5.17. [18] *The free left Ehresmann monoid on X is isomorphic to some $\mathcal{P}(X^*, \mathcal{Y})$.*

Kambites [26] has also determined the structure of the free left Ehresmann monoid, using a completely different approach. He also determines the structure of the free Ehresmann monoid [27]. In both cases he uses labelled trees, in what can be seen as an analogue of Munn's approach [38] to the structure of free inverse semigroups.

A strong M-triple with $\mathcal{X} = \mathcal{Y}$ gives rise to a proper left restriction semigroup with the normal form property. We therefore anticipate that our class of monoids of the form $\mathcal{P}(T, \mathcal{Y})$ are a subclass of a class of monoids of the form $\mathcal{P}(T, \mathcal{X}, \mathcal{Y})$, where $\mathcal{P}(T, \mathcal{X}, \mathcal{Y})$ is constructed from a monoid T acting by order preserving maps on a partially ordered set $\mathcal{X}$ containing $\mathcal{Y}$ as semilattice. We also anticipate that that this class can be abstractly characterised by some notion of 'proper', which will of necessity be a little more technical than that for left restriction semigroups.

A suggested notion, that of 'T-proper' appears in [3]. The author and her colleagues and students are working on semigroups of the form $\mathcal{P}(T, \mathcal{X}, \mathcal{Y})$, and, indeed, the corresponding approaches in the two-sided case.

Acknowledgement The author would like to thank the organisers of ICA 2010, and in particular, Professor K.P. Shum, for the opportunity to participate in this meeting.

References

[1] J. Araújo and M. Kinyon, 'An elegant 3-basis for inverse semigroups', *arXiv:1003.4028v3.*

[2] S. Armstrong, 'The structure of type A semigroups', *Semigroup Forum* **29** (1984), 319–336.

[3] M. Branco, G.M.S. Gomes and V. Gould, 'Left adequate and left Ehresmann monoids', *I.J.A.C.*, to appear; see also *preprint at http://www-users.york.ac.uk/~varg1.*

[4] M. Branco, G.M.S. Gomes and V. Gould, 'Extensions and covers for semigroups whose idempotents form a left regular band', *Semigroup Forum*, **81** (2010), 51–70.

[5] J. R. Cockett and S. Lack, Restriction categories I: categories of partial maps, *Theoretical Computer Science*, **270** (2002), 223-259.

[6] C. Cornock, *Restriction Semigroups: Structure, Varieties and Presentations* PhD thesis, in preparation.

[7] C. Cornock and V. Gould, 'Proper restriction semigroups' *preprint at http://www-users.york.ac.uk/~varg1.*

[8] C. Ehresmann, *Oevres complètes et commentées*, (A. C. Ehresmann, Ed.) Suppl. Cahiers Top. Géom. Diff., Amiens, 1980–1984.

[9] A. El-Qallali, *Structure Theory for Abundant and Related Semigroups*, PhD Thesis, York 1980.

[10] J. Fountain, 'A class of right *PP* monoids', *Quart. J. Math. Oxford (2)* **28** (1977), 285-300.

[11] J. Fountain, 'Adequate semigroups', *Proc. Edinburgh Math. Soc. (2)* **22** (1979), 113-125.

[12] J. Fountain, 'Abundant semigroups', *Proc. London Math. Soc.* **44** (1982), 103–129.

[13] J. Fountain, G.M.S. Gomes and V. Gould, ' A Munn type representation for a class of E-semiadequate semigroups', *J. Algebra* **218** (1999), 693-714.

[14] J. Fountain, G.M.S. Gomes and V. Gould, 'Free ample monoids' *I.J.A.C.* **19** (2009), 527–554.

[15] G. M. S. Gomes and V. Gould, 'Proper weakly left ample semigroups' *I.J.A.C.* **9** (1999), 721–739.

[16] G.M.S. Gomes and V. Gould, Graph expansions of unipotent monoids, *Communications in Algebra* **28** (2000), 447-463.

[17] G.M.S. Gomes and V. Gould, 'Fundamental Ehresmann Semigroups', *Semigroup Forum* **63** (2001), 11–33.

[18] G. Gomes and V. Gould, 'Left adequate and left Ehresmann monoids II', *preprint at http://maths.york.ac.uk/www/varg1.*

[19] V. Gould and M. Kambites, 'Faithful functors from cancellative categories to cancellative monoids, with an application to ample semigroups', *I.J.A.C.* **15** (2005), 683–698.

[20] V. Gould and C. Hollings, 'Restriction semigroups and inductive constellations' *Communications in Algebra* **38** (2010), 261–287.

[21] V. Gould 'Notes on restriction semigroups and related structures', *notes available at http://www-users.york.ac.uk/~varg1.*

[22] J. M. Howie, 'The maximum idempotent separating congruence on an inverse semigroup', *Proc. Edinburgh Math. Soc.* **14** (1964), 71–79.

[23] J. M. Howie, *Fundamentals of semigroup theory*, Oxford University Press, (1995).

[24] M. Jackson and T. Stokes, 'An invitation to C-semigroups', *Semigroup Forum* **62** (2001), 279-310.

[25] P.R. Jones, *private communication.*

[26] M. Kambites, 'Retracts of trees and free left adequate semigroups', *Proc. Edinburgh Math. Soc.*, to appear; see also *arXiv:0904.0916.*

[27] M. Kambites, 'Free adequate semigroups', *arXiv:0902.0297.*

[28] M.V. Lawson, 'The Structure of Type A Semigroups', *Quart. J. Math. Oxford*, **37 (2)** (1986), 279–298.

[29] M.V. Lawson, 'Semigroups and ordered categories I. The reduced case', *J. Algebra* **141** (1991), 422-462.

[30] M.V. Lawson, *Inverse semigroups, the theory of partial symmetries*, World Scientific 1998.

[31] E. Manes, 'Guarded and banded semigroups', *Semigroup Forum* **72** (2006), 94-120.

[32] D. B. McAlister, 'Groups, Semilattices and Inverse Semigroups', *Trans. American Math. Soc.* **192** (1974), 227–244.

[33] D. B. McAlister, 'Groups, Semilattices and Inverse Semigroups II', *Trans. American Math. Soc.* **196** (1974), 351–370.

[34] W.D. Munn, 'A class of irreducible matrix representations of an arbitrary inverse semigroup', *Proc. Glasgow Math. Ass.* **5** (1961), 41–48.

[35] W.D. Munn, 'A certain sublattice of the lattice of congruences of a regular semigroup', *Proc. Cambridge Phil. Soc.* **60** (1964), 385–391.

[36] W.D. Munn, 'Fundamental inverse semigroups', *Quart. J. Math. Oxford* **21** (1970), 157–170.

[37] W.D. Munn, '0-bisimple inverse semigroups', *J. Algebra* **15** (1970) 570–588.

[38] W.D. Munn 'Free inverse semigroups', *Proc. London Math. Soc.* **29** (1974), 385–404.

[39] W.D. Munn, A note on E-unitary inverse semigroups', *Bull. London Math. Soc.* **8** (1976), 71–76.

[40] K.S.S. Nambooripad, *Structure of Regular Semigroups I', Memoirs of the American Mathematical Society* **224** 1979.

[41] M. Petrich and N. R. Reilly, 'A Representation of E-unitary Inverse Semigroups', *Quart. J. Math. Oxford* **30** (1979), 339–350.

[42] B.M. Schein, 'On the theory of generalized grouds and generalized heaps', in *The theory of semigroups and its applications I*, University of Saratov, Saratov, 1965, 286–324 (Russian).

[43] B. M. Schein, 'Relation algebras and function semigroups', *Semigroup Forum* **1** (1970), 1–62.

[44] B.M. Schein, 'On the theory of inverse semigroups and generalised grouds', *Amer. Math. Soc. Translations*, **113** (1979), 89–122.

[45] V. S. Trokhimenko, 'Menger's function systems', *Izv. Vysš. Učebn. Zaved. Matematika* **11**(138) (1973), 71-78 (Russian).

On Some Constructions and Results of the Theory of Partially Soluble Finite Groups

Wenbin Guo*
Department of Mathematics, University of Science and Technology of China,
Hefei 230026, P. R. China; and
Department of Mathematics, Xuzhou Normal University,
Xuzhou, 221116, P.R. China
E-mail: wbguo@xznu.edu.cn

Vasilii G. Safonov
Department of Mathematics, Belarus State University,
Minsk 220010, Belarus
E-mail: safonov@minedu.unibel.by

Alexander N. Skiba
Department of Mathematics, Francisk Skorina Gomel State University,
Gomel 246019, Belarus
E-mail: alexander.skiba49@gmail.com

Dedicated to Professor Shum Kar-Ping on the occasion of his 70th birthday

Abstract

In this paper we discuss some constructions and recent results of the theory of soluble and partially soluble finite groups.

1 Introduction

$G^{\mathfrak{X}}$-residual, verieties of groups, formations. Let $\mathfrak{X}$ be any non-empty class of groups. Then we use $G^{\mathfrak{X}}$ to denote the intersection of all normal subgroups N of the group G with $G/N \in \mathfrak{X}$.

*Research of the first author is supported by a NNSF grant of China (Grant #10771180)

Keywords: finite group, $\mathfrak{F}$-hypercentre, soluble group, supersoluble group, nilpotent group, quasisoluble group, quasisupersoluble group, quasinilpotent group, saturated formation, Baer-local formation, the generalized Fitting subgroup.

Mathematics Subject Classification (2000): 20D10, 20D15, 20D20

If, for example, $\mathfrak{X} = \mathcal{A}$ is the class of all abelian groups, then $G^{\mathcal{A}} = G'$ is the commutator subgroup of G; if $\mathfrak{X} = \mathfrak{N}$ is the class of all nilpotent groups and G is a finite group, then $G^{\mathfrak{N}}$ is the smallest member of the lower central series of G.

A class $\mathfrak{X}$ of groups is said to be a variety of groups if $\mathfrak{X}$ is the class of all groups in which some given system of laws hold. In general, $G/G^{\mathfrak{X}}$ does not belong to $\mathfrak{X}$. But also a situation is possible when not only $G/G^{\mathfrak{X}}$ belongs to $\mathfrak{X}$ but also and every homomorphic image of $G/G^{\mathfrak{X}}$ belongs to $\mathfrak{X}$: *A class $\mathfrak{X}$ of groups is a variety of groups if and only if for any group G, every homomorphic image of $G/G^{\mathfrak{X}}$ belongs to $\mathfrak{X}$.*

This observation is a corollary of the well known Kagalowskiy Theorem [1, Chapter 1, Section 5] on varieties of groups.

In the universe of all finite groups, analogously one can defined a formation of groups: *A class $\mathfrak{F}$ of finite groups is said to be a formation if for every finite group G, every homomorphic image of $G/G^{\mathfrak{F}}$ belongs to $\mathfrak{F}$.*

The Gaschütz's product of the group classes $\mathfrak{M}$ and $\mathfrak{H}$ is the class

$$\mathfrak{M}\mathfrak{H} = \{ \text{ is a group}|\ G^{\mathfrak{H}} \in \mathfrak{M}\}.$$

Such an operator defines a semigroup on the set of all varieties and a semigroup on the set of all formations. The first of these semigroups is isomorphically embedded into the second one.

Theorem 1.1 (A.N. Skiba, unpublished). *Let $\mathbf{V}$ be the semigroup of all varieties, p a prime and $\mathbf{P}$ the semigroup of all formations of finite p-groups. Then $\mathbf{V} \simeq \mathbf{P}$.*

Moreover, we have the following

Theorem 1.2 [2, Chapter 2]. *Let for any variety $\mathfrak{V}$, $\mathbf{fin}\mathfrak{V}$ is the class of all finite groups in $\mathfrak{V}$. Then*

(1) **fin** *isomorphically maps the semigroup of all locally finite varieties into the semigroup of all formations of finite groups.*

(2) **fin** *isomorphically maps the lattice of all locally finite varieties into the lattice of all formations of finite groups.*

Theorem 1.2 shows, in particular, that the lattice of all formations of finite groups is not distributive. In this regard, a large number of papers on the theory of formations has been associated with the determination of distributive lattices formations. Among recent works in this direction we mention here only to papers of V.G. Safonov [3, 4], where the distributivity of the lattice of all totally local formations was proved and the paper of W. Guo and K.P. Shum [5] where the wide classes of formations with boolean lattice of subformations were found.

The set of all idempotents of the semigroup of all formations of finite groups is continual. The formation of all finite p-groups is an example of an idempotent. There are also other examples of idempotents of this semigroup (see [6, 7, 8, 9]). Nevertheless, there is still an open

Question (A.I. Starostin, 1984). *Describe all the idempotents of the semigroup of all formations of finite groups.*

Saturated and Baer-local formations. In all that follows the term "a group" means "a finite group".

Recall that a formation $\mathfrak{F}$ is said to be saturated if $G \in \mathfrak{F}$ whenever $G^{\mathfrak{F}} \leq \Phi(N)$ for some normal subgroup N of G.

A formation $\mathfrak{F}$ is said to be composition, solubly saturated, or Baer-local formation if $G \in \mathfrak{F}$ whenever $G^{\mathfrak{F}} \leq \Phi(N)$ for some *soluble* normal subgroup N of G.

These two classes of formations are useful in various applications of the formation theory (see [10, 11, 2, 69, 13, 14]).

The most famous examples of saturated formations are the classes of all soluble $\mathfrak{S}$, supersoluble $\mathfrak{U}$, nilpotent $\mathfrak{N}$, p-soluble, p-supersoluble, p-nilpotent groups, dispersive (in sense Ore) groups and so on.

$\mathfrak{F}$-hypercentre. If N is a normal subgroup of a group G and H/K is a chief factor of G below N, then we say that H/K is a G-chief factor of N.

Let $\mathfrak{X}$ be a class of groups. A chief factor H/K of a group G is called $\mathfrak{X}$-*central* provided $(H/K) \leftthreetimes (G/C_G(H/K)) \in \mathfrak{X}$. Otherwise, it is called $\mathfrak{X}$-*eccentric.*

For example, If $\mathfrak{X} = \mathfrak{N}$ the class of all nilpotent groups, then an $\mathfrak{N}$-central chief factor H/K of G is precisely a central chief factor, i.e. $C_G(H/K) = G$. In the case when $\mathfrak{X} = \mathfrak{U}$ is the class of all supersoluble groups, then a $\mathfrak{U}$-central chief factor is exactly a cyclic chief factor.

The product of all normal subgroups N of G whose all G-chief factors are $\mathfrak{X}$-central is called the $\mathfrak{X}$-hypercentre of G and denoted by $Z_{\mathfrak{X}}(G)$ [69, p. 389].

Thus, $Z_{\mathfrak{N}}(G) = Z_{\infty}(G)$ is the hypercentre of G; $Z_{\mathfrak{U}}(G)$ is the largest normal subgroup of G whose all G-chief factors are cyclic.

The hypercentre already becomes a classic tool in the study of groups.

The general theory of the $\mathfrak{X}$-hypercentre was developed in the works of Baer, Huppert, P. Schmid, and L.A. Shemetkov (see [11, 69, 13]). In particular, as a result of common efforts, the following deep results have been proved.

Theorem 1.3. *Let G be a group.*

(1) *If $\mathfrak{F}$ is a Baer-local formation, then $[G^{\mathfrak{F}}, Z_{\mathfrak{F}}(G)] = 1$.*

(2) *If $\mathfrak{F}$ is a saturated formation, then $Z_{\mathfrak{F}}(G) = C_U(G^{\mathfrak{F}})$ for any maximal $\mathfrak{F}$-subgroup U of G such that $G = UG^{\mathfrak{F}}$.*

(3) $C_G(G^{\mathfrak{F}}) = Z_{\mathfrak{F}}(G) \times (Z(G^{\mathfrak{F}} \cap E_{\mathfrak{F}}(G)))$.

In this theorem $E_{\mathfrak{F}}(G)$ is the product of all normal subgroups N of G whose all G-chief factors are $\mathfrak{F}$-eccentric.

2 Some applications of the $\mathfrak{U}$-hypercentre

Introduction. Recall that a subgroup A of a group G is said to be quasinormal (Ore) or permutable (Stonehewer) in G if $AB = BA$ for all subgroups B of G; A is said to be S-quasinormal, S-permutable, or $\pi(G)$-permutable in G (Kegel [15]) if $AP = PA$ for all Sylow subgroups P of G.

The subgroups $Z_{\mathfrak{N}}(G)$ and $Z_{\mathfrak{U}}(G)$ are closely related to quasinormal and S-quasinormal subgroups. For instance, it was proved in [16] that if $A_G = 1$ and A is a quasinormal subgroup of G, then $A \leq Z_{\mathfrak{N}}(G)$; if $A_G = 1$ and A is a modular element (in the sense of Kurosh, see [17, p. 43]) of the subgroup lattice of G, then $A \leq Z_{\mathfrak{U}}(G)$ [17, Theorem 5.2.5]).

Here A_G is the largest normal subgroup of G contained in A.

The essential influence of the subgroups $Z_{\mathfrak{N}}(G)$ and $Z_{\mathfrak{U}}(G)$ on the structure of G makes them useful to describe some important classes of groups. For example, if all subgroups of G of prime order and order 4 are contained in $Z_{\mathfrak{N}}(G)$, then G is nilpotent (N. Itô). If all these subgroups are contained in $Z_{\mathfrak{U}}(G)$, then G is supersoluble (Huppert, Doerk). If all subgroups of G of prime order are contained in $Z_{\mathfrak{U}}(G)$, then G is soluble (Gaschütz). A group G is quasinilpotent if and only if $G/Z_{\mathfrak{N}}(G)$ is semisimple [18, X, Theorem 13.6].

Mukherjee proved in [19] that the $\mathfrak{U}$-hypercentre $Z_{\mathfrak{U}}(G)$ of G coincides with the largest term of the chain of subgroups

$$1 = Q_0 \leq Q_1 \leq Q_2 \leq \cdots$$

where $Q_i(G)/Q_{i-1}(G)$ is the subgroup of $G/Q_{i-1}(G)$ generated by the set of all cyclic quasinormal subgroups of $G/Q_{i-1}(G)$. A characterization of the $\mathfrak{U}$-hypercentre related to S-quasinormal subgroups was obtained in the paper [20].

The first applications of the subgroup $Z_{\mathfrak{U}}(G)$ back to the work of Baer [21].

Currently, this subgroup can be found in studies of many other authors (see, for example, [22-34] and Chapter 1 in the book [35]).

On two problems of L.A. Shemetkov. Recall that $F^*(G)$ denotes the generalized Fitting subgroup of G, that is, the product of all normal *quasinilpotent* subgroups of E [18, Chapter X].

The subgroup A of G is said to be c-normal in G (Wang [36]) if G has a normal subgroup T such that $AT = G$ and $A \cap T \leq A_G$. A is said to be c-supplemented in G (Ballester-Bolinches, Wang and Guo [37]) if G has a subgroup T such that $AT = G$ and $A \cap T \leq A_G$, the largest normal subgroup of G contained in A. Buckley [38] obtained a description of nilpotent groups of odd order all of whose subgroups of prime order are normal. As a consequence, he also proved that a group of odd order is supersoluble if all its subgroups of prime order are normal [38]. Applying Huppert-Doerk's description of minimal non-supersoluble groups [39, 40], we can go further and prove that a group is supersoluble if all its cyclic subgroups of prime order and order 4 are normal. Later, Srinivasan [41] proved that a group G is supersoluble if every maximal subgroup of every Sylow subgroup of G is normal in G. These results have been developed in various directions, especially in the framework of formation theory.

Recall that a formation $\mathfrak{F}$ is a class of groups which is closed under taking homomorphic images and such that each group G has the smallest normal subgroup (denoted by $G^{\mathfrak{F}}$) whose quotient is in $\mathfrak{F}$. A formation $\mathfrak{F}$ is said to be saturated if $G \in \mathfrak{F}$ for any group G with $G/\Phi(G) \in \mathfrak{F}$. If $\mathfrak{F}$ is a saturated formation containing all supersoluble groups and G is a group with a normal subgroup E, then the following results are true.

(1) If $G/E \in \mathfrak{F}$ and every cyclic subgroup of E of prime order and order 4 is either S-quasinormal (Ballester-Bolinches and Pedraza-Aguilera [42], Asaad and Csörgő [43]) or c-normal (Ballester-Bolinches and Wang [44]) or c-supplemented (Ballester-Bolinches, Wang and Guo [37], Wang and Li [45]) in G, then $G \in \mathfrak{F}$.

(2) If $G/E \in \mathfrak{F}$ and every cyclic subgroup of every Sylow subgroup of $F^*(E)$ of prime order and order 4 is either S-quasinormal (Li and Wang [46]) or c-normal (Wei, Wang and Li [47]) or c-supplemented (Wang, Wei and Li [48], Wei, Wang and Li [49]) in G, then $G \in \mathfrak{F}$.

(3) If $G/E \in \mathfrak{F}$ and every maximal subgroup of every Sylow subgroup of E is either S-quasinormal (Asaad [50]) or c-normal (Wei [51]) or c-supplemented (Ballester-Bolinches and Guo [52]) in G, then $G/E \in \mathfrak{F}$.

(4) If $G/E \in \mathfrak{F}$ and every maximal subgroup of every Sylow subgroup of $F^*(E)$ is either S-quasinormal (Li and Wang [53]) or c-normal (Wei, Wang and Li [47]) or c-supplemented (Wei, Wang and Li [48]) in G, then $G \in \mathfrak{F}$.

In these results $F^*(E)$ denotes the generalized Fitting subgroup of E, that is, the product of all normal quasinilpotent subgroups of E [18, Chapter X].

Bearing in mind the above results Prof. L.A. Shemetkov asked in 2004 at Gomel Algebraic Seminar the following two questions:

(I) *Is it true that all the abovementioned results can be strengthened by proving that every G-chief factor below E is cyclic?*

(II) *Is it true that the conclusion about the cyclic character of the G-chief factors below E still holds if we omit the condition "$G/E \in \mathfrak{F}$ "?*

A partial solution of these problems was obtained in [28, Theorem 1.4]. A complete answer to the questions was obtained in [33].

Recall that a subgroup H of a group G is said to be S-supplemented in G [54] if G has a subgroup T such that $G = HT$ and $T \cap H \leq H_{sG}$, where H_{sG} is the subgroup generated by all subgroups of H which are S-quasinormal in G.

Theorem 2.1 [33, Theorem A]. *Let E be a normal subgroup of a group G. Suppose that for every non-cyclic Sylow subgroup P of E, every maximal subgroup of P or every cyclic subgroup of P of prime order and order 4 is S-supplemented in G. Then each G-chief factor below E is cyclic.*

Theorem 2.2 [33, Theorem B]. *Let $\mathfrak{F}$ be any formation and G a group. If E is a normal subgroup of G and $F^*(E) \leq Z_{\mathfrak{F}}(G)$, then $E \leq Z_{\mathfrak{F}}(G)$.*

Corollary 2.3 [33, Corollary 1.2]. *Let E be a normal subgroup of a group G. If every G-chief factor below $F^*(E)$ is cyclic, then every G-chief factor below E is cyclic.*

It is rather clear that if $\mathfrak{F}$ is a saturated formation containing all supersoluble groups and G is a group with a cyclic normal subgroup E such that $G/E \in \mathfrak{F}$, then $G \in \mathfrak{F}$. Hence Theorem 2.1 and Corollary 2.3 allow us to give affirmative answers to Questions I and II. Finally, in view of Corollary 2.3, Theorem 2.1 not only generalizes all the results in [36-38, 41-53] mentioned above but also gives shorter proofs of many of them.

3 On factorizations of groups with $\mathfrak{F}$-hypercentral intersections of the factors

One of the highlights of the proof of Theorem 2.1 is the following result allowing to carry out inductive reasonings.

Theorem 3.1 [33, Lemma 2.4]. *Let A, B and E be normal subgroups of a group G. Suppose that $G = AB$ and $E \leq Z_{\mathfrak{U}}(A) \cap Z_{\mathfrak{U}}(B)$. If $(|G : A|, |G : B|) = 1$, then $E \leq Z_{\mathfrak{U}}(G)$.*

But in fact, this theorem is a generalization of the following well-known result of the theory of supersolvable groups.

Corollary 3.2 (Friesen). *Let $G = AB$ where A, B are normal supersoluble subgroups of G. If $(|G : A|, |G : B|) = 1$, then G is supersoluble.*

This important observation led us to the following general problem:

Let $G = AB$ be the product of two subgroups A and B of G. What we can say about the structure of G if $A \cap B \leq Z_{\mathfrak{F}}(A) \cap Z_{\mathfrak{F}}(B)$ for some class of groups $\mathfrak{F}$?

The paper [33] is devoted to the analysis of this topic.

In particular the following facts were proved.

Theorem 3.3 [34, Theorem 3.4]. *Suppose that G has three subgroups A_1, A_2 and A_3 whose indices $|G : A_1|$, $|G : A_2|$, $|G : A_3|$ are pairwise coprime. If $A_i \cap A_j \leq Z_{\mathfrak{S}}(A_i) \cap Z_{\mathfrak{S}}(A_j)$ for all $i \neq j$, then G is soluble.*

In this theorem $\mathfrak{S}$ denotes the class of all soluble groups.

Corollary 3.4 (H. Wielandt [55]). *If G has three soluble subgroups A_1, A_2 and A_3 whose indices $|G : A_1|$, $|G : A_2|$, $|G : A_3|$ are pairwise coprime, then G is itself soluble.*

In the following theorem, $c(G)$ denotes the nilpotent class of a nilpotent group G.

Theorem 3.5 [34, Theorem 3.7]. *Suppose that G has three subgroups A_1, A_2 and A_3 whose indices $|G : A_1|$, $|G : A_2|$, $|G : A_3|$ are pairwise coprime. Let p be a prime. Then:*

(1) If $A_i \cap A_j \le Z_{\mathfrak{F}}(A_i) \cap Z_{\mathfrak{F}}(A_j)$ for all $i \neq j$, where $\mathfrak{F}$ is the class of all p-closed groups, then G is p-closed.

(2) If $A_i \cap A_j \le Z_{\mathfrak{F}}(A_i) \cap Z_{\mathfrak{F}}(A_j)$ for all $i \neq j$, where $\mathfrak{F}$ is the class of all p-decomposable groups, then G is p-decomposable.

(3) If $A_i \cap A_j \le Z_n(Z_\infty(A_i)) \cap Z_n(Z_\infty(A_j))$ for all $i \neq j$, then G is nilpotent and $c(G) \le n$.

The following corollaries are well known.

Corollary 3.6 (O. Kegel). *If G has three nilpotent subgroups A_1, A_2 and A_3 whose indices $|G : A_1|$, $|G : A_2|$, $|G : A_3|$ are pairwise coprime, then G is itself nilpotent.*

Corollary 3.7 (K. Doerk). *If G has three abelian subgroups A_1, A_2 and A_3 whose indices $|G : A_1|$, $|G : A_2|$, $|G : A_3|$ are pairwise coprime, then G is itself abelian.*

Theorem 3.8 [34, Theorem 3.10]. *Suppose that G has four subgroups A_1, A_2, A_3 and A_4 whose indices $|G : A_1|$, $|G : A_2|$, $|G : A_3|$, $|G : A_4|$ are pairwise coprime. If $A_i \cap A_j \le Z_{\mathfrak{U}}(A_i) \cap Z_{\mathfrak{U}}(A_j)$ for all $i \neq j$, then G is supersoluble.*

Corollary 3.9 (K. Doerk). *If G has four supersoluble subgroups A_1, A_2, A_3 and A_4 whose indices $|G : A_1|$, $|G : A_2|$, $|G : A_3|$, $|G : A_4|$ are pairwise coprime, then G is supersoluble.*

Recall that a subgroup H of G is said to be abnormal if $x \in \langle H, H^x \rangle$. It is clear that if H is a abnormal in G, then $N_G(H) = H$.

Theorem 3.10 [34, Theorem 3.13]. *Suppose that G has three abnormal subgroups A_1, A_2 and A_3 whose indices $|G : A_1|$, $|G : A_2|$, $|G : A_3|$ are pairwise coprime.*

(1) If $A_i \cap A_j \le Z_{\mathfrak{F}}(A_i) \cap Z_{\mathfrak{F}}(A_j)$ for all $i \neq j$, where $\mathfrak{F}$ is the class of all metanilpotent groups, then G is metanilpotent.

(2) If $A_i \cap A_j \le Z_{\mathfrak{U}}(A_i) \cap Z_{\mathfrak{U}}(A_j)$ for all $i \neq j$, then G is supersoluble.

Corollary 3.11 (A.F. Vasilyev [62]). *If G has three abnormal supersoluble subgroups A_1, A_2 and A_3 whose indices $|G : A_1|$, $|G : A_2|$, $|G : A_3|$ are pairwise coprime, then G is itself supersoluble.*

Finally, we mention the following result.

Theorem 3.12 [34, Theorem 3.16]. *A group G is supersoluble if and only if every maximal subgroup V of every Sylow subgroup of G either is normal or has a supplement T in G such that $V \cap T \le Z_{\mathfrak{U}}(T)$.*

Corollary 3.13 (W. Guo, K.P. Shum, A.N.Skiba [71]). *A group G is supersoluble if and only if every maximal subgroup of every Sylow subgroup of G either is normal or has a supersoluble supplement in G.*

4 Quasi-$\mathfrak{F}$-groups

Now we consider some applications of the $\mathfrak{X}$-hypercentre in the theory of generalized quasinilpotent groups.

A group G is said to be *quasinilpotent* if for every its chief factor H/K and every $x \in G$, x induces an inner automorphism on H/K [18, Chapter X].

Note that since for every central chief factor H/K every element of G induces trivial automorphism on H/K, one can say that a group G is quasinilpotent if for every its *non-central* chief factor H/K and every $x \in G$, x induces an inner automorphism on H/K. This obvious observation allows us to consider the following generalization of quasisnilpotent groups.

Definition 4.1 [30, 31]. Let $\mathfrak{F}$ be a class of groups and G a group. We say that G *is a quasi-$\mathfrak{F}$-group* if for every $\mathfrak{F}$-eccentric chief factor H/K of G, every automorphism of H/K induced by an element of G is inner.

In particular, we say that G is a *quasisupersoluble group* if for every *non-cyclic* chief factor H/K of G, every automorphism of H/K induced by an element of G is inner.

A group G is called semisimple if G is either the unit group or the direct product of non-abelian simple group. In particular any non-abelian simple group is semisimple.

The theory of quasinilpotent groups is well represented in the book [18]. A key result of this theory was the following structure theorem.

Theorem 4.2 [18, Chapter X, Theorem 3.6]. *A group G is quasinilpotent if and only if $G/Z_{\infty}(G)$ is semisimple.*

The first question that arises when we consider the quasisupersoluble groups or the Quasi-$\mathfrak{F}$-groups, in general, is the following:

What can we say about the structure of the Quasi-$\mathfrak{F}$-groups?

The following theorem gives a complete answer to this question in the case of quasisupersoluble groups.

Theorem 4.3 [32, Theorem C]. *A group G is quasisupersoluble if and only if $G/Z_{\mathfrak{U}}(G)$ is semisimple.*

In general case, we have

Theorem 4.4 [31, Theorem C]. *Let $\mathfrak{F}$ be a saturated normally hereditary formation. Then a group G is a quasi-$\mathfrak{F}$-group if and only if $G/Z_{\mathfrak{F}}(G)$ is semisimple.*

Here we call, following [56], that a class $\mathfrak{F}$ of groups *(normally) hereditary*, if $\mathfrak{F}$ contains every (normal) subgroup of every its group.

Surprising similarities in the structure of the quasinilpotent groups and the quasi-$\mathfrak{F}$-groups makes a real suggestion that the quasi-$\mathfrak{F}$-groups inherit many other interesting properties of quasinilpotent groups. This assumption was confirmed in the above-mentioned papers [30, 31, 32].

My immediate goal is to discuss some of the results of these papers.

The books [11, 2, 69, 13, 14] contains numerous applications of Baer-local formations. Nevertheless, it has long remained an open question how wide is the class of Baer-local formations.

It is well known that the class $\mathfrak{F}$ of all nilpotent groups is saturated. L.A. Shemetkov [57] the first to show that the class $\mathfrak{N}^*$ (we here use the notation in [18]) of all quasinilpotent groups is a Baer-local formation. Perhaps, the class $\mathfrak{N}^*$ was the only know classic example of the Baer-local formation which is not saturated.

Following [64], a group G is said to be an SC-group if every chief factor of G is a simple group. We note in passing that the groups every chief factor of which is a simple group were named in [65] c-supersoluble.

SC-Groups have many interesting properties [64, 65]. In particular, the class of all such groups is a new example of the Baer-local formation [66].

By above Theorem 4.3, we see that every quasisupersoluble group is an SC-group.

These observations are a motivation for attempts to find new series of Baer-local formations among classes of quasi-$\mathfrak{F}$-groups.

Theorem 4.5 [32, Theorem A]. *The class $\mathfrak{U}^*$ of all quasisupersoluble groups is a normally hereditary Baer-local formation.*

In general case, we have

Theorem 4.6 [31, Theorem A]. *Suppose that $\mathfrak{F}$ is a saturated formation containing all nilpotent groups. Then:*

(1) *$\mathfrak{F}^*$ is a Baer-local formation.*

(2) *If $\mathfrak{F}$ is normally hereditary, then $\mathfrak{F}^*$ is normally hereditary.*

(3) *If $\mathfrak{F}$ is a Fitting class, then $\mathfrak{F}^*$ is a Fitting class.*

A class $\mathfrak{F}$ of groups is called a Fitting class if $\mathfrak{F}$ is normally hereditary and $\mathfrak{F}$ contains every group $G = AB$, where A and B are normal in G and $A, B \in \mathfrak{F}$.

On the base of Theorem 4.3, one can easily obtain examples of quasisupersoluble groups. For example, let $A = C_7 \leftthreetimes \langle\alpha\rangle$, where $|C_7| = 7$ and α is an automorphism of C_7 with $|\alpha| = 3$. Let $B = A \times A_7$. Then by Theorem 4.2, B is quasisupersoluble and not quasinilpotent. The group $C = [B]\langle\beta\rangle$, where β is an inner automorphism of A_7 with $|\beta| = 2$ and α acts trivially on A, is an SC-group but not a quasisupersoluble group.

5 G-covering systems

If P is a Sylow p-subgroup of G, then we write Σ_p to denote a set of subgroups which contains at least one supplement to G of each maximal subgroup of P.

Let $\mathfrak{F}$ be a class of groups. Following [71], we call a set Σ of subgroups of G a *G-covering subgroup system for the class $\mathfrak{F}$* if $G \in \mathfrak{F}$ whenever $\Sigma \subseteq \mathfrak{F}$.

If P is a Sylow p-subgroup of G and p is odd, then the set $\{N_G(J(P)), C_G(Z(G))\}$ (Thompson) and the one-element set $\{N_G(Z(J(P))\}$ (Glauberman-Thompson) are G-covering subgroup systems for the class of all p-nilpotent groups. Now let Σ_i be any set of subgroups $H_1, H_2, \ldots, H_i$ of G whose indices $|G : H_1|, |G : H_2|, \ldots, |G : H_i|$ are pairwise coprime. Then Σ_3 is a G-covering subgroup system for the classes of all soluble (Wielandt), all nilpotent (Kegel) and all abelian (Doerk) groups. It is not difficult to find examples which show that Σ_3 is not a G-covering subgroup system for the classes of all supersoluble and all metanilpotent groups. However, Σ_4 is a G-covering subgroup system for both these classes (Doerk, Kramer). According to [68], the set Σ of all normalizers of all Sylow subgroups of G is a G-covering subgroup system for the class of all nilpotent groups and, by Fedri and Serens [70], Σ is not a G-covering subgroup system for the class of all supersoluble groups. In the paper [73] the following theorem in this trend was proved.

Theorem 5.1 [73, Theorem A]. *The set $\Sigma_p \cup \Sigma_q$ is a G-covering subgroup system for the classes of all metanilpotent, all supersoluble, all p-supersoluble, all p-nilpotent and all abelian groups.*

Corollary 5.2 (W. Guo, K.P. Shum, A.N.Skiba [71]). *Let Σ be a set of subgroups which contains at least one supplement of each maximal subgroup of every Sylow subgroup of G. Then Σ is a G-covering subgroup system for the classes of all nilpotent and all supersoluble groups.*

Corollary 5.3 (W. Guo, K.P. Shum, A.N.Skiba [71]). *Let Σ be a set of subgroups which contains at least one supplement of each maximal subgroup of every Sylow subgroup of G. If G is p-soluble, then Σ is a G-covering subgroup system for the classes of all p-nilpotent and all p-supersoluble groups.*

The proof of Theorem 5.1 consists of many steps and the following theorem is one of them.

Theorem 5.4 [73, Theorem B]. *Let $\{p_1, p_2, \ldots, p_t\} \subseteq \pi(G)$ and r be a prime.*

(1) *The set Σ_{p_1} is a G-covering subgroup system for the classes of all r-decomposable and all nilpotent groups.*

(2) *The set $\Sigma_{p_1} \cup \Sigma_{p_2} \cup \ldots \cup \Sigma_{p_t}$ is a G-covering subgroup system for the class $\mathfrak{N}^t$.*

It this theorem $\mathfrak{N}^t$ denotes the class of all soluble groups of nilpotent length at most t [69, p. 358].

6 X-permutable subgroups

Recall that a subgroup A of a group G is said to permute with a subgroup B if $AB = BA$. A subgroup A is said to be a permutable or a quasinormal subgroup of G if A is permutable with all subgroups of G.

Often we meet the situation $AB \neq BA$, nevertheless there exists an element $x \in G$ such that $AB^x = B^xA$, for instance, we have the following cases:

1) Let $G = AB$ be a group. If A_p and B_p are Sylow p–subgroups of A and of B respectively, then $A_pB_p \neq B_pA_p$ in general, but G has an element x such that $A_pB_p^x = B_p^xA_p$.

2) If A and B are Hall subgroups of a soluble group G, then there exists an element $x \in G$ such that $AB^x = B^xA$ (cf. [69, Chapter I, 4.11]).

3) If A and B are normally embedded subgroups (see Definition 7.1 in [69, Chapter I]) of a soluble group, then A is permutable with some conjugate of B (cf. [69, Chapter I, 4.17]).

The above examples and some other examples of such kind, the following definitions are inspired:

Definition [74, 75, 76]. Let A, B be subgroups of a group G and X a non-empty subset of G. Then we say that A *X-permutes with* B if there exists some $x \in X$ such that $AB^x = B^xA$.

The importance of the concept of X-permutable subgroups is connected first of all to the observation that many important classes of groups can be described in terms of X-permutable subgroups [74, 75, 76, 77, 78, 79, 80] and so on. In particular the following results in this trend were proved.

Theorem 6.1 ([75, Theorem 3.6]). *Let $X = G^{\mathfrak{N}^*}$ and $Y = X_{\mathfrak{S}}$ is the soluble radical of X. Then the following are equivalent:*

(1) G is soluble.

(2) G has a soluble maximal subgroup M and every Sylow subgroup of G X-permutes with M and with all Sylow subgroups of G.

(3) Every maximal subgroup of G Y-permutes with all Sylow subgroups of G.

Theorem 6.2 ([75, Theorem 3.7]). *Let $X = F(G)$ be the commutator subgroup of G. Then G is supersoluble if and only if every maximal subgroup of G X-permutes with all subgroups of G.*

Theorem 6.3 ([75, Theorem 3.17]). *A group G is nilpotent if and only if it has a nilpotent abnormal subgroup X such that every two Sylow subgroups of G are X-permutable.*

Corollary 6.4. *Let X be a Carter subgroup of a soluble group G. Then G is nilpotent if and only if every two Sylow subgroups of G are X-permutable.*

7 New criteria of existence and conjugacy of Hall subgroups of finite groups

The famous Schur-Zassenhaus Theorem asserts that: *If G has a normal Hall π-subgroup A, then G is a $E_{\pi'}$-group (that is, G has a Hall π'-group). Moreover, if either A or G/A is soluble, then A is a $C_{\pi'}$-subgroup (that is, any two Hall π'-subgroups of G are conjugate).*

In 1928, Hall [81] proved that: *A finite soluble group has a Hall π-subgroup and any two Hall π-subgroups are conjugate in G.*

In 1949, Čunihin developed further the Schur-Zassenhaus and Hall's theorem and proved the following classical result.

Theorem (S. A. Čunihin [82]). *If G is π-separable, then G is a E_π-group and a $E_{\pi'}$-group. Moreover, if G is π-soluble, then G is a C_π-group and a $C_{\pi'}$-group.*

Note that a group G is said to be π-separable if G has a normal series

$$1 = G_0 \le G_1 \le \ldots \le G_{t-1} \le G_t = G, \tag{$*$}$$

where each index $|G_i : G_{i-1}|$ is either a π-number or a π'-number. A group G is said to be π-soluble if each index $|G_i : G_{i-1}|$ of Series (*) is either a π-prime power (that is, a power of some prime in π) or a π'-number.

The example of the group $PSL(2,7)$ shows that the condition of normality for the members of Series (*) could not be omitted.

It is well known that the above Schur-Zassenhaus theorem, Hall theorem and Čunihin theorem are truly fundamental results of group theory. In connection with these important results, the following two problems have naturally arisen:

Problem I. Whether the conclusion of the Schur-Zassenhaus Theorem holds if the Hall subgroup A of G is not normal ? In other words, can we weaken the condition of normality for the Hall subgroup A of G so that the conclusion of the Schur-Zassenhaus Theorem is still true ?

Problem II. Whether we can replace the condition of normality for the members of Series (*) by some weaker condition, for example, by permutability of the members of Series (*) with some systems of subgroups of G.

Some results about Problem I had been obtained in [84, 85]. In the paper [86], the following further generalization of Schur-Zassenhaus Theorem was proved.

Theorem 7.1 [86, Theorem A]. *Let A be a Hall π-subgroup of G. Let $G = AT$ for some subgroup T of G, and let q be a prime. If A permutes with every Sylow p-subgroup of T, for all primes $p \neq q$, and either A or T is soluble, then T contains a complement of A in G and any two complements of A in G are conjugate.*

By using the Feit-Thompson theorem, we obtain the following stronger version of Theorem 7.1.

Theorem 7.2 [86, Theorem A^*]. *Let A be a Hall π-subgroup of G. Let $G = AT$ for some subgroup T of G, and let q be a prime. If A permutes with every Sylow p-subgroup of T for all primes $p \neq q$, then T contains a complement of A in G and any two complements of A in G are conjugate.*

Recall that a subgroup H of G is said to be a supplement of a subgroup A in G if $AH = G$. Let

$$1 = H_0 \leq H_1 \leq \ldots \leq H_{t-1} \leq H_t = G \qquad (**)$$

be some subgroup series of G. We say that a subgroup series

$$1 = T_t \leq T_{t-1} \leq \ldots \leq T_1 \leq T_0 = G$$

is a *supplement of Series* $(**)$ *in* G if T_i is a supplement of H_i in G for all $i = 0, 1, \ldots, t$.

The another purpose of the paper [86] was to give the positive answer to Problem II.

Theorem 7.3 [86, Theorem B]. *Suppose that G has a subgroup series*

$$1 = H_0 \leq H_1 \leq \ldots \leq H_{t-1} \leq H_t = G$$

and a supplement

$$1 = T_t \leq T_{t-1} \leq \ldots \leq T_1 \leq T_0 = G$$

of this series in G such that H_i permutes with every Sylow subgroup of T_i for all $i = 1, 2, \ldots, t$. If each index $|H_{i+1} : H_i|$ is either a π-number or a π'-number, then G is a E_π-group and a $E_{\pi'}$-group. Moreover, if each π-index $|H_{i+1} : H_i|$ is a prime power, then G has a soluble Hall π-subgroup.

Corollary 7.4. *Suppose that G has a subgroup series*

$$1 = H_0 \leq H_1 \leq \ldots \leq H_{t-1} \leq H_t = G$$

and a supplement

$$1 = T_t \leq T_{t-1} \leq \ldots \leq T_1 \leq T_0 = G$$

of this series in G such that H_i permutes with all Sylow subgroups of T_i for all $i = 1, 2, \ldots, t$. If each index $|H_{i+1} : H_i|$ $(i = 0, 1, \cdots, t-1)$ is a prime power, then G is a E_π-group, for any set π of primes.

Corollary 7.5. *Suppose that G has a subgroup series*

$$1 = H_0 \leq H_1 \leq \ldots \leq H_{t-1} \leq H_t = G$$

and a supplement

$$1 = T_t \leq T_{t-1} \leq \ldots \leq T_1 \leq T_0 = G$$

of this series in G such that H_i permutes with all Sylow subgroups of T_i for all $i = 1, 2, \ldots, t$. If each index $|H_{i+1} : H_i|$ is a prime power, then G is soluble.

Theorem 7.6 [86, Theorem C]. *Suppose that G has a subgroup series*

$$1 = H_0 < H_1 \leq \ldots \leq H_{t-1} \leq H_t = G$$

and a subgroup T such that $G = H_1T$ and H_i permutes with all subgroups of T for all $i = 1, 2, \ldots, t$. If each index $|H_{i+1} : H_i|$ is either a π-number or a π'-number, then G is a C_π-group and a $C_{\pi'}$-group.

The following example shows that, under conditions of Theorems 7.1, 7.4 or 7.6 the group G is not necessary π-separable.

Example. Let $G = A_5 \times C_7$, where C_7 is a group of order 7 and A_5 is the alternating group of degree 5. Let C_5 be a Sylow 5-subgroup of A_5. Consider the subgroup series

$$1 = H_0 < H_1 < H_2 < H_3 = G, \qquad (***)$$

where $H_1 = A_4$ and $H_2 = A_5$. Then the series $1 = T_3 < T_2 < T_1 < T_0 = G$, where $T_2 = C_7$ and $T_1 = C_5 \times C_7$, is a supplement of Series (***) in G. It is clear also that H_i permutes with all subgroups of T_i, for all i. Let $\pi = \{5, 7\}$. Then every index of Series (***) is either a π-number or a π'-number. However, G is not π-separable.

References

[1] Hanna Neumann, *Varieties of Groups*, Springer-Verlag, Berlin-Heidelberg-NewYork, 1967.

[2] L. A. Shemetkov, A. N. Skiba, *Formations of algebraic systems*, Nauka, Moscow, Main Editorial Board for Physical and Mathematical Literature, 1989.

[3] V.G. Safonov, On a question of the theory of totally saturated formations of finite groups, *Algebra Colloquium*, **15**(1), 2008, 119-128.

[4] V.G. Safonov, On modularity of the lattice of totally saturated formations, *Comm. Algebra*, **35**(11), 2007, 3495-3502.

[5] W. Guo, K.P. Shum, On totally local formations of finite groups, *Comm. Algebra*, **30**(5), 2002, 2117-2131.

[6] S.F. Kamornikov, L.A. Shemetkov, Decomposable idempotents of the semigroup of formations of finite groups, *Dokl. AN BSSR*, **35**(10), 1991, 869-871.

[7] O.V. Melnikov, Indecomposability of minimal idempotents of the semigroup of formations, *Questions in Algebra*, **9**, 1996, 42-47.

[8] A. Ballester-Bolinches, M.D.Perez-Ramos, Some questions of the Kourovka notebook concerning formation products, *Comm. Algebra*, **26**(5), 1998, 1581-1587.

[9] A. Ballester-Bolinches, C. Calvo, and R. Esteban-Romero, A question from the Kourovka Notebook, *Bull. Austral. Math. Soc.*, **68**(3), 2003, 461-470.

[10] B. Huppert, *Endliche Gruppen I.* Springer-Verlag, Berlin-Heidelberg-New York, 1967.

[11] L. A. Shemetkov, *Formations of finite Groups*, Moscow, Nauka, Main Editorial Board for Physical and Mathematical Literature, 1978.

[12] K. Doerk, T. Hawkes, *Finite Soluble Groups*, Walter de Gruyter, Berlin–New York, 1992.

[13] W. Guo, *The Theory of Classes of Groups*, Science Press-Kluwer Academic Publishers, Beijing-New York-Dordrecht-Boston-London, 2000.

[14] A. Ballester-Bolinches, L. M. Ezquerro, *Classes of Finite groups*, Springer, Dordrecht, 2006.

[15] O. Kegel, Sylow-Gruppen and Subnormalteiler endlicher Gruppen, *Math. Z.*, **78**, 1962, 205–221.

[16] R. Maier, P. Schmid, The embedding of permutable subgroups in finite groups, *Math. Z.*, **131**, 1973, 269–272.

[17] R. Schmidt, *Subgroup Lattices of Groups*, de Gruyter, Berlin, 1994.

[18] B. Huppert, N. Blackburn, *Finite Groups III*, Springer-Verlag, Berlin-New-York, 1982.

[19] N. P. Mukherjee, The hyperquasicenter of a finite groups, *Proc. Amer. Math. Soc.*, **32**, 1972, 24-28.

[20] Alexander N. Skiba, A characterization of the hypercyclically embedded subgroups of finite groups, *Journal of Pure and Applied Algebra*, **315**(3), 2011, 257-261.

[21] R. Baer, Supersoluble immersion, *Canad. J. Math.*, **11**, 1959, 353-369.

[22] K.A. Al-Sharo, E.A. Molokova, L.A. Shemetkov, Factorizable groups and formations, *Acta Applicantae Mathematicae*, **85** (1-3), 2005, 3-10.

[23] K.A. Al-Sharo, L.A. Shemetkov, On subgroups of prime order in a finite group, *Ukr. Math. J.*, **54**(6), 2002, 915-923.

[24] O.L. Shemetkova, Finite groups with a system of generalized central elements, *Algebra and Discrete mathematics*, **4**, 2004, 59-71.

[25] K.A. Al-Sharo, O.L. Shemetkova, An application of the concept of a generalized central element, *Algebra and Discrete mathematics*, **4**, 2007, 1-10.

[26] A.A. Rodionov, L.A. Shemetkov, On p-locally N-closed formations of finite groups, *Proc. of Mathematics Institute*, **18**(1), 2010, 92-98.

[27] O.L. Shemetkova, On finite groups with Q-central elements of prime order, *Proc. of Mathematics Institute*, **16**(1), 2008, 97-99.

[28] L. A. Shemetkov and A. N. Skiba. On the $\mathfrak{X}\Phi$-hypercentre of finite groups, *J. Algebra*, **322**, 2009, 2106–2117.

[29] Vladimir O. Lukyanenko, Alexander N. Skiba, Finite groups in which τ-normality is a transitive Relation, *Rend. Sem. Mat. Padova*, **124**, 2010, 97-106.

[30] W. Guo, A.N. Skiba, On finite quasi-$\mathfrak{F}$-groups, *Comm. Algebra*, **37**, 2009, 470-481.

[31] Wenbin Guo, A.N. Skiba, On some classes of finite quasi-$\mathfrak{F}$-groups, *J. Group Theory*, **12**, 2009, 407-417.

[32] Wenbin Guo, A.N. Skiba, On quasisupersoluble and p-quasisupersoluble finite groups, *Algebra Collocuium*, **17**(4), 2010, 549-556.

[33] A.N. Skiba, On two questions of L.A. Shemetkov concerning hypercyclically embedded subgroups of finite groups, *J. Group Theory*, **(13)**(6), 2010, 841-950.

[34] Wenbin Guo, A.N. Skiba, On factorizations of finite groups with $\mathfrak{F}$-hypercentral intersections of the factors, *J. Group Theory* (in Press).

[35] M. Weinstein, Between Nilpotent and Solvable, Polygonal Publishing House, 1982.

[36] Y. Wang. c-normality of groups and its properties, *J. Algebra*, **180**, 1996, 954–965.

[37] A. Ballester-Bolinches, Y. Wang and X. Y. Guo. c-supplemented subgroups of finite groups, *Glasgow Math. J.*, **42**, 2000, 383–389.

[38] J. Buckley. Finite groups whose minimal subgroups are normal. *Math. Z.* **15** (1970), 15–17.

[39] B. Huppert. Normalteiler and maximale Untergruppen endlicher Gruppen, *Math. Z.* **60**, 1954, 409–434.

[40] K. Doerk. Minimal nicht uberauflosbare endlicher Gruppen, *Math. Z.*, **91**, 1966, 198–205.

[41] S. Srinivasan. Two sufficient conditions for supersolvability of finite groups, *Israel J. Math.*, **35**, 1980, 210–214.

[42] A. Ballester-Bolinches and M. C. Pedraza-Aguilera, On minimal subgroups of finite groups, *Acta Math. Hungar.*, **73**, 1996, 335–342.

[43] M. Asaad and P. Csörgő, Influence of minimal subgroups on the structure of finite group, *Arch. Math. (Basel)*, **72**, 1999, 401–404.

[44] A. Ballester-Bolinches and Y. Wang. Finite groups with some C-normal minimal subgroups. *J. Pure Appl. Algebra*, **153**, 2000, 121–127.

[45] Y. Wang, Y. Li and J. Wang. Finite groups with c-supplemented minimal subgroups. *Algebra Colloquium*, **10(3)**, 2003, 413–425.

[46] Y. Li and Y. Wang. The influence of minimal subgroups on the structure of a finite group. *Proc. Amer. Math. Soc.*, **131**, 2002, 337–341.

[47] H. Wei, Y. Wang and Y. Li. On c-normal maximal and minimal subgroups of Sylow subgroups of finite groups, II. *Comm. Algebra*, **31**, 2003, 4807–4816.

[48] Y. Wang, H. Wei and Y. Li. A generalization of Kramer's theorem and its applications. *Bull. Australian Math. Soc.* **65**, 2002, 467–475.

[49] H. Wei, Y. Wang and Y. Li. On c-supplemented maximal and minimal subgroups of Sylow subgroups of finite groups. *Proc. Amer. Math. Soc.* **132(8)**, 2004, 2197–2204.

[50] M. Asaad. On maximal subgroups of finite group. *Comm. Algebra*, **26**, 1998, 3647–3652.

[51] H. Wei. On c-normal maximal and minimal subgroups of Sylow subgroups of finite groups. *Comm. Algebra*, **29**, 2001, 2193–2200.

[52] A. Ballester-Bolinches and X. Y. Guo. On complemented subgroups of finite groups. *Arch. Math.*, **72**, 1999, 161–166.

[53] Y. Li and Y. Wang. The influence of π-quasinormality of some subgroups of a finite group. *Arch. Math. (Basel)*, **81**, 2003, 245–252.

[54] A. N. Skiba, On weakly s-permutable subgroups of finite groups, J. Algebra, **315**(1), 2007, 192-209.

[55] H. Wielandt, Uber die Normalstrukture von mehrfach faktorisierbaren Gruppen, *B. Austral Math. Soc.*, **1**, 1960, 143-146.

[56] A. I. Mal'cev, *Algebraic Systems*, Nauka, Moscow, Main Editorial Board for Physical and Mathematical Literature, 1970.

[57] L. A. Shemetkov, Composition formations and radicals of finite groups, Ukrainsk. Math. Zh., **40**(3), 1988, 369–375.

[58] D. Gorenstein, *Finite Groups*. Harper & Row Publishers, New York-Evanston-London, 1968.

[59] R. Baer, Classes of finite groups and their properties, *Illinois J. Math.*, **1**, 1957, 115-187.

[60] D. Friesen, Products of normal supersoluble subgroups, *Proc. Amer. Math. Soc.*, **30**, 1971, 46–48.

[61] O. H. Kegel, Zur Struktur mehrafach faktorisierbarer endlicher Gruppen, Math. Z., **87**, 1965, 409-434.

[62] A. F. Vasilyev, Formations and their recognazing, *Proc. Gomel State University*, **2(41)**, 2007, 23-29.

[63] W. Guo, K. P. Shum, A. N. Skiba, G-covering subgroup systems for the classes of supersoluble and nilpotent groups, *Israel J. Math.*, **138**, 2003, 125–138.

[64] D. J. S. Robinson, The structure of finite groups in which permutability is a transitive relation, *J. Austral Math. Soc.*, **70**, 2001, 143–159'

[65] V.A. Vedernikov, On some classes of finite groups, *Dokl. Akad. nauk BSSR*, **32**(10), 872-875.

[66] A. F. Vasil'yev, T.I. Vasil'yeva, On finite groups whose principial factors are simple groups, *Izvestiya VUZ. Matematika*, **41**(11),19979, 10-14.

[67] R. Laue, Dualization for saturation for locally defined formations, *J. Algebra*, **52**, 1978, 347–353.

[68] M. Bianchi, A.G. Mauri and P. Hauck, On finite groups with nilpotent Sylow-normalizers, *Arch. Math.*, **47**, 1986, 193-197.

[69] K. Doerk, T. Hawkes, *Finite Soluble Groups*, Berlin–New York: Walter de Gruyter, 1992.

[70] V. Fedri and L. Serens, Finite soluble groups with supersoluble Sylow normalizers, *Arch. Math.*, **50**, 1988, 11-18.

[71] W. Guo, K. P. Shum, A. N. Skiba, G-covering subgroup systems for the classes of supersoluble and nilpotent groups, *Israel J. Math.*, **138**, 2003, 125–138.

[72] W. Guo, K. P. Shum, A. N. Skiba, G-covering subgroup systems for the classes of p-supersoluble and p-nilpotent finite groups, *Siberian Math. J.*, **45**(3), 2004, 453-442.

[73] W. Guo, A. N. Skiba, On G-covering subgroup systems of finite groups, *Acta Math. Hungar.*, (in Press).

[74] W. Guo, K.P. Shum and A.N. Skiba, Conditionally Permutable Subgroups and Supersolubility of Finite Groups, *Southeast Asian Bull Math.*, **29**, 2005, 493-510.

[75] W.Guo, A.N. Skiba, and K.P. Shum, X-permutable subgroups, *Siberian Mathematical Journal*, **48(4)**, 2007, 742-759.

[76] W.Guo, K.P. Shum and A.N. Skiba, X-semipermutable Subgroups of Finite Groups, *J. Algebra*, **315**, 2007, 31–41.

[77] W.Guo, K.P. Shum and A.N. Skiba, Criterions of Supersolubility for Products of Supersoluble Groups, *Publ. Math. Debrecen*, **68/3-4**, 2006, 433–449.

[78] W.Guo, K.P. Shum and A. N. Skiba, X-permutable maximal subgroups of Sylow subgroups of finite groups, *Ukrain. Matem. J.*, **58(10)**, 2006, 1299-1309.

[79] Bao-jun Li, A.N. Skiba, New characterizations of finite supersoluble groups, *Science in China Serias A: Mathematics*, **50**(1), 2008.

[80] W. Guo, K.P. Shum and A.N. Skiba, X_m-semipermutable Subgroups of Finite Groups, *Math. Nachr.*, **283**(11), 2010, 1603-1612.

[81] P. Hall, A note on soluble groups, J. London Math. Soc., 1928, **3**, 98–105.

[82] S. A. Čunihin, On theorems of Sylow's type, Doklady Akad. Nauk. SSSR, 1949, **66**, 65-68.

[83] W. Guo, A.N. Skiba, Criterions of Existence of Hall Subgroups in Non-soluble Finite Groups,*Acta Math. Sin.*, **26**(2), 2010, 295-304.

[84] Wenbin Guo, K. P. Shum and Alexander N. Skiba, X-semipermutable subgroups of finite groups, J. Algebra, 2007, **315**, 31-41.

[85] Wenbin Guo, K.P. Shum and Alexander N. Skiba, Schur-Zassenhaus Theorem for X-permutable subgroups, Algebra Colloquium, 2008, **15**(2), 185-192.

[86] W. Guo, A. N. Skiba, New criterions of existence and conjugacy of Hall subgroups of finite groups, *Proc. Amer. Math. Soc.*, (in Press).

Right *pp* Semigroups with Right Adequate Transversals

Xiaojiang Guo
Department of Mathematics, Jiangxi Normal University, Nanchang,
Jiangxi 330022, People's Republic of China
E-mail: xjguo1967@sohu.com; xjguo@jxnu.edu.cn

Yaohua Hu
Zhuhai City polytechnic, Zhuhai,
Guangdong 519090, People's Republic of China
E-mail: huyaohua2005@126.com

Dedicated to Professor Yuqi Guo and Professor K.P. Shum on the occasion of their 70th birthday

Abstract

In this paper, we consider rpp semigroups with quasi-ideal right adequate tranversals. After obtaining some characterizations of such semigroups, we establish the structure of this class of *rpp* semigroups in terms of right regular partial bands and right adequate semigroups. In addition, some special cases are considered.

Key words: *rpp* semigroup; right adequate semigroup; (quasi-ideal) right adequate transversal; right regular partial band.
Mathematical Subject Classification (2000): 20M10

1 Introduction

A semigroup S is called *right principal projective*, in short, *rpp*, if for all $a \in S$, the principal right ideal aS^1, regarded as a right S^1-system, is projective. Equivalently, a semigroup S is a rpp semigroup if and only if each $\mathcal{L}^*$-class of S contains an idempotent. Dually, we can define *left principal projective* (in short, *lpp*) *semigroups*. A semigroup S is called *abundant* if each $\mathcal{L}^*$-class and each $\mathcal{R}^*$-class of S contains at least one idempotent. Obviously, an abundant semigroup is just a semigroup which is both lpp and rpp. Moreover, an abundant semigroup S is called *adequate* if the set $E(S)$ of idempotents of S forms a semilattice. Regular semigroups are abundant semigroups and inverse semigroups are adequate semigroups.

In 1982, Blyth and McFadden [1] introduced the concept of inverse transversals of regular semigroups. Since then many authors have investigated various kinds of regular semigroups with inverse transversals. As an analogue of inverse transversals in the range of abundant semigroups, El-Qallali [3] defined adequate transversals and further established the structure of abundant semigroups satisfying the regularity condition and with a multiplicative type-A transversal. Guo [6] obtained the structure of general abundant semigroups with a multiplicative adequate transversal. Chen [2], Guo-Shum [8], Guo-Wang [9] and Luo [13] considered various kinds of abundant semigroups with quasi-ideal adequate transversals. Guo-Xie [10] probed a class of naturally ordered abundant semigroups each of which is indeed an abundant semigroup with some kind of transversals. Recently, Guo-Peng [7] researched split quasi-adequate semigroups.

Right adequate semigroups are analogue of adequate semigroups in the theory of rpp semigroups (see [4]), it is natural to ask how to define the concept of right adequate transversals in a rpp semigroup. This is the aim of this paper.

2 Preliminaries

Throughout this paper we shall use the notions and notations of [5] and [12]. Other undefined terms can be found in [11]. Here we provided some known results repeatedly used without mention in the sequel. First, we recall some of the basic facts about the relations $\mathcal{L}^*$ and the dual for $\mathcal{R}^*$.

Lemma 2.1 [5] *Let S be a semigroup and $a, b \in S$. Then the following statements are equivalent:*

(1) $a\mathcal{L}^*b$.
(2) *for all $x, y \in S^1$, $ax = ay$ if and only if $bx = by$.*

As an easy and useful consequence, we have

Corollary 2.2 [5] *Let $a, e = e^2 \in S$. Then the following statements are equivalent:*

(1) $a\mathcal{L}^*e$.
(2) *$a = ae$ and for all $x, y \in S^1$, $ax = ay$ implies $ex = ey$.*

It is well known that $\mathcal{L} \subseteq \mathcal{L}^*$ and $\mathcal{R} \subseteq \mathcal{R}^*$. In particular, if a, b are regular elements, $a\mathcal{L}^*b$ ($a\mathcal{R}^*b$) if and only if $a\mathcal{L}b$ ($a\mathcal{R}b$). For convenience, if $a \in S$, a^* denotes a typical idempotent related to a by $\mathcal{L}^*$ and a^+ one of those related to a by $\mathcal{R}^*$. If $\mathcal{K}^*$ is one of the relations $\mathcal{L}^*, \mathcal{R}^*, \mathcal{D}^*, \mathcal{J}^*$ and $\mathcal{H}^*$, we shall denote by K_a^* the $\mathcal{K}^*$-class of S containing a.

Following Fountain [4], a rpp semigroup S is called a *right adequate semigroup* if $E(S)$ (the set of idempotents of S) is a semilattice. It is easy to see that each $\mathcal{L}^*$-class of a right adequate semigroup contains exactly one idempotent. Moreover, we have

Lemma 2.3 [4] *If S is a right adequate semigroup, then $(ab)^* = (a^*b)^*$ for all $a, b \in S$.*

Let S be a *rpp* semigroup and U a *rpp* subsemigroup of S. We call U a $*-$*subsemigroup* of S if for any $a \in S$, there exists $e \in E(U)$ such that $a\mathcal{L}^*(S)e$.

By a *quasi-ideal* I of a semigroup S, we mean a subset I of S which satisfies $ISI \subseteq I$.

As in Guo-Xie [10], we call a nonempty set M a *groupoid* if there is a partial operation on M. A groupoid $(M, \circ)$ is said to be a *partial semigroup* if, for all $x, y, z \in M$, if xy, yz and one of $(xy)z$ and $x(yz)$ are defined, then the other one of $(xy)z$ and $x(yz)$ is defined and $x(yz) = (xy)z$. If, in addition, there exists a unipotent semigroup S (that is, a semigroup in which each $\mathcal{L}$-class and each $\mathcal{R}$-class contains at most one idempotent) such that $E(S) = M$, then M is called a *partial semilattice*. A partial semigroup M is called a *left (right) regular partial band* if M is the disjoint union of left (right) zero rectangular bands M_α with $\alpha \in Y$, where Y is a partial semilattice, satisfying the following conditions:

(PB1) for all $\alpha, \beta \in Y$ and $x \in M_\alpha, y \in M_\beta$, if $\alpha \preceq \beta$, then xy and yx are defined for all $x \in M_\alpha, y \in M_\beta$;

(PB2) for all $x \in M$ and $y \in M_\alpha$, if xy and yx are defined and such that $xy = x$ and $yx = y$ ($xy = y$ and $yx = x$), then $x \in M_\alpha$.

In this case, we call Y the *structure partial semilattice* of the left (right) regular partial band M. Moreover, a subset N of the left (right) regular partial band $(M, \circ) = \bigcup_{\alpha \in Y} M_\alpha$ with structure partial semilattice Y is called a *skeleton* if $(N, \circ)$ is partial semilattice isomorphic to Y and $|N \bigcap M_\alpha| = 1$ for all $\alpha \in Y$. In what follows, by a left (right) regular partial band M with semilattice skeleton Y, we mean that M is a left (right) regular partial band M in which Y is a skeleton of M and Y is a semilattice. Note that each M_α is a left (right) zero band, thus we obtain that

- $x \circ x$ is defined for a left (right) partial regular band M and any $x \in M$.

Example 2.4 Let I, J be left zero semigroups and 0 be a symbol. Put $S = I \cup J \cup \{0\}$. Define a multiplication $\otimes$ on S by

$$a \otimes b = \begin{cases} a & a, b \in I \text{ or } a, b \in J; \\ 0 & \text{Otherwise} \end{cases}$$

By a routine calculation, $(S, \otimes)$ is a left regular band. Of cause, $(S, \otimes)$ is a partial semigroup. Now let $\alpha \in I$ and $\beta \in J$. Then $Y := \{\alpha, \beta, 0\}$ is a skeleton of S. It can be easily checked that $(S, \otimes)$ is a left partial band with semilattice skeleton Y.

Next, we give a left regular partial band with semilattice skeleton which is not a left regular band. The following example is essentially due to [8, Example 2.4].

Example 2.5 Let $M = \{\alpha, \beta, 0, x\}$. Define a partial multiplication by

$\circ$	α	β	x	0
α	α	0	not defined	0
β	0	β	β	0
x	0	x	x	0
0	0	0	0	0

It is easy to check that $(M, \circ)$ is a partial semigroup. If we let $Y = \{\alpha, \beta, 0\}$, then $(Y, \circ)$ is a semilattice. Obviously, $I_\alpha = \{\alpha\}, I_\beta = \{\beta\}, I_0 = \{0\}$ and $I_\beta = \{\beta, x\}$ are all left zero rectangular bands in the partial semigroup $(M, \circ)$. By routine computing, $(M, \circ)$ is a left regular partial band with semilattice skeleton Y.

Lemma 2.6 *Let $(M, \circ)$ be a left regular partial band with semilattice skeleton Y, and assume M is the disjoint union of left zero bands M_α with $\alpha \in Y$. If $(Y, \circ)$ is still a semilattice, then $x \circ u$ is defined for any $x \in M$ and $u \in Y$.*

Proof. Assume $x \in M_v$ with $v \in Y$. Note that M_v is a left zero band, $x \circ v$ is defined and $x \circ v = x$. However, $v \circ u$ is defined and $v \circ u \leq u, v$ since $(Y, \circ)$ is a semilattice. Now, by Condition $(PB1)$, $x \circ (v \circ u)$ is defined. Since M is a partial semigroup, this shows that $(x \circ v) \circ u$ is defined and $(x \circ u =)\ (x \circ v) \circ u = x \circ (v \circ u)$. We complete the proof. □

Lemma 2.7 *Let S be a rpp semigroup and $x, y \in S$. If there exists $f \in E(S)$ such that $x = yf$ and $f\mathcal{R}^* y^*$, then $f\mathcal{L}^* x$.*

Proof. Clearly $x = xf$. Let $s, t \in S^1$. By the arguments after Corollary 2.2, $f\mathcal{R}^* y^*$ implies that $f\mathcal{R} y^*$. Then

$$\begin{aligned} xs = xt \quad &\Longrightarrow yfs = yft \\ &\Longrightarrow y^* fs = y^* ft \quad \text{(since } y\mathcal{L}^* y^*\text{)} \\ &\Longrightarrow fs = ft \qquad\quad \text{(since } f\mathcal{R} y^*\text{, giving } y^* f = f\text{)} \end{aligned}$$

Now, by Corollary 2.2, we have $f\mathcal{L}^* x$. □

Let S be a *rpp* semigroup with a set of idempotent E. Let S° be a right adequate $*$-subsemigroup of S and E° be the semilattice of idempotents of S°. The semigroup S° is called a *right adequate transversal* for S if for each element x in S, there are a unique element x° in S° and idempotent f in E such that $x = x^\circ f$, where $f\mathcal{R}(x^\circ)^*$ and $(x^\circ)^*$ in E°. In this case, we have

(1) *f is uniquely determined by x.* Indeed, if $x^\circ f = x = x^\circ f_1$, for some $f_1 \mathcal{R} x^{\circ *}$, then $f_1 \mathcal{R} x^{\circ *} \mathcal{R} f$, and by Lemma 2.7, $f_1 \mathcal{L}^* x \mathcal{L}^* f$. This shows $f_1 \mathcal{H} f$, and thereby $f_1 = f$. We shall denote f by f_x and write $\Lambda = \{f_x : x \in S\}$.

(2) $\forall x \in S$, $(x^\circ)^* = f_x (x^\circ)^*$ *and* $f_x = (x^\circ)^* f_x$. This is since $f_x \mathcal{R} (x^\circ)^*$.

(3) $\forall x \in S$, $x^\circ = x^\circ (x^\circ)^* = x^\circ f_x (x^\circ)^* = x (x^\circ)^*$. For also $f_x \mathcal{R} (x^\circ)^*$.

Also, we say that the right adequate transversal S° is *quasi-ideal* if $f_x a \in S^\circ$ for all $x \in S$ and $a \in S^\circ$. Equivalently, S° is quasi-ideal if and only if S° is a quasi-ideal of S. In fact, if S° is a quasi-ideal right adequate transversal for S, a routine checking shows that S° is a quasi-ideal of S; conversely, if S° is a quasi-ideal of S, by noting that $(x^\circ)^* f_x = f_x$ since $(x^\circ)^* \mathcal{R} f_x$, we immediately have $f_x a = (x^\circ)^* f_x a \in S^\circ$, for any $a \in S^\circ$.

The following proposition gives some properties of right adequate transversals.

Proposition 2.8 *Let S° be a right adequate transversal for a rpp semigroup S. Then*

(1) $\Lambda = \{f \in E : (\exists h \in E^\circ) f \mathcal{R} h\}$. *Moreover, Λ is a right regular partial band with semilattice skeleton E°.*

(2) $E^\circ \Lambda \subseteq \Lambda$.

(3) *For any $x \in S$, $(x^\circ)^\circ = x^\circ$ and $f_{x^\circ} = (f_x)^\circ = x^{\circ *}$.*

(4) *For any $x \in \Lambda$, $x = f_x$.*

(5) *For any $x, y \in S$, $f_{xy} = f_{xy} f_y$.*

Proof. (1) If $f \in \Lambda$, then there exists $x \in S$ such that $f_x = f$ and $f \mathcal{R} x^{\circ *}$, where $x^{\circ *} \in E^\circ$. Conversely, if $f \in E$ and there exists $h \in E^\circ$ such that $h \mathcal{R} f$, then $f = hf$, and this implies $f = f_f \in \Lambda$. The rest is straightforwards.

(2) Take $e \in E^\circ$ and $f \in \Lambda$, by part (1), there exists $h \in E^\circ$ such that $h \mathcal{R} f$, hence $eh \mathcal{R} ef$. Note that

$$\begin{aligned} efef &= efehf \\ &= efhef \quad (\text{since } h, e \in E^\circ) \\ &= ehef \quad (\text{since } h \mathcal{R} f) \\ &= ef, \end{aligned}$$

that is, $ef \in E$. Again from (1) and $eh \mathcal{R} ef$ (where $eh \in E^\circ$), we can deduce that $ef \in \Lambda$.

(3) For any $x \in S$, since $x^\circ = x^\circ x^{\circ *}$ and $x^{\circ *} \mathcal{R} x^{\circ *}$, by the uniqueness of $(x^\circ)^\circ$, we have $(x^\circ)^\circ = x^\circ$ and $f_{x^\circ} = x^{\circ *}$. On the other hand, since $f_x \mathcal{R} x^{\circ *}$, we have $f_x = x^{\circ *} f_x$ and so $(f_x)^\circ = x^{\circ *}$.

(4) Assume $x \in \Lambda$. Then by (3), $x = f_a$ for some $a \in S$, so that $x^\circ = a^{\circ *}$. Notice that $f_a \mathcal{R} a^{\circ *}$, we can obtain that $x = f_a = a^{\circ *} f_a = x^\circ x$, which yields that $x = f_a = f_x$.

(5) Assume $x, y \in S$. Since $f_y \mathcal{L} y^{\circ *}$, we have $y f_y = y$ and so $xy f_y = xy$. This shows that $f_{xy} = f_{xy} f_y$ since $f_{xy} \mathcal{L}^* xy$. □

3 Quasi-ideal right adequate transversals

In this section, we establish some characterizations of rpp semigroups with quasi-ideal right adequate transversals.

Theorem 3.1 *Let S° be a right adequate transversal for a rpp semigroup S. Then the following statements are equivalent:*

(1) *S° is a quasi-ideal right adequate transversal for S.*

(2) *For all $x, y \in S$, $(xy)^\circ = xy^\circ$.*

(3) *For all $x \in S$ and $e \in E^\circ$, $f_x e \in S^\circ$.*

(4) *For all $x, y \in S$, $f_{xy} = (xy)^{\circ *} f_y$.*

Proof. (1) $\Rightarrow$ (2) Suppose that S° is a quasi-ideal right adequate transversal for S. Then for all $x, y \in S$, $xy^\circ = x^\circ(f_x y^\circ) \in S^\circ$ since S° is a quasi-ideal of S. Note that $\mathcal{R}$ is a left congruence. We have $(xy^\circ)^*(y^\circ)^* \mathcal{R} (xy^\circ)^* f_y$. On the other hand, since $xy^\circ y^{\circ *} = xy^\circ$, we have $(xy^\circ)^* = (xy^\circ)^*(y^\circ)^*$. Thus

$$\begin{aligned} ((xy^\circ)^* f_y)^2 &= (xy^\circ)^* f_y (xy^\circ)^* f_y = (xy^\circ)^* f_y (xy^\circ)^* (y^\circ)^* f_y \\ &= (xy^\circ)^* f_y (y^\circ)^* (xy^\circ)^* f_y = (xy^\circ)^* (y^\circ)^* (xy^\circ)^* f_y \\ &= (xy^\circ)^* f_y \end{aligned}$$

since E° is a semilattice and $(xy^\circ)^*, (y^\circ)^* \in E^\circ$, giving $(xy^\circ)^*(y^\circ)^* = (y^\circ)^*(xy^\circ)^*$, and since $f_y(y^\circ)^* = (y^\circ)^*$ (the forgoing arguments before Proposition 2.8). Now, $(xy^\circ)^* \mathcal{R} (xy^\circ)^* f_y$. But $xy = (xy^\circ) f_y = (xy^\circ)(xy^\circ)^* f_y$, therefore $(xy)^\circ = xy^\circ$ since S° is a right adequate transversal for S. This proves (2).

(2) $\Rightarrow$ (3) It is trivial if note that $e^\circ = e$.

(3) $\Rightarrow$ (4) Assume that (3) holds. Then $f_{xy}(y^\circ)^* \in S^\circ$. By Proposition 2.8(5), we have

$$(*) \qquad \begin{aligned} (xy)^{\circ *} f_{xy} &= f_{xy} = f_{xy} f_y = f_{xy} (y^\circ)^* f_y = (f_{xy}(y^\circ)^*) f_y \\ &= (f_{xy}(y^\circ)^*)(f_{xy}(y^\circ)^*)^* f_y. \end{aligned}$$

Because $f_{xy}(y^\circ)^* = (f_{xy}(y^\circ)^*)(y^\circ)^*$, we have $(f_{xy}(y^\circ)^*)^* = (f_{xy}(y^\circ)^*)^*(y^\circ)^*$, so that

$$(f_{xy}(y^\circ)^*)^* = (f_{xy}(y^\circ)^*)^*(y^\circ)^* \mathcal{R} (f_{xy}(y^\circ)^*)^* f_y$$

by $(y^\circ)^* \mathcal{R} f_y$ and since $\mathcal{R}$ is a left congruence. Now, since S° is a right adequate transversal for S, and by Equ. $(*)$, we have $(xy)^{\circ *} = f_{xy} y^{\circ *}$ and $f_{xy} = (f_{xy} y^{\circ *})^* f_y = ((xy)^{\circ *})^* f_y = (xy)^{\circ *} f_y$ by noting that $(xy)^{\circ *} \mathcal{L}^* ((xy)^{\circ *})^*$ and that each $\mathcal{L}^*$-class of a right adequate semigroup contains precisely one idempotent, giving $(xy)^{\circ *} = ((xy)^{\circ *})^*$.

(4) $\Rightarrow$ (1) Suppose (4) is satisfied. Let $x \in S$ and $a \in S^\circ$. Then by (4), we have

$$\begin{aligned} f_x a &= (f_x a)^\circ f_{f_x a} = (f_x a)^\circ (f_x a)^{\circ *} f_a = (f_x a)^\circ a^* \\ &\in S^\circ \end{aligned}$$

and whence (1) holds. □

Proposition 3.2 *Let S° be a quasi-ideal right adequate transversal for the rpp semigroup S. Then for all $e \in E(S)$,*

(1) *$E(Se)$ is a left regular band.*

(2) *eSe is a right adequate subsemigroup of S.*

Proof. (1) Let $x, y \in E(Se)$. Then by Theorem 3.1 (4), $x = xe = (xe)^\circ f_e = x^\circ f_e$ and $(x^\circ)^* \mathcal{R} (x^\circ)^* f_e$. Thus

$$(f_e x^\circ)^2 = f_e x^\circ f_e x^\circ = f_e(x x^\circ) = f_e(xx)^\circ = f_e x^\circ \in E^\circ$$

and similarly, $f_e y^\circ \in E^\circ$. Therefore

$$\begin{aligned} xy &= (xy)^2 = x^\circ(f_e y^\circ)(f_e x^\circ)(f_e y^\circ) f_e = x^\circ(f_e y^\circ)(f_e x^\circ) f_e \\ &= xyx \end{aligned}$$

and $E(Se)$ is a band. Moreover, $E(Se)$ is a left regular band.

(2) Let $x \in eSe$, and $f \in E(S)$ with $x\mathcal{L}^*(S)f$. Then $x = xe$ and so $f = fe$ (by $x\mathcal{L}^* f$), which implies that $ef \in E(S)$, giving $ef \in E(eSe)$, and $ef\mathcal{L}(S)f$. On the other hand, by Lemma 2.1, $\mathcal{L}^* \cap (eSe \times eSe) \subseteq \mathcal{L}^*(eSe)$. Thus $x\mathcal{L}^*(eSe)ef$ and whence eSe is a rpp semigroup for any $e \in E(S)$.

Now, it remains to verify that $E(eSe)$ is a semilattice. By (1), $E(eSe)$ is a left regular band. Thus for all $x, y \in E(eSe)$, by the proof of (1), we have

$$\begin{aligned} xy &= (xy)^2 = e^\circ(f_e x^\circ)(f_e y^\circ) f_e = e^\circ(f_e y^\circ)(f_e x^\circ) f_e = eyx \\ &= yx \end{aligned}$$

and $E(eSe)$ is a semilattice, as required. □

Corollary 3.3 *Let S° be a quasi-ideal right adequate transversal for the abundant semigroup S. Then for all $e \in E(S)$, eSe is an adequate semigroup.*

Proof. It follows from Proposition 3.2. □

Definition 3.4 *Let S be a rpp semigroup and $\overline{S}$ an adequate semigroup. If $\overline{S}$ is a right adequate transversal for S, then we call $\overline{S}$ an* ***adequate transversal of type-R*** *for S.*

In the theorem below, we give a characterization of quasi-ideal adequate transversals of type-R.

Proposition 3.5 *Let S° be a quasi-ideal right adequate transversal for the rpp semigroup S. If S° is an adequate transversal of type-R for S, then S is an abundant semigroup.*

Proof. Assume that S° is an adequate transversal of type-R for S. Let $x \in S$. Then there exists $e \in E^\circ$ such that $e\mathcal{R}^* x^\circ$, since S° is an adequate semigroup. By the dual of Corollary 2.2, this implies that $ex^\circ = x^\circ$ and $ex = ex^\circ f_x = x^\circ f_x = x$. If $u, v \in S^1$ and $ux = vx$, then since S° is a quasi-ideal right adequate transversal for S, we have $ux^\circ, vx^\circ \in S^\circ$, and thus

$$\begin{aligned} ux^\circ &= ux^\circ(x^\circ)^* = ux^\circ f_x (x^\circ)^* = ux(x^\circ)^* = vx(x^\circ)^* \\ &= vx^\circ, \end{aligned}$$

which implies that $ue = ve$, since S° is a $*$-subsemigroup of S so that $x^\circ \mathcal{R}^*(S^\circ)e$ implies $x^\circ \mathcal{R}^* e$. Therefore by the dual of Corollary 2.2, $x\mathcal{R}^*(S)e$. Consequently, S is an lpp semigroup whence S is an abundant semigroup. □

Theorem 3.6 *Let S° be a quasi-ideal right adequate transversal for the rpp semigroup S. If $x^\circ \in E^\circ$ for all $x \in E(S)$, then the following statements are equivalent:*

(1) *S° is an adequate transversal of type-R for S.*

(2) *S is an abundant semigroup.*

Proof. By Proposition 3.5, we need only to prove (2) $\Rightarrow$ (1). Now, suppose that S is an abundant semigroup. We need only to prove that S is an lpp semigroup. Let $a \in S^\circ$. Then there exists $x \in E(S)$ such that $x\mathcal{R}^*(S)a$. Since $x \in E(S)$, we have

$$(**)\qquad \begin{aligned} x^\circ f_x x^\circ &= x^\circ f_x x^\circ (x^\circ)^* = x^\circ f_x x^\circ (f_x (x^\circ)^*) = x^\circ f_x x^\circ f_x (x^\circ)^* \\ &= x^2 (x^\circ)^* = x(x^\circ)^* = x^\circ (f_x (x^\circ)^*) = x^\circ (x^\circ)^* \\ &= x^\circ. \end{aligned}$$

Note that $x\mathcal{L}^* f_x$. We have $x\mathcal{L} f_x$ and $f_x x = f_x$, so that $f_x x^\circ f_x = f_x$. Together with Equ. (**), x° is an inverse of f_x, and we have $f_x x^\circ \in E^\circ$ and $f_x x^\circ \mathcal{L} x^\circ$. But S° is a right adequate semigroup, E° is a semilattice. This implies that $f_x x^\circ = x^\circ$ since $f_x x^\circ \mathcal{L} x^\circ$. It follows that $x = x^\circ f_x = (f_x x^\circ) f_x = f_x$. Hence $x^\circ = (x^\circ)^* \mathcal{R} f_x = x$ since $x^\circ \in E^\circ$ giving $x^\circ = (x^\circ)^*$, and thus $x^\circ x = x$ and $x x^\circ = x^\circ$. On one hand, by $x\mathcal{R}^*(S)a$, we have $xa = a$ and $x^\circ a = x^\circ x a = xa = a$. On the other hand, if $u, v \in S^{\circ 1}$ and $ua = va$, then $ux = vx$ since $a\mathcal{R}^*(S)x$. Now, by the forgoing proof, x° is an inverse of f_x $(= x)$ and $ux^\circ = (ux)x^\circ x = (vx)x^\circ = vx^\circ$. Summing up, by the dual of Corollary 2.2, $a\mathcal{R}^*(S^\circ)x^\circ$ and thereby S° is an lpp semigroup. We complete the proof. □

Let us turn back to the proof of Theorem 3.6. When we prove that $x = f_x$ for any $x \in E(S)$, we need use the property that $x^\circ \in E^\circ$ for $x \in E(S)$. Thus that $x^\circ \in E^\circ$ for $x \in E(S)$ is a sufficient condition that $x^\circ \mathcal{R} x$, giving $E(S) = \Lambda$. But the reverse implication is clear. Therefore we indeed prove the following proposition.

Proposition 3.7 *Let S° be a quasi-ideal right adequate transversal for the rpp semigroup S. Then the following statements are equivalent:*

(1) *$x^\circ \in E^\circ$ for all $x \in E(S)$.*

(2) *$E(S) = \Lambda$.*

In what follows, we call S° a *right adequate transversal of the point property* for the rpp semigroup if S° is a right adequate transversal for S, and $e^\circ \in E^\circ$ for any $e \in E(S)$.

4 The structure theorems

In this section, we shall establish the structure of rpp semigroups with quasi-ideal right adequate transversals.

Let T be a right adequate semigroup with semilattice Y of idempotents and $\Lambda = \bigcup_{\alpha \in Y} \Lambda_\alpha$ a right regular partial band with semilattice skeleton Y. Define a mapping by

$$\langle , \rangle : \Lambda \times T \to T, \quad (x,t) \to \langle x,t \rangle$$

and put

$$QRT = QRT(T, \Lambda, \langle , \rangle) = \{(a,x) \in T \times \Lambda : x \in \Lambda_{a^*}\}.$$

The above triple $(T, \Lambda, \langle , \rangle)$ is called a *QRT-system* provided

($QRT1$) For all $x \in \Lambda, y \in Y$ and $s,t \in T$, if yx is defined in Λ, then $\langle yx, st \rangle = y\langle x, s \rangle t$.

($QRT2$) Let $x \in \Lambda, u \in T$. If x is in Y, then $\langle x, u \rangle = xu$ where xu is the product of x and u in T.

($QRT3$) For any $(y,s) \in QRT$, $\langle y, s^* \rangle = s^*$.

For a QRT-system $(T, \Lambda, \langle , \rangle)$, we have

(QRT4) For all $s,t,w \in T$ and $y,n \in \Lambda$ with $(s,y) \in QRT$, by ($QRT1$), we have

$$s\langle y, t\langle n,w\rangle\rangle = s\langle s^*y, t\langle n,w\rangle\rangle = ss^*\langle y,t\rangle\langle n,w\rangle = s\langle y,t\rangle\langle n,w\rangle.$$

(QRT5) For all $x \in \Lambda, t \in T$, by ($QRT1$), we immediately have $\langle x,t\rangle t^* = \langle x,t\rangle$.

Given a QRT-system $(T, \Lambda, \langle , \rangle)$. Define a multiplication on QRT by

$$(s,x) \circ (t,y) = (u, u^*y),$$

where $u = s\langle x,t\rangle$. Then $\circ$ is well defined. In fact, by ($QRT5$), we have $\langle x,t\rangle t^* = \langle x,t\rangle$ and so $ut^* = u$. It follows that $u^*t^* = u^*$ and $u^*\omega t^*$ (where ω is a naturally partial order on the set of idempotents Y). So, u^*y is defined and $u^*y \in \Lambda_{u^*}$. Thus $(u, u^*y) \in QRT(T, \Lambda, \langle , \rangle)$. It is to say that $\circ$ is well defined. Thereby $\circ$ is closed.

Lemma 4.1 *$(QRT, \circ)$ is a semigroup.*

Proof. It suffices to show that $(QRT, \circ)$ satisfies the associative law. For this, let $(s,y),(t,n),(w,v) \in QRT(T, \Lambda, \langle , \rangle)$. Compute

$$\begin{aligned}((s,y)\circ(t,n))\circ(w,v) &= (s\langle y,t\rangle, (s\langle y,t\rangle)^*n)\circ(w,v)\\ &= (s\langle y,t\rangle\langle (s\langle y,t\rangle)^*n, w\rangle, (s\langle y,t\rangle\langle (s\langle y,t\rangle)^*n,w\rangle)^*v)\\ &= (s\langle y,t\rangle(s\langle y,t\rangle)^*\langle n,w\rangle, (s\langle y,t\rangle(s\langle y,t\rangle)^*\langle n,w\rangle)^*v)\\ &= (s\langle y,t\rangle\langle n,w\rangle, (s\langle y,t\rangle\langle n,w\rangle)^*v).\end{aligned}$$

On the other hand, since $s\langle y,t\rangle\langle n,w\rangle = s\langle y,t\rangle\langle n,w\rangle\langle n,w\rangle^*$, we have $(s\langle y,t\rangle\langle n,w\rangle)^* = (s\langle y,t\rangle\langle n,w\rangle)^*\langle n,w\rangle^*$ and

$$\begin{aligned}(s,y)\circ((t,n)\circ(w,v)) &= (s,y)\circ(t\langle n,w\rangle, (t\langle n,w\rangle)^*v)\\ &= (s\langle y, t\langle n,w\rangle\rangle, (s\langle y,t\langle n,w\rangle\rangle)^*(t\langle n,w\rangle)^*v)\\ &= (s\langle y,t\rangle\langle n,w\rangle, (s\langle y,t\rangle\langle n,w\rangle)^*(t\langle n,w\rangle)^*v)\\ &= (s\langle y,t\rangle\langle n,w\rangle, (s\langle y,t\rangle\langle n,w\rangle)^*v),\end{aligned}$$

where the third "=" is by $(QRT4)$. Thus $(s,y)\circ((t,n)\circ(w,v)) = ((s,y)\circ(t,n))\circ(w,v)$ and whence $\circ$ is associative, as required. $\square$

Lemma 4.2 *For all* $(s,y),(t,n)\in QRT(T,\Lambda,\langle,\rangle)$,

(1) $(s,y)\in E(QRT(T,\Lambda,\langle,\rangle))$ *if and only if* $s\langle y,s\rangle = s$.
(2) $QRT(T,\Lambda,\langle,\rangle)$ *is a rpp semigroup.*
(3) $(s,y)\mathcal{L}^*(t,n)$ *if and only if* $s\mathcal{L}^*t$ *and* $y=n$.
(4) $QRT^\circ = \{(t,t^*)\in QRT : t\in T\}$ *is a quasi-ideal right adequate transversal for* QRT.

Proof. (1) It is a routine calculation.

(2) Let $(s,y)\in QRT$. Then $(s^*,y)\in QRT$. By (1) and (QRT3), $(s^*,y)\in E(QRT)$. By (QRT5), we have

$$\begin{aligned}(s,y)(s^*,y) &= (s\langle y,s^*\rangle,(s\langle y,s^*\rangle)^*y)\\ &= (ss^*\langle y,s^*\rangle,(ss^*\langle y,s^*\rangle)^*y) = (ss^*,(ss^*)^*y)\\ &= (s,y).\end{aligned}$$

On the other hand, if $(u,z),(v,w)\in QRT^1$ and $(s,y)(u,z)=(s,y)(v,w)$, then by comparing the components, we have $s\langle y,u\rangle = s\langle y,v\rangle$ and $(s\langle y,u\rangle)^*z = (s\langle y,v\rangle)^*w$. By the prior equality, $s^*\langle y,u\rangle = s^*\langle y,v\rangle$. By the latter one, we have $(s^*\langle y,u\rangle)^*z = (s^*\langle y,v\rangle)^*w$ by using Lemma 2.3. Thus

$$\begin{aligned}(s^*,y)(u,z) &= (s^*\langle y,u\rangle),(s^*\langle y,u\rangle)^*z)\\ &= (s^*\langle y,v\rangle,(s^*\langle y,v\rangle)^*w)\\ &= (s^*,y)(v,w).\end{aligned}$$

Now, by Corollary 2.2, $(s,y)\mathcal{L}^*(s^*,y)$. Therefore QRT is a rpp semigroup.

(3) By the proof of (2), we have

$$\begin{aligned}(s,y)\mathcal{L}^*(t,n) &\Leftrightarrow (s^*,y)\mathcal{L}(t^*,n)\\ &\Leftrightarrow (s^*,y)\circ(t^*,n)=(s^*,y),(t^*,n)\circ(s^*,y)=(t^*,n)\\ &\Leftrightarrow s\mathcal{L}^*t, y=n.\end{aligned}$$

(4) By (QRT2), it is easy to check that QRT° is a subsemigroup of QRT, and by (1) and (3), QRT° is a right adequate *-subsemigroup of QRT. Let $(s,y)\in QRT$. Clearly, $(s,y)=(s,s^*)(s^*,y)$. Since

$$(s^*,s^*)(s^*,y) = (s^*\langle s^*,s^*\rangle, s^*\langle s^*,s^*\rangle y) = (s^*,y)$$

and

$$(s^*,y)(s^*,s^*) = (s^*\langle y,s^*\rangle,(s^*\langle y,s^*\rangle)^*s^*) = (s^*s^*,(s^*s^*)^*s^*) = (s^*,s^*),$$

we have $(s^*,s^*)\mathcal{R}(s^*,y)$. On the other hand, if $(\alpha,z)\in E(QRT)$, $(s^*,s^*)\mathcal{R}(\alpha,z)$ and $(s,y)=(s,s^*)(\alpha,z)$, then by comparing the components of the last equality, $s=s\langle s^*,\alpha\rangle$ and $y=s^*z$. Hence $s=ss^*\alpha=s\alpha$. By $(s^*,s^*)\mathcal{R}(\alpha,z)$, we have $(s^*,s^*)=(\alpha,z)(s^*,s^*)$ and $(\alpha,z)=(s^*,s^*)(\alpha,z)$, hence $s^*=\alpha s^*$ and $\alpha s^*=s^*$, thus $\alpha = s^*$. Now, $y = s^*z = \alpha z = z$. Therefore $(s^*, y) = (\alpha, z)$. Consequently, QRT° is a right adequate transversal for QRT.

Compute

$$\begin{aligned}(t,t^*)(s,y)(u,u^*) &= (t\langle t^*, s\rangle, (t\langle t^*, s\rangle)^*y)(u,u^*) \\ &= (ts, (ts)^*y)(u,u^*) \\ &= (ts\langle (ts)^*y, u\rangle, (ts\langle (ts)^*y, u\rangle)^*u^*) \\ &= (ts\langle y, u\rangle, (ts\langle y, u\rangle)^*u^*) \\ &= (ts\langle y, u\rangle, (ts\langle y, u\rangle)^*) \\ &\in QRT^\circ.\end{aligned}$$

The last "=" is since $ts\langle (ts)^*y, u\rangle = ts\langle (ts)^*y, u\rangle u^*$, giving $(ts\langle (ts)^*y, u\rangle)^* = (ts\langle (ts)^*y, u\rangle)^*u^*$. Thus QRT° is a quasi-ideal of QRT.

We have now proved that QRT° is a quasi-ideal right adequate transversal for QRT. □

We arrive at the structure theorem of rpp semigroups with a quasi-ideal right adequate tranversal.

Theorem 4.3 *Let $(T, \Lambda; \langle,\rangle)$ be a QRT-system. Then $QRT(T, \Lambda; \langle,\rangle)$ is a rpp semigroup with the quasi-ideal right adequate transversal QRT°, isomorphic to T. Conversely, any rpp semigroup with a quasi-ideal right adequate transversal can be constructed in this way.*

Proof. By Lemma 4.2, we need only to prove the converse part. Now, assume that S is a rpp semigroup with the quasi-ideal right adequate transversal S°. Then by Proposition 2.8, Λ is a right regular partial band having E° as its semilattice skeleton. Define

$$\langle,\rangle : \Lambda \times S^\circ \to S^\circ;\ (y,t) \mapsto \langle y, t\rangle = yt.$$

Since S° is a quasi-ideal right adequate transversal for S, it is clear that $\langle,\rangle$ is well defined. And, it is easy to check that the triple $(S^\circ, \Lambda; \langle,\rangle)$ satisfies Conditions (QRT1), (QRT2) and (QRT3). So, $(S^\circ, \Lambda; \langle,\rangle)$ is a QRT-system. Now, we can form the semigroup $QRT(S^\circ, \Lambda; \langle,\rangle)$.

It remains to prove that the mapping

$$\theta : QRT(S^\circ, \Lambda; \langle,\rangle) \to S;\ (s,x) \mapsto sx$$

is a semigroup isomorphism. Since S° is a right adequate transversal for S, it is easy to see that θ is a bijection. And, that θ is a homomorphism follows from the following computation:

$$\begin{aligned}((s,x)(t,y))\theta &= (s\langle x,t\rangle, (s\langle x,t\rangle)^*y)\theta = sxt(sxt)^*y = (sx)(ty) \\ &= (s,x)\theta \bullet (t,y)\theta.\end{aligned}$$

Thus θ is a semigroup isomorphism. We complete the proof. □

By Lemma 4.2 (1) and Theorem 4.3, the following corollary is immediate.

Corollary 4.4 *Let $(T, \Lambda; \langle , \rangle)$ be a QRT-system. If for all $(s, y) \in QRT$, $s\langle y, s\rangle = s$ implies that $s \in E(T)$, then $QRT(T, \Lambda; \langle , \rangle)$ is a rpp semigroup with the quasi-ideal right adequate transversal QRT° of the point property, which is isomorphic to T. Conversely, any rpp semigroup with a quasi-ideal right adequate transversal of the point property can be constructed in this way.*

Acknowledgments The authors would like to thank the referees who pointed out several mistakes in the manuscript. This work is supported by the NNSF of China (grant: 10961014); the NSF of Jiangxi Province; the SF of Education Department of Jiangxi Province and the SF of Jiangxi Normal University, China.

References

[1] Blyth T. S. and McFadden R., Regular semigroups with a multiplicative inverse transversals, Proc. Roy. Soc. Edinburgh 92A(1982), 253-270.

[2] Chen J. F., Abundant semigroups with adequate transversals, Semigroup Forum 60(2000), 67-79.

[3] El-Qallali A., Abundant semigroups with a multiplicative type A transversal, Semigroup Forum 47(1993), 327-340.

[4] Fountain J. B., Adequate Semigroups, Proc. Edinburgh Math. Soc. 22(1979), 113-125.

[5] Fountain J. B., Abundant Semigroups, Proc. London Math. Soc. 44(1982), 103-129.

[6] Guo X. J., Abundant semigroups with a multiplicative adequate transversal, Acta Mathematica Sinica (new ser.) 18(2002), 229-244.

[7] Guo X. J. and Peng T. T., Split quasi-adequate semigroups, To appear in "Acta Math. Sinica (new ser.)".

[8] Guo X. J. and Shum K. P., Abundant semigroups with Q-adequate transversals and some of their special cases, Algebra Colloquium 14(2007), 687-704.

[9] Guo X. J. and Wang L. M., Idempotent-connected abundant semigroups which are disjoint unions of quasi-ideal adequate transversals, Communications in Algebra 30(2002), 1779-1800.

[10] Guo X. J. and Xie X.-Y., Naturally ordered abundant semigroups in which each idempotent has a greatest inverse, Communications in Algebra 35(2007), 2324-2339.

[11] Guo Y. Q., Shum K. P. and Zhu P. Y., The structure of left C-rpp semigroups, Semigroup Forum 50(1995), 9-23.

[12] Howie J. M., An introduction to semigroup theory, London: Academic Press 1976.

[13] Luo M. X., Relationship between the quasi-ideal with adequate transversal of an abundant semigroup, Semigroup Forum 67(2003), 411-418.

Power Automorphisms and Induced Automorphisms in Finite Groups*

Xiuyun Guo

Department of Mathematics, Shanghai University,
Shanghai, P R China
**E-mail: xyguo@shu.edu.cn*

Dedicated to Professor K P Shum of his 70th birthday

An automorphism of a group G is said to be a power automorphism if it leaves every subgroup of G invariant. In this paper, we collect our recently work about power automorphisms and induced automorphisms.

Keywords: Power automorphisms; Norm; Induced automorphisms.

1. Introduction

All groups considered in this paper are finite.

An automorphism α of a group G is called fixed-point-free if $C_G(\alpha) = \{g \in G | g^\alpha = g\} = 1$. A typical result about fixed-point-free automorphisms is the following Thompson theorem.

Theorem 1.1 (Thompson, 1959). *If a group G has a fixed-point-free automorphism of prime order, then G is nilpotent.*

In this line many authors continue to investigate the relationship between the structure of groups and the fixed-point-free automorphisms (see Refs. 2–21).

Now recall that an automorphism of a group G is said to be a power automorphism if it leaves every subgroup of G invariant. It is interesting to investigate the relation between the structure of groups and the power automorphisms. The following example illustrate that we can not say anything if a group has a power automorphism of prime order.

*The research of the work was partially supported by the National Natural Science Foundation of China(11071155), SRFDP(200802800011), and Shanghai Leading Academic Discipline Project(J50101).

Example 1.1. *Let H be a group and p a prime such that $|H| < p$. Then* $\alpha\colon (xy) \longmapsto (xy^{|H|})$ *(for $x \in H$ and $y \in \langle g \rangle$)* *is a power automorphism of a group $G = H \times \langle g \rangle$, where $\langle g \rangle$ is a cyclic group of order p.*

However, It is easy to see the following result is true.

Theorem 1.2. *Let G be a non-abelian simple group and α be a power automorphisms of G. Then α must be the identity automorphism of G.*

So the structure of groups may be influenced by its power automorphism group—the set of all power automorphisms. We will collect some results about this topic.

2. On norms

In order to investigate the relationship between the structure of groups and its power automorphism groups, it is better to investigate the relationship between the structure of groups and the intersection of power automorphism groups and the inner automorphism groups. Now we recall the norm of groups which was introduced by Baer in 1934.[22]

Definition 2.1. *Let G be a group. Then the intersection of normalizers of all subgroups in G ia called the norm of G, denoted by $N(G)$.*

It is clear that every element in $N(G)$ induces a power automorphism of G. It is also clear that $N(G)$ is a characteristic subgroup of G and the center $\zeta(G)$ of G is contained in $N(G)$. However, the converse is not true.

Example 2.1. *Let p be an odd prime and $G = \langle a, b \big| a^{p^n} = b^p = 1, a^b = a^{1+p^{n-1}}, n \geqslant 2\rangle$. Then $Z(G) = \langle a^p \rangle, N(G) = \langle a^p \rangle \times \langle b \rangle$.*

For a long time, many authors have investigated both the properties of $N(G)$ and the relation between $N(G)$ and the structure of G.[22–29] For example, Baer in Ref. 25 showed that $Z(G) = 1$ if and only if $N(G) = 1$ for any group G. Thus a group G is nilpotent if and only if $N(G/X) \neq 1$ whenever X is a proper normal subgroup of G. So how to characterize the p-nilpotent groups by using the norm becomes a interesting question. In 2007 Wang and Guo[30] first prove the following result:

Theorem 2.1 (Wang and Guo, 2007). *Let p be a prime dividing the order of a finite group G and P a Sylow p-subgroup of G. Assume that*

(1) $\Omega_1(P) \leq N(N_G(P))$;

(2) when $p = 2$, *in addition,* $\Omega_2(P) = \langle \Omega_1(P), x \mid x \in P$ *is quasi-central in* $N_G(P)$ *and* $o(x) = 4\rangle$.

Then G *is* p*-nilpotent.*

Theorem 2.1 is a generalization of both Burnside Theorem for p-nilpotent groups and Itô Lemma for p-nilpotent groups. Applying Theorem 2.1, the authors can give a sufficient condition for a group to be nilpotent. In fact, the authors obtain a more general result in Theorem 2.2 by using the formation theory.

Theorem 2.2 (Wang and Guo, 2007). *Suppose* $\mathcal{F}$ *is a saturated formation containing the class of nilpotent groups. Let* G *be a group with a normal subgroup* N *such that* $G/N \in \mathcal{F}$. *Assume that* $\Omega_1(P) \leq N(N_G(P))$ *and when* $p = 2$, *in addition,* $\Omega_2(P) = \langle \Omega_1(P), x \mid x \in P$ *is quasi-central in* $N_G(P)$ *and* $o(x) = 4\rangle$ *for every prime* p *dividing the order of* N *and every Sylow* p*-subgroup* P *of* N. *Then* $G \in \mathcal{F}$.

Remark 2.1. *Recall that a subgroup* H *of a group* G *is said to be quasi-normal in* G *if* $HK = KH$ *for any subgroup* K *of* G. *An element* x *of* G *is said to be quasi-central in* G *if* $\langle x \rangle$ *is quasi-normal in* G. *Obviously, if* $x \in N(G)$, *then* x *is quasi-central in* G. *In general, the inverse is not true. However, it is easy to see that* x *is a quasi-central element of a finite group* G *if and only if* $x \in N(G)$ *if* p *is the smallest prime dividing the order of* G *and* x *is of order* p. *Hence the above theorem generalizes the main results in [8, Theorem 1] and [9, Theorem 1].*

The authors also give a result to characterize the central chief factors in a group.

Theorem 2.3 (Wang and Guo, 2007). *Let* N *be a normal subgroup of a finite group* G. *Assume for every Sylow subgroup* P *of* N *that* $\Omega_1(P) \leq N(N_G(P))$. *In addition, If* P *is a Sylow 2-subgroup of* N, *assume that* $\Omega_2(P) \leq N(N_G(P))$. *Then all chief factors of* G *contained in* N *are centralized by* G.

Recall that a group G is called a Dedekind group if every subgroup of G is normal in G. It is clear that a group G is a Dedekind group if and only if $G = N(G)$. What about the structure of a finite p-group whose norm is exactly of index p? The following result gives an answer.

Theorem 2.4 (Wang and Guo, 2007). *A finite p-group P satisfies $|P : N(P)| = p$ if and only if $P = R \times A$, where*

$$R = \langle x, u \mid x^{p^n} = u^{p^m} = 1, x^u = x^{1+p^{n-1}}, 1 \leq m < n, \text{and } n \geq 3 \text{ if } p = 2\rangle,$$

and A is an abelian p-group with $exp(A) < p^n$.

For generalization case we have the following:

Theorem 2.5 (Wang and Guo, 2009). *Let $\pi = \{p, q\}$. Then $|G : N(G)| = pq$ if and only if $G = H \times D$, H is a π-group, D a π'-group, and $|H : N(H)| = pq$, D is a Dedekind-group.*

Theorem 2.6 (Wang and Guo, 2009). *Let $\pi = \{p, q\}$ and $p > q$. Then π-group H satisfies $|H : N(H)| = pq$ if and only if H is of one of the following types:*

(1) $H = P \times Q$, $P \in Syl_p(H)$, $Q \in Syl_p(H)$ and $|P : N(P)| = p, |Q : N(Q)| = q$;
(2) $H = A_p \times A_q \times L$, where $L = \langle a\rangle \rtimes \langle x\rangle$, $|a| = p, |x| = q$ and x induces a automorphism of $\langle a\rangle$ with order q;
(3) $H = A_p \times E_{2^e} \times L$, where $L = Q_8 \rtimes \langle a\rangle$, $|a| = p$ and $|Q_8/C_{Q_8}(a)| = 2$;
(4) $H = A_p \times E_{2^e} \times Q_8 \times L$, where L is the dihedral group of order 2p;
(5) $H = A_p \times L$, where $L = Q \rtimes \langle a\rangle, |a| = p$, Q is a q-group, $|Q : N(Q)| = q$ and $C_Q(a) = N(Q)$.

Recall that a group G is a Hamiltonian-group if and only if $G = Q_8 \times A_2 \times A_0$, where Q_8 is the quaternion group of order 8, A_2 an elementary abelian 2-group and A_0 an abelian group of odd order. For this kind of groups, we have the following:

Theorem 2.7 (Wang and Guo, 2009). *If G is a 2-group and N(G) is a Hamiltonian-group, then G is also a Hamiltonian-group.*

Another interesting question is: How about the structure of finite groups if either $\langle x\rangle \trianglelefteq G$ or $x \in N(G)$ $\forall x \in G($ it is called a N-group)? Related groups is J-groups which is investigated by A. Mann etc in Ref. 33. A group is called a J-group if $\forall x \in G$, either $\langle x\rangle \trianglelefteq G$ or $\langle x, x^g\rangle \trianglelefteq G$ for any $g \in G$ but $g \notin N_G(\langle x\rangle)$. There are examples which are N-groups but not J-groups, also there are examples which are J-groups but not N-groups.

Theorem 2.8 (Guo and Wang, 2009). *Let G be a group. Then G is a N-group if and only if G is a nilpotent group with all of its Sylow-subgroups being N-groups, and at most one of them is not a Dedekind group.*

So it is important to investigate the N-groups whose order is a power of a prime.

Theorem 2.9 (Guo and Wang, 2009). *Let P be a p-group and a N-group. Then*

(1) $P' \leq N(P)$;

(2) $cl(P) \leqslant 2$;

(3) $P/N(P)$ is cyclic.

Theorem 2.10 (Guo and Wang, 2009). *Let P be a p-group and a N-group, but not Dedekind group. Then $P = \langle B, g\rangle \ltimes \langle x\rangle, \langle B, g\rangle$ is abelian, and*

(1) $o(x) = p^{m+n}, o(g) = p^r, 1 \leqslant m \leqslant r \leqslant n+1-m, exp(B) \leqslant p^{n+1-m}$, when $p = 2$, $n + m \geqslant 3$;

(2) $x^g = x^{1+p^n}, [B, x] = 1$.

Now we recall a result about the norm: $Z(G) \leq N(G) \leq Z_2(G)$ for every group G, where $Z(G)$ is the center of G and $Z_2(G)$ is defined by $Z_2(G)/Z(G) = Z(G/Z(G))$.

Question 2.1. *How about the structure of G with $N(G) = Z(G)$?*

Question 2.2. *How about the structure of G with $N(G) = Z_2(G)$?*

3. On power automorphisms

For convenience, we use $A(G)$ to denote the automorphism group of a group G and $P(G)$ to denote the power automorphism group of G. Although we have Example 1.1, Theorem 1.2 encourages us to investigate the properties of power automorphism groups and the structure of groups whenever power automorphism groups are given. Cooper ever investigate the properties of Power automorphism.[34] First we list the following fact about abelian groups.

Theorem 3.1. *Let A be an abelian group and $\alpha \in P(G)$.*

(1) If there is an element with infinite order in A, then $a^\alpha = a^{-1}, \forall a \in A$;

(2) If A is a p-group of finite exponent, then there is a positive integer l such that $a^\alpha = a^l \forall a \in A$. Furthermore, if α nontrivial and has order prime to p, then α is fixed-point-free;

(3) If A is a finite abelian group, then there is a positive integer l such that $a^\alpha = a^l \forall a \in A$.

Now we turn now to the determination of groups with its power automorphism groups having small indices. For convenience, we shall call a group G a J_i-group($i = 1, 2$) if $|Aut(G) : P(G)|$ is a product of i distinct primes. Also we call a group G a J_0-group if $A(G) = P(G)$.

Theorem 3.2 (Wang and Guo, 2009). *A group G is a J_0-group if and only if G is cyclic.*

Lemma 3.1 (Wang and Guo, 2009). *A group A can not be a direct product factor of a J_1-group if A is one of the following type groups*

(1) A is a non-cyclic abelian p-group;

(2) $A = Q_8$;

(3) $A = \langle x, u \mid x^{p^n} = u^{p^m} = 1, x^u = x^{1+p^{n-1}}, 1 \leq m < n \rangle$, where p be an odd prime;

(4) $A = \langle x, u \mid x^{2^n} = u^{2^m} = 1, x^u = x^{1+2^{n-1}}, 1 \leq m < n$ and $n \geq 3 \rangle$.

By using Lemma 3.1 we can prove the following:

Theorem 3.3 (Wang and Guo, 2009). *There exists no group G such that $|A(G) : P(G)|$ is a prime.*

Lemma 3.2 (Wang and Guo, 2009). *Suppose that a group $L = \langle a, x \mid a^p = x^{q^n} = 1, a^x = a^r \rangle$, where $r \not\equiv 1 (mod\ p)$, and $r^q \equiv 1 (mod\ p)$. Then*

(i) $|Aut(L)| = p(p-1)q^{n-1}$, $|P(L)| = q^{n-1}$:

(ii) L is a J_2-group if and only if $p = 3$ and $q = 2$;

(iii) $\gamma \in P(L)$ for any central automorphism γ of L.

Lemma 3.3 (Wang and Guo, 2009). *Let H be a group such that $H = \langle t \rangle \times L$, where $o(t) = 3^e > 1$ and $L = \langle a, x \mid a^3 = x^{2^n} = 1, a^x = a^{-1} \rangle$. Then H is not a J_2-group.*

Lemma 3.4 (Wang and Guo, 2009). *Let H be a group such that $H = \langle t \rangle \times L$, where $o(t) = 2^e > 1$ and $L = \langle a, x \mid a^3 = x^{2^n} = 1, a^x = a^{-1} \rangle$. Then H is not a J_2-group.*

Theorem 3.4 (Wang and Guo, 2009). *Let G be a group. Then $|A(G) : P(G)|$ is a product of two distinct primes if and only if $G = A \times L$ is a direct product of two subgroups A and L of relatively prime order, where A*

is cyclic, and L is isomorphic with K_4*,* Q_8 *or the group* $\langle a, x | a^3 = x^{2^n} = 1, a^x = a^{-1}, n \geq 1\rangle$.

Now a surprising consequence of Theorem 3.4 is the following Corollary 3.1.

Corollary 3.1 (Wang and Guo, 2009). *A group* G *is a* J_2*-group if and only if* $Aut(G)/P(G) \simeq S_3$.

Corollary asserts that if a group G satisfies $|Aut(G) : P(G)|$ is a product of two distinct primes, then these two distinct primes can only be 2 and 3.

Corollary 3.2 (Wang and Guo, 2009).
Let G *be a group. If* $|Aut(G)| = pq$*, where* p, q *are primes with* $p > q$*, then*

(i) $q = 2$*, and* $pq + 1$ *is still a prime;*
(ii) $G \simeq K_4, S_3, C_9, C_{18}, C_{pq+1}$ *or* $C_{2(pq+1)}$.

Notice that $x^\alpha = x$ for every element x of order 2 if α be a power automorphism of a group G. So we state the following

Definition 3.1. *A power automorphism* θ *of a finite group* G *is called to be pre-fixed-point-free if* $C_G(\theta)$ *is an elementary abelian 2-group.*

Question 3.1. *How about the structure of a group if it has a pre-fixed-point free automorphism?*

Question 3.2. *How about the structure of a group if it has a power automorphism* α *such that* $C_G(\theta)$ *is an abelian 2-group?*

In fact, we already answer the Question 3.1.

4. On induced automorphisms

For any group G a natural fact is: $N_G(H)/C_G(H) \lesssim \mathrm{A}(H)$ for every $H \leq G$. So a interesting question is:

Question 4.1. *How about the structure of a group* G *if* $N_G(H)/C_G(H)$ *is known for some subgroups* H *of* G*?*

There are many important results related with this question. For example:

Theorem 4.1 (Zassenhaus). *If for every abelian subgroup A, $N_G(A)/C_G(A) = 1$, then G is abelian.*

Theorem 4.2 (Burnside). *Let $P \in Syl_p(G)$. If $N_G(P)/C_G(P) = 1$, then G is p-nilpotent.*

Theorem 4.3 (Frobenius). *Let $P \in Syl_p(G)$. If for every $P_1 \leq P$, $N_G(P_1)/C_G(P_1)$ is a p-group, then G is p-nilpotent.*

Theorem 4.4 (Guo). *Let G be a p-solvable group and $P \in Syl_p(G)$. If for every $P_1 \leq P$, $N_G(P_1)/C_G(P_1)$ is a strong p-closed group, then G is p-supersolvable.*

For convenience, we write $\mathrm{A}_G(H) = N_G(H)/C_G(H)$. It is clear that $\mathrm{In}(H) \leq \mathrm{A}_G(H) \leq \mathrm{A}(H)$. So the natural questions are the following:

Question 4.2. *How about the structure of a group G if $A_G(H) = In(H)$ for some subgroups H?*

Question 4.3. *If $A_G(H) = A(H)$ for some subgroups H in a group G, how about the structure of G?*

For the question 4.3, Lennox and Wiegold give the following answer.

Theorem 4.5 (Lennox and Wiegold, 1993). *If for every $H \leq G$, $A_G(H) = A(H)$, then $G \simeq S_n$, $n = 1, 2, 3$.*

Before closing this section, we list the following results which are the generalization of the above results:

Theorem 4.6 (Li and Guo, 2007). *Let G be a 2-groupthen every abelian subgroup of G is a TI-subgroup if and only if G is following one*

(1) G is an abelian 2-group;

(2) G is a Hamiltonian 2-group;

(3) G is the center product of Q_8 and D_8;

(4) $G(2^n, 2) = \langle a, b | a^{2^n} = b^2 = 1, b^{-1}ab = a^{1+2^{n-1}}, n > 1\rangle$;

(5) $G(2, 2, 2^n) = \langle a, b, c | a^2 = b^2 = c^{2^n} = 1, [a, b] = c^{2^{n-1}}, [a, c] = [b, c] = 1, n > 0\rangle$.

Theorem 4.7 (Li and Guo, 2007). *Let p be an odd prime and G a p-group. Then every abelian subgroup of G is a TI-subgroup if and only if G is the following one*

(1) G is an abelian p-group;

(2) $G(p^n, p) = \langle a, b | a^{p^n} = b^p = 1, b^{-1}ab = a^{1+p^{n-1}}, n > 1\rangle$;

(3) $G(p, p, p^n) = \langle a, b, c | a^p = b^p = c^{p^n} = 1, [a, b] = c^{p^{n-1}}, [a, c] = [b, c] = 1, n > 0\rangle$.

Theorem 4.8 (Guo, Li and Flavell, 2007). *Let G be a group. If every abelian subgroup of G is a TI-subgroup, then one of the following is true*

(1) G nilpotent;

(2) G is a solvable Frobenius group. If K is the Frobenius kernel and H a Frobenius complement, then

(a) every H-chief factor of K is cyclic.

(b) K is isomorphic to $Z_p \times Z_p$ for some prime p, H acts irreducibly on K and either H is cyclic or the direct product of a cyclic group of odd order and Q_8.

(3) G is isomorphic to S_4;

(4) G is isomorphic to A_5 or $PSL(2, 7)$ or $PSL(2, 9)$.

References

1. J. G. Thompson, Finite groups with fixed-point-free automorphisms of prime order, *Proc. Nat. Acad. Sci. U.S.A.*, **45**(1959), 578–581.
2. Shult, Nilpotence of the commutator subgroup in groups admitting fixed point free operator groups, *Pacific Journal of Math.*, **17**(1966), 323-347.
3. R.P. Martineau, Elementary abelian fixed-point-free automorphism groups, *Quart. J. Math.* , Oxford(2), **23**(1972), 205-212.
4. R.P. Martineau, Solubility of groups admitting certain fixed-point-free automorphism groups, *Math. Z.*, **130**(1973), 143-147.
5. T.R. Berger, Nilpotent fixed-poinautomorphism groups of solvable groups, *Math. Z.*, **131**(1973), 305, 305-312.
6. M.R. Pettet, Fixed-point-free automorphism groups of square-free exponent, *Proc. London Math. Soc.*, **33**(1976), 361-384.
7. E.W. Ralston, Sovability of finite groups admitting fixed-point-free automorphism of order rs, *J. Algebra*, **23**(1972) 164-180.
8. P.J. Rowley, Solubility of finite groups admitting a fixed-point-free abelian automorphism group of square-free exponent rs, *Proc. London Math. Soc.*, **37**(1978), 385-421.
9. Peter Rowley, Solubility of finite groups admitting a fixed-point-free abelian automorphism of order rst, *Pacific Journal of Math.*, **93**(1981), 201-235.
10. Gulin Ercan and I.S. Guloglu, Finite groups admitting a fixed-point-free abelian automorphism of order rst, *J. Group Theory*, **7**(2004), 437-446.
11. J. N. Ward, On finite solvable groups and the fixed-point groups of automorphisms, Bull. Austral. Math. Soc. **5**(1971), 375-378.

12. Paul Flavell and Geoffrey R. Robinson, Fixed points of coprime automorphisms and generalizations of Glauberman's Z^*-theorem, *J. Algebra*, **226**(2000), 714–718.
13. Paul Flavell, The fixed points of coprime action, *Arch. Math. (Basel)*, **75**(2000), 173–177.
14. G. Higman , Groups and rings which have automorphisms without non-trivial fixed elements, *J. London Math. Soc.* **32**(1957), 321–334.
15. A. Shalev A., On almost fixed point free automorphisms, *J. Algebra*, **157**(1993), 271–282.
16. A. Shalev, Automorphisms of finite groups of bounded rank, *Israel J. Math.*, **82**(1993), 395–404.
17. E. I. Khukhro, Almost regular automorphisms of finite groups of bounded rank, *Siberian Math. J.*, **37**(1996), 1237–1241.
18. A. Jaikin-Zapirain, Finite groups of bounded rank with an almost regular automorphism, *Israel J. Math.*, **129**(2002), 209-220.
19. A. Jaikin-Zapirain, On the almost regular automorphisms of finite p- groups, *Advances in Mathematics* **153**(2000), 391–402.
20. E. I. Khukhro, Finite groups of bounded rank with an almost regular automorphism of prime order.(Russian *Sibirsk. Mat. Zh.* **43** (2002), 1182–1191. translation in *Siberian Math. J.*, **43**(2002), 955–962.
21. E.I. Khukhro and V.D. Mazurov, Finite groups with an automorphism of prime order whose centralizer has small rank, *J. Algebra*, **301**(2006), 474–492.
22. R. Baer, 'Der Kern, eine charakteristiche Untergruppe', *Compositio Math.* **1**(1934), 254–283.
23. R. Baer, 'Zentrum und Kern von Gruppen mit Elementen unendlicher Ordnung', *Compositio Math.* **2**(1935), 247–249.
24. R. Baer, 'Groups with abelian norm quotient group', *Amer.J.Math.* **61**(1939), 700–708.
25. R. Baer, 'Norm and hypernorm', *Publ.Math.Debrecen* **4**(1956), 347–350.
26. R. A. Bryce and J. Cossey, 'The Wielandt subgroup of a finite soluble group', *J. London Math. Soc.* **2**(1989), 244–256.
27. J. C. Beidleman, H. Heineken and M. Newell, Center and norm, *Bull. Austral. Math. Soc.*, **69(3)**(2004), 457-464.
28. E.A.Ormerod, On the Wielandt length of matabelian p-groups, *Arch.Math(Basel)*, **57(3)**(1991), 212-215.
29. E.Schenkman, On the norm of a group, *Illinois J.Math.*, **4**(1960),150-152.
30. Junxin Wang and Xiuyun Guo, On the norm of finite groups, *Algebra Colloquium*, **14**(2007), 605-612.
31. Xiuyun Guo and Junxin Wang, On generalized Dedekind Groups, *Acta Math. Hungar.*, **122**(2009), 37-44.
32. Junxin Wang and Xiuyun Guo, Finite groups with its power automorphism groups having small indeces, *Acta Mathematica Scnica(English Series)*, **25**(2009), 1097-1108.
33. M.Herzom, P.Longobardi, M. Maj, A. Mann, On generalized Dedekind groups and Tarski super Monsters, *J.Algebra*, **226**(2000), 690-713.

34. C.D.H.Cooper, Power automorphisms of a group, *Math.Z.*, **107**(1968), 335-356.
35. J.Lennox and J. Wiegold, On a question of Deaconescu about automorphisms, *Rend. Sem. Mat. Univ. Padova*, **89**(1993), 83-86.
36. Shirong Li and Xiuyun Guo, Finite p-groups whose abelian subgroups have a trivial intersection, *Acta Mathematica Scnica(English Series)*, **23**(2007), 731-734
37. Xiuyun Guo, Shirong Li and Paul Flavell, Finite groups whose abelian subgroups are TI-subgroups, *Journal of Algebra*, **307**(2007), 565-569

Arrangements of Hyperplanes, Lower Central Series, Chen Lie Algebras and Resonance Varieties

Michel Jambu

For Professor Kar Pin Shum, on the occasion of his seventieth birthday

1 Introduction

In its simplest manifestation, an arrangement is merely a finite collection of lines in the real plane whose the complement consists of a finite number of polygonal bounded and unbounded regions. Determining the numbers of these regions turns out to be purely combinatorial problem which can easily solve by a recursion whose solution is given by formulas involving only the number of lines and the number of lines through each intersection point. These formulas generalize to collection of hyperplanes of $\mathbb{R}^l$ where the recursive formulas are satisfied by an evaluation of the characteristic polynomials forms of the (reverse-ordered) poset of intersections. The study of characteristic polynomials forms the backbone of the combinatorial, and much of the algebraic theory of arrangements.
From the topological standpoint, a richer situation is presented by arrangements of complex hyperplanes (in $\mathbb{C}^l$ or $\mathbb{P}^l$). In this case, the complement is connected, and its topology, as reflected in the fundamental group or the cohomology ring for instance, is much more interesting.
The motivation and many of the applications of the topological theory arose initially from the connection with braids.
Let $\mathcal{A}_l = \{z_i = z_j\}_{1 \leq i < j \leq l} = \{H_{ij}\}_{1 \leq i < j \leq l}$ be the arrangement of diagonal hyperplanes in $\mathbb{C}^l$ with complement $M_l = \mathbb{C}^l - \cup H_{ij}$ the configuration space of l-ordered points in $\mathbb{C}^l$. We called such an arrangement, braid arrangement.

- Fox and Neuwirth (1962): $\pi_1(M_l) \simeq P_l$ (pure braid group with l-strands).

- Fadell and Neuwith (1962): M_l is aspherical.
- V.I. Arnold (1969): $H^*(M_l;\mathbb{C})$. It is the beginning of a very active period of researches in this area.
 Then, extension to more general arrangements.
- E. Brieskorn (1970): $H^*(M;\mathbb{C})$ for any arrangements where M is the complement.
- P. Deligne (1972): M is aspherical where M is the complement of a complexified simplicial arrangement.
- P. Orlik and L. Solomon (1980): $H^*(M;\mathbb{Z}) \simeq A^*_{\mathbb{Z}}(\mathcal{A}) := E_{\mathbb{Z}}/\mathcal{J}$ (which is a combinatorial presentation of the cohomology algebra) and $P_M(t) = \sum_{X\in L(\mathcal{A})} \mu(X)(-t)^{codim(X)}$.
- Rybnikov (1994): $\pi_1(M)$ is not combinatorially defined.

Although topological invariants of the complement $M(\mathcal{A})$ are closely connected to the combinatorics of the arrangement, it is not *a priori* enough to determine the fundamental group $\pi_1(M_{\mathcal{A}}) = G(\mathcal{A})$ and, as such, it is not easy to handle. According to a classical construction of W. Magnus, the associated graded Lie algebra $\mathrm{gr}(G(\mathcal{A}))$ defined by the *lower central series* of $G(\mathcal{A})$ reflects many properties of $G(\mathcal{A})$. The ranks of the abelian groups $\mathrm{gr}_k(G(\mathcal{A}))$, called LCS ranks, are important numerical invariants of $G(\mathcal{A})$. As shown by Kohno (based on foundational work by Sullivan and Morgan), the associated graded Lie algebra $\mathrm{gr}(G(\mathcal{A}))$ and the *holonomy Lie algebra* $\mathcal{H}(\mathcal{A})$ which is determined by the intersection lattice $\mathcal{L}(\mathcal{A})$ are rationally isomorphic and he proved an explicit formula dor the LCS ranks. However, for the larger class of *hypersolvable* arrangements defined by Jambu and Papadima, we have an isomorphism $\mathrm{gr}(G(\mathcal{A})) \cong \mathcal{H}(\mathcal{A})$ and an explicit formula for the LCS ranks is known. However, both the direct product and the semi-direct product of free groups, may be realized as the fundamental groups of the complements of (different) arrangements and their LCS ranks are equal. These groups cannot be distinguished by means of their associated Lie algebra.
K.T. Chen introduced a more manageable approximation to the LCS ranks. The *Chen groups* of a group are the lower central series quotients of its maximal metabelian quotient. The direct sum of the Chen groups is a graded Lie algebra. Papadima and Suciu proved that the rational Chen Lie Algebra is combinatorially determined.
In this paper, after a short introduction to arrangements of hyperplanes.

Then we proceed by showing different examples. The main example is the famous braid groups. In the following sections, we introduce some generalizations in terms of hyperplane arrangements, fiber-type and hypersolvable ones. Then we recall the Magnus theory relating groups theory and Lie algebras theory. Then, we introduce the Chen groups and following Cohen and Suciu, we give some examples showing that Chen groups allow to distinguish non isomorphic groups which cannot be distinguish by means of lower central series. Finally, we introduce the resonance varieties and we give several examples which illustrate the Suciu's conjectures about LCS formulas.

2 Arrangements of Hyperplanes

2.1 Generalities

Let $V = \mathbf{C}^l$ and $\mathcal{A} = \{H_1, \cdots, H_n\}$ be an arrangement of n hyperplanes of V [9] and $M = V - \bigcup_{H \in \mathcal{A}} H$. Let $\mathcal{L}(\mathcal{A})$ be the lattice of the intersections of hyperplanes, ordered by reverse inclusion.
Then

$$H^*(M) \cong A^*(\mathcal{A})$$

with $A^*(\mathcal{A})$ is the *Orlik-Solomon* algebra defined combinatorially.

$$A^*(\mathcal{A}) := \bigwedge(e_1, \cdots, e_n)/\mathcal{J}$$

where $\mathcal{J}$ is an ideal defined by the dependence relations between hyperplanes of $\mathcal{A}$.

$$P_M(t) = \sum_{X \in L(\mathcal{A})} \mu(X)(-t)^{codim(X)}.$$

2.2 Product of arrangements

Let $(\mathcal{A}_1, V_1)$ and $(\mathcal{A}_2, V_2)$ be two arrangements of hyperplanes and $V = V_1 \oplus V_2$. Define the **product** arrangement $(\mathcal{A}_1 \times \mathcal{A}_2)$ by

$$(\mathcal{A}_1 \times \mathcal{A}_2) = \{H_1 \oplus V_2 \mid H_1 \in \mathcal{A}_1\} \cup \{V_1 \oplus H_2 \mid H_2 \in \mathcal{A}_2\}$$

Define $M_1 := V_1 - \bigcup_{H \in \mathcal{A}_1} H$, $M_2 := V_2 - \bigcup_{H \in \mathcal{A}_2} H$ and $M := V - \bigcup_{H \in \mathcal{A}_1 \times \mathcal{A}_2} H$ then

$$P_M(t) = P_{M_1}(t) \times P_{M_2}(t)$$

$$\pi_1(M) = \pi_1(M_1) \times \pi_1(M_2)$$

which is the proof of the previous remark.

3 Braid Groups

The *Braid group with l strands* denoted B_l admits the following presentation:

- generators: $\sigma_1, \cdots, \sigma_{l-1}$
- relations: $\sigma_i\sigma_j = \sigma_j\sigma_i$ for $|i-j|> 1$ and
  $$\sigma_{i+1}\sigma_i\sigma_{i+1} = \sigma_i\sigma_{i+1}\sigma_i \text{ for } i \leq l-2$$

Let $\pi : B_l \longrightarrow \mathbf{S}_l$ be the natural homomorphism where $\mathcal{S}_l$ is the symmetric group on l and define $P_l := \ker\pi$ as the *Pure braid group.*

The group P_l can be realized as the fundamental group of the complement M_l of the diagonal hyperplanes H_{ij} of $\mathbb{C}^l$ defined by $z_i = z_j$ for $1 \leq i < j \leq l$,

$$M_l := \mathbb{C}^l - \bigcup_{1\leq i<j\leq l} H_{ij}$$

$$P_l \cong \pi_1(M_l)$$

$$\mathbf{C} - \{\text{p} - 1 \text{ points}\} \hookrightarrow M_p \longrightarrow M_{p-1}$$

is a linear fibration where $M_p \longrightarrow M_{p-1}$ is the restriction of the map $\mathbb{C}^{p+1} \longrightarrow \mathbb{C}^p$ which forgets the last coordinate.

Properties:

- M_l is a $K(\pi, 1)$-space
- $\pi_1(M_l) \cong \mathbf{F}_{l-1} \rtimes \mathbf{F}_{l-2} \rtimes \cdots \times \mathbf{F}_2 \rtimes \mathbf{F}_1$ (iterated semi-direct product of free groups)
- $H^*(M_l) \cong \bigoplus_{k=1}^{l-1} H^1(\bigvee_k S^1)$ and $P_{M_l}(t) = \prod_{k=1}^{l-1}(1+kt)$

3.1 Fiber-type arrangements

This is a natural generalization of the braid arrangements.

Definition 3.1 *[5] $\mathcal{A}$ is a* **fiber-type arrangement** *if there exists a sequence of subarrangements:*

$$\mathcal{A}_1 \subset \mathcal{A}_2 \subset \cdots \subset \mathcal{A}_l = \mathcal{A}$$

such that $|\mathcal{A}_1|= 1$ and $M(\mathcal{A}_{i+1}) \longrightarrow M(\mathcal{A}_i)$ is a locally trivial fibration with fiber $\mathbf{C} - \{|\mathcal{A}_{i+1} - \mathcal{A}_i| \text{ points}\}$.

$\{d_{i+1} :=| \mathcal{A}_{i+1} - \mathcal{A}_i |, 0 \leq i < l-1\}$ where $\mathcal{A}_0 = \emptyset$, is called the set of the *exponents* of $\mathcal{A}$ and $P_M(t) = \prod_i (1 + d_i t)$ for all exponents d_i.

Theorem 3.2 [5]*[Falk, Randell] The fiber-type arrangements satisfy the LCS formula:*

$$\prod_{k \geq 1} (1 - t^k)^{\phi_k(M)} = \prod_{i=1}^{l} (1 - d_i t)$$

Notice that Π_l and P_l may be realized as the fundamental groups of different fiber-type arrangements with same exponents $\{1, 2, \cdots, l-1\}$.

3.2 Hypersolvable arrangements

The class of hypersolvable arrangements [6] contains both fiber-type and *generic* arrangements and many others.

Definition 3.3 *An arrangement is called* **generic** *iff it is a cone over a general position arrangement.*

Notice that all the fiber-type arrangements are $K(\pi, 1)$ which means that the complement is $K(\pi, 1)$ and the generic arrangements are never $K(\pi, 1)$. Let $\mathcal{A} = \{H_1, \ldots, H_n\}$ be a central arrangement in the complex vector space V. We also denote $\mathcal{A} = \{\alpha_1, \ldots, \alpha_n\} \subset \mathbb{P}(V^*)$ its set of defining equations, viewed as points in the dual projective space. Let $\mathcal{B} \subset \mathcal{A}$ be a proper, non-empty sub-arrangement, and set $\overline{\mathcal{B}} = \mathcal{A} - \mathcal{B}$. We say that $(\mathcal{A}, \mathcal{B})$ is a *solvable extension* if the following conditions are satisfied.

- (i) No points $a \in \overline{\mathcal{B}}$ sits on a projective line determined by $\alpha, \beta \in \mathcal{B}$.

- (ii) For every $a, b \in \overline{\mathcal{B}}$, there exists a point $\alpha \in \mathcal{B}$ on the line passing through a and b. (In the presence of condition (i), this point is uniquely determined, and will be denoted by $f(a,b)$.
- (iii) For every distinct points $a, b, c \in \overline{\mathcal{B}}$, the three points $f(a,b), f(a,c)$ and $f(b,c)$ are either equal or collinear.

Note that only two possibilities may occur: either $\text{rank}(\mathcal{A}) = \text{rank}(\mathcal{B}) + 1$ (*fibered case*), or $\text{rank}(\mathcal{A}) = \text{rank}(\mathcal{B})$ (*singular case*).

Definition 3.4 *[6] [Jambu-Papadima] The arrangement $\mathcal{A}$ is called* **hypersolvable** *if it has a hypersolvable composition series, i.e., an ascending chain of sub-arrangements, $\mathcal{A}_1 \subset \cdots \subset \mathcal{A}_i \subset \mathcal{A}_{i+1} \subset \cdots \subset \mathcal{A}_l = \mathcal{A}$, where rank $\mathcal{A}_1 = 1$, and each extension $(\mathcal{A}_{i+1}, \mathcal{A}_i)$ is solvable.*

It is shown that there is a "*deformation*" of every hypersolvable arrangement to a fiber-type one with the same intersections up to codimension 2 and therefore with the same fundamental group.

Definition 3.5 *The* **quadratic** *Orlik-Solomon algebra is defined by*

$$\overline{A}^*(\mathcal{A}) := \bigwedge(e_1, \cdots, e_n)/J$$

where J is the homogeneous ideal generated by

$$(\mathcal{R}_{\mathcal{A}}) \qquad e_{i_1} \wedge e_{i_2} - e_{i_1} \wedge e_{i_3} + e_{i_2} \wedge e_{i_3}$$

with codim$(H_{i_1} \cap H_{i_2} \cap H_{i_3}) = 2$.

Definition 3.6 *The* **quadratic** *Poincaré polynomial $\overline{P}_M(t)$ is the Poincaré polynomial of $\overline{A}^*(\mathcal{A})$.*

Theorem 3.7 [6][7]*[Jambu-Papadima] Let $\mathcal{A}$ be a hypersolvable arrangement. Then $\prod_{k \geq 1} (1-t^k)^{\phi_k(\mathcal{A})} = \overline{P}_M(-t)$ (called the generalized LCS formula).*

Notice that if $\mathcal{A}$ is fiber-type, then $\overline{A}^*(\mathcal{A}) = A^*(\mathcal{A})$ and $\overline{P}_M(t) = P_M(t)$.

4 The Magnus Theory

There is a very strong analogy between the theory of *groups* and the theory of *Lie algebras*. The most well-known is the one between Lie groups and Lie algebras; every finite dimensional (complex) Lie algebra is the Lie algebra

of some (complex) Lie group.
Another connection is due to Magnus and developed by Lazard and we will consider it in the following.
Let G be an arbitrary group. How *"far"* is it from an abelian group?
Let G^{ab} be the *abelianization* of G, that is $G^{\mathrm{ab}} = G/[G,G]$ where $[G,G]$ is the subgroup of commutators.
If G is abelian, then $[G,G] = 0$ and $G^{\mathrm{ab}} = G$.
If G is *perfect*, then $[G,G] = G$ and $G^{\mathrm{ab}} = 0$.
Let us consider the *Lower Central Series* of G which is denoted by $(\Gamma_n G)_{n\geq 1}$ where:

- $\Gamma_1 G = G$
- $\Gamma_{n+1} G = [G, \Gamma_n G]$

Properties:

- $\Gamma_{n+1} G$ is a subgroup of $\Gamma_n G$
- $\Gamma_n G/\Gamma_{n+1} G$ is an abelian group which is finitely generated if G is finitely generated.
- $[\Gamma_m G, \Gamma_n G] \subset \Gamma_{m+n} G$.

Define

$$\mathrm{gr}_n G = \Gamma_n G/\Gamma_{n+1} G$$

which is an abelian group for any $n \geq 1$ and

$$\mathrm{gr} G = \bigoplus_{n\geq 1} \mathrm{gr}_n G$$

There is a natural structure of Lie algebra on $\mathrm{gr}G$ over $\mathbb{Z}$ where the Lie bracket $[x,y]$ is induced from the group commutator $(x,y) = xyx^{-1}y^{-1}$.
Let denote $\phi_n(G) = \mathrm{rank}(\mathrm{gr}_n G)$. They are important numerical invariants of G. Although they may be very difficult to determine, many properties of the group G are reflected into properties of its associated Lie algebra $\mathrm{gr}G$.
Then a natural question is, given the group G, to determine the Lie algebra $\mathrm{gr}G$ and to compute $\phi_n(G)$ for every n.

4.1 Some examples: Free groups

Let $G = \mathbf{F}_l$ be a free group of rank l. Magnus showed that

$$\mathrm{gr}(\mathbf{F}_l) = \mathbf{L}_l$$

where $\mathbf{L}_l$ is the free Lie algebra on l generators, whose ranks were computed by Witt.
Let R be a commutative ring with unit and let $R\!<\!A\!>$ be the free associative algebra over the set A of l elements. The product $[x, y] = xy - yx$ turns $R\!<\!A\!>$ into a Lie algebra. The Lie subalgebra $\mathbf{L}_A(R)$ generated by A is called the *free Lie algebra* over A. Notice that $R\!<\!A\!>$ is the universal enveloping algebra of $\mathbf{L}_A(R)$.
When $R = \mathbb{Z}$, we denote $\mathbf{L}_A(R)$ by $\mathbf{L}_A$. So, $\mathbf{L}_l$ is the free Lie algebra on l generators over $\mathbb{Z}$.

Theorem 4.1 (Witt) *Let $\mathbf{F}_l$ be the free group of rank l. Then*

$$\phi_k(\mathbf{F}_l) = \mathrm{rank}(\mathbf{L}_l)_k = \frac{1}{k}\sum_{d|k} \mu(d) l^{k/d}$$

where $(\mathbf{L}_l)_k$ is the homogeneous component of rank k of the free Lie algebra $\mathbf{L}_l$ and μ is the classical Möbius function.

In fact, we will consider the following equalities coming from the proof of the Witt theorem, which is called *Witt formula* or *LCS formula*:

$$\prod_{k\geq 1}(1 - t^k)^{\phi_k(\mathbf{F}_l)} = 1 - lt$$

Let us now consider a direct product of free groups $G = \mathbf{F}_{i_1} \times \mathbf{F}_{i_2} \times \ldots \times \mathbf{F}_{i_n}$. Then, we get the LCS formula for the direct product of free groups.

$$\prod_{k\geq 1}(1 - t^k)^{\phi_k(G)} = \prod_{k\geq 1}(1 - t^k)^{\phi_k(\mathbf{F}_{i_1})} \ldots \prod_{k\geq 1}(1 - t^k)^{\phi_k(\mathbf{F}_{i_n})} = \prod_{j=1}^{n}(1 - i_j t).$$

4.2 Braid arrangements

Theorem 4.2 [8] *[Kohno] Let $\mathcal{A}_l$ be the braid arrangement, H_{ij} of diagonal hyperplanes of $\mathbb{C}^l$ defined by $z_i = z_j$ for $1 \leq i < j \leq l$,*

$$\prod_{k \geq 1} (1 - t^k)^{\phi_k(M_l)} = \prod_{1 \leq k \leq l-1} (1 - kt)$$

Remark: Both the semi-direct product P_n and the direct product $\Pi_n := \mathbf{F}_{n-1} \times \cdots \times \mathbf{F}_2 \times \mathbf{F}_1$, may be realized as the fundamental groups of the complements of (distinct) arrangements of hyperplanes. Neither homology nor the lower central series can distinguish between Π_n and P_n.

5 Chen Groups

The *Chen groups* [3] of a group G are the lower central series quotients of G modulo its second commutator subgroup G''. Recall that $G^{(i+1)} = [G^{(i)}, G^{(i)}]$, then $G' = [G, G]$ and $G'' = [\Gamma_2 G, \Gamma_2 G]$.
The group G/G'' is metabelian. It is finitely-generated if G is finitely-generated. It fits into the exact sequence:

$$0 \longrightarrow G'/G'' \longrightarrow G/G'' \longrightarrow G/G' \longrightarrow 0$$

The k**-th Chen** group of G is, by definition, $\mathrm{gr}_k(G/G'')$. Let $\theta_k(G) = \phi_k(G/G'')$ be its rank. The projection $G \longrightarrow G/G''$ induces surjections $\mathrm{gr}_k G \longrightarrow \mathrm{gr}_k(G/G'')$. Thus $\phi_k \geq \theta_k$ for all k and $\phi_k = \theta_k$ for $k \leq 3$.
Assume $G/G' \cong \mathbb{Z}^n$. The Chen groups of G can be determined from the *Alexander invariant* $B = G'/G''$ (viewed as a module over $\mathbb{Z}[G/G']$). Then $\mathrm{gr}_k(G/G'') = \mathrm{gr}_{k-2} B$, for $k \geq 2$ and we have

$$\sum_{k \geq 0} \theta_{k+2} t^k = \mathrm{Hilb}(\mathrm{gr} B)$$

where $\mathrm{gr} B = \bigoplus_{k \geq 0} \mathrm{gr}_k B$ (viewed as a module over $\mathrm{gr}\mathbb{Z}[G/G']$)and $\mathrm{Hilb}(\mathrm{gr} B)$ is the Hilbert series of the graded module $\mathrm{gr} B$. A presentation for $\mathrm{gr} B$ can be obtained from a presentation for B via the well-known Gröbner basis algorithm for finding the tangent cone to a variety.

Theorem 5.1 (Chen, Murasugi) *Let $G = \mathbf{F}_l$. Then*

$$\theta_k(\mathbf{F}_l) = (k-1).\binom{l+k-2}{k}, \quad k \geq 2$$

Theorem 5.2 [**3**]*[Cohen, Suciu] 1. Let* $G = \pi_1(M_1 \times M_2)$, $\pi_1(M_1) = G_1$, $\pi_1(M_2) = G_2$, *then*

$$\theta_k(G) = \theta_k(G_1) + \theta_k(G_2)$$

2. Let $G = \mathbf{F}_{d_1} \times \cdots \times \mathbf{F}_{d_l}$ *be a direct product of free groups, then the Chen groups of* G *are free abelian and*

$$\theta_1(G) = \sum_{i=1}^{l} d_i$$

$$\theta_k(G) = (k-1)\sum_{i=1}^{l} \binom{k+d_i-2}{k} \text{ for } k \geq 2$$

In particular, let $\Pi_l = \mathbf{F}_{l-1} \times \cdots \times \mathbf{F}_1$ *then*

$$\theta_1(\Pi_l) = \binom{l}{2}$$

$$\theta_k(\Pi_l) = (k-1)\binom{k+l-2}{k+1} \text{ for } k \geq 2$$

3. The Chen groups of the pure braid group P_l *are free abelian. The rank* θ_k *are given by*

$$\theta_1 = \binom{l}{2}, \qquad \theta_2 = \binom{l}{3}$$

and

$$\theta_k = (k-1).\binom{l+1}{4} \text{ for } k \geq 3$$

Corollary 5.3 *For* $l \geq 4$, *the groups* P_l/P_l'' *and* Π_l/Π_l'' *are not isomorphic. For* $l \geq 4$, *the groups* P_l *and* Π_l *are not isomorphic.*

Remark 5.4 $P_2 \cong \mathbf{F}_1$ *and* $P_3 \cong \mathbf{F}_2 \times \mathbf{F}_1$.

Example 5.5 *Let be the two arrangements* $\mathcal{A}$ *and* $\mathcal{B}$ *defined by:*

$$Q(\mathcal{A}) = xyz(x-y)(x-z)(x-2z)(x-3z)(y-z)(x-y-z)$$

and

$$Q(\mathcal{B}) = xyz(x-y)(x-z)(x-2z)(x-3z)(y-z)(x-y-2z)$$

They are fiber-type with the same exponents $(1, 4, 4)$. The sequences are:

$$\{x = 0\} \subset \{x = 0, z = 0, x - z = 0, x - 2z = 0, x - 3z = 0\} \subset \mathcal{A}$$

$$\{x = 0\} \subset \{x = 0, z = 0, x - z = 0, x - 2z = 0, x - 3z = 0\} \subset \mathcal{B}$$

Therefore homology groups and lower central series quotients are isomorphic. However,

$$\theta_1(G(\mathcal{A})) = \theta_1(G(\mathcal{B})) = 9$$

$$\theta_2(G(\mathcal{A})) = \theta_2(G(\mathcal{B})) = 12$$

$$\theta_3(G(\mathcal{A})) = \theta_3(G(\mathcal{B})) = 40$$

and for $k \geq 4$

$$\theta_k(G(\mathcal{A})) = \frac{1}{2}(k-1)(k^2 + 3k + 24)$$

$$\theta_k(G(\mathcal{B})) = \frac{1}{2}(k-1)(k^2 + 3k + 22)$$

then

$$G(\mathcal{A}) \ncong G(\mathcal{B})$$

Notice that the groups $G(\mathcal{A})$ and $G(\mathcal{B})$ cannot be distinguished by means of the LCS formula.

Theorem 5.6 **[10]***[Papadima, Suciu] For any complex arrangement* $\mathcal{A}$,

$$\mathrm{gr}(G(\mathcal{A})/G''(\mathcal{A})) \otimes \mathbb{Q} \cong (\mathcal{H}_M/\mathcal{H}''_M) \otimes \mathbb{Q}$$

In particular, the rational holonomy Lie algebra of the arrangement is combinatorially determined by the level 2 of the intersection lattice, $\mathcal{L}_2(\mathcal{A})$. Hence, the Chen ranks are combinatorially determined.
There is a conjecture (*Suciu*) which makes this combinatorial dependence explicit.
This is related to the cohomology of the Orlik-Solomon algebra which figures in the Aomoto-Gelfand theory of generalized hypergeometric functions, and in solutions of the Knizhnik-Zamolodchikov equations of conformal field theory.

6 Resonance Varieties

For each $a_\lambda = \sum \lambda_i a_i \in A^1(\mathcal{A}$, the Orlik-Solomon algebra can be turned into a cochain complex $(A(\mathcal{A}), a_\lambda)$, with i-th term the degree i graded piece of $(A(\mathcal{A})$ and with differential given by multiplication by a_λ.

$$0 \longrightarrow A^0(\mathcal{A}) \longrightarrow A^1(\mathcal{A}) \longrightarrow \cdots \longrightarrow A^{l-1}(\mathcal{A}) \longrightarrow A^l(\mathcal{A}) \longrightarrow 0$$

The cohomology of this complex determines a stratification of the parameter $\mathbb{C}^n$. The cohomology $H^*(A(\mathcal{A}), a_\lambda)$ is isomorphic to the cohomology of the complement $M(\mathcal{A})$ with coefficients in a local system determined by a_λ, when a_λ satisfies some *non-resonance* conditions dependent only on $M(\mathcal{A})$. The p-th **resonance variety** is defined as

$$R_p(\mathcal{A}) = \{\lambda \in \mathbb{C}^n \mid H^p(A(\mathcal{A}], a\lambda) \neq 0\}$$

Then Falk [4] showed that $H^1(A(\mathcal{A}), a_\lambda) \neq 0$ precisely when a_λ belongs to an affine variety called the *resonance variety* $R_1(\mathcal{A})$ of the arrangement $\mathcal{A}$. He showed that $R_1(\mathcal{A})$ is a linear subspace of $\mathbf{C}^n$ which is a union of subspaces of dimension at least 2, as follows.
A partition $\mathsf{P} = (p_1 \mid \cdots \mid p_q)$ of $\mathcal{A}$ is called *neighborly* if

$$p_j \cap I \geq \mid I \mid -1 \Rightarrow I \subset p_j \text{ for all } I \in \mathcal{L}_2(\mathcal{A})$$

To a neighborly partition corresponds an irreducible subvariety of $R_1(\mathcal{A})$

$$L_{\mathsf{P}} = \Delta_n \cap \bigcap_{\{I \in \mathcal{L}_2(\mathcal{A}) \mid I \not\subset p_j, \text{ any } j\}} \{\lambda \mid \sum_{i \in I} \lambda_i = 0\}$$

where $\Delta_n = \{\lambda \in \mathbb{C}^n \mid \sum_{i=0}^{n} \lambda_i = 0\}$.
Conversely, all components of $R_1(\mathcal{A}]$ arise from neighborly partitions of sub-arrangements of $\mathcal{A}$.
Then $R_1(\mathcal{A}) = \bigcup L_i$ and for any $r \geq 0$ let $h_r = \mid \{L_i \mid \dim L_i = r\} \mid$ be the number of components of $R_1(\mathcal{A})$ of dimension r.
h_r can be computed directly from the lattice $\mathcal{L}(\mathcal{A})$ by computing neighborly partitions of sub-arrangements of $\mathcal{A}$ and finding $\dim L_{\mathsf{P}}$.
This is the conjecture [11] (*Suciu*): Let $R_1(\mathcal{A}) = \bigcup_{i=1}^{v} L_i$ be the decomposition of $R_1(\mathcal{A})$ into linear components. Then if $\phi_4(\mathcal{A}) = \theta_4(\mathcal{A})$

$$\prod_{k \geq 1} (1 - t^k)^{\phi_k(\mathcal{A})} = \prod_{i=1}^{v} (1 - (\dim L_i)t).$$

6.1 Some examples

- **1. Braid arrangements $\mathcal{A}_4$.**
  Let $Q(\mathcal{A}_4) = xyz(x-y)(x-z)(y-z)$ be the defining polynomial of $\mathcal{A}_4$.
  $R_1(\mathcal{A}_4) = L_{124} \cup L_{135} \cup L_{236} \cup L_{456} \cup L_{16|25|34}$ but $\phi_4(\mathcal{A}_4) = 21$ and $\theta_4(\mathcal{A}_4)$ so the conjecture does not work. This arrangement is fiber-type and the LCS ranks $\phi_k(\mathcal{A}_4)$ can be computed by the LCS formula.

- **2. Arrangement $\mathcal{A}_{X_3}$.**
  Let $Q(\mathcal{A}_{X_3}) = xyz(y+z)(x-z)(2x+y)$ be the defining polynomial of $\mathcal{A}_{X_3}$.
  $R_1(\mathcal{A}_{X_3}) = L_{135} \cup L_{236} \cup L_{456}$ and $\dim L_i = 2$ for any L_i. Moreover $\phi_k(\mathcal{A}_{X_3}) = \theta_4(\mathcal{A}_{X_3}) = 9$, so

  $$\prod_{k \geq 1} (1-t^k)^{\phi_k(\mathcal{A}_{X_3})} = \prod_{i=1}^{3} (1-2t) = (1-2t)^3$$

  Notice that $\mathcal{A}_{X_3}$ is not hypersolvable.

- **3. Arrangement $\mathcal{A}_{X_2}$.**
  Let $Q(\mathcal{A}_{X_2}) = xyz(x+y)(x-z)(y-z)(x+y-2z)$ be the defining polynomial of $\mathcal{A}_{X_2}$.
  $R_1(\mathcal{A}_{X_2}) = L_{136} \cup L_{245} \cup L_{127} \cup L_{237} \cup L_{567}$ and $\dim L_i = 2$ for any L_i. Moreover $\phi_k(\mathcal{A}_{X_2}) = \theta_4(\mathcal{A}_{X_2}) = 15$; so

  $$\prod_{k \geq 1} (1-t^k)^{\phi_k(\mathcal{A}_{X_2})} = \prod_{i=1}^{5} (1-2t) = (1-2t)^5$$

  Notice that $\mathcal{A}_{X_2}$ is not hypersolvable.

- **4. Non-Fano arrangement $\mathcal{A}$.**
  Let $Q(\mathcal{A}) = xyz(x-y)(x-z)(y-z)(x+y-z)$ be the defining polynomial of $\mathcal{A}$. $P_M(t) = (1+t)(1+3t)^2$; $\mathcal{A}$ is not hypersolvable (it is called factored); $\phi_k(\mathcal{A}) = 42$ and $\theta_4(\mathcal{A}) = 27$ so the conjecture does not work.

References

[1] D. Cohen, Cohomology and intersection cohomology of complex hyperplane arrangements, Adv. Math. **97**, 1993, 231 – 266.

[2] D. Cohen, P. Orlik, Arrangements and Local Systems, Mathematical Research Letters **7**, 2000, 299 – 316.

[3] D. Cohen, A. Suciu, The Chen groups of the pure braid group, Contemp. Math. **181**, 1995, 45 – 64.

[4] M. Falk, Arrangements and Cohomology, Annals of Combinatorics **1**, 1997, 135 – 157.

[5] M. Falk, R. Randell, The lower central series of a fiber-type arrangement, Invent. Math. **82**, 1985, 77 – 88.

[6] M. Jambu, S. Papadima, A generalization of fiber-tyme arrangments and a new deformation method, Topology, **37**, 1998, 1135 – 1164.

[7] M. Jambu, S. Papadima, Deformations of hypersolvable arrangements, Topology and its Appl. **118**, 2002, 103 – 111.

[8] T. Kohno, Srie de Poincar-Koszul associe aux groupes d etresses pures, Invent. Math. **82**, 1985, 57 – 75.

[9] P. Orlik, H. Terao, Arrangements of Hyperplanes, Grundlehren Math. Wiss., Vol **300**, Springer-Verlag, Berlin, 1992.

[10] S. Papadima, A. Suciu, Chen Lie Algebras,

[11] A. Suciu, Fundamental goups of line arrangments, Enumerative aspects, Contemp. Math.

[12] S. Yuzvinsky, Cohomology of the Brieskorn-Orlik-Solomon algebras, Comm. Algebra **23** 1995, 5339 – 5354.

Laboratoire J.A. Dieudonné, UMR 6621, Université de Nice, France
E-mail address: jambu@unice.fr

Remarks on p-Adic Cantor Series Expansions

N. R. Kanasri*

Department of Mathematics, Faculty of Science
Khon Kaen University, Khon Kaen 40002, Thailand
**E-mail: rompurk@hotmail.com*

V. Laohakosol

Department of Mathematics, Faculty of Science, Kasetsart University,
Bangkok 10900, Thailand and Centre of Excellence in Mathematics,
CHE, Si Ayutthaya Road, Bangkok 10400, Thailand
E-mail: fscivil@ku.ac.th

The problem of characterizing rational numbers by their p-adic Cantor series expansions is described. Some partial characterization and quantitative results are derived. In particular, it is shown that a rational number whose numerator fall within a favorable range can be characterized by its p-adic Cantor series expansion.

Keywords: p-adic Cantor series expansion; Rational numbers.

1. Introduction

The Knopfmachers' algorithm,,[2] provides us with a construction of several series expansion representations in the p-adic number field such as the Engel, Lüroth, Sylvester and Cantor series expansions. Arising from these works is the problem of characterizing rational numbers by such expansions. In,[1] a complete characterization of rational numbers via its so-called Engel-type expansion was given, while that of Lüroth-type expansion was given in[4] and that of Sylvester-type expansion in.[5] However, the characterization through p-adic Cantor series expansions is still open. The objectives of this article are first to establish a characterization of one particular class of rational numbers via their p-adic Cantor series expansions and second to derive some related quantitative results.

Let p be a fixed prime, $\mathbb{Q}_p$ the field of p-adic numbers equipped with the p-adic valuation $|.|_p$, so normalized that $|p|_p = p^{-1}$, and let $\mathbb{Z}_p$ be the ring of p-adic integers. For $A \in \mathbb{Q}_p$, define the *order* $v_p(A)$ of A by

$$|A|_p = p^{-v_p(A)} \text{ and } v_p(0) = +\infty.$$

Each $A \in \mathbb{Q}_p$ is uniquely representable in the form

$$A = \sum_{n=v_p(A)}^{\infty} c_n p^n, \quad c_n \in \mathbb{F}_p := \{0, 1, ..., p-1\}, \ c_{v_p(A)} \neq 0.$$

The *fractional part*, $\langle A \rangle$, of A is defined as the finite series

$$\langle A \rangle = \begin{cases} \sum_{n=v_p(A)}^{0} c_n p^n & \text{if } v_p(A) \leq 0, \\ 0 & \text{if } v_p(A) > 0. \end{cases}$$

Denote the set of all fractional parts by

$$S_p := \{\langle A \rangle \, ; A \in \mathbb{Q}_p\} \quad \subset \quad \mathbb{Q}.$$

The Knopfmachers' series expansion algorithm proceeds as follows: for $A \in \mathbb{Q}_p$, let $a_0 := \langle A \rangle \in S_p$. Define $A_1 := A - a_0$. If $A_n \neq 0 \ \ (n \geq 1)$ is already defined, put $a_n = \left\langle \frac{1}{A_n} \right\rangle \in S_p$, and

$$A_{n+1} = \left(A_n - \frac{1}{a_n} \right) \frac{s_n}{r_n}, \tag{1}$$

if $a_n \neq 0$, where r_n, s_n are non-zero rational numbers which may depend on $a_1, ..., a_n$. Then

$$\begin{aligned} A = a_0 + A_1 = a_0 + \frac{1}{a_1} + \frac{r_1}{s_1} A_2 = ... = a_0 + \frac{1}{a_1} + \frac{r_1}{s_1} \frac{1}{a_2} \\ + \cdots + \frac{r_1 \cdots r_{n-1}}{s_1 \cdots s_{n-1}} \frac{1}{a_n} + \frac{r_1 \cdots r_n}{s_1 \cdots s_n} A_{n+1}. \end{aligned}$$

The process ends in a finite expansion if some $A_{n+1} = 0$. If some $a_n = 0$, then A_{n+1} is not defined. This is ruled out by imposing the condition

$$v_p(s_n) - v_p(r_n) \geq 2v_p(a_n) - 1.$$

Thus

$$A = a_0 + \frac{1}{a_1} + \sum_{n=1}^{\infty} \frac{r_1 \cdots r_n}{s_1 \cdots s_n} \cdot \frac{1}{a_{n+1}}.$$

We are interested here in the case where

$$r_n = a_n + 1, \quad s_n = a_n.$$

In this case, the algorithm yields a well-defined (with respect to the p-adic valuation) and unique series expansion called p-adic Cantor series expansion of the form

$$A = a_0 + \frac{1}{a_1} + \sum_{n=2}^{\infty} \left(1 + \frac{1}{a_1}\right) \cdots \left(1 + \frac{1}{a_{n-1}}\right) \frac{1}{a_n},$$

where the digits a_n are, by the algorithm, subject to the restrictions

$$a_0 = \langle A \rangle, \tag{2}$$

$$a_n = \frac{c_{\nu_n}}{p^{\nu_n}} + \ldots + \frac{c_1}{p} + c_0 \in S_p \; ; \; \nu_n = -v_p(a_n); \; c_i \in \mathbb{F}_p; \; c_{\nu_n} \neq 0 \quad (n \geq 1), \tag{3}$$

$$\nu_{n+1} \geq 2\nu_n + 1 \quad (n \geq 1). \tag{4}$$

For $\alpha \in \mathbb{N}$, define

$$D_\alpha := \{b_\alpha := c_\alpha + c_{\alpha-1}p + \ldots + c_0 p^\alpha; \; c_i \in \mathbb{F}_p \; (i = 0, 1, \ldots, \alpha), \; c_\alpha \neq 0\}.$$

Note that for each $b_\alpha \in D_\alpha$, we have the bounds

$$1 \leq b_\alpha \leq (p-1)(1 + p + \ldots + p^\alpha) = p^{\alpha+1} - 1 < p^{\alpha+1}. \tag{5}$$

2. Results

Here and throughout, the symbol $|\cdot|$ denotes the usual absolute value. Let

$$A_1 = A = p^\nu \frac{\alpha}{\beta} = p^{\nu_1} \frac{\alpha_1}{\beta_1} \in \mathbb{Q}.$$

By the recurrence relation (1), it follows that $A_n \in \mathbb{Q}$ for each $n \in \mathbb{N}$. Observe that for each $a_n \in S_p$, we have

$$0 < a_n < \sum_{k=0}^{\infty} \frac{p-1}{p^k} = p,$$

and so we can write $a_n = \frac{b_{\nu_n}}{p^{\nu_n}}$, where $b_{\nu_n} \in D_{\nu_n}$, and

$$0 < b_{\nu_n} \leq p^{\nu_n+1} - 1 < p^{\nu_n+1}.$$

Furthermore, each A_n can be represented in the form

$$A_n = p^{\nu_n} \frac{\alpha_n}{\beta_n},$$

where $\alpha_n \in \mathbb{Z}, \; \beta_n \in \mathbb{N}, \; \gcd(\alpha_n, \beta_n) = 1, p \nmid \alpha_n \beta_n$. The recurrence relation (1) becomes

$$(b_{\nu_n} + p^{\nu_n}) \, \alpha_{n+1} p^{\nu_{n+1} - \nu_n} \beta_n = (b_{\nu_n} \alpha_n - \beta_n) \, \beta_{n+1}. \tag{6}$$

Since $\gcd(\alpha_{n+1}p^{\nu_{n+1}-\nu_n}, \beta_{n+1}) = 1$, then $\beta_{n+1} \mid (b_{\nu_n} + p^{\nu_n})\beta_n$ and so

$$\left|\alpha_{n+1}p^{\nu_{n+1}-\nu_n}\right| \leq \left|\frac{(b_{\nu_n} + p^{\nu_n})\,\alpha_{n+1}p^{\nu_{n+1}-\nu_n}\beta_n}{\beta_{n+1}}\right| = |b_{\nu_n}\alpha_n - \beta_n|. \tag{7}$$

Since

$$\nu_{n+1} \geq 2\nu_n + 1 \geq \nu_n + 2\nu_{n-1} + 2 \geq \ldots \geq \nu_n + \nu_{n-1} + \ldots + 2\nu_1 + n, \tag{8}$$

we get

$$|\alpha_{n+1}| < |\alpha_n| + \frac{\beta_n}{p^{\nu_{n+1}-\nu_n}} \leq |\alpha_n| + \frac{\beta_n}{p^{\nu_n+1}}. \tag{9}$$

From $\beta_{n+1} \mid (b_{\nu_n} + p^{\nu_n})\beta_n$, we deduce

$$0 < \beta_{n+1} \leq (b_{\nu_n} + p^{\nu_n})\beta_n \leq \ldots \leq (b_{\nu_n} + p^{\nu_n})\cdots(b_{\nu_1} + p^{\nu_1})\beta_1 \tag{10}$$

and so

$$\frac{\beta_n}{p^{\nu_{n+1}-\nu_n}} \leq \frac{b_{\nu_{n-1}} + p^{\nu_{n-1}}}{p^{\nu_{n-1}}} \cdots \frac{b_{\nu_1} + p^{\nu_1}}{p^{\nu_1}} \frac{\beta_1}{p^{n+\nu_1}}. \tag{11}$$

Claim. If α_n and α_{n+1} are both > 0, then $\alpha_{n+1} < \alpha_n$.
Proof of Claim. From the recurrence relation (6), we have

$$p^{\nu_{n+1}-\nu_n}\frac{\alpha_{n+1}}{\beta_{n+1}} = \frac{\alpha_n b_{\nu_n} - \beta_n}{(b_{\nu_n} + p^{\nu_n})\beta_n},$$

since the left-hand fraction is in lowest form, we get

$$p^{\nu_{n+1}-\nu_n}\alpha_{n+1} \leq \alpha_n b_{\nu_n} - \beta_n < \alpha_n b_{\nu_n},$$

so $\alpha_{n+1} < \alpha_n$.

From the Claim, if we start off with $A = A_1 = p^{\nu}\frac{\alpha}{\beta} = p^{\nu_1}\frac{\alpha_1}{\beta_1} > 0$, then $\alpha_1 > 0$ and being integers eventually for some $m \in \mathbb{N}$, we must have either $\alpha_m = 0$ which yields a terminating expansion, or $\alpha_m < 0$ which yields all remaining $\alpha_k < 0$ $(k \geq m)$ resulting in an infinite expansion.

Henceforth, without loss of generality, *we assume that all* $\alpha_n < 0$ $(n \geq 1)$. If $A < -1$ (referred to as **Case L**), then $1 < |A| = |A_1| = p^{\nu_1}\frac{|\alpha_1|}{\beta_1}$, we get $\beta_1 < p^{\nu_1}|\alpha_1|$, while if $-1 < A < 0$, (referred to as **Case G**), then $1 > |A| = |A_1| = p^{\nu_1}\frac{|\alpha_1|}{\beta_1}$, and so $\beta_1 > p^{\nu_1}|\alpha_1|$.

We now state a number of preliminary results whose proofs are left as exercises.

Lemma 1. *There are two disjoint cases.*
I. ***Case G:*** *We have*

(1) $\beta_n > p^{\nu_n}|\alpha_n|$ *for all* $n \geq 1$*;*

(2) $1 > \ldots > p^{\nu_n}\frac{|\alpha_n|}{\beta_n} > \ldots > p^{\nu_1}\frac{|\alpha_1|}{\beta_1} > 0 \quad (n > 1)$, *so* $\lim_{n\to\infty} p^{\nu_n}\frac{|\alpha_n|}{\beta_n}$ *exists; let it be* $L_1 \leq 1$;

(3) If there exists $n_0 \in \mathbb{N}$ *such that* $|\alpha_{n_0}| \leq p-1$ *and* $\epsilon_n \leq \frac{1}{p-1} \quad (n \geq n_0)$, *then* $|\alpha_{n+1}| \leq p-1 \quad (n \geq n_0)$.

II. ***Case L:*** *We have*

(1) $\beta_n < p^{\nu_n}|\alpha_n|$ *for all* $n \geq 1$.

(2) $p^{\nu_1}\frac{|\alpha_1|}{\beta_1} > \ldots > p^{\nu_n}\frac{|\alpha_n|}{\beta_n} > \ldots > 1 \quad (n > 1)$, *so* $\lim_{n\to\infty} p^{\nu_n}\frac{|\alpha_n|}{\beta_n}$ *exists; let it be* $L_2 \geq 1$;

(3) If $|\alpha_n| \leq p-1$, *then* $|\alpha_{n+1}| \leq p-1 \quad (n \in \mathbb{N})$.

III. *We have* $L_1 = L_2 = 1$.

IV. *For* $n \in \mathbb{N}$, *we have* $\gcd(\alpha_n, \alpha_{n+1}) = 1$.

To pursue our analysis, let

$$\beta_n = (1+\epsilon_n)p^{\nu_n}|\alpha_n|, \tag{12}$$

where, by Lemma 1(III), ϵ_n is a rational number $\to 0$ as $n \to \infty$. Note also that for all $n \in \mathbb{N}$, we always have:

- $\epsilon_1 > \epsilon_2 > \ldots > \epsilon_{n-1} > \epsilon_n > \ldots > 0$ in Case G, but
- $0 > \ldots > \epsilon_n > \epsilon_{n-1} > \ldots > \epsilon_2 > \epsilon_1 > -1$ in Case L.

Lemma 2. ***Case G:*** *If there exists* $N \in \mathbb{N}$ *such that* $|\alpha_N| \leq 2p-1$ *and* $\epsilon_n \leq \frac{1}{2p-1} \quad (n \geq N)$, *then* $|\alpha_{n+1}| \leq 2p-1 \quad (n \geq N)$.
Case L: *If* $|\alpha_n| \leq 2p-1$, *then* $|\alpha_{n+1}| \leq 2p-1$ $(n \in \mathbb{N})$.

Lemma 3. *Let* $r, \beta \in \mathbb{N}$ *with* $p \nmid \beta$, $\beta < p^{r+1}$. *Then we have the following unique p-adic Cantor series expansion*

$$\begin{aligned}-\frac{p^r}{\beta} =& \frac{1}{a_1} + \left(1+\frac{1}{a_1}\right)\frac{1}{a_2} + \cdots \\ &+ \left(1+\frac{1}{a_1}\right)\left(1+\frac{1}{a_2}\right)\cdots\left(1+\frac{1}{a_{n-1}}\right)\frac{1}{a_n} - \frac{p^{r(n+1)}}{b_{r(1)}\cdots b_{r(n)}\beta},\end{aligned}$$

where $r(1) = r$, $b_r = p^{r+1} - \beta$, *and for* $i \geq 1$,

$$\begin{aligned}&r(i+1) = 2r(i)+1,\ a_i = \frac{b_{r(i)}}{p^{r(i)}}, \\ &b_{r(i+1)} = p^{r(i+1)+1} - \left(b_{r(1)} + p^{r(1)}\right)\left(b_{r(2)} + p^{r(2)}\right)\cdots\left(b_{r(i)} + p^{r(i)}\right)\beta.\end{aligned}$$

Our main result reads :

Theorem 4. *Let* $A = p^{\nu}\alpha/\beta \in \mathbb{Q} \cap p\mathbb{Z}_p \setminus \{0\}$ *with* $\nu \in \mathbb{N},\ \alpha \in \mathbb{Z},\ \beta \in \mathbb{N},\ \gcd(\alpha, \beta) = 1$.

Case L: *Let* $0 < \delta < \frac{p}{2p-1}$.

If $|\alpha| \le 2p-1$, *and there exists* $N_1 \in \mathbb{N}$, *such that* $b_{\nu_n} \ge (p-1-\delta)p^{\nu_n}\quad (n \ge N_1)$, *then the p-adic Cantor series expansion of A is of the form*

$$A = \frac{1}{a_1} + \left(1 + \frac{1}{a_1}\right)\frac{1}{a_2} + \ldots + \left(1 + \frac{1}{a_1}\right)\cdots\left(1 + \frac{1}{a_{n-1}}\right)\frac{1}{a_n}$$
$$- \left(1 + \frac{1}{a_1}\right)\cdots\left(1 + \frac{1}{a_n}\right)\frac{p^{\nu_{n+1}}}{\beta_{n+1}} \qquad (n \in \mathbb{N}),$$

where $\beta_{n+1} \in \mathbb{N},\ p \nmid \beta_{n+1}$, *and* $\beta_{n+1} < p^{\nu_{n+1}+1}$; *i.e., its tail is as specified in Lemma 3.*

Case G: *Let* $0 < \Delta < \frac{p}{2p-1}$.

If there exists $N_2 \in \mathbb{N}$, *such that* $|\alpha_{N_2}| \le 2p-1$, *and* $\epsilon_n < \frac{1}{2p-1}$, $b_{\nu_n} \ge (p-1-\Delta)p^{\nu_n}\quad (n \ge N_2)$, *then the same conclusion as in the previous case L holds.*

Proof. For Case L, if $|\alpha| \le 2p-1$, by Lemma 2 all the remaining numerators α_n satisfy $|\alpha_n| \le 2p-1$. Consider two possibility depending on $\beta_{n+1} = (b_{\nu_n} + p^{\nu_n})\beta_n$. Possibility 1 : there exists a positive integer N such that $\beta_{N+1} \ne (b_{\nu_N} + p^{\nu_N})\beta_N$, then $2\beta_{N+1} \le (b_{\nu_N} + p^{\nu_N})\beta_N$, since $\beta_{N+1} \mid (b_{\nu_N} + p^{\nu_N})\beta_N$. By (6) and Lemma 1 (II,1), we have

$$|\alpha_{N+1}| \le \frac{b_{\nu_N}|\alpha_N| + \beta_N}{2p^{\nu_{N+1}-\nu_N}} < \frac{|\alpha_N|}{2} + \frac{\beta_N}{2p^{\nu_N+1}} \le \frac{p+1}{2p}|\alpha_N| \le p.$$

By Lemma 1 (II,3), $|\alpha_n| < p$ for all $n \ge N+1$. Lemma 1 (II,1) then implies that $\beta_n < p^{\nu_n+1}$ for all $n \ge N+1$, and together with (9), we deduce that $|\alpha_{n+1}| \le |\alpha_n|\quad (n \ge N+1)$. Thus for all sufficiently large n, we must have $|\alpha_{n+1}| = |\alpha_n|$, and Lemma 1(IV) shows that $\alpha_n = \alpha_{n+1} = -1$. Then by Lemma 1(III) there must exist such an n satisfying also that $\beta_n < p^{\nu_n+1}$. From such an n onwards, the tail of A is as specified in Lemma 3.

Possibility 2 : $\beta_{n+1} = (b_{\nu_n} + p^{\nu_n})\beta_n$ for all $n \ge 1$. By (6), we have

$$p^{\nu_{n+1}-\nu_n}|\alpha_{n+1}| = b_{\nu_n}|\alpha_n| + \beta_n.$$

Since $0 > \epsilon_n \to 0$ as $n \to \infty$, there exists $N_\delta\ (\ge N_1) \in \mathbb{N}$, such that

$$-\epsilon_n < \frac{p}{2p-1} - \delta \quad (n \ge N_\delta).$$

For $n \geq N_\delta$, since $b_{\nu_n} \geq (p-1-\delta)p^{\nu_n}$, writing $b_{\nu_n} = p^{\nu_n+1} - x_n$, we have $1 \leq x_n \leq (1+\delta)p^{\nu_n}$. Then

$$p^{\nu_{n+1}-\nu_n} |\alpha_{n+1}| = p^{\nu_n+1} |\alpha_n| + (\beta_n - x_n |\alpha_n|),$$

which shows that $p^{\nu_n+1} \mid (\beta_n - x_n |\alpha_n|)$. From

$$\beta_n - x_n |\alpha_n| = \{(1+\epsilon_n)p^{\nu_n} - x_n\} |\alpha_n|,$$

we have

$$-1 < -\frac{\delta - \epsilon_n}{p}(2p-1) \leq \frac{\beta_n - x_n |\alpha_n|}{p^{\nu_n+1}} \leq \frac{(1+\epsilon_n)p^{\nu_n} - 1}{p^{\nu_n+1}}(2p-1) < 2.$$

This forces $\frac{\beta_n - x_n|\alpha_n|}{p^{\nu_n+1}} = 0$ or 1.
If $\frac{\beta_n - x_n|\alpha_n|}{p^{\nu_n+1}} = 1$, then

$$\left\{(1+\epsilon_n)p^{\nu_n} + b_{\nu_n} - p^{\nu_n+1}\right\} |\alpha_n| = \beta_n - x_n |\alpha_n| = p^{\nu_n+1},$$

which implies that $p \mid b_{\nu_n}$, a contradiction. Thus $\frac{\beta_n - x_n|\alpha_n|}{p^{\nu_n+1}} = 0$, then $p^{\nu_{n+1}-\nu_n} |\alpha_{n+1}| = p^{\nu_n+1} |\alpha_n|$ and so $|\alpha_{n+1}| = |\alpha_n|$ for all n ($\geq N_\delta$) sufficiently large. The result follows by the same arguments as in the last possibility.

In Case G, assume that there exists $N_2 \in \mathbb{N}$, such that $|\alpha_{N_2}| \leq 2p-1$, and $\epsilon_n < \frac{1}{2p-1}$, $b_{\nu_n} \geq (p-1-\Delta)p^{\nu_n}$ $(n \geq N_2)$. By Lemma 2, we have $|\alpha_n| \leq 2p-1$ $(n \geq N_2)$. Consider two possibility depending on $\beta_{n+1} = (b_{\nu_n} + p^{\nu_n})\beta_n$. Possibility 1 : If there exists a positive integer $N'(\geq N_2)$ such that $\beta_{N'+1} \neq \left(b_{\nu_{N'}} + p^{\nu_{N'}}\right)\beta_{N'}$, then $2\beta_{N'+1} \leq \left(b_{\nu_{N'}} + p^{\nu_{N'}}\right)\beta_{N'}$, since $\beta_{N'+1} \mid \left(b_{\nu_{N'}} + p^{\nu_{N'}}\right)\beta_{N'}$. By (6), we have

$$|\alpha_{N'+1}| \leq \frac{b_{\nu_{N'}} |\alpha_{N'}| + \beta_{N'}}{2p^{\nu_{N'+1}-\nu_{N'}}} < \left(\frac{1}{2} + \frac{1+\epsilon_{N'}}{2p}\right) |\alpha_{N'}| \leq p + \frac{1}{2} + \frac{1}{2p^{\nu_n+1}} < p+1.$$

Since $p \nmid \alpha_{N'+1}$, we have $|\alpha_{N'+1}| \leq p-1$.

By Lemma 1 (I,3), all the remaining numerators α_n satisfy $|\alpha_n| \leq p-1$ for all $n \geq N'+1$. For $n \geq N'+1$, using (12), and $\epsilon_n < \frac{1}{2p-1} < \frac{1}{p-1}$, we have $\frac{\beta_n}{p^{\nu_n+1}} \leq 1$ and so (9) yields $|\alpha_{n+1}| \leq |\alpha_n|$. Thus for all sufficiently large n, we must have $|\alpha_{n+1}| = |\alpha_n|$. The result follows by the same arguments as in Cases L.

Possibility 2 : $\beta_{n+1} = (b_{\nu_n} + p^{\nu_n})\beta_n$ for all $n \geq 1$. By (6), we have

$$p^{\nu_{n+1}-\nu_n} |\alpha_{n+1}| = b_{\nu_n} |\alpha_n| + \beta_n.$$

For $n \geq N_2$, since $b_{\nu_n} \geq (p-1-\Delta)p^{\nu_n}$, writing $b_{\nu_n} = p^{\nu_n+1} - x_n$, we have $1 \leq x_n \leq (1+\Delta)p^{\nu_n}$. Then

$$p^{\nu_{n+1}-\nu_n} |\alpha_{n+1}| = p^{\nu_n+1} |\alpha_n| + (\beta_n - x_n |\alpha_n|),$$

which shows that $p^{\nu_n+1} \mid (\beta_n - x_n |\alpha_n|)$. From

$$\beta_n - x_n |\alpha_n| = \{(1+\epsilon_n)p^{\nu_n} - x_n\} |\alpha_n|,$$

we have

$$-1 < -\frac{\Delta - \epsilon_n}{p}(2p-1) \le \frac{\beta_n - x_n |\alpha_n|}{p^{\nu_n+1}} \le \frac{(1+\epsilon_n)p^{\nu_n} - 1}{p^{\nu_n+1}}(2p-1) < 2,$$

since $\epsilon_n > \Delta - \frac{p}{2p-1}$. It follows that $\frac{\beta_n - x_n|\alpha_n|}{p^{\nu_n+1}} = 0$ or 1.
If $\frac{\beta_n - x_n|\alpha_n|}{p^{\nu_n+1}} = 1$, then

$$\left\{(1+\epsilon_n)p^{\nu_n} + b_{\nu_n} - p^{\nu_n+1}\right\} |\alpha_n| = \beta_n - x_n |\alpha_n| = p^{\nu_n+1},$$

which implies that $p \mid b_{\nu_n}$, a contradiction. Thus $\frac{\beta_n - x_n|\alpha_n|}{p^{\nu_n+1}} = 0$, and so $p^{\nu_{n+1}-\nu_n} |\alpha_{n+1}| = p^{\nu_n+1} |\alpha_n|$. It follows that $|\alpha_{n+1}| = |\alpha_n|$ for all n $(\ge N_\Delta)$ sufficiently large. The result follows by the same arguments as in the last possibility. □

3. Further Observation

Since the main result depend on the range of $|\alpha|$, this such condition do not hold further consideration are needed. In this section we describe some further observation.

Prop 3.1. (i) Let $\epsilon \in (0,1)$.
Case G: If there exists $N_\epsilon \in \mathbb{N}$ such that $b_{\nu_n} \le (p-1-\epsilon)p^{\nu_n}$ and $\epsilon_n < \epsilon$ $(n \ge N_\epsilon)$, then $|\alpha_{n+1}| < |\alpha_n|$ for all $n \ge N_\epsilon$.
Case L: For all $n \in \mathbb{N}$, if $b_{\nu_n} \le (p-1)p^{\nu_n}$, then $|\alpha_{n+1}| < |\alpha_n|$.

(ii) Let $l \in \mathbb{N},\ 0 < \Delta < \frac{p}{p^l-1},\ 0 < \delta < \frac{p}{p^l-1}$.
Case G: If there exists $N_\Delta \in \mathbb{N}$, depending only on Δ but not on δ, such that

$$0 < \epsilon_n < \frac{p}{p^l-1} - \Delta, \beta_{n+1} = \left(b_{\nu_n} + p^{\nu_n}\right)\beta_n,$$
$$(p-1+\Delta)\, p^{\nu_n} \ge b_{\nu_n} \ge (p-1-\delta)p^{\nu_n} \quad (n \ge N_\Delta)$$

then either $\alpha_n = -1$ or $|\alpha_n| \ge p^l + 1$ $\ (n \ge N_\Delta)$.
Case L: If there exists $N_\delta \in \mathbb{N}$ depending on δ but not on Δ, such that

$$\delta - \frac{p}{p^l-1} < \epsilon_n < 0, \beta_{n+1} = \left(b_{\nu_n} + p^{\nu_n}\right)\beta_n,$$
$$(p-1+\Delta)\, p^{\nu_n} \ge b_{\nu_n} \ge (p-1-\delta)p^{\nu_n} \quad (n \ge N_\delta),$$

then either $\alpha_n = -1$ or $|\alpha_n| \ge p^l + 1$ $\ (n \ge N_\delta)$.

Proof. (i) For Case G, by (7), we have

$$|\alpha_{n+1}| \leq \frac{b_{\nu_n}|\alpha_n| + \beta_n}{p^{\nu_{n+1}-\nu_n}} \leq \frac{\{(p-1-\epsilon)p^{\nu_n} + p^{\nu_n}(1+\epsilon_n)\}|\alpha_n|}{p^{\nu_n+1}}$$
$$= \left(1 - \frac{\epsilon - \epsilon_n}{p}\right)|\alpha_n|.$$

In Case G, we have the result by the existence of N_ϵ. For Case L, this is immediate from the previous discussion by taking ϵ to be 0 and noting that in this case $\epsilon_n < 0$.

(ii) For Case G, since $(p-1+\Delta)p^{\nu_n} \geq b_{\nu_n} \geq (p-1-\delta)p^{\nu_n}$, writing $b_{\nu_n} = p^{\nu_n+1} - x_n$, we have $x_n \in \mathbb{N}$, $(1-\Delta)p^{\nu_n} \leq x_n \leq (1+\delta)p^{\nu_n}$. Since $\beta_{n+1} = (b_{\nu_n} + p^{\nu_n})\beta_n$, by (6)

$$p^{\nu_{n+1}-\nu_n}|\alpha_{n+1}| = b_{\nu_n}|\alpha_n| + \beta_n = p^{\nu_n+1}|\alpha_n| + (\beta_n - x_n|\alpha_n|),$$

which shows that $p^{\nu_n+1} \mid (\beta_n - x_n|\alpha_n|)$. If $|\alpha_n| \leq p^l$, then $1 \leq |\alpha_n| \leq p^l - 1$. Let $N_\Delta \in \mathbb{N}$ be such that $0 < \epsilon_n < \frac{p}{p^l-1} - \Delta$ for all $n \geq N_\Delta$. From

$$\beta_n - x_n|\alpha_n| = \{(1+\epsilon_n)p^{\nu_n} - x_n\}|\alpha_n|,$$

we have, for all $n \geq N_\Delta$,

$$-p^{\nu_n+1} < -(\delta - \epsilon_n)p^{\nu_n}|\alpha_n| \leq \beta_n - x_n|\alpha_n| \leq (\Delta + \epsilon_n)p^{\nu_n}|\alpha_n| < p^{\nu_n+1}.$$

This forces $\beta_n - x_n|\alpha_n| = 0$, implying that $\alpha_n = -1$. For Case L, this is immediate from the previous discussion by choosing $N_\delta \in \mathbb{N}$ so that $\delta - \frac{p}{p^l-1} < \epsilon_n < 0$ for all $n \geq N_\delta$. □

Prop 3.2. Let $p > \varepsilon > 0$, $\delta > 1$, and let L_n be the number of indices $i \in \{1, 2, \ldots, n\}$ for which $b_{\nu_i} > (p-1-\varepsilon)p^{\nu_i}$. For n sufficiently large, we have

$$L_n > \frac{n \log\left(1 - \frac{\varepsilon}{p}\right)^{-1} + \log\left(\frac{p^{\nu_1}}{\delta\beta_1}\right)}{\log\left(1 + \frac{1}{p}\right) - \log\left(1 - \frac{\varepsilon}{p}\right)}.$$

Proof. By (10) and (8), we have

$$|A_{n+1}| = \frac{|\alpha_{n+1}|\,p^{\nu_{n+1}}}{\beta_{n+1}} \geq \frac{|\alpha_{n+1}|\,p^{(\nu_n+1)+\cdots+(\nu_1+1)+\nu_1}}{(b_{\nu_n} + p^{\nu_n})\cdots(b_{\nu_1} + p^{\nu_1})\beta_1}$$
$$> \frac{|\alpha_{n+1}|\,p^{\nu_1}}{\beta_1\left(1 + \frac{1}{p}\right)^{L_n}\left(1 - \frac{\varepsilon}{p}\right)^{n-L_n}}.$$

For n sufficiently large, if $L_n \leq \frac{n\log\left(1-\frac{\varepsilon}{p}\right)^{-1}+\log\left(\frac{p^{\nu_1}}{\delta\beta_1}\right)}{\log\left(1+\frac{1}{p}\right)-\log\left(1-\frac{\varepsilon}{p}\right)}$, then

$$\frac{p^{\nu_1}}{\beta_1\left(1+\frac{1}{p}\right)^{L_n}\left(1-\frac{\varepsilon}{p}\right)^{n-L_n}} \geq \delta,$$

and so $|A_{n+1}| > \delta\,|\alpha_{n+1}| \geq \delta$, contradicting Lemma 1 (III). □

Acknowledgments

This research is supported by the Thailand Research Funds RTA 5180005 and the Centre of Excellence in Mathematics, CHE, Thailand.

References

1. P. J. Grabner and A. Knopfmacher, *Arithmetic and metric properties of p-adic Engel series expansions*, Publ. Math. Debrecen 63(2003), 363-377.
2. A. Knopfmacher and J. Knopfmacher, *Series expansions in p-adic and other non-archimedean fields*, J. Number Theory 32(1989), 297-306.
3. A. Knopfmacher and J. Knopfmacher, *Infinite series expansions for p-adic numbers*, J. Number Theory 41(1992), 131-145.
4. V. Laohakosol and N. Rompurk, *A characterization of rational elements by Lüroth type series expansions in the p-adic number field and in the field of Laurent series over a finite field*, Acta Arith. 122.2(2006), 195-205.
5. V. Laohakosol and N. Rompurk Kanasri, *A characterization of rational numbers by p-adic Sylvester series expansions*, Acta Arith. 130.4(2007), 389-402.
6. I. Niven, *Irrational Numbers*, The Carus Math. Monographs No. 11, Mathematical Assoc. of America, 1967.
7. O. Perron, *Irrationalzahlen*, Chelsea, New York, 1951.

On k-Regular Ternary Semirings

S. Kar* **and K. Das**†‡

Department Of Mathematics, Jadavpur University.

188, Raja S. C. Mallick Road, Kolkata — 700032; India.

Abstract

In this paper we characterize k-regular ternary semirings and study some properties of k-regular ternary semirings. We also show that for a ternary semigroup S, the power ternary semiring P(S) is a k-regular ternary semiring if and only if S is a regular ternary semigroup.

AMS Mathematics Subject Classification (2000) : $16Y30$.

Keywords : Ternary Semiring, Power Ternary Semiring, k-regular Ternary Semiring, Ternary Semigroup, Regular Ternary Semigroup.

1 Introduction

The notion of ternary semirings was first introduced by T. K. Dutta and S. Kar [3] in 2003 as a generalization of ternary ring introduced by W. G. Lister [16] in 1971. The notion of regularity in ring was studied by J. von Neumaan [17] in 1936. The notion of regular ternary ring was introduced by T. Vasile [18] in 1987. Subsequently, the notion of regular semiring was also introduced and studied as a generalization of regular ring. S. Bourne [1] generalize the notion of regularity in semirings. If a semiring S happens to be a ring then the von

*karsukhendu@yahoo.co.in

†koushikdas.maths@gmail.com

‡The author is thankful to UGC, India for financial assistance.

Neumaan regularity and Bourne regularity are equivalent. But this is not true in a semiring in general. To distinguish the notion of Bourne regularity of a semiring from the notion of von Neumaan regularity it was renamed as k-regularity. The notion of k-regular ternary semiring was introduced by T. K. Dutta and S. Kar [3]. In this paper we characterize k-regular ternary semirings and obtain some important results regarding k-regularity in ternary semirings.

2 Preliminaries

Definition 2.1 [3] A non-empty set S together with a binary operation, called addition and a ternary multiplication, denoted by juxtaposition, is said to be a ternary semiring if S is an additive commutative semigroup satisfying the following conditions :

(i) $(abc)de = a(bcd)e = ab(cde)$,

(ii) $(a+b)cd = acd + bcd$,

(iii) $a(b+c)d = abd + acd$,

(iv) $ab(c+d) = abc + abd$ for all $a, b, c, d, e \in S$.

Definition 2.2 Let S be a ternary semiring. If there exists an element $0 \in S$ such that $0 + x = x$ and $0xy = x0y = xy0 = 0$ for all $x, y \in S$ then '0' is called the zero element or simply the zero of the ternary semiring S. In this case we say that S is a ternary semiring with zero.

Example 2.3 Let $\mathbb{Z}_0^-$ be the set of all negative integers with zero. Then with the usual binary addition and ternary multiplication, $\mathbb{Z}_0^-$ forms a ternary semiring with zero.

By a ternary semiring S we shall always mean a ternary semiring S with zero, unless otherwise stated.

Definition 2.4 A ternary semiring S is said to be commutative if $x_1x_2x_3 = x_{\sigma(1)}x_{\sigma(2)}x_{\sigma(3)}$ for every permutation σ of $\{1, 2, 3\}$ and $x_1, x_2, x_3 \in S$.

Let A, B, C be three subsets of a ternary semiring S. Then by ABC, we mean the set of all finite sums of the form $\sum_{i=1}^{n} a_ib_ic_i$, where $a_i \in A$, $b_i \in B$, $c_i \in C$ and $n \in \mathbb{N}$ (set of all natural numbers).

Definition 2.5 An additive subsemigroup I of a ternary semiring S is called

(i) a left ideal of S if $SSI \subseteq I$,

(ii) a lateral ideal of S if $SIS \subseteq I$,

(iii) a right ideal of S if $ISS \subseteq I$,

(iv) a two-sided ideal of S if I is both a left and a right ideal of S,

(v) an ideal of S if I is a left, a right and a lateral ideal of S.

Definition 2.6 An ideal I of a ternary semiring S is called a proper ideal of S if $I \neq S$.

Proposition 2.7 Let S be a ternary semiring and $a \in S$. Then the principal

(i) left ideal generated by 'a' is given by $< a >_l = SSa + na$,

(ii) right ideal generated by 'a' is given by $< a >_r = aSS + na$,

(iii) two-sided ideal generated by 'a' is given by $< a >_t = SSa + aSS + SSaSS + na$,

(iv) lateral ideal generated by 'a' is given by $< a >_m = SaS + SSaSS + na$,

(v) ideal generated by 'a' is given by $< a > = SSa + aSS + SaS + SSaSS + na$,

where $n \in \mathbb{Z}_0^+$ (set of all positive integers with zero).

Definition 2.8 A non-empty subset A of a ternary semiring S is called a k-subset of S if $x + y \in A$; $x \in S$, $y \in A$ imply that $x \in A$.

Definition 2.9 A (left, right, lateral) ideal I of a ternary semiring S is called a (left, right, lateral) k-ideal if $x + y \in I$; $x \in S$, $y \in I$ imply that $x \in I$.

Definition 2.10 Let A be a non-empty subset of a ternary semiring S. Then the k-closure of A, denoted by $\overline{A}$, is defined by $\overline{A} = \{a \in S : a + b = c \text{ for some } b, c \in A\}$.

We note that for non-empty subsets A, B of a ternary semiring S, $A \subseteq B \Longrightarrow \overline{A} \subseteq \overline{B}$.

A non-empty subset A of a ternary semiring S is a k-subset of S if and only if $A = \overline{A}$.

Proposition 2.11 *Let A be a (left, right, lateral) ideal of a ternary semiring S. Then the k-closure of A, denoted by $\overline{A}$, is defined by $\overline{A} = \{a \in S : a + b = c$ for some $b, c \in A\}$.*

We can easily show that (i) $A \subseteq \overline{A}$, (ii) $A \subseteq B \Longrightarrow \overline{A} \subseteq \overline{B}$, (iii) $\overline{\overline{A}} = \overline{A}$.

We note that a (left, right, lateral) ideal A of a ternary semiring S is a k-ideal of S if and only if $A = \overline{A}$.

Remark 2.12 *Let A be a (left, right, lateral) ideal of a ternary semiring S. Then the k-closure of A i.e. $\overline{A}$ is a (left, right, lateral) k-ideal of S. Also note that if A, B, C are three k-ideals of S, then $A \cap B \cap C$ is also a k-ideal of S.*

Proposition 2.13 Let S be a ternary semiring and $a \in S$. Then the principal

(i) left k-ideal generated by 'a' is given by $\overline{<a>_l} = \overline{SSa + na}$,

(ii) right k-ideal generated by 'a' is given by $\overline{<a>_r} = \overline{aSS + na}$,

(iii) lateral k-ideal generated by 'a' is given by $\overline{<a>_m} = \overline{SaS + SSaSS + na}$,

(iv) k-ideal generated by 'a' is given by $\overline{<a>} = \overline{SSa + aSS + SaS + SSaSS + na}$,

where $n \in \mathbb{Z}_0^+$ (set of all positive integers with zero).

Definition 2.14 Let I be a proper ideal of a ternary semiring S. Then the congruence on S, denoted by ρ_I and defined by $s\rho_I s'$ if and only if $s + a = s' + a'$ for some $a, a' \in I$, is called the Bourne congruence on S defined by the ideal I. We denote the Bourne congruence class of an element r of S by r/I and denote the set of all such congruence classes of S by S/I. If the Bourne congruence ρ_I defined by the ideal I is proper i.e. $0/I \neq S/I$, then we define addition and ternary multiplication on S by $a/I + b/I = (a + b)/I$ and $(a/I)(b/I)(c/I) = (abc)/I$ for all $a, b, c \in S$.

With these two operations, S/I forms a ternary semiring and is called the Bourne factor ternary semiring or simply the factor ternary semiring.

Definition 2.15 Let S and T be two ternary semirings and $f : S \to T$ be a mapping. Then the mapping $f : S \to T$ is called a homomorphism of S into T if

(i) $f(a + b) = f(a) + f(b)$ and

(ii) $f(abc) = f(a)f(b)f(c)$ for all $a, b, c \in S$.

Definition 2.16 A ternary semiring S is said to be additively idempotent if $a + a = a$ for all $a \in S$.

Definition 2.17 An element a of a ternary semiring S is called an idempotent element of S if $a^3 = a$. A ternary semiring S is called an idempotent ternary semiring if every element of S is idempotent.

Definition 2.18 A non-empty set S together with a ternary multiplication, denoted by juxtaposition, is said to be a ternary semigroup if S satisfies ternary associative law — $(abc)de = a(bcd)e = ab(cde)$ for all $a, b, c, d, e \in S$.

Definition 2.19 An element a in a ternary semiring (semigroup) S is called regular if there exists $x \in S$ such that $axa = a$. A ternary semiring (semigroup) S is called regular if all of its elements are regular.

Definition 2.20 [6] A proper ideal Q of a ternary semiring S is called a semiprime ideal of S if $I^3 \subseteq Q$ implies that $I \subseteq Q$ for any ideal I of S.

Definition 2.21 [6] A proper ideal Q of a ternary semiring S is called a completely semiprime ideal of S if $a^3 \in Q$ implies that $a \in Q$ for any $a \in S$.

3 k-regular Ternary Semirings

Definition 3.1 Let S be a ternary semiring and $a \in S$. Then a is called k-regular if there exist $x, y \in S$ such that $a + aya = axa$.

If all the elements of S are k-regular then S is called a k-regular ternary semiring.

Note 3.2 *If a ternary semiring S is additively idempotent, then S is k-regular if and only if for every $a \in S$ there exists $x \in S$ such that $a + axa = axa$.*

Since S is k-regular, for any $a \in S$ there exist $x, y \in S$ such that $a + aya = axa$.

Now by adding both sides by $(axa + aya)$ we get,

$a+aya+(axa+aya) = axa+(axa+aya) \Longrightarrow a+(aya+aya)+axa = (axa+axa)+aya \Longrightarrow$ $a+aya+axa = axa+aya$; *(since S is additively idempotent)* $\Longrightarrow a+a(x+y)a = a(x+y)a \Longrightarrow$ $a + aza = aza$, *where* $z = x + y$.

The converse is obvious.

Example 3.3 Let $S = \mathbb{Z}_0^-$, the set of all negative integers with zero. Then with the binary addition defined by $a + b = min\{a, b\}$ and usual ternary multiplication S forms a k-regular ternary semiring.

Example 3.4 Let $S' = \{r\sqrt{2} : r \in \mathbb{Q}$ (set of all rational numbers)$\}$. Then with the usual binary addition and ternary multiplication S' forms a k-regular ternary semiring.

Remark 3.5 *Each regular ternary semiring is also a k-regular ternary semiring but the converse is not true. Let S be a regular ternary semiring. Then for any $a \in S$ there exists $x \in S$ such that $a = axa$. Now $a = axa$ implies that $a + aya = axa + aya$ for all $y \in S$. Then $a + aya = a(x+y)a$ i.e. $a + aya = aza$, where $z = (x+y)$. So S is k-regular.*

But the converse is not necessarily true in general. For example, we consider the ternary semiring $S = \mathbb{Z}_0^-$, where $\mathbb{Z}_0^-$ is the set of all negative integers with zero with the binary addition, $a + b = min\{a, b\}$ and usual ternary multiplication. Now we have $a + a(-1)a = a(-1)a$ for all $a \in S$ which implies that S is a k-regular ternary semiring. But we note that for every $a\ (\neq 0, -1) \in \mathbb{Z}_0^-$ there exists no $x \in \mathbb{Z}_0^-$ such that $axa = a$. This implies that $S = \mathbb{Z}_0^-$ is not a regular ternary semiring.

We note that every left and right ideal of a k-regular ternary semiring may not be a k-regular ternary semiring; however for a lateral ideal of a k-regular ternary semiring, we have the following result :

Proposition 3.6 *Let S be a k-regular ternary semiring and I be a lateral ideal of S. Then I is a k-regular ternary semiring, considering I as a ternary semiring.*

Proof : Let S be a k-regular ternary semiring and I be a lateral ideal of S. Let $a \in I$. Then there exist $x, y \in S$ such that $a + aya = axa$. This implies that

$$a + ay(a + aya) + ax(aya) = ax(a + aya) + ay(aya)$$
$$\Longrightarrow a + ay(axa) + ax(aya) = ax(axa) + ay(aya)$$
$$\Longrightarrow a + a(yax + xay)a = a(xax + yay)a.$$

Since I is a lateral ideal of S and $a \in I$, we find that $yax, xay, xax, yay \in I$. This implies that $(yax + xay) \in I$ and $(xax + yay) \in I$. Consequently a is k-regular in I and hence I is a k-regular ternary semiring.

Note 3.7 Every ideal of a k-regular ternary semiring is a k-regular ternary semiring.

Theorem 3.8 *Let f be a homomorphism from a k-regular ternary semiring S onto a ternary semiring T. Then T is a k-regular ternary semiring i.e. homomorphic image of a k-regular ternary semiring is also a k-regular ternary semiring.*

Proof : Let S be a k-regular ternary semiring and T be a ternary semiring such that there is an onto homomorphism $f : S \longrightarrow T$. Let $a' \in T$. Then $a' = f(a)$ for some $a \in S$. Since S is k-regular there exist $x, y \in S$ such that $a + aya = axa$. Since f is a homomorphism we have, $f(a + aya) = f(axa) \Longrightarrow f(a) + f(aya) = f(axa) \Longrightarrow f(a) + f(a)f(y)f(a) = f(a)f(x)f(a)$. Thus it follows that $a' + a'y'a' = a'x'a'$ (where $x' = f(x) \in T$ and $y' = f(y) \in T$). So T is a k-regular ternary semiring.

Corollary 3.9 *Let S be a k-regular ternary semiring and I be a k-ideal of S. Then the factor ternary semiring S/I is also a k-regular ternary semiring.*

Proof : Since $f : S \longrightarrow S/I$ defined by $f(a) = a/I$ for all $a \in S$ is a homomorphism of S onto S/I, we find that S/I is a homomorphic image of S. Again, since S is k-regular, by Theorem 3.8, it follows that S/I is also a k-regular ternary semiring.

Lemma 3.10 *Let S be a ternary semiring and $a \in S$. Then $\overline{\overline{<a>_r}\ \overline{<a>_m}\ \overline{<a>_l}} \subseteq \overline{aSa}$.*

Proof : First we prove that $\overline{\overline{<a>_r}\ S\ \overline{<a>_l}} \subseteq \overline{aSa}$.

Let $x \in \overline{\overline{<a>_r}\ S\ \overline{<a>_l}}$. Then $x + \sum\limits_{finite} p_i q_i r_i = \sum\limits_{finite} x_i y_i z_i$, where $p_i, x_i \in \overline{<a>_r} = \overline{(aSS + na)}$, $q_i, y_i \in S$ and $r_i, z_i \in \overline{<a>_l} = \overline{(SSa + na)}$ and $n \in \mathbb{Z}_0^+$.

This implies that

$$p_i + \sum_{finite} (as_i t_i) + n_1 a = \sum_{finite} (as_i' t_i') + n_2 a,$$
$$x_i + \sum_{finite} (ad_i e_i) + n_3 a = \sum_{finite} (ad_i' e_i') + n_4 a,$$
$$r_i + \sum_{finite} (b_i c_i a) + n_5 a = \sum_{finite} (b_i' c_i' a) + n_6 a$$

and $z_i + \sum\limits_{finite} (u_i v_i a) + n_7 a = \sum\limits_{finite} (u_i' v_i' a) + n_8 a$,

where $s_i, t_i, s_i', t_i', d_i, e_i, d_i', e_i', b_i, c_i, b_i', c_i', u_i, v_i, u_i'$ *and* $v_i' \in S$ and $n_1, n_2, ..., n_8 \in \mathbb{Z}_0^+$.

Now we find that

$$x + \sum_{finite}((p_i + \sum_{finite}(as_it_i) + n_1a)q_ir_i) + \sum_{finite}((\sum_{finite}(ad_ie_i) + n_3a)y_iz_i)$$
$$= \sum_{finite}((x_i + \sum_{finite}(ad_ie_i) + n_3a)y_iz_i) + \sum_{finite}((\sum_{finite}(as_it_i) + n_1a)q_ir_i)$$

$$\Longrightarrow x + \sum_{finite}((\sum_{finite}(as'_it'_i) + n_2a)q_ir_i) + \sum_{finite}((\sum_{finite}(ad_ie_i) + n_3a)y_iz_i)$$
$$= \sum_{finite}((\sum_{finite}(ad'_ie'_i) + n_4a)y_iz_i) + \sum_{finite}((\sum_{finite}(as_it_i) + n_1a)q_ir_i)$$

$$\Longrightarrow x + \sum_{finite}((\sum_{finite}(as'_it'_i) + n_2a)q_i(r_i + \sum_{finite}(b_ic_ia) + n_5a))$$
$$+ \sum_{finite}((\sum_{finite}(ad_ie_i) + n_3a)y_i(z_i + \sum_{finite}(u_iv_ia) + n_7a))$$
$$+ \sum_{finite}((\sum_{finite}(ad'_ie'_i) + n_4a)y_i(\sum_{finite}(u_iv_ia) + n_7a)) + \sum_{finite}((\sum_{finite}(as_it_i) + n_1a)q_i(\sum_{finite}(b_ic_ia) + n_5a))$$
$$= \sum_{finite}((\sum_{finite}(ad'_ie'_i) + n_4a)y_i(z_i + \sum_{finite}(u_iv_ia) + n_7a))$$
$$+ \sum_{finite}((\sum_{finite}(as_it_i) + n_1a)q_i(r_i + \sum_{finite}(b_ic_ia) + n_5a))$$
$$+ \sum_{finite}((\sum_{finite}(as'_it'_i) + n_2a)q_i(\sum_{finite}(b_ic_ia) + n_5a)) + \sum_{finite}((\sum_{finite}(ad_ie_i) + n_3a)y_i(\sum_{finite}(u_iv_ia) + n_7a))$$

$$\Longrightarrow x + \sum_{finite}((\sum_{finite}(as'_it'_i) + n_2a)q_i(\sum_{finite}(b'_ic'_ia) + n_6a))$$
$$+ \sum_{finite}((\sum_{finite}(ad_ie_i) + n_3a)y_i(\sum_{finite}(u'_iv'_ia) + n_8a)) + \sum_{finite}((\sum_{finite}(ad'_ie'_i) + n_4a)y_i(\sum_{finite}(u_iv_ia) + n_7a))$$
$$+ \sum_{finite}((\sum_{finite}(as_it_i) + n_1a)q_i(\sum_{finite}(b_ic_ia) + n_5a))$$
$$= \sum_{finite}((\sum_{finite}(ad'_ie'_i) + n_4a)y_i(\sum_{finite}(u'_iv'_ia) + n_8a))$$
$$+ \sum_{finite}((\sum_{finite}(as_it_i) + n_1a)q_i(\sum_{finite}(b'_ic'_ia) + n_6a))$$
$$+ \sum_{finite}((\sum_{finite}(as'_it'_i) + n_2a)q_i(\sum_{finite}(b_ic_ia) + n_5a)) + \sum_{finite}((\sum_{finite}(ad_ie_i) + n_3a)y_i(\sum_{finite}(u_iv_ia) + n_7a)).$$

This implies that $x \in \overline{aSa}$.

Thus it follows that $\overline{\overline{<a>_r S <a>_l}} \subseteq \overline{aSa}$. Again $\overline{\overline{<a>_r <a>_m <a>_l}} \subseteq \overline{\overline{<a>_r S <a>_l}}$.

So we find that $\overline{\overline{<a>_r <a>_m <a>_l}} \subseteq \overline{aSa}$.

Now we have the following characterization theorem for k-regular ternary semiring :

Theorem 3.11 *The following conditions in a ternary semiring S are equivalent :*

(i) S is k-regular.

(ii) For any right k-ideal R, lateral k-ideal M and left k-ideal L of S, $R \cap M \cap L = \overline{RML}$.

(iii) For $a, b, c \in S$; $\overline{<a>_r} \cap \overline{<b>_m} \cap \overline{<c>_l} = \overline{\overline{<a>_r}\ \overline{<b>_m}\ \overline{<c>_l}}$.

(iv) For $a \in S$; $\overline{<a>_r} \cap \overline{<a>_m} \cap \overline{<a>_l} = \overline{\overline{<a>_r}\ \overline{<a>_m}\ \overline{<a>_l}}$.

Proof : $(i) \Longrightarrow (ii)$.

Suppose S is a k-regular ternary semiring and R be a right k-ideal, M be a lateral k-ideal and L be a left k-ideal L of S. Since R is a right k-ideal of S, we have $RML \subseteq RSS \subseteq R$. Now $RML \subseteq R \Longrightarrow \overline{RML} \subseteq \overline{R} = R$. Similarly, we have $\overline{RML} \subseteq M$ and $\overline{RML} \subseteq L$. Then $\overline{RML} \subseteq R \cap M \cap L$.

Conversely, let $a \in R \cap M \cap L$. Since S is k-regular, there exist $x, y \in S$ such that $a + aya = axa$ which implies that $a + ay(a + aya) + ax(aya) = ax(a + aya) + ay(aya)$ i.e. $a + ay(axa) + ax(aya) = ax(axa) + ay(aya)$. So $a + a(yax + xay)a = a(xax + yay)a$. Now since M is a lateral ideal and $a \in M$, we find that $yax, xay, xax, yay \in M$. So $(yax + xay) \in M$ and $(xax + yay) \in M$ and hence $a(yax + xay)a \in RML$ and $a(xax + yay)a \in RML$. This implies that $a \in \overline{RML}$. So we find that $R \cap M \cap L \subseteq \overline{RML}$. Consequently, $R \cap M \cap L = \overline{RML}$.

$(ii) \Longrightarrow (iii)$ and $(iii) \Longrightarrow (iv)$ are obvious.

Thus it remains to show that $(iv) \Longrightarrow (i)$.

Let $a \in S$. Clearly, $a \in \overline{<a>_r} \cap \overline{<a>_m} \cap \overline{<a>_l} = \overline{\overline{<a>_r}\ \overline{<a>_m}\ \overline{<a>_l}}$. Again we have from Lemma 3.10, $\overline{\overline{<a>_r}\ \overline{<a>_m}\ \overline{<a>_l}} \subseteq \overline{aSa}$. So we find that $a \in \overline{aSa}$. This implies that there exist $x, y \in S$ such that $a + aya = axa$. Thus a is k-regular for all $a \in S$ and hence S is a k-regular ternary semiring.

Corollary 3.12 *The following conditions in a ternary semiring S are equivalent :*

(i) S is k-regular.

(ii) For any right k-ideal A and left k-ideal B of S, $A \cap B = \overline{ASB}$.

(iii) For $a, b \in S$; $\overline{<a>_r} \cap \overline{<b>_l} = \overline{\overline{<a>_r}\ S\ \overline{<b>_l}}$.

(iv) For $a \in S$; $\overline{<a>_r} \cap \overline{<a>_l} = \overline{\overline{<a>_r}\ S\ \overline{<a>_l}}$.

Definition 3.13 An ideal I of a ternary semiring S is called k-idempotent if $\overline{I^3} = I$.

Remark 3.14 *Since for a k-ideal I of a ternary semiring S; $I = \overline{I}$, it follows that a k-ideal I of a ternary semiring S is k-idempotent if $\overline{I^3} = \overline{I}$.*

Also note that an idempotent k-ideal of a ternary semiring is k-idempotent but the converse is not true.

Theorem 3.15 *A commutative ternary semiring S is k-regular if and only if every k-ideal of S is k-idempotent.*

Proof : Let S be a commutative k-regular ternary semiring and I be any k-ideal of S. Then from Theorem 3.11, we find that $I \cap I \cap I = \overline{III}$ i.e. $I = \overline{I^3}$. This implies that I is k-idempotent. Since I is an arbitrary k-ideal of S, it follows that every k-ideal of S is k-idempotent.

Conversely, suppose that every k-ideal of a commutative ternary semiring S is k-idempotent. Let $a \in S$. Then $\overline{<a>}$ is a k-ideal of S. So it is k-idempotent i.e. $\overline{\overline{<a>}\,\overline{<a>}\,\overline{<a>}} = \overline{<a>}$. Now $a \in \overline{<a>} \implies a \in \overline{\overline{<a>}\,\overline{<a>}\,\overline{<a>}}$. Again from Lemma 3.10, we have $\overline{\overline{<a>}\,\overline{<a>}\,\overline{<a>}} \subseteq \overline{aSa}$. Therefore $a \in \overline{aSa}$ and hence there exist $x, y \in S$ such that $a + aya = axa$. Since a is an arbitrary element of S, so S is a k-regular ternary semiring.

Definition 3.16 A proper ideal Q of a ternary semiring S is called a k-semiprime ideal of S if $I^3 \subseteq Q$ implies that $I \subseteq Q$ for any k-ideal I of S.

Note 3.17 Every semiprime ideal of a ternary semiring S is also a k-semiprime ideal of S.

Theorem 3.18 *A commutative ternary semiring S is k-regular if and only if every k-ideal of S is k-semiprime.*

Proof : Let S be a commutative k-regular ternary semiring and Q be any k-ideal of S such that $A^3 \subseteq Q$ for any k-ideal A of S. Now since S is k-regular, from Theorem 3.11, we have $\overline{A^3} = A$. Now $A^3 \subseteq Q$ implies that $\overline{A^3} \subseteq \overline{Q} = Q$. So we find that $A \subseteq Q$. Hence Q is a k-semiprime ideal of S.

Conversely, suppose that every k-ideal of a commutative ternary semiring S is k-semiprime. Let $a \in S$. Since S is commutative, $\overline{aSa}$ is a k-ideal of S. So $\overline{aSa}$ is k-semiprime. Now we have from Lemma 3.10, $\overline{<a>}\,\overline{<a>}\,\overline{<a>} \subseteq \overline{\overline{<a>}\,\overline{<a>}\,\overline{<a>}} \subseteq \overline{aSa}$. Since $\overline{aSa}$ is k-semiprime and $\overline{<a>}\,\overline{<a>}\,\overline{<a>} \subseteq \overline{aSa}$, so $\overline{<a>} \subseteq \overline{aSa}$. Now $a \in \overline{<a>} \Longrightarrow a \in \overline{aSa}$. Therefore there exist $x, y \in S$ such that $a + aya = axa$. Since $a \in S$ is arbitrary, so S is a k-regular ternary semiring.

Theorem 3.19 *A commutative ternary semiring S is k-regular if and only if every k-ideal of S is complelely semiprime.*

Proof : Let S be a commutative k-regular ternary semiring and I be any k-ideal of S. Let $a^3 \in I$ for some $a \in S$. Since S is k-regular, there exist $x, y \in S$ such that $a+aya = axa$. Again, since S is commutative, we have $aya = yaa$ and $axa = xaa$. Now $a+aya = axa \Longrightarrow a+yaa = xaa \Longrightarrow a + y(a + yaa)a + xy(aaa) = x(a + yaa)a + yy(aaa) \Longrightarrow a + y(xaa)a + xy(aaa) = x(xaa)a + yy(aaa) \Longrightarrow a + yx(aaa) + xy(aaa) = xx(aaa) + yy(aaa)$. Since I is an k-ideal of S and $aaa \in I$, $yx(aaa) + xy(aaa) \in I$ and $xx(aaa) + yy(aaa) \in I$. So we find that $a \in \overline{I} = I$. Hence I is a complelely semiprime ideal of S.

Conversely, suppose that every k-ideal of a commutative ternary semiring S is complelely semiprime and $a \in S$. Since S is commutative, $\overline{aSa}$ is a k-ideal of S. So $\overline{aSa}$ is a complelely semiprime ideal of S. Now $a^3 \in aSa \subseteq \overline{aSa}$. So $a \in \overline{aSa}$. Thus there exist $x, y \in S$ such that $a + aya = axa$. Since $a \in S$ is arbitrary, so S is a k-regular ternary semiring.

The notion of ternary semiring comes out from the notion of ternary semigroup as follows:

Let S be a ternary semigroup and $P(S)$ be the set of all non-empty subsets of S.

We now define binary addition and ternary multiplication on $P(S)$ by :

$U + V = U \cup V$ and $UVW = \{abc \in S : a \in U, b \in V, c \in W\}$ for all $U, V, W \in P(S)$.

It can be easily verified that together with this binary addition and ternary multiplication, $P(S)$ forms a ternary semiring. We call this ternary semiring $P(S)$ the power ternary semiring of S.

Now we have the following result :

Theorem 3.20 *Let S be a ternary semigroup. Then the power ternary semiring $P(S)$ is a k-regular ternary semiring if and only if S is a regular ternary semigroup.*

Proof : Let $P(S)$ be a k-regular ternary semiring and $a \in S$. Then A=$\{$a$\}$ $\in P(S)$. Since $P(S)$ is k-regular, there exist $X, Y \in P(S)$ such that $A + AYA = AXA$. This implies that $A \subseteq AXA$. So there exist $x \in X \subseteq S$ such that $a = axa$. Hence S is a regular ternary semigroup.

Conversely, suppose that S is a regular ternary semigroup and $A \in P(S)$ i.e. $A \subseteq S$. Then for each $a \in A$, there exists $x_a \in S$ such that $a = ax_aa$. Now we can consider the set $X = \{x_a \in S : a \in A\}$. Then $X \subseteq S$ and $A \subseteq AXA$. This implies that $X \in P(S)$ and $A + AXA = AXA$. Hence $P(S)$ is a k-regular ternary semiring.

Remark 3.21 *This is an interesting problem to study the power ternary semiring $P(S)$ with the help of the properties of ternary semigroup S and vice-versa.*

References

[1] Bourne, S. : *The Jacobson Radical of a Semiring*; Proc. Nat. Acad. Sci. USA; 37(1951), 163-170.

[2] Bourne, S. : *On homomorphism theorem for Semirings*; Proc. Nat. Acad. Sci. USA; 38(1952), 118-119.

[3] Dutta, T. K. and Kar, S. : *On Regular Ternary Semirings*; Advances in Algebra, Proceedings of the ICM Satellite Conference in Algebra and Related Topics, World Scientific (2003), 343 - 355.

[4] Dutta, T. K. and Kar, S. : *A Note On Regular Ternary Semirings*; Kyungpook Mathematical Journal, Vo. 46, No. 3 (2006), 357 - 365.

[5] Dutta, T. K. and Kar, S. : *On Prime Ideals And Prime Radical Of Ternary Semirings*; Bull. Cal. Math. Soc., Vol. 97, No. 5 (2005), 445 - 454.

[6] Dutta, T. K. and Kar, S. : *On Semiprime Ideals And Irreducible Ideals Of Ternary Semirings*; Bull. Cal. Math. Soc., Vol. 97, No. 5 (2005), 467 - 476.

[7] Dutta, T. K. and Kar, S. : *On Ternary Semifields*; Discussiones Mathematicae - General Algebra and Applications, Vol. 24, No. 2 (2004), 185 - 198.

[8] Dutta, T. K. and Kar, S. : *On The Jacobson Radical Of A Ternary Semiring*; Southeast Asian Bulletin of Mathematics, Vol. 28, No. 1 (2004), 1 - 13.

[9] Dutta, T. K. and Kar, S. : *A Note On The Jacobson Radical Of A Ternary Semiring*; Southeast Asian Bulletin of Mathematics, Vol. 29, No. 2 (2005), 321 - 331.

[10] Dutta, T. K. and Kar, S. : *Two Types Of Jacobson Radicals Of Ternary Semirings*; Southeast Asian Bulletin of Mathematics, Vol. 29, No. 4 (2005), 677 - 687.

[11] Dutta, T. K. and Kar, S. : *On Matrix Ternary Semirings*; International Journal of Mathematics and Analysis, Vol. 1, No. 1 (2006), 81 - 95.

[12] Kar, S. : *On Quasi-ideals And Bi-ideals Of Ternary Semirings*; International Journal of Mathematics and Mathematical Sciences; Vol. 2005, Issue 18 (2005), 3015 - 3023.

[13] Kar, S. : *On Structure Space Of Ternary Semirings*; Southeast Asian Bulletin Of Mathematics, Vol. 31, No. 3 (2007), 537 - 545.

[14] Kar, S. and Bhunia, S. : *A Characterization Of Ternary Semiring*; International Journal Of Algebra, Number Theory And Applications, Vol. 1, No. 1 (2009), 43 - 51.

[15] Kar, S., Sircar, J. and Mandal, S. : *On Right Strongly Prime Ternary Semirings*; East - West Journal Of Mathematics; Vol. 12, No. 1 (2010); 59 - 68.

[16] Lister, W. G. : *Ternary Rings*; Trans. Amer. Math. Soc. 154 (1971), 37-55.

[17] Neumann, von J. : *On regular rings*; Proc. Nat. Acad. Sci. USA 22 (1936), 707-713.

[18] Vasile, Tamas. : *Regular ternary rings*; An. Stiin. Univ. Al. I. Cuza. Ia si Sec. Ia Mat. 33 (1987), no. 2, 89-92.

Dual Right Serial Algebras

D. Kariman*[1,2], Risnawita[1] and I. M. Maris[1,3]

[1] *Algebra Research Division, Faculty of Mathematics and Natural Sciences, Institut Teknologi Bandung, Indonesia*
**E-mail : delsi.kariman@yahoo.co.id*
[2] *Jurusan Matematika STKIP PGRI Sumatera Barat, Indonesia*
[3] *Program Studi Pendidikan Matematika STAIN Batusangkar, Indonesia*

Algebras and coalgebras formed through their quiver are called path algebras and path coalgebras. The quiver of path coalgebras (algebras) can be obtained from the quiver of path algebras (coalgebras) by reversing its arrows. We apply these facts to show that dual right serial algebras are left serial coalgebras.

Keywords: Path Algebra; Path Coalgebra; Quiver; Right Serial Algebra.

1. Introduction

In the paper of Muchtadi and Garminia [2], it is shown that the quiver of the dual bound path algebra $D(KQ/I)$ can be obtained from the quiver Q by reversing arrows and in Assem[1] it is shown that a right serial algebra A can be associated with a quiver with property: for every point a, there exists at most one arrow of source a. In this paper we show that the quiver of dual basic right serial K-algebra A has this property for every point a, there exists at most the sink of one arrow. In fact, this coalgebra is a right serial coalgebra.

This paper is organized as follows: in section 2, we will give the definition of path algebras and path coalgebras. In section 3, we will explain about the quiver with relations. In section 4, we will give the definition of right serial algebra. Finally in the last section, we will explain the main theorem about the quiver of dual basic right serial K-algebra.

2. Path Algebras and Path Coalgebras

A *quiver* $Q = (Q_0, Q_1, s, t)$ is a quadruple consisting of two sets: Q_0 (whose elements are called points, or vertices) and Q_1 (whose elements are called arrows), and two maps $s, t : Q_1 \to Q_0$ which associate to each arrow $\alpha \in Q_1$

its source $s(\alpha) \in Q_0$ and its target $t(\alpha) \in Q_0$, respectively, see Muchtadi and Garminia.[2] If $a, b \in Q_0$, a *path* of length $l \geq 1$ from i to j is a formal composition of arrows

$$p = \alpha_1 \alpha_2 ... \alpha_l,$$

where $s(\alpha_1) = i$, $t(\alpha_l) = j$ and $t(\alpha_{k-1}) = s(\alpha_k)$, for $k = 2, \cdots, l$. To any vertex $a \in Q_0$ we attach a trivial path of length 0, say e_i, starting and ending at a such that for any arrow α (resp. β) such that $s(\alpha) = i$ (resp. $t(\beta) = i$) then $e_i \alpha = \alpha$ (resp. $\beta e_i = \beta$). We identify the set of vertices and the set of trivial paths.

Let KQ be the K-vector space generated by the set of all paths in Q. Then KQ can be endowed with a structure of K-algebra with multiplication induced by concatenation of path.KQ is called the path algebra of the quiver Q. The algebra KQ can be graded by

$$KQ = KQ_0 \oplus KQ_1 \oplus \cdots \oplus KQ_m \oplus \cdots,$$

where Q_m is the set of all paths of lengths m.

Definition 2.1. Let Q be a finite connected quiver. Then, the ideal of path algebra KQ generated by arrows of Q is called the arrow ideal and is denoted by R_Q.

Definition 2.2. Let Q be a finite quiver and R_Q be the arrow ideal of path algebra KQ generated by arrows of Q.Then it is called the arrow ideal in the path algebra KQ. An ideal I in KQ is admissible if there exists $m > 2$ such that

$$R_Q^m \subseteq I \subseteq R_Q^2.$$

If I is an admissible ideal in $KQ, (Q, I)$,then I is called the bound quiver. The quotient algebra KQ/I is called he bound path algebra.

The path algebra KQ can be viewed as a graded K-coalgebra with comultiplication induced by the decomposition of paths, that is, if $p = \alpha_1 \cdots \alpha_m$ is a path from the vertex i to the vertex j, thenthe following statwement holds:

$$\triangle(p) = e_i \otimes p + p \otimes e_j + \textstyle\sum_{i=1}^{m-1} \alpha_1 \cdots \alpha_i \otimes \alpha_i + 1 \cdots \alpha_m = \textstyle\sum_{\tau v = p} \tau \otimes v$$

and for a trivial path e_i, we have $\triangle(e_i) = e_i \otimes e_i$. The counit of KQ is defined by the formula $\epsilon(\alpha) = 1$ if $\alpha \in Q_0$ and $\epsilon(\alpha) = 0$ if α is a path of length ≥ 1. The coalgebra $(KQ, \triangle, \epsilon)$ is the *path coalgebra* of the quiver Q. For the convenience we denoted by KQ the path algebra of Q and by CQ the path coalgebra of Q.

3. Quiver with Relations

Let $v \in CQ$, the *orthogonal subspace* to v is the set $v^{\perp} = f \in D(CQ) : f(v) = 0$. More generally, for any subset $S \subseteq CQ$, we define the *orthogonal subspace* to S to be the space

$$S^{\perp} = \{f \in D(CQ) : f(S) = 0\}.$$

Since KQ^{op} can be embedded in $D(CQ)$, we may consider the orthogonal suspace to S in KQ^{op}

$$S^{\perp_{KQ^{op}}} = S^{\perp} \cap KQ^{op} = \{w \in KQ^{op} : \langle S, w\rangle = 0\}.$$

Reciprocally, for any subset $T \subseteq D(CQ)$, the *orthogonal subspace* to T in CQ is defined by the formula

$$T^{\perp_{CQ}} = \{v \in CQ : f(v) = 0 \text{ for all } f \in T\},$$

and if $T \subseteq KQ^{op}$, then we can write $T^{\perp_{CQ}} = \{v \in CQ : \langle v, w\rangle = 0$ for all $w \in T\}$. For simplicity we write $\perp$ instead of $\perp_{CQ}$.

Definition 3.1. Let (Q^{op}, I) be a quiver with relations. Then, the bound path coalgebra of (Q^{op}, I) is defined by the subspace of CQ,

$$C(Q, I) = \{a \in CQ : \langle a, I\rangle = 0\}.$$

Lemma 3.1. *Let Q be a quiver and C a relation subcoalgebra of CQ. Then, $C^{\perp KQ^{op}}$ is an admissible ideal of KQ.*

Now, we will state the relationship between the path algebra and the path coalgebra. The proof of the following theorem and corollaries can be seen in the paper of Muchtadi and Garminia [2].

Theorem 3.1. *Let Q be a finite quiver. Then the following statements hold:*
1. If I is an admissible ideal of KQ^{op} then $D(KQ^{op}/I) \cong I^{\perp} = C(Q, I)$.
2. If C is a finite dimensional relation subcoalgebra of CQ then $D(C) \cong KQ^{op}/C^{\perp KQ^{op}}$

Corollary 3.1. *Let C be a relation subcoalgebra of CQ and $D(C)$ be its dual. Then $D(C)$ is isomorphic to the bound path algebra of quiver (Q^{op}, I),* *i.e., the quiver of dual path algebra $D(C)$ can be obtained from the quiver Q by reversing the arrows.*

Corollary 3.2. *Let KQ/I be a bound path algebra where I is an admissible ideal of KQ, and $D(KQ)$ be its dual. Then $D(KQ/I)$ is isomorphic to the bound path coalgebra of quiver (Q^{op}, I) i.e., the quiver of dual bound path algebra $D(KQ/I)$ can be obtained from the quiver Q by reversing arrows.*

4. Right Serial Algebra

Definition 4.1. A right module M_A is said to be **uniserial** if it has a unique composition series.

In other word, M is uniserial if and only if its submodule lattice is a chain. Clearly, if M is uniserial, then so is every submodule of M, and every quotient of M. Moreover, the dual DM of M is a uniserial left A-module. Because a uniserial module M necessarily has a simple top (and a simple socle), it must be indecomposable.

We now describe those algebras that have the property that every indecomposable projective module is uniserial.

Definition 4.2. An algebra A is called right serial if every indecomposable projective right A-module is uniserial.

The shape of the ordinary quiver of a right serial algebra follows easily from the next lemma.

Lemma 4.1. *An algebra A is right serial if and only if for every indecomposable projective right module P the module $radP/rad^2P$ is simple or zero.*

Proof. See Assem [1]. □

Theorem 4.1. *A basic K-algebra A is right serial if and only if, for every point a of its quiver Q_A, there exists at most one arrow of source a.*

Proof. See Assem in [1]. □

5. Left Serial Coalgebra

A finite dimensional coalgebra C is called *pointed* if each simple subcoalgebra is of dimension one (see Muchtadi and Garminia [2]).

Lemma 5.1. *A pointed coalgebra C is left serial if and only if each vertex in the quiver Q_C is the sink of at most one arrow.*

Proof. See Gomez-Torrecillas and Navarro [3]. □

By applying the Corollary 3.2 to Theorem 4.1, we obtain the following theorem.

Theorem 5.1. *The quiver of dual basic right serial K-algebra A has the following property: for every point a of its quiver Q_A, there exists at most one arrow of sink a. Thus, dual of a basic right serial K-algebra is a left serial K-coalgebra; and dual of a relation left serial K-coalgebra is a basic right serial K-algebra.*

For further research, we could study the representation theory of serial coalgebras via serial algebras.

Acknowledgement. The authors would like to thank Intan Muchtadi and Hanni Garminia for their various discussion and suggestions contributed to this paper. The authors are supported by Hibah Penelitian Ilmu Pengetahuan Terapan DIPA DIKTI ITB based on SK Dekan FMIPA No.14a/ SK/ K01.7/ KP/ 2010.

References

1. I.Assem, A. Skowronski,and D. Simson, *Elements of Reprensentation Theory of Associative Algebras*, London Mathematical Society Student Texts 65, (Cambridge University Press, London, 2006).
2. I. Muchtadi-Alamsyah and H. Garminia, *Quiver of Bound Path Algebras and Bound Path Coalgebras, ITB J. Sci.* **42 A No 2**, 153-162, (2010).
3. J.Gomez-Torrecillas and G.Navarro, *Serial coalgebras and their valued Gabriel quivers J Algebra* **319**, 5039-5059, (2008).

δ-Derivations of Algebras and Superalgebras

I. Kaygorodov

Sobolev Inst. of Mathematics
Novosibirsk, Russia
**E-mail: kib@math.nsc.ru*
www.math.nsc.ru

This article about δ-derivations of Lie, alternative, Malcev, Jordan, Filippov and ternary Malcev algebras, δ-derivations and δ-superderivations of Lie and Jordan superalgebras.

Keywords: δ-derivation, δ-superderivation, Jordan algebra, Jordan superalgebra, Lie superalgebra, Lie algebra, Filippov algebra, ternary Malcev algebra.

1. δ-derivations of algebras

The notion of a derivation of an algebra was generalized by many mathematicians in a number of different directions. Thus, in [Fil1], we can find the definition of a δ-derivation of an algebra. Recall that with $\delta \in F$ fixed, a *δ-derivation* of an algebra A is conceived of as a linear map ϕ satisfying the condition

$$\phi(xy) = \delta(\phi(x)y + x\phi(y))$$

for arbitrary elements $x, y \in A$.

A definition for a 1-derivation coincides with the usual definition of a derivation. A 0-derivation is any endomorphism ϕ of A such that $\phi(A^2) = 0$. A *nontrivial δ-derivation* is a nonzero δ-derivation ϕ which is not a 1- or 0-derivation, nor a $\frac{1}{2}$-derivation such that $\phi \in \Gamma(A)$.

In [Fil1], also, $\frac{1}{2}$-derivations are described for an arbitrary primary Lie Φ-algebra A ($\frac{1}{6} \in \Phi$) with a nondegenerate symmetric invariant bilinear form. Namely, it was proved that the linear map $\phi : A \to A$ is a $\frac{1}{2}$-derivation iff $\phi \in \Gamma(A)$, where $\Gamma(A)$ is the centroid of an algebra A. This implies that if A is a central simple Lie algebra having a nondegenerate symmetric invariant bilinear form over a field of characteristic $p \neq 2, 3$, then any $\frac{1}{2}$-derivation ϕ is represented as $\phi(x) = \alpha x$, $\alpha \in \Phi$. In [Fil2], it was proved that every primary Lie Φ-algebra does not have a nonzero δ-derivation if $\delta \neq -1, 0, \frac{1}{2}, 1$, and that every primary Lie Φ-algebra A ($\frac{1}{6} \in \Phi$) with a nonzero antiderivation satisfies the identity $[(yz)(tx)]x + [(yx)(zx)]t = 0$ and is a three-dimensional central simple algebra over a field of quotients of the center $Z_R(A)$ of its right multiplication algebra $R(A)$.

In [Hop], we can find an example oa a nontrivial (-1)-derivations (antiderivations) for a Lie algebra sl_2.

Example 1 [Hop]. $sl_2 \cong k^3$ with the product defined by

$$\begin{bmatrix} a \\ b \\ c \end{bmatrix} \begin{bmatrix} x \\ y \\ z \end{bmatrix} = \begin{bmatrix} bx - cy \\ 2ay - 2bx \\ 2cx - 2az \end{bmatrix}, \text{ where } \begin{bmatrix} a & b \\ c & -a \end{bmatrix} \mapsto \begin{bmatrix} a \\ b \\ c \end{bmatrix}.$$

Then it easy to show that $Antider(sl_2) = \left\{ \begin{bmatrix} -2a & b & c \\ 2c & a & d \\ 2b & e & a \end{bmatrix} | a, b, c, d, e \in k \right\}.$

In [Fil2], we can also find an example of a nontrivial $\frac{1}{2}$-derivation for the Witt algebra W_1, i.e., a $\frac{1}{2}$-derivation which is not an element of the centroid of W_1.

Example 2 [Fil2]. It is well-known that $W_1 = Der(K[x])$. Map $R : y \to ys(x_1, x_2, x_3)$, where $ys(x_1, x_2, x_3) = \sum_{\sigma \in S_3} (-1)^\sigma y x_{\sigma(1)} x_{\sigma(2)} x_{\sigma(3)}$ for some $x_1, x_2, x_3 \in W_1$ is a nontrivial $\frac{1}{2}$-derivation.

Filippov also noted that δ-derivations of a prime Lie algebra form a commutative ring which contains the centroid of the algebra, and posed the question: is it true that this ring does not contain zero divisors?

In [Zus], P. Zusmanovich compute δ-derivations of the modular Zassenhaus algebra. It turns out that Zassenhaus algebra has δ- derivations which are zero divisors, what provides a negative answer to the Filippovs question. Further examples can be obtained by tensoring Zassenhaus algebra with the reduced polynomial algebra and adding a "tail" of derivations a construction typical in the structure theory of modular Lie algebras. Note that the property to possess nontrivial δ-derivations distinguishes Zassenhaus algebra.

Example 3 [Zus]. Recall that divided powers algebra $O_1(n)$ is a p^n-dimensional associative commutative algebra with basis $\{x^i | 0 \leq i \leq p^n\}$ and multiplication

$$x_i x_j = \binom{i+j}{j} x^{i+j}.$$

$W_1(n)$ can be viewed as a Lie algebra of derivations $O_1(n)\partial$, where $\partial(x^i) = x^{i-1}$. In such realization, it possesses a basis $\{e_i = x^{i+1}\partial | -1 \leq i \leq p^n - 2\}$ with multiplication

$$[e_i, e_j] = \left(\binom{i+j+1}{j} - \binom{i+j+1}{i}\right) e_{i+j}.$$

Map $D(e_i) = (1 - \delta_{i,p-2})(i+2)e_{i+1}$, where $\delta_{i,p-2} = 0$, if $i \neq p-2$ and $\delta_{i,p-2} = 1$, if $i = p-2$ is a nilpotent $\frac{1}{2}$-derivation of a Lie algebra $W_1(1)$.

δ-Derivations of primary alternative and non-Lie Mal'tsev Φ-algebras with restrictions on an operator ring Φ were described in [Fil3]. It turns out that algebras in these classes have no nonzero δ-derivation if $\delta \neq 0, \frac{1}{2}, 1$.

Also, in [Zus] we note a restriction on δ which should satisfy a non-nilpotent δ-derivation of a complex perfect finite-dimensional Lie algebra and we see a simple formula expressing δ-derivations of current Lie algebras in terms of its tensor factors, similar to the known formula for the ordinary derivations.

Later in [Kay3], we considered δ-derivations of semisimple finite-dimensional Jordan algebras and simple finite-dimensional non-commutative Jordan algebras with rank $k > 1$. It turns out that algebras in these classes have no nontrivial δ-derivations. Also we considered δ-derivations for unital flexible algebras over a field of characteristic not 2.

In [Kay6] we described δ-derivations of ternary Malcev algebra M_8 over field of characteristic not 2,3 and simple finite-dimensional Filippov algebras over algebraic closed field of zero characteristic. We obtained some new examples of nontrivial (-1)-derivations for Filippov algebras.

A linear mapping $\phi : A \to A$ is called a n-ary *δ-derivation* if

$$\phi([x_1, \dots, x_n]) = \delta \sum_{i=1}^{n} [x_1, \dots, \phi(x_i), \dots, x_n].$$

for two arbitrary elements $x_1, \dots, x_n \in A$.

By the *centroid* $\Gamma(A)$ of a n-ary algebra A we mean a set of all linear mappings $\chi : A \to A$ satisfying the condition

$$\chi([x_1, \dots, x_n]) = [x_1, \dots, \chi(x_i), \dots, x_n]$$

for two arbitrary elements $x_1, \dots, x_n \in A$. If ϕ is a $\frac{1}{n}$-derivation of n-ary algebra A, than $\phi \in \Gamma(A)$.

A definition for a 1-derivation coincides with the usual definition of a n-ary derivation. A 0-derivation is an arbitrary endomorphism ϕ of A such that $\phi([A, \dots, A]) = 0$. A *nontrivial δ-derivation* of n-ary algebra A is a nonzero δ-derivation which is not a 1- or 0-derivation, nor an element of the centroid.

Example 4 [Kay6]. Let A_{n+1} be a n-ary algebra with bases elements $\{e_1, \ldots, e_{n+1}\}$ and n-ary anticommutative multiplication $[x_1, \ldots, x_n]$, such that

$$[e_1, \ldots, \hat{e}_i, \ldots, e_{n+1}] = (-1)^{n+i+1} e_i.$$

Then every map D, such that $D(e_i) = \sum_{i=1}^{n+1} \alpha_{ij} e_i$, where $\sum_{i=1}^{n+1} a_{ii} = 0$ and $a_{ij} = a_{ji}$, is a nontrivial (-1)-derivation.

In [Kay6], too, we can find a new examples of a nontrivial $\frac{1}{2}$-derivation and (-1)-derivation for a Lie algebras.

Example 5 [Kay6]. Consider a direct sum of vector spaces $J = A + Ax$. We shall define multiplication on J. For arbitrary elements $a, b \in A$, their product in J is the product ab in A and

$$a(bx) = (ab)x, (ax)b = (ab)x, (ax)(bx) = ab.$$

We define map $\psi : \psi(ax) = \phi(a), \psi(a) = \phi(a)x$.

If ϕ is a map from example 1 and $A = sl_2$, than ψ is a nontrivial (-1)-derivation of J.

If ϕ is a map from example 2 and $A = W_1$, than ψ is a nontrivial $\frac{1}{2}$-derivation of J.

2. δ-derivations and δ-superderivations of superalgebras

Let Γ be the Grassmann algebra over F generated by elements $1, e_1, \ldots, e_n, \ldots$ and defined by relations $e_i^2 = 0$ and $e_i e_j = -e_j e_i$. Products $1, e_{i_1} e_{i_2} \ldots e_{i_k}$, $i_1 < i_2 < \ldots < i_k$, form a basis for the algebra Γ over F. Denote by $\Gamma_{\bar{0}}$ and $\Gamma_{\bar{1}}$ subspaces generated by products of, respectively, even and odd lengths. Then Γ is representable as a direct sum of those subspaces (written $\Gamma = \Gamma_{\bar{0}} \oplus \Gamma_{\bar{1}}$), and the following relations hold: $\Gamma_{\bar{i}} \Gamma_{\bar{j}} \subseteq \Gamma_{\bar{i}+\bar{j}\,(\mathrm{mod}2)}$, $i, j = 0, 1$. In other words, Γ is a $\mathbb{Z}_2$-graded algebra (or superalgebra) over F.

Now let $A = A_{\bar{0}} \oplus A_{\bar{1}}$ be an arbitrary superalgebra over F. Consider the tensor product $\Gamma \otimes A$ of F-algebras. Its subalgebra

$$\Gamma(A) = \Gamma_{\bar{0}} \otimes A_{\bar{0}} + \Gamma_{\bar{1}} \otimes A_{\bar{1}}$$

is called the *Grassmann envelope* of a superalgebra A.

Let Ω be some variety of algebras over F. A superalgebra $A = A_{\bar{0}} \oplus A_{\bar{1}}$ is called a *Ω-superalgebra* if its Grassmann envelope $\Gamma(A)$ is an algebra in Ω.

Let A be a superalgebra. By a *superspace* we mean a $\mathbb{Z}_2$-graded space. A homogeneous element ψ of an endomorphism superspace $A \to A$ is called a *superderivation* if

$$\psi(xy) = \psi(x)y + (-1)^{p(x)p(\psi)}x\psi(y).$$

Suppose $\delta \in F$. A linear mapping $\phi : A \to A$ is called a *δ-superderivation* if

$$\phi(xy) = \delta(\phi(x)y + (-1)^{p(x)p(\phi)}x\phi(y)).$$

Consider a Lie superalgebra $G = G_{\bar{0}} + G_{\bar{1}}$ and fix an element $x \in G_{\bar{1}}$. Then $ad_x : y \to xy$ is a superderivation of G having parity $p(ad_x) = 1$. Obviously, for any superalgebra, multiplication by an element of the base field F is an even $\frac{1}{2}$-superderivation.

By the *supercentroid* $\Gamma_s(A)$ of a superalgebra A we mean a set of all homogeneous linear mappings $\chi : A \to A$ satisfying the condition

$$\chi(ab) = \chi(a)b = (-1)^{p(a)p(\chi)}a\chi(b)$$

for two arbitrary homogeneous elements $a, b \in A$.

A definition for a 1-superderivation coincides with the usual definition of a superderivation. A 0-superderivation is an arbitrary endomorphism ϕ of A such that $\phi(A^2) = 0$. A *nontrivial δ-superderivation* is a nonzero δ-superderivation which is not a 1- or 0-derivation, nor an element of the centroid.

In [Kay1], δ-derivations of simple finite-dimensional Jordan superalgebras over an algebraically closed field of zero characteristic were characterized. δ-Derivations of classical Lie superalgebras were described in [Kay2]. In [Kay3], we considered δ-derivations of Cartan-type Lie superalgebras, and also δ-superderivations of simple finite-dimensional Lie superalgebras, Jordan superalgebras and non-commutative Jordan superalgebras with rank $k > 1$ over an algebraically closed field of characteristic 0. It turns out that superalgebras in these classes have no nontrivial δ-derivations and δ-superderivations .

Also δ-derivations (and δ-superderivations) of prime Lie superalgebras considered P. Zusmanovich [Zus]. He was proved that every primary Lie superalgebra does not have a nonzero δ-derivation if $\delta \neq -1, 0, \frac{1}{2}, 1$.

In [KayZh], δ-derivations of simple unital superalgebras of Jordan brackets over an field of characteristic other than 2 were characterized.

Example 6 [KayZh]. Let Γ be an associative commutative algebra with a derivation D. Consider the direct sum of vector spaces $J(\Gamma, D) = \Gamma + \Gamma x$. We shal define multiplication on J. For arbitrary elements $a, b \in \Gamma$, their product in J is the product ab in Γ and

$$a(bx) = (ab)x, (ax)b = (ab)x, (ax)(bx) = D(a)b - aD(b).$$

We define map $\phi(a) = az$, where $D(z) \neq 0$. If J is a simple superalgebra, than ϕ ia a nontrivial $\frac{1}{2}$-derivation.

In [Kay5], δ-derivations of semisimple finite-dimensional Jordan superalgebras over an algebraically closed field of characteristic other than 2 were characterized. It turns out that superalgebras in these classes have nontrivial $\frac{1}{2}$-derivations and $\frac{1}{2}$-superderivations .

Example 7 [Kay5]. Let Z be an associative commutative algebra with a derivation D over field F with characteristic $p \neq 2$, such that

(i) Z had no proper D-invariant ideals,

(ii) the only constants of the derivation D in Z are scalars $\alpha \cdot 1, \alpha \in F$.

Consider the direct sum of vector spaces $V_{1/2}(Z, D) = Z + Zx$. We shal define multiplication on $V_{1/2}(Z, D)$. For arbitrary elements $a, b \in Z$, their product in $V_{1/2}(Z, D)$ is the product ab in Γ and

$$a(bx) = \frac{1}{2}(ab)x, (ax)b = \frac{1}{2}(ab)x, (ax)(bx) = D(a)b - aD(b).$$

We define map ϕ: $\phi(y) = (1 + p(y))z \cdot y$ for fix $z \in A \setminus \{Fe\}$. Than ϕ is nontrivial $\frac{1}{2}$-derivation of Jordan superalgebra $V_{1/2}(Z, D)$.

Example 8 [Kay5]. If $p \neq 3$ and $\phi : A \to A$ is a map, such that $\phi(A) = 0, \phi(ax) = az$ for fixed $z \in A \setminus \{0\}$, then ϕ is a nontrivial odd $\frac{1}{2}$-superderivation of Jordan superalgebra $V_{1/2}(Z, D)$.

If $p = 3$ and map ϕ, such that $\phi(a) = (\alpha D(a))x, \phi(ax) = D(\alpha D(a)) + az$ for $z \in A, \alpha \in F$, such that $\alpha z \neq 0$, than ϕ is a nontrivial odd $\frac{1}{2}$-superderivation of Jordan superalgebra $V_{1/2}(Z, D)$.

Also, in [Kay4] and [Kay6], we considered δ-superderivations of Generalized Kantor Double with a prime even part.

Main result is the following theorem.

Theorem. *If semisimple finite-dimensional Jordan superalgebra J over algebraic closed field of characteristic $p \neq 2$ has a nontrivial δ-derivation, than $J = J_* \oplus J^*$, where $J^* \cong V_{1/2}(Z, D)$ or $J^* \cong J(\Gamma, D)$.*

The author would like to express profound gratitude to P. S. Kolesnikov for immeasurable help and assistance.

References

[Fil1] Filippov V. T., *δ-Derivations of Lie algebras,* Sib. Mat. Zh., **39**, No. 6, 1998, 1409-1422.

[Fil2] Filippov V. T., *δ-Derivations of prime Lie algebras,* Sib. Mat. Zh., **40**, No. 1, 1999, 201-213.

[Fil3] Filippov V. T., *δ-Derivations of prime alternative and Mal'tsev algebras,* Algebra Logika, **39**, No. 5, 2000, 618-625.

[Kay1] Kaigorodov I. B., *δ-Derivations of simple finite-dimensional Jordan superalgebras,* Algebra and Logic, **46**, 5, 2007, 318-329.

[Kay2] Kaygorodov I. B., *δ-Derivations of classical Lie superalgebras,* Sib. Math. J., **50**, 3, 2009, 434-449.

[Kay3] Kaigorodov I. B., *δ-Superderivations of simple finite-dimensional Jordan and Lie superalgebra,* Algebra and Logic, **49**, No. 2, 2010, 130-144.

[Kay4] Kaygorodov I. B., *Generalized Kantor Double*, Vestnik Samarskogo gos. universiteta, **78**, No. 4, 2010, 42-50 [in Russian].

[Kay5] Kaygorodov I. B., *δ-Derivations of semisimple finite-dimensional Jordan superalgebras*, Mathematical Notes, **88**, (2011), to appear.

[Kay6] Kaygorodov I. B., *Generalized δ-derivations and δ-derivations of n-ary algebras*, Sib. Math. J., to appear.

[KayZh] Kaygorodov I. B., Zhelyabin V. N., *δ-Superderivations of simple superalgebras of Jordan brackets,* St. Peterburg Math. Journal, **23**, (2011), to appear.

[Zus] Zusmanovich P., *On δ-derivations of Lie algebras and superalgebras*, J. of Algebra, **324**, No. 12, 2010, 3470–3486.

Decomposition of Some Types of Ordered Semigroups

Niovi Kehayopulu, Michael Tsingelis

Department of Mathematics, University of Athens
157 84 Panepistimiopolis, Athens, Greece

Dedicated to Professor K. P. Shum on the occasion of his 70th birthday

Abstract. We are interested in ordered semigroups S which inter their properties in their σ-classes, where σ is a complete semilattice congruence on S. This gives information about the structure of ordered semigroups referring to the decomposition of these semigroups into components of the same type.

AMS 2000 Subject Classification: 06F05 (20M10).

Keywords: Regular, intra-regular, left (right) regular, completely regular, left (right) quasi-regular, k-regular, intra-k-regular, left (right) k-regular, archimedean, weakly commutative, left (right) simple, simple, left (right) strongly simple ordered semigroup, semilattice (complete semilattice) of semigroups of type $\mathcal{K}$ or $\mathcal{C}$

1. Introduction and prerequisites

If S is an ordered semigroup and σ a complete semilattice congruence on S, then S is left quasi-regular if and only if the σ-class $(a)_\sigma$ of S containing a is a left quasi-regular subsemigroup of S for every $a \in S$ [5]. As a consequence, an ordered semigroup is a semilattice (resp. complete semilattice) of left strongly simple semigroups if and only if it is a union of left strongly simple semigroups. In the present paper we prove that the regular, intra-regular, left (right) regular, completely regular, k-regular, intra-k-regular, left (right) k-regular, completely k-regular, left (right) quasi-k-regular ordered semigroups are such semigroups. Denote by $\mathcal{K}$ the class of ordered semigroups which are regular, intra-regular, left (right) regular, completely regular, left (right) quasi-regular, k-regular, intra-k-regular, left (right) k-regular, completely k-regular, left (right) quasi-k-regular, respectively; and by $\mathcal{C}$ the class of ordered semigroups which are archimedean, weakly commutative, left (right) simple, simple, left (right) strongly simple, respectively. We prove that an ordered semigroup S is of type $\mathcal{K}$ if and only if it is a complete semilattice of semigroups of type $\mathcal{K}$, equivalently it is a semilattice of semigroups of type $\mathcal{K}$ and this is equivalent to say that it is a union of semigroups of type $\mathcal{K}$. If an ordered semigroup S is of type $\mathcal{C}$, then it is a complete semilattice of semigroups of type $\mathcal{C}$, and so a semilattice of semigroups of type $\mathcal{C}$, and a union of semigroups of type $\mathcal{C}$, as well. In particular, if $\{S_i \mid i \in I\}$ is a family of semigroups of type $\mathcal{C}$ at the same time a chain under the inclusion relation and $S = \bigcup\{S_i \mid i \in I\}$, then S is of type $\mathcal{C}$, as well.

A subsemigroup F of an ordered semigroup S is called a *filter* of S if (1) $a, b \in S$ such that $ab \in F$ implies $a \in F$ and $b \in F$ and (2) if $a \in F$ and $b \in S$ such that $b \geq a$, then $b \in F$. For an element a of S, we denote by $N(a)$ the filter of S generated by a, and by $\mathcal{N}$ the equivalence relation on S defined by $\mathcal{N} = \{(a, b) \mid N(a) = N(b)\}$. For an ordered semigroup S and a subset H of S we denote by $(H]$ the subset of S defined by

$$(H] = \{t \in S \mid t \leq h \text{ for some } h \in H\}.$$

An ordered semigroup S is called *regular* if for every $a \in S$ there exists $x \in S$ such that $a \leq axa$. It is called *intra-regular* if for every $a \in S$ there exist $x, y \in S$ such that $a \leq xa^2y$. S is called *left* (resp. *right*) *regular* if for every $a \in S$ there exists $x \in S$ (resp. $y \in S$) such that $a \leq xa^2$ (resp. $a \leq a^2y$). It is called *completely regular* if it is at the same time regular, left regular and right regular. As one can easily see, an ordered semigroup S is regular if and only if $a \in (aSa]$ for every $a \in S$, equivalently if $A \subseteq (ASA]$ for every $A \subseteq S$. S is intra-regular if and only if $a \in (Sa^2S]$ for every $a \in S$, equivalently if $A \subseteq (SA^2S]$ for every $A \subseteq S$. S is left regular if and only if $a \in (Sa^2]$ for every $a \in S$, equivalently if $A \subseteq (SA^2]$ for every $A \subseteq S$. S is right regular if and only if $a \in (a^2S]$ for every $a \in S$, equivalently if $A \subseteq (A^2S]$ for every $A \subseteq S$. An ordered semigroup S is completely regular if and only if $a \in (a^2Sa^2]$ holds for every $a \in S$, equivalently if $A \subseteq (A^2SA^2]$ for every $A \subseteq S$. An ordered semigroup S is called *archimedean* [2] if, for every $a, b \in S$ there exists $n \in N$ such that $a^n \in (SbS]$. It is called *weakly commutative* [2] if for every $a, b \in S$ there exists $n \in N$ such that $(ab)^n \in (bSa]$. It is called *left* (resp. *right*) *simple*, if S is the only left (resp. right) ideal of S, that is, if for every left (resp. right) ideal T of S, we have $T = S$. One can easily prove that S is left (resp. right) simple if and only if $S = (Sa]$ (resp. $S = (aS]$) for every $a \in S$. S is called *simple* if S itself is the only ideal of S. An ordered semigroup S is simple if and only if $S = (SaS]$ for every $a \in S$ (i.e. for every $a, b \in S$ there exist $x, y \in S$ such that $b \leq xay$). An ordered semigroup S is called *left* (resp. *right*) *quasi-regular* [5] if $a \in (SaSa]$ (resp. $a \in (aSaS]$) for every $a \in S$. It is called *left* (resp. *right*) *strongly simple* [5] if it is both simple and left (resp. right) quasi-regular. An ordered semigroup S is called *k-regular* if for every $a \in S$ we have $a^k \in (a^k S a^k]$. It is called *intra-k-regular* if $a^k \in (Sa^{2k}S]$ for every $a \in S$. It is called *left (resp. right) k-regular* if $a^k \in (Sa^{k+1}]$ (resp. $a^k \in (a^{k+1}S]$) for every $a \in S$. The concepts of k-regular, intra-k-regular, left (resp. right) k-regular, left (right) quasi-regular, left (right) quasi-k-regular, left (right) strongly simple ordered semigroups extend the corresponding once for semigroups, and we keep the same terminology as in [6]. A subsemigroup T of an ordered semigroup $(S, ., \leq)$ is called regular, intra-regular, k-regular etc., if the set T with the multiplication "." and the order "$\leq$" of S is so.

Let S be an ordered semigroup. An equivalence relation σ on S is called *congruence* if $(a, b) \in \sigma$ implies $(ac, bc) \in \sigma$ and $(ca, cb) \in \sigma$ for every $c \in S$. A congruence σ on S is called *semilattice congruence* if $(a^2, a) \in \sigma$ and $(ab, ba) \in \sigma$ for every $a, b \in S$.

A semilattice congruence σ on S is called *complete* if $a \leq b$ implies $(a, ab) \in \sigma$ [3]. The relation $\mathcal{N}$ mentioned above is the least complete semilattice congruence on S [3]. If σ is a semilattice congruence on S, then the σ-class $(a)_\sigma$ containing a ($a \in S$) is a subsemigroup of S for every $a \in S$. An ordered semigroup is called a *semilattice* (resp. *complete semilattice*) *of semigroups of a given type, say* $\mathcal{T}$, if there exists a semilattice congruence (resp. complete semilattice congruence) σ on S such that the σ-class $(a)_\sigma$ is a subsemigroup of S of type $\mathcal{T}$ for every $a \in S$. An ordered semigroup S is a semilattice of semigroups of type $\mathcal{T}$ if and only if there exists a semilattice Y and a family $\{S_\alpha \mid \alpha \in Y\}$ of subsemigroups of S of type $\mathcal{T}$ such that

(1) $S = \bigcup_{\alpha \in Y} S_\alpha$

(2) $S_\alpha \cap S_\beta \neq \emptyset$ for every $\alpha \in Y$.

(3) $S_\alpha S_\beta \subseteq S_{\alpha\beta}$ for every for every $\alpha, \beta \in Y$ (cf., for example, [1]).

It is a complete semilattice of semigroups of type $\mathcal{T}$ if, in addition, $S_\beta \cap (S_\alpha] \neq \emptyset$ implies $\beta = \alpha\beta$ [4]. The first definition is very useful for applications, especially when we use computer. The second one shows its importance referring to the decomposition of semigroups.

Regular, intra-regular, left (right)-regular, completely regular, left (right) quasi-regular ordered semigroups play an essential role in studying the structure of ordered semigroups. So we deal with such semigroups though our results hold for k-regular, intra-k-regular, left (right) k-regular, left (right) quasi-k-regular ordered semigroups, where k is an arbitrary natural number. The results for the k-case can be proved similarly in the way indicated in Proposition 2.7.

2. On ordered semigroups of type $\mathcal{K}$

Denote by $\mathcal{K}$ the class of ordered semigroups which are regular, intra-regular, left (right) regular, completely regular, left (right) quasi-regular, k-regular, intra-k-regular, left (right) k-regular, completely k-regular, left (right) quasi- k-regular, respectively. We get information about the decomposition of semigroups of type $\mathcal{K}$ into their components of the same type. In this respect, we prove that if an ordered semigroup S is of type $\mathcal{K}$ and σ a complete semilattice congruence on S, then the σ-class $(a)_\sigma$ of S containing a ($a \in S$) is also so. As a consequence, for an ordered semigroup S, the following are equivalent: (1) S is of type $\mathcal{K}$. (2) S is a complete semilattice of semigroups of type $\mathcal{K}$. (3) S is a semilattice of semigroups of type $\mathcal{K}$. (4) S is a union of semigroups of type $\mathcal{K}$.

Proposition 2.1. *Let S be a regular ordered semigroup and σ a complete semilattice congruence on S. Then the σ-class $(a)_\sigma$ is a regular subsemigroup of S for every $a \in S$.*

Proof. Let $a \in S$ and $b \in (a)_\sigma$. Then there exists $y \in (a)_\sigma$ such that $b \leq byb$. In fact: Since $b \in S$ and S is regular, we have $b \leq bxb$ for some $x \in S$. Then $b \leq bx(bxb) = b(xbx)b$. On the other hand, $xbx \in (a)_\sigma$. Indeed: Since $b \leq bxb$ and σ is complete semilattice congruence on S, we have $(b, b^2xb) \in \sigma$. Since σ is a semilattice congruence on S, we have $(xb, bx) \in \sigma$, $(b^2xb, b^3x) \in \sigma$, $(b^3, b) \in \sigma$, $(b^3x, bx) \in \sigma$, so $(b^2xb, bx) \in \sigma$. Thus we have $(b, bx) \in \sigma$, $(b, xb) \in \sigma$, and $(bx, xbx) \in \sigma$. Since $(b, b^2xb) \in \sigma$, $(b^2xb, xb) \in \sigma$ and $(bx, xbx) \in \sigma$, we have $(b, xbx) \in \sigma$ and $xbx \in (b)_\sigma = (a)_\sigma$. □

Proposition 2.2. *Let S be an intra-regular ordered semigroup and σ a complete semilattice congruence on S. Then the σ-class $(a)_\sigma$ is an intra-regular subsemigroup of S for every $a \in S$.*

Proof. Let $a \in S$ and $b \in (a)_\sigma$. Then there exist $z, w \in (a)_\sigma$ such that $b \leq zb^2w$. In fact: Since S is intra-regular, we have $b \leq xb^2y$ for some $x, y \in S$. Then

$$\begin{aligned} b &\leq x(xb^2y)(xb^2y)y \leq x^2b^2yx(xb^2y)(xb^2y)y^2 \\ &= (x^2b^2yx^2)b^2(yxb^2y^3) \end{aligned}$$

On the other hand, $x^2b^2yx^2 \in (a)_\sigma$ and $yxb^2y^3 \in (a)_\sigma$. In fact: Since σ is complete, we have $(b, bxb^2y) \in \sigma$. Since σ is a semilattice congruence, we have $(bxb^2y, x^2b^2yx^2) \in \sigma$. Then $(b, x^2b^2yx^2) \in \sigma$, and $x^2b^2yx^2 \in (b)_\sigma = (a)_\sigma$. In a similar way we get $yxb^2y^3 \in (a)_\sigma$. □

Proposition 2.3. *Let S be a left (resp. right) regular ordered semigroup and σ a complete semilattice congruence on S. Then $(a)_\sigma$ is a left (resp. right) regular subsemigroup of S for every $a \in S$.*

Proof. Let S be left regular, $a \in S$ and $b \in (a)_\sigma$. Then there exists $z \in (a)_\sigma$ such that $b \leq zb^2$. In fact: Since $b \in S$ and S is left regular, we have $b \leq xb^2$ fir some $x \in S$. Then we have $b \leq xb(xb^2) = (xbx)b^2$. On the other hand, $xbx \in (a)_\sigma$. Indeed: Since σ is complete, we have $(b, bxb^2) \in \sigma$. Since σ is a semilattice congruence, we have $(bxb^2, xbx) \in \sigma$. Thus we have $xbx \in (b)_\sigma = (a)_\sigma$. If S is right regular, the proof is similar. □

By Propositions 2.1 and 2.3 we have the following proposition. An independent proof is the following:

Proposition 2.4. *Let S be a completely regular ordered semigroup and σ a complete semilattice congruence on S. Then $(a)_\sigma$ is a completely regular subsemigroup of S for every $a \in S$.*

Proof. Let $a \in S$ and $b \in (a)_\sigma$. Then there exists $z \in (a)_\sigma$ such that $b \leq b^2zb^2$. In fact: Since $b \in S$ and S is completely regular, we have $b \leq b^2xb^2$ for some $x \in S$. Then we have

$$b \leq b(b^2xb^2)x(b^2xb^2)b = b^2(bxb^2xb^2xb)b^2.$$

On the other hand, $bxb^2xb^2xb \in (a)_\sigma$. In fact: Since σ is complete, we have $(b, b^3xb^2) \in \sigma$. Since σ is a semilattice congruence, we have $(b^3xb^2, bxb^2xb^2xb) \in \sigma$. Thus we have $(b, bxb^2xb^2xb) \in \sigma$, and $bxb^2xb^2xb \in (b)_\sigma = (a)_\sigma$. □

Proposition 2.5. [5] *Let S be a left (resp. right) quasi-regular ordered semigroup and σ a complete semilattice congruence on S. Then $(a)_\sigma$ is a left (resp. right) quasi-regular subsemigroup of S for every $a \in S$.*

Definition 2.6. An ordered semigroup S is called *k-regular* if $a^k \in (a^kSa^k]$ for every $a \in S$. It is called *intra-k-regular* if $a^k \in (Sa^{2k}S]$ for every $a \in S$. It is called *left k-regular* if $a^k \in (Sa^{k+1}]$ for every $a \in S$, *right k-regular* if $a^k \in (a^{k+1}S]$ for every $a \in S$, *completely k-regular* if it is at the same tine k-regular, left k-regular and right k-regular. S is called *left quasi-k-regular* if $a^k \in (Sa^kSa^k]$ for every $a \in S$ and *right quasi-k-regular* if $a^k \in (a^kSa^kS]$ for every $a \in S$.

Proposition 2.7. *Let S be a k-regular ordered semigroup and σ a complete* *semilattice congruence on S. Then $(a)_\sigma$ is a k-regular subsemigroup of S for every $a \in S$.*

Proof. Let $a \in S$ and $b \in (a)_\sigma$. Then there exists $z \in (a)_\sigma$ such that $b^k \leq b^k z b^k$. In fact: Since S is k-regular, there exists $x \in S$ such that $b^k \leq b^k x b^k$. Then we have

$$b^k \leq b^k x(b^k x b^k) = b^k(x b^k x) b^k.$$

Since σ is complete, we have $(b^k, b^k b^k x b^k) \in \sigma$. Since σ is a semilattice congruence, $(b^k, b^k b^k x b^k) \in \sigma$ implies $(b, x b^k x) \in \sigma$. Thus we have $x b^k x \in (b)_\sigma = (a)_\sigma$. □

Proposition 2.8. *Let S be an intra-k-regular ordered semigroup and σ a complete semilattice congruence on S. Then the σ-class $(a)_\sigma$ is an intra-k-regular subsemigroup of S for every $a \in S$.*

Proposition 2.9. *Let S be a left (resp. right) k-regular ordered semigroup and σ a complete semilattice congruence on S. Then $(a)_\sigma$ is a left (resp. right) k-regular subsemigroup of S for every $a \in S$.*

Proposition 2.10. *Let S be a completely k-regular ordered semigroup and σ a complete semilattice congruence on S. Then $(a)_\sigma$ is a completely k-regular subsemigroup of S for every $a \in S$.*

Proposition 2.11. *Let S be a left (resp. right) quasi-k-regular ordered semigroup and σ a complete semilattice congruence on S. Then $(a)_\sigma$ is a left (resp. right) quasi-k-regular subsemigroup of S for every $a \in S$.*

Remark 2.12. The converse statement in Propositions 2.1–2.11 also holds: If S is an ordered semigroup, σ a semilattice congruence on S and $(a)_\sigma$ is a regular subsemigroup of S for every $a \in S$, then S is regular. The same holds if we replace the word "regular" by "intra-regular", " left (right) regular", "completely regular", "k-regular", "k-intra-regular", "completely k-regular", left (right) quasi-k-regular, respectively.

Remark 2.13. If S is an ordered semigroup, $\{S_i \mid i \in I\}$ a family of subsemigroups of S of type $\mathcal{K}$ and $S = \bigcup\{S_i \mid i \in I\}$, then S is of type $\mathcal{K}$ as well.

Theorem 2.14. *Let S be an ordered semigroup. The following are equivalent:*

(1) *S is of type $\mathcal{K}$.*
(2) *S is a complete semilattice of semigroups of type $\mathcal{K}$.*
(3) *S is a semilattice of semigroups of type $\mathcal{K}$.*

(4) *S is a union of semigroups of type $\mathcal{K}$.*

Proof. (1) $\Longrightarrow$ (2). Let S be regular. The relation $\mathcal{N}$ is a complete semilattice congruence on S [3]. By Proposition 2.1, the class $(a)_{\mathcal{N}}$ is a regular subsemigroup of S for every $a \in S$. Thus S is a complete semilattice of regular semigroups. The implications (2) $\Longrightarrow$ (3) and (3) $\Longrightarrow$ (4) are obvious, and one can easily see that (4) $\Longrightarrow$ (1). □

3. On ordered semigroups of type $\mathcal{C}$

Denote by $\mathcal{C}$ the class of ordered semigroups which are archimedean, weakly commutative, left (right) simple, simple, left (right) strongly simple, respectively. In this paragraph we prove that if an ordered semigroup S is of type $\mathcal{C}$ and σ a complete semilattice congruence on S, then the σ-class $(a)_\sigma$ of S containing a ($a \in S$) is also so. As a consequence, if an ordered semigroup S is of type $\mathcal{C}$, then it is a complete semilattice of semigroups of type $\mathcal{C}$ which, in turn, is a semilattice of semigroups of type $\mathcal{C}$, and so a union of semigroups of type $\mathcal{C}$. In particular, if $\{S_i \mid i \in I\}$ is a family of semigroups of type $\mathcal{C}$ at the same time a chain under the inclusion relation and $S = \bigcup\{S_i \mid i \in I\}$, then S is of type $\mathcal{C}$, as well.

Proposition 3.1. *Let S be an archimedean ordered semigroup and σ a complete semilattice congruence on S. Then $(a)_\sigma$ is an archimedean subsemigroup of S for every $a \in S$.*

Proof. Let $a \in S$ and $b, c \in (a)_\sigma$. Then there exist $m \in N$ and $x, y \in (a)_\sigma$ such that $b^m \leq xcy$. In fact: Since $b, c \in S$ and S is archimedean, there exist $n \in N$ and $s, t \in S$ such that $b^n \leq sct$. Since σ is complete, we have $(b^n, b^n sct) \in \sigma$. Then, since σ is a semilattice congruence, we have $(b, bsct) \in \sigma$. Moreover we have

$$\begin{aligned} b^{3n+2} &= bb^n b^n b^n b \leq b(sct)(sct)(sct)b \\ &= (bscts)c(tsctb) \end{aligned}$$

Since $(b, bsct) \in \sigma$ and σ is a semilattice congruence, we have $(b, bscts) \in \sigma$ and $(b, tsctb) \in \sigma$, so $bscts \in (b)_\sigma = (a)_\sigma$ and $tsctb \in (b)_\sigma = (a)_\sigma$. □

Proposition 3.2. *Let S be a weakly commutative ordered semigroup and σ a complete semilattice congruence on S. Then $(a)_\sigma$ is a weakly commutative subsemigroup of S for every $a \in S$.*

Proof. Let $a \in S$, $x, y \in (a)_\sigma$. Then there exist $m \in N$ and $s \in (a)_\sigma$ such that $(xy)^m \leq ysx$. In fact: Since $x, y \in S$ and S is weakly commutative, there exist $n \in N$ and $t \in S$ such that $(xy)^n \leq ytx$. Since σ is complete, we have $((xy)^n, (xy)^n ytx) \in \sigma$. Since σ is a semilattice congruence, we get $(xy, txyt) \in \sigma$. Moreover we have

$$(xy)^{2n} = (xy)^n (xy)^n \leq (ytx)(ytx) = y(txyt)x.$$

On the other hand, $txyt \in (a)_\sigma$. Indeed: Since $x, y \in (a)_\sigma$ and $(a)_\sigma$ is a subsemigroup of S, we have $xy \in (a)_\sigma$, so $(a, xy) \in \sigma$. Since $(a, xy) \in \sigma$ and $(xy, txyt) \in \sigma$, we have $(a, txyt) \in \sigma$, and $txyt \in (a)_\sigma$. □

Proposition 3.3. *Let S be a left (resp. right) simple ordered semigroup and σ a complete semilattice congruence on S. Then $(a)_\sigma$ is a left (resp. right) simple subsemigroup of S for every $a \in S$.*

Proof. Let S be left simple, $a \in S$ and $b, c \in (a)_\sigma$. Then there exists $y \in (a)_\sigma$ such that $c \leq yb$. In fact: Since $b, c \in S$ and S is left simple, there exists $x \in S$ such that $c \leq xb$. Since $cb, b \in S$ and S is left simple, there exists $z \in S$ such that $b \leq z(cb)$. Then we have

$$c \leq x(zcb) = (xzc)b.$$

Moreover, $xzc \in (a)_\sigma$. Indeed: Since σ is complete, we have $(c, cxzcb) \in \sigma$. Since σ is a semilattice congruence, $(c, cxzcb) \in \sigma$ implies $(c, xzcb) \in \sigma$. Since $b, c \in (a)_\sigma$ and $(a)_\sigma$ is a subsemigroup of S, we have $cb \in (a)_\sigma$, so $(cb, a) \in \sigma$. Since $(cb, a) \in \sigma$ and $(a, c) \in \sigma$, we have $(cb, c) \in \sigma$ then, since σ is a semilattice congruence, we have $(xzcb, xzc) \in \sigma$. Since $(c, xzcb) \in \sigma$ and $(xzcb, xzc) \in \sigma$, we have $(c, xzc) \in \sigma$. Then $xzc \in (c)_\sigma = (a)_\sigma$. The rest of the proof is similar. □

Proposition 3.4. *Let S be a simple ordered semigroup and σ a complete semilattice congruence on S. Then $(a)_\sigma$ is a simple subsemigroup of S for every $a \in S$.*

Proof. Let $a \in S$ and $b, c \in (a)_\sigma$. Then there exist $z, w \in (a)_\sigma$ such that $c \leq zbw$. In fact: Since $b, c \in S$ and S is simple, there exist $x, y \in S$ such that $c \leq xby$. Since $cbc, b \in S$ and S is simple, there exist $s, t \in S$ such that $b \leq s(cbc)t$. Then we have

$$c \leq x(scbct)y = (xsc)b(cty).$$

Moreover, $xsc, cty \in (a)_\sigma$. Indeed: Since $c \leq xby$, we have $xsc \leq xsxby$. Since σ is complete, we have $(xsc, xscxsxby) \in \sigma$. Since σ is a semilattice congruence, $(xsc, xscxsxby) \in \sigma$ implies $(xsc, cxsby) \in \sigma$. In a similar way, from $b \leq scbct$, we have $sb \leq sscbct$, $(sb, sbsscbct) \in \sigma$, $(sbsscbct, bscbct) \in \sigma$, $(sb, bscbct) \in \sigma$. From $b \leq scbct$, we get $(b, bscbct) \in \sigma$, then $(b, sb) \in$

σ, and $(cxby, cxsby) \in \sigma$. From $c \le xby$, we have $(c, cxby) \in \sigma$, then $(c, cxsby) \in \sigma$. Since $(c, cxsby) \in \sigma$ and $(xsc, cxsby) \in \sigma$, we have $(c, xsc) \in \sigma$, and $xsc \in (c)_\sigma = (a)_\sigma$. In a similar way we prove that $cty \in (a)_\sigma$. □

By Propositions 2.5 and 3.4 we have the following proposition

Proposition 3.5. *Let S be a left (resp. right) strongly simple ordered semigroup and σ a complete semilattice congruence on S. Then $(a)_\sigma$ is a left (resp. right) strongly simple subsemigroup of S for every $a \in S$.*

Remark 3.6. The converse statement in Propositions 3.1–3.5 also holds: If S is an ordered semigroup, σ a semilattice congruence on S and $(a)_\sigma$ is an archimedean subsemigroup of S for every $a \in S$, then S is also archimedean. The same holds if we replace the word "archimedean" by "weakly commutative", "left (right) simple", "simple".

Remark 3.7. If S is an ordered semigroup, $\{S_i \mid i \in I\}$ a family of subsemigroups of S of type $\mathcal{C}$, chain under inclusion relation, and $S = \bigcup\{S_i \mid i \in I\}$, then S is of type $\mathcal{C}$ as well.
As a result, we have the following:

Theorem 3.8. *Let S be an ordered semigroup. We consider the statements:*

(1) *S is of type $\mathcal{C}$.*
(2) *S is a complete semilattice of semigroups of type $\mathcal{C}$.*
(3) *S is a semilattice of semigroups of type $\mathcal{C}$.*
(4) *S is a union of semigroups of type $\mathcal{C}$.*

Then (1) $\Longrightarrow$ (2) $\Longrightarrow$ (3) $\Longrightarrow$ (4). In particular, if $\{S_i \mid i \in I\}$ is a family of semigroups of type $\mathcal{C}$ at the same time a chain under the inclusion relation and $S = \bigcup\{S_i \mid i \in I\}$, then S is of type $\mathcal{C}$, as well.

References

1. N. Kehayopulu, *On right regular and right duo ordered semigroups*, Math. Japon. **36**, no. 2 (1991), 201–206.
2. N. Kehayopulu, *On weakly commutative ordered semigroups*, Semigroup Forum **56**, no. 1 (1998), 32–35.
3. N. Kehayopulu, M. Tsingelis, *Remark on ordered semigroups.* In: Partitions and holomorphic mappings of semigroups (Russian), 50–55, Obrazovanie, St. Petersburg, 1992.
4. N. Kehayopulu, M. Tsingelis, *On ordered semigroups which are complete chains of semigroups*, Sci. Math. Jpn., to appear.
5. N. Kehayopulu, M. Tsingelis, *On ordered semigroups which are semilattices of left strongly simple semigroups*, preprint, to appear.
6. M. Mitrović, *Semilattices of Archimedean Semigroups*, University of Nis, Faculty of Mechanical Engineering, Nis, 2003. xiv+160 pp. ISBN: 86-80587-32-X

On Representations of Dialgebras and Conformal Algebras

P. S. Kolesnikov

Sobolev Institute of Mathematics,
Novosibirsk, 630090, Russia
E-mail: pavelsk@math.nsc.ru

In this note, we observe a relation between dialgebras (in particular, Leibniz algebras) and conformal algebras. The purpose is to show how the methods of conformal algebras help solving problems on dialgebras, and, conversely, how the ideas of dialgebras work for conformal algebras.

Keywords: Leibniz algebra; Dialgebra; Vertex operator algebra; Conformal algebra.

1. Conformal Algebras

The notion of a conformal algebra was introduced in [1] (in [2], a similar notion appeared under the name of a vertex Lie algebra). This notion is an important tool for studying vertex operator algebras. The latter came into algebra from mathematical physics (namely, from the 2-dimensional conformal field theory, what explains the name "conformal algebra"), that was initiated by [3]. The algebraic essence of vertex operator structures was extracted in [4] and later developed in a series of works, e.g., [5–7]. The relations between vertex and conformal algebras are very much similar to the relations between ordinary associative and Lie algebras.

In conformal field theory, the operator product expansion (OPE) describes the commutator of two fields. Let V be a (complex) space of states, and let $Y : V \to \operatorname{End} V[[z, z^{-1}]]$, $Y : b \mapsto Y(b, z)$, be a state-field correspondence of a vertex algebra. Then the commutator of two fields can be expressed as a finite distribution

$$[Y(a,w), Y(b,z)] = \sum_{n\geq 0} \frac{1}{n!} Y(c_n, z) \frac{\partial^n \delta(w-z)}{\partial z^n}, \quad a, b \in V,$$

where $c_n \in V$, $\delta(w-z) = \sum\limits_{m\in\mathbb{Z}} w^m z^{-m-1}$ is the formal delta-function. The formal Fourier transformation

$$[Y(a,z)_\lambda Y(b,z)] = \mathrm{Res}_{w=0} \exp\{\lambda(w-z)\}[Y(a,w), Y(b,z)] \qquad (1)$$

is called the λ-bracket on the space of fields $\{Y(a,z) \mid a \in V\}$. Here λ is a new formal variable, and $\mathrm{Res}_{w=0}F(w,z)$ means the residue at $w = 0$, i.e., the formal series in z that is a coefficient of $F(w,z)$ at w^{-1}.

The algebraic properties of the λ-bracket (1) lead to the formal definition of a conformal algebra over a field $\Bbbk$ of characteristic 0.

Definition 1.1 ([1]). *A conformal algebra is a left (unital) module C over the polynomial algebra $H = \Bbbk[T]$ endowed with a binary $\Bbbk$-linear operation*

$$(\cdot_\lambda\cdot) : C \otimes C \to C[\lambda], \qquad (2)$$

such that $(Ta_\lambda b) = -\lambda(a_\lambda b)$, $(a_\lambda Db) = (T+\lambda)(a_\lambda b)$.

In terms of fields, T is just the ordinary derivation with respect to z.

Every conformal algebra can be represented by formal distributions over an ordinary algebra. Let C be an object described by Definition 1.1. Consider the space of Laurent polynomials $\Bbbk[t,t^{-1}]$ as a right H-module with respect to the following action: $f(t)T = -f'(t)$. Then

$$\mathcal{A}(C) = \Bbbk[t,t^{-1}] \otimes_H C$$

carries the natural algebra structure:

$$(f \otimes_H a) \cdot (g \otimes_H b) = (g \otimes_H 1)(f \otimes_H (a_{-T}b)), \quad a,b \in C, \ f,g \in \Bbbk[t,t^{-1}].$$

The space of formal distributions $\mathcal{A}(C)[[z,z^{-1}]]$ that consists of all series

$$Y(a,z) = \sum_{n\in\mathbb{Z}} (t^n \otimes_H a) z^{-n-1}, \quad a \in C,$$

can be endowed with the action of $T = d/dz$ and with a λ-bracket $(\cdot_\lambda\cdot)$ similar to (1), where the commutator is replaced with the ordinary product of distributions. Then

$$(Y(a,z)_\lambda Y(b,z)) = Y((a_\lambda b), z), \quad a,b \in C,$$

i.e., C is isomorphic to a formal distribution conformal algebra over $\mathcal{A}(C)$.

The algebra $\mathcal{A}(C)$ is called the coefficient algebra[8] of C, or annihilation algebra.[1]

Definition 1.2 ([8]). *Let $\mathcal{V}$ be a variety of algebras (associative, alternative, Lie, etc.). Then a conformal algebra C is said to be $\mathcal{V}$-conformal algebra if $\mathcal{A}(C)$ belongs to $\mathcal{V}$.*

Associative and Lie conformal algebras, their representations, and cohomologies have been studied in a series of papers, e.g., [9–15]. In particular, associative conformal algebras naturally appear in the study of representations of Lie conformal algebras.

Example 1.1. Consider one of the simplest (though important) examples of conformal algebras. Suppose A is an ordinary algebra (not necessarily associative or Lie). Then the free H-module

$$\operatorname{Cur} A = H \otimes A$$

generated by the space A endowed with the λ-bracket $(f(T) \otimes a)_\lambda (g(T) \otimes b) = f(-\lambda) g(T+\lambda) \otimes ab$, is called the current conformal algebra.

If A belongs to a variety $\mathcal{V}$ defined by a family of polylinear identities then $\operatorname{Cur} A$ is a $\mathcal{V}$-conformal algebra.

Certainly, current conformal algebras and their subalgebras do not exhaust the entire class of conformal algebras. For example, $W = \Bbbk[T, x]$ with respect to the operation

$$(f(T,x)_\lambda g(T,x)) = f(-\lambda, T) g(T+\lambda, x+\lambda)$$

is an associative conformal algebra (called Weyl conformal algebra[8]), and $\mathrm{Vir} = \Bbbk[T]$ with respect to

$$(f(T)_\lambda g(T)) = f(-\lambda) g(T+\lambda)(T+2\lambda)$$

is a Lie conformal algebra (called Virasoro conformal algebra[1]).

Conformal algebra is said to be finite if it is a finitely generated H-module.

2. Dialgebras

The following notion appears naturally from a certain noncommutative analogue of Lie homology theory.

Definition 2.1 ([16]). *A (left) Leibniz algebra is a linear space L with a bilinear operation $[\cdot,\cdot]$ such that*

$$[x,[y,z]] = [[x,y],z] + [y,[x,z]], \quad x,y,z \in L.$$

The defining identity means that the operator of left multiplication $[x,\cdot]$ is a derivation of L. Leibniz algebras are the most popular noncommutative generalizations of Lie algebras. The following structures play the role of associative enveloping algebras for Leibniz algebras.

Definition 2.2 ([17]). *An (associative) dialgebra is a linear space D endowed with two bilinear operations $(\cdot \dashv \cdot)$, $(\cdot \vdash \cdot)$ such that*

$$x \dashv (y \vdash z) = x \dashv (y \dashv z), \quad (x \dashv y) \vdash z = (x \vdash y) \vdash z, \tag{3}$$

$$x \vdash (y \vdash z) = (x \vdash y) \vdash z, \tag{4}$$

$$x \dashv (y \dashv z) = (x \dashv y) \dashv z, \tag{5}$$

$$x \vdash (y \dashv z) = (x \vdash y) \dashv z, \tag{6}$$

for all $x, y, z \in D$.

In particular, the operation $[a, b] = a \vdash b - b \dashv a$, $a, b \in D$, turns an associative dialgebra D into a Leibniz algebra denoted by $D^{(-)}$.

A systematical study of associative dialgebras was performed in [18]. Also, in [19] and [20] the notions of alternative and commutative dialgebras were introduced. These definitions also appear in the general categorical approach using the language of operads.[21]

Shortly speaking, an operad A is a collection of spaces $A(n)$, $n \geq 1$, such that a composition rule $A(n) \otimes A(m_1) \otimes \cdots \otimes A(m_n) \to A(m_1 + \cdots + m_n)$ and an action of a symmetric group are defined in such a way that some natural axioms hold (associativity of a composition, existence of a unit in $A(1)$, and equivariance of the composition with respect to the symmetric group action).

A linear space A over a field $\Bbbk$ may be considered as an operad (see, e.g., [22] as a general reference), where $A(n) = \operatorname{Hom}(A^{\otimes n}, A)$. In the free operad denoted by Alg, the spaces Alg(n) are spanned by (planar) binary trees with n leaves.

An algebra structure on a linear space A is just a functor of operads Alg $\to A$. If $\mathcal{V}$ is a variety of algebras defined by polylinear identities then there exists a free $\mathcal{V}$-operad $\mathcal{V}$-Alg built on polylinear polynomials of the free $\mathcal{V}$-algebra. There exists a canonical functor Alg $\to \mathcal{V}$-Alg, and it is clear that an algebra A belongs to $\mathcal{V}$ if and only if there exists a functor $\mathcal{V}$-Alg $\to A$ such that the following diagram is commutative.

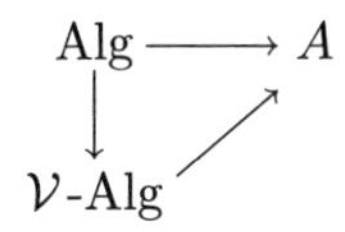

A very similar definition works for dialgebras. An operad Dialg whose spaces are spanned by planar binary trees with 2-colored vertices (colors 1 and 2 stand for $\vdash$ and $\dashv$, respectively) has an image equivalent to the product Alg$\otimes\mathcal{E}$, where $\mathcal{E}$ is the free $\mathcal{V}_c$-operad corresponding to the variety

$\mathcal{V}_c$ of associative and commutative algebras, $\dim \mathcal{E}(n) = n$ (see [21] for details).

Suppose $\mathcal{V}$ is a variety of algebras defined by polylinear identities. For a linear space D, a functor Dialg $\to D$ defines two bilinear operations $\vdash$ and $\dashv$ on D. Conversely, any system $(D, \vdash, \dashv)$ may be considered as a functor Dialg $\to D$.

Definition 2.3 ([21]). *A linear space D with two bilinear operations $\vdash$ and $\dashv$ is said to be $\mathcal{V}$-dialgebra if there exists a functor $\mathcal{V}$-Alg $\otimes\, \mathcal{E} \to D$ such that the following diagram is commutative.*

$$\begin{array}{ccc} \text{Dialg} & \longrightarrow & D \\ \downarrow & & \downarrow \\ \text{Alg} \otimes \mathcal{E} & \longrightarrow & \mathcal{V}\text{-Alg} \otimes \mathcal{E} \end{array}$$

The last definition is easy to translate into the language of identities. First, identify Alg(n) with the space of polylinear non-associative polynomials in $x_1, \dots, x_n$; for Dialg we have a similar interpretation. Next, consider the following linear maps $\Psi_k : \text{Alg}(n) \to \text{Dialg}(n)$, $k = 1, \dots, n$:

$$\Psi_k : (x_{j_1} \dots x_k \dots x_{j_n}) \mapsto (x_{j_1} \vdash \dots \vdash x_k \dashv \dots \dashv x_{j_n}),$$

preserving the bracketing scheme. Then we have

Theorem 2.1 ([21]). *Assume $\{f_i \mid i \in I\}$ is the family of polylinear defining identities of a variety $\mathcal{V}$. Then D is a $\mathcal{V}$-dialgebra if and only if D satisfies the identities $\Psi_k(f_i) = 0$ for all $i \in I$, $k = 1, \dots, \deg f_i$.*

If $\mathcal{V}$ is the variety of Lie algebras then $f = x_1x_2 + x_2x_1$ is one of its defining identities. Since $\Psi_1(f) = x_1 \dashv x_2 + x_2 \vdash x_1$, we can describe Lie dialgebras in terms of single operation, say, $[a, b] = a \vdash b$. Then the class of Lie dialgebras coincides with the class of Leibniz algebras.

Note that all $\mathcal{V}$-dialgebras satisfy the relations (3), called 0-identities.[21] The following approach to the definition of varieties of dialgebras was proposed in [23].

Let D be a dialgebra that satisfies 0-identities. Then the linear span D_0 of all elements $a \vdash b - a \dashv b$, $a, b \in D$, is an ideal of D. The quotient $\bar{D} = D/D_0$ is an ordinary algebra with a single operation. Moreover, the following actions are well-defined:

$$\begin{array}{ll} \bar{D} \otimes D \to D, & D \otimes \bar{D} \to D, \\ (a + D_0) \otimes b \mapsto a \vdash b, & a \otimes (b + D_0) \mapsto a \dashv b. \end{array}$$

Denote by $\hat{D}$ the split null extension $\bar{D} \oplus D$, assuming $D^2 = 0$.

Theorem 2.2 ([23]). *Suppose $\mathcal{V}$ is a variety of algebras with polylinear defining identities. Then D is a $\mathcal{V}$-dialgebra if and only if D satisfies the 0-identities and $\hat{D}$ is an algebra from $\mathcal{V}$.*

A curious relation between conformal algebras and dialgebras was noted in [21]. It turns out that if C is a $\mathcal{V}$-conformal algebra in the sense of Definition 1.2 then the same linear space endowed with just two operations

$$a \vdash b = (a_\lambda b)|_{\lambda=0}, \quad a \dashv b = (a_\lambda b)|_{\lambda=-T}, \quad a, b \in C,$$

is a $\mathcal{V}$-dialgebra denoted by $C^{(0)}$. Conversely, every $\mathcal{V}$-dialgebra can be embedded into an appropriate $\mathcal{V}$-conformal algebra. The last statement easily follows from Theorem 2.2 and

Theorem 2.3 (c.f. [24]). *Let D be a dialgebra satisfying the 0-identities. Then the map $D \to H \otimes \hat{D}$, $a \mapsto 1 \otimes (a + D_0) + T \otimes a$, $a \in D$, is an injective homomorphism of dialgebras $D \to (\mathrm{Cur}\, \hat{D})^{(0)}$. Therefore, D is a $\mathcal{V}$-dialgebra if and only if there exists a $\mathcal{V}$-algebra A such that $D \subseteq (\mathrm{Cur}\, A)^{(0)}$.*

Thus, there are three equivalent definitions of what is a dialgebra of a given variety provided by Theorems 2.1, 2.2, and 2.3.

3. Some Classical Theorems for Leibniz Algebras

Since Leibniz algebras are just Lie dialgebras in the sense of Definition 2.3, we may use Theorem 2.3 to get natural generalizations of some classical statements on Lie algebras to the class of Leibniz algebras. These are: the Engel Theorem, the Poincaré—Birkhoff—Witt (PBW) Theorem, and the Ado Theorem.

We will need the following statement (c.f. Theorem 3 in [25]).

Theorem 3.1. *Let L be a Leibniz algebra, and let V be a module over the Lie algebra $\bar{L}$. Then there exists an injective homomorphism $\rho : L \to (\mathrm{Cur}\, \mathrm{gl}(V \oplus (L \otimes V)))^{(0)}$ of Leibniz algebras.*

Proof. For every $x \in L$, denote $\bar{x} = x + L_0 \in \bar{L}$ and define $\rho(x) \in H \otimes \mathrm{gl}(V \oplus (L \otimes V))$ as follows:

$$\rho(x) = 1 \otimes \rho_0(x) + T \otimes \rho_1(x), \quad \rho_i(x) \in \mathrm{gl}(V \oplus (L \otimes V)),$$

where

$$\rho_0(x) : v \mapsto \bar{x}v, \quad \rho_0(x) : a \otimes v \mapsto a \otimes \bar{x}v + [x, a] \otimes v,$$
$$\rho_1(x) : v \mapsto x \otimes v, \quad \rho_1(x) : a \otimes v \mapsto 0$$

for all $a \in L$, $v \in V$. It is clear that ρ is injective ($\rho_1(x) \neq 0$ for $x \neq 0$). Let us check that ρ is a homomorphism of Leibniz algebras. First, $(\rho(x)_\lambda \rho(y)) = [1 \otimes \rho_0(x) - 1 \otimes \lambda\rho_1(x), 1 \otimes \rho_0(y) + (T + \lambda)\rho_1(y)]$. for all $x, y \in L$. Hence, $[\rho(x), \rho(y)] = 1 \otimes [\rho_0(x), \rho_0(y)] + T \otimes [\rho_0(x), \rho_1(y)]$. Next, it is straightforward to compute

$$[\rho_0(x), \rho_0(y)] : v + (a \otimes w) \mapsto [\bar{x}, \bar{y}]v + a \otimes [\bar{x}, \bar{y}]w + [[x, y], a] \otimes w,$$
$$[\rho_0(x), \rho_1(y)] : v + (a \otimes w) \mapsto [x, y] \otimes v$$

for all $v, w \in V$, $a \in L$. Therefore, $[\rho_0(x), \rho_0(y)] = \rho_0([x, y])$, $[\rho_0(x), \rho_1(y)] = \rho_1([x, y])$, i.e., $[\rho(x), \rho(y)] = \rho([x, y])$. □

The following statement immediately follows from Theorem 3.1 applied to $V = L$.

Theorem 3.2 ([26–28]). *Let L be a finite-dimensional Leibniz algebra such that all operators $[x, \cdot] \in \operatorname{End} L$ are nilpotent. Then L itself is a nilpotent Leibniz algebra.*

Recall that for a Lie algebra L the classical PBW Theorem states that the universal enveloping associative algebra $U(L)$ is isomorphic (as a linear space) to the symmetric algebra $S(L)$. For Leibniz algebras, the role of associative envelopes belongs to associative dialgebras.

Theorem 3.3 ([18,29]). *The universal enveloping associative dialgebra $Ud(L)$ of a Leibniz algebra L is isomorphic (as a linear space) to $U(\bar{L}) \otimes L$.*

As in the case of Lie algebras, the main technical difficulty in the proof of the PBW Theorem for Leibniz algebras is to show that "normal" monomials are linearly independent. In [30], another proof of this independence was obtained by making use of Gröbner—Shirshov bases theory for associative dialgebras. However, one may just apply Theorem 3.1 to $V = U(\bar{L})$, see [25] for details.

Another interesting question is similar to the Ado Theorem: Whether a finite-dimensional Leibniz algebra can be embedded into a finite-dimensional associative dialgebra? It turns out, the answer is positive. Indeed, it is enough to apply Theorem 3.1 to $V = \Bbbk$, a trivial 1-dimensional module over $\bar{L}$. In particular, we may conclude that an n-dimensional Leibniz algebra can be embedded into an associative dialgebra D such that $\dim D \leq 2(n + 1)^2$.

4. Jordan Dialgebras

An associative dialgebra D turns into a Leibniz algebra $D^{(-)}$ if we define the bracket $[x,y] = x \vdash y - y \dashv x$. This is natural to expect that if we define new operation

$$x \circ y = x \vdash y + y \dashv x, \quad x, y \in D,$$

then the algebra $D^{(+)} = (D, \circ)$ obtained would be a noncommutative analogue of a Jordan algebra. Roughly speaking, it relates to Jordan algebras in the same way as Leibniz algebras relate to Lie algebras.

This is indeed a Jordan dialgebra; the commutativity identity turns into $\Psi_1(x_1x_2 - x_2x_1) = x_1 \dashv x_2 - x_2 \vdash x_1$, so we may describe the dialgebra with only one operation. Objects of this type appeared also in [31,32].

Definition 4.1. A Jordan dialgebra is a linear space with a bilinear product satisfying the following identities:

$$\begin{gathered}[x_1, x_2]x_3 = 0, \\ (x_1^2, x_2, x_3) = 2(x_1, x_2, x_1x_3), \quad x_1(x_1^2x_2) = x_1^2(x_1x_2).\end{gathered} \tag{7}$$

Here $[a,b]$ and (a,b,c) stand for the commutator $ab - ba$ and associator $(ab)c - a(bc)$, respectively. The first identity in (7) comes from the 0-identities, the second and third appear from the Jordan identity. In [33] these algebras were called semi-special quasi-Jordan algebras.

Recall that a Jordan algebra J is said to be special if there exists an associative algebra A such that $J \subseteq A^{(+)}$. The class of all homomorphic images of all special Jordan algebras is a variety denoted by SJ. This is well-known that SJ does not coincide with the variety of all Jordan algebras. Those defining identities of SJ that do not hold in all Jordan algebras are called special identities (or s-identities, for short).

I was shown in [34] that the minimal degree of an s-identity is equal to 8. However, the description of all s-identities is still an open problem.

For Jordan dialgebras, the same theory makes sense.

Definition 4.2 ([33]). *A Jordan dialgebra J is said to be special if there exists an associative dialgebra D such that $J \subseteq D^{(+)}$.*

It is clear that the class of all homomorphic images of all special Jordan dialgebras is a variety. Let us denote this variety by DiSJ. The notion of an s-identity for Jordan dialgebras is a natural generalization of s-identities for Jordan algebras.

The following statement was proved in [33] by making use of computer algebra methods.

Theorem 4.1 ([33]). 1. *For Jordan dialgebras, there are no s-identities of degree* ≤ 7*;*
2. *There exists an identity of degree 8 that holds on all special Jordan dialgebras and on all Jordan algebras, but does not hold on all Jordan dialgebras.*

On the other hand, the variety SJ leads to the notion of a SJ-dialgebra by Definition 2.3. It turns out that these two different approaches lead to the same class of dialgebras.

Theorem 4.2 ([35]). *The variety of* SJ*-dialgebras coincides with* DiSJ.

This fact allows to deduce a correspondence between s-identities for Jordan algebras and dialgebras.

Theorem 4.3 ([35]). *Let* $f(x_1, \ldots, x_n)$ *be a polylinear s-identity for Jordan algebras. Then* $\Psi_k f$, $k = 1, \ldots, n$, *is an s-identity for Jordan dialgebras. Conversely, if* $g(x_1, \ldots, x_n)$ *is an s-identity for Jordan dialgebras then*

$$g(x_1, \ldots, x_n) = \sum_{k=1}^{n} g_k, \quad g_k = \Psi_k(f_k)$$

for some nonassociative polynomials $f_k(x_1, \ldots, x_n)$*, and at least one of* f_k *is an s-identity for Jordan algebras.*

Note that Theorem 4.3 works for polylinear identities only, so it says nothing about the identity from Theorem 4.1(2).

A series of classical results for special Jordan algebras can be transferred to dialgebras. In particular, the Shirshov—Cohn Theorem states that every 2-generated Jordan algebra is special. It turns out that the free 2-generated Jordan dialgebra is special, but its homomorphic image may not be special.[35] However, Theorem 2.3 implies that every 1-generated Jordan dialgebra is special.

Another problem on Jordan dialgebras concerns their relation to Leibniz algebras. The classical Tits—Kantor—Koecher construction allows to build a Lie algebra $T(J)$ for a given Jordan algebra J in such a way that structure of J is closely related with the structure of $T(J)$. This is natural to expect[31] that a similar construction for a Jordan dialgebra should lead to Lie dialgebra, i.e., Leibniz algebra.

Conformal algebras allow to solve this problem.

Let J be a Jordan dialgebra, and let $\hat{J}$ stands for the split null extension $\bar{J} \oplus J$ (see Theorem 2.2). This is a Jordan algebra, and it follows from Theorem 2.3 that $J \subseteq (\operatorname{Cur} \hat{J})^{(0)}$.

Denote by

$$T(\hat{J}) = \hat{J}^+ \oplus S(\hat{J}) \oplus \hat{J}^-$$

the Tits—Kantor—Koecher construction[36–38] for $\hat{J}$. Here $\hat{J}^\pm$ are linear spaces isomorphic to $\hat{J}$, and $S(\hat{J}) \subseteq \operatorname{End}\hat{J} \oplus \operatorname{Der}\hat{J}$ is spanned by

$$U_{a,b} = L_{ab} + [L_a, L_b], \quad a, b \in \hat{J},$$

where L_x denotes the operator of left multiplication: $L_x(y) = xy$. The images of $a \in J$ in the isomorphic copies $J^\pm$ are denoted by $a^\pm$. This is a Lie algebra, J^+ and J^- are its abelian subalgebras, and $[a^-, b^+] = L_{ab} + [L_a, L_b]$ for $a, b \in J$. Therefore, $\mathcal{L}(J) = (\operatorname{Cur} T(\hat{J}))^{(0)}$ is a Leibniz algebra. Then the elements $1 \otimes (a + J_0)^\pm + T \otimes a^\pm \in \mathcal{L}(J)$, $a \in J$, generate a Leibniz algebra $T(J)$

Theorem 4.4 ([24]). *Let J be a Jordan dialgebra. Then $T(J)$ is a solvable Leibniz algebra if and only if J is a Penico solvable;*[39] *$T(J)$ is nilpotent if and only if so is J.*

5. On Embedding of Conformal Algebras

One of the basic facts about associative algebras states that every finite-dimensional associative algebra A can be presented by matrices. Indeed, even if A does not contain a unit element, we may consider $A^\# = A \oplus \Bbbk 1$, and then there is a faithful representation $L : A \to mathrmEnd\, A^\#$, $L(a) : x \to ax$.

For a conformal algebra C of rank n over H, the role of $\operatorname{End} A$ belongs to $\operatorname{Cend} C$, which is isomorphic to the conformal algebra of $n \times n$ matrices over the Weyl conformal algebra. The following properties define an analogue of the unit element for conformal algebras.

Definition 5.1 ([14]). *Suppose C is a conformal algebra. An element $e \in C$ is said to be a (conformal) unit in C if $(e_\lambda x)|_{\lambda=0} = x$ for all $x \in C$ and $e_\lambda e = e$.*

Associative conformal algebra with a unit has a very natural structure.

Proposition 5.1 ([14]). *Let C be a semisimple associative conformal algebra with a unit. Then there exists an associative algebra A with a locally nilpotent derivation ∂ such that $C \simeq H \otimes A$ with respect to the operation*

$$(f(T) \otimes a)_\lambda (g(T) \otimes b) = f(-\lambda) g(T + \lambda) \otimes a \exp\{\lambda \partial\} b, \quad a, b \in A, \ f, g \in H.$$

Remark 5.1. In particular, if $\partial = 0$ then such C is just the current algebra $\operatorname{Cur} A$; if $A = \Bbbk[x]$ and $\partial = d/dx$ then C is the Weyl conformal algebra.

This is the reason why the following problem[40] makes sense: Is it possible to join a conformal unit to an associative conformal algebra. Moreover, a finite associative conformal algebra can be embedded into matrices over the conformal Weyl algebra if and only if one can join a unit to this conformal algebra. The following statement answers positively to this question.

Theorem 5.1 ([41]). *If C is a finite associative conformal algebra which is a torsion-free H-module then there exists an associative conformal algebra C_e with a unit such that $C \subseteq C_e$.*

The last Theorem does not hold for all conformal algebras. For example, consider the free H-module C generated by the space $W = \Bbbk[x] \oplus \Bbbk w$ with the following operation on generators:

$$W_\lambda w = 0, \quad f(x)_\lambda g(x) = f(x - T - \lambda)g(x), \quad w_\lambda f(x) = f(T)w,$$

for $f, g \in \Bbbk[x]$. This is an associative conformal algebra. Assume there exists a unital conformal algebra C_e with unit e such that $C \subseteq C_e$. Then associativity implies

$$((e_\lambda w)_0 x^n) = (e_\lambda (w_{-\lambda} x^n)) = e_\lambda T^n w = (T + \lambda)^n (e_\lambda w), \; n \geq 1.$$

This is impossible since $(e_\lambda w)$ is a polynomial in λ, and its degree does not depend on the choice of n, except for $(e_\lambda w) = 0$. But $(e_0 w) = w$ by the definition of a conformal unit. The contradiction obtained shows that C_e does not exist.

Another problem on embeddings of conformal algebras concerns the following observation. If C is an associative conformal algebra with operations $T : C \to C$ and $(\cdot_\lambda \cdot) : C \otimes C \to C[\lambda]$ then the same module over $H = \Bbbk[T]$ with respect to the new operation

$$[x_\lambda y] = (x_\lambda y) - (y_\mu x)|_{\mu = -T - \lambda}$$

is a Lie conformal algebra.[1] It is natural to denote this conformal algebra by $C^{(-)}$.

In contrast to the case of ordinary algebras, there exist Lie conformal algebras that can not be embedded into associative conformal algebras.[12] However, it is unknown whether the following statement is true.

Conjecture 5.1. *Suppose L is a finite Lie conformal algebra which is a torsion-free H-module. Then L can be embedded into an associative conformal algebra with unit.*

It was shown in [10] that every such L has a maximal solvable ideal R, so L/R is semisimple.

The conjecture obviously holds for semisimple conformal algebras. In [42], it was shown that a nilpotent Lie conformal algebra can be embedded into a nilpotent associative conformal algebra with the same index of nilpotency. This proves Conjecture 5.1 for nilpotent algebras, but it actually holds in much more general class of Lie conformal algebras that includes finite torsion-free solvable algebras.

The idea comes from the construction of a conformal representation for a Leibniz algebra in Theorem 3.1.

Let us first consider a Lie conformal algebra of the type $L = \operatorname{Cur}\mathfrak{g}$, where $\mathfrak{g}$ is a Lie algebra, $\dim \mathfrak{g} < \infty$. This is straightforward to check that the embedding built in the proof of Theorem 3.1 is in fact a homomorphism of conformal algebras. This proves Conjecture 5.1 without a reference to the classical Ado Theorem.

In the more general case, the Lie Theorem for conformal algebras proved in [10] allows to deduce the following fact.

Theorem 5.2 ([41]). *Suppose L is a Lie conformal algebra which is a semi-direct product of a current conformal algebra* $\operatorname{Cur}\mathfrak{g}$ *($\dim \mathfrak{g} < \infty$) and a finite torsion-free solvable Lie conformal algebra R. Then there exists a finite-dimensional associative algebra A such that $L \subseteq (\operatorname{Cur} A)^{(-)}$.*

Acknowledgments

This work was partially supported by RFBR 09-01-00157, SSc 3669.2010.1, SB RAS Integration project N 97, Federal Target Grants 02.740.11.0429, 02.740.11.5191, 14.740.11.0346, and by the ADTP Grant 2.1.1.10726.

References

1. V. Kac, *Vertex algebras for beginners. 2nd ed.* (Providence, RI: American Mathematical Society, 1998).
2. M. Primc, *J. Pure Appl. Algebra* **135**, 253 (1999).
3. A.A. Belavin, A.M. Polyakov, A.B. Zamolodchikov, *Nucl. Phys. B* **241**, 333 (1984).
4. R.E. Borcherds, *Proc. Natl. Acad. Sci. USA* **84**, 3068 (1986).
5. I. Frenkel, J. Lepowsky, A. Meurman, *Vertex operator algebras and the monster.* (Boston etc.: Academic Press, Inc., 1988).
6. C. Dong, J. Lepowsky, *Generalized vertex algebras and relative vertex operators.* (Basel: Birkhäuser, 1993).
7. H.-S. Li, *J. Pure Appl. Algebra* **109**, 143 (1996).

8. M. Roitman, *J. Algebra* **217**, 496 (1999).
9. B. Bakalov, V.G. Kac, A.A. Voronov, *Commun. Math. Phys.* **200**, 561 (1999).
10. A. D'Andrea, V.G. Kac, *Sel. Math., New Ser.* **4**, 377 (1998).
11. S.-J. Cheng, V.G. Kac, *Asian J. Math.* **1**, 181 (1997).
12. M. Roitman, *Sel. Math., New Ser.* **6**, 319 (2000).
13. C. Boyallian, V.G. Kac, J.I. Liberati, *J. Algebra* **260**, 32 (2003).
14. A. Retakh, *J. Algebra* **237**, 769 (2001).
15. V.G. Kac, A. Retakh, *J. Algebra Appl.* **7**, 517 (2008).
16. J.-L. Loday, *Enseign. Math., II. Sér.* **39**, 269 (1993).
17. J.-L. Loday, *C. R. Acad. Sci., Paris, Sér. I* **321**, 141 (1995).
18. J.-L. Loday, Dialgebras., in *Dialgebras and Related Operads.*, ed. J.-L. Loday et al. (Berlin: Springer, 2001) pp. 7–66.
19. D. Liu, *J. Algebra* **283**, 199 (2005).
20. F. Chapoton, An endofunctor in the category of operads. (Un endofoncteur de la catégorie des opérades.), in *Dialgebras and Related Operads.*, ed. J.-L. Loday et al. (Berlin: Springer, 2001) pp. 105–110.
21. P. Kolesnikov, *Sib. Mat. Zh.* **49**, 322 (2008).
22. T. Leinster, *Higher operads, higher categories.* (Cambridge: Cambridge University Press, 2004).
23. A. Pozhidaev, 0-dialgebras with bar-unity, Rota-Baxter and 3-Leibniz algebras., in *Groups, Rings and Group Rings.*, ed. A. Giambruno et al. (Providence, RI: American Mathematical Society (AMS), 2009) pp. 245–256.
24. V. Gubarev, P. Kolesnikov, Conformal representations of Leibniz algebras. Comm. Algebra, to appear.
25. P. Kolesnikov, *Sib. Mat. Zh.* **49**, 540 (2008).
26. Sh.A. Ayupov, B.A. Omirov, On Leibniz algebras., in *Algebra and operators theory*, ed. Y. Khakimdjanov et al. (Dordrecht: Kluwer Academic Publishers, 1998) pp. 1–13.
27. A. Patsourakos, *Commun. Algebra* **35**, 3828 (2007).
28. D. W. Barnes, On Engel's Theorem for Leibniz algebras. (2010), arXiv: 1012.0608v1.
29. M. Aymon, P.-P. Grivel, *Commun. Algebra* **31**, 527 (2003).
30. L.A. Bokut, Y. Chen, C. Liu, *Int. J. Algebra Comput.* **20**, 391 (2010).
31. R. Velásquez, R. Felipe, *Commun. Algebra* **36**, 1580 (2008).
32. M. Bremner, On the definition of quasi-jordan algebra. Comm. Algebra, to appear. ArXiv:1008.2009.
33. M. Bremner, L.A. Peresi, Special identities for quasi-Jordan algebras. Comm. Algebra, to appear. ArXiv:1008.2723.
34. C.M. Glennie, Identities in Jordan algebras., in *Computational Problems in Abstract Algebra. Proc. Conf., Oxford, 1967*, ed. J. Leech (Oxford: Pergamon, 1970) pp. 307–313.
35. V.Yu. Voronin, Special and exceptional Jordan dialgebras. (2010), arXiv:1011.3683.
36. I. Kantor, *Sov. Math., Dokl.* **5**, 1404 (1964).
37. M. Koecher, *Am. J. Math.* **89**, 787 (1967).
38. J. Tits, *Nederl. Akad. Wet., Proc., Ser. A* **65**, 530 (1962).

39. N. Jacobson, *Structure and representations of Jordan algebras.* (Providence, RI: American Mathematical Society, 1968).
40. A. Retakh, *J. Algebra* **304**, 543 (2006).
41. P. Kolesnikov, On finite representations of conformal algebras J. Algebra, to appear. ArXiv:1005.3805.
42. M. Roitman, *J. Lie Theory* **15**, 575 (2005).

On Some Results of Finite Solvable Groups*

Xianhua Li
School of Mathematical Science, Soochow University,
Suzhou Jiangsu 215006, People's Republic of China
E-Mail: xhli@suda.edu.cn

Dedicated to Professor Shum Kar-Ping on the occasion of his 70th birthday

Abstract

In this paper, our recently several work and main results are introduced, including the motivation of these works.

1 Introduction

All groups considered are finite.

Given a group G, let $H < G$ denote that H is a proper subgroup of G and let G_p denote a Sylow p-subgroup of G. If $A \leq G$ then we write A_G for the core $\cap_{x \in G} A_x$ of A in G. Let $\pi(|G|)$ stand for the set of prime divisors of $|G|$. The rest of terminology and notation is standard.

The relation between the properties of subgroups of a group and its structure is always a question of particular interest in the theory of groups. Of the various families of subgroups that can influence on the structure of the group, much have been done. In this paper, we would like to introduce two recent works and some results, which are our join works with my co-workers and have been published in several papers (see [9], [10], [11], [12] and [18]).

*Supported by the National NSF of China (Grant N. 10571128, 10871032) and NSF of Jiangsu Province (Grant N. BK2008156) and Suzhou City Senior Talent Supporting Project.

2 Theta pairs and the structure of finite groups

It is well-known that each maximal subgroup of a soluble group is a complement of a chief factor of G. Taking this elementary fact as starting point, Deskins [5] and Mukherjee and Bhattacharya [13] introduced the interesting concepts of normal index, completions and θ-pairs, respectively. All of them are associated with a maximal subgroup and turned out to be useful in studying the normal structure of a group (see [2], [13], [20],[19] and their references). More recently, the concepts of c-normality and c-supplementation are introduced, respectively, also contribute to a better understanding of the normal structure of the groups(see [17], [3], [6] and [7]). We extend the concept of the θ-pairs from maximal subgroups to proper subgroups of a group such that all these terms are integrated into a uniform concept.

Definition 2.1. *(see Definition 1 [10]) Given a proper subgroup H of a group G, we call (A, B) a θ-pair of H in G if he following conditions hold:*

(i) $A \leq G$, $\langle H, A\rangle = G$ and $B = (A \cap H)_G$;

(ii) if A_1/B is a proper subgroup of A/B and $A_1/B \lhd G/B$ then $G \neq \langle H, A_1\rangle$.

Obviously, the pair (G, H_G) satisfies the condition (i). Hence the set

$$\theta = \{A \mid H_G \leq A \unlhd G,\ G = HA\}$$

is non-empty. Let A be an element of $\mathcal{S}$ of minimal order. It is clear that (A, H_G) is a θ-pair for H in G. Thus we have proved the following proposition:

Proposition 2.1. *For every subgroup H of a group G, the set $\theta(H)$ of all θ-pairs for H in G is non-empty.*

Given $H < G$, denote by $\theta(H)$ the family of all θ-pairs of H. It is easy to verify that θ(H) is the same as the set of θ-pairs of maximal subgroups on finite groups as defined in [13] when H is a maximal subgroup of G.

If $(C, D) \in \theta(H)$ and C is normal and subnormal in G, then we call (C, D) a normal and a subnormal θ-pair of H, respectively. A partial order is defined in $\theta(H)$ by means of $(A, B) \leq (C, D)$ if and only if $A \leq C$. In this case, $B \leq D$ also. Hence,it is clear that (A, B) means a maximal θ-pair for H. The maximal elements in $\theta(H)$ do exist. We here call a maximal element with respect to $\leq$ a maximal θ-pair. The elementary properties of θ-pairs of general proper subgroups of finite groups are similar to these of θ-pairs of maximal subgroups. We list them by some lemmas because they are fundamental.

Lemma 2.1. *(see Lemma 1, 2 [10]) Let H be a proper subgroup of G. Then:*

(1) If (A, B), $(C, D) \in \theta(H)$, and $(C, D) \leq (A, B)$ then $D \leq B$.

(2) There exists a normal θ-pair of H, and if (C, D) is a normal θ-pair of H, then there exists a normal maximal θ-pair (A, B) of H such that $A/B \cong C/D$ and $(C, D) \leq (A, B)$.

(3) If (C, D) is a normal maximal θ-pair of H then $D = H_G$.

Lemma 2.2. *(see Lemma 3 in [10]) Suppose that $H < G$ (1) If $N \lhd G$ and $N \leq D$ then (C, D) is a θ-pair of H in G if and only if $(C/N, D/N)$ is a θ-pair of H/N in G/N.*

(2) Let $N \lhd G$ and $N \leq H$, but $N \not\leq D$. If (C, D) is a maximal θ-pair of H, then there exists a normal θ-pair (A, B) of H such that N/B and A/B is a section of CN/DN.

Recall that a subgroup H of a group G is said to be c-normal (weak c-normal and c-supplemented, respectively) in G if there exists a normal subgroup (a subnormal subgroup and a subgroup, respectively) K of G such that $G = HK$ and $H \cap K \leq H_G$.

Lemma 2.3. *(see Lemma 4 in [10]) Let G be a group and let H be a subgroup of G. Then*

(1) If H is c-supplemented in G, $H \leq M \leq G$, then H is c-supplemented in M;

(2) Let $N \lhd G$ and $N \leq H$. Then H is c-supplemented in G if and only if H/N is c-supplemented in G/N;

(3) Let π be a set of primes. Let N be a normal π'-subgroup and let H be a π-subgroup of G. If H is c-supplemented in G then HN/N is c-supplemented in G/N;

(4) Let $H \leq G$ and $L \leq \Phi(H)$. If L is c-supplemented in G then $L \lhd G$ and $L \leq \Phi(G)$.

Theorem 2.1. *(see Theorem 1 in [10]) Let G be a group and let H be a subgroup of G. Then the following statements hold:*

(1) H is c-normal in G if and only if there is a normal s-pair (A, B) for H in G such that $H \cap A = B$.

(2) H is weak c-normal in G if and only if there is a subnormal θ-pair (A, B) for H in G such that $H \cap A = B$.

(3) H is c-supplemented in G if and only if there is an s-pair (A, B) of H in G such that $H \cap A = B$.

The above Theorem is a characterization theorem of c-normality, weak c-normality and c-supplementation of a group by using the θ-pairs.

By Theorem 2.1,the definitions of c-normality etc can be developed by imposing some conditions on the θ-pairs. Hence all results which are obtained by using c-normal subgroup, weak c-normal subgroup and c-supplemented subgroup of G will have the form in $\theta-$pairs of subgroups of G.

In the following theorems, by using the concept of θ-pair of subgroups of G, we can give some characterization theorems for the solvability, supersolvability and nilpotence of a group. It is well-known that if every Sylow subgroup of a group G is cyclic then G is metacyclic. For the group G containing some non-cyclic Sylow subgroups, we get the following theorem.

Theorem 2.2. *(see Theorem 3, 4 in [9]) Let G be a group with at least a non-cyclic Sylow subgroup. Then G is solvable (supersolvable, nilpotent, respectively) if and only if for every maximal subgroup H of any non-cyclic Sylow subgroup of G, there exists* $(A, B) \in \theta(H)$ *such that* A/B *is solvable (supersolvable, nilpotent, respectively).*

Considering the generalization of Theorem 2.2, we can prove the following result.

Theorem 2.3. *(see Theorem 3.1 in [11]) Assume that for every maximal subgroup of any Sylow subgroup of a normal subgroup N of G there is a normal* θ*-pair* (C, D) *such that* C/D *is supersolvable. Then G is supersolvable.*

Corollary 2.1. *(see Corollary 3.3 in [11]) Suppose that for each maximal subgroup of each Sylow subgroup of* G'*, there exists a normal* θ*-pair* (C, D) *such that* C/D *is supersolvable, then G is supersolvable.*

Theorem 2.4. *(see Theorem 3.4 in [11]) Assume that for every maximal subgroup of any Sylow subgroup of a normal subgroup N of G there is a normal* θ*-pair* (C, D) *such that* C/D *is nilpotent. Then G is nilpotent.*

3 On the normal indices of proper subgroups of finite groups

In 1954, Huppert obtained the following famous theorem: a group G is supersolvable if and only if the index of every maximal subgroup of G is prime. After that, Deskins in [5] introduced the normal index of a maximal subgroup M of a group G. The normal index of a maximal subgroup M of a group G, denoted by $\eta(G : M)$, is the order of a chief factor H/K of G, where H is a minimal supplement of M in G. Several authors have already given the characterization theorems of solvablity, supersolvablity, et al. of a group G by using the normal indices of some maximal subgroups(see [5], [4], [14], [15] and [1]). This approach provides some well quantitative versions and ways of characterizations of a group G. However, the definition of the normal index is confined to the maximal subgroups only and the maximal subgroups are some special subgroups, we can not go further. In fact, Guo and Li in [8] have pointed out that the concept of c-normality provides a better tool for us and it makes the whole matters easier and simpler than using the normal indices alone. We now extend the concept of the normal index from maximal subgroups to some proper subgroups of a group. We prove that if the normal index of a subgroup H of G is equal to the index $|G : H|$, then H is c-normal in G. Hence, we obtain a quantitative characterization of c-normality of subgroups by using the normal index. It means that the c-normality of a subgroup H is equivalent to that H has the special normal index. On the other hand, the normal index is a number, which provides a new tool and a new approach to study the properties of groups by using the quantitative information of themselves. By considering the normal index of a proper subgroup, we are able to obtain the characterizations of a group to be solvable, supersolvable and nilpotent.

3.1 Definitions and preliminary results

Definition 3.1. *(see Definition 2.2 in [18]) Let H be a proper subgroup of a group G and K a normal subgroup of G. Then $|K/K \cap H_G|$ is called the normal index of H in G, denoted by $\eta^*(G : H)$, if K satisfies the following conditions:*

(1) $G = KH$;

(2) if $G = TH$ for any normal subgroup T of G, then we have $|T/T \cap H_G| \geq |K/K \cap H_G|$.

Let G be a group and H be a proper subgroup of G. Let K_1 and K_2 be normal subgroups of G such that K_1 satisfies the conditions of Definition 3.1 and K_2 satisfies $H_G \leq K_2$, $G = K_2H$ and $|K_2| \leq |T|$ for any normal subgroup T of G such that $H_G \leq T$ and $G = TH$. Then $G = K_1H = K_2H$, $|K_2/H_G| = |K_2/K_2 \cap H_G| \geq |K_1/K_1 \cap H_G|$ by Definition 3.1. On the other hand, by $G = (K_1H_G)H$ and the hypothesis of K_2, we have $|K_1/K_1 \cap H_G| = |K_1H_G/H_G| \geq |K_2/H_G|$. Hence, $|K_1/K_1 \cap H_G| = |K_2/H_G|$. Thus, we have shown that Definition 3.1 is equivalent to the following definition 3.2.

Definition 3.2. *(see Definition 2.2 in [12]) Let H be a proper subgroup of a group G and K a normal subgroup of G containing H_G. Then $|K/H_G|$ is called the normal index of H in G, if K satisfies*

(1) $G = KH$;

(2) $|K| \leq |T|$ for any normal subgroup T of G such that $H_G \leq T$ and $G = TH$.

Theorem 3.1. *(see Proposition 2.1 and Lemma 2.2 in [18]) Let G be a group. Then the following statements hold:*

(1) If H is a maximal subgroup of G, then $\eta^*(G:H)=\eta(G:H)$;

(2) If H is a normal subgroup of G, then $\eta^*(G:H)=|G:H|$.

(3) If $N \trianglelefteq G$ and H is a subgroup of G such that $N \leq H$, then $\eta^*(G/N:H/N)=\eta^*(G:H)$.

The part(1) in the above Theorem shows that the normal index is a generalization of the concept defined in [5].

3.2 Normal index theorems and properties of a group

In the theory of groups, some index theorems have fundamental importance, for example, Lagrange's theorem, et.al.. We are interested in what will happen if we replace the index by the normal index in some index theorems. That is, if the normal index theorem is true, what kind of properties does G has? Let H and K be subgroups of G. Firstly, because $|G:H| \mid \eta^*(G:H)$, if $(\eta^*(G:H),\eta^*(G:K))=1$, then $(|G:H|,|G:K|)=1$ and so $G=HK$. Secondly, as an analogy of Lagrange's theorem, we consider the case $|G|=|H|\eta^*(G:H)$. This is equivalent to $|G:H|=\eta^*(G:H)$.

Theorem 3.2. *(see Theorem 3.1 in [12]) Let H be a proper subgroup of a group G. Then $\eta^*(G:H)=|G:H|$ if and only if H is c-normal in G.*

Theorem 3.2 gives a quantitative characterization of the c-normality of subgroups. By replacing the condition H is c-normal in G by $\eta^*(G:H)=|G:H|$, we can obtain a quantitative version of all results obtained by using the c-normal subgroups. For instance, Theorem 3.2 in [8], Theorem 4.1 in [17] and Theorem 4.2 in [16] can be written in the following forms, respectively.

(1) Let G_1 be a class of Frobenius groups with elementary abelian kernel N and cyclic complement M, such that $|M|$ is square-free and the order of M is not prime. If G is G_1-free and $\eta^*(G:L)=|G:L|$ for each 2-maximal subgroup L of G, then G is supersolvable.

(2) Let G be a group. Suppose $\eta^*(G:P_1)=|G:P_1|$ for every Sylow subgroup P of G and every maximal subgroup P_1 of P. Then G is supersolvable.

(3) Let G be a group and p be the smallest prime divisor of $|G|$. Assume that G is A_4-free and $\eta^*(G:P_2)=|G:P_2|$ for every 2-maximal subgroup P_2 of the Sylow p-subgroups of G. Then $G/O_p(G)$ is p-nilpotent.

Secondly, to be a parallel theorem as the index theorem $|G|=|G:K||K:H|$ for $H \leq K \leq G$, we consider the question: if $\eta^*(G:H)=\eta^*(G:K)\eta^*(K:H)$ for any $H \leq K \leq G$, what properties does G have?

Definition 3.3. *(see Definition 3.2 in [12]) Let G be a group. Then G is called an NIT-group if $\eta^*(G:H)=\eta^*(G:K)\eta^*(K:H)$ for any $H \leq K \leq G$.*

Theorem 3.3. *(see Theorem 3.4 in [12]) Let G be a group. Then the following statements are equivalent.*

(1) G is an NIT-group.

(2) G is a supersolvable NIT-group.

(3) $\eta^*(G : H) = |G : H|$ for any subgroup H of G.

(4) H is c-normal in G for any subgroup H of G.

Theorem 3.4. *(see Theorem 3.5 in [12]) Let G be a group. If for every maximal subgroup M of G and every maximal subgroup H of M, $\eta^*(G : H) = \eta^*(G : M)\eta^*(M : H)$, then G is solvable and every 2-maximal subgroup of G is c-normal in G.*

Corollary 3.1. *(see Corollary 3.6 in [12]) Let G be a group and $\mathcal{F}$ be the class of Frobenius groups with elementary abelian kernel N and cyclic complement K, such that $|K|$ is square-free and the order of K is not prime. If G is $\mathcal{F}$-free and $\eta^*(G : H) = \eta^*(G : M)\eta^*(M : H)$ for every maximal subgroup M of G and every maximal subgroup H of M, then G is supersolvable.*

3.3 The normal indices of some special subgroups

First we consider the influence of the normal indices of some 2-maximal subgroups of a group G on the structure of G.

Theorem 3.5. *(see Theorem 3.1 in [18]) Let G be a group. Then G is solvable if and only if there exists a solvable 2-maximal subgroup L of G such that $\eta^*(G : L) = |G : L|$.*

Theorem 3.6. *(see Theorem 3.2 in [18]) Let G be a group. Then G is solvable if $\eta^*(G : L) = |G : L|$ for every 2-maximal subgroup L.*

Theorem 3.7. *(see Theorem 3.3 in [18]) Let p be a prime dividing the order of a group G. If $\eta^*(G : L)$ is the power of some prime or p'-number for every 2-maximal subgroup L of G, then G is $p-$solvable.*

Theorem 3.8. *(see Theorem 3.4 in [18][19]) Let p be a prime dividing the order of a group G. If $\eta^*(G : L)_p = 1$ for every 2-maximal subgroup L of G, then G is $p-$nilpotent.*

Theorem 3.9. *(see Theorem 4.6 in [12]) Let G be a group. Suppose that G has a maximal subgroup M such that $\eta^*(G : L) = |G : L|$ for each maximal subgroup L of M. Then*

(1) G is solvable.

(2) M/M_G is a cyclic group of square-free order.

(3) Assume that $M_G < M$. Then $Z(G/M_G) = 1$, $(G/M_G)' = F(G/M_G)$ is a Sylow subgroup of G/M_G and a minimal normal complement subgroup of M/M_G in G/M_G.

(4) If $M_G = 1$, then M is a cyclic group of square-free order, $F(G)$ is a Sylow subgroup of G and a minimal normal complement subgroup of M in G. Furthermore, for each maximal subgroup K of G, if K is not normal in G, then there exist $g \in G$ such that $K = M^g$ and $K \cap M = 1$ if $g \notin M$.

Corollary 3.2. *(see Corollary 4.7 in [12]) Let G be a group. Then G is solvable if $\eta^*(G : L) = |G : L|$ for every 2-maximal subgroup L of G.*

Theorem 3.10. *(see Theorem 4.8 in [12]) A group G is solvable if and only if there exists a solvable 2-maximal subgroup L of G such that $|\pi(\eta^*(G : L))| \leq 2$.*

Theorem 3.11. *(see Theorem 4.9 in [12]) Let G be a group. If G has a maximal subgroup M satisfying one of the following conditions, then G is supersolvable:*

(1) $\eta^*(G : L)$ is either square-free or the square of a prime for any maximal subgroup L of M.

(2) M is supersolvable, $M_G = 1$ and M has a maximal subgroup L such that $\eta^*(G : L)$ is either square-free or the square of a prime.

(3) M is supersolvable , $M_G = 1$ and $\eta(G : M)$ is square-free.

(4) M is supersolvable, $F(G) \nleq M$ and M has a maximal subgroup L such that $\eta^*(G : L)$ is either square-free or the square of a prime.

(5) M is supersolvable, $F(G) \nleq M$ and $\eta(G : M)$ is square-free.

(6) M is cyclic and $\eta(G : M)$ is square-free.

(7) M has a cyclic maximal subgroup L such that $\eta^*(G : L)$ is either square-free or the square of a prime.

Corollary 3.3. *(see Corollary 4.10 in [12]) Let G be a group. Then G is supersolvable if G satisfies one of the following conditions:*

(1) $\eta^*(G : L)$ is either square-free or the square of a prime for every 2-maximal subgroup L of G.

(2) $\eta^*(G : L)$ is square-free for every 2-maximal subgroup L of G.

Theorem 3.12. *(see Theorem 3.6 in [18]) Let G be a group. Then G is supersolvable if $\eta^*(G : L)$ is square free for every 2-maximal subgroup L.*

Now we consider the influence of the normal indices of some Sylow subgroups of a group G on the structure of G.

Theorem 3.13. *(see Theorem 4.11 in [12]) Let p be a prime dividing the order of a group G and P a Sylow p-subgroup of G. Assume that $\eta^*(G:P) = p^\alpha|G:P|$ where $\alpha \leqslant 2$. Then the following conclusions are true:*

(1) If $p=2$ and G is A_4-free, then $G/O_2(G)$ is 2-nilpotent. Furthermore, if $|O_2(G)| \leqslant 2^2$, then G is 2-nilpotent.

(2) If $p \neq 2$ and $(|G|, p^2-1) = 1$, then $G/O_p(G)$ is p-nilpotent. Furthermore, if $|O_p(G)| \leqslant p^2$, then G is p-nilpotent.

Theorem 3.14. *(see Theorem 4.12 in [12]) Let G be a group and $p \in \pi(G)$. Assume that there exists a Sylow p-subgroup P of G such that $\eta^*(G:P) = p^\alpha|G:P|$, where $\alpha \leqslant 1$. If either $p = min\pi(G)$ or $(|G|, p-1) = 1$, then $G/O_p(G)$ is p-nilpotent. Specifically, G is solvable.*

Theorem 3.15. *(see Theorem 3.5 in [18]) Let p be a prime dividing the order of a group G and P a Sylow $p-$subgroup of G. If $\eta^*(G:P) = |G:P|$, then G is $p-$solvable.*

Corollary 3.4. *(see Corollary 3.1 in [18][19]) Let G be a group. Then G is solvable if $\eta^*(G:P) = |G:P|$ for every Sylow $p-$subgroup P.*

Corollary 3.5. *(see Corollary 3.2 in [18]) Let P be a Sylow $2-$subgroup of a group G. If $\eta^*(G:P) = |G:P|$, then G is solvable.*

A 2-maximal subgroup H of a group G is called c-2-maximal if $H < \cdot M < \cdot G$ and $|G:M|$ is composite.

Theorem 3.16. *(see Theorem 3.7 in [18]) Let G be a group. Then G is solvable if $\eta^*(G:L) = |G:L|$ for every c-2-maximal subgroup L.*

Theorem 3.17. *(see Theorem 3.7 in [18]) Let G be a group. Then G is supersolvable if $\eta^*(G:L)$ is square free for every c-2-maximal subgroup L.*

References

[1] A. Ballester-Bolinches, On the normal index of maximal subgroups in finite groups, Journal of Pure and Applied Algebra, 64(1990), 113-118.

[2] A. Ballester-Bolinches, Saturated formations, theta pairs and completions in finite groups, Siberian Math. J., 37(2)(1996), 207-212

[3] A. Ballester-Bolinches, Xiuyun Guo and Yanming Wang, c-supplemented subgroups of finite groups, Glasgow Math. J., 42(3)(2000), 383-389

[4] J. C. Beidleman and A. E. Spencer, The normal index of maximal subgroups in finite groups, Illinois J. Math., 16(1972), 95-101.

[5] W. E. Deskins, On maximal subgroups, Proc. Symp. Pure Math., Amer. Math. Soc., 1(1959), 100-104

[6] Xiuyun Guo, Xiuping Sun and K.P. Shum, On the solvability of certain c-supplemented finite groups. Southeast Asian Bull. Math. 28 (2004), no. 6, 1029V1040.

[7] Xiuyun Guo and K. P. Shum, Finite p-nilpotent groups with some subgroups c-supplemented, J. Aust. Math. Soc. 78 (2005), no. 3, 429V43.

[8] Duyu Li, Xiuyun Guo, The influence of c-normality of subgroups on the structure of finite groups, Journal of Pure and Applied Algebra 150(2000), 53-60.

[9] Xianhua Li and A. Ballester-Bolinches, On supplements of subgroups of finite groups, Bollettino U.M.I, (8)9-B(2006), 567-574

[10] Xianhua Li and Shiheng Li, Theta pairs and the structure of finite groups, Siberian Mathematical Journal, 45(3)(2004), 557-561

[11] Xianhua Li and Ailing Nan, The structure of finite groups and the θ-pairs of subgroups, CONTRIBUTIONS TO GENERAL ALGEBRA 19 Proceedings of the Olomouc Conference 2010 (AAA79 + CYA25) Verlag Johannes Heyn, Klagenfurt 2010, 155-158

[12] Xianhua Li and Xinjian Zhang, On the normal indices of proper subgroups of finite groups, Journal of Pure and Applied Algebra, 214(2010), 1285-1290

[13] N. P. Mukherjee and P. Bhattacharya, On Theta Pairs for A Maximal Subgroup, Proc. Amer. Math. Soc., 109(3)(1990), 589-596

[14] N. P. Mukherjee, Prabir Bhattacharya, The normal index of a finite group, Pacific J. Math., 132(1)(1988), 143-148.

[15] N. P. Mukherjee and Prabir Bhattacharya, The normal index of a maximal subgroup of a finite group, Proc. Amer. Math. Soc., 106(1)(1989), 25-32.

[16] Yanming Wang, Finite groups with some subgroups of Sylow subgroups c-supplemented, Journal of Algebra, 224(2000),467-478.

[17] Yanming Wang, C-Normality of Groups and its properties, Journal of Algebra, 180(1996), 954-965

[18] Xia Yin, Xianhua Li, The normal index of subgroups in finite groups, Asian-European Journal of Mathematics, 3(3)(2010), 511-519

[19] Yaoqing Zhao and Shirong Li, On Theta Pairs for Maximal Subgroups, J. Pure and Applied Algebra, 147 (2000), 133-142

[20] Yaoqing Zhao, On the Deskins completions, theta completions and theta Pairs for maximal subgroups, Comm. in Algebra, 26(10)(1998), 3141-3153

A Note on the Dual of Some Special Biserial Algebras

I. M. Maris*[1,2], D. Kariman[1,3], Risnawita[1], and I. Muchtadi-Alamsyah[1]

[1] *Algebra Research Division, Faculty of Mathematics and Natural Sciences, Institut Teknologi Bandung, Indonesia,*
* *E-mail: ikametizamaris@yahoo.com*
[2] *Program Studi Pendidikan Matematika Stain Batusangkar, Indonesia,*
[3] *Jurusan Matematika STKIP PGRI Sumatera Barat, Indonesia*

It is known that quivers can be used to represent algebras and coalgebras. These algebras and coalgebras are called the path algebras and path coalgebras. Thus, the quiver of a path coalgebra (resp. algebra) can be obtained from the quiver of a path algebra (resp. coalgebra) by reversing its arrows.We will apply this fact to show that the dual special biserial algebras are still special biserial coalgebras.

Keywords: Quiver, Path Algebra, Path Coalgebra, Special Biserial Algebras

1. Introduction

Algebras and coalgebras can be represented as quiver (directed graph). In A. Assem [1] and W. Chin [2], it is explained how to construct algebras (resp. coalgebras) from quiver which are now called path algebras (resp. path coalgebras). In Muchtadi and Garminia [7], it is shown that the dual of bound path algebras (resp. coalgebras) are bound path coalgebras (resp. algebras). In this paper, we will apply these results to study some special biserial algebras.

The defining property of biserial algebras was first mentioned by Tachikawa in [9].The name "biserial" was first given by Fuller in [4]. A biserial algebra is an algebra which satisfies the radical of every indecomposable left or right module can be decomposed into the sum of two uniserial modules U and V, where $l(U \cap V) = 1$. A special biserial algebra is a special type of biserial algebra that can be represented as $A \cong kQ/I$ with (Q, I) a bound quiver satisfying: each vertex of Q is start-point or end-point of at most two arrows and for an arrow α, there is at most one arrow β such that $\alpha\beta \notin I$ and at most one arrow γ such that $\gamma\alpha \notin I$ (see Skowronski and Waschbusch in [8]). Several special biserial coalgebras can be defined dually.

In this note, we will discuss about some special biserial algebras and we will show that the dual of a special biserial algebra is still a special biserial coalgebra. The result is the similar as the recent result given in W.Chin [3] but we use a different approach.

This paper is organized as follows: in section 2, we will review some results on quivers of bound path algebras and bound path coalgebras. In section 3, we will show that dual of special biserial algebra is special biserial coalgebra. For basic notions on path algebras and path coalgebras, The readers are referred to A. Assem [1], Muchtadi [5] , Muchtadi and Garminia [6].

2. Quiver of Bound Path Algebras and Bound Path Coalgebras

For a quiver Q with a set of vertices $\{1, 2, ..., n\}$, we denote by Q^{op} the quiver having same set of vertices, and its arrows are obtained by reversing the arrows of the quiver Q. We denote by kQ the path algebra and CQ the path coalgebra.

In the following, we first cite some results on the bound path algebras and bound path coalgebras which are taken from Muchtadi and Garminia [7].We will just state the results without giving the proofs.

Definition 2.1. Let I be an admissible ideal of kQ^{op} and (Q^{op}, I) be a quiver with relations. Then, the bound path coalgebra of (Q^{op}, I) is defined by the subspace of CQ so that $C(Q, I) = \{a \in CQ : \langle a, I\rangle = 0\}$.

Lemma 2.1. *Let Q be a finite quiver and C a finite dimensional relation subcoalgebra of CQ, Then,there exists an admissible ideal I of KQ^{op} such that $C = C(Q, I)$.*

Theorem 2.1. *Let Q be a finite quiver.*

(1) If I is an admissible ideal of KQ^{op} then $D(KQ^{op}/I) \cong C(Q, I)$.
(2) If C is a finite dimensional relation subcoalgebra of CQ then $D(C) \cong KQ^{op}/C^{\perp KQ^{op}}$ where $C^{\perp KQ^{op}} = \{w \in KQ^{op} : \langle C, w\rangle = 0\}$.

Corollary 2.1. *Let C be a relation subcoalgebra of CQ and $D(C)$ be its dual. Then, $D(C)$ is isomorphic to the bound path algebra of quiver (Q^{op}, I),* *i.e., the quiver of dual path algebra $D(C)$ can be obtained from the quiver Q by reversing arrows.*

Corollary 2.2. *Let KQ/I be a bound path algebra, where I is an admissible ideal of KQ and $D(KQ)$ be its dual. Then, $D(KQ/I)$ is isomorphic to the bound path coalgebra of quiver (Q^{op}, I) i.e., the quiver of dual bound path algebra $D(KQ/I)$ can be obtained from the quiver Q by reversing the arrows.*

3. The Dual of a Special Biserial Algebra

Definition 3.1. An algebra A is called special biserial if $A \cong kQ/I$ with (Q, I) a bound quiver satisfying the following conditions:

(1) Each vertex of Q is start-point or end-point of at most two arrows.
(2) For an arrow α, there is at most one arrow β such that $\alpha\beta \notin I$ and at most one arrow γ such that $\gamma\alpha \notin I$.

Let Q be the quiver with vertex set Q_0 dan arrow set Q_1. Let $C \subset CQ$ be a finite dimensional relation subcoalgebra. If $c = \sum a_i p_i \in C$ is a k-linier combination of distinct paths p_i with $a_i \neq 0$ for all i, then, c is said to be a *reduced element* and is denoted by $supp(c) = \{p_i\}$. We also write $p_i \vdash C$ (we say p_i appears in C) if $p_i \in supp(c)$ for some $c \in C$.

Definition 3.2. A finite dimensional relation subcoalgebra C of CQ is called a special biserial coalgebra if it satisfies the following conditions:

(1) Every vertex of Q is the start of at the most two arrows and the end of at most two arrows.
(2) Given any arrow β, there is at most one arrow α such that $\alpha\beta \vdash B$ and at most one arrow γ such that $\beta\gamma \vdash B$.

In W. Chin [2], it is shown that for a relational finite dimensional coalgebra C, C is a special biserial coalgebra if and only if $D(C)$ is a (basic) special biserial algebra. Now, we state below our Main Theorem.

Theorem 3.1. *Let A be a basic finite dimensional algebra. Then, A is special biserial if and only if $D(A)$ is a relational special biserial coalgebra.*

Before giving the proof, we need to establish the following lemma:

Lemma 3.1. *Let Q be a quiver and I an admissible ideal. Then, for every path p, $p \notin I$ if and only if $p^* \vdash C(Q, I)$.*

Proof. If $p \notin I$ then $p^* \in D(KQ/I) = C(Q,I)$, hence $p^* \vdash C(Q,I)$. Conversely, if $p \in I$, then $p^* = 0$ and $p^* \notin suppc$, for all $c \in C(Q,I)$. □

Proof. (proof of the Main Theorem) Based on Corollary 2.2, we can get that by reversing the arrows of the quiver of the special biserial algebra so that we will get the arrows of the quiver of special biserial coalgebra. Hence, the first conditions are equivalent. By Lemma 3.1, it is clear that the second conditions are equivalent as well. □

Acknowledgments

The research of the authors are supported by Hibah Penelitian Ilmu Pengetahuan Terapan DIPA DIKTI ITB based on SK Dekan FMIPA No.14a/ SK/ K01.7/ KP/ 2010.

References

1. I.Assem, A. Skowronski,and D. Simson, *Elements of Reprensentation Theory of Associative Algebras*, London Mathematical Society Student Texts 65, (Cambridge University Press, London, 2006).
2. W.Chin, *A brief introduction to coalgebra representation theory in Hopf Algebra*, Lecture Notes in Pure and Appl.Math 237, 109-131, (Marcel Dekker, New York, 2004).
3. W.Chin, *Special biserial coalgebras and representations of quantum SL(2)*,(2010) arXiv:math/0609370v4 [math.QA].
4. K.R.Fuller, *Biserial Ring*, Lecture Notes in Mathematics 734, 64-90 (Springer, 1979).
5. I.Muchtadi-Alamsyah, *Algebras and quivers*, Proceeding ISSM Paris, (2005).
6. I. Muchtadi-Alamsyah and H. Garminia, *Quiver of Path Algebras and Path Coalgebras,* Proceeding IndoMS International Conference on Mathematics and Its Applications, 23-28, (2009).
7. I. Muchtadi-Alamsyah and H. Garminia, *Quiver of Bound Path Algebras and Bound Path Coalgebras, ITB J. Sci.* **42 A No 2**, 153-162, (2010).
8. A. Skowronski and J. Waschbusch, *Representation-finite biserial algebras, Journal fur Reine und Angewandte Mathematik*, **345** 172-181, (1983).
9. H. Tachikawa, *On algebra of which every indecomposable reresentation has an irreducible one as the top or the bottom Loewy constituent, Mathematische Zeitschrift* **75**, 215-227, (1961).

On Recognizability of Finite Groups by Spectrum

V. D. Mazurov

Algebra Department, Sobolev Institute of Mathematics,
Siberian Branch of Russian Academy of Sciences,
Novosibirsk, 630090, Russia
E-mail: mazurov@math.nsc.ru

For a finite group G, denote by $\mathcal{G}$ the set of all nonisomorphic finite groups having the same set of element orders as G. We prove that $\mathcal{G}$ is infinite if and only if there exists $H \in \mathcal{G}$ with a non-trivial soluble normal subgroup.

Keywords: spectrum; recognizable group; isospectral groups.

1. Introduction

For a finite group G, denote by $\pi(G)$ the set of prime divisors of $|G|$ and by $\omega(G)$ *the spectrum* of G, i.e. the set of element orders of G. This set is a subset of the set of natural numbers which is closed under divisibility condition and hence is uniquely defined by the set $\mu(G)$ of its maximal under divisibility elements. A group G is said to be *recognizable* by spectrum $\omega(G)$ (shortly, *recognizable*), if every finite group H with $\omega(H) = \omega(G)$ is isomorphic to G. In other words, G is recognizable if $h(G) = 1$ where $h(G)$ is the number of pairwise non-isomorphic groups H which are *isospectral* to G, that is having the same spectrum as G. A group G is said to be *irrecognizable* if $h(G)$ is infinite.

The goal of this paper is a proof of the following

Theorem 1.1. *A finite group G is irrecognizable if and only there exists a finite group H isospectral to G with a non-trivial soluble normal subgroup.*

2. Proofs

Let G be a finite group.

2.1. *Sufficiency*

Suppose first that $H = H_0$ is a finite group with $\omega(H) = \omega(G)$ which contains a non-trivial soluble normal subgroup W. Without loss of the generality, we can assume that W is an elementary abelian p-group for some prime p. By induction on n prove that, for every $n = 0, 1, ...$, there exists a group H_n of order $|W|^n|H|$ which is isospectral to G and contains a normal subgroup W_n of order $|W|$. This is obvious for $n = 0$. Suppose that, for $n > 0$ there exist such a group H_{n-1}. Denote by H_n the natural semidirect product of $V = W_{n-1}$ by $K = H_{n-1}$ where the action of K on V is defined by conjugation in K. It is obvious that $|H_n| = |W||H_{n-1}| = |W|^n|H|$.

We claim that $\omega(H_n) = \omega(G)$. Clearly, $\omega(G) = \omega(K) \subseteq \omega(H_n)$. Let $(g, v) \in H_n$, $g \in K$, $v \in V$ and m be the order of Vg in G/V. If $V \ni u = g^m \neq 1$ then the order of $(g, v)^m = (u, *)$ is equal to p, so the order of (g, v) is equal to the order of g. If $g^n = 1$ and $1 \neq (g, v)^n = (1, vv^g \cdots v^{g^{n-1}})$ then $(gv)^n = vv^g \cdots v^{g^{n-1}} \neq 1$, which implies that orders of (g, v) and gv are equal.

Thus, we have an infinite sequence $H_0, H_1, ...$ of groups of increasing orders every of which is isospectral to G, and hence G is irrecognizable.

2.2. *Necessity*

We need the following claims.

Lemma 2.1. *For every natural n, there are only a finite number of finite simple groups of period n.*

Here, we say that G is of period n if $x^n = 1$ for all $x \in G$.

Proof. This follows from the Classification of the Finite Simple Groups (see[1] or[2]). □

Lemma 2.2. *For given natural n, k and a prime p, there exist only finitely many finite groups of period p^n with k generators.*

Proof. This was proven by A.I. Kostrikin[3] in the case $n = 1$ and later by E.I. Zelmanov[4,5] for arbitrary n. □

As has been proved already in 1956 by Ph. Hall and G. Higman,[6] validity of Lemmas 2.1 and 2.2 implies

Lemma 2.3. *For given natural n and s, there are only a finite number of finite groups of period n with s generators.*

Let $k(n,s)$ be the maximal order of a finite group of period n with s generators.

Lemma 2.4. *Let ω be a set of natural numbers. Then there exists a natural $k = k(\omega)$ such that every finite group G with $\omega(G) = \omega$ contains a subgroup H for which $\omega(H) = \omega(G)$ and $|H| \leq k$.*

Proof. Let $\mu(G) = \{m_1, ..., m_s\}$ and n be the lowest common multiple of $m_1, ..., m_s$. Then $x^n = 1$ for every $x \in G$. For every $i = 1, ..., s$, choose $g_i \in G$ of order m_i and set $H = \langle g_1, ..., g_s \rangle$. Clearly, $\mu(H) = \mu(G)$, so H is isospectral to G. By Lemma 2.3, $|H| \leq k(n,s)$. □

Now, suppose that G is irrecognizable. Let $\mathcal{G}$ be the set of all non-isomorphic finite groups which are isospectral to G. Clearly, $\mathcal{G}$ is infinite and so contains a group of arbitrary large order.

Suppose that, contrary to the conclusion of Theorem 1.1, every member of $\mathcal{G}$ has no non-trivial soluble normal subgroup. For every $X \in \mathcal{G}$, denote by $S(X)$ the socle of X, i.e. the subgroup of X generated by all minimal normal subgroups of X. Then X is isomorphic to an automorphism group of $S(X)$ and hence $\mathcal{G}$ contains a group X with $S(X)$ of arbitrary large order.

Since $S(X) = L_1 \times ... \times L_t$ where $L_i, i = 1, ..t$, is a simple non-abelian group of period dividing the period of G, Lemma 2.1 implies that there exists $X \in \mathcal{G}$ such that $t > k^2$ where $k = k(\omega)$. Without loss of the generality, we can suppose that G coincides with this X.

By Lemma 2.4, there exists a subgroup H of G such that H is isospectral to G and $|H| \leq k$. Since $\omega(SH) = \omega(G)$ where $S = S(G)$, we can suppose that $G = SH$.

Obviously, G acts on the set $\mathcal{S} = \{L_1, ..., L_t\}$ by conjugation and the size of each orbit is at most $|H| \leq k$. Thus, G has at least $t/k > k$ orbits on $\mathcal{S}$ and hence $S = M_1 \times ... \times M_l$ where $M_i, i = 1, ..., l$, is a nontrivial normal in G subgroup equal to the direct product of L_j-s from one and the same orbit, and $l > k \geq |H|$.

Clearly, there exists some M_i with $M_i \cap H = 1$. Let T be a non-trivial Sylow subgroup of this M_i and $N = N_G(T)$. By Frattini argument, $M_i N = G$ and G/M_i is isomorphic to $N/N \cap M_i$. Since $H \cap M_i = 1, H$ is a section of G/M_i an hence is a section of N. It follows that $\omega(G) = \omega(H) \subseteq \omega(N) \subseteq \omega(G)$ and hence N is isospectral to G. Since N contains a non-trivial soluble normal subgroup T, Theorem 1.1 is proved.

Note that the first part of proof follows the arguments from.[7]

Acknowledgements

We thank the Russian Foundation of Basic Researches for its support through the grants 10-01-90007 and 11-01-00456. Our work is supported also by the program "Development of scientific potential of the higher school" of Russian Ministry of Science and Education (project 2.1.1.10726) and the Federal Target Grant "Scientific and educational personnel of innovation Russia" for 2009-2013 (government contracts 02.740.11.5191 and 14.740.11.0346).

References

1. G. A. Jones, Varieties and simple groups, *J. Austral. Math. Soc.* **17**, 163-173 (1974).
2. M. Aschbacher, P. B. Kleidman, M. W. Liebeck, Exponents of almost simple groups and an application to the restricted Burnside problem, *Math. Z.* **208**, 401-409 (1991).
3. A. I. Kostrikin, *Vokrug Bernsajda* (Russian), Moskva: Izdatel'stvo Nauka. Glavnaya Redaktsiya Fiziko- Matematicheskoj Literatury, 1986. English translation: A. I. Kostrikin, *Around Burnside.* Ergebnisse der Mathematik und ihrer Grenzgebiete, 3. Folge, **20**. Berlin etc.: Springer-Verlag, 1990.
4. E. I. Zel'manov, Solution of the restricted Burnside problem for groups of odd exponent (Russian), *Izv. AN SSSR, Ser. Matem.* **54**, 41-60 (1990). Translated in: *Mathematics of the USSR-Izvestiya*, **36**, 41-60 (1991).
5. E. I. Zel'manov, Solution of the restricted Burnside problem for 2-groups (Russian), *Matem. sb.* **182**, 568-592 (1991). Translated in: *Mathematics of the USSR-Sbornik* **72**, 543-565 (1992).
6. P. Hall, G. Higman, On the p-length of p-soluble groups and reduction theorems for Burnside's problem. *Proc. Lond. Math. Soc., III. Ser.* **6**, 1-42 (1956).
7. N. Chigira, W. J. Shi. More on the set of elements orders in finite groups. *Northeast Math. J.* **12**, 257-260 (1996).

Recognizability of Finite Groups by Order and Degree Pattern

A. R. Moghaddamfar

Department of Mathematics, K. N. Toosi University of Technology, P. O. Box 16315-1618, Tehran, Iran.

E-mails: moghadam@kntu.ac.ir and moghadam@mail.ipm.ir

Dedicated to Professor K. P. Shum on the occasion of his 70-th birthday

Abstract

The Gruenberg-Kegel graph $\mathrm{GK}(G) = (V_G, E_G)$ or prime graph of a finite group G is a simple graph with $V_G = \pi(G)$, the set of all primes dividing the order of G, and such that two vertices p and q are joined by an edge, $\{p, q\} \in E_G$, if G contains an element of order pq. The degree $\deg(p)$ of a vertex $p \in V_G$, is the number of edges incident on p. Let $\pi(G) = \{p_1, p_2, \ldots, p_k\}$ with $p_1 < p_2 < \cdots < p_k$. Now, we consider the k-tuple $D(G) = (\deg(p_1), \deg(p_2), \ldots, \deg(p_k))$, which is called the degree pattern of G. The group G is called k-fold OD-recognizable if there exist exactly k non-isomorphic groups H satisfying condition $(|H|, D(H)) = (|G|, D(G))$. Moreover, a 1-fold OD-recognizable group is simply called OD-recognizable. In this article, we first give a survey to all finite groups which have already been recognized by order and degree pattern. Then we prove that the simple group $C_3(4)$ is OD-recognizable.

1 Introduction

Let G be a finite group, $\pi(G)$ the set of all prime divisors of its order and $\omega(G)$ be the spectrum of G, that is the set of its element orders. The *Gruenberg-Kegel graph* $\mathrm{GK}(G)$ or *prime graph of* G is a simple graph with vertex set $\pi(G)$ in which two vertices p and q are joined by an edge (and we write $p \sim q$) if and only if $pq \in \omega(G)$. Let $t(G)$ be the number of connected components of $\mathrm{GK}(G)$. The ith connected component is denoted by $\pi_i = \pi_i(G)$ for each i. If $2 \in \pi(G)$, then we assume that $2 \in \pi_1(G)$. The classification of finite simple groups with disconnected Gruenberg-Kegel graph is obtained by Williams [27] and Kondrat'ev [17]. An corrected list of these groups can be found in [18].

We recall that a *clique* in a graph is a set of pairwise adjacent vertices. An independence set in a graph is a set of pairwise non-adjacent vertices. It is worth noting that if S is a simple group with disconnected prime graph, then all connected components $\pi_i(S)$ for $2 \leq i \leq t(S)$ are clique, for instance, see [17], [25] and [27].

AMS subject Classification 2010: 20D05, 20D06, 20D08.

Keyword and phrases: prime graph, degree pattern, OD-recognition of a finite group.

The *degree* $\deg(p)$ *of a vertex* $p \in \pi(G)$ is the number of edges incident on p. If $\pi(G) = \{p_1, p_2, \ldots, p_k\}$ with $p_1 < p_2 < \cdots < p_k$, then we define

$$\mathrm{D}(G) := \big(\deg(p_1), \deg(p_2), \ldots, \deg(p_k)\big),$$

which is called the *degree pattern of* G. Set

$$\Omega_n(G) = \{p \in \pi(G) | \deg(p) = n\},$$

for $n = 0, 1, 2, \ldots, k-1$. Clearly,

$$\pi(G) = \bigcup_{n=0}^{k-1} \Omega_n(G).$$

Moreover, since $\deg(p) = 0$ if and only if $\{p\}$ is a connected component of $\mathrm{GK}(G)$, we have $|\Omega_0(G)| \leq s(G) \leq 6$ (see [27]). A group G is called a $C_{p,p}-$group if $p \in \Omega_0(G)$.

Given a finite group M, denote by $h_{\mathrm{OD}}(M)$ the number of isomorphism classes of finite groups G such that $|G| = |M|$ and $\mathrm{D}(G) = \mathrm{D}(M)$. In terms of the function h_{OD}, groups M are classified as follows:

Definition 1 *A finite group* M *is called* k-fold OD-recognizable *if* $h_{\mathrm{OD}}(M) = k$. *Usually, a* 1*-fold OD-recognizable group is simply called* OD-recognizable.

It is well-known that for each positive integer n there are only *finitely* many non-isomorphic groups of order n normally denoted by $\nu(n)$. In fact, according to Cayley's theorem a group of order n can be embedded into the symmetric group S_n on n letters. Hence we observe that

$$\nu(n) \leq \binom{n!}{n}.$$

For a sharper upper bound, we have $\nu(n) \leq (n!)^{\log_2 n}$(see [24], page 109). Now since

$$1 \leq h_{\mathrm{OD}}(G) \leq \nu(|G|),$$

the following result follows immediately.

Theorem 1.1 *All finite groups of order* n *are* k*-fold* OD*-recognizable for some* k.

In order to formulate the obtained results, we need some notation and definitions. Throughout the paper, we assume that q is a prime power. We write $L_n(q)$ instead of the projective special linear group $\mathrm{PSL}(n, q)$ and write $U_n(q)$ instead of the projective special unitary group $\mathrm{PSU}(n, q)$. We use $B_n(q)$ and $C_n(q)$ to denote the simple orthogonal and symplectic groups, respectively. (In Atlas [6] notation, these are the groups $O_{2n+1}(q)$ and $S_{2n}(q)$, respectively.)

For recent results concerning the simple groups which are k-fold OD-recognizable, for $k \geq 2$, it was shown in [2], [19] and [22] that each of the following pairs $\{K_1, K_2\}$ of groups:

$\{A_{10},\ \mathbb{Z}_3 \times J_2\}$,

$\{B_3(5),\ C_3(5)\}$,

$\{B_m(q),\ C_m(q)\}$, $\quad m = 2^f \geq 2$, $|\pi\big((q^m+1)/2\big)| = 1$, q is an odd prime power,

$\{B_p(3),\ C_p(3)\}$, $\quad |\pi\big((3^p-1)/2\big)| = 1$, p is an odd prime, satisfy $|K_1| = |K_2|$ and $\mathrm{D}(K_1) = \mathrm{D}(K_2)$, and $h_{\mathrm{OD}}(K_i) = 2$. In general, for simple groups $B_m(q)$ and $C_m(q)$ we have $|B_m(q)| = |C_m(q)|$ and $\mathrm{D}(B_m(q)) = \mathrm{D}(C_m(q))$ (see [26, Proposition 7.5]). Hence, if $B_m(q)$ and $C_m(q)$ are non-isomorphic groups, then it follows that $h_{\mathrm{OD}}(B_m(q)) = h_{\mathrm{OD}}(C_m(q)) \geq 2$.

Until recently, we do not know if there exists a *simple* group which is k-fold OD-recognizable for $k \geq 3$. Therefore, one may ask the following question as an open problem:

Problem 1. *Is there a simple group S for which $h_{\mathrm{OD}}(S) \geq 3$?*

Table 1 lists finite simple groups which are currently known to be OD-recognizable or 2-fold OD-recognizable.

Table 1.

S	Conditions on S	h_{OD}	Refs.
A_n	$n = p, p+1, p+2$ (p a prime)	1	[19], [20]
	$n = p+3,\ p \in \pi(100!) \setminus \{7\}$	1	[10], [21], [23], [31]
	$n = 10$	2	[22]
$L_2(q)$	$q \neq 2, 3$	1	[19], [20], [35]
$L_3(q)$	$\lvert\pi(\frac{q^2+q+1}{d})\rvert = 1,\ d = (3, q-1)$	1	[19]
$U_3(q)$	$\lvert\pi(\frac{q^2-q+1}{d})\rvert = 1,\ d = (3, q+1), q > 5$	1	[19]
$L_4(q)$	$q = 5, 7$	1	[1]
$L_3(9)$		1	[33]
$U_3(5)$		1	[34]
$U_4(7)$		1	[1]
$L_n(2)$	$n = p$ or $p+1$, for which $2^p - 1$ is a prime	1	[1]
$L_9(2)$		1	[16]
$R(q)$	$\lvert\pi(q \pm \sqrt{3q} + 1)\rvert = 1,\ q = 3^{2m+1},\ m \geq 1$	1	[19]
$\mathrm{Sz}(q)$	$q = 2^{2n+1} \geq 8$	1	[19], [20]
$B_m(q), C_m(q)$	$m = 2^f \geq 4,\ \lvert\pi\big((q^m+1)/2\big)\rvert = 1,$	2	[2]
$B_2(q) \cong C_2(q)$	$\lvert\pi\big((q^2+1)/2\big)\rvert = 1,\ q \neq 3$	1	[2]
$B_m(q) \cong C_m(q)$	$m = 2^f \geq 2,\ 2\vert q,\ \lvert\pi\big(q^m+1\big)\rvert = 1,\ (m, q) \neq (2, 2)$	1	[2]
$B_p(3), C_p(3)$	$\lvert\pi\big((3^p-1)/2\big)\rvert = 1$, p is an odd prime	2	[2], [19]
$B_3(5), C_3(5)$		2	[2]
S	A sporadic simple group	1	[19]
S	A simple group with $\lvert\pi(S)\rvert = 4,\ \ S \neq A_{10}$	1	[32]
S	A simple group with $\lvert S\rvert \leq 10^8,\ \ S \neq A_{10},\ U_4(2)$	1	[29]
S	A simple $C_{2,2}$- group	1	[20]

However, among non-simple groups, there are some groups which are k-fold OD-recognizable for $k \geq 3$. The following are some examples:

Example 1. Let G be a *nilpotent* group with

$$|G| = p_1^{n_1} p_2^{n_2} \cdots p_k^{n_k},$$

where k, $n_1, \ldots, n_k$ are positive integers and $p_1, p_2, \ldots, p_s$ distinct primes. Then G is a direct product of its Sylow p-subgroups:

$$G = S_{p_1} \times S_{p_2} \times \cdots \times S_{p_k},$$

from which we conclude that

$$Z(G) = Z(S_{p_1}) \times Z(S_{p_2}) \times \cdots \times Z(S_{p_k}).$$

Therefore, we deduce that

$$\pi(Z(G)) = \pi(G) = \{p_1, p_2, \ldots, p_k\},$$

and so the prime graph $\mathrm{GK}(G)$ is a clique. In this case, $\mathrm{D}(G) = (d, d, \ldots, d)$ where $d = k - 1$. Therefore, if G is not a cyclic group, then

$$h_{\mathrm{OD}}(G) \geq \nu_{\mathrm{nil}}(|G|) \geq 2,$$

where $\nu_{\mathrm{nil}}(n)$ denotes the number of isomorphism classes of nilpotent groups of order n.

Example 2. Assume that G_1 and G_2 are two non-isomorphic finite groups with

$$|G_1| = p_1^{\alpha_1} p_2^{\alpha_2} \cdots p_r^{\alpha_r} \quad \text{and} \quad |G_2| = q_1^{\beta_1} q_2^{\beta_2} \cdots q_s^{\beta_s}.$$

we set $m := p_1 p_2 \cdots p_r$ and $n := q_1 q_2 \cdots q_s$. Now, we construct the following groups:

$$G_2 \times \mathbb{Z}_{mn} \times H_1 \quad \text{and} \quad G_1 \times \mathbb{Z}_{mn} \times H_2,$$

where H_i $(i = 1, 2)$ is a group of order $|G_i|$. Evidently, the prime graph of these groups are complete. Therefore, they have the same degree pattern, namely, the $(s + r)$-tuple $(l, l, \ldots, l)$ where $l = s + r - 1$. Also, it is easy to see that these groups have the same order. This example shows that the constructed groups are k-fold OD-recognizable for $k \geq \nu(|G_1|) + \nu(|G_2|)$.

Example 3. Considering the following groups

$$G_1 = U_4(2) \times \mathbb{Z}_6, \quad G_2 = A_5 \times \mathbb{Z}_{2592}, \quad G_3 = A_6 \times \mathbb{Z}_{432},$$

we observe that $|G_i| = 2^7 \cdot 3^5 \cdot 5$ and $\mathrm{D}(G_i) = (2, 2, 2)$ for $1 \leq i \leq 3$. Hence, these groups are k-fold OD-recognizable with $k \geq 3$.

Example 4. One might prefer to construct some groups with the same order and degree pattern and non-complete prime graph. Let G be a finite group with non-regular prime graph $\mathrm{GK}(G)$. If there exists a prime $p \in \pi(G)$ such that $\deg(p) = |\pi(G)| - 1$, then for an arbitrary p-group, say P, we have $\Gamma(P \times M) = \Gamma(M)$ and so $\mathrm{D}(P \times G) = \mathrm{D}(G)$. Let us fix an arbitrary p-group P of order p^n where $n \geq 2$. Then, it is clear that

$$h_{\mathrm{OD}}(P \times G) \geq \nu(p^n) \cdot h_{\mathrm{OD}}(G) \geq \mathrm{Par}(n) \cdot h_{\mathrm{OD}}(G) \geq 2 \cdot h_{\mathrm{OD}}(G),$$

where $\mathrm{Par}(n)$ denotes the number of partitions of n. For instance, if $G = A_{10}$, then we know that $\mathrm{D}(G) = (2,3,2,1)$ and $h_{\mathrm{OD}}(A_{10}) = 2$. Indeed, as we mentioned before, the groups A_{10} and $\mathbb{Z}_3 \times J_2$ are only groups with the same order and degree pattern. Now assume that P is a 3-group of order 27. Then $\nu(27) = 5$. In fact, P is isomorphic to one of the following groups:

- $\mathbb{Z}_3 \times \mathbb{Z}_3 \times \mathbb{Z}_3$, $\mathbb{Z}_9 \times \mathbb{Z}_3$ or $\mathbb{Z}_{27}$.
- An extra-special 3-group of order 27 and exponent 3.
- An extra-special 3-group of order 27 and exponent 9.

Therefore, it is easy to see that $h_{\mathrm{OD}}(P \times G) \geq 5 \cdot h_{\mathrm{OD}}(G) = 10$.

In general, if $|\pi(G)| = k > 1$, $\Omega_{k-1}(G) \neq \emptyset$ and H is a $\Omega_{k-1}(G)$-group of order $|H|$, then we have

$$h_{\mathrm{OD}}(H \times G) \geq \nu(|H|) \cdot h_{\mathrm{OD}}(G).$$

If n is a positive integer, then $\pi(n)$ denotes the set of prime divisors of n. Given a finite group G, the order of G can be expressed as a product of some coprime positive integers m_i, $i = 1, 2, \ldots, t(G)$, with $\pi(m_i) = \pi_i$. These integers m_i's are called the *order components of* G. Let $\mathrm{OC}(G) = \{m_1, m_2, \ldots, m_{t(G)}\}$ be the set of all order components of G. The order components of simple groups with disconnected prime graphs are obtained in Tables 1-4 in [3].

Given a finite group M, define $h_{\mathrm{OC}}(M)$ to be the number of isomorphism classes of finite groups with the same set $\mathrm{OC}(M)$ of order components. In terms of the function h_{OC}, groups M are classified as follows:

Definition 2 *A finite group M is called k-*fold OC-recognizable *if $h_{\mathrm{OC}}(M) = k$. Usually, a 1-fold OC-recognizable group is simply called* OC-recognizable.

It is clear that $1 \leq h_{\mathrm{OC}}(M) < \infty$ for any finite group M. There are scattered results in the literature showing that some almost simple groups are k-fold OC-recognizable where $k \in \{1, 2\}$. For example, the following simple groups are OC-characterizable: All the sporadic simple groups [3], Suzuki-Ree groups [4], $L_2(q)$ [5], $L_p(q)$ [11], $L_{p+1}(q)$ [12], $U_p(q)$ [13], $U_{p+1}(q)$ [14]. In addition, the following simple groups are 2-fold OC-recognizable: $G \cong B_p(3)$ and $C_p(3)$ [8], $B_n(q)$ and $C_n(q)$ where $n = 2^m \geq 4$ and q is an odd prime power [15]. On the other hand, a simple group S with connected prime graph is not OC-recognizable, because $h_{\mathrm{OC}}(S) \geq \nu_{\mathrm{nil}}(|S|) \geq 2$.

Note that, the values of the functions h_{OD} and h_{OC} may be different. For example, there are only four non-isomorphic groups of order 30, which we listed in Table 2. Now, it can be easily seen that $h_{\mathrm{OD}}(\mathbb{Z}_{30}) = h_{\mathrm{OD}}(\mathbb{Z}_3 \times D_{10}) = h_{\mathrm{OD}}(\mathbb{Z}_5 \times D_6) = 1$, while $h_{\mathrm{OC}}(\mathbb{Z}_{30}) = h_{\mathrm{OC}}(\mathbb{Z}_3 \times D_{10}) = h_{\mathrm{OC}}(\mathbb{Z}_5 \times D_6) = 3$.

Table 2. The groups of order 30.

G	$\mathrm{GK}(G)$	$t(G)$	$\mathrm{OC}(G)$	$D(G)$	h_{OD}	h_{OC}
$\mathbb{Z}_{30}$	$2 \sim 3 \sim 5 \sim 2$	1	$\{30\}$	$(2,2,2)$	1	3
$\mathbb{Z}_3 \times D_{10}$	$2 \sim 3 \sim 5$	1	$\{30\}$	$(1,2,1)$	1	3
$\mathbb{Z}_5 \times D_6$	$2 \sim 5 \sim 3$	1	$\{30\}$	$(1,1,2)$	1	3
D_{30}	$2,\ 3 \sim 5$	2	$\{2, 15\}$	$(0,1,1)$	1	1

We denote by $H : K$ (resp. $H \cdot K$) a split extension (resp. a non-split extension) of a normal subgroup H by another subgroup K. Note that, split extensions are the same as semi-direct products.

Table 3 lists finite non-solvable groups which are currently known to be OD-recognizable or k-fold OD-recognizable with $k \geq 2$.

Table 3.

G	Conditions on G	$h_{\rm OD}$	Refs
Aut(M)	M is a sporadic group $\neq J_2, M^cL$	1	[20]
S_n	$n = p,\ p+1$ ($p \geq 5$ is a prime)	1	[20]
M	$M \in \mathcal{C}_1$	2	[22]
M	$M \in \mathcal{C}_2$	2	[19]
M	$M \in \mathcal{C}_3$	8	[22]
M	$M \in \mathcal{C}_4$	3	[21, 10, 23]
M	$M \in \mathcal{C}_5$	2	[22]
M	$M \in \mathcal{C}_6$	3	[22]
M	$M \in \mathcal{C}_7$	6	[23]
M	$M \in \mathcal{C}_8$	1	[30]
M	$M \in \mathcal{C}_9$	9	[30]
M	$M \in \mathcal{C}_{10}$	1	[34]
M	$M \in \mathcal{C}_{11}$	3	[34]
M	$M \in \mathcal{C}_{12}$	6	[34]

$$\begin{aligned}
\mathcal{C}_1 &= \{A_{10}, J_2 \times \mathbb{Z}_3\} \\
\mathcal{C}_2 &= \{S_6(3), O_7(3)\} \\
\mathcal{C}_3 &= \{S_{10},\ \mathbb{Z}_2 \times A_{10},\ \mathbb{Z}_2 \cdot A_{10},\ \mathbb{Z}_6 \times J_2,\ S_3 \times J_2,\ \mathbb{Z}_3 \times (\mathbb{Z}_2 \cdot J_2), \\
&\qquad (\mathbb{Z}_3 \times J_2) \cdot \mathbb{Z}_2,\ \mathbb{Z}_3 \times \mathrm{Aut}(J_2)\}. \\
\mathcal{C}_4 &= \{S_{p+3},\ \mathbb{Z}_2 \cdot A_{p+3},\ \mathbb{Z}_2 \times A_{p+3}\}, \\
&\qquad \text{where } p \in \pi(100!) \setminus \{7\} \text{ and } p+2 \text{ is not a prime.} \\
\mathcal{C}_5 &= \{\mathrm{Aut}(M^cL),\ \mathbb{Z}_2 \times M^cL\}. \\
\mathcal{C}_6 &= \{\mathrm{Aut}(J_2),\ \mathbb{Z}_2 \times J_2,\ \mathbb{Z}_2 \cdot J_2\}. \\
\mathcal{C}_7 &= \{\mathrm{Aut}(S_6(3)),\ \mathbb{Z}_2 \times S_6(3),\ \mathbb{Z}_2 \cdot S_6(3),\ \mathbb{Z}_2 \times O_7(3),\ \mathbb{Z}_2 \cdot O_7(3),\ \mathrm{Aut}(O_7(3))\}. \\
\mathcal{C}_8 &= \{L_2(49) : 2_1,\ L_2(49) : 2_2,\ L_2(49) : 2_3\}. \\
\mathcal{C}_9 &= \{L \cdot 2^2,\ \mathbb{Z}_2 \times (L : 2_1),\ \mathbb{Z}_2 \times (L : 2_2),\ \mathbb{Z}_2 \times (L \cdot 2_3),\ \mathbb{Z}_2 \cdot (L : 2_1), \\
&\qquad \mathbb{Z}_2 \cdot (L : 2_2),\ \mathbb{Z}_2 \cdot (L \cdot 2_3),\ \mathbb{Z}_4 \times L,\ (\mathbb{Z}_2 \times \mathbb{Z}_2) \times L\}, \text{ where } L = L_2(49). \\
\mathcal{C}_{10} &= \{U_3(5),\ U_3(5) : 2\} \\
\mathcal{C}_{11} &= \{U_3(5) : 3,\ \mathbb{Z}_3 \times U_3(5),\ \mathbb{Z}_3 \cdot U_3(5)\} \\
\mathcal{C}_{12} &= \{L : S_3,\ \mathbb{Z}_2 \cdot (L : 3),\ \mathbb{Z}_3 \times (L : 2),\ \mathbb{Z}_3 \cdot (L : 2),\ (\mathbb{Z}_2 \times L) \cdot \mathbb{Z}_2, \\
&\qquad (\mathbb{Z}_3 \cdot L) \cdot \mathbb{Z}_2\}, \text{ where } L = U_3(5).
\end{aligned}$$

We conclude the introduction with notation to be used in the rest of this article. The *socle* of a group G is the subgroup generated by the set of all minimal normal subgroups of G; it is denoted by $\mathrm{Soc}(G)$. We denote by $\mathrm{Syl}_p(G)$ the set of all Sylow p-subgroups of G, where $p \in \pi(G)$. Moreover G_p denotes a Sylow p-subgroup of G for $p \in \pi(G)$. If H is a subgroup of G, then $C_G(H)$ and $N_G(H)$ are, respectively, the centralizer and the normalizer of H in G.

2 OD-Recognizability of the Symplectic Group $C_3(4)$

We start with the following lemma which is easily obtained from [6] and [9]:

Lemma 1 *Let $M = C_3(4)$. The following hold for M.*

(1) $|M| = 2^{18} \cdot 3^4 \cdot 5^3 \cdot 7 \cdot 13 \cdot 17$.
(2) $\mu(M) = \{8, 12, 20, 30, 34, 51, 63, 65, 85\}$.
(3) $\mathrm{D}(M) = (3, 4, 4, 1, 1, 3)$,
(4) $|\mathrm{Out}(M)| = 2$.
(5) *The prime graph of M is depicted in Fig. 1.*

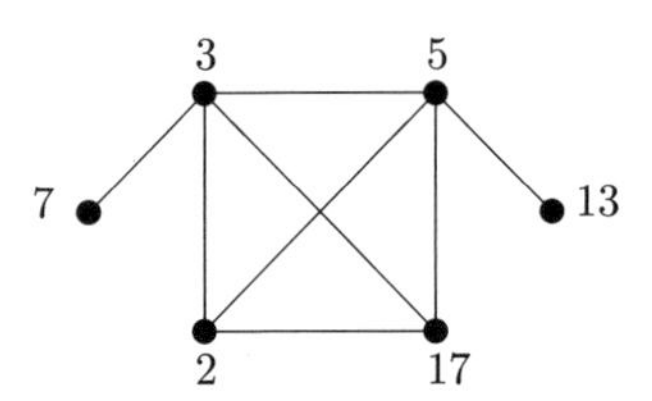

Fig. 1. $\mathrm{GK}(C_3(4))$

We shall need the following lemma of Zavarnitsin [28].

Lemma 2 *Let $S = P_1 \times \cdots \times P_t$, where P_i's are isomorphic non-Abelian simple groups. Then*

$$\mathrm{Aut}(S) \cong \Big(\mathrm{Aut}(P_1) \times \cdots \times \mathrm{Aut}(P_t)\Big) \rtimes S_t.$$

In particular, $|\mathrm{Aut}(S)| = \prod_{i=1}^{t} |\mathrm{Aut}(P_i)| \cdot t!$.

Lemma 3 *([23]) Let S be a simple group such that $\pi(S) \subseteq \{2, 3, 5, 7, 13, 17\}$. Then S is isomorphic to one of the simple groups listed in Table 4.*

Table 4. *The simple groups S with $\pi(S) \subseteq \{2,3,5,7,13,17\}$*

S	$\|S\|$	s	S	$\|S\|$	s
A_5	$2^2 \cdot 3 \cdot 5$	2	$L_4(3)$	$2^7 \cdot 3^6 \cdot 5 \cdot 13$	4
A_6	$2^3 \cdot 3^2 \cdot 5$	4	$B_3(3)$	$2^9 \cdot 3^9 \cdot 5 \cdot 7 \cdot 13$	2
$U_4(2)$	$2^6 \cdot 3^4 \cdot 5$	2	$O_8^+(3)$	$2^{12} \cdot 3^{12} \cdot 5^2 \cdot 7 \cdot 13$	24
A_7	$2^3 \cdot 3^2 \cdot 5 \cdot 7$	2	$G_2(3)$	$2^6 \cdot 3^6 \cdot 7 \cdot 13$	2
A_8	$2^6 \cdot 3^2 \cdot 5 \cdot 7$	2	$C_3(3)$	$2^9 \cdot 3^9 \cdot 5 \cdot 7 \cdot 13$	2
A_9	$2^6 \cdot 3^4 \cdot 5 \cdot 7$	2	$L_3(3^2)$	$2^7 \cdot 3^6 \cdot 5 \cdot 7 \cdot 13$	4
A_{10}	$2^7 \cdot 3^4 \cdot 5^2 \cdot 7$	2	$L_2(3^3)$	$2^2 \cdot 3^3 \cdot 7 \cdot 13$	6
$B_3(2)$	$2^9 \cdot 3^4 \cdot 5 \cdot 7$	1	$U_4(5)$	$2^7 \cdot 3^4 \cdot 5^6 \cdot 7 \cdot 13$	4
$O_8^+(2)$	$2^{12} \cdot 3^5 \cdot 5^2 \cdot 7$	6	$B_2(5)$	$2^6 \cdot 3^2 \cdot 5^4 \cdot 13$	2
$L_3(2^2)$	$2^6 \cdot 3^2 \cdot 5 \cdot 7$	12	$L_2(5^2)$	$2^3 \cdot 3 \cdot 5^2 \cdot 13$	4
$L_2(2^3)$	$2^3 \cdot 3^2 \cdot 7$	3	$L_2(13)$	$2^2 \cdot 3 \cdot 7 \cdot 13$	2
$U_3(3)$	$2^5 \cdot 3^3 \cdot 7$	2	$C_4(2)$	$2^{16} \cdot 3^5 \cdot 5^2 \cdot 7 \cdot 17$	1
$U_4(3)$	$2^7 \cdot 3^6 \cdot 5 \cdot 7$	8	$O_8^-(2)$	$2^{12} \cdot 3^4 \cdot 5 \cdot 7 \cdot 17$	2
$U_3(5)$	$2^4 \cdot 3^2 \cdot 5^3 \cdot 7$	6	$F_4(2)$	$2^{24} \cdot 3^6 \cdot 5^2 \cdot 7^2 \cdot 13 \cdot 17$	2
$L_2(7)$	$2^3 \cdot 3 \cdot 7$	2	$L_4(2^2)$	$2^{12} \cdot 3^4 \cdot 5^2 \cdot 7 \cdot 17$	4
$B_2(7)$	$2^8 \cdot 3^2 \cdot 5^2 \cdot 7^4$	2	$U_4(2^2)$	$2^{12} \cdot 3^2 \cdot 5^3 \cdot 13 \cdot 17$	4
$L_2(7^2)$	$2^4 \cdot 3 \cdot 5^2 \cdot 7^2$	4	$C_2(2^2)$	$2^8 \cdot 3^2 \cdot 5^2 \cdot 17$	4
J_2	$2^7 \cdot 3^3 \cdot 5^2 \cdot 7$	2	$C_3(2^2)$	$2^{18} \cdot 3^4 \cdot 5^3 \cdot 7 \cdot 13 \cdot 17$	2
${}^3D_4(2)$	$2^{12} \cdot 3^4 \cdot 7^2 \cdot 13$	3	$O_8^+(2^2)$	$2^{24} \cdot 3^5 \cdot 5^4 \cdot 7 \cdot 13 \cdot 17^2$	12
${}^2F_4(2)'$	$2^{11} \cdot 3^3 \cdot 5^2 \cdot 13$	2	$L_2(2^4)$	$2^4 \cdot 3 \cdot 5 \cdot 17$	4
$U_3(2^2)$	$2^6 \cdot 3 \cdot 5^2 \cdot 13$	4	$L_3(2^4)$	$2^{12} \cdot 3^2 \cdot 5^2 \cdot 7 \cdot 13 \cdot 17$	24
$G_2(2^2)$	$2^{12} \cdot 3^3 \cdot 5^2 \cdot 7 \cdot 13$	2	$C_2(13)$	$2^6 \cdot 3^2 \cdot 5 \cdot 7^2 \cdot 13^4 \cdot 17$	2
$B_2(2^3)$	$2^{12} \cdot 3^4 \cdot 5 \cdot 7^2 \cdot 13$	6	$L_2(13^2)$	$2^3 \cdot 3 \cdot 5 \cdot 7 \cdot 13^2 \cdot 17$	4
$\mathrm{Sz}(2^3)$	$2^6 \cdot 5 \cdot 7 \cdot 13$	3	$L_2(17)$	$2^4 \cdot 3^2 \cdot 17$	2
$L_2(2^6)$	$2^6 \cdot 3^2 \cdot 5 \cdot 7 \cdot 13$	6	$U_3(17)$	$2^6 \cdot 3^4 \cdot 7 \cdot 13 \cdot 17^3$	6
$L_3(3)$	$2^4 \cdot 3^3 \cdot 13$	2	He	$2^{10} \cdot 3^3 \cdot 5^2 \cdot 7^3 \cdot 17$	2

Now we are ready to state and prove the main theorem of this section.

Theorem 2.1 *The symplectic group $C_3(4)$ is OD-recognizable.*

Proof. Let $M = C_3(4)$. Suppose G is a finite group, such that

$$|G| = |M| = 2^{18} \cdot 3^4 \cdot 5^3 \cdot 7 \cdot 13 \cdot 17 \quad \text{and} \quad \mathrm{D}(G) = \mathrm{D}(M) = (3,4,4,1,1,3),$$

(see Lemma 1). We have to show that G is isomorphic to M. It is evident that the prime graph of G is connected, since $\deg(3) = \deg(5) = 4$. Moreover, by hypothesis, we immediately conclude that the only possibilities for the prime graph $\mathrm{GK}(G)$ of G are:

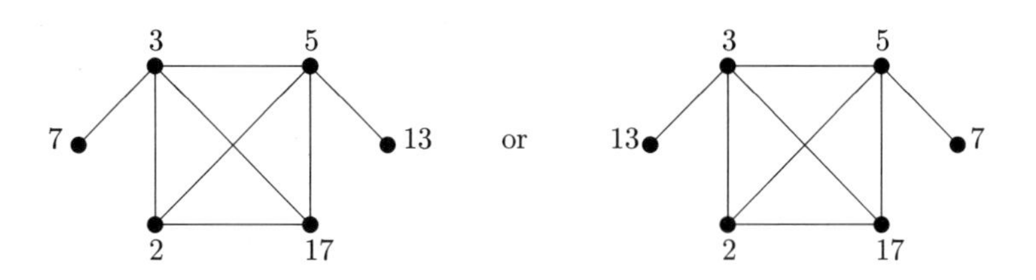

Therefore, we conclude that $\{2,3,5,6,7,10,13,15,34,51,85\} \subseteq \omega(G)$, and the subsets $\{2,7,13\}$ and $\{7,13,17\}$ of vertices are independent sets of $\mathrm{GK}(G)$. In the sequel, we break up the proof into a sequence of lemmas. Let K be the maximal normal solvable subgroup of G.

Lemma 4 *K is a $\{2,3,5\}$-group. In particular, G is non-solvable.*

Proof. First, we show that K is a $13'$-group. Assume the contrary and let 13 divide the order of K. In this case K possesses an element x of order 13. We set $C := C_G(x)$ and $N := N_G(\langle x\rangle)$. By the structure of $\mathrm{D}(G)$, it follows that C is a $\{p,13\}$-group where $p \in \{3,5\}$. Now using (N/C)-Theorem the factor group N/C is embedded in $\mathrm{Aut}(\langle x\rangle) \cong \mathbb{Z}_{12}$. Hence, N is a $\{2,3,5,13\}$-group. Now, by Frattini argument $G = KN$. This implies that $\{7,17\} \subseteq \pi(K)$. Since K is solvable, it possesses a Hall $\{7,17\}$-subgroup L of order $7 \cdot 17$. Clearly L is cyclic and hence $7 \sim 17$, which is a contradiction.

Next, we show that K is a p'-group for $p \in \{7,17\}$. Let $p \in \pi(K)$, $K_p \in \mathrm{Syl}_p(K)$ and $N = N_G(K_p)$. Again, by Frattini argument $G = KN$ and hence 13 divides the order of N. Let L be a subgroup of N of order 13. Since L normalizes K_p, G contains a subgroup of order $13 \cdot p$ and this leads to a contradiction as before, since $p \nmid 13-1$. Therefore K is a $\{2,3,5\}$-group.

In addition, since $K \neq G$, it follows that G is non-solvable. This completes the proof. □

Lemma 5 *The factor group G/K is an almost simple group. In fact, $M \leq G/K \leq \mathrm{Aut}(M)$.*

Proof. Let $H := G/K$ and $S := \mathrm{Soc}(H)$. Evidently, $S = P_1 \times \cdots \times P_m$, where P_i's are non-Abelian simple groups. This implies that $Z(S) = 1$, or equivalently $C_H(S) \cap S = 1$. But then $C_H(S) = 1$, since otherwise $C_H(S)$ would contain minimal normal subgroups of H disjoint from S, which is a contradiction. Consequently, we get

$$H \cong \frac{N_H(S)}{C_H(S)} \hookrightarrow \mathrm{Aut}(S).$$

In what follows, we will show that $m = 1$ and $P_1 \cong M$.

Suppose that $m \geq 2$. In this case, it is easy to see that $\{7,13\} \cap \pi(S) = \emptyset$, since otherwise $\deg(7) \geq 2$ or $\deg(13) \geq 2$, which is a contradiction. On the other hand, by

Lemma 4, we observe that $13 \in \pi(H) \subseteq \pi(\mathrm{Aut}(S))$. Thus, we may assume that 13 divides the order of $\mathrm{Out}(S)$. But

$$\mathrm{Out}(S) = \mathrm{Out}(S_1) \times \cdots \times \mathrm{Out}(S_r),$$

where the groups S_j are direct products of isomorphic P_i's such that

$$S \cong S_1 \times \cdots \times S_r.$$

Therefore, for some j, 13 divides the order of an outer automorphism group of a direct product S_j of t isomorphic simple groups P_i. Since $\max \pi(P_i) = 17$, it follows that $|\mathrm{Out}(P_i)|$ is not divisible by 13, see [23, Table 4]. Now, by Lemma 2, we obtain $|\mathrm{Aut}(S_j)| = |\mathrm{Aut}(P_i)|^t \cdot t!$. Therefore, $t \geq 13$ and so 2^{26} must divide the order of G, which is a contradiction. Therefore $m = 1$ and $S = P_1$.

Now, from Lemma 4, we easily conclude that

$$|S| = 2^a \cdot 3^b \cdot 5^c \cdot 7 \cdot 13 \cdot 17,$$

where $2 \leq a \leq 18$, $0 \leq b \leq 4$ and $0 \leq c \leq 3$. Using collected results contained in Table 2, we deduce that $S \cong C_3(4)$ or $L_3(2^4)$. But in the case that $S \cong L_3(2^4)$, we have $7 \cdot 13 \in \omega(S)$ (see [7]), which is a contradiction. This completes the proof of lemma. □

Lemma 6 *G is isomorphic to M.*

Proof. By Lemma 5, $C_3(4) \leq G/K \leq \mathrm{Aut}(C_3(4))$. Since $|\mathrm{Out}(C_3(4)| = 2$ (see Lemma 1), we conclude that $G/K \cong C_3(4)$ or $G/K \cong \mathrm{Aut}(C_3(4))$. However, by order consideration we deduce that $|K| = 1$ and $G \cong C_3(4)$, as desired. □

The proof of the theorem is now complete. □

References

[1] M. Akbari, A. R. Moghaddamfar and S. Rahbariyan, *A characterization of some finite simple groups through their orders and degree patterns*, to appear in Algebra Colloquium.

[2] M. Akbari and A. R. Moghaddamfar, *Simple groups which are 2-fold OD-characterizable*, to appear in Bulletin of the Malaysian Mathematical Sciences Society.

[3] G. Y. Chen, *A new characterization of sporadic simple groups*, Algebra Colloq., 3(1)(1996), 49-58.

[4] G. Y. Chen. A new characterization of Suzuki-Ree groups. *Sci. in Chinese (Ser. A)* 27(5) (1997), 430-433.

[5] G. Y. Chen. A new characterization of $PSL(2, q)$. *Southeast Asian Bull. Math.*, 22(1998), 257-263.

[6] J. H. Conway, R. T. Curtis, S. P. Norton, R. A. Parker, R. A. Wilson, *Atlas of Finite Groups*, Clarendon Press, oxford, 1985.

[7] M. R. Darafsheh, A. R. Moghaddamfar, A. R. Zokayi, *A recognition of simple groups* $\mathrm{PSL}(3, q)$ *by their element orders*, Acta Math. Sci. Ser. B Engl. Ed., 24(1)(2004), 45-51.

[8] M. R. Darafsheh, *On non-isomorphic groups with the same set of order components*, J. Korean Math. Soc., 45(1)(2008), 137-150.

[9] I. B. Gorshkov, *Recognition of finite simple groups with orders having prime divisors at most* 17 *by the spectrum*, (Russian) Sib. Elektron. Mat. Izv., 7 (2010), 14-20.

[10] A. A. Hoseini and A. R. Moghaddamfar, *Recognizing alternating groups* A_{p+3} *for certain primes* p *by their orders and degree patterns*, to appear in Frontiers of Mathematics in China.

[11] A. Khosravi and B. Khosravi, *A new characterization of* $\mathrm{PSL}(p, q)$, Comm. Algebra 32(6) (2004), 2325-2339.

[12] Behrooz Khosravi, Bahman Khosravi and Behnam Khosravi, *Characterizability of* $\mathrm{PSL}(p+1, q)$ *by its order component(s)*, Houston J. Math., 32(3) (2006), 683-700.

[13] Bahman Khosravi, Behnam Khosravi and Behrooz Khosravi, *A new characterization of* $\mathrm{PSU}(p, q)$, Acta Math. Hungar. 107(3) (2005), 235-252.

[14] A. Khosravi and B. Khosravi, *Characterizability of* $\mathrm{PSU}(p + 1, q)$ *by its order component(s)*, Rocky Mountain J. Math. 36(5) (2006), 1555-1575.

[15] A. Khosravi and B. Khosravi, *r-Recognizability of $B_n(q)$ and $C_n(q)$ where $n = 2^m \geq 4$*, J. Pure Appl. Algebra, 199(1-3) (2005), 149-165.

[16] B. Khosravi, *Some characterizations of $L_9(2)$ related to its prime graph*, Publicationes Mathematicae Debrecen, Tomus 75, Fasc. 3-4, (2009).

[17] A. S. Kondratév, *Prime graph components of finite simple groups*, Math. Sb., 180(6)(1989), 787-797.

[18] A. S. Kondrat'ev and V. D. Mazurov, *Recognition of alternating groups of prime degree from their element orders*, Siberian Mathematical Journal, 41(2)(2000), 294-302.

[19] A. R. Moghaddamfar, A. R. Zokayi and M. R. Darafsheh, *A characterization of finite simple groups by the degrees of vertices of their prime graphs*, Algebra Colloquium, 12 (3)(2005), 431-442.

[20] A. R. Moghaddamfar and A. R. Zokayi, *Recognizing finite groups through order and degree pattern*, Algebra Colloquium, 15(3)(2008), 449-456.

[21] A. R. Moghaddamfar and A. R. Zokayi, OD-*Characterization of alternating and symmetric groups of degrees* 16 *and* 22, Front. Math. China, 4(4)(2009), 669-680.

[22] A. R. Moghaddamfar and A. R. Zokayi, OD-*Characterization of certain finite groups having connected prime graphs*, Algebra Colloquium, 17(1)(2010), 121-130.

[23] A. R. Moghaddamfar and S. Rahbariyan, *More on the* OD-*characterizability of a finite group*, to appear in Algebra Colloquium.

[24] J. S. Rose, *A Course on Group Theory*, Cambridge University Press, Cambridge, 1978.

[25] M. Suzuki, *On the prime graph of a finite simple group–an application of the method of Feit-Thompson-Bender-Glauberman*, Groups and combinatorics–in memory of Michio Suzuki, Adv. Stud. Pure Math., 32 (Math. Soc. Japan, Tokyo, 2001), 41-207.

[26] A. V. Vasiliev and E. P. Vdovin, *An adjacency criterion in the prime graph of a finite simple group*, Algebra and Logic 44(6)(2005), 381-406.

[27] J. S. Williams, *Prime graph components of finite groups*, J. Algebra 69(2)(1981), 487-513.

[28] A. V. Zavarnitsin, *Recognition of alternating groups of degrees* $r+1$ *and* $r+2$ *for prime* r *and the group of degree* 16 *by their element order sets*, Algebra and Logic, 39 (6)(2000), 370-377.

[29] L. C. Zhang and W. J. Shi, *OD-Characterization of all simple groups whose orders are less than* 10^8, Front. Math. China, 3(3)(2008), 461-474.

[30] L.C. Zhang, W.J. Shi, OD-Characterization of almost simple groups related to $L_2(49)$, *Arch. Math.* (Brno), 44(3)(2008), 191-199.

[31] L. C. Zhang, W. J. Shi, L. L. Wang and C. G. Shao, OD-*Characterization of* A_{16}, Journal of Suzhou University (Natural Science Edition), 24(2)(2008), 7-10.

[32] L. C. Zhang and W. J. Shi, *OD-Characterization of simple* K_4*-groups*, Algebra Colloquium, 16 (2) (2009) 275-282.

[33] L. C. Zhang, W.J. Shi, C. G. Shao and L. L. Wang, OD-*Characterization of the simple group* $L_3(9)$, Journal of Guangxi University (Natural Science Edition), 34(1)(2009), 120-122.

[34] L. C. Zhang and W.J. Shi, OD-*Characterization of almost simple groups related to* $U_3(5)$, Acta Mathematica Sinica (English Series), 26(1), 2010, 161-168.

[35] L. C. Zhang and W. J. Shi, OD-*Characterization of the projective special linear groups* $L_2(q)$, to appear in Algebra Colloquium.

Regular Elements of Generalized Order-Preserving Transformation Semigroups

W. Mora[1] and Y. Kemprasit[2]

[1]Department of Mathematics, Faculty of Science
Prince of Songkla University, Songkhla 90112, Thailand
k_winita@hotmail.com
[2]Department of Mathematics, Faculty of Science
Chulalongkorn University, Bangkok 10330, Thailand
yupaporn.k@chula.ac.th

Abstract

For a chain X, let $OT(X), OP(X)$ and $OI(X)$ be the order-preserving full transformation semigroup, the order-preserving partial transformation semigroup and the order-preserving 1-1 partial transformation semigroup on X, respectively. It is known that $OP(X)$ and $OI(X)$ are regular semigroups. In addition, the regular elements of $OT(X)$ have been characterized. To extend these results, the following semigroups are introduced. For chains X and Y, denote by $OT(X,Y)$ the set of all order-preserving transformations from X into Y. For $\theta \in OT(Y,X)$, let $(OT(X,Y),\theta)$ stand for the semigroup $(OT(X,Y),*)$ where $\alpha * \beta = \alpha\theta\beta$ for all $\alpha, \beta \in OT(X,Y)$. We define the semigroups $(OP(X,Y),\theta)$ with $\theta \in OP(Y,X)$ and $(OI(X,Y),\theta)$ with $\theta \in OI(Y,X)$ analogously. The semigroups $(OT(X,Y),\theta)$, $(OP(X,Y),\theta)$ and $(OI(X,Y),\theta)$ can be considered as generalizations of $OT(X), OP(X)$ and $OI(X)$, respectively. In this paper, we characterize the regular elements of these generalized order-preserving transformation semigroups.

Mathematics Subject Classification: 20M17, 20M20.
Keywords: Regular element of a semigroup, generalized order-preserving transformation semigroup.

1 Introduction

For a mapping α, the domain and the range of α are denoted by $\operatorname{dom}\alpha$ and $\operatorname{ran}\alpha$, respectively. For $A \subseteq \operatorname{dom}\alpha$, denote by $\alpha_{|A}$ the restriction of α to A. The value of α at $x \in \operatorname{dom}\alpha$ is written as $x\alpha$. The identity mapping on a set X will be denoted by 1_X.

An element a of a semigroup S is called *regular* if $a = axa$ for some $x \in S$ and let $\operatorname{Reg}(S)$ denote the set of all regular elements of S. We call S a *regular semigroup* if $\operatorname{Reg}(S) = S$.

For nonempty subsets A and B of a partially ordered set (poset)X, let $A < B$ mean that $a < b$ for all $a \in A$ and $b \in B$.

For posets X and Y, the mapping $\alpha : X \to Y$ is said to be *order-preserving* if for any $x_1, x_2 \in \operatorname{dom}\alpha$, $x_1 \leq x_2$ in X implies $x_1\alpha \leq x_2\alpha$ in Y. X and Y are said to be *order-isomorphic* if there is a bijection φ from X onto Y such that φ and φ^{-1}

are order-preserving. A mapping $\alpha : X \to Y$ is called *order-reversing* if for any $x_1, x_2 \in \operatorname{dom}\alpha$, $x_1 \leq x_2$ in X implies $x_1\alpha \geq x_2\alpha$ in Y and we call X and Y are *anti-order-isomorphic* if there is a bijection φ from X onto Y such that φ and φ^{-1} are order-reversing.

For a poset X, let $OT(X), OP(X)$ and $OI(X)$ denote the order-preserving full transformation semigroup, the order-preserving partial transformation semigroup and the order-preserving 1-1 partial transformation semigroup on X, respectively. If X is a finite chain, $OT(X), OP(X)$ and $OI(X)$ have been counted respectively by Howie [4], Gomes and Howie [2] and Fernandes [1]. A well-known isomorphism theorem was provided in [10, p. 222-223] as follows : For posets X and Y, $OT(X) \cong OT(Y)$ as semigroups if and only if X and Y are either order-isomorphic or anti-order-isomorphic. The analogous results have been provided in [8] that for posets X and Y, $OP(X) \cong OP(Y)[OI(X) \cong OI(Y)]$ as semigroups if and only if X and Y are either order-isomorphic or anti-order-isomorphic. It is known from [3, p. 203] that $OT(X)$ is a regular semigroup if X is a finite chain. This result was extended in [6] that $OT(X)$ is regular for any chain X order-isomorphic to a subset of $\mathbb{Z}$, the set of integers with their natural order. In fact, Kim and Kozhukhov [9] characterized a countable chain X such that $OT(X)$ is a regular semigroup. It was also proved in [6] that for an interval X in $\mathbb{R}$, the set of real numbers with usual order, $OT(X)$ is a regular semigroup if and only if X is closed and bounded. In addition, it was provided in [6] that for any chain X, $OP(X)$ and $OI(X)$ are always regular. In general, $OT(X)$ need not be regular for any chain X. Then we have given in [11] a characterization determining when an element of $OT(X)$ is regular where X is a chain as follows: For a chain X and $\alpha \in OT(X)$, $\alpha \in \operatorname{Reg}(OT(X))$ if and only if the following conditions hold.

(i) If $\operatorname{ran}\alpha$ has an upper bound in X, then $\operatorname{ran}\alpha$ has a maximum.
(ii) If $\operatorname{ran}\alpha$ has a lower bound in X, then $\operatorname{ran}\alpha$ has a minimum.
(iii) If $a \in X \setminus \operatorname{ran}\alpha$ is neither an upper bound nor a lower bound of $\operatorname{ran}\alpha$, then $\{x \in \operatorname{ran}\alpha \mid x < a\}$ has a maximum or $\{x \in \operatorname{ran}\alpha \mid a < x\}$ has a minimum.

To generalize $OT(X), OP(X)$ and $OI(X)$, the following semigroups are introduced as follows: For posets X and Y, let $OT(X,Y), OP(X,Y)$ and $OI(X,Y)$ be the set of all order-preserving transformations from X into Y, the set of all order-preserving transformations from subsets of X into Y and the set of all 1-1 order-preserving transformations from subsets of X into Y, respectively. For $\theta \in OT(Y,X)$, let $(OT(X,Y),\theta)$ be the semigroup $(OT(X,Y),*)$ where $\alpha*\beta = \alpha\theta\beta$ for all $\alpha, \beta \in OT(X,Y)$. The semigroups $(OP(X,Y),\theta)$ with $\theta \in OP(Y,X)$ and $(OI(X,Y),\theta)$ with $\theta \in OI(Y,X)$ are defined similarly. Notice that $(OT(X,X),1_X)$ $= OT(X)$, $(OP(X,X),1_X) = OP(X)$ and $(OI(X,X),1_X) = OI(X)$. In [7], the authors characterized when $(OT(X,Y),\theta)$ is a regular semigroup where X and Y are chains. In addition, the regularity of $(OP(X,Y),\theta)$ and $(OI(X,Y),\theta)$ was determined in [5].

The purpose of this paper is to give necessary and sufficient conditions for the elements of $(OT(X,Y),\theta)$, $(OP(X,Y),\theta)$ and $(OI(X,Y),\theta)$ to be regular where X and Y are chains. The regularity of the elements in $(OT(X,Y),\theta)$ is given in terms of the regularity of some elements in $OT(X)$. Note that the characterization of the regularity of the elements in $OT(X)$ is mentioned above.

2 Main Results

Throughout this section, let X and Y be any chains.

The following lemmas will be used for our work. The proof of first lemma is simple and it is omitted.

Lemma 2.1. *If $\alpha : X \to Y$ is order-preserving and $a, b \in \operatorname{ran}\alpha$ such that $a < b$, then $a\alpha^{-1} < b\alpha^{-1}$.*

Lemma 2.2. *Let A and B be nonempty sets. If α and β are mappings from a subset of A into B and γ is a mapping from a subset of B into A such that $\alpha = \alpha\gamma\beta\gamma\alpha$, then the following conditions hold.*

(i) $\operatorname{ran}\alpha = \operatorname{ran}(\gamma\alpha)$.
(ii) $\operatorname{ran}\alpha \subseteq \operatorname{dom}\gamma$.
(iii) γ *is 1-1 on* $\operatorname{ran}\alpha$.

Proof. Since $\operatorname{ran}\alpha = \operatorname{ran}(\alpha\gamma\beta\gamma\alpha) \subseteq \operatorname{ran}(\gamma\alpha) \subseteq \operatorname{ran}\alpha$, we obtain (i). Also, we have

$$\begin{aligned}\operatorname{ran}\alpha = (\operatorname{dom}\alpha)\alpha = \big(\operatorname{dom}(\alpha\gamma\beta\gamma\alpha)\big)\alpha &\subseteq \big(\operatorname{dom}(\alpha\gamma)\big)\alpha \\ &= \big((\operatorname{ran}\alpha \cap \operatorname{dom}\gamma)\alpha^{-1}\big)\alpha \\ &= \operatorname{ran}\alpha \cap \operatorname{dom}\gamma \subseteq \operatorname{dom}\gamma.\end{aligned}$$

This verifies (ii). Since $\alpha = \alpha\gamma\beta\gamma\alpha$, it follows that $z = z\gamma\beta\gamma\alpha$ for all $z \in \operatorname{ran}\alpha$, or in other words, $\gamma\beta\gamma\alpha$ is the identity on $\operatorname{ran}\alpha$. If $y_1, y_2 \in \operatorname{ran}\alpha$ are such that $y_1\gamma = y_2\gamma$, then $y_1 = y_1\gamma\beta\gamma\alpha = y_2\gamma\beta\gamma\alpha = y_2$, so (iii) follows. □

First, we characterize the regular elements of $(OT(X,Y),\theta)$ with $\theta \in OT(Y,X)$.

Theorem 2.3. *For $\theta \in OT(Y,X)$ and $\alpha \in OT(X,Y)$, $\alpha \in \operatorname{Reg}((OT(X,Y),\theta))$ if and only if the following conditions hold.*

(i) $\alpha\theta \in \operatorname{Reg}(OT(X))$.
(ii) $\operatorname{ran}\alpha = \operatorname{ran}(\theta\alpha)$.
(iii) θ *is 1-1 on* $\operatorname{ran}\alpha$.

Proof. Assume that $\alpha \in \operatorname{Reg}((OT(X,Y),\theta))$. Then there exists $\beta \in OT(X,Y)$ such that $\alpha = \alpha\theta\beta\theta\alpha$. Thus $\alpha\theta, \beta\theta \in OT(X)$ and $\alpha\theta = (\alpha\theta)(\beta\theta)(\alpha\theta)$. This verifies (i) and, of course, (ii) and (iii) follow immediately from Lemma 2.2.

For the converse, assume that (i), (ii) and (iii) hold. Let $\beta \in OT(X)$ be such that $\alpha\theta = (\alpha\theta)\beta(\alpha\theta)$. Then $\alpha\big(\theta_{|\operatorname{ran}\alpha}\big) = \alpha\theta\beta\alpha\big(\theta_{|\operatorname{ran}\alpha}\big)$. Since $\theta|_{\operatorname{ran}\alpha}$ is 1-1, it follows that $\alpha = \alpha\theta\beta\alpha$. Then $\operatorname{ran}\alpha = \operatorname{ran}(\alpha\theta\beta\alpha) \subseteq \operatorname{ran}(\beta\alpha) \subseteq \operatorname{ran}\alpha$, so $\operatorname{ran}\alpha = \operatorname{ran}(\beta\alpha)$. Hence $\operatorname{ran}(\beta\alpha) = \operatorname{ran}\alpha = \operatorname{ran}(\theta\alpha)$. For each $y \in \operatorname{ran}(\beta\alpha) = \operatorname{ran}(\theta\alpha)$, choose an element $d_y \in y(\theta\alpha)^{-1}$. Then $d_y \in Y$ and $d_y(\theta\alpha) = y$ for all $y \in \operatorname{ran}(\beta\alpha)$. Note that $X = \bigcup_{y \in \operatorname{ran}(\beta\alpha)} y(\beta\alpha)^{-1}$ which is a disjoint union. Define $\beta' : X \to Y$ by a bracket notation as follows:

$$\beta' = \begin{pmatrix} y(\beta\alpha)^{-1} \\ d_y \end{pmatrix}_{y \in \operatorname{ran}(\beta\alpha)},$$

that is, $x\beta' = d_y$ for all $x \in y(\beta\alpha)^{-1}$ and $y \in \operatorname{ran}(\beta\alpha)$. If $x \in X$, then $x\alpha \in \operatorname{ran}\alpha = \operatorname{ran}(\beta\alpha)$ and $x\alpha = x\alpha\theta\beta\alpha = (x\alpha\theta)\beta\alpha$, so $x\alpha\theta \in (x\alpha)(\beta\alpha)^{-1}$ which implies that

$x\alpha\theta\beta'\theta\alpha = (x\alpha\theta)\beta'\theta\alpha = d_{x\alpha}(\theta\alpha) = x\alpha$. Hence $\alpha = \alpha\theta\beta'\theta\alpha$. To show that β' is order-preserving, let $x_1, x_2 \in X$ be such that $x_1 < x_2$. Then $x_1\beta\alpha \leq x_2\beta\alpha$. If $x_1\beta\alpha = x_2\beta\alpha$, then $x_1, x_2 \in (x_1\beta\alpha)(\beta\alpha)^{-1}$, so $x_1\beta' = d_{x_1\beta\alpha} = x_2\beta'$. Assume that $x_1\beta\alpha < x_2\beta\alpha$. Since $\operatorname{ran}(\beta\alpha) = \operatorname{ran}(\theta\alpha)$, we get $x_1\beta\alpha, x_2\beta\alpha \in \operatorname{ran}(\theta\alpha)$. Since $\theta\alpha \in OT(Y)$, it follows from Lemma 2.1 that $(x_1\beta\alpha)(\theta\alpha)^{-1} < (x_2\beta\alpha)(\theta\alpha)^{-1}$. It follows that $d_{x_1\beta\alpha} < d_{x_2\beta\alpha}$. Since $((x_1\beta\alpha)(\beta\alpha)^{-1})\beta' = \{d_{x_1\beta\alpha}\}$ and $((x_2\beta\alpha)(\beta\alpha)^{-1})\beta' = \{d_{x_2\beta\alpha}\}$, we have that $x_1\beta' = d_{x_1\beta\alpha} < d_{x_2\beta\alpha} = x_2\beta'$.

The proof is thereby completed. □

Next, we characterize the elements of $\operatorname{Reg}((OP(X,Y),\theta))$ with $\theta \in OP(Y,X)$ and $\operatorname{Reg}((OI(X,Y),\theta))$ with $\theta \in OI(Y,X)$.

Theorem 2.4. *For $\theta \in OP(Y,X)$ and $\alpha \in OP(X,Y)$,$\alpha \in \operatorname{Reg}((OP(X,Y),\theta))$ if and only if the following conditions hold.*

(i) $\operatorname{ran}\alpha = \operatorname{ran}(\theta\alpha)$.
(ii) $\operatorname{ran}\alpha \subseteq \operatorname{dom}\theta$.
(iii) *θ is 1-1 on* $\operatorname{ran}\alpha$.

Proof. It is immediate from Lemma 2.2 that if $\alpha \in \operatorname{Reg}((OP(X,Y),\theta))$, then conditions (i), (ii) and (iii) hold.

Now suppose, conversely, that conditions (i), (ii) and (iii) hold. Then $\operatorname{ran}(\alpha\theta) = (\operatorname{ran}\alpha \cap \operatorname{dom}\theta)\theta = (\operatorname{ran}\alpha)\theta$. Since $\operatorname{ran}\alpha = \operatorname{ran}(\theta\alpha)$, we get $y(\theta\alpha)^{-1} \neq \emptyset$ for every $y \in \operatorname{ran}\alpha$. For each $y \in \operatorname{ran}\alpha$, choose an element $d_y \in y(\theta\alpha)^{-1}$. Then $d_y \in Y$ and $d_y(\theta\alpha) = y$ for all $y \in \operatorname{ran}\alpha$. Define $\beta : \operatorname{ran}(\alpha\theta)(= (\operatorname{ran}\alpha)\theta) \to Y$ by

$$\beta = \begin{pmatrix} y\theta \\ d_y \end{pmatrix}_{y\in\operatorname{ran}\alpha}.$$

The mapping β is well-defined by (iii). To show that β is order-preserving, let $y_1, y_2 \in \operatorname{ran}\alpha$ be such that $y_1\theta < y_2\theta$. Since θ is order-preserving, it follows from Lemma 2.1 that $y_1 < y_2$ since $y_1 \in (y_1\theta)\theta^{-1}$ and $y_2 \in (y_2\theta)\theta^{-1}$. Since $\theta\alpha \in OP(Y)$ and $y_1, y_2 \in \operatorname{ran}\alpha = \operatorname{ran}(\theta\alpha)$, by Lemma 2.1, $y_1(\theta\alpha)^{-1} < y_2(\theta\alpha)^{-1}$. But $d_{y_1} \in y_1(\theta\alpha)^{-1}$ and $d_{y_2} \in y_2(\theta\alpha)^{-1}$, so $d_{y_1} < d_{y_2}$. Then $(y_1\theta)\beta = d_{y_1} < d_{y_2} = (y_2\theta)\beta$. Hence $\beta \in OP(X,Y)$. It remains to show that $\alpha = \alpha\theta\beta\theta\alpha$. Since for $x \in \operatorname{dom}\alpha, x\alpha\theta \in \operatorname{dom}\beta$ and $x\alpha\theta\beta = d_{x\alpha} \in \operatorname{dom}(\theta\alpha)$, this implies that $\operatorname{dom}(\alpha\theta\beta\theta\alpha) = \operatorname{dom}\alpha$. If $x \in \operatorname{dom}\alpha$, then $x\alpha\theta\beta\theta\alpha = (x\alpha\theta)\beta\theta\alpha = d_{x\alpha}(\theta\alpha) = x\alpha$. Hence $\alpha = \alpha\theta\beta\theta\alpha$. This shows that α is regular in $(OP(X,Y),\theta)$ and the verification is complete. □

Theorem 2.5. *For $\theta \in OI(Y,X)$ and $\alpha \in OI(X,Y)$, $\alpha \in \operatorname{Reg}((OI(X,Y),\theta))$ if and only if the following conditions hold.*

(i) $\operatorname{dom}\alpha \subseteq \operatorname{ran}\theta$.
(ii) $\operatorname{ran}\alpha \subseteq \operatorname{dom}\theta$.

Proof. Assume that α is a regular element of $(OI(X,Y),\theta)$. Then there is $\beta \in OI(X,Y)$ such that $\alpha = \alpha\theta\beta\theta\alpha$. It follows from Lemma 2.2 that $\operatorname{ran}\alpha = \operatorname{ran}(\theta\alpha)$ and $\operatorname{ran}\alpha \subseteq \operatorname{dom}\theta$. Then $(\operatorname{dom}\alpha)\alpha = \operatorname{ran}\alpha = \operatorname{ran}(\theta\alpha) = (\operatorname{ran}\theta \cap \operatorname{dom}\alpha)\alpha$, so $\operatorname{dom}\alpha = \operatorname{ran}\theta \cap \operatorname{dom}\alpha$ because α is 1-1. Hence $\operatorname{dom}\alpha \subseteq \operatorname{ran}\theta$.

Conversely, assume that (i) and (ii) hold. Then $\operatorname{ran}(\theta\alpha) = (\operatorname{ran}\theta \cap \operatorname{dom}\alpha)\alpha = (\operatorname{dom}\alpha)\alpha = \operatorname{ran}\alpha$ and $\operatorname{dom}(\alpha\theta) = (\operatorname{ran}\alpha \cap \operatorname{dom}\theta)\alpha^{-1} = (\operatorname{ran}\alpha)\alpha^{-1} = \operatorname{dom}\alpha$. Define $\beta = (\alpha\theta)^{-1}\alpha(\theta\alpha)^{-1}$. It is evident that $\beta \in OI(X,Y)$. We also have that $\alpha\theta\beta\theta\alpha = \alpha\theta(\alpha\theta)^{-1}\alpha(\theta\alpha)^{-1}\theta\alpha = 1_{\operatorname{dom}(\alpha\theta)}\alpha 1_{\operatorname{ran}(\theta\alpha)} = 1_{\operatorname{dom}\alpha}\alpha 1_{\operatorname{ran}\alpha} = \alpha$, so $\alpha \in \operatorname{Reg}((OI(X,Y),\theta))$, as desired. $\square$

References

[1] V.H. Fernandes, Semigroups of order preserving mappings on a finite chians: A new class of divisor, *Semigroup Forum* **54**(1997), 230-236.

[2] G.M.S. Gomes and J.M. Howie, On the ranks of certain semigroups of order-preserving transformations, *Semigroup Forum* **45**(1992), 272-282.

[3] P. M. Higgins, *Techniques of semigroup theory*, New York, Oxford University Press, 1992.

[4] J. M. Howie, Product of idempotents in certain semigroups of transformations, *Proc. Edinburgh Math Soc.* **17**(1971), 223-236.

[5] S. Jaidee, *Order-preserving generalized transformation semigroups*, Master Thesis, Chulalongkorn University, 2003.

[6] Y. Kemprasit and T. Changphas, Regular order-preserving transformation semigroups, *Bull. Austral. Math. Soc.* **62**(2000), 511-524.

[7] Y. Kemprasit and S. Jaidee, Regularity and isomorphism theorems of generalized order-preserving transformation semigroups, *Vietnam J. Math.* **33**(2000), 253-260.

[8] Y. Kemprasit, W. Mora and T. Rungratgasame, Isomorphism theorems for semigroups of order-preserving partial transformations, *Int. J. Algebra* **4** (17)(2010), 799-808.

[9] V. I. Kim and I.B. Kozhukhov, Regularity conditions for semigroups of isotone transformations of countable chains, *J. Math. Sci.* **152**(2)(2008), 203-208.

[10] E.S. Lyapin, *Semigroups*, Translations of Mathematical Monographs Vol. 3 Providence, R.I., Amer. Math. Soc., 1974.

[11] W. Mora and Y. Kemprasit, Regular elements of some order-preserving transformation semigroups, *Int. J. Algebra* **4**(13)(2010), 631-641.

Finite Field Basis Conversion and Normal Basis in Characteristic Three

I. Muchtadi-Alamsyah*, F. Yuliawan and A. Muchlis

Algebra Research Division, Faculty of Mathematics and Natural Sciences, Institut Teknologi Bandung, Jl. Ganesha 10, Bandung 40132, Indonesia
*E-mail: *ntan@math.itb.ac.id, fajar.yuliawan@math.itb.ac.id, muchlis@math.itb.ac.id*

We propose a storage efficient basis conversion from Gaussian normal basis to polynomial basis and vice versa with operations on the ground field in characteristic three, some common hardware technique (such as shifting and wiring) and multiplications on polynomial basis.

Keywords: Finite fields; Polynomial basis, Gaussian normal basis, Storage-efficient basis conversion

1. Introduction

Arithmetic in finite fields of characteristic three has several applications in cryptosystem based on pairing (e.g. Tate pairing and Weil pairing) such as identity based cryptosystems. Elements in a finite field can be represented in many ways, depending on the choice of the basis. Thus, conversion between different choices of basis would be very useful, for example to provide interoperability between cryptographic applications and to implement some efficient arithmetic in one basis using another basis. In special-purpose hardware devices having small storage such as hardware for cryptographic applications, conversion in a storage-efficient manner is required.

Two common choices of basis are polynomial basis and normal basis (Granger et al[2]). In general, multiplication on polynomial basis is faster than that on normal basis. A special type of normal basis called Gaussian normal basis can be used to speed up multiplication. In this paper, we propose a storage efficient basis conversion from Gaussian normal basis to polynomial basis and vice versa with operations on the ground field in characteristic three, some common hardware technique (such as shifting and wiring) and multiplications on polynomial basis.

The paper is organized as follows: in section 2 we will give mathematical background. In section 3 we will give the basic techniques for the conversions and the conversion algorithms. In last section we will give conclusion and suggestions for further research.

2. Mathematical Background

Recall that the characteristic of a field is the least positive integer n such that

$$\underbrace{1+1+\cdots+1}_{n \text{ times}} = 0.$$

The characteristic of a field is always a prime number. For a finite field having characteristic p, the number of its elements is a power of p. Conversely, for each prime power p^m, there exists a finite field with p^m elements. We will denote the finite field with p^m elements as $\mathbb{F}_{p^m}$.

The finite field $\mathbb{F}_{p^m}$ is a vector space over the ground field $\mathbb{F}_p$ of dimension m. If $N = \{\alpha_0, \alpha_1, \alpha_2, \ldots, \alpha_{m-1}\}$ is a basis, then each element w of $\mathbb{F}_{p^m}$ can be represented uniquely as

$$w = w_0\alpha_0 + w_1\alpha_1 + \cdots + w_{m-1}\alpha_{m-1}.$$

In this case, the m-tuple $(w_0, w_1, \ldots, w_{m-1})$ is called the representation of w in basis N.

After a basis for is chosen, the algorithm for finite field operation in terms of finite field representation can be derived. For example, addition is a pointwise addition in the ground field, but multiplication and inversion are more complicated depending on the basis chosen. Two common choices of basis are polynomial basis and normal basis.

A polynomial basis of $\mathbb{F}_{p^m}$ is a basis of the form $\{\alpha^0, \alpha^1, \alpha^2, \ldots, \alpha^{m-1}\}$. Polynomial basis can be obtained taking α to be a root of any irreducible polynomial of degree m over $\mathbb{F}_p$. In this case, the element α is called the generator of the polynomial basis N. If α is a root of an irreducible polynomial $f(X)$, this representation is basically the same as the quotient ring $\mathbb{F}_p/\langle f(X)\rangle$. Multiplication of two field elements in this representation is done by multiplying two polynomial modulo the irreducible polynomial chosen and inversion can be done using the Extended Euclidean Algorithm.

A normal basis is a basis of the form $\{\alpha^{p^0}, \alpha^{p^1}, \alpha^{p^2}, \ldots, \alpha^{p^{m-1}}\}$ and α is again called generator of the normal basis. Multiplication and inversion in normal basis representation are generally more complicated than that of polynomial basis. The advantage of using normal basis is that the p-th power an element is just a cyclic shift of its representation i.e. if the representation of w is $(w_0, w_1, \ldots, w_{m-1})$, then the representation of w^p is $(w_{m-1}, w_0, w_1, \ldots, w_{m-2})$ since

$$\left(\sum_{i=0}^{m-1} w_i \alpha^{p^i}\right)^p = \sum_{i=0}^{m-1} w_i \alpha^{p^{i+1}} = \sum_{i=0}^{m-1} w_{i-1} \alpha^{p^i}$$

where $w_{-1} = w_{m-1}$. It also follows that the p-th root of w is represented as $(w_1, \ldots, w_{m-1}, w_0)$. In particular, in characteristic three, the cube and the cube root of an element can be easily computed using normal bases. This is useful in the implementation of Tate pairing in pairing-based cryptography.

It is known that every finite field has a normal basis. A normal basis can be constructed using Gauss periods. Gauss periods were introduced by Gauss to study the construction of a regular polygon using ruler and compass.

Definition 2.1. Let $r = mk + 1$ be a prime different from p, and $\mathcal{K}$ the unique subgrup of order k of the multiplicative group $\mathbb{Z}_r^\times$. Let $\mathcal{K}_0, \mathcal{K}_1, \ldots, \mathcal{K}_{m-1}$ be the cosets of $\mathcal{K}$ in $\mathbb{Z}_r^\times$. Let β be a primitive r-th root of unity in $\mathbb{F}_{p^{mk}}$. For all $0 \leq i \leq m-1$, define

$$\alpha_i = \sum_{\gamma \in \mathcal{K}_i} \beta^\gamma \in \mathbb{F}_{p^m}.$$

Then $\alpha_0, \alpha_1, \ldots, \alpha_{m-1}$ are called Gauss periods of type (m, k) over $\mathbb{F}_p$.

Wassermann gives the exact condition for a gauss period of type (m, k) to be a generator of a normal basis for $\mathbb{F}_{p^m}$. We have the following theorem.

Theorem 2.1. *Let $r = mk + 1$ be a prime different from p and $\mathcal{K}$ the unique subgrup of order k of the multiplicative group $\mathbb{Z}_r^\times$ and β a primitive r-th root of unity in $\mathbb{F}_{p^{mk}}$. Then the Gauss period*

$$\alpha = \sum_{\gamma \in \mathcal{K}} \beta^\gamma \in \mathbb{F}_{p^m}$$

is a generator of a normal basis in $\mathbb{F}_{p^m}$ over $\mathbb{F}_p$ if and only if $\gcd(e, m) = 1$, where e is the index of the subgrup generated by p in $\mathbb{Z}_r^\times$. In particular, α is a generator if p is a generator of the cyclic group $\mathbb{Z}_r^\times$.

In the above theorem, the normal basis generated by α is called a Gaussian normal basis of type k, or type k normal basis in short. For use in cryptographic applications, k is chosen to be small. Muchlis et al discussed

Table 1. The values of $100 \leq m \leq 300$ such that $\mathbb{F}_{3^m}$ has type 2 normal basis

m	r	m	r	m	r	m	r
105	211	141	283	194	389	*251	503
111	223	146	293	200	401	254	509
*113	227	155	311	209	419	260	521
116	233	158	317	221	443	278	557
119	239	165	331	224	449	*281	563
125	251	*173	347	230	461	284	569
128	257	176	353	231	463	285	571
*131	263	*179	359	*233	467	*293	587
134	269	189	379	*239	479	296	593
140	281	*191	383	243	487	299	599

Note: * denotes the values of m which are primes.

a conversion algorithm between type 1 normal basis and polynomial basis and also between type 2 normal basis and polynomial basis in binary finite fields (Muchlis et al[3]). In this paper, we will focus mainly on type 2 normal basis in finite fields of chraracteristic three.

Let $m > 1$ be such that $r = 2m + 1$ is a prime and let β be a primitive r-th root of unity in the finite fields $\mathbb{F}_{3^{2m}}$. By Theorem (2.1), the element $\alpha = \beta + \beta^{-1} \in \mathbb{F}_{3^m}$ is a generator of a type 2 normal basis in $\mathbb{F}_{3^m}$ over $\mathbb{F}_3$ if and only if $\gcd(e, m) = 1$, where e is the index of the subgrup generated by 3 in $\mathbb{Z}_{2m+1}^{\times}$. Type 2 normal basis seems to exist for a reasonably dense set of values of m, e.g., for 21.4% of all $m \leq 500$ (there are 107 values of $m \leq 500$ such that $\mathbb{F}_{3^m}$ has type 2 normal basis) and 18.9% of all $m \leq 1000$. If m is also required to be a prime, for example for security reasons on the cryptographic applications, then there are 37 values of $m \leq 1000$.

If α generate a type 2 normal basis in $\mathbb{F}_{3^m}$, the minimal polynomial of α is known to be the irreducible polynomial (Blake et al[1])

$$f(X) = \sum_{i=0}^{\lfloor m/2 \rfloor} \binom{m-i}{i} (-1)^i X^{m-2i} + \sum_{i=0}^{\lfloor (m-1)/2 \rfloor} \binom{m-1-i}{i} (-1)^i X^{m-(2i+1)}.$$

Now, since $\alpha^{3^i} = \left(\beta + \beta^{-1}\right)^{3^i} = \beta^{3^i} + \beta^{-3^i}$, the type 2 normal basis generated by α is

$$\{\beta + \beta^{-1}, \beta^3 + \beta^{-3}, \beta^{3^2} + \beta^{-3^2}, \ldots, \beta^{3^{m-1}} + \beta^{-3^{m-1}}\}.$$

Using the condition in type 2 normal basis characterization, we can obtain a new basis which is a permutation of the normal basis (Blake et al,[1] Sunar and Koc[4]).

Lemma 2.1. *Let* $\alpha = \beta + \beta^{-1}$ *as in above. The set*

$$\{\beta + \beta^{-1}, \beta^2 + \beta^{-2}, \beta^3 + \beta^{-3}, \ldots, \beta^m + \beta^{-m}\}$$

is a permutation of the type 2 normal basis generated by α.

Proof. We will divide it into two cases, depending on the parity of m.

If m is even, since $\gcd(e, m) = 1$, the order of 3 in $\mathbb{Z}_{2m+1}^{\times}$ is $2m$. If $3^i = \pm 3^j$ in $\mathbb{Z}_{2m+1}$ for some $0 \le i < j \le m-1$, then $3^{j-i} = \pm 1$ and hence $3^{2(j-i)} = 1$ which is impossible since $0 < 2(j-i) < 2m$.

Now if m is odd, the order 3 in $\mathbb{Z}_{2m+1}^{\times}$ is m or $2m$. If $3^i = \pm 3^j$ $\mathbb{Z}_{2m+1}$ for some $0 \le i < j \le m-1$, then $3^{j-i} = \pm 1$ and hence $3^{2(j-i)} = 1$ which is also impossible since $2(j-i)$ is not divisible by m.

We conlude that $\{3^0, 3^1, 3^2, \ldots, 3^{m-1}\} = \{\pm 1, \pm 2, \ldots, \pm m\}$. The lemma follows immediately since β a primitive $(2m+1)$-th root of unity. □

We will call the new basis, a type 2 permuted normal basis generated by α. This basis plays a key role in our conversion algorithm as an intermediate basis.

3. Conversion Algorithm

In this section, we present our conversion algorithms between type 2 normal basis and polynomial basis. The algorithms consists of two main steps. One of these is a conversion between type 2 normal basis and type 2 permuted normal basis. Since the later is a permutation of the former, hardware implementation of this conversion can be simply done by rewiring. Thus, we only need to consider the conversion between type 2 permuted normal basis and polynomial basis.

3.1. *Basic techniques for conversion*

Let $\alpha = \beta + \beta^{-1}$ be the generator of a type 2 permuted normal bases in $\mathbb{F}_{3^m}$. Hence, it also generate a polynomial basis. For convenience, let $\alpha_i = \beta^i + \beta^{-i}$ for all $1 \le i \le m$ and $\alpha_0 = \beta^0 + \beta^{-0} = -1$. Observe that $\alpha_1 = \alpha$ and

$$\alpha\alpha_i = \left(\beta^1 + \beta^{-1}\right)\left(\beta^i + \beta^{-i}\right) = \beta^{i+1} + \beta^{-(i+1)} + \beta^{i-1} + \beta^{-(i-1)}.$$

Hence, $\alpha\alpha_i = \alpha_{i+1} + \alpha_{i-1}$ for all $1 \le i \le m-1$. Since $\beta^{m+1} = \beta^{-m}$, we also have $\alpha\alpha_m = \alpha_m + \alpha_{m-1}$. Using these identites, we have the following lemmas which give some relations between type 2 permuted normal basis and polynomial basis with the same generator.

Lemma 3.1. *The representation of* $1 \in \mathbb{F}_{3^m}$ *in type 2 permuted normal basis is* $(-1, -1, \ldots, -1)$ *and if* $(w_0, w_1, \ldots, w_{m-1})$ *is a representation of* w *in type 2 permuted normal basis generated by* α, *then the representasion* $(v_0, v_1, \ldots, v_{m-1})$ *of* $v = \alpha w$ *(in the same basis) satisfies the relation*

$$v_0 = w_0 + w_1, \quad v_{m-1} = w_0 + w_{m-2} + w_{m-1}$$

and

$$v_i = w_0 + w_{i-1} + w_{i+1} \quad \text{for all } 1 \leq i \leq m-2.$$

Consequently, the representation $(x_0, x_1, \ldots, x_{m-1})$ *of* $x = \alpha^{-1} w$ *can also be obtained by solving the system*

$$\begin{aligned} x_0 + x_1 &= w_0; \\ x_0 + x_{i-1} + x_{i+1} &= w_i \quad \text{for all } 1 \leq i \leq m-2; \\ x_0 + x_{m-2} + x_{m-1} &= w_{m-1}. \end{aligned}$$

Proof. Since $\beta^{2m+1} = 1$ and $\beta \neq 1$, then $\beta^i = \beta^{i-2m-1}$ and

$$1 + \beta + \beta^2 + \cdots + \beta^{2m} = \frac{\beta^{2m+1} - 1}{\beta - 1} = 0.$$

Hence

$$1 = \sum_{i=1}^{m} -\beta^i + \sum_{i=m+1}^{2m} -\beta^{i-2m-1} = \sum_{i=1}^{m} -\left(\beta^i + \beta^{-i}\right) = \sum_{i=1}^{m} -\alpha_i$$

and the first part of the lemma is proved.

For the second part, we observe that

$$\begin{aligned} \sum_{i=0}^{m-1} v_i \alpha_{i+1} = \alpha \sum_{i=0}^{m-1} w_i \alpha_{i+1} &= \sum_{i=0}^{m-2} w_i \alpha \alpha_{i+1} + w_{m-1} \alpha \alpha_m \\ &= \sum_{i=0}^{m-2} w_i (\alpha_{i+2} + \alpha_i) + w_{m-1}(\alpha_m + \alpha_{m-1}) \\ &= w_0 \alpha_0 + w_1 \alpha_1 + \sum_{i=2}^{m-1} (w_{i-2} + w_i) \alpha_i + (w_{m-2} + w_{m-1}) \alpha_m. \end{aligned}$$

The result follows from the identity $\alpha_0 = -1 = \alpha_1 + \alpha_2 + \cdots + \alpha_m$ and by comparing coefficients of $\alpha_1, \alpha_2, \ldots, \alpha_m$. □

The second part of Lemma (3.1) above shows that we can obtain the representation of polynomial basis elements $1, \alpha, \alpha^2, \ldots, \alpha^{m-1}$ in type 2 normal basis iteratively: having the representation of α^i for some $0 \leq i \leq m-2$, we can obtain the representation of α^{i+1}.

Lemma 3.2. *Let $1 \leq i \leq m-2$. If $(b_0, b_1, \ldots, b_{m-1})$ is the representation of α_i in polynomial basis generated by α, then*

$$b_{i+1} = b_{i+2} = \cdots = b_{m-1} = 0.$$

Furthermore, if $2 \leq i \leq m-2$ and $(a_0, a_1, \ldots, a_{m-1})$, $(c_0, c_1, \ldots, c_{m-1})$ are the representation of α_{i-1} and α_{i+1}, respectively in polynomial basis generated by α then $c_0 = -a_0$ and

$$c_j = -a_j + b_{j-1} \quad \textit{for all } 1 \leq j \leq m-1.$$

Proof. We will prove the first part using induction on i. Since $\alpha_1 = \alpha$ and $\alpha_2 = \alpha\alpha_1 - \alpha_0 = \alpha^2 + 1$, the representations of α_1 and α_2 in polynomial basis generated by α are $(0, 1, 0, 0, \ldots, 0)$ and $(1, 0, 1, 0, 0, \ldots, 0)$. Hence, the statement is true for $i = 1, 2$. Assuming that $(w_0, w_1, \ldots, w_{j-1}, 0, 0, \ldots, 0)$ and $(v_0, v_1, \ldots, v_j, 0, 0, \ldots, 0)$ is the representation of α_{j-1} and α_j for some $2 \leq j \leq m-3$, the representation of α_{j+1} is then

$$(w_0, w_1 + v_0, w_2 + v_1, \ldots, w_{j-1} + v_{j-2}, v_{j-1}, v_j, 0, 0, \ldots, 0)$$

since $\alpha_{j+1} = \alpha\alpha_j - \alpha_{j-1}$ and the statement is also true for $i = j+1$.

For the second part of the lemma, since $\alpha_{i+1} = \alpha\alpha_i - \alpha_{i-1}$ and $b_{m-1} = 0$, we have

$$\sum_{k=0}^{m-1} c_k \alpha^k = \alpha \sum_{k=0}^{m-1} b_k \alpha^k - \sum_{k=0}^{m-1} a_k \alpha^k = -a_0 + \sum_{k=1}^{m-1} (b_{k-1} - a_k) \alpha^k$$

and the result follows by comparing coefficients. □

Almost similar to the previous lemma, we can use Lemma (3.2) above to obtain the representation of type 2 permuted normal basis elements in polynomial basis, except the last elements: for $2 \leq i \leq m-2$, if the representation of α_{i-1} and α_i in polynomial basis generated by α are known, then the representation of α_{i+1} can also be obtained. We do not need the representation of α_m in polynomial basis since if $(w_0, w_1, \ldots, w_{m-1})$ is the representation of w in type 2 permuted normal basis, then

$$w = w_{m-1}\alpha_m + \sum_{i=1}^{m-1} w_{i-1}\alpha_i = -w_{m-1} + \sum_{i=1}^{m-1} (w_{i-1} - w_{m-1})\alpha_i$$

since we have the identity $\alpha_m = -(1 + \alpha_1 + \alpha_2 + \cdots + \alpha_{m-1})$.

3.2. *The conversion algorithms*

In this section, we will present our conversion algorithms between type 2 permuted normal basis and polynomial basis. First, we present the conversion from polynomial basis generated by α to type 2 permuted normal basis (also generated by α). The following algorithm which is based on Lemma (3.1) can be considered as a special case of the Generate-Accumulate method that is proposed by Kaliski and Liskov.

Algorithm 3.1. A conversion algorithm from polynomial basis representation to type 2 permuted normal basis representation.
Input: The representation $(a_0, a_1, \ldots, a_{m-1})$ of an element a in polynomial basis.
Output: The representation $(w_0, w_1, \ldots, w_{m-1})$ of a in type 2 permuted normal basis.
Procedure:
$v := (-1, -1, \ldots, -1)$
$w := (0, 0, \ldots, 0)$
for i **from** 0 **to** $m - 1$ **do**
  $w := w + a_i \times v$
  $v := (-1, -1, \ldots, -1) + \text{RSHIFT}(v) + \text{LSHIFTLL}(v)$

In the above algorithm, RSHIFT is the usual right shift function, that is RSHIFT of $(v_0, v_1, v_2, \ldots, v_{m-2}, v_{m-1})$ is $(0, v_0, v_1, \ldots, v_{m-3}, v_{m-2})$ and LSHIFTLL is a left shift function that leave the last coefficient unchanged, that is LSHIFTLL of $(v_0, v_1, v_2, \ldots, v_{m-2}, v_{m-1})$ is $(v_1, v_2, v_3, \ldots, v_{m-2}, v_{m-1}, v_{m-1})$. Based on Lemma (3.1), entering the first iteration, v is the representation of 1 in type 2 permuted normal basis, and entering the i-th loop, it is the representation of α^{i-1}, which is the elements of the input basis.

Next, we propose a conversion algorithm from type 2 permuted normal basis representation to polynomial basis representation based on Lemma (3.2).

Algorithm 3.2. A conversion algorithm from type 2 permuted normal basis representation to polynomial basis representation.
Input: The representation $(w_0, w_1, \ldots, w_{m-1})$ of an element a in type 2 permuted normal basis.
Output: The representation $(a_0, a_1, \ldots, a_{m-1})$ of a in polynomial basis.
Procedure:

$b := (0, 1, 0, 0, \ldots, 0)$
$c := (1, 0, 1, 0, 0, \ldots, 0)$
$a := (-w_{m-1}, w_0 - w_{m-1}, 0, 0, \ldots, 0)$
for i **from** 1 **to** $m - 1$ **do**
  $a := a + (w_i - w_{m-1}) \times c$
  $d := b$
  $b := c$
  $c := \text{RSHIFT}(c) - d$

4. Conclusion

We have described in this paper several new algorithms for storage-efficient conversion between Gaussian normal basis and polynomial basis. The use of some permutation of Gaussian normal basis and polynomial basis with the same generator enable us to replace multiplications on optimal normal basis with simpler operations and some multiplications on polynomial basis. For further research, one may apply these algorithms for use in pairing-based cryptosystems.

Acknowledgement

The authors are supported by Hibah Penelitian Ilmu Pengetahuan Terapan DIPA DIKTI ITB based on SK Dekan FMIPA No.14a/ SK/ K01.7/ KP/ 2010.

References

1. I. Blake, S.Gao, A.Menezes, R. Mullin, S. Vanstone, and T. Yaghoobian, *Applications of Finite Fields*,(Kluwer Academic Publishers, 1993).
2. R.Granger, D.Page, M.Stam, *Hardware and Software Normal Basis Arithmetic for Pairing-Based Cryptography in Characteristic Three*, *IEEE Transactions on Computers* **Volume 54 Issue 7**, 852 - 860, (2005).
3. A. Muchlis, M.H.Khusyairi, F.Yuliawan, *Storage Efficient Finite Field Basis Conversion via Efficient Multiplication*, preprint, (2010).
4. B. Sunar and C. K. Koc, *An efficient optimal normal basis type II multiplier*, *IEEE Transactions on Computers.* **50(1)**, 83-87, (2001).

The Maximal Separative Homomorphic Image of an LA**-Semigroup

Qaiser Mushtaq and Muhammad Inam

ABSTRACT. In this paper a congruence relation ρ has been defined on an LA**-semigroup S by $a\rho b$ if and only if $a^n b = a^{n+1}$ and $b^n a = b^{n+1}$, where n is a positive integer. It has been proved also that S/ρ is a maximal separative homomorphic image of S.

1. INTRODUCTION

T. Tamura and N. Kimura in [8], proved that any commutative semigroup S is uniquely expressible as a semilattice of archimedean semigroups. Later E. Hewitt and H.S. Zuckerman in [1], proved the following mutually equivalent conditions: S is separative, the archimedean components of S are cancellative, and S can be embedded in a union of groups. Q. Mushtaq and S.M. Yusuf in [5], have extended their results to a locally associative LA-semigroup S. In this paper we have extended these results to an LA**-semigroup. Specifically, a congruence relation ρ on LA**-semigroup S has been defined by $a\rho b$ if and only if $a^n b = a^{n+1}$ and $b^n a = b^{n+1}$, where n is a positive integer. It has been proved that S/ρ is a maximal separative homomorphic image of S.

A groupoid S is called a left almost semigroup (abbreviated as an LA-semigroup) [3] if its elements satisfy the left invertive law

$$(ab)c = (cb)a, \text{ for all } a, b, c \in S. \tag{1}$$

It is also called an Abel-Grassmann's groupoid (abbreviated as an AG-groupoid) in [7]. Several examples and interesting properties of LA-semigroups can be found in [5] and [6]. It is also known [3] that in an LA-semigroup S, the medial law

$$(ab)(cd) = (ac)(bd) \tag{2}$$

holds for all $a, b, c, d \in S$. An LA-semigroup S is called an LA**-semigroup, if for all $a, b, c \in S$ the following condition is satisfied.

$$a\,(bc) = b\,(ac)\,. \tag{3}$$

2000 *Mathematics Subject Classification.* 20M10 and 20N99.

Key words and phrases. LA-semigroups, LA**-semigroups, Abel Grassmann's groupoid, Medial law, Left invertive law.

For example, if $S = \{a, b, c, d, e\}$, then S forms an LA**-semigroup with respect to the multiplication defined by the following Cayley table.

.	a	b	c	d	e
a	d	d	d	d	d
b	d	d	d	d	e
c	d	d	d	d	e
d	d	d	d	d	d
e	d	a	a	d	d

If a belongs to S, then we define a^m for every positive integer m as follows:
(*i*) $a^m = (((aa)\,a)\,a)\,a\ ...\ m$ times. It implies that $a^2a = a^3$, $a^3a = a^4$, ...
(*ii*) $a^{m-1}a = a^m$, $a^m a = a^{m+1}$.

Lemma 1. *If a belongs to S, then for every positive integer m,*

(*i*) $a^m = a^{m-1}a = a^{m-3}a^3 = a^{m-5}a^5 = a^{m-7}a^7 = ...$
(*ii*) $a^m = a^2a^{m-2} = a^4a^{m-4} = a^6a^{m-6} = ...$

Proof. It is straight forward to see that $a^m = a^{m-1}a = \left(a^{m-2}a\right)a = (aa)\,a^{m-2} = a^2a^{m-2} = a^2\left(a^{m-3}a\right) = a^{m-3}\left(a^2a\right) = a^{m-3}a^3 = \left(a^{m-4}a\right)a^3 = \left(a^3a\right)a^{m-4} = a^4a^{m-4}$
$= a^4\left(a^{m-5}a\right) = a^{m-5}\left(a^4a\right) = a^{m-5}\left(a^5\right)$. Hence the proof follows immediately. □

The following results from [2] are since extensively used in this paper, we therefore reproduce them here.
(*i*). If $a \in S$, then for every positive integer n, $a^{2n} = a^{2n-1}a = aa^{2n-1}$.
(*ii*). For $m = 2n+1$, $a^{2n+1} = a^{2n}a = a^2a^{2n-1}$.
(*iii*). If $a \in S$, then for every positive integer m and n, $a^m a^{2n-1} = a^{m+2n-1}$.
The above result can easily be proved by using induction on m and by result (*i*), $aa^{2n-1} = a^{2n}$. The rest of the proof is same as in [[2], Theorem 4.7].
(*iv*). If $a \in S$, then for every positive integer m and n, $a^{2n}a^m = a^{2n+m}$.
(*v*). For every a, $b \in S$ and positive integer n, $(ab)^n = a^n b^n$.
(*vi*). For any $a \in S$ and positive integers k and m, $\left(a^k\right)^m = a^{km}$.

Theorem 1. *In S,*

$$(ab)\,(cd) = (ac)\,(bd) = (dc)\,(ba) = (db)\,(ca)$$

for all a, b, c, $d \in S$.

Proof. Available in [7]. □

An element $a \in S$ is called a left associative element of S, if for all b and $c \in S$

$$a\,(bc) = (ab)\,c.$$

Theorem 2. *$H = \left\{a^2 \in G :\ for\ all\ a \in S\right\}$ is a commutative subsemigroup of S and it contains all the left associative elements of S.*

Proof. Let a^2 and b^2 belong to H. Then

$$a^2b^2 = (aa)\,(bb) = (bb)\,(aa) = b^2a^2$$

Again, for any $a^2 \in H$, and b, $c \in S$, we have

$$a^2\,(bc) = (aa)\,(bc) = (cb)\,(aa) = (cb)\,a^2 = \left(a^2b\right)c.$$ □

Lemma 2. *If $a \in S$, then for every positive integer m and n, $a^{2n}a^m = a^{m+2n} = a^{m+1}a^{2n-1} = a^{2n-2}a^{m+2} = a^{m+3}a^{2n-3} = \ldots$*

Proof.

(i)
$$a^{m+2n} = a^{2n}a^m = \left(a^{2n-1}a\right)a^m = \left(a^m a\right)a^{2n-1}$$

(ii)
$$a^{2n}a^m = a^{m+1}a^{2n} = a^{m+1}\left(a^{2n-2}a\right) = a^{2n-2}\left(a^{m+1}a\right) = a^{2n-2}a^{m+2}$$

(iii)
$$a^{2n}a^m = \left(a^{2n-3}a\right)a^{m+2} = \left(a^{m+2}a\right)a^{2n-3} = a^{m+3}a^{2n-3}$$

and so on. □

Define a relation ρ on a S as follows:

$$a\rho b \text{ if and only if } b^n a = b^{n+1} \text{ and } a^n b = a^{n+1}.$$

Theorem 3. *If there exist positive integers m and n such that $b^m a = b^{m+1}$ and $a^n b = b^{n+1}$ then $a\rho b$, where a and $b \in S$.*

Proof. Without any loss of generality we suppose that $m < n$. Then two cases arise here.

(i) When m and n are both odd or are both even. Then $n-m$ is even and so by definition $b^m a = b^{m+1}$.This implies that $b^{n-m}\left(b^m a\right) = b^{n-m}b^{m+1}$, since $b^{n-m} \in H$. So we can write $\left(b^{n-m}b^m\right)a = b^{n-m}b^{m+1}$, which implies that $\left(b^{n-m}b^m\right)a = b^{n+1}$ and hence $b^{n-m+m}a = b^{n+1} = b^n a$.

(ii) When one of m and n is even and the other is odd. Suppose m is even and n is odd and that $m < n$. Then $n-m$ is odd. Now $b^m a = b^{m+1}$ implies that $b^{n-m}\left(b^m a\right) = b^{n-m}b^{m+1} = b^{n+1}$ and $b^m\left(b^{n-m}a\right) = b^{n+1}$. Since $b^m \in H$, so $\left(b^m b^{n-m}\right)a = b^{n+1}$ and hence $b^n a = b^{n+1}$. □

Theorem 4. *The relation ρ defined on S is a congruence relation.*

Proof. Now $a\rho b$ implies that $b^k a = b^{k+1}$, $a^k b = a^{k+1}$. The relation ρ is clearly reflexive and symmetric. For transitivity we may proceed as follows.

(i) If k is odd, let $a\rho b$ and $b\rho c$ so that there exist positive integers n and m such that,

$$\begin{aligned} b^{2n-1}a &= b^{2n},\ a^{2n-1}b = a^{2n} \\ c^{2m-1}b &= c^{2m},\ b^{2m-1}c = b^{2m} \end{aligned}$$

Suppose that l is odd, and $l = (2n)(2m) - 1 = (2n-1)(2m) + (2m-1)$.Then

$$c^l a = c^{(2n-1)(2m)+(2m-1)}a = \left(c^{(2n-1)(2m)}c^{2m-1}\right)a$$
$$= \left(\left(c^{2m}\right)^{2n-1}c^{2m-1}\right)a = \left(\left(c^{2m-1}b\right)^{2n-1}c^{2m-1}\right)a$$
$$= \left(\left(c^{(am-1)(2n-1)}b^{2n-1}\right)c^{2m-1}\right)a = \left(ac^{2m-1}\right)\left(c^{(am-1)(2n-1)}b^{2n-1}\right)$$
$$= \left(a\left(c^{2m-2}c\right)\right)\left(c^{(am-1)(2n-1)}b^{2n-1}\right) = \left(c\left(c^{2m-2}a\right)\right)\left(c^{(am-1)(2n-1)}b^{2n-1}\right)$$
$$= \left(cc^{(2m-1)(2n-1)}\right)\left(\left(c^{2m-2}a\right)b^{2n-1}\right) = \left(cc^{(2m-1)(2n-1)}\right)\left(b^{2n-1}a\right)c^{2m-2}$$
$$= \left(c^{(2m-1)(2n-1)+1}\right)\left(\left(b^{2n-1}a\right)c^{2m-2}\right) = \left(c^{(2m-1)(2n-1)+1}\right)\left(b^{2n}c^{2m-2}\right)$$

$$= b^{2n}\left(c^{(2m-1)(2n-1)+1}c^{2m-2}\right) = b^{2n}c^{(2m-1)(2n-1)+1+2m-2}$$

$$= b^{2n}\left(c^{2n(2m-1)-(2m-1)+1+2m-2}\right) = b^{2n}\left(c^{2n(2m-1)}\right)$$

$$= b^{2n}\left(c^{2m-1}\right)^{2n} = \left(c^{2m-1}\right)^{2n}b^{2n} = \left(c^{2m-1}b\right)^{2n}$$

$$= \left(c^{2m}\right)^{2n} = c^{(2m)(2n)} = c^{l+1}.$$

(ii) If k is even, let $a\rho b$ and $b\rho c$ so that there exist positive integers n and m such that,

$$\begin{aligned} b^{2n}a &= b^{2n+1},\ a^{2n}b = a^{2n+1} \\ c^{2m}b &= c^{2m+1},\ b^{2m}c = b^{2m+1} \end{aligned}$$

Suppose that l is even and $l = 2n(2m+1) + 2n$. Therefore,

$$\begin{aligned} c^{l}a &= c^{2n(2m+1)+2m}a = \left(c^{2n(2m+1)}c^{2m}\right)a \\ &= \left(\left(c^{2m}b\right)^{2n}c^{2m}\right)a = \left(ac^{2m}\right)\left(c^{2m}b\right)^{2n} \\ &= \left(ac^{2m}\right)\left(\left(c^{2m}\right)^{2n}b^{2n}\right) = \left(a\left(c^{2m-1}c\right)\right)\left(\left(c^{2m}\right)^{2n}b^{2n}\right) \\ &= \left(c^{2m-1}(ac)\right)\left(c^{(2m)(2n)}b^{2n}\right) = \left(c^{2m-1}(ac)\right)\left(c^{(2m)(2n)}b^{2n}\right) \\ &= \left(c^{2m-1}c^{(2m)(2n)}\right)\left((ac)\,b^{2n}\right) = \left(c^{2m-1}c^{(2m)(2n)}\right)\left(\left(b^{2n}c\right)a\right) \\ &= \left(c^{2m-1}c^{(2m)(2n)}\right)\left(b^{2n}(ca)\right) = \left(c^{2m-1}c^{(2m)(2n)}\right)\left(c\left(b^{2n}a\right)\right) \\ &= \left(c^{2m-1}c^{(2m)(2n)}\right)\left(cb^{2n+1}\right) = \left(c^{2m-1}c\right)\left(c^{(2m)(2n)}b^{2n+1}\right) \\ &= c^{2m}\left(c^{(2m)(2n)}b^{2n+1}\right) = \left(c^{2m}c^{(2m)(2n)}\right)b^{2n+1} \\ &= \left(c^{2m+(2m)(2n)}\right)b^{2n+1} = c^{(2m)(2n+1)}b^{2n+1} = \left(c^{2m}\right)^{2n+1}b^{2n+1} \\ &= \left(c^{2m}b\right)^{2n+1} = c^{(2m+1)(2n+1)} = c^{2n(2m+1)+2m+1} = c^{l+1} \end{aligned}$$

Hence the relation ρ is an equivalence relation. For compatible of ρ suppose that $a\rho b$ and $c \in S$. Then,

$$a^{k}b = a^{k+1} \text{ and } b^{k}a = b^{k+1}.$$

Also

$$(ac)^{k}(bc) = \left(a^{k}c^{k}\right)(bc) = \left(a^{k}b\right)\left(c^{k}c\right) = a^{k+1}c^{k+1} = (ac)^{k+1}.$$

Similarly, $(bc)^{k}(ac) = (bc)^{k+1}$. This implies that $ac\rho bc$. Now,

$$(ca)^{k}(cb) = \left(c^{k}a^{k}\right)(cb) = \left(c^{k}c\right)\left(a^{k}b\right) = c^{k+1}a^{k+1} = (ca)^{k+1}$$

implies that $(cb)^{k}(ca) = (cb)^{k+1}$, that is, $ca\rho cb$. Hence ρ is a congruence relation on S. □

A relation σ on S is called separative if and only if $a^2\sigma ab$ and $b^2\sigma ab$ implies that $a\sigma b$.

Theorem 5. *The relation ρ is separative.*

Proof. Let $a^2 \rho ab$ and $b^2 \sigma ab$. Then there exists positive integers m and n such that

(i, ii) $$\left(a^2\right)^n (ab) = \left(a^2\right)^{n+1}, \ (ab)^n a^2 = (ab)^{n+1}$$

(iii, iv) $$(ab)^m b^2 = (ab)^{m+1}, \ \left(b^2\right)^m (ab) = \left(b^2\right)^{m+1}$$

Then (i) implies that

(v) $$a^{2n} (ab) = a^{2(n+1)}$$

Also,

$$\begin{aligned} a^{2n}(ab) &= a\left(a^{2n}b\right) = a\left(\left(aa^{2n-1}\right)b\right) = a\left(\left(ba^{2n-1}\right)a\right) \\ &= \left(ba^{2n-1}\right)a^2 = \left(a^2a^{2n-1}\right)b = a^{2+2n-1}b \\ &= a^{2n+1}b \end{aligned}$$

So (v) implies that $a^{2n+1}b = a^{2(n+1)}$. Similarly, $b^{2m+1}a = b^{2(m+1)}$. Then by Theorem 3, we get $a\rho b$. □

Theorem 6. *S/ρ is the maximal separative homomorphic image of S.*

Proof. Since ρ is separative, therefore S/ρ is separative. We now show that ρ is contained in every separative congruence relation σ on S. Let $a\rho b$ so that there exists a positive integer n such that

$$b^n a = b^{n+1} \text{ and } a^n b = a^{n+1}.$$

We show that $a\sigma b$, where σ is a separative congruence on S. Let k be any positive integer such that

(i) $$b^k a \sigma b^{k+1}, a^k k \sigma a^{k+1}$$

an suppose that $k > 2$.Then $\left(b^{k-1}a\right)^2 = \left(b^{k-1}\right)^2 a^2 = b^{2k-2}a^2 = b^{2k-2}(aa) = \left(b^k b^{k-2}\right)(aa) = \left(b^k a\right)\left(b^{k-2}a\right)$ where k is odd or even. But, $b^k a \sigma b^{k+1}$ implies that

(ii) $$\left(b^k a\right)\left(b^{k-2}a\right)\sigma b^{k+1}\left(b^{k-2}a\right)$$

Here we consider the two case separately, that is, when k is even and when k is odd.

Case 1.

Let k be odd. Then because $k+1$ is even, $b^{k+1} \in H$. Therefore by the left invertive law and the fact that $b^{k-1} \in H$,

$$b^{k+1}\left(b^{k-2}a\right) = \left(b^{k+1}b^{k-2}\right)a = \left(\left(b^k b\right)b^{k-2}\right)a = \left(b^{k-2}b\right)b^k$$

(iii) $$= \left(b^{k-1}b^k\right)a = b^{k-1}\left(b^k a\right) = b^k\left(b^{k-1}a\right).$$

Now (i) implies that $\left(b^k a\right)\left(b^{k-2}a\right)\sigma b^k\left(b^{k-1}a\right)$. It further implies that $\left(b^k b^{k-2}\right)(aa)\sigma b^k\left(b^{k-1}a\right)$, yielding $b^{k+k-2}a^2\sigma b^k\left(b^{k-1}a\right)$.

Therefore $b^{2k-2}a^2\sigma b^k\left(b^{k-1}a\right)$ and $\left(b^{k-1}\right)^2 a^2\sigma b^k\left(b^{k-1}a\right)$. Hence

(iv) $$\left(b^{k-1}a\right)^2\sigma b^k\left(b^{k-1}a\right)$$

Also, $b^{k+1}\sigma b^k a$, and $b^{k-1}b^{k+1}\sigma b^{k-1}\left(b^k a\right)$ this implies that $b^{k-1+k+1}\sigma b^{k-1}\left(b^k a\right)$.It further means that $b^{2k}\sigma b^{k-1}\left(b^k a\right)$.Hence

(v) $$b^{2k}\sigma b^k\left(b^{k-1}a\right)$$

Now from (iv) and (v), we conclude that $\left(b^{k-1}a\right)^2\sigma b^k\left(b^{k-1}a\right)$ and $b^{2k}\sigma b^k\left(b^{k-1}a\right)$.

Case 2.

Let k be even. Then

$$\begin{aligned} b^{k+1}\left(b^{k-2}a\right) &= b^{k-2}\left(b^{k+1}a\right) = \left(b^{k-2}b^{k+1}\right)a = b^{k-2+k+1}a = b^{k-1+k}a \\ &= \left(b^{k}b^{k-1}\right)a = b^{k}\left(b^{k-1}a\right). \end{aligned}$$

Hence

(vi) $$b^{k-2}\left(b^{k+1}a\right) = b^{k-1}\left(b^{k}a\right)$$

Thus by using (vi), we have

(vii) $$\left(b^{k-1}a\right)^{2}\sigma b^{k}\left(b^{k-1}a\right)$$

Now, $b^{k+1}\sigma b^{k}a$ implies that $b^{k-1}b^{k+1}\sigma b^{k-1}\left(b^{k}a\right)$. Since k is even, so $k-1$ and $k+1$ are odd, therefore by result (iii) of [2], $b^{k-1+k+1}\sigma b^{k-1}\left(b^{k}a\right)$. This implies that

(viii) $$b^{2k}\sigma b^{k}\left(b^{k-1}a\right).$$

Thus we conclude from (vii) and $(viii)$, that $\left(b^{k}\right)^{2}\sigma\left(b^{k}\right)\left(b^{k-1}a\right)$ and $\left(b^{k}\right)\left(b^{k-1}a\right)\sigma\left(b^{k-1}a\right)^{2}$. Hence the relation is true whether k is odd or even. Let $X = b^{k}$ and $Y = b^{k-1}a$. Then from $\left(b^{k}\right)^{2}\sigma b^{k}\left(b^{k-1}a\right)$ and $\left(b^{k-1}a\right)^{2}\sigma b^{k}\left(b^{k-1}a\right)$ we have $X^{2}\sigma XY$ and $Y^{2}\sigma XY$. This implies that $X\sigma Y$, that is, $b^{k}\sigma b^{k-1}a$. Similarly, $a^{k}b\sigma a^{k-1}$ Therefore, if

(ix) $$a^{k}b\sigma a^{k-1} \text{ and } b^{k}\sigma b^{k-1}a$$

holds for k, it follows that (ix) holds for $k = 1$. Therefore $ba\sigma b^{2}$ and $ab\sigma a^{2}$. Since σ is separative, therefore $a\sigma b$. This implies that $\rho \subseteq \sigma$. Hence S/ρ is a maximal separative homomorphic image of S. □

Theorem 7. *Let ρ and σ be separative congruences on S. If $\rho \cap (S \times S) \subseteq \sigma \cap (S \times S)$, then $\rho \subseteq \sigma$.*

Proof. First note that $\left(a^{2}(ab)\right)\left(a^{2}b^{2}\right) = \left(a^{2}a^{2}\right)\left((ab)b^{2}\right) = a^{4}\left(\left(bb^{2}\right)a\right) = \left(bb^{2}\right)\left(a^{4}a\right) = b^{3}a^{5}$. Then the rest of the proof is same as in [2]. □

References

[1] Hewitt, E and H. S. Zuckerman, *The irreducible representation of Semigroup related to the semigroup*, 111. J. Math. , 1(1957), 188-213.

[2] Kamran, M. S, *Conditions for LA-senigroups to resemble associative structures*, Ph. D. Thesis, Quaid-i-Azam University, Islamabad, 1993.

[3] Kazim, M. A and M. Naseeruddin, *On almost semigroups*, Alig. Bull. Math., 2(1972), 1-7.

[4] Mushtaq, Q and S. M. Yusuf, *On LA-semigroups*, The Alig. Bull. Math., 8(1978), 65-70.

[5] Mushtaq, Q and S. M. Yusuf, *On locally associative LA-semigroup*, J. Nat. Sci. Math., 1, 19(1979), 57-62.

[6] Mushtaq, Q and Q. Iqbal, *Decomposition of locally associative LA-semigroup*, Semigroup Forum, 41(1991), 155-164.

[7] Protic, P. V and Milan Božinović, *Some congruences on an AG**-groupoid*, FILOMAT(Niš), Algebra, Logic and Dis. Math. , 9, 3(1995), 879-886.

[8] Tamura, T and N. Kimura, *On decomposition of a commutative semigroup*, Kodai Math. Sem. Rep. , 1954, 102-112.

Department of Mathematics, Quaid-i-Azam University, Islamabad, Pakistan

E-mail address: qmushtaq@isb.apollo.net.pk

A Note on Frobenius-Schur Indicators

Siu-Hung Ng

ABSTRACT. This exposition concerns two different notions of Frobenius-Schur indicators for finite-dimensional Hopf algebras. These two versions of indicators coincide when the underlying Hopf algebra is semisimple. We are particularly interested in the family of pivotal finite-dimensional Hopf algebras with unique pivotal element; both indicators are gauge invariants of this family of Hopf algebras. We obtain a formula for the (pivotal) Frobenius-Schur indicators for the regular representation of a pivotal Hopf algebra. In particular, we use this formula for the four dimensional Sweedler algebra and demonstrate the difference of these two indicators.

1. INTRODUCTION

The second Frobenius-Schur indicator of a representation over $\mathbb{C}$ of a finite group was introduced more than a century ago [6]. This indicator $\nu_2(V)$ of an irreducible representation V of a finite group G can only be 0, 1 or -1 which is respectively determined by whether V is complex, real or pseudo-real representation of G. Moreover, the indicator $\nu_2(V)$ can be computed by the formula:

$$\nu_2(V) = \frac{1}{|G|}\sum_{g\in G}\chi_V(g^2), \tag{1}$$

where χ_V is the character afforded by V. In general, the n-th Frobenius-Schur indicator of a representation of V of G is given by

$$\nu_n(V) = \frac{1}{|G|}\sum_{g\in G}\chi_V(g^n)\,. \tag{2}$$

The higher indicators appear to be more obscure than the second indicator but they certainly carries the arithmetic properties of the group G. If one defines $\theta_n(g)$ for each $g \in G$ as the number of solutions $x \in G$ such that $x^n = g$, then θ_n is a class function on G. Moreover,

$$\theta_n = \sum_{V\in \mathrm{Irr}(G)} \nu_n(V)\chi_V$$

where $\mathrm{Irr}(G)$ denotes a complete set of non-isomorphic irreducible representations of G. The reader are referred to some basic references (such as [2, 8, 19]) on representation theory of finite groups for these basic facts.

A *bialgebra* H over a field $\Bbbk$ is a $\Bbbk$-algebra equipped with two algebra maps $\Delta : H \to H\otimes H$ and $\epsilon : H \to \Bbbk$ which satisfy the conditions:

$$(\mathrm{id}\otimes\Delta)\Delta = (\Delta\otimes\mathrm{id})\Delta, \quad (\epsilon\otimes\mathrm{id})\Delta = (\mathrm{id}\otimes\epsilon)\Delta = \mathrm{id}\,. \tag{3}$$

The Sweedler notation is often used to denote $\Delta(h) = h_{(1)} \otimes h_{(2)}$ with the summation suppressed. In the sequel, we will use the notation: $\Delta^{(1)} = \mathrm{id}_H$, $\Delta^{(2)} = \Delta$ and

$$\Delta^{(n+1)} = (\Delta\otimes\mathrm{id}_H)\Delta^{(n)} \quad \text{for all integer } n \geq 2.$$

In Sweedler's notation, $\Delta^{(n)}(h) = h_{(1)} \otimes \cdots \otimes h_{(n)}$ for all integer $n \geq 2$. The n-th Sweedler power $h^{[n]}$ of $h \in H$ is defined as

$$h^{[n]} = h_{(1)} \cdot \cdots \cdot h_{(n)},$$

and the assignment $P_n : H \to H$, $h \mapsto h^{[n]}$ will be called the n-th Sweedler power map of H.

The bialgebra H is called a *Hopf algebra* if H admits an *antipode* which is a $\Bbbk$-linear map $S : H \to H$ such that

$$m(S \otimes \mathrm{id})(h) = \epsilon(h) = m(S \otimes \mathrm{id})(h)$$

for all $h \in H$ where m denotes the multiplication on H. A *left (resp. right) integral* of H is a non-zero element $\Lambda \in H$ such that $h\Lambda = \epsilon(h)\Lambda$ (resp. $\Lambda h = \epsilon(h)\Lambda$). An element of H is called a *two-sided integral* if it is both left and right integral. We refer the readers to [13, 20] for the basic facts and notations for Hopf algebras.

The left (or right) integrals of a finite-dimensional Hopf algebra H span a 1-dimensional ideal. Moreover, H is semisimple if, and only if, $\epsilon(\Lambda)$ for some left integral Λ. In this case, Λ is a two sided integral and the normalized (two-sided) integral Λ, that is $\epsilon(\Lambda) = 1$, of H is uniquely determined by H.

In the remainder of this note, we simply consider finite-dimensional Hopf algebras over the field of complex numbers $\mathbb{C}$.

The group algebra $\mathbb{C}[G]$ is a semisimple Hopf algebra with the comultiplication map $\Delta : \mathbb{C}[G] \to \mathbb{C}[G] \otimes \mathbb{C}[G]$, the counit map $\epsilon : \mathbb{C}[G] \to \mathbb{C}$ and the antipode S given by

$$\Delta(g) = g \otimes g, \quad \epsilon(g) = 1, \quad S(g) = g^{-1} \tag{4}$$

for $g \in G$. The element $\Lambda = \frac{1}{|G|}\sum_{g\in G} g$ is a normalized integral of $\mathbb{C}[G]$, i.e. $\epsilon(\Lambda) = 1$.

Linchenko and Montgomery [11] generalized the notion of the n-th Frobenius-Schur indicator $\nu_n(V)$ for the representation V of semisimple Hopf algebra H over $\mathbb{C}$ by using uniqueness of normalized integral semisimple Hopf algebras, namely for each integer $n \geq 2$,

$$\nu_n(V) = \chi_V(\Lambda^{[n]}) . \tag{5}$$

where χ_V is the character afforded by V.

The Frobenius-Schur theorem for the second indicator in relation to the existence of non-degenerate H-invariant symmetric or skew-symmetric bilinear form on a irreducible representation of a complex semisimple Hopf algebra was discovered in [11]. The arithmetic properties of the higher indicators were extensively studied in [10]. One of the remarkable consequences of these arithmetic properties is the Cauchy theorem [10] for complex semisimple Hopf algebras H which asserts that $\dim H$ and $\exp(H)$ have common prime factor, where $\exp(H)$ can be defined as the order of Drinfeld element in the Drinfeld double $D(H)$ (cf. [3]). The result also answered the questions raised by Etingof and Gelaki in [3].

Before the inception of [11], some notions of Frobenius-Schur (FS) indicator were also considered in rational conformal field theory (RCFT). Bantay introduced a version of the 2nd FS indicator for RCFT as a formula in terms of the modular data [1], and a categorical version of the 2nd FS indicator was studied by Fuchs, Ganchev, Szlachnyi, and Vescernys in [7]. These apparently unrelated notions of Frobenius-Schur indicators suggested a possibility that they are related and are invariant of the underlying monoidal categories. The author in collaboration Mason [12] introduced a definition for the representations of semisimple quasi-Hopf algebras H which is a generalization of the definition of Linchenko-Montgomery.

Moreover, if H and K are complex semisimple quasi-Hopf algebra with monoidally equivalent representation categories via the monoidal equivalence $\mathcal{F} : \text{Rep}(H) \to \text{Rep}(K)$, then $\nu_2(V) = \nu_2(\mathcal{F}(V))$.

When V is a simple H-module, $\nu_2(V)$ can only be 0, 1, or -1. A version of Frobenius-Schur Theorem is proved for semisimple quasi-Hopf algebra with this generalized notion of indicator. The canonical pivotal structure discovered in [5] has played an very important role in [12]. This fact was also recognized by Schauenburg, and he reintroduced the definition of FS indicator for a representation of a complex semisimple quasi-Hopf algebra using the canonical pivotal structure in [18]. This indicator is shown to be identical to that of [12]. The discovery initiated a generalized definition of the n-th Frobenius-Schur indicator for each object in a pivotal tensor category [15]. By using the canonical pivotal structures in a modular category and in the representation category of a complex semisimple quasi-Hopf algebra, the Bantay's 2-nd indicator formula and the indicators for semisimple (quasi-)Hopf algebras are recovered from this notion of (pivotal) Frobenius-Schur indicators (cf. [16, 14]). These new developments on indicators have yielded a Cauchy Theorem for semisimple quasi-Hopf algebras[14, Theorem 8.4], and two Congruence Subgroup Theorems for modular categories [17, Theorems 6.7, 6.8].

2. Pivotal Hopf algebras and beyond

Let H be a finite-dimensional Hopf algebra over $\mathbb{C}$ but not necessarily semisimple. Then $\text{Rep}(H)$ is a pivotal category if, and only if, the square of the antipode S of H is a conjugation by a group-like element g, i.e.$S^2(h) = ghg^{-1}$. Such group-like element is called a *pivotal element* and H is called a *pivotal Hopf algebra.* It is clear that if g is a pivotal element, then the coset $gZG(H)$ is the set of all pivotal elements of H where $ZG(H)$ is the group of central group-like elements of H. Unlike the semisimple case, there is no canonical choice of pivotal element in general. It is important to note that the pivotal Frobenius-Schur indicators are invariant of pivotal categories. If one would like distinguish only the monoidal structures of the representation categories of two non-isomorphic pivotal Hopf algebras, using the higher pivotal Frobenius-Schur indicators for these pivotal categories might not be straightforward.

This suggests a need of invariants for the monoidal structure of $\text{Rep}(H)$ for any finite-dimensional Hopf algebras H. By [16, Theorem 2.2], if H and K are finite-dimensional Hopf algebra over $\mathbb{C}$ such that $\text{Rep}(H)$ and $\text{Rep}(K)$ are $\mathbb{C}$-linearly equivalent monoidal categories, then $H \cong K^F$ where K^F is a Drinfeld twist by a gauge transformation F on H which satisfies some 2-cocycle conditions. We simply call H and K are *gauge equivalent* Hopf algebras in this case.

Let $\mathfrak{C}$ be a collection of finite-dimensional Hopf algebras over $\mathbb{C}$ which is closed under gauge equivalence classes. Following the terminology of [9], a quantity $f(H)$ defined for each $H \in \mathfrak{C}$ is called a gauge invariant for the collection $\mathfrak{C}$ if $f(K) = f(H)$ for all Hopf algebras K gauge equivalent to H. Typical examples of gauge invariants of H are $\dim(H)$ and quasi-exponent $\text{qexp}(H)$ defined by Etingof and Gelaki [4], and they are gauge invariants for the collection $\mathfrak{H}$ of all the finite-dimensional Hopf algebras over $\mathbb{C}$.

In [9], the author, in collaboration with Kashina and Montgomery, has introduced a $\mathbb{C}$-linear recursive sequence $\nu(H) = \{\nu_n(H)\}_{n \geq 1}$ which is given by

$$\nu_1(H) = 1, \quad \nu_n(H) = \text{Tr}(S \circ P_{n-1}) \quad \text{for } n \geq 2,$$

where S, P_m are respectively the antipode and the m-th Sweedler power of H. It has been shown in [9, Theorem 2.2] that the sequence $\nu(H)$ is a gauge invariant for $\mathfrak{H}$. Moreover, by [9, Corollary 2.6], we have

$$\nu_n(H) = \lambda(\Lambda^{[n]}) \tag{6}$$

where λ and Λ are both left (or both right) integrals of H^* and H respectively satisfying the condition $\lambda(\Lambda) = 1$. When H is semisimple, it follows by [10] that $\nu_n(H)$ coincides with the n-th Frobenius-Schur indicator of the regular representation of a semisimple Hopf algebra defined in [11].

Now, we come back to consider the collection $\mathfrak{P}$ of pivotal Hopf algebras with a unique pivotal element such as the Taft algebras or the small quantum group $U_q(\mathfrak{sl}_2)$. Since the pivotal element is unique for $H \in \mathfrak{P}$, $\text{Rep}(H)$ has exactly one pivotal structure. If K is a Hopf algebra gauge equivalent to H, then $\text{Rep}(K)$ also has a unique pivotal structure and so K has a unique pivotal element. Therefore, $\mathfrak{P}$ is a collection closed under gauge equivalence. Moreover, for any monoidal equivalence $\mathcal{F} : \text{Rep}(H) \to \text{Rep}(K)$ for some Hopf algebras $H, K \in \mathfrak{P}$, $\mathcal{F}$ is automatically a pivotal equivalence. Therefore, by the preceding remark,

$$\nu_n^p(H) = \nu_n^p(\mathcal{F}(H)) = \nu_n^p(K)$$

for all $n \in \mathbb{N}$, where $\nu_n^p(V)$ denotes the n-th pivotal Frobenius-Schur indicator of V in the pivotal category $\text{Rep}(H)$. In particular, the sequence $\nu^p(H) = \{\nu_n^p(H)\}_{n\geq 1}$ is a gauge invariant for $\mathfrak{P}$.

One natural question is how these gauge invariants $\nu(H)$ and $\nu^p(H)$ are related. In the following section, we will demonstrate that these two sequences are different, in general, for $H \in \mathfrak{P}$.

3. Pivotal Frobenius-Schur indicators for pivotal Hopf algebras

Let H be a pivotal Hopf algebra with a pivotal element g. For any left H-module V, the map $j : V \to V^{\vee\vee}$ given by

$$j(x) = g\hat{x}$$

is a natural isomorphism of H-modules, where $\hat{x}(f) = f(x)$ for all $f \in H^*$. Since g is group-like, $j : Id \to (-)^{\vee\vee}$ is an isomorphism of tensor functors. Hence, j defines a pivotal structure on $\text{Rep}(H)$. Consider the pivotal category $(\text{Rep}(H), j)$. Let λ and Λ be a right integral of H^* and a left integral of H, respectively, such that $\lambda(\Lambda) = 1$. For any $V \in \text{Rep}(H)$, $\theta : H \otimes {}^{\circ}V \to H \otimes V$ defined by $\theta(h \otimes v) = h_{(1)} \otimes h_{(2)}v$ is an H-module isomorphisms, where ${}^{\circ}V$ denotes the trivial left H-module with the underlying space V. Hence,

$$(H \otimes V)^H = \Lambda_{(1)} \otimes \Lambda_{(2)}V.$$

In particular, the linear map $\alpha : V \to (H \otimes V)^H$, $v \mapsto \Lambda_{(1)} \otimes \Lambda_{(2)}v$, is an isomorphism with

$$\alpha^{-1}(\sum_i h_i \otimes v_i) = \sum_i \lambda(h_i)v_i \,.$$

Thus that map $\overline{\alpha} : V \to \text{Hom}_H(\mathbb{C}, H \otimes V)$, defined by

$$\overline{\alpha}(v)(1) = \Lambda_{(1)} \otimes \Lambda_{(2)}v,$$

is a $\mathbb{C}$-linear isomorphism.

By [15], the map $E_{H,V} : \mathrm{Hom}_H(\mathbb{C}, H \otimes V) \to \mathrm{Hom}_H(\mathbb{C}, V \otimes H)$ is defined by

$$E_{H,V}(f) := \left(\mathbb{C} \xrightarrow{\mathrm{db}} H^\vee \otimes H^{\vee\vee} \xrightarrow{H^\vee \otimes f \otimes H^{\vee\vee}} H^\vee \otimes H \otimes V \otimes H^{\vee\vee} \xrightarrow{\mathrm{ev} \otimes j^{-1}} V \otimes H\right),$$

where $\mathrm{db} : \mathbb{C} \to H^\vee \otimes H^{\vee\vee}$ is the dual basis map. If $f \in \mathrm{Hom}_H(\mathbb{C}, H \otimes V)$ and $f(1) = \sum_i h_i \otimes v_i$, then $E_{H,V}(f)(1) = \sum_i v_i \otimes g^{-1}h_i$.

Now, we take $V = H^{\otimes(m-1)}$ for some integer $m \geq 1$. Note that $H^{\otimes 0} = \mathbb{C}$ by convention. The isomorphism $\overline{\alpha} : V \to \mathrm{Hom}_H(\mathbb{C}, H \otimes V)$ determines a unique endomorphism $\overline{E}^{(m)} = \overline{\alpha}^{-1} \circ E_{H,H^{\otimes(m-1)}} \circ \overline{\alpha}$ on V. Since $E_{H,\mathbb{C}} = \mathrm{id}$, and so is $\overline{E}^{(1)}$. For $m \geq 2$, $h \in H$ and $v \in H^{\otimes(m-1)}$,

$$\overline{E}^{(m)}(h \otimes v) = \lambda(\Lambda_{(2)}h)\Lambda_{(3)}v \otimes g^{-1}\Lambda_{(1)}. \tag{7}$$

The m-th pivotal Frobenius-Schur indicator $\nu_m^p(H)$ of the regular representation of H is defined as $\mathrm{Tr}(E_{H,H^{\otimes(m-1)}})$, which is also equal to $\mathrm{Tr}(\overline{E}^{(m)})$. The following lemma is useful for simplifying $\mathrm{Tr}(\overline{E}^{(m)})$.

Lemma 3.1. *Let A be an algebra over a field $\Bbbk$, $a_1, \ldots, a_n \in A$, $f \in A^*$. Let $T : A^{\otimes n} \to A^{\otimes n}$ be the $\Bbbk$-linear map given by*

$$T(v_1 \otimes \cdots \otimes v_n) = f(v_1)a_1v_2 \otimes \cdots \otimes a_{n-1}v_n \otimes a_n, \quad \textit{for } v_1 \otimes \cdots v_n \in A^{\otimes n}.$$

Then $\mathrm{Tr}(T) = f(a_1 \cdots a_{n-1}a_n)$.

Proof. Let $\{b^i\}$, $\{b_i\}$ be dual bases for A^* and A respectively. Then we have

$$\begin{aligned}
\mathrm{Tr}(T) &= \sum_{i_1,\cdots,i_n} f(b_{i_1})\langle b^{i_1}, a_1b_{i_2}\rangle \cdots \langle b^{i_{n-1}}, a_{n-1}b_{i_n}\rangle\langle b^{i_n}, a_n\rangle \\
&= \sum_{i_1,\cdots,i_n} f(b_{i_1})\langle b^{i_1}, a_1b_{i_2}\rangle \cdots \langle b^{i_{n-1}}, a_{n-1}a_n\rangle \\
&= \sum_{i_1} f(b_{i_1})\langle b^{i_1}, a_1 \cdots a_{n-1}a_n\rangle \\
&= f(a_1 \cdots a_{n-1}a_n). \quad \square
\end{aligned}$$

Therefore, we have

Proposition 3.2. *Let H be a finite-dimensional pivotal Hopf algebra over $\mathbb{C}$. With respect to the pivotal element $g \in H$, $\nu_1^p(H) = 1$ and the m-th pivotal Frobenius-Schur indicator $\nu_m^p(H)$ of H is given by*

$$\nu_m^p(H) = \lambda(\Lambda_{(2)}^{[m-1]}g^{-1}\Lambda_{(1)}).$$

for all integers $m \geq 2$.

Proof. By definition, $\nu_1^p(H) = \mathrm{Tr}(E_{H,\mathbb{C}}) = \mathrm{Tr}\left(\mathrm{id}_{\mathrm{Hom}_H(\mathbb{C},H)}\right)$. Since $\dim \mathrm{Hom}_H(\mathbb{C}, H) = 1$, $\nu_1^p(H) = 1$.

We now assume $m \geq 2$. By the preceding remark, $\nu_m^p(H) = \mathrm{Tr}(\overline{E}^{(m)})$. In view of (7) and Lemma 3.1,

$$\nu_m^p(H) = \lambda(\Lambda_{(2)}\Lambda_{(3)} \cdots \Lambda_{(m)}g^{-1}\Lambda_{(1)}) = \lambda(\Lambda_{(2)}^{[m-1]}g^{-1}\Lambda_{(1)}). \quad \square$$

We can now use the proposition to compute the pivotal Frobenius-Schur indicators of a Taft algebra [21] which has exactly one pivotal element. Hence, they are Hopf algebras in $\mathfrak{P}$. For simplicity, we consider the Taft algebra of dimension 4, which is also called the Sweedler algebra H_4.

A Taft algebra $T_n(\omega)$ is a $\mathbb{C}$-algebra generated by x, g subject to the relations:

$$xg = \omega gx, \quad x^n = 0, \quad g^n = 1$$

where ω is a primitive n-th root of unity. The comultiplication Δ, the counit ϵ and the antipode of $T_n(\omega)$ are given by

$$\Delta(g) = g \otimes g, \quad \Delta(x) = x \otimes 1 + g \otimes x, \quad \epsilon(g) = 1, \quad \epsilon(x) = 0,$$
$$S(g) = g^{-1} \quad \text{and} \quad S(x) = -gx\,.$$

In particular, $S^2(h) = ghg^{-1}$ for all $h \in T_n(\omega)$. Therefore, g is a pivotal element. Since there is no non-trivial central group-like element in $T_n(\omega)$, g is the unique pivotal element of $T_n(\omega)$, i.e. $T_n(\omega) \in \mathfrak{P}$.

For the Sweedler algebra H_4 or $T_2(-1)$, $\Lambda = x + gx$ is a left integral and $\lambda \in H^*$, defined by $\lambda(g^i x^j) = \delta_{i,0}\delta_{j,1}$ for $0 \le i, j \le 1$, is a right integral such that $\lambda(\Lambda) = 1$. Note that

$$\Delta^{(m)}(g) = g^{\otimes m}, \quad \Delta^{(m)}(x) = x \otimes 1^{\otimes(m-1)} + g \otimes x \otimes 1^{\otimes(m-2)} + \cdots + g^{\otimes(m-1)} \otimes x$$

for all integer $m \ge 2$. Therefore,

$$g^{[m]} = g^m, \quad x^{[m]} = \sum_{i=0}^{m-1} g^i x, \quad (gx)^{[m]} = \sum_{i=0}^{m-1} gxg^{m-i-1}.$$

For any integer $m \ge 2$,

$$\begin{aligned}
\nu_m^p(H_4) &= \lambda(\Lambda_{(2)}^{[m-1]} g^{-1} \Lambda_{(1)}) \\
&= \lambda(1^{[m-1]} gx + x^{[m-1]} + g^{[m-1]} x + (gx)^{[m-1]} g) \\
&= \lambda\left(\sum_{i=0}^{m-2} g^i x + g^{m-1} x + \sum_{i=0}^{m-2} (-1)^{m-i-1} g^{m-i} x\right) \\
&= \#\{i \mid 0 \le i \le m-1, i \text{ even}\} - \#\{i \mid 0 \le i \le m-2,\ m-i \text{ even}\} \\
&= \begin{cases} 0 & \text{if } m \text{ is even,} \\ 1 & \text{if } m \text{ is odd.} \end{cases}
\end{aligned}$$

Therefore, $\nu^p(H_4) = \{1, 0, 1, 0, \ldots\}$.

Another version of gauge invariant $\nu(H_4)$, simply called the FS-indicators of H_4, was computed in [9] and it is given by

$$\nu(H_4) = \{1, 2, 3, 4, \ldots\}.$$

Obviously, $\nu^p(H_4) \ne \nu(H_4)$ are both invariant of the monoidal category $\text{Rep}(H_4)$.

The example H_4 suggests the notion of Frobenius-Schur indicators defined in [9] and the pivotal Frobenius-Schur indicators defined in [15] are intrinsically different. One natural question is whether there exists a gauge invariants $\kappa(H)$ for $H \in \mathfrak{H}$ such that $\kappa(H) = \nu^p(H)$ for all $H \in \mathfrak{P}$. It is highly unclear whether there is any natural relationships (in terms of some structures of $\text{Rep}(H)$) between $\nu^p(H)$ and $\nu(H)$ for $H \in \mathfrak{P}$. Such relations could reveal more interesting gauge invariants for the complete collection $\mathfrak{H}$ of finite-dimensional Hopf algebras.

Acknowledgments: The author would like to acknowledge the partial supported by the Shanghai Leading Academic Discipline Project(J50101) for his visit to Shanghai University in June 2010.

References

[1] Peter Bantay. The Frobenius-Schur indicator in conformal field theory. *Phys. Lett. B*, 394(1-2):87–88, 1997.

[2] Charles W. Curtis and Irving Reiner. *Methods of representation theory. Vol. I.* Wiley Classics Library. John Wiley & Sons Inc., New York, 1990. With applications to finite groups and orders, Reprint of the 1981 original, A Wiley-Interscience Publication.

[3] Pavel Etingof and Shlomo Gelaki. On the exponent of finite-dimensional Hopf algebras. *Math. Res. Lett.*, 6(2):131–140, 1999.

[4] Pavel Etingof and Shlomo Gelaki. On the quasi-exponent of finite-dimensional Hopf algebras. *Math. Res. Lett.*, 9(2-3):277–287, 2002.

[5] Pavel Etingof, Dmitri Nikshych, and Viktor Ostrik. On fusion categories. *Ann. of Math. (2)*, 162(2):581–642, 2005.

[6] F. G. Frobenius and I. Schur. Uber die reellen darstellungen der endlichen gruppen. *Sitzungsber. Akad. Wiss. Berlin*, pages 186–208, 1906.

[7] J. Fuchs, A. Ch. Ganchev, K. Szlachányi, and P. Vecsernyés. S_4 symmetry of $6j$ symbols and Frobenius-Schur indicators in rigid monoidal C^* categories. *J. Math. Phys.*, 40(1):408–426, 1999.

[8] I. Martin Isaacs. *Character theory of finite groups.* Dover Publications Inc., New York, 1994. Corrected reprint of the 1976 original [Academic Press, New York; MR **57** #417].

[9] Yevgenia Kashina, Susan Montgomery, and Siu-Hung Ng. On the trace of the antipode and higher indicators. *preprint*, arXiv:0910.1628.

[10] Yevgenia Kashina, Yorck Sommerhäuser, and Yongchang Zhu. On higher Frobenius-Schur indicators. *Mem. Amer. Math. Soc.*, 181(855):viii+65, 2006.

[11] V. Linchenko and S. Montgomery. A Frobenius-Schur theorem for Hopf algebras. *Algebr. Represent. Theory*, 3(4):347–355, 2000. Special issue dedicated to Klaus Roggenkamp on the occasion of his 60th birthday.

[12] Geoffrey Mason and Siu-Hung Ng. Central invariants and Frobenius-Schur indicators for semisimple quasi-Hopf algebras. *Adv. Math.*, 190(1):161–195, 2005.

[13] Susan Montgomery. *Hopf algebras and their actions on rings*, volume 82 of *CBMS Regional Conference Series in Mathematics.* Published for the Conference Board of the Mathematical Sciences, Washington, DC, 1993.

[14] Siu-Hung Ng and Peter Schauenburg. Frobenius-Schur indicators and exponents of spherical categories. *Adv. Math.*, 211(1):34–71, 2007.

[15] Siu-Hung Ng and Peter Schauenburg. Higher Frobenius-Schur indicators for pivotal categories. In *Hopf algebras and generalizations*, volume 441 of *Contemp. Math.*, pages 63–90. Amer. Math. Soc., Providence, RI, 2007.

[16] Siu-Hung Ng and Peter Schauenburg. Central invariants and higher indicators for semisimple quasi-Hopf algebras. *Trans. Amer. Math. Soc.*, 360(4):1839–1860, 2008.

[17] Siu-Hung Ng and Peter Schauenburg. Congruence subgroups and generalized frobenius-schur indicators. *Comm. Math. Phys.*, 300(1):1–46, 2011.

[18] Peter Schauenburg. On the Frobenius-Schur indicators for quasi-Hopf algebras. *J. Algebra*, 282(1):129–139, 2004.

[19] Jean-Pierre Serre. *Linear representations of finite groups.* Springer-Verlag, New York, 1977. Translated from the second French edition by Leonard L. Scott, Graduate Texts in Mathematics, Vol. 42.

[20] Moss E. Sweedler. *Hopf algebras.* W. A. Benjamin, Inc., New York, 1969. Mathematics Lecture Note Series.

[21] Earl J. Taft. The order of the antipode of finite-dimensional Hopf algebra. *Proc. Nat. Acad. Sci. U.S.A.*, 68:2631–2633, 1971.

Department of Mathematics, Iowa State University
E-mail address: rng@iastate.edu

Zariski Topology of Prime Spectrum of a Module

N. V. Sanh, L. P. Thao, N. F. A. Al-Mayahi and K. P. Shum

Nguyen Van Sanh*, Le Phuong Thao**
Noori F. A. Al-Mayahi† and Kar Ping Shum‡
*Department of Mathematics, Faculty of Science
Mahidol University, Bangkok 10400, Thailand
frnvs@mahidol.ac.th
**Department of Mathematics, Cantho University
Can Tho city, Vietnam
lpthao@ctu.edu.vn
† College of Computer Science & Mathematics
University of Al-Qadisiya, Iraq
nafm2005@yahoo.com
‡ Department of Mathematics, The University of Hong Kong
Hong Kong SAR, RPC
kpshum@maths.hku.hk

We introduce and investigate the Zariski topology of the Prime Spectrum of a module M over an arbitrary associative ring R.

1. Introduction

Throughout this paper, all rings are associative rings with an identity and all modules are unitary right R-modules. Let M be a right R-module and $S = \mathrm{End}_R(M)$, its endomorphism ring. A submodule X of M is called a *fully invariant* submodule of M if for any $s \in S$, we have $s(X) \subset X$. By the definition, the class of all fully invariant submodules of M is non-empty and closed under intersections and sums. Especially, a right ideal of R is a fully invariant submodule of R_R if it is a two-sided ideal of R.

Following [9], a fully invariant proper submodule X of a right R-module M is said to be a *prime submodule* of M (we say that X is prime in M) if for any ideal I of S, and any fully invariant submodule U of $M, I(U) \subset X$ implies $I(M) \subset X$ or $U \subset X$. Especially, an ideal P of R is a *prime ideal* if for any ideals I, J of $R, IJ \subset P$ implies $I \subset P$ or $J \subset P$. A right R-module M is called a prime module if 0 is prime in M.

A fully invariant submodule X of a right R-module M is said to be a *semiprime submodule* if it is an intersection of prime submodules of M. A right R-module M is called a *semiprime module* if 0 is a semiprime submodule of M.

For a subset $X \subset M$, denote $I_X = \{f \in S | f(M) \subset X\}$. If X is a submodule of M, then it is clear that I_X is a right ideal of S, and if X is a fully submodule of M, then I_X is a two sided ideal of S. Following,[9] if X is a prime (resp. semiprime) submodule of M, then I_X is is a prime (resp. semiprime) ideal of S. The converse statement is true if M is a self-generator. By a self-generator, we mean that M generates all its submodules. For a right R-module M, we denote $P(M)$, the intersection of all prime submodules of M, and we call $P(M)$ the prime radical of the right R-module M. If M is quasi-projective, then $P(M/P(M)) = 0$. The notations and terminologies that we used in this paper are standard.For basic knowledge of Rings and Modules, the reader is referred to the monogrphies 1-6], [8], [12-13].

The following results have been appeared in [9, Theorem 1.2] and we present here to check the primeness.

Theorem 1.1 *Let X be a proper fully invariant submodule of M. Then the following conditions are equivalent:*

(1) X is a prime submodule of M;
(2) For any right ideal I of S, any submodule U of M, if $I(U) \subset X$, then either $I(M) \subset X$ or $U \subset X$;
(3) For any $\varphi \in S$ and fully invariant submodule U of M, if $\varphi(U) \subset X$, then either $\varphi(M) \subset X$ or $U \subset X$;
(4) For any left ideal I of S and subset A of M, if $IS(A) \subset X$, then either $I(M) \subset X$ or $A \subset X$;
(5) For any $\varphi \in S$ and for any $m \in M$, if $\varphi(S(m)) \subset X$, then either $\varphi(M) \subset X$ or $m \in X$.
Moreover, if M is quasi-projective, then the above conditions are equivalent to:
(6) M/X is a prime module.
In addition, if M is quasi-projective and a self-generator, then the above conditions are equivalent to:
(7) If I is an ideal of S and U a fully invariant submodule of M such that $I(M)$ and U properly contain P, then $IU \not\subset P$.

Theorem 1.2 *Let M be a right R-module, $S = \mathrm{End}(M_R)$ and X a fully invariant submodule of M. If X is a prime submodule of M, then I_X is a prime ideal of S. The converse ofthe above statement is true if M is a self-generator.*

2. Zariski Topology

For the description of classical Zariski Topology, the reader is referred to [14]. It is noted that a problem proposed by J Ohm related to sepectrals in posets was considered in [8].We now use the usual symbol Spec (M) (or X^M for short) to denote the set of all prime submodules of M, and we call *Spec*(M) *the prime spectrum* of M. Let N be a fully invariant submodule of M, Denote $V^\star(N) = \{P \in Spec(M) \mid N \subset P\}$ and $V(N) = \{P \in Spec(M) \mid I_N \subset I_P\}$. The following Lemma is straightforward.

Lemma 2.1

The set $V^\star(N)$ has the following properties:

(a) $V^\star(0) = X^M$ *and* $V^\star(M) = \phi$;
(b) *For any family of fully invariant submodules* $\{N_i\}_{i\in J}$ *of* M, *we have* $\bigcap_{i\in J} V^\star(N_i) = V^\star(\sum_{i\in J}(N_i))$;
(c) *For any fully invariant submodules* N, L *of* M, *we have* $V^\star(N) \cup V^\star(L) \subset V^\star(N \cap L)$.

Some properties of $V(N)$ are given in the following lemma.

Lemma 2.2

The set $V(N)$ has the following properties:

(i) $V(0) = X^M$ *and* $V(M) = \phi$;
(ii) *For any family of fully invariant submodules* $\{N_i\}_{i\in J}$ *of* M, *we have* $\bigcap_{i\in J} V(N_i) = V(\sum_{i\in J} I_{N_i}(M))$;
(iii) *For any fully invariant submodules* N, L *of* M, *we have* $V(N)\cup V(L) = V(N \cap L)$.

Proof.

(i) This part is clear from the definition. (ii) Let $P \in Spec(M)$. Then $P \in \bigcap_{i \in J} V(N_i)$, and this implies that $I_{N_i} \subset I_P$, for all $i \in J$. Hence, $I_{N_i}(M) \subset P$, for all $i \in J$. It follows that $\sum_{i \in J} I_{N_i}(M) \subset P$, showing that $I_{\sum_{i \in J} I_{N_i}(M)} \subset I_P$. Thus, $P \in V(\sum_{i \in J} I_{N_i}(M))$. Conversely, let $P \in V(\sum_{i \in J} I_{N_i}(M))$. Then $I_{\sum_{i \in J} I_{N_i}(M)} \subset I_P$, and this implies that $I_{(\sum_{i \in J} I_{N_i})(M)} \subset I_P$, so $\sum_{i \in J} I_{N_i} \subset I_P$, and hence $I_{N_i} \subset I_P$, for all $i \in J$. This shows that $P \in \bigcap_{i \in J} V(N_i)$. (iii) Let $P \in Spec(M)$. If $P \in V(N) \cup V(L)$, then $I_N \subset I_P$ *or* $I_L \subset I_P$. This implies $I_{N \cap L} = I_N \cap I_L \subset I_P$ and hence $P \in V(N \cap L)$. Conversely, if $P \in V(N \cap L)$, then $I_{N \cap L} \subset I_P$, showing that $I_N.I_L \subset I_N \cap I_L \subset I_P$. Since $P \in X^M$ we have $I_P \in X^S$. So $I_N \subset I_P$ *or* $I_L \subset I_P$. Thus, $P \in V(N)$ or $P \in V(L)$, that is, $P \in V(N) \cup V(L)$. Therefore $V(N) \cup V(L) = V(N \cap L)$. □

We now put

$$\Gamma^\star(M) = \{V^\star(N) \mid N \textit{ is a fully invariant submodule of } M\}$$

$$\Gamma(M) = \{V(N) \mid N \textit{ is a fully invariant submodule of } M\}$$

Note that, from Lemma 2.1, there exists a topology on X^M, having $\Gamma^\star(M)$ as the set of all closed sets if and only if $\Gamma^\star(M)$ is closed under finite union.

From Lemma 2.2, there exists a topology, say Γ, on X^M, having $\Gamma(M)$ as the family of all closed sets. The topology Γ is called the *Zariski topology* on $Spec(M)$.

Proposition 2.3 *Let N and L be fully invariant submodules of M. Then, we have the following statements.*

(1) If $I_N = I_L$, then $V(N) = V(L)$. The converse statement is true if both N and L are prime in M;

(2) $V(N) = V(I_N(M)) = V^\star(I_N(M))$. In particular, we have $V(PM) = V^\star(¶M)$ for any ideal ¶ of the ring S.

Proof. (1) It is clear that $V(N) = V(L)$ if $I_N = I_L$. Conversely, suppose that N and L are prime submodules of M and $V(N) = V(L)$. Then $I_N \subset I_L$ and $I_L \subset I_N$, showing that $I_N = I_L$. (2) Firstly, we will show that $V(N) = V(I_N(M))$. For each $P \in V(N)$, we have $I_N \subset I_P$, implying that $I_N(M) \subset I_P(M) \subset P$. Then $I_{I_N(M)} \subset I_P$. It follows that $P \in V(I_N(M))$.

Conversely, let $P \in V(I_N(M))$. Then, $I_{I_N(M)} \subset I_P$, that is , $I_N \subset I_P$, showing that $P \in V(N)$. Hence, $V(N) = V(I_N(M))$.

We now show that $V(I_N(M)) = V^\star(I_N(M))$. For any $P \in V(I_N(M))$, we have $I_{I_N(M)} \subset I_P$. It would imply that $I_N \subset I_P$. Hence, $I_N(M) \subset I_P(M) \subset P$. It means that $P \in V^\star(I_N(M))$. Conversely, let $P \in V^\star(I_N(M))$. Then, $I_N(M) \subset P$, and therefore ,$I_{I_N(M)} \subset I_P$, proving that $P \in V(I_N(M))$. Thus, $V(I_N(M)) = V^\star(I_N(M))$.

Let ¶ be any ideal of S. We finally show that $V(¶M) = V^\star(¶M)$. Let $Q \in V(¶M)$. Then $I_{¶M} \subset I_Q$. This leads to $I_{¶M}(M) \subset I_Q(M) \subset Q$. Since $¶ \subset I_{¶M}$, we get $¶M \subset I_{¶M}(M) \subset Q$, and hence $Q \in V^\star(¶M)$. Conversely, let $Q \in V^\star(¶M)$. Then $¶M \subset Q$, and we have $I_{¶M} \subset I_Q$. This shows that $Q \in V(¶M.)$ Hence $V(¶M) = V^\star(¶M)$. □

Theorem 2.4 *Let M be a right R-module which is a self-generator. Then $V^\star(N) \cup V^\star(L) = V^\star(N \cap L)$ for any fully invariant submodules N, L of M. In this case, we have $\Gamma^\star(M) = \Gamma(M)$.*

Proof. Let $P \in Spec(M)$. We have $P \in V^\star(N) \cup V^\star(L)$. It is equivalent to $N \subset P$ *or* $L \subset P$, or equivalently, $I_N \subset I_P$ *or* $I_L \subset I_P$. This means that $I_N \cap I_L \subset I_P$, i.e., $I_{N\cap L} \subset I_P$. It means if and only if $N \cap L \subset P$, or equivalently, $P \in V^\star(N \cap L)$. Thus, $V^\star(N) \cup V^\star(L) = V^\star(N \cap L)$. By Proposition 3, we have $V(N) = V(I_N(M)) = V^\star(I_N(M)) = V^\star(N)$. □

Note that, by Theorem 2.4, when M is a self-generator, then there exists a topology, say $\Gamma^\star$, on X^M, having $\Gamma^\star(M)$ as the collection of all closed sets. Moreover the topology $\Gamma^\star$ concides with the Zariski topology of X^M.

3. The natural maps

Suppose that $Spec(M) \neq \emptyset$. Then $Spec(S) \neq \emptyset$. Define $\psi : Spec(M) \to Spec(S)$, $P \mapsto I_P$. The map ψ is well-defined and ψ is called the *natural map* of $Spec(M)$. Next, we will investigate some properties of this natural map. Note that, if M is a self-generator, then ψ is surjective, by Theorem 1.2.

Proposition 3.1 *The natural map ψ of $Spec(M)$ is continuous. More precisely, $\psi^{-1}(V^S(¶)) = V(¶M)$ for any ideal ¶ of S.*

Proof. Let U be a closed subset in $Spec(S)$. Then we can easily deduce that $U = V^S(¶)$, for some ideal ¶ of S. Take any $N \in Spec(M)$. Then, $N \in \psi^{-1}(V^S(¶))$, that is, $\psi(N) = I_N \in V^S(¶)$, or equivalently, $¶ \subset I_N$, that is $¶M \subset N$. This means that $I_{¶M} \subset I_N$, or equivalently, $N \in V(¶M)$. Thus, $\psi^{-1}(V^S(¶)) = V(¶M)$ which is a closed set in $Spec(M)$, proving that ψ is continuous. □

For any prime ideal $\mathcal{P}$ of S, we denote $Spec_{\mathcal{P}}(M) = \{P \in Spec(M) \mid I_P = \mathcal{P}\}$ Then we have the following Proposition.

Proposition 3.2

Let M be a right R-module and $P,\ Q \in Spec(M)$. Then the following conditions are equivalent:

(1) The natural map $\psi : Spec(M) \to Spec(S)$ is injective;
(2) If $V(P) = V(Q)$, then $P = Q$;
(3) $|Spec_{\mathcal{P}}(M)| \leq 1$ for any $\mathcal{P} \in Spec(S)$.

Proof. (1) $\Rightarrow$ (2). Assume that $V(P) = V(Q)$. Since $P,\ Q \in X^M$, we have $I_P = I_Q$, and this implies that $\psi(P) = \psi(Q)$. Therefore, $P = Q$ since ψ is an injective map.

(2) $\Rightarrow$ (3). If $Spec_{\mathcal{P}}(M) = \emptyset$, then $|Spec_{\mathcal{P}}(M)| = 0$. Assume that $Spec_{\mathcal{P}}(M) \neq \emptyset$ and $P,\ Q \in Spec_{\mathcal{P}}(M)$. Then $I_P = I_Q = \mathcal{P}$. This shows that $V(P) = V(Q)$, and hence $P = Q$ by (2). Thus $|Spec_{\mathcal{P}}(M)| \leq 1$.

(3) $\Rightarrow$ (1). Assume that $\psi(P) = \psi(Q)$, where $P,\ Q \in Spec(M)$. Then $I_P = I_Q = \mathcal{P}$, is a prime ideal of S. Thus $P,\ Q \in Spec_{\mathcal{P}}(M)$. By (3), we have $P = Q$. □

Theorem 3.3 *Let M be a right R-module and $\psi : X^M \to X^S$ be the natural map of X^M. If ψ is surjective, then ψ is open and closed.*

Proof. By Proposition 3.1, ψ is continuous map such that $\psi^{-1}(V^S(\P)) = V(PM)$ for any ideal $\P$ of S. Let N be a fully invariant submodule of M. Then we have $\psi^{-1}(V^S(I_N)) = V(I_N(M)) = V(N)$. Since ψ is surjective, $\psi(V(N)) = V^S(I_N)$ which is closed in $Spec(S)$. Thus, ψ is a closed map. Similarly, $\psi(X^M \setminus V(N)) = \psi(\psi^{-1}(X^S) \setminus \psi^{-1}(V^S(I_N))) = \psi(\psi^{-1}(X^S \setminus V^S(I_N)) = X^S \setminus V^S(I_N)$ which is open in $Spec(S)$. Hence, ψ is an open map. □

For each $f \in S$, denote $X_f^M = X^M \setminus V(Sf(M))$ and $X_f^S = X^S \setminus V^S(SfS)$. Then X_f^M is an open set of X^M and it is clear that $X_0^M = \emptyset$ and $X_{1_S}^M = X^M$.

Proposition 3.4 *Let M be a right R-module with the natural map $\psi : X^M \to X^S$ and $f \in S$. Then, we have the following facts:*

(1) $\psi^{-1}(X_f^S) = X_f^M$;
(2) $\psi(X_f^M) \subset X_f^S$. The equality holds if ψ is surjective.

Proof. (1) We have $\psi^{-1}(X_f^S) = \psi^{-1}(X^S \setminus V^S(SfS)) = X^M \setminus \psi^{-1}(V^S(SfS)) = X^M \setminus V(SfS(M)) = X^M \setminus V(Sf(M)) = X_f^M$.

(2) From $X_f^M = \psi^{-1}(X_f^S)$, we get $\psi(X_f^M) \subset X_f^S$. If ψ is a surjective map then it is clear that we have the equality. □

Theorem 3.5 *Let M be a right R-module. Then the set $\mathcal{B} = \{X_f^M \mid f \in S\}$ forms a basis for the Zariski topology on X^M.*

Proof. If $X^M = \emptyset$. Then $\mathcal{B} = \emptyset$. In this case, the theorem is true. Assume now that $X^M \neq \emptyset$ and U is an open set of X^M. Then $U = X^M \setminus V(N)$, where N is a fully invariant submodule of M. Recall that $V(N) = V(I_N(M)) = V^\star(I_N(M))$, so $V(N) = V^\star(I_N(M)) = V^\star(\sum_{f_i \in I_N} Sf_iS(M)) = \bigcap_{f_i \in I_N} V^\star(Sf_iS(M)) = \bigcap_{f_i \in I_N} V(Sf_iS(M)) = \bigcap_{f_i \in I_N} V(Sf_i(M))$. Thus $U = X^M \setminus \bigcap_{f_i \in I_N} V(Sf_i(M)) = \bigcup_{f_i \in I_N} (X^M \setminus V(Sf_i(M)) = \bigcup_{f_i \in I_N} X_{f_i}^M$. Therefore, $\mathcal{B}$ is a basis for the Zariski topology of X^M. □

We note that our definition of Zariski topology on a right R-module M can be considered as a generaliztion of that of rings in.[14] For a ring R, the set of all prime ideals $Spec(R)$ (or X^R) is called the prime spectrum of the ring R. Let I be any ideal of R, and denote $V^R(I) = \{P \in Spec(R) \mid I \subset P\}$. Then

(1) $V^R(0) = X^R$ and $V^R(R) = \phi$;
(2) If $\{E_i\}_{i \in J}$ is any family of ideals of R, then $\bigcap_{i \in J} V^R(E_i) = V^R(\bigcup_{i \in J}(E_i)$;
(3) If I, J are ideals of R, then $V^R(I) \cup V^R(J) = V^R(IJ) = V^R(I \cap J)$.

Let $\Gamma(R) = \{V^R(I) \mid I$ is an ideal of $R\}$. From $(1) - (3)$, there exists a topology, say Γ^R, on $Spec(R)$ having $\Gamma(R)$ as the family of all closed sets. This topology is called the *Zariski topology* on $Spec(R)$. The following lemma is from,[14] and we use it to prove the Theorem 3.7.

Lemma 3.6 *Let R be a ring. Then, X^R is compact.*

Applying this Lemma, we can prove the following Theorem.

Theorem 3.7 *Let M be a right R-module. If the natural map ψ of X^M is surjective, then X^M is compact. In particular, if M is a self-generator, then X^M is compact.*

Proof. Since $\mathcal{B} = \{X_f^M \mid f \in S\}$ forms a basis for the Zariski topology on X^M, for any open cover of X^M, we can write $X^M = \bigcup_{i \in \Omega} X_{f_i}^M$. Then $X^S = \psi(X^M) = \psi(\bigcup_{i \in \Omega} X_{f_i}^M) = \bigcup_{i \in \Omega} \psi(X_{f_i}^M) = \bigcup_{i \in \Omega} X_{f_i}^S$. By Lemma 3.6, X^S is compact, therefore there exists a finite set $J \subset \Omega$ such that $X^S = \bigcup_{i \in J} X_{f_i}^S$.

Thus $X^M = \psi^{-1}(X^S) = \psi^{-1}(\bigcup_{i\in J} X^S_{f_i}) = \bigcup_{i\in J} \psi^{-1}(X^S_{f_i}) = \bigcup_{i\in J} X^M_{f_i}$, proving that X^M is compact. □

4. Closed sets in Spec(M)

Let Y be a subset of X^M. We denote the intersection of all elements in Y by $J(Y)$, and the closure of Y in X^M by $cl(Y)$. The following Theorem gives a characterization of closed sets in X^M.

Theorem 4.1 *Let M be a right R-module and Y be a subset of X^M. Then $V(J(Y)) = cl(Y)$. Therefore, Y is closed in X^M if and only if $V(J(Y)) = Y$.*

Proof. Let $P \in Y$. Then $J(Y) \subset P$. This implies $I_{J(Y)} \subset I_P$, and hence $P \in V(J(Y))$. Thus $Y \subset V(J(Y))$. Now, let $V(N)$ be any closed subset of X^M such that $Y \subset V(N)$. Then $I_N \subset I_P$, for any $P \in Y$. Thus $I_N \subset \bigcap_{P\in Y} I_P = I_{\bigcap_{P\in Y} P} = I_{J(Y)}$. Take any $Q \in V(J(Y))$. Then $I_Q \supset I_{J(Y)} \supset I_N$, and thus $Q \in V(N)$. This gives $V(J(Y)) \subset V(N)$. It means that $V(J(Y))$ is the smallest closed subset of X^M containing Y. Hence $V(J(Y)) = cl(Y)$. □

Proposition 4.2 *Let M be a right R-module and $P \in X^M$. Then we have the following:*

(1) $cl(\{P\}) = V(P)$;
(2) For any $Q \in X^M$, $Q \in cl(\{P\})$ if and only if $I_P \subset I_Q$ if and only if $V(Q) \subset V(P)$;
(3) If M is a self-generator, then the set $\{P\}$ is closed in X^M if and only if P is a maximal prime submodule of M.

Proof. (1) This part is clear from Proposition 4.1.

(2) We see that $Q \in cl(\{P\})$ means $Q \in V(P)$ (by (1)). It is equivalent to say that $I_P \subset I_Q$, i.e., $V(Q) \subset V(P)$.

(3) Suppose that $\{P\}$ is closed. Then, $V(P) = cl(\{P\}) = \{P\}$. Let Q be a prime submodule of M and $Q \supset P$. Then $I_Q \supset I_P$, that is, $Q \in V(P) = \{P\}$, and hence $Q = P$. Thus, P is a maximal prime submodule of M.

Conversely, let $Q \in cl(\{P\})$. Then $Q \in V(P)$ and therefore $I_P \subset I_Q$. Since M is a self-generator, we have $P \subset Q$. Since P is a maximal prime, $P = Q$. Thus, $cl(\{P\}) = \{P\}$ proving that $\{P\}$ is closed. □

Recall that a topological space A is called *irreducible* if for any decomposition $A = B \cup C$ with $B,\ C$ are closed sets of A, we have $A = B$ or $A = C$. A subset B of A is called *irreducible* if it is irreducible as a subspace of A. It is equivalent to saying that if $B_1,\ B_2$ are closed sets in A such that

$B \subset B_1 \cup B_2$, then $B \subset B_1$ or $B \subset B_2$. A topological space X is called a *T_0-space* if for any two distinct points of X, there is an open set which contains only one of these points.

Corollary 4.3 *Let M be a right R-module. For any $P \in X^M$, $V(P)$ is an irreducible closed subset of X^M.*
Proof. Since $\{P\}$ is irreducible, $cl(\{P\})$ is irreducible. But then $cl(\{P\}) = V(P)$, and we see that $V(P)$ is irreducible. □

Theorem 4.4 *Let M be a right R-module with the Zariski topology on $Spec(M)$. Then the following conditions are equivalent:*

(1) X^M is a T_0 space;
(2) The natural map $\psi : X^M \to X^S$ is injective;
(3) For $P,\ Q \in X^M$, if $V(P) = V(Q)$, then $P = Q$;
(4) $|Spec_{\mathcal{P}}(M)| \leq 1$ for any $\mathcal{P} \in Spec(S)$.

Proof. (1) $\Rightarrow$ (3). Suppose that X^M is a T_0 space. Let $P,\ Q \in X^M$ such that $V(P) = V(Q)$. Then $cl(\{P\}) = cl(\{Q\})$. If $P \neq Q$, then there exists an open set U of X^M such that $P \in U$ but $Q \notin U$. Then $X^M \setminus U$ is a closed set of X^M such that $Q \in X^M \setminus U$ and this gives $cl(\{Q\}) \subset X^M \setminus U$. It follows that $cl(\{P\}) \subset X^M \setminus U$, showing that $P \notin U$, a contradiction. Thus $P = Q$.

(3) $\Rightarrow$ (1). Let $P,\ Q$ be two distinct points of X^M. By (3), $V(P) \neq V(Q)$. If every open set of X^M which contains one point will contain the other point, then $P \in cl(\{Q\})$ and $Q \in cl(\{P\})$. Thus $P \in V(Q)$ and $Q \in V(P)$, and so $V(P) = V(Q)$, a contradiction.

(2) $\Leftrightarrow$ (3) $\Leftrightarrow$ (4) by Proposition 3.3. □

Corollary 4.5 *Let M be a right R-module which is a self-generator. Then X^M is a T_0-space with the Zariski topology.*
Proof. Let $P,\ Q \in X^M$. If $V(P) = V(Q)$, then $I_P = I_Q$. Since M is a self-generator, we have $P = Q$. Thus X^M is a T_0-space. □

Let X be a topological space and $x,\ y$ be distinct points in X. We say that x and y can be *separated* if each point lies in an open set which does not contain the other point. X is called a *T_1-space* if any two distinct points in X can be separated. A topological space X is a T_1-space if and only if any one point set of X is closed in X.

Theorem 4.6 *Let M be a right R-module which is a self-generator. Then X^M is a T_1 space if and only if $Maxp(M) = Spec(M)$, where $Maxp(M)$ is the set of all maximal prime submodules of M.*

Proof. Applying Proposition 4.2. □

Theorem 4.7 *Let M be a right R-module with surjective natural map $\psi : X^M \to X^S$. Then the following conditions are equivalent:*

(1) X^M is connected;
(2) X^S is connected.

Proof. (1) $\Rightarrow$ (2). Since ψ is surjective, continuous and X^M is connected, X^S is connected.

(2) $\Rightarrow$ (1). Assume that X^S is connected. If X^M is disconnected, then there exists a nonempty proper subset Y of X^M which is both open and closed. By Theorem 3.3, $\psi(Y)$ is both open and closed subset of X^S. It is clear that $\psi(Y) \neq \emptyset$. We will show that $\psi(Y)$ is a proper subset of X^S. Since Y is open in X^M, we have $Y = X^M \setminus V(N)$, where N is a fully invariant submodule of M. Then by Theorem 3.3 again, $\psi(Y) = X^S \setminus V^S(I_N)$. If $\psi(Y) = X^S$, then $X^S \setminus V^S(I_N) = X^S$.It follows that $V^S(I_N) = \emptyset$ and thus $\psi^{-1}(V^S(I_N)) = \emptyset$ and we get $V(I_N(M)) = \emptyset$. Thus $Y = X^M \setminus V(N) = X^M \setminus V(I_N(M)) = X^M \setminus \emptyset = X^M$, a contradiction. Thus X^M is connected.
□

Theorem 4.8 *Let M be a right R-module and Y, a subset of X^M. If $J(Y)$ is a prime submodule of M, then Y is irreducible. Conversely, if Y is irreducible, then $T = \{I_P \mid P \in Y\}$ is an irreducible subset of X^S, i.e, $J^\star(T) = I_{J(Y)}$ is a prime ideal of S.*

Proof. Suppose that $J(Y) = Q$ is a prime submodule of M. By Corollary 4.3, $V(Q) = V(J(Y)) = cl(Y)$ is irreducible. Thus, Y is irreducible.

Conversely, if Y is irreducible, then $\psi(Y) = \{I_P \mid P \in Y\} = T$ is an irreducible subset of X^S since the natural map ψ of X^M is continuous. By Lemma 3.6, $J^\star(T) = \bigcap_{P \in Y} I_P = I_{J(Y)}$ is a prime ideal of S. □

Theorem 4.9 *Let M be a right R-module with the surjective natural map $\psi : X^M \to X^S$ and $Y \subset X^M$. Then Y is an irreducible closed subset of X^M if and only if $Y = V(P)$, for some $P \in X^M$.*

Proof. If $Y = V(P)$, $P \in X^M$, then Y is an irreducible closed subset of X^M, by Corollary 4.3. Conversely, if Y is an irreducible closed subset of X^M, then $Y = V(N)$, with N is a fully invariant submodule of M. Since Y is closed, $Y = V(N) = V(J(Y))$ by Proposition 4.1. Note that Y is irreducible, $I_{J(Y)} = \mathcal{P}$ is a prime ideal of S, by Theorem 4.8. By hypothesis, ψ is a surjective map, there exists $P \in X^M$ such that $\mathcal{P} = \psi(P) = I_P$. Thus, $I_P = I_{J(Y)} = \mathcal{P}$, proving that $Y = V(J(Y)) = V(P)$. □

Proposition 4.10 *Let M be a nonzero right R-module. Then the following conditions are equivalent:*

(1) For any fully invariant submodules N, L of M, $V^\star(N) = V^\star(L) \Rightarrow N = L$;

(2) Every proper fully invariant submodule of M is a semiprime submodule.

Proof. (1) $\Rightarrow$ (2). Let N be a proper fully invariant submodule of M. If $V^\star(N) = \emptyset$, then $V^\star(N) = V^\star(M)$. It follows that $N = M$, a contradiction. Thus, $V^\star(N) \neq \emptyset$. Put $L = \bigcap_{P \in V^\star(N)} P$. Then, it is easy to see that $N \subset L$, and so $V^\star(L) \subset V^\star(N)$. Now, let $Q \in V^\star(N)$. Then $Q \supset L$, and hence $Q \in V^\star(L)$. This implies that $V^\star(N) \subset V^\star(L)$. Thus, $V^\star(N) = V^\star(L)$ and hence $N = L$, showing that N is a semiprime submodule of M.

(2) $\Rightarrow$ (1). Suppose that N is a semiprime submodule of M. Then, it is clear that $N = \bigcap_{P \in V^\star(N)} P$. Hence, if $V^\star(N) = V^\star(L)$, then $N = L$. □

5. Prime dimension

Definition 5.1 Let M be a right R-module. We first suppose that every prime submodule of M is contained in a maximal prime submodule. We now define, by transfinite induction, the set X_α of prime submodules of M. Firstly, put $X_{-1} = \emptyset$. For an ordinal $\alpha \geq 0$, if X_β has been defined for all ordinal $\beta < \alpha$, let X_α be the set of prime submodules P of M such that all prime submodules, properly containing P, must belong to $\bigcup_{\beta<\alpha} X_\beta$. In particular, X_0 is the set of all maximal prime submodules of M. If there exists an ordinal γ such that $X_\gamma = Spec(M)$, then we say that $\dim M$ exists and if γ is the smallest ordinal such that $X_\gamma = Spec(M)$, then we put $\dim M = \gamma$. Now,we call $\dim M$ the *prime dimension* of M. We remark here that the some new results related with the prime and semiprime Goldie modules have been recently established by Van Sanh and his associates in [10], and also Van Sanh has given a generalized version of Hopins and Liviski theorems for modules in [11].

Theorem 5.2 *Let M be a right R-module. Then X^M is a T_1 space if and only if $dim M \leq 0$.*

Proof. Assume that X^M is a T_1 space. If $X^M = \emptyset$, then $\dim M = -1$. Let $X^M \neq \emptyset$. We claim that every prime submodule of M is a maximal prime submodule. Let $P \in X^M$. If P is not maximal prime, then there exists $Q \in X^M$ such that P is properly contained in Q. Since $\{P\}$ is closed in

X^M, we have $\{P\} = cl(\{P\}) = V(P)$. From $P \subset Q$ we get $I_P \subset I_Q$, and then $Q \in V(P) = \{P\}$, that is , $Q = P$, a contradiction. Thus every prime submodule of M is maximal prime, and so $\dim M = 0$.

Conversely, assume that $\dim M \leq 0$. If $\dim M = -1$, then $X^M = \emptyset$ and so it is a T_1 space. Now, suppose that $\dim M = 0$. Then $X^M \neq \emptyset$ and every prime submodule is a maximal prime submodule. Let $P \in X^M$. Then $cl(\{P\}) = V(P) = \{P\}$, hence $\{P\}$ is closed in X^M. Thus, X^M is a T_1-space. □

Corollary 5.3 *Let M be a nonzero right R-module with* $\dim M = 0$. *Then the following conditions are equivalent:*

(1) For any two fully invariant submodules N, L of M, $V^\star(N) = V^\star(L) \Rightarrow N = L$;
(2) Every proper fully invariant submodule of M is a semiprime submodule.
(3) Every proper fully invariant submodule of M is an intersection of maximal prime submodules.

Proof. (1) $\Rightarrow$ (2). By Proposition 4.10.

(2) $\Rightarrow$ (3). Since $\dim M = 0$, we have $X^M \neq \emptyset$ and any prime submodule of M is maximal prime and the result follows.

(3) $\Rightarrow$ (1). By Proposition 4.10. □

Proposition 5.4 *Let M be a finitely generated right R-module. Then every proper fully invariant submodule of M is contained in a maximal prime submodule of M.*

Proof. Suppose that $M = x_1R + ... + x_kR$. Let N be a proper fully invariant submodule of M. Then the set

$$\Omega = \{B \mid B \text{ is a proper fully invariant submodule of M containing } N\}$$

is nonempty since $N \in \Omega$. The set Ω is ordered by inclusion. Let Γ be any nonempty chain in Ω and $C = \bigcup_{P \in \Gamma} P$. Then C is a fully invariant submodule of M containing N. If $C = M$, then there exists $P \in \Gamma$ such that $\{x_1, ..., x_k\} \subset P$. Thus, $P = M$, a contradiction. Hence C is an upper bound of Γ in Ω. By Zorn's Lemma, there exists a maximal element X in Ω. We claim that X is a prime submodule of M. Let $\varphi \in S$ and U, a fully invariant submodule of M such that $\varphi(U) \subset X$ and $U \not\subset X$. Then $M = U + X$, so $\varphi(M) = \varphi(U) + \varphi(X) \subset X$, proving that X is prime in M, by Theorem 1.1. Let Y be any prime submodule of M such that $X \subset Y$. Then $Y \in \Omega$. By the maximality of X, we have $X = Y$. Therefore, X is maximal prime submodule of M. □

If we let $N = 0$ in the proof of Proposition 5.4, then we have the following corollary:

Corollary 5.5 *Every finitely generated module $M \neq 0$ has a maximal prime submodule.*

Proposition 5.6 *Let $M \neq 0$ be a finitely generated right R-module. Then X^M is a T_1-space if and only if* $\dim M = 0$.

Proof. ($\Rightarrow$) Suppose that X^M is a T_1-space. Then $\dim M \leq 0$, by Theorem 5.2. Since M is a finitely generated module, M contains a maximal prime submodule, by Corollary 5.5. Thus $X^M \neq \emptyset$, and so $\dim M = 0$.

($\Leftarrow$) By Theorem 5.2. □

Let X be a topological space. X is called a *Hausdorff space* if for any two distinct points x, y of X, there exist an open set U containing x and an open set V containing y such that $U \cap V = \emptyset$.

Let M be a right R-module with $|Spec(M)| \leq 1$. Then $Spec(M)$ is a Hausdorff space. The following proposition is the case when $|Spec(M)| \geq 2$.

Theorem 5.6 *Let M be a right R-module with $|Spec(M)| \geq 2$. If $Spec(M)$ is a Hausdorff space, then* $\dim M = 0$.

Proof. Let P, $Q \in Spec(M)$ such that $P \neq Q$. Since $Spec(M)$ is a Hausdorff space, there exists open sets U, V of X^M such that $P \in U$, $Q \in V$ and $U \cap V = \emptyset$. Since U, V are open, there exist fully invariant submodules N, L of M such that $U = X^M \setminus V(N)$, $V = X^M \setminus V(L)$. Since $P \notin V$, we have $P \in V(L)$ and so $I_L \subset I_P$. If $P \subset Q$, then $I_P \subset I_Q$. Thus $I_L \subset I_Q$, so $Q \in V(L)$, a contradiction. Thus $P \not\subset Q$. Similarly, $Q \not\subset P$, showing that $\dim M = 0$. □

6. Prime radical and m-system

Definition 6.1 Let M be a right R-module. A non-empty set $X \subset M \setminus \{0\}$ is called an *m-system* if for any ideal $\mathcal{A}$ of S, and for all submodules K, L of M, if $(K+L) \cap X \neq \emptyset$ and $(K+\mathcal{A}M) \cap X \neq \emptyset$, then $(K+\mathcal{A}L) \cap X \neq \emptyset$.

Lemma 6.2 *Let M be a right R-module and P, a proper fully invariant submodule of M. Then P is a prime submodule of M if and only if $X := M \setminus P$ is an m-system.*

Proof. ($\Rightarrow$) Suppose that P is a prime submodule of M. Let $\mathcal{A}$ be an ideal of S and K, L be submodules of M such that $(K+L) \cap X \neq \emptyset$ and $(K+\mathcal{A}M) \cap X \neq \emptyset$. If $(K+\mathcal{A}L) \cap X = \emptyset$, then $(K+\mathcal{A}L) \subset P$. Then $K \subset P$ and $\mathcal{A}L \subset P$. Since P is prime in M, we have $\mathcal{A}M \subset P$ or $L \subset P$. Thus $(K+L) \cap X = \emptyset$ or $(K+\mathcal{A}M) \cap X = \emptyset$, a contradiction. Therefore,

X is an m-system in M.

($\Leftarrow$) Suppose that X is an m-system in M. Let $\mathcal{A}$ be an ideal of S and L be a fully invariant submodule of M such that $\mathcal{A}L \subset P$. If $L \not\subset P$ and $\mathcal{A}M \not\subset P$, then $L \cap X \neq \emptyset$ and $\mathcal{A}M \cap X \neq \emptyset$. Thus $\mathcal{A}L \cap X \neq \emptyset$, a contradiction. Hence, P is a prime submodule of M. □

Proposition 6.3 *Let M be a right R-module, $X \subset M$ be an m-system and P, a fully invariant submodule of M maximal with respect to the property that $P \cap X = \emptyset$. Then, P is prime in M.*

Proof. Let $\mathcal{A}$ be an ideal of S and L, a fully invariant submodule of M such that $\mathcal{A}L \subset P$. If $\mathcal{A}M \not\subset P$ and $L \not\subset P$, then $(P + L) \cap X \neq \emptyset$ and $(P + \mathcal{A}M) \cap X \neq \emptyset$ by the maximal property of P. Thus $(P + \mathcal{A}L) \cap X \neq \emptyset$ and hence $P \cap X \neq \emptyset$, a contradiction. It shows that P is prime in M. □

Definition 6.4 Let M be a right R-module. For a fully invariant submodule N of M, if there is a prime submodule containing N, then we define:

$r(N) := \{m \in M \mid \text{every m} - \text{system containing m meets } N\}$

If there is no prime submodule containing N, then we define $r(N) = M$.

Theorem 6.5 *Let M be a right R-module and N, a fully invariant submodule of M. Then either $r(N) = M$ or $r(N) = J(V^\star(N))$.*

Proof. Suppose that $r(N) \neq M$. It means that the set $V^\star(N) \neq \emptyset$. We first show that $r(N) \subset J(V^\star(N))$. Let $m \in r(N)$ and $P \in V^\star(N)$. Then $M \setminus P$ is an m-system in M. If $M \setminus P$ contains m, then $M \setminus P$ meets N and hence also P, a contradiction. Therefore $M \setminus P$ does not contain m, and so $m \in P$. Thus $r(N) \subset J(V^\star(N))$. Conversely, suppose $m \notin r(N)$. Then there exist an m-system X containing m and does not meet N. By Zorn's Lemma, there exists a fully invariant submodule $P \supset N$ which is maximal with respect to the property that $P \cap X = \emptyset$. By Proposition 6.3, P is a prime submodule of M and we have $m \notin P$. Therefore we have $r(N) = J(V^\star(N))$. □

Note that $r(0)$ is the prime radical $P(M)$ of M.

Let I be an ideal of a ring R. Recall that, if there exists a prime ideal of R containing I, then we define $\sqrt{I}$ is the intersection of all prime ideals of R containing I. If there are no prime ideals containing I, we put $\sqrt{I} = R$.

We now formulate the following definition.

Definition 6.6 An ideal I of a ring R is called *radical ideal* if $\sqrt{I} = I$. A fully invariant submodule X of a right R-module M is said to be a *radical submodule* if $r(X) = X$.

Theorem 6.7 *Let M be a right R-module which is a self-generator and X, a radical submodule of M. Then I_X is a radical ideal of S.*

Proof. Since M is a self-generator, X has no prime submodule if and only if S has no prime ideal. Therefore, if $X = r(X) = M$, then $I_X = S = \sqrt{S}$. For the case $r(X) \neq M$, we have $X = r(X) = \bigcap_{P \in V^\star(X)} P$. Thus $I_X = \bigcap_{P \in V^\star(X)} I_{P_i}$. Hence I_X is a radical ideal of S. □

Proposition 6.8 *Let M be a right R-module and N, a fully invariant submodule of M. Then $\sqrt{I_N}(M) \subset r(N)$.*

Proof. If $r(N) = M$, then the result is immediate. Otherwise, if X is any prime submodule of M that contains N, then I_X is a prime ideal of S and $I_X \supset I_N$. Thus $\sqrt{I_N} \subset I_X$, and so $\sqrt{I_N}(M) \subset I_X(M) \subset X$. Since X is an arbitrary prime submodule of M containing N, we have $\sqrt{I_N}(M) \subset r(N)$. □

Proposition 6.9 *Let M be a quasi-projective and finitely generated right R-module which is a self-generator. Let N be a fully invariant submodule of M. Then $\sqrt{I_N}(M) = r(N)$.*

Proof. From Proposition 6.3, we have $\sqrt{I_N}(M) \subset r(N)$. Now, we write $r(N) = I_{r(N)}(M)$ and we will show that $I_{r(N)} \subset \sqrt{I_N}$. Let ¶ be a prime ideal of S such that $I_N \subset$ ¶. Then ¶M is a prime submodule of M and ¶$M \supset I_N(M) = N$ and so ¶$M \supset r(N)$. Since $I_{r(N)} = Hom(M, r(N)) \subset Hom(M,$ ¶$M) =$ ¶ (by [13, 18.4]), we have $I_{r(N)} \subset \sqrt{I_N}$. It follows that $r(N) \subset \sqrt{I_N}(M)$. □

Proposition 6.10 *Let M be a right R-module which is a self-generator. Let Y be a subset of X^M. If Y is irreducible, then $J(Y)$ is a prime submodule of M.*

Proof. By Theorem 2.4, we have $V(N) = V^\star(N)$ for any fully invariant submodule of M. It is clear that $J(Y)$ is a proper fully invariant submodule of M. Clearly, $Y \subset V^\star(J(Y))$. Let $\mathcal{I}$ be an ideal of S and U, a fully invariant submodule of M such that $\mathcal{I}U \subset J(Y)$. Then $Y \subset V^\star(J(Y)) \subset V^\star(\mathcal{I}U) \subset V^\star(\mathcal{I}M) \cup V^\star(U)$. Since $V^\star(\mathcal{I}M)$, $V^\star(U)$ are closed sets of X^M and Y is irreducible, we have $Y \subset V^\star(\mathcal{I}M)$ or $Y \subset V^\star(U)$. If $Y \subset V^\star(\mathcal{I}M)$, then $P \supset \mathcal{I}M$, for all $P \in Y$, and hence $\mathcal{I}M \subset J(Y)$. If $Y \subset V^\star(U)$,then $P \supset U$, for all $P \in Y$, and therefore $U \subset J(Y)$. Thus $\mathcal{I}M \subset J(Y)$ or $U \subset J(Y)$, proving that $J(Y)$ is a prime submodule of M by Theorem 1.1. □

Corollary 6.11 *Let M be a right R-module which is a self-generator. Let N be a fully invariant submodule of M. Then $V^\star(N)$ is irreducible if and only if $r(N)$ is prime in M. Consequently, X^M is irreducible if and only if*

$P(M)$ is a prime submodule of M.
Proof. ($\Rightarrow$) by Proposition 6.10.
($\Leftarrow$) by Theorem 4.8. $\square$

Proposition 6.12 *Let M be a prime module. Then $Spec(M)$ is a T_1-space if and only if 0 is the only prime submodule of M.*
Proof. ($\Rightarrow$) Since $Spec(M)$ is a T_1 space, $dimM \leq 0$ (by Theorem 5.2). Since $Spec(M) \neq \emptyset$, $\dim M = 0$, and so 0 is the maximal prime submodule of M. Therefore, 0 is the only prime submodule of M.
($\Leftarrow$) Since 0 is the only prime submodule of M, it is the only maximal prime submodule of M. Thus $\dim M = 0$, and so $Spec(M)$ is a T_1-space. $\square$

Lemma 6.13 *Let M be a right R-module. Let $Y = \{P_1, ..., P_n\}$ be a finite subset of X^M. Then $cl(Y) = \bigcup_{i=1}^{n} V(P_i)$.*

Proof. By Proposition 4.1, $cl(Y) = V(J(Y)) = V(\bigcap_{i=1}^{n} P_i)$. We have $V(P_i) \subset V(\bigcap_{i=1}^{n} P_i)$, for all $i = 1, ..., n$; and so $\bigcup_{i=1}^{n} V(P_i) \subset V(\bigcap_{i=1}^{n} P_i)$. Conversely, let $Q \in V(\bigcap_{i=1}^{n} P_i)$. Then $I_{\bigcap_{i=1}^{n} P_i} \subset I_Q$ It follows that $I_{P_1}...I_{P_n} \subset I_Q$. Since Q is a prime submodule of M, I_Q is a prime ideal of S. Thus $I_{P_k} \subset I_Q$, for some $k = 1, ..., n$. Thus $Q \in V(P_k)$, and so $Q \in \bigcup_{i=1}^{n} V(P_i)$. Hence $V(\bigcap_{i=1}^{n} P_i) \subset \bigcup_{i=1}^{n} V(P_i)$. $\square$

Proposition 6.14 *Let M be a right R-module. Let Y be a finite irreducible closed subset of X^M. Then $Y = V(P)$ for some $P \in X^M$.*
Proof. Suppose that $Y = \{P_1, ..., P_n\}$ is an irreducible closed subset of X^M. Then $Y = cl(Y) = \bigcup_{i=1}^{n} V(P_i)$, by Lemma 6.13. Since Y is irreducible, $Y = V(P_i)$ for some $i = 1, ..., n$. $\square$

References

1. F. W. Anderson and K. R. Fuller, "Rings and Categories of Modules", Graduate Texts in Math., No 13, Springer-Verlag, Berlin - Heidelberg - New York, 1992.
2. M. Behboodi and M. R. Haddadi, *Classical Zariski Topology of Modules and Spectral spaces I,* International Electronic Journal of Algebra, **4**(2008), 104-130.
3. A. W. Chatters, C. R. Hajarnavis, "Rings With Chain Conditions", Pitman Advanced Publishing Program, 1980.

4. K. R. Goodearl and R. B. Warfield, Jr., "An Introduction to Noncommutative Noetherian Ring", Cambridge Univ. Press, Cambridge, UK, 1989.
5. F. Kasch, "Modules and Rings", London Math. Soc., 1982.
6. T.Y Lam, "A First Course in Noncommutative Rings", Springer-Verlag, New York, 1991.
7. H. C. Li and K. P. Shum, *On a problem of spectral posets,* J. Appl. Algebra Discrete Struct. **1**(3)(2003), 203-209.
8. Donald S. Passman, "A Course in Ring Theory", AMS Chelsea Publishing, Amer. Math. Soc., Providence, Rhode Island, 2004.
9. N. V. Sanh, N. Anh Vu, S. Asawasamrit, K.F. Amed and L.P. Thao, Primeness in Module Category, Asian-European J. Math, 3(1), 2010, 145-154.
10. N.V.Sanh, S. Aswasamrit,K.F. Amed and L.P. Thao, *On prime and semiprime Goldie modules,* Asian-European J. Math, to appear.
11. N. V. Sanh and L. P. Thao, *A generalization of Hopkin -Levizki Theorem,* Southeast-Asian Bull. of Mathematics, to appear.
12. B. Stenström, "Rings of Quotients", Springer-Verlag, Berlin - Heidelberg - New York, 1975.
13. R. Wisbauer, "Foundations of Module and Ring Theory" , Gordon and Breach, Tokyo, e.a., 1991.
14. Ünsal Tekir, *The Zariski Topology on the Prime Spectrum of a Module over Noncommutative Rings,* Algebra Colloquium, **16:4**(2009), 691-698.

On the Subgroups of the Triangle Group $*2p\infty$

E. D. B. Provido* and M. L. A. N. de las Penas**

Mathematics Department, Ateneo de Manila University,
Loyola Heights, Quezon City, 1108 Philippines
E-mail: *eprovido@ateneo.edu, **mlp@mathsci.math.admu.edu.ph

M. C. B. Decena

Department of Mathematics and Physics, College of Science,
University of Santo Tomas, Espana, Manila, Philippines
E-mail: mcebd@yahoo.com

In this work, we discuss the low index subgroups of the triangle group $*2p\infty$ (also called the extended Hecke group) which were derived using tools in color symmetry theory. Moreover, we present the low index subgroups of the index 2 subgroups of the extended modular group $*23\infty$.

Keywords: triangle groups, extended Hecke groups, extended modular groups, color symmetry

1. Introduction

The group generated by the reflections P, Q and R in the sides of a triangle Δ with interior angles $\pi/p, \pi/q$ and π/r is called a *triangle group* denoted by $*pqr$. The elements of $*pqr$ are symmetries of the triangle tiling $\mathcal{T}$ obtained by repeatedly reflecting the triangle Δ in its sides. The tiling $\mathcal{T}$ is either in the spherical ($\mathbb{S}^2$), Euclidean ($\mathbb{E}^2$), or hyperbolic ($\mathbb{H}^2$) plane according as $1/p+1/q+1/r$ is larger, equal, or smaller than 1, respectively. The *extended Hecke group* $*2p\infty$ and the *extended modular group* $*23\infty$ are some of the well known examples of triangle groups. These groups and their subgroups have been studied for many aspects in the literature [1, 2, 3, 4, 5, 6]. The *modular group*, $23\infty \cong PSL(2, Z)$, for example, is an index 2 subgroup of the extended modular group and has especially been of great interest in many fields of mathematics, such as number theory, automorphic function theory, group theory and graph theory [7, 8, 9, 10]. Moreover, the modular group, plays a crucial role in the uniformization of algebraic curves and in the construction of subgroups of larger matrix groups (e.g, $SL(n, Z)$ and $SL(n, R)$).

The main focus of this paper would be to determine the low index subgroups of the extended Hecke group, which will be derived using a geometric method based on tools in color symmetry theory. The approach will allow for a characterization of each subgroup by type of symmetries by looking at corresponding colored hyperbolic tilings by ideal triangles. The richness of $\mathbb{H}^2$ symmetries compared with the Euclidean plane is overwhelming. One of the motivations for this detailed exploration of symmetries in hyperbolic groups lies in its relevance to structures in condensed chemical systems [11]. In [11, 12, 13], the important role of hyperbolic triangle groups and their subgroups appears in connection with the construction of $3D$-nets from hyperbolic tilings which are wrapped into appropriate triply periodic minimal surfaces.

In this work, we will also illustrate the process of arriving at the low index subgroups of the index 2 subgroups of the the extended modular group. Morever, we include a short discussion on the derivation of the low index subgroups of the modular group. The basis for our methods will be discussed in the next section.

2. An approach in determining subgroups of two dimensional symmetry groups

In our work, the determination of the subgroups of the extended Hecke groups is facilitated by the link of group theory and color symmetry theory. More particularly, the derivation of the subgroups is realized by a correspondence that arises between a subgroup of a group M of symmetries of a given tiling and M–transitive colorings of the tiling. This idea is encapsulated in the following result which was established in [14]:

Theorem 2.1. *Let $\mathcal{T}$ be a tiling of the spherical, Euclidean or hyperbolic plane by copies of a triangle t which fill the plane with no gaps and overlaps. Let H be the group generated by reflections in the sides of t. Let M be a subgroup of H and $\mathcal{O} = \{mt \mid m \in M\}$ be the M–orbit of t.*

(i) Suppose L is a subgroup of M of index n. Let $\{m_1, m_2, \ldots, m_n\}$ be a complete set of left coset representatives of L in M and $C = \{c_1, c_2, \ldots, c_n\}$ a set of n colors. The assignment $m_i Lt \rightarrow c_i$ defines an n–coloring of $\mathcal{O}$ which is M–transitive.

(ii) In an M–transitive n–coloring of $\mathcal{O}$, the elements of M which fix a specific color in the colored set $\mathcal{O}$ form a subgroup of M of index n.

In the above theorem we assume that the tiling $\mathcal{T}$ is the H–orbit of t, that is $\mathcal{T} = \{ht \mid h \in H\}$; t forms a fundamental region for H and $Stab_H(t) = \{h \in H \mid ht = t\}$ is the trivial group $\{e\}$. [15]

Moreover, for a subgroup M of H, $Stab_M(t) = \{e\}$ and M acts transitively on $\mathcal{O} = \{mt \mid m \in M\}$. Consequently, there is a one-to-one correspondence between M and $\mathcal{O}$ given by $m \to mt$, $m \in M$. The action of M on $\mathcal{O}$ is regular, where $m' \in M$ acts on $mt \in \mathcal{O}$ by sending it to its image under m'. We look at M–transitive n–colorings of $\mathcal{O}$, that is, n–colorings of $\mathcal{O}$ for which the group M has a transitive action on the set C of n colors. For a discussion on how to arrive at such a coloring and the proof of the preceding theorem, the reader may refer to [14].

In this work we will show that the index n subgroups of the extended Hecke group H is derived using n–colorings of $\mathcal{O} = Ht$ (in this case, $M = H$). An n–coloring of $\mathcal{O}$ is an onto function $f : \mathcal{O} \to C$, that is, to each $mt \in \mathcal{O}$ is assigned a color in C. The coloring determines a partition $\mathcal{P} = \{f^{-1}(c_i) \mid c_i \in C\}$ where $f^{-1}(c_i)$ is the set of elements of $\mathcal{O}$ assigned color c_i. Equivalently, we may think of the coloring as a partition of $\mathcal{O}$.

Remark: Note that given a subgroup L of M of index n and a set of n colors $\{c_1, c_2, \ldots, c_n\}$, then there correspond $(n-1)!$ M–transitive n–colorings of $\mathcal{O}$ with L fixing c_1. In an M–transitive n–coloring of $\mathcal{O}$ with L fixing c_1, the set of triangles Lt is assigned c_1 and the remaining $n-1$ colors are distributed among the $m_i Lt$, $m_i \notin L$.

3. Low index subgroups of $*2p\infty$

In this section, the low index subgroups of the extended Hecke group $*2p\infty$ are derived using Theorem 2.1 given the setting described as follows. Consider a hyperbolic triangle Δ with interior angles $\pi/2, \pi/p$ and zero. Repeatedly reflecting Δ in its sides results in a tiling $\mathcal{T}$ of $\mathbb{H}^2$ by copies of Δ. Let P, Q, R denote reflections along the sides of Δ. The group H generated by P, Q, R is called the extended Hecke group and is denoted by $*2p\infty$. The generators P, Q, R satisfy the relations $P^2 = Q^2 = R^2 = (RP)^2 = (QR)^p = e$.

The elements of H are symmetries of $\mathcal{T}$ and the triangle Δ forms a fundamental region for H. This means that if $h \in H$ and $h\Delta$ is the image of Δ under h then the union of these images as h varies over the elements of H is the entire plane and, moreover, if $h_1, h_2 \in H$, $h_1 \neq h_2$, the respective interiors of $h_1\Delta$ and $h_2\Delta$ are disjoint. Each triangle in the tiling is the image of Δ under a uniquely determined $h \in H$. Note that $Stab_H(\Delta)$ is $\{e\}$.

In Fig. 1(a) we present an example of a tiling in the upper half plane model of $\mathbb{H}^2$ by a triangle Δ with interior angles $\pi/2, \pi/4$ and zero. The axes of reflections P, Q, R and centers of rotations QR and RP are labeled as shown. The zero angle edges are parallel to each other and meet at infinity. The group H generated by P, Q, R is the group $*24\infty$.

In this work we will use Conway's orbifold notation [16] to describe the subgroups of the extended Hecke group. Orbifold symbols provide a compact label for all two dimensional symmetry groups. The symbol describes the type of symmetries within a fundamental domain of the group: $*$ indicates a mirror reflection, $\times$ a glide reflection, and a number n indicates a rotation of order n. Moreover, if a number n comes after the $*$, the center of the corresponding rotation lies on mirror lines, so the symmetry there is dihedral of order $2n$.

The following theorem discusses the index 2, 3, 4 and 5 subgroups of the extended Hecke group. Each subgroup, (up to conjugacy in H) corresponding to a coloring of $\mathcal{T}$ is described in Tables 1, 2 and 3 through its orbifold symbol. The corresponding fundamental region is given together with the generators of each subgroup.

Theorem 3.1. *The extended Hecke group* $H = *2p\infty$ *has*

(i) 7 index 2 subgroups if p is even (Table 1 nos. 1-7).
(ii) 3 index 3 subgroups up to conjugacy in H if p is divisible by both 2 and 3 (Table 1 nos. 8-10).
(iii) 22 index 4 subgroups up to conjugacy in H if p is divisible by 2, 3 and 4 (Table 2 nos. 11-32).
(iv) 13 index 5 subgroups up to conjugacy in H if p is divisible by 2, 3, 4 and 5 (Table 3 nos. 33-45).

Proof. To arrive at the index 2, 3, 4 and 5 subgroups of the extended Hecke group H, we construct respectively, 2–, 3–, 4– and 5–colorings of $\mathcal{T}$ where all elements of H effect color permutations and H acts transitively on the set of colors. For such an n–coloring of $\mathcal{T}$, $n \in \{2, 3, 4, 5\}$, a homomorphism $\Pi : H \to S_n$ is defined, where S_n is the symmetric group of order $n!$. The group H is generated by P, Q, and R, thus Π is completely determined when $\Pi(P)$, $\Pi(Q)$, and $\Pi(R)$ are specified. To construct a coloring, we consider a fundamental region for H and assign to it color c_1. The arguments to obtain 2–, 3–, 4–, 5–colorings of $\mathcal{T}$ where each of the generators P, Q, R permutes the colors, that is, the associated color permutation to a generator is of order 1 or 2, are as follows:

(i) In constructing 2–colorings of $\mathcal{T}$, each of P, Q, R may be assigned either the identity permutation (1) or the permutation (12). The possible 2–colorings of $\mathcal{T}$ are listed in Table 1 line nos. 1–7.
The coloring shown in Fig. 1(a) is a result of the color scheme suggested in Table 1 line no. 1, where each of the reflections P, Q, R interchanges colors c_1 and c_2. As a consequence of this assignment of colors, the p–fold rotation QR together with the 2–fold rotation RP and the ∞–fold rotation QP generate the subgroup fixing color c_1. The orbifold notation for the group is $2p\infty$. This group is also known as the *Hecke group*.
Consider the coloring shown in Fig. 1(b) corresponding to the color assignment given in Table 1 line no. 2 where P interchanges colors c_1 and c_2 while both Q and R fix the colors c_1 and c_2. The reflections Q, R, PQP together generate the subgroup fixing color c_1 in the coloring. Note that the products of pairs of these reflections, namely: the p–fold rotations QR and $PQRP$, and the ∞–fold rotation $(QP)^2$ also fix color c_1. The orbifold notation for the given group is $*pp\infty$. Note that the color scheme given in Table 1 line nos. 3 and 4 will result in colorings where the respective subgroups fixing color c_1 are groups of similar type.
Shown in Fig. 1(c) is the coloring resulting from the color assignment given in Table 1 line no. 5 where P fixes the colors and each of Q and R interchanges colors c_1 and c_2. Resulting from this color scheme, the reflection P as well as the p–fold rotation QR generate the subgroup fixing color c_1 in the coloring. This group in orbifold notation is $p * \infty$. Groups of similar type result from color assignments given in Table 1 line nos. 6 and 7. Considering the colorings that will yield distinct subgroups there will be 7 subgroups of index 2 in $*2p\infty$.

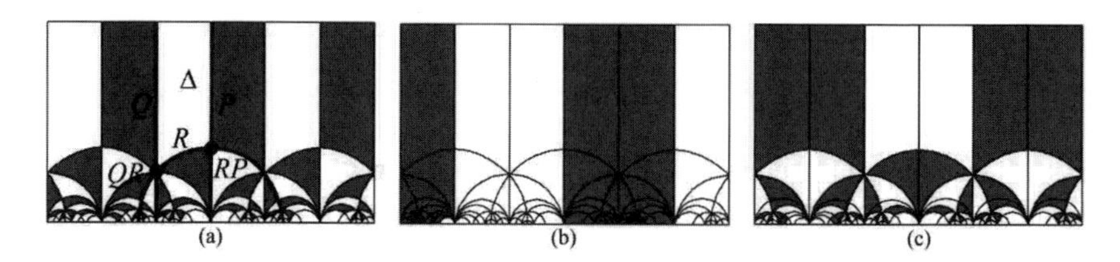

Fig. 1. (Color online) The colorings of the tiling by triangles with angles $\pi/4, \pi/2$ and zero resulting from the color assignments given in Table 1: (a) 1 (b) 2 (c) 5. The colors white and red are used to represent, respectively, colors c_1 and c_2.

Table 1. Colorings that will give rise to index 2 and 3 subgroups of $H = *2p\infty$.

no	index in H	P	Q	R	Generators	a fundamental region	orbifold notation
1	2	(12)	(12)	(12)	QR, RP, QP	$\Delta \cup Q(\Delta)$	$2p\infty$
2	2	(12)	(1)	(1)	R, Q, PQP	$\Delta \cup P(\Delta)$	$*pp\infty$
3	2	(1)	(12)	(1)	P, R, QRQ, QPQ	$\Delta \cup Q(\Delta)$	$*(p/2)\infty\infty$
4	2	(1)	(1)	(12)	RQR, P, Q	$\Delta \cup R(\Delta)$	$*22(p/2)\infty$
5	2	(1)	(12)	(12)	RQ, P	$\Delta \cup R(\Delta)$	$p * \infty$
6	2	(12)	(1)	(12)	RP, Q, PQP	$\Delta \cup P(\Delta)$	$\infty * (p/2)$
7	2	(12)	(12)	(1)	QP, R, QRQ	$\Delta \cup Q(\Delta)$	$2 * (p/2)\infty$
8	3	(13)	(12)	(13)	QPQ, QRQ, RP, RQR	$\Delta \cup Q(\Delta) \cup R(\Delta)$	$2 * 2(p/3)\infty$
9	3	(1)	(12)	(13)	QRQ, RQR, P, QPQ	$\Delta \cup Q(\Delta) \cup R(\Delta)$	$*2(p/3)\infty\infty$
10	3	(12)	(13)	(1)	PQP, QPQ, QRQ, R	$\Delta \cup P(\Delta) \cup Q(\Delta)$	$*2(p/2)p\infty$

(ii) In arriving at 3–colorings of $\mathcal{T}$, this time each of P, Q, R may be assigned either the identity permutation or either of the permutations (12), (13) or (23). Now, since we assume H acts transitively on the set of 3 colors, we consider the assignments that bring forth a $\Pi(H)$ which is a transitive subgroup of S_3. Note that there are two transitive subgroups of S_3, namely S_3 and the cyclic group Z_3. It is not possible for $\Pi(H)$ to be Z_3 since this group cannot be generated by elements of order two. Thus we only consider the case when $\Pi(H)$ is S_3. In order to generate S_3, two of the three permutations (12), (13) or (23) must be associated with two from among the generators P, Q, and R. The remaining third generator is associated either with the identity, with one of the two already assigned earlier or with the third permutation of order 2. The possible 3–colorings that will yield distinct subgroups of index 3 distinct up to conjugacy in H are given in Table 1 line nos. 8-10.

The color scheme suggested in Table 1 line no. 8 gives rise to the coloring shown in Fig. 2(a) where the reflection Q interchanges c_1 and c_2 and the reflections P and R interchange c_1 and c_3. As a consequence of this assignment of colors, the reflections QPQ, QRQ, RQR together with the 2–fold rotation RP generate the subgroup fixing c_1. The $(p/3)$–fold rotation $(QR)^3$, the 2–fold rotation $QRPQ$ and the ∞–fold rotation $(QP)^3$ also fix c_1. The orbifold notation for the group is $2 * 2(p/3)\infty$.

The coloring given in Fig. 2(b) corresponding to the color assignment given in Table 1 line no. 9 has P fixing all the colors, Q interchanging c_1 and c_2 and R interchanging c_1 and c_3. The reflections QRQ, RQR, P and QPQ generate the subgroup fixing c_1 of the given coloring; where the region bounded by the axes of these reflections serves as a fundamental region. Note that the products of pairs of these reflections, namely: the $(p/3)$–fold rotation $(QR)^3$, the ∞–fold rotations $RPQR$ and $(PQ)^2$, and the 2–fold rotation $QRPQ$ also fix c_1. In orbifold notation, the group is $*2(p/3)\infty\infty$. Note that the color assignment given in Table 1 line no. 10 will result in a coloring where the respective subgroup fixing c_1 is a group of similar type.

Fig. 2. (Color online) The colorings of the tiling by triangles with angles $\pi/2, \pi/3$ and zero resulting from the color assignments given in Table 1: (a) 8 (b) 9. The colors white, red and blue are used to represent, respectively, colors c_1, c_2 and c_3.

(iii) In constructing 4–colorings of $\mathcal{T}$ that will yield index 4 subgroups of H, each of P, Q, R is assigned either the identity permutation; a 2–cycle or a product of two disjoint 2–cycles: (12), (13), (14), (23), (24), (34), (12)(34), (13)(24) or (14)(23). We only look at the situations that arise when $\Pi(H)$ is a transitive subgroup of S_4. In particular, we consider the cases when $\Pi(H)$ is either S_4, D_4 or $V = \{e, (12)(34), (13)(24), (14)(23)\}$, a Klein 4 group. The possible H–transitive 4–colorings of $\mathcal{T}$ that can be constructed when $\Pi(H) \cong V, D_4$ and S_4 are given in Table 2 line nos. 11-32. These colorings will give rise to subgroups of index 4 distinct up to conjugacy in H.

Consider the coloring given in Fig. 3(a) corresponding to the color assignment given in Table 2 line no. 11 where the reflections P, Q, R respectively interchange c_2 and c_4; c_1 and c_2; c_1 and c_3. Consequently, the reflections $P, QPQPQ, QRQ, RQR$ generate the subgroup fixing c_1 of the given coloring; where the region bounded by the axes of reflections serves as a fundamental region. Note that the products of pairs of these reflections, namely: the $(p/3)$–fold rotation $(QR)^3$, the ∞–fold rotation $(PQ)^3$, the p–fold rotation $QPQRPQ$ and the ∞–

fold rotation $RQPR$ also fix c_1. In orbifold notation, the group is $*p(p/3)\infty\infty$. Similar groups will fix c_1 of the colorings resulting from the color assignments from Table 2 line nos. 14-17.
For the given coloring appearing in Fig. 3(b), corresponding to the color assignment in Table 2 line no. 12, the reflection R interchanges c_1 and c_2; c_3 and c_4; whereas P interchanges the colors c_1 and c_2; and Q interchanges the colors c_1 and c_3. As a result, the reflections $RQR, QRQRQ, QPQ$ together with the 2–fold rotation RP generate the subgroup fixing c_1 of the given coloring. The $(p/4)$–fold rotation $(QR)^4$, the ∞–fold rotations $QRPQRQ$ and $(PQ)^3$ also fix c_1. The orbifold notation for the given group is $2*(p/4)\infty\infty$. Note that the color assignments given in Table 2 line nos. 13, 18, and 27-29 will give rise to groups of similar type.
Now, other types of subgroups fixing color c_1 can be derived from the remaining color assignments given in Table 2. The coloring shown in Fig. 3(c) results from the color scheme presented in Table 2 line no. 19 where c_1 and c_3; c_2 and c_4 are at the same time interchanged by the reflection Q while the reflections R and P interchange c_1 and c_2; c_3 and c_4, respectively. As a consequence of this assignment of colors, the reflections P and QRQ fix c_1 in the given coloring. The translation $RQPQ$ also fixes c_1. Together they generate the subgroup fixing c_1 of the corresponding coloring. In addition, the ∞–fold rotation $(PQ)^4$ and the $(p/4)$–fold rotation $(QR)^4$ also fix c_1 of the given coloring. In orbifold notation, this subgroup is $*\infty*(p/4)$.

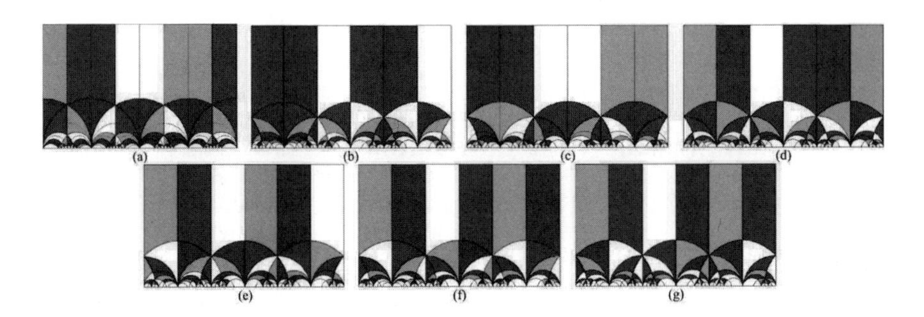

Fig. 3. (Color online) The colorings of the tiling by triangles with angles (a) $\pi/2, \pi/3$ and zero; (b)-(g) $\pi/2, \pi/4$ and zero resulting from the color assignments given in Table 2: (a) 11; (b) 12; (c) 19; (d) 20; (e) 25; (f) 26; (g) 30. The colors white, red, blue and green are used to represent, respectively, colors c_1, c_2, c_3 and c_4.

Table 2. Colorings that will give rise to index 4 subgroups of $H = *2p\infty$.

no	index in H	P	Q	R	Generators	a fundamental region	orbifold notation
11	4	(24)	(12)	(13)	$P, QPQPQ, QRQ, RQR$	$\Delta \cup R(\Delta) \cup Q(\Delta) \cup QP(\Delta)$	$*p(p/3)\infty\infty$
12	4	(12)	(13)	(12)(34)	$RP, QRQRQ, RQR, QPQ$	$\Delta \cup Q(\Delta) \cup R(\Delta) \cup QR(\Delta)$	$2*(p/4)\infty\infty$
13	4	(13)(24)	(12)	(13)	$RP, QPQPQ, PQP, QRQ$	$\Delta \cup Q(\Delta) \cup P(\Delta) \cup QP(\Delta)$	$2*p(p/3)\infty$
14	4	(1)	(12)	(13)(24)	$P, QRQRQ, RQR, QPQ$	$\Delta \cup Q(\Delta) \cup R(\Delta) \cup QR(\Delta)$	$*(p/4)\infty\infty\infty$
15	4	(12)	(13)(24)	(1)	$R, PQPQP, QPQ, QRQ, PQRQP$	$\Delta \cup P(\Delta) \cup Q(\Delta) \cup PQ(\Delta)$	$*\infty 2(p/2)(p/2$
16	4	(1)	(13)(24)	(12)	$P, RQRQR, QRQ, QPQ, RQPQR$	$\Delta \cup R(\Delta) \cup Q(\Delta) \cup RQ(\Delta)$	$*2(p/4)2\infty\infty$
17	4	(13)(24)	(12)	(1)	$R, QPQPQ, PQP, QRQ$	$\Delta \cup Q(\Delta) \cup P(\Delta) \cup QP(\Delta)$	$*p(p/2)p\infty$
18	4	(12)	(13)(24)	(12)	$QRQ, QPQ, RP, PQRQP, PQPQP,$	$\Delta \cup P(\Delta) \cup Q(\Delta) \cup PQ(\Delta)$	$2*2(p/4)2\infty$
19	4	(34)	(13)(24)	(12)	$P, QRQ, RQPQ$	$\Delta \cup Q(\Delta) \cup R(\Delta) \cup QP(\Delta)$	$*\infty*(p/4)$
20	4	(12)	(13)(24)	(12)(34)	$QRQR, QPQ, RP, QRPRQ$	$\Delta \cup Q(\Delta) \cup R(\Delta) \cup QR(\Delta)$	$(p/2)2*\infty$
21	4	(12)(34)	(12)	(13)(24)	$PQ, RQR, RPQPR$	$\Delta \cup R(\Delta) \cup P(\Delta) \cup RP(\Delta)$	$\infty*(p/4)\infty$
22	4	(13)(24)	(12)	(12)(34)	$QR, PQP, PRQRP$	$\Delta \cup P(\Delta) \cup R(\Delta) \cup PR(\Delta)$	$p*(p/2)\infty$
23	4	(12)(34)	(13)(24)	(12)	$RP, QRQ, PQPQ$	$\Delta \cup Q(\Delta) \cup P(\Delta) \cup QP(\Delta)$	$2\infty*(p/4)$
24	4	(13)(24)	(12)	(13)(24)	$QRPQ, PQP, RP, QPQPQ$	$\Delta \cup Q(\Delta) \cup P(\Delta) \cup QP(\Delta)$	$22*(p/4)\infty$
25	4	(14)(23)	(12)	(13)(24)	RQR, PQP, QRP	$\Delta \cup P(\Delta) \cup R(\Delta) \cup PR(\Delta)$	$*(p/4)\infty\times$
26	4	(12)(34)	(13)(24)	(14)(23)	PQR, PRQ, QPR	$\Delta \cup P(\Delta) \cup Q(\Delta) \cup R(\Delta)$	$(p/2)\infty\times$
27	4	(1)	(12)(34)	(13)(24)	$P, QPQ, QRQR$	$\Delta \cup Q(\Delta) \cup R(\Delta) \cup QR(\Delta)$	$(p/2)*\infty\infty$
28	4	(13)(24)	(1)	(12)(34)	$RQR, PQP, Q, PRQRP$	$\Delta \cup P(\Delta) \cup R(\Delta) \cup PR(\Delta)$	$*(p/2)\infty(p/2)$
29	4	(12)(34)	(13)(24)	(1)	$PQPQ, QRQ, R, PQRQP$	$\Delta \cup P(\Delta) \cup Q(\Delta) \cup PQ(\Delta)$	$\infty*(p/2)(p/2$
30	4	(13)(24)	(12)(34)	(12)(34)	$QR, PQRP, PQPQ$	$\Delta \cup P(\Delta) \cup Q(\Delta) \cup PQ(\Delta)$	$pp\infty$
31	4	(12)(34)	(13)(24)	(12)(34)	$RP, QRPQ, PQPQ$	$\Delta \cup Q(\Delta) \cup P(\Delta) \cup QP(\Delta)$	$2(p/2)2\infty$
32	4	(12)(34)	(12)(34)	(13)(24)	$PQ, RPQR$	$\Delta \cup R(\Delta) \cup P(\Delta) \cup RP(\Delta)$	$(p/2)\infty\infty$

Given in Fig. 3(d) is a coloring arising from the color scheme in Table 2 line no. 20, where the reflection R interchanges c_1 and c_2; c_3 and c_4; whereas P interchanges the colors c_1 and c_2; and Q interchanges the colors c_1 and c_3; c_2 and c_4. The reflections QPQ and $QRPRQ$ together with the 2–fold rotation RP and the $(p/2)$–fold rotation $(QR)^2$ generate the subgroup fixing c_1 of the given coloring. Also fixing c_1 is the ∞–fold rotation $(QP)^4$. The orbifold notation for the given group is $(p/2)2 * \infty$. Note that the color assignments given in Table 2 line nos. 21-24 will give rise to groups of similar type.

Figure 3(e), on the other hand, shows a coloring of the tiling obtained from the color assignment given in Table 2 line no. 25 where c_1 and c_4; c_2 and c_3 are interchanged by the reflection P while the reflection Q interchanges c_1 and c_2; R interchanges c_1 and c_3; c_2 and c_4. As a consequence of this assignment of colors, the reflections RQR and PQP together with the glide reflection QPR generate the subgroup fixing c_1 of the given coloring. The $(p/4)$–fold rotation $(QR)^4$ and the ∞–fold rotation $(PQ)^4$ also fix c_1. In orbifold notation, this group is $*(p/4)\infty\times$.

The color scheme suggested in line no. 26, Table 2, gives rise to the coloring in Fig. 3(f), where the subgroup fixing c_1 is generated by three glide reflections, namely: PQR, PRQ, QPR. Note that products of pairs of glide reflections result in the following rotations fixing c_1: the $(p/2)$–fold rotation $(QR)^2$ and the ∞–fold rotation $(PQ)^2$. The orbifold notation for this subgroup is $(p/2)\infty\times$.

Finally, consider the coloring in Fig. 3(g) resulting from the color assignment given in Table 2 line no. 30. In this coloring, three rotations generate the subgroup fixing c_1. These are the two p-fold rotations $QR, PQRP$ and the ∞–fold rotation $(PQ)^2$. This subgroup in orbifold notation is $pp\infty$. Groups of similar type arise from the color assignment given in Table 2 line nos. 31-32. This gives the derivation of the 22 index 4 subgroups of H up to conjugacy in H.

(iv) To arrive at 5–colorings of $\mathcal{T}$ that will give rise to index 5 subgroups of H such that H is to permute the colors in the resulting coloring, each of P, Q, R is associated either the identity permutation or a permutation of order 2 coming from S_5, that is, a 2–cycle, or a product of two disjoint 2–cycles. Considering the fact that H acts transitively on the set of 5 colors, we look at the situation that arises when $\Pi(H)$ is a transitive subgroup of S_5. Note that there are 5 transitive subgroups of S_5 [17], namely, S_5, the cyclic group Z_5, the alternating group A_5, the

Table 3. Colorings that will give rise to index 5 subgroups of $H = *2p\infty$.

no	index in H	P	Q	R	Generators fixing 1	a fundamental region	orbifold notation
33	5	(1)	(23)(45)	(13)(25)	$P, Q, RQPQR,$ $RQRQPQRQR$	$\Delta \cup R(\Delta)$ $\cup RQ(\Delta)$ $\cup RQR(\Delta)$ $\cup RQRQ(\Delta)$	$*(p/5)\infty\infty\infty\infty2$
34	5	(23)(45)	(13)(25)	(1)	$P, R, QRQ,$ $QPRPQ,$ $QPQRQPQ,$ $QPQPQPQPQ,$ $QPQPRPQPQ$	$\Delta \cup Q(\Delta)$ $\cup QP(\Delta)$ $\cup QPQ(\Delta)$ $\cup QPQP(\Delta)$	$*(p/2)(p/2)p\infty2$
35	5	(23)(45)	(13)(25)	(23)(45)	$P, R, QRPQ,$ $QPQRPQPQ$ $QPQPQPQPQ$	$\Delta \cup Q(\Delta)$ $\cup QP(\Delta)$ $\cup QPQ(\Delta)$ $\cup QPQP(\Delta)$	$22*\infty2(p/5)$
36	5	(23)	(15)(24)	(45)	$P, R, QPQ,$ $QRQRQRQ,$ $QRQPQPQRQ$	$\Delta \cup Q(\Delta)$ $\cup QR(\Delta)$ $\cup QRQ(\Delta)$ $\cup QRQP(\Delta)$	$*(p/4)p\infty\infty2$
37	5	(23)	(15)(25)	(23)(45)	$P, R, QPQ,$ $QRQRPQRQ,$ $QRQPQPQRQ$	$\Delta \cup Q(\Delta)$ $\cup QR(\Delta)$ $\cup QRQ(\Delta)$ $\cup QRQP(\Delta)$	$2*2(p/5)\infty\infty$
38	5	(23)	(24)(35)	(15)(23)	$P, Q, RQRPQR,$ $RQPQPQPQR,$ $RQPQRQPQR$	$\Delta \cup R(\Delta)$ $\cup RQ(\Delta)$ $\cup RQP(\Delta)$ $\cup RQPQ(\Delta)$	$2*2(p/5)\infty\infty$
39	5	(23)	(24)(35)	(15)	$P, Q, RQRQR,$ $RQPQPQPQR,$ $RQPQRQPQR$	$\Delta \cup R(\Delta)$ $\cup RQ(\Delta)$ $\cup RQP(\Delta)$ $\cup RQPQ(\Delta)$	$*(p/3)(p/2)2\infty\infty$
40	5	(24)(35)	(13)	(25)(34)	$P, QPQPQ,$ $R, QRQRQ,$ $QPRQRPQ$	$\Delta \cup Q(\Delta)$ $\cup QR(\Delta)$ $\cup QP(\Delta)$ $\cup QPR(\Delta)$	$*(p/3)(p/2)2\infty\infty$
41	5	(24)(35)	(13)(45)	(35)	$P, R, QRPQ,$ $QPQRQPQ,$ $QPQPQPQPQ$ $QPQPRPQPQ$	$\Delta \cup Q(\Delta)$ $\cup QP(\Delta)$ $\cup QPQ(\Delta)$ $\cup QPQP(\Delta)$	$2*p\infty2(p/4)$
42	5	(24)(35)	(13)(45)	(24)	$P, R, QRQ,$ $QPQRPQPQ$ $QPQPQPQPQ$	$\Delta \cup Q(\Delta)$ $\cup QP(\Delta)$ $\cup QPQ(\Delta)$ $\cup QPQP(\Delta)$	$2*(p/3)(p/2)2\infty$
43	5	(23)(45)	(15)(23)	(24)(35)	$R, QPQPQ,$ $P, QRQRPQ$	$\Delta \cup Q(\Delta)$ $\cup R(\Delta)$ $\cup QP(\Delta)$ $\cup RP(\Delta)$	$*(p/5)2\infty$
44	5	(23)(45)	(13)(25)	(24)(35)	$P, R, QPQRQ,$ $QPRQRPQ$	$\Delta \cup Q(\Delta)$ $\cup QP(\Delta)$ $\cup QPQ(\Delta)$ $\cup QPR(\Delta)$	$*(p/5)2\infty\times$
45	5	(23)(45)	(13)(25)	(25)(34)	$P, R, QRQRQ,$ $QPRQPQ,$ $QRPQPQ$	$\Delta \cup Q(\Delta)$ $\cup QR(\Delta)$ $\cup QP(\Delta)$ $\cup QPQP(\Delta)$	$p*(p/3)2\infty$

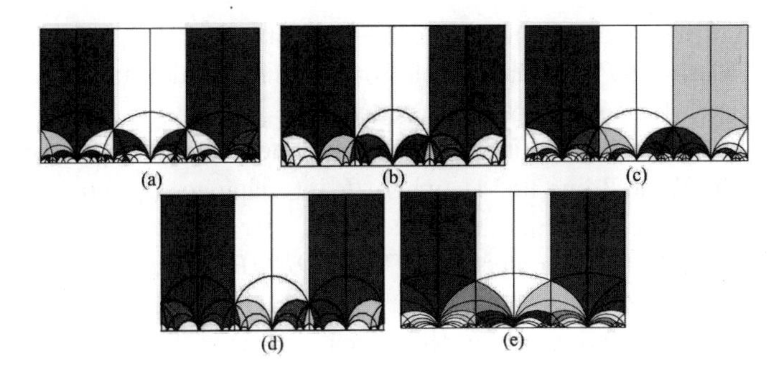

Fig. 4. (Color online) The colorings of the tiling by triangles with angles (a), (c) $\pi/2, \pi/4$ and zero; (b), (d) $\pi/2, \pi/5$ and zero; (e) $\pi/2, \pi/3$ and zero resulting from the color assignments given in Table 3: (a) 34; (b) 35; (c) 41; (d) 44; (e) 45. The colors white, red, blue, green and yellow are used to represent, respectively, colors c_1, c_2, c_3, c_4 and c_5.

dihedral group D_5 and the group F_5, a group of order 20 isomorphic to $AGL_1(5)$, a 1-dimensional affine group over a field F of order 5. However, the groups Z_5 and F_5 cannot be generated by elements of order 2 which means $\Pi(H)$ cannot be Z_5 or F_5. Thus we only consider the case when $\Pi(H)$ is either S_5, A_5 or D_5. The list of possible H–transitive 5–colorings of $\mathcal{T}$ that can be constructed when $\Pi(H) \cong S_5, A_5$ or D_5 are given in Table 1 line nos. 33-45. These colorings will give rise to subgroups of index 5 distinct up to conjugacy in H.

The coloring shown in Fig. 4(a) corresponds to the color assignment given in Table 3 line no. 34 where R fixes the colors, P interchanges c_2 and c_3, c_4 and c_5; and Q interchanges c_1 and c_3, c_2 and c_5. As a result of this color assignment, the reflections $P, R, QRQ, QPRPQ, QPQRQPQ, QPQPQPQPQ, QPQPRPQPQ$ generate the subgroup fixing c_1 in the given coloring. The product of pairs of these reflections, such as the $(p/2)$–fold rotations $(QR)^2$ and $QP(QR)^2PQ$, the p–fold rotation $QPQPQRPQPQ$, the ∞–fold rotation $(PQ)^5$ and the 2–fold rotation RP also fix c_1. In orbifold notation, the group is $*(p/2)(p/2)p\infty 2$. Similar groups will fix c_1 of the colorings resulting from the color assignments from Table 3 line nos. 33, 36, 39, 40 and 43.

The color assignment given in Table 3 line no. 35 associated to the coloring shown in Fig. 4(b) in which both the reflections P, R interchange c_2 and c_3, c_4 and c_5; while Q interchanges c_1 and c_3, c_2 and c_5, gives rise to the coloring in which the reflections $P, R, QPQPQPQPQ$ together with the 2–fold rotations $QRPQ, QPQRPQPQ$ generate the subgroup fixing c_1 in the coloring. The $(p/5)$–fold rotation $QP(QR)^5PQ$, the 2–fold rotation RP and the ∞–fold rotation $(PQ)^5$ also fix c_1. In orbifold notation, this subgroup is $22*\infty 2(p/5)$.

The coloring given in Fig. 4(c) corresponds to the color assignment given in Table 3 line no. 41, where R interchanges c_3 and c_5; while P interchanges c_2 and c_4, c_3 and c_5; and Q interchanges c_1 and c_3, c_4 and c_5. As a result the reflections P, R, $QPQRQPQ$, $QPQPQPQPQ$, $QPQPRPQPQ$ together with the 2–fold rotation $QRPQ$ generate the subgroup fixing c_1 in the given coloring. The p–fold rotation $QPQPQRPQPQ$, the ∞–fold rotation $(PQ)^5$, the 2–fold rotation RP, and the $(p/4)$–fold rotation $(QR)^4$, also fix c_1. The orbifold notation for this group is $2*p\infty 2(p/4)$. The group arising from the color assignment given in Table 3 line nos. 37, 38 and 42 are of similar type.

The coloring given in Fig. 4(d) is a result of the color assignment given in Table 3 line no. 44, where P interchanges c_2 and c_3, c_4 and c_5; Q interchanges c_1 and c_3, c_2 and c_5; and R interchanges c_2 and c_4, c_3 and c_5. This gives a subgroup generated by three reflections $P, R, QPRQRPQ$ and a glide reflection $QPQRQ$ which fix c_1 in the given coloring. The $(p/5)$–fold rotation $QP(QR)^5PQ$, the 2–fold rotation PR and the ∞–fold rotation $QR(QP)^5RQ$ also fix c_1. The orbifold notation for the group is $*(p/5)2\infty\times$.

Lastly, the color scheme given in Table 3 line no. 45, corresponding to Fig. 4(e) gives rise to a subgroup generated by reflections $P, R, QRQRQ$, a p–fold rotation $QPRQPQ$ (with center not lying on a reflection axis), and a translation $QRPQPQ$. The $(p/3)$–fold rotation $(QR)^3$, the 2–fold rotation RP and the ∞–fold rotation $(QP)^5$ are among the symmetries fixing c_1. In orbifold notation this group is $p*(p/3)2\infty$. This gives the complete list of the 13 index 5 subgroups in H fixing c_1 for the 5–colorings of $\mathcal{T}$ presented in Table 3. □

4. Low index subgroups of the index 2 subgroups of $*23\infty$

The approach discussed in the previous section in obtaining the low index subgroups of $*2p\infty$ may be applied to also determine the low index subgroups of the index 2 subgroups of $*2p\infty$. To illustrate the method, the low index subgroups of the index 2 subgroups of the extended modular group $*23\infty$ will be considered.

The extended modular group is generated by P, Q, R which are reflections along the sides of a hyperbolic triangle Δ with angles $\pi/2, \pi/3$ and 0. Based on the results given in Table 1 there are 3 index 2 subgroups of

$*23\infty$, namely, the modular group $K_1 = \langle RP, QP \rangle$, $K_2 = \langle Q, R, PQP \rangle$, and $K_3 = \langle P, RQ \rangle$. In orbifold symbol, these groups are 23∞, $*33\infty$ and $3*\infty$, corresponding respectively to line nos 1, 2 and 5 of Table 1. Since p is not even ($p = 3$), then there are only three index 2 subgroups of $*23\infty$. Note that the subgroups listed in Table 1 line nos. 3, 4, 6 and 7 require p to be even.

In applying Theorem 2.1 to derive the low index subgroups of an index 2 subgroup K_i ($i = 1, 2, 3$) of $*23\infty$, we consider the K_i–orbit of Δ, $\mathcal{O}_i = K_i\Delta = \{k\Delta | k \in K_i\}$ and construct K_i–transitive colorings of $\mathcal{O}_i$. In determining respectively, the index 2, 3, 4 and 5 subgroups of K_i, we construct 2–, 3–, 4– and 5–colorings of $\mathcal{O}_i$. For an n–coloring of $\mathcal{O}_i$, $n \in \{2, 3, 4, 5\}$, a homomorphism $\Pi_i : K_i \to S_n$ is defined. The group K_1 for instance is generated by RP and QP, thus Π_1 is completely determined when $\Pi_1(RP)$ and $\Pi_1(QP)$ are specified. On the other hand, K_2 is generated by Q, R, PQP so Π_2 is determined when $\Pi_2(Q), \Pi_2(R)$ and $\Pi_2(PQP)$ are specified. Finally, since the generators of K_3 are P, RQ so Π_3 is determined when $\Pi_3(P), \Pi_3(RQ)$ are also given.

Construction of tables similar to Tables 1, 2 and 3, this time pertaining to assignments of permutations to RP, QP (Table 4) ; Q, R, PQP (Table 5) and P, RQ (Table 6) will yield transitive 2–, 3–, 4–, and 5–colorings of $\mathcal{O}_1, \mathcal{O}_2$ and $\mathcal{O}_3$, respectively. Moreover, we note that if $y \in M \leq K_i$ ($i = 1, 2, 3$), then the order of $\Pi_i(y)$ must divide the order of y. That is, to obtain 2–, 3–, 4– and 5–colorings of $\mathcal{O}_i$ where each of the generators of K_i permutes the colors, the orders of the color permutations associated to the generators of K_i should be divisors respectively of the orders of the given generators.

The index 2, 3, 4 and 5 subgroups of 23∞, $*33\infty$ and $3*\infty$, arising respectively from the K_i–transitive colorings of $\mathcal{O}_i$ ($i = 1, 2, 3$) are described using orbifold symbols and given in Tables 4, 5 and 6. We discuss the derivation for the subgroups of the modular group 23∞ below. Similar methods are employed to determine the subgroups of K_2 and K_3 which correspond to the colorings in Tables 5 and 6, respectively.

In determining respectively, the index 2, 3, 4 and 5 subgroups of the modular group K_1, we construct 2–, 3–, 4– and 5–colorings of $\mathcal{O}_1$. To obtain 2–, 3–, 4–, or 5-colorings of $\mathcal{O}_1$ where each of the generators RP, QP of the modular group permutes the colors, the orders of the color permutations associated to RP and QP should be divisors, respectively of the orders of RP and QP. Note that RP is of order two and QP is of infinite order.

In constructing 2–colorings of $\mathcal{O}_1$ that will yield index 2 subgroups of

Table 4. Colorings corresponding to index 2, 3, 4 and 5 subgroups of $K_1 = 23\infty$.

no	index in H	$a = RP$	$b = QP$	Generators fixing 1	orbifold notation
1	2	(12)	(12)	$ba^{-1}, b^{-1}a^{-1}$	33∞
2	3	(1)	(123)	a, bab^{-1}, b^3	222∞
3	3	(23)	(12)	a, b^{-2}	$2\infty\infty$
4	4	(12)(34)	(234)	b, ab^3a^{-1}	$22\infty\infty$
5	4	(34)	(1234)	a, bab^{-1}, b^{-4}	223∞
6	5	(23)(45)	(12345)	$a, b^2a^{-1}b^{-1}, b^{-2}a^{-1}b$	233∞

Table 5. Colorings corresponding to index 2, 3, 4 and 5 subgroups of $K_2 = *33\infty$.

no	index in H	$a = Q$	$b = R$	$c = PQP$	Generators fixing 1	orbifold notation
1	2	(12)	(12)	(12)	ba, ca	33∞
2	3	(23)	(23)	(12)	$a, b, cbac$	$\infty * \infty$
3	3	(23)	(12)	(23)	$a, c, bcab$	$3 * \infty 3$
4	3	(23)	(12)	(12)	$a, cb, bacab$	$3 * 3\infty$
5	3	(23)	(12)	(13)	a, bac	$*\infty\times$
6	4	(34)	(23)	(12)	$a, b, cac, cbacabc$	$*3\infty3\infty$
7	4	(34)	(23)	(13)	$a, cabac, b, cacac$	$*\infty3\infty3$
8	5	(23)(45)	(12)(45)	(12)(34)	$a, cb, bacbacab$	$33 * \infty$

Table 6. Colorings corresponding to index 2, 3, 4 and 5 subgroups of $K_3 = 3 * \infty$.

no	index in H	$a = P$	$b = RQ$	Generators fixing 1	orbifold notation
1	2	(12)	(1)	b, aba	$33\times$
2	3	(1)	(123)	$a, bab^{-1}, b^{-1}ab$	$\infty\infty*$
3	3	(23)	(123)	a, bab	$3 * \times$
4	4	(12)(34)	(234)	$b, ababa$	$3\infty\times$
5	4	(34)	(123)	$a, bab^{-1}, b^{-1}abab$	$3*$
6	5	(23)(45)	(124)	$a, babab^{-1}, b^{-1}abab$	$33 * \times$

K_1, we assign to RP and QP either the identity permutation (1) or the 2–cycle (12). The consequent color assignments will give rise to one index 2 subgroup which is precisely the index 4 subgroup of the extended modular group $*23\infty$ (Table 2 no. 30) generated by rotations. This group in orbifold notation is 33∞.

In arriving at 3–colorings of $\mathcal{O}_1$ that will give rise to index 3 subgroups of K_1, we assign to RP and QP color permutations that will generate a transitive subgroup of S_3. We will only consider the permutation assignments that will result in $\Pi_1(K_1)$ isomorphic to Z_3 or S_3. The color assignments that give rise to index 3 subgroups of K_1 are listed in Table 4 line nos. 2-3. Associated with the color scheme in line no. 2 is the index 3 subgroup fixing color c_1 generated by three 2–fold rotations and one ∞-fold rotation given, respectively, by: RP, $QRPQ$, $PQRPQP$ and $(QP)^3$. The group in orbifold notation is 222∞. Resulting from the color scheme suggested in line no. 3 is the subgroup fixing color c_1 generated by one 2–fold rotation and two ∞–fold rotations. These rotations are $RP, QRQP, (QP)^3$, respectively. The three rotations generate the index 3 subgroup of K_1 with orbifold notation $2\infty\infty$.

In determining the 4–colorings of $\mathcal{O}_1$ that will correspond to index 4 subgroups of K_1, we assign to QP and RP color permutations that will yield any transitive subgroup of S_4. We will consider the permutation assignments that bring forth a $\Pi_1(K_1)$ isomorphic to $V = \{e, (12)(34), (13)(24), (14)(23)\}$, Z_4, D_4, A_4, or S_4. The possible K_1–transitive 4–colorings of $\mathcal{O}_1$ that can be constructed when $\Pi_1(K_1)$ is isomorphic to these groups are given in Table 4 line nos. 4-5. These colorings will give rise to subgroups of index 4 distinct up to conjugacy in K_1 which we characterize in terms of the Conway orbifold notation as follows: The color scheme given in line no. 4 gives rise to the coloring of $\mathcal{O}_1$ where c_1 is fixed by two 2–fold rotations $RQRPQR, PRQRPQRP$ and two ∞–fold rotations $RQPQPQRP, QP$. In orbifold notation, the group is $22\infty\infty$. The color assignment in line no. 5 corresponds to the subgroup fixing c_1 in the coloring generated by two 2–fold rotations RP, QR, a 3–fold rotation $PQPRQPQP$ and ∞–fold rotation $QPQRQP$. The group in orbifold notation is 223∞.

For the derivation of the 5-colorings of $\mathcal{O}_1$ that will correspond to index 5 subgroups of K_1, we consider the permutation assignments that bring forth a $\Pi_1(K_1)$ isomorphic to either one of the following transitive subgroups of S_5: S_5, Z_5, A_5, D_5 and F_5. We arrive at one index 5 subgroup of K_1 up to conjugacy in K_1. The permutation assignments given in line no. 6 gives the coloring of $\mathcal{O}_1$ where c_1 is fixed by a 2–fold rotation RP, 3–fold rotations $QPQRPQ$ and $RQPQPRQP$, and the ∞–fold rotation $QPRQPQPRPQ$. In orbifold notation, this group is 233∞ . This concludes the derivation of the low index subgroups of $*32\,\mathrm{inf}$.

5. Outlook

There is still a lot to be investigated on the subgroup structure of the extended Hecke groups and the extended modular groups. Studies have been carried out on commutator subgroups, principal congruence subgroups and power subgroups of the extended Hecke group and extended modular groups, see for instance [2, 4, 5, 6]. The conjugacy classes of torsion elements of the extended Hecke group have also been investigated in [3], and this information has been used to find some normal subgroups and Fuchsian subgroups. It would be worthwhile to verify and generalize these results using the methods found in this paper.

The Picard group or the Gaussian modular group $PSL(2, Z[i])$, is the three dimensional counterpart of the modular group, consisting of linear fractional transformations with coefficients Gaussian integers. There are existing studies on subgroups of the Picard group using other methods [3, 18, 19, 20]. It would be interesting to explore the subgroup structure of the Picard group using a similar approach provided in this paper. In deriving its subgroups, the Picard group will be viewed as a group of symmetries of a three dimensional tiling by a hyperbolic tetrahedron with dihedral angles $\pi/4, \pi/4, \pi/3, \pi/2, \pi/2$ and $\pi/2$, having symmetry group generated by reflections in the faces of the tetrahedron. The Picard group contains the modular group as a subgroup. It will be also useful to identify the embeddings of the Hecke groups in the Picard group [21].

Another point of investigation would be to look at the subgroup structure of the higher dimensional analogues of the modular group and Picard group. In [22], Ahlfor's description of hyperbolic isometries via Clifford algebras $\mathcal{C}_n[1, 2]$ gives rise to a natural generalization of the modular and Picard groups in higher dimensions, such as $PSL(2, Z[i, j])$. Studying these analogues of the modular and Picard groups using tools from color symmetry theory would be noteworthy. Future work would also include the study of quaternionic modular groups, which occur as subgroups of groups generated by reflections with axes along the facets of four- and five-dimensional hyperbolic Coxeter polytopes [23, 24].

Acknowledgments

Eden Provido and Louise De Las Peñas would like to acknowledge funding support of the Ateneo de Manila University Loyola Schools through the Scholarly Work Faculty Grant. Ma. Carlota Decena is grateful for the support from the Research Center for Natural Sciences, University of Santo Tomas.

References

1. G. A. Jones and J. Thornton, *J London Math Soc.* **34**, 26 (1986).
2. O. Koruoğlu, R. Sahin and S. Ikikardes, *Bull. Braz Math Soc. New* **37**, 1 (2006).
3. N. Ozgur and R. Sahin, *Turk J. of Math.* **27**, 473 (2003).
4. R. Sahin and O. Bizim, *Mathematica Acta Scientia* **23**, 497 (2003).
5. R. Sahin, S. Ikikardes and O. Koruoglu, *Rocky Mount. J Math.* **63**, 1033 (2006).
6. R. Sahin, S. Ikikardes and O. Koruoglu, *Turk. J. Math.* **28**, 143 (2004).
7. M. Lang, *J. Algebra* **248**, 202 (2002).
8. Q. Mushtaq and U. Hayat, *Indian J. Pure Appl. Math.* **38**, 345 (2007).
9. M. Newman, *Illinois J. Math.* **6**, 480 (1962).
10. M. Newman, *Illinois J. Math.* **8**, 262 (1964).
11. S. Hyde, A. Larsson, T. Di Matteo, S. Ramsden and V. Robins, *Austr. J. Chem.* **56**, 981 (2003).
12. S. Hyde, S. Ramsden and V. Robins, *Acta Crystallogr.* **A65**, 81 (2009).
13. V. Robins, S. Ramsden and S. Hyde, *Eur. Phys. J. B.* **39** (2004).
14. M. L. De Las Peñas, R. Felix and G. Laigo, *Z. Kristallogr.* **225** (2010).
15. E. Vinberg, *Russian Math. Surveys* **40**, 31 (1985).
16. J. H. Conway and D. Huson, *Structural Chem.* **13** (2002).
17. J. H. Conway, A. Hulpke and J. Mckay, *J. Comput. Math* **1** (1998).
18. B. Fine and M. Newman, *Trans. American Math. Soc.* **302**, 769 (1987).
19. N. Ozgur, *Contr. Alg. and Geom.* **8**, 262 (1964).
20. H. Waldinger, *Proc. Amer. Math. Soc.* **16**, 1373 (1965).
21. B. Fine, *Can. J. Math.* **28**, 481 (1976).
22. C. Maclachlan, P. Waterman and N. Wielenberg, *Trans. Amer. Math. Soc.* **312**, 739 (1989).
23. N. Johnson and A. Weiss, *Linear Alg. and its Appl.* **295**, 159 (1999).
24. A. Weiss, Tessellations and related modular groups, in *European Women in Mathematics: Proceedings of the Tenth General Meeting*, (Singapore, 2001).

Relation Between Weak Entwining Structures and Weak Corings

Nikken Prima Puspita

Department of Mathematics,
Faculty of Mathematics and Natural Sciences,Diponegoro University,
Jl. Prof. Soedharto, Kampus Undip Tembalang, Semarang, Indonesia 50275
Email : nikkenprima@yahoo.com

Abstract. Given a commutative ring R with unit, R-algebra A and R-coalgebra C. Triple (A,C,ψ) is called (weak) entwining structure if there is R-linear map $\psi: C\otimes_R A \to A\otimes_R C$ that fulfil some axioms. In the other hand, from algebra A and coalgebra C we can consider $A\otimes_R C$ as a left A-module canonically such that (A,C,ψ) is entwined structure if only if $A\otimes_R C$ is a A-coring. In particular, we obtain that (A,C,ψ) is a weak entwined structure if only if $A\otimes_R C$ is a weak A-coring.
Keywords : algebra, coalgebra, coring, entwining structure.

1. Introduction

In this paper we assume that R is a commutative ring with unit. In Brzeziński and Wisbauer [3] R-algebra (A,μ,ι) and R-coalgebra (C,Δ,ε) is called entwined and (A,C,ψ) is said to be an entwining structure if there exists a R-linear map $\psi: C\otimes_R A \to A\otimes_R C$ such that fulfil some axioms. It is described by Brzeziński [2] on a *bow-tie* diagram.

R-algebra A dan R-coalgebra C can be considered as R-module. From A and C as R-module, we can construct tensor product $A\otimes_R C$. Moreover, from right A-action $\alpha:(A\otimes_R C)\otimes_R A \to A\otimes_R C$, $\alpha((a\otimes b)\otimes c) = a\psi(c\otimes b)$, $A\otimes_R C$ is a (A,A)-bimodule and we obtain $A\otimes_R C$ is a weak coring. From Brzeziński [4] we have relation between weak coring and weak entwining structure , i. e. (A,C,ψ) is an entwining structure if only if $A\otimes_R C$ has an A-coring structure given by the comultiplication

$$\underline{\Delta} := I_A \otimes \Delta : A\otimes_R C \to A\otimes_R C\otimes_R C \simeq (A\otimes_R C)\otimes_A (A\otimes_R C),$$

and counit $\underline{\varepsilon} := I_A \otimes \varepsilon : A \otimes_R C \to A.$ Weak coring is a structure like coring but weak coring is obtained from non-unital bimodule (see Puspita [7], Wisbauer [9]). We will see relation between coring and entwining structures can be used on weak coring $A \otimes_R C.$

In section 2 we give definitions of corings and weak corings. Those are generalization from coalgebra (see Brzeziński [3], Puspita [7] and Wisbauer [9]). In the next section from Brzeziński [4] given definitions of entwining structures and weak entwining structures. In section 4 finally we have relation between weak entwining structures and weak corings, i.e $A \otimes_R C$ is a weak coring if only if $A \otimes_R C$ is an entwining structure.

2. Corings and Weak Corings

In 1960 Sweedler Introduced the study of coalgebras and comodules over field. A vector space C over field F with comultiplication $\Delta : C \to C \otimes_F C$ and counit $\varepsilon : C \to F$ is called F -coalgebra. The study of coalgebras over commutative rings and noncommutative rings are presented in Brzeziński and Wisbauer [3]. In this section, we are given basic information of corings and weak corings (see Brzeziński [3], Puspita [7] and Wisbauer [9]). Throughout A will be an assosiative ring with unit.

Definition 2.1. *Let C be an (A, A) non-unital bimodule.*

(i). An (A, A)-bilinear map $\underline{\Delta} : C \to C \otimes_A A \otimes_A C,$ i.e $(\forall c \in C)\underline{\Delta}(c) = \sum c_{\underline{1}} \otimes 1 \otimes c_{\underline{2}}$ is called a weak comultiplication.

(ii). An (A, A)-bilinear map $\underline{\varepsilon} : C \to A$ is called weak counit for $\underline{\Delta}$ provided we have a commutative diagram on figure 1.

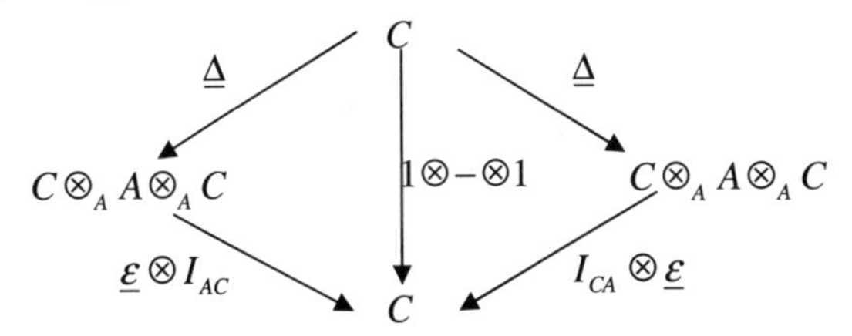

figure 1. *Weak counit diagram*

Figure 1 is commutative for $c \in C,$ $\sum \underline{\varepsilon}(c_{\underline{1}})c_{\underline{2}} = 1c1 = \sum c_{\underline{1}}\underline{\varepsilon}(c_{\underline{2}}).$

Definition 2.2. *An (A, A)-non-unital bimodule C is called weak coring provided it has weak comultiplication $\underline{\Delta}$ and weak counit $\underline{\varepsilon}$.*

Definition 2.3. *Let* $(C,\underline{\Delta},\underline{\varepsilon})$ *be an weak* A*-coring. If* C *is an* (A,A)*-unital bimodule with left or right unital, then* C *is called* ***pre-coring.*** *If* C *is an* (A,A)*-unital, then* C *is an* A*-* ***coring****.*

Based on Definition 2.3., we conclude that every A-coring are a weak A-coring. A weak A-coring is an A-coring if only if C is an (A,A)-unital bimodule.

3. Entwining Structures

Entwining structure introduced by Brzeziński and Majid [1]. Some authors have presented their observation in the same object in various text books as well see Brzeziński [4] and Brzeziński and Wisbauer[3].

Definition 3.1. *Let* (A,μ,ι) *be a* R*-algebra and* (C,Δ,ε) *be a* R*-coalgebra. Triple* (A,C,ψ) *is called entwining structure provided there exist* R*-linear map* $\psi : C\otimes_R A \to A\otimes_R C$ *such that*

(1). $\psi\circ(I_C\otimes\mu)=(\mu\otimes I_C)\circ(I_A\otimes\psi)\circ(\psi\otimes I_A)$,

(2). $(I_A\otimes\Delta)\circ\psi=(\psi\otimes I_C)\circ(I_C\otimes\psi)\circ(\Delta\otimes I_A)$,

(3). $\psi\circ(I_C\otimes\iota)=\iota\otimes I_C$,

(4). $(I_A\otimes\varepsilon)\circ\psi=\varepsilon\otimes I_A$.

The axioms in Definition 3.1. are described on **bow-tie** diagram (see Brzeziński [2], Brzeziński and Wisbauer[3]) as follow :

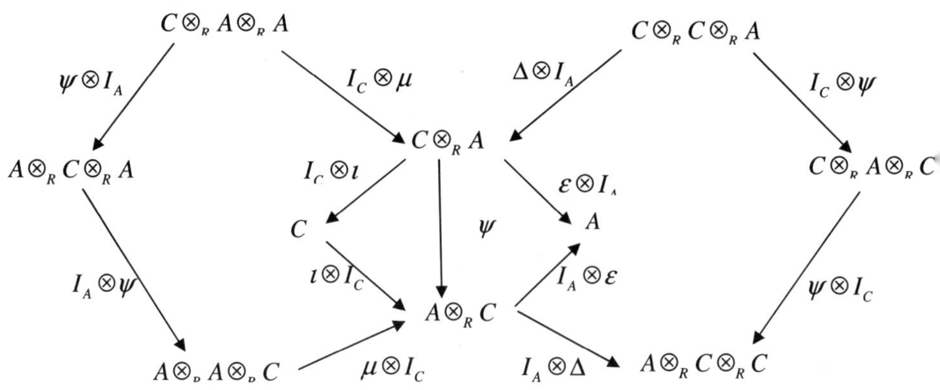

Figure 2. *bow-tie* commutative diagram

Defined a map $\psi : C \otimes_R A \rightarrow A \otimes_R C$, $\psi(c \otimes a) = \sum a_\psi \otimes c^\psi$, for $a_\psi \in A, c^\psi \in C$. Figure 2 is commutative, it means that for any $c \otimes a_1 \otimes a_2 \in C \otimes_R A \otimes_R A$, $c \otimes a \in C \otimes_R A$,

1. $\sum (a_1 a_2)_\psi c^\psi = \sum a_{1_\psi} a_{2_\varphi} c^{\psi\varphi}$
2. $\sum a_\psi \otimes c_{\underline{1}}^\psi \otimes c_{\underline{2}}^\psi = \sum a_{\psi\varphi} c_{\underline{1}}^\varphi c_{\underline{2}}^\psi$
3. $\sum 1_\psi \otimes c^\psi = 1 \otimes c$
4. $\sum a_\psi \varepsilon(c^\psi) = \varepsilon(c) a.$

Definition for weak entwining structures analog with Definition 3.1. The differences are caused by $A \otimes_R C$ as a non unital module so the conditions that need to be fulfiled are still involved an element unit 1. The following definition are presented in Hungerford [6] and Wisbauer [9].

Definisi 3.2. *Let* (A, μ, ι) *be a* R *- algebra and* (C, Δ, ε) *is a* R *-coalgebra. Triple* (A, C, ψ) *is called weak entwining structure provided there exist a* R *-linear maps* $\psi : C \otimes_R A \rightarrow A \otimes_R C$, $\psi(a \otimes c) = \sum a_\psi c^\psi$ *for* $a_\psi \in A$ *and* $c^\psi \in C$ *such that :*

(1). $\sum (ab)_\psi c^\psi = \sum a_\psi b_\varphi c^{\psi\varphi}$

(2). $\sum a_\psi (c^\psi{}_{\underline{1}} \otimes 1) \otimes c^\psi{}_{\underline{2}} = \sum a_{\psi\varphi} \otimes c_{\underline{1}}^\varphi \otimes c_{\underline{2}}^\psi$

(3). $\sum a_\psi \varepsilon(c^\psi) = \sum \varepsilon(c^\psi) 1_\psi a$

(4). $\sum 1_\psi \otimes c^\psi = \sum \varepsilon(c_{\underline{1}}^\psi) 1_\psi \otimes c_{\underline{2}}.$

4. Weak Entwining Structures and Weak Corings

As a R-module, product tensor between R-algebra A and R-coalgebra C is denoted by $A \otimes_R C$. In this section it will be explained the relation between (weak) entwining structures and (weak) corings) $A \otimes_R C$. We are now proving our main theorem.

Theorem 4.1. *Let* (A, μ, ι) *be a* R *-algebra and* (C, Δ, ε) *be a* R *-coalgebra. Triple* (A, C, ψ) *is an entwining structure if only if* $A \otimes_R C$ *is an* A *-coring.*

PROOF.

$(\Leftarrow)$ Assume that $A\otimes_R C$ is an A-coring over comultiplication and counit

$$\underline{\Delta}: A\otimes_R C \xrightarrow{I_A\otimes\Delta} (A\otimes_R C)\otimes_A (A\otimes_R C) \simeq (A\otimes_R C).1\otimes_R C,$$

$$a\otimes c \;\mapsto\; \sum (a\otimes c_{\underline{1}})\otimes_A (1\otimes c_{\underline{2}}) \mapsto \sum (a\otimes c_{\underline{1}}).(1\otimes c_{\underline{2}}),$$

$$\underline{\varepsilon}: A\otimes_R C \to (A\otimes_R C).1 \xrightarrow{I_A\otimes\varepsilon} A,$$

$$a\otimes c \;\mapsto\; (a\otimes c).1 \quad\mapsto\; a\varepsilon(c).$$

The following R-linear map is defined by right A-action $A\otimes_R C.$

$$\psi: C\otimes_R A \to A\otimes_R C,\; c\otimes a \mapsto (1\otimes c).a$$

$\psi(c\otimes a)=\sum a_\psi c^\psi$, for $a_\psi \in A, c^\psi \in C.$ We will show that (A,C,ψ) is an entwining structure by ψ. For any $a,b\in A$ and $c\in C$

(i). by associative properties from right action

$$\sum (ab)_\psi \otimes c^\psi = (1\otimes c).a.b = ((1\otimes c).a).b = \left(\sum a_\psi c^\psi\right).b = \left(1\otimes \sum a_\psi c^\psi\right).b = \sum a_\psi b_\varphi c^{\psi\varphi}.$$

(ii). By comultiplication in $A\otimes_R C$ we have

$$\begin{aligned}
\underline{\Delta}(1\otimes c).a &= \underline{\Delta}\left(\sum a_\psi c^\psi\right)\\
&= \sum a_\psi \underline{\Delta}(c^\psi)\\
&= \sum a_\psi \left(c^\psi{}_{\underline{1}} \otimes c^\psi{}_{\underline{2}}\right)\\
&= \sum a_\psi \otimes c^\psi{}_{\underline{1}} \otimes c^\psi{}_{\underline{2}}
\end{aligned}$$

$$\begin{aligned}
\underline{\Delta}((1\otimes c).a) &= \underline{\Delta}(1\otimes c).a\\
&= \left(\sum (1\otimes c_{\underline{1}})\otimes_A (1\otimes c_{\underline{2}})\right).a\\
&= \sum 1\otimes c_{\underline{1}}\left(\sum a_\psi (1\otimes c_{\underline{2}})^\psi\right)\\
&= \sum 1\otimes c_{\underline{1}}\left(\sum (a_\psi 1) c_{\underline{2}}^\psi\right)\\
&= \sum (1\otimes c_{\underline{1}}).a_\psi \otimes c_{\underline{2}}^\psi\\
&= \sum a_{\psi\varphi} (1\otimes c_{\underline{1}})^\varphi \otimes c_{\underline{2}}^\psi\\
&= \sum a_{\psi\varphi} \otimes c_{\underline{1}}^\varphi \otimes c_{\underline{2}}^\psi
\end{aligned}$$

(iii). R -linear map $\underline{\varepsilon}$ is a module homorphism, so that

$$\begin{aligned}\sum a_\psi \varepsilon(c^\psi) &= (I_A \otimes \varepsilon)\sum a_\psi c^\psi \\ &= (I_A \otimes \varepsilon)\circ \psi(c\otimes a) \\ &= (I_A \otimes \varepsilon)((1\otimes c).a) \\ &= \varepsilon(c)a \\ &= \varepsilon \otimes I_A(c\otimes a)\end{aligned}$$

(iv). As an A -coring, $A\otimes_R C$ is a right unital A -module, so from unital properties we have $1\otimes c = (1\otimes c).1 = \sum 1_\psi c^\psi$.

By (i) – (iv) (A,C,ψ) is an entwining structure.

$(\Rightarrow)$ Suppose that (A,C,ψ) is an entwining structure, to show $A\otimes_R C$ is an A -coring, so in the first step we have to show that $A\otimes_R C$ is an (A,A)-unital bimodule. Left A - action for $A\otimes_R C$ is trivial. By R -linear map $\psi : A\otimes_R C \to C\otimes_R A,$ defined right A - action in $A\otimes_R C$

$$(A\otimes_R C)\otimes_R A \to A\otimes_R C,\ (a\otimes b)\otimes c \mapsto a\psi(c\otimes b)$$

by right A -action above, $A\otimes_R C$ is an (A,A)-unital bimodule.

For any $(a\otimes c),(a'\otimes c')\in A\otimes_R C$ and $r,s\in A$

(i). $$\begin{aligned}(a\otimes c).x+(a'\otimes c').x &= (a\otimes c)\otimes x+(a'\otimes c')\otimes x \\ &= ((a\otimes c)+(a'\otimes c'))\otimes x \\ &= ((a\otimes c)+(a'\otimes c')).x\end{aligned}$$

(ii). $$\begin{aligned}(a\otimes c).(x+y) &= a\psi(c\otimes(x+y)) \\ &= a\psi(c\otimes x+c\otimes y) \\ &= a\psi(c\otimes x)+a\psi(c\otimes y) \\ &= (a\otimes c).x+(a\otimes c).y\end{aligned}$$

(iii). $((a\otimes c).x).y=(a\psi(c\otimes x)).y$

$$=a(1\otimes c).xy$$
$$=a\psi(c\otimes xy)$$
$$=(a\otimes c)(xy)$$

(iv). $(a\otimes c).1=a\otimes\psi(c\otimes 1)$

$$=a\otimes\left(\sum 1_{\psi}c^{\psi}\right)$$
$$=a\otimes(1\otimes c) \text{ (see Definition 3.1 (4))}$$
$$=a\otimes c$$

Furthermore we define a map

$$\underline{\Delta}: A\otimes_R C\xrightarrow{I_A\otimes\Delta}(A\otimes_R C)\otimes_A(A\otimes_R C)\simeq(A\otimes_R C).1\otimes_R C,$$
$$a\otimes c \;\mapsto\; \sum(a\otimes c_{\underline{1}})\otimes_A(1\otimes c_{\underline{2}})\mapsto\sum(a\otimes c_{\underline{1}}).(1\otimes c_{\underline{2}}),$$
$$\underline{\varepsilon}: A\otimes_R C\to(A\otimes_R C).1\xrightarrow{I_A\otimes\varepsilon}A,$$
$$a\otimes c\;\mapsto\;(a\otimes c).1\;\mapsto\; a\varepsilon(c).$$

R-linear map $\underline{\Delta}$ and $\underline{\varepsilon}$ above sequentially are a comultiplication and counit for an (A,A)-bimodule $A\otimes_R C$

$$(I_{A\otimes_R C}\otimes\underline{\varepsilon})\circ\underline{\Delta}(a\otimes c)=(I_{A\otimes_R C}\otimes\underline{\varepsilon})\sum(a\otimes c_{\underline{1}}).1\otimes c_{\underline{2}}$$
$$=\sum a\otimes c_{\underline{1}}\varepsilon(c_{\underline{2}})$$
$$=a\sum c_{\underline{1}}\varepsilon(c_{\underline{2}})$$
$$=a\otimes c \text{ (by counital as a coalgebra). } \square$$

Theorem 4.2. *Let (A,μ,ι) be a R- algebra and (C,Δ,ε) be a R-coalgebra. Triple (A,C,ψ) is an entwining structure if only if $A\otimes_R C$ is a weak A-coring.*

PROOF.

$(\Leftarrow)$ We have that $A\otimes_R C$ is a weak A- coring over weak comultiplication $\underline{\Delta}$ and weak counit $\underline{\varepsilon}$.

$$\underline{\Delta}: A\otimes_R C\xrightarrow{I_A\otimes\Delta}(A\otimes_R C)\otimes_A A\otimes_A(A\otimes_R C)\simeq(A\otimes_R C)\otimes_A(A\otimes_R C)\simeq(A\otimes_R C).1\otimes_R$$

$$a\otimes c \;\mapsto\; \sum(a\otimes c_{\underline{1}})\otimes_A(1\otimes c_{\underline{2}})\mapsto\sum(a\otimes c_{\underline{1}}).(1\otimes c_{\underline{2}}),$$

$$\underline{\varepsilon}: A\otimes_R C\to(A\otimes_R C).1\xrightarrow{I_A\otimes\varepsilon}A,$$

$$a\otimes c\;\mapsto\;(a\otimes c).1\;\;\mapsto\; a\varepsilon(c).$$

By right A-action of $A\otimes_R C,$ we defined a R-linear map

$$\psi: C\otimes_R A\to A\otimes_R C,\; c\otimes a\mapsto(1\otimes c).a$$

$\psi(c\otimes a)=\sum a_\psi c^\psi$, for $a_\psi\in A, c^\psi\in C$. We must show that (A,C,ψ) fulfil Definition 3.2. For 3.2 (1) analog with Theorem 4.2 (i). For any $a,b\in A$ and $c\in C$

(i). by weak comultiplication definition of $A\otimes_R C$

$$\begin{aligned}\underline{\Delta}(1\otimes c).a&=\underline{\Delta}\left(\sum a_\psi c^\psi\right)\\&=\sum a_\psi\underline{\Delta}(c^\psi)\\&=\sum(a_\psi\otimes c^\psi{}_{\underline{1}})\otimes(1\otimes c^\psi{}_{\underline{2}})\\&=\sum(a_\psi\otimes 1\otimes c^\psi{}_{\underline{1}})\otimes(1\otimes c^\psi{}_{\underline{2}})\\&=\sum a_\psi\otimes((1\otimes c^\psi{}_{\underline{1}}).1)\otimes c^\psi{}_{\underline{2}}\\&=\sum a_\psi\psi(c^\psi{}_{\underline{1}}\otimes 1)\otimes c^\psi{}_{\underline{2}}\end{aligned}$$

$$\begin{aligned}\underline{\Delta}((1\otimes c).a)&=\underline{\Delta}(1\otimes c).a\\&=\sum(1\otimes c_{\underline{1}})\otimes_A(1\otimes c_{\underline{2}}).a\\&=\sum 1\otimes c_{\underline{1}}\left(\sum(a_\psi 1)c_{\underline{2}}^\psi\right)\\&=\sum(1\otimes c_{\underline{1}}).a_\psi\otimes c_{\underline{2}}^\psi\\&=\sum a_{\psi\varphi}(1\otimes c_{\underline{1}})^\varphi\otimes c_{\underline{2}}^\psi\\&=\sum a_{\psi\varphi}\otimes c_{\underline{1}}^\varphi\otimes c_{\underline{2}}^\psi\end{aligned}$$

becomes $\sum a_\psi\psi(c^\psi{}_{\underline{1}}\otimes 1)\otimes c^\psi{}_{\underline{2}}=\sum a_{\psi\varphi}\otimes c_{\underline{1}}^\varphi\otimes c_{\underline{2}}^\psi$.

(ii). Morphism $\underline{\varepsilon}$ is an A-module homomorphism so that

$$\begin{aligned}\sum a_{\psi}\varepsilon\left(c^{\psi}\right) &= \left(I_A \otimes \varepsilon\right)\sum a_{\psi}c^{\psi} \\ &= \left(I_A \otimes \varepsilon\right)\circ\psi\left(c \otimes a\right) \\ &= \left(I_A \otimes \varepsilon\right)\left(1 \otimes c\right).a \\ &= \left(I_A \otimes \varepsilon\right)\left(1 \otimes c\right).1.a \\ &= \left(I_A \otimes \varepsilon\right)\left(\sum 1_{\psi}c^{\psi}\right).a \\ &= \sum \varepsilon\left(c^{\psi}\right)1_{\psi}a\end{aligned}$$

(iii). By weak counital we have

$$\begin{aligned}\sum 1_{\psi}c^{\psi} &= \left(1 \otimes c\right).1 \\ &= \left(\underline{\varepsilon} \otimes I_A\right)\circ\underline{\Delta}\left(1 \otimes c\right) \text{ (by weak counital)} \\ &= \left(\underline{\varepsilon} \otimes I_A\right)\left(\sum\left(1 \otimes c_{\underline{1}}\right).1 \otimes c_{\underline{2}}\right) \\ &= \left(\underline{\varepsilon} \otimes I_A\right)\left(\sum 1_{\psi}c_{\underline{1}}^{\psi} \otimes c_{\underline{2}}\right) \\ &= \sum 1_{\psi}\underline{\varepsilon}\left(c_{\underline{1}}^{\psi}\right) \otimes c_{\underline{2}} \\ &= \sum \underline{\varepsilon}\left(c_{\underline{1}}^{\psi}\right)1_{\psi} \otimes c_{\underline{2}}\end{aligned}$$

(i) – (iii) showed that (A, C, ψ) is a weak entwining structure. Conversely is analog with Theorem 4.2. □

References

[1] Brzeziński, T., Majid, Sh. (1998), *Coalgebra Bundles*, Comm. Math. Phys, 191 : 467-492.

[2] Brzeziński, T. (2001) *The cohomology structure of an algebra entwined with coalgebra*, Jurnal of Algebra 235 : 176-202.

[3] Brzeziński, T., Wisbauer, R. (2003) *Coring and comodules*, Germany.

[4] Brzeziński, T. *The Structures of Corings*, Alg. Rep Theory, to appear.

[5] Caenepeel, S., de Groot, E. (2000) Modules over weak entwining structures, Contemporary Mathematics 267 : 32-54.

[6] Hungerford, T.W. (1974) *Algebra, Graduate text in Mathematics*, Springer-Verlag, Berlin.

[7] Puspita, N. P. (2009) *Koring Lemah*, Thesis, Gadjah Mada University, Yogyakarta.

[8] Wisbauer, R. (1991) *Foundation of Module and Ring Theory*, Gordon and Breach Science Publishers, Germany.

[9] Wisbauer, R., (2001) *Weak Coring*, Jurnal of Algebra 245 : 123 – 160.

The Development of the Theory of Almost Distributive Lattices

G. C. Rao

Department of Mathematics, Andhra Universitry,
Visakhapatnam, Andhra Pradesh, INDIA
E-mail: gcraomaths@yahoo.co.in

The concept of an Almost Distributive Lattice (ADL) has come into being in 1981. The development of the theory of Almost Distributive Lattices since its inception is given. The concepts of $S-$ ideal in ADLs, normal ADLs, Topological representation of normal ADLs, conjunctively $S-$regular ADLs, $S-$normal ADLs, Heyting ADLs, and $L-$ADLs are studied.

Keywords: Almost Distributive Lattice (ADL), ideal, filter, Prime ideal, Prime filter, normal ADL, $S-$normal ADL, $S-$relatively normal ADL, uni subADL, $S-$ideal, $S-$filter, annihilator ideal, $\alpha-$ideal, Heyting ADL, $L-$ADL.

1. Introduction

There are several lattice structures that arise through generalizations of Boolean algebras. Since a Boolean algebra is a complemented distributive lattice, significant generalizations can be achieved either by relaxing the distributivity and retaining complementation (which is then no longer unique), or by retaining the distributivity and relaxing the complementation. These include the ring theoretic generalizations like regular rings, p-rings, biregular rings etc. on one hand and the lattice theoretic generalizations like distributive lattices, Heyting Algebras etc. on the other.

For this reason, in 1981, Swamy and Rao[38] introduced the concept of an Almost Distributive Lattice(ADL) as a common abstraction of almost all the existing ring theoretic generalizations of a Boolean algebra like $p-$rings(Mc Coy and Montgomery[9]), regular rings (Von-Neumann[46]), biregular rings (Arens and Kaplansky[1]), associate rings (Sussman[37]), p_1-rings (Subrahmanyam[34]), triple systems(Subrahmanyam[36]), Baer-ings (Meuborn[10]) and $m-$domain rings(Subrahmanyam[35]) on one hand and the class of distributive lattices on the other.

The main aim of this talk is to explain the development of the theory of

Almost Distributive Lattices since their inception in 1981. Most of the results are stated without proofs. For detailed proofs one can go through the references given.

2. Almost Distributive Lattice (ADL)

In [38], an Almost Distributive Lattice was originated as a bi-product while obtaining an algebraic characterization to the set of all global sections with compact supports of a sheaf of sets over a locally Boolean Space.

Definition 2.1. An Almost Distributive Lattice is an algebra $(R, \vee, \wedge, 0)$ or type $(2, 2, 0)$ satisfying the following identities.
1. $a \vee 0 = a$
2. $0 \wedge a = 0$
3. $(a \vee b) \wedge c = (a \wedge c) \vee (b \wedge c)$
4. $a \wedge (b \vee c) = (a \wedge b) \vee (a \wedge c)$
5. $a \vee (b \wedge c) = (a \vee b) \wedge (a \vee c)$
6. $(a \vee b) \wedge b = b$.

It can be proved that an ADL satisfies almost all the properties of of distributive lattice except possibly the commutativity of $\vee$, the commutativity of $\wedge$ and the right distributivity of $\vee$ over $\wedge$. Each of these implies all the other and make the structure a distributive lattice with zero.

Example 2.1. Let $(R, +, \cdot, 0, 1)$ be a commutative regular ring and if we define, for any $a, b \in R$,

$$a \vee b = a + b + a_0 b \quad \text{and} \quad a \wedge b = a_0 b,$$

where a_0 is the unique idempotent such that $aR = a_0R$, then the system $(R, \vee, \wedge, 0)$ is an ADL.

Example 2.2. Every non-empty set X can be regarded as an ADL as follows. Let $x_0 \in X$. Define the binary operations $\vee, \wedge$ on X by

$$x \vee y = \begin{cases} x & \text{if } x \neq x_0 \\ y & \text{if } x = x_0 \end{cases} \qquad x \wedge y = \begin{cases} y & \text{if } x \neq x_0 \\ x_0 & \text{if } x = x_0. \end{cases}$$

Then $(X, \vee, \wedge, x_0)$ is an ADL (where x_0 is the zero) and is called the discrete ADL.

From now on words R stands for an ADL $(R, \vee, \wedge, 0)$.

Definition 2.2. For any $a, b \in R$, we say that a less than or equal to b and write $a \leq b$ if $a = a \wedge b$ or equivalently, $a \vee b = b$. It can be easily seen that the above definition " $\leq$ " is a partial order on R.

Theorem 2.1. *Let $(R, \vee, \wedge, 0)$ be an ADL with 0. Then the following are equivalent:*
1. $(R, \vee, \wedge, 0)$ *is a distributive lattice with a smallest element* 0
2. *The poset* $(R, \leq)$ *is directed above*
3. $\vee$ *is commutative in* R
4. $\wedge$ *is commutative in* R
5. $\vee$ *is right distributive over* $\wedge$ *in* R
6. *The relation* $\theta = \{(a, b) \in R \times R \mid b \wedge a = a\}$ *is symmetric.*

Definition 2.3. A non-empty sub set I of an ADL R is called an ideal of R if $a \vee b \in I$ and $a \wedge x \in I$ for any $a, b \in I$ and $x \in R$. Also, a non-empty subset F of R is said to be a filter of R if $a \wedge b \in F$ and $x \vee a \in F$ for any $a, b \in F$ and $x \in R$.

The set $I(R)$ of all ideals of R is a bounded distributive lattice with least element $\{0\}$ and greatest element R under set inclusion in which, for any $I, J \in I(R)$, $I \cap J$ is the infimum of I and J while the supremum is given by $I \vee J := \{a \vee b \mid a \in I, b \in J\}$. A proper ideal P of R is called a prime ideal if, for any $x, y \in R$, $x \wedge y \in P \Rightarrow x \in P$ or $y \in P$. A proper ideal M of R is said to be maximal if it is not properly contained in any proper ideal of R. For any subset S of R the smallest ideal containing S is given by $(S] := \{(\bigvee_{i=1}^{n} s_i) \wedge x \mid s_i \in S, x \in R \text{ and } n \in N\}$. If $S = \{s\}$, we write $(s]$ instead of $(S]$ and it is called Principal ideal of R. Similarly, for any $S \subseteq R$, $[S) := \{x \vee (\bigwedge_{i=1}^{n} s_i) \mid s_i \in S, x \in R \text{ and } n \in N\}$ is the smallest filter of R containing S.. If $S = \{s\}$, we write $[s)$ instead of $[S)$.
The set $PI(R)$ of all principal ideals of R forms a distributive lattice under the operations $\vee$ and $\wedge$. This enables us to extend many concepts that exists in the class of distributive lattices to the class of ADLs.

3. Normal ADLs

W.H.Cornish [5]introduced the concept of normal lattices and relatively normal lattices and studied their properties. On these lines we defined, in [24], an ADL to be normal if its principal ideal lattice is a normal lattice. We characterized the normal ADLs in terms of its prime ideals, minimal prime ideals and annihilator ideals. We also defined the concept of sectionally normal and relatively normal ADLs and proved that every normal ADL is sectionally normal and conversely.

Definition 3.1. Let P be an ideal in R. Then define

$O(P) = \{x \in R \mid x \wedge a = 0, \text{ for some } a \in R \setminus P\}$

If P is a prime ideal then, from the definition, we get that $O(P) \subseteq P$. Also observe that $x \in O(P)$ if and only if $(x)^*$ is not contained in P where $x^* = \{a \in R \mid x \wedge a = 0\}$. In general $O(P)$ is not an ideal of R but we have the following.

Lemma 3.1. *For any prime ideal P of R, $O(P)$ is an ideal of R.*

Definition 3.2. An ADL R is said to be normal if the principal ideal lattice $PI(R)$ is a normal lattice.

The following theorem can be easily proved.

Theorem 3.1. *An ADL R is normal if and only if every prime ideal of R contains a unique minimal prime ideal of R.*

The following theorem gives equivalent conditions for an ADL to be normal.

Theorem 3.2. *Let $(R, \vee, \wedge, 0)$ be an ADL. Then the following conditions are equivalent:*
1. *Any two distinct minimal prime ideals are co-maximal.*
2. *R is normal.*
3. *For each prime ideal P of R, $O(P)$ is a prime ideal of R.*
4. *For any $x, y \in R$, $x \wedge y = 0 \Rightarrow (x)^*$ and $(y)^*$ are co-maximal.*
5. *For any $x, y \in R$, $(x \wedge y)^* = (x)^* \vee (y)^*$.*

For any $a, b \in R$ with $a < b$, the set $[a, b] = \{x \in R \mid a \leq x \leq b\}$ is said to be an interval in R. An ADL R is called a relatively complemented ADL if each interval in R is a complemented lattice.

Theorem 3.3. *Every relatively complemented ADL R is normal.*

Definition 3.3. (1) R is said to be sectionally normal if, for any $x > 0$ in R, each interval $[0, x]$ is a normal lattice.
(2) R is said to be relatively normal if, for any $x, y \in R$ with $x < y$, each interval $[x, y]$ is a normal lattice.

Now, we have the following.

Theorem 3.4. *Let $(R, \vee, \wedge, 0)$ be an ADL with 0. Then the following statements are equivalent:*
1. *R is normal.*
2. *Each ideal $J \neq R$ is normal as a subADL of R.*
3. *R is sectionally normal.*

4. $S-$ideals in ADLs

The concept of $S-$ideals in an ADL was introduced in [18] as follows.

Definition 4.1. Let S be a subADL of R. An ideal I of R is called an S-ideal of R if $I = (I \cap S]$, the ideal generated by $I \cap S$. An $S-$ideal I is called an $S-$prime ideal of R if $I \cap S$ is a prime ideal of S and $S-$maximal ideal if $I \cap S$ is a maximal ideal of S. It can be easily seen that every $S-$maximal ideal is an $S-$prime ideal.

The concept of $S-$ ideal in a distributive lattice was due to Cignoli [4].

Definition 4.2. If R has maximal elements, then a subADL S of R is said to be a uni subADL if it contains all maximal elements of R.

Theorem 4.1. *Let R be an ADL with maximal elements and S a uni subADL of R. Every proper $S-$ideal is an intersection of $S-$prime ideals.*

In the following theorem we characterize an $S-$maximal ideal.

Theorem 4.2. *Let R be an ADL with maximal elements, S a uni subADL of R and M an $S-$ideal of R. Then M is an $S-$maximal ideal of R if and only if M is maximal among all proper $S-$ideals of R.*

In order to find a necessary and sufficient condition for a prime ideal of R to become a $S-$prime ideal of R, we introduced the concept of conjunctively $S-$regular ADL as follows.

Definition 4.3. Let R be an ADL and S a subADL of R. Then R is called conjunctively $S-$regular, if $PI(R)$ is conjunctively $PI(S)-$regular. That is, for $x, y \in R$ and $a \in S$ such that $(x] \cap (y] \subseteq (a]$, there exist $(b], (c] \in PI(S)$ such that $(x] \subseteq (b]$, $(y] \subseteq (c]$ and $(b \wedge c] \subseteq (a]$.

The following theorem gives elementwise characterization of conjunctively $S-$regular ADL.

Theorem 4.3. *Let R be an ADL and S a subADL of R. Then R is conjunctively $S-$regular if and only if, for any $x, y \in R$ and $a \in S$ such that $a \wedge x \wedge y = x \wedge y$, there exist elements $b, c \in S$ such that $b \wedge x = x$, $c \wedge y = y$ and $a \wedge b \wedge c = b \wedge c$.*

As a generalization of the concept of normal ADL, we defined $S-$normal ADL as follows.

Definition 4.4. Let R be an ADL with maximal elements and S a uni subADL of R. R is called $S-$normal if $PI(R)$ is $PI(S)-$normal. That is, for $x, y \in R$ such that $(x] \cap (y] = (0]$, there exist $(a]$, $(b] \in PI(S)$ such that $(x] \cap (a] = (0]$, $(y] \cap (b] = (0]$ and $(a] \vee (b] = PI(R)$
A normal ADL is also an $S-$normal ADL where $S = R$.

In the following theorem, an $S-$normal ADL is characterized elementwise.

Theorem 4.4. *Let R be an ADL with maximal elements and S a uni subADL of R. Then R is $S-$normal if and only if, for any $x, y \in R$ such that $x \wedge y = 0$, there exist elements $a, b \in S$ such that $x \wedge a = 0 = y \wedge b$ and $a \vee b$ is a maximal element.*

Definition 4.5. Let R be an ADL with maximal elements. Then
$B = \{a \in R \mid$ there exists $b \in R$ such that $a \wedge b = 0$ and $a \vee b$ is maximal$\}$ is called the Birkhoff centre of R and $(B, \vee, \wedge)$ is a uni subADL of R which is also a relatively complemented ADL.

If S is the Birkhoff centre B of R, then we have the following characterization of a $B-$normal ADL.

Theorem 4.5. *Let R be an ADL with maximal elements and B the Birkhoff centre of R. Then the following conditions are equivalent:*
1. R is conjunctively $B-$regular
2. R is $B-$normal
3. Any $B-$maximal ideal of R is a prime ideal of R
4. Any $B-$maximal filter of R is contained in a unique maximal filter of R
5. Any $B-$maximal ideal of R is a minimal prime ideal of R.

5. Topological characterization of normal ADLs

Topological characterization of a normal ADL was given in [19]. We begin with the following. Let R be an ADL with at least two elements and $Spec_F R$ denote the set of all prime filters of R. For any $A \subseteq R$, let $H'(A) = \{P \in Spec_F R \mid A \nsubseteq P\}$ and for any $x \in R$, we write $H'(x) = H'(\{x\})$. Then we have the following.

Lemma 5.1. *For any $x, y \in R$, the following hold:*

1. $\bigcup_{x \in R} H'(x) = Spec_F R$
2. $H'(x) \cap H'(y) = H'(x \vee y)$
3. $H'(x) \cup H'(y) = H'(x \wedge y)$
4. $H'(x) = \emptyset$ *iff x is a maximal element.*

By the above lemma, we get that the collection $\{H'(x) \mid x \in R\}$ forms a base for a topology on $Spec_F R$. This topology is known as hull-kernel topology.
The following theorem gives topological characterization of a normal ADL.

Theorem 5.1. *Let R be an ADL with maximal elements in which every dense element is a maximal element. Then the following conditions are equivalent:*

1. *R is normal*
2. *$Max_F R$ is a Hausdorff space*
3. *Every prime filter is contained in a unique maximal filter.*

Theorem 5.2. *Let R be an ADL with maximal elements. Then the following conditions are equivalent:*

1. *$Spec_F R$ is a T_1-space*
2. *$Spec_F R = Max_F R$*
3. *Every prime ideal of R is a minimal prime ideal of R.*

Next theorem characterizes $B-$normal ADLs topologically.

Theorem 5.3. *Let R be an ADL with maximal elements in which every dense element is a maximal element and B the Birkhoff centre of R. Then the following conditions are equivqlent:*

1. *R is $B-$normal*
2. *$Max_F R$ is a Boolean space*
3. *$Max_F R$ homeomorphic to $Spec_F B$.*

6. $S-$relatively normal ADLs

In [17], we introduced the concept of $S-$relatively normal ADLs. First we give the following.

Definition 6.1. Let $x, y \in R$. Define $\lfloor x, y \rfloor_S = \{a \in S \mid y \wedge a \wedge x = a \wedge x\}$. We call $\lfloor x, y \rfloor_S$ an $S-$relative annihilator. It can be observed that $a \in \lfloor x, y \rfloor_S$ iff $y = y \vee (a \wedge x)$. Clearly $\lfloor x, y \rfloor_S$ is an ideal of S.

In the following theorem an $S-$normal ADL is charcterized in terms of $S-$relative annihilators of R.

Theorem 6.1. *Let R be an ADL with maximal elements and S a uni subADL of R. Then the following conditions are equivalent:*

(1) R is $S-$normal

(2) $\lfloor x, y \rfloor_S \vee \lfloor y, x \rfloor_S = S$, *for any* $x, y \in R$ *with* $x \wedge y = 0$

(3) For any prime filter F *of* S *and for any* $x, y \in R$ *with* $x \wedge y = 0$, *there exists* $a \in F$ *such that* $x \wedge a$ *and* $y \wedge a$ *are comparable.*

The following theorem gives a number of characterizations for an ADL to become an $S-$relatively normal ADL.

Theorem 6.2. *Let* R *be an ADL with maximal elements and* S *a uni subADL of* R. *Then the following conditions are equivalent:*

(1) R *is* $S-$*relatively normal*

(2) For each pair $x, y \in R$, *there is no proper ideal of* S *contain both* $\lfloor x, y \rfloor_S$ *and* $\lfloor y, x \rfloor_S$

(3) The set of all filters of R *that contain a given* $S-$*prime filter of* R *form a chain*

(4) The set of all prime filters of R *that contain a given* $S-$*prime filter of* R *form a chain*

(5) Any proper filter of R *that contain a given* $S-$*prime filter of* R *is prime.*

Since the lattice theoretic duality principle does not hold good in ADLs. We introduced the concept of dual $S-$relative normality in an ADL. First we give dual $S-$normality the following.

Definition 6.2. Let R be an ADL, S a uni subADL of R and $x, y \in R$. We define $\lceil x, y \rceil_S = \{a \in S \mid (x \vee a) \vee y = x \vee a\}$. We call $\lceil x, y \rceil_S$ an $S-$relative dual annihilator. It can be observed that $a \in \lceil x, y \rceil_S$ iff $y = (x \vee a) \wedge y$. Clearly $\lceil x, y \rceil_S$ is a filter of S.

Definition 6.3. Let R be an ADL with maximal elements and S a uni subADL of R. Then R is called dually $S-$normal if, for any $x, y \in R$ with $x \vee y$ is a maximal element in R, there exist $a, b \in R$ such that $x \vee a,\ y \vee b$ are maximal elements and $a \wedge b = 0$.

Theorem 6.3. *Let* R *be an ADL with maximal elements and* S *a uni subADL of* R. *Then the following are equivalent:*

(1) R *is dually* $S-$*normal*

(2) $\lceil x, y \rceil_S \vee \lceil y, x \rceil_S = S$, *for any* $x, y \in R$ *with* $x \vee y$ *is a maximal element.*

In case S is equal to the Birkhoff centre B of R, we have the following.

Theorem 6.4. *Let* R *be an ADL with maximal elements in which every dual dense element is zero element and* B *the Birkhoff centre of* R. *Then the following are equivalent:*

(1). R is dually B−normal
(2). The set MaxR of all maximal ideals of R is a Boolean space
(3). MaxR is homeomorphic to SpecB, the set of all prime ideals of B.

The concept of dually $S-$relative normal ADL is given in the following.

Definition 6.4. Let R be an ADL with maximal elements and S a uni subADL of R. R is called dually $S-$relatively normal if $PI(R)$ is dually $PI(S)-$completely normal lattice. That is, for any $x, y \in R$, there exist $a, b \in S$ such that $(y] \subseteq (x] \vee (a]$, $(x] \subseteq (y] \vee (b]$ and $(a] \cap (b] = (0]$.

The following theorem gives elementwise characterization of a dually $S-$relatively normal ADL.

Theorem 6.5. *Let R be an ADL with maximal elements and S a uni subADL of R. Then R is dually $S-$relatively normal if and only if for any $x, y \in R$, there exist $a, b \in R$ such that $(x \vee a) \vee y = x \vee a$, $(y \vee b) \vee x = y \vee b$ and $a \wedge b = 0$.*

Theorem 6.6. *Let R be an ADL with maximal elements and S a uni subADL of R. Then the following conditions are equivalent:*

(1) R is dually $S-$relatively normal
(2) For each pair $x, y \in R$, there is no proper filter of S containing both $\lceil x, y \rceil_S$ and $\lceil y, x \rceil_S$
(3) The set of all ideals of R that contain a given $S-$prime ideal of R form a chain
(4) The set of all prime ideals of R that contain a given $S-$prime ideal of R form a chain
(5) Any proper ideal of R that contain a given $S-$prime ideal of R is a prime.

Combining relative normality and dually relative normality of an ADL and give the following.

Definition 6.5. Let R be an ADL with maximal elements. Then R is called a linear ADL if R is both relatively normal and dually relatively normal. If S is a uni subADL of R, then R is called an $S-$linear ADL if R is both $S-$relatively normal and dually $S-$relatively normal.

In case S is the Birkhoff centre B of R, the two concepts $B-$relative normality and dually $B-$relative normality are equivalent as shown in the following.

Theorem 6.7. *Let R be an ADL with maximal elements and B the Birkhoff centre of R. Then the following conditions are equivalent:*

1. *R is $B-$relatively normal*
2. *R is dually $B-$relatively normal*
3. *Given $x, y \in R$, there is $a \in B$ and a complement a' of a such that $y \wedge a \wedge x = a \wedge x$ and $x \wedge a' \wedge y = a' \wedge y$*
4. *Given $x, y \in R$, there is $a \in B$ a complement a' of a such that $x \vee a \vee y = x \vee a$ and $y \vee a' \vee x = y \vee a'$*
5. *R is a linear ADL and the minimal prime ideals of R are $B-$maximal ideals of R*
6. *R is linear ADL and the minimal prime filters of R are $B-$maximal filters of R.*

7. Annihilator ideals and $\alpha-$ideals in ADLs

The concept of an annihilator ideal in an ADL was given in [27] as follows.

Definition 7.1. For any non-empty subset A of an ADL R with 0, define

$$A^* = \{x \in R / a \wedge x = 0, \text{ for all } a \in A\}$$

then A^* is an ideal of R it is called the annihilator of A. For any $a \in R$, we have $\{a\}^* = (a]^*$, where $(a]$ is the principal ideal generated by a.For any $\emptyset \neq A \subseteq R$, we have clearly $A \cap A^* = (0]$.

Now we prove the following.

Theorem 7.1. *Let R be an ADL with 0. Let $\mathcal{A}(R)$ be the set of all annihilator ideals of R. Then $(A(R), \underline{\vee}, \wedge)$ is a complete Boolean Algebra, in which for any $I, J \in A(R), I \underline{\vee} J = (I^* \cap J^*)^*$ and $I \wedge J = I \cap J$.*

The concept of an $\alpha-$ideal in an ADL was given in [26] as follows.

Definition 7.2. Let R be an ADL with 0. An ideal I of R is called an α-ideal if $\quad (x]^{**} \subseteq I$ for all $x \in I$.

Example 7.1. Let $A = \{0, a\}$ and $B = \{0, b_1, b_2\}$ be two discrete ADLs. Write $R = A \times B = \{(0,0), (0,b_1), (0,b_2), (a,0), (a,b_1), (a,b_2)\}$. Then $(R, \vee, \wedge, 0')$ is an ADL where $0' = (0,0)$, under point-wise operations.
Take $I = \{(0,0), (0,b_1), (0,b_2)\}$. Clearly I is an ideal of R.
Clearly $((0,0)]^{**} = \{(0,0)\} \subset I$. Also $((0,b_1)]^* = ((0,b_2)]^* = \{(0,0), (a,0)\}$. So $((0,b_1)]^{**} = ((0,b_2)]^{**} = ((0,0)]^* \cap ((a,0)]^* = R \cap \{(0,0), (0,b_1), (0,b_2)\} = I$. Thus I is an α-ideal of R.

In the following, we define an extension of an ideal I, which leads to a useful characterization of $\alpha-$ideals.

Definition 7.3. Let R be an ADL with 0. For any ideal I of R, define

$I^e = \{ x \in R \mid (a]^* \subseteq (x]^* \text{ for some } a \in I \}$. Then I^e is the smallest α-ideal of R containing I.

Lemma 7.1. *Let R be an ADL with 0. Then for any ideals I, J of R, we have the following :*

(1). $I \subseteq I^e$
(2). $I \subseteq J \Rightarrow I^e \subseteq J^e$
(3). $(I \cap J)^e = I^e \cap J^e$
(4). $I^e \vee J^e \subseteq (I \vee J)^e$
(5). $(I^e)^e = I^e$
(6). $(I^*)^e = I^*$.

If R is an ADL with 0, then the set $\mathcal{I}_\alpha(R)$ of all $\alpha-$ideals of R forms a distributive lattice.

Theorem 7.2. *Let R be an ADL with 0. Then for any ideal I of R, the following are equivalent:*

(1). *I is an α-ideal*
(2). $I = I^e$
(3). *for $x, y \in R, (x]^* = (y]^*$ and $x \in I$ imply $y \in I$*
(4). $I = \bigcup_{x \in I}(x]^{**}$
(5). *for $x, y \in R, h(x) = h(y)$ and $x \in I$ imply $y \in I$, where $h(x) = \{P \in \mathcal{P} / \ x \in P\}$, $\mathcal{P}$ is the set of all minimal prime ideals of R.*

In [40], we have defined the concept of a $*-$ADL as an ADL R in which, for any $x \in R$, $(x)^{**} = (x')^*$ for some $x' \in R$.

In the following theorem $*-$ADL is characterized in terms of $\alpha-$ideals of R.

Theorem 7.3. *Let R be an ADL with 0 in which the set D of dense elements of R is non-empty. Then the following conditions are equivalent:*

(1). *R is $\star$-ADL and $I^* \neq (0]$ for each proper α-ideal I of R*
(2). *$I \cap D \neq \emptyset$ for each $I \in \mathcal{DI}(R)$, the set of all dense ideals of R*
(3). *Every α-ideal is an annihilator ideal*
(4). *$\mathcal{I}_\alpha(R)$ is semicomplemented*
(5). *$\mathcal{I}_\alpha(R)$ has a unique dense element.*

8. Heyting ADLs

Heyting algebra is a relatively pseudo-complemented distributive lattice. It arises from non-classical logic and was first investigated by Skolem T. in 1920 [45]. It is named as Heyting algebra after the Dutch mathematician Arend Heyting. It was introduced by Birkhoff G. under a different name Brouwerian lattice and with a different notation. We begin with the following.

Definition 8.1. Let $(H, \vee, \wedge, 0, m)$ be an ADL with 0 and a maximal element m. Suppose $\longrightarrow$ is a binary operation on H satisfies the following conditions: for all $a, b.c \in H$,
(1). $a \longrightarrow a = m$
(2). $(a \longrightarrow b) \wedge b = b$
(3). $a \wedge (a \longrightarrow b) = a \wedge b \wedge m$
(4). $a \longrightarrow (b \wedge c) = (a \longrightarrow b) \wedge (a \longrightarrow c)$
(5). $(a \vee b) \longrightarrow c = (a \longrightarrow c) \wedge (b \longrightarrow c)$.
Then $(H, \vee, \wedge, 0, m)$ is called a Heyting Almost Distributive Lattice (HADL).

An HADL becomes a Heyting algebra once it becomes a lattice. Therefore we give a number of equivalent conditions for an HADL to become a Heyting algebra.

Theorem 8.1. *Let H be an HADL with 0 and a maximal element m. Then the following are equivalent:*
(1) . H is a Heyting algebra
(2). H is a distributive lattice
(3). The poset $(H, \leq)$ is directed above
(4). $\vee$ is commmutative
(5). $\wedge$ is commutative
(6). $\vee$ is right distributive over $\wedge$
(7). $\theta := \{(a, b) \in H \times H \mid b \wedge a = a\}$ is antisymmetric
(8). For any $a, b, c \in H$, $a \wedge c \leq b \Leftrightarrow c \leq a \longrightarrow b$
(9). $b \leq a \longrightarrow b$ for all $a, b \in H$
(10). $\theta_a := \{(x, y) \in H \times H \mid (x \longrightarrow y) \wedge (y \longrightarrow x) \geq a\}$ is a congruence relation on H, for all $a \in H$.

In the following we give another characterization of an HADL

Theorem 8.2. *Let $(H, \vee, \wedge, 0, m)$ be an HADL. Then the following are equivalent:*

(1). H is a Heyting algebra
(2). For any $a, b, c \in H$, $a \wedge c \leq b \Leftrightarrow c \leq a \longrightarrow b$
(3). $b \leq a \longrightarrow b$ for all $a, b \in H$.

Theorem 8.3. *Let H be an ADL with 0 and a maximal element m. Then the following are equivalent:*
(1). H is an HADL
(2). $[0, a]$ is a Heyting algebra for all $a \in H$
(3). $[0, m]$ is a Heyting algebra.

The following theorem gives justification for the definition of HADL.

Theorem 8.4. *Let H be an ADL with 0 and a maximal element m. Then H is an HADL if and only if $PI(H)$ is a Heyting algebra.*

The concept of $L-$ADL is given in [11] as a generalization of $L-$algebra as follows.

Definition 8.2. An HADL $(L, \vee, \wedge, \longrightarrow, 0, m)$ is called an $L-$Almost Distributive Lattice(LADL) if $[(a \longrightarrow b) \vee (b \longrightarrow a)] \wedge m = m$, for all $a, b \in L$

Now we give the following characterization of an ADL to become an LADL.

Theorem 8.5. *Let L be an ADL with 0 and a maximal element m, then the following are equivalent:*
(1) L is an LADL
(2) $[0, a]$ is an $L-$algebra for all $a \in L$
(3) $[0, m]$ is an $L-$algebra.

Theorem 8.6. *Let L be an HADL, then L is an LADL if and only if $PI(L)$ is an $L-$algebra.*

An LADL becomes an $L-$algebra once it becomes a lattice. Therefore, we give a number of equivalent conditions for an LADL to become an L-algebra in the following.

Theorem 8.7. *Let $(L, \vee, \wedge, \longrightarrow, 0, m)$ be an LADL. Then the following are equivalent:*
(1) L is an $L-$algebra
(2) L is a Heyting algebra
(3) L is a distributive lattice
(4) The poset $(L, \leq)$ is directed above
(5) $\vee$ is commutative

(6) $\wedge$ is commutative
(7) $\vee$ is right distributive over $\wedge$
(8) $\theta := \{(a,b) \in L \times L \mid b \wedge a = a\}$ is antisymmetric
(9) For $a, b, c \in L,\ a \wedge c \leq b \Leftrightarrow c \leq a \longrightarrow b$
(10) $b \leq a \longrightarrow b$ for all $a, b \in L$
(11) $\theta := \{(x,y) \in L \times L \mid (x \longrightarrow y) \wedge (y \longrightarrow x) \geq a\}$ is a congruence relation on L, for all $a \in L$.

The concept of an implicative filter in an Heyting ADL was given in [15] as follows.

Definition 8.3. Let $(H, \vee, \wedge, \longrightarrow, 0, m)$ be an HADL. A non-empty subset F of L is said to be an implicative filter if
(1). $a, b \in F \Rightarrow a \wedge b \in F$
(2). $a \in F,\ b \in L \Rightarrow b \longrightarrow a \in F$.

Theorem 8.8. *Let $(H, \vee, \wedge, \longrightarrow, 0, m)$ and $(H', \vee, \wedge, \longrightarrow, 0', m')$ be two HADLs and $\alpha : H \longrightarrow H'$ be a homomorphism from H into H'. Then we define the kernel of α as ker $\alpha = \{x \in H \mid \alpha(x) = m'\}$. Then $Ker\alpha$ is right filter as well as an implicative filter.*

9. Closed elements and dense elements in Heyting ADLs

The concept of a closed elements in a Heyting ADL was given in [12]as follows.

Definition 9.1. Let $(H, \vee, \wedge, \longrightarrow, 0, m)$ be an HADL. Define, for any $x \in H,\ x^* = x \longrightarrow 0$ and $H^* = \{x^* \mid x \in H\}$. Then the operation is a pseudo-complementation in H. The elements of H^* are called closed elements of H. Also, for $x, y \in H^*$ we define $x \underline{\vee} y = (x^* \wedge y^*)^*$.

we give the following theorem.

Theorem 9.1. *Let $(H, \vee, \wedge, \longrightarrow, 0, m)$ be an HADL. Then H^* is a bounded implicative semilattice.*

Dense elements in a Heyting ADL were extensively studied in [13]. First we give the definition of a dense element in a Heyting ADL H in the following.

Definition 9.2. Let $(H, \vee, \wedge, \longrightarrow, 0, m)$ be an HADL. Define $D_H = \{x \in H \mid x^* = 0\}$. Then the elements of D_H are called a dense elements in H. It can be proved that D_H is an implicative filter of H.

In the following we define the concept of an admissible map on HADLs.

Definition 9.3. Let $(H, \vee, \wedge, \longrightarrow, 0, m)$ be an HADL and $f : H^* \times H \longrightarrow H$ be a map such that:
(1). $f_a : H \longrightarrow H$ defined by $f_a(d) = f(a, d)$ is an epimorphiasm
(2). $f(a \wedge b, d) = f(a, f(b, d))$
(3). $a \leq b \Rightarrow f(b, d) \leq f(a, d)$
(4). $f(0, d) = m, f(m, d) = d$
for any $a, b \in H^*$. Then f is said to be an admissible map.

Now we give the following.

Theorem 9.2. *Let* $(L, \vee, \wedge, \longrightarrow, 0, m)$ *be an HADL and define, for any* $[a, d], [b, e] \in \frac{H^* \times H}{\theta}$, $[a, d] \overline{\wedge} [b, e] = [a \wedge b, d \wedge e]$ *and* $[a, d] \longleftarrow [b, e] = [a^* \underline{\vee} b, f(a, (d \longrightarrow e) \wedge m)]$. *Then* $(\frac{H^* \times H}{\theta}, \overline{\wedge}, \longleftarrow .[0, 0], [m, m])$ *is a bounded implicative semilattice.*

Theorem 9.3. *Let* $(L, \vee, \wedge, \longrightarrow, 0, m)$ *be an HADL, D be set of all dense elements of H and* $F = \{[a, b] \in \frac{H^* \times H}{\theta} \mid [a, d]^* = [0, 0]\}$, *then* $\gamma : D \longrightarrow F$, $\gamma(d) = [m, d]$ *is an epimorphism.*

We have further generalized the concept of an ADL and introduced the concept of generalized ADL and studied its properties extensively in [20 and 23].

References

1. Arens, R.F. and Kaplansky, I.: Topological representation of algebras, Trans. Amer.Math.Soc., 63(1948), 457-481.
2. Birkhoff, G.: Lattice Theorey, Vol.25,American Mathematical Society Colloquium Publication,1967.
3. Burris, S, Sankappanavar, H.P.: A course in universal algebra, Springer Verlag.1981.
4. Cignoli, R.: Boolean Multiplicative Closures. I & II. Proc. Japan Acad., 42, (1966), P. 1168-1174
5. Cornish,W.H.: Normal lattices, J.Aust.Math.Soc., Vol.14(1971), 200-215.
6. Frink,O: Pseudo-complements in semilattices, Duke Math. Journal,20, 505-515(1962).
7. Gratzer, G.: A generalization on Stone's representation theorem for Boolean algebra,Duke. Math.Journal 30,469-474((1963).

8. Gratzer, G: Lattice Theorey: First Concepts and Distributive Lattices, W.H.Freeman and Company, Sanfransisco (1971).
9. McCoy, N.H. and Mantgomery, D.: A representation of generalized Boolean rings, Duke. Math.J., 3(1937), 455-459.
10. Meuborn, A.A.: Regular rings and Baer rings, Math.Z., 121(1971), 211-219.
11. Rao.G.C, Berhanu Assaye and Prasad, R.: L-Almost Distributive Lattices, Accepted for publication in Asian-European Journal of Mathematics.
12. Rao.G.C, and Berhanu Assaye.: Closed elements in Heyting Almost Distributive Lattices, International Journal of Computational Cognition, Vol.9, No.1, March 2011.
13. Rao.G.C, and Berhanu Assaye.: Dense elements in Heyting Almost Distributive Lattices, International Journal of Computational Cognition, Vol.9, No.1, March 2011.
14. Rao.G.C, Berhanu Assaye and Ratna Mani, M.V.: Heyting Almost Distributive Lattices, International Journal of Computational Cognition, Vol.8, No.3, September 2010 (85-89).
15. Rao.G.C and Berhanu Assaye.: Implicative Filters in Almost Distributive Lattices, International Journal of Computational Cognition, Vol.8, No.4, December 2010, (55-59).
16. Rao.G.C, Rafi, N and Ravi Kumar Bandaru.: Closure Operators in Almost Distributive Lattices, International Mathematical Forum, 5, 2010, No.19, 929-935.
17. Rao.G.C, Rafi, N and Ravi Kumar Bandaru.: S-Linear Almost Distributive Lattices, European J. Pure and Applied Math., Vol. 3, 2010, No. 4, 704-716.
18. Rao.G.C, Rafi, N and Ravi Kumar Bandaru.: S-ideals in Almost Distributive Lattices, Accepted for publication in Southeast Asian Bulletin of Mathematics.
19. Rao.G.C, Rafi, N and Ravi Kumar Bandaru.: On Prime Filters in an Normal Almost Distributive Lattice, Accepted for publication in Southeast Asian Bulletin of Mathematics.
20. Rao.G.C, Ravi Kumar Bandaru and Rafi, N.: Generalized Almost Distributive Lattices-I, Southeast Asian Bulletin of Mathematics,Vol.33,2009, 1175-1188.
21. Rao.G.C, Ravi Kumar Bandaru and Rafi, N.: A characterization of Congruence Kernels in Pseudo-complemented Generalized Almost Distributive Lattices, International Journal of Algebra, Vol.3, 2009, No.16, 785-791.
22. Rao.G.C, Ravi Kumar Bandaru and Rafi, N.: Pseudo-complementation on Generalized Almost Distributive Lattices, Asian-European Journal of Mathematics, Vol. 3, No. 2 (2010) 335-346.
23. Rao.G.C, Ravi Kumar Bandaru and Rafi, N.: Generalized Almost Distributive Lattices-II, Accepted for publication in Southeast Asian Bulletin of Mathematics.
24. Rao.G.C, and Ravi Kumar, S.: Normal Almost Distributive Lattices, Southeast Asian Bulletin of Mathematics, Vol 32(5), 2008, 831- 841.
25. Rao.G.C, and Ravi Kumar, S.: Minimal prime ideals in almost distributive lattices , Int. Journal of Contemp. Math. Sciences, Vol. 4, 2009, no. 9-12,

475-484.
26. Rao.G.C, and Sambasiva Rao, M.: Alpha-ideals in Almost Distributive Lattices, Int. Journal of Contemp. Math. Sciences, Vol. 4, 2009, no. 9-12, 457-466.
27. Rao.G.C, and Sambasiva Rao, M.: Annihilator ideals in Almost Distributive Lattices, International Mathematical Forum, Vol 4, 2009, No. 15, 733-746.
28. Rao.G.C, and Sambasiva Rao, M.: Alpha-ideals and Prime ideals in Almost Distributive Lattices, International Journal of Algebra, Vol. 3, 2009, no. 5-8, 221-229.
29. Rao.G.C, and Sambasiva Rao, M.: Alpha-ideals in Generalized Stone Almost Distributive Lattices, Accepted for publication in Southeast Asian Bulletin of Mathematics.
30. Rao.G.C, and Sambasiva Rao, M.: Congruences and alpha-ideals in Almost Distributive Lattices, Accepted for publication in Southeast Asian Bulletin of Mathematics.
31. Rao.G.C, and Sambasiva Rao, M.: Annulets in Almost distributive lattices, European J. Pure and Applied Math., 2 (2009), no.1, 58-72.
32. Rao.G.C, and Sambasiva Rao, M.: Generalized Alpha-Stone Almost Distributive Lattices and Prime Alpha-ideals, International Journal of Computational Cognition,Vol.8, No.4, December 2010, (44-49) .
33. Rao.G.C, and Sambasiva Rao, M.: Prime Alpha-ideal characterization of sectionally *- Almost Distributive Lattices, International Journal of Computational Cognition, Vol.8, No.4, December 2010, (75-79).
34. Subrahmanyam, N.V.: Lattice theory for certain classes of rings, Math.Ann., 139(1960),275-286.
35. Subrahmanyam, N.V.: Structure theory for a generalized Boolean ring, Math.Ann., 141(1960),297-310.
36. Subrahmanyam, N.V.: An extension of Boolean lattice theory, Math.Ann., 151(1963),332-345.
37. Sussman,I.: A generalization for Boolean rings, Math.Ann., 136(1958),326-338.
38. Swamy, U.M. and Rao.G.C.: Almost Distributive Lattices, Journal of Australian Math.Soc,(Series A) 31, 77-91(1981).
39. Swamy, U.M., Rao.G.C., and Nanaji Rao.G.:Pseudo-Complementation on Almost Distributive Lattices, Southeast Asian Bulletin of Mathematics, 24:95-104(2000).
40. Swamy, U.M., Rao.G.C.:and Nanaji Rao.G.:Stone Almost Distributive Lattices, Southeast Asian Bulletin of Mathematics, 27: 513-526(2003).
41. Swamy, U.M., Rao.G.C.: and Nanaji Rao.G.:Dense Elements in Almost Distributive Lattices, Southeast Asian Bulletin of Mathematics, 27: 1081-1088(2004).
42. Swamy, U.M., Rao.G.C.: and Nanaji Rao.G.:On Characterizations of Stone Almost Distributive Lattices, Southeast Asian Bulletin of Mathematics, 32, 1167-1176(2008).
43. Swamy, U.M., Rao.G.C.: and Pragathi, Ch Boolean Centre of Almost Distributive Lattices, Southeast Asian Bulletin of mathematics, (2008), Vol 32: 985-994.

44. Swamy, U.M., Rama Rao.V.V.: Triple and sheaf representation of Stone Lattices, Algebra Universalis 5, 104-113 (1975).
45. Thoralf Skolem.: Logico-combinational investigations in the satisfiability or probability of mathematical propositions, Harvard University press, Cambridge, (1920).
46. Von Neumann, J.: On Regular Rings, Proc. Nat. Acad. Sci., 22 (1936), 707-713, U.S.A.

$S-$Relatively Normal Almost Distributive Lattices

G. C. Rao and N. Rafi

Department of Mathematics, Andhra University,
Visakhapatnam, Andhra Pradesh, INDIA
E-mail: gcraomaths@yahoo.co.in
E-mail: rafimaths@gmail.com

The concept of $S-$filters in an ADL is introduced and its properties are studied. It is proved that the set of all $S-$filters of an ADL R is isomorphic to the set of all filters of S, where S is a subADL of R. We characterized $S-$relatively normal ADL in terms of its $S-$relative annihilators.

Keywords: Almost Distributive Lattice (ADL), $S-$normal ADL, uni subADL, $S-$filter, Prime ideal, Prime filter, $S-$relative annihilators, $S-$relatively normal ADL.

1. Introduction

The concept of an Almost Distributive Lattice (ADL) was introduced by Swamy and Rao[13] as a common abstraction of the existing ring theoretic and lattice theoretic generalizations of a Boolean algebra. The concept of an ideal in an ADL was introduced in[13] analogous to that in a distributive lattice and it was observed that the set $PI(R)$ of all principal ideals of an ADL R forms a distributive lattice. This enables us to extend many existing concepts from the class of distributive lattices to the class of ADLs. In our paper,[11] we introduced the concept of an $S-$normal ADL R, where S is a uni subADL of R. The concept of $S-$filters in a distributive lattice was given by Cignoli.[3] In this paper, we extend the concept of an $S-$filters to an ADL and study their properties. We characterize $S-$normal ADL in terms of its $S-$prime filters. We define the concept of a $S-$relatively normal ADL and characterize it in terms of its $S-$relative annihilators. Also we give another characterization in terms of a congurences relations in R.

2. Preliminaries

Definition 2.1 ([13]). *An Almost Distributive Lattice with zero or simply ADL is an algebra* $(R, \vee, \wedge, 0)$ *of type* $(2, 2, 0)$ *satisfying*

1. $(x \vee y) \wedge z = (x \wedge z) \vee (y \wedge z)$
2. $x \wedge (y \vee z) = (x \wedge y) \vee (x \wedge z)$
3. $(x \vee y) \wedge y = y$
4. $(x \vee y) \wedge x = x$
5. $x \vee (x \wedge y) = x$
6. $0 \wedge x = 0$
7. $x \vee 0 = x$.

Every non-empty set X can be regarded as an ADL as follows. Let $x_0 \in X$. Define the binary operations $\vee, \wedge$ on X by

$$x \vee y = \begin{cases} x & \text{if } x \neq x_0 \\ y & \text{if } x = x_0 \end{cases} \qquad x \wedge y = \begin{cases} y & \text{if } x \neq x_0 \\ x_0 & \text{if } x = x_0. \end{cases}$$

Then $(X, \vee, \wedge, x_0)$ is an ADL (where x_0 is the zero) and is called a discrete ADL. If $(R, \vee, \wedge, 0)$ is an ADL, for any $a, b \in R$, define $a \leq b$ if and only if $a = a \wedge b$ (or equivalently, $a \vee b = b$), then $\leq$ is a partial ordering on R.

Theorem 2.1 ([13]). *If* $(R, \vee, \wedge, 0)$ *is an ADL, then for any* $a, b, c \in R$, *we have the following:*

(1). $a \vee b = a \Leftrightarrow a \wedge b = b$
(2). $a \vee b = b \Leftrightarrow a \wedge b = a$
(3). $\wedge$ *is associative in* R
(4). $a \wedge b \wedge c = b \wedge a \wedge c$
(5). $(a \vee b) \wedge c = (b \vee a) \wedge c$
(6). $a \wedge b = 0 \Leftrightarrow b \wedge a = 0$
(7). $a \vee (b \wedge c) = (a \vee b) \wedge (a \vee c)$
(8). $a \wedge (a \vee b) = a$, $\quad (a \wedge b) \vee b = b$ *and* $a \vee (b \wedge a) = a$
(9). $a \leq a \vee b$ *and* $a \wedge b \leq b$
(10). $a \wedge a = a$ *and* $a \vee a = a$
(11). $0 \vee a = a$ *and* $a \wedge 0 = 0$
(12). *If* $a \leq c$, $\ b \leq c$ *then* $a \wedge b = b \wedge a$ *and* $a \vee b = b \vee a$
(13). $a \vee b = (a \vee b) \vee a$.

It can be observed that an ADL R satisfies almost all the properties of a distributive lattice except the right distributivity of $\vee$ over $\wedge$, commutativity of $\vee$, commutativity of $\wedge$. Any one of these properties make an ADL

R a distributive lattice as given in the following:

Theorem 2.2 ([13]). *Let $(R, \vee, \wedge, 0)$ be an ADL with 0. Then the following are equivalent:*

1). $(R, \vee, \wedge, 0)$ *is a distributive lattice with* 0
2). $a \vee b = b \vee a$, *for all* $a, b \in R$
3). $a \wedge b = b \wedge a$, *for all* $a, b \in R$
4). $(a \wedge b) \vee c = (a \vee c) \wedge (b \vee c)$, *for all* $a, b, c \in R$.

As usual, an element $m \in R$ is called maximal if it is a maximal element in the partially ordered set $(R, \leq)$. That is, for any $a \in R$, $m \leq a \Rightarrow m = a$.

Theorem 2.3 ([13]). *Let R be an ADL and $m \in R$. Then the following are equivalent:*

1). m *is maximal with respect to* $\leq$
2). $m \vee a = m$, *for all* $a \in R$
3). $m \wedge a = a$, *for all* $a \in R$
4). $a \vee m$ *is maximal, for all* $a \in R$.

As in distributive lattices,[1] a non-empty sub set I of an ADL R is called an ideal of R if $a \vee b \in I$ and $a \wedge x \in I$ for any $a, b \in I$ and $x \in R$. Also, a non-empty subset F of R is said to be a filter of R if $a \wedge b \in F$ and $x \vee a \in F$ for any $a, b \in F$ and $x \in R$.

The set $I(R)$ of all ideals of R is a bounded distributive lattice with least element $\{0\}$ and greatest element R under set inclusion in which, for any $I, J \in I(R)$, $I \cap J$ is the infimum of I and J while the supremum is given by $I \vee J := \{a \vee b \mid a \in I, b \in J\}$. A proper ideal P of R is called a prime ideal if, for any $x, y \in R$, $x \wedge y \in P \Rightarrow x \in P$ or $y \in P$. A proper ideal M of R is said to be maximal if it is not properly contained in any proper ideal of R. It can be observed that every maximal ideal of R is a prime ideal. Every proper ideal of R is contained in a maximal ideal. For any subset S of R the smallest ideal containing S is given by $(S] := \{(\bigvee_{i=1}^{n} s_i) \wedge x \mid s_i \in S, x \in R \text{ and } n \in N\}$. If $S = \{s\}$, we write $(s]$ instead of $(S]$. Similarly, for any $S \subseteq R$, $[S) := \{x \vee (\bigwedge_{i=1}^{n} s_i) \mid s_i \in S, x \in R \text{ and } n \in N\}$ is the smallest filter of R containing S. If $S = \{s\}$, we write $[s)$ instead of $[S)$.

Theorem 2.4 ([13]). *For any x, y in R the following are equivalent:*

1). $(x] \subseteq (y]$
2). $y \wedge x = x$
3). $y \vee x = y$
4). $[y) \subseteq [x)$.

For any $x, y \in R$, it can be verified that $(x] \vee (y] = (x \vee y]$ and $(x] \wedge (y] = (x \wedge y]$. Hence the set $PI(R)$ of all principal ideals of R is a sublattice of the distributive lattice $I(R)$ of ideals of R. This enables us to extend many concepts from distributive lattices to the class of ADLs.

3. $S-$filters in ADLs

The concept $S-$filters in a distributive lattice was given by Cignoli.[3] In this section, we define the concept of an $S-$filter in an ADL R. A subADL of an ADL with 0 carries the usual meaning where 0 is treated as a nullary operation and it is called a uni subADL if it contains all maximal elements of R. If S is a subADL of R, then $F \cap S$ is a filter of S for any filter F of R. Through out this paper R represents an ADL and S stands for a subADL of R.

We begin with the following.

Definition 3.1. Let R be an ADL with maximal elements and S a uni subADL of R. A filter F of R is called an $S-$filter of R if F is generated by $F \cap S$. An $S-$filter F is called an $S-$prime($S-$maximal) filter of R if $F \cap S$ is a prime(maximal) filter of S. It can be observed that every $S-$maximal filter of R is an $S-$prime filter.

Example 3.1. Let $A = \{0, a\}$ and $B = \{0, b_1, b_2\}$ be two discrete ADLs. Then $R = A \times B = \{(0,0), (a,0), (0,b_1), (0,b_2), (a,b_1), (a,b_2)\}$ is an ADL under point-wise operations. Take $F = \{(0,b_1), (0,b_2), (a,b_1), (a,b_2)\}$ and $S = \{(0,0), (0,b_1), (a,b_1)\}$. Clearly, F is a filter of R and S is a subADL of R. Now $F \cap S = \{(0,b_1), (a,b_1)\}$, $(0,b_2) \vee (0,b_1) = (0,b_2)$ and $(a,b_2) \vee (a,b_1) = (a,b_2)$, we get that F is an $S-$filter of R. Also, it can be observed that F is an $S-$maximal filter of R.

Now, the following result can be verified easily.

Theorem 3.1. *A filter F of R is an $S-$filter of R if and only if for any $x \in F$, there exists $s \in F \cap S$ such that $x \vee s = x$.*

The set $F(R)$ of all filters of an ADL R is a complete distributive lattice, in which the glb and lub of any family $\{F_\alpha \mid \alpha \in \Delta\}$ are given by $\bigcap_{\alpha\in\Delta} F_\alpha$ and $\bigvee_{\alpha\in\Delta} F_\alpha = \{x \vee (\bigwedge_{i=1}^{n} s_i) \mid s_i \in \bigcup_{\alpha\in\Delta} F_\alpha \text{ for } 1 \leq i \leq n,\ n \in N \text{ and } x \in R\}$ respectively.[7] Regarding $S-$filters of R we have the following results.

Theorem 3.2. *If F_1, F_2 are two $S-$filters of R, then $F_1 \cap F_2$ is an $S-$filter of R.*

Proof. Let $x \in F_1 \cap F_2$. That implies $x \in F_1$ and $x \in F_2$. Since F_1, F_2 are S-filters of R, then there exist $s_1 \in F_1 \cap S$ and $s_2 \in F_2 \cap S$ such that $x \vee s_1 = x$ and $x \vee s_2 = x$. That implies $s_1 \vee s_2 \in (F_1 \cap F_2) \cap S$. Now $x \vee (s_1 \vee s_2) = x$. Therefore $F_1 \cap F_2$ is an $S-$filter of R. □

Theorem 3.3. *If $\{F_\alpha\}_{\alpha\in\Delta}$ is a family of $S-$filters of R, then $\bigvee_{\alpha\in\Delta} F_\alpha$ is also an $S-$filter of R.*

Proof. Let $\{F_\alpha\}_{\alpha\in\Delta}$ be a family of $S-$filters of R. Then $\bigvee_{\alpha\in\Delta} F_\alpha$ is a filter of R . Let $t \in \bigvee_{\alpha\in\Delta} F_\alpha$. Then $t = x \vee (\bigwedge_{i=1}^{n} t_i)$, where $t_i \in F_{\alpha_i}$ for $1 \leq i \leq n$ and $x \in R$. Since each F_{α_i} is an $S-$filter of R, there exists $s_i \in F_{\alpha_i} \cap S$ such that $t_i \vee s_i = t_i$. Write $s = \bigwedge_{i=1}^{n} s_i$, then $s \in (\bigvee F_\alpha) \cap S$.

Now,

$$
\begin{aligned}
t \wedge s &= (x \vee (\bigwedge_{i=1}^{n} t_i)) \wedge (\bigwedge_{i=1}^{n} s_i) \\
&= \bigwedge_{i=1}^{n} (t_i \wedge s_i) \\
&= (x \vee (\bigwedge_{i=1}^{n} t_i)) \wedge (\bigwedge_{i=1}^{n} t_i) \wedge (\bigwedge_{i=1}^{n} s_i) \\
&= (\bigwedge_{i=1}^{n} t_i) \wedge (\bigwedge_{i=1}^{n} s_i) \\
&= (\bigwedge_{i=1}^{n} s_i) = s.
\end{aligned}
$$

Hence $t \vee s = t$. Therefore $\bigvee_{\alpha\in\Delta} F_\alpha$ is an $S-$filter of R. □

Lemma 3.1. *If $y \in S$ then the principal filter $(y]$ in R generated by y is an $S-$filter of R.*

Proof. Let $t \in [y)$. Then $y \in S \cap [y)$ and $t = t \vee y$. Therefore $[y)$ is an $S-$filter of R. □

If F is a filter of S, then $F^e = \{a \vee x \mid a \in R \text{ and } x \in F\}$ is a filter of R. Now we prove the following lemma.

Lemma 3.2. *Let R be an ADL and S a subADL of R. If F is a filter of S, then $F = F^e \cap S$.*

Proof. Let $x \in F^e \cap S$. Then $x = t \vee a$, for some $t \in R$ and $a \in F$. Now $x \wedge a = (t \vee a) \wedge a = a$. Then $x = x \vee a$. Since F is filter of S, we get that $x \in F$. Therefore $F^e \cap S \subseteq F$. Clearly $F \subseteq F^e \cap S$. Hence $F = F^e \cap S$. □

Theorem 3.4. *Let R be an ADL and S a subADL of R. If F is a filter of S, then F^e is an $S-$filter of R.*

Proof. Let $t \in F^e$. Then $t = a \vee x$ for some $a \in R$ and $x \in F$. Since $x \in F$, we have $x \in F^e \cap S$ and $t \vee x = (a \vee x) \vee x = a \vee x = t$. Therefore F^e is an $S-$filter of R. □

Lemma 3.3. *Let R be an ADL and S a subADL of R. If F is an $S-$filter of R, then $(F \cap S)^e = F$.*

Proof. Since $F \cap S \subseteq F$ and F is a filter of R, we get that $(F \cap S)^e \subseteq F$. Now, let $x \in F$. Then there exists $s \in F \cap S$ such that $x \vee s = x$. Now, $s \in F \cap S$ and $x \in R$, implies that $x \vee s \in (F \cap S)^e$. Therefore $x \in (F \cap S)^e$ and hence $F \subseteq (F \cap S)^e$. Thus $(F \cap S)^e = F$. □

We denote the set of all $S-$filters of R by $SF(R)$ and the set of all filters of S by $F(S.)$ In the above lemma, we have proved that every filter F of S gives rise to an $S-$filter F^e of R. In fact this correspondence is an isomorphism as shown in the following theorem.

Theorem 3.5. *Let R be an ADL and S a subADL of R. Then the mapping $f : F(S) \longrightarrow SF(R)$ defined by $f(F) = F^e$ is a lattice isomorphism.*

Proof. By the theorem 3.4, for each $F \in F(S)$, $F^e \in SF(R)$. Then f is well defined. Let $F_1, F_2 \in F(S)$. If $F_1 \subseteq F_2$, then clearly $F_1^e \subseteq F_2^e$. Now suppose $F_1^e \subseteq F_2^e$. Then $F_1^e \cap S \subseteq F_2^e \cap S$. So that, by lemma 3.2 , $F_1 \subseteq F_2$. By lemma 3.3, we get that f is onto. □

The following corollary follows from the above theorem.

Corollary 3.1. *$SF(R)$ is a complete sublattice of $F(R)$.*

Now, we prove the following

Lemma 3.4. *Let R an ADL with maximal elements, S a uni subADL of R and P a prime filter of R. If F is an $S-$filter of R contained in P, then there exists an $S-$prime filter $\overline{P}$ of R such that $F \subseteq \overline{P} \subseteq P$.*

Proof. Since P is a prime filter of R and S is a uni subADL of R, then $P \cap S$ is a prime filter of S. Therefore, by theorem 3.4, $(P \cap S)^e$ is an $S-$filter of R and $(P \cap S)^e \cap S = P \cap S$. Thus we get that $(P \cap S)^e \cap S$ is a prime filter of S. Therefore $(P \cap S)^e$ is an $S-$prime filter of R. Now, we prove that $F \subseteq (P \cap S)^e \subseteq P$. Since $F \subseteq P$, we get that $F \cap S \subseteq P \cap S$ and hence $(F \cap S)^e \subseteq (P \cap S)^e$. Therefore $F \subseteq (P \cap S)^e$. Clearly, $P \cap S \subseteq P$ and hence $(P \cap S)^e \subseteq P$. Then $F \subseteq (P \cap S)^e \subseteq P$. □

Theorem 3.6. *Let R be an ADL with maximal elements, S a uni subADL of R, F an $S-$filter of R and K a non-empty subset of R, which is closed under the operation join such that $F \cap K = \emptyset$. Then there exists an $S-$prime filter P of R such that $F \subseteq P$ and $P \cap K = \emptyset$.*

Proof. Write $\mathfrak{F} = \{J \mid J \text{ is an } S-\text{ filter of } R,\ F \subseteq J \text{ and } J \cap K = \emptyset\}$. Clearly $\mathfrak{F} \neq \emptyset$ since $F \in \mathfrak{F}$ and $(\mathfrak{F}, \subseteq)$ is a poset in which for every chain $\{I_\alpha\}$, $\bigcup I_\alpha$ is an upper bound. By the Zorn's lemma, $\mathfrak{F}$ has a maximal element P (say). We prove that $P \cap S$ is a prime filter of S. Let $a, b \in S$ such that $a \vee b \in P \cap S$. Suppose $a \notin P \cap S$ and $b \notin P \cap S$. Then $a \notin P$ and $b \notin P$. So that $P \vee [a)$ and $P \vee [b)$ are $S-$filters of R contains P properly. By the maximality of P, we get that $(P \vee [a)) \cap K \neq \emptyset$ and $(P \vee [b)) \cap K \neq \emptyset$. Choose $x \in (P \vee [a)) \cap K$ and $y \in (P \vee [b)) \cap K$. Now $x \vee y \in ((P \vee [a)) \cap (P \vee [b))) \cap K = (P \vee [a \vee b)) \cap K$. Hence $(P \vee [a \vee b)) \cap K \neq \emptyset$. This implies that $a \vee b \notin P$, which is a contradiction. Therefore either $a \in P \cap S$ or $b \in P \cap S$. Hence P is an $S-$prime filter of R. □

Corollary 3.2. *Let R be an ADL with maximal elements, S a uni subADL of R, F an $S-$filter of R and $a \in R$ such that $a \notin F$. Then there exists an $S-$prime filter P of R such that $F \subseteq P$ and $a \notin P$.*

Proof. Follows from above theorem by taking $K = \{a\}$. □

Theorem 3.7. *Let R be an ADL with maximal elements, S a uni subADL of R. Then every proper $S-$filter of R is the intersection of all S-prime filters of R containing it.*

Proof. Let F be a proper $S-$filter of R.
Write $J = \bigcap\{P \mid P \text{ is an } S - \text{prime filter and } F \subseteq P\}$. Suppose $x \notin F$. Then, by above theorem, there exists an $S-$prime filter P of R such that $F \subseteq P$ and $x \notin P$. Thus $x \notin J$ and hence $J \subseteq F$. Clearly $F \subseteq J$. Therefore $F = J$. □

Corollary 3.3. *Let R be an ADL with maximal elements, S a uni subADL of R. Then intersection of all $S-$prime filters of R is equal to the set of all maximal elements of R.*

Corollary 3.4. *Let R be an ADL with maximal elements, S a uni subADL of R and x a non-maximal element in R. Then there exists an $S-$prime filter P of R such that $x \notin P$.*

Corollary 3.5. *Let R be an ADL with maximal elements, S a uni subADL of R, I an $S-$ideal of R and F an $S-$filter of R such that $I \cap F = \emptyset$. Then there exists an $S-$prime ideal(filter) P of R such that $I \subseteq P$ and $P \cap F = \emptyset$ ($F \subseteq P$ and $P \cap I = \emptyset$).*

Corollary 3.6. *Let R be an ADL with maximal elements, S a uni subADL of R. If $s \in S$, $x \in R$ such that $x \vee s \neq x$, then there exists an $S-$prime filter P of R such that $[s) \subseteq P$ and $x \notin P$.*

Proof. Since $s \in S$ and by lemma 3.1, we get that $[s)$ is an $S-$filter and $x \notin [s)$. Hence by the theorem 3.7, there exists an $S-$prime filter P of R such that $[s) \subseteq P$ and $x \notin P$. □

Theorem 3.8. *Let R be an ADL with maximal elements, S a uni subADL of R and M an $S-$filter of R. Then M is an $S-$maximal filter of R if and only if M is maximal among all proper $S-$filters of R.*

Proof. Assume that M is an $S-$maximal filter of R. Clearly, M is a proper $S-$filter of R. Let N be any proper $S-$filter of R such that $M \subseteq N$. Since M is an $S-$maximal filter, we get that $M \cap S$ is a maximal filter of S. Now $M \cap S \subseteq N \cap S$ implies that $M \cap S = N \cap S$. Hence $M = (M \cap S)^e = (N \cap S)^e = N$ (by lemma 3.3). Therefore M is maximal among all proper $S-$filters of R. Conversely, assume that M is maximal among all proper $S-$filters of R. Clearly, $M \cap S$ is a filter of S and $0 \notin M \cap S$. Hence $M \cap S$ is a proper filter of S. Let N be any proper filter of S such that $M \cap S \subseteq N$. Then $(M \cap S)^e \subseteq N^e$. That is $M \subseteq N^e$ and N^e is a proper $S-$filter of R. Then $M = N^e$. So that $M \cap S = N^e \cap S = N$. □

Definition 3.2.[11] Let R be an ADL with maximal elements and S a uni subADL of R. R is called $S-$normal if for any $x, y \in R$ such that $x \wedge y = 0$ then there exist elements $a, b \in S$ such that $x \wedge a = 0 = y \wedge b$ and $a \vee b$ is a maximal element.

Now, we have the following result.

Theorem 3.9. *If R is an ADL with maximal elements and S is a uni subADL of R, then R is $S-$normal if and only if any $S-$prime filter of R contained in a unique maximal filter of R.*

4. $S-$relatively Normal ADLs

In this section, we have studied about $S-$normality in an ADL R, where S is a subADL of R with 0. In,[8] Ravi Kumar has characterized Relative Normal Almost Distributive Lattice in terms of its relative annihilators. R. Cignoli gave the concept of $S-$completely normal lattice.[4] In this section, we introduce the concept of the $S-$relative normality in an ADL R through its principal ideal lattice $PI(R)$, define $S-$relative annihilator and characterize $S-$relative normal Almost Distributive Lattice in terms of its $S-$relative annihilators and $S-$prime filters.
We begin with the following.

Definition 4.1. Let R be an ADL, S a subADL of R and $x, y \in R$. Then we define $\lfloor x, y \rfloor_S = \{a \in S \mid y \wedge a \wedge x = a \wedge x\}$.

We observe that $a \in \lfloor x, y \rfloor_S$ iff $y = y \vee (a \wedge x)$.

Lemma 4.1. *For any $x, y \in R$, $\lfloor x, y \rfloor_S$ is an ideal of S.*

Definition 4.2. For any $x, y \in R$, the ideal $\lfloor x, y \rfloor_S$ is called an $S-$relative annihilator of x with respect to y.

Now, the following result can be verified easily.

Lemma 4.2. *Let $x, y \in R$. Then for any $a \in S$, $a \in \lfloor x, y \rfloor_S$ iff $x \wedge a \leq y \wedge a$.*

Lemma 4.3. *Let F be any prime filter of S. For any $x, y \in R$, if $y \in F \vee [x)$, then $F \cap \lfloor x, y \rfloor_S$ is non-empty.*

In the following we prove some important properties of $S-$relative annihilators that will be useful for proving results in this section.

Lemma 4.4. *Let R be an ADL with maximal elements and S a uni subADL of R. For any $a, b, c \in R$, we have the following:*

(1) $s \in \lfloor a, b \rfloor_S \;\Rightarrow\; \lfloor a, s \rfloor_S \subseteq \lfloor a, b \rfloor_S$
(2) *If* $a \leq b$, *then for any* $c \in R$, $\lfloor b, c \rfloor_S \subseteq \lfloor a, c \rfloor_S$ *and* $\lfloor c, a \rfloor_S \subseteq \lfloor c, b \rfloor_S$
(3) $\lfloor a, b \rfloor_S = \lfloor a, a \wedge b \rfloor_S = \lfloor a, b \wedge a \rfloor_S = \lfloor a \vee b, a \rfloor_S = \lfloor b \vee a, a \rfloor_S = \lfloor a \vee b, a \wedge b \rfloor_S$
(4) $S = \lfloor 0, a \rfloor_S = \lfloor a, a \rfloor_S = \lfloor a, a \vee b \rfloor_S = \lfloor a, b \vee a \rfloor_S = \lfloor a \wedge b, a \rfloor_S = \lfloor b \wedge a, a \rfloor_S = \lfloor a \wedge b, a \vee b \rfloor_S$
(5) *For any* $a, b, c \in R$, $\lfloor a \vee b, c \rfloor_S = \lfloor b \vee a, c \rfloor_S$, $\lfloor a \wedge b, c \rfloor_S = \lfloor b \wedge a, c \rfloor_S$, $\lfloor c, a \wedge b \rfloor_S = \lfloor c, b \wedge a \rfloor_S$ *and* $\lfloor c, a \vee b \rfloor_S = \lfloor c, b \vee a \rfloor_S$
(6) *For any* $a, b, c \in R$,
  (i). $\lfloor a, c \rfloor_S \vee \lfloor b, c \rfloor_S \subseteq \lfloor a \wedge b, c \rfloor_S$
  (ii). $\lfloor a, b \rfloor_S \vee \lfloor a, c \rfloor_S \subseteq \lfloor a, b \vee c \rfloor_S$
  (iii). $\lfloor a \vee b, c \rfloor_S = \lfloor a, c \rfloor_S \cap \lfloor b, c \rfloor_S$
  (iv). $\lfloor a, b \wedge c \rfloor_S = \lfloor a, b \rfloor_S \cap \lfloor a, c \rfloor_S$.

Definition 4.3.[4] Let L be a bounded distributive lattice and S a sublattice of L containing 0 and 1. L is called $S-$completely normal if for any $x, y \in L$, there exist $s, t \in S$ such that $x \wedge s \leq y$, $y \wedge t \leq x$ and $s \vee t = 1$.

Now we define the concept of an $S-$relatively normal ADL in the following.

Definition 4.4. Let R be an ADL with maximal elements and S a uni subADL of R. R is called $S-$relatively normal if the lattice $PI(R)$ is a $PI(S)-$completely normal lattice.

Theorem 4.1. *Let R be an ADL with maximal elements and S a uni subADL of R. Then R is $S-$relatively normal if and only for any $x, y \in R$, there exist $a, b \in S$ such that $y \wedge a \wedge x = a \wedge x$, $x \wedge b \wedge y = b \wedge y$ and $a \vee b$ is a maximal element.*

Corollary 4.1. *Let R be an ADL with maximal elements and S a uni subADL of R. Then R is $S-$relatively normal if and only if for each pair $x, y \in R$, $\lfloor x, y \rfloor_S \vee \lfloor y, x \rfloor_S = S$.*

Let R be an ADL and F a filter in R. Then the relation $\Psi(F) = \{(x, y) \in R \times R \mid x \wedge t = y \wedge t$, for some $t \in F\}$ is a congruence relation on R. Then the set $R/\Psi(F) = \{x/\Psi(F) \mid x \in R\}$ is an ADL. Let $\prod$ be the natural homomorphism from R onto $R/\psi(F)$ defined by $\prod(x) = x/\psi(F)$, for all $x \in R$.

Theorem 4.2. *Let R be an ADL with maximal elements and S a uni subADL of R. Then R is $S-$relatively normal if and only if $R/\psi(F)$ is* *chain, for each prime filter F of S.*

The following result follows directly from the above theorem.

Theorem 4.3. *Each $S-$relatively normal ADL is a subdirect product of the bounded chains R/P, where P runs through the set of the prime ideals of S.*

References

1. Birkhoff, G.: Lattice Theory. *Amer. Math. Soc. Colloq. Publ. XXV, Providence* (1967), U.S.A.
2. Cignoli, R.: Boolean Multiplicative Closures. I & II. *Proc. Japan Acad.*, **42** (1966), P. 1168-1174
3. Cignoli, R.: Stone filters and stone ideals in distributive lattices . *Bull. Math. soc. Sci. Math. R. S. Roumanie.*, **15** (1971), P. 131-137.
4. Cignoli, R.: The lattice of global sections of sheaves of chains over Boolean spaces. *Algebra Univesalis*, **8** (1978), 357-373.
5. Cornish, W.H. : Normal Lattices. *J. Aust. Math. Soc.* **16** (1972), 200-215.
6. Gratzer,G., *Lattice theory : first concepts and Distributive lattices*, W.H.Freeman and Company, San Fransisco, 1971.
7. Rao, G.C.: Almost Distributive Lattices. Doctoral Thesis (1980), Dept. of Mathematics, Andhra University, Visakhapatnam.
8. Rao, G.C. and Ravi Kumar, S.: Normal Almost Distributive Lattices. *Southeast Asian Bullettin of Mathematics*, **32** (2008), 831-841.
9. Rao, G.C. Rafi, N. and Ravi Kumar Bandaru.: Closure Operators in Almost Distributive Lattices. *International Mathematical Forum*, **5** (2010), no. 19, 929-935.
10. Rao, G.C. Rafi, N. and Ravi Kumar Bandaru.: $S-$Linear Almost Distributive Lattices. European J. Pure and Applied Math., **3**, 2010, no. 4, 704-716.
11. Rao, G.C. Rafi, N. and Ravi Kumar Bandaru.: $S-$ideals in Almost Distributive Lattices. *Accepted for publication in Southeast Asian Bulletin of Mathematics.*
12. Rao, G.C. Rafi, N. and Ravi Kumar Bandaru.: On Prime Filters in a Normal Almost Distributive Lattice. *Accepted for publication in Southeast Asian Bulletin of Mathematics.*
13. Swamy, U.M. and Rao, G.C.: Almost Distributive Lattices. *J. Aust. Math. Soc. (Series A)*, **31** (1981), 77-91.
14. Swamy, U.M. and Ramesh, S.: Birkhoff Centre of ADL, *Int. J. Algebra*, **3** (2009), 539-546.

α-Normal Almost Distributive Lattices

G. C. Rao[1] and M. Sambasiva Rao[2]

[1] *Department of Mathematics, Andhra University,*
Visakhapatnam-530003, INDIA
gcraomaths@yahoo.co.in

[2] *Department of Mathematics, MVGR College of Engineering,*
Vizianagaram-535005, INDIA
mssraomaths35@rediffmail.com

The concept of an α-normal Almost Distributive Lattice(ADL) is introduced. This class of ADLs are characterized in terms of prime α-ideals and annulets. It is proved that every $\star$-ADL is an α-normal ADL. Finally, a sufficient condition is derived for every $\star$-ADL to become a normal ADL.

Keywords: Almost Distributive Lattice(ADL), normal ADL, α-normal ADL, $\star$-ADL, prime α-ideal, minimal prime ideal.

1. Introduction

In,[7] U.M. Swamy and G.C. Rao introduced the concept of an Almost Distributive Lattice(ADL) and made a detailed study of algebraic properties of the ADL. The notion of $\star$-ADLs was introduced by U.M. Swamy, G.C. Rao and G. Nanaji Rao.[8] G.C. Rao and S. Ravikumar[2,3] introduced the concepts of minimal prime ideals and normal ADLs and extensively studied their properties. Recently in the year 2009, the concept of α-ideals was introduced and extensively studied by the authors in.[5] In this paper, the concept of an α-normal ADL is introduced as a generalization of normal ADLs in terms of prime α-ideals. Though every normal ADL is an α-normal ADL, the converse is not true. A sufficient condition for an α-normal ADL to become a normal ADL is derived. The class of α-normal ADLs are characterized in terms of prime α-ideals. Finally, it is observed that every $\star$-ADL is an α-normal ADL but not a normal ADL and hence a sufficient condition is derived for a $\star$-ADL to become a normal ADL.

2. Preliminaries

In this section, we recall certain definitions and important results mostly from,[3],[4],[5],[7],[8] those will be required in the text of the paper.

Definition 2.1.[7] An Almost Distributive Lattice (ADL)with zero is an algebra $(L, \vee, \wedge, 0)$ of type (2,2,0) satisfying

1). $(x \vee y) \wedge z = (x \wedge z) \vee (y \wedge z)$
2). $x \wedge (y \vee z) = (x \wedge y) \vee (x \wedge z)$
3). $(x \vee y) \wedge y = y$
4). $(x \vee y) \wedge x = x$
5). $x \vee (x \wedge y) = x$
6). $0 \wedge x = 0$ for any $x, y, z \in L$

Every non-empty set X can be regarded as an ADL as follows. Let $x_0 \in X$. Define two binary operations $\vee, \wedge$ on X by:

$$x \vee y = \begin{cases} x \text{ if } x \neq x_0 \\ y \text{ if } x = x_0 \end{cases} \qquad x \wedge y = \begin{cases} y \text{ if } x \neq x_0 \\ x_0 \text{ if } x = x_0 \end{cases}$$

Then $(X, \vee, \wedge, x_0)$ is an ADL and is called a discrete ADL. Throughout this paper L stands for an ADL $(L, \vee, \wedge, 0)$ with zero. For any $a, b \in L$, define $a \leq b$ if and only if $a = a \wedge b$ (or equivalently, $a \vee b = b$), then $\leq$ is a partial ordering on L.

Theorem 2.1.[7] *For any $a, b, c \in L$, we have the following*

1). $a \vee b = a \Leftrightarrow a \wedge b = b$
2). $a \vee b = b \Leftrightarrow a \wedge b = a$
3). $a \wedge b = b \wedge a$ whenever $a \leq b$
4). $\wedge$ is associative in L
5). $a \wedge b \wedge c = b \wedge a \wedge c$
6). $(a \vee b) \wedge c = (b \vee a) \wedge c$
7). $a \wedge b = 0 \Leftrightarrow b \wedge a = 0$
8). $a \vee b = b \vee a$ whenever $a \wedge b = 0$
9). $a \vee (b \wedge c) = (a \vee b) \wedge (a \vee c)$
10). $a \wedge (a \vee b) = a, (a \wedge b) \vee b = b$ and $a \vee (b \wedge a) = a$
11). $a \leq a \vee b$ and $a \wedge b \leq b$
12). $a \wedge a = a$ and $a \vee a = a$
13). $0 \vee a = a$ and $a \wedge 0 = 0$

14). If $a \leq c$ *and* $b \leq c$ *then* $a \wedge b = b \wedge a$ *and* $a \vee b = b \vee a$
15). $a \vee b = a \vee b \vee a$.

A non-empty subset I of L is called an ideal(filter) of L if $a \vee b \in I (a \wedge b \in I)$ and $a \wedge x \in I (x \vee a \in I)$ whenever $a, b \in I$ and $x \in L$. If I is an ideal of L and $a, b \in L$, then $a \wedge b \in I \Leftrightarrow b \wedge a \in I$. The set $\mathcal{I}(L)$ of all ideals of L is a complete distributive lattice with the least element $\{0\}$ and the greatest element L under set inclusion in which, for any $I, J \in \mathcal{I}(L)$, $I \cap J$ is the infimum of I, J and the supremum is given by $I \vee J = \{ i \vee j \mid i \in I, j \in J \}$. For any $a \in L$, $(a] = \{a \wedge x \mid x \in L\}$ is the principal ideal generated by a. The set $\mathcal{PI}(L)$ of all principal ideals of L is a sublattice of $\mathcal{I}(L)$. A proper ideal(filter) P of L is said to be prime if for any $x, y \in L$, $x \wedge y \in P (x \vee y \in P) \Rightarrow x \in P$ or $y \in P$. It is clear that a subset P of L is a prime ideal if and only if $L - P$ is a prime filter. A prime ideal P of L is called minimal if there is no prime ideal Q of L such that $Q \subset P$. For any prime ideal P of L, $O(P) = \{ x \in R \mid x \wedge y = 0 \text{ for some } y \notin P \}$ is an ideal of L such that $O(P) \subseteq P$.

Theorem 2.2.[2] *For any prime ideal* P *of* L, $O(P)$ *is the intersection of all minimal prime ideals contained in* P.

Theorem 2.3.[7] *Let* I *be an ideal and* F *a filter of an ADL* L *such that* $I \cap F = \emptyset$. *Then there exists a prime ideal* P *such that* $I \subseteq P$ *and* $P \cap F = \emptyset$.

Theorem 2.4.[3] *Let* P *be a prime ideal of an ADL* L. *Then* P *is a minimal prime ideal if and only if for each* $x \in P$ *there exists* $y \notin P$ *such that* $x \wedge y = 0$.

For any $A \subseteq L$, $A^* = \{ x \in L \mid a \wedge x = 0 \text{ for all } a \in A\}$ is an ideal of L. We write $(a]^*$ for $\{a\}^*$ and this is called an annulet of L.

Lemma 2.1.[4] *For any two ideals* I *and* J *of* L, *we have the following:*

1). If $I \subseteq J$, *then* $J^* \subseteq I^*$
2). $I \subseteq I^{**}$
3). $I^{***} = I^*$
4). $I \cap J = (0] \Leftrightarrow I \subseteq J^*$
5). $(I \vee J)^* = I^* \cap J^*$.

An element $a \in L$ is called dense[9] if $(a]^* = (0]$. An ideal I of L is called dense if $I^* = (0]$. An ADL L is called a $\star$-ADL,[8] if to each $x \in L$, $(x]^{**} = (x']^*$ for some $x' \in L$. An ADL L is a $\star$-ADL if and only if to each $x \in L$, there exists $y \in L$ such that $x \wedge y = 0$ and $x \vee y$ is dense. An ADL L is called normal[2] if every prime ideal contains a unique minimal prime ideal. An ADL L is normal if and only if $(x]^* \vee (y]^* = (x \wedge y]^*$ for all $x, y \in L$.

An ideal I of an ADL L is called an α-ideal[5] if $(x]^{**} \subseteq I$, for any $x \in I$. For any ideal I of L, the set $I^e = \{x \in L \mid (a]^* \subseteq (x]^* \text{ for some } a \in I\}$ is the smallest α-ideal of L such that $I \subseteq I^e$. An α-ideal I of L is called proper if $I \neq L$.

Lemma 2.2.[5] *Let L be an ADL. Then for any ideals I, J of L, we have:*

1). $I \subseteq J \Rightarrow I^e \subseteq J^e$
2). $(I \cap J)^e = I^e \cap J^e$
3). $(I^e)^e = I^e$.

Theorem 2.5.[5] *Let I be an ideal of an ADL L. Then I is an α-ideal if and only if for $x, y \in L, (x]^* = (y]^*$ and $x \in I$ imply that $y \in I$.*

The set $I_\alpha(L)$ of all α-ideals of L forms a complete distributive lattice with respect to the operations $\wedge$ and $\tilde{\vee}$ given by $I \wedge J = I \cap J$ and $I \,\tilde{\vee}\, J = (I \vee J)^e$.

3. α-normal ADLs

In this section, the concept of α-normal ADLs is introduced and some properties of this class of ADLs are studied. Finally, certain necessary and sufficient conditions for an ADL to become an α-normal ADL are derived. Throughout this section L stands for an ADL with 0, unless otherwise mentioned.

We begin this section with the definition of an α-normal ADL.

Definition 3.1. An ADL L is said to be an α-normal ADL if every prime α-ideal contains a unique minimal prime ideal.

Example 3.1. Let $A = \{0, a\}$ and $B = \{0, b_1, b_2\}$ be two discrete ADLs. Write $L = A \times B = \{(0,0), (0, b_1), (0, b_2), (a, 0), (a, b_1), (a, b_2)\}$. Then $(L, \vee, \wedge, 0')$ is an ADL under point-wise operations, in which the zero element is $0' = (0, 0)$.

Consider the ideals $I = \{(0,0),(a,0)\}$ and $J = \{(0,0),(0,b_1),(0,b_2)\}$. It can be easily observed that I and J are prime α-ideals. Since $\{(0,0)\}$ is not a prime ideal, we get that I and J are containing unique minimal prime ideals I and J respectively. Therefore L is an α-normal ADL.

Example 3.2. Let $L = \{0, a, b, c\}$ and define $\vee$ and $\wedge$ on L as follows:

$\vee$	0	a	b	c
0	0	a	b	c
a	a	a	b	b
b	b	b	b	b
c	c	b	b	c

$\wedge$	0	a	b	c
0	0	0	0	0
a	0	a	a	0
b	0	a	b	c
c	0	0	c	c

Then clearly $(L, \vee, \wedge, 0)$ is an ADL. Consider the ideals $I = \{0, a\}$ and $J = \{0, c\}$. Then it can be easily observed that I and J are prime α-ideals in L. Since $\{0\}$ is not a prime ideal, we get that I and J are containing unique minimal prime ideals respectively I and J. Therefore L is an α-normal ADL.

Lemma 3.1. *Every normal ADL is an α-normal ADL.*

PROOF. Let L be a normal ADL. Then every prime ideal of L contains a unique minimal prime ideal. Hence every prime α-ideal of L contains a unique minimal prime ideal. Therefore L is an α-normal ADL.

In general, the converse of the above lemma is not true. That is, every α-normal ADL need not be a normal ADL. In fact, it is not true even in a distributive lattice. It may be observed in the following example:

Example 3.3. Let $L = \{0, a, b, c, 1\}$ be a distributive lattice whose Hasse diagram is given in the following figure:

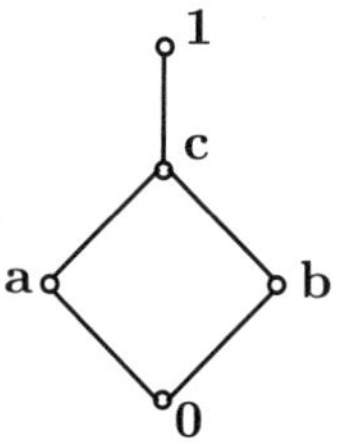

Consider the ideals $I = \{0, a\}$ and $J = \{0, b\}$ of L.
Clearly I and J are prime α-ideals of L.
Also I, J are minimal prime ideals of L.
Therefore L is an α-normal.
But the prime ideal $P = \{0, a, b, c\}$ of L is containing two distinct minimal prime ideals I and J.
Therefore L is an α-normal but not normal.

1
c
a
b
0

Though every α-normal ADL need not be a normal, in the following, we derive a sufficient condition for every α-normal ADL to become normal.

Theorem 3.1. *An α-normal ADL in which every prime ideal is non-dense is a normal ADL.*

PROOF. Let L be an α-normal ADL in which every prime ideal is non-dense. Let P be a prime ideal of L. Choose $x, y \in L$ such that $(x]^* = (y]^*$ and $x \in P$. Since P is non-dense, there exists $0 \neq x' \in L$ such that $x' \in P^*$. Hence $x \wedge x' = 0$. Thus $x' \wedge y = 0 \in P$. Thus $y \in P$, because $x' \in P^*$. Hence P is a prime α-ideal. Since L is an α-normal, P contains a unique minimal prime ideal. Therefore L is normal.

We now characterize an α-normal ADL in terms of its prime α-ideals.

Theorem 3.2. *The following conditions are equivalent in an ADL L :*

(1). L is α-normal
(2). For each prime α-ideal P, $O(P)$ is prime
(3). For all $x, y \in L$, $x \wedge y = 0$ implies $(x]^* \tilde{\vee} (y]^* = L$
(4). For all $x, y \in L$, $(x]^* \tilde{\vee} (y]^* = (x \wedge y]^*$.

PROOF. (1) $\Rightarrow$ (2): Assume that L is an α-normal ADL. Let P be a prime α-ideal of L. Then P contains a unique minimal prime ideal, say Q. Then by Theorem 2.3, $O(P) = Q$. Therefore $O(P)$ is a prime ideal of L.

(2) $\Rightarrow$ (3): Assume the condition (2). Let $x, y \in L$ such that $x \wedge y = 0$. Suppose $(x]^* \tilde{\vee} (y]^* \neq L$. Then $(x]^* \tilde{\vee} (y]^*$ is a proper α-ideal. Hence $\{(x]^* \tilde{\vee} (y]^*\} \cap D = \emptyset$. Hence there exists a prime α-ideal P such that $(x]^* \tilde{\vee} (y]^* \subseteq P$. Now

$$\begin{aligned} (x]^* \tilde{\vee} (y]^* \subseteq P &\Rightarrow (x]^* \subseteq P \text{ and } (y]^* \subseteq P \\ &\Rightarrow x \notin O(P) \text{ and } y \notin O(P) \\ &\Rightarrow x \wedge y \notin O(P) \quad \text{(since } O(P) \text{ is prime)} \\ &\Rightarrow 0 \notin O(P) \end{aligned}$$

Which is a contradiction. Therefore $(x]^* \tilde{\vee} (y]^* = L$.

(3) $\Rightarrow$ (4): Assume the condition (3). Let $x, y \in L$ be two arbitrary elements. We have always that $(x]^* \tilde{\vee} (y]^* \subseteq (x \wedge y]^*$.

$$\text{Now} \quad \begin{aligned} a \in (x \wedge y]^* &\Rightarrow a \wedge x \wedge y = 0. \\ &\Rightarrow (a \wedge x) \wedge (a \wedge y) = 0 \\ &\Rightarrow (a \wedge x]^* \tilde{\vee} (a \wedge y]^* = R \\ &\Rightarrow a \in (a \wedge x]^* \tilde{\vee} (a \wedge y]^* \end{aligned}$$

$$\Rightarrow\ a \in \{(a \wedge x]^* \vee (a \wedge y]^*\}^e$$
$$\Rightarrow\ (s \vee t]^* \subseteq (a]^* \text{ for some } s \in (a \wedge x]^*,\ t \in (a \wedge y]^*$$

Since $s \in (a \wedge x]^*$ and $t \in (a \wedge y]^*$, we get that $a \wedge s \in (x]^*$ and $a \wedge t \in (y]^*$.

Hence
$$\begin{aligned}(s \vee t]^* \subseteq (a]^* &\Rightarrow (a]^{**} \subseteq (s \vee t]^{**}.\\ &\Rightarrow (a]^{**} = (a]^{**} \cap (s \vee t]^{**}\\ &\Rightarrow (a]^{**} = (a \wedge (s \vee t)]^{**}\\ &\Rightarrow (a]^{**} = ((a \wedge s) \vee (a \wedge t)]^{**}\\ &\Rightarrow ((a \wedge s) \vee (a \wedge t)]^* = (a]^*\\ &\Rightarrow a \in \{(x]^* \vee (y]^*\}^e\\ &\Rightarrow a \in (x]^* \tilde{\vee} (y]^*\end{aligned}$$

Thus $(x \wedge y]^* \subseteq (x]^* \tilde{\vee} (y]^*$. Therefore $(x]^* \tilde{\vee} (y]^* = (x \wedge y]^*$.

$(4) \Rightarrow (1)$: Assume the condition (4). Let P be a prime α-ideal of L. Suppose that P contains two distinct minimal prime ideals, say Q_1, Q_2. Clearly Q_1 and Q_2 are α-ideals in L. Choose $x \in Q_1 - Q_2$ and $y \in Q_2 - Q_1$. Since Q_1, Q_2 are minimal primes, there exist $x_1 \notin Q_1$ and $y_1 \notin Q_2$ such that $x \wedge x_1 = y \wedge y_1 = 0$. Now

$$\begin{aligned}(y \wedge x_1) \wedge (x \wedge y_1) &= y \wedge x \wedge x_1 \wedge y_1\\ &= x \wedge x_1 \wedge y \wedge y_1\\ &= 0 \wedge 0\\ &= 0\end{aligned}$$

Since $y \wedge x_1 \notin Q_1$ and Q_1 is minimal prime, we get that$(y \wedge x_1]^* \subseteq Q_1$. Since $x \wedge y_1 \notin Q_2$ and Q_2 is minimal prime, we get that $(x \wedge y_1]^* \subseteq Q_2$. Hence

$$\begin{aligned}L = (0]^* &= (\,(y \wedge x_1) \wedge (x \wedge y_1)\,]^*\\ &= (y \wedge x_1]^* \tilde{\vee} (x \wedge y_1]^*\\ &\subseteq Q_1 \tilde{\vee} Q_2\\ &\subseteq P\end{aligned}$$

Which is a contradiction. Hence L is α-normal. Thus the proof is completed.

In general a $\star$-ADL need not be a normal ADL. It may be seen in the example 1.5. We have already observed that L is not a normal ADL. Now $(0]^{**} = (c]^*$, $(a]^{**} = (b]^*$, $(b]^{**} = (a]^*$, $(c]^{**} = (0]^*$ and $(1]^{**} = (c]^*$. Hence it is a $\star$-ADL but not normal. However, we prove that the class of $\star$-ADLs contained in the class of α-normal ADLs. For this, we need the following.

Lemma 3.2. *Let L be a $\star$-ADL. Then every proper α-ideal of L is contained in a minimal prime ideal.*

PROOF. Let I be a proper α-ideal of L. Then clearly $I \cap D = \emptyset$. Hence there exists a prime α-ideal P of L such that $I \subseteq P$ and $P \cap D = \emptyset$. We now show that P is minimal. Let $x \in P$. Since L is a $\star$-ADL, there exists $y \in L$ such that $x \wedge y = 0$ and $x \vee y \in D$. Since $P \cap D = \emptyset$, we get $x \vee y \notin P$. Hence $y \notin P$. Therefore P is a minimal prime ideal.

Theorem 3.3. *Every $\star$-ADL is an α-normal ADL.*

PROOF. Let L be a $\star$-ADL. Let $x, y \in L$ be such that $x \wedge y = 0$. Suppose $(x]^* \tilde{\vee} (y]^* \neq L$. Since $(x]^*$ and $(y]^*$ are α-ideals, we get that $(x]^* \tilde{\vee} (y]^*$ is a proper α-ideal of L. Hence by above Lemma, there exists a minimal prime ideal, say P such that $(x]^* \tilde{\vee} (y]^* \subseteq P$. Hence $(x]^* \subseteq P$ and $(y]^* \subseteq P$. Since P is minimal, we get $x \notin P$ and $y \notin P$. Therefore $0 = x \wedge y \notin P$. Which is a contradiction. Hence $(x]^* \tilde{\vee} (y]^* = L$. Therefore by Theorem 3.7, L is an α -normal ADL. The following corollary is a direct consequence of Theorems 3.6 and 3.9.

Corollary 3.1. *A $\star$-ADL in which every prime ideal is non-dense is a normal ADL.*

References

1. Birkhoff, G.: Lattice Theory. *Amer. Math. Soc. Colloq. Publ. XXV, Providence* (1967), U.S.A.
2. G.C. Rao and S. Ravikumar: "Normal Almost Distributive Lattices", *Southeast Asian Bulletin of Mathematics* **32**(2008), 831 - 841.
3. G.C. Rao and S. Ravikumar, "Minimal Prime Ideals in Almost Distributive Lattices", *Int. J. Contemp. Math. Sci.* **4**(2009), no.10, 475 - 784.
4. G.C. Rao and M. Sambasiva Rao, "Annihilator Ideals in Almost Distributive Lattices", *International Mathematical Forum* **4**(2009), no.15, 733 - 746.
5. G.C. Rao and M. Sambasiva Rao, "α-Ideals in Almost Distributive Lattices", *Inter. J. Contemp. Math. Sci.* **4**(2009), no.10, 457 - 466.
6. G.C. Rao, *Almost Distributive Lattices*, Doctoral Thesis, Andhra University, Waltair, 1980.
7. U.M. Swamy and G.C. Rao, "Almost Distributive Lattices", *J. Austral. Math. Soc. (Series A)* **31** (1981), 77 - 91.
8. U.M. Swamy, G.C. Rao and G. Nanaji Rao, "Stone Almost Distributive Lattices", *Southeast Asian Bulletin of Mathematics* **27** (2003), 513 - 526.
9. U.M. Swamy, G.C. Rao and G. Nanaji Rao, "Dense Elements in Almost Distributive Lattices", *Southeast Asian Bulletin of Mathematics* **27** (2004), 1081 - 1088.

A Lecture on Pseudo-Complemented Almost Distributive Lattices

G. Nanaji Rao

Department of Mathematics, Andhra University,
Visakhapatnam, Andhra Pradesh, INDIA
E-mail: nani6us@yahoo.com

The concept of pseudo-complementation $*$ on an Almost Distributive Lattice (ADL) with 0 is introduced and it is proved that it is equationally definable. A one to one correspondence between the pseudo-complementation on an ADL L with 0 and maximal elements of L is obtained. It is also proved that the set $L^* = \{a^* \mid a \in L\}$ is a Boolean algebra which is independent (up to isomorphism) of the pseudo-complementation $*$ on L. The concept of $*-$Almost Distributive Lattice, Stone Almost Distributive Lattice are introduced. Some necessary and sufficient conditions for an ADL to become a $*-$ ADL (Stone ADL) in topological and algebraic terms are proved. Characterization of $*-$ADLs and Stone ADLs in terms of prime ideals and minimal prime ideals are established.

Keywords: Almost Distributive Lattice (ADL), Prime ideal, Prime filter, Pseudo-complementation, Boolean algebra, maximal element, dense element, Hull-kernel topology, $*-$ADL, Stone ADL .

1. Introduction

A pseudo-complemented lattice is a lattice L with 0 such that to each $a \in L$, there exists $a^* \in L$ such that for all $x \in L$, $a \wedge x = 0$ if and only if $x \leq a^*$. Here a^* is called the pseudo-complement of a. For each element a of a pseudo-complemented lattice L, a^* is uniquely determined by a, so that " $*$ " can be regarded as a unary operation on L. Moreover, each pseudo-complemented lattice contains the unit element 0^* ($0 \wedge x = 0 \Leftrightarrow x \leq 0^*$ for all x). It follows that every pseudo-complemented lattice L can be regarded as an algebra $(L, \vee, \wedge, *, 0, 1)$ of type $(2, 2, 1, 0, 0)$. The fact that class of pseudo-complemented distributive lattices is equationally definable was first observed by P. Ribenboin [9]. Also, K.B. Lee [6]proved that the class of pseudo-complemented distributed lattice is generated by its finite members and a complete description of the lattice of equational classes of pseudo-complemented distributive lattices is given. G. Gratzer [5] studied about the pseudo-complemented semi-lattice L, and proved that the set $L^* = \{a^* \mid a \in L\}$, where $*$ is a pseudo-complementation on L, becomes a Boolean algebra.

After the Boole's aximitization of the two valued propositional calculus, into a Boolean algebra, many generalizations of the Boolean algebras have come into being. The concept of an Almost Distributive Lattice (ADL) was introduced by Swamy and Rao [14]as a common abstraction of almost all the lattice theoretic and ring theoretic generalization of a Boolean algebra like, $P-$rings, bi-regular rings, associative rings, p_1-rings triple system, Bair-rings and $m-$domains rings.

In [15], we introduced the concept of pseudo-complementation on Almost Distributive Lattice (ADL). It is well known that in the case of distributive lattice, the pseudo-complementation $*$ if exists, is unique. But, in the case of ADLs, there can be several pseudo-complementations. In fact, if there is a pseudo-complementation on an ADL L, then we prove that each maximal element of L, gives rise to a pseudo-complementation and that this correspondence between the set of maximal elements of L, and the set of pseudo-complementations on L is one-to-one. Also, we prove that if $*$ is a pseudo-complementation on an ADL L, then the set $L^* = \{a^* \mid a \in L\}$ becomes a Boolean algebra and if $*$ and $\perp$ are two pseudo-complementations on L, then we prove that the corresponding Boolean algebras L^* and $L^\perp$ are isomorphic. In other words, the Boolean algebra L^* is independent (up to isomorphism) of the pseudo-complementation $*$. Also, we prove that the pseudo-complementation $*$ on an ADL is equationally definable.

The problem of the characterizing pseudo-complemented distributive lattices satisfying $x^* \vee x^{**} = 1$ (called Stone lattice) has been studied by several mathematicians like O. Frink [3], G, Gratzer [4], T.P.Speed [11], etc,. In [15], we have proved that if $*$ and $\perp$ are two pseudo-complementations on an ADL L, then $a^* \vee a^{**} = 0^* \Leftrightarrow a^\perp \vee a^{\perp\perp} = 0^\perp$. for all $a \in L$. This motivates us, in [16], we have introduced a more generalized class of ADLs like $*-$ADLs and Stone ADLs which properly contains the class of ADLs with pseudo-complementation as a generalization of results of T.P. Speed [11]. Also, we have characterized the Stone ADLs, both algebraically and topologically in terms of their prime ideals and minimal prime ideals with hull-kernel topology and dual hull-kernel topology. Also, the $*-$ADLs was characterized by means of dense of elements. Most of the results are stated without proofs. For detailed proofs one can go through the references given.

2. Preliminaries

First we recall certain definitions and properties of Almost Distributive Lattice from [7],[14], that are required in the text.

Definition 2.1. An algebra $(L, \vee, \wedge, 0)$ of type (2,2,0) is called an Almost Distributive Lattice(ADL) if , for any $x, y, z \in L$,
(1) $x \wedge (y \vee z) = (x \wedge y) \vee (x \wedge z)$
(2) $(x \vee y) \wedge z = (x \wedge z) \vee (y \wedge z)$
(3) $(x \vee y) \wedge x = x$
(4) $(x \vee y) \wedge y = y$
(5) $x \vee (x \wedge y) = x$
(6) $0 \wedge x = 0$

It is observed that in [14], that any non-empty set A can be regarded as an ADL with some arbitrary fixed element $a_0 \in A$ as zero where the operation $\vee$ and $\wedge$ are defined as follows.

Definition 2.2. Let A be a non-empty set and $a_0 \in A$. For any $a, b \in A$, define,

$$(a \vee b) = \begin{cases} a & \text{if } a \neq a_0 \\ b & \text{if } a = a_0 \end{cases}$$

$$(a \wedge b) = \begin{cases} b & \text{if } a \neq a_0 \\ a_0 & \text{if } a = a_0 \end{cases}$$

This is called a discrete ADL with 0.

Lemma 2.1. *Let L be an ADL. Then we have the following:*
(1) $a \vee b = a \Leftrightarrow a \wedge b = b$
(2) $a \vee b = b \Leftrightarrow a \wedge b = a$
(3) $a \wedge b = b \wedge a$ *whenever* $a \leq b$
(4) $\wedge$ *is associative in* L
(5) $a \wedge b \wedge c = b \wedge a \wedge c$
(6) $(a \vee b) \wedge c = (b \vee a) \wedge c$

(7) $a \wedge b = 0 \Leftrightarrow b \wedge a = 0$
(8) $a \vee (b \wedge c) = (a \vee b) \wedge (a \vee c)$
(9) $a \wedge (a \vee c) = a,\ (a \wedge b) \vee b = b$ *and* $a \vee (b \wedge a) = a$
(10) $a \leq a \vee b$ *and* $a \wedge b \leq b$
(11) $a \wedge a = a$ *and* $a \vee a = a$
(12) $a \vee 0 = a = 0 \vee a$ *and* $a \wedge 0 = 0$
(13) *if* $a \leq c, b \leq c$ *for some* $c \in L$, *then* $a \wedge b = b \wedge a$ *and* $a \vee b = b \vee a$.

Definition 2.3. A homomorphism between ADLs L and L' is a mapping $f : L \to L'$ satisfying the following:
(1) $f(a \vee b) = f(a) \vee f(b)$
(2) $f(a \wedge b) = f(a) \wedge f(b)$
(3) $f(0) = 0$.

A nonempty subset I of L is called an ideal of L if $x \vee y \in I$ and $x \wedge a \in I$ whenever $x, y \in I$ and $a \in L$. For any $A \subseteq L$, the ideal generated by A is $(A] = \{(\bigvee_{i=1}^{n} a_i) \wedge x \mid a_i \in A, x \in L, n \in Z^+\}$. If $A = \{a\}$,then we write (a] for (A]. The set of all principal ideals of L is a distributive lattice and it is denoted by PI(L). A proper ideal P of L is called prime if for any $x, y \in L, x \wedge y \in P$ then $x \in P$ or $y \in P$. For any $x, y \in L$,define $x \leq y$ if and only if $x = x \wedge y$ or equivalently $x \vee y = y$. Then $\leq$ is a partial ordering on L in which 0 is the least element. For any $x, y \in L$ with $x \leq y, [x, y] = \{t \in L \mid x \leq t \leq y\}$ is a bounded distributive lattice with respect to the operations in L. If, in addition, $[x, y]$ is a Boolean algebra, then L is relatively complemented ADL and in this case, the operation $\vee$ is associative. An element m is maximal in $(L, \leq)$ if and only if $m \wedge x = x$ for all $x \in L$.For any $A \subseteq L, A^* = \{x \in L \mid a \wedge x = 0\ \forall\ a \in A\}$ is an ideal of L. We write $[a]^*$ for $\{a\}^*$. We don't know, so far , whether $\vee$ is associative in an ADL or not. In this talk, L denotes an ADL in which $\vee$ is associative.

Lemma 2.2. *Let L be an ADL and* $a \in L$. *Then* $(a] = L$ *if and only if a is a maximal element.*

Lemma 2.3. *Let L be an ADL and I an ideal of L. Then, for any* $a, b \in L$, *we have the following:*
(1) $(a] = \{a \wedge x \mid x \in L\}$
(2) $a \in (b] \Leftrightarrow b \wedge a = a$
(3) $a \wedge b \in I \Leftrightarrow b \wedge a \in I$
(4) $(a] \cap (b] = (a \wedge b] = (b \wedge a]$
(5) $(a] \vee (b] = (a \vee b] = (b \vee a]$.

3. Pseudo-Complementation

In this section, we give the definition of pseudo-complementation on an almost distributive lattice with 0 and prove some basic properties of an ADL with pseudo-complementation and we prove that an ADL with Pseudo-complementation is equationally definable.

Definition 3.1. Let$(L, \vee, \wedge, 0)$ be an ADL with 0. Then a unary operation $a \mapsto a^*$ on L is called a pseudo-complementation on L if, for any $a, b \in L$, it satisfies the following conditions:
(P_1) $a \wedge b = 0 \Rightarrow a^* \wedge b = b$;
(P_2) $a \wedge a^* = 0$;
(P_3) $(a \vee b)^* = a^* \wedge b^*$.

We observe that p_1, P_2 and P_3 are independent.

Example 3.1. Let X be a discrete ADL with 0 and with at least two elements, say $1, 2$ other than 0. Then $(X^3, \vee, \wedge, 0)$ is an ADL with 0, where $\vee, \wedge$ are defined coordinatewise. Now, for any $x \in X^3$, we write $\mid x \mid$ for the number of non-zero entries in x. Define $*$ on x^3 as follows: For any $x \in X^3$ and $i = 1, 2, 3$

$$x_i^* = \begin{cases} 0 & \text{if } x_i \neq 0 \\ 1 & \text{if } x_i = 0 \text{ and } |x| = 1 \text{ and } 0^* = (2,2,2) \\ 2 & \text{if } x_i = 0 \text{ and } |x| > 1 \end{cases}$$

then $(X^3, \vee, \wedge, 0)$ is an ADL with $(0, 0, 0)$ as 0 which satisfies P_1 and P_2 but not (P_3) (if $a = (1, 0, 0)$ and $b = (0, 1, 0)$, then $a^* = (0, 1, 1), b^* = (1, 0, 1)$, and $a \vee b = (1, 1, 0)$, so $(a \vee b)^* = (0, 0, 2)$ and $a^* \wedge b^* = (0, 0, 1)$)

Example 3.2. Let L be an ADL with 0 with at least two elements. Define $a^* = 0$ for all $a \in L$. Then L satisfies (P_2) and (P_3) but not (P_1) (if $0 \neq b \in L$, then $0 \wedge b = 0$ and $0^* \wedge b = 0 \neq b$.)

Example 3.3. Let L be a bounded distributive lattice with bounds $0 \neq 1$. Now for any $a \in L$, define $a^* = 1$ for all $a \in L$. Then L satisfies (P_1) and (P_3) but not (P_2).

Now, we give certain examples of pseudo-complementation on ADLs.

Example 3.4. Let $(R, +, \cdot, 0)$ be a commutative regular ring. To each $a \in R$, Let a^0 be the unique idempotent element in R such that $aR = a^0R$. Define, for any $a, b \in R$,
(1) $a \wedge b = a^0 b$;
(2) $a \vee b = a + (1 - a^0)b$;
(3) $a^* = 1 - a^0$;
then $(R, \vee, \wedge, 0)$ is an ADL with 0 and $*$ is a pseudo-complementation on R.

In the case of a distributive lattice with 0, it is well known that the element a^* satisfying the properties (P_1) and(P_2) is unique (if it exists) and that (P_3) is a consequence of (P_1) and (P_2) and hence, there can be at most one pseudo-complementation. However, in an ADL with 0, there can be several pseudo-complementations; for, consider the following examples.

Example 3.5. Let $(X, \vee, \wedge, 0)$ be a discrete ADL with 0. For any $x \neq 0$ in X, define

$$a^* = \begin{cases} 0 & \text{if } a \neq 0 \\ x & \text{if } a = 0 \end{cases}$$

then $*$ is a pseudo-complementation on X. Here, with each $x \neq 0$ in X, we obtain a pseudo-complementation on $(X, \vee, \wedge, 0)$. More generally, we have the following:

Example 3.6. Let X be a non-empty set with at least two elements and let Y be any set and $f_0 \in X^Y$. Now, for any $a, b \in X^Y$, define

$$(1) \qquad (a \vee b)(y) = \begin{cases} a(y) & \text{if } a(y) \neq f_0(y) \\ b(y) & \text{if } a(y) = f_0(y); \end{cases}$$

$$(2) \qquad (a \wedge b)(y) = \begin{cases} b(y) & \text{if } a(y) \neq f_0(y) \\ f_0(y) & \text{if } a(y) = f_0(y). \end{cases}$$

Then $(X^Y, \vee, \wedge, f_0)$ is an ADL with f_0 as zero element. Now, let $f \in X^Y$ such that $f(y) \neq f_0(y)$ for all $y \in Y$. For any $a, b \in X^Y$, define

$$a^f(y) = \begin{cases} f_0(y) & \text{if } a(y) \neq f_0(y) \\ f(y) & \text{if } a(y) = f_0(y) \end{cases}$$

Then $a \mapsto a^f$ is a pseudo-complementation on X^Y and, conversely, if $a \mapsto a^*$ is a pseudo-complementation on X^Y, then there exists $f \in X^Y$

such that $f(y) \neq f_0(y)$ for all $y \in Y$ and $a^* = a^f$ for all $a \in X^Y$ (take $f = f_0^*$)

In the following, we prove that any relatively complemented ADL with 0 is pseudo-complemented,and hence, the examples given in Examples $3.4 - 3.6$ are special cases of this.

Theorem 3.1. *Let L be a relatively complemented ADL with 0 and with a maximal element m_0. For any $a \in L$, define a^* to be the complement of a in $[0, a \vee m_0]$. Then $*$ is a pseudo-complementation on L.*

Now, we give the some properties of pseudo-complementation in the following.

Lemma 3.1. *Let L be an ADL with 0 and $*$ a pseudo-complementation on L. Then, for any $a, b \in L$, we have the following:*
(1) 0^* *is maximal;*
(2) *if a is maximal, then $a^* = 0$;*
(3) $0^{**} = 0$;
(4) $a^* \wedge a = 0$;
(5) $a^{**} \wedge a = a$;
(6) $a^* = a^{***}$;
(7) $a^* = 0 \Leftrightarrow a^{**}$ *is maximal;*
(8) $a^* \leq 0^*$;
(9) $a^* \wedge b^* = b^* \wedge a^*$;
(10) $a \leq b \Rightarrow b^* \leq a^*$;
(11) $a^* \leq (a \wedge b)^*$ *and* $b^* \leq (a \wedge b)^*$;
(13) $a = 0 \Leftrightarrow a^{**} = 0$;

Lemma 3.2. *Let L be an ADL with 0 and $*$ a pseudo-complementation on L. Then for any $a, b \in L$, the following are equivalent:*
(1) $a \wedge b = 0$;
(2) $a^{**} \wedge b = 0$;
(3) $a^{**} \wedge b^{**} = 0$;
(4) $a \wedge b^{**} = 0$;

Lemma 3.3. *Let L be an ADL with 0 and $*$ a pseudo-complementation on L. Then, for any $a, b \in L$, the following hold:*

(1) $(a \wedge b)^{**} = a^{**} \wedge b^{**}$;
(2) $(a \wedge b)^{*} = (b \wedge a)^{*}$;
(3) $(a \vee b)^{*} = (b \vee a)^{*}$;

In [6], Lee derived certain sets of identities characterizing pseudo-complemented distributive lattice. In the following theorems, we obtain a set of of identities which characterize a pseudo-complementation on an ADL with 0.

Theorem 3.2. *Let* $(L, \vee, \wedge, 0)$ *be an ADL with* 0. *Then a unary operation* $* : L \longrightarrow L$ *is a pseudo-complementation on* L *if and only if it satisfies the following equations:*
(1) $a \wedge a^{*} = 0$;
(2) $a^{**} \vee a = a^{**}$;
(3) $(a \vee b)^{*} = a^{*} \wedge b^{*}$;
(4) $(a \wedge b)^{**} = a^{**} \wedge b^{**}$;
(5) $0^{*} \wedge a = a$;

Theorem 3.3. *Let* $(L, \vee, \wedge, 0)$ *be an ADL with* 0. *Then a unary operation* $* : L \longrightarrow L$ *is a pseudo-complementatin on* L *if and only if it satisfies the following equations:*
(1) $a^{*} \wedge b = (a \wedge b)^{*} \wedge b$;
(2) $0^{*} \wedge a = a$;
(3) $0^{**} = 0$;
(4) $(a \vee b)^{*} = a^{*} \wedge b^{*}$

4. One-to-One Correspondence

Here, we prove that, for any ADL with 0 and with pseudo-complementation $*$, the set $L^{*} = \{a^{*} | a \in L\}$ becomes a Boolean algebra. In Sec. 3, it is remarked that an ADL with 0 can have more than one pseudo-complementation and examples were given to this effect. In fact, if L is an ADL with a pseudo-complementation $*$, then to each maximal element $m \in L$, we obtain a pseudo-complementation $*_m$. we prove that this correspondence between the maximal elements of L and pseudo-complementations on L is one-to-one and that the Boolean algebra L^{*} is independent (upto isomorphism) of the pseudo-complementation $*$.

Theorem 4.1. *Let L be an ADL with 0 and $*$ a pseudo-complementation on L. For any $a^*, b^* \in L^*$, define $a^* \leq b^*$ if and only if $a^* \wedge b^* = a^*$. Then $(L^*, \leq)$ is a Boolean algebra, in which $(a^{**} \wedge b^{**})^*$ is the least upper bound (l.u.b) and $(a^* \wedge b^*)$ is the greatest lower bound (g.l.b) of a^* and b^*.*

We remarked that an ADL with 0 can have more than one pseudo-complementation. Now, we prove that for any two pseudo-complementations $*$, $\perp$ on L, the Boolean algebras L^* and $L^\perp$ are isomorphic. First we prove the following lemma.

Lemma 4.1. *Let L be an ADL with 0 and let $*$ and $\perp$ be two pseudo-complementations on L. Then for any $a, b \in L$, we have the following:*
(1) $a^* \wedge a^\perp$ *and* $a^* \vee a^\perp = a^*$;
(2) $a^{*\perp} = a^{\perp\perp}$;
(3) $a^* = b^* \Leftrightarrow a^\perp = b^\perp$;
(4) $a^* = 0 \Leftrightarrow a^\perp = 0 \Leftrightarrow (a \wedge b = 0 \Rightarrow b = 0)$;
(5) $a^\perp = a^* \wedge 0^\perp$;
(6) $a^* \vee a^{**} = 0^* \Leftrightarrow a^\perp \vee a^{\perp\perp} = 0^\perp$;

Theorem 4.2. *Let L be an ADL with 0 and $*$ a pseudo-complementation on L. Let M be the set of all maximal elements in L and let $PC(L)$ be the set of all pseudo-complementations on L. For any $m \in M$, define $*_m : L \longrightarrow L$ by $a^{*_m} = a^* \wedge m$ for all $a \in L$. Then $m \mapsto *_m$ is a bijection of M onto $PC(L)$.*

In the following theorem, we prove that the Boolean algebra L^* is independent, upto isomorphism of the pseudo-complementations $*$ on L

Theorem 4.3. *Let L be an ADL with 0 and let $*, \perp$ be two pseudo-complementation on L .Then the map $f : L^* \longrightarrow L^\perp$ defined by $f(a^*) = a^\perp$ is an isomorphism of Boolean algebras.*

5. $*$-Almost Distributive Lattices

In this section we introduce $*-$ADLs which properly contains the class of ADLs with pseudo-complementation. Also, we give topological characterization of a $*-$ADLs in terms of hull-kernel topology as well as dual hull-kernel topology on the set of minimal prime ideals of an ADL L. First we prove the following lemma.

Lemma 5.1. *Let L be an ADL and $x, y \in L$. Then the following statements hold:*
(1) $[x \vee y]^* = [x]^* \cap [y]^*$
(2) $[x \wedge y]^* = [y \wedge x]^*$
(3) $[x]^{***} = [x]^*$
(4) $x \leq y \Rightarrow [y]^* \subseteq [x]^*$
(5) $[x \wedge y]^{**} = [x]^* \cap [y]^*$

If L is a bounded distributive lattice and X is the set of all prime ideals of L, then the hull-kernel topology $\Im_h^X$ on X is the topology on X for which $\{X_a \mid a \in L\}$ is a basis, where for any $a \in L, X_a = \{P \in X \mid a \notin P\}$. In this topology, for any subset F of X, the closure $\overline{F}$ of F is given by $\overline{F} = \{Q \in X \mid \bigcap_{P \in F} P \subseteq Q\}$. The dual hull-kernel topology $\Im_d^X$ on X is the topology for which $\{h_X(a) \mid a \in L\}$ is a basis, where $h_X(a) = \{P \in X \mid a \in P\}$.For any $A \subseteq L$, write $h_X(A) = \{P \in X \mid A \subseteq P\}$ and it can be observed that $h_X(A) = \bigcap_{a \in A} h_X(a)$.

All these concepts can be analogously defined in case of ADLs also. Let L be an ADL and $Y(M)$ denote the set of all prime ideals (minimal prime ideals) of L. The hull-kernel topology on Y is denoted by $\Im_h^Y$. In this topology, for any $a \in L$, the corresponding basic open set is denoted by Y_a. We write $\Im_h^M$ for the topology on M induced from $\Im_h^Y$. In this topology, the basic open sets are $\{M_x \mid x \in L\}$ where $M_x = M \cap Y_x$. The dual hull-kernel topology on $Y(M)$ is denoted by $\Im_d^Y(\Im_d^M)$. In this topology, for any $a \in L$, the basic open sets are denoted by $h_Y(a)(h_M(a))$.Now, we state the following lemma whose proof is straight forward.

Lemma 5.2. *Let L be an ADL and $x, y \in L$. Then the following statements hold:*
(1) *A prime ideal P of L in minimal if and only if $[x]^* - P \neq \phi$ for all $x \in P$,*
(2) $M_x = h_M([x]^*)$,
(3) $h_M(x) = h_M([x]^{**})$,
(4) $[x]^* \subseteq [y]^* \Leftrightarrow h_M(x) \subseteq h_M(y)$,
(5) $[x]^* \subseteq [y]^* \Leftrightarrow M_y \subseteq M_x$,
(6) $[z]^* = [x]^* \cap [y]^* \Leftrightarrow h_M(z) = h_M(x) \cap h_M(y)$,
(7) $[x]^{**} = [y]^* \Leftrightarrow h_M(x) = h_M(y)$,

Now we introduce a new class of ADLs called $*-$ADLs which includes a class of ADLs with pseudo-complementation.

Definition 5.1. An ADL $(L, \vee, \wedge, 0)$ is called a $*-$ ADLs, if , to each $x \in L, [x]^{**} = [x^{'}]^{*}$ for some $x' \in L$.

Example 5.1. Let $A = \{0, a\}$, $B = \{0, b_1, b_2\}$ be two discrete ADLs. Define $\vee$ and $\wedge$ on $A \times B$ under pointwise. Then $A \times B$ is a $*-$ADL.

We now give topological characterization of $*-$ ADLs in the following theorem.

Theorem 5.1. *Let L be an ADL. Then the following statements are equivalent:*
(1) *L is a $*-$ ADL*
(2) $\Im_h^M = \Im_d^M$,
(3) *M is compact in Hull kerneal topology,*

If we define $\theta = \{(x, y) \in L \times L \mid [x]^* = [y]^*\}$, then θ is a congruence on L and L/θ is distributive lattice with 0. An element x of L is called dense if $[x]^* = \{0\}$. The filter of all dense elements in L is denoted by D.Now we give an algebraic characterization of $*-$ ADLs in the following theorem.

Theorem 5.2. *Let L be an ADL. Then the following statements are equivalent:*
(1)*L is a $*-$ ADL.*
(2) *L/θ is a Boolean algebra.*
(3) *for any $x \in L$, there is $x' \in L$ such that $x \wedge x' = 0, x \vee x' \in D$.*
(4) *For any ideal I of L with $I \cap D = \phi$, there is a minimal prime ideal $P \supseteq I$.*

6. Stone Almost Distributive Lattices

Now we introduce the concept of a Stone ADL, analogous to that of a Stone lattice, and charecterize the Stone ADLs in terms of the hull-kernel topology on the set Y. If m is maximal element in L, then Y_a is a clopen subset of Y for all $a \in B[0, m]$ where $B[0, m]$ stands for the Boolean algebra of all complemented elements of the bounded distributive lattice $[0, m]$. It can be easily proved that if m_1 and m_2 are maximal elements in L,then the map $x \longmapsto x \wedge m_2$ is an isomorphism between the bounded distributive lattices $[0, m_1]$ and $[0, m_2]$.

Definition 6.1. Let L be an ADL and $*$ a pseudo-complementation on L.Then L is called Stone ADL if, for any $x \in L, x^* \vee x^{**} = 0^*$.

It can be easily observed that examples (3.4) and (3.6) are Stone ADLs. Now we give a topological characterization of Stone ADLs in the following theorem.

Theorem 6.1. *Let L be an ADL with a maximal element m. Then the following statements are equivalent:*
(1) *L is a Stone ADL.*
(2) *For any $x \in L, [x]^* \vee [x]^{**} = L$.*
(3) *For any $x \in L, \overline{Y_x}$ is open in the Hull kernel topology on Y.*

Next theorem gives a characterization of Stone ADLs in terms of minimal prime ideals of L. First we prove the following lemma's.

Lemma 6.1. *For any non-empty set F of L which is closed under $\wedge$, define $\theta_F = \{(x, y) \in L \times L \mid x \wedge d = y \wedge d \ for \ some \ d \in F\}$.Then θ_F is a congruence on L*

Lemma 6.2. *Let E be the set of all prime ideals of L contained in $L - F$ and G be the set of all prime ideals of L/θ_F. Then the map f defined by $f(P) = \overline{P} = \{x/\theta_F \in L/\theta_F \mid x \in P\}$ is an isomorphism of E onto G.*

Remark 6.1. By above lemma, if $L = F$ then both sets E and F are empty.

Now we prove the following theorem.

Theorem 6.2. *A $*-$ ADL L with a maximal element is a Stone ADL if and only if for any two distinct minimal prime ideals P, Q of L, $P \vee Q = L$*

For any ADL L,we denote the set of ideals of L by $I(L)$.Then $(I(L), \vee, \cap)$ is a distributive lattice where for any $I, J \in I(L), I \vee J = \{a \vee b \mid a \in I, b \in J\}$

and $\cap$ is set intersection [7]. Now we give some important equivalent conditions for a $*-$ ADL to become a Stone ADL

Theorem 6.3. *Let L be a $*-$ ADL and m be a maximal element in L.Then the following conditions are equivalent:*
(1) *L is a Stone ADL.*
(2) *$[x \wedge y]^* = [x]^* \vee [y]^*$ for all $x, y \in L$.*
(3) *L^{**} is a sublattice of the lattice $(I(L), \vee, \cap)$ where $L^{**} = \{[x]^{**} \mid x \in L\}$.*
(4) *For any $x \in L$, $[x]^* = [a]^*$ for some $a \in B[0, m]$.*
(5) *If $x \wedge y = 0$ then there exists $a \in B[0, m]$ such that $x \in (a]$ and $y \in (a']$, where a' is the complement of a in $B[0, m]$.*

Now, we give another characterization of Stone ADLs in terms of minimal prime ideals of L.For this, we state the following lemma.

Lemma 6.3. *Let m be a maximal element in L. If $P \in Y$, then $P^c = P \cap B[0, m]$ is a prime ideal of $B[0, m]$.*

Theorem 6.4. *Let L be a Stone ADL and Q be a prime ideal of $B[0, 0^*]$. Then $Q^e = \{a \wedge x \mid a \in Q, x \in L\}$ is a minimal prime ideal of L.*

Note. In general, for any ADL L with a maximal element m, we get that Q^e is the smallest ideal of L containing Q

Now we prove the following theorem.

Theorem 6.5. *Let be a $*-$ ADL with a maximal element m. Then L is Stone ADL if and only if $P = (P \cap B[0, m])^e$ for all $P \in M$*

7. Prime Ideal Characterization of Stone ADLs

In this section , we give the prime ideal characterization of Stone ADLs. Let m be a maximal element in L and X denote the Boolean space of all prime ideals of the Boolean algebra $B[0, m]$ with hull-kernel topology ;that is, the topology for which $\{X_(a) \mid a \in B[0, m]\}$ is a basis, where for any $a \in B[0, m]$, $X_a = \{P \in X \mid a \notin P\}$. If $P \in Y$ then P^c is a prime ideal of $B[0, m]$ (by lemma 6.3) and it can be observed that the map$f : Y \longrightarrow X$

defined by $f(P) = P^c$ for all $P \in Y$, is continuous. Then we have the following.

Lemma 7.1. *If $P^e \in Y$ for all $P \in X$,then the map $g : X \longrightarrow Y$ defined by $g(P) = P^e$, has the following properties:*
(1) $f \circ g = id_x$;that is, $P^{e^c} = P$ for all $P \in X$,
(2) $g^{-1}(Y_x)$ is closed in X for all $x \in L$.

Now we prove the following theorem.

Theorem 7.1. *Suppose $P^e \in Y$ for all $P \in X$ and m be a maximal element in L. Then the following statements are equivalent:*
(1) L is a Stone ADL.
(2) For any $x \in L$, there exists largest element $a \in B[0, m]$ such that $x \wedge a = 0$.
(3) The map g is continuous.

A subspace S of a topological space T is said to be retract of T, if there exists a continuous map $\theta : T \longrightarrow S$ such that $\theta(a) = a$ for all $a \in S$. Finally, we characterize Stone ADLs in terms of its prime ideals. Before going to the theorem, we need the following lemma.

Lemma 7.2. *Let L be an ADL with a maximal element $m, x \in L$ and $a \in B[0, m]$.Then $a \wedge x = x$ if and only if $M_x \subseteq M_a$.*

Theorem 7.2. *Let L be an ADL with a maximal element m. Then the following conditions are equivalent:*
(1) L is a Stone ADL.
(2) M is retract of Y.
(3) $f/M : M \longrightarrow X$ is a homeomorphism.

References

1. Birkhoff, G.: Lattice Theorey, Vol.25,American Mathematical Society Colloquium Publication,1967.

2. Burris, S, Sankappanavar, H.P.: A course in universal algebra, Springer Verlag.1981.
3. Frink,O: Pseudo-complements in semilattices, Duke Math. Journal,20, 505-515(1962).
4. Gratzer, G.: A generalization on Stone's representation theorem for Boolean algebra,Duke. Math.Journal 30,469-474((1963).
5. Gratzer, G: Lattice Theorey: First Concepts and Distributive Lattices, W.H.Freeman and Company, Sanfransisco (1971).
6. Lee, K.B.: Equational class of distributive pseudo-complemented lattice, Cand. J.Math., 22, 881-891(1970)
7. Rao, G.C.: Almost distributive lattices, Doctoral Thesis, Andhra University,Waltair, 1980.
8. Rao. G.C.: and S. Ravi Kumar.:Minimal prime ideals in almost distributive lattices , Int. Journal of Contemp. Math. Sciences, Vol. 4, no. 9-12, 475-484(2009).
9. Ribenboim, P.: Characterization of the pseudo-compliment in distributive lattice with least element, Summa Brasil. Math., 2(1949), No.4, 43-49.MR 11,75.
10. Speed, T.P: Some remarks on a class of distributive lattices, J.Austral.Math. Soc.9,289-296(1969).
11. Speed, T.P: On stone lattices, J.Austral.Math. Soc.9,297-307(1969).
12. Swamy, U.M., Mankyamba, P.: Prime ideal characterization of stone lattices, Maths Seminar Note 7, Kobe University, 25-31(1979).
13. Swamy,U.M., Murti, G.S.: Boolean centre of a universal algebra, Algebra Universals 13,202-205(1981)
14. Swamy, U.M. and Rao.G.C.: Almost Distributive Lattices, Journal of Australian Math.Soc,(Series A) 31, 77-91(1981).
15. Swamy, U.M., Rao.G.C.: and Nanaji Rao.G.:Pseudo-Complementation on Almost Distributive Lattices, Southeast Asian Bulletin of Mathematics, 24:95-104(2000).
16. Swamy, U.M., Rao.G.C.:and Nanaji Rao.G.:Stone Almost Distributive Lattices, Southeast Asian Bulletin of Mathematics, 27: 513-526(2003).
17. Swamy, U.M., Rao.G.C.: and Nanaji Rao.G.:Dense Elements in Almost Distributive Lattices, Southeast Asian Bulletin of Mathematics, 27: 1081-1088(2004).
11. Swamy, U.M., Rao.G.C.: and Nanaji Rao.G.:On Characterizations of Stone Almost Distributive Lattices, Southeast Asian Bulletin of Mathematics, 32, 1167-1176(2008).
18. Swamy, U.M., Rama Rao.V.V.: Triple and sheaf representation of Stone Lattices, Algebra Universalis 5, 104-113(1975).

Iteration of Total and Partial Unary Operations

C. Ratanaprasert*, K. Denecke

Dedicated to Prof. K. P. Shum's 70th birthday

Abstract

Iterating a unary operation f defined on an n-element set A one obtains the descending chain

$$A \supseteq Imf \supseteq Imf^2 \supseteq \cdots \supseteq Imf^m = Imf^{m+1}$$

of images. The least integer $\lambda(f)$ with $Imf^{\lambda(f)} = Imf^{\lambda(f)+1}$ is called the *pre-period* of f. The pre-period of $f$is an integer between 0 and $n-1$. If $\lambda(f) = n-1$ for $n \geq 1$, then f is called a *long-tailed* (LT)-operation and if $\lambda(f) = n-2$ for $n \geq 2$, f is said to be an LT_1-operation. LT- and LT_1-operations and their invariant equivalence relations are characterized in [3]. In [4] these results are extended to partial operations and in [5] to n-ary operations. In this paper we characterize unary total and partial operations for arbitrary $\lambda(f) = n-t, 0 < t < n$.
Key words: iteration, unary total and partial operation, pre-period.
AMS Mathematics Subject Classification: 03B50, 08A30, 08A55.

1 Preliminaries

Iterating a unary operation f defined on the finite set A one obtains the descending chain

$$A \supseteq Imf \supseteq Imf^2 \supseteq \cdots \supseteq Imf^m = Imf^{m+1}.$$

The least integer $\lambda(f)$ with $Imf^{\lambda(f)} = Imf^{\lambda(f)+1}$ is called the *pre-period* of f. The pre-period of f is an integer between 0 and $n-1$. If $\lambda(f) = n-1$ for $n \geq 1$, then f is called a *long-tailed* (LT)-operation and if $\lambda(f) = n-2$ for $n \geq 2$, f is said to be a LT_1-operation. LT- and LT_1-operations and their invariant equivalence relations are characterized in [3]. Let O_A^1 be the set of all unary operations defined on the set A. We summarize the results from [3].

*Research of the first author supported by the Thailand Research Fund.

Theorem 1.1 ([3]) *Let $f \in O_A^1$ be a total unary operation and $|A| = n \geq 2$. Then the following propositions are satisfied:*

(i) $\lambda(f) = n-1$ *if and only if there exists an element $d \in A$ such that*

$$A = \{d, f(d), f^2(d), \ldots, f^{n-1}(d) = f^n(d)\} \text{(see e.g. [2])} .$$

(ii) *Assume now that $n \geq 3$. Then $\lambda(f) = n-2$ and $|Imf^{n-2}| = 1$ if and only if there are distinct elements $u, v \in A$ such that $A = \{u, v, f(v), \ldots, f^{n-2}(v)\}$ and such that there is an exponent k with $0 \leq k \leq n-2$ with $f(u) = f^{k+1}(v)$ and there is an integer m with $m + k = n-2$ such that $f^{m+1}(u) = f^m(u)$.*

(iii) *We have $\lambda(f) = n-2$ and $|Imf^{n-2}| = 2$ if and only if there are different elements $u, v \in A$ such that either*

a) $A = \{v, u, f(u), \ldots, f^{n-2}(u)\}$ *with* $v = f(v)$ *and* $f^{n-1}(u) = f^{n-2}(u)$, *or*

b) $A = \{u, f(u), f^2(u) \ldots, v = f^{n-2}(u), f^{n-1}(u)\}$ *where* $v = f^n(u) = f^{n-2}(u)$.

In [4] these results are extended to partial operations. Let P_A^1 be the set of all partial unary operations defined on a set A and let id_A be the identity mapping on A. If $domf \subset A$, then f is called *proper partial.*

Let $f : A \multimap A$ be a partial unary operation defined on a finite set A with $|A| = n$ and with domain $domf$. Let ∞ be an element which does not belong to A and assume that $B := A \cup \{\infty\}$. Then we define a total operation $f^+ \in O_B^1$ as follows:

$$f^+(b) = \begin{cases} f(b) & \text{if } b \in domf \\ \infty & \text{otherwise} \end{cases} .$$

f^+ is said to be the *one-point extension* of f. Clearly, $(id_A)^+ = id_B$. Let $o^+ = c_\infty$ be the constant operation on B with value ∞. For $C \subseteq P_A^1$ we define $C^+ := \{f^+ \mid f \in P_A^1\}$. It is well-known and easy to prove (see e.g. [4]) that $(P_A^1; \circ, id_A)$ is isomorphic to $((P_A^1)^+; \circ, id_B)$ with respect to the mapping $+ : P_A^1 \to (P_A^1)^+$ defined by $f \mapsto f^+$ for every $f \in P_A^1$. A consequence of this isomorphism is

Lemma 1.2 ([4]) *Let $|A| = n \geq 2$ and let $f \in P_A^1$ be a partial unary operation defined on A. Then $|Im(f^+)^l| = |Imf^l| + 1$ for any $l \geq 0$. For the pre-periods we have $\lambda(f) = \lambda(f^+)$.*

For partial operations we have the following characterizations.

Theorem 1.3 *Let $|A| \geq 2$ and $f \in P_A^1 \setminus O_A^1$ and assume that f is not bijective on $domf$. Then $\lambda(f) = n$ if and only if there exists an element $d \in domf$ such that $A = \{d, f(d), \ldots, f^{n-1}(d)\}$ and $f^{n-1}(d) \notin domf$.*

Theorem 1.4 *Let A be a set with $|A| = n \geq 2$ and let $f : A \multimap\!\!\rightarrow A$ be a unary proper partial operation. Then $\lambda(f) = n - 1$ if and only if there are distinct elements $u, v \in A$ such that either*

(i) $A = \{u, v, f(v), \ldots, f^{n-2}(v)\}$ *where* $f^{n-2}(v) \notin domf$ *and there is an integer* k *with* $0 \leq k \leq n-2$ *such that* $f(u) = f^{k+1}(v)$*; or*

(ii) $A = \{u, v, f(v), \ldots, f^{n-2}(v)\}$ *where* $u = f(u)$ *and* $f^{n-2}(v) \notin domf$*; or*

(iii) $A = \{u, v, f(v), \ldots, f^{n-2}(v)\}$ *where* $\{u, f^{n-2}(v)\} \cap domf = \emptyset$.

2 Iteration of Total Operations

Let A be an n-element set and let $f : A \to A$ be an operation. If $\lambda(f) = 0$, then f is a permutation and conversely. If $\lambda(f) = n - 1$, then f is a long-tailed operation. We want to characterize arbitrary unary operations with $\lambda(f) = n - t$ for $0 < t < n$.

Proposition 2.1 *Let A be a finite set with $|A| = n$ and let $f : A \to A$ be a unary operation with $\lambda(f) = n - t$ for $0 < t < n$. Then*

(i) $C := Imf^{\lambda(f)}$ *is the largest subset of A such that $f|C$ is a permutation,*

(ii) *for* $D := A \setminus Imf$ *we have* $A \setminus C = \bigcup\limits_{v \in D} \{v, f(v), \ldots, f^{m_v}(v)\}$ *where* $f^{m_v+1}(v) \in C$ *and* $|\{v, f(v), \ldots, f^{m_v}(v)\}| \leq n - t$*, for all* $v \in D$ *and*

(iii) *there is an element* $u \in D$ *such that* $m_u + 1 = n - t$.

Proof: (i) Since $Imf^{\lambda(f)} = Imf^{\lambda(f)+1}$, the mapping $f^{\lambda(f)} : Imf^{\lambda(f)} \to Imf^{\lambda(f)}$ is a bijection. Therefore, $f|Imf^{\lambda(f)}$ is a permutation. If X is a subset of A such that $f|X$ is a permutation, then $X \subseteq Imf^{\lambda(f)}$. (We note that $|C| = |Imf^{\lambda(f)}| \leq t$.)

(ii) Since $\lambda(f) \neq 0$, we have $C \subseteq Imf \subset A$ and then $D := A \setminus Imf \subseteq A \setminus C$. If $v \in D$, then $v \in A \setminus C$ and then there is a largest non-negative integer m_v such that $B_v := \{v, f(v), \ldots, f^{m_v}(v)\}$ is a subset of $A \setminus C$ and $f^{m_v+1}(v) \in C$. Let now $a \in A \setminus C$. If $a \notin Imf$, then $a \in D$; hence $a \in B_a \subseteq \bigcup\limits_{v \in D} B_v$. If $a \in Imf$, then there is an element $b \in A \setminus C$ with $a = f(b)$ since otherwise, i.e. if $b \in C$, then $a = f(b) \in C$ since $f|C$ is a permutation. Now, if $b \notin Imf$, then we obtain $b \in B_b \subseteq \bigcup\limits_{v \in D} B_v$, otherwise we continue as before. Since A is finite, this procedure stops with an element $w \in D$ and a positive integer m such that $a = f^m(w) \in B_w \subseteq \bigcup\limits_{v \in D} B_v$. Therefore, $A \setminus C \subseteq \bigcup\limits_{v \in D} B_v$.

The inclusion $\bigcup_{v\in D} B_v \subseteq A\setminus C$ is clear since $B_v \subseteq A\setminus C$ for all $v \in D$. Altogether we have equality. The proof shows also that $f^{m_v+1}(v) \in C$.
Because of $f|B_v(f^{m_v}(v)) \in C$ the mapping $f|B_v : B_v \to B_v$ is proper partial and by Theorem 1.2 we get $\lambda(f|B_v) = m_v + 1 = |B_v|$. Since $B_v \subseteq A\setminus C$, we have $\lambda(f|B_v) \leq \lambda(f|(A\setminus C))$ and altogether

$$|B_v| = \lambda(f|B_v) \leq \lambda(f|(A\setminus C)) = \lambda(f) = n - t.$$

(iii) Since D is finite, there is an element $u \in D$ such that $\lambda(f|B_u) = max\{\lambda(f|B_v) \mid v \in D\}$. Let $q := \lambda(f|B_u)(= m_u + 1)$. We prove that $Imf^q = Imf^{q+1}$. Clearly, $Imf^{q+1} \subseteq Imf^q$. Assume that $b \in Imf^q$. Then there is an element $a \in A$ such that $f^q(a) = b$. If $a \in C$, then $f^m(a) \in C$ for all $m \geq 1$; hence, $b \in C \subseteq Imf^{q+1}$. Assume that $a \notin C$. Then $b \notin C$, since $f(a) = b \in C$ iff $a \in C$. There are an element $v \in D$ and an integer p with $0 \leq p \leq m_v$ such that $a = f^p(v)$. Since $b \notin C$, we have $b \in B_v$ and $b = f^{q+p}|B_v(a)$. Now we consider the one-point extension of the partial operation $f|B_v : B_v \to B_v$ and use that $\lambda(f|B_v) = \lambda((f|B_v)^+) = m_v + 1 \leq q$. This gives $Im((f|B_v)^+)^{q+1} = Im((f|B_v)^+)^{q+p}$. From $Im(f|B_v)^{q+p} \subseteq Im((f|B_v)^+)^{q+p}$ we obtain $b = f^{q+p}(v) \in Im((f|B_v)^+)^{q+1}$. Since $b \in B_v$, we have $b \in Imf^{q+1}$. This shows $Imf^q \subseteq Imf^{q+1}$ and altogether we have equality. Since $\lambda(f)$ is the least integer such that $Imf^{\lambda(f)} = Imf^{\lambda(f)+1}$, we get $q \geq \lambda(f) = n - t$. Together with $n - t \geq q$ we get $m_u + 1 = \lambda(f|B_u) = q = n - t$. ■

In the following proposition we use also the denotation $C := Imf^{\lambda(f)}$.

Proposition 2.2 *Let A be a finite set and let $f : A \to A$ be a unary operation satisfying (i), (ii), (iii) of Proposition 2.1. Let $C := Imf^{\lambda(f)}$. Then there is a subset $\emptyset \neq D \subseteq A\setminus Imf$ such that*

$$A = \bigcup_{v\in D}\{v, f(v), \ldots, f^{m_v}(v), \ldots, f^{p_v}(v)\} \cup \bigcup_{c\in C}\{c, f(c), \ldots, f^{q_c}(c)\}$$

where

a) $0 \leq m_v < n - t$ *and* $f^{p_v+1}(v) = f^{m_v+1}(v)$ *for all* $v \in D$,

b) *there is an element* $u \in D$ *such that* $m_u + 1 = n - t$,

c) $f^{q_c+1}(c) = c$ *for all* $c \in C$.

Proof: Let $f : A \to A$ be a unary operation satisfying (i), (ii), (iii) of Proposition 2.1. The permutation $f|C$ can be written as $f|C = \alpha_1\alpha_2\cdots\alpha_k$, where $\alpha_1, \ldots, \alpha_k$ are disjoint cycles and where $C_1, \ldots, C_k$ are pairwise disjoint sets corresponding to

$\alpha_1, \ldots, \alpha_k$. Then $C_i = \{c_i, f(c_i), \ldots, f^{m_i}(c_i)\}$, where $f^{m_i+1}(c_i) = c_i$ for all $1 \leq i \leq k$ and $C = \bigcup_i C_i$. By (ii), there is a non-empty subset $D \subseteq A \setminus Imf$ such that $A \setminus C = \bigcup_{v \in D} \{v, f(v), \ldots, f^{m_v}(v)\}$, where $f^{m_v+1}(v) \in C$ and $0 \leq m_v < n-t$ for all $v \in D$. For each $v \in D$, let $f^{m_v+1}(v) \in C_v \in \{C_1, \ldots, C_k\}$. Then $A \setminus C = \bigcup_{v \in D} \{v, f(v), \ldots, f^{m_v}(v), f^{m_v+1}(v), \ldots, f^{p_v}(v)\}$ where $f^{p_v+1}(v) = f^{m_v+1}(v)$ for all $v \in D$. Hence $A = \bigcup_{v \in D} \{v, f(v), \ldots, f^{m_v}(v), \ldots, f^{p_v}(v)\} \cup \bigcup_{c \in C} \{c, f(c), \ldots, f^{q_c}(c)\}$ and conditions a) and b) are satisfied. Condition c) follows from condition (iii). ■

The converse of Proposition 2.2 is also satisfied and we have:

Proposition 2.3 *Let A be a finite set and let $f : A \to A$ be a unary operation. If there are sets C and D with $\emptyset \neq C \subseteq Imf^{\lambda(f)}, \emptyset \neq D \subseteq A \setminus Imf$ such that*

$$A = \bigcup_{v \in D} \{v, f(v), \ldots, f^{m_v}(v), \ldots, f^{p_v}(v)\} \cup \bigcup_{c \in C} \{c, f(c), \ldots, f^{q_c}(c)\}$$

and such that there is an integer t, for $1 \leq t \leq n$ which satisfies conditions a) *and* b) *from Proposition 2.2 and assume that condition* c) *from Proposition 2.2 is also satisfies, then $\lambda(f) = n - t$.*

Proof: To prove that $\lambda(f) = n - t$ we have to show that $Imf^{n-t} \subseteq Imf^{n-t+1}$ since the opposite inclusion is always satisfied. Let $b \in Imf^{n-t}$. Then there is an $a \in A$ such that $b = f^{n-t}(a)$. If $a \in C$, then $b \in C \subseteq Imf^m$ for all $m \geq 1$; so, $b \in Imf^{n-t+1}$. If $a \in A \setminus C$, then we use the ideas from the proof of Proposition 2.1 and consider the case that $a \in \bigcup_{v \in D} B_v$ where $B_v = \{v, f(v), \ldots, f^{m_v}(v)\}$ and $f^{m_v+1}(v) \in C$ for all $v \in D$. Then there is an element $v \in D$ and an integer $0 \leq s \leq m_v$ such that $b = f^{n-t+s}(v)$. Since $f|B_v$ is a partial operation on B_v with $\lambda(f|B_v) = m_v + 1 \leq n - t$, for the one-point extension of f we have $Im(f^+|B_v)^{m_v+1} = Im(f^+|B_v)^{n-t} = Im(f^+|B_v)^{n-t+s}$ for all $s \geq 0$. Therefore, $f^{n-t+s}(v) = (f^+|B_v)^{n-t+s}(v) \in Im(f^+|B_v)^{n-t+s} = Im(f^+|B_v)^{m_v+1}$ which implies that $b \notin B_v$ (since $f|B_v$ is a partial LT-operation with $\lambda(f|B_v) = |B_v|$). Thus $b \in Imf^{\lambda(f)} \subseteq Imf^{n-t+1}$ which implies $\lambda(f) \leq n - t$. But condition b) implies $n - t \leq \lambda(t)$. Altogether we have $\lambda(f) = n - t$. ■

Combining the last three propositions we get the following characterization of unary operations f on an n-element set with $\lambda(f) = n - t$ for $1 \leq t \leq n - 1$:

Theorem 2.4 *Let A be a finite set with $|A| = n$ and let $f : A \to A$ be a unary operation. Then $\lambda(f) = n - t$ for $0 < t < n$ if and only if the following conditions are satisfied:*

(i) *$C := Imf^{\lambda(f)}$ is the largest subset of A such that $f|C$ is a permutation,*

(ii) *for $D := A \setminus Imf$ we have $A \setminus C = \bigcup_{v \in D} \{v, f(v), \ldots, f^{m_v}(v)\}$ where $f^{m_v+1}(v) \in C$ and $|\{v, f(v), \ldots, f^{m_v}(v)\}| \leq n - t$, for all $v \in D$ and*

(iii) *there is an element $u \in D$ such that $m_u + 1 = n - t$.*

Remark:

1. If $t = 1$, i.e. if $\lambda(f) = n - 1$, then $|Imf^{\lambda(f)}| = 1$, and there is an element $u \in A \setminus Imf$ such that $B = \{u, f(u), \ldots, f^{n-2}(u)\}$ and $f^{n-1}(u) \in C$. Hence, $A = B \cup C = \{u, f(u), \ldots, f^{n-1}(u)\}$ and we obtain condition (i) from Theorem 1.1.

2. If $t = 2$, i.e. if $\lambda(f) = n - 2$, then $|Imf^{\lambda(f)}| = 1$ or $|Imf^{\lambda(f)}| = 2$.
case 1: $|Imf^{\lambda(f)}| = 1$. Then there is an element $u \in A \setminus Imf$ such that $B = \{u, f(u), \ldots, f^{n-3}(u)\}$ and $f^{n-2}(u) \in C$. Hence $|A \setminus (B \cup C)| = 1$ and there exists an element $v \in D \setminus \{u\}$ such that $A = \{v, u, f(u), \ldots, f^{n-2}(u)\}$ where $f^{n-2}(u) = f^{n-1}(u)$ and $f(v) \in A \setminus \{u, v\}$ (i.e. there is an integer k such that $f(v) = f^{k+1}(u)$).
case 2: $|Imf^{\lambda(f)}| = 2$. Then there is an element $u \in A \setminus Imf$ such that $B = \{u, f(u), \ldots, f^{n-3}(u)\}$ and $f^{n-2}(u) \in C$. Therefore either $Imf^{\lambda(f)} = \{f^{n-2}(u), f^{n-1}(u)\}$ where $f^n(u) = f^{n-2}(u)$ or $Imf^{\lambda(f)} = \{v, f^{n-2}(u)\}$ where $f^{n-1}(u) = f^{n-2}(u)$ and $f(v) = v$. Altogether in the second case we have either $A = \{u, f(u), \ldots, v = f^{n-2}(u), f^{n-1}(u)\}$ where $f^n(u) = v$; or $A = \{v, u, f(u), \ldots, f^{n-2}(u)\}$ where $f(v) = v$ and $f^{n-1}(u) = f^{n-2}(u)$. Therefore we obtain conditions (ii) and (iii) from Theorem 1.1.

For $\lambda(f) = n - 3$ we get:

Proposition 2.5 *Let A be a finite set with $|A| = n > 3$ and let $f : A \to A$ be a unary operation with $\lambda(f) = n - 3$. Then there is a maximal subset C of A ($C = Imf^{n-3}$) such that $f|C$ is a permutation and either*

(i) *there is an element $u \in A \setminus C$ such that $A \setminus C = \{u, f(u), \ldots, f^{n-4}(u)\}$ where $f^{n-3}(u) \in C$, or*

(ii) *there are distinct elements $u, v \in A \setminus C$ such that $A \setminus C := \{u, v, f(v), \ldots, f^{n-4}(v)\}$ where $f(u) \in A \setminus \{u, v\}$ and $f^{n-3}(v) \in C$, or*

(iii) *there are distinct elements $u, v, w \in A \setminus C$ such that $A = \{u, v, w, f(w), \ldots, f^{n-3}(w)\}$ where $f(u), f(v) \in A \setminus \{u, v, w\}$ and $C = \{f^{n-3}(w)\}$.*

3 Iteration of Partial Operations

Theorem 1.3 and Theorem 1.4 characterize partial unary operations with $\lambda(f) = n$ and with $\lambda(f) = n - 1$, respectively. Let A be a finite set with $|A| = n$ and let $f : A \multimap A$ be a proper partial unary operation which is not bijective on $domf$. Then $D := domf \setminus Imf \neq \emptyset$. Let C be the largest subset of A such that $f|C$ is a permutation (i.e. $C = Imf^{\lambda(f)}$ if it is non-empty).

Lemma 3.1 *For each $v \in D$, there are $m_v \geq 2$ and a subset $B_v := \{v, f(v), \ldots, f^{m_v-2}(v)\}$ such that $f^{m_v-1}(v) \notin domf$ or $f^{m_v-1}(v) \in C$.*

Proof: Let $v \in D$. If $f(v) \notin domf$ or $f(v) \in C$, then $B_v = \{v\}$ with $m_v = 2$. If $f(v) \in domf \setminus C$, we consider $f^2(v)$. Assume that $v, f(v), \ldots, f^{k-1}(v) \in domf \setminus C$ for $k \geq 1$. If $f^k(v) \notin domf$ or $f^k(v) \in C$, then $B_v = \{v, f(v), \ldots, f^{k-1}(v)\}$ and $m_v = k + 1$. If $f^k(v) \in domf \setminus C$, we consider $f^{k+1}(v)$ and conclude in a similar way. Since A is finite, there is a positive integer m_v such that $B_v = \{v, f(v), \ldots, f^{m_v-2}(v)\}$ is the required subset. ■

Lemma 3.2 *Let $D' := \{f^{m_v-1}(v) \mid v \in D\} \setminus C$ and let $B := domf \cup D'$. Define $g : B \to B$ by*

$$g(x) = \begin{cases} x & \text{if } x \in D' \\ f(x) & \text{if } x \in domf \end{cases}.$$

Then

(i) *g is a total unary operation and $\overline{C} := C \,\dot{\cup}\, D'$ is the largest subset of B such that $g|\overline{C}$ is a permutation.*

(ii) *$D \cap Img = \emptyset$.*

(iii) *$Img = Imf \cup D' = Imf$.*

(iv) *$Img^m = Imf^m \cup D'$ for all $m \geq 1$.*

(v) *$\lambda(f) = \lambda(g)$.*

Proof: (i) This is obvious.
(ii) Suppose $v \in D \cap Img$. Then $v \in D = domf \setminus Imf$ implies $v \notin Imf$ and then $v \neq f^m(t)$ for all $t \in domf$ and for all $m \geq 1$; hence $v \notin D'$ and then $v \notin D'$ and $v \notin Imf$ implies that $v \notin Img(= Imf \cup D'$, see (iv)), a contradiction.
(iii) This is clear.
(iv) Let $m \geq 1$ and let $t \in Imf^m$. Then there is an element $a \in domf$ such that

$t = f^m(a)$. Hence, $f^k(a) \in domf$ for all $0 \leq k < m$. So, $t = f^m(a) = g^m(a) \in Img^m$. Therefore, $Imf^m \subseteq Img^m$ and then $Imf^m \cup D' \subseteq Img^m$.
Conversely, let $t \in Img^m$. Then there is an element $a \in B$ such that $t = g^m(a)$. If $g^k(a) \in domf$ for all $0 \leq k < m$, then $t = g^m(a) = f^m(a) \in Imf^m$. If there is a least integer k with $0 \leq k < m$ such that $g^k(a) \notin domf$, then $g^s(a) = f^s(a)$ for all $0 \leq s \leq k$; so, $g^s(a) \in Imf$, but $g^s(a) \notin C$ implies that $g^s(a) \in D'$. Now we have $t = g^m(a) = g^{m-s}(g^s(a)) = g^s(a) \in D'$. Therefore, $t \in Imf^m \cup D'$.
(v) From $Img^{\lambda(f)} = Imf^{\lambda(f)} \cup D' = Imf^{\lambda(f)+1} \cup D' = Img^{\lambda(f)+1}$ we obtain $\lambda(g) \leq \lambda(f)$. Similarly, $Img^{\lambda(g)} = Img^{\lambda(g)} \setminus D' = Img^{\lambda(g)+1} \setminus D' = Imf^{\lambda(g)+1}$ implies $\lambda(f) \leq \lambda(g)$. Altogether we have $\lambda(f) = \lambda(g)$. ■

Using these propositions we obtain the following characterization of unary partial operations f with $\lambda(f) = n - t$ for $0 < t < n$.

Theorem 3.3 *Let $f : A \multimap A$ be a proper partial unary operation. Then $\lambda(f) = n - t$ for $0 < t < n$ if and only if $domf = C \cup \bigcup_{v \in D} B_v$ where $m_v \leq n - t$ for all $v \in D$ and there is an element $u \in D$ such that $m_u = n - t$.*

Proof: Let $D' := \{f^{m_v - 1}(v) \mid v \in D\} \setminus C$.
$\Rightarrow$ Assume that $\lambda(f) = n - t$ for $0 < t < n$ and let g be the total operation as defined in Lemma 3.2. Then by Lemma 3.2 (v) we have $\lambda(f) = \lambda(g) = n - t$. Therefore we can apply the results from the total case (see Theorem 2.4) and obtain:

(i) $\overline{C} := Img^{\lambda(g)} \neq \emptyset$ is the largest subset of A such that $g|\overline{C}$ is a permutation. Let $C := Img^{\lambda(g)} \setminus D'$ (the case that $C = \emptyset$, i.e. that $\overline{C} = D'$ is possible). Then $C = (Imf^{\lambda(f)} \cup D') \setminus D' = Imf^{\lambda(f)}$ by Lemma 3.2 (iv). If $C \neq \emptyset$, then $C \subseteq B \setminus D' = domf$ and $g|C = f|C$ is a permutation.

(ii) There is a non-empty subset $\overline{D}$ with $\overline{D} \cap Img = \emptyset$ such that $B \setminus \overline{C} = \bigcup_{v \in \overline{D}} B_v$ where $B_v = \{v, g(v), \ldots, g^{m_v - 2}(v)\}, g^{m_v - 1}(v) \in \overline{C}$ and $m_v \leq n - t$ for all $v \in \overline{D}$ and there is an element $u \in \overline{D}$ such that $m_u = n - t$. Clearly, $D = \overline{D}$. Since $B \setminus \overline{C} = (domf \cup D') \setminus (Imf^{\lambda(f)} \cup D') = domf \setminus Imf^{\lambda(f)} = domf \setminus C$, from $x \in B \setminus \overline{C}$ we obtain that $f^s(x) = g^s(x)$ for all $s \geq 0$, and so $domf \setminus C = \bigcup_{v \in D} B_v$ where $B_v = \{v, f(v), \ldots, f^{m_v - 2}(v)\}$ with $f^{m_v - 1}(v) \notin domf$ or $f^{m_v - 1}(v) \in C$ because of

$$x \in \overline{C} = C \cup D' \Rightarrow x \in C \text{ or } x \in D' \Rightarrow x \in C \text{ or } x \notin domf.$$

Therefore, $domf = C \cup \bigcup_{v \in D} B_v$ where $m_v \leq n - t$ for all $v \in D$ and there is $u \in D$ such that $m_u = n - t$.

$\Leftarrow$ Let $f : A \multimap\!\!\rightarrow A$ be a proper partial unary operation such that $domf = C \cup \bigcup_{v \in D} B_v$ where $B_v = \{v, f(v), \ldots, f^{m_v-2}(v)\}, f^{m_v-1}(v) \notin domf$ or $f^{m_v-1}(v) \in C$; and $m_v \leq n-t$ for all $v \in D$ and assume that there is an element $u \in D$ such that $m_u = n-t$. Let g be the total unary operation as defined in Lemma 3.2. Then Lemma 3.2 (i) implies that $\overline{C} = C \cup D'$ is the largest subset of $domg$ such that $g|C$ is a permutation. By assumption we have $domf \setminus C = \bigcup_{v \in D} B_v$. Moreover, $B \setminus \overline{C} = (domf \cup D') \setminus (Imf^{\lambda(f)} \cup D') = domf \setminus Imf^{\lambda(f)} = domf \setminus C = \bigcup_{v \in D} B_v$. So, g satisfies conditions (i), (ii), (iii) of Theorem 2.4 and then $\lambda(g) = n-t$. By Lemma 3.2 (v) we have $\lambda(f) = \lambda(g)$. Therefore, $\lambda(f) = n - t$. ■

References

[1] K. Denecke, S. L. Wismath, *Universal Algebra and Coalgebra*, World Scientific, 2009.

[2] K. Denecke, R. Pöschel, *The characterization of primal algebras by hyperidentities*, Contributions to General Algebra 6, Verlag Hölder -Pichler-Tempsky, Wien, Verlag B.G. Teubner, Stuttgart, 1988, 67-88.

[3] C. Ratanaprasert, K. Denecke, *Unary Operations with Long Pre-Periods*, Discrete Mathematics with its Application in Information Science and Related Topics, Conference, Beibei, Chongqing, (20/10/2004) in Discrete Mathematics 2008, vol. 308, no 21 (180 p.) pp. 4998-5005.

[4] C. Ratanaprasert, K. Denecke, *Partial Unary Operations with Long Pre-Periods*, Journal of Applied Algebra and Discrete Structures, Vol. 5 (2007), No. 3, pp. 133-145.

[5] C. Ratanaprasert, K. Denecke, *N-ary Operations with Long Pre-periods*, preprint 2006, submitted to Journal of Multiple-valued Logic and Soft Computation.

Author's addresses:

C. Ratanaprasert, Department of Mathematics, Faculty of Science, Silpakorn University, Nakhon Pathom, Thailand, e-mail: ratach@su.ac.th

K. Denecke, Universität Potsdam, Institut für Mathematik, Potsdam, Germany, e-mail: kdenecke@rz.uni-potsdam.de

Inverse Semigroups and Their Generalizations

X. M. Ren*

Department of Mathematics,
Xi'an University of Architecture and Technology,
Xi'an 710055, China
E-mail: xmren@xauat.edu.cn

K. P. Shum†

Department of Mathematics,
The University of Hong Kong,
Pokfulam Road, Hong Kong (SAR), China
E-mail: kpshum@maths.hku.hk

As generalizations of groups, inverse semigroups were first studied by V. V. Vagner (1952) and independently by G. B. Preston (1954). The theory of inverse semigroups play a major rule in development of the algebraic theory of semigroups. The main purpose of this paper is to give a brief survey of inverse semigroups and their generalizations in the class of regular semigroups, in the class of abundant semigroups and within the class of U-abundant semigroups. In particular, we exhibit some basic properties and structures of inverse semigroups and their generalizations, for example, left inverse semigroups, quasi-inverse semigroups, orthodox semigroups, adequate semigroups, type A semigroups, $\mathcal{L}^*$-inverse semigroups, $\mathcal{Q}^*$-inverse semigroups, type W semigroups, Ehresmann semigroups and U-orthodox semigroups and others. We establish a corresponding hierarchy of some important semigroups which generalize the inverse semigroups for regular semigroups, abundant semigroups and weakly abundant semigroups, respectively.

Keywords: Inverse semigroups; Orthodox semigroups; Type A semigroups; Type W semigroups; Ehresmann semigroups; U-orthodox semigroups.

*The research is supported by by National Natural Science Foundation of China Grant (No:10971160) and a NSF grant of Shaanxi Province (No: SJ08A06).
†The research is partially supported by a grant of Wu Jiehyee Charitable Foundation, Hong Kong, 2007/09.

Introduction

Recall that an element a of a semigroup S is said to be *regular* if there exists an element $x \in S$ such that $axa = a$. An element a' of a semigroup S is said to be an inverse element of the element $a \in S$ if

$$aa'a = a, \qquad a'aa' = a. \tag{1}$$

A semigroup S is said to be *regular* if every element of S is regular. A semigroup S is called an inverse semigroup if every a in S possesses a unique inverse, i.e. if there exists a unique inverse element a^{-1} in S such that

$$aa^{-1}a = a, \qquad a^{-1}aa^{-1} = a. \tag{2}$$

Inverse semigroups were first studied by V. V. Vagner[1,2] and independently by G. B. Preston[3,4] . Vagner called the inverse semigroups *the generalized groups.* It is probably better to regard an inverse semigroup as a generalized group or a generalization of a group rather than a specialization of a semigroup because the theory of inverse semigroups has many features in common with the theory of groups (see Ref. 5 and 6). In 1961 A. H. Clifford and G. B. Preston offered the opinion that inverse semigroups were the most promising class of semigroups for future study, and the intervening years have amply justified their forecast.

The theory of inverse semigroups and their generalizations play a major rule in development of the theory of semigroups. In this paper, we give a survey on inverse semigroups and its some generalizations. In Section 2, we recall some basic concepts and results for inverse semigroups and its some generalizations in the class of regular semigroups, for example, left inverse semigroups, quasi-inverse semigroups and orthodox semigroups. In Section 3 the study of some abundant semigroups, such as, adequate semigroups, type A semigroups, $\mathcal{L}^*$-inverse semigroups, $\mathcal{Q}^*$-inverse semigroups and type W semigroups, which are some kinds of generalizations of inverse semigroups for the class of abundant semigroups is resumed. In Section 4 some concepts and structure theorems of semigroups which generalizes inverse semigroups in the class of weakly abundant semigroups are given.

For notations and terminologies not given in this paper, the reader is referred to Ref. 5,7–12.

1. Basic definitions and results

In 1951 J. A. Green[13] introduced the following five equivalence relations $\mathcal{L}, \mathcal{R}, \mathcal{H}, \mathcal{D}$ and $\mathcal{J}$ on a semigroup S. These Green's equivalences are especially significant in the study of regular semigroups. We adopt here a usual notation for the above Green's relations as following:

$$\begin{aligned}
\mathcal{L} &= \{(a,b) \in S \times S \mid (\exists x, y \in S^1) xa = b, yb = a\}, \\
\mathcal{R} &= \{(a,b) \in S \times S \mid (\exists x, y \in S^1) ax = b, by = a\}, \\
\mathcal{H} &= \mathcal{L} \wedge \mathcal{R}, \\
\mathcal{D} &= \mathcal{L} \vee \mathcal{R}, \\
\mathcal{J} &= \{(a,b) \in S \times S \mid (\exists x, y, u, v \in S^1) xay = b, ubv = a\}.
\end{aligned}$$

It is easy to see that all these five relations can be reduced to the universal relation if S is a group.

By using the Green's relations and the idempotents of a semigroup, we can give some alternative descriptions for inverse semigroups.

Theorem 1.1 [Ref. 5, V. Th. 1.2]. *The following statements about a semigroup S are equivalent:*

(i) *S is an inverse semigroup;*
(ii) *S is regular and idempotents commute;*
(iii) *each $\mathcal{L}$-class and each $\mathcal{R}$-class of S contains a unique idempotent;*
(iv) *each principal left ideal and each principal right ideal of S contains a unique idempotent generator.*

As a special case of inverse semigroups, we need to notice Clifford semigroups. By a Clifford semigroup we mean a regular semigroup S in which the idempotents are *central*, i.e. in which $ex = xe$ for every idempotent e and every x in S. The following statements for a Clifford semigroup hinted that the structure of a semigroup S as a whole can be deeply influenced by the properties of the idempotents of S.

Theorem 1.2 [Ref. 5, IV. Th. 2.1]. *If S is a semigroup with set E of idempotents, then the following statements are equivalent:*

(i) *S is a Clifford semigroup;*
(ii) *S is regular and $\mathcal{D}^S \cap (E \times E) = 1_E$;*
(iii) *S is a semilattice of groups;*
(iv) *S is a strong semilattice of groups.*

To extend the class of inverse semigroups, G. L. Bailes[14] , P. S. Venkatesan[15] and M. Yamada[16] introduced and investigated left (right) inverse semigroups. According to M. Yamada,[16] a left inverse semigroup S is a regular semigroup in which the set of all idempotents of S is a left regular band, i.e. for any $e, f \in E(S)$, the identity $efe = ef$ holds in S; A quasi-inverse semigroup S is a regular semigroup in which the set of all idempotents of S is a regular band, i.e. for any $e, f, g \in E(S)$, the identity $efge = efege$ holds in S. In fact, a regular semigroup S is an inverse semigroup if and only if the set of all idempotents of S is a semillatice, i.e. for any $e, f \in E(S)$, the identity $ef = fe$ holds in S. Hence, left inverse semigroups and quasi-inverse semigroup are two generalizations of inverse semigroups in the class of regular semigroups.

In 1974 P. S. Venkatesan given the following descriptions for left inverse semigroups.

Theorem 1.3 [Ref.15, Th. 1]. *The following conditions on a regular semigroup S are equivalent:*
(i) *S is a left inverse semigroup;*
(ii) *$efe = ef$ for any idempotents e, f of S;*
(iii) *If a' and a'' are inverses of the element a of S then $aa' = aa''$;*
(iv) *$Se \cap Sf = Sef$ for any two idempotents e, f of S.*

For quasi-inverse semigroups, the following result was given by M. Yamada.[16]

Theorem 1.4 [Ref.16, Th. 3]. *A regular semigroup is a quasi-inverse semigroup if and only if it is isomorphic a subdirect product of a left inverse semigroup and a right inverse semigroup.*

Furthermore, M. Yamada[16] established a structure theorem of a left inverse semigroup. We first restate below the definition of the left half-direct product of semigroups given by M. Yamada.[16]

Definition 1.5 [Ref. 16]. *Let I be an inverse semigroup and Y the semilattice of idempotents of I. Let $B = \bigcup_{\alpha \in Y} B_\alpha$ be a left regular band, where each B_α is a left zero band. Suppose that φ is a mapping of I into the endomorphism semigroup $End(B)$ of B given by $\gamma \mapsto \sigma_\gamma$ for $\gamma \in I$ such that the family $\{\sigma_\gamma : \gamma \in I\}$ satisfies the following conditions:*

(C1) *for each $\gamma \in I, B_\alpha \sigma_\gamma \subseteq B_{\gamma\alpha(\gamma\alpha)^{-1}}$. In particular, if $\gamma \in Y, \sigma_\gamma$ is an inner endomorphism on B such that $e^{\sigma_\gamma} = fe$ for some $f \in B_\gamma$ and all $e \in B$.*

(C2) *for any $\alpha, \beta \in I$ and $f \in B_{\alpha\beta(\alpha\beta)^{-1}}, \sigma_\beta \sigma_\alpha \delta_f = \sigma_{\alpha\beta}\delta_f$, where δ_f is the inner endomorphism induced by f on B satisfying $h^{\delta_f} = fh = fhf$, for all $h \in B$.*

For the set $B \bowtie I = \{(e,\omega) \mid \omega \in I$ *and* $e \in B_{\omega\omega^{-1}}\}$, *we define a multiplication on* $B \bowtie I$ *as follows:*

$$(e,\omega)(f,\tau) = (ef^{\sigma_\omega}, \omega\tau),$$

where $f^{\sigma_\omega} = f\sigma_\omega$.

It can be easily verified that $B \bowtie I$ *is a semigroup under the above multiplication. We call this semigroup a left half-direct product of* B *and* I, *determined by* φ, *and we now denote this semigroup by* $B \bowtie_\varphi I$.

Theorem 1.6 [Ref. 16, Th. 2]. *A left inverse semigroup is isomorphic to a left half-direct product of a left regular band and an inverse semigroup.*

Conversely, a left half-direct product of a left regular band and an inverse semigroup is a left inverse semigroup.

For the structure of quasi-inverse semigroups, M. Yamada in Ref. 17 first introduced the concept of $H.D.$-product of semigroups and established the following construction.

Theorem 1.7 [Ref. 17, Th. 6]. *Let* S *be a quasi-inverse semigroup whose set of idempotents forms a regular band* E. *Let* δ *be the smallest inverse congruence on* S *such that* $\Gamma = S/\delta$ *is the greatest inverse semigroup induced by* δ *and let* Y *be the semilattice of* Γ. *Define the congruences* η_1, η_2 *on* E *by* $e\eta_1 f$ *if and only if* $e\mathcal{R}f$; $e\eta_2 f$ *if and only if* $e\mathcal{L}f$, *respectively. For* $X \subseteq E$, *write* $\widetilde{X} = \{\tilde{e} \mid e \in X\}$ *and* $\widehat{X} = \{\hat{e} \mid e \in X\}$, *where* $\tilde{e}$ *and* $\hat{e}$ *are the* η_1*-class and the* η_2*-class containing* $e \in X$, *respectively. Then the following statements hold :*

(i) $E/\eta_1 = \widetilde{E}$ *is a left regular band such that* $\widetilde{E} = \bigcup_{\alpha\in Y} \widetilde{E_\alpha}$, *where every* $\widetilde{E_\alpha}$ *is a left zero band;* $E/\eta_2 = \widehat{E}$ *is a right regular band such that* $\widehat{E} = \bigcup_{\alpha\in Y} \widehat{E_\alpha}$, *where each* $\widehat{E_\alpha}$ *is a right zero band, for every* $\alpha \in Y$.

(ii) S *is isomorphic to an H.D.-product of* $\widetilde{E}, \Gamma$ *and* $\widehat{E}$ *with respect to the mappings* φ' *and* ψ', *respectively. Conversely, any H.D.-product of a left regular band* $I = \bigcup_{\alpha\in Y} \Lambda_\alpha$, *an inverse semigroup* Γ *and a right regular band* $\Lambda = \bigcup_{\alpha\in Y} \Lambda_\alpha$ *with respect to the mappings* φ' *and* ψ' *is a quasi-inverse semigroup* S, *where* Γ *is the greatest inverse semigroup homomorphic image of* S *and* Y *is the semilattice of idempotents of* Γ.

By Theorem 1.1, a regular semigroup S is an inverse semigroup if and only if the set of all idempotents of S become a semilattice, i.e. the identity $ef = fe$ holds for any $e, f \in E(S)$. An orthodox semigroup is defined as a regular semigroup in which the set of idempotents forms a subsemigroup. Thus, an orthodox semigroup can be regarded as a generalization of an inverse semigroup.

N. Reilly and H. E. Scheiblich[18] in 1967 given an important observation for orthodox semigroups:

Theorem 1.8 [Ref. 18]. *If S is a regular semigroup, then the following statements are equivalent:*

(i) *S is orthodox ;*
(ii) *for any a, b in S, if a' is an inverse of a and b' is an inverse of b, then $b'a'$ is an inverse of ab;*
(iii) *if e is idempotent then every inverse of e is idempotent.*

Theorem 1.9 [Ref. 5]. *A regular semigroup S is orthodox if and only if*

$$(\forall a, b \in S) \quad [V(a) \cap V(b) \neq \emptyset \Rightarrow V(a) = V(b)].$$

The following important result for an orthodox semigroup was given by T. E. Hall in 1969.

Theorem 1.10 [Ref. 5]. *If S is an orthodox semigroup then the relation γ defined by*

$$\gamma = \{(a, b) \in S \times S \mid V(a) = V(b)\}$$

is a congruence on S. It is the smallest inverse semigroup congruence on S.

In the structure theory of orthodox semigroups, we have the following well known Hall-Yamada theorem of orthodox semigroups. This is very graceful and interesting result.

Theorem 1.11 [Hall-Yamada, Ref. 5]. *Let B be a band and T an inverse semigroup whose semilattice of idempotents is isomorphic to B/ε. Let γ_1 be the minimum inverse semigroup congruence on the Hall semigroup W_B and ψ an idempotent-separating homomorphism from T into W_B/γ_1 whose range contains all the idempotents of W_B/γ_1. Then the Hall-Yamada semigroup $S = \mathcal{H}(B, T, \psi)$ is an orthodox semigroup whose band of idempotents is isomorphic to B. If γ is the minimum inverse semigroup congruence on S, then $S/\gamma \simeq T$.*

Conversely, if S is an orthodox semigroup whose band of idempotents is B then there exists an idempotent-separating homomorphism $\theta : S/\gamma \to W_B/\gamma_1$ whose range contains all the idempotents of W_B/γ_1 such that S is isomorphic to $\mathcal{H}(B, S/\gamma, \theta)$.

2. Abundant semigroups

In this section, we will restate some properties and results of abundant semigroups which can be regarded as generalizations of inverse semigroups and regular semigroups mentioned in Section 1.

In 1982 an abundant semigroup was first introduced and studied by J. B. Fountain.[9] To show the definition of an abundant semigroup, we first cite a set of relations called Green's star relations on a semigroup S as follows:

$$\begin{aligned}
\mathcal{L}^* &= \{(a,b) \in S \times S \mid (\forall x, y \in S^1) ax = ay \Leftrightarrow bx = by\},\\
\mathcal{R}^* &= \{(a,b) \in S \times S \mid (\forall x, y \in S^1) xa = ya \Leftrightarrow xa = yb\},\\
\mathcal{H}^* &= \mathcal{L}^* \wedge \mathcal{R}^*,\\
\mathcal{D}^* &= \mathcal{L}^* \vee \mathcal{R}^*,\\
\mathcal{J}^* &= \{(a,b) \in S \times S \mid J^*(a) = J^*(b)\},
\end{aligned}$$

where $J^*(a)$ denote the principal *-ideal generated by the element a in S (see Ref. 9).

We have the following basic lemma for the Green's star relations.

Lemma 2.1 [Ref. 9, Lemma 1.1]. *Let S be a semigroup and let $a, b \in S$. Then the following conditions are equivalent:*

(i) $(a, b) \in \mathcal{L}^*$;

(ii) *there is an S^1-isomorphism $\varphi : aS^1 \to bS^1$ with $a\varphi = b$.*

Clearly, on any semigroup S we have $\mathcal{L} \subseteq \mathcal{L}^*$ and $\mathcal{R} \subseteq \mathcal{R}^*$. It is easy to see that for regular elements $a, b \in S$, $(a, b) \in \mathcal{L}^*$ if and only if $(a, b) \in \mathcal{L}$. For any result concerning $\mathcal{L}^*$ there is a dual result for $\mathcal{R}^*$. Moreover, we can easily see that $\mathcal{L}^*$ is a right congruence and $\mathcal{R}^*$ is a left congruence on S, respectively.

An abundant semigroup S is a semigroup in which each $\mathcal{L}^*$-class and each $\mathcal{R}^*$-class contains an idempotent. It is clear that a regular semigroup is abundant. As a generalization of an inverse semigroup, J. B. Fountain[19] first introduced and studied an adequate semigroup in the class of abundant semigroups. *An adequate semigroup* is an abundant semigroup S in which the set of idempotents of S forms a semilattice.

Theorem 2.2 [Ref. 19, Proposition 1.3]. *Let S be a semigroup, E its set of idempotents and T the set of regular elements in S. Then the following conditions are equivalent:*

(i) *S is adequate;*
(ii) *T is an inverse subsemigroup of S and E has non-empty intersection with each $\mathcal{L}^*$-class and each $\mathcal{R}^*$-class of S ;*
(iii) *T is an inverse subsemigroup of S and T has non-empty intersection with each $\mathcal{L}^*$-class and each $\mathcal{R}^*$-class of S ;*
(iv) *each $\mathcal{L}^*$-class and each $\mathcal{R}^*$-class of S contains a unique idempotent and the subsemigroup generated by E is regular.*

From Theorem 2.2 we see that if S is an adequate semigroup, then each $\mathcal{L}^*$-class and each $\mathcal{R}^*$-class of S contains a unique idempotent.For an element a of such a semigroup, the idempotent in the $\mathcal{L}^*$-class and the $\mathcal{R}^*$-class containing a will be denoted by a^* and $a^\dagger$, respectively. An adequate semigroup S is called *type A* if $ea = a(ea)^*$ and $ae = (ae)^\dagger a$ for all elements a in S and all idempotents e in S.

Theorem 2.3 [Ref. 19, Proposition 1.5]. *Let S be an adequate semigroup. Then the following conditions are equivalent:*
(i) *S is a type A semigroup;*
(ii) *for every idempotent e in S and every element a in S, $eS^1 \cap aS^1 = eaS^1$ and $S^1e \cap S^1a = S^1ae$;*
(iii) *there are embeddings $\phi_1 : S \to S_1, \phi_2 : S \to S_2$ where S_1, S_2 are inverse semigroups and $a^*\phi_1 = (a\phi_1)^* = (a\phi_1)^{-1}a\phi_1, a^\dagger\phi_2 = (a\phi_2)^\dagger = (a\phi_2)(a\phi_1)^{-1}$.*

Theorem 2.4 [Ref. 19, Proposition 1.6]. *If S is an adequate semigroup with semilattice of idempotents E, then*
(i) *for all $a, b \in S$, $a\mathcal{L}^*b$ if and only if $a^* = b^*$; $a\mathcal{R}^*b$ if and only if $a^\dagger = b^\dagger$;*
(ii) *for all $a, b \in S$, $(ab)^* = (a^*b)^*$ and $(ab)^\dagger = (ab^\dagger)^\dagger$;*
(iii) *for all $a, b \in S$, $(ab)^* \leqslant b^*$ and $(ab)^\dagger \leqslant a^\dagger$ where $\leqslant$ is the usual ordering on E.*

From the Definition and Theorem above, we can see that both adequate semigroups and type A semigroups are generalizations of inverse semigroups in the class of abundant semigroups.

Now we consider the following special adequate semigroups, namely, C-a semigroups. By a C-a semigroup we mean an abundant semigroup S whose idempotents are central in S.

Theorem 2.5 [Ref. 19, Proposition 2.9]. *Let S be an adequate semigroup whose set of idempotents E is a semilattice. Then the following conditions are equivalent :*
(i) *S is a C-a semigroup ;*
(ii) *every $\mathcal{H}^*$-class of S contains an idempotent ;*
(iii) *for all elements a of S, $a^* = a^+$;*
(iv) $\mathcal{L}^* = \mathcal{R}^* = \mathcal{H}^*$;
(v) *S is a strong semilattice of cancellative monoids.*

Although many properties of abundant semigroups are hereditary from the properties of regular semigroups, but there exist some remarkable differences between these two kinds of semigroups, both in structures and properties. We now list some of these differences below:

(i) The homomorphic image of a regular semigroup is still regular, but the homomorphic image of an abundant semigroup is not necessarily abundant[20] . In other words, if ρ is a congruence on an abundant semigroup S then its quotient semigroup S/ρ is not necessarily abundant.

(ii) The semilattice of regular semigroups is obviously regular, but it is not true for a semilattice of abundant semigroups.

(iii) In regular semigroups, it is well-known that the Lallement Lemma holds, that is, if ρ is a congruence on a regular semigroup S and $a\rho$ is an idempotent of S/ρ, then there exists an idempotent $e \in S$ such that $e\rho = a\rho$. However, in abundant semigroups, the Lallement Lemma may not hold.

Definition 2.6 [Ref. 20]. *A semigroup homomorphism $\varphi : S \to T$ is called a good homomorphism if for any $a, b \in S$, $a\mathcal{L}^*(S)b$ implies that $a\varphi\mathcal{L}^*(T)b\varphi$ and $a\mathcal{R}^*(S)b$ implies that $a\varphi\mathcal{R}^*(T)b\varphi$.*

Correspondingly, we call a semigroup congruence ρ a *good congruence* if its natural homomorphism $\varphi^\sharp : S \to S/\rho$ is a good homomorphism.

Equipped with the above definition, we can see immediately that the good homomorphic image of an abundant semigroup is still abundant.

Definition 2.7 [Ref. 20]. *A semigroup S is called an idempotent-connected (in brevity IC) semigroup if for every $a \in S$, with its corresponding idempotents $a^\dagger \in R_a^*(S)$ and $a^* \in L_a^*(S)$, respectively, there exists a bijection mapping $\alpha : \langle a^\dagger \rangle \to \langle a^* \rangle$ satisfying $xa = a(x\alpha)$, where $x \in \langle a^\dagger \rangle$.*

Thus, every regular semigroup S is an IC semigroup. This is because for any $a \in S$ and any $a' \in V(a)$, if we define a mapping $\alpha : \langle aa' \rangle \to \langle a'a \rangle$ such that $x\alpha = a'xa$ for $x \in \langle aa' \rangle$, then we can easily observe that such mapping α is the required mapping that makes the regular semigroup S an IC semigroup.

The following characterization for type A semigroups was given by A. El-Qallali and J. B. Fountain in Ref. 20.

Theorem 2.8 [Ref. 20, Proposition 3.1]. *An adequate semigroup S with semilattice of idempotents E satisfies IC condition if and only if S is a type A semigroup.*

The structure theory of adequate semigroups and type A semigroups have been extensively investigated by many authors. In particular, both Armstrong[21] and Lawson[22] studied the structure of type A semigroups and obtained several nice structure theorems for such semigroups.

In 1984 Armstrong[21] first defined the following partial order $\leqslant_l$ (partial order $\leqslant_r$) on an adequate semigroup S by

$$a \leqslant_l b \ (a \leqslant_r b) \text{ if and only if } a = be \ (a = eb)$$

for some $e^2 = e \in S$. Later on, Lawson[22] considered the partial order $\leqslant = \leqslant_r \cap \leqslant_l$ on an adequate semigroup. He obtained the following theorems for type A semigroups.

Theorem 2.9 [Ref. 22, Proposition 1.5]. *An adequate semigroup S is a type A semigroup if and only if $\leqslant = \leqslant_r = \leqslant_l$.*

Theorem 2.10 [Ref. 22, Proposition 1.6]. *Let S is a type A semigroup. Then the following statements hold :*

(i) *the partial order $\leqslant$ is compatible with the multiplication on S;*

(ii) *the restriction of $\leqslant$ to the idempotents of S is the natural partial order on the idempotents of S.*

(iii) *if* $e^2 = e \in S$ *and* a *is an element with* $a \leqslant e$, *then* a *is an idempotent.*
(iv) *if* $a, b \in S$, $a \leqslant b$ *and* $a\mathcal{L}^*b$ $(a\mathcal{R}^*b)$, *then* $a = b$.

On the other hand, S. Armstrong[21] established a structure theorem of type A semigroup by using the concept of a weak Croisot groupoid.

S. Armstrong[21] first introduced a concept of a weak Croisot groupoid (for details, see Ref. 21). If the weak Croisot groupoid G has a semilattice of idempotents $(E, \leqslant)$, then for any $e, f \in E$ with $f \leqslant e$, there is a pair of $\mathcal{R}^*$-classes R_e^* and R_f^* in G. In this case, we can define a mapping $\phi_{e,f} : R_e^* \to R_f^*$ by $a \mapsto fa$. Similarly, for the $\mathcal{L}^*$-classes L_e^* and L_f^* with $f \leqslant e$ in G, we also have $\psi_{e,f} : L_e^* \to L_f^*$ by $a \mapsto af$. The families of the mappings $\{\phi_{e,f} : R_e^* \to R_f^* \mid e \geqslant f\}$ and $\{\psi_{e,f} : L_e^* \to L_f^* \mid e \geqslant f\}$ are called the *structure mappings* of G if they satisfy the following conditions:

(C_1) $\phi_{e,e}$ is the identity mapping on R_e^* for all $e \in E$;

(C_2) if $e, f, g \in E$ and $g \leqslant f \leqslant e$, then $\phi_{e,g} = \phi_{e,f}\phi_{f,g}$;

(C_3) if $e, f \in E$ with $f \leqslant e$, then $e\phi_{e,f} = f$;

(C_4) for every $a \in G$ there exists a bijection $\theta_a : a^*E \to a^\dagger E$ such that for $f \leqslant a$, $f\theta_a \leqslant a^\dagger$ and $a\psi_{a^*,f} = a\phi_{a^\dagger, f\theta_a}$;

(C_5) if $a = b\phi_{b^\dagger, a^\dagger}$ then $a^* \leqslant b^*$ and $b\phi_{b^\dagger,a^\dagger}c\phi_{b^*,a^*} = (bc)\phi_{b^\dagger,a^\dagger}$ for $c \in R_{b^*}^*$.

Now, define a multiplication on G by

$$ab = (a\psi_{a^*, a^*b^\dagger})(b\phi_{b^\dagger, a^*b^\dagger}).$$

Then, one can easily verify that the groupoid G endowed with such multiplication is a semigroup, and is denoted by $G(\psi, \phi)$.

Equipped with the above concepts, the following structure theorem for type A semigroups can be given.

Theorem 2.11 [Ref. 21, Theorem 4.2]. *Given a weak Croisot groupoid* $(G, \cdot)$ *and families* ψ, ϕ *of mappings between* $\mathcal{L}^*$*-classes and* $\mathcal{R}^*$*-classes respectively of* G, *satisfying conditions* (C_1)-(C_5). *Then* $G(\psi, \phi)$ *is a type* A *semigroup and is the unique type semigroup with* *-*trace* $(G, \cdot)$ *and families of structure mappings* ψ, ϕ.

Conversely, every type A *semigroup* S *is isomorphic to some* $G(\psi, \phi)$ *for some suitable* G, ψ *and* ϕ.

By the *-trace of a type A semigroup S, we mean the partial groupoid $tr^*(S) = (S, \cdot)$ with a partial binary operation "·" defined by

$$a \cdot b = \begin{cases} ab & L_a^* \cap R_b^* \cap E \neq \emptyset, \\ \text{undefined} & \text{otherwise.} \end{cases}$$

Thus, in Theorem above, a weak Croisot groupoid $(G, \cdot)$ is indeed the *-trace of the type A semigroup S.

In order to find analogue of right (left) inverse semigroups in the class of abundant semigroups, A. El-Qallali[23] introduced the concept of $\mathcal{L}^*$–unipotent semigroup. He called an abundant semigroup S an $\mathcal{L}^*$-unipotent semigroup if its idempotents form a subsemigroup of S and every $\mathcal{L}^*$-class of S contains a unique idempotent.

Theorem 2.12 [Ref. 23, Lemma 1.1]. *Let S be a quasi-adequate semigroup which is abundant and in which the set E of idempotents becomes a subsemigroup. Then the following conditions are equivalent :*

(i) *S is $\mathcal{L}^*$-unipotent ;*
(ii) *for any $a \in S$, there exists a unique idempotent $e \in E$ such that $L^*(a) = Se$;*
(iii) *every principal left *-ideal has a unique idempotent generator ;*
(iv) *$fef = ef$ for any $e, f \in E$, i.e. the set of idempotents of S forms a right regular band.*

For $\mathcal{L}^*$-unipotent semigroups, A. El-Qallali proved the following result.

Theorem 2.13 [Ref. 23, Proposition 1.2]. *Let S be a semigroup with a set of idempotents E. If T is the set of regular elements of S then the following conditions are equivalent :*

(i) *S is $\mathcal{L}^*$-unipotent ;*
(ii) *T is a left inverse subsemigroup of S and E has non-empty intersection with each $\mathcal{L}^*$-class and each $\mathcal{R}^*$-class of S ;*
(iii) *T is a left inverse subsemigroup of S and T has non-empty intersection with each $\mathcal{L}^*$-class and each $\mathcal{R}^*$-class of S ;*
(iv) *$|R_a{}^* \cap E| \geqslant 1$ and $|L_a{}^* \cap E| = 1$ for all $a \in S$ and the subsemigroup generated by E is regular.*

Theorem 2.14 [Ref. 23, Proposition 1.3]. *Let S be a quasi-adequate semigroup with band of idempotent E. Then the following conditions are equivalent :*

(i) *S is $\mathcal{L}^*$-unipotent* ;
(ii) *$eS \cap fS = efS$ for any $e, f \in E$* ;
(iii) *$ef\mathcal{R}fe$ for any $e, f \in E$* ;
(iv) *the Green's relations $\mathcal{R}$ and $\mathcal{J}$ coincide on E* ;
(v) *$a^\dagger ea = ea$ for all $e \in E$, $a \in S$, $a^\dagger \in R_a{}^* \cap E$.*

As inspired by the definition of type A semigroups, we defined in Ref. 24 the *$\mathcal{L}^*$-inverse semigroups.* An abundant semigroup S is called an $\mathcal{L}^*$-inverse semigroup if S is an IC semigroup whose idempotents form a left regular band[24] .

Remark 1. As pointed out by G. L. Bailes in Ref. 14 , a semigroup S is a right inverse semigroup if and only if every $\mathcal{R}$-class of S contains a unique idempotent. Based on this fact, we can easily deduce that a regular semigroup S is a right inverse semigroup if and only if its set of idempotent forms a left regular band. On the other hand, M. Yamada[16] called a regular semigroup a *right inverse semigroup* if its set of idempotents forms a right regular band. Hence, there are discrepancy between G. L. Bailes and M. Yamada on the definition of left (right) inverse semigroups. Our definition in Ref. 24 for $\mathcal{L}^*$-inverse semigroups above is a generalization of left inverse semigroups, in the sense of M. Yamada.

The authors[24] have considered the following equivalence on an $\mathcal{L}^*$-inverse semigroup :

$$\delta = \{(a, b) \in S \times S \mid b = b^\dagger a, a = a^\dagger b \text{ where } \ b^\dagger \in E(a^\dagger), a^\dagger \in E(b^\dagger)\}.$$

Theorem 2.15 [Ref. 24, Theorem 2.8]. *The following statements hold for an $\mathcal{L}^*$-inverse semigroup S :*
(i) *the relation δ on S is the minimum type A good congruence, i.e. S/δ is a type A semigroup;*
(ii) $\delta \cap \mathcal{R}^* = 1$.

If B is the set of all idempotents of an $\mathcal{L}^*$-inverse semigroup S, then B is obviously a left regular band. According to the above theorem, one can easily see that S/δ is a type A semigroup. This gives us the clue to construct an $\mathcal{L}^*$-inverse semigroup by using the ingredients: type A semigroup S/δ and the left regular band B.

Let Γ be a type A semigroup with semilattice Y of idempotents. Let $B = \cup_{\alpha \in Y} B_\alpha$ be a semilattice decomposition of a left regular band B into left zero bands B_α.

Because the type A semigroup Γ is abundant, we can always identify the element $\gamma \in \Gamma$ by its corresponding idempotent $\gamma^{\dagger} \in R^*_{\gamma}(\Gamma) \cap E$ or by $\gamma^* \in L^*_{\gamma}(\Gamma) \cap E$, respectively. Moreover, since the type A semigroup Γ is also an IC abundant semigroup, there is a connecting isomorphism $\eta : \langle \omega^{\dagger} \rangle \to \langle \omega^* \rangle$ such that $\alpha\omega = \omega(\alpha\eta)$ for any $\alpha \in \langle \omega^{\dagger} \rangle$ and $\omega \in \Gamma$.

Now, we form the set $B \bowtie \Gamma = \{(e, \gamma) | \ e \in B_{\gamma^{\dagger}}, \gamma \in \Gamma\}$. In order to make this set $B \bowtie \Gamma$ a semigroup, we need to introduce a multiplication "$*$" defined on the set $B \bowtie \Gamma$ by the following mapping. Firstly, we define a mapping $\varphi : \Gamma \to End(B)$ by $\gamma \mapsto \sigma_{\gamma}$, where $\sigma_{\gamma} \in End(B)$ which is the endomorphism semigroup on B. This mapping satisfies following properties:

(P1) Absorbing: for each $\gamma \in \Gamma$ and $\alpha \in Y$, we have $B_{\alpha}\sigma_{\gamma} \subseteq B_{(\gamma\alpha)^{\dagger}}$. In particular, if $\gamma \in Y$, then σ_{γ} is an inner endomorphism on B such that $e^{\sigma_{\gamma}} = fe$ for some $f \in B_{\gamma}$ and all $e \in B$.

(P2) Focusing: for $\alpha, \beta \in \Gamma$ and $f \in B_{(\alpha\beta)^{\dagger}}$, we have $\sigma_{\beta}\sigma_{\alpha}\delta_f = \sigma_{\alpha\beta}\delta_f$, where δ_f is an inner endomorphism induced by f on B satisfying $h^{\delta_f} = fh$ for all $h \in B$.

(P3) Homogenizing: for $e \in B_{\omega^{\dagger}}, g \in B_{\tau^{\dagger}}$ and $h \in B_{\xi^{\dagger}}$, if $\omega\tau = \omega\xi$ and $eg^{\sigma_{\omega}} = eh^{\sigma_{\omega}}$, then $fg^{\sigma_{\omega^*}} = fh^{\sigma_{\omega^*}}$, for any $f \in B_{\omega^*}$.

(P4) Idempotent connecting: assume that for any $\omega \in \Gamma, \eta$ is the connecting isomorphism which maps $\langle \omega^{\dagger} \rangle$ to $\langle \omega^* \rangle$ by $\alpha \mapsto \alpha\eta$. If $(e, \omega^{\dagger})$ and $(f, \omega^*) \in B \bowtie \Gamma$, then there is a bijection $\theta : \langle e \rangle \to \langle f \rangle$ such that

(i) $e\theta = f$ and $g = e(g\theta)^{\sigma_{\omega}}$, for $g \in \langle e \rangle$;

(ii) for $g \in \langle e \rangle$ and $\alpha \in \langle \omega^{\dagger} \rangle, (g, \alpha) \in B \bowtie \Gamma$ if and only if $(g\theta, \alpha\eta) \in B \bowtie \Gamma$.

Equipped with the above mapping φ, we now define a multiplication "$*$" on $B \bowtie \Gamma$ by

$$(e, \omega) * (f, \tau) = (ef^{\sigma_{\omega}}, \omega\tau)$$

for any $(e, \omega), (f, \tau) \in B \bowtie \Gamma$, where $f^{\sigma_{\omega}} = f\sigma_{\omega}$.

It can be verified that the multiplication "$*$" defined above for the set $B \bowtie \Gamma$ is associative. We call the semigroup a left wreath product of a left regular band B and a type A semigroup Γ under a mapping φ, denoted by $B \bowtie_{\varphi} \Gamma$.

We are now going to establish a structure theorem for $\mathcal{L}^*$-inverse semigroups (The structure theorem for $\mathcal{R}^*$-inverse semigroups can be dually established).

Theorem 2.16 [Ref. 24, Theorem 4.1]. *A semigroup S is an $\mathcal{L}^*$-inverse semigroup if and only if S is a left wreath product of a left regular band B and a type A semigroup Γ.*

We observe here that if the conditions (P3) and (P4) in Theorem 2.16 are removed and the element $\gamma^\dagger$ is replaced by $\gamma\gamma^{-1}$, and also replace the type A semigroup by an inverse semigroup Γ, then the left wreath product $B \bowtie_\varphi \Gamma$ above will become the left half-direct product of a left regular band and an inverse semigroup which turns out to be a left inverse semigroup. Thus, as a special case of our Theorem 2.16, we can immediately deduce Theorem 1.6 of Yamada in 1973 for left inverse semigroups.

We now construct an example of a non-trivial $\mathcal{L}^*$-inverse semigroup from[24] by using the left wreath product of semigroups introduced above.

Example 1. We first let $Y = \{\alpha, \beta\}$ be a basic semilattice with $\alpha\beta = \beta$. Let $\Gamma_1 = \{\alpha, \alpha_n = 2^n A \mid n \in N\}$, where A denotes the matrix $\begin{pmatrix} 1 & 0 \\ 1 & 0 \end{pmatrix}$. Then we can check that Γ_1 forms an infinite monoid, under the matrix multiplication, where the identity of Γ_1 is α and $\alpha = A$.

Again, we let $\Gamma_2 = \{\beta, \beta_1, \beta_2, \cdots, \beta_5\}$ be the symmetric group S_3,where β is the identity of S_3. Then, we can verify that $\Gamma = \Gamma_1 \cup \Gamma_2$ forms a semigroup with the following Table 1, where $\alpha_n, \alpha_m \in \Gamma_1, n, m = 1, 2, \cdots$.

$*$	α	β	β_1	β_2	β_3	β_4	β_5	α_n
α	α	β	β_1	β_2	β_3	β_4	β_5	α_n
β	β	β	β_1	β_2	β_3	β_4	β_5	β
β_1	β_1	β_1	β	β_4	β_5	β_2	β_3	β_1
β_2	β_2	β_2	β_5	β	β_4	β_3	β_1	β_2
β_3	β_3	β_3	β_4	β_5	β	β_1	β_2	β_3
β_4	β_4	β_4	β_3	β_1	β_2	β_5	β	β_4
β_5	β_5	β_5	β_2	β_3	β_1	β	β_4	β_5
α_m	α_m	β	β_1	β_2	β_3	β_4	β_5	α_{m+n}

Table 1

By definition, we can easily check that $\Gamma = \Gamma_1 \cup \Gamma_2$ is an abundant semigroup satisfying the IC condition and hence Γ is a type A semigroup since the idempotents $\{\alpha, \beta\}$ of Γ forms a semilattice.

To construct a left regular band, we let $B = B_\alpha \cup B_\beta = \{e_1, e_2, e_3, e_4\}$, where $B_\alpha = \{e_1, e_2\}, B_\beta = \{e_3, e_4\}$ are respectively the left zero bands.

Then $B = \bigcup_{\alpha \in Y} B_\alpha$ is a left regular band with the following Table 2.

$*$	e_1	e_2	e_3	e_4
e_1	e_1	e_1	e_4	e_4
e_2	e_2	e_2	e_4	e_4
e_3	e_3	e_3	e_3	e_3
e_4	e_4	e_4	e_4	e_4

Table 2

By using the above constructed type A semigroup Γ and the left regular band B, we form the set product $S = B \bowtie \Gamma = \{a, b, c, d, e, f, g, h, u, v, w, x, y, z, a_n, b_m \mid n, m \in N\}$, where the elements of $S = B \bowtie T$ are the following elements:

$$\begin{array}{llll} a = (e_1, \alpha), & b = (e_2, \alpha), & c = (e_3, \beta), & d = (e_4, \beta) \\ e = (e_3, \beta_1), & f = (e_4, \beta_1), & g = (e_3, \beta_2), & h = (e_4, \beta_2) \\ u = (e_3, \beta_3), & v = (e_4, \beta_3), & w = (e_3, \beta_4), & x = (e_4, \beta_4) \\ y = (e_3, \beta_5), & z = (e_4, \beta_5), & a_n = (e_1, \alpha_n), & b_m = (e_2, \alpha_m). \end{array}$$

By applying the definition of left wreath product of B and Γ, as described above, we can link up the semigroups B and Γ by using a mapping ϕ. Now we give the mapping $\phi : \Gamma \to End(B)$ by $\gamma \mapsto \sigma_\gamma$ as follows:

$$\sigma_\alpha = \begin{pmatrix} e_1 & e_2 & e_3 & e_4 \\ e_1 & e_1 & e_4 & e_4 \end{pmatrix}, \sigma_{\alpha_n} = \begin{pmatrix} e_1 & e_2 & e_3 & e_4 \\ e_2 & e_2 & e_4 & e_4 \end{pmatrix}, \sigma_\beta = \begin{pmatrix} e_1 & e_2 & e_3 & e_4 \\ e_3 & e_3 & e_3 & e_3 \end{pmatrix},$$

$$\sigma_{\beta_1} = \begin{pmatrix} e_1 & e_2 & e_3 & e_4 \\ e_4 & e_4 & e_4 & e_4 \end{pmatrix}, \sigma_{\beta_2} = \begin{pmatrix} e_1 & e_2 & e_3 & e_4 \\ e_3 & e_4 & e_4 & e_4 \end{pmatrix}, \sigma_{\beta_3} = \begin{pmatrix} e_1 & e_2 & e_3 & e_4 \\ e_3 & e_3 & e_3 & e_4 \end{pmatrix},$$

and $\sigma_{\beta_3} = \sigma_{\beta_4} = \sigma_{\beta_5}$. The mapping ϕ maps the type A semigroup Γ into the semigroup $End(B)$ of endomorphisms on B. The multiplication of the semigroup $S = B \bowtie_\phi \Gamma$ is given by $(s, \omega) * (t, \tau) = (st^{\sigma_\omega}, \omega\tau)$, for any element $(s, \omega), (t, \tau) \in S = B \bowtie_\phi \Gamma$, for example if we let $a = (e_1, \alpha), f = (e_4, \beta_1) \in B \bowtie_\phi \Gamma$, then we define $a * f = (e_1 e_4^{\sigma_\alpha}, \alpha\beta_1) = (e_1 e_4, \beta_1) = (e_4, \beta_1) = f$. Similarly, we can define the multiplication on S for other cases.

By using the above defined multiplication "$*$", we obtain the Table 3 for the semigroup $S = B \bowtie_\phi \Gamma$. where $n, m, i, j = 1, 2, 3, \cdots$.

In Table 3, we can easily see that $E(S) = \{a, b, c, d\}$ is the set of all idempotents of the semigroup S. Hence, we can check that the only $\mathcal{L}^*$-classes of S are the sets $\{a, b, a_n, b_i \mid n, i \in N\}$ and $\{c, d, e, f, g, h, u, v, w, x, y, z\}$. The $\mathcal{R}^*$-classes of S are the sets $\{a, a_n \mid n \in N\}$; $\{b, b_i \mid i \in N\}$; $\{c, e, g, u, w, y\}$ and $\{d, f, h, v, x, z\}$, respectively. Clearly, each $\mathcal{R}^*$-class and each $\mathcal{L}^*$-class of S contains an idempotent and hence S is an abundant semigroup.

$*$	a	b	c	d	e	f	g	h	u	v	w	x	y	z	a_n	b_j
a	a	a	d	d	f	f	h	h	v	v	x	x	z	z	a_n	a_j
b	b	b	d	d	f	f	h	h	v	v	x	x	z	z	b_n	b_j
c	c	c	c	c	e	e	g	g	u	u	w	w	y	y	c	c
d	d	d	d	d	f	f	h	h	v	v	x	x	z	z	d	d
e	e	e	e	e	c	c	w	w	y	y	g	g	u	u	e	e
f	f	f	f	f	d	d	x	x	z	z	h	h	v	v	f	f
g	g	g	g	g	y	y	c	c	w	w	u	u	e	e	g	g
h	h	h	h	h	z	z	d	d	x	x	v	v	f	f	h	h
u	u	u	u	u	w	w	y	y	c	c	e	e	g	g	u	u
v	v	v	v	v	x	x	z	z	d	d	f	f	h	h	v	v
w	w	w	w	w	u	u	e	e	g	g	y	y	c	c	w	w
x	x	x	x	x	v	v	f	f	h	h	z	z	d	d	x	x
y	y	y	y	y	g	g	u	u	e	e	c	c	w	w	y	y
z	z	z	z	z	h	h	v	v	f	f	d	d	x	x	z	z
a_m	a_m	a_m	d	d	f	f	h	h	v	v	x	x	z	z	a_{m+n}	a_{m+j}
b_i	b_i	b_i	d	d	f	f	h	h	v	v	x	x	z	z	b_{i+n}	b_{i+j}

Table 3

Furthermore, we can easily observe that $E(S)$, the set of all idempotents of the semigroup S, forms a left regular band and $S_1 = \{a, b, c, d, e, f, g, h, u, v, w, x, y, z\}$ is a left inverse subsemigroup of S. Let $\langle v \rangle$ denotes the subsemigroup of S generated by all $x \in E(S)$ with $x \leqslant v$. Then, we immediately see that $\langle a \rangle = \{a, d\}$, $\langle b \rangle = \{b, d\}$, $\langle c \rangle = c$ and $\langle d \rangle = d$ in S. In this way, we can check from the above Table 3 that S is indeed an IC abundant semigroup and by our Definition, S is an $\mathcal{L}^*$-inverse semigroup. Because every element of $S \setminus S_1$ is a non-regular element, S is of course not a left inverse semigroup. Also, since the idempotents of S do not commute with each other, S is not a type A semigroup. This example illustrates that the class of type A semigroups and the class of left inverse semigroups are two proper subclasses of the class of $\mathcal{L}^*$-inverse semigroups. Thus, the $\mathcal{L}^*$-inverse semigroups are indeed the generalizations of both type A semigroups and left inverse semigroups.

An $\mathcal{L}^*$-inverse semigroup can be regarded as a natural generalization of a left inverse semigroup in the class of regular semigroups. Now, we furthermore consider $\mathcal{Q}^*$-inverse semigroups, which are IC abundant semigroups whose set of idempotents forms a regular band. This class of semigroups not only contains the $\mathcal{L}^*$-inverse semigroups as its special subclass but also they can be regarded as an analogue of the quasi-inverse semigroups considered by M. Yamada in Ref. 16 within the class of regular semigroups.

Thus, we can establish a corresponding hierarchy between the class of regular semigroups and the class of abundant semigroups, starting from inverse semigroups, left (right) inverse semigroups, quasi-inverse semigroups and orthodox semigroups to type A semigroups, $\mathcal{L}^*$-inverse ($\mathcal{R}^*$-inverse) semigroups, $\mathcal{Q}^*$-inverse semigroups and type W semigroups.

We call an IC abundant semigroup S a *$\mathcal{Q}^*$-inverse semigroup* if the set of its idempotents E forms a regular band, i.e. E satisfies the identity $efege = efge$, for all e, f and g in E.

We assume that S is a $\mathcal{Q}^*$-inverse semigroup whose set of idempotents E forms a regular band. Denote the $\mathcal{J}$-class containing the element $e \in E$ by $E(e)$. We first have the following result.

Theorem 2.17 [Ref. 25, Theorem 3.2]. *If an equivalence relation δ on S is defined by $a\delta b$ if and only if $b = eaf$ and $a = gbh$ for some $e \in E(a^+), f \in E(a^*), g \in E(b^+)$ and $h \in E(b^*)$, then the equivalence relation δ is the smallest type A good congruence on S.*

Let S be a $\mathcal{Q}^*$-inverse semigroup with a regular band of idempotents E. Define relations μ_l and μ_r on S as follows:

$$(a, b) \in \mu_l \Leftrightarrow (xa, xb) \in \mathcal{L}^* \quad (x \in E), \tag{3}$$

$$(a, b) \in \mu_r \Leftrightarrow (ax, bx) \in \mathcal{R}^* \quad (x \in E). \tag{4}$$

Put $\rho_1 = \delta \cap \mu_r$ and $\rho_2 = \delta \cap \mu_l$ on S (see Ref. 25). We are now able to establish the following theorem for $\mathcal{Q}^*$-inverse semigroups.

Theorem 2.18 [Ref. 25, Theorem 3.7 and Proposition 3.8]. *Let S be a $\mathcal{Q}^*$-inverse semigroup and ρ_1, ρ_2 be the congruences given above on S. Then the following statements hold:*

(i) *ρ_1 is an $\mathcal{L}^*$-inverse semigroup good congruence on S, i.e. S/ρ_1 is an $\mathcal{L}^*$-inverse semigroup.*

(ii) *ρ_2 is an $\mathcal{R}^*$-inverse semigroup good congruence on S, i.e. S/ρ_2 is an $\mathcal{R}^*$-inverse semigroup.*

(iii) *$\rho_1 \cap \rho_2 = 1_S$, where 1_S is the identity relation on S.*

To obtain structure theory for $\mathcal{Q}^*$-inverse semigroups, the concept of the *wreath product* of semigroups was introduced by Ren and Shum in Ref. 25 as follows:

In the *wreath product* of semigroups, we need the following ingredients :

(a) Y: a semilattice.

(b) Γ: a type A semigroup whose set of idempotents is the semilattice Y.

(c) I: a left regular band such that $I = \bigcup_{\alpha \in Y} I_\alpha$, where I_α is a left zero band for all $\alpha \in Y$.

(d) Λ: a right regular band such that $\Lambda = \bigcup_{\alpha \in Y} \Lambda_\alpha$,where Λ_α is a right zero band for all $\alpha \in Y$.

We now form the following sets:

$$I \bowtie \Gamma = \{(e, \omega) |\ \omega \in \Gamma, e \in I_{\omega^+}\},$$

$$\Gamma \bowtie \Lambda = \{(\omega, i) |\ \omega \in \Gamma, i \in \Lambda_{\omega^*}\},$$

and

$$I \bowtie \Gamma \bowtie \Lambda = \{(e, \omega, i) |\ \omega \in \Gamma, e \in I_{\omega^+} \text{ and } i \in \Lambda_{\omega^*}\}.$$

Since $\omega \in \Gamma$ and Γ is a type A semigroup, there are some idempotents $\omega^\dagger \in R^*_\omega(\Gamma) \cap E(\Gamma)$ and $\omega^* \in L^*_\omega(\Gamma) \cap E(\Gamma)$. Also since the set of idempotents of Γ forms a semilattice, $\omega^\dagger$ and ω^* are in Y. This illustrates that the sets $I \bowtie \Gamma, \Gamma \bowtie \Lambda$, and $I \bowtie \Gamma \bowtie \Lambda$ are well-defined. We only need to define an associative multiplication on the set $I \bowtie \Gamma \bowtie \Lambda$ so that the set $I \bowtie \Gamma \bowtie \Lambda$ under the multiplication turns out to be a semigroup.

Before we define a multiplication on $I \bowtie \Gamma \bowtie \Lambda$, we need to give a description for the structure mappings.

Define a mapping $\varphi : \Gamma \to End(I)$ by $\gamma \mapsto \sigma_\gamma$ for $\gamma \in \Gamma$ and $\sigma_\gamma \in End(I)$ satisfying the following conditions:

(P1) For each $\gamma \in \Gamma$ and $\alpha \in Y$, we have $I_\alpha \sigma_\gamma \subseteq I_{(\gamma\alpha)^\dagger}$. In particular, if $\gamma \in Y$ then σ_γ is an inner endomorphism on I such that there exists $g \in I_\gamma$ with $e^{\sigma_\gamma} = ge$, for all $e \in I$, where e^{σ_γ} denotes $e\sigma_\gamma$.

(P2) For $\alpha, \beta \in \Gamma$ and $f \in I_{(\alpha\beta)^\dagger}$, we have $\sigma_\beta \sigma_\alpha \delta_f = \sigma_{\alpha\beta}\delta_f$, where δ_f is an inner endomorphism induced by f on I satisfying $h^{\delta_f} = fh = fhf$, for all $h \in I$.

(P3) For $e \in I_{\omega^\dagger}, g \in I_{\tau^\dagger}$ and $h \in I_{\xi^\dagger}$, if $\omega\tau = \omega\xi$ and $eg^{\sigma_\omega} = eh^{\sigma_\omega}$, then $fg^{\sigma_{\omega^*}} = fh^{\sigma_{\omega^*}}$, for all $f \in I_{\omega^*}$.

(P4) Assume that for any $\omega \in \Gamma, \eta$ is the connecting isomorphism which maps $\langle \omega^\dagger \rangle$ to $\langle \omega^* \rangle$ by $\alpha \mapsto \alpha\eta$. If $(e, \omega^\dagger)$ and $(f, \omega^*) \in I \bowtie \Gamma$, then there is a bijection $\theta : \langle e \rangle \to \langle f \rangle$ such that

(i) $e\theta = f$ and $ge^{\sigma_\alpha} = e(g\theta)^{\sigma_\omega}$ for any $g \in \langle e \rangle$ and $\alpha \in \langle \omega^\dagger \rangle$.

(ii) For any $g \in \langle e \rangle$ and $\alpha \in \langle \omega^\dagger \rangle, (g, \alpha) \in I \bowtie \Gamma$ if and only if $(g\theta, \alpha\eta) \in I \bowtie \Gamma$.

Similarly, define a mapping $\psi : \Gamma \to End(\Lambda)$ by $\gamma \mapsto \rho_\gamma$ for $\gamma \in \Gamma$ and $\rho_\gamma \in End(\Lambda)$ satisfying the following conditions:

(P1)$'$ For each $\gamma \in \Gamma$ and $\alpha \in Y$, we have $\Lambda_\alpha \rho_\gamma \subseteq \Lambda_{(\alpha\gamma)^*}$. In particular, if $\gamma \in Y$, then ρ_γ is an inner endomorphism on Λ such that there exists $i \in \Lambda_\gamma$ with $j^{\rho_\gamma} = ji$ for all $j \in \Lambda$, where j^{ρ_γ} denotes $j\rho_\gamma$.

(P2)$'$ For $\alpha, \beta \in \Gamma$ and $i \in \Lambda_{(\alpha\beta)^*}$, we have $\rho_\alpha \rho_\beta \varepsilon_i = \rho_{\alpha\beta} \varepsilon_i$, where ε_i is an inner endomorphism induced by i on Λ such that $j^{\varepsilon_i} = ji = iji$ for any $j \in \Lambda$.

(P3)$'$ For $i \in \Lambda_{\omega^*}, j \in \Lambda_{\tau^*}$ and $k \in \Lambda_{\xi^*}$, if $\tau\omega = \xi\omega$ and $j^{\rho_\omega} i = k^{\rho_\omega} i$, then $j^{\rho_{\omega^\dagger}} m = k^{\rho_{\omega^\dagger}} m$ for all $m \in \Lambda_{\omega^\dagger}$.

(P4)$'$ Assume that for any $\omega \in \Gamma, \eta$ is the connecting isomorphism which maps $\langle \omega^\dagger \rangle$ to $\langle \omega^* \rangle$ by $\alpha \mapsto \alpha\eta$. If $(\omega^\dagger, j)$ and $(\omega^*, i) \in \Gamma \bowtie \Lambda$, Then there is a bijection $\theta' : \langle i \rangle \to \langle j \rangle$ such that the following conditions hold:

(i) $j\theta' = i, k^{\rho_\omega} i = i^{\rho_{\alpha\eta}}(k\theta')$, for any $k \in \langle j \rangle$ and $\alpha \in \langle \omega^\dagger \rangle$;

(ii) For any $k \in \langle i \rangle$ and $\alpha \in \langle \omega^\dagger \rangle, (\alpha, k) \in \Gamma \bowtie \Lambda$ if and only if $(\alpha\eta, k\theta') \in \Gamma \bowtie \Lambda$.

After gluing up the above components I, Γ and Λ together with the mappings φ and ψ, we now define a multiplication on the set $I \bowtie \Gamma \bowtie \Lambda$ by

$$(e, \omega, i) * (f, \tau, j) = (ef^{\sigma_\omega}, \omega\tau, i^{\rho_\tau} j), \tag{5}$$

for any $(e, \omega, i), (f, \tau, j) \in I \bowtie \Gamma \bowtie \Lambda$, where $f^{\sigma_\omega} = f\sigma_\omega$ and $i^{\rho_\tau} = i\rho_\tau$.

Using the properties $(P1), (P2), (P1)'$ and $(P2)'$, we can easily verify that the above multiplication " $*$" on $I \bowtie \Gamma \bowtie \Lambda$ is associative. We now call the above constructed semigroup *the wreath product* of I, Γ and Λ with respect to φ and ψ, and denote it by $Q = I \bowtie_\varphi \Gamma \bowtie_\psi \Lambda$.

Theorem 2.19 [Ref. 25, Theorem 4.4]. *The wreath product $I \bowtie_\varphi \Gamma \bowtie_\psi \Lambda$ of a left regular band I, a type A semigroup Γ and a right regular band Λ with respect to the mappings φ and ψ is a $\mathcal{Q}^*$-inverse semigroup.*

Conversely, every $\mathcal{Q}^*$-inverse semigroup S can be expressed by a wreath *product of* $I \bowtie_{\varphi} \Gamma \bowtie_{\psi} \Lambda$.

Remark 2. The class of $\mathcal{Q}^*$-inverse semigroups contains several interesting classes of semigroups as its special subclasses. We only discuss some of these special subclasses as follows.

(a) $\mathcal{L}^*$-inverse semigroups and $\mathcal{R}^*$-inverse semigroups

By Theorem 2.19, a $\mathcal{Q}^*$-inverse semigroup S can be expressed as a wreath product $I \bowtie_{\varphi} \Gamma \bowtie_{\psi} \Lambda$ of I, Γ and Λ with respect to the mappings φ and ψ, where Γ is a type A semigroup, I and Λ are respectively a left regular band and a right regular band. In Theorem 2.19, if $\Lambda = \emptyset$, then $S = I \bowtie_{\varphi} \Gamma$, which is an $\mathcal{L}^*$-inverse semigroup. Similarly, if we let $I = \emptyset$, then $\Gamma \bowtie_{\psi} \Lambda$ becomes an $\mathcal{R}^*$-inverse semigroup. Thus, the class of $\mathcal{L}^*$-inverse semigroups and the class of $\mathcal{R}^*$-inverse semigroups are two special subclasses of the class of $\mathcal{Q}^*$-inverse semigroups. In this case, we can easily reobtain Theorem 2.16 for structure of $\mathcal{L}^*$-inverse semigroups, as a corollary of Theorem 2.19.

(b) Quasi-inverse semigroups

We know that a quasi-inverse semigroup is a regular semigroup whose set of idempotents forms a regular band. It is clear that a quasi-inverse semigroup is a special $\mathcal{Q}^*$-inverse semigroup.

When S is a quasi-inverse semigroup, we can define a relation δ on S by $a\delta b$ if and only if $b = eaf$ for some $e \in E(aa')$ and $f \in E(a'a)$, where a' is an inverse element of a. It can be immediately seen from [20] that δ is the smallest inverse semigroup congruence on S and so $\Gamma = S/\delta$ is the greatest inverse semigroup homomorphism image of S. Obviously, the inverse semigroup $\Gamma = S/\delta$ must be a type A semigroup whose set of idempotents forms a semilattice. As a result, a wreath product $I \bowtie_{\varphi} \Gamma \bowtie_{\psi} \Lambda$ of S, regarded as a Q^*-inverse semigroup, can be simplified by using the so called *half-direct product* (in brevity, *H.D.*-product) of a quasi-inverse semigroup given by M. Yamada in Ref. 17 as described in Theorem 1.7.

In 1958, Kimura first considered[5] the spined product of semigroups as follows: if S and T are two semigroups having a common homomorphic image H, and if $\phi : S \to H$ and $\psi : T \to H$ are homomorphisms onto H, then the spined product of S and T with respect to H, ϕ and ψ is defined by

$$Y = \{(s,t) \in S \times T \mid s\phi = t\psi\}.$$

For $\mathcal{Q}^*$-inverse semigroups, we have also the following another constructions.

Theorem 2.20 [Ref. 25, Theorem 5.1]. *A semigroup S is a $\mathcal{Q}^*$-inverse semigroup if and only if S is a spined product of an $\mathcal{L}^*$-inverse semigroup $S_1 = I \bowtie_\varphi \Gamma$ and an $\mathcal{R}^*$-inverse semigroup $S_2 = \Gamma \bowtie_\psi \Lambda$ with respect to a type A semigroup Γ.*

We now give a constructive example of a $\mathcal{Q}^*$-inverse semigroup which from Ref. 25.

Example 2. Let

$$a = \begin{pmatrix} 1 & 1 & 0 \\ 0 & 0 & 0 \\ 0 & 0 & 0 \end{pmatrix}, \quad b = \begin{pmatrix} 1 & 1 & 1 \\ 0 & 0 & 0 \\ 0 & 0 & 0 \end{pmatrix}$$

$$c = \begin{pmatrix} 0 & 0 & 0 \\ 1 & 1 & 0 \\ 0 & 0 & 0 \end{pmatrix}, \quad d = \begin{pmatrix} 0 & 0 & 0 \\ 1 & 1 & 1 \\ 0 & 0 & 0 \end{pmatrix}.$$

Also, let $a_n = 2^n a$, $b_n = 2^n b$, $c_n = 2^n c$ and $d_n = 2^n d$ for any $n \geqslant 1$. Then, we can easily verify that the set $S_1 = \{a, b, c, d, a_n, b_n, c_n, d_n \mid n \geqslant 1\}$ under the usual matrix multiplication forms a non-regular semigroup. Now we construct a semigroup $S = \{a, b, c, d, e, f, g, h, i, j, u, v, w, x, y, z,$ $a_n, b_n, c_n, d_n \mid n \geqslant 1\}$ with the following multiplication Table 4 which includes S_1 as its subsemigroup.

In Table 4, one can easily see that $\{a, b, c, d, e, f, g, h\}$ is the set of all idempotents of the semigroup S and $T = \{a, b, c, d, e, f, g, h, i, j, u, v, w, x,$ $y, z\}$ forms a regular subsemigroup of S. Moreover, it can be easily checked that $E(S) = \{a, b, c, d, e, f, g, h\}$ is a regular band, that is, $xyxzx = xyzx$ for any $x, y, z \in E(S)$. Hence, T is indeed a quasi-inverse subsemigroup of S.

By definition, it can be checked that the $\mathcal{L}^*$-classes of S are the sets $\{a, c, a_n, c_p\}$, $\{b, d, b_m, d_q\}$, $\{e, g, i, u, w, y\}$ and $\{f, h, j, v, x, z\}$ and the $\mathcal{R}^*$-classes of S are the sets $\{a, b, a_n, b_m\}, \{c, d, c_p, d_q\}$, $\{e, f, i, j, w, x\}$ and $\{g, h, u, v, y, z\}$ for any $n, m, p, q \geqslant 1$. Thus, S is an abundant semigroup since each $\mathcal{L}^*$-class and each $\mathcal{R}^*$-class of S contains an idempotent.

According to Table 4, we can observe that $\langle x \rangle = \langle x, g, h \rangle$ for any $x \in \{a, b, c, d\}$ and $\langle x \rangle = x$ for any $x \in \{e, f, g, h\}$, where $\langle x \rangle$ is used to denote the subsemigroup of S generated by all $y \in E(S)$ with $y \leqslant x$.

As a consequence, we can easily see that S satisfies the IC condition as mentioned above, and hence S is an IC abundant semigroup. Now, by Definition, S is a $\mathcal{Q}^*$-inverse semigroup because $E(S)$ is a regular band. Clearly, the class of quasi-inverse semigroups is a proper subclass of the class of $\mathcal{Q}^*$-inverse semigroups. Moreover, since $E(S)$ is not a left regular band, S is not an $\mathcal{L}^*$-inverse semigroup. In fact, an $\mathcal{L}^*$-inverse semigroup is a special $\mathcal{Q}^*$-inverse semigroups.

$*$	a	b	c	d	e	f	g	h	i	j	u	v	w	x	y	z	a_n	b_m	c_p	d_q
a	a	b	a	b	g	h	g	h	u	v	u	v	y	z	y	z	a_n	b_m	a_p	b_q
b	a	b	a	b	g	h	g	h	u	v	u	v	y	z	y	z	a_n	b_m	a_p	b_q
c	c	d	c	d	g	h	g	h	u	v	u	v	y	z	y	z	c_n	d_m	c_p	d_q
d	c	d	c	d	g	h	g	h	u	v	u	v	y	z	y	z	c_n	d_m	c_p	d_q
e	e	e	e	e	e	f	e	f	i	j	i	j	w	x	w	x	e	e	e	e
f	f	f	f	f	e	f	e	f	i	j	i	j	w	x	w	x	f	f	f	f
g	g	g	g	g	g	h	g	h	u	v	u	v	y	z	y	z	g	g	g	g
h	h	h	h	h	g	h	g	h	u	v	u	v	y	z	y	z	h	h	h	h
i	i	i	i	i	i	j	i	j	w	x	w	x	e	f	e	f	i	i	i	i
j	j	j	j	j	i	j	i	j	w	x	w	x	e	f	e	f	j	j	j	j
u	u	u	u	u	u	v	u	v	y	z	y	z	g	h	g	h	u	u	u	u
v	v	v	v	v	u	v	u	v	y	z	y	z	g	h	g	h	v	v	v	v
w	w	w	w	w	w	x	w	x	e	f	e	f	i	j	i	j	w	w	w	w
x	x	x	x	x	w	x	w	x	e	f	e	f	i	j	i	j	x	x	x	x
y	y	y	y	y	y	z	y	z	g	h	g	h	u	v	u	v	y	y	y	y
z	z	z	z	z	y	z	y	z	g	h	g	h	u	v	u	v	z	z	z	z
a_l	a_l	b_l	a_l	b_l	g	h	g	h	u	v	u	v	y	z	y	z	a_{l+n}	b_{l+m}	a_{l+p}	b_{l+q}
b_k	a_k	b_k	a_k	b_k	g	h	g	h	u	v	u	v	y	z	y	z	a_{k+n}	b_{k+m}	a_{k+p}	b_{k+q}
c_r	c_r	d_r	c_r	d_r	g	h	g	h	u	v	u	v	y	z	y	z	c_{r+n}	d_{r+m}	c_{r+p}	d_{r+q}
d_s	c_s	d_s	c_s	d_s	g	h	g	h	u	v	u	v	y	z	y	z	c_{s+n}	d_{s+m}	c_{s+p}	d_{s+q}

Table 4

Naturally, one would ask what will be the corresponding analogue of orthodox semigroups in the class of abundant semigroups ? The answer is quasi-adequate semigroups or type W semigroups.

Quasi-adequate semigroup and type W semigroups were firstly initiated by El-Qallali and Fountain in Ref. 26. They first called an abundant semigroup whose idempotents form a subsemigroup an quasi-adequate semigroup. Then, they defined a binary relation on such semigroups by

$$a \delta b \quad \text{if and only if} \quad E(a^+) a E(a^*) = E(b^+) b E(b^*)$$

for some idempotents a^+, a^*, b^+, b^* related to a and b, respectively. If the above binary relation δ is a congruence on the quasi-adequate semigroup S

and it is IC, then El-Qallali and Fountain called the semigroup a type W semigroup[26] . They asked: Is an IC quasi-adequate semigroup a type W semigroup ? This open question has now been answered by X. J. Guo in Ref. 27. In fact, X. J. Guo proved the following crucial result.

Theorem 2.21 [Ref. 27, Theorem 4.1.3]. *If S is an IC quasi-adequate semigroup then the above binary relation δ is a congruence on S.*

This result not only answers the question of El-Qallila and Fountain [6], but also pointed out that a quasi-adequate semigroup S is a type W semigroup if and only if S is IC.

It is obvious that both quasi-adequate semigroups and type W semigroups in the class of abundant semigroups are naturally generalizations of orthodox semigroups in the class of regular semigroups.

We now cite some properties of quasi-adequate semigroups obtained by A. El-Qallali and J. B. Fountain[26] .

Theorem 2.22 [Ref. 26, Proposition 1.3]. *Let S be a semigroup and T be the set of all regular elements in S. Then the following conditions are equivalent:*
(i) *S is quasi-adequate ;*
(ii) *S is abundant and T is an orthodox semigroup ;*
(iii) *T is an orthodox subsemigroup which has a non-empty intersection with every $\mathcal{L}^*$-class and every $\mathcal{R}^*$-class of S.*

By A. El-Qallali and J. B. Fountain[26] , we can obtain a construction of type W semigroups. The method is briefly sketched below :

First, we give a band B and a type A semigroup T with semilattice Y of idempotents isomorphic to $B/\mathcal{J}$. Let $\psi : T \to W_B/\gamma$ be an idempotent-separating good homomorphism whose range contains all the idempotents of W_B/γ, where γ is the minimum inverse semigroup congruence on W_B which is the *Hall* semigroup of the band B(see Ref. 5). Now, we form the spined product under ψ by $S = K(B,T,\psi) = \{\,(x,t) \in W_B \times T \mid x\gamma = t\psi\,\}$. Then, it can be checked that S is a type W semigroup. In fact, we have the following construction theorem.

Theorem 2.23 [Ref. 26, Theorem 3.3 and Theorem 3.5]. *The spined product $S = K(B,T,\psi) = \{\,(x,t) \in W_B \times T \mid x\gamma = t\psi\,\}$ is a type W semigroup* *with band of idempotents isomorphic to B and $S/\delta \cong T$.*

Conversely, any type W semigroup S is isomorphic to $K(B, S/\delta, \psi)$.

3. Weakly abundant semigroups

As a further generalization of Green's relations $\mathcal{L}, \mathcal{R}, \mathcal{H}, \mathcal{D}$ and $\mathcal{J}$, a new set of relations from (see Ref. 28,29,31,32) on a semigroup S was introduced in the following way. Let S be a semigroup and E be a set of all idempotents of S. Let $U \subseteq E$ be a non-empty subset. Then the five relations $\widetilde{\mathcal{L}}^U, \widetilde{\mathcal{R}}^U, \widetilde{\mathcal{H}}^U, \widetilde{\mathcal{D}}^U$ and $\widetilde{\mathcal{J}}^U$ based on U are defined as follows:

$$\begin{aligned}
\widetilde{\mathcal{L}}^U &= \{(a,b) \in S \times S \mid (\forall e \in U)\ ae = a \Leftrightarrow be = b\}, \\
\widetilde{\mathcal{R}}^U &= \{(a,b) \in S \times S \mid (\forall e \in U)\ ea = a \Leftrightarrow eb = b\}, \\
\widetilde{\mathcal{H}}^U &= \widetilde{\mathcal{L}}^U \wedge \widetilde{\mathcal{R}}^U, \\
\widetilde{\mathcal{D}}^U &= \widetilde{\mathcal{L}}^U \vee \widetilde{\mathcal{R}}^U, \\
\widetilde{\mathcal{J}}^U &= \{(a,b) \in S \times S \mid \widetilde{J}(a) = \widetilde{J}(b)\},
\end{aligned}$$

where we use $\widetilde{J}(a)$ to denote the principal U-admissible ideal generated by the element a in S (in detail, see Ref. 31).

It is easy to show that $\mathcal{L} \subseteq \mathcal{L}^* \subseteq \widetilde{\mathcal{L}}^U$. In particular, if S is abundant and $U = E(S)$ then $\widetilde{\mathcal{L}}^U = \mathcal{L}^*$; if S is regular and $U = E(S)$ then $\widetilde{\mathcal{L}}^U = \mathcal{L}^* = \mathcal{L}$. Of course, there is a dual result for $\widetilde{\mathcal{R}}^U$.

A semigroup S is called a U-semiabundant semigroup if each $\widetilde{\mathcal{L}}^U$-class and each $\widetilde{\mathcal{R}}^U$-class of S contains an element from U, denoted by (S, U). In the case, we call U a set of projections of S and the elements of U are called projections. From now on we will write (S, U) to mean that S is a U-semiabundant semigroup.

It is well-known from Ref. 33, see Example 3.6 of Ref. 33 for a counterexample, that $\widetilde{\mathcal{L}}^U$ is not necessarily a right congruence on S and $\widetilde{\mathcal{R}}^U$ is not necessarily a left congruence on S. We say that a U-semiabundant semigroup (S, U) is *weakly abundant*(in Ref. 30, called U-abundant) if (S, U) satisfies the *congruence conditions*, i.e. $\widetilde{\mathcal{L}}^U$ is a a right congruence and $\widetilde{\mathcal{R}}^U$ is a left congruence on S. As usual, we use $\widetilde{L}_a$ and $\widetilde{R}_a$ to denote the $\widetilde{\mathcal{L}}^U$-class and the $\widetilde{\mathcal{R}}^U$-class of (S, U) containing the element a, respectively. Moreover,

a^* and $a^\dagger$ are used to denote a typical projection of $\widetilde{L}_a \cap U$ and a typical projection of $\widetilde{R}_a \cap U$, respectively. It is important to note that in general the projections in the $\widetilde{L}_a$ and the $\widetilde{R}_a$ of weakly abundant semigroups are not unique.

Following the terminologies above, both regular semigroups and abundant semigroups are weakly abundant semigroups when $U = E(S)$.

Example 3. Let $S_\alpha = \{u, v, s, t\}$ be a semigroup with the following Table 5.

	u	v	s	t
u	u	v	s	t
v	v	v	s	t
s	s	s	t	v
t	t	t	v	s

Table 5

Let $S_\beta = \{a, a^2, a^3, ...\}$ be an infinite monogenic generated by a such that $S_\alpha \cap S_\beta = \emptyset$.

Consider the set $S = S_\alpha \cup S_\beta$ with the binary operation which extends the multiplications on S_α and S_β given by $xy = yx = y$ for $x \in S_\alpha$ and $y \in S_\beta$.

Thus, S forms a semigroup and the set of all idempotents of S is $E(S) = \{u, v\}$.

It can be checked that the $\mathcal{L}$-classes of S are the sets $\{u\}, \{v, s, t\}, \{a\}$, $\{a^2\}, \{a^3\}, \cdots$. It is clear that the semigroup S is not a regular semigroup.

For the elements $a, u \in S$, we have that $as = at = a$ for $s, t \in S$. But, $s = us \neq ut = t$. By definition of $\mathcal{L}^*$, we have that $(a, u) \notin \mathcal{L}^*(S)$. Similarly, $(a, v) \notin \mathcal{R}^*(S)$. Then, S is also not an abundant semigroup.

However, if we take $U = E(S) = \{u, v\}$, regard it as the set of projections of S, we have that $\widetilde{\mathcal{L}}^U = \widetilde{\mathcal{R}}^U$ on S and also $\widetilde{\mathcal{L}}^U$-classes of S are the sets $\{u\}$ and $\{v, s, t, a, a^2, \cdots\}$. Thus, (S, U) is indeed a weakly abundant semigroup.

This example illustrates that the class of weakly abundant semigroups contains both the class of abundant semigroups and the class of regular semigroups as its proper subclasses.

In 1991 M. V. Lawson[33] introduced first and studied an Ehresmann semigroup. By an *Ehresmann semigroup* we mean a weakly abundant semigroup in which the set of projections U is a semilattice. Clearly, inverse semigroups are Ehresmann semigroups and so they are adequate semigroups.

Definition 3.1 [Ref. 33 and 30]. *A semigroup homomorphism* $\phi : (S,U) \to (T,V)$ *is said to be admissible if* $\phi|_U : U \to V, a\widetilde{\mathcal{L}}^U b$ *implies* $a\phi\widetilde{\mathcal{L}}^V b\phi$ *and* $a\widetilde{\mathcal{R}}^U b$ *implies* $a\phi\widetilde{\mathcal{R}}^V b\phi$ *for all* $a, b \in (S,U)$. *Naturally, a congruence* ρ *on* (S,U) *is said to be admissible if the natural homomorphism* $\rho^\natural : (S,U) \to ((S,U)/\rho, U/\rho)$ *is admissible.*

In studying the Ehresmann semigroups, Lawson obtained the following very nice result.

Theorem 3.2 [Ref. 33, Theorem 4.24]. *The category of Ehresmann semigroups and admissible homomorphisms is isomorphic to the category of Ehresmann categories and strongly ordered functors.*

By the definition of Ehresmann semigroups (S,U), it is easy to show that every $\widetilde{\mathcal{L}}^U$-class and every $\widetilde{\mathcal{R}}^U$-class of (S,U) contains a unique projection from U. We denote by a^* $(a^\dagger)$ the unique projection in the $\widetilde{\mathcal{L}}^U$-class ($\widetilde{\mathcal{R}}^U$-class) of a. Thus, we have the following

Theorem 3.3 [Ref. 34]. *Let* (S,U) *be an Ehresmann semigroup.*
(i) *for all* $a, b \in S$, $(ab)^* = (a^*b)^*$ *and* $(ab)^\dagger = (ab^\dagger)^\dagger$;
(ii) *for all* $a \in S$ *and* $e \in U$, $(ae)^* = a^*e$ *and* $(ea)^\dagger = ea^\dagger$.

G. M. S. Gomes and V. Gould in 2001 studied a special subclass of the class of Ehresmann semigroups named *fundamental Ehresmann semigroups* (in detail, see Ref. 35).

In fact, Ehresmann semigroups in the class of weakly abundant semigroups can be regarded as a natural generalization of inverse semigroups in the class of regular semigroups.

In Ref. 30, we discussed the *U-ample semigroups*, which are generalization of type A semigroup in the class of abundant semigroups.

Definition 3.4 [Ref. 30, Definition 5.1]. *An Ehresmann semigroup (S, U) is called U-ample if for each element a of (S, U) and each projection e in U, (S, U) satisfies the following conditions:*

(i) $ea = a(ea)^*$;
(ii) $ae = (ae)^\dagger a$.

Definition 3.5 [Ref. 30, Definition 4.1]. *A U-semiabundant semigroup (S, U) is said to be projection-connected (for brevity, PC condition) if for every $a \in (S, U)$ and any $a^\dagger \in \widetilde{R}_a \cap U$, $a^* \in \widetilde{L}_a \cap U$, there exists an isomorphism $\alpha : \langle a^\dagger \rangle \to \langle a^* \rangle$ such that $xa = a(x\alpha)$ for any $x \in \langle a^\dagger \rangle$.*

Theorem 3.6 [Ref. 30, Proposition 5.5]. *An Ehresmann semigroup (S, U) is PC if and only if it is a U-ample semigroup.*

Definition 3.7 [Ref. 30, Definition 6.1]. *A weakly abundant semigroup (S, U) is said to be U-orthodox if (S, U) is PC and the set of projections U forms a subsemigroup.*

A structure theorem of U-orthodox semigroups have been given by authors in Ref. 30 as follows.

Theorem 3.8 [Ref. 30, Theorem 8.7]. *Let U be a band and (T, V) a V-ample semigroup whose semilattice of projections is isomorphic to $U/\mathcal{J}$. Let γ be the smallest inverse semigroup congruence on the Hall semigroup W_U and let ψ be a projection-separating admissible homomorphism from (T, V) into W_U/γ whose range contains all the idempotents of W_U/γ. Then $(S, \bar{U}) = \mathcal{M}(U, (T, V), \psi)$ is a $\bar{U}$-orthodox semigroup whose band of projections is isomorphic to U. If δ is the minimum ample semigroup congruence on $(S, \bar{U})$, then $((S, \bar{U})/\delta, \bar{U}/\delta) \simeq (T, V)$.*

Conversely, if (S, U) is a U-orthodox semigroup with band of projections U, then there exists a projection-separating admissible homomorphism $\theta : ((S, U)/\delta, U/\delta) \to W_U/\gamma$ whose range contains all the idempotents of W_U/γ such that (S, U) is isomorphic to $\mathcal{M}(U, ((S, U)/\delta, U/\delta), \theta)$.

Thus Theorem 2.23 of type W semigroups given by El-Qallali and Fountain in Ref. 26 and Theorem 1.11 of orthodox semigroups given by Yamada-Hall follows from Theorem 3.8 directly.

In summary, we establish a corresponding table of hierarchy of some important semigroups for the class of regular semigroups, the class of abundant semigroups and the class of weakly abundant semigroups as follows:

Regular semigroups	**Abundant semigroups**	**Weakly abundant semigroups**
Clifford semigroups	C-a semigroups	
Inverse semigroups	Adequate semigroups Type A semigroups	Ehresmann semigroups U-ample semigroups
Left inverse semigroups	$\mathcal{R}^*$-unipotent semigroups $\mathcal{L}^*$-inverse semigroups	
Quasi-inverse semigroups	$\mathcal{Q}^*$-inverse semigroups	
Orthodox semigroups	Quasi-adequate semigroups Type W semigroups	U-orthodox semigroups

Table 6

Acknowledgements

The authors cordially thank two anonymous referees for their valuable comments which lead to the improvement of this paper.

References

1. V. V. Vagner, Generalized groups, *Dokl. Akad. Nauk SSSR* **84** (1952), 1119–1122 (Russian).
2. V. V. Vagner, Theory of generalized heaps and generalized groups, *Mat. Sbornik (N.S.)*, **32** (1953), 545–632(Russian).
3. G. B. Preston, Inverse semi-groups, *J. London Math. Soc.*, **29** (1954), 396–403.
4. G. B. Preston, Inverse semi-groups with minimal right ideals, *J. London Math. Soc.*, **29** (1954), 404–411.
5. J. M. Howie, An introduction to semigroup theory, Academic Press, 1976.
6. M. Petrich, Inverse semigroups, John Wiley & Sons, New York, Chichester, Brisbane, Toronto and Singapore, 1984.
7. T. E. Hall, On regular semigroups, *J. Algebra,* **24** (1973), 1–24.
8. A. H. Clifford and G. B. Preston, *The algebraic theory of semigroups, Amer. Math. Soc. I* (1961), II(1967).
9. J. B. Fountain, Abundant semigroups, *Proc. London Math. Soc.*, **44** (1982), 103–129.
10. X. M. Ren and K. P. Shum, The structure of superabundant semigroups, *Science in China Ser. A*, (5) **47** (2004), 756–771.
11. X. M. Ren and K. P. Shum, On generalized orthogroups, *Comm. Algebra,* (6) **29** (2001), 2341–2361.

12. K. P. Shum, X. M. Ren and Y. Q. Guo, On C^*-quasiregular semigroups, *Comm. Algebra,* (9) **27** (1999), 4251–4274.
13. J. A. Green, On the structure of semigroups, *Ann. of Math.,* **54** (1951), 163–172.
14. G. L. Bailes, Right inverse semigroup, *J. Algebra,* **26** (1973), 492–507.
15. P. S. Venkatesan, Right (left) inverse semigroups, *J. Algebra,* **31** (1974), 209–217.
16. M. Yamada, Orthodox semigroups whose idempotents satisfy a certain identity, *Semigroup Forum* **6** (1973), 113–128.
17. M. Yamada, Note on a certain class of orthodox semigroups, *Semigroup Forum,* **6** (1973), 180–188.
18. N. R. Reilly and H. E. Scheiblich, Congruences on regular semigroups, *Pacific J. Math.,* **23** (1967), 349–360.
19. J. B. Fountain, Adequate semigroups, *Proc. Edinburgh Math. Soc.,* **22** (1979), 113–125.
20. A. El-Qallali and J. B. Fountain, Idempotent-connected abundant semigroups, *Proc. Roy. Soc. Edinburgh Sect. A,* **91** (1981), 79–90.
21. S. Armstrong, The structure of type A semigroups, *Semigroup Forum,* **29** (1984), 319–336.
22. M. V. Lawson, The structure of type A semigroups, *Quart. J. Math. Oxford,* (2) **37** (1986), 279–298.
23. A. El-Qallali, $\mathcal{L}^*$-unipotent semigroups, *J. Pure Appl. Algebra,* **62** (1989), 19–33.
24. X. M. Ren and K. P. Shum, The structure of $\mathcal{L}^*$-inverse semigroups, *Science in China Ser. A,* (8) **49** (2006), 1065–1081.
25. X. M. Ren and K. P. Shum, The structure of $\mathcal{Q}^*$-inverse semigroups, *J. Algebra,* **325** (2011), 1-17.
26. A. El-Qallali and J. B. Fountain, Quasi-adequate semigroups, *Proc. Roy. Soc. Edinburgh Sect. A,* **91** (1981), 91–99.
27. X. J. Guo, *Some studies on left pp semigroups,* Ph. D. Thesis, Lanzhou University, 1995.
28. A. El-Qallali, *Structure theory for abundant and related semigroups,* Ph. D. Thesis, York, 1980.
29. M. V. Lawson, Rees matrix semigroups, *Proc. Edinburgh Math. Soc.,* **33** (1990), 23–37.
30. X. M. Ren, Y. H. Wang and K. P. Shum, On U-orthodox semigroups, *Science in China Ser. A,* (2) **52** (2009), 329–350.
31. X. M. Ren, K. P. Shum and Y. Q. Guo, A generalized Clifford theorem of semigroups, *Science in China Ser. A,* (4) **53** (2010), 1097–1101.
32. Y. Q. Guo, C. M. Gong and X. M. Ren, A survey on the origin and developments of Green's relations on semigroups, *Journal of Shandong Unversity (Natural Science),* (8) **45** (2010), 1–18.
33. M. V. Lawson, Semigroups and ordered categories, I. the reduced case, *J. Algebra,* **141** (1991), 422–462.
34. J. B. Fountain, G. M. S. Gomes and V. Gould, A Munn type representation for a class of E-semiadequate semigroups, *J. Algebra,* **218** (1999), 693–714.

35. G. M. S. Gomes and V. Gould, Fundamental Ehresmann semigroups, *Semigroup Forum,* **63** (2001), 11–33.
36. Y. Q. Guo, K. P. Shum and P. Y. Zhu, The structure of left C-*rpp* semigroups, *Semigroup Forum,* **50** (1995), 9–23.
37. Y. Q. Guo, K. P. Shum and C. M. Gong, On $(*, \sim)$-Green's relations and ortho-lc-monoids, *to appear in Communication in Algebra.*
38. X. M. Ren, K. P. Shum and Y. Q. Guo, On super R^*-unipotent semigroups, *Southeast Asian Bulletin of Mathematics,* **22** (1998), 199–208.
39. X. M. Ren, K. P. Shum and Y. Q. Guo, On spined products of quasi-rectangular groups, *Algebra Colloquium,* **4:2** (1997), 187–194.
40. K. P. Shum and X. M. Ren, Abundant semigroups with left central idempotents, *Pure Math. Appl.,* (1) **10** (1999), 109–113.
41. K. P. Shum and Y. Q. Guo, Regular semigroups and their generalizations, *rings, groups and algebras* 181–226, Lecture notes in Pure and Appl. Math., 181, Dekker, New York, 1996.

Correction of Two-Dimensional Solid Burst Errors in LRTJ-Spaces

Sapna Jain*
Department of Mathematics
University of Delhi
Delhi 110 007
India
E-mail: sapnajain@gmx.com

Abstract. In this paper, we obtain a lower bound for two-dimensional array codes correcting solid burst array errors with weight constraint in LRTJ-spaces.

AMS Subject Classification (2000): 94B05

Keywords: Linear array codes, Lee metric, RT metric, solid burst errors

1. Introduction

The choice of a metric for a given parallel communication system plays an important role as the channel model should match the metric d to be employed for developing a suitable array code, and hence for a communication system to operate reliably. Thus, given a modulation scheme, one metric may be better suited than another. In [4], the author introduced a new pseudo-metric on $Mat_{m\times s}(\mathbf{Z}_q)$, the module space of all $m \times s$ matrices over the finite ring $\mathbf{Z}_q$, generalizing the classical Lee metric [7] and array RT-metric [9] and named this pseudo-metric as the Generalized-Lee-RT-Pseudo-Metric (the GRTP-Metric) which further appeared as LRTJ-metric (Lee-Rosenbloom-Tfassman-Jain-metric) in [1].

Again in [5], the author introduced the notion of two-dimensional solid burst errors for the array coding and devised the array codes correcting and detecting solid burst array errors in RT-spaces. The objective of this paper is to establish a lower bound on the parameters of array codes equipped with the LRTJ-metric for the correction of two-dimensional solid burst errors with weight constraint.

2. Definitions and notations

Let $\mathbf{Z}_q$ be the ring of integers modulo q. Let $Mat_{m\times s}(\mathbf{Z}_q)$ be the set of all $m \times s$ matrices with entries from $\mathbf{Z}_q$. Then $Mat_{m\times s}(\mathbf{Z}_q)$ is a module over $\mathbf{Z}_q$. Let V be a $\mathbf{Z}_q$-submodule of the module $Mat_{m\times s}(\mathbf{Z}_q)$. Then V is called an array code (In fact, linear array code). For q prime, $\mathbf{Z}_q$ becomes a field and correspondingly $Mat_{m\times s}(\mathbf{Z}_q)$ and V become the vector space and a sub space respectively over the field $\mathbf{Z}_q$. We note that the space $Mat_{m\times s}(\mathbf{Z}_q)$ is identifiable with the space $\mathbf{Z}_q^{ms}$. Every matrix in $Mat_{m\times s}(\mathbf{Z}_q)$ can be represented as an $1 \times ms$ vector by writing the first row of matrix followed by second row and so forth. Similarly, very vector in $\mathbf{Z}_q^{ms}$ can be represented as an $m \times s$ matrix in $Mat_{m\times s}(\mathbf{Z}_q)$ by separating the co-ordinates of the vector into m groups of s-coordinates. Also, we define the modular

*Address for Communication: 32, Uttranchal 5, I.P. Extension, Delhi 110092, India

value $|a|$ of an element $a \in \mathbf{Z}_q$ by

$$|a| = \begin{cases} a & \text{if} \quad 0 \leq a \leq q/2 \\ q-a & \text{if} \quad q/2 < a \leq q-1. \end{cases}$$

We note that the non-zero modular value $|a|$ can be obtained by two different elements a and $q-a$ of $\mathbf{Z}_q$ provided $\{q$ is odd$\}$ or $\{q$ is even and $a \neq [q/2]\}$, i.e.

$$|a| = |q-a| \quad \text{if} \quad \begin{cases} q \quad \text{is odd} \\ \text{or} \\ q \text{ is even and} \quad a \neq q/2. \end{cases}$$

If q is even and $a = [q/2]$ or if $a = 0$, then $|a|$ is obtained in only one way viz., $|a| = a$.

Thus, there may be one or two equivalent values of $|a|$ which we shall refer to as repetitive equivalent values of a. The number of repetitive equivalent values of a will be denoted by e_a, where

$$e_a = \begin{cases} 1 & \text{if } \{\, q \text{ is even and } a = q/2\} \text{ or } \{a = 0\} \\ 2 & \text{if } \{\, q \text{ is odd and } a \neq 0\} \text{ or } \{q \text{ is even, } a \neq 0 \text{ and } a \neq q/2\}. \end{cases}$$

We now define the LRTJ-metric as follows [1, 4]):

Let $Y \in Mat_{1\times s}(\mathbf{Z}_q)$ with $Y = (y_1, y_2, \cdots, y_s)$.

Define the row-weight of Y as

$$wt_\rho(Y) = \begin{cases} \max_{j=1}^{s}|y_j| + \max_{j=1}^{s}\{j-1 \mid y_j \neq 0\} \text{ if } Y \neq 0 \\ \\ 0 \qquad\qquad\qquad\qquad\qquad\qquad \text{if } Y = 0. \end{cases}$$

Then $0 \leq wt_\rho(Y) \leq [q/2] + s - 1$. Extending the definition of the row-weight to the class of all $m \times s$ matrices as

$$wt\rho(A) = \sum_{i=1}^{m} wt\rho(R_i)$$

where $A = \begin{bmatrix} R_1 \\ R_2 \\ \cdots \\ R_m \end{bmatrix} \in Mat_{m\times s}(\mathbf{Z}_q)$ and R_i denotes the ith row of A. Then wt_ρ satisfies $0 \leq wt_\rho(A) \leq m([q/2]+s-1)\ \forall\, A \in Mat_{m\times s}(\mathbf{Z}_q)$ and determines a pseudo-metric on $Mat_{m\times s}(\mathbf{Z}_q)$ if we set $d(A, A') = wt_\rho(A - A')\ \forall\ A, A' \in Mat_{m\times s}(\mathbf{Z}_q)$. We name this pseudo-metric as Lee-Rosenbloom-Tfassman-Jain-metric (or LRTJ-metric) because of the following observations:

1. For $s = 1$, it is just the classical Lee metric [7].
2. For $q = 2, 3$, new pseudo-metric reduces to the RT metric [9].

Remarks.:

1. For $q > 3$,
$$\text{LRTJ-wt } (A) > \text{RT-wt } (A) \ \forall \ A \in Mat_{m\times s}(\mathbf{Z}_q)$$

2. For $s = 1$ and $q = 2, 3$, LRTJ-metric reduces to the Hamming metric [8].

Also, we shall use the following notations:

1. $[x] =$ The largest integer less than or equal to x.

2. Q_i will denote the sum of repetitive equivalent values up to i i.e.,
$$Q_i = e_0 + e_1 + \cdots + e_i$$
where e_i denotes the repetitive equivalent value of i.

3. Lower bound for array codes correcting solid burst array errors with weight constraint in LRTJ-spaces

We begin with the definition of two-dimensional solid bursts for the array coding [5]:

Definition 3.1. A solid burst of order pr(or $p \times r$)$(1 \le p \le m, 1 \le r \le s)$ in the space $\text{Mat}_{m\times s}(F_q)$ is an $m \times s$ matrix A having a $p \times r$ submatrix B with all entries in B nonzero and remaining entries in the $m \times s$ matrix A as zero.

Note. For $p = 1$, Definition 3.1 reduces to the definition of solid-burst for classical codes [10].

Example 3.2. Consider the linear space $\text{Mat}_{3\times 3}(F_2)$. Then all the solid-bursts of order 2×2 are given by:

$$\begin{pmatrix} 1 & 1 & 0 \\ 1 & 1 & 0 \\ 0 & 0 & 0 \end{pmatrix}, \begin{pmatrix} 0 & 1 & 1 \\ 0 & 1 & 1 \\ 0 & 0 & 0 \end{pmatrix}, \begin{pmatrix} 0 & 0 & 0 \\ 1 & 1 & 0 \\ 1 & 1 & 0 \end{pmatrix}, \begin{pmatrix} 0 & 0 & 0 \\ 0 & 1 & 1 \\ 0 & 1 & 1 \end{pmatrix}.$$

Now we obtain a lower bound on the number of parity check digits required to correct all weighted soild bursts of order pr $(1 \le p \le m, 1 \le r \le s)$ in $Mat_{m\times s}(\mathbf{Z}_q)$ having LRTJ-weight w or less $(1 \le w \le m([q/2] + s - 1))$. For this, we first prove a lemma that enumerates the number of solid bursts of order $pr(1 \le p \le m, 1 \le r \le s)$ having LRTJ-weight w or less.

Lemma 3.3. *The number of solid bursts of order* $pr(1 \le p \le m, 1 \le r \le s)$ *in*

$Mat_{m\times s}(\mathbf{Z}_q)$ having LRTJ-weight w or less $(1 \le w \le m([q/2]+s-1))$ is given by

$$S^{p\times r}_{m\times s}(\mathbf{Z}_q, w) = \begin{cases} m \displaystyle\sum_{j=1}^{min(w,s)} (Q_{w-(j-1)} - 1) \quad \text{if } p = r = 1, \\ m \displaystyle\sum_{j=1}^{min(w-r+1,s-r+1)} (Q_{w-(j+r-2)} - 1)^r \qquad \text{if } p = 1, r \ge 2, \\ (m-p+1)\displaystyle\sum_{j=1}^{L}\Bigg(\sum_{r_l^{(j)}} \frac{p!}{\prod_{l=1}^{[q/2]} r_l^{(j)}} \times \\ \times \displaystyle\prod_{l=1}^{[q/2]} \left(e_l (Q_l - 1)^{r-2}\Big((Q_l - 1) + (r-1)(Q_{l-1} - 1)\Big)\right)^{r_l^{(j)}}\Bigg) \\ \qquad\qquad \text{if } p \ge 2, r \ge 1, \end{cases} \tag{1}$$

where for a fixed j, $r_l^{(j)}$ $(1 \le l \le [q/2])$ are non-negative integers satisfying

$$\begin{aligned} & r_l^{(j)} \ge 0 \text{ for all } 1 \le l \le [q/2], \\ & \sum_{l=1}^{[q/2]} (l + (j+r-2)) r_l^{(j)} \le w, \\ & \sum_{l=1}^{[q/2]} r_l^{(j)} = p, \end{aligned} \tag{2}$$

and

$$L = \max\Big\{0, \min\Big\{[w/p] - r + 1, s - r + 1\Big\}\Big\}. \tag{3}$$

Proof. Consider a solid burst $A = \begin{bmatrix} A_1 \\ A_2 \\ \cdots \\ A_m \end{bmatrix}$ where $A_i = (a_{i_1}, a_{i_2}, \cdots, a_{i_s})$, of order $pr(1 \le p \le m, 1 \le r \le s)$ having LRTJ-weight w or less $(1 \le w \le m([q/2]+s-1))$. Let B be the $p \times r$ nonzero submatrix of A such that all the entries in B are nonzero and entries in A outside B are zero. There are three cases depending upon the values of p and r.

Case 1. When $p = 1,\ r = 1$.

In this case, the 1×1 nonzero submatrix B can have (i, j) as its starting position in $m \times s$ matrix A where i can take values from 1 to m and j can take values from 1 to min (w, s). With (i, j) as the starting position of 1×1 nonzero submatrix B, entry in B can be filled in $e_1 + e_2 + \cdots + e_{w-(j-1)} = Q_{w-(j-1)} - 1$ ways as any nonzero element of $\mathbf{Z}_q$ having modular value $w - (j-1)$ or less can be filled in that position. Therefore, number of solid bursts of order 1×1 having LRTJ-weight w or less in $Mat_{m\times s}(\mathbf{Z}_q)$ is given by

$$\begin{aligned} S^{1\times 1}_{m\times s}(\mathbf{Z}_q, w) &= m \sum_{j=1}^{min(w,s)} (e_1 + e_2 + \cdots + e_{w-(j-1)}) \\ &= m \sum_{j=1}^{min(w,s)} (Q_{w-(j-1)} - 1). \end{aligned}$$

Case 2. When $p = 1,\ r \geq 2$.

In this case, the number of starting positions for the $1 \times r$ nonzero submatrix B in $m \times s$ matrix A is $m \times \min(w - r + 1, s - r + 1)$ and entries in the $1 \times r$ submatrix B can be selected in $(Q_{w-(j+r-2)} - 1)^r$ ways. Therefore, the number of solid-bursts of order $1 \times r$ having LRTJ-weight w or less in $\mathrm{Mat}_{m\times s}(\mathbf{Z}_q)$ is given by

$$S^{1\times r}_{m\times s}(\mathbf{Z}_q, w) = m \sum_{j=1}^{min(w-r+1,s-r+1)} (Q_{w-(j+r-2)} - 1)^r.$$

Case 3. When $p \geq 2,\ r \geq 1$.

In this case, let the $p \times r$ nonzero submatrix B starts at the $(i,j)^{th}$ position in A where $1 \leq i \leq (m-p+1)$ and $1 \leq j \leq L$ and L is given by (3). Out of p rows of B (where submatrix B starts at the jth column in matrix A), let $r_l^{(j)}$ $(1 \leq l \leq [q/2])$ be the number of rows of B (and hence that of matrix A) having LRTJ-weight equal to $l + (j + r - 2)$. Then $r_l^{(j)} \geq 0$ for all $1 \leq l \leq [q/2]$. The number of ways in which p rows of B can be selected is given by

$$\sum_{r_l^{(j)}} \frac{p!}{\prod_{l=1}^{[q/2]} r_l^{(j)}} \prod_{l=1}^{[q/2]} \left(e_l (Q_l - 1)^{r-2} \Big((Q_l - 1) + (r-1)(Q_{l-1} - 1) \Big) \right)^{r_l^{(j)}} \tag{4}$$

where for a fixed j, $r_l^{(j)}$ $(1 \leq l \leq [q/2])$ are nonnegative integers satisfying (2). Summing (4) over i and j $(1 \leq i \leq m-p+1, 1 \leq j \leq L)$, we get the number of solid bursts of order pr $(2 \leq p \leq m, 1 \leq r \leq s)$ having LRTJ-weight w or less $(2 \leq w \leq m([q/2] + s - 1))$ and is given by

$$\begin{aligned} S^{p\times r}_{m\times s}(\mathbf{Z}_q, w) &= \sum_{i=1}^{(m-p+1)} \sum_{j=1}^{L} \Bigg(\sum_{r_l^{(j)}} \frac{p!}{\prod_{l=1}^{[q/2]} r_l^{(j)}} \times \\ &\quad \times \prod_{l=1}^{[q/2]} \left(e_l (Q_l - 1)^{r-2} \Big((Q_l - 1) + (r-1)(Q_{l-1} - 1) \Big) \right)^{r_l^{(j)}} \Bigg) \\ &= (m-p+1) \sum_{j=1}^{L} \Bigg(\sum_{r_l^{(j)}} \frac{p!}{\prod_{l=1}^{[q/2]} r_l^{(j)}} \times \end{aligned}$$

$$\times \prod_{l=1}^{[q/2]} \left(e_l (Q_l - 1)^{r-2} \Big((Q_l - 1) + (r-1)(Q_{l-1} - 1) \Big) \right)^{r_l^{(j)}} \Bigg)$$

where L is given by (3) and $r_l^{(j)}$ satisfy the constraints (2). □

Example 3.4. Take $m = p = 2, r = s = 1, q = 5$ and $w = 3$. Then $S_{2\times1}^{2\times1}(\mathbf{Z}_5, 3)$ is given as follows using Lemma 3.3:

$$\begin{aligned} S_{2\times1}^{2\times1}(\mathbf{Z}_5, 3) &= \sum_{r_1^{(1)}, r_2^{(1)}} \frac{2!}{r_1^{(1)}! r_2^{(1)}!} (e_1 (Q_1 - 1)^{-1} (Q_1 - 1))^{r_1^{(1)}} \times \\ &\quad \times (e_2 (Q_2 - 1)^{-1} (Q_2 - 1))^{r_2^{(1)}} \\ &= \sum_{r_1^{(1)}, r_2^{(1)}} \frac{2!}{r_1^{(1)}! r_2^{(1)}!} 2^{r_1^{(1)}} 2^{r_2^{(1)}} \quad (\text{Note that } e_1 = e_2 = 2 \text{ over } \mathbf{Z}_5) \end{aligned} \tag{5}$$

and $r_1^{(1)}, r_2^{(1)}$ are non-negative integers satisfying the following constraints:

$$\begin{aligned} r_1^{(1)} + 2r_2^{(1)} &\leq 3, \\ r_1^{(1)} + r_2^{(1)} &= 2. \end{aligned} \tag{6}$$

The feasible solutions for $(r_1^{(1)}, r_2^{(1)})$ satisfying the constraints (6) are given by

$$(r_1^{(1)}, r_2^{(1)}) = (1, 1), (2, 0).$$

Therefore, from (5) we have

$$\begin{aligned} S_{2\times1}^{2\times1}(\mathbf{Z}_5, 3) &= \frac{2!}{1!1!} \times 2^1 \times 2^1 + \frac{2!}{2!0!} \times 2^2 \times 2^0 \\ &= 8 + 4 = 12. \end{aligned}$$

These 12 solid bursts of order 2×1 having LRTJ-weight 3 or less in $\mathrm{Mat}_{2\times1}(\mathbf{Z}_5)$ are given as follows:

$$\begin{pmatrix}1\\1\end{pmatrix}, \begin{pmatrix}1\\2\end{pmatrix}, \begin{pmatrix}1\\3\end{pmatrix}, \begin{pmatrix}1\\4\end{pmatrix}, \begin{pmatrix}2\\1\end{pmatrix}, \begin{pmatrix}2\\4\end{pmatrix}, \begin{pmatrix}3\\1\end{pmatrix}, \begin{pmatrix}3\\4\end{pmatrix}, \begin{pmatrix}4\\1\end{pmatrix} \begin{pmatrix}4\\2\end{pmatrix}, \begin{pmatrix}4\\3\end{pmatrix}, \begin{pmatrix}4\\4\end{pmatrix}.$$

Now, we obtain a lower bound on the number of parity check digits for the correction of solid bursts of the order $pr(1 \leq p \leq m, 1 \leq r \leq s)$ having LRTJ-weight w or less $(1 \leq w \leq m([q/2] + s - 1))$.

Theorem 3.5. *An (n, k) linear LRTJ-metric array code $V \subseteq Mat_{m\times s}(\mathbf{Z}_q)$ (where $n = ms$) that corrects all solid bursts of order pr $(1 \leq p \leq m, 1 \leq r \leq s)$ having LRTJ-weight w or less $(1 \leq w \leq m([q/2] + s - 1))$ must satisfy*

$$q^{n-k} \geq 1 + S_{m\times s}^{p\times r}(\mathbf{Z}_q, w)$$

where $S_{m\times s}^{p\times r}(\mathbf{Z}_q, w)$ *is given by (1) in Lemma 3.3.*

Proof. The proof follows from the fact that the number of available cosets must be greater than or equal to the number of correctable error arrays including the null array. □

Acknowledgment. The author would like to thank her husband Dr. Arihant Jain for his constant support and encouragement for pursuing her research work.

References

[1] E. Deza and M.M. Deza, Encyclopedia of Distances, Elsevier, 2008, p.270.

[2] S. Jain, *Bursts in m-Metric Array Codes*, Linear Algebra and Its Applications, 418 (2006), 130-141.

[3] S. Jain, *Campopiano-Type Bounds in Non-Hamming Array Coding*, Linear Algebra and Its Applications, 420 (2007), 135-159.

[4] S. Jain, *Array Codes in Generalized-Lee-RT-Pseudo-Metric (GLRTP-Metric)*, Algebra Colloquium, 17 (Spec1) (2010), 727-740.

[5] S. Jain, *Solid Bursts-From Hamming to RT-Spaces*, to appear in Ars Combinatoria.

[6] S. Jain and K.P. Shum, *Correction of CT Burst Array Errors in Generalized-LEE-RT Spaces*, Acta Mathematica Sinica, English series, Vol. 26 (2010), 1475-1484.

[7] C. Y. Lee, *Some properties of non-binary error correcting codes*, IEEE Trans. Information Theory, IT-4 (1958), 77-82.

[8] W. W. Peterson and E.J. Weldon, Jr., Error Correcting Codes, 2nd Edition, MIT Press, Cambridge, Massachusetts, 1972.

[9] M. Yu. Rosenbloom and M.A. Tsfasman, *Codes for m-metric*, Problems of Information Transmission, 33 (1997), 45-52.

[10] S.G.S. Shiva and C.L. Sheng, *Multiple solid burst-error correcting binary codes*, IEEE. Trans. Info. Theory, Vol. IT-15 (1959), 188-189.

Recent Developments of Semirings

M. K. Sen[a] and A. K. Bhuniya[b]

a) Department of Pure Mathematics, University of Calcutta, Kolkata, India

b) Department of Mathematics, Visva-Bharati, Santiniketan-731235, India.

E-mail addresses: senmk6@yahoo.com(M. K. Sen), anjankbhuniya@gmail.com(A. K. Bhuniya).

Dedicated to Prof. K. P. Shum on his 70th birthday

Abstract

Here we give an account of recent developments of completely regular semirings, Clifford semirings, k-regular semirings and the variety of idempotent semirings.

Key words and phrases: Completely regular semirings, Clifford semirings, idempotent semirings, k-regular semirings.

2000 Mathematics subject Classification: 16Y60.

1 Introduction

In the year 1934, Vandiver [36] published a paper entitled "Note on a simple type of algebra in which cancellation law of addition does not hold" in the Bulletin of American Mathematical Society. This structure, in its most general form, is known as "semiring". Though it was first appeared formally in thirties but it's real appreciation was begun only at the beginning of seventies, when the mathematics community realized that this is not merely a generalization of rings but also a generalization of distributive lattices and that structurally semirings are more similar to semigroups than to rings. Semirings which are subdirect products of a ring and a distributive lattice, distributive lattices of rings, idempotent semirings etc. are result of the influence of the extensive developments of the semigroup theory during this period starting at sixties. Now we have many papers on semirings covering diverse aspects of this theory. So there is enough material to be surveyed, rather it is not possible to cover arbitrarily many directions. Here we give an account of recent developments of completely regular semirings, its different subvarieties including the variety of idempotent semirings and k-regular semirings only.

2 Preliminaries

A *semiring* $(S, +, \cdot)$ is an algebra with two binary operations $+$ and $\cdot$ such that both the *additive reduct* $(S, +)$ and the *multiplicative reduct* $(S, \cdot)$ are semigroups and such that the following distributive laws hold:

$$x(y + z) = xy + xz \text{ and } (x + y)z = xz + yz.$$

The set all additive idempotents of a semiring S will be denoted by $E^+(S)$. Thus $E^+(S) = \{a \in S \mid a + a = a\}$. Similarly the set of all multiplicative idempotents are denoted by $E^{\cdot}(S)$. If $E^+(S) = E^{\cdot}(S) = S$, then S is called an idempotent semiring. A semiring S is called a *b-lattice* if S is an idempotent semiring such that the additive reduct $(S, +)$ is a semilattice.

An ideal I of a semiring S is called a *k-ideal* if for any $x, y \in S$, $x \in I$ and either $x + y \in I$ or $y + x \in I$ imply $y \in I$ (Golan [9] called such ideals subtractive). Let S and T be two semirings. Then $f : S \longrightarrow T$ is said to be a homomorphism if $f(x + y) = f(x) + f(y)$ and $f(xy) = f(x)f(y)$ for all $x, y \in S$. We denote the set of all congruences on a semiring S by $C(S)$, the identity relation by Δ and the universal relation by ∇. $(C(S), \vee, \wedge)$ is a lattice, where $\rho \wedge \sigma = \rho \cap \sigma$ and $\rho \vee \sigma$ is the congruence generated by $\rho \cup \sigma$ for all $\rho, \sigma \in C(S)$. Let $\mathcal{C}$ be the class of all C-semirings. A congruence ρ on a semiring S is said to be a C-congruence if $S/\rho \in \mathcal{C}$. Thus a congruence ρ on S is said to be a distributive lattice congruence if S/ρ is a distributive lattice. A semiring S is said to be an idempotent semiring (distributive lattice, b-lattice) of C-semirings if there exists an idempotent semiring (distributive lattice, b-lattice) congruence ρ on S such that each ρ-class is a member of $\mathcal{C}$.

For a semiring $(S, +, \cdot)$ the Green relations $\mathcal{L}, \mathcal{R}, \mathcal{H}$ and $\mathcal{D}$ on the additive(multiplicative) reduct $(S, +)((S, \cdot))$ will be denoted by $\mathcal{L}^+, \mathcal{R}^+, \mathcal{H}^+$ and $\mathcal{D}^+[\mathcal{L}^{\bullet}, \mathcal{R}^{\bullet}, \mathcal{H}^{\bullet}, \mathcal{D}^{\bullet}]$.

A class $\mathbf{V}$ of semirings is said to form a variety if $\mathbf{V}$ is closed under homomorphic images, taking subsemirings and direct products. If $\mathbf{V}$ is a variety of semirings then a variety $\mathbf{U}$ of semirings is said to be a subvariety of $\mathbf{V}$ if $\mathbf{U} \subseteq \mathbf{V}$. Let $L(\mathbf{V})$ be the collection of all subvarieties of a variety $\mathbf{V}$. Then $(L(\mathbf{V}), \vee, \wedge)$ is a lattice, where $\mathbf{U} \vee \mathbf{W} = \{S \in \mathbf{V} \mid S$ is a subdirect product of some $S_1 \in \mathbf{U}$ and $S_2 \in \mathbf{W}\}$ and $\mathbf{U} \wedge \mathbf{W} = \mathbf{U} \cap \mathbf{W}$. According to the famous Birkhoff Theorem, a class $\mathbf{V}$ of semirings is a variety if and only if **V** is an equational class. For example, the class **I** of all idempotent semirings is a variety determined by the additional identities

$$x \cdot x \approx x \text{ and } x + x \approx x.$$

For any two varieties **V** and **W** of semirings, the *Mal'cev product* $\mathbf{V} \circ \mathbf{W}$ of **V** and **W** is the class of all semirings S on which there exists a congruence ρ such that $S/\rho \in \mathbf{W}$ and such that the ρ-classes belong to **V**.

For notions and terminologies not given in this paper, the readers are referred to Howie [15], Petrich and Reilly [26] for a background on semigroup theory, Hebisch and Weinert [13], Golan [9] for semiring theory and McKenzie, McNulty and Taylor [16] for information concerning universal algebra.

3 Completely Regular Semirings

Clifford [5] defined an element a of a semigroup S to have a relative inverse if there exists $e \in S$ such that $ae = ea = a$ and there exists an element a' in S such that $aa' = a'a = e$. If S is a semigroup in which every element has a relative inverse then S is a union of groups. Such semigroups are presently known as completely regular semigroups, now, which have become a major and core content of the theory of semigroups and which have been studied extensively [15], [26].

Sen, Maity and Shum [34] proposed an axiomatic formulation for the completely regular semirings as follows:

Definition 3.1 *A semiring S is said to be completely regular if for each $a \in S$ there exists $x \in S$ such that*

$$a = a + x + a, \ a + x = x + a, \ \textit{and}\ a(a + x) = a + x.$$

First two conditions are nothing but the assumption that the additive reduct $(S, +)$ is a completely regular semigroup, where as the third makes the additive identity of the maximal subgroup H_a of the completely regular semigroup $(S, +)$ an absorbing zero in H_a, this is in fact, an expected interaction of the zero of the additive subgroup H_a with multiplication.The algebraic structure which posses all the properties of a ring except additive commutativity are known as skew-rings, the name due to Grillet [11].

Theorem 3.2 *The following conditions on a semiring S are equivalent:*

(i) S is completely regular;

(ii) every $\mathcal{H}^+$-class is a skew-ring;

(iii) S is a union (disjoint) of skew-rings.

If S is a completely regular semiring then $\mathcal{J}^+$ is the least b-lattice congruence on S. A completely regular semiring S is called *completely simple* if $\mathcal{J}^+ = S \times S$. The following theorem is an analogue of the famous Clifford theorem describing semilattice decompositions of the completely regular semigroups into completely simple semigroups.

Theorem 3.3 *A semiring S is completely regular if and only if it is a b-lattice of completely simple semirings.*

The variety $\mathbf{S}$ of all semirings which are union of rings were studied by Guo and Pastijn in [12]. Interestingly, the distributive laws force $\mathcal{H}^+$ in an utterly intrinsic way to be a congruence on every $S \in \mathbf{S}$ [12], which immediately implies that:

Theorem 3.4 *[12] A semiring $S \in \mathbf{S}$ if and only if S is an idempotent semiring of rings.*

There are semirings $S \in \mathbf{S}$ such that $E^+(S)$, the set of all additive idempotents, is not a subsemiring of S [12]. Since, for every $S \in \mathbf{S}$, $(E^+(S), \cdot)$ is a band, it means that $E^+(S)$ is not additive closed, i.e. $(S, +)$ is not orthodox. Semirings in $\mathbf{S}$ for which the set $E^+(S)$ of all additive idempotents is a subsemiring of S are called *orthodox*. Rings as well as distributive lattices are orthodox.

Theorem 3.5 *If a semiring S is a b-lattice of rings then S is orthodox.*

Let S be an orthodox semiring. Then $E^+(S)$ is an idempotent semiring. So every subvariety of the variety $\mathbf{I}$ of all idempotent semirings will correspond to some subvariety of the variety $\mathbf{O}$ of all orthodox semirings. Let $\mathbf{R}i$ be the variety of all rings.

Theorem 3.6 *For every subvariety $\mathbf{V}$ of $\mathbf{I}$, $\mathbf{R}i \circ \mathbf{V} \subseteq \mathbf{S}$.*

Theorem 3.7 *Let $\mathbf{V}$ be a subvariety of $\mathbf{I}$. Then $\mathbf{V} \vee \mathbf{R}i$ is the subvariety of $\mathbf{O}$ of all orthodox semirings S such that $E^+(S) \in \mathbf{V}$. In particular, $\mathbf{O} = \mathbf{I} \vee \mathbf{R}i$*

Theorem 3.8 *The mappings*

$$[\mathbf{R}i, \mathbf{O}] \longrightarrow L(\mathbf{I}), \text{ defined by } \mathbf{V} \longrightarrow \mathbf{V} \cap \mathbf{I},$$

and

$$L(\mathbf{I}) \longrightarrow [\mathbf{R}i, \mathbf{O}], \text{ defined by } \mathbf{U} \longrightarrow \mathbf{U} \vee \mathbf{R}i$$

are pairwise inverse isomorphisms and hence the interval $[\mathbf{R}i, \mathbf{O}]$ *in* $L(\mathbf{O})$ *is isomorphic to the lattice* $L(\mathbf{I})$ *of all subvarieties of the variety* $\mathbf{I}$ *of all idempotent semirings.*

3.1 Semirings which are subdirect products of a ring and a distributive lattice

Both rings and distributive lattices form subvarieties of the variety of semirings, more specifically, they are subvarieties of the variety of all semirings with commutative regular addition. Then it is natural to ask for the join of the variety of all rings and the variety of all distributive lattices. Bandelt and Petrich [2], starting with this quest, characterized the semirings with commutative regular addition which are subdirect products of rings and distributive lattices.

If S is a semiring such that $(S,+)$ is commutative and regular then the additive reduct $(S,+)$ is a semilattice of groups. For each element a of S, let a' be the unique additive inverse of a and $a^o = a + a'$ be the identity of the maximal subgroup of S, which contains a. Bandelt and Petrich proved that:

Theorem 3.9 *[2] If* S *is a semiring such that* $(S,+)$ *is commutative and regular then* S *is a distributive lattice of rings if and only if for all* $a, b \in S$,

$$aa^o = a^o, \; ab^o = b^oa \text{ and } a + a^ob = a. \tag{3.1}$$

Again if a semiring S is a subdirect product of a distributive lattice and a ring then $E^+(S)$ is a k-subset of S which is equivalent to the condition that for all $a, b \in S$,

$$a + b = a \text{ implies that } b \in E^+(S). \tag{3.2}$$

Theorem 3.10 *Let* S *be a semiring whose additive reduct* $(S,+)$ *is regular. Then* S *is a subdirect product of a ring and a distributive lattice if and only if its additive reduct* $(S,+)$ *is commutative and* S *satisfies (3.1) and (3.2).*

Now, since the additive reduct of such semirings are strong semilattices of groups, we expect for a construction analogous to the construction of a strong semilattices of semigroups. So far, we have two such constructions: one due to Bandelt and Petrich [2] and other is its modified form due to S. Ghosh [7].

Let D be a distributive lattice and $\{S_\alpha : \alpha \in D\}$ be a family of pairwise disjoint semirings. For each $\alpha \leq \beta$ in D, let $\phi_{\alpha,\beta} : S_\alpha \longrightarrow S_\beta$ be a semiring monomorphism satisfying

$$\phi_{\alpha,\alpha} = I_{S_\alpha}, \text{ the identity mapping on } S_\alpha \tag{3.3}$$

$$\phi_{\alpha,\beta}\phi_{\beta,\gamma} = \phi_{\alpha,\gamma} \text{ if } \alpha \leq \beta \leq \gamma \tag{3.4}$$

and an additional condition:

$$S_\alpha\phi_{\alpha,\gamma}S_\beta\phi_{\beta,\gamma} \subseteq S_{\alpha\beta}\phi_{\alpha\beta,\gamma} \text{ if } \alpha + \beta \leq \gamma, \tag{3.5}$$

Then $S = \bigcup_{\alpha \in T} S_\alpha$ can be made a semiring with addition and multiplication defined by: for $a \in S_\alpha, b \in S_\beta$,

$$a + b = a\phi_{\alpha,\alpha+\beta} + b\phi_{\beta,\alpha+\beta} \tag{3.6}$$

$$\text{and } a \cdot b = c \in S_{\alpha\beta} \text{ if } c\phi_{\alpha\beta,\alpha+\beta} = a\phi_{\alpha,\alpha+\beta} \cdot b\phi_{\beta,\alpha+\beta}. \tag{3.7}$$

This semiring S is denoted by $S =< D; S_\alpha, \phi_{\alpha,\beta} >$.

Bandelt and Petrich [2] characterized the semirings which are subdirect product of a ring and a distributive lattice within the class of all regular semirings with commutative addition. But a subsemigroup T of a regular semigroup S need not be a regular semigroup, from which it follows that though both rings and distributive lattices are regular semirings with commutative addition, still their subdirect product may not be additive regular. Ghosh [7] characterized the class of all semirings which are subdirect products of a distributive lattice and a ring. To do this he has introduced E-inversive semirings.

The 0 of a semiring S (if it exists) is called absorbing if $a0 = 0a = 0$ for all $a \in S$. An additive commutative semiring S with an absorbing zero is called a halfring if it is additive cancellative i.e. for all $a, b, c \in S$, $a + c = b + c$ implies that $a = b$. An additive commutative semiring S is called E-inversive if its additive reduct $(S, +)$ is an E-inversive semigroup i.e. for each $a \in S$, there is $x \in S$ such that $a + x \in E^+(S)$.

Theorem 3.11 *[?] A semiring S is a subdirect product of a distributive lattice and a ring if and only if it is an E-inversive strong distributive lattice of halfrings.*

Mukhopadhyay et. al. generalized the result on subdirect product of a distributive lattice and a ring by Bandelt and Petrich in a dual sense: one that they did it for the additive inverse semirings whose additive reduct is not necessarily commutative and another that they characterized the semirings which are subdirect products of a ring and a (commutative) idempotent semiring, b-lattice, distributive quasi lattice each of which is a generalization of distributive lattice.

Theorem 3.12 *An inverse semiring S is a subdirect product of a ring and an additive idempotent semiring if and only if $E^+(S)$ is a k-ideal of S.*

Inverse semirings S which are subdirect product of a ring and an idempotent semiring are precisely those which satisfy (3.2) and $aa^o = a^o$ for all $a \in S$.

Following theorem is one from which Sen et. al. [17], [18] initiated the idea of p-regularity of a semiring.

Theorem 3.13 *An inverse semiring S is a subdirect product of a distributive lattice and a ring if and only if it satisfies (3.1) and (3.2) and for each $a \in S$ there is $b \in S$ such that*

$$a + aba = 2a. \tag{3.8}$$

Note that in a semiring S, for $a, b \in S$, $a + aba = 2a$ implies that $na + aba = (n+1)a$ for all $n \in \mathbb{N}$ but the converse is not true in general. Again if S happens to be a ring then $a + aba = 2a$ implies the von Neumann regularity of the ring S.

Definition 3.14 *Let S be a semiring. Then S is said to be p-regular (p stands for 'periodic') if for each $a \in S$, there is $b \in S$ such that $na + aba = (n+1)a$ for some $n \in \mathbb{N}$.*

An inverse semiring S is p-regular if and only if for each $a \in S$ there is $b \in S$ such that $a + aba = 2a$. They have extended this idea of p-regularity and have introduced p-simple semirings, p-ideals, p-idempotents, p-semifields etc..

3.2 Clifford and generalized Clifford semirings

Ghosh [7] introduced Clifford semirings as the semirings which are strong distributive lattices of rings. Additive reduct $(S, +)$ of such a semiring S being strong semilattice of commutative groups, is a commutative Clifford semigroup.

Definition 3.15 *A semiring S is called a Clifford semiring if it is an additively commutative inverse semiring such that $E^+(S)$ is a sublattice as well as a k-ideal of S.*

Now, since the additive reduct $(S,+)$ is commutative, $E^+(S)$ is an ideal of S. Thus for $E^+(S)$ to be a k-ideal in an additively commutative semiring S, it is sufficient that for all $a,b \in S$,

$$a + b = b \text{ implies that } a \in E^+(S). \tag{3.9}$$

As a natural generalization of strong semilattices of groups, strong distributive lattices of rings comes first, but on the other hand this generalization pre-assumes condition like additive commutativity. Sen, Maity and Shum [33] characterized the semirings which are distributive lattices of skew-rings. We expect that the readers will not be confused to call such semirings Clifford semirings, in fact Sen, Maity and Shum introduced such semirings with this name.

Definition 3.16 *A semiring S is called a Clifford semiring if it is additive inverse semiring such that for every $a \in S$ its inverse a' satisfies*

$$a + a' = a' + a \text{ and } a(a + a') = a + a'$$

and $E^+(S)$ is a distributive sublattice as well as a k-ideal of S.

As immediate characterizations of such semirings we have:

Theorem 3.17 *Let S be a semiring. Then the following conditions are equivalent:*

(i) S is a Clifford semiring;

(ii) S is an additive inverse semiring satisfying (3.9) and for all $a,b \in S$,

$$a + a' = a' + a, \; a(a + a') = a + a', \; (a + a')b = b(a + a') \text{ and } a + (a + a')b = a,$$

where a' is the additive inverse of a.

(iii) S is a strong distributive lattice of skew-rings.

If B is a b-lattice, $\{S_\alpha : \alpha \in B\}$ be a family of mutually disjoint semirings and for all $\alpha \leq \beta$ in B, $\phi_{\alpha,\beta} : S_\alpha \longrightarrow S_\beta$ is a semiring monomorphism satisfying (3.3), (3.4) and (3.5), then $S = \cup_{\alpha \in B} S_\alpha$ is a semiring with respect to the addition and multiplication as defined in (3.6) and (3.7) respectively [33]. This semiring S is called strong b-lattice of semirings and is denoted by $S =< B; S_\alpha, \phi_{\alpha,\beta} >$. As a generalization of strong distributive lattices of skew-rings Sen, Maity and Shum introduced strong b-lattices of skew-rings which they named generalized Clifford semirings.

Definition 3.18 *A semiring S is called a generalized Clifford semiring if it is additive inverse semiring such that for every $a \in S$ its inverse a' satisfies*

$$a + a' = a' + a \text{ and } a(a + a') = a + a'$$

and $E^+(S)$ is a k-ideal of S.

Following theorem demonstrates structure of the generalized Clifford semirings.

Theorem 3.19 *Let S be a semiring. Then the following conditions are equivalent:*

(i) S is a generalized Clifford semiring;

(ii) S is an additive inverse semiring satisfying (3.9) and for all $a, b \in S$,

$$a + a' = a' + a \text{ and } a(a + a') = a + a',$$

where a' is the additive inverse of a;

(iii) S is a strong b-lattice of skew-rings;

(iv) S is additive inverse subdirect product of a b-lattice and a skew-ring.

4 k-Regular Semirings

A ring R is regular (due to von Neumann) if the multiplicative reduct $(R, \cdot)$ is a regular semigroup. Bourne [4] extended the notion of regularity to semirings and defined a semiring S to be regular if for each $a \in S$ there exist $x, y \in S$ such that $a + axa = aya$. If a semiring S happens to be a ring then the Neumann regularity and the Bourne regularity are equivalent. This is not true in a semiring in general [27]. Adhikari, Sen and Weinert [1] renamed the Bourne regularity of a semiring as k-regularity to distinguish from the notion of von Neumann regularity. Here we continue our discussion with this new name.

Recently k-regular have been studied extensively by Sen, Weinert, Adhikari, Mukhopadhyay and Bhuniya [1], [27] – [30], [35]. There are many examples of k-regular semirings .

Sen and Mukhopadhyay characterized the k-regular semirings, mainly commutative k-regular semirings by their (left, right) k-ideals.

Theorem 4.1 *Let S be a semiring. Then S is k-regular if and only if $\overline{RL} = R \cap L$ for any right k-ideal R and left k-ideal L of S.*

Sen and Bhuniya, being motivated by the following results, studied the k-regular semirings whose additive reduct is a semilattice i.e. the k-regular semirings in $\mathbb{SL}^+$.

Theorem 4.2 *[30] Every semiring whose additive reduct is a semilattice can be embedded into a k-regular semiring of endomorphisms of a semilattice.*

Let F be a semigroup. Then the set $P_f(F)$ of all finite subsets of F is a semiring, where

$$A + B = A \cup B$$
$$\text{and} \quad AB = \{ab \mid a \in A, \ b \in B\} \quad \text{for all} \quad A, B \in P_f(F).$$

The following result is also an enthuistic addition in the series of natural examples of k-regular semirings.

Theorem 4.3 *[30] Let F be a semigroup. Then the semiring $P_f(F)$ is k-regular if and only if F is a regular semigroup.*

Also there is a fair correspondence between k-ideals of a semiring and ideals of a semigroup. For any semigroup F, J is a k-ideal (left, right) of the semiring $P_f(F)$ of all finite subsets of F if and only if $J = P_f(I)$ for some ideal (left, right) I of F. This result is an inspiring one for introducing Green's equivalences in $\mathbb{SL}^+$ [27].

Note that a semiring $S \in \mathbb{SL}^+$ is k-regular if and only if for each $a \in S$ there exists $x \in S$ such that $a + axa = axa$. Then $ax + (ax)^2 = (ax)^2$, which indicates the presence of some elements e in the semirings in $\mathbb{SL}^+$ such that $e + e^2 = e^2$. Such elements are called almost idempotents. If a semiring $S \in \mathbb{SL}^+$ is such that $a + a^2 = a^2$ for all $a \in S$ then S is called an almost idempotent semiring [31]. The class of all almost idempotent semirings is a variety which we denote by **AI**. Following proposition describes the free almost idempotent semiring over a band F. Let **B** be the variety of all bands.

Proposition 4.4 *Let F be a semigroup. Then the semiring $P_f(F) \in$ **AI** if and only if $F \in$ **B**.*

Let X be a nonempty set and F_X be the free semigroup on X. Let ρ be the congruence on $P_f(F_X)$ generated by $\{(U + U^2, \ U^2) \mid U \in P_f(F_X)\}$. We define $i : X \longrightarrow P_f(F_X)/\rho$ by $i(x) = \{x\rho\}$.

Theorem 4.5 *Let X be a nonempty set, $S \in \mathbf{AI}$, and $\phi : X \longrightarrow S$ be any mapping. Then there exists a unique semiring homomorphism $\bar{\psi} : P_f(F_X)/\rho \longrightarrow S$ such that $\bar{\psi} oi = \phi$.*

Thus we see that the variety of all almost idempotent semirings corresponds the variety of all bands via the semiring $P_f(F)$. In search of the almost idempotent semirings corresponding to the different subvarieties of bands and in view of the result that every band is a semilattice of rectangular bands, we should look first for the least distributive lattice congruence on an almost idempotent semiring.

Let S be an almost idempotent semiring. Then the binary relation η on S defined by: for $a, b \in S$,

$a\eta b$ if there exists $x \in S$ such that

$$a + axbxa = axbxa \text{ and } b + bxaxb = bxaxb$$

is the least distributive lattice congruence on S.

Thus rectangular almost idempotent semirings can be defined as follows:

Definition 4.6 *An almost idempotent semiring S is said to be rectangular if for all $a, b \in S$ there exists $x \in S$ such that*

$$a + axbxa = axbxa.$$

In the following theorem we show that the rectangular almost idempotent semirings really correspond to the rectangular bands.

Theorem 4.7 *Let F be a semigroup. Then the semiring $P_f(F)$ is rectangular almost idempotent semiring if and only if F is a rectangular band.*

The following result is an analogue of the semilattice decomposition of bands into rectangular bands.

Theorem 4.8 *Every almost idempotent semiring is a distributive lattice of rectangular almost idempotent semirings.*

In [31], the authors also characterize left (right) zero, weakly commutative, (left, right) normal almost idempotent semirings. The authors also characterize completely k-regular semirings [28] and k-Clifford semirings [29]. Distributive lattice decompositions of the semirings in $\mathbb{SL}^+$ into Archimedean components can be found in [3] and another important addition in the series of papers on the semirings in $\mathbb{SL}^+$ is [6].

5 Variety of Idempotent Semirings

The Green's relations on both the additive reduct $(S,+)$ and multiplicative reduct $(S,\cdot)$ of an idempotent semiring S have been proved to be very useful in characterizing the variety $\mathbf{I}$ of idempotent semirings. The equivalence relations $\mathcal{D}^+$ and $\mathcal{D}^\bullet$ on an idempotent semiring S are the least semilattice congruences on the reducts $(S,+)$ and $(S,\cdot)$ respectively. Again for any idempotent semiring S the least distributive lattice congruence on S exists. If η is the least distributive lattice congruence on an idempotent semiring S then both $(S/\eta,+)$ and $(S/\eta,\cdot)$ are semilattices and so $\mathcal{D}^+ \subseteq \eta$ and $\mathcal{D}^\bullet \subseteq \eta$. In [32], Sen, Guo and Shum ask for a description of the idempotent semirings for which η coincides with either one of $\mathcal{D}^+$ and $\mathcal{D}^\bullet$.

If $\mathcal{D}^+$ is the least distributive lattice congruence on an idempotent semiring S, then for any two $x,y \in S$, $(x+xy)\mathcal{D}^+x$ which implies that

$$x + xy + x = x.$$

Similarly $x+yx+x = x$ for all $x,y \in S$. Sen et. al. shown these two conditions to be sufficient such that $\mathcal{D}^+$ is the least distributive lattice congruence on S.

A *band semiring* is an idempotent semiring which satisfies the identities

$$x + xy + x \approx x \tag{5.1}$$

$$x + yx + x \approx x \tag{5.2}$$

The class of all band semirings is a subvariety of the variety of all idempotent semirings. We will denote the variety of all band semirings by $\mathbf{BI}$. Let $\mathbf{S}l^+$ be the variety of idempotent semirings S whose additive reduct $(S,+)$ is a semilattice i.e. the variety of all b-lattices. Every band semiring S satisfies the following identity

$$xy + yx + xy \approx xy. \tag{5.3}$$

Thus $\mathbf{BI} \cap \mathbf{S}l^+ = \mathbf{D}$, the variety of all distributive lattices.

We denote the variety of all idempotent semirings whose additive reduct is a rectangular band by $\mathbf{R}^+$. Note that $\mathbf{R}^+\mathbf{BI}$.

If S is an idempotent semiring, then two partial orders $\leq_+$ and $\leq_.$ can be defined on S by the following: for $a,b \in S$,

$$a \leq_+ b \Leftrightarrow a+b = b+a = b,$$

$$a \leq_. b \Leftrightarrow ab = ba = a.$$

Theorem 5.1 *[32] Let S be an idempotent semiring. Then the following conditions are equivalent.*

1. *$\mathcal{D}^+$ is the least distributive lattice congruence on S;*

2. *$S \in \mathbf{BI}$;*

3. *S satisfies (5.1) and (5.3);*

4. *S satisfies the identity*

$$x \approx (x+y+x)x(x+y+x); \tag{5.4}$$

5. *$S \in \mathbf{R}^+ \circ \mathbf{D}$;*

6. *$\leq_+ \subseteq \leq_\cdot$.*

Wang, Zhou and Guo [37] proved that the additive reduct of a band semiring is a regular band. We can think the following identity

$$x \approx xyx + x + xyx \tag{5.5}$$

as the dual of (5.4). As we expect from the dual nature of this identity is fulfilled in the following theorem.

Theorem 5.2 *Let S be an idempotent semiring. Then $\leq_\cdot \subseteq \leq_+$ if and only if S satisfies (5.5).*

Then Pastijn and Zhao [23] proved that the idempotent semirings S which satisfy the equivalent conditions of this theorem are precisely those for which $\mathcal{D}^\bullet$ is the least distributive lattice congruence.

Let $\mathbf{N}$ be the variety of idempotent semirings which satisfies the identity:

$$x + xyx + x \approx x. \tag{5.6}$$

This is known as the generalized absorptive law [22] and is equivalent to

$$x(x+y+x)x \approx x.$$

The Green's relation $\mathcal{D}^\bullet$ is a congruence on every $S \in \mathbf{N}$. Also we have

Theorem 5.3 *[23] Let S be an idempotent semiring. Then $S \in \mathbf{N}$ if and only if $\mathcal{D}^+ \vee \mathcal{D}^\bullet$ is the least distributive lattice congruence on S.*

Thus for $S \in \mathbf{N}$, $\mathcal{D}^\bullet$ is the least distributive lattice congruence if and only if $\mathcal{D}^+ \subseteq \mathcal{D}^\bullet$. Pastijn and Zhao characterized the variety of all idempotent semirings S such that $\mathcal{D}^\bullet$ is the least distributive lattice congruence.

Theorem 5.4 *[23] Let S be an idempotent semiring. Then the following conditions are equivalent:*

1. *$\mathcal{D}^\bullet$ is the least distributive lattice congruence on S;*

2. *$S \in \mathbf{N}$ and $\mathcal{D}^+ \subseteq \mathcal{D}^\bullet$;*

3. *S satisfies the identity (5.5).*

Zhao, Shum and Guo [38], [39] characterized the variety of idempotent semirings by the Green's relations on both the reducts. In [39], the authors characterized the variety of idempotent semirings such that $\mathcal{L}^\bullet \cap \xi$ is a congruence, $\mathcal{L}^\bullet \cap \xi = \Delta$ and $\mathcal{L}^\bullet \cap \xi = \nabla$, where $\xi \in \{\mathcal{L}^+, \mathcal{R}^+, \mathcal{D}^+, \mathcal{D}^\bullet\}$ as equational classes.

Theorem 5.5 *Let $S \in \mathbf{I}$. Then $\mathcal{L}^\bullet \cap \mathcal{D}^+$ is a congruence on S if and only if S satisfies the following identities:*

$$(yx + xyx + z)(xyx + yx + z) \approx yx + xyx + z,$$
$$(z + yx + xyx)(z + xyx + yx) \approx z + yx + xyx,$$
$$z(yx + xyx)z(xyx + yx) \approx z(yx + xyx).$$

Similar results for the varieties of idempotent semirings such that $\mathcal{L}^\bullet \cap \xi$ is a congruence, where $\xi \in \{\mathcal{L}^+, \mathcal{R}^+, \mathcal{D}^\bullet\}$ can be found in [39].

Both the classes of all idempotent semirings such that $\mathcal{L}^\bullet \cap \xi = \Delta$ and $\mathcal{L}^\bullet \cap \xi = \nabla$, where $\xi \in \{\mathcal{L}^+, \mathcal{R}^+, \mathcal{D}^+, \mathcal{D}^\bullet\}$ are subvarieties of $\mathbf{I}$. Let us denote these subvarieties by:

$$\mathbf{L}_d = \{S \in \mathbf{I} \mid \mathcal{L}^\bullet \cap \mathcal{D}^+ \in C(S)\},\ \mathbf{L}_l = \{S \in \mathbf{I} \mid \mathcal{L}^\bullet \cap \mathcal{L}^+ \in C(S)\},\ \mathbf{L}_r = \{S \in \mathbf{I} \mid \mathcal{L}^\bullet \cap \mathcal{R}^+ \in C(S)\},$$
$$\mathbf{L}_d^* = \{S \in \mathbf{I} \mid \mathcal{L}^\bullet \cap \mathcal{D}^\bullet \in C(S)\},\ \mathbf{L}_{d_0} = \{S \in \mathbf{I} \mid \mathcal{L}^\bullet \cap \mathcal{D}^+ = \Delta\},\ \mathbf{L}_{d_1} = \{S \in \mathbf{I} \mid \mathcal{L}^\bullet \cap \mathcal{D}^\bullet = \nabla\}.$$

Similarly we define $\mathbf{L}_{l_0}$, $\mathbf{L}_{l_1}$, $\mathbf{L}_{r_0}$, $\mathbf{L}_{r_1}$, $\mathbf{L}_{d_0}^*$ and $\mathbf{L}_{d_1}^*$.

Theorem 5.6

1. *$\mathbf{L}_d = \mathbf{L}_{d_1} \circ \mathbf{L}_{d_0}$,*

2. *$\mathbf{L}_l = \mathbf{L}_{l_1} \circ \mathbf{L}_{l_0}$,*

3. *$\mathbf{L}_r = \mathbf{L}_{r_1} \circ \mathbf{L}_{r_0}$,*

4. *$\mathbf{L}_d^* = \mathbf{L}_{d_1}^* \circ \mathbf{L}_{d_0}^*$.*

Zhao et. al. also characterized similarly all the varieties of idempotent semirings determined by the dual criterions i.e. with $\mathcal{L}^{\bullet} \vee \xi$, $\xi \in \{\mathcal{L}^{+}, \mathcal{R}^{+}, \mathcal{D}^{+}, \mathcal{D}^{\bullet}\}$.

In the paper [20] on the weak commutativity in idempotent semirings F. Pastijn characterized the variety $\mathbf{U}$ of all idempotent semiring such that $xy+yx \approx yx+xy$. Given any idempotent semiring S, $\mathcal{D}^{+} \cap \mathcal{D}^{\bullet}$ is the least $\mathbf{U}$-congruence on S. Though it is not at all unexpected, but really interesting that:

Theorem 5.7 *[20]* $\mathbf{U} = \mathbf{S}l^{+} \vee \mathbf{S}l^{\bullet}$, *where* $\mathbf{S}l^{\bullet}$ *is the variety of all idempotent semirings whose multiplicative reduct is a semilattice.*

Another interesting feature of this subvariety is that for all $S \in \mathbf{U}$, both the additive reduct and the multiplicative reduct of S are regular bands.

Also the lattice $L(\mathbf{U})$ of all subvarieties of $\mathbf{U}$ is a subdirect product of the lattices $L(\mathbf{S}l^{+})$ and $L(\mathbf{S}l^{\bullet})$. In fact,

Theorem 5.8 *[20] Let* $L = \{(\mathbf{W_1}, \mathbf{W_2}) \in L(\mathbf{S}l^{+}) \times L(\mathbf{S}l^{\bullet}) \mid \mathbf{W_1} \cap \mathbf{Bi} = \mathbf{W_2} \cap \mathbf{Bi}\}$. *Then L is a subdirect product of $L(\mathbf{S}l^{+})$ and $L(\mathbf{S}l^{\bullet})$. The mappings*

$$\phi : L(\mathbf{U}) \longrightarrow L, \mathbf{W} \longrightarrow (\mathbf{W} \cap \mathbf{S}l^{+}, \mathbf{W} \cap \mathbf{S}l^{\bullet})$$

$$\text{and } \psi : L \longrightarrow L(\mathbf{U}), (\mathbf{W_1}, \mathbf{W_2}) \longrightarrow \mathbf{W_1} \vee \mathbf{W_2}$$

are pairwise inverse isomorphisms.

References

[1] Adhikari, M. R., Sen, M. K. and Weinert, H. J., On k-regular semirings, *Bull. Cal. Math. Soc.* **88**(1996),141–144.

[2] Bandelt, H. J. and Petrich, M., Subdirect products of rings and distributive lattices, *Proc. Edin. Math. Soc.* **25** (1982), 155-171.

[3] Bhuniya, A. K. and Mondal, T. K., Distributive lattice decompositions of semirings with a semillatice additive reduct, *Semigroup Forum* **80**(2010), 293–301.

[4] Bourne, S., The Jacobson radical of a semiring, *Proc. Nat. Acad. Sci. U.S.A.* **37**(1951), 163.

[5] Clifford, A. H., Semigroups admitting relative inverses, *Annals of Math.* **42**(1941), 1037–1049.

[6] Damljanović, N., Ćirić, M. and Bogdanović, S., Congruence openings of additive Greens relations on a semiring, *Semigroup Forum*

[7] Ghose, S., A characterisation of semirings which are subdirect products of a distributive lattice and a ring, *Semigroup Forum* **59**(1999), 106-120.

[8] Ghosh, S., Pastijn, F. and Zhao, X. Z., Varieties generated by ordered bands I, *Order***22(2)**(2005), 109-128.

[9] Golan, J. S., *Semirings and Their Applications.* Kluwer Academic Publishers, Dordrecht, 1999.

[10] Green, J. A., On the structure of semigroups, *Annals of Mathematics* **54**(1951), 163–172.

[11] Grillet, M. P., Semirings with a completely simple additive Semigroup, *J. Austral. Math. Soc.*; **20(A)**(1975), 257-267.

[12] Guo, Y. Q. and Pastijn, F., Semirings which are unions of rings, *Science in China (Series A)* **45(2)**(2002), 172–195.

[13] Hebisch, U. and Weinert, H. J., *Semirings: Algebraic Theory and Applications in Computer Science*, World Scientific, Singapore (1998).

[14] Henricksen, M., Ideals in semirings with commutative addition, *Amer. Math. Soc. Notices* **6**, 321 (1958).

[15] Howie, J. M., *Fundamentals of Semigroup Theory*, Clarendon Press, Oxford (1995).

[16] McKenzie, R. N., McNulty, G. F. and Taylor, W. F., *Algebras, Lattices, Vareities*, Vol-I, Wadsworth and Brooks, Monterey 1987.

[17] Mukhopadhyay, P. and Ghosh, S., A New Class of Ideals in Semirings, *Southeast Asian Bull. of Math.* **23(2)**(1999) 253264.

[18] Mukhopadhyay, P., Sen, M. K. and Ghosh, S., p-ideals in p-regular semirings, *Southeast Asian Bulletin of Mathematics* **26**(2002), 439452.

[19] Pastijn, F., Varieties generated by ordered bands II, *Order*

[20] Pastijn, F., Weak commutativity in idempotent semirings, *Semigroup Forum* **72**(2006), 283-311.

[21] Pastijn, F. and Guo, Y. Q., The lattice of idempotent distributive semiring varieties, *Science in China(Series A)* **29**(1999), 391-407.

[22] Pastijn, F. and Romanowska, A., Idempotent distributive semirings I, *Acta Sci. Math.(Szeged)* **44**(1982), 239-253.

[23] Pastijn, F. and Zhao, Z. X., Green's $\mathcal{D}$-relation for the multiplicative reduct of an idempotent semiring, *Archivum Mathematicum(Brno)* **36**(2000), 77-93.

[24] Pastijn, F. and Zhao, Z. X., Varieties of idempotent semirings with commutative addition, *Algebra Universalis* **54**(2005), 301-321.

[25] Petrich, M., *Lectures in Semigroups*, John Wiley & Sons (1977).

[26] Petrich, M. and N. Reilly, *Completely Regular Semigroups*, Wiley IEEE, 1999.

[27] Sen, M. K. and Bhuniya, A. K., On additive idempotent k-regular semirings. *Bull. Cal. Math. Soc.* **93**(2001), 371-384.

[28] Sen, M. K. and Bhuniya, A. K., Completely k-regular semirings, *Bull. Cal. Math. Soc.* **97**(2005), 455–466.

[29] Sen, M. K. and Bhuniya, A. K., On additive idempotent k-Clifford semirings, *Southeast Asian Bulletin of Mathematics* **32**(2008), 1149–1159.

[30] Sen, M. K. and Bhuniya, A. K., On semirings whose additive reduct is a semilattice. *Semigroup Forum*, DOI 10.1007/s00233-010-9271-9.

[31] Sen, M. K. and Bhuniya, A. K., The structure of almost idempotent semirings, *Algebra Colloquium* **17**(Spec 1) (2010) 851-864.

[32] Sen, M. K., Guo, Y. Q. and Shum, K. P., A class of idempotent semirings, *Semigroup Forum* **60**(2000), 351-367.

[33] Sen, M. K., Maity, S. K. and Shum, K. P., Clifford semirings and generalized Clifford semirings, *Taiwanese Journal of Mathematics* **9(3)** (2005), 433–444.

[34] Sen, M. K., Maity, S. K. and Shum, K. P., On Completely Regular Semirings, *Bull. Cal. Math. Soc.* **98(4)**(2006), 329–328.

[35] Sen, M. K. and Mukhopadhyay, P., von Neumannian regularity in semirings, *Kyungpook Math. J.* **35(2)**(1995), 249-258.

[36] Vandiver, H. S., Note on a simple type of algebra in which the cancellation law of addition does not hold, *Bull. Amer. Math. Soc.* **40**(1934), 914-920.

[37] Wang, Z. P., Zhou, Y. L. and Guo, Y. Q., A note on band semirings, *Semigroup Forum* **71**(2005), 439-442.

[38] Zhao, X. Z., Guo, Y. Q. and Shum, K. P., $\mathcal{D}$-subvarieties of the variety of idempotent semirings, Algebra Colloquium, 9(2002), 15-28.

[39] Zhao, X. Z., K. P. Shum and Y. Q. Guo: $\mathcal{L}$-subvarieties of the variety of idempotent semirings, Algebra Universalis, 46(2001), 75-96.

Local Definitions of Group Formations

L. A. SHEMETKOV

Department of Mathematics, Francisk Skorina Gomel State University, Gomel 246019, Belarus
E-mail: shemetkov@gsu.by

This article analyses relations between local definitions of group formations: local and ω-local satellites, composition and $\mathfrak{L}$-composition satellites, $\mathfrak{X}$-formation functions of Förster. It is proved that a non-empty formation of finite groups is $\mathfrak{X}$-local if and only if it has an $\mathfrak{X}$-composition satellite.

AMS Mathematics Subject Classification (2000): 20D10

Keywords: finite groups, saturated formation

1. Introduction

In the class $\mathfrak{E}$ of finite groups, the idea of local definitions of saturated formations was realized for the first time by W. Gaschütz.[1] Later local definitions of partially saturated formations were introduced and investigated in Refs. 2–9. Following Ref. 9, we formulate a general approach.

A local definition is a map $f{:}\mathfrak{E} \to \{\text{formations}\}$ together with a f-rule which decide whether a chief factor is f-central or f-eccentric in a group. In addition, we follow the agreement that the local definition f does not distinguish between non-identity groups with the same (up to isomorphism) set of composition factors. Therefore, for any prime p, f is not distinguish between any two non-identity p-groups; we will denote through $f(p)$ a value of f on non-identity p-groups.

If a class $\mathfrak{F}$ coincides with the class of all groups all of whose chief factors are f-central, we say that f is a local definition of $\mathfrak{F}$.

In this paper we analyse relations between local definitions of different types. We consider only finite groups.

2. Preliminaries

We use standard notations and definitions.[10] We say that a map f does not distinguish between $\mathfrak{H}$-groups if $f(A) = f(B)$ for any two groups A and

B in $\mathfrak{H}$. Recall that a formation is a class of groups which is closed under taking homomorphic images and subdirect products. Following Gaschütz, the $\mathfrak{F}$-residual $G^{\mathfrak{F}}$ of a group G is the least normal subgroup with quotient in $\mathfrak{F}$. The Gaschütz product $\mathfrak{F} \circ \mathfrak{H}$ of formations $\mathfrak{F}$ and $\mathfrak{H}$ is defined as the class of all groups G such that $G^{\mathfrak{H}} \in \mathfrak{F}$.

Char($\mathfrak{X}$) is the set of orders of all groups of prime order in $\mathfrak{X}$. A group G is called a pd-group if its order is divisible by a prime p; an ωd-group (a chief ωd-factor) is a group (a chief factor) being pd-group for some $p \in \omega$; $G_{\omega d}$ is the largest normal subgroup all of whose G-chief factors are ωd-groups ($G_{\omega d} = 1$ if all minimal normal subgroups in G are ω'-groups). If $\mathfrak{H}$ is a class of groups, then $\mathfrak{H}_\omega$ is the class of all ω-groups in $\mathfrak{H}$. A chief factor H/K of G is called a chief $\mathfrak{H}$-factor if $H/K \in \mathfrak{H}$.

$O_\omega(G)$ is the greatest normal ω-subgroup in G; $\pi(G)$ is the set of primes dividing the order of a group G; $\pi(\mathfrak{F}) = \cup_{G \in \mathfrak{F}} \pi(G)$; $\mathfrak{N}$ is the class of all nilpotent groups; $\mathfrak{A}$ is the class of all abelian groups; (G) is the class of all groups isomorphic to G; $\mathfrak{J}$ is the class of simple (abelian and non-abelian) groups; if $\mathfrak{L}$ is a subclass in $\mathfrak{J}$, then $\mathfrak{L}' = \mathfrak{J} \setminus \mathfrak{L}$; $\mathfrak{L}^+$ is the set of all abelian groups in $\mathfrak{L}$; $\mathfrak{L}^- = \mathfrak{L} \setminus \mathfrak{L}^+$. $\mathrm{E}\mathfrak{H}$ is the class of all groups having a subnormal series all of whose factors are $\mathfrak{H}$-groups; $G_{\mathrm{E}\mathfrak{H}}$ is the $\mathrm{E}\mathfrak{H}$-radical of G, the largest normal $\mathrm{E}\mathfrak{H}$-subgroups in G. If $S \in \mathfrak{J}$, then $C^S(G)$ is the intersection of centralizers of all chief $\mathrm{E}(S)$-factors of G; Com(G) is the set of all groups that are isomorphic to composition factors of a group G; $\mathrm{Com}(\mathfrak{F}) = \cup_{G \in \mathfrak{X}} \mathrm{Com}(G)$; $\mathrm{Com}^+(\mathfrak{F}) = \mathrm{Com}(\mathfrak{F}) \cap \mathfrak{J}^+$.

3. Local and ω-local satellites

The following type of a local definition was proposed by W. Gaschütz.[1]

Definition 3.1. Let f be a local definition such that $f(A) = \bigcap_{p \in \pi(A)} f(p)$ for any group $A \neq 1$. Let an f-rule be defined as follows: a chief factor H/K of a group G is f-central if $G/C_G(H/K) \in f(H/K)$. Then f is called *a local satellite.*

The convenient notation $LF(f)$ for a group class with a local satellite f was introduced by Doerk and Hawkes.[10] Clearly, $LF(f)$ is a non-empty formation.

The following proposition is evident.

Proposition 3.1. *Let f be a local satellite and $\pi = \{p \in \mathbb{P} \mid f(p) \neq \varnothing\}$. Then $LF(f)$ consists precisely of groups G satisfying the following condition: $G/O_{p',p}(G) \in f(p)$ for any $p \in \pi \cap \pi(G)$. Thus, if $\pi = \varnothing$, we have*

$LF(f) = (1)$. *If* $\pi \neq \varnothing$, *we have that* $LF(f) = \mathfrak{E}_\pi \bigcap(\bigcap_{p\in\pi}(\mathfrak{E}_{p'} \circ \mathfrak{E}_p \circ f(p)))$.

Following Gaschütz, a formation $\mathfrak{F}$ is called saturated if $G/\Phi(G) \in \mathfrak{F}$ always implies $G \in \mathfrak{F}$. By definition, the empty set is a saturated formation. The following remarkable result is known as the Gaschütz–Lubeseder–Schmid theorem, see Ref. 10, Theorem IV.4.6.

Theorem 3.1. *A non-empty formation has a local satellite if and only if it is saturated.*

Evidently, the formation $\mathfrak{A}_p \times \mathfrak{N}_{p'}$ of all nilpotent groups with an abelian Sylow p-subgroup is not saturated, but for any prime $q \neq p$, $G/(\Phi(G) \cap O_q(G)) \in \mathfrak{A}_p \times \mathfrak{N}_{p'}$ always implies $G \in \mathfrak{A}_p \times \mathfrak{N}_{p'}$. One more fact of the same sort is the following. Consider a saturated formation of the form $\mathfrak{M} \circ \mathfrak{H}$. Here $\mathfrak{H}$ can be non-saturated, but for every prime $p \in \mathbb{P} \setminus \pi(\mathfrak{M})$, $G/(\Phi(G) \cap O_p(G)) \in \mathfrak{H}$ always implies $G \in \mathfrak{H}$. The facts of such kind lead to the concept of a ω-saturated formation.[4]

Definition 3.2. Let ω be a set of primes. A formation $\mathfrak{F}$ is called ω-*saturated* if for every prime $p \in \omega$, $G/(\Phi(G) \cap O_p(G)) \in \mathfrak{F}$ always implies $G \in \mathfrak{F}$.

The problem of finding of local definitions of ω-saturated formations was considered in Refs. 7 and 9. While solving this problem the following concept of small centralizer was useful (see Ref. 6).

Definition 3.3. Let H/K be a chief factor of a group G. The *small centralizer* $c_G(H/K)$ of H/K in G is the subgroup generated by all normal subgroups N of G such that $\mathrm{Com}(NK/K) \cap \mathrm{Com}(H/K) = \varnothing$.

Using Definition 3.2 we can introduce the concept 'ω-saturated satellite' as follows.

Definition 3.4. Let ω be a set of primes, and f a local definition which does not distinguish between non-identity ω'-groups; if $\omega' \neq \varnothing$, we denote through $f(\omega')$ a value of f on non-identity ω'-groups. In addition, we assume that $f(A) = \bigcap_{p\in\pi(A)\cap\omega} f(p)$ for any ωd-group A. Let an f-rule be defined by the following way: a chief factor H/K of G is f-central in G if either H/K is an ωd-group and $G/C_G(H/K) \in f(H/K)$ or else H/K is an ω'-group and $G/c_G(H/K) \in f(\omega')$. Then f is called an ω-*local satellite*. We denote by $LF_\omega(f)$ the class of all groups all of whose chief factors are f-central.

Clearly, if $\omega = \mathbb{P}$, then an ω-local satellite f is a local satellite, and $LF_\omega(f) = LF(f)$. If $\omega \neq \mathbb{P}$ and $f(\omega') = \varnothing$, then $LF_\omega(f) = LF(h)$ where $h(p) = f(p)$ if $p \in \omega$, and $h(p) = \varnothing$ if $p \in \omega'$.

We note that in Ref. 7 an agreement $f(1) = f(\omega')$ was used; in the present paper we do not follow this agreement. Therefore, we have always $1 \in LF_\omega(f)$.

Lemma 3.1 (see Ref. 9, Lemma 1). *Let $\mathfrak{L}$ be a subclass in $\mathfrak{J}$, and $\{S_i \mid i \in I\}$ be the set of all chief* $\mathrm{E}\mathfrak{L}$*-factors of G. Then $\bigcap_{i\in I} c_G(S_i) = G_{\mathrm{E}(\mathfrak{L}')}$ is the* $\mathrm{E}(\mathfrak{L}')$*-radical of G.*

Remark 3.1. In Lemma 1 the set $\{c_G(S_i) \mid i \in I\}$ can be empty. We always follow the agreement that the intersection of an empty set of subgroups of G coincides with G.

The following proposition is similar to Proposition 3.1.

Proposition 3.2. *Let f be an ω-local satellite, and ω a proper subset in $\mathbb{P}$. Let $\pi = \{p \in \omega \mid f(p) \neq \varnothing\}$. Then:*

(1) if $\pi = \varnothing$ and $f(\omega') = \varnothing$, then $LF_\omega(f) = (1)$;

(2) if $\pi = \varnothing$ and $f(\omega') \neq \varnothing$, then $LF_\omega(f) = \mathfrak{E}_{\omega'} \cap f(\omega')$;

(3) if $f(\omega') \neq \varnothing$, then $LF_\omega(f)$ consists precisely of groups G such that $G/G_{\omega d} \in f(\omega')$ and $G/O_{p',p}(G) \in f(p)$ for any $p \in \pi(G) \cap \omega$.

Proof. Statements (1) and (2) are evident.

Prove (3). Assume that $f(\omega') \neq \varnothing$, and let $G \in LF_\omega(f)$. Let $\mathfrak{T}$ be the set of all chief ω'-factors in G. If a chief factor H/K of G is an ω'-group, then $G/c_G(H/K) \in f(\omega')$. Therefore, $G/\bigcap_{H/K\in\mathfrak{T}} c_G(H/K) \in f(\omega')$. By Lemma 3.1, $\bigcap_{H/K\in\mathfrak{T}} c_G(H/K) = G_{\omega d}$. So, $G/G_{\omega d} \in f(\omega')$. If $p \in \omega$ and H/K is an chief pd-factor, then $G/C_G(H/K) \in f(p)$, and we have $G/O_{p',p}(G) \in f(p)$.

Conversely, let G be a group such that $G/G_{\omega d} \in f(\omega')$ and $G/O_{p',p}(G) \in f(p)$ for any $p \in \pi(G) \cap \omega$. Clearly, we have that all G-chief ωd-factors are f-central. Let H/K be a G-chief ω'-factor of G. Then $G_{\omega d}K/K \subseteq c_G(H/K)$, and $G/G_{\omega d} \in f(\omega')$ implies $G/c_G(H/K) \in f(\omega')$. □

The following result extends Theorem 3.1 to ω-saturated formations.

Theorem 3.2 (see Ref. 7, Theorem 1). *Let ω be a set of primes. A non-empty formation has a ω-local satellite if and only if it is ω-saturated.*

4. Composition and $\mathfrak{L}$-composition satellites

Gaschütz's main idea[1] was to study groups modulo p-groups, and he implemented it through local satellites of soluble formations. While considering non-soluble formations, we have to follow the following principle: study groups modulo p-groups and simple groups. That approach was proposed in my lecture at the conference in Sverdlovsk (Russia) in 1973; in that lecture[11] composition satellites were considered under the name 'primarily homogeneous screens'.

Definition 4.1. Let f be a local definition, and let an f-rule be defined as follows: a chief factor H/K of a group G is f-central if $G/C_G(H/K) \in f(H/K)$. Then f is called *a composition satellite.* We denote by $CF(f)$ the class of all groups all of whose chief factors are f-central.

As an example, we consider the class $\mathfrak{N}^*$ of all quasinilpotent groups (for the definition of a quasinilpotent group, see Ref. 12, Definition X.13.2). It is easy to check that $\mathfrak{N}^* = CF(f)$ where f is a composition satellite such that $f(p) = (1)$ for every prime p, and $f(S) = form(S)$ for every non-abelian simple group S. Here $form(S)$ is a least formation containing S; it consists of all groups of the form $A_1 \times \cdots \times A_n$ with $A_i \simeq S$ for any i. The formation $\mathfrak{N}^*$ is non-saturated, but it is solubly saturated in the following sence.

Definition 4.2. A formation $\mathfrak{F}$ is called *solubly saturated* if $G/(\Phi(G_{G_\mathfrak{S}}) \in \mathfrak{F}$ always implies $G \in \mathfrak{F}$. Here $G_\mathfrak{S}$ is the soluble radical of G.

As pointed out in Ref. 10, formations with composition satellites were also considered—in different terminology—by R. Baer in his unpublished manuscript. The question of the coincidense of the family of non-empty solubly saturated formations and the family of formations having composition satellites was solved by the following result due to R. Baer.

Theorem 4.1 (see Ref. 10, Theorem IV.4.17). *A non-empty formation has a composition satellite if and only if it is solubly saturated.*

A composition satellite h is called integrated if $h(S) \subseteq CF(h)$ for any simple group S. If $\mathfrak{F} = CF(f)$, then $\mathfrak{F} = CF(h)$ where $h(S) = f(S) \cap \mathfrak{F}$ for any simple group S. Thus, if a formation has a composition satellite, then it has an integrated composition satellite.

The following theorem proved independently in Ref. 13 and Ref. 14 was the first important result on composition formations.

Theorem 4.2. *Let f be an integrated composition satellite. Let A be a group of automorphisms of a group G such that the following condition holds: $A/C_A(H/K) \in f(H/K)$ for every A-composition factor H/K of G. Then $A \in CF(f)$.*

Applying Theorem 4.2 to the formation $\mathfrak{U}$ of all supersoluble groups, I proved in 1974 the following theorem.

Theorem 4.3 (see Ref. 13, Theorem 2.4). *Let A be a group of automorphisms of a group G. Assume that there exists a chain of A-invariant subgroups*

$$G = G_0 > G_1 > \cdots > G_n = 1$$

with prime indices $|G_{i-1} : G_i|$. Then A is supersoluble.

In 1968 S.A. Syskin tried to prove Theorem 4.2 in the soluble universe, but his proof[15] is false.

In Ref. 5 there has been begun studying of ω-solubly saturated formations.

Definition 4.3. Let ω be a set of primes. A formation $\mathfrak{F}$ is called:

(1) *ω-solubly saturated* if $G/O_\omega(\Phi(\text{the}\,\omega\text{-soluble radical of}\,G) \in \mathfrak{F}$ always implies $G \in \mathfrak{F}$;
(2) *$\mathfrak{N}_p$-saturated* if $G/\Phi(O_p(G)) \in \mathfrak{F}$ always implies $G \in \mathfrak{F}$;
(3) *$\mathfrak{N}_\omega$-saturated* if it is $\mathfrak{N}_p$-saturated for any $p \in \omega$.

It is proved[5] that a formation is ω-solubly saturated if and only if it is $\mathfrak{N}_\omega$-saturated. Articles Refs. 8,9 gives a description of local definitions of $\mathfrak{N}_\omega$-saturated formations.

Definition 4.4. Let $\mathfrak{L}$ be a class of simple groups. Let f be a local definition which does not distinguish between all non-identity E($\mathfrak{L}'$)-groups; if $\mathfrak{L}' \neq \varnothing$, we denote by $f(\mathfrak{L}')$ an value of f on non-identity E($\mathfrak{L}'$)-groups. Let f-rule be defined as follows: a chief factor H/K of a group G is f-central in G if either H/K is an E$\mathfrak{L}$-group and $G/C_G(H/K) \in f(H/K)$ or H/K is a E($\mathfrak{L}'$)-group and $G/c_G(H/K) \in f(\mathfrak{L}')$. Then f is called an *$\mathfrak{L}$-composition satellite.* We denote by $CF_{\mathfrak{L}}(f)$ the class of all groups all of whose chief factors are f-central.

Clearly, if $\mathfrak{L} = \mathfrak{J}$, then an $\mathfrak{L}$-composition satellite f is a composition satellite, and $CF_{\mathfrak{L}}(f) = CF(f)$. If $\mathfrak{L} \neq \mathfrak{J}$ and $f(\mathfrak{L}') = \varnothing$, then $CF_{\mathfrak{L}}(f) = CF(h)$ where $h(S) = f(S)$ if $S \in \mathfrak{L}$, and $h(S) = \varnothing$ if $S \in \mathfrak{L}'$.

We note that in Ref. 8 an agreement $f(1) = f(\mathfrak{L}')$ was used; in the present paper we do not follow this agreement. Therefore, we have always $1 \in CF_{\mathfrak{L}}(f)$.

Proposition 4.1. *Let $\mathfrak{L}$ be a class of simple groups, and f an $\mathfrak{L}$-composition satellite. Let $\mathfrak{K} = \{S \in \mathfrak{L} \mid f(S) \neq \varnothing\}$. Then:*

(1) if $\mathfrak{K} = \varnothing$ and $f(\mathfrak{L}') = \varnothing$, then $CF_{\mathfrak{L}}(f) = (1)$;
(2) if $\mathfrak{K} = \varnothing$ and $f(\mathfrak{L}') \neq \varnothing$, then $CF_{\mathfrak{L}}(f) = \mathrm{E}(\mathfrak{L}') \cap f(\mathfrak{L}')$;
(3) if $f(\mathfrak{L}') \neq \varnothing$, then $CF_{\mathfrak{L}}(f)$ consists precisely of groups G such that $G/G_{\mathrm{E}\mathfrak{L}} \in f(\mathfrak{L}')$ and $G/C^S(G) \in f(S)$ for every $S \in \mathrm{Com}(G) \cap \mathfrak{L}$.

Proof. Assertions (1) and (2) are evident.

Prove (3). Assume that $f(\mathfrak{L}') \neq \varnothing$, and let $G \in CF_{\mathfrak{L}}(f)$. Let $\mathfrak{T}$ be the set of all chief $\mathrm{E}(\mathfrak{L}')$-factors in G. If a chief factor H/K of G is an $\mathrm{E}(\mathfrak{L}')$-group, then $G/c_G(H/K) \in f(\mathfrak{L}')$. Therefore, $G/\bigcap_{H/K \in \mathfrak{T}} c_G(H/K) \in f(\mathfrak{L}')$. By Lemma 3.1, $\bigcap_{H/K \in \mathfrak{T}} c_G(H/K) = G_{\mathrm{E}(\mathfrak{L})}$. So, $G/G_{\mathrm{E}(\mathfrak{L})} \in f(\mathfrak{L}')$. If $S \in \mathfrak{L}$ and H/K is an chief $\mathrm{E}(\mathfrak{L})$-factor, then $G/C_G(H/K) \in f(S)$, and we have $G/C^S(G) \in f(S)$.

Conversely, let G be a group such that $G/G_{\mathrm{E}\mathfrak{L}} \in f(\mathfrak{L}')$ and $G/C^S(G) \in f(S)$ for every $S \in \mathrm{Com}(G) \cap \mathfrak{L}$. Clearly, we have that all G-chief $\mathrm{E}\mathfrak{L}$-factors are f-central. Let H/K be a G-chief $\mathrm{E}(\mathfrak{L}')$-factor of G. Then $G_{\mathrm{E}\mathfrak{L}} \subseteq c_G(H/K)$, and therefore $G/G_{\mathrm{E}\mathfrak{L}} \in f(\mathfrak{L}')$ implies $G/c_G(H/K) \in f(\mathfrak{L}')$. □

Theorem 4.4 (see Ref. 8, Theorem 1). *Let $\mathfrak{F}$ be a non-empty formation, $\mathfrak{L}$ a class of simple groups, $\omega = \pi(\mathfrak{L}^+)$. The following conditions are equivalent:*

(1) $\mathfrak{F}$ is ω-solubly saturated;
(2) $\mathfrak{F}$ has an $\mathfrak{L}$-composition satellite;
(3) $\mathfrak{F}$ has an $\mathfrak{L}^+$-composition satellite;
(4) $\mathfrak{F} = CF_{\mathfrak{L}}(f)$ where $f(S) = \mathfrak{F}$ if $S \in \mathfrak{L}' \cup \mathfrak{L}^-$, and $f(p) = \mathfrak{N}_p \circ f(p)$ if $p \in \omega$.

Remark 4.1. Clearly, Theorem 4.1 follows from Theorem 4.4 in the case $\mathfrak{L} = \mathfrak{J}$.

Remark 4.2. It follows from Theorem 4.4 that every non-empty formation $\mathfrak{F}$ with the property $\mathrm{Com}^+(\mathfrak{F}) \cap \mathfrak{L} = \varnothing$ has an $\mathfrak{L}$-composition satellite.

Remark 4.3. When $\mathfrak{L} = \mathfrak{L}^+$ and $\omega = \pi(\mathfrak{L})$, we usually use the term 'ω-composition satellite' and the notations $CF_\omega(f)$, $f(\omega')$ in place of the term '$\mathfrak{L}$-composition satellite' and the notations $CF_{\mathfrak{L}}(f)$, $f(\mathfrak{L}')$.

5. $\mathfrak{X}$-local formations

In 1985 Förster[3] introduced the concept '$\mathfrak{X}$-local formation' in order to obtain a common extension of Theorem 3.1 and 4.1.

Definition 5.1. Let $\mathfrak{X}$ be a class of simple groups such that $\mathrm{Char}(\mathfrak{X}) = \pi(\mathfrak{X})$. Consider a map

$$f : \pi(\mathfrak{X}) \cup \mathfrak{X}' \longrightarrow \{\text{formations}\}$$

which does not distinguish between any two non-identity isomorphic groups. Denote through $LF_{\mathfrak{X}}(f)$ the class of all groups G satisfying the following conditions:

(1) if H/K is a chief $\mathrm{E}\mathfrak{X}$-factor of a group G, then $G/C_G(H/K)$ belongs to $f(p)$ for any $p \in \pi(H/K)$;
(2) if G/L is a monolithic quotient of G and $\mathrm{Soc}(G/L) \in \mathrm{E}(\mathfrak{X}')$, then $G/L \in f(S)$ where $S \in \mathrm{Com}(\mathrm{Soc}(G/L))$.

The class $LF_{\mathfrak{X}}(f)$ is a formation; it is called an $\mathfrak{X}$-*local formation.*

$\mathfrak{X}$-local formations were investigated in Refs. 16–20. In Ref. 21 it was proved with help of some lemmas in Ref. 19 that every $\mathfrak{X}$-local formation has an $\mathfrak{X}^+$-composition satellite. Now we give a direct proof of that fact.

Theorem 5.1. *Let $\mathfrak{X}$ be a class of simple groups such that* $\mathrm{Char}(\mathfrak{X}) = \pi(\mathfrak{X})$. *Let $\mathfrak{L}$ be a class of simple groups such that $\mathfrak{L}^+ = \mathfrak{X}^+$. Then the following statements hold:*

(1) if a non-empty formation has an $\mathfrak{L}$-composition satellite, then it is an $\mathfrak{X}^+$-local formation;
(2) if a non-empty formation is an $\mathfrak{X}$-local formation, then it has an $\mathfrak{L}$-composition satellite.

Proof. Statement (1) is proved in Ref. 20, p. 152.

Prove (2). Set $\omega = \mathrm{Char}(\mathfrak{X})$. Evidently, $\mathfrak{L}^- \cup \mathfrak{L}' = \mathfrak{X}^- \cup \mathfrak{X}' = (\mathfrak{L}^+)' = (\mathfrak{X}^+)'$.

Let $\mathfrak{F}$ be a $\mathfrak{X}$-local formation, $\mathfrak{F} = LF_{\mathfrak{X}}(f)$. Consider an $\mathfrak{L}$-composition satellite h such that $h(p) = f(p)$ if $p \in \omega$, and $h(S) = \mathfrak{F}$ if $S \in \mathfrak{L}^- \cup \mathfrak{L}'$. We will prove that $\mathfrak{F} = CF_{\mathfrak{L}}(h)$.

Suppose that $\mathfrak{F} \not\subseteq CF_{\mathfrak{L}}(h)$. Let G be a group of the least order in $\mathfrak{F} \setminus CF_{\mathfrak{L}}(h)$. Then G is monolithic, and $G/M \in CF_{\mathfrak{L}}(h)$ where M is the socle of G. Evidently, $c_G(M) = 1$; if M is non-abelian, then $C_G(M) = 1$. Clearly, M is the $CF_{\mathfrak{L}}(h)$-residual of G, and every chief factor between G

and M is h-central. If M is $\mathrm{E}(\mathfrak{L}^- \cup \mathfrak{L}')$-group, then $G \in \mathfrak{F} = h(M)$, and we have that M is h-central in G. Assume now that M is a p-group, $p \in \omega$. Since $G \in \mathfrak{F}$, we have $G/C_G(M) \in f(p) = h(p)$, i. e., M is h-central. We see that $G \in CF_{\mathfrak{L}}(h)$, a contradiction. Thus, $\mathfrak{F} \subseteq CF_{\mathfrak{L}}(h)$.

Suppose now that $CF_{\mathfrak{L}}(h) \not\subseteq \mathfrak{F}$. Choose a group G of the least order in $CF_{\mathfrak{L}}(h) \setminus \mathfrak{F}$. Then G is monolithic, and $G/M \in \mathfrak{F}$ where $M = G^{\mathfrak{F}}$ is the socle of G. Assume that M is an $\mathrm{E}(\mathfrak{L}^- \cup \mathfrak{L}')$-group. Then from h-centrality of M it follows that $G \in \mathfrak{F}$. Assume that M is a p-group, $p \in \omega$. Then $G/C_G(M) \in h(p) = f(p)$. We see that all the chief factors and all the quotients of G satisfies conditions (1) and (2) of Definition 5.1. So, $G \in \mathfrak{F}$, a contradiction. Thus, $\mathfrak{F} = CF_{\mathfrak{L}}(h)$. □

Corollary 5.1. *Let $\mathfrak{X}$ be a class of simple groups such that* $\mathrm{Char}(\mathfrak{X}) = \pi(\mathfrak{X})$. *If a non-empty formation $\mathfrak{F}$ is an $\mathfrak{X}$-local formation then it has an $\mathfrak{X}^+$-composition satellite.*

Corollary 5.2. *Let $\mathfrak{X}$ be a class of simple groups such that* $\mathrm{Char}(\mathfrak{X}) = \pi(\mathfrak{X})$. *If a non-empty formation $\mathfrak{F}$ is an $\mathfrak{X}$-local formation then it is $\pi(\mathfrak{X})$-solubly saturated.*

References

1. W. Gaschütz, Zur Theorie der endlichen auflösbaren Gruppen, *Math. Z.* **80** (1963), 300–305.
2. L. A. Shemetkov, On the product of formations,*Dokl. Akad. Nauk BSSR*,**28**:2 (1984), 101–103.
3. P. Förster, Projective Klassen endlicher Gruppen IIa. Gesättigte Formationen: ein allgemeiner Satz von Gaschütz–Lubeseder–Baer Typ, *Publ. Sec. Mat. Univ. Autònoma Barcelona*, **29**:2–3 (1985), 39–76.
4. L. A. Shemetkov and A. N. Skiba, On partially local formations, *Dokl. Akad. Nauk Belarus*, **39**:3 (1995), 9–11.
5. L. A. Shemetkov, Frattini extensions of finite groups and formations, *Comm. Algebra* **25**(3) (1997), 955–964.
6. A. Ballester-Bolinches, L. A. Shemetkov, On lattices of p-local formations of finite groups, *Math. Nachr.*, **186** (1997), 57–65.
7. L. A. Shemetkov and A. N. Skiba, Multiply ω-local formations and Fitting classes of finite groups, *Siberian Advances in Mathematics*, **10**(2) (2000), 112–141.
8. L. A. Shemetkov and A. N. Skiba, Multiply $\mathfrak{L}$-composition formations of finite groups, *Ukrainian Math. Journal*, **52** (6) (2000) 898–913.
9. L. A. Shemetkov, On partially saturated formations and residuals of finite groups, *Comm. Algebra*, **29**:9 (2001), 4125–4137.
10. K. Doerk and T. Hawkes, *Finite Soluble Groups* (Walter de Gruyter, Berlin–New York, 1992).

11. L. A. Shemetkov, Two directions in the development of the theory of non-simple finite groups (a lecture delivered at the Twelth All-Union Algebra Colloquium held in Sverdlovsk in September,1973) *Russian Math. Surveys*, **30**:2 (1975), 186–206).
12. B. Huppert and N. Blackburn, *Finite Groups* II (Springer-Verlag, Berlin–Heidelberg–New York, 1982).
13. L. A. Shemetkov, Graduated formations of finite groups, *Math. USSR Sbornik*, **23**:4 (1974), 593–611; translated from *Matem. Sbornik*, **94**:4 (1974), 628–648.
14. P. Schmid, Locale Formationen endlicher Gruppen, *Math. Z.*, 137 (1974), 31–48.
15. S. A. Syskin, Over-solvable groups of automorphisms of finite solvable groups, *Algebra and Logic*, **7** (1968), 193–194; translated from *Algebra Logika*, **7**:3 (1968), 105–107.
16. A. Ballester, Remarks on formations, *Isr. J. Math.*, **73**:1 (1991), 97–106.
17. A. Ballester-Bolinches, C. Calvo and R. Esteban-Romero, $\mathfrak{X}$-saturated formations of finite groups, *Comm. Algebra*, **33**:4 (2005), 1053–1064.
18. A. Ballester-Bolinches, C. Calvo and R. Esteban-Romero, Products of formations of finite groups, *J. Algebra*, **299** (2006), 602–615.
19. A. Ballester-Bolinches, C. Calvo and L.A. Shemetkov, On partially saturated formations of finite groups, *Sbornik: Mathematics*, **198**:6 (2007), 757–775; translated from *Matem. Sbornik*, **198**:6 (2007), 3–24.
20. A. Ballester-Bolinches and L. M. Ezquerro, *Classes of finite groups* (Springer, Dordrecht, 2006).
21. L. A. Shemetkov, A note on $\mathfrak{X}$-local formations, *Problems in Physics, Mathematics and Technics* **4**(5) (2010), 64–65.

Sheaves Over Boolean Spaces

Invited talk on Oct 8, 2010 at ICA-2010, Yogyakarta, Indonesia

by

U. M. Swamy
(e-mail: umswamy@yahoo.com)

It is well known that the power set $\mathcal{P}(X)$ of any nonempty set X forms a Boolean algebra under the usual set theoritic operations and that $\mathcal{P}(X)$ is isomorphic to $\mathbf{2}^X$, where $\mathbf{2}$ is the two element algebra $\{0, 1\}$. M.H. Stone [7] established that any Boolean algebra B is isomorphic to a subalgebra of $\mathbf{2}^X$ for a suitable set X . Here X is taken as the set of maximal ideals of a given Boolean algebra B and, for each $a \in B$, $f_a : X \longrightarrow \mathbf{2}$ is defined by

$$f_a(M) = \begin{cases} 1 & \text{if } a \notin M \\ 0 & \text{if } a \in M \end{cases}$$

and it is proved that $a \longmapsto f_a$ is an isomorphism of B onto the subalgebra $B_1 = \{ f_a : a \in B \}$ of $\mathbf{2}^X$. In order to identify B_1 in $\mathbf{2}^X$, X is topologised in such a way that $B_1 = \mathcal{C}(X, \mathbf{2})$, the set of all continuous mappings of X into the discrete two element space $\mathbf{2}$.

In general, for any topological space X, the set $\mathcal{C}(X, \mathbf{2})$ of continuous functions of X into the discrete $\mathbf{2}$ is a Boolean algebra under the pointwise operations; note that $\mathcal{C}(X, \mathbf{2})$ is isomorphic with the algebra of clopen (closed and open) subsets of X. On the other hand, if B is any Boolean algebra and X is the set of maximal ideals of B, then the class

$$\{ X_a : a \in B\}, \text{ where } X_a = \{ M \in X : a \notin M\},$$

is a base for a topology on X with respect to which X is a compact Hausdorff and totally disconnected space; X together with this topology is called the Stone space of B or the spectrum of B and is denoted by $SpecB$. Further $B \cong \mathcal{C}(SpecB, \mathbf{2})$. However, for any topological space X, X is homeomorphic to $Spec\ \mathcal{C}(X, \mathbf{2})$ if and only if X is a compact Hausdorff and totally disconnected space. Infact $B \longmapsto SpecB$ and $X \longmapsto \mathcal{C}(X, \mathbf{2})$ are one-to-one correspondences, which are inverses to each other, between Boolean algebras and compact Hausdorff and totally disconnected topological spaces and, for this reason, such topological spaces are called Boolean spaces. The correspondences $B \longmapsto SpecB$ and $X \longmapsto \mathcal{C}(X, \mathbf{2})$ are called Stone duality between Boolean algebras and Boolean spaces. This is a very strong duality in the sense that it is a contravarient equivalence between the category of Boolean algebras and the category of Boolean spaces. For any Boolean algebras B_1 and B_2 and the spaces $X_1 = SpecB_1$ and $X_2 = SpecB_2$, we have the following corresepondences.

$$\text{Hom } (B_1, B_2) \approx \mathcal{C}(X_2, X_1)$$

$$\text{Monomorphisms } (B_1, B_2) \approx \text{Surjective } \mathcal{C}(X_2, X_1)$$

$$\text{Epimorphisms } (B_1, B_2) \approx \text{Injective } \mathcal{C}(X_2, X_1)$$

$$\text{Isomorphisms of } B_1 \text{ onto } B_2 \approx \text{Homeomorphisms of } X_2 \text{ onto } X_1$$

$$\text{Ideals of } B_1 \approx \text{Open sets in } X_1$$

$$\text{Maxiaml ideals of } B_1 \approx \text{Points of } X_1$$

Note that , for any Boolean space X, we have

$$\mathcal{C}(X, \mathbf{2}) \subseteq \mathbf{2}^X \subseteq \mathcal{P}(X \times \mathbf{2}),$$

$$X \times \mathbf{2} = \text{the disjoint union of } \{x\} \times \mathbf{2},\ x \in X \text{ and } \{x\} \times \mathbf{2} \cong \mathbf{2}$$

and hence $\{x\} \times \mathbf{2}$ can be viewed as the two element Boolean algebra glued at x on X and any element of $\mathcal{C}(X, \mathbf{2})$ is a continuous section which cuts each $\{x\} \times \mathbf{2}$, $x \in X$ at exactly one point.

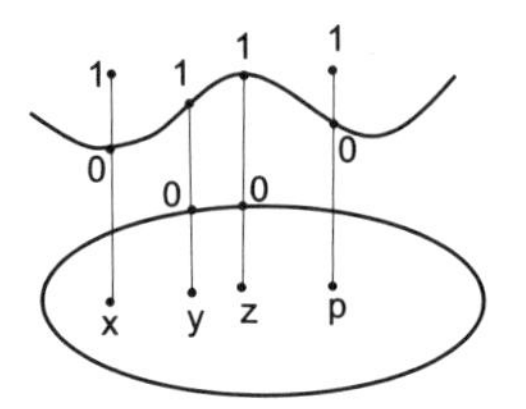

1. ALMOST BOOLEAN ALGEBRAS

The above discussion is abstracted in the following.

Definition 1.1: A triple (S, Π, X) is called a **sheaf of sets over X** if S and X are topological spaces and $\Pi : S \longrightarrow X$ is a local homeomorphism, in the sense that Π is a surjection and, for each $s \in S$, there exist neighbourhoods U and Y of s and $\Pi(s)$ in S and X respectively such that the restriction Π/U of Π to U is a homeomorphism.

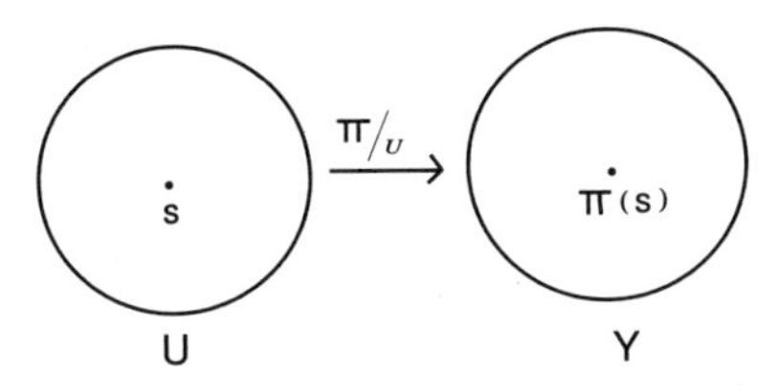

X is called the **base space**, S the **sheaf space** and Π the **projection** . For each $p \in X$, the set

$$S_p = \Pi^{-1}(p) = \{ s \in S : \Pi(s) = p \}$$

is called the **stalk** at p.

Definition 1.2: Let (S, Π, X) be a sheaf of sets over X. For any $Y \subseteq X$, a continuous mapping $f : Y \longrightarrow S$ such that $\Pi \circ f = Id_Y$ (that is, $f(p) \in S_p$ for all $p \in Y$) is called a **section on Y**. Sections on X are called **global sections**.

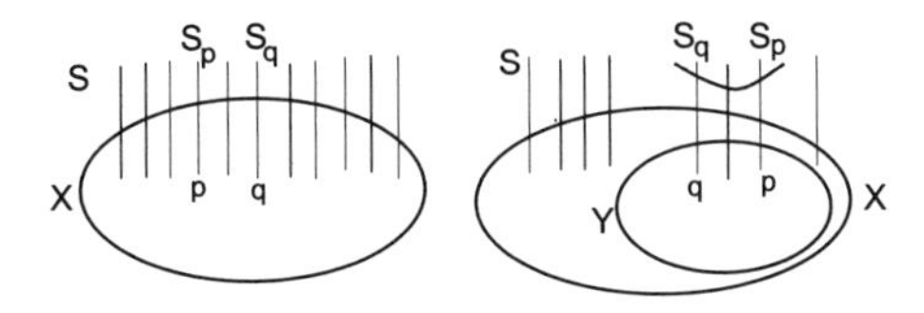

Let us recall the following elementary properties of sheaves.

Theorem 1.3: The following hold in any sheaf (S, Π, X) of sets over any topological space X.

(1) Any section is an injective map.

(2) Any section carries open sets onto open sets; that is, it is an open map. In fact , the class $\{ f(U) \mid U$ is open on X and f is a section on $U \}$ is a base for open sets in the sheaf space S.

(3) If $Y \subseteq X$, f and g are sections on Y and $p \in Y$ such that $f(p) = g(p)$, then there exists a neighbourhood U of p such that $f/U = g/U$. That is the set $\{ p \in Y \mid f(p) = g(p) \}$ is open in Y.

(4) The set $\langle f, g \rangle = \{ p \in Y \mid f(p) \neq g(p) \}$ is closed in Y and is called the **relative support** of f with g.

(5) For any $s \in S$, there exist a neighbourhood U of $\Pi(s)$ and a section f on U such that $s = f(\Pi(s))$.

(6) The subspace topology on each stalk of S_p is discrete.

We are mainly considering sheaves over Boolean spaces; that is , the sheaves where the base space is a Boolean space (a compact Hausdorff and totally disconnected topological space). In the following we obtain an algebraic structure satisfying all the axioms of a Boolean algebra, except possibly the commutativity of the operations $\wedge$ and $\vee$.

Theorem 1.4: Let (S, Π, X) be a sheaf of sets over a Boolean space X and consider the set

$$B = \{ (U, f) : U \text{ is open in } X \text{ and } f \text{ is a section on } U \}$$

For any $a = (U, f)$ and $b = (W, g)$ in B, define

$$a \wedge b = (U \cap W,\ g/U \cap W) \quad \text{and} \quad a \vee b = (U \cup W,\ f \vee g)$$

where $f \vee g : U \cup W \longrightarrow S$ is defined by

$$(f \vee g)(p) = \begin{cases} f(p) & \text{if } p \in U \\ g(p) & \text{if } p \in W - U \end{cases}$$

Also, let $o = (\phi, \phi)$ and $l = (X, m)$, where m is a global section. Then $(B, \vee, \wedge, o, l)$ is an algebraic structure satisfying all the axioms of a Boolean algebra, except possibly the commutativity of $\wedge$ and $\vee$.

An algebraic structure $(B, \vee, \wedge, o, l)$ of the above type is called an almost Boolean algebra. In the following, we give a precise definition.

Definition 1.5: A system $(B, \vee, \wedge, o, m)$ is called an **Almost Boolean Algebra (ABA)** if B is a nonempty set, $\vee$ and $\wedge$ are binary operations on B and o and m are distinguished distinct elements of B satisfying the following for any a, b and $c \in B$.

(1) $a \vee 0 = a$

(2) $0 \wedge a = 0$

(3) $a \vee (b \wedge c) = (a \vee b) \wedge (a \vee c)$

(4) $a \wedge (b \vee c) = (a \wedge b) \vee (a \wedge c)$

(5) $(a \vee b) \wedge c = (a \wedge c) \vee (b \wedge c)$

(6) $(a \vee b) \wedge b = b$

(7) For any a and $b \in B$, there exists $x \in B$ such that $a \vee x = a \vee b$ and $a \wedge x = 0$

(8) $m \wedge a = a$

Examples 1.6

(1) The system $(B, \vee, \wedge, o, l)$ obtained from a sheaf of sets over a Boolean space, as in theorem 2 above, is an Almost Boolean Algebra.

(2) Let $(R, +, ., 0, 1)$ be a commutative regular ring with identity (that is, a commutative ring with identity in which, for each $a \in R$, there exists unique idempotent a_0 in R such that $aR = a_0R$). For any a and b in R, define

$$a \wedge b = a_0 b$$

$$\text{and} \quad a \vee b = a_0 + b - a_0 b$$

Then $(R, \vee, \wedge, 0, 1)$ is an Almost Boolean Algebra

(3) Let X be any set with at least two elements. Choose arbitrary elements $0 \neq m$ in X. For any x and y in X, define

$$x \wedge y = \begin{cases} 0 & \text{if } x = 0 \\ y & \text{if } x \neq 0 \end{cases}$$

$$\text{and } x \vee y = \begin{cases} y & \text{if } x = 0 \\ x & \text{if } x \neq 0 \end{cases}$$

Then $(X, \vee, \wedge, o, m)$ is an Almost Boolean Algebra.

The following can be proved using the axioms in the definition 3 of an Almost Boolean Algebra

Theorem 1.7: The following hold in any Almost Boolean Algebra $(B, \vee, \wedge, o, m)$ for any elements a, b and c in B.

(1) $a \wedge a = a = a \vee a$

(2) $(a \wedge b) \wedge c = a \wedge (b \wedge c)$

(3) $(a \vee b) \vee c = a \vee (b \vee c)$

(4) $(a \wedge b) \wedge c = (b \wedge a) \wedge c$

(5) $(a \vee b) \wedge c = (b \vee a) \wedge c$

(6) $\{ x \wedge a \ : \ x \in B \}$ is a Boolean algebra under the induced operations with 0 as the smallest element and a as the largest element

(7) For any x and $y \in B$, define

$$x \leq y \iff x = x \wedge y \quad (\iff x \vee y = y)$$

Then $\leq$ is a partial order on B.

(8) The element x such that $a \vee x = a \vee b$ and $a \wedge x = 0$ is unique and is denoted by a^b

In the following we give several equivalent conditions for an Almost Boolean Algebra to become a Boolean algebra.

Theorem 1.8: An Almost Boolean Algebra $(B, \vee, \wedge, o, m)$ is a Boolean algebra if and only if any one (and hence all) of the following conditions are satisfied for all elements a, b and c in B.

(1) $a \wedge b = b \wedge a$

(2) $a \vee b = b \vee a$

(3) $(a \wedge b) \vee a = a$

(4) $a \wedge (b \vee a) = a$

(5) $(a \wedge b) \vee c = (a \vee c) \wedge (b \vee c)$

(6) The partially ordered set $(B, \leq)$ is directed above

(7) The infimum of a and b exists and is equal to $a \wedge b$

(8) The supremum of a and b exists and is equal to $a \vee b$

(9) $a \vee m = m$

(10) $(B, \leq)$ has a unique maximal element.

2. CONSTRUCTION OF A SHEAF OF SETS OVER A BOOLEAN SPACE

We have discussed a procedure in theorem 1.4 to construct an Almost Boolean Algebra from a given sheaf of sets over a Boolean space. In the following, we give a method of constructing a sheaf of sets over a Boolean space from a given ABA . First we derive a Boolean space from an ABA.

Definition 2.1: Let $\mathbf{B} = (B, \vee, \wedge, o, m)$ be an ABA. A nonempty subset I of B is called an **ideal** of $\mathbf{B}$ if

$$a \text{ and } b \in I \implies a \vee b \in I \text{ and } a \wedge x \in I \text{ for all } x \in B$$

(Note that $x \wedge a = (x \wedge a) \wedge a = (a \wedge x) \wedge a = a \wedge (x \wedge a)$ and hence $x \wedge a \in I$ for all $a \in I$ and $x \in B$, so that I is actually a two sided ideal in the usual sense)

The ideals of any ABA form a complete distributive lattice, with respect to the inclusion order. Being algebraic, it satisfies the infinite meet distributivity also and therefore it is a complete Browerian algebra. As usual, a proper ideal is said to be a **maximal ideal** if it is maximal among all proper ideals.

Theorem 2.2: Let I be a proper ideal of an Almost Boolean Algebra $\mathbf{B}$. Then the following are equivalent to each other.

(1) I is a maximal ideal of $\mathbf{B}$

(2) For any a and $b \in \mathbf{B}$, $a \wedge b \in I \implies a \in I$ or $b \in I$

(3) For any ideals J and K of $\mathbf{B}$, $J \cap K \subseteq I \implies J \subseteq I$ or $K \subseteq I$.

Definition 2.3: Let $\mathbf{B}$ be an ABA and X the set of all maximal ideals of $\mathbf{B}$. For any $a \in B$, let

$$X_a = \{ M \in X : a \notin M \}$$

Then the class $\{ X_a : a \in B \}$ forms a base for a topology on X. We denote X together with this topology by $Spec\mathbf{B}$ and call it the spectrum of $\mathbf{B}$ or Stone space of $\mathbf{B}$.

Theorem 2.4: The following hold for any Almost Boolean Algebra $\mathbf{B}$.

(1) $Spec\mathbf{B}$ is a Boolean space

(2) $Spec\mathbf{B} \cong SpecB_m$ for any maximal element m in B, where B_m is the Boolean algebra $\{ x \wedge m : x \in B \}$.

Let $\mathbf{B} = (B, \vee, \wedge, o, m)$ be an ABA and $X = Spec\mathbf{B}$. For any $a \in B$, let

$$\theta_a = \{ (x, y) \in B \times B : a \wedge x = a \wedge y \}$$

Then θ_a is a congruence relation on $\mathbf{B}$. For any $P \in X$, let

$$\theta_p = \bigcup_{a \in B-P} \theta_a .$$

Then θ_P is a congruence relation on $\mathbf{B}$ and let

$$S_p = B/\theta_P - \{ \theta_P(0) \} \text{ and } S = \bigsqcup_{P \in X} S_P .$$

For any $a \in B$, let $\widehat{a} : X_a \longrightarrow S$ be defined by

$$\widehat{a}(P) = \theta_P(a) \text{ for all } P \in X_a .$$

Note that $\theta_P(a) = \theta_P(0) \iff a \in P \iff P \notin X_a$.
Topologise S with the largest topology with respect to which each $\widehat{a}$, $a \in B$, is continuous. Finally define

$$\Pi : S \longrightarrow X \text{ by } \Pi(s) = P \text{ if } s \in S_P .$$

Then we have the following.
Theorem 2.5: (S, Π, X) constructed above is a sheaf of sets over the Boolean space X.

Theorem 2.6: Let $\mathbf{B}$ be an ABA and (S, Π, X) be the sheaf obtained above from $\mathbf{B}$. Then

$$a \mapsto (X_a, \widehat{a})$$

is an isomorphism of B onto the ABA constructed from the sheaf (S, Π, X) as in theorem 1.4 .

Definition 2.7

Two sheaves of sets (S, Π, X) and (T, η, Y) are said to be isomorphic if there exist homeomorphisms $\alpha : S \longrightarrow T$ and $\beta : X \longrightarrow Y$ such that the following diagram is commutative; that is, $\eta \circ \alpha = \beta \circ \Pi$

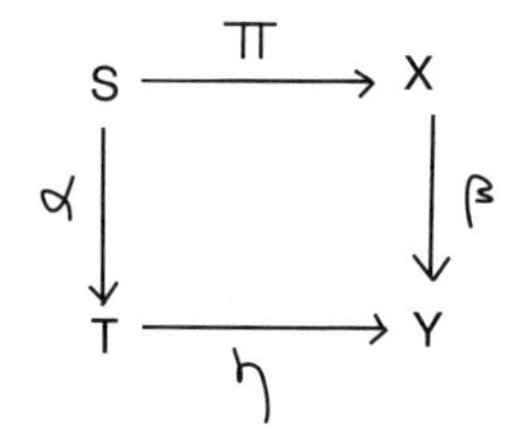

Theorem 2.8: Let (S, Π, X) be a sheaf of sets over a Boolean space X and B be the ABA obtained from (S, Π, X) as in theorem 1.4. Then (S, Π, X) is isomorphic to the sheaf obtained from B as in theorem 2.5

3. ALMOST BOOLEAN RINGS

We define the notion of an Almost Boolean ring and obtain a one-to-one correspondence between the structures of Almost Boolean algebras and those of an Almost Boolean ring on any given set which is analogous to that between the structures of Boolean algebras and of Boolean rings.

Definition 3.1: A system $(R, +, ., o, m)$ is called an **Almost Boolean Ring** (ABR) if R is a nonempty set, $+$ and $.$ are binary operations on R and o and m are two distinct distinguished elements in R satisfying the following for any a, b and c in R.

(1) $a + 0 \; = \; 0$

(2) $a + a \; = \; 0$

(3) $(ab)c \; = \; a(bc)$

(4) $aa \; = \; a$

(5) $a(b + c) \; = \; ab + ac$

(6) $(a + b)c \; = \; ac + bc$

(7) $(a + (b + c))d \; = \; ((a + b) + c)d$

(8) $ma \; = \; a$

Theorem 3.2: The following are equivalent to each other for any Almost Boolean ring $\mathbf{R} = (R, +, ., o, m)$

(1) $\mathbf{R}$ is a Boolean ring (that is, a non trivial ring with unity in which every element is an idempotent)

(2) The operation . is commutative

(3) The operation + is associative

(4) The operation + is commutative .

Theorem 3.3: Let $\mathbf{B} = (B, \vee, \wedge, o, m)$ be an Almost Boolean Algebra. For any a and $b \in B$, define

$$a + b \; = \; a^b \vee b^a$$

where a^b is the unique element in B such that $a \vee a^b = a \vee b$ and $a \wedge a^b = 0$. Then $(B, +, \wedge, o, m)$ is an Almost Boolean Ring and is denoted by $\mathbf{R(B)}$.

Theorem 3.4: Let $\mathbf{R} = (R, +, ., o, m)$ be an Almost Boolean Ring. For any a and $b \in R$, define

$$a \vee b = a + (b + ab) .$$

Then $(R, \vee, ., o, m)$ is an Almost Boolean Algebra and is denoted by $\mathbf{B(R)}$.

Theorem 3.5: For any Almost Boolean Algebra $\mathbf{B}$,

$$\mathbf{B(R(B))} = \mathbf{B}$$

and, for any Almost Boolean ring $\mathbf{R}$,

$$\mathbf{R(B(R))} = \mathbf{R}$$

Theorem 3.6: $\mathbf{B} \mapsto \mathbf{R(B)}$ and $\mathbf{R} \mapsto \mathbf{B(R)}$ establish a duality between Almost Boolean Algebras and Almost Boolean Rings.

4. CERTAIN GENERALIZATIONS

Several generalizations of Almost Boolean Algebras have come up in recent times and much work is done on these. In the following we present a few of these generalizations. For details one can go through the works mentioned in the references.

4.1 Almost distributive lattices

A system $(L, \vee, \wedge, o)$ is called an **Almost distributive lattice**(ADL) if L is a nonempty set, $\vee$ and $\wedge$ are binary operations on L and o is a distinguished element of L satisfying the following for any a, b and c in L.

(1) $a \vee 0 = a$

(2) $0 \wedge a = 0$

(3) $(a \vee b) \wedge c = (a \wedge c) \vee (b \wedge c)$

(4) $a \wedge (b \vee c) = (a \wedge b) \vee (a \wedge c)$

(5) $a \vee (b \wedge c) = (a \vee b) \wedge (a \vee c)$

(6) $(a \vee b) \wedge b = b$

(7) $a \vee (b \vee c) = (a \vee b) \vee c$

In an ADL L one can define a partial order by

$$a \leq b \iff a = a \wedge b \ (\iff a \vee b = b) .$$

There are several equivalent conditions , as in Theorem 1.8, on the partial order $\leq$ and the operations $\vee$ and $\wedge$ for an ADL to become a distributive lattice with zero.

4.2 Pseudo complemented ADL's

An ADL $(L, \vee, \wedge, 0)$ is called **pseudo complemented** if there is an unary operation $*$ on L satisfying the following for any a and b in L.

(1) $a \wedge b = 0 \iff a^* \wedge b = b$

(2) $(a \vee b)^* = a^* \wedge b^*$.

The unary operation $*$ is called a **pseudo complementation** on L.

If $*$ is a pseudo complementation on L, then 0^* is a maximal element in L and $* \mapsto 0^*$ is a one-to-one correspondence between the pseudo complementations on L and the maximal elements in L. In the case of lattices, there can be atmost one maximal element and hence there can be atmost one pseudo complementation.

4.3 Stone ADL's

A pseudo complemented ADL L is called a **Stone ADL** if $x^* \vee x^{**} = 0^*$ for all $x \in L$.

The notion of a Stone ADL is independent of the pseudo complementation, since for any pseudo complementations $*$ and $\perp$ on an ADL L and for any $x \in L$, we have

$$x^* \vee x^{**} = 0^\star \text{ if and only if } x^\perp \vee x^{\perp\perp} = 0^\perp .$$

If L is Stone ADL, then $L^* = \{ x^* \ : \ x \in L \}$ is a Boolean algebra which is independent of the pseudo complementation, in the sense that, for any pseudo complementations $*$ and $\perp$ on L, L^* is isomorphic to $L^\perp$. It is known that L is a Stone ADL if and only if L is isomorphic to the ADL of global sections of a sheaf (S, Π, X) of dense ADL's over a Boolean space X (an ADL is called dense if, for any nonzero elements x and y, $x \wedge y$ is again nonzero).

4.4 Normal ADL's

An ADL L is called **normal** if any one of the following equivalent conditions is satisfied

(1) Any two distinct minimal prime ideals of L are comaximal, in the sense that their join is the whole of L

(2) Any prime ideal of L contains a unique minimal prime ideal

(3) For any x and $y \in L$, $x \wedge y = 0 \implies (x)^* \vee (y)^* = L$, where $(x)^* = \{ a \in L : a \wedge x = 0 \}$

It can be observed that, for any elements a and b in an ADL L with $a \leq b$, the interval $[a, b]$ is a bounded distributive lattice under the operations $\vee$ and $\wedge$ induced by those in L. An ADL L is said to be **relatively normal** if $[a, b]$ is a normal lattice for any a and b in L with $a \leq b$, which is equivalent to saying that any two incomparable prime ideals of L are comaximal.

References

[1] Birkhoff,G., Lattice Theory, Amer. Math.Soc.colloquium publication, 1967.

[2] Comer,S.D., Representation of algebras by sheaves over Boolean spaces, Pacific Jour.Math, 38(1974),29-38.

[3] Cornish,W.H., Normal lattices, Jour. Australian Math.Soc., 14(1971),200-215.

[4] Ramesh,S., Amicable sets in ADL's, Jour. of AP Society of Math. Sciences, 1(2008),207-216.

[5] Rao,G.C., and Ravi Kumar,S., Normal ADL's, South East Asian Bull. Math., 32(2008),831-841.

[6] Speed,T.P., On Stone lattices, Jour.Australian Math.Soc., 9(1969), 297-307.

[7] Stone,M.H., Theory of Representations for Boolean algebras, Trans.Amer.Math.Soc,40(1936),40-111.

[8] Subrahmanyam,N.V., An extension of Boolean lattice theory, Math.Ann, 151(1963),332-345.

[9] Swamy,U.M., Representation of Universal Algebras by sheaves, Proc.Amer.Math.Soc., 45(1974),53-58.

[10] Swamy,U.M., and Rama Rao,V.V., Triple and sheaf representations of Stone lattices, Algebra Universatis,5(1975),104-113.

[11] Swamy,U.M., and Manikyamba,P., Prime ideal characterization of Stone lattices, Math. Seminar Notes,7(1979),25-31.

[12] Swamy,U.M., and Rao,G.C., Almost Distributive Lattices, Jour.Australian Math.Soc(Series A), 31(1981),77-91.

[13] Swamy,U.M., and Murthy,G.S.N., Boolean centre of an Universal algebra, Algebra Universatis, 13(1981),202-205.

[14] Swamy,U.M., Rao,G.C., and Rao,G.N., Pseudo-complementations on ADL's, South East Asian Bull.Math, 24(2000),95-104.

[15] Swamy,U.M., Rao,G.C., and Rao,G.N., Stone ADL's, South East Asian Bull.Math, 27(2003),513-526.

[16] Swamy.U.M, Rao.G.C and Pragathi,Ch., Boolean centre of an ADL,South East Asian Bull.Math,27(2003).

[17] Swamy,U.M., and Ramesh,S., Birkhoff centre of an ADL, International Jour. Algebra, 3(2009),539-546.

[18] Swamy,U.M., Ramesh,S., and Raj,Ch.S.S., Prime ideal characterization of Stone ADL's, Asian-Europian Jour. of Math., 3(2010),357-367.

Some Properties of Semirings

T. Vasanthi
Department of Applied Mathematics
Yogi Vemana University
Kadapa - 516003, A.P., India.
E-mail: vasanthi_thatimakula@yahoo.co.in

Abstract

Additive and multiplicative structures play an important role in determining the structure of semirings. In this paper we study the properties of semirings and ordered semirings. We examine whether the algebraic structure of $(S, \cdot)$ may determine the order structure of $(S, +)$ and vice -versa. We also study the additive and multiplicative structures of semirings. A semiring is said to be a Positive Rational Domain (PRD) if $(S, \cdot)$ is an abelian group. We discuss properties of semirings in which $ab = a + b + ab$, a, b in S. We prove that in a PRD $(S, +, \cdot)$, $(S, +)$ is a commutative semigroup if $(S, +)$ is cancellative. We establish that in a semiring with integral multiple property (IMP), $(S, +)$ is O-Archimedean if and only if $(S, \cdot)$ is O-Archimedean. We also study the properties of totally ordered semirings with Integral Multiple property (IMP).

Keywords: Non-negatively ordered; Non-positively ordered; Positively totally ordered; Negatively totally ordered; IMP;

Preliminaries: A triple $(S, +, \cdot)$ is called a semiring if $(S, +)$ is a semigroup; $(S, \cdot)$ is a semigroup; $a(b + c) = ab + ac$ and $(b + c)a = ba + ca$ for every $a, b, c \in S$. A semiring $(S, +, \cdot)$ is said to be totally ordered if there exists a full ordering on S under which $(S, +)$ and $(S, \cdot)$ are totally ordered semigroups. A totally ordered (t.o.) semigroup $(S, \cdot)$ is non - negatively (non - positively) ordered if $x^2 \geq x(x^2 \leq x)$ for every x in S; $(S, \cdot)$ is positively ordered in strict sense (negatively ordered in strict sense) if $xy \geq x$ and $xy \geq y(xy \leq x$ and $xy \leq y)$ for every $x, y \in S$. $E[+]$ denotes the set of all additive idempotents. A semiring $(S, +, \cdot)$ is said to be a positive rational domain (PRD) if and only if $(S, \cdot)$ is an abelian group. A semigroup $(S, \cdot)$ is said to be quasi commutative if $ab = b^m a$ for some integer $m \geq 1$. In [4], the structure of semirings in which $(S, \cdot)$ contains the multiplicative identity or $(S, \cdot)$ is cancellative are studied. This motivated the author to study the properties of semirings when S satisfies IMP, PRD or $ab = a + b + ab$ for all $a, b \in S$.

If a totally ordered semiring contains multiplicative identity, then $(S, +)$ is either non-negatively ordered or non-positively ordered; This is observed in proposition 1[4]. The converse of 1[4] is not necessarily true. This is evident from the following examples:

(i) Let $S = 2, 4, 6, \cdots$ with usual addition, multiplication and ordering. Here $(S, +)$ is non-negatively ordered. If we take the dual order, then $(S, +)$ is non-positively ordered. Here S does not contain multiplicative identity 1.

(ii) (a)

$+$	$\frac{1}{3}$	$\frac{2}{3}$	1
$\frac{1}{3}$	$\frac{2}{3}$	1	1
$\frac{2}{3}$	1	1	1
1	1	1	1

$\cdot$	$\frac{1}{3}$	$\frac{2}{3}$	1
$\frac{1}{3}$	$\frac{2}{3}$	1	1
$\frac{2}{3}$	1	1	1
1	1	1	1

(b)

$+$	$\frac{1}{4}$	$\frac{2}{4}$	$\frac{3}{4}$	1
$\frac{1}{4}$	$\frac{3}{4}$	$\frac{3}{4}$	1	1
$\frac{2}{4}$	$\frac{3}{4}$	1	1	1
$\frac{3}{4}$	1	1	1	1
1	1	1	1	1

$\cdot$	$\frac{1}{4}$	$\frac{2}{4}$	$\frac{3}{4}$	1
$\frac{1}{4}$	$\frac{2}{4}$	$\frac{3}{4}$	1	1
$\frac{2}{4}$	$\frac{3}{4}$	1	1	1
$\frac{3}{4}$	1	1	1	1
1	1	1	1	1

(c)

$+$	$\frac{1}{5}$	$\frac{2}{5}$	$\frac{3}{5}$	$\frac{4}{5}$	1
$\frac{1}{5}$	$\frac{3}{5}$	$\frac{4}{5}$	1	1	1
$\frac{2}{5}$	$\frac{4}{5}$	$\frac{4}{5}$	1	1	1
$\frac{3}{5}$	1	1	1	1	1
$\frac{4}{5}$	1	1	1	1	1
1	1	1	1	1	1

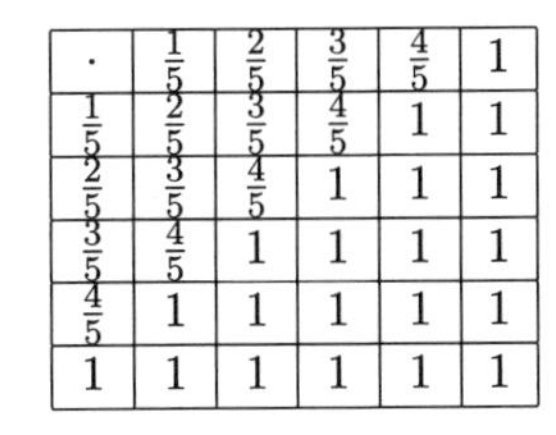

$\cdot$	$\frac{1}{5}$	$\frac{2}{5}$	$\frac{3}{5}$	$\frac{4}{5}$	1
$\frac{1}{5}$	$\frac{2}{5}$	$\frac{3}{5}$	$\frac{4}{5}$	1	1
$\frac{2}{5}$	$\frac{3}{5}$	$\frac{4}{5}$	1	1	1
$\frac{3}{5}$	$\frac{4}{5}$	1	1	1	1
$\frac{4}{5}$	1	1	1	1	1
1	1	1	1	1	1

THEOREM 1: Let $(S, +, \cdot)$ be a t.o.s.r. in which $E(+) = \phi$. Then $(S, +)$ is strictly non-negatively or strictly non-positively ordered.
PROOF: Suppose $(S, +)$ is not strictly non-positively ordered.
Then there exists an x such that $x < 2x$.
If $y \in S$ such that $2y < y$, then xy and yx are in E[+] using Theorem 11[3] which is a contradiction to our assumption that $E(+) = \phi$.
Therefore $2y > y$. Hence $(S, +)$ is strictly non-negatively ordered.

THEOREM 2: Let $(S, +, \cdot)$ be a totally ordered semiring and satisfy $ab = a + b + ab$ for all $a, b \in S$. If $(S, +)$ is p.t.o, then $(S, \cdot)$ is p.t.o.
PROOF: $ab = a + b + ab \geq a$ and b (Since, $(S, +)$ is p.t.o.)
$\Rightarrow ab \geq a$ and b
$\Rightarrow (S, \cdot)$ is p.t.o.

THEOREM 3: Let $(S, +, \cdot)$ be a semiring. If S contains multiplicative identity which is also additive identity, then $a + b + ab = ab$ for all $a, b \in S$.
PROOF: $a + b + ab = a + 1.b + ab$
$= a + (1 + a)b$
$= a + ab$
$= a.1 + ab$

$= a(1 + b)$
$= ab$
$\therefore a + b + ab = ab$ for all $a, b \in S$.

THEOREM 4: Let $(S, +, \cdot)$ be a semiring satisfying the condition $a + b + ab = ab$ for all $a, b \in S$. If S contains the multiplicative identity which is also an absorbing element with respect to $+$, then $a+b = ab$(mono semiring) for all $a, b \in R$.
PROOF: $a + b + ab = ab$ for all $a, b \in S$
$a + 1.b + ab = ab$ for all $a, b \in S$
$\Rightarrow a + (1 + a)b = ab$ for all $a, b \in S$
$\Rightarrow a + 1 \cdot b = ab$ for all $a, b \in S$.
$\Rightarrow a + b = ab$

THEOREM 5: Let $(S, +, \cdot)$ be a semiring with multiplicative identity which is also additive idempotent. Then $(S, +)$ is a band.
PROOF: By hypothesis $1 + 1 = 1$
$\Rightarrow a.(1 + 1) = a.1$ for all $a \in S$
$\Rightarrow a.1 + a.1 = a.1$
$\Rightarrow a + a = a$ for all $a \in S$
$\therefore (S, +)$ is a band.

THEOREM 6: Let $(S, +, \cdot)$ be a PRD. If $(S, +)$ is cancellative, then $(S, +)$ is commutative.
PROOF: Since $(S, +, \cdot)$ is a PRD, S contains multiplicative identity. Now $x + y + x + y = (x + y)(1 + 1) = x + x + y + y$. Since $(S, +)$ is cancellative, $y + x = x + y$.

THEOREM 7: Let $(S, +, \cdot)$ be a PRD semiring. Assume that $\mid S \mid > 1$. Then the following are true.
(1) $S = T_1 \bigcup T_2, T_1 \bigcap T_2 = \{e\}$, where $(T_1, \cdot)$ is a non-negatively ordered semigroup, $(T_2, \cdot)$ is a non-positively ordered semigroup and e is the multiplicative identity.
(2) $(T_1, \cdot) \cong (T_2, \cdot)$
(3) S is an infinite set, T_1 and T_2 are infinite subsets of S.
(4) For any two elements $x, y \in S$,
(i) $xy \geq x(xy \geq y)$ if and only if $y(x)$ is non-negatively ordered.
(ii) $xy \leq x(xy \leq y)$ if and only if $y(x)$ is non-positively ordered.
PROOF:(1) Let $T_1 = \{x \in S : x^2 \geq x\}, T_2 = \{x \in S : x^2 \leq x\}$
Let $x, y \in T_1$. Then $x^2 \geq x$ and $y^2 \geq y$. $(xy)^2 = (xy)(xy) = x^2y^2 \geq xy(\because (S, \cdot)$ is commutative). So $xy \in T_1$.
$\therefore (T_1, \cdot)$ is a semigroup. Similarly we can prove that $(T_2, \cdot)$ is a semigroup.
Let $x \in T_1$. Then $x^2 \geq x \Rightarrow x^{-1}x^2 \geq x^{-1}x \Rightarrow x \geq e$
If $y \in T_2$. Then $y^2 \leq y \Rightarrow y^{-1}y^2 \leq y^{-1}y \Rightarrow y \leq e$.

(2) $(T_1, \cdot) \cong (T_2, \cdot)$ by the map $\phi : x \to x^{-1}$ for all $x \in T_1$
(3) Since there is no finite PRD of order > 1, the proof follows.
(4) (i) Suppose y is non-negatively ordered i.e., $y^2 \geq y$
Now $xy^2y^{-1} \geq xyy^{-1} \Leftrightarrow xy \geq x$
(ii) is exactly the dual of (i).

THEOREM 8: Let $(S, +, \cdot)$ be a t.o. PRD. Then $(S, +)$ is non-negatively ordered or non-positively ordered.
PROOF: Since $(S, +, \cdot)$ is a PRD, it contains multiplicative identity.
Now using proposition 1[4], $(S, +)$ is either non-negatively ordered or non - positively ordered.

THEOREM 9: Let $(S, +, \cdot)$ be a semiring with IMP. If $(S, \cdot)$ is periodic, then $(S, +)$ is periodic.
PROOF: Let $x \in S$, Since S satisfies IMP, $x^2 = kx$
Also $x^n = x^m$, since $(S, \cdot)$ is periodic.
$\Rightarrow x^{n-2}.x^2 = x^{m-2}.x^2$
$\Rightarrow x^{n-2}.kx = x^{m-2}.kx$
Applying $x^2 = kx$ repeatly we finally obtain that $lx = px$
$\Rightarrow x$ is a periodic element w.r.to $'+'$.
$\therefore (S, +)$ is periodic.

THEOREM 10: Let $(S, +, \cdot)$ be a semiring with IMP in which $(S, \cdot)$ is semisimple and contains the multiplicative identity. Then $(S, +)$ is periodic.
PROOF: $a = 1.a.1.a.1 = a^2$
Since S satisfies IMP, $a^2 = na$
Therefore $a = na$.
$\Rightarrow (S, +)$ is periodic.

THEOREM 11: Let $(S, +, \cdot)$ be a t.o.s.r. with IMP. Then $(S, \cdot)$ is O-Archimedean if and only if $(S, +)$ is O-Archimedean.
PROOF: Suppose that $(S, +)$ is O-Archimedean.
Suppose if possible there exists an element x in S such that $a^n < x$ for every natural number n.
Since, a is additively O-Archimedean there exists a positive integer m such that $ma \geq x$.
Now $a^3 = a.a^2 = a(ta) = ta^2 = t(ta) = t^2a$
$a^4 = a.a^3 = a(t^2a) = a^2t^2 = t^2(ta) = t^3a$
$a^{r+1} = t^ra$
$t^ma = a^{m+1} < x \leq ma < t^ma$, which is a contradiction.
Therefore, $(S, \cdot)$ is O-Archimedean. Conversely, Suppose that $(S, \cdot)$ is O-Archimedean
Let $a \in S$ with $a < b$. Suppose that $ma^2 < b$ for every m.
If $m = 1, a^2 < b$ and $a^3 = a^2.a = ta.a = ta^2 < b$.

then, $a^3 < b$
Continuing in this way, we deduce that $a^n < b$ for every positive integer n, which contradicts the fact that $(S, \cdot)$ is O-Archimedean.
Therefore, $4ma^2 \geq b$ and thus, for every a and b in $(S, \cdot)$, there exists a positive integer m such that $ma^2 \geq b$. Now,$a^2 = ta \Rightarrow b \leq ma^2 = m(ta) = mt)a$. This proves that $(S, +)$ is O-Archimedean.

THEOREM 12: Let $(S, +, \cdot)$ be a semiring with IMP. If $(S, +)$ is a band, then the following statements hold:
(i) $(S, \cdot)$ is a band
(ii) If $(S, +, \cdot)$ is a Positive Rational Domain (PRD), then S reduces to a singleton set.
(iii) If $(S, \cdot)$ is quasi commutative, then $(S, \cdot)$ is commutative.
PROOF:(i) Since $(S, +)$ is a band, for every $a \in S, a = a + a = 2a = 3a = 4a = 5a = \cdots = na \cdots$ for some positive integer n. But $a^2 = na \Rightarrow a^2 = a$ for every $a \in S$. $\therefore (S, \cdot)$ is a band.
(ii) Let $(S, +, \cdot)$ be a PRD. Then by the definition of PRD, $(S, \cdot)$ is a group. By (i), $(S, \cdot)$ is a band i.e., $a^2 = a$ for every $a \in S$. Since $(S, \cdot)$ is a group, by cancellation, $a = e$. This implies $\mid S \mid = 1$.
(iii) Suppose that $(S, \cdot)$ is quasi commutative i.e., $ab = b^m a$ for some integer $m \geq 1$. But every element in $(S, \cdot)$ is an idempotent, using (i), $\therefore b^m = b$ for some positive integer $m \geq 1$. Hence, $ab = ba$ for every $a, b \in S$. Therefore, $(S, \cdot)$ is commutative.

References

[1] M. Satyanarayana and T. Vasanthi, Structure of Certain Classes of Semirings, Proceedings of the First International Symposium on Algebraic Structures and Number Theory, Hong Kong, 1988, 132-147.

[2] M. Satyanarayana and C. Srihari Nagore, Integrally Ordered Semigroups. Semigroup Forum 17(1979), 101-111.

[3] M. Satyanarayana, On the Additive Semigroup Structure of Semirings. Semigroup Forum 23(1981), 7-14.

[4] M.Satyanarayana, On the Additive Semigroup Structure of Ordered Semirings. Semigroup Forum 31(1985), 193-199.

On $CN-$Groups and $CT-$Groups[1]

Lifang Wang

School of Mathematics, Shanxi Normal University, Linfen 041004, China
Email: lfwang2003@yahoo.com.cn

Yanming Wang[2]

Lingnan College, Sun Yatsen Univ., Guangzhou, 510275, China
Email: stswym@mail.sysu.edu.cn

Abstract

A finite group is called a $CN-$*group* if every subgroup of it is $c-$normal. A finite group G is called a $CT-$*group* if $c-$normality is transitive in G. In this paper, we determine the structure of $CN-$groups and CT-groups and prove that solvable $CT-$groups are exactly the same as $CN-$groups.

2000 Mathematics Subject Classification: 20D10, 20D20
Keywords: $c-$normal subgroups, $CN-$groups, transitive, $CT-$groups.

1 Introduction

As a well known result, Dedekind and Bare [7, 5.3.7] determined the structure of groups with every subgroup normal. In fact, they proved that every subgroup of a group G is normal if and only if G is abelian or the direct product of a quaternion

[1]Project supported in part by NSF 10871210 of China

[2]The corresponding author

group of order 8, an elementary abelian 2-group and an abelian group of odd order. The group is also called Hamitonion group. Gheorghe [5] determined the structure of groups whose subgroups are quasinormal. It is clear that every subgroup of a group G is $s-$quasinormal if and only if G is nilpotent.

Let G be a finite group. A subgroup H of G is called $c-$ normal in G if there exists a normal subgroup N of G such that $G = HN$ and $H \cap N \leq H_G = Core_G(H)$, where H_G is the largest normal subgroup of G that is contained in H. It was introduced the definition in [8]. Recently, many authors investigated the influence of $c-$normality of subgroups on the structure of groups [2,6,9,10].

A group G is said to be a $T-$*group* (resp. $PT-$*group*, $PST-$*group*), if normality (resp. permutability, $s-$quasinormality) is a transitive relation, i.e. if D is a normal subgroup H and H is a normal subgroup K, then D is a normal subgroup of K (quasinormal or s-quasinormal respectively). Soluble $T-$group, $PT-$group and $PST-$group have been studied by many authors. The structure of soluble $T-$groups was determined by Gaschütz[4] in 1957. He showed that these are exactly the groups with an abelian normal Hall subgroup L of odd order such that G/L is a Dedekind group and the elements of G induce power automorphisms in L. The corresponding theorem for soluble $PT-$group (resp. $PST-$group) is due to Zacher [11] (resp. Agrawal [1]): here one has to replace "Dedekind" in Gaschütz's theorem by "nilpotent modular"(resp. "nilpotent").

$C-$normality, like normality, permutability and $s-$quasinormality, is not a transitive relation. For example, $\langle(12)\rangle$ is $c-$normal in S_3, S_3 is $c-$normal in S_4 but $\langle(12)\rangle$ is not $c-$normal in S_4.

Similar to the Hamiltonion group, a natural question is as follows:

Question 1: What is the structure of a finite group with every subgroup $c-$normal?

The question was actually asked by John Cossey, Ballester-Bolinches, Wang and some other people.

It is natural to introduce the following definitions.

Definition 1. A finite group G is called a $CN-$*group* if every subgroup of G is $c-$normal in G.

Definition 2. A finite group G is called a $CT-group$ if $c-$normality is transitive in G, that is, if A is a $c-$ normal subgroup of B and B is a $c-$normal subgroup of a group H, then A is $c-$normal in H.

We also have interest to ask the following question

Question 2: What is the structure of a finite $CT-group$?

In this paper, we try to determine the structure of CN-group and CT-group so to answer the question 1 and question 2. We are interested in finding that the solvable $CT-$groups are exactly $CN-$groups in Theorem 3.2.

We introduce the definitions.

Definition 3. A finite group G is called an elementary abelian group if G is abelian and its Sylow subgroup are elementary abelian groups.

Our notation is standard and it can be referred [3].

2 $CN-$groups

In this section, we investigate the structure of $CN-$group. First, for a $CN-$ $p-$group, we have the following result.

Theorem 2.1. *Let G be a $p-$group. Then G is a $CN-$group if and only if every subgroup of $\Phi(G)$ is normal in G.*

Proof. Assume that G is a $CN-$group. Let A be a subgroup of $\Phi(G)$. Then A is $c-$normal in G. By definition, there exists a normal subgroup N of G such that $G = AN$ and $A \cap N \leq A_G$. Since $A \leq \Phi(G)$, it implies that $G = AN = N$ and $A = A \cap N = A_G$, hence A is normal in G.

Conversely, assume that every subgroup of $\Phi(G)$ is normal in G. Take any subgroup A of G. If $A \leq \Phi(G)$, then A is normal in G by hypothesis. If $A \not\leq \Phi(G)$, then, since $G/\Phi(G)$ is elementary abelian, there exists a normal subgroup N of G such that $G/\Phi(G) = A\Phi(G) \times N/\Phi(G)$. Hence, we have that $G = AN$ and $A \cap N \leq \Phi(G)$. By hypothesis, $A \cap H \trianglelefteq G$. Therefore, $G = AN$, where $N \trianglelefteq G$ and $A \cap N \leq A_G$. So, A is $c-$normal in G.

The proof is complete.

For general case, we have the following theorem about the structure of finite $CN-$groups.

Theorem 2.2. *Let G be a finite group. Then the following statements are equivalent:*

(1) G is a $CN-$group;

(2) there exists a normal subgroup H of G such that G/H is elementary abelian and every element of G induces a power automorphism of H by conjugate.

In order to prove the above theorem. we first prove the following Lemma:

Lemma 2.1. *Let G be a finite $CN-$group, then the following statements holds:*

(i) $\Phi(A) \trianglelefteq G$ and $\Phi(A) \leq \Phi(G)$, for any subgroup A of G;

(ii) every subgroup of $\Phi(G)G'$ is normal in G.

Proof. (i) By hypothesis, G is a CN-group, so we have that $\Phi(A)$ is $c-$normal in G. Then there exists a normal subgroup N of G such that $G = N\Phi(A)$ and $N \cap \Phi(A) \leq (\Phi(A))_G$. Now, $A = A \cap G = \Phi(A)(A \cap N) = A \cap N$ by modular law. Therefore $A \leq N$ and $\Phi(A) = N \cap \Phi(A) \leq (\Phi(A))_G$. Hence, $\Phi(A) \trianglelefteq G$.

If $\Phi(A) \not\leq \Phi(G)$, then there exists a maximal subgroup M of G such that $G = M\Phi(A)$. Now $A = A \cap M\Phi(A) = (A \cap M)\Phi(A) = A \cap M \leq M$. Therefore, $\Phi(A) \leq M$, a contradiction. Hence, $\Phi(A) \leq \Phi(G)$.

(ii) First, we prove that every cyclic subgroup $\langle x \rangle$ of $\Phi(G)G'$ is normal in G.

Since $\langle x \rangle$ is $c-$normal in G, there exists a normal subgroup N of G such that $G = \langle x \rangle N$ and $\langle x \rangle \cap N \leq \langle x \rangle_G$. If $\langle x \rangle \ntrianglelefteq G$, then N is a proper subgroup of G. Hence there exists a maximal subgroup M of G such that $N \leq M$. Since $G/N \cong \langle x \rangle / \langle x \rangle \cap N$ is abelian, it follows that $G' \leq N$. Therefore, $G = \langle x \rangle N \leq (\Phi(G)G')N = \Phi(G)N \leq M < G$, a contradiction. Hence, $\langle x \rangle \trianglelefteq G$.

Finally, since every cyclic subgroup of $\Phi(G)G'$ is normal in G and every subgroup is generated by some cyclic subgroups, it implies that every subgroup of $\Phi(G)G'$ is normal in G.

Proof of Theorem 2.2

(i)$\Rightarrow$(ii). Let $H = \Phi(G)G'$. Then G/H is abelian since $G' \leq H$. Since $\Phi(P) \leq \Phi(G)$ for every Sylow subgroup P of G, we have that every Sylow subgroup of G/H is elementary abelian. Hence G/H is elementary abelian. By Lemma 2.1(ii), we have that every cyclic subgroup of H is normal in G. Then, every element of G induces a power automorphism of H by conjugate.

(ii)$\Rightarrow$(i). clearly, every subgroup of H is normal in G. Let A be any subgroup of G. If $A \leq H$, then $A \trianglelefteq H$ by hypothesis. If $A \leq H$, then there exists a normal subgroup N of G such that $G/H = AH/H \times N/H$. Hence $G = AN$ and $A \cap N \leq H$. By hypothesis, we have that $A \cap N \trianglelefteq G$. Therefore, $G = AN$ and $A \cap N \leq A_G$. Hence A is $c-$normal in G. Hence G is a $CN-$group.

Since an odd Dedekind group is abelian, from Theorem 2.2, we have the following result.

Corollary 2.1. *If G is an odd $CN-$group, then there exists a abelian normal subgroup H of G such that G/H is elementary abelian. In particular, G is a metabelian group.*

3 $CT-$groups

Now, we study the $CT-$groups. First, we give some lemmas which will be useful in the proof of our theorems.

Lemma 3.1. [8, Lemma 2.1] *Let G be a group. Then*

(i) If H is normal in G, then H is $c-$normal in G;

(ii) If H is $c-$normal in G, $H \leq K \leq G$, then H is $c-$normal in K;

(iii) Let $K \trianglelefteq G$, $K \leq H$. Then H is $c-$normal in G if and only if H/K is $c-$normal in G/K.

Lemma 3.2. [2, Lemma 2.2] *Let G be a finite group and p a prime number. If P is a minimal normal $p-$subgroup of G and $\langle x \rangle \leq P$ is $c-$normal in G, then $P = \langle x \rangle$.*

Lemma 3.3. *If G is a finite $CT-$group and $N \trianglelefteq G$, then G/N is a $CT-$group.*

Proof. Suppose that A/N is $c-$normal in B/N and B/N is $c-$normal in G/N. Then A is $c-$normal in B and B is $c-$normal in G by Lemma 3.1. Since G is a $CT-$group, it follows that A is $c-$normal in G. Again by Lemma 3.1, G/N is $c-$normal in G/N and G/N is a $CT-$group.

Theorem 3.1. *If G is a finite solvable $CT-$group, then*

(i) every subnormal subgroup of G is $c-$normal;

(ii) every minimal normal subgroup of G is a cyclic subgroup of G of prime order;

(iii) G is supersolvable;

(iv) every subgroup of G' is normal in G.

Proof. (i) Since a normal subgroup of G is $c-$normal in G, it follows that every subnormal subgroup of G is $c-$normal.

(ii) Let N be a minimal normal subgroup of G. Then N is elementary abelian since G is solvable. Hence every subgroup of N of prime order is $c-$normal in G by (i). It follows from Lemma 3.2 that N is a cyclic group of prime order.

(iii) Let N be a minimal normal subgroup of G. Then G/N is a $CT-$group by Lemma 3.3. By induction hypothesis, we have that G/N is supersolvable. Since N is cyclic, we have that G is supersolvable.

(iv) By (iii), we have that G' is nilpotent. Since every subgroup of G' is subnormal in G, by (i), we have that every subgroup of G' is $c-$normal in G.

Let $\langle x\rangle$ be any cyclic subgroup of G'. Then there exists a normal subgroup N of G such that $G = \langle x\rangle N$ and $\langle x\rangle \cap N \leq \langle x\rangle_G$. Since $G/N \cong \langle x\rangle/\langle x\rangle \cap N$ is abelian, then $G' \leq N$ and $\langle x\rangle \leq N$. Hence $\langle x\rangle \trianglelefteq G$. Thus every cyclic subgroup of G' is normal in G. It is clear that every subgroup of G' is normal in G.

From the above theorem, we have the following result.

Theorem 3.2. *Let G be a finite solvable group. Then G is a $CT-$group if and only if G is a $CN-$group.*

Proof. If G is a $CN-$group, then every subgroup of G is $c-$normal in G. It is clear that G is a $CT-$ group.

Conversely, we assume that G is a $CT-$group. Let A be any subgroup of G. By Theorem 3.1(iv), we have that $A\cap G' \trianglelefteq G$ and $A\cap G' \leq A_H \leq A_G$. Hence A is $c-$normal in $H = AG'$. Since $H \trianglelefteq G$ and G is a $CT-$group, it follows that A is $c-$normal in G. Therefore, any subgroup A of G is $c-$normal in G and G is a $CN-$group.

Corollary 3.1. *Let G be a finite solvable group. Then G is a $CT-$group if and only if there exists a normal subgroup H of G such that G/H is elementary abelian and every element of G induces a power automorphism of H by conjugate.*

For non-solvable $CT-$groups, we have the following result.

Theorem 3.3. *Let G be a finite group. If G is a $CT-$group and N is a minimal normal subgroup , then N is simple.*

Proof. Let $N = N_1 \times N_2 \times \cdots \times N_s$, where $N_1, N_2, \cdots, N_s$ are isomorphic simple groups. If $s > 1$, then N_i is $c-$normal in G by Theorem 3.1. Therefore, there exists a normal subgroup K of G such that $G = N_iK$ and $N_i \cap K \leq (N_i)_G = 1$. By modular law, we have that $N = N_i(N \cap K)$. Since N is a minimal normal subgroup of G and $K \trianglelefteq G$, it follows that $N \cap K = 1$ or N. If $N \cap K = 1$, then $N = N_i$, a contradiction. If $N \cap K = N$, then $N_1 \leq N \leq K$, $K = G$ and $N_1 \trianglelefteq G$, a contradiction. Hence, $s = 1$ and N is simple.

Remark. From Theorem 3.3 and Lemma 3.3, we see that every chief factor of a finite $CT-$group is simple.

References

[1] R. K. Agrawal, Finite groups whose subnormal subgroups permute with all Sylow subgroups, Proc. Amer. Soc. 47(1975), 77-83.

[2] A. Ballester-Bolinches; Yanming Wang, Finite groups with some minimal $c-$normal subgroups, J. Pure Appl. Alg., 153(2000), 121-127.

[3] K. Doerk and T. Hawkes, Finite solvable groups, Walter De Gruyter(Berlin-New York), 1992.

[4] W. Gaschütz, Gruppen, in dennen das Normalteilersein transitiv ist, J. Reine Angew. Math. 198(1957), 87-92.

[5] Pic, Georges Sur une quation fondamentale relative aux groupes finis de substitutions linaires. (French) C. R. Acad. Sci. Paris 228, (1949). 1268–1270.

[6] Xiuyun Guo; K. P. Shum, On c-normal maximal and minimal subgroups of Sylow p-subgroups of finite groups. Arch. Math. (Basel) 80 (2003), no. 6, 561–569.

[7] D.J.S.Robinson, A course in the theory of groups. New York-Berlin 1993.

[8] Y. Wang, $C-$normality of gorups and its properties, J. Alg., 180(1996), 954-965.

[9] Wei, Huaquan, On c-normal maximal and minimal subgroups of Sylow subgroups of finite groups. Comm. Algebra 29 (2001), no. 5, 2193–2200.

[10] Wei, Huaquan; Wang, Yanming; Li, Yangming, On c-normal maximal and minimal subgroups of Sylow subgroups of finite groups. II. Comm. Algebra 31 (2003), no. 10, 4807–4816.

[11] G.Zacher, I gruppi risolubli in cui i sottogruppi di composizione coincidono con i sottogrupi quasi-normali, Atti Accad. Naz. Lincei Rend. cl. Sci. Fis. Mat. Natur. (8)37(1964), 150-154.

The Solvability of a System of Matrix Equations over an Arbitrary Division Ring[1]

Qing-Wen Wang*,[a], K. P. Shum[b]

[a] Department of Mathematics, Shanghai University, 99 Shangda Road, Shanghai 200444, China, E-mail: wqw858@yahoo.com.cn

[b] Department of Mathematics, The University of Hong Kong,Pokfulam, Hong Kong, China, Email: kpshum@maths.hku.hk

Abstract: In this paper,we give a necessary and sufficient condition for the solvability to a system of linear matrix equations $A_1X_1 = C_1, X_1B_1 = C_2, A_2X_2 = C_3, X_2B_2 = C_4, A_3X_3 = C_5, X_3B_3 = C_6, A_4X_1B_4+A_5X_2B_5+A_6X_3B_6 = C_7$ over an arbitrary division ring.The findings of this paper extend some known results in the literature.

Keywords: System of linear matrix equations; Rank; Division ring; Generalized inverse

2000 AMS Subject Classifications 15A24, 15A09, 15A03, 15A30

1. Introduction

Throughout this paper, we denote an arbitrary division ring by $\mathbb{R}$, the set of all $m \times n$ matrices over $\mathbb{R}$ by $\mathbb{R}^{m\times n}$, the identity matrix with the appropriate size by I, the column right space, the row left space of a matrix A over $\mathbb{R}$ by $\mathcal{R}(A)$, $\mathcal{N}(A)$, respectively, the dimension of $\mathcal{R}(A)$ by $\dim \mathcal{R}(A)$, a generalized inverse of a matrix A by A^- which satisfies $AA^-A = A$. Moreover, two matrices L_A and R_A stand for the two projectors $L_A = I - A^-A$, $R_A = I - AA^-$ induced by A. By the text of Hungerford [4], for a matrix A over $\mathbb{R}$, $\dim \mathcal{R}(A) = \dim \mathcal{N}(A)$, which is called the rank of A and denoted by $r(A)$.

The aim of this paper is to discuss the solability to the following system of linear matrix equations over $\mathbb{R}$:

$$\begin{cases} A_1X_1 = C_1 \\ X_1B_1 = C_2 \\ A_2X_2 = C_3 \\ X_2B_2 = C_4 \\ A_3X_3 = C_5 \\ X_3B_3 = C_6 \\ A_4X_1B_4 + A_5X_2B_5 + A_6X_3B_6 = C_7 \end{cases} . \tag{1.1}$$

[1]This research was supported by the grants from the Ph.D. Programs Foundation of Ministry of Education of China (20093108110001), the Scientific Research Innovation Foundation of Shanghai Municipal Education Commission (09YZ13), and Shanghai Leading Academic Discipline Project (J50101)

* Corresponding author

This research is motivated by a several well-known linear matrix equations in the literature, especially for the following matrix equations

$$A_1X_1 = C_1, X_1B_1 = C_2, \tag{1.2}$$

$$A_1X_1B_1 + A_2X_2B_2 = C, \tag{1.3}$$

$$A_1X_1 = C_1, X_1B_1 = C_2, A_4X_1B_4 = C_7,$$

$$A_4X_1B_4 + A_5X_2B_5 + A_6X_3B_6 = C_7. \tag{1.4}$$

The classical matrix equations mentioned above have been widely investigated (see, e.g. [2]–[11]).

2. The necessary and sufficient condition for the solvability of (1.1)

We begin with the following lemma whose proof is just like the one over the complex field.

Lemma 2.1. *Let $A_1 \in \mathbb{R}^{m_1\times p_1}$, $B_1 \in \mathbb{R}^{q_1\times n_1}$, $C_1 \in \mathbb{R}^{m_1\times q_1}$, $C_2 \in \mathbb{R}^{p_1\times n_1}$ be known and $X_1 \in \mathbb{R}^{p_1\times q_1}$ unknown. Then the following statements are equivalent:*
(1) System (1.2) is consistent.
(2)

$$A_1A_1^-C_1 = C_1,\ C_2B_1^-B_1 = C_2,\ A_1C_2 = C_1B_1.$$

(3)

$$A_1C_2 = C_1B_1,\ r\left[A_1, C_1\right] = r\left(A_1\right),\ r\begin{bmatrix} B_1 \\ C_2 \end{bmatrix} = r\left(B_1\right).$$

In this case, the general solution of (1.2) is

$$X_1 = A_1^-C_1 + L_{A_1}C_2B_1^- + L_{A_1}Y_1R_{B_1},$$

where Y_1 is an arbitrary matrix over $\mathbb{R}$ with appropriate dimension.

The next Lemma due to Marsglia and Styan [6] can be generalized to $\mathbb{R}$.

Lemma 2.2. *Let $A \in \mathbb{R}^{m\times n}$, $B \in \mathbb{R}^{m\times k}$ and $C \in \mathbb{R}^{l\times n}$. Then they satisfy the following rank equalities*
(a) $r\begin{bmatrix} A & B \end{bmatrix} = r(A) + r(R_AB) = r(B) + r(R_BA)$.
(b) $r\begin{bmatrix} A \\ C \end{bmatrix} = r(A) + r(CL_A) = r(C) + r(AL_C)$.
(c) $r\begin{bmatrix} A & B \\ C & 0 \end{bmatrix} = r(B) + r(C) + r(R_BAL_C)$.

By using Lemma 2.2, we can easily deduce the following lemma.

Lemma 2.3. *Let $A \in \mathbb{R}^{m\times n}$, $B \in \mathbb{R}^{m\times k}$, $C \in \mathbb{R}^{l\times n}$, $D \in \mathbb{R}^{j\times k}$ and $E \in \mathbb{R}^{l\times i}$. Then they satisfy the following rank equalities*

(*d*) $r(CL_A) = r\begin{bmatrix} A \\ C \end{bmatrix} - r(A)$.

(*e*) $r\begin{bmatrix} B & AL_C \end{bmatrix} = r\begin{bmatrix} B & A \\ 0 & C \end{bmatrix} - r(C)$.

(*f*) $r\begin{bmatrix} C \\ R_B A \end{bmatrix} = r\begin{bmatrix} C & 0 \\ A & B \end{bmatrix} - r(B)$.

(*g*) $r\begin{bmatrix} A & BL_D \\ R_E C & 0 \end{bmatrix} = r\begin{bmatrix} A & B & 0 \\ C & 0 & E \\ 0 & D & 0 \end{bmatrix} - r(D) - r(E)$.

The following Lemma was due to Tian [9] in 2000 which can also be generalized to $\mathbb{R}$.

Lemma 2.4. *Let $A_1 \in \mathbb{R}^{m_1\times n_1}$, $A_2 \in \mathbb{R}^{m_2\times n_2}$, $A_3 \in \mathbb{R}^{m_3\times n_3}$, $A_4 \in \mathbb{R}^{m_4\times n_4}$, $B_1 \in \mathbb{R}^{m_1\times p}$, $B_2 \in \mathbb{R}^{m_2\times p}$, $B_3 \in \mathbb{R}^{m_3\times p}$, $B_4 \in \mathbb{R}^{m_4\times p}$, $C_1 \in \mathbb{R}^{q\times n_1}$, $C_2 \in \mathbb{R}^{q\times n_2}$, $C_3 \in \mathbb{R}^{q\times n_3}$, $C_4 \in \mathbb{R}^{q\times n_4}$ be known and the quadruple matrix equations*

$$B_1XC_1 = A_1, B_2XC_2 = A_2, B_3XC_3 = A_3, B_4XC_4 = A_4$$

satisfy the four conditions

$$\mathcal{N}(B_1) \subseteq \mathcal{N}(B_2), \mathcal{R}(C_2) \subseteq \mathcal{R}(C_1), \mathcal{N}(B_3) \subseteq \mathcal{N}(B_4), \mathcal{R}(C_4) \subseteq \mathcal{R}(C_3).$$

Then they have a common solution if and only if the following fourteen rank equalities are all satisfied

$$r\,[B_i, A_i] = r\,(B_i),$$

$$r\begin{bmatrix} C_i \\ A_i \end{bmatrix} = r\,(C_i), i = 1,2,3,4,$$

$$r\begin{bmatrix} A_1 & 0 & B_1 \\ 0 & -A_2 & B_2 \\ C_1 & C_2 & 0 \end{bmatrix} = r\begin{bmatrix} B_1 \\ B_2 \end{bmatrix} + r\,[C_1, C_2],$$

$$r\begin{bmatrix} A_3 & 0 & B_3 \\ 0 & -A_4 & B_4 \\ C_3 & C_4 & 0 \end{bmatrix} = r\begin{bmatrix} B_3 \\ B_4 \end{bmatrix} + r\,[C_3, C_4],$$

$$r\begin{bmatrix} A_i & 0 & B_i \\ 0 & -A_j & B_j \\ C_i & C_j & 0 \end{bmatrix} = r\begin{bmatrix} B_i \\ B_j \end{bmatrix} + r\,[C_i, C_j], i = 1,2, j = 3,4.$$

Lemma 2.5. (*See* [8], [10]) *Let* $A_1 \in \mathbb{R}^{m\times p}$, $A_2 \in \mathbb{R}^{m\times s}$, $B_1 \in \mathbb{R}^{q\times n}$, $B_2 \in \mathbb{R}^{t\times n}$ *and* $C \in \mathbb{R}^{m\times n}$ *be given. Then system (1.3) is consistent if and only if*

$$r\left[C, A_1, A_2\right] = r\left[A_1, A_2\right], r\begin{bmatrix} C \\ B_1 \\ B_2 \end{bmatrix} = r\begin{bmatrix} B_1 \\ B_2 \end{bmatrix},$$

$$r\begin{bmatrix} C & A_1 \\ B_2 & 0 \end{bmatrix} = r(A_1) + r(B_2), r\begin{bmatrix} C & A_2 \\ B_1 & 0 \end{bmatrix} = r(A_2) + r(B_1).$$

We now give the main theorem of this paper.

Theorem 2.6. *Let* $A_1 \in \mathbb{R}^{m_1\times p_1}$, $A_2 \in \mathbb{R}^{m_2\times p_2}$, $A_3 \in \mathbb{R}^{m_3\times p_3}$, $A_4 \in \mathbb{R}^{m\times p_1}$, $A_5 \in \mathbb{R}^{m\times p_2}$, $A_6 \in \mathbb{R}^{m\times p_3}$, $B_1 \in \mathbb{R}^{q_1\times n_1}$, $B_2 \in \mathbb{R}^{q_2\times n_2}$, $B_3 \in \mathbb{R}^{q_3\times n_3}$, $B_4 \in \mathbb{R}^{q_1\times n}$, $B_5 \in \mathbb{R}^{q_2\times n}$, $B_6 \in \mathbb{R}^{q_3\times n}$, $C_1 \in \mathbb{R}^{m_1\times q_1}$, $C_2 \in \mathbb{R}^{p_1\times n_1}$, $C_3 \in \mathbb{R}^{m_2\times q_2}$, $C_4 \in \mathbb{R}^{p_2\times n_2}$, $C_5 \in \mathbb{R}^{m_3\times q_3}$, $C_6 \in \mathbb{R}^{p_3\times n_3}$, $C_7 \in \mathbb{R}^{m\times n}$ *be known and* $X_1 \in \mathbb{R}^{p_1\times q_1}$, $X_2 \in \mathbb{R}^{p_2\times q_2}$, $X_3 \in \mathbb{R}^{p_3\times q_3}$ *unknown. Then system (1.1) is consistent if and only if the following equalities are all satisfied*

$$A_1C_2 = C_1B_1,\ r\left[A_1, C_1\right] = r\left(A_1\right),\ r\begin{bmatrix} B_1 \\ C_2 \end{bmatrix} = r\left(B_1\right), \tag{2.1}$$

$$A_2C_4 = C_3B_2,\ r\left[A_2, C_3\right] = r\left(A_2\right),\ r\begin{bmatrix} B_2 \\ C_4 \end{bmatrix} = r\left(B_2\right), \tag{2.2}$$

$$A_3C_6 = C_5B_3,\ r\left[A_3, C_5\right] = r\left(A_3\right),\ r\begin{bmatrix} B_3 \\ C_6 \end{bmatrix} = r\left(B_3\right), \tag{2.3}$$

$$r\begin{bmatrix} A_1 & 0 & 0 & C_1B_4 \\ 0 & A_2 & 0 & C_3B_5 \\ 0 & 0 & A_3 & C_5B_6 \\ A_4 & A_5 & A_6 & C_7 \end{bmatrix} = r\begin{bmatrix} A_1 & 0 & 0 \\ 0 & A_2 & 0 \\ 0 & 0 & A_3 \\ A_4 & A_5 & A_6 \end{bmatrix}, \tag{2.4}$$

$$r\begin{bmatrix} B_1 & 0 & 0 & B_4 \\ 0 & B_2 & 0 & B_5 \\ 0 & 0 & B_3 & B_6 \\ A_4C_2 & A_5C_4 & A_6C_6 & C_7 \end{bmatrix} = r\begin{bmatrix} B_1 & 0 & 0 & B_4 \\ 0 & B_2 & 0 & B_5 \\ 0 & 0 & B_3 & B_6 \end{bmatrix}, \tag{2.5}$$

$$r\begin{bmatrix} A_1 & 0 & 0 & C_1B_4 \\ 0 & A_3 & 0 & C_5B_6 \\ 0 & 0 & B_2 & B_5 \\ A_4 & A_6 & A_5C_4 & C_7 \end{bmatrix} = r\begin{bmatrix} A_1 & 0 \\ 0 & A_3 \\ A_5 & A_6 \end{bmatrix} + r\left[B_2, B_5\right], \tag{2.6}$$

$$r\begin{bmatrix} A_2 & 0 & 0 & C_3B_5 \\ 0 & A_3 & 0 & C_5B_6 \\ 0 & 0 & B_1 & B_4 \\ A_5 & A_6 & A_4C_2 & C_7 \end{bmatrix} = r\begin{bmatrix} A_2 & 0 \\ 0 & A_3 \\ A_4 & A_6 \end{bmatrix} + r\left[B_1, B_4\right], \tag{2.7}$$

$$r\begin{bmatrix} A_1 & 0 & 0 & C_1B_4 \\ 0 & A_2 & 0 & C_3B_5 \\ 0 & 0 & B_3 & B_6 \\ A_4 & A_5 & A_6C_6 & C_7 \end{bmatrix} = r\begin{bmatrix} A_1 & 0 \\ 0 & A_2 \\ A_4 & A_5 \end{bmatrix} + r\left[B_3, B_6\right], \tag{2.8}$$

$$r\begin{bmatrix} A_6 & A_4C_2 & A_5C_4 & C_7 \\ A_3 & 0 & 0 & C_5B_6 \\ 0 & B_1 & 0 & B_4 \\ 0 & 0 & B_2 & B_5 \end{bmatrix} = r\begin{bmatrix} B_1 & 0 & B_4 \\ 0 & B_2 & B_5 \end{bmatrix} + r\begin{bmatrix} A_3 \\ A_6 \end{bmatrix}, \tag{2.9}$$

$$r\begin{bmatrix} A_4 & A_5C_4 & A_6C_6 & C_7 \\ A_1 & 0 & 0 & C_1B_4 \\ 0 & B_2 & 0 & B_5 \\ 0 & 0 & B_3 & B_6 \end{bmatrix} = r\begin{bmatrix} B_2 & 0 & B_5 \\ 0 & B_3 & B_6 \end{bmatrix} + r\begin{bmatrix} A_1 \\ A_4 \end{bmatrix}, \tag{2.10}$$

$$r\begin{bmatrix} A_5 & A_4C_2 & A_6C_6 & C_7 \\ A_2 & 0 & 0 & C_3B_5 \\ 0 & B_1 & 0 & B_4 \\ 0 & 0 & B_3 & B_6 \end{bmatrix} = r\begin{bmatrix} B_1 & 0 & B_4 \\ 0 & B_3 & B_6 \end{bmatrix} + r\begin{bmatrix} A_2 \\ A_5 \end{bmatrix}, \tag{2.11}$$

$$r\begin{bmatrix} C_7 & 0 & 0 & A_5C_4 & A_6C_6 & A_4 & 0 & A_6 \\ 0 & -C_7 & -A_4C_2 & 0 & 0 & 0 & A_5 & A_6 \\ C_1B_4 & 0 & 0 & 0 & 0 & A_1 & 0 & 0 \\ 0 & -C_3B_5 & 0 & 0 & 0 & 0 & A_2 & 0 \\ 0 & -C_5B_6 & 0 & 0 & 0 & 0 & 0 & A_3 \\ 0 & B_4 & B_1 & 0 & 0 & 0 & 0 & 0 \\ B_5 & 0 & 0 & B_2 & 0 & 0 & 0 & 0 \\ B_6 & B_6 & 0 & 0 & B_3 & 0 & 0 & 0 \end{bmatrix}$$

$$= r\begin{bmatrix} A_4 & 0 & A_6 \\ 0 & A_5 & A_6 \\ A_1 & 0 & 0 \\ 0 & A_2 & 0 \\ 0 & 0 & A_3 \end{bmatrix} + r\begin{bmatrix} 0 & B_4 & B_1 & 0 & 0 \\ B_5 & 0 & 0 & B_2 & 0 \\ B_6 & B_6 & 0 & 0 & B_3 \end{bmatrix}. \tag{2.12}$$

Proof. Suppose that the system (1.1) is consistent, and X_{01}, X_{02}, X_{03} are the solutions to the system (1.1). Then it is easy to verify (2.1)-(2.3). Hence (2.4) follows from

$$P\begin{bmatrix} A_1 & 0 & 0 & C_1B_4 \\ 0 & A_2 & 0 & C_3B_5 \\ 0 & 0 & A_3 & C_5B_6 \\ A_4 & A_5 & A_6 & C_7 \end{bmatrix} Q = \begin{bmatrix} A_1 & 0 & 0 \\ 0 & A_2 & 0 \\ 0 & 0 & A_3 \\ A_4 & A_5 & A_6 \end{bmatrix},$$

where

$$P=\begin{bmatrix} I & 0 & 0 & 0 \\ 0 & I & 0 & 0 \\ 0 & 0 & I & 0 \\ 0 & 0 & 0 & I \end{bmatrix}, Q=\begin{bmatrix} I & 0 & 0 & -X_{01}B_4 \\ 0 & I & 0 & -X_{02}B_5 \\ 0 & 0 & I & -X_{03}B_6 \\ 0 & 0 & 0 & I \end{bmatrix}.$$

Similarly, we can easily show that the other equalities of (2.5)-(2.12) hold.
Conversely, if (2.1)-(2.12) hold. Then by Lemma 2.1 and (2.1)-(2.3), we know that equations $A_1X_1 = C_1$, $X_1B_1 = C_2$; $A_2X_2 = C_3$, $X_2B_2 = C_4$ and $A_3X_3 = C_5$, $X_3B_3 = C_6$ are all consistent, and the general solutions of the above equations can be expressed, respectively, as the following

$$X_1 = A_1^- C_1 + L_{A_1}C_2B_1^- + L_{A_1}Y_1R_{B_1}, \tag{2.13}$$

$$X_2 = A_2^- C_3 + L_{A_2}C_4B_2^- + L_{A_2}Y_2R_{B_2}, \tag{2.14}$$

and

$$X_3 = A_3^- C_5 + L_{A_3}C_6B_3^- + L_{A_3}Y_3R_{B_3}, \tag{2.15}$$

where Y_1, Y_2 and Y_3 are arbitrary matrices over $\mathbb{R}$ with appropriate sizes. Putting (2.13), (2.14) and (2.15) into equation $A_4X_1B_4 + A_5X_2B_5 + A_6X_3B_6 = C_7$ yields

$$D_1Y_1E_1 + D_2Y_2E_2 + D_3Y_3E_3 = F, \tag{2.16}$$

which can be written as

$$D_1Y_1E_1 + D_2Y_2E_2 = F - D_3Y_3E_3, \tag{2.17}$$

where $D_1 = A_4L_{A_1}, D_2 = A_5L_{A_2}, D_3 = A_6L_{A_3}, E_1 = R_{B_1}B_4, E_2 = R_{B_2}B_5, E_3 = R_{B_3}B_6$,

$$F = C_7 - A_4(A_1^- C_1 + L_{A_1}C_2B_1^-)B_4 - A_5(A_2^- C_3 + L_{A_2}C_4B_2^-)B_5 - A_6(A_3^- C_5 + L_{A_3}C_6B_3^-)B_6.$$

By Lemma 2.5, we know that equation (2.17) is consistent if and only if there exists an Y_3 satisfying the following four equalities

$$r\,[D_1, D_2, F - D_3Y_3E_3] = r\,[D_1, D_2], \tag{2.18}$$

$$r\begin{bmatrix} E_1 \\ E_2 \\ F - D_3Y_3E_3 \end{bmatrix} = r\begin{bmatrix} E_1 \\ E_2 \end{bmatrix}, \tag{2.19}$$

$$r\begin{bmatrix} F - D_3Y_3E_3 & D_1 \\ E_2 & 0 \end{bmatrix} = r(D_1) + r(E_2), \tag{2.20}$$

$$r\begin{bmatrix} F - D_3Y_3E_3 & D_2 \\ E_1 & 0 \end{bmatrix} = r(D_2) + r(E_1). \tag{2.21}$$

It follows from Lemma 2.2 that (2.18)-(2.21) are equivalent to the following four matrix equations, respectively,

$$R_PD_3Y_3E_3 = R_PF, R_{D_1}D_3Y_3E_3L_{E_2} = R_{D_1}FL_{E_2}, R_{D_2}D_3Y_3E_3L_{E_1} = R_{D_2}FL_{E_1}, D_3Y_3E_3L_Q = FL_Q,$$

which can be simply written as

$$G_1Y_3H_1 = K_1, G_2Y_3H_2 = K_2, G_3Y_3H_3 = K_3, G_4Y_3H_4 = K_4, \tag{2.22}$$

where

$$G_1 = R_P D_3, G_2 = R_{D_1} D_3, G_3 = R_{D_2} D_3,$$
$$G_4 = D_3, H_1 = E_3, H_2 = E_3 L_{E_2}, H_3 = E_3 L_{E_1},$$
$$H_4 = E_3 L_Q, K_1 = R_P F, K_2 = R_{D_1} F L_{E_2}, K_3 = R_{D_2} F L_{E_1}, K_4 = F L_Q.$$

It is not difficult to deduce that the given matrices in (2.22) satisfy the following four inclusions

$$\mathcal{N}(G_1) \subseteq \mathcal{N}(G_2), \mathcal{R}(H_2) \subseteq \mathcal{R}(H_1), \mathcal{N}(G_3) \subseteq \mathcal{N}(G_4), \mathcal{R}(H_4) \subseteq \mathcal{R}(H_3).$$

By Lemma 2.4, the system (2.22) is consistent if and only if the following fourteen rank equalities hold

$$r[G_i, K_i] = r(G_i), \; r\begin{bmatrix} H_i \\ K_i \end{bmatrix} = r(H_i), i = 1,2,3,4. \tag{2.23}$$

$$r\begin{bmatrix} K_1 & 0 & G_1 \\ 0 & -K_i & G_i \\ H_1 & H_i & 0 \end{bmatrix} = r\begin{bmatrix} G_1 \\ G_i \end{bmatrix} + r[H_1, H_i], i = 2,3,4. \tag{2.24}$$

$$r\begin{bmatrix} K_i & 0 & G_i \\ 0 & -K_4 & G_4 \\ H_i & H_4 & 0 \end{bmatrix} = r\begin{bmatrix} G_i \\ G_4 \end{bmatrix} + r[H_i, H_4], i = 2,3. \tag{2.25}$$

$$r\begin{bmatrix} K_2 & 0 & G_2 \\ 0 & -K_3 & G_3 \\ H_2 & H_3 & 0 \end{bmatrix} = r\begin{bmatrix} G_2 \\ G_3 \end{bmatrix} + r[H_2, H_3]. \tag{2.26}$$

Substituting the explicit expressions of G_i, H_i and K_i $(i = 1,2,3,4)$ into the eight rank equalities in (2.23) and simplifying them by Lemma 2.1, Lemma 2.2, Lemma 2.3 and Gaussian block operations, we know that they are equivalent to the eight rank equalities in (2.4)-(2.11), respectively. Taking the explicit expressions of G_i, H_i and K_i $(i = 1,2,3,4)$ into the five rank equalities in (2.24) and (2.25) and simplifying them by Lemma 2.1, Lemma 2.2, Lemma 2.3 and Gaussian block operations, we also know that they are equivalent to the eight rank equalities in (2.4)-(2.11), respectively. But the processes are much tedious, we omit them here. Similarly, we can obtain (2.12) by simplifying (2.26). From the above, we know that (2.16) is consistent if and only if the nine rank equalities (2.4)-(2.12) hold. Hence, there exist matrices Y_{01}, Y_{02} and Y_{03} satisfying

$$D_1 Y_{01} E_1 + D_2 Y_{02} E_2 + D_3 Y_{03} E_3 = F,$$

i.e., Y_{01}, Y_{02} and Y_{03} are solutions of (2.16).

Let

$$\alpha_0 = A_1^- C_1 + L_{A_1} C_2 B_1^- + L_{A_1} Y_{01} R_{B_1}, \tag{2.27}$$

$$\beta_0 = A_2^- C_3 + L_{A_2} C_4 B_2^- + L_{A_2} Y_{02} R_{B_2}, \tag{2.28}$$

$$\gamma_0 = A_3^- C_5 + L_{A_3} C_6 B_3^- + L_{A_3} Y_{03} R_{B_3}. \tag{2.29}$$

Then it is easy to verify that

$$A_4\alpha_0 B_4 + A_5\beta_0 B_5 + A_6\gamma_0 B_6 = C_7.$$

Obviously, (2.27)-(2.29) are also solutions of equations

$$\begin{matrix} A_1X_1 = C_1 \\ X_1B_1 = C_2 \end{matrix}, \quad \begin{matrix} A_2X_2 = C_3 \\ X_2B_2 = C_4 \end{matrix}, \quad \begin{matrix} A_3X_3 = C_5 \\ X_3B_3 = C_6 \end{matrix},$$

respectively. Therefore, if (2.1)-(2.12) hold, then system (1.1) is consistent. The proof is completed. □

3. Solvable conditions to some special cases of system (1.1)

In this section, we consider the special cases of Theorem 2.6.

Corollary 3.1. *Let $A_4 \in \mathbb{R}^{m\times p_1}$, $A_5 \in \mathbb{R}^{m\times p_2}$, $A_6 \in \mathbb{R}^{m\times p_3}$, $B_4 \in \mathbb{R}^{q_1\times n}$, $B_5 \in \mathbb{R}^{q_2\times n}$, $B_6 \in \mathbb{R}^{q_3\times n}$, $C_7 \in \mathbb{R}^{m\times n}$ be known and $X_1 \in \mathbb{R}^{p_1\times q_1}$, $X_2 \in \mathbb{R}^{p_2\times q_2}$, $X_3 \in \mathbb{R}^{p_3\times q_3}$ unknown. Then equation (1.4) is solvable if and only if the following rank equalities are all satisfied*

$$r\left[A_4, A_5, A_6, C_7\right] = r\left[A_4, A_5, A_6\right], r\begin{bmatrix} B_4 \\ B_5 \\ B_6 \\ C_7 \end{bmatrix} = r\begin{bmatrix} B_4 \\ B_5 \\ B_6 \end{bmatrix}, \tag{3.1}$$

$$r\begin{bmatrix} C_7 & A_4 & A_5 \\ B_6 & 0 & 0 \end{bmatrix} = r\left[A_4, A_5\right] + r\left(B_6\right), r\begin{bmatrix} C_7 & A_4 & A_6 \\ B_5 & 0 & 0 \end{bmatrix} = r\left[A_4, A_6\right] + r\left(B_5\right), \tag{3.2}$$

$$r\begin{bmatrix} C_7 & A_5 & A_6 \\ B_4 & 0 & 0 \end{bmatrix} = r\left[A_5, A_6\right] + r\left(B_4\right), r\begin{bmatrix} C_7 & A_6 \\ B_4 & 0 \\ B_5 & 0 \end{bmatrix} = r\begin{bmatrix} B_4 \\ B_5 \end{bmatrix} + r\left(A_6\right), \tag{3.3}$$

$$r\begin{bmatrix} C_7 & A_5 \\ B_4 & 0 \\ B_6 & 0 \end{bmatrix} = r\begin{bmatrix} B_4 \\ B_6 \end{bmatrix} + r\left(A_5\right), r\begin{bmatrix} C_7 & A_4 \\ B_5 & 0 \\ B_6 & 0 \end{bmatrix} = r\begin{bmatrix} B_5 \\ B_6 \end{bmatrix} + r\left(A_4\right), \tag{3.4}$$

$$r\begin{bmatrix} C_7 & 0 & A_4 & 0 & A_6 \\ 0 & -C_7 & 0 & A_5 & A_6 \\ B_5 & 0 & 0 & 0 & 0 \\ 0 & B_4 & 0 & 0 & 0 \\ B_6 & B_6 & 0 & 0 & 0 \end{bmatrix} = r\begin{bmatrix} B_5 & 0 \\ 0 & B_4 \\ B_6 & B_6 \end{bmatrix} + r\begin{bmatrix} A_4 & 0 & A_6 \\ 0 & A_5 & A_6 \end{bmatrix}. \tag{3.5}$$

Proof. In Theorem 2.6, let $A_1, A_2, A_3, B_1, B_2, B_3, C_1, C_2, C_3, C_4, C_5$and C_6 vanish. Then simplifying (2.1)-(2.12) yields (3.1)-(3.5). □

Remark 3.1. *Corollary 3.1 is one of the main theorems of* [9].

Similarly, using Theorem 2.6, we can obtain the following.

Corollary 3.2. *Suppose* $A_1 \in \mathbb{R}^{m_1 \times p_1}$, $A_2 \in \mathbb{R}^{m_2 \times p_2}$, $A_4 \in \mathbb{R}^{m \times p_1}$, $A_5 \in \mathbb{R}^{m \times p_2}$, $B_4 \in \mathbb{R}^{q_1 \times n}$, $B_5 \in \mathbb{R}^{q_2 \times n}$, $C_1 \in \mathbb{R}^{m_1 \times q_1}$, $C_3 \in \mathbb{R}^{m_2 \times q_2}$, $C_7 \in \mathbb{R}^{m \times n}$ *be known and* $X_1 \in \mathbb{R}^{p_1 \times q_1}$, $X_2 \in \mathbb{R}^{p_2 \times q_2}$ *unknown. Then system*

$$A_1X_1 = C_1, A_2X_2 = C_3, A_4X_1B_4 + A_5X_2B_5 = C_7$$

is consistent if and only if the following rank equalities are all satisfied

$$r\,[A_1, C_1] = r\,(A_1)\,,\ r\,[A_2, C_3] = r\,(A_2)\,,$$

$$r\begin{bmatrix} A_1 & C_1B_4 \\ A_4 & C_7 \\ 0 & B_5 \end{bmatrix} = r\begin{bmatrix} A_1 \\ A_4 \end{bmatrix} + r\,(B_5)\,, r\begin{bmatrix} B_4 \\ B_5 \\ C_7 \end{bmatrix} = r\begin{bmatrix} B_4 \\ B_5 \end{bmatrix},$$

$$r\begin{bmatrix} A_2 & C_3B_5 \\ A_5 & C_7 \\ 0 & B_4 \end{bmatrix} = r\begin{bmatrix} A_2 \\ A_5 \end{bmatrix} + r\,(B_4)\,, r\begin{bmatrix} C_1B_4 & A_1 & 0 \\ C_3B_5 & 0 & A_2 \\ C_7 & A_4 & A_5 \end{bmatrix} = r\begin{bmatrix} A_1 & 0 \\ 0 & A_2 \\ A_4 & A_5 \end{bmatrix}.$$

Corollary 3.3. *Let* $A_1 \in \mathbb{R}^{m_1 \times p_1}$, $A_2 \in \mathbb{R}^{m_2 \times p_2}$, $A_4 \in \mathbb{R}^{m \times p_1}$, $A_5 \in \mathbb{R}^{m \times p_2}$, $B_1 \in \mathbb{R}^{q_1 \times n_1}$, $B_2 \in \mathbb{R}^{q_2 \times n_2}$, $B_4 \in \mathbb{R}^{q_1 \times n}$, $B_5 \in \mathbb{R}^{q_2 \times n}$, $C_1 \in \mathbb{R}^{m_1 \times q_1}$, $C_2 \in \mathbb{R}^{p_1 \times n_1}$, $C_3 \in \mathbb{R}^{m_2 \times q_2}$, $C_4 \in \mathbb{R}^{p_2 \times n_2}$, $C_7 \in \mathbb{R}^{m \times n}$ *be known and* $X_1 \in \mathbb{R}^{p_1 \times q_1}$, $X_2 \in \mathbb{R}^{p_2 \times q_2}$ *unknown. Then system*

$$A_1X_1 = C_1, X_1B_1 = C_2, A_2X_2 = C_3, X_2B_2 = C_4, A_4X_1B_4 + A_5X_2B_5 = C_7$$

is consistent if and only if

$$A_1C_2 = C_1B_1,\ A_2C_4 = C_3B_2, r\,[A_1, C_1] = r(A_1),$$

$$r\,[A_2, C_3] = r(A_2), r\begin{bmatrix} B_1 \\ C_2 \end{bmatrix} = r(B_1),\ r\begin{bmatrix} B_2 \\ C_4 \end{bmatrix} = r(B_2),$$

$$r\begin{bmatrix} A_1 & 0 & C_1B_4 \\ 0 & A_2 & C_3B_5 \\ A_4 & A_5 & C_7 \end{bmatrix} = r\begin{bmatrix} A_1 & 0 \\ 0 & A_2 \\ A_4 & A_5 \end{bmatrix}, r\begin{bmatrix} B_1 & 0 & B_4 \\ 0 & B_2 & B_5 \\ A_4C_2 & A_5C_4 & C_7 \end{bmatrix} = r\begin{bmatrix} B_1 & 0 & B_4 \\ 0 & B_2 & B_5 \end{bmatrix},$$

$$r\begin{bmatrix} A_1 & 0 & C_1B_4 \\ A_4 & A_5C_4 & C_7 \\ 0 & B_2 & B_5 \end{bmatrix} = r\begin{bmatrix} A_1 \\ A_4 \end{bmatrix} + r\,[B_2, B_5]\,, r\begin{bmatrix} A_2 & 0 & C_3B_5 \\ A_5 & A_4C_2 & C_7 \\ 0 & B_1 & B_4 \end{bmatrix} = r\begin{bmatrix} A_2 \\ A_5 \end{bmatrix} + r\,[B_1, B_4]\,.$$

Corollary 3.4. *Let* $A_5 \in \mathbb{R}^{m \times p_1}$, $A_6 \in \mathbb{R}^{m \times p_2}$, $B_1 \in \mathbb{R}^{q_1 \times n_1}$, $B_2 \in \mathbb{R}^{q_1 \times n_2}$, $B_3 \in \mathbb{R}^{q_2 \times n_3}$, $B_4 \in \mathbb{R}^{q_2 \times n_4}$, $B_5 \in \mathbb{R}^{q_1 \times n}$, $B_6 \in \mathbb{R}^{q_2 \times n}$, $C_1 \in \mathbb{R}^{p_1 \times n_1}$, $C_2 \in \mathbb{R}^{p_1 \times n_2}$, $C_3 \in \mathbb{R}^{p_2 \times n_3}$, $C_4 \in \mathbb{R}^{p_2 \times n_4}$, $C_7 \in \mathbb{R}^{m \times n}$ *be known and* $X_1 \in \mathbb{R}^{p_1 \times q_1}$, $X_2 \in \mathbb{R}^{p_2 \times q_2}$ *unknown. Then system*

$$X_1B_1 = C_1, X_1B_2 = C_2, X_2B_3 = C_3, X_2B_4 = C_4, A_5X_1B_5 + A_6X_2B_6 = C_7$$

is consistent if and only if the following rank equalities are all satisfied

$$r\begin{bmatrix} C_i \\ B_i \end{bmatrix} = r\left(B_i\right), i = 1, 2, 3, 4,$$

$$r\left[A_5, A_6, C_7\right] = r\left[A_5, A_6\right],$$

$$r\begin{bmatrix} C_1 & C_2 \\ B_1 & B_2 \end{bmatrix} = r\left[B_1, B_2\right], r\begin{bmatrix} C_3 & C_4 \\ B_3 & B_4 \end{bmatrix} = r\left[B_3, B_4\right],$$

$$r\begin{bmatrix} A_6 & A_5C_1 & A_5C_2 & C_7 \\ 0 & B_1 & B_2 & B_5 \end{bmatrix} = r\left(A_6\right) + r\left[B_1, B_2, B_5\right],$$

$$r\begin{bmatrix} A_5 & A_6C_3 & A_6C_4 & C_7 \\ 0 & B_3 & B_4 & B_6 \end{bmatrix} = r\left(A_5\right) + r\left[B_3, B_4, B_6\right],$$

$$r\begin{bmatrix} A_6C_3 & A_5C_1 & A_6C_4 & A_5C_2 & C_7 \\ B_3 & 0 & B_4 & 0 & B_6 \\ 0 & B_1 & 0 & B_2 & B_5 \end{bmatrix} = r\begin{bmatrix} B_3 & 0 & B_4 & 0 & B_6 \\ 0 & B_1 & 0 & B_2 & B_5 \end{bmatrix}.$$

4. Conclusion

In this paper, we have derived a necessary and sufficient condition for the system (1.1) has a solution. Some known results can be regarded as the special cases of our results. In closing this paper, we propose an open problem: Could we give an expression of the general solution of the system (1.1) ?

References

[1] J. K. Baksalary and P. Kala, The matrix equation $AXB + CYD = E$, Linear Algebra Appl. 30 (1980) 141-147.

[2] P. Bhimasankaram, Common solutions to the linear matrix equations $AX = B, CX = D$, and $EXF = G$, *Sankhya Ser. A*, 38 (1976): 404-409.

[3] L. Huang and Q. Zeng, The solvability of matrix equation $AXB + CYD = E$ over a simple Artinian ring, Linear and Multilinear Algebra, 38 (1995) 225-232.

[4] T. W. Hungerford, *Algebra*, Spring-Verlag New York Inc., 1980.

[5] A. P. Liao, Z. Z Bai, Y. Lei, Best approximate solution of matrix equation $AXB + CYD = E$, SIAM J. Matrix Anal. Appl. 27 (3) (2006) 675-688.

[6] G. Marsaglia and G. P. H. Styan, Equalities and inequalities for ranks of matrices, *Linear and Multilinear Algebra* 2 (1974) 269-292.

[7] S. K. Mitra, The matrix equations $AX = C, XB = D$, *Linear Algebra Appl.* 59 (1984) 171-181.

[8] A. B. Özgüle, The matrix equation $AXB + CYD = E$ over a principal ideal domain, SIAM J. Matrix Anal. Appl. 12 (1991) 581-591.

[9] Y. Tian, The solvability of two linear matrix equations, Linear and Multilinear Algebra 48 (2000) 123-147.

[10] Q.W. Wang, A system of matrix equations and a linear matrix equation over arbitrary regular rings with identity, Linear Algebra Appl. 384 (2004) 43-54.

[11] Q.W. Wang, H. S. Zhang, On solutions to the quaternion matrix equation $AXB + CYD = E$, Electronic Journal of Linear Algebra 17 (2008) 343-358.

Arithmetic Properties of a Curious Arithmetic Function*

Tingting Wang
Mathematical College, Sichuan University, Chengdu 610064, P.R. China
email: vip529529@163.com

Qianrong Tan
School of Computer Science and Technology, Panzhihua University, Panzhihua 617000, P.R. China
emails: tqrmei6@126.com

Shaofang Hong
Mathematical College, Sichuan University, Chengdu 610064, P.R. China
email: sfhong@scu.edu.cn, s-f.hong@tom.com, hongsf02@yahoo.com

Abstract

In this paper, we introduce the arithmetic function $f_3 : \mathbb{N}^* \to \mathbb{Q}^+$ defined by $f_3(3^k l) = l^{1-k}$ for any integers $k \geq 0$ and $l \geq 1$ with $3 \nmid l$. We show several important arithmetic properties about this function and then use them to establish some curious results concerning the 3-adic valuation.

Keywords: arithmetic function; least common multiple; 3-adic valuation.

1 Introduction

Let $\mathbb{N}^* = \mathbb{N} \setminus \{0\}$ and $\mathbb{Q}^+ = \{x \in \mathbb{Q} | x > 0\}$ denote the set of positive integers and the set of positive rational numbers, respectively. Arithmetic functions are classical topic in number theory (see, for example, [1], [5] and [10]). Recently, some kind of periodic functions are introduced by some authors for the investigation of least common multiple of integer sequences. See, for instance, [2], [4], [6]-[9]. In this paper, we introduce a curious arithmetic function $f_3 : \mathbb{N}^* \to \mathbb{Q}^+$ defined by

$$f_3\left(3^k l\right) = l^{1-k}\,(\forall k, l \in \mathbb{N}, (3, l) = 1).$$

It is clear that $f_3(n)$ is not always an integer, because we have, for example, $f_3(1) = 1, f_3(3) = 1, f_3(18) = \frac{1}{2}, f_3(36) = \frac{1}{4}$ etc. But we will find surprisingly the product

*S. Hong is the corresponding author and was supported partially by National Science Foundation of China Grant #10971145 and by the Ph.D. Programs Foundation of Ministry of Education of China Grant #20100181110073.

$F_3(n) := \prod_{k=1}^{n} f_3(k)$ is always an integer for any positive integer n, and it is in fact a multiple of all prime number not equal to 3 and not exceeding n. Finally, we establish several curious results involving the 3-adic valuations about this function. We note that the arithmetic function $f_2(2^k l) = l^{1-k}$ was introduced by Farhi in [3], who also obtained similar results presented in this paper for the even prime case.

To study f_3, we need two auxiliary arithmetic functions $g : \mathbb{Q}^+ \to \mathbb{N}^*$ and $h_3 : \mathbb{N}^* \to \mathbb{Q}^+$, defined respectively by

$$g(x) := \begin{cases} x & \text{if } x \in \mathbb{N} \\ 1 & \text{else,} \end{cases} \tag{1.1}$$

and

$$h_3(r) := \frac{r}{\prod_{i=1}^{\infty} g\left(\frac{r}{3^i}\right)} \, (\forall r \in \mathbb{N}). \tag{1.2}$$

Notice that the product in the denominator of the right-hand side of (1.2) is actually finite because $g\left(\frac{r}{3^i}\right) = 1$ for any sufficiently large i, which means that h_3 is well defined.

As usual, we let v_3 denote the 3-adic valuation, that is, for any positive rational number x, there are two positive integers u and v such that $x = 3^{v_3(x)} \frac{u}{v}$, $3 \nmid u$, $3 \nmid v$. Denote the number $\frac{u}{v}$ by $R_3(x)$. Then we have $x = 3^{v_3(x)} R_3(x)$, and so $f_3(x) = (R_3(x))^{1-v_3(x)}, \forall x \in \mathbb{N}$. By $\lfloor \cdot \rfloor$ we mean the integer-part function. The following well-known facts will be used freely throughout this paper:

$$\left\lfloor \frac{\lfloor \frac{x}{a} \rfloor}{b} \right\rfloor = \left\lfloor \frac{x}{ab} \right\rfloor, \qquad a, b \in \mathbb{N}^*, x \in \mathbb{R}; \qquad v_p(n!) = \sum_{i=1}^{\infty} \left\lfloor \frac{n}{p^i} \right\rfloor, \qquad n \in \mathbb{N};$$

$$\binom{n}{k_1, ..., k_s} = \prod_{i=1}^{s} \binom{k_1 + ... + k_i}{k_i}, \; k_1 + ... + k_s = n, \; k_1, ..., k_s, n \in \mathbb{N}.$$

2 The main results and proofs

In this section, we give the main results of this paper. Let $H_3(n) := \prod_{r=1}^{n} h_3(r)$, and $L_n := \operatorname{lcm}(1, ..., n)$ for any positive integer n. Let's begin with the following result.

Theorem 2.1. *Let n be a positive integer. Then $F_3(n)$ is an integer.*

Proof. First, by the definition of g, we have

$$\prod_{i=1}^{\infty} g\left(\frac{r}{3^i}\right) = \prod_{i=1}^{\infty} g\left(3^{v_3(r)-i} R_3(r)\right) = \prod_{i=1}^{v_3(r)} g\left(3^{v_3(r)-i} R_3(r)\right)$$

$$= \prod_{i=1}^{v_3(r)} 3^{v_3(r)-i} R_3(r) = \prod_{i=0}^{v_3(r)-1} 3^i R_3(r) = 3^{\frac{v_3(r)(v_3(r)-1)}{2}} (R_3(r))^{v_3(r)}.$$

Then

$$h_3(r) = \frac{r}{\prod_{i=1}^{\infty} g\left(\frac{r}{3^i}\right)} = 3^{\frac{v_3(r)(3-v_3(r))}{2}} \left(R_3(r)\right)^{1-v_3(r)} = 3^{\frac{v_3(r)(3-v_3(r))}{2}} f_3(r).$$

It follows that $f_3(r) = 3^{\frac{v_3(r)(v_3(r)-3)}{2}} h_3(r)$. Thus, for all $n \in \mathbb{N}^*$, we have

$$F_3(n) = 3^{\sum_{r=1}^{n} \frac{v_3(r)(v_3(r)-3)}{2}} H_3(n). \tag{2.1}$$

Now, taking R_3 of each of the two sides of this identity, we get

$$F_3(n) = R_3\left(H_3(n)\right). \tag{2.2}$$

Therefore, in order to prove Theorem 2.1, it is sufficient to prove that $H_3(n)$ is an integer for any $n \in \mathbb{N}$. For this purpose, we note that $\prod_{k=1}^{r} g\left(\frac{k}{a}\right) = \left\lfloor\frac{r}{a}\right\rfloor! \,(\forall r, a \in \mathbb{N})$. Thus

$$H_3(n) = \prod_{r=1}^{n} \frac{r}{\prod_{i=1}^{\infty} g\left(\frac{r}{3^i}\right)} = \frac{n!}{\prod_{i=1}^{\infty}\prod_{r=1}^{n} g\left(\frac{r}{3^i}\right)} = \frac{n!}{\prod_{i=1}^{\infty}\left\lfloor\frac{n}{3^i}\right\rfloor!}. \tag{2.3}$$

(Remark that the product in the denominator of the right-hand side of (2.3) is actually finite because $\left\lfloor\frac{n}{3^i}\right\rfloor = 0$ for any sufficiently large i).

But $\sum_{i=1}^{\infty}\left\lfloor\frac{n}{3^i}\right\rfloor \leq \sum_{i=1}^{\infty}\frac{n}{3^i} = \frac{n}{2} < n$. Hence $\frac{n!}{\prod_{i=1}^{\infty}\left\lfloor\frac{n}{3^i}\right\rfloor!}$ is a multiple of the multinomial coefficient $\binom{\left\lfloor\frac{n}{3}\right\rfloor + \left\lfloor\frac{n}{3^2}\right\rfloor + \cdots}{\left\lfloor\frac{n}{3}\right\rfloor, \left\lfloor\frac{n}{3^2}\right\rfloor, \cdots}$ which is an integer. In other words, $\frac{n!}{\prod_{i=1}^{\infty}\left\lfloor\frac{n}{3^i}\right\rfloor!}$ is an integer. This completes the proof of Theorem 2.1. □

Consequently, we discuss the divisibility of $F_3(n)$ by $R_3(L_n)$.

Theorem 2.2. *For any positive integer n, $F_3(n)$ is a multiple of $R_3\left(L_n\right)$. In particular, $F_3(n)$ is a multiple of all prime numbers not equal to 3 and not exceeding n.*

Proof. According to (2.2) and (2.3), we just need to show that $\frac{n!}{\prod_{i=1}^{\infty}\left\lfloor\frac{n}{3^i}\right\rfloor!}$ is a multiple of L_n. In other words, it suffices to prove that for all prime numbers p, we have

$$v_p\left(\frac{n!}{\prod_{i=1}^{\infty}\left\lfloor\frac{n}{3^i}\right\rfloor!}\right) \geq \alpha_p, \tag{2.4}$$

where $\alpha_p = v_p\left(L_n\right)$, which is the greatest power of p not exceeding n. Since for any given prime p (see, for example, [5]),

$$v_p\left(\frac{n!}{\prod_{i=1}^{\infty}\left\lfloor\frac{n}{3^i}\right\rfloor!}\right) = \sum_{i=1}^{\infty}\left\lfloor\frac{n}{p^i}\right\rfloor - \sum_{i=1}^{\infty}\sum_{j=1}^{\infty}\left\lfloor\frac{\left\lfloor\frac{n}{p^i}\right\rfloor}{3^j}\right\rfloor$$

$$=\sum_{i=1}^{\infty}\left(\left\lfloor\frac{n}{p^i}\right\rfloor-\sum_{j=1}^{\infty}\left\lfloor\frac{\left\lfloor\frac{n}{p^i}\right\rfloor}{3^j}\right\rfloor\right)=\sum_{i=1}^{\alpha_p}\left(\left\lfloor\frac{n}{p^i}\right\rfloor-\sum_{j=1}^{\infty}\left\lfloor\frac{\left\lfloor\frac{n}{p^i}\right\rfloor}{3^j}\right\rfloor\right).$$

Since $p \le n$, it follows that for $1 \le i \le \alpha_p$, we have

$$\left\lfloor\frac{n}{p^i}\right\rfloor-\sum_{j=1}^{\infty}\left\lfloor\frac{\left\lfloor\frac{n}{p^i}\right\rfloor}{3^j}\right\rfloor \ge \left\lfloor\frac{n}{p^i}\right\rfloor-\sum_{j=1}^{\infty}\frac{\left\lfloor\frac{n}{p^i}\right\rfloor}{3^j}=\left\lfloor\frac{n}{p^i}\right\rfloor\left(1-\sum_{j=1}^{\infty}\frac{1}{3^j}\right)=\frac{1}{2}\left\lfloor\frac{n}{p^i}\right\rfloor>0.$$

But $\left\lfloor\frac{n}{p^i}\right\rfloor-\sum_{j=1}^{\infty}\left\lfloor\frac{\left\lfloor\frac{n}{p^i}\right\rfloor}{3^j}\right\rfloor$ is an integer. This implies that $\left\lfloor\frac{n}{p^i}\right\rfloor-\sum_{j=1}^{\infty}\left\lfloor\frac{\left\lfloor\frac{n}{p^i}\right\rfloor}{3^j}\right\rfloor \ge 1$. Hence

$$v_p\left(\frac{n!}{\prod_{i=1}^{\infty}\left\lfloor\frac{n}{p^i}\right\rfloor!}\right) \ge \sum_{i=1}^{\alpha_p} 1 = \alpha_p,$$

which confirms (2.4) and so the proof of Theorem 2.2 is complete. □

Let's now consider the properties of p-adic valuations.

Theorem 2.3. *For all positive integers n and all primes $p \neq 3$, we have:*

$$\sum_{r=1}^{n} v_3(r)v_p(r) \le \sum_{r=1}^{n} v_p(r) - \left\lfloor \log_p n \right\rfloor.$$

Proof. By Theorem 2.2, we have

$$\sum_{r=1}^{n} v_p\left(f_3(r)\right) = v_p\left(F_3(n)\right) \ge v_p\left(R_3(L_n)\right) = v_p\left(L_n\right) = \left\lfloor \log_p n \right\rfloor.$$

But for $r \ge 1$, we have

$$v_p\left(f_3(r)\right) = v_p((R_3(r))^{1-v_3(r)}) = \left(1-v_3(r)\right)v_p(R_3(r)) = \left(1-v_3(r)\right)v_p(r).$$

We then derive that

$$\sum_{r=1}^{n}\left(1-v_3(r)\right)v_p(r) \ge \left\lfloor \log_p n \right\rfloor,$$

and the desired result follows immediately. This completes the proof of Theorem 2.3. □

Finally we give the last main result as the conclusion of this paper.

Theorem 2.4. *Let n be a positive integer, and let $a_0 + a_1 \times 3 + a_2 \times 3^2 + \dots + a_s \times 3^s$ be the 3-adic representation of n. Then we have*

$$\sum_{r=1}^{n} v_3(r)\left(3-v_3(r)\right) = \frac{1}{2}\sum_{i=1}^{s}(3^i-1)a_i + \sum_{i=1}^{s} i a_i. \tag{2.5}$$

In particular, for any $m \in \mathbb{N}$, we have

$$\sum_{r=1}^{3^m} v_3(r)\,(3 - v_3(r)) = \frac{1}{2}(3^m - 1) + m. \tag{2.6}$$

Proof. First, it is easy to check that (2.5) holds when $n = 1$ and 2. In what follows, we let $n \geq 3$. Then taking the 3-adic valuation on the both sides of the identity (2.1) and using (2.3), we obtain

$$\sum_{r=1}^{n} \frac{v_3(r)\,(3 - v_3(r))}{2} = v_3\left(H_3(n)\right) = v_3\left(\frac{n!}{\prod_{i=1}^{\infty}\left\lfloor \frac{n}{3^i}\right\rfloor !}\right)$$
$$= \sum_{i=1}^{\infty}\left\lfloor \frac{n}{3^i}\right\rfloor - \sum_{i=1}^{\infty}\sum_{j=1}^{\infty}\left\lfloor \frac{n}{3^{i+j}}\right\rfloor$$

Notice that $\#\{(i,j) \in \mathbb{N}^{*2} | i + j = k\} = k - 1$. So we have

$$\sum_{i=1}^{\infty}\sum_{j=1}^{\infty}\left\lfloor \frac{n}{3^{i+j}}\right\rfloor = \sum_{k=2}^{\infty}\sum_{i+j=k}\left\lfloor \frac{n}{3^k}\right\rfloor = \sum_{k=2}^{\infty}(k-1)\left\lfloor \frac{n}{3^k}\right\rfloor.$$

Thus

$$\sum_{r=1}^{n} \frac{v_3(r)\,(3 - v_3(r))}{2} = \sum_{i=1}^{\infty}\left\lfloor \frac{n}{3^i}\right\rfloor - \sum_{i=1}^{\infty} i\left\lfloor \frac{n}{3^{i+1}}\right\rfloor.$$

Since

$$2\sum_{i=1}^{\infty}\left((i-1)\left\lfloor \frac{n}{3^i}\right\rfloor - i\left\lfloor \frac{n}{3^{i+1}}\right\rfloor\right) = 0,$$

it follows that

$$\sum_{i=1}^{\infty}\left\lfloor \frac{n}{3^i}\right\rfloor - \sum_{i=1}^{\infty} i\left\lfloor \frac{n}{3^{i+1}}\right\rfloor = \sum_{i=1}^{\infty}\left((2i-1)\left\lfloor \frac{n}{3^i}\right\rfloor - 3i\left\lfloor \frac{n}{3^{i+1}}\right\rfloor\right).$$

However, according to the 3–adic representation $n = a_0 + a_1 \times 3 + a_2 \times 3^2 + ... + a_s \times 3^s$, we infer that

$$(2i-1)\left\lfloor \frac{n}{3^i}\right\rfloor - 3i\left\lfloor \frac{n}{3^{i+1}}\right\rfloor = \begin{cases} ia_i + (i-1)\left\lfloor \frac{n}{3^i}\right\rfloor & \text{if } i = 1, ..., s, \\ 0 & \text{if } i > s. \end{cases}$$

Hence

$$\sum_{i=1}^{\infty}\left\lfloor \frac{n}{3^i}\right\rfloor - \sum_{i=1}^{\infty} i\left\lfloor \frac{n}{3^{i+1}}\right\rfloor = \sum_{i=1}^{s} ia_i + \sum_{i=1}^{\infty}(i-1)\left\lfloor \frac{n}{3^i}\right\rfloor = \sum_{i=1}^{s} ia_i + \sum_{i=1}^{\infty} i\left\lfloor \frac{n}{3^{i+1}}\right\rfloor.$$

It follows that

$$\sum_{i=1}^{\infty} i\left\lfloor \frac{n}{3^{i+1}}\right\rfloor = \frac{1}{2}\sum_{i=1}^{\infty}\left\lfloor \frac{n}{3^i}\right\rfloor - \frac{1}{2}\sum_{i=1}^{s} ia_i.$$

Therefore

$$\sum_{i=1}^{\infty}\left\lfloor\frac{n}{3^i}\right\rfloor-\sum_{i=1}^{\infty}i\left\lfloor\frac{n}{3^{i+1}}\right\rfloor=\frac{1}{2}\sum_{i=1}^{\infty}\left\lfloor\frac{n}{3^i}\right\rfloor+\frac{1}{2}\sum_{i=1}^{s}ia_i.$$

But it is well known that

$$\sum_{i=1}^{\infty}\left\lfloor\frac{n}{3^i}\right\rfloor=\frac{n-\sum_{i=0}^{s}a_i}{2}(=v_3(n!)).$$

Thereby,

$$\begin{aligned}\sum_{r=1}^{n}\frac{v_3(r)\,(3-v_3(r))}{2}&=\frac{n-\sum_{i=0}^{s}a_i}{4}+\frac{1}{2}\sum_{i=1}^{s}ia_i\\&=\frac{\sum_{i=1}^{s}\left(3^i-1\right)a_i}{4}+\frac{1}{2}\sum_{i=1}^{s}ia_i.\end{aligned}$$

So (2.5) follows immediately.

Now let $n=3^m$. Then there is a unique $i=m$ in the numerator of (2.5), and we have

$$\sum_{r=1}^{3^m}\frac{v_p(r)\,(3-v_3(r))}{2}=\frac{3^m-1}{4}+\frac{m}{2}.$$

Thus (2.6) follows immediately. The proof of Theorem 2.4 is complete. □

References

[1] T.M. Apostol, *Introduction to analytic number theory,* Springer-Verlag, New York, 1976.

[2] B. Farhi, Nontrivial lower bounds for the least common multiple of some finite sequences of integers, *J. Number Theory* 125 (2007), 393-411.

[3] B. Farhi, A study of a curious arithmetic function, to appear. arXiv:1004.2269.

[4] B. Farhi and D. Kane, New results on the least common multiple of consecutive integers, *Proc. Amer. Math. Soc.* 137 (2009), 1933-1939.

[5] G.H. Hardy and E.M. Wright, *The Throry of numbers*, Fifth Ed., Oxford Univ. Press, London, 1979.

[6] S. Hong and G. Qian, The least common multiple of consecutive terms in arithmetic progressions, *Proc. Edinburgh Math. Soc.*, doi:10.1017/S0013091509000431.

[7] S. Hong, G. Qian and Q. Tan, The least common multiple of sequences associated with Euler function, submitted.

[8] S. Hong, G. Qian and Q. Tan, The least common multiple of consecutive jumping terms in the sequences associated with integer-valued polynomials, submitted.

[9] S. Hong and Y. Yang, On the periodicity of an arithmetical function, *C.R. Acad. Sci. Paris, Ser. I* 346 (2008), 717-721.

[10] L.-K. Hua, *Introduction to number theory,* Springer-Verlag, Berlin Heidelberg, 1982.

Operator Ideals Arising from Generating Sequences

Ngai-Ching Wong

Dedicated to Professor Kar-Ping Shum in honor of his seventieth birthday

ABSTRACT. In this note, we will discuss how to relate an operator ideal on Banach spaces to the sequential structures it defines. Concrete examples of ideals of compact, weakly compact, completely continuous, Banach-Saks and weakly Banach-Saks operators will be demonstrated.

1. INTRODUCTION

Let $T : E \to F$ be a linear operator between Banach spaces. Let U_E, U_F be the closed unit balls of E, F, respectively. Note that closed unit balls serve simultaneously the basic model for open sets and bounded sets in Banach spaces. The usual way to describe T is to state either the bornological property, via TU_E, or the topological property, via $T^{-1}U_F$, of T. An other way classifying T is through the sequential structures T preserves. Here are some well-known examples.

Examples 1.1. (1) T is *bounded* (i.e., TU_E is a bounded subset of F)

$\Leftrightarrow$ T is *continuous* (i.e., $T^{-1}U_F$ is a 0-neighborhood of E in the norm topology)

$\Leftrightarrow$ T is *sequentially bounded* (i.e., T sends bounded sequences to bounded sequences);

(2) T is *of finite rank* (i.e., TU_E spans a finite dimensional subspace of F)

$\Leftrightarrow$ T is *weak-norm continuous* (i.e., $T^{-1}U_F$ is a 0-neighborhood of E in the weak topology)

$\Leftrightarrow$ T sends bounded sequences to sequences spanning finite dimensional subspaces of F;

Date: January 11, 2011; for the Proceedings of ICA 2010 (World Scientific).
2000 *Mathematics Subject Classification.* 47L20, 47B10 46A11, 46A17.
Key words and phrases. operator ideals, generating sequences, generating bornologies.
Partially supported by a Taiwan National Science Council grant 99-2115-M-110-007-MY3.

(3) T is *compact* (i.e., TU_E is totally bounded in F)

$\Leftrightarrow T$ is continuous in the topology of uniform convergence on norm compact subsets of E' (i.e., $T^{-1}U_F \supseteq K^\circ$, the polar of a norm compact subset K of the dual space E' of E)

$\Leftrightarrow T$ is *sequentially compact* (i.e., T sends bounded sequences to sequences with norm convergent subsequences); and

(4) T is *weakly compact* (i.e., TU_E is relatively weakly compact in F)

$\Leftrightarrow T$ is continuous in the topology of uniform convergence on weakly compact subsets of E' (i.e., $T^{-1}U_F \supseteq K^\circ$, the polar of a weakly compact subset K of E')

$\Leftrightarrow T$ is *sequentially weakly compact* (i.e., T sends bounded sequences to sequences with weakly convergent subsequences).

In [18, 17, 15], we investigate the duality of the topological and bornological characters of operators demonstrated in the above examples, and applied them in the study of operator ideal theory in the sense of Pietsch [6]. In [16], we use these concepts in classifying locally convex spaces. They are also used by other authors in operator algebra theory in [13, 2].

After giving a brief account of the equivalence among the notions of operator ideals, generating topologies, and generating bornologies developed in [18, 17, 15] in Section 2, we shall explore into the other option of using sequential structures in Section 3. The theory is initialed by Stephani [11, 12]. We provide a variance here. In particular, we will show that the notions of operator ideals, generating bornologies and generating sequences are equivalent in the context of Banach spaces. However, we note that the sequential methodology does not seem to be appropriate in the context of locally convex spaces, as successfully as in [18, 16]. Some examples of popular operator ideals are treated in our new ways as demonstrations in Section 4.

The author would like to take this opportunity to thank Professor Kar-Ping Shum. He has learned a lot from Professor Shum since he was a student in the Chinese University of Hong Kong in 1980's. With about 300 publications, Professor Shum has served as a good model for the author and other fellows since then. The current work is based partially on the thesis of the author finished when he was studying with Professor Shum.

2. The triangle of operators, topologies and bornologies

Let X be a vector space over $\mathbb{K} = \mathbb{R}$ or $\mathbb{C}$. Following Hogbe-Nlend [4], by a *vector bornology* on X we mean a family $\mathcal{B}$ of subsets of X satisfies the following conditions:

(VB_1) $X = \cup\mathcal{B}$;
(VB_2) if $B \in \mathcal{B}$ and $A \subseteq B$ then $A \in \mathcal{B}$;
(VB_3) $B_1 + B_2 \in \mathcal{B}$ whenever $B_1, B_2 \in \mathcal{B}$;
(VB_4) $\lambda B \in \mathcal{B}$ whenever $\lambda \in \mathbb{K}$ and $B \in \mathcal{B}$;
(VB_5) the circle hull $\mathrm{ch}\, B$ of any B in $\mathcal{B}$ belongs to $\mathcal{B}$.

Elements in $\mathcal{B}$ are called $\mathcal{B}$-bounded sets in X. A vector bornology $\mathcal{B}$ on X is called a *convex bornology* if $\Gamma B \in \mathcal{B}$ for all B in $\mathcal{B}$, where ΓB is the absolutely convex hull of B. The pair $(X, \mathcal{B})$ is called a *convex bornological space* which is denoted by $X^{\mathcal{B}}$.

Let $(X, \mathcal{T})$ be a locally convex space. For any subset B of X, the *σ-disked hull* of B is defined to be

$$\Gamma_\sigma B = \{\textstyle\sum_{n=1}^{\infty} \lambda_n b_n : \sum_{n=1}^{\infty} |\lambda_n| \leq 1, \lambda_n \in \mathbb{K}, b_n \in B, n = 1, 2, \ldots\}.$$

A set B in X is said to be *σ-disked* if $B = \Gamma_\sigma B$. An absolutely convex, bounded subset B of X is said to be *infracomplete* (or a *Banach disk*) if the normed space $(X(B), r_B)$ is complete, where

$$X(B) = \bigcup_{n \geq 1} nB \quad \text{and} \quad r_B(x) := \inf\{\lambda > 0 : x \in \lambda B\}$$

is the gauge of B defined on $X(B)$.

Lemma 2.1 ([18]). *Let $(X, \mathcal{P})$ be a locally convex space and B an absolutely convex bounded subset of X.*

(a) If B is σ-disked then B is infracomplete.
(b) If B is infracomplete and closed then B is σ-disked.

Let X, Y be locally convex spaces. Denote by $\sigma(X, X')$ the weak topology of X with respect to its dual space X', while $\mathcal{P}_{\mathrm{ori}}(X)$ is the original topology of X. We employ the notion $\mathcal{M}_{\mathrm{fin}}(Y)$ for the *finite dimensional bornology* of Y which has a basis consisting of all convex hulls of finite sets. On the other hand, $\mathcal{M}_{\text{von}}(Y)$ is used for the *von Neumann bornology* of Y which consists of all $\mathcal{P}_{\text{ori}}(Y)$-bounded subsets of Y. Ordering of topologies and bornologies are induced by set-theoretical inclusion, as usual. Moreover, we write briefly $X_{\mathcal{P}}$ for a vector space X equipped with a locally convex topology $\mathcal{P}$ and $Y^{\mathcal{M}}$ for a vector space Y equipped with a convex vector bornology $\mathcal{M}$.

Let X, Y be locally convex spaces. We denote by $\mathfrak{L}(X,Y)$ and $L^{\times}(X,Y)$ the collection of all linear operators from X into Y which are continuous and (locally) bounded (i.e., sending bounded sets to bounded sets), respectively.

Definition 2.2 (see [16]). Let $\mathcal{C}$ be a subcategory of locally convex spaces.

(1) (**"Operators"**) A family $\mathfrak{A} = \{\mathfrak{A}(X,Y) : X, Y \in \mathcal{C}\}$ of algebras of operators associated to each pair of spaces X and Y in $\mathcal{C}$ is called an *operator ideal* if

OI$_1$: $\mathfrak{A}(X,Y)$ is a nonzero vector subspace of $\mathfrak{L}(X,Y)$ for all X, Y in $\mathcal{C}$; and

OI$_2$: $RTS \in \mathfrak{A}(X_0,Y_0)$ whenever $R \in \mathfrak{L}(Y,Y_0)$, $T \in \mathfrak{A}(X,Y)$ and $S \in \mathfrak{L}(X_0,X)$ for any X_0, X, Y and Y_0 in $\mathcal{C}$.

(2) (**"Topologies"**) A family $\mathcal{P} = \{\mathcal{P}(X) : X \in \mathcal{C}\}$ of locally convex topologies associated to each space X in $\mathcal{C}$ is called a *generating topology* if

GT$_1$: $\sigma(X,X') \subseteq \mathcal{P}(X) \subseteq \mathcal{P}_{\text{ori}}(X)$ for all X in $\mathcal{C}$; and

GT$_2$: $\mathfrak{L}(X,Y) \subseteq \mathfrak{L}(X_{\mathcal{P}}, Y_{\mathcal{P}})$ for all X and Y in $\mathcal{C}$.

(3) (**"Bornologies"**) A family $\mathcal{M} = \{\mathcal{M}(Y) : Y \in \mathcal{C}\}$ of convex vector bornologies associated to each space Y in $\mathcal{C}$ is called a *generating bornology* if

GB$_1$: $\mathcal{M}_{\text{fin}}(Y) \subseteq \mathcal{M}(Y) \subseteq \mathcal{M}_{\text{von}}(Y)$ for all Y in $\mathcal{C}$; and

GB$_2$: $\mathfrak{L}(X,Y) \subseteq L^{\times}(X^{\mathcal{M}}, Y^{\mathcal{M}})$ for all X and Y in $\mathcal{C}$.

Definition 2.3. Let $\mathfrak{A}$ be an operator ideal, $\mathcal{P}$ a generating topology and $\mathcal{M}$ a generating bornology on $\mathcal{C}$.

(1) (**"Operators"** $\to$ **"Topologies"**) For each X_0 in $\mathcal{C}$, the $\mathfrak{A}$*–topology* of X_0, denoted by $\mathcal{T}(\mathfrak{A})(X_0)$, is the projective topology of X_0 with respect to the family

$$\{T \in \mathfrak{A}(X_0,Y) : Y \in \mathcal{C}\}.$$

In other words, a seminorm p of X_0 is $\mathcal{T}(\mathfrak{A})(X_0)$–continuous if and only if there is a T in $\mathfrak{A}(X_0, Y)$ for some Y in $\mathcal{C}$ and a continuous seminorm q of Y such that

$$p(x) \le q(Tx), \quad \forall x \in X_0.$$

In this case, we call p an $\mathfrak{A}$*–seminorm* of X_0.

(2) (**"Operators"** $\to$ **"Bornologies"**) For each Y_0 in $\mathcal{C}$, the $\mathfrak{A}$*–bornology* of Y_0, denoted by $\mathcal{B}(\mathfrak{A})(Y_0)$, is the inductive bornology of Y_0 with respect to the family

$$\{T \in \mathfrak{A}(X, Y_0) : X \in \mathcal{C}\}.$$

In other words, a subset B of Y_0 is $\mathcal{B}(\mathfrak{A})(Y_0)$–bounded if and only if there is a T in $\mathfrak{A}(X, Y_0)$ for some X in $\mathcal{C}$ and a topologically bounded subset A of X such that

$$B \subseteq TA.$$

In this case, we call B an $\mathfrak{A}$*–bounded subset* of Y_0.

(3) (**"Topologies"** $\to$ **"Operators"**) For X, Y in $\mathcal{C}$, let

$$\mathcal{O}(\mathcal{P})(X, Y) = \mathfrak{L}(X_{\mathcal{P}}, Y)$$

be the vector space of all continuous operators from X into Y which is still continuous with respect to the $\mathcal{P}(X)$–topology.

(4) (**"Bornologies"** $\to$ **"Operators"**) For X, Y in $\mathcal{C}$, let

$$\mathcal{O}(\mathcal{M})(X, Y) = \mathfrak{L}(X, Y) \cap L^{\times}(X, Y^{\mathcal{M}})$$

be the vector space of all continuous operators from X into Y which send bounded sets to $\mathcal{M}(Y)$–bounded sets.

(5) (**"Topologies"** $\leftrightarrow$ **"Bornologies"**) For X, Y in $\mathcal{C}$, the $\mathcal{P}^{\circ}(Y)$*–bornology* of Y (resp. $\mathcal{M}^{\circ}(X)$*–topology* of X) is defined to be the bornology (resp. topology) polar to $\mathcal{P}(X)$ (resp. $\mathcal{M}(Y)$). More precisely,

- a bounded subset A of Y is $\mathcal{P}^{\circ}(Y)$–bounded if and only if its polar A° is a $\mathcal{P}(Y'_{\beta})$–neighborhood of zero; and
- a neighborhood V of zero of X is a $\mathcal{M}^{\circ}(X)$–neighborhood of zero if and only if V° is $\mathcal{M}(X'_{\beta})$–bounded.

Here are two examples: the ideals $\mathfrak{K}$ of compact operators and $\mathfrak{P}$ of absolutely summing operators (see e.g. [6]), the generating systems $\mathcal{P}_{pc}$ of precompact topologies (see e.g. [7]) and $\mathcal{P}_{pn}$ of prenuclear topologies (see e.g. [8, p. 90]), and the generating systems $\mathcal{M}_{pc}$ of precompact bornologies and $\mathcal{M}_{pn}$ of prenuclear bornologies (see e.g. [5]), respectively.

Definition 2.4. An operator ideal $\mathfrak{A}$ on Banach spaces is said to be

(1) *injective* if $S \in \mathfrak{A}(E, F_0)$ infers $T \in \mathfrak{A}(E, F)$, whenever $T \in \mathfrak{L}(E, F)$ and $\|Tx\| \leq \|Sx\|, \forall x \in E$;
(2) *surjective* if $S \in \mathfrak{A}(E_0, F)$ infers $T \in \mathfrak{A}(E, F)$, whenever $T \in \mathfrak{L}(E, F)$ and $TU_E \subseteq SU_{E_0}$.

The *injective hull* $\mathfrak{A}^{\text{inj}}$ and the *surjective hull* $\mathfrak{A}^{\text{sur}}$, of $\mathfrak{A}$ is the intersection of all injective and surjective operator ideals containing $\mathfrak{A}$, respectively. For example, $\mathfrak{K} = \mathfrak{K}^{\text{inj}} = \mathfrak{K}^{\text{sur}}$ and $\mathfrak{P} = \mathfrak{P}^{\text{inj}} \subsetneq \mathfrak{P}^{\text{sur}}$ (see, e.g., [6]).

In the following, we see that the notions of operator ideals, generating topologies, and generating bornologies are equivalent (see also [13] for the operator algebra version).

Theorem 2.5 ([9, 10, 18, 16]). *Let $\mathfrak{A}$ be an operator ideal, $\mathcal{P}$ a generating topology and $\mathcal{M}$ a generating bornology on Banach spaces. We have*

(1) $\mathcal{T}(\mathfrak{A}) = \{\mathcal{T}(\mathfrak{A})(X) : X \in \mathcal{C}\}$ *is a generating topology.*
(2) $\mathcal{B}(\mathfrak{A}) = \{\mathcal{B}(\mathfrak{A})(Y) : Y \in \mathcal{C}\}$ *is a generating bornology.*
(3) $\mathcal{O}(\mathcal{P}) = \{\mathcal{O}(\mathcal{P})(X, Y) : X, Y \in \mathcal{C}\}$ *is an operator ideal.*
(4) $\mathcal{O}(\mathcal{M}) = \{\mathcal{O}(\mathcal{M})(X, Y) : X, Y \in \mathcal{C}\}$ *is an operator ideal.*
(5) $\mathcal{P}^\circ = \{\mathcal{P}^\circ(Y) : Y \in \mathcal{C}\}$ *is a generating bornology.*
(6) $\mathcal{M}^\circ = \{\mathcal{M}^\circ(Y) : Y \in \mathcal{C}\}$ *is a generating topology.*
(7) $\mathcal{O}(\mathcal{T}(\mathfrak{A})) = \mathfrak{A}^{\text{inj}}$.
(8) $\mathcal{O}(\mathcal{B}(\mathfrak{A})) = \mathfrak{A}^{\text{sur}}$.
(9) $\mathcal{T}(\mathcal{O}(\mathcal{P})) = \mathcal{P}$.
(10) $\mathcal{B}(\mathcal{O}(\mathcal{M})) = \mathcal{M}$.

For generating topologies $\mathcal{P}$ and $\mathcal{P}_1$, and generating bornologies $\mathcal{M}$ and $\mathcal{M}_1$ on Banach spaces, we can also associate to them the operator ideals with components

$$\begin{aligned}
\mathcal{O}(\mathcal{P}/\mathcal{P}_1)(X,Y) &= \mathfrak{L}(X_{\mathcal{P}}, Y_{\mathcal{P}_1}),\\
\mathcal{O}(\mathcal{M}/\mathcal{M}_1)(X,Y) &= L^{\times}(X^{\mathcal{M}_1}, Y^{\mathcal{M}}) \cap \mathfrak{L}(X,Y)\\
\mathcal{O}(\mathcal{P}/\mathcal{M})(X,Y) &= \mathfrak{L}(X_{\mathcal{P}}, Y^{\mathcal{M}}).
\end{aligned}$$

These give rise to

$$\begin{aligned}
\mathcal{O}(\mathcal{P}/\mathcal{P}_1) &= \mathcal{O}(\mathcal{P}_1)^{-1} \circ \mathcal{O}(\mathcal{P}) \quad [11],\\
\mathcal{O}(\mathcal{M}/\mathcal{M}_1) &= \mathcal{O}(\mathcal{M}) \circ \mathcal{O}(\mathcal{M}_1)^{-1} \quad [12],\\
\mathcal{O}(\mathcal{P}/\mathcal{M}) &= \mathcal{O}(\mathcal{M}) \circ \mathcal{O}(\mathcal{P}) = \mathcal{O}(\mathcal{M}) \circ \mathcal{O}(\mathcal{P}) \quad [18].
\end{aligned}$$

Here, the product $\mathfrak{A} \circ \mathfrak{B}$ consists of operators of the form ST with $S \in \mathfrak{A}$ and $T \in \mathfrak{B}$, while the quotient $\mathfrak{A} \circ \mathfrak{B}^{-1}$ consists of those S such that $ST \in \mathfrak{A}$ whenever $T \in \mathfrak{B}$. Readers are referred to Pietsch's classic [6] for information regarding quotients and products of operator ideals.

3. Generating sequences

The following is based on a version in [11].

Definition 3.1. A family $\Phi(E)$ of bounded sequences in a Banach space E is called a *vector sequential structure* if

- $\{ax_n + by_n\} \in \Phi(E)$ whenever $\{x_n\}, \{y_n\} \in \Phi(E)$ and a, b are scalars;
- every subsequence of a sequence in $\Phi(E)$ belongs to $\Phi(E)$.

Clearly, the biggest vector sequential structure on a Banach space E is the family $\Phi_b(E)$ of all bounded sequences in E. On the other hand, we let $\Phi_{\text{fin}}(E)$ to be the vector sequential structure on E of all bounded sequences $\{x_n\}$ with finite dimensional ranges, i.e., there is a finite dimensional subspace of F containing all x_n's.

Given two Banach spaces E, F with vector sequential structures Φ, Ψ, we let $\mathfrak{L}(E^{\Phi}, F^{\Psi})$ be the vector space of all continuous linear operators sending a bounded sequence in Φ to a bounded sequence in F with a subsequence in Ψ.

Definition 3.2. A family $\Phi := \{\Phi(F) : F \text{ is a Banach space}\}$ of (bounded) vector sequential structures on each Banach space is called a *generating sequential structure* if

> **GS$_1$:** $\Phi_{\text{fin}}(F) \subseteq \Phi(F) \subseteq \Phi_b(F)$ for every Banach space F.
>
> **GS$_2$:** $\mathfrak{L}(E, F) \subseteq \mathfrak{L}(E^\Phi, F^\Phi)$ for all Banach spaces E, F.

Definition 3.3. Let $\mathfrak{A}$ be an operator ideal on Banach spaces. Call a bounded sequence $\{y_n\}$ in a Banach space F an $\mathfrak{A}$-*sequence* if there is a bounded sequence $\{x_n\}$ in a Banach space E and a continuous linear operator T in $\mathfrak{A}(E, F)$ such that $Tx_n = y_n$, $n = 1, 2, \ldots$. Denote by $\mathcal{S}(\mathfrak{A})(F)$ the family of all $\mathfrak{A}$-sequences in F.

Lemma 3.4. *Let $\mathfrak{A}$ be an operator ideal on Banach spaces. Then*

$$\mathcal{S}(\mathfrak{A}) := \{\mathcal{S}(\mathfrak{A})(F) : F \textit{ is a Banach space}\}$$

is a generating sequential structure.

Proof. It is easy to see that all $\mathcal{S}(\mathfrak{A})(F)$ is a bounded vector sequential structure. Since the ideal $\mathfrak{F}$ of continuous linear operators of finite rank is the smallest operator ideal, while the ideal $\mathfrak{L}$ of all continuous linear operators is the biggest, we see that (GS$_1$) is satisfied. On the other hand, (GS$_2$) follows from (OI$_2$) directly. □

We now work on the converse of Lemma 3.4. Let Φ be a generating sequential structure. Denote by $\mathcal{M}^\Phi_{\text{base}}(E)$ the family of all bounded sets M in a Banach space E such that every sequence $\{x_n\}$ in M has a subsequence $\{x_{n_k}\}$ in $\Phi(E)$.

Lemma 3.5. *The family $\mathcal{M}^\Phi(E)$ of all (bounded) sets with circle hull in $\mathcal{M}^\Phi_{base}(E)$ in a Banach space E forms a generating bornology $\mathcal{M}^\Phi$.*

Proof. Let $M, N \in \mathcal{M}^\Phi(E)$, let λ be a scalar, and let $W \subseteq M$. It is easy to see that the circle hull $\text{ch}\, M$, λM and W are all in $\mathcal{M}^\Phi(E)$. For the sum $M + N$, we first notice that both $\text{ch}\, M$ and $\text{ch}\, N$ belong to $\mathcal{M}^\Phi_{\text{base}}(E)$. Now

$$M + N \subseteq \text{ch}\, M + \text{ch}\, N \in \mathcal{M}^\Phi_{\text{base}}(E)$$

and

$$\text{ch}\,(M + N) \subseteq \text{ch}\, M + \text{ch}\, N$$

give $M + N \in \mathcal{M}^{\Phi}(E)$. All these together say that $\mathcal{M}^{\Phi}(E)$ is a vector bornology on the Banach space E.

The condition (GB_1) follows directly from (GS_1). For (GB_2), let $T \in \mathfrak{L}(E,F)$ and $M \in \mathcal{M}^{\Phi}(E)$. Then for every sequence $\{Tx_n\}$ in TM, we will have a subsequence of $\{x_n\}$ in $\Phi(E)$. By (GS_2), we have a further subsequence $\{Tx_{n_k}\}$ belonging to $\mathcal{M}^{\Phi}(F)$. $\square$

Definition 3.6. A generating sequential structure Φ is said to be *normal* if for every $\{x_n\}$ in $\Phi(E)$ and every bounded scalar sequence (λ_n) in ℓ^{∞} of norm not greater than 1, we have $\{\lambda_n x_n\} \in \Phi(E)$.

It is clear that $\mathcal{S}(\mathfrak{A})$ is normal for any operator ideal $\mathfrak{A}$.

Lemma 3.7. *For every normal generating sequential structure Φ, we have*

$$\mathcal{M}^{\Phi}_{\text{base}} = \mathcal{M}^{\Phi}.$$

Proof. It suffices to show that every M in $\mathcal{M}^{\Phi}_{\text{base}}(E)$ has its circle hull $\operatorname{ch} M$ in $\mathcal{M}^{\Phi}_{\text{base}}(E)$ for each Banach space E, i.e., $\mathcal{M}^{\Phi}_{\text{base}}$ is itself circled. To this end, let $\{x_n\}$ be from $\operatorname{ch} M$. Then there is a (λ_n) from ℓ^{∞} of norm not greater than 1 such that

$$x_n = \lambda_n y_n, \quad n = 1, 2, \ldots$$

for some $\{y_n\}$ in M. Now, there is a subsequence $\{y_{n_k}\}$ in $\Phi(E)$ ensuring that $\{x_{n_k}\} = \{\lambda_{n_k} y_{n_k}\} \in \Phi(E)$ for Φ being normal. $\square$

Definition 3.8. Let Φ be a generating sequential structure on Banach spaces. Let E, F be Banach spaces. Denote by $\mathcal{O}(\Phi)(E,F)$ the family of all continuous linear operators T in $\mathfrak{L}(E,F)$ sending every bounded sequence $\{x_n\}$ in E to a bounded sequence in F with a subsequence $\{Tx_n\}$ in $\Phi(F)$. In other words, $\mathcal{O}(\Phi)(E,F) = \mathfrak{L}(E^{\Phi_b}, F^{\Phi})$.

Theorem 3.9. *Let Φ be a generating sequential structure on Banach spaces. Then $\mathcal{O}(\Phi)$ is a surjective operator ideal. More precisely,*

$$\mathcal{O}(\Phi) = \mathcal{O}(\mathcal{M}^{\Phi}).$$

Proof. The ideal property of $\mathcal{O}(\Phi)$ is trivial. For the surjectivity of $\mathcal{O}(\Phi)$, we let $S \in \mathcal{O}(\Phi)(E_0, F)$ and $T \in \mathfrak{L}(E,F)$ such that $TU_E \subseteq SU_{E_0}$. Here, U_E, U_{E_0} are the closed unit balls of E, E_0, respectively. We need to show that $T \in \mathcal{O}(\Phi)(E,F)$. In fact, for any bounded sequence $\{x_n\}$ in E, we have

$$Tx_n = Sy_n, \quad n = 1, 2, \ldots,$$

with some bounded sequence $\{y_n\}$ in F. Now, the fact $\{Sy_n\}$ has a subsequence $\{Sy_{n_k}\} \in \Phi(F)$ implies that $T \in \mathcal{O}(\Phi)(E,F)$, as asserted.

Finally, we check the representation. The inclusion $\mathcal{O}(\Phi) \subseteq \mathcal{O}(\mathcal{M}^\Phi)$ is an immediate consequence of the definition of $\mathcal{M}^\Phi$. Conversely, let $T \in \mathcal{O}(\mathcal{M}^\Phi)(E,F)$, i.e., $TU_E \in \mathcal{M}^\Phi(F)$. Let $\{x_n\}$ be a bounded sequence in E. We can assume $\|x_n\| \leq 1$ for all n. Then $\{Tx_n\} \subseteq TU_E$ and hence possesses a subsequence $\{Tx_{n_k}\} \in \Phi(F)$. In other words, $T \in \mathcal{O}(\Phi)(E,F)$. □

Corollary 3.10. *If a generating sequential structure Φ is normal then*

$$\mathcal{S}(\mathcal{O}(\Phi)) = \Phi.$$

Conversely, if $\mathfrak{A}$ is an operator ideal on Banach spaces then

$$\mathcal{O}(\mathcal{S}(\mathfrak{A})) = \mathfrak{A}^{\text{sur}}.$$

Theorem 3.11. *Let Φ, Ψ be two generating sequential structures. Let*

$$\mathcal{O}(\Phi/\Psi)(E,F) := \mathfrak{L}(E^\Psi, F^\Phi),$$

i.e., $T \in \mathcal{O}(\Phi/\Psi)(E,F)$ if and only if the continuous linear operator T sends each bounded sequence $\{x_n\}$ in $\Psi(E)$ to a sequence in F with a subsequence $\{Tx_{n_k}\}$ in $\Phi(F)$. Then $\mathcal{O}(\Phi/\Psi)(E,F)$ is an operator ideal on Banach spaces. Moreover, if Ψ is normal, then

$$\mathcal{O}(\Phi/\Psi) = \mathcal{O}(\mathcal{M}^\Phi/\mathcal{M}^\Psi) = \mathcal{O}(\mathcal{M}^\Phi) \circ \mathcal{O}(\mathcal{M}^\Psi)^{-1}.$$

Proof. Conditions (OI_1) and (OI_2) follow from Conditions (GS_1) and (GS_2), respectively.

Assume Ψ is normal. Let $T \in \mathcal{O}(\Phi/\Psi)(E,F)$ and $M \in \mathcal{M}^\Psi(E)$ be circled. Suppose, on contrary, $TM \notin \mathcal{M}^\Phi(F)$. By definition, there is a sequence $\{Tx_n\}$ in TM having no subsequence $\{Tx_{n_k}\}$ in $\Phi(F)$. But since $\{x_n\} \subseteq M$, it has a subsequence $\{x_{n_k}\}$ in $\Psi(E)$. And thus $\{Tx_{n_k}\}$ has a further subsequence in $\Phi(F)$, a contradiction.

Conversely, let $T \in \mathcal{O}(\mathcal{M}^\Phi/\mathcal{M}^\Psi)(E,F)$. It is easy to see that the range of any sequence $\{x_n\}$ in $\Psi(E)$ is a bounded set in $\mathcal{M}^\psi(E)$, by noting that Ψ is normal. The range of the sequence $\{Tx_n\}$ is bounded in $\mathcal{M}^\Phi(F)$. Therefore, there is a subsequence $\{Tx_{n_k}\}$ of $\{Tx_n\}$ belongs to $\Phi(F)$. This ensures $T \in \mathcal{O}(\Phi/\Psi)(E,F)$. □

Definition 3.12. Given two generating sequential structures Φ and Ψ. We say that $\Psi(E)$ is *coordinated to* $\Phi(E)$ if the following conditions holds. Suppose $\{x_n\} \in \Psi(E)$. The bounded sequence $\{x_n\} \notin \Phi(E)$ if and only if $\{x_n\}$ has a subsequence $\{x_{n_k}\}$ which has no further subsequence belonging to $\Phi(E)$. We write, in this case, $\Phi(E) \ll \Psi(E)$. If $\Phi(E) \ll \Psi(E)$ for all Banach spaces E then we write $\Phi \ll \Psi$.

Theorem 3.13. *Assume Φ, Ψ be two generating sequential structures such that $\Phi(F) \ll \Psi(F)$ for some Banach space F. Then $\mathcal{O}(\Phi/\Psi)(E,F)$ consists of exactly those continuous linear operators sending bounded sequences in $\Psi(E)$ to bounded sequences in $\Phi(F)$ for any Banach space E.*

Proof. Assume $T \in \mathcal{O}(\Phi/\Psi)(E,F)$ and $\{x_n\} \in \Psi(E)$. Now, suppose $\{Tx_n\} \notin \Phi(F)$. Since $\{Tx_n\} \in \Psi(F)$ by (GS_2), we can verify the condition in Definition 3.12. Therefore, we might have a subsequence $\{Tx_{n_k}\}$ having no further subsequence in $\Phi(F)$. However, by the facts that $\{x_{n_k}\} \in \Psi(E)$ and $T \in \mathcal{O}(\Phi/\Psi)(E,F)$, we arrive at the asserted contradiction. □

Corollary 3.14. *Let Φ, Ψ be two generating sequential structures such that $\Phi \ll \Psi$ and Ψ is normal. Then, the operator ideal $\mathcal{O}(\mathcal{M}^\Phi, \mathcal{M}^\Psi)$ consists of exactly those operators sending sequences in Ψ to sequences in Φ.*

4. Examples

We begin with some useful generating bornologies.

Examples 4.1. A set W in a Banach space F is called

(1) *nuclear* (see, e.g., [11]) if

$$W \subseteq \{y \in F : y = \sum_{n=1}^{\infty} \lambda_n y_n, |\lambda_n| \leq 1\}$$

for an absolutely summable sequence $\{y_n\}$ in F;

(2) *unconditionally summable* (see, e.g., [11]) if

$$W \subseteq \{y \in F : y = \sum_{n=1}^{\infty} \lambda_n y_n \text{ in norm}, |\lambda_n| \leq 1\}$$

for an unconditionally summable series $\sum_{n=1}^{\infty} y_n$ in F;

(3) *weakly unconditionally summable* (see, e.g., [11]) if

$$W \subseteq \{y \in F : y = \sum_{n=1}^{\infty} \lambda_n y_n \text{ weakly}, |\lambda_n| \leq 1\}$$

for a weakly unconditionally summable series $\sum_{n=1}^{\infty} y_n$ in F;

(4) *limited* ([1]) if

$$\lim_{n\to\infty} \sup_{a \in W} |y'_n(a)| = 0$$

for any $\sigma(F', F)$-null sequence $\{y'_n\}$ in F'.

The above defines generating bornologies $\mathcal{M}_\nu$, $\mathcal{M}_{uc}$, $\mathcal{M}_{wuc}$, and $\mathcal{M}_{\lim}$, respectively.

Another way to obtain generating bornologies is through generating sequential structures.

Examples 4.2. A bounded sequence $\{x_n\}$ in a Banach space F is called

(1) δ_0-*fundamental* ([11]) if

$$t_n = x_{n+1} - x_n, \quad n = 1, 2, \ldots,$$

forms a weakly unconditionally summable series;

(2) *limited* ([1]) if

$$\langle x_n, x'_n \rangle \to 0 \text{ as } n \to \infty$$

for any $\sigma(F', F)$-null sequence $\{x'_n\}$ in F';

(3) *Banach-Saks* if

$$\lim_{n\to\infty} \frac{x_1 + x_2 + \cdots + x_n}{n} = x_0$$

for some x_0 in F in norm.

Examples 4.3. The following are all generating sequential structures.

(1) Φ_{δ_0} of all δ_0-fundamental sequences.
(2) $\Phi_{\lim}$ of all limited sequences.
(3) Φ_{BS} of all Banach-Saks sequences.

(4) Φ_c of all convergent sequences.

(5) Φ_{wc} of all weakly convergence sequences.

(6) Φ_{wCa} of all weakly Cauchy sequences.

Examples 4.4 ([11]). The following generating bornologies can be induced by the corresponding generating sequential structures as in Lemma 3.5.

(1) The δ_0-compact bornology $\mathcal{M}_{\delta_0}$ is defined by Φ_{δ_0}.

(2) The compact bornology $\mathcal{M}_c$ is defined by Φ_c.

(3) The weakly compact bornology $\mathcal{M}_{wc}$ is defined by Φ_{wc}.

(4) The Rosenthal compact bornology $\mathcal{M}_{wCa}$ is defined by Φ_{wCa}.

(5) The Banach-Saks bornology $\mathcal{M}_{\mathrm{BS}}$ is defined by Φ_{BS}.

Proposition 4.5. *The generating bornology induced by $\Phi_{\lim}$ is $\mathcal{M}_{\lim}$.*

Proof. First, we notice that $\Phi_{\lim}$ is normal. Hence, $\mathcal{M}^{\Phi_{\lim}}_{\mathrm{base}} = \mathcal{M}^{\Phi_{\lim}}$ by Lemma 3.7. Let $W \in \mathcal{M}_{\lim}(F)$. Then, by definition,

$$\lim_{n\to\infty} \sup_{a\in W} |\langle a, y'_n\rangle| = 0$$

for any $\sigma(F', F)$-null sequence $\{y'_n\}$ in F' and any sequence $\{y_n\}$ in W. Hence, $W \in \mathcal{M}^{\Phi_{\lim}}(F)$.

Conversely, if W is a bounded set in F such that every sequence $\{y_n\}$ in W has a limited subsequence $\{y_{n_k}\}$, we need to check that $W \in \mathcal{M}_{\lim}$. Assume, on contrary, that there were some $\sigma(F', F)$-null sequence $\{y'_n\}$ in F' and some sequence $\{a_n\}$ in W such that

$$\lim_{n\to\infty} |\langle a_{n_k}, y'_{n_k}\rangle| = 0.$$

This is a contradiction and thus W is a member of $\mathcal{M}_{\lim}(F)$. □

Proposition 4.6. *The generating bornologies $\mathcal{M}_c$, $\mathcal{M}_{wc}$, $\mathcal{M}_{wCa}$, $\mathcal{M}_\nu$, $\mathcal{M}_{uc}$, $\mathcal{M}_{wuc}$ and $\mathcal{M}_{\lim}$ are all σ-disked.*

Proof. For the case of $\mathcal{M}_{\lim}$, see [1]. For all others, see [11]. □

We are now ready to give a number of examples.

Examples 4.7. Using Definition 2.3 and Theorem 3.9, we can obtain the following operator ideals on Banach spaces through the associated generating bornologies.

(1) The ideal $\mathfrak{N}^{\text{sur}}$ *of co-nuclear operators* for the nuclear bornology $\mathcal{M}_\nu$.
(2) The ideal $\mathfrak{L}_{lim}$ *of limited operators* for the limited bornology $\mathcal{M}_{\text{lim}}$.
(3) The ideal $\mathfrak{BS}$ *of Banach-Saks operators* for the Banach-Saks bornology $\mathcal{M}_{\text{BS}}$.
(4) The ideal $\mathfrak{K}$ *of compact operators* for the compact bornology $\mathcal{M}_c$.
(5) The ideal $\mathfrak{W}$ *of weakly compact operators* for the weakly compact bornology $\mathcal{M}_{wc}$.
(6) The ideal $\mathfrak{R}$ *of Rosenthal compact operators* for the Rosenthal compact bornology $\mathcal{M}_{wCa}$.
(7) The ideal $\mathfrak{D}_0$ *of δ_0-compact operators* for the δ_0-compact bornology $\mathcal{M}_{\delta_0}$.

For a proof of (a), see [6, p. 112]. For (b), see [3]. For (c), see Theorem 3.9. For others, see [11].

Concerning Theorem 3.13, we have

Examples 4.8. (1) $\Phi_c \ll \Phi_{wc}$.
(2) $\Phi_c \ll \Phi_{wCa}$.
(3) $\Phi_{wc} \ll \Phi_{wCa}$.
(4) $\Phi_c \ll \Phi_{\delta_0}$.
(5) $\Phi_c \ll \Phi_{\text{lim}}$.

The first four can be found in [11]. For the last one, we observe the fact that each limited sequence is weakly Cauchy (see [1]) and (c).

Examples 4.9. The following operator ideals have desirable representations.

(1) The ideal $\mathfrak{U}$ of unconditionally summing operators

$$\begin{aligned}\mathfrak{U} &= \mathcal{O}(\mathcal{M}_{uc}/\mathcal{M}_{wuc}) = \mathcal{O}(\mathcal{M}_{uc}) \circ \mathcal{O}(\mathcal{M}_{wuc})^{-1} \\ &= \mathcal{O}(\mathcal{M}_c/\mathcal{M}_{\delta_0}) = \mathcal{O}(\mathcal{M}_c) \circ \mathcal{O}(\mathcal{M}_{\delta_0})^{-1} = \mathfrak{K} \circ \mathfrak{D}_0^{-1}.\end{aligned}$$

(2) The ideal $\mathfrak{V}$ of completely continuous operators

$$\begin{aligned}\mathfrak{V} &= \mathcal{O}(\Phi_c/\Phi_{wc}) = \mathcal{O}(\mathcal{M}_c/\mathcal{M}_{wc}) = \mathcal{O}(\mathcal{M}_c) \circ \mathcal{O}(\mathcal{M}_{wc})^{-1} = \mathfrak{K} \circ \mathfrak{W}^{-1} \\ &= \mathcal{O}(\Phi_c/\Phi_{wCa}) = \mathcal{O}(\mathcal{M}_c/\mathcal{M}_{wCa}) = \mathcal{O}(\mathcal{M}_c) \circ \mathcal{O}(\mathcal{M}_{wCa})^{-1} = \mathfrak{K} \circ \mathfrak{R}^{-1}.\end{aligned}$$

(3) The ideal $w\mathfrak{SC}$ of weakly sequentially complete operators

$$s\mathfrak{SC} = \mathcal{O}(\Phi_{wc}/\Phi_{wCa}) = \mathcal{O}(\mathcal{M}_{wc}/\mathcal{M}_{wCa}) = \mathcal{O}(\mathcal{M}_{wc}) \circ \mathcal{O}(\mathcal{M}_{wCa})^{-1} = \mathfrak{W} \circ \mathfrak{R}^{-1}.$$

(4) The ideal $\mathfrak{GP}$ of Gelfand-Phillips operators

$$\mathfrak{GP} = \mathcal{O}(\Phi_c/\Phi_{\lim}) = \mathcal{O}(\mathcal{M}_c/\mathcal{M}_{\lim}) = \mathcal{O}(\mathcal{M}_c) \circ \mathcal{O}(\mathcal{M}_{\lim})^{-1} = \mathfrak{K} \circ \mathfrak{L}_{\lim}^{-1}.$$

(5) The ideal $w\mathfrak{BS}$ of weakly Banach-Saks operators

$$w\mathfrak{BS} = \mathcal{O}(\Phi_{\mathrm{BS}}/\Phi_{wc}) = \mathcal{O}(\mathcal{M}_{\mathrm{BS}}/\mathcal{M}_{wc}) = \mathcal{O}(\mathcal{M}_{\mathrm{BS}}) \circ \mathcal{O}(\mathcal{M}_{wc})^{-1} = \mathfrak{BS} \circ \mathfrak{W}^{-1}.$$

See [11] for a proof of (a), (b) and (c). See [3] for (d). For (e), we simply recall that a T in $\mathfrak{L}(E, F)$ is called *weakly Banach-Saks* if T sends each weakly convergent sequence to a sequence possessing a Banach-Saks subsequence, by definition.

Proposition 4.10 ([11]). *Let $\mathfrak{A}$ be a surjective operator ideal such that $\mathfrak{A}$ is idempotent, i.e., $\mathfrak{A}^2 = \mathfrak{A}$. Then for any operator ideal $\mathfrak{B} \supseteq \mathfrak{A}$, there exists an operator ideal $\mathfrak{C}$ such that*

$$\mathfrak{A} = \mathfrak{C} \circ \mathfrak{B}^{\mathrm{sur}},$$

and $\mathfrak{C}$ can be chosen to be $\mathfrak{A} \circ (\mathfrak{B}^{\mathrm{sur}})^{-1}$.

Proof. We present a proof here for completeness. Denote

$$\mathcal{M}_{\mathfrak{A}} := \mathcal{M}(\mathfrak{A}), \quad \mathcal{M}_{\mathfrak{B}} := \mathcal{M}(\mathfrak{B})$$

and set

$$\mathfrak{C} := \mathcal{O}(\mathcal{M}_{\mathfrak{A}}/\mathcal{M}_{\mathfrak{B}}).$$

Now the facts $\mathfrak{A} \subseteq \mathfrak{B}$ and

$$\mathfrak{A} \subseteq \mathfrak{C} = \mathcal{O}(\mathcal{M}_{\mathfrak{A}}) \circ \mathcal{O}(\mathcal{M}_{\mathfrak{B}})^{-1} = \mathfrak{A} \circ (\mathfrak{B}^{\mathrm{sur}})^{-1}$$

implies that

$$\mathfrak{A} = \mathfrak{A}^2 \subseteq \mathfrak{C} \circ \mathfrak{B} \subseteq \mathfrak{C} \circ \mathfrak{B}^{\mathrm{sur}}.$$

On the other hand, if $T \in \mathfrak{C} \circ \mathfrak{B}^{\mathrm{sur}}(E, F)$ then $T = RS$ for some $R \in \mathfrak{C}(G, F) = L^{\times}(G^{\mathcal{M}_{\mathfrak{B}}}, F^{\mathcal{M}_{\mathfrak{A}}}) \cap \mathfrak{L}(G, F)$ and $S \in \mathfrak{B}^{\mathrm{sur}}(E, G) = L^{\times}(E, G^{\mathcal{M}_{\mathfrak{B}}}) \cap \mathfrak{L}(E, G)$ with some Banach space G. Hence,

$$T \in L^{\times}(E, F^{\mathcal{M}_{\mathfrak{A}}}) \cap \mathfrak{L}(E, F) = \mathfrak{A}(E, F). \qquad \square$$

Examples 4.11 ([11]). (1) Since the surjective ideal $\mathfrak{K}$ of compact operators is idempotent and contained in the surjective ideal $\mathfrak{W}$, $\mathfrak{R}$, and $\mathfrak{D}_0$ of weakly compact operators, Rosenthal compact operators, and δ_0-compact operators, respectively, we have

(a) $\mathfrak{K} = \mathfrak{V} \circ \mathfrak{W}$ since $\mathfrak{V} = \mathcal{O}(\mathcal{M}_c/\mathcal{M}_{wc}) = \mathfrak{K} \circ \mathfrak{W}^{-1}$;

(b) $\mathfrak{K} = \mathfrak{V} \circ \mathfrak{R}$ since $\mathfrak{V} = \mathcal{O}(\mathcal{M}_c/\mathcal{M}_{wcA}) = \mathfrak{K} \circ \mathfrak{R}^{-1}$;

(c) $\mathfrak{K} = \mathfrak{U} \circ \mathfrak{D}_0$ since $\mathfrak{U} = \mathcal{O}(\mathcal{M}_c/\mathcal{M}_{\delta_0}) = \mathfrak{K} \circ {\mathfrak{D}_0}^{-1}$.

(2) Since the ideal $\mathfrak{W}$ of weakly compact operators is idempotent and contained in the surjective ideal $\mathfrak{R}$ of Rosenthal compact operators, we have

$$\mathfrak{W} = w\mathfrak{S}\mathfrak{C} \circ \mathfrak{R},$$

where $w\mathfrak{S}\mathfrak{C} = \mathcal{O}(\mathcal{M}_{wc}/\mathcal{M}_{wCa}) = \mathfrak{W} \circ \mathfrak{R}^{-1}$ is the ideal of weakly sequentially complete operators.

References

[1] J. Bourgain and J. Diestel, "Limited operators and strictly cosingularity", *Math. Nachr.*, **119** (1984), 55–58.

[2] J. Conradie and G. West, "Topological and bornological characterisations of ideals in von Neumann algebras: II", *Integral Equations Operator Theory*, **23** (1995), no. 1, 49–60.

[3] L. Drewnowski, "On Banach spaces with the Gelfand-Phillips property", *Math. Z.*, **93** (1986), 405–411.

[4] H. Hogbe-Nlend, *Bornologies and functional analysis*, Math. Studies **26**, North–Holland, Amsterdam, 1977.

[5] H. Hogbe-Nlend, *Nuclear and co-nuclear spaces*, Math. Studies **52**, North–Holland, Amsterdam, 1981.

[6] A. Pietsch, *Operator Ideals*, North–Holland, Amsterdam, 1980.

[7] D. Randtke, "Characterizations of precompact maps, Schwartz spaces and nuclear spaces", *Trans. Amer. Math. Soc.*, **165** (1972), 87–101.

[8] H. H. Schaefer, *Topological Vector Spaces*, Springer–Verlag, Berlin–Heidelberg–New York, 1971.

[9] I. Stephani, "Injektive operatorenideale über der gesamtheit aller Banachräume und ihre topologische erzeugung", *Studia Math.*, **38** (1970), 105–124.

[10] I. Stephani, "Surjektive operatorenideale über der gesamtheit aller Banachräume und ihre Erzeugung", *Beiträge Analysis*, . **5** (1973), 75–89.

[11] I. Stephani, "Generating system of sets and quotients of surjective operator ideals", *Math. Nachr.*, **99** (1980), 13–27.

[12] I. Stephani, "Generating topologies and quotients of injective operator ideals", in "Banach Space Theory and Its Application (proceedings, Bucharest 1981)", *Lecture Notes in Math.*, **991**, Springer–Verlag, Berlin–Heidelberg–New York, 1983, 239–255.

[13] G. West, "Topological and bornological characterisations of ideals in von Neumann algebras: I", *Integral Equations Operator Theory*, **22** (1995), no. 3, 352–359.

[14] N.-C. Wong, *Operator ideals on locally convex spaces*, Master Thesis, The Chinese University of Hong Kong, 1987.

[15] N.-C. Wong, "Topologies and bornologies determined by operator ideals, II", *Studia Math.* **111(2)** (1994), 153–162.

[16] N.-C. Wong, "The triangle of operators, topologies, bornologies", in *Third International Congress of Chinese Mathematicians. Part 1, 2*, 395–421, AMS/IP Stud. Adv. Math., **42**, Part 1, 2, Amer. Math. Soc., Providence, RI, 2008. arXiv:math/0506183v1 [math.FA]
[17] N.-C. Wong and Y.-C. Wong, "The bornologically surjective hull of an operator ideal on locally convex spaces", *Math. Narch.*, **160** (1993), 265–275.
[18] Y.-C. Wong and N.-C. Wong, "Topologies and bornologies determined by operator ideals", *Math. Ann.*, **282** (1988), 587–614.

Department of Applied Mathematics, National Sun Yat-sen University, and National Center for Theoretical Sciences, Kaohsiung, 80424, Taiwan, R.O.C.

E-mail address: `wong@math.nsysu.edu.tw`

OD-Characterization of Alternating and Symmetric Groups of Degree 106 and 112*

[1,2]Yanxiong Yan and [1]Guiyun Chen†
1. School of Mathematics and Statistics,
Southwest University,
Beibei, Chongqing 400715,
People's Republic of China;
2. Department of Mathematics,
Chongqing Education College,
Nanping, Chongqing 400067,
People's Republic of China.

Abstract

The degree pattern of a finite group G associated to its prime graph has been introduced in [1] and it is proved that the following simple groups are uniquely determined by their degree patterns and orders: all sporadic simple groups, alternating groups A_p ($p \geq 5$ is a twin prime) and some simple groups of Lie type. In the present paper, we continue this investigation. In particular, we show that the alternating groups A_{106} and A_{112} are OD-characterizable. We will also show that the symmetric groups S_{106} and S_{112} are 3-fold OD-characterizable.

Keywords: *prime graph, degree pattern, degree of a vertex, OD-characterization of a finite group.*
2000 Mathematics Subject Classification: 20D05

1 Introduction

Throughout this paper, all the groups under consideration are finite, and simple groups are non-abelian. For any group G, we use $\pi_e(G)$ to denote the set of orders of its elements and $\pi(G)$ the set of prime divisors of $|G|$. One of the well-known simple graphs associated to G is the prime graph (or Gruenberg-Kegel graph) denoted as $\Gamma(G)$. This graph is constructed as follows: The vertex set of this graph is $\pi(G)$, and two distinct vertices p, q are joined by an edge if and only if $pq \in \pi_e(G)$. In this case, we write $p \sim q$. The number of connected components of $\Gamma(G)$ is denoted as $t(G)$ and the connected components of $\Gamma(G)$ as $\pi_i = \pi_i(G)$ $(i = 1, 2, \ldots, t(G))$. When $|G|$ is even, we suppose that $2 \in \pi_1(G)$. We also denote by $\pi(n)$ the set of all primes dividing n, where n is a natural number.

In this article, we also use the following notations. Given a finite group G, denote by $Soc(G)$ the socle of G which is the subgroup generated by the set of all minimal normal subgroups of G. $Syl_p(G)$ denotes the set of all Sylow p-subgroups of G, where $p \in \pi(G)$. And P_r denotes a Sylow r-subgroup of G for $r \in \pi(G)$. All further unexplained notations are standard and can be found for instance in [2].

*Project supported by the NSFC(Grant No.10771172), Excellent Teachers Supporting Project and Keystone Project of Ministry of Education of China.
†Corresponding author.
E-mail address: 2003yyx@163.com (Y. X. Yan), gychen_99@yahoo.com (G. Y. Chen).

Definition 1.1. *(see[1])Let G be a finite group and $|G| = p_1^{\alpha_1} p_2^{\alpha_2} \cdots p_k^{\alpha_k}$, where p_i are primes and α_i are integers. For $p \in \pi(G)$, let $deg(p) := |\{q \in \pi(G) | p \sim q\}|$, called the* degree *of p. We also define $D(G) := (deg(p_1), deg(p_2), \ldots, deg(p_k))$, where $p_1 < p_2 < \cdots < p_k$. We call $D(G)$ the* degree pattern *of G.*

Definition 1.2. *(see[1]) A group M is called k-fold* OD-characterizable *if there exist exactly k non-isomorphic groups G such that $|G| = |M|$ and $D(G) = D(M)$. Moreover, a 1-fold OD-characterizable group is simply called an OD-*characterizable group.

Definition 1.3. *A group G is said to be an almost simple related to S if and only if $S \trianglelefteq G \leq Aut(S)$ for some non-abelian simple group S.*

In a series of articles (Ref. to [1, 3, 4, 5, 6, 7, 8]), it was shown that many finite almost simple groups are OD-characterizable, which are included in the following proposition.

Proposition. A finite group G is OD-characterizable if G is one of the following groups:

(1) All sporadic simple groups and their automorphism groups except $Aut(J_2)$ and $Aut(M^cL)$.

(2) The alternating groups A_p, A_{p+1}, A_{p+2} and the symmetric groups S_p and S_{p+1}, where p is a prime.

(3) All finite simple K_4-groups except A_{10}.

(4) The simple groups of Lie type $L_2(q)$, $L_3(q)$, $U_3(q)$. ${}^2B_2(q)$ and ${}^2G_2(q)$ for certain prime power q.

(5) All finite simple $C_{2,2}$-groups.

(6) The alternating groups A_{p+3} $(7 \neq p \in \pi(100!))$.

(7) The almost simple groups of $Aut(F_4(2))$, $Aut(O_{10}^+(2))$ and $Aut(O_{10}^-(2))$.

(8) All finite almost simple K_3-groups except $Aut(U_4(2))$.

In this article we will show that the alternating groups A_{106} and A_{112} are OD-characterizable. Indeed, we will prove the following theorem.

Theorem A. The alternating simple groups A_{106} and A_{112} are OD-characterizable.

As we mentioned already, the alternating groups A_p, A_{p+1} and A_{p+2}, where p is a prime number, are OD-characterizable (see Proposition (1)). In fact, A_{10} is the first alternating group which does not appear in the list. It is interesting to point out that the prime graph of A_p, A_{p+1} and A_{p+2} for all primes $p \geq 5$ are disconnected while the alternating group A_{10} has a connected prime graph. On the other hand, proposition (6) says that all alternating groups $A_{p+3}(7 \neq p \in \pi(100!))$ are OD-characterizable. In fact, Theorem A and Proposition (1) and (6) imply the following corollary.

Corollary 1. All alternating groups A_n, where $10 \neq n \leq 112$, are OD-characterizable.

In this paper, we will also prove the following theorem

Theorem B. The symmetric groups S_{106} and S_{112} are 3-fold OD-characterizable.

2 Preliminaries

In this section, we consider some results which will be applied for our further investigations.

Lemma 2.1. *(see [9]) The group S_n (or A_n) has an element of order $m = p_1^{\alpha_1} \cdot p_2^{\alpha_2} \cdots p_s^{\alpha_s}$, where $p_1, p_2, \cdots, p_s$ are distinct primes and $\alpha_1, \alpha_2, \cdots \alpha_s$ are natural numbers, if and only if $p_1^{\alpha_1}+p_2^{\alpha_2}+\cdots+p_s^{\alpha_s} \leq n$ (or $p_1^{\alpha_1}+p_2^{\alpha_2}+\cdots+p_s^{\alpha_s} \leq n$ for m odd, and $p_1^{\alpha_1}+p_2^{\alpha_2}+\cdots+p_s^{\alpha_s} \leq n-2$ for m even).*

Lemma 2.2. *Let A_n (or S_n) be an alternating group (or a symmetric group) of degree n. Then the following assertions hold.*

(1) Let $p, q \in \pi(A_n)$ be odd primes. Then $p \cdot q \in \pi_e(A_n)$ if and only if $p + q \leq n$.

(2) Let $p \in \pi(A_n)$ be an odd prime. Then $2 \cdot p \in \pi_e(A_n)$ if and only if $p + 4 \leq n$.

(3)Let $p, q \in \pi(S_n)$. Then $p \cdot q \in \pi_e(S_n)$ if and only if $p + q \leq n$.

Proof. This is a direct consequence of Lemma 2.1.

Lemma 2.3. *Let A_{p+3} be an alternating group, where p is a prime and $p+2$ is a composite number. Suppose $|\pi(A_{p+3})| = d$. Then the following assertions hold.*

(1) *$deg(2) = d - 2$. In particular, $2 \cdot r \in \pi_e(A_{p+3})$ for each $r \in \pi(A_{p+3})\backslash\{p\}$.*

(2) *$deg(3) = d - 1$, i.e., $3 \cdot r \in \pi_e(A_{p+3})$ for each $r \in \pi(A_{p+3})$.*

(3) *$deg(p) = 1$. In other words, $p \cdot r \in \pi_e(A_{p+3})$, where $r \in \pi(A_{p+3})$, if and only if $r = 3$.*

(4) *$Exp(|A_{p+3}|, 2) < p$.*

(5) *$Exp(|A_{p+3}|, r) < \frac{p-1}{2}$, where $5 \leq r \in \pi(A_{p+3})$. particularly, if $r > [\frac{p+3}{2}]$, then $Exp(|A_{p+3}|, r) = 1$.*

Remark 1. The exponent of the largest power of a prime p in the factorization of a positive integer $m(> 1)$ is denoted as $Exp(m, p)$. For example, if [ ] denotes the Gauss's integer function x, then $Exp(n!, p) = [\frac{n}{p}] + [\frac{n}{p^2}] + [\frac{n}{p^3}] + \cdots$.

Proof. By Lemma 2.2, we have $2 \cdot p \in \pi_e(A_{p+3})$. Obviously, $r + 4 \leq p + 3$ for each $r \in \pi(A_{p+3})\backslash\{p\}$, it follows that $2 \cdot r \in \pi_e(A_{p+3})$ and so $deg(2) = d - 2$. By the same reason, we have $deg(3) = d - 1$. For $r \in \pi(A_{p+3})\backslash\{2, p\}$, by Lemma 2.2, it is easy to see that $p \cdot r \in \pi_e(A_{p+3})$ if and only if $p + r \leq p + 3$. Hence $r = 3$ and $deg(p) = 1$.

Till now we have proved that (1), (2) and (3) hold. Next, we prove the remaining (4) and (5) hold.

Since $Exp(|A_{p+3}|, 2) = \frac{1}{2}([\frac{p+3}{2}] + [\frac{p+3}{2^2}] + [\frac{p+3}{2^3}] + \cdots) \leq \frac{1}{2}(\frac{p+3}{2} + \frac{p+3}{2^2} + \frac{p+3}{2^3} + \cdots)$ $= \frac{p+3}{2}(\frac{1}{2} + \frac{1}{2^2} + \frac{1}{2^3} + \cdots) = \frac{p+3}{2} < \frac{p+p}{2} = p$, hence, $Exp(|A_{p+3}|, 2) < p$.

By the same reason as above we can prove $Exp(|A_{p+3}|, r) < \frac{p-1}{2}$, where $5 \leq r \in \pi(A_{p+3})$. Clearly, if $r > [\frac{p+3}{2}]$, then we have $Exp(|A_{p+3}|, r) = 1$. This completes the proof of Lemma 2.3.

Lemma 2.4. *(see [10]) Let S be a finite non-abelian simple group with order having prime divisors at most 109. Then S is isomorphic to one of the following simple groups listed in Table 1-3 in [10]. In particular, if $|\pi(Out(G))| \neq 1$, then $\pi(Out(G)) \subseteq \{2, 3, 5\}$.*

Remark 2. Given a prime p, we use $\mathcal{F}_p$ to denote the set of nonabelian finite simple groups G such that $p \in \pi(G) \subseteq \{2, 3, 5, \cdots, p\}$. By [10], the members of $\mathcal{F}_{109}$ are ordered according to the size of their prime spectrum (listed in Table 1-3). The number of groups in each set $\mathcal{F}_{109}$ is given after the symbol "♯". For each group, we also know that the prime decomposition of the order. However, since the members of $\mathcal{F}_{109}$ are too many and the order decompositions occupy too much space, the detailed Table 1-3 are omitted. In the latter case, i.e, if $|\pi(Out(G))| \neq 1$, using [2], it is easy to see that the statement of the lemma is correct by checking each choice of p.

Lemma 2.5. *(see [11]) Let $S = P_1 \times P_2 \times \cdots \times P_r$, where P_i are isomorphic non-Abelian simple groups. Then $Aut(S) = (Aut(P_1) \times Aut(P_2) \times \cdots \times Aut(P_r)) \rtimes \mathbb{S}_r$*

3 OD-Characterizable of the Alternating Groups

We again recall that all the alternating groups A_p, A_{p+1} and A_{p+2}, where p is a prime number, are OD-characterizable (see [3]). On the other hand, it has been shown that all the A_{p+3} for any $p \in \pi(100!)\backslash\{7\}$ are OD-characterizable in [4]. So far there is no alternating group not OD-characterizable is found. Hence, the authors in [5] put forward the following conjecture:

Conjecture. All alternating groups A_{p+3} with $p \neq 7$ are OD-characterizable.

In this section, we are going to give an affirmative answer to this question for another two alternating groups and prove alternating groups A_{106} and A_{112} are OD-characterizable.

Theorem 1. The alternating group A_{106} is OD-characterizable.

Proof. Let G be a finite group satisfying

(1) $|G| = |A_{106}| = 2^{101} \cdot 3^{50} \cdot 5^{25} \cdot 7^{17} \cdot 11^{9} \cdot 13^{8} \cdot 17^{6} \cdot 19^{5} \cdot 23^{4} \cdot 29^{3} \cdot 31^{3} \cdot 37^{2} \cdot 41^{2} \cdot 43^{2} \cdot 47^{2} \cdot 53^{2} \cdot 59 \cdot 61 \cdot 67 \cdot 71 \cdot 73 \cdot 79 \cdot 83 \cdot 89 \cdot 97 \cdot 101 \cdot 103$

and

(2) $D(G) = D(A_{106}) = (25, 26, 26, 24, 23, 23, 23, 22, 22, 20, 20, 18, 17, 17, 16, , 15, 15, 14, 12, 11, 11, 9, 9, 7, 4, 3, 1)$. We have to show that $G \cong A_{106}$. It is evident that $\{2, p\} \cup \{pq | p+q \leq 106\} \cup \{2p | p+4 \leq 106\} \subseteq \pi_e(G)$ and $(\{2p | p+4 > 106\} \cup \{pq | p+q > 106\}) \cap \pi_e(G) = \varnothing$, where $2 \neq p, q \in \pi(G)$. Clearly, the prime graph of G is connected , because the vertex 3 is joint to all other vertices. Moreover, it is easy to see that $\Gamma(G) = \Gamma(A_{106})$. in the following, we write the proof in a number of separate lemmas.

Lemma 3.1. Let K be the maximal normal solvable subgroup of G. Then K is a $\{2, 3\}$-group. In particular, G is nonsolvable.

Proof. We first prove that K is a $103'$-group. Indeed, if not, then K would contain an element x of order 103. Set $C = C_G(x)$ and $N = N_G(\langle x \rangle)$. By the structure of $D(G)$, it follows that C is a $\{3, 103\}$-group. Now using n-c Theorem the factor group N/C is isomorphic to a subgroup of $Aut(\langle x \rangle) \cong \mathbb{Z}_2 \times \mathbb{Z}_3 \times \mathbb{Z}_{17}$. Hence, N is a $\{2, 3, 17, 103\}$-group. By Frattini argument we have that $G = KN$. This implies that $r \in \pi(K)$ for each $r \in \pi(G)\backslash\{2, 3, 17, 103\}$. Since K is solvable, it possesses a Hall $\{r, 103\}$-subgroup. In particular, K possesses a Hall $\{101, 103\}$-subgroup L , which is a cyclic subgroup of order $101 \cdot 103$. Hence $101 \sim 103$, which is a contradiction to Lemma 2.3 (3).

Next, we prove that K is a p'-group for each $p \in \pi(G)\backslash\{2, 3\}$. Let $p \in \pi(K)$, $P \in Syl_p(K)$ and $N = N_G(P)$. Again, by Frattini argument $G = KN_G(P)$ and hence 103 divides the order of N. Let T be a subgroup of N of order 103. Since T normalizes P, by n-c Theorem, we have that $N_G(P)/C_G(P) \lesssim Aut(P)$. It is easy to see that $103 \nmid | Aut(P) |$, Thus $T \leq C_G(Q)$. Hence, $103 \cdot p \in \pi_e(G)$ and so $deg(103) \geq 2$, which leads to a contradiction as above. Therefore K is a $\{2, 3\}$-group. Since $K \neq G$, it follows at once that G is nonsolvable. This completes the proof.

Lemma 3.2. The quotient group G/K is an almost simple group. In fact, $S \lesssim G/K \lesssim Aut(S)$, where S is a simple group.

Proof. Let $\overline{G} := G/K$ and $S := Soc(\overline{G})$. Then $S = B_1 \times B_2 \times \cdots \times B_m$, where B_i $(1 \leq i \leq m)$ are nonabelian simple groups and $S \lesssim \overline{G} \lesssim Aut(S)$. In what follows, we will show that $m = 1$.

Suppose that $m \geq 2$. We assert that 103 does not divide the order of S. Otherwise $2 \cdot 103 \in \pi_e(G)$, which is impossible for $\Gamma(G) = \Gamma(A_{106})$. Hence, for every i we have $B_i \in \mathcal{F}_p$, where p is a prime and $p < 103$. On the other hand, by Lemma 2.4, we observe that $103 \in \pi(\overline{G}) \subseteq \pi(Aut(S))$. Thus, we may assume that 103 divides the order of $Out(S)$. But

$$Out(S) = Out(S_1) \times Out(S_2) \times \cdots \times Out(S_r),$$

where the groups S_j are direct products of isomorphic $B_i^{'}s$ such that

$$S = S_1 \times S_2 \times \cdots \times S_r.$$

Therefore for some j, 103 divides the order of an outer automorphism group of a direct product S_j of t isomorphic simple groups B_i for some $1 \leq i \leq m$. Since $B_i \in \mathcal{F}_p$, it follows that $|Out(B_i)|$ is not divided by 103 by Lemma 2.4. Now, by Lemma 2.5, we obtain $|Aut(S_j)| = |Aut(B_i)|^t \cdot t!$. Therefore $t \geq 103$ and so 2^{206} must divide the order of G. However, $Exp(|A_{106}|, 2) = Exp(|G|, 2) < 103$ by Lemma 2.3 (4), which is a contradiction. Thus $m = 1$ and $S = B_1$. This completes the proof of Lemma 3.2.

Lemma 3.3. G is isomorphic to the alternating group A_{106}.

Proof. By Lemma 2.4 and 3.1, it is evident that $|S| = 2^a \cdot 3^b \cdot 5^{25} \cdot 7^{17} \cdot 11^9 \cdot 13^8 \cdot 17^6 \cdot 19^5 \cdot 23^4 \cdot 29^3 \cdot 31^3 \cdot 37^2 \cdot 41^2 \cdot 43^2 \cdot 47^2 \cdot 53^2 \cdot 59 \cdot 61 \cdot 67 \cdot 71 \cdot 73 \cdot 79 \cdot 83 \cdot 89 \cdot 97 \cdot 101 \cdot 103$, where $2 \leq a \leq 101, 1 \leq b \leq 50$. Using Table 1-3 in [10], we see that S can only be isomorphic to one of the following simple groups: A_{103}, A_{104}, A_{105} and A_{106}.

If $S \cong A_{103}$, then $A_{103} \lesssim G/K \lesssim Aut(A_{103}) \cong S_{103}$, and so it follows that $G/K \cong S_{103}$ or A_{103}. In the case $G/K \cong S_{103}$, it is easy to see that $3 \cdot 103 \in \pi_e(\overline{G}) \backslash \pi_e(S_{103})$, which is a contradiction. In the latter case, $G/K \cong A_{103}$ by comparing orders, we deduce that $7||K|$, a contradiction to Lemma 3.1.

Similarly, we see that S cannot be isomorphic to the alternating groups A_{104} and A_{105}. Therefore, $S \cong A_{106}$. According to Lemma 3.2, we have that $A_{106} \lesssim G/K \lesssim Aut(A_{106}) \cong S_{106}$. Again, by comparing orders we see that G/K can only be isomorphic to A_{106}. Hence, we have $K = 1$ and $G \cong A_{106}$, This completes the proof of lemma, which concludes the theorem.

Theorem 2. The alternating group A_{112} is OD-characterizable.

Proof. Let G be a finite group satisfying

(1) $|G| = |A_{112}| = 2^{108} \cdot 3^{54} \cdot 5^{26} \cdot 7^{18} \cdot 11^{10} \cdot 13^8 \cdot 17^6 \cdot 19^5 \cdot 23^4 \cdot 29^3 \cdot 31^3 \cdot 37^2 \cdot 41^2 \cdot 43^2 \cdot 47^2 \cdot 53^2 \cdot 59 \cdot 61 \cdot 67 \cdot 71 \cdot 73 \cdot 79 \cdot 83 \cdot 89 \cdot 97 \cdot 101 \cdot 103 \cdot 107 \cdot 109$

and

(2) $D(G) = D(A_{106}) = (27, 28, 27, 26, 25, 24, 23, 23, 23, 22, 21, 20, 19, 18, 17, , 16, 16, 15, 14, 13, 12, 11, 10, 9, 6, 5, 4, 3, 1)$.

Clearly, the prime graph of G is connected since $deg(3) = 28$ and $|\pi(G)| = 29$. Furthermore, it is easy to see that $\Gamma(G) = \Gamma(A_{112})$.

Let K denote the maximal normal solvable subgroup of G. By the same reason as the proof of Theorem 1, K is a $\{2,3\}$-group and $A_{112} \lesssim G/K \lesssim Aut(A_{112}) \cong S_{112}$. Hence $G/K \cong A_{112}$ or S_{112}. In the case that $G/K \cong A_{112}$, by order consideration, we deduce that $K = 1$ and $G \cong A_{112}$ and the desired conclusion follows in this case. In the latter case we see that $2^{109}||G|$, a contradiction, which proves Theorem 2.

Therefore, the proof of Theorem A follows from Theorem 1 and Theorem 2.

As an immediately corollarly, we have

Corollary 2. If G is a finite group such that $|G| = |A_{p+3}|$ and $\pi_e(G) = \pi_e(A_{p+3})$, where $7 \neq p \in \pi(112!)$. Then $G \cong A_{p+3}$.

Proof. If $\pi_e(G) = \pi_e(A_{p+3})$, then G and A_{p+3} have the same degree pattern. Hence the result follows from Theorem A.

4 OD-Characterizable of the Symmetric Groups

In this section, we will discuss the automorphism groups of A_{106} and A_{112}, i.e., S_{106} and S_{112}. At first glance, it seems that there is not much difference in the degree pattern and order between these simple groups and their automorphism groups. Clearly, the index of these simple groups in their automorphism groups is 2. Furthermore, in the prime graph $\Gamma(S_{106})$ (resp. $\Gamma(S_{112})$), the degree of vertex 2 and the degree of the largest prime divisor, that is, vertex 103 in $\Gamma(S_{106})$ (resp. 109 in $\Gamma(S_{112})$), increase by one, and the degree of other vertices remain as in $\Gamma(S_{106})$ (resp. $\Gamma(S_{112})$). In spite of these slight changes, the symmetric groups S_{106} and S_{112} are not OD-Characterizable. According to Theorem A, we can see that A_{106} and A_{112} are OD-Characterizable. in the following part, we will prove that the symmetric groups S_{106} and S_{112} are 3-fold OD-Characterizable.

Theorem 3. The symmetric groups S_{106} and S_{112} are 3-fold OD-characterizable.

Proof. Let G be a finite group satisfying

(1) $|G| = |S_{106}| = 2^{102} \cdot 3^{50} \cdot 5^{25} \cdot 7^{17} \cdot 11^{9} \cdot 13^{8} \cdot 17^{6} \cdot 19^{5} \cdot 23^{4} \cdot 29^{3} \cdot 31^{3} \cdot 37^{2} \cdot 41^{2} \cdot 43^{2} \cdot 47^{2} \cdot 53^{2} \cdot 59 \cdot 61 \cdot 67 \cdot 71 \cdot 73 \cdot 79 \cdot 83 \cdot 89 \cdot 97 \cdot 101 \cdot 103$

and

(2) $D(G) = D(S_{106}) = (26, 26, 26, 24, 23, 23, 23, 22, 22, 20, 20, 18, 17, 17, 16, , 15, 15, 14, 12, 11, 11, 9, 9, 7, 4, 3, 2)$.

It is evident that $\{pq|p+q \leq 106\} \subseteq \pi_e(G)$ and $\{pq|p+q > 106\} \cap \pi_e(G) = \varnothing$, where $p, q \in \pi(G)$. Clearly, the prime graph of G is connected , because the vertex 2 is joint to all other vertices. Moreover, it is easy to see that $\Gamma(G) = \Gamma(S_{106})$.

Suppose that K is the maximal normal solvable subgroup of G. Arguing as in the proof of Theorem 1, K is a $\{2,3\}$-group and $A_{106} \lesssim G/K \lesssim Aut(A_{106}) \cong S_{106}$, and so it follows that $G/K \cong S_{106}$ or A_{106}. In the case that $G/K \cong S_{106}$, by comparing orders, we deduce that $K = 1$ and $G \cong S_{106}$. In the latter case we have $|K| = 2$ and so $K \leq Z(G)$. Therefore G is a central extension of $\mathbb{Z}_2$ by A_{106} and G is isomorphic to one of the following groups:

$$2 \cdot A_{106}(\text{non-split extension of } \mathbb{Z}_2 \text{ by } A_{106});$$

$$2 : A_{106} \cong \mathbb{Z}_2 \times A_{106}(\text{split extension of } \mathbb{Z}_2 \text{ by } A_{106}).$$

It is easy to know that either G is isomorphic to $2 \cdot A_{106}$ or $2 : A_{106} \cong \mathbb{Z}_2 \times A_{106}$, which satisfies the conditions $|G| = |S_{106}|$ and $D(G) = D(S_{106})$. Hence, S_{106} is 3-fold OD-characterizable.

We omit the details for S_{112} because the arguments are quite similar as those for S_{106}. We only mention that the non-isomorphic groups $2 \cdot A_{112}$ (non-split extension of $\mathbb{Z}_2$ by A_{112}) and $2 : A_{112} \cong \mathbb{Z}_2 \times A_{112}$ (split extension of $\mathbb{Z}_2$ by A_{112}) have the same order and degree pattern as S_{112}. Hence, S_{112} is 3-fold OD-characterizable and so the theorem is proved.

References

[1] A. R. Moghaddamfar, A. R. Zokayi, M. R. Darafsheh, *A characterization of finite simple groups by the degrees of vertices of their prime graphs*, Algebra Colloquium, **12** (3) (2005) 431-442.

[2] J. H. Conway, R. T. Curtis, S. P. Norton, R. A. Parker, R. A. Wilson, *Atlas of Finite Groups*, Oxford University Press, London / New York, (1985).

[3] A. R. Moghaddamfar, A. R. Zokayi, *Recognizing finite groups through order and degree pattern*, Algebra Colloquium, **15** (3) (2008) 449-456.

[4] A. A. Hoseini, A. R. Moghaddamfar, *Recognizing alternating groups A_{p+3} for certain primes p by their orders and degree patterns*, Frontiers of Mathematics in China, **5** (3) (2010) 541-553.

[5] A. R. Moghaddamfar, A. R. Zokayi, *OD-Characterization of alternating and symmetric groups of degrees* 16 *and* 22, Frontiers of Mathematics in China, **4** (4) (2009) 669-680.

[6] A. R. Moghaddamfar, A. R. Zokayi, *OD-Characterizability of certain finite groups having connected prime graphs*, Algebra Colloquium **17** (1) (2010) 121-130.

[7] L. C. Zhang, W. J. Shi, *OD-characterization of simple K_4-groups*, Algebra Colloquium **16** (2) (2009) 275-282.

[8] Y. X. Yan, G. Y. Chen, *A new characterization of almost Simple K_3- groups* , to appear.

[9] A. Zavarnitsine and V. D. Mazurov, *Element orders in covering of symmetric and alternating groups*, Algebra and Logic **38** (3) (1999) 159-170.

[10] A. V. Zavarnitsine, *Finite simple groups with narrow prime spectrum*, Siberian Electronic Mathematical Reports, **6** (2009) 1-12.

[11] A. V. Zavarnitsin, *Recognition of alternating groups of degrees $r+1$ and $r+2$ for prime r and the group of degree 16 by their element order sets,* Algebra and Logic, **39** (6) (2000) 370-477.

[12] D. Gorenstein, *Finite Groups*, Harper and Row, New York, (1980).

[13] G. Y. Chen, *On Struture of Frobenius and 2-Frobenius group*, Journal of Southwest China Normal University, **20** (5) (1995) 485-487 (in Chinese).

[14] J. S. Williams, *Prime graph components of finite groups*, Journal of Algebra, **69** (2) (1981) 487-513.

[15] D. Passman, *Permutation Groups*, Benjamin Inc., New York, (1968).

[16] W. J. Shi, *A new characterization of some simple groups of Lie type*, Contemporary Math., **82** (1989) 171-180.

Cotorsion Pairs of Complexes

Xiaoyan Yang

Dedicated to the 70-year-old birthday of Professor K. P. Shum

Abstract: Let C be a complex of right R-modules. In this article, we will see that the functor $C \otimes -$ gives rise to a complete hereditary cotorsion pair. We also introduce the notion of a duality pair and demonstrate how the left half of such a pair is "often" covering and preenveloping.

2000 *Mathematics Subject Classification*: 13D05; 18G35.

Keywords: duality pair; cotorsion theory; complex.

1 Preliminaries

In this paper, $\mathcal{C}$ will be the abelian category of complexes of R-modules. This category has enough projectives and injectives. This can be seen from the fact that any complex of the form

$$\cdots \longrightarrow 0 \longrightarrow M \xrightarrow{\text{id}} M \longrightarrow 0 \longrightarrow \cdots$$

with M projective (injective) is projective (injective). For objects C and D of $\mathcal{C}$, $\mathrm{Hom}(C, D)$ is the abelian group of morphisms from C to D in $\mathcal{C}$ and $\mathrm{Ext}^i(C, D)$ for $i \geq 0$ will denote the groups we get from the right derived functors of Hom.

In this paper a complex

$$\cdots \longrightarrow C^{-1} \xrightarrow{\delta^{-1}} C^0 \xrightarrow{\delta^0} C^1 \xrightarrow{\delta^1} \cdots$$

will be denoted C. We will use subscripts to distinguish complexes. So if $\{C_i\}_{i \in I}$ is a family of complexes, C_i will be

$$\cdots \longrightarrow C_i^{-1} \xrightarrow{\delta_i^{-1}} C_i^0 \xrightarrow{\delta_i^0} C_i^1 \xrightarrow{\delta_i^1} \cdots$$

Given M a left R-module, we will denote by $\overline{M}$ the complex

$$\cdots \longrightarrow 0 \longrightarrow M \xrightarrow{\text{id}} M \longrightarrow 0 \longrightarrow \cdots$$

with the first M in the -1-th position. Also we mean by $\underline{M}$ the complex with M in the 0-th place and 0 in the other places. Given a complex C and an integer m, $C[m]$ denotes the complex such that $C[m]^n = C^{m+n}$ and whose boundary operators are $(-1)^m \delta^{m+n}$. The nth cycle module is defined as $\mathrm{Ker}\delta^n$ and is denoted $\mathrm{Z}^n C$. The nth boundary module is defined as $\mathrm{Im}\delta^{n-1}$ and is denoted $\mathrm{B}^n C$.

If C and D are complexes and let $\mathrm{Hom}^{\cdot}(C,D)$ be the complex of abelian groups with

$$\mathrm{Hom}^{\cdot}(C,D) = \prod_{t\in\mathbb{Z}} \mathrm{Hom}(C^t, D^{n+t})$$

and such that if $f \in \mathrm{Hom}^{\cdot}(C,D)$ then

$$(\delta^n f)^m = \delta_D^{n+m} f^m - (-1)^n f^{m+1}\delta_C^m.$$

Then $\mathrm{Z}^0\mathrm{Hom}^{\cdot}(C,D)$ will be the group $\mathrm{Hom}_{\mathcal{C}}(C,D)$ of morphisms from C to D.

Given two complexes C and D. Let $\underline{\mathrm{Hom}}(C,D) = \mathrm{Z}(\mathrm{Hom}^{\cdot}(C,D))$. We then see that $\underline{\mathrm{Hom}}(C,D)$ can be made into a complex with $\underline{\mathrm{Hom}}(C,D)^m$ the abelian group of morphisms from C to $D[m]$ and with boundary operator given by $f \in \underline{\mathrm{Hom}}(C,D)^m$, then $\delta^m(f) : C \to D[m+1]$ with $\delta^m(f)^n = (-1)^m \delta_D^{n+m} f^n$ for any $n \in \mathbb{Z}$. Let C be a complex of right R-modules and D be a complex of left R-modules. We define $C \otimes D$ to be $\frac{(C\otimes^{\cdot} D)}{\mathrm{B}(C\otimes^{\cdot} D)}$. Then with the maps

$$\frac{(C\otimes^{\cdot} D)^n}{\mathrm{B}^n(C\otimes^{\cdot} D)} \longrightarrow \frac{(C\otimes^{\cdot} D)^{n+1}}{\mathrm{B}^{n+1}(C\otimes^{\cdot} D)}, \quad x\otimes y \longmapsto \delta_C(x)\otimes y$$

where $x \otimes y$ is used to denote the coset in $\frac{(C\otimes^{\cdot} D)^n}{\mathrm{B}^n(C\otimes^{\cdot} D)}$, we get a complex. We note that the new functor $\underline{\mathrm{Hom}}(C,D)$ will have right derived functors whose values will be complexes. These values should certainly be denoted $\underline{\mathrm{Ext}}^i(C,D)$. It is not hard to see that $\underline{\mathrm{Ext}}^i(C,D)$ is the complex

$$\cdots \longrightarrow \mathrm{Ext}^i(C,D[n-1]) \longrightarrow \mathrm{Ext}^i(C,D[n]) \longrightarrow \mathrm{Ext}^i(C,D[n+1]) \longrightarrow \cdots$$

with boundary operator induced by the boundary operator of D. For C a complex of left R-modules we have two functors $-\otimes C : \mathcal{C}_R \to \mathcal{C}_{\mathbb{Z}}$ and $\underline{\mathrm{Hom}}(C,-) : {}_R\mathcal{C} \to \mathcal{C}_{\mathbb{Z}}$, where $\mathcal{C}_R$ (resp., ${}_R\mathcal{C}$) denotes the category of complexes of right R-modules (resp., left R-modules). Since $-\otimes C : \mathcal{C}_R \to \mathcal{C}_{\mathbb{Z}}$ is a right exact functor, we can construct left derived functors which we denote by $\mathrm{Tor}_1(-,C)$.

Let $0 \to C \to D \to D/C \to 0$ be a short exact sequence of complexes. We say the sequence is pure if for any complex X, the sequence $0 \to X\otimes C \to X \otimes D \to X \otimes (D/C) \to 0$ is exact.

By $\mathcal{E}$ denote the class of all exact complexes. For any complex C, the character complex $\underline{\mathrm{Hom}}(C,\overline{Q/Z})$ will be denote by C^+.

2 Cotorsion pairs of complexes from left adjoints

Let $\mathcal{A}$ be an abelian category. Given a class of objects $\mathcal{F}$, an $\mathcal{F}$-precover of an object M is a morphism $f : F \to M$ with $F \in \mathcal{F}$, such that $\mathrm{Hom}(F',F) \to \mathrm{Hom}(F',M) \to 0$ is exact for any $F' \in \mathcal{F}$. If Moreover, $f \circ g = f$ implies that g is an isomorphism whenever $g \in \mathrm{End}(F)$, then $f : F \to M$ is an $\mathcal{F}$-cover. $\mathcal{F}$-preenvelopes and $\mathcal{F}$-envelopes are defined dually. We say that a class $\mathcal{F}$ is (pre)covering if every object has an $\mathcal{F}$-(pre)cover. A cotorsion theory in $\mathcal{A}$ is defined as a pair $(\mathcal{F},\mathcal{C})$ of classes of $\mathrm{Ob}(\mathcal{A})$ such that $\mathcal{F}^\perp = \mathcal{C}$ and ${}^\perp\mathcal{C} = \mathcal{F}$. Recall that given a class of objects $\mathcal{F}$, its orthogonal class $\mathcal{F}^\perp$ (${}^\perp\mathcal{F}$) is defined as the class of objects C such that $\mathrm{Ext}^1(F,C)=0$ ($\mathrm{Ext}^1(C,F)=0$) for all $F \in \mathcal{F}$. A cotorsion theory is said to be perfect if every R-module has an $\mathcal{F}$-cover and an $\mathcal{C}$-envelope.

Let C be a fixed complex of right R-modules.

$$\mathcal{A}_1 = \{D|\mathrm{Tor}_1(C,D)=0\}, \quad \mathcal{A} = \{D|\mathrm{Tor}_k(C,D)=0 \text{ for all } k \geq 1\}.$$

We now point out some properties of $\mathcal{A}_1$ that will also be used to show $(\mathcal{A}, \mathcal{A}^{\perp})$ is a cotorsion pair. We start with an alternate description of the class $\mathcal{A}_1$.

Lemma 2.1 $\mathcal{A}_1$ equals the class of all complexes D for which $0 \to C \otimes A \to C \otimes B \to C \otimes D \to 0$ is exact whenever $0 \to A \to B \to D \to 0$ is exact.

Proof. If $D \in \mathcal{A}_1$, then it is clear that D is in the class described since $\mathrm{Tor}_1(C,D)=0$. Conversely, suppose D is a complex for which $0 \to C \otimes A \to C \otimes B \to C \otimes D \to 0$ is exact whenever $0 \to A \to B \to D \to 0$ is exact. Let $0 \to K \to F \to D \to 0$ be exact with F flat. Apply $C \otimes -$, and conclude $\mathrm{Tor}_1(C,D)=0$.

Theorem 2.2 $(\mathcal{A}_1, \mathcal{A}_1^{\perp})$ is a cotorsion pair.

Proof. Let $X \in {}^{\perp}(\mathcal{A}_1^{\perp})$. We need to show $X \in \mathcal{A}_1$. This means we needs to show $\mathrm{Tor}_1(C,X)=0$. Suppose $A \in \mathcal{A}_1$. Then $\underline{\mathrm{Ext}}^1(A,C^+) \cong \mathrm{Tor}_1(C,A)^+$, and so

$$0 = \underline{\mathrm{Ext}}^1(A,C^+)^{nm} = [\underline{\mathrm{Ext}}^1(A,C^+)^n]^m \cong \mathrm{Ext}^1(A[-n],C^+[m])$$

for all n,m, which gives that $\mathrm{Tor}_1(C,A[-n])^+ \cong \underline{\mathrm{Ext}}^1(A[-n],C^+)=0$. Thus $A[-n] \in \mathcal{A}_1$ for all n. Let $Y \in \mathcal{A}_1^{\perp}$. Then $\mathrm{Ext}^1(A,Y)=0$ for all $A \in \mathcal{A}_1$, and so $\mathrm{Ext}^1(A,Y[n]) \cong \mathrm{Ext}^1(A[-n],Y)=0$ for all n. Hence $Y[n] \in \mathcal{A}_1^{\perp}$ for all n. Since $C^+ \in \mathcal{A}_1^{\perp}$, then $C^+[n] \in \mathcal{A}_1^{\perp}$. So $\mathrm{Ext}^1(X,C^+[n])=0$ for all n. That is, $\mathrm{Tor}_1(C,X)^+ \cong \underline{\mathrm{Ext}}^1(X,C^+)=0$. Thus $X \in \mathcal{A}_1$.

Proposition 2.3 Let $M \in$ Mod-R and $C = \overline{M}$. Then $D \in \mathcal{A}_1$ if and only if $\mathrm{Tor}_1(M,D^n)=0$ for all n.

Proof. ($\Longrightarrow$) Let $0 \to A \to P \to D \to 0$ be exact with P a projective complex. Then we have the following commutative diagram with exact rows:

$$\begin{array}{ccccccccc}
0 & \longrightarrow & \overline{M} \otimes A & \longrightarrow & \overline{M} \otimes P & \longrightarrow & \overline{M} \otimes D & \longrightarrow & 0 \\
 & & \cong\downarrow & & \cong\downarrow & & \cong\downarrow & & \\
0 & \longrightarrow & M \otimes_R A & \longrightarrow & M \otimes_R P & \longrightarrow & M \otimes_R D & \longrightarrow & 0
\end{array}$$

So $0 \to M \otimes_R A^n \to M \otimes_R P^n \to M \otimes_R D^n \to 0$ is exact for all n. Thus $\mathrm{Tor}_1^R(M,D^n)=0$.

($\Longleftarrow$) Let $0 \to A \to B \to D \to 0$ be exact. Then $0 \to M \otimes_R A^n \to M \otimes_R B^n \to M \otimes_R D^n \to 0$ is exact for all n and we have the following commutative diagram:

$$\begin{array}{ccccccccc}
0 & \longrightarrow & M \otimes_R A & \longrightarrow & M \otimes_R B & \longrightarrow & M \otimes_R D & \longrightarrow & 0 \\
 & & \cong\downarrow & & \cong\downarrow & & \cong\downarrow & & \\
 & & \overline{M} \otimes A & \longrightarrow & \overline{M} \otimes B & \longrightarrow & \overline{M} \otimes D & \longrightarrow & 0
\end{array}$$

with the upper row exact. Then the lower row exact. So $D \in \mathcal{A}_1$.

Proposition 2.4 Let $N \in R$-Mod and $\mathrm{Tor}_1^R(C^n,N)=0$ for all n. Then $\overline{N} \in \mathcal{A}_1$.

Proof. Since $\mathrm{Ext}^1(C, \overline{N^+}[n]) \cong \mathrm{Ext}^1_R(C^{-n-2}, N^+) \cong \mathrm{Tor}_1^R(C^{-n-2}, N)^+ = 0$ by [6, Lemma 3.1] for all n, then $\mathrm{Tor}_1(C, \overline{N})^+ \cong \underline{\mathrm{Ext}}^1(C, \overline{N}^+) = 0$. Thus $\overline{N} \in \mathcal{A}_1$.

Lemma 2.5 $\mathcal{A}$ is a Kaplansky class.

Proof. Let $0 \to D' \to D \to D'' \to 0$ be pure exact with $D \in \mathcal{A}$. Then $0 \to D''^+ \to D^+ \to D'^+ \to 0$ is split. Since $\underline{\mathrm{Ext}}^k(C, D^+) \cong \mathrm{Tor}_k(C, D)^+$ for all $k \geq 1$, then $\mathrm{Tor}_k(C, D') = \mathrm{Tor}_k(C, D'') = 0$, and so $D', D'' \in \mathcal{A}$.

Theorem 2.6 $(\mathcal{A}, \mathcal{A}^\perp)$ is a perfect hereditary cotorsion pair.

Proof. Using Lemma 2.5 and Definition of $\mathcal{A}$.

Example 2.7 (1) If C be any DG-flat complex. Then $\mathcal{A}$ is the class of all exact complexes. So the associated cotorsion pair is $(\mathcal{E}, \mathcal{E}^\perp)$.

(2) Let R be an n-Gorenstein ring, and let $\{I_j\}_{j \in J}$ be the set of (representatives of) all indecomposable injective right R-modules. Define $I = \oplus_{j \in J} I_j$ and $C = \overline{I}$. We claim that $D \in \mathcal{A}$ if and only if D is Gorenstein flat. If D is Gorenstein flat, then $\mathrm{Tor}_k(\overline{I}, D) = 0$ for all $k \geq 1$. On the other hand, suppose $D \in \mathcal{A}$. Since $\mathrm{Tor}_k^R(-, D^n) = 0$ commutes with direct sums, we must have $\mathrm{Tor}_k^R(I_j, D^n) = 0$ for all indecomposable I_j, all $k \geq 1$ and all n. Now if E is any injective right R-module, then since R is right noetherian, it must be a direct limit of indecomposable injectives. Since $\mathrm{Tor}_k^R(-, D^n) = 0$ commutes with direct limits, we must have $\mathrm{Tor}_k^R(E, D^n) = 0$ for all $k \geq 1$. So D^n is a Gorenstein flat left R-module, and hence D is Gorenstein flat. Thus $(\mathcal{GF}, \mathcal{GF}^\perp)$ is a perfect hereditary cotorsion pair.

Let $\mathcal{F}$ be a class of complexes. $\mathcal{F}$ is called a Kaplansky class if there exists a regular cardinal $\mathcal{N}$ with the following property: For every $F \in \mathcal{F}$ and $x \in F$, there exists a subcomplex $P \subseteq F$ such that $x \in P$ and $P, F/P \in \mathcal{F}$ and $|P| < \mathcal{N}$.

Lemma 2.8 Let $M : \cdots \to M^{-2} \to M^{-1} \to M^0 \to 0$ be an exact complex with each M^i finitely presented. Then $(\prod_{i \in I} F_i) \otimes M \cong \prod_{i \in I}(F_i \otimes M)$ for any family $(F_i)_{i \in I}$ of complexes.

Proof. Since M^i is finitely generated, there is an exact sequence $R^{n_i} \to M^i \to 0$ for some $n_i \geq 0$. Set $P_0 = \bigoplus_{i=-\infty}^{0} \overline{R^{n_i}}[-i-1]$. Then $P_0 \to M \to 0$ is exact. Set $K = \mathrm{Ker}(P_0 \to M)$. Then every K^i is finitely generated and $K^m = 0$ for all $m > 1$. Similarly, we have an exact sequence $P_1 \to K \to 0$ and $P_1 = \bigoplus_{i=-\infty}^{1} \overline{R^{m_i}}[-i-1]$. Since

$$
\begin{aligned}
(\prod_{i \in I} F_i) \otimes P_0 &\cong \bigoplus_{i=-\infty}^{0} ((\prod_{i \in I} F_i) \otimes \overline{R^{n_i}}[-i-1]) \\
&\cong \bigoplus_{i=-\infty}^{0} ((\prod_{i \in I} F_i)[-i-1] \otimes_R R^{n_i}) \\
&\cong \bigoplus_{i=-\infty}^{0} \prod_{i \in I} (F_i[-i-1] \otimes_R R^{n_i}) \\
&\cong \bigoplus_{i=-\infty}^{0} \prod_{i \in I} (F_i \otimes \overline{R^{n_i}}[-i-1]) \\
&\cong \prod_{i \in I} (F_i \otimes P_0)
\end{aligned}
$$

and $(\prod_{i\in I} F_i)\otimes P_1 \cong \prod_{i\in I}(F_i\otimes P_1)$ by [2, Proposition 2.1], then we have the following commutative diagram with exact rows:

$$\begin{array}{ccccccc} (\prod_{i\in I} F_i)\bigotimes P_1 & \longrightarrow & (\prod_{i\in I} F_i)\bigotimes P_0 & \longrightarrow & (\prod_{i\in I} F_i)\bigotimes M & \longrightarrow & 0 \\ \alpha\downarrow & & \beta\downarrow & & \gamma\downarrow & & \\ \prod_{i\in I}(F_i\bigotimes P_1) & \longrightarrow & \prod_{i\in I}(F_i\bigotimes P_0) & \longrightarrow & \prod_{i\in I}(F_i\bigotimes M) & \longrightarrow & 0 \end{array}$$

Since α and β are isomorphisms, then γ is an isomorphism.

Theorem 2.9 Let R be right coherent and C a finitely presented complex. Then every complex of left R-modules has an $\mathcal{A}$-preenvelope.

Proof. Let $A \in {}_R\mathcal{C}$ and $D \in \mathcal{A}$. Since $\mathcal{A}$ is a Kaplansky class, for every morphism $A \xrightarrow{f} D$, there exists $D' \in \mathcal{A}$ and a cardinal $\mathcal{N}$ such that $|D'| < \mathcal{N}$ and f factors $A \to D' \to D$. Then we say that any two morphisms $A \to D$ and $A \to D'$ with $D, D' \in \mathcal{A}$ and $|D|, |D'| < \mathcal{N}$ are equivalent if and only if any diagram

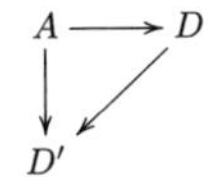

can be completed by an isomorphism. Now if we take X a set of representatives of such $A \to D$, then $A \to \prod_X D$ is an $\mathcal{A}$-preenvelope.

3 Cotorsion pairs induced by duality pairs

In this section we define duality pairs and give several examples. We also prove how suitable duality pairs induce cotorsion pairs.

Definition 3.1 A duality pair over R is a pair $(\mathcal{M},\mathcal{N})$, where $\mathcal{M}$ is a class of complexes of left (right) R-modules and $\mathcal{N}$ is a class of complexes of right (left) R-modules, which are closed under isomorphisms, subject to the following conditions:

(1) For any complex M, one has $M \in \mathcal{M}$ if and only if $M^+ \in \mathcal{N}$.

(2) $\mathcal{N}$ is closed under direct summands and finite sums.

A duality pair $(\mathcal{M},\mathcal{N})$ is called (co)product-closed if the class $\mathcal{M}$ is closed under (co)products in the category of all complexes.

A duality pair $(\mathcal{M},\mathcal{N})$ is called perfect if it is coproduct-closed, if $\mathcal{M}$ is closed under extensions, and if $\overline{R}$ belongs to $\mathcal{M}$.

Let R be a ring and let $R' = R[x]/(x^2)$. Then R' may be viewed as a graded ring, with a copy of R (generated by 1) in degree 0 and a copy of R (generated by x) in degree -1. One can check that the category R'-Mod is isomorphic to the category of unbounded R-chain complexes.

Theorem 3.2 Let $(\mathcal{M},\mathcal{N})$ be a duality pair. Then $\mathcal{M}$ is closed under pure subcomplexes, pure quotients and pure extensions. Furthermore, the following hold:

(1) If $(\mathcal{M},\mathcal{N})$ is product-closed, then $\mathcal{M}$ is preenveloping.

(2) If $(\mathcal{M},\mathcal{N})$ is coproduct-closed, then $\mathcal{M}$ is covering.

(3) If $(\mathcal{M},\mathcal{N})$ is perfect, then $(\mathcal{M},\mathcal{M}^{\perp})$ is a perfect cotorsion pair.

Proof. First we prove that $\mathcal{M}$ is closed pure subcomplexes, pure quotients and pure extensions, that is, given a pure exact sequence of complexes $0 \to M' \to M \to M'' \to 0$, then $M \in \mathcal{M}$ if and only if $M', M'' \in \mathcal{M}$. Apply $\underline{\mathrm{Hom}}(-,\overline{Q/Z})$ to the sequence above, we get a split exact sequence. By definition, it follows that $M^{+} \in \mathcal{N}$ if and only if $M'^{+}, M''^{+} \in \mathcal{N}$, and so $M \in \mathcal{M}$ if and only if $M', M' \in \mathcal{M}$.

(1) Let $C \in {}_R\mathcal{C}$ and $M \in \mathcal{M}$. Since $\mathcal{M}$ is closed under pure subcomplexes, for any morphism $C \xrightarrow{f} M$ there exist $M' \in \mathcal{M}$ and a cardinal $\mathcal{N}$ such that $|M'| < \mathcal{N}$ and factors $C \to M' \to M$ by [5, Lemma 5.2.1]. By analogy with the proof of Theorem 2.9, we have $\mathcal{M}$ is preenveloping since $\mathcal{M}$ is closed under products.

(2) Let $C \in {}_R\mathcal{C}$ and $M \in \mathcal{M}$. Let $M \xrightarrow{f} C$ be any morphism, we want to show that there are $M' \in \mathcal{M}$ and a cardinal $\mathcal{N}$ such that f can be factors through M' with $|M'| < \mathcal{N}$. If $|M| < \mathcal{N}$ let $M' = M$. So suppose $|M| \geq \mathcal{N}$. Consider a subcomplex $S \subseteq M$ maximal with respect to the two properties that S is pure in M and that $S \subseteq \mathrm{Ker}(M \to C)$. Let $M' = M/S$. Then $M' \in \mathcal{M}$ by hypothesis. We want to argue that $|M'| < \mathcal{N}$. Let K be the kernel of $M' \to C$. Then $|M'/K| \leq |C|$. So if $|M'| \geq \mathcal{N}$, there is a nonzero pure submodule T/S of M' contained in K as an R'-modules. Since $\otimes_{R'} = \otimes_{\mathcal{C}}$, then T is a pure subcomplex of M and is contained in the kernel of $M \to C$. This contradicts the choice of S. Let X be a set with cardinality $\mathcal{N}$ such that any morphism $M \to C$ with $M \in \mathcal{M}$ has a factorization $M \to M' \to C$ for some $M' \in \mathcal{M}$ with $M' \in X$ (as sets). Let $(\varphi_i)_{i\in I}$ give all such morphisms $\varphi_i : M_i \to C$ with $M_i \in X$ (as sets). Then $\bigoplus_{i\in I} M_i \to C$ is an $\mathcal{M}$-precover. Let P be a finitely presented complex. Then $P[-n]$ is finitely presented for all n. Let $\{M_i\}$ be a system in $\mathcal{M}$ indexed by a partially ordered set. Then the diagram is commutative

$$\begin{array}{ccccc} \mathrm{Hom}(P[-n],\bigoplus M_i) & \longrightarrow & \mathrm{Hom}(P[-n],\varinjlim M_i) & & \\ {\scriptstyle\varphi}\downarrow & & {\scriptstyle\varphi'}\downarrow & & \\ \bigoplus \mathrm{Hom}(P[-n],M_i) & \longrightarrow & \varinjlim \mathrm{Hom}(P[-n],M_i) & \longrightarrow & 0 \end{array}$$

with the lower row exact. Since φ, φ' are isomorphisms, $\mathrm{Hom}(P,(\bigoplus M_i)[n]) \to \mathrm{Hom}(P,(\varinjlim M_i)[n]) \to 0$ is exact for all n. So $\underline{\mathrm{Hom}}(P,\bigoplus M_i) \to \underline{\mathrm{Hom}}(P,\varinjlim M_i) \to 0$ is exact, which implies that $\bigoplus M_i \to \varinjlim M_i \to 0$ is pure exact. Thus $\varinjlim M_i \in \mathcal{M}$ by hypothesis and $\mathcal{M}$ is covering.

(3) By (1) and (2).

Example 3.3 Consider for each integer $n \geq 0$ the following complexes classes:

(1) Let $\mathcal{F}_n = \{C \in {}_R\mathcal{C} | \mathrm{fdim} C \leq n\}$, $\mathcal{I}_n = \{D \in \mathcal{C}_R | \mathrm{idim} D \leq n\}$.

Then $(\mathcal{F}_n, \mathcal{I}_n)$ is a perfect duality pair, and so $(\mathcal{F}_n, \mathcal{F}_n^{\perp})$ is a perfect cotorsion pair.

(2) Let $\mathrm{dg}\mathcal{F} = \{F \in {}_R\mathcal{C} | F \text{ is DG-flat}\}$, $\mathrm{dg}\mathcal{I} = \{I \in \mathcal{C}_R | I \text{ is DG-injective}\}$.

Then $(\mathrm{dg}\mathcal{F}, \mathrm{dg}\mathcal{I})$ is a perfect cotorsion pair. So $(\mathrm{dg}\mathcal{F}, \mathrm{dg}\mathcal{F}^{\perp})$ is a perfect cotorsion pair.

(3) Let $\mathcal{E}_l = \{E \in {}_R\mathcal{C} | E \text{ is exact}\}$, $\mathcal{E}_r = \{E \in \mathcal{C}_R | E \text{ is exact}\}$.

Then $(\mathcal{E}_l, \mathcal{E}_r)$ is a perfect duality pair. Thus $(\mathcal{E}_l, \mathcal{E}_l^{\perp})$ is a perfect cotorsion pair.

(4) Let R be an n-Gorenstein and $\mathcal{GF} = \{M \in {}_R\mathcal{C} | M \text{ is G-flat}\}$, $\mathcal{GI} = \{N \in \mathcal{C}_R | N \text{ is G-injective}\}$.

Then $(\mathcal{GF}, \mathcal{GI})$ is a product-closed perfect duality pair. Let $0 \to M' \to M \to M'' \to 0$ be pure exact with $M \in \mathcal{GF}$. Then $0 \to M''^{+} \to M^{+} \to M'^{+} \to 0$ is split, and so $M'^{+}, M''^{+} \in \mathcal{GI}$, which gives $M', M'' \in \mathcal{GF}$. Thus $(\mathcal{GF}, \mathcal{GF}^{\perp})$ is a perfect cotorsion pair.

References

[1] E.E. Enochs, O.M.G. Jenda and J. Xu, Orthogonality in the category of complexes, Math J. Okayama Univ 38 (1996) 25-46.

[2] E.E. Enochs and J.R. Garcĺa-Rozas, Tensor Products of Complexes, Math J. Okayama Univ 39 (1997) 17-39.

[3] E.E. Enochs and O.M.G. Jenda, Relative Homological Algebra, de Gruyter Exp. Math. Vol. 30, Walter de Gruyter and Co., Berlin, 2000.

[4] E.E. Enochs and J.A. López-Ramos, Kaplansky classes, Rend. Sem. Math. Univ. Padova 107 (2002) 67–79.

[5] J.R. García-Rozas, Covers and Envelopes in the Category of Complexes, Research Notes in Mathematics, Vol. 407, Chapman and Hall/CRC, Nem York, 1999.

[6] J. Gillespie, The flat model structure on Ch(R), Trans. Amer. Math. Soc. 356 (2004) 3369–3390.

[7] J. Gillespie, Cotorsion pairs of modules from left adjoints, Comm. Algebra 37 (2009) 2570-2582

[8] H. Holm and P. Jørgensen, Covers, preenvelopes and purity, Illinois J. Math. to appear..

[9] H. Holm and P. Jørgensen, Cotorsion pairs induced by duality pairs, to appear.

Quasirecognition by Prime Graph of the Simple Group $E_7(q)$*

Qingliang Zhang[1], Wujie Shi[2,†], Rulin Shen[3]

1. School of Mathematical Sciences, Suzhou University, 215006 Jiangsu, P.R. China
2. Department of Mathematics and Statistics, Chongqing University of Arts and Sciences, Chongqing, 402160 P.R. China
3. Department of Mathematics, Hubei Institute for Nationalities, Hubei, 445000 P.R. China

Abstract: Let G be a finite group. The main result of this paper is as follows: the simple group $E_7(q)$ with disconnected prime graph is quasirecognizable by prime graph. Hence we generalize some known results of $E_7(q)$.

Keywords: quasirecognition, finite simple group, prime graph

1. Introduction

If n is an integer, then we denote by $\pi(n)$ the set of all prime divisors of n. If G is a finite group, then $\pi(|G|)$ is denoted by $\pi(G)$. We construct the prime graph of G which is denoted by $\Gamma(G)$ as follows: the vertex set is $\pi(G)$ and two distinct primes p and q are joined by an edge (we write $p \sim q$) if and only if G contains an element of order pq. Let $t(G)$ be the number of connected components of $\Gamma(G)$ and let $\pi_1, \pi_2, \cdots, \pi_{t(G)}$ be the connected components of $\Gamma(G)$. If $2 \in \pi(G)$, we always suppose that $2 \in \pi_1$. And we denote by (a, b) the greatest common divisor of positive integers a and b. The spectrum of a finite group G which is denoted by $\omega(G)$, is the set of its element orders. Obviously, $\omega(G)$ is partially ordered by divisibility. Therefore it is uniquely determined by $\mu(G)$, the subset of its maximal elements. The set of those numbers $n \in \mu(G)$ whose all prime divisors belong to $\pi_i(G)$, is denoted by $\mu_i(G)$.

A finite group G is said to be recognizable by spectrum if the equality $\omega(H) = \omega(G)$ implies that $H \cong G$. A finite simple nonabelian group G is said to be quasirecognizable by spectrum if

*Project supported by the NNSF of China (Grant No. 10871032), the SRFDP of China (Grant No. 20060285002), (complex integration project 2008.01.02).

†Corresponding author.

E-mail address: qingliangstudent@163.com (Qingliang Zhang), wjshi@suda.edu.cn (Wujie Shi), shenrulin@hotmail.com (Rulin Shen).

each finite group H with $\omega(H) = \omega(G)$ has a composition factor isomorphic to G (see [2]).

A finite group G is said to be recognizable by prime graph, if the equality $\Gamma(H) = \Gamma(G)$ implies that $H \cong G$. A finite simple nonabelian group G is said to be quasirecognizable by prime graph if each finite group H with $\Gamma(H) = \Gamma(G)$ has a unique nonabelian composition factor isomorphic to G (see [9]). In [2] it is proved that every finite simple group with at least three connected components (except A_6) is quasirecognizable by spectrum. However, we do not know whether or not every finite simple group with at least three connected components is quasirecognizable by prime graph.

Hagie in [6] determined finite groups G satisfying $\Gamma(G) = \Gamma(S)$, where S is a sporadic simple group. In [10] and [13] with the same prime graph as a CIT simple group and $PSL(2,q)$ where $q = p^\alpha < 100$ are determined. It is proved that if $q = 3^{2n+1}(n > 0)$, then the simple group ${}^2G_2(q)$ is uniquely determined by its prime graph [9, 20]. Also in [11] its proved that $L_2(p)$, where $p > 11$ is a prime number and $p \neq 12k + 1$ for some $k > 0$, is recognizable by prime graph. In [7] and [12] groups with the same prime graph as $L_{16}(2)$ and $L_2(q)$ are determined. Recently it is proved that the simple group ${}^2F_4(q)$, where $q = 2^{2m+1}(m \geq 1)$, is quasirecognizable by prime graph in [1].

Note that a finite group is recognizable by spectrum need not be recognizable by prime graph. For example, A_5 is recognizable by spectrum , but is not recognizable by prime graph. And the reason is as follows. Let $G \cong A_6$ or $G = [N]Q$, where $Q \cong A_5$ and N is an elementary abelian 2-group and a direct sum of natural $SL(2,4)$-modules. Here $[N]Q$ is denoted the semidirect product of N and Q, where $N \trianglelefteq [N]Q$. Its evident that $\Gamma(A_5) = \Gamma(G)$, thus A_5 is not recognizable by prime graph. Also a finite group is recognizable by spectrum need not be even quasirecognizable by prime graph. On the other hand, the recognizability by prime graph implies the recognizability by spectrum, also the quasirecognizability by prime graph implies the quasirecognizability by spectrum.

In this paper we show that the simple group $E_7(q)$ with disconnected prime graph is quasirecognizable by prime graph, and also we give a generalization of some known results of W.J. Shi for $E_7(q)$. And the main result is as follows:

Main Theorem. The simple group $E_7(q)$ with disconnected prime graph is quasirecognizable by prime graph.

The second author of this paper, W.J. Shi, put forward the following conjecture in [14]:

Conjecture. Let G be a finite group and M be a finite simple group. Then $G \cong M$ if and only if (i) $|G| = |M|$ and (ii) $\omega(G) = \omega(H)$.

A series of papers proved that this conjecture is valid for most finite simple groups (see a survey in [16]). In fact the characterization of finite simple groups by spectrum or prime graph is a generalization of the conjecture. Throughout the proof we use the classification of finite simple groups and we use the tables 1a-1c in [18]. In this paper, all groups are finite and by simple groups we mean nonabelian simple groups. All further unexplained notations are standard and refer to [3], for example.

2. Preliminaries

First we give an easy remark:

Remark 2.1. Let K be a subgroup of G. Let N be a normal subgroup of K and $p \sim q$ in $\Gamma(K/N)$. Then $p \sim q$ in $\Gamma(G)$. In fact if $xN \in K/N$ has order pq, then there is a power of x which has order pq.

Definition 2.2 ([5]). A finite group G is called 2-Frobenius group if it has a normal series $1 \trianglelefteq H \trianglelefteq K \trianglelefteq G$, where K and G/H are Frobenius groups with kernels H and K/H, respectively.

Lemma 2.3 ([19, Theorem A)]. If G is a finite group with $t(G) > 1$, then one of the following statements holds:

(a) G is a Frobenius group or a 2-Frobenius group and $t(G) = 2$;

(b) there exists a nonabelian simple group S such that $S \leq \overline{G} = G/K \leq Aut(S)$, where K is the maximal normal soluble subgroup of G; furthermore, K and $\overline{G}/S$ are $\pi_1(G)$-groups, $\Gamma(S)$ is disconnected, $t(S) \geq t(G)$, and for every i, $2 \leq i \leq t(G)$,there is j, $2 \leq j \leq t(S)$, such that $\mu_i(G) = \mu_j(S)$.

Lemma 2.4. (i) If G is a finite group and $\Gamma(G) = \Gamma(E_7(2))$, then G has a normal series $1 \trianglelefteq H \trianglelefteq K \trianglelefteq G$ such that $G/K = 1$ or G/K is a π_1-group, $H = 1$ or a nilpotent π_1-group, and K/H is a nonabelian simple group with $t(K/H) \geq 3$, and $G/K \leq Out(K/H)$. Also if $j \in \{2, 3\}$, then there exists an $i \geq 2$ such that $\pi_j(E_7(2)) = \pi_i(K/H)$.

(ii) If G is a finite group and $\Gamma(G) = \Gamma(E_7(3))$, then G has a normal series $1 \trianglelefteq H \trianglelefteq K \trianglelefteq G$ such that $G/K = 1$ or G/K is a π_1-group, $H = 1$ or a nilpotent π_1-group, and K/H is a non-abelian simple group with $t(K/H) \geq 3$, and $G/K \leq Out(K/H)$. Also if $j \in \{2, 3\}$, then there exists an $i \geq 2$ such that $\pi_j(E_7(3)) = \pi_i(K/H)$.

Proof. (i) Since $t(G) = 3$, by using Lemma 2.3, it follows that G is neither a Frobenius group nor a 2-Frobenius group. Therefore by Lemma 2.3, G has a normal series $1 \trianglelefteq H \trianglelefteq K \trianglelefteq G$ such that $G/K = 1$ or G/K is a π_1-group, and $H = 1$ or a nilpotent π_1-group by Lemma 2.4, also K/H is a nonabelian simple group. We know that a π_1-group is a group of even order with one component. So by using Remark 2.1, it follows that $t(K/H) \geq t(G) = 3$. By assumption, $K/H \trianglelefteq G/H$ and hence $N_{G/H}(K/H) = G/H$. We claim that $C_{G/H}(K/H) = 1$. Otherwise let $xH \in C_{G/H}(K/H)$ and p_0 be a prime divisor of $|xH|$. As we mentioned above, $\pi_2 \bigcup \pi_3 \subseteq \pi(K/H)$ and by the definition of prime graph it follows that every prime divisor of $|K/H|$ is connected to p_0, which is a contradiction, since $t(G) = 3$. Therefore $C_{G/H}(K/H) = 1$, and so G/H is isomorphic to a subgroup of $Aut(K/H)$. On the other hand, K/H is a nonabelian simple group and so $K/H \cong Inn(K/H)$, which implies that $G/K \leq Out(K/H)$.

(ii) we can prove the truth of (ii) in the same way as in (i). □

Lemma 2.5 ([4, 8]). With the exceptions of the relations $239^2 - 2 \cdot 13^4 = -1$ and $3^5 - 2 \cdot 11^2 = 1$ every solution of the equation $p^m - 2q^n = \pm 1$; p, q prime; m, $n > 1$, has exponents $m = n = 2$.

Lemma 2.6 ([8]). The only solution of the diophantine equation $p^m - q^n = 1$; p, q prime; and m, $n > 1$ is $3^2 - 2^3 = 1$.

In the sequel we recall the concept of quadratic residue and the Legendre symbol from number theory.

Remark 2.7 ([17]). Let $(k, n) = 1$. If there is an integer x such that $x^2 \equiv k \pmod{n}$, then k is called a quadratic residue (mod n). Otherwise k is called a quadratic nonresidue (mod n).

Let p be an odd prime. The symbol (a/p) will have the value 1 if a is a quadratic residue (mod p), -1 if a is a quadratic nonresidue (mod p), and zero if $p \mid a$. The symbol (a/p) is called the Legendre symbol.

Let p be a prime number and $(a, p) = 1$. Let $k \geq 1$ be the smallest positive integer such that $a^k \equiv 1 \pmod{p}$. Then k is called the order of a with respect p and we denote it by $ord_p(a)$. Obviously, by the Fermat little theorem it follows that $ord_p(a) \mid (p-1)$. Also if $a^n \equiv 1 \pmod{p}$, then $ord_p(a) \mid n$.

Lemma 2.8 ([17]). Let p be an odd prime, then $(-1/p) = (-1)^{(p-1)/2}$.

Lemma 2.9 ([21]). Let p be a prime and n be a positive integer. Then one of the following holds:

(i) there is a primitive prime p' for $p^n - 1$, that is, $p' \mid (p^n - 1)$ but $p' \nmid (p^m - 1)$, for every $1 \leq m < n$;

(ii) $p = 2$, $n = 1$ or 6;

(iii) p is a Mersenne prime and $n = 2$.

3. Proof of the main results

By [18, Tables 1a-1c], we know that if $t(E_7(q)) > 1$, then $q = 2$ or $q = 3$. Consequently we will consider two cases in the following, namely, we will show the quasirecognizability by prime graph of $E_7(2)$ and $E_7(3)$ separately. Of course, the proof for $E_7(2)$ is different from the proof for $E_7(3)$ in many places.

Theorem 3.1. *Let G be a finite group such that $\Gamma(G) = \Gamma(E_7(2))$, then G has a unique* nonabelian composition factor $K/H \cong E_7(2)$ and $K = G$.

Proof. By [18, Tables 1a-1c], we have $t(G) = 3$ and $\pi_1(G) = \{2,3,5,7,11,13,17,19,31,43\}$. Also the odd components are $\pi_2(G) = \{73\}$ and $\pi_3(G) = \{127\}$.

By Lemma 2.4, if G is a finite group and $\Gamma(G) = \Gamma(E_7(2))$, then G has a normal series $1 \trianglelefteq H \trianglelefteq K \trianglelefteq G$ such that $G/K = 1$ or G/K is a π_1-group, $H = 1$ or a nilpotent π_1-group, and K/H is a nonabelian simple group with $t(K/H) \geq 3$, and $G/K \leq Out(K/H)$. Now by using the classification of finite simple groups and according to [18, Tables 1a-1c], we consider the following cases:

Case 1. Let $K/H \cong A_p$, where p and $p-2$ are primes. Then the odd components of K/H are $\{p\}$ and $\{p-2\}$ according to [18, Tables 1a-1c]. We know that $\pi_2(K/H) \bigcup \pi_3(K/H) = \pi_2(G) \bigcup \pi_3(G)$, thus $\{p, p-2\} = \{73, 127\}$, which is a contradiction.

Case 2. Let $K/H \cong A_1(q)$, where $q = p^\alpha$ and $4 \mid q+1$. Then the odd components of K/H are $\{p\}$ and $\pi(\frac{p^\alpha-1}{2})$.

(2.1) If $p = 73$, then $\pi(\frac{p^\alpha-1}{2}) = \{127\}$. Therefore $p^\alpha - 1 = 2 \cdot 127^\beta$ for some $\beta > 0$ and so by using Lemma 2.5, it follows that $\alpha = 1$ or $\beta = 1$ or $\alpha = \beta = 2$. If $\alpha = 1$, then $72 = 2 \cdot 127^\beta$ has no solution. If $\beta = 1$, then we have $73^\alpha = 255$, which is impossible. If $\alpha = \beta = 2$, it follows that $73^2 - 1 = 2 \cdot 127^2$ and this is a contradiction.

(2.2) If $p = 127$, then we can get a contradiction similarly.

Case 3. Let $K/H \cong A_1(q)$, where $q = p^\alpha$ and $4 \mid q-1$. Then we can get a contradiction in the way as in Case 2.

Case 4. Let $K/H \cong A_1(q)$, where $q = 2^\alpha$ and $\alpha > 1$. We know that $\pi_2(K/H) \bigcup \pi_3(K/H) = \pi_2(G) \bigcup \pi_3(G)$, thus $\pi(2^{2\alpha} - 1) = \{73, 127\}$. Obviously $3 \in \pi(2^{2\alpha} - 1)$ and so $3 \in \{73, 127\}$, which is a contradiction.

Case 5. Let $K/H \cong {}^2A_5(2)$. Hence the odd components of K/H are $\{7\}$ and $\{11\}$, and it follows that$\{7, 11\}$=$\{73, 127\}$, which is a contradiction.

Similarly $K/H \ncong S$, where S is a sporadic simple group in [18, Tables 1a-1c].

Case 6. Let $K/H \cong {}^2D_p(3)$, where $p = 2^m + 1 \geq 3$. Then the odd components of K/H are $\pi(\frac{3^{p-1}+1}{2})$ and $\pi(\frac{3^p+1}{4})$.

(6.1) If $\pi(\frac{3^{p-1}+1}{2}) = \{73\}$, then we have $3^{p-1} + 1 = 2 \cdot 73^\alpha$ and so by using Lemma 2.5, it follows that $p-1 = 1$ or $\alpha = 1$ or $p-1 = \alpha = 2$. If $p-1 = 1$, then $p = 2$, which is a contradiction. If $\alpha = 1$, then $3^{p-1} = 145$, which is impossible. If $p-1 = \alpha = 2$, it follows that $10 = 2 \cdot 73^2$ and this is a contradiction.

(6.2) If $\pi(\frac{3^{p-1}+1}{2}) = \{127\}$, then we can get a contradiction similarly.

Case 7. Let $K/H \cong G_2(q)$, where $q = 3^\alpha$. Then the odd components of K/H are $\pi(q^2-q+1)$ and $\pi(q^2+q+1)$. Since $\pi_2(K/H) \bigcup \pi_3(K/H) = \pi_2(G) \bigcup \pi_3(G)$, we have $\pi(q^4+q^2+1) = \{73, 127\}$, and it follows that $q^4 + q^2 + 1 = 73^\beta \cdot 127^\gamma$ for some $\beta, \gamma > 0$, namely, $3^{4\alpha} + 3^{2\alpha} + 1 = 73^\beta \cdot 127^\gamma$. Therefore $3 \equiv (-1)^\gamma \pmod 8$, which is a contradiction.

Case 8. Let $K/H \cong F_4(q)$, where $q = 2^\alpha$. Then the odd components of K/H are $\pi(q^4+1)$ and $\pi(q^4 - q^2 + 1)$.

(8.1) If $\pi(2^{4\alpha} + 1) = \{73\}$, it follows that $2^{4\alpha} + 1 = 73^\beta$ for some $\beta > 0$. We know that $2^{4\alpha} + 1 \equiv 2 \pmod 3$ and $73^\beta \equiv 1 \pmod 3$, which is a contradiction.

(8.2) If $\pi(2^{4\alpha} + 1) = \{127\}$, similarly we can get a contradiction.

Case 9. Let $K/H \cong {}^2B_2(q)$, where $q = 2^{2m+1} > 2$. Then the odd components of K/H are $\pi_2(K/H) = \pi(q-1)$ and $\pi_3(K/H) = \pi(q + \sqrt{2q} + 1)$ and $\pi_4(K/H) = \pi(q - \sqrt{2q} + 1)$.

(9.1) If $\{73\} = \pi(q-1)$, then $2^{2m+1} - 1 = 73^\alpha$, and it follows that $\alpha = 1$ by using Lemma 2.6. Therefore $2^{2m+1} = 74$, which is impossible.

(9.2) If $\{73\} = \pi(2^{2m+1} + 2^{m+1} + 1)$, then $2^{2m+1} + 2^{m+1} = 73^k - 1$ for some $k > 0$, so $2^{2m+1} + 2^{m+1} \equiv 0 \pmod 3$, namely $(-1) + (-1)^{m+1} \equiv 0 \pmod 3$, and it follows that m is an odd number. We know that $\{127\} = \pi(q-1)$ or $\pi(2^{2m+1} - 2^{m+1} + 1)$. If $\{127\} = \pi(q-1)$, then $127^\alpha = 2^{2m+1} - 1$ for some $\alpha > 0$. And by using Lemma 2.6 we have $\alpha = 1$. Hence $m = 3$, and it follows that $2^7 + 2^4 = 73^k - 1$, namely, $73^k = 145$, which is impossible. Thus $\{127\} = \pi(2^{2m+1} - 2^{m+1} + 1)$, which implies that $2^{2m+1} - 2^{m+1} + 1 = 127^\alpha$ for some $\alpha > 0$. Therefore $2^{2m+1} - 2^{m+1} \equiv 0 \pmod 3$ and it follows that $(-1) + (-1)^m \equiv 0 \pmod 3$, and so m is an even number, which is a contradiction since we have shown that m is an odd number.

(9.3) If $\{73\} = \pi(2^{2m+1} - 2^{m+1} + 1)$, similarly we can get a contradiction.

Case 10. Let $K/H \cong {}^2F_4(q)$, where $q = 2^{2m+1} > 2$. We know that the odd components of K/H are $\pi_2(K/H) = \pi(q^2 + \sqrt{2q^3} + q + \sqrt{2q} + 1)$ and $\pi_3(K/H) = \pi(q^2 - \sqrt{2q^3} + q - \sqrt{2q} + 1)$. Therefore $\pi_2(K/H) \bigcup \pi_3(K/H) = \pi(q^4 - q^2 + 1) = \{73, 127\} = \pi_2(G) \bigcup \pi_3(G)$.

Let $r \in \pi_2(K/H) \bigcup \pi_3(K/H)$. By assumption, $r \mid (q^2 + \sqrt{2q^3} + q + \sqrt{2q} + 1)(q^2 - \sqrt{2q^3} + q - \sqrt{2q} + 1)$. Hence $r \mid (q^4 - q^2 + 1)$, which implies that $r \mid q^6 + 1$. Consequently $q^6 \equiv -1 \pmod r$ and so $(-1/r) = 1$. Therefore $r \equiv 1 \pmod 4$ by Lemma 2.8, which is a contradiction, since $127 \equiv 3 \pmod 4$.

Case 11. Let $K/H \cong {}^2G_2(q)$, where $q = 3^{2m+1} > 3$. We know that the odd components of K/H are $\pi_2(K/H) = \pi(q - \sqrt{3q} + 1)$ and $\pi_3(K/H) = \pi(q + \sqrt{3q} + 1)$. Therefore $\pi_2(K/H) \bigcup \pi_3(K/H) = \pi(q^2 - q + 1) = \{73, 127\} = \pi_2(G) \bigcup \pi_3(G)$. Thus $127 \in \pi(q^2 - q + 1) \subseteq \pi(q^3 + 1)$ and it follows that $q^3 \equiv -1 \pmod{127}$, and so $(-1/127) = 1$, which implies $127 \equiv 1 \pmod 4$ by Lemma 2.8 and this is a contradiction, since $127 \equiv 3 \pmod 4$.

Case 12. Let $K/H \cong E_8(q)$, where $q = p^\alpha$ and $q \equiv 2, 3 \pmod 5$. Then the odd components of K/H are $\pi_2(K/H) = \pi(\frac{q^{10}-q^5+1}{q^2-q+1})$, $\pi_3(K/H) = \pi(\frac{q^{10}+q^5+1}{q^2-q+1})$, and $\pi_4(K/H) = \pi(q^8 - q^4 + 1)$, also the even component is $\pi(q(q^8-1)(q^{12}-1)(q^{14}-1)(q^{18}-1)(q^{20}-1))$.

(12.1) If $\pi(q^8 - q^4 + 1) = \{127\}$, then $127 \in \pi(q^{12}+1)$, which implies that $q^{12} \equiv -1 \pmod{127}$, and it follows that $(-1/127) = 1$, so $127 \equiv 1 \pmod 4$ by using Lemma 2.8 and this is a contradiction.

(12.2) If $\pi(\frac{q^{10}-q^5+1}{q^2-q+1}) = \{127\}$, similarly we can get a contradiction.

(12.3) If $\pi(\frac{q^{10}+q^5+1}{q^2+q+1}) = \{127\}$, then $127 \in \pi(q^{15}-1)$. Consequently $q^{15} \equiv 1 \pmod{127}$. If $q \equiv 1 \pmod{127}$ or $q^3 \equiv 1 \pmod{127}$ or $q^5 \equiv 1 \pmod{127}$, then $127 \mid (q^{12}-1)$ or $127 \mid (q^{20}-1)$, and it follows that $127 \in \pi_1(K/H) \subseteq \pi_1(G)$, which is a contradiction. Thus 127 is a primitive prime of $q^{15}-1$ by Lemma 2.9 and consequently $15 \mid 127 - 1$, which is a contradiction.

Similarly we can show that $K/H \not\cong E_8(q)$, where $q = p^\alpha$ and $q \equiv 0, 1, 4 \pmod 5$.

Case 13. Let $K/H \cong E_7(2)$. Now we prove that $K = G$. By Lemma 2.4, we know that $G/K \cong Out(E_7(2))$ and $Out(E_7(2)) = 1$. It follows that $G = K$.

Now the proof of Theorem 3.1 is complete. □

Corollary 3.1. Let G be a finite group satisfying $|G| = |E_7(2)|$ and $\Gamma(G) = \Gamma(E_7(2))$, then $G \cong E_7(2)$.

Proof. By Theorem 3.1, G has a normal subgroup H such that $G/H \cong E_7(2)$, which implies $H = 1$ since $|G| = |E_7(2)|$. □

Theorem 3.2. *Let G be a finite group such that $\Gamma(G) = \Gamma(E_7(3))$, then G has a unique nonabelian composition factor $K/H \cong E_7(3)$.*

Proof. In the following we only show some cases whose proof are different from the corresponding ones in Theorem 3.1. And all the other cases are similar to Theorem 3.1, so we omit the proof for convenience.

By [18, Tables 1a-1c], we have $t(G) = 3$ and $\pi_1(G) = \{2, 3, 5, 7, 11, 13, 19, 37, 41, 61, 73, 547\}$. Also the odd components are $\pi_2(G) = \{757\}$ and $\pi_3(G) = \{1093\}$.

By Lemma 2.4, if G is a finite group and $\Gamma(G) = \Gamma(E_7(3))$, then G has a normal series $1 \trianglelefteq H \trianglelefteq K \trianglelefteq G$ such that $G/K = 1$ or G/K is a π_1-group, $H = 1$ or a nilpotent π_1-group, and K/H is a nonabelian simple group with $t(K/H) \geq 3$, and $G/K \leq Out(K/H)$. Now by using the classification of finite simple groups and according to [18, Tables 1a-1c], we consider the following cases:

Case 1. Let $K/H \cong G_2(q)$, where $q = 3^\alpha$. Then the odd components of K/H are $\pi(q^2-q+1)$ and $\pi(q^2+q+1)$. Since $\pi_2(K/H)\bigcup\pi_3(K/H) = \pi_2(G)\bigcup\pi_3(G)$, we have $\pi(q^4+q^2+1) = \{757, 1093\}$, and it follows that $q^4+q^2+1 = 757^\beta\cdot 1093^\gamma$ for some $\beta, \gamma > 0$, namely, $3^{4\alpha}+3^{2\alpha}+1 = 757^\beta \cdot 1093^\gamma$. We know that $3^{4\alpha}+3^{2\alpha}+1 \equiv 3 \pmod 4$ and $757^\beta \cdot 1093^\gamma \equiv 1 \pmod 4$, which is a contradiction.

Case 2. Let $K/H \cong {}^2G_2(q)$, where $q = 3^{2m+1} > 3$. We know that the odd components of K/H are $\pi_2(K/H) = \pi(q-\sqrt{3q}+1)$ and $\pi_3(K/H) = \pi(q+\sqrt{3q}+1)$.

(2.1) If $\pi(q-\sqrt{3q}+1) = \{757\}$, then $\pi(3^{2m+1}+3^{m+1}+1) = \{1093\}$. Hence $3^{m+1}(3^m-1) = 757^\alpha - 1$ and $3^{m+1}(3^m+1) = 1093^\beta - 1$ for some $\alpha, \beta > 0$. We know that $4 \mid 757^\alpha - 1$ and $4 \mid 1093^\beta - 1$. Thus $3^m - 1 \equiv 0 \pmod 4$ and $3^m + 1 \equiv 0 \pmod 4$ and it follows that $4 \mid 2$, which is a contradiction.

(2.2) If $\pi(q+\sqrt{3q}+1) = \{757\}$, similarly we can get a contradiction.

Case 3. Let $K/H \cong E_8(q)$, where $q = p^\alpha$ and $q \equiv 2, 3 \pmod 5$. Then the odd components of K/H are $\pi_2(K/H) = \pi(\frac{q^{10}-q^5+1}{q^2-q+1})$, $\pi_3(K/H) = \pi(\frac{q^{10}+q^5+1}{q^2-q+1})$and $\pi_4(K/H) = \pi(q^8-q^4+1)$, also the even component is $\pi(q(q^8-1)(q^{12}-1)(q^{14}-1)(q^{18}-1)(q^{20}-1))$.

(3.1) If $\pi(\frac{q^{10}-q^5+1}{q^2-q+1}) = \{757\}$, then $757 \in \pi(q^{10}-q^5+1) \subseteq q^{15}+1$, which implies that $q^{30} \equiv 1 \pmod{757}$. If $q^i \equiv 1 \pmod{757}$, where $i \in \{1, 3, 5, 6, 10\}$, then $757 \in \pi_1(K/H)$, since $\pi_1(K/H) = \pi(q(q^8-1)(q^{12}-1)(q^{14}-1)(q^{18}-1)(q^{20}-1))$. We know that $\pi_1(K/H) \subseteq \pi_1(G)$, so $757 \in \pi_1(G)$, which is a contradiction. Thus 757 is a primitive prime of $q^{30}-1$ by Lemma 2.9 and consequently $30 \mid 757-1$, which is a contradiction.

(3.2) If $\pi(\frac{q^{10}+q^5+1}{q^2+q+1}) = \{757\}$, then similarly we can show that 757 is a primitive prime of $q^{15}-1$ and consequently $15 \mid 757-1$, which is a contradiction.

(3.3) If $\pi(q^8-q^4+1) = \{757\}$, also we can prove that 757 is a primitive prime of $q^{24}-1$ and consequently $24 \mid 757-1$, which is a contradiction.

Similarly we can show that $G \not\cong E_8(q)$, where $q = p^\alpha$ and $q \equiv 0, 1, 4 \pmod 5$.

At the beginning of the proof of Theorem 3.2, we have said that all the other cases are similar to the corresponding ones in Theorem 3.1, and we omit the proof for convenience.

Now the proof of Theorem 3.2 is complete. □

Corollary 3.2. Let G be a finite group satisfying $|G| = |E_7(3)|$ and $\Gamma(G) = \Gamma(E_7(3))$, then $G \cong E_7(3)$.

Proof. By Theorem 3.2, G has a nonabelian composition factor K/H such that $K/H \cong E_7(3)$, which implies $H = 1$ and $K = G$ since $|G| = |E_7(3)|$. □

Note that the conjecture of the second author was proved to be right for $E_7(q)$ in [15], thus Corollary 3.1 and Corollary 3.2 generalize the relative results.

References

[1] Z. Akhlaghi, M. Khatami and B. Khosravi, Quasirecognition by prime graph of the finite simple group ${}^2F_4(q)$, Acta. Math. Hungar., 122(4) (2009), 387-397.

[2] O.A. Alekseeva and A.S. Kondra'ev, Quasirecognition of one class of finite simple groups by the set of element orders, Siberian Math. J., 44 (2003), 195-207.

[3] J.H. Conway, R.T. Curtis, S.P. Norton, R.A. Parker, R.A. Wilson, Atlas of Finite Groups, Clarendon Press, Oxford: Clarendon Press, 1985.

[4] P. Crescenzo, A diophantine equation which arises in the theory of finite groups, Advances Math. 17 (1975), 25-29.

[5] K.W. Gruenberg and K.W. Rogggenkamp, Decomposition of the augmentation ideal and of the relation modules of a finite group, Proc. London Math. Soc., 31(2) (1975), 149-166.

[6] M. Hagie, The prime graph of a sporadic simple group, Comm. Algebra, 31 (2003), 4405-4424.

[7] B. Khosravi, n-recognition by prime graph of the simple group $PSL(2,q)$, J. Algebra Appl., 7(6) (2008), 735-748.

[8] A. Khosravi and B. Khosravi, A new characterization of some alternating and symmetric groups (II), Houston J. Math., 30(4) (2004), 465-478.

[9] A. Khosravi and B. Khosravi, Quasirecognition by prime graph of the finite simple group ${}^2G_2(q)$, Siberian Math. J., 48 (2007), 570-577.

[10] B. Khosravi and B. Khosravi and B. Khosravi, groups with the same prime graph as a CIT simple group, Houston J. Math., 33 (2007), 967-977.

[11] B. Khosravi, B. Khosravi and B. Khosravi, On the prime graph of $PSL(2,p)$ where $p > 3$ is a prime number, Acta. Math., Hungar., 116 (2007), 295-307.

[12] B. Khosravi and B. Khosravi and B. Khosravi, A characterization of the the finite simple group $L_{16}(2)$by its prime graph, Manuscripta Math., 126 (2008), 49-58.

[13] B. Khosravi and S. Salehi Amiri, On the prime graph of $L_2(q)$ where $q = p^\alpha < 100$, Quasigroups and Related Systems, 14 (2006), 179-190.

[14] W.J. Shi, A new characterization of the sporadic simple groups, Group Theory, Proceedings of the 1987 Singapore Group Theory Conference, Walter de Gruyter, Berlin New York, (1989), 531-540.

[15] W.J. Shi, The pure quantitative characterization of finite groups (I), Progress in Nature Sci. China, 4(3) (1994), 316-326.

[16] W.J. Shi, Pure quantitative characterization of finite groups, Front. Math. China, 2(1) (2007),123-125.

[17] W. Sierpinski, Elementary Theory of Number, Panstowe Wydawnictwo Naukowe, Warsaw, (Monografie Matematyczne, vol. 42) (1964).

[18] A.V. Vasil'ev and M. A. Grechkoseeva, On recognition of the finite simple orthogonal group of dimension 2^m, $2^m + 1$ and $2^m + 2$ over a field of characteristic 2, Siberian Math. J., 45 (2004), 420-432.

[19] J.S. Williams, Prime graph components of finite groups, J. Algebra, 69(2) (1981), 487-513.

[20] A.V. Zavarnitsin, Recognition of finite groups by the prime graph, Algebra Logic, 43 (2006), 220-231.

[21] K. Zsigmondy, Zur Theorie der Potenzreste, Monatsh. Math., und Phys., 3 (1892), 265-284.

A Characterization of Metacyclic p-Groups by Counting Subgroups

Qinhai Zhang

Department of Mathematics, Shanxi Normal University,
Linfen, Shanxi 041004, China
E-mail: zhangqh@dns.sxnu.edu.cn zhangqhcn@hotmail.com
www.sxnu.edu.cn

Assume G is a group of order p^n. $s_k(G)$ and $n_k(G)$ denote the number of subgroups and normal subgroups of order p^k of G, respectively. In this paper, we obtained the following result: Let G be a metacyclic p-group of order odd prime and H a finite p-group. If $s_k(G) = s_k(H)$ and $n_k(G) = n_k(H)$ for arbitrary a natural number k, then $G \cong H$.

Keywords: Metacyclic p-groups; Counting subgroups; Regular p-groups, p^s-abelian p-groups.

1. Introduction

For convenience, we introduce the following symbols. Assume G is a finite p-group. We use $s_k(G)$ to denote the number of subgroups of order p^k of G, $n_k(G)$ the number of normal subgroups of order p^k of G, $a_k(G)$ the number of abelian subgroups of order p^k of G and $c_k(G)$ the number of cyclic subgroups of order p^k of G.

Counting subgroups is important in the study of finite p-groups since the structure of some p-groups can be characterized by their number of subgroups. Some old and well-known results are

Theorem 1.1. *Assume G is a group of order p^n. Then*

(1) If $s_1(G) = 1$, then G is cyclic for $p > 2$, and G is a cyclic or the generalized quaternion group for $p = 2$;

(2) If $s_m(G) = 1$ for $1 < m < n$, then G is cyclic;

(3) If $s_m(G) = c_m(G)$ for $1 < m \leq n$, that is, every subgroup of order p^m is cyclic, then G is cyclic, or G is the generalized quaternion group for $p^m = 4$.

Recently, some further results are obtained. For example,

Theorem 1.2.[2] *Assume G is a group of order p^n and H is an elementary abelian p-group of order p^n, $n \geq 1$. Then*

(1) $s_1(G) \leq s_1(H)$, the equality holds if and only if $\exp(G) = p$;

(2) for $2 \leq m < n$, $s_m(G) \leq s_m(H)$, the equality holds if and only if $G \cong H$.

Theorem 1.3.[3] *Assume G is a group of order p^n, $p > 2$ and $n \geq 5$. Then for arbitrary natural number k with $0 \leq k \leq n$, $s_k(G) \leq 1 + p + 2p^2$ if and only if G is a group of order p^n with a cyclic subgroup of index p^2.*

General speaking, it is not enough to judge two groups are isomorphic upon condition that the number of subgroups of the the two groups is equal. The groups of order p^3 and exponent p^2 are such examples. Based on such observation, M.Y.Xu and H.P.Qu in their book[4] proposed the following

Problem 12.1.12 Assume G and H are groups of order p^n satisfy, for $m = 0, 1, \cdots, n$,

$$s_m(G) = s_m(H), c_m(G) = c_m(H), a_m(G) = a_m(H), n_m(G) = n_m(H).$$

Is G isomorphic to H?

It is pity that the answer is still negative.

Example 1.1. Assume $p \geq 5$, $G = \langle a, b, c \mid a^{p^2} = b^p = c^p = 1, [a, b] = c, [c, a] = 1, [c, b] = a^p \rangle$, $H = \langle a, b, c \mid a^{p^2} = b^p = c^p = 1, [a, b] = c, [c, a] = 1, [c, b] = a^{\sigma p} \rangle$, where σ is a fixed quadratic non-residue(mod p). It is easy to know that $G \ncong H$ by checking the list of groups of order p^4. But G and H satisfy the all conditions of Problem 12.1.12.

In this paper, we obtained a characterization of matacyclic p-groups by counting subgroups.

Throughout this paper p will an odd prime. For undefined notation and terminology, the reader is referred to book[1] .

2. Preliminaries

For convenience and self-containing, we introduce the following concepts.

Definition 2.1. Assume G is a group of order p^n. G is called *metacyclic* if G have a cyclic normal subgroup $\langle a \rangle$ such that $G/\langle a \rangle$ is cyclic; G is called *maximal class* if the nilpotency class $c(G) = n - 1$; G is called *regular* if $(ab)^p = a^p b^p c_3{}^p \ldots c_m{}^p$ for arbitrary $a, b \in G$, where $c_i \in \langle a, b\rangle'$.

Definition 2.2. Assume G is a group of order p^n and $\exp(G) = p^e$. For any integer s with $0 \leq s \leq e$, we define

$$\Omega_s(G) = \langle a \in G \mid a^{p^s} = 1\rangle, \quad \mho_s(G) = \langle a^{p^s} \mid a \in G\rangle.$$

Now, set $|\Omega_i(G)/\Omega_{i-1}(G)| = p^{\omega_i}$, $1 \leq i \leq e$. It is easy to prove that $|\mho_{i-1}(G)/\mho_i(G)| = p^{\omega_i}$ and $\omega = \omega_1 \geq \omega_2 \geq \cdots \geq \omega_e > 0$. We call $(\omega_1, \ldots, \omega_e)$ the *ω-invariants* of G.

For $1 \leq j \leq \omega = \omega_1$, we use e_j to denote the number of such that $\omega_i \geq j$. Then $e = e_1 \geq e_2 \geq \cdots \geq e_\omega \geq 1$. We call $(e_1, \ldots, e_\omega)$ the *e-invariants* or *type invariants* of G. If G is abelian, then $p^{e_1}, \ldots, p^{e_\omega}$ are the orders of basis elements.

Definition 2.3. Assume G is a group of order p^n and C_i denotes subgroups $C_G(G_i/G_{i+2}$ of G for $i = 2, 3, \cdots, n-2$, where G_i is the ith term of the lower central series of G. An element s is called *uniform* if $s \notin \bigcup_{i=2}^{n-2} C^{(i)}$.

3. The number of subgroups of metacyclic p-groups of order odd prime

The following lemma is fundamental in counting subgroups.

Lemma 3.1 (P.Hall's enumeration principle). *Assume G is a finite p-group. Then*

$$s(G) - \sum_{M \in \mathcal{S}_1} s(M) + p \sum_{M \in \mathcal{S}_2} s(M) - \cdots + (-1)^d p^{\binom{d}{2}} s(\Phi(G)) = 0.$$

Where $\Phi(G)$ is the Frattini subgroup of G. The subgroup of G containing $\Phi(G)$ is called a large subgroup. For $i = 0, 1, \cdots, d = d(G)$, $\mathcal{S}_i$ denotes the set of large subgroups of index p^i, $\mathfrak{S}$ a set consisting of proper subgroups of G and $s(M)$ the number of subgroups of $\mathfrak{S}$ contained in M.

For metacyclic p-groups, Newman and Xu in their jointed paper[5] classified such groups.

Theorem 3.1.[5] *Any metacyclic p-group G, p odd, has the following presentation:*

$$G = \langle a, b : a^{p^{r+s+u}} = 1,\ b^{p^{r+s+t}} = a^{p^{r+s}},\ a^b = a^{1+p^r}\rangle$$

where r, s, t, u are non-negative integers with $r \geq 1$ and $u \leq r$. Different values of the parameters r, s, t and u with the above conditions give non-isomorphic metacyclic p-groups. We use $< r, s, t, u; p >$ to denote the group.

Furthermore, G is split if and only if $stu = 0$.

It is easy to see that G is a regular p-group with the type invariant $(r+s+t+u, r+s)$ and $\mho_{s+u}(G) \leq Z(G)$.

Assume G is a metacyclic p-groups of order odd prime. Then G is a regular p-group with $\omega \leq 2$. Assume the type invariant of G is (e_1, e_2). Obviously,

$$|\Omega_k(G)| = \begin{cases} p^{2k}, & 1 \leq k \leq e_2; \\ p^{k+e_2}, & e_2 < k \leq e_1. \end{cases}$$

We observed the fact: $c_k(G) = (|\Omega_k(G)| - |\Omega_{k-1}(G)|)/(p^k - p^{k-1})$. Thus we have the following

Theorem 3.2. *Assume G is a metacyclic p-group with the type invariant (e_1, e_2). Then*

$$c_k(G) = \begin{cases} p^{k-1}(1+p), & 1 \leq k \leq e_2; \\ p^{e_2}, & e_2 < k \leq e_1. \end{cases}$$

For $s_k(G)$, we have the following

Theorem 3.3. *Assume G is a metacyclic p-group with the type invariant (e_1, e_2). Then*

$$s_k(G) = \begin{cases} 1+p+\cdots+p^k, & 1 \leq k \leq e_2; \\ 1+p+\cdots+p^{e_2}, & e_2 < k < e_1; \\ 1+p+\cdots+p^{e_1+e_2-k}, & e_1 \leq k \leq e_1+e_2. \end{cases}$$

Proof. We use induction on $|G|$. If $|G| \leq p^2$, then the conclusion is obvious. Assume $|G| \geq p^3$ and the conclusion is true for the groups of order less than $|G|$.

If $1 \leq k < e_1$, then $\Omega_k(G) < G$. We observed $s_k(G) = s_k(\Omega_k(G))$. Thus, by induction hypothesis, the Theorem is true.

If $e_1 \leq k \leq e_1 + e_2$, we take a maximal subgroup of G with the type invariant (e_1', e_2'), then $e_1' + e_2' = e_1 + e_2 - 1$ and $e_1' \leq e_1 \leq k$. Assume the type invariant of $\Phi(G)$ is (e_1'', e_2''). Then $e_1''+e_2'' = e_1+e_2-2$ and $e_1'' \leq e_1 \leq k$. By Lemma 3.1 and induction hypothesis we have $s_k(G) = (1+p)(1+p+\cdots+p^{e_1+e_2-1-k}) - p(1+p+\cdots+p^{e_1+e_2-2-k}) = 1+p+\cdots+p^{e_1+e_2-k}$. □

Theorem 3.4.

Assume $G \cong < r, s, t, u; p >$. *If* G *is non-abelian, then*

(1) the order of non-normal subgroups of G *with the smallest order is* p^{r-u+1};

(2) If $s \neq 0$, *then the order of non-normal subgroups of* G *with the largest order is* $p^{r+2s+u+t-1}$;

(3) If $s = 0$, *the order of non-normal subgroups of* G *with the largest order is* p^{r+u-1}.

Proof. Let $x = ab^{-p^t}$. Then $o(x) = p^{r+s}$.

(1) We consider subgroup $H_1 = \langle x^{p^{s+u-1}} \rangle$. Then $|H_1| = p^{r-u+1}$. By calculation we have the minimal subgroup of H_1 is $\langle a^{p^{r+s-1}} b^{-p^{r+s+t-1}} \rangle$. It follows that $[x^{p^{s+u-1}}, b] = a^{p^{r+s+u-1}} \notin H_1$. Thus H_1 is non-normal. So we need only to prove that all subgroups of G whose order $\leq p^{r-u}$ are normal. Assume $M < G$ and $|M| \leq p^{r-u}$, Then $M \leq \Omega_{r-u}(G) = \langle x^{p^{s+u}}, a^{p^{s+2u}} \rangle \leq \mho_{s+u}(G) \leq Z(G)$. This implies $M \trianglelefteq G$.

(2) If $s \neq 0$, we consider subgroup $H_2 = \langle b, a^{p^{r+1}} \rangle$. Then $|H_2| = p^{r+2s+t+u-1}$. Since $[b, a] = a^{p^r} \notin H_2$, H_2 is not normal. So we need only to prove that all subgroups of G whose order $\leq p^{r+2s+t+u}$ are normal. Assume $M < G$ and $|M| \geq p^{r+2s+t+u}$. Then $|\langle a \rangle : M \cap \langle a \rangle| \leq |G : M| \leq p^r$ and $|\langle a \rangle : G'| = p^r$. Thus $G' \leq M \cap \langle a \rangle \leq M$. This implies $M \trianglelefteq G$.

(3) If $s = 0$, since G is non-abelian, we have $u \geq 1$, and $G = \langle b \rangle \rtimes \langle x \rangle$. We consider subgroup $H_3 = \langle b^{p^{r+t+1}}, x \rangle$. Then $|H_2| = p^{r+u-1}$. Since $[b, x] = b^{p^{r+t}} \notin H_3$, H_3 is not normal. So we need only to prove that all subgroups of G whose order $\leq p^{p^{r+u}}$ are normal. Assume $M < G$ and $|M| \geq p^{r+u}$. Then $|\langle b \rangle : M \cap \langle b \rangle| \leq |G : M| \leq p^{r+t}$ and $|\langle b \rangle : G'| = p^{r+t}$. Thus $G' \leq M \cap \langle b \rangle \leq M$. This implies $M \trianglelefteq G$. □

4. A characterization of metacyclic p-groups by counting subgroups

First, we give some lemmas.

Lemma 4.1. *Assume* G *and* H *are finite abelian* p*-groups, and* $s_k(G) = s_k(H)$ *for arbitrary a natural number* k. *Then* $G \cong H$.

Proof. First, we prove by induction that $\Omega_i(G) \cong \Omega_i(H)$ for arbitrary a natural number i.

Since G and H are abelian, $s_i(G) = c_i(G) + s_i(\Omega_{i-1}(G))$ and $s_i(H) = c_i(H) + s_i(\Omega_{i-1}(H))$. By induction we have $\Omega_s(G) \cong \Omega_s(H)$ if $s < i$. In particular, $\Omega_{i-1}(G) \cong \Omega_{i-1}(H)$. Thus $c_i(G) = c_i(H)$.

Since $c_i(G) = \frac{|\Omega_i(G)|-|\Omega_{i-1}(G)|}{p^i-p^{i-1}}$ and $c_i(H) = \frac{|\Omega_i(H)|-|\Omega_{i-1}(H)|}{p^i-p^{i-1}}$, $|\Omega_i(G)| = |\Omega_i(H)|$. It follows that $|\Omega_s(G)| = |\Omega_s(H)|$ if $s \leq i$. Obviously, $\Omega_i(G)$ and $\Omega_i(H)$ have the same ω-invariant and type invariant. Moreover, since $\Omega_i(G)$ and $\Omega_i(H)$ are abelian, $\Omega_i(G) \cong \Omega_i(H)$.

Finally, assume $p^e = \exp(G) \geq \exp(H)$ without loss of generality. Then $G = \Omega_e(G)$ and $H = \Omega_e(H)$. Thus $G \cong H$. □

Lemma 4.2. *Assume G and H are metacyclic p-groups of order odd prime. If $s_k(G) = s_k(H)$ and $n_k(G) = n_k(H)$ for arbitrary a natural number k. Then $G \cong H$.*

Proof. We observed a p-group L of order odd is abelian if and only if $s_k(L) = n_k(L)$. Thus we get G and H are abelian or non-abelian in the same time.

If G and H are abelian, then, by Lemma 4.1, the conclusion is true.

If G and H are not abelian, then we can assume $G \cong < r_1, s_1, t_1, u_1; p >$, $H \cong < r_2, s_2, t_2, u_2; p >$ by Theorem 3.1.

If $s_1 s_2 \neq 0$, then, by Theorem 3.3 and Lemma 3.4, we have

$$\begin{cases} r_1 + s_1 = r_2 + s_2, & (1) \\ r_1 + s_1 + t_1 + u_1 = r_2 + s_2 + t_2 + u_2, & (2) \\ r_1 + 2s_1 + u_1 + t_1 - 1 = r_2 + 2s_2 + u_2 + t_2 - 1, & (3) \\ r_1 - u_1 + 1 = r_2 - u_2 + 1. & (4) \end{cases}$$

By solving above equation system we have $r_1 = r_2, s_1 = s_2, t_1 = t_2, u_1 = u_2$. It follows by Theorem 3.1 that $G \cong H$.

If $s_1 = s_2 = 0$, then by (1), (2) and (4) we get $G \cong H$. So we assume that $s_1 = 0, s_2 \neq 0$.

By Theorem 3.3 and Lemma 3.4 we have

$$\begin{cases} r_1 = r_2 + s_2, & (1') \\ r_1 + t_1 + u_1 = r_2 + s_2 + t_2 + u_2, & (2') \\ r_1 + u_1 - 1 = r_2 + 2s_2 + u_2 + t_2 - 1, & (3') \\ r_1 - u_1 + 1 = r_2 - u_2 + 1. & (4') \end{cases}$$

By $(1')$ and $(4')$, we have $s_2 = u_1 - u_2$. From $(1'), (3')$ and $(4')$ we get $t_2 = 0$. By $(2')$, $(4')$ and $s_2 = u_1 - u_2$, we have $t_1 = u_2 - u_1$. Thus $t_1 = -s_2 < 0$. This contradicts Theorem 3.1. □

Lemma 4.3.[6] *Assume G is a group of order p^n. If G has not any normal and elementary abelian subgroup of order p^3, Then one of the following is true:*

(1) *G is a metacyclic p-group;*

(2) *G is a 3-group of maximal class;*

(3) *$G = EM$, where $E = \Omega_1(G)$ is a non-abelian group of order p^3 and exponent p, M is a cyclic group of order p^{n-2}, Moreover, $Z(G)$ is cyclic, $|G/Z(G)| \leq p^3$, and $|M : C_G(E) \cap M| \leq p$.*

By Lemma 4.3 we get immediately the following

Lemma 4.4. *Assume G is a finite p-group. If $p \geq 5$, Then G is metacyclic p-group if and only if $s_1(G) \leq 1 + p$.*

Remark. If $p = 3$, Lemma 4.4 is not true. For example, $\langle a, b, c | b^9 = c^3 = 1, [c, b] = 1, a^3 = b^{-3}, [b, a] = c, [c, a] = b^{-3} \rangle$ is a counterexample. But we have the following

Lemma 4.5. *Assume G is a metacyclic 3-group and H is a finite p-group. If $|G| = |H| \neq 3^4$, $s_1(G) = s_1(H)$ and $s_2(G) = s_2(H)$, then H is also a metacyclic 3-group.*

Proof. If $|G| = |H| < 3^4$, Obviously, the conclusion ie true. Assume that $|G| = |H| > 3^4$.

Since G is a metacyclic 3-group, $s_1(H) = s_1(G) \leq 4$. By Lemma 4.3 we have H is metacyclic or a 3-group of maximal class.

If H is a 3-group of maximal class, then by $s_1(H) = 4$ we know H is a 3-group of maximal class whose uniform elements are of order 9. It follows that $s_2(H) > c_2(H) \geq (2|H|/3)/(9-3) = |H|/9 \geq 3^3$. On the other hand, by Theorem 3.3 we get $s_2(G) \leq 1 + 3 + 3^2$, this is a contradiction. So H is also a metacyclic 3-group. □

Remark. Lemma 4.5 is not true for groups of order 3^4. For example, let $G = \langle a, b \; | a^9 = b^9 = 1, a^b = a^4 \rangle$, $H = \langle a, b, c \; | a^9 = c^3 = 1, b^3 = a^3, [a, b] = c, [c, a] = 1, [c, b] = a^{-3} \rangle$. By calculation we get $s_1(G) = s_1(H) = 4, s_2(G) = s_2(H) = 13$. But H is a 3-group of maximal class.

Finally, we obtained the main theorem in this paper:

Theorem 4.1. *Let G be a metacyclic p-group of order odd prime and H a finite p-group. If $s_k(G) = s_k(H)$ and $n_k(G) = n_k(H)$ for arbitrary a natural number k, then $G \cong H$.*

Proof. If $|G| = |H| \leq 3^4$, then, by checking the list of groups of order $\leq 3^4$, the conclusion is true. If $|G| = |H| \neq 3^4$, then , by Lemma 4.4 and

Lemma 4.5, we have H is also a metacyclic 3-group. The conclusion follows by Lemma 4.2. □

Acknowledgments

This work was supported by NSFC (Grant No. 11071150) and NSF of Shanxi Province (Grant No. 2008012001).

References

1. Huppert, B. *Endliche Gruppen I*, Springer-Verlag, 1967.
2. Y.Fan, A characterization of elementary abelian p-groups by counting subgroups, *Math. Practice Theory* 1 (1988), 63–65. (in Chinese)
3. H.P. Qu, Y. Sun and Q.H. Zhang, Finite p-groups in which the number of subgroups of possible order are less than or equal p^3, *Chin. Ann. Math.*, 31B:4(2010), 497-506.
4. M.Y. Xu and H.P. Qu, *Finite p-groups*, Beijing University Press, Beijing, 2010. (in Chinese)
5. M. F. Newman and M.Y. Xu, Metacyclic groups of prime-power order (Research announcement), *Adv. Math.(Beijing)* 17(1988), 106-107.
6. N.Blackburn, Generalization of centain elementary theorems on p groups, *Proc.London Math.Soc.* 11:3(1961), 1–22.

The Variety Generated by All Non-Permutative and Non-Idempotent Semigroups of Order Four*†

Wen Ting Zhang and Yan Feng Luo
Department of Mathematics, Lanzhou University,
Lanzhou, Gansu, 730000, P. R. of China
zhangwt@lzu.edu.cn, luoyf@lzu.edu.cn

Abstract

Denote by $\mathbf{S}_n$ the variety generated by all semigroups of order n. The varieties $\mathbf{S}_2$ and $\mathbf{S}_3$ are finitely based, and Volkov proved that the variety $\mathbf{S}_n$ is non-finitely based for all $n \geq 5$. However, the finite basis problem for $\mathbf{S}_4$ is still open. Let $\mathbf{I}_4$ (resp. $\mathbf{P}_4$) be the variety generated by all idempotent (resp. permutative) semigroups of order four and $\mathbf{S}'_4$ be the variety generated by all non-permutative and non-idempotent semigroups of order four. Clearly, $\mathbf{S}_4 = \mathbf{I}_4 \vee \mathbf{P}_4 \vee \mathbf{S}'_4$ and $\mathbf{I}_4$ (resp. $\mathbf{P}_4$) is finitely based. In this paper, it is shown that the variety $\mathbf{S}'_4$ is finitely based and a finite identity basis for $\mathbf{S}'_4$ is given.

Keywords: Semigroups; Varieties; Finite basis.

Mathematics Subject Classification 2000: 20M07.

1 Introduction

In the theory of semigroup varieties, semigroups of small order play crucial roles. The presence or absence of certain small semigroups in a variety often controls important structural and equational properties satisfied by the variety. For instance, any locally finitely variety containing Perkins's semigroup B_2^1 of order six is non-finitely based [16], while any variety that does not contain the null semigroup of order two consists entirely of completely regular semigroups. Refer to [19, Table 1] for a list of other properties satisfied by a variety that excludes a combination of other semigroups of order two.

In recent years, small semigroups have been studied in several ways. Most notably, all semigroups of order five [20, 21] or less [4, 15] had been shown to be finitely based. On the other hand, a semigroup with as few as four elements can generate a variety that is *finitely universal* [12] in the sense that its subvariety lattice contains an isomorphic copy of any finite lattice. (In contrast, the variety generated by

*Dedicated to Professor K. P. Shum on the Occasion of his 70th birthday.

†This research was partially supported by the National Natural Science Foundation of China (No. 10571077, 10971086).

any finite group [14] or finite associative ring [9] is finitely based and contains only finitely many subvarieties.) Refer to the surveys [17, 19, 18] for more information on semigroup varieties and the lattices they constitute.

Small semigroups have also been studied collectively. For each integer $n \geq 2$, let $\mathbf{S}_n$ be the variety generated by all semigroups of order n. The variety $\mathbf{S}_2$ is contained in the variety which is defined by the identities set $\{x^2 \approx x^4, xyzt \approx xzyt\}$ (see [8]) and so $\mathbf{S}_2$ is finitely based. The finite basis property for the variety $\mathbf{S}_3$ was first announced in [18], and a finite identity basis was given in [13]. In [22], Volkov has shown that for each integer $n \geq 5$, the variety $\mathbf{S}_n$ is non-finitely based. These results naturally led to question whether or not the variety $\mathbf{S}_4$ is finitely based. Up to isomorphism and anti-isomorphism, there are 126 semigroups of order four, and so directly to solve the finitely basis problem for the variety $\mathbf{S}_4$ is quite difficult.

Note that the finite basis property for semigroups of order four was verified by Edmunds in [4]. Since a finite semigroup is finitely based if it is either idempotent [2, 5, 6] or permutative [15], Edmunds essentially proved that all non-permutative and non-idempotent semigroups of order four are finitely based. There are ten such semigroups up to isomorphism and anti-isomorphism. All subvarieties of varieties generated by these ten semigroups have been described in [7, 10, 11, 23, 24]

Let

(a) $\mathbf{I}_4$ the variety generated by all idempotent semigroups of order four;

(b) $\mathbf{P}_4$ the variety generated by all permutative semigroups of order four;

(c) $\mathbf{S}_4'$ the variety generated by all non-permutative and non-idempotent semigroups of order four.

Then $\mathbf{S}_4 = \mathbf{I}_4 \vee \mathbf{P}_4 \vee \mathbf{S}_4'$. It is well known that both any idempotent variety and any periodic variety that satisfies some permutation identity are finitely based. Hence $\mathbf{I}_4$ and $\mathbf{P}_4$ are finitely based.

In this paper, we take our main attention to the finite basis problem for the variety $\mathbf{S}_4'$. It is shown that the variety $\mathbf{S}_4'$ is finitely based and a finite identity basis is given.

This paper contains three sections. Notation and some background material are given in Section 2. In Section 3, a finite identity basis for $\mathbf{S}_4'$ is obtained, and so $\mathbf{S}_4'$ is finitely based.

2 Preliminaries

Most of the natation and background material of this paper are given in this section. Refer to the monograph [1] for any undefined terminology.

2.1 Letters and words

Let $\mathcal{X}$ be a fixed countably infinite alphabet throughout. For any subset $\mathcal{A}$ of $\mathcal{X}$, denote by $\mathcal{A}^*$ the free monoid over $\mathcal{A}$. Elements of $\mathcal{X}$ and $\mathcal{X}^*$ are referred to as *letters* and *words* respectively.

Let x, y be any distinct letters and $\mathbf{w}$ be any word. Then

- the *content* of $\mathbf{w}$, denoted by $\mathsf{C}(\mathbf{w})$, is the set of letters occurring in $\mathbf{w}$;
- the *head* of $\mathbf{w}$, denoted by $\mathsf{h}(\mathbf{w})$, is the first letter occurring in $\mathbf{w}$;
- the *tail* of $\mathbf{w}$, denoted by $\mathsf{t}(\mathbf{w})$, is the last letter occurring in $\mathbf{w}$;
- the *initial part* of $\mathbf{w}$, denoted by $\mathsf{ip}(\mathbf{w})$, is the word obtained from $\mathbf{w}$ by retaining the first occurrence of each letter;
- the *final part* of $\mathbf{w}$, denoted by $\mathsf{fp}(\mathbf{w})$, is the word obtained from $\mathbf{w}$ by retaining the last occurrence of each letter;
- The *multiplicity* of x in $\mathbf{w}$, denoted by $\mathsf{m}(x, \mathbf{w})$, is the number of times x occurs in $\mathbf{w}$;
- The *multiplicity* of x before the first occurrence of y in $\mathbf{w}$, denoted by $\overrightarrow{\mathsf{m}}_y(x, \mathbf{w})$, is the number of times x occurs before the first occurrence of y in $\mathbf{w}$;
- The *multiplicity* of x after the last occurrence of y in $\mathbf{w}$, denoted by $\overleftarrow{\mathsf{m}}_y(x, \mathbf{w})$, is the number of times x occurs after the last occurrence of y in $\mathbf{w}$;
- The *multiplicity* of x between the first and the last occurrence of y in $\mathbf{w}$, denoted by $\overleftrightarrow{\mathsf{m}}_y(x, \mathbf{w})$, is the number of times x occurs between the first and the last occurrence of y in $\mathbf{w}$;
- x is *simple* in $\mathbf{w}$ if $\mathsf{m}(x, \mathbf{w}) = 1$.

A word is said to be *simple* if all of its letters are simple in it.

Example 2.1 *Suppose that* $\mathbf{w} = x^2zxy^2xtzx^3t^4$. *Then*

(1) $\mathsf{C}(\mathbf{w}) = \{x, y, z, t\}$, $\mathsf{ip}(\mathbf{w}) = xzyt$ *and* $\mathsf{fp}(\mathbf{w}) = yzxt$;

(2) $\mathsf{m}(x, \mathbf{w}) = 7,\ \mathsf{m}(y, \mathbf{w}) = \mathsf{m}(z, \mathbf{w}) = 2,$ *and* $\mathsf{m}(t, \mathbf{w}) = 5$;

(3) $\overrightarrow{\mathsf{m}}_y(x, \mathbf{w}) = 3,\ \overleftrightarrow{\mathsf{m}}_y(x, \mathbf{w}) = 0$ *and* $\overleftarrow{\mathsf{m}}_y(x, \mathbf{w}) = 4$;

(4) $\overrightarrow{\mathsf{m}}_x(y, \mathbf{w}) = 0,\ \overleftrightarrow{\mathsf{m}}_x(y, \mathbf{w}) = 2$ *and* $\overleftarrow{\mathsf{m}}_x(y, \mathbf{w}) = 0$.

Let $\mathbf{w}$ be a word and $\mathsf{C}(\mathbf{w}) = \{x_1, x_2, \ldots, x_n\}$. Let $\{x_{i_1}, x_{i_2}, \ldots, x_{i_m}\}$ be a subset of $\mathsf{C}(\mathbf{w})$, we denote by $\mathbf{w}(x_{i_1}, \ldots, x_{i_m})$ the word obtained from $\mathbf{w}$ by deleting every occurrence of the letters in $\mathsf{C}(\mathbf{w})\backslash\{x_{i_1}, \ldots, x_{i_m}\}$ from $\mathbf{w}$.

2.2 Identities and varieties

An identity is typically written as $\mathbf{u} \approx \mathbf{v}$ where $\mathbf{u}$ and $\mathbf{v}$ are nonempty words. Let Π be any set of identities. The deducibility of an identity $\mathbf{u} \approx \mathbf{v}$ from Π is indicated by $\Pi \vdash \mathbf{u} \approx \mathbf{v}$ or $\mathbf{u} \overset{\Pi}{\approx} \mathbf{v}$. Let $\mathbf{V}$ be a semigroup variety. An identity $\mathbf{u} \approx \mathbf{v}$ is said to be *satisfied* by $\mathbf{V}$, denoted by $\mathbf{V} \vDash \mathbf{u} \approx \mathbf{v}$, if it is satisfied by all semigroups in $\mathbf{V}$. The variety $\mathbf{V}$ *defined* by Π is the class of all semigroups that satisfy all identities in Π; in this case, Π is said to be an identity *basis* for the variety. A variety $\mathbf{V}$ is said to be *finitely based* if it possesses a finite basis, otherwise, $\mathbf{V}$ is called *non-finitely based.*

A set of identities Π is said to be *closed under deleting simple letters* if Π contains every identity which is deduced from any identity of Π by setting some subset of the simple letters in the identity equal to 1. We denote by $\Pi^\star$ the closure of Π under deleting simple letters.

2.3 Identities of some small semigroups

Let Y_2 be the semilattice of order two and N_2^1 be the null semigroup of order two with an identity element adjoined. These semigroups are given by the following presentations:

$$Y_2 =< 0, 1 >,$$
$$N_2^1 =< 0, 1, a : a^2 = 0 > .$$

The variety generated by a semigroup S is denoted by $\operatorname{var} S$. For any $\mathsf{x} \in \{\mathsf{C}, \mathsf{ip}, \mathsf{fp}\}$, an identity $\mathbf{u} \approx \mathbf{v}$ is said to be x-*compliant* if $\mathsf{x}(\mathbf{u}) = \mathsf{x}(\mathbf{v})$.

Lemma 2.2 *(1) An identity holds in the semilattice Y_2 if and only if it is* C-*compliant.*

(2) An identity $\mathbf{u} \approx \mathbf{v}$ holds in the monoid N_2^1 if and only if for each $x \in \mathcal{X}$, either $\mathsf{m}(x, \mathbf{u}) = \mathsf{m}(x, \mathbf{v}) = 1$ or $\mathsf{m}(x, \mathbf{u}), \mathsf{m}(x, \mathbf{v}) \geq 2$.

Proof. These results are well-known and easy to verify. See, for example, [1]. □

In [4], Edmunds has listed all non-permutative and non-idempotent semigroups of order four up to isomorphism and anti-isomorphism, and an identity basis for each variety generated by these semigroups was given. There are 10 such semigroups and 9 such varieties. It turns out that the following three semigroups

A	a	b	c	d
a	a	a	a	a
b	a	b	c	d
c	c	c	c	c
d	c	d	a	b

B	a	b	c	d
a	a	a	a	a
b	a	a	a	b
c	c	c	c	c
d	a	b	c	d

C	a	b	c	d
a	a	a	a	a
b	a	a	a	b
c	a	b	c	b
d	a	a	a	d

and the dual semigroup A^* of A are required in this paper. Note that in [4], the semigroups A, B and C are denoted by $S(6, 25)$, $S(5, 2)$ and $S(4, 22)$ respectively.

Lemma 2.3 *([4, Proposition])*

(1) $\{x \approx x^3, xyx^2 \approx xy, xy^2x \approx xyxy\}$ *is an identity basis for* var A;

(2) $\{x^2 \approx x^3, xyx \approx x^2y\}$ *is an identity basis for* var B;

(3) $\{x^2 \approx x^3, xyx \approx x^2yx \approx xy^2x \approx xyx^2 \approx xyxy \approx yxy\}$ *is an identity basis for* var C.

Lemma 2.4 *Let* var $A \models \mathbf{u} \approx \mathbf{v}$. *Then*

(1) $\mathsf{ip}(\mathbf{u}) = \mathsf{ip}(\mathbf{v})$;

(2) $\mathsf{m}(x, \mathbf{u}) \equiv \mathsf{m}(x, \mathbf{v}) \pmod 2$ *for all* $x \in \mathcal{X}$;

(3) $\overrightarrow{\mathsf{m}}_y(x, \mathbf{u}) \equiv \overrightarrow{\mathsf{m}}_y(x, \mathbf{v}) \pmod 2$ *for all* $x, y \in \mathcal{X}$.

Proof. It follows from Lemma 4.5 of [3] after noting that the semigroup M_{31} of [3] is just A with a zero adjoined. □

Dually, we have

Lemma 2.5 *Let* var $A^* \models \mathbf{u} \approx \mathbf{v}$. *Then*

(1) $\mathsf{fp}(\mathbf{u}) = \mathsf{fp}(\mathbf{v})$;

(2) $\mathsf{m}(x, \mathbf{u}) \equiv \mathsf{m}(x, \mathbf{v}) \pmod 2$ *for all* $x \in \mathcal{X}$;

(3) $\overleftarrow{\mathsf{m}}_y(x, \mathbf{u}) \equiv \overleftarrow{\mathsf{m}}_y(x, \mathbf{v}) \pmod 2$ *for all* $x, y \in \mathcal{X}$.

Corollary 2.6 *Let* var $A \vee$ var $A^* \models \mathbf{u} \approx \mathbf{v}$. *Then* $\overleftrightarrow{\mathsf{m}}_y(x, \mathbf{u}) \equiv \overleftrightarrow{\mathsf{m}}_y(x, \mathbf{v}) \pmod 2$ *for all* $x, y \in \mathcal{X}$.

Proof. Since

$$\begin{aligned} \mathsf{m}(x, \mathbf{u}) &= \overrightarrow{\mathsf{m}}_y(x, \mathbf{u}) + \overleftrightarrow{\mathsf{m}}_y(x, \mathbf{u}) + \overleftarrow{\mathsf{m}}_y(x, \mathbf{u}), \\ \mathsf{m}(x, \mathbf{v}) &= \overrightarrow{\mathsf{m}}_y(x, \mathbf{v}) + \overleftrightarrow{\mathsf{m}}_y(x, \mathbf{v}) + \overleftarrow{\mathsf{m}}_y(x, \mathbf{v}), \end{aligned}$$

it follows directly from Lemmas 2.4 and 2.5. □

Lemma 2.7 *Let* var $B \models \mathbf{u} \approx \mathbf{v}$. *Then either* $\mathsf{m}(x, \mathbf{u}) = \mathsf{m}(x, \mathbf{v}) = 1$ *or* $\mathsf{m}(x, \mathbf{u})$, $\mathsf{m}(x, \mathbf{v}) \geq 2$ *for all* $x \in \mathcal{X}$.

Proof. Note that N_2^1 is a subsemigroup of B. It follows directly from Lemma 2.2(2). □

3 A finite basis for $\mathbf{S}_4'$

In this section, we shall prove that the variety $\mathbf{S}_4'$ is finitely based and a finite identity basis will be given.

Let

$$xyx^2zx \approx xyzx, \tag{I}$$
$$xz_1yz_2xyz_3xz_4y \approx xz_1yz_2yxz_3xz_4y, \tag{II}$$
$$xz_1yz_2xyz_3yz_4x \approx xz_1yz_2yxz_3yz_4x, \tag{III}$$
$$tz_1xz_2x^2y^2z_3yz_4t \approx tz_1xz_2y^2x^2z_3yz_4t, \tag{IV}$$
$$tz_1xz_2x^2y^2z_3tz_4y \approx tz_1xz_2y^2x^2z_3tz_4y, \tag{V}$$
$$xz_1tz_2x^2y^2z_3yz_4t \approx xz_1tz_2y^2x^2z_3yz_4t, \tag{VI}$$
$$xz_1tz_2x^2y^2z_3tz_4y \approx xz_1tz_2y^2x^2z_3tz_4y \tag{VII}$$

and $\Gamma = \{\text{I}, \text{II}, \text{III}, \text{IV}, \text{V}, \text{VI}, \text{VII}\}$.

From these identities it is easy to see that

Lemma 3.1 *(1)* $\{\text{I}\}^\star$ *implies that* $x^2 \approx x^4$ *and* $x^3yx \approx xyx \approx xyx^3$*;*

(2) $\{\text{II}, \text{III}\}^\star$ *implies that* $\mathbf{w}_1xy\mathbf{w}_2 \approx \mathbf{w}_1yx\mathbf{w}_2$ *for* $\mathbf{w}_1, \mathbf{w}_2 \in \mathcal{X}^+$*, and* $x, y \in \mathsf{C}(\mathbf{w}_1) \cap \mathsf{C}(\mathbf{w}_2)$*;*

(3) $\{\text{IV}, \text{V}, \text{VI}, \text{VII}\}^\star$ *implies that* $\mathbf{w}_1x^2y^2\mathbf{w}_2 \approx \mathbf{w}_1y^2x^2\mathbf{w}_2$ *for* $\mathbf{w}_1, \mathbf{w}_2 \in \mathcal{X}^+$, $\mathsf{C}(\mathbf{w}_1) \cap \mathsf{C}(\mathbf{w}_2) \neq \emptyset$*,* $x \in \mathsf{C}(\mathbf{w}_1)$ *and* $y \in \mathsf{C}(\mathbf{w}_2)$*.*

Lemma 3.2 *Let* $\mathbf{w}_1, \mathbf{w}_2, \mathbf{w} \in \mathcal{X}^+$ *with* $\mathsf{C}(\mathbf{w}) \subseteq \mathsf{C}(\mathbf{w}_1) \cap \mathsf{C}(\mathbf{w}_2)$*. Then*

(1) $\{\text{II}, \text{III}\}^\star$ *implies that* $\mathbf{w}_1\mathbf{w}\mathbf{w}_2 \approx \mathbf{w}_1\bar{\mathbf{w}}\mathbf{w}_2$*, where* $\bar{\mathbf{w}}$ *is a word obtained from* $\mathbf{w}$ *by any permutation of letters in* $\mathbf{w}$*;*

(2) $\{\text{I}, \text{II}, \text{III}\}^\star$ *implies that* $\mathbf{w}_1\mathbf{w}^2\mathbf{w}_2 \approx \mathbf{w}_1\mathbf{w}_2$*.*

Proof. Part (1) follows from Lemma 3.1(2), and part (2) follows from (1) and the identity I. □

Lemma 3.3 $\Gamma^\star$ *implies that*

$$\mathbf{w}_1x\mathbf{w}y\mathbf{w}_2 \approx \mathbf{w}_1y\bar{\mathbf{w}}y\bar{\mathbf{w}}x\bar{\mathbf{w}}x\mathbf{w}_2,$$

where

(1) $\mathbf{w}_1, \mathbf{w}_2, \mathbf{w}, \bar{\mathbf{w}} \in \mathcal{X}^+$ *and* $\mathsf{C}(\mathbf{w}_1) \cap \mathsf{C}(\mathbf{w}_2) \neq \emptyset$*;*

(2) $x \in \mathsf{C}(\mathbf{w}_2) \cap \mathsf{C}(\mathbf{w})$ *and* $y \in \mathsf{C}(\mathbf{w}_1) \cap \mathsf{C}(\mathbf{w})$*;*

(3) both $\mathbf{w}$ *and* $\bar{\mathbf{w}}$ *are simple words and* $\mathbf{w}$ *and* $x\bar{\mathbf{w}}y$ *can be obtained from one another by permutation of letters;*

(4) $\mathsf{C}(\mathbf{w})\backslash\{x,y\} = \mathsf{C}(\bar{\mathbf{w}}) \subseteq \mathsf{C}(\mathbf{w}_1) \cap \mathsf{C}(\mathbf{w}_2)$.

Proof. It follows that

$$
\begin{array}{lll}
& \mathbf{w}_1 x\mathbf{w} y\mathbf{w}_2 & \\
\approx & \mathbf{w}_1 x^2\bar{\mathbf{w}}y^2\mathbf{w}_2 & \text{by } \mathsf{C}(\mathbf{w}) \subseteq \mathsf{C}(\mathbf{w}_1 x) \cap \mathsf{C}(y\mathbf{w}_2) \text{ and Lemma 3.2(1)} \\
\approx & \mathbf{w}_1 x^2 y\bar{\mathbf{w}}y\mathbf{w}_2 & \text{by } \mathsf{C}(\bar{\mathbf{w}}y) \subseteq \mathsf{C}(\mathbf{w}_1 x^2) \cap \mathsf{C}(y\mathbf{w}_2) \text{ and Lemma 3.2(1)} \\
\approx & \mathbf{w}_1 x^2 y\bar{\mathbf{w}}y\bar{\mathbf{w}}^2\mathbf{w}_2 & \text{by } \mathsf{C}(\bar{\mathbf{w}}) \subseteq \mathsf{C}(\mathbf{w}_1) \cap \mathsf{C}(\mathbf{w}_2) \text{ and Lemma 3.2(2)} \\
\approx & \mathbf{w}_1 x^2 (y\bar{\mathbf{w}})^2\bar{\mathbf{w}}\mathbf{w}_2 & \\
\approx & \mathbf{w}_1 (y\bar{\mathbf{w}})^2 x^2\bar{\mathbf{w}}\mathbf{w}_2 & \text{by Lemma 3.1(3)} \\
\approx & \mathbf{w}_1 y\bar{\mathbf{w}}y\bar{\mathbf{w}}x\bar{\mathbf{w}}x\mathbf{w}_2 & \text{by } \mathsf{C}(x\bar{\mathbf{w}}) \subseteq \mathsf{C}(\mathbf{w}_1(y\bar{\mathbf{w}})^2 x) \cap \mathsf{C}(\mathbf{w}_2) \text{ and Lemma 3.2(1)}
\end{array}
$$

□

Any word $\mathbf{w}$ can always be uniquely written in the form

$$\mathbf{w} = x_1^{e_1} x_2^{e_2} \cdots x_k^{e_k}$$

where $e_i \geq 1$ and $x_i \neq x_{i+1}$ for all i. The sequence $x_1 x_2 \cdots x_k$ is said to be the *appearance sequence* of $\mathbf{w}$. The number of times a letter x appears in the sequence $x_1 x_2 \cdots x_k$ is denoted by $\mathsf{a}(x,\mathbf{w})$. The expression ${}_ix$ in $\mathbf{w}$ means the ith occurrence of a letter x in the appearance sequence.

Example 3.4 *Suppose that* $\mathbf{w} = x^2 zxy^2 xtzx^3 t^4$*. Then*

(1) the appearance sequence of $\mathbf{w}$ *is* $xzxyxtzxt$*;*

(2) $\mathsf{a}(x,\mathbf{w}) = 4$, $\mathsf{a}(y,\mathbf{w}) = 1$, *and* $\mathsf{a}(z,\mathbf{w}) = 2 = \mathsf{a}(t,\mathbf{w})$*;*

(3) $\mathsf{m}({}_1x,\mathbf{w}) = 2$, $\mathsf{m}({}_2x,\mathbf{w}) = \mathsf{m}({}_3x,\mathbf{w}) = 1$, $\mathsf{m}({}_4x,\mathbf{w}) = 3$ *and* $\mathsf{m}(x,\mathbf{w}) = 7$.

A word $\mathbf{u}$ is said to be in *canonical form* if all of the following conditions hold:

(CF1) if $\mathsf{a}(x,\mathbf{u}) = 1$ for some $x \in \mathcal{X}$, then $\mathsf{m}(x,\mathbf{u}) \in \{1,2,3\}$;

(CF2) if $\mathsf{a}(x,\mathbf{u}) = n \geq 2$ for some $x \in \mathcal{X}$, then $\mathsf{m}({}_1x,\mathbf{u}), \mathsf{m}({}_nx,\mathbf{u}) \in \{1,2\}$ and $\mathsf{m}({}_ix,\mathbf{u}) = 1$ for all $i \in \{2,\ldots,n-1\}$;

(CF3) if $\mathbf{u} = \mathbf{w}_1 xy\mathbf{w}_2$ for some $\mathbf{w}_1, \mathbf{w}_2 \in \mathcal{X}^+$ with $x, y \in \mathsf{C}(\mathbf{w}_1) \cap \mathsf{C}(\mathbf{w}_2)$, then ${}_1x$ occurs before ${}_1y$ in $\mathbf{u}$;

(CF4) if $\mathbf{u} = \mathbf{w}_1 x\mathbf{w} y\mathbf{w}_2$ for some $\mathbf{w}_1, \mathbf{w}_2, \mathbf{w} \in \mathcal{X}^+$, $x, y \in \mathsf{C}(\mathbf{w})$, $\mathsf{C}(\mathbf{w})\backslash\{x,y\} \subseteq \mathsf{C}(\mathbf{w}_1) \cap \mathsf{C}(\mathbf{w}_2) \neq \emptyset$, and either $x \in \mathsf{C}(\mathbf{w}_1)\backslash\mathsf{C}(\mathbf{w}_2)$, $y \in \mathsf{C}(\mathbf{w}_2)\backslash\mathsf{C}(\mathbf{w}_1)$ or $x \in \mathsf{C}(\mathbf{w}_2)\backslash\mathsf{C}(\mathbf{w}_1)$, $y \in \mathsf{C}(\mathbf{w}_1)\backslash\mathsf{C}(\mathbf{w}_2)$, then $x \in \mathsf{C}(\mathbf{w}_1)\backslash\mathsf{C}(\mathbf{w}_2)$ and $y \in \mathsf{C}(\mathbf{w}_2)\backslash\mathsf{C}(\mathbf{w}_1)$.

Remark 3.5 *By conditions (CF2), (CF3) and Corollary 3.2(2), it is easy to see that if* $\mathbf{w}_1\mathbf{w}\mathbf{w}_2$ *is a word in canonical form with* $\mathbf{w}_1, \mathbf{w}_2, \mathbf{w} \in \mathcal{X}^+$ *and* $\mathsf{C}(\mathbf{w}) \subseteq \mathsf{C}(\mathbf{w}_1) \cap \mathsf{C}(\mathbf{w}_2)$*, then* $\mathbf{w}$ *is a simple word.*

Proposition 3.6 *Each word is $\Gamma^\star$-equivalent to a word in canonical form.*

Proof. It suffices to apply the identities $\Gamma^\star$ to a word $\mathbf{u}$ until it satisfies all of the conditions (CF1)-(CF4). First, Conditions (CF1) and (CF2) are satisfied by Lemma 3.1(1) and $\text{I} \in \Gamma^\star$. Then condition (CF3) is satisfied by Lemma 3.1(2). Finally, condition (CF4) is satisfied by Lemma 3.3 and Remark 3.5. □

Let $\mathbf{w}$ be any word. Denote by $\widehat{\mathbf{w}}$ the word obtained from $\mathbf{w}$ by retaining the first and the last occurrences of each letter in $\mathbf{w}$. For example, if $\mathbf{w} = x^2zxy^2xtzx^3t^4$, then $\widehat{\mathbf{w}} = xzyytzxt$. It is clear that $\mathsf{C}(\widehat{\mathbf{w}}) = \mathsf{C}(\mathbf{w})$ and

$$\mathsf{m}(x, \widehat{\mathbf{w}}) = \begin{cases} 1 & \text{if } \mathsf{a}(x, \mathbf{w}) = 1, \\ 2 & \text{if } \mathsf{a}(x, \mathbf{w}) \geq 2. \end{cases} \tag{3.1}$$

Now let $\mathbf{w}$ be a word in canonical form and $\widehat{\mathbf{w}} = x_1x_2\cdots x_n$ where $x_1, x_2, \ldots x_n$ are possibly equal letters. Then $\mathbf{w}$ can always be uniquely written in the form

$$\mathbf{w} = (\prod_{i=1}^{n-1} x_i^{e_i}\mathbf{w}_i)x_n^{e_n} \tag{3.2}$$

where

(1) $e_i \geq 1$ for $i = 1, 2, \ldots, n$;

(2) $\mathbf{w}_i$ is a (possibly empty) simple word with $\mathsf{C}(\mathbf{w}_i) \subseteq \{x_1, \ldots, x_i\} \cap \{x_{i+1}, \ldots, x_n\}$, $\mathsf{h}(\mathbf{w}_i) \neq x_i$ and $\mathsf{t}(\mathbf{w}_i) \neq x_{i+1}$ for $i = 1, 2, \ldots, n-1$. In particular, if $\mathbf{w}_i$ is the empty word, then $x_i \neq x_{i+1}$.

It follows from (3.1) that $n = \sum_{x \in \mathsf{C}(\mathbf{w})} \mathsf{m}(x, \widehat{\mathbf{w}})$.

Now let $\mathbf{u}$ and $\mathbf{v}$ be words in canonical form. Then we can rewrite $\mathbf{u}, \mathbf{v}$ in the form of (3.2), that is

$$\mathbf{u} = (\prod_{i=1}^{k-1} x_i^{e_i}\mathbf{u}_i)x_k^{e_k} \qquad \text{and} \qquad \mathbf{v} = (\prod_{j=1}^{l-1} y_j^{f_j}\mathbf{v}_j)y_l^{f_l}. \tag{3.3}$$

In the following four lemmas, we shall show that if $\mathbf{S}_4' \models \mathbf{u} \approx \mathbf{v}$, then $k = l$, $\widehat{\mathbf{u}} = \widehat{\mathbf{v}}$, $e_i = f_i$ and $\mathbf{u}_i = \mathbf{v}_i$ for $i = 1, 2, \ldots, k$, and hence $\mathbf{u} = \mathbf{v}$.

Lemma 3.7 *If $\mathbf{S}_4' \models \mathbf{u} \approx \mathbf{v}$, then*

(1) $\mathsf{a}(x, \mathbf{u}) = \mathsf{a}(x, \mathbf{v}) = 1$ *or* $\mathsf{a}(x, \mathbf{u}), \mathsf{a}(x, \mathbf{v}) \geq 2$ *for each* $x \in \mathsf{C}(\mathbf{u}) = \mathsf{C}(\mathbf{v})$*;*

(2) $k = l$.

Proof. Since A, B, C and A^* are subsemigroups of $\mathbf{S}_4'$, the identity $\mathbf{u} \approx \mathbf{v}$ satisfies all properties listed in Lemmas 2.4, 2.5 and 2.7 and Corollary 2.6. In particular,

$\mathsf{C}(\mathbf{u}) = \mathsf{C}(\mathbf{v})$, $\mathsf{ip}(\mathbf{u}) = \mathsf{ip}(\mathbf{v})$, $\mathsf{fp}(\mathbf{u}) = \mathsf{fp}(\mathbf{v})$, and $\overleftrightarrow{\mathsf{m}}_y(x, \mathbf{u}) \equiv \overleftrightarrow{\mathsf{m}}_y(x, \mathbf{v}) \pmod 2$ for all $x, y \in \mathcal{X}$. Suppose that $\mathsf{a}(x, \mathbf{u}) = 1$ and $\mathsf{a}(x, \mathbf{v}) \geq 2$ for some $x \in \mathcal{X}$. Let

$$\mathbf{u} = \mathbf{u}_1 x^e \mathbf{u}_2, \qquad \mathbf{v} = \mathbf{v}_1 x \mathbf{v}_2 x \mathbf{v}_3,$$

for some $\mathbf{u}_1, \mathbf{u}_2, \mathbf{v}_1, \mathbf{v}_3 \in (\mathcal{X}\backslash\{x\})^*$, $\mathbf{v}_2 \in \mathcal{X}^+$ with $\mathsf{C}(\mathbf{v}_2)\backslash\{x\}$ is nonempty and $e \geq 1$.

Claim 1. $\mathsf{C}(\mathbf{v}_2)\backslash\{x\} \subseteq \mathsf{C}(\mathbf{v}_1) \cup \mathsf{C}(\mathbf{v}_3)$.

In fact, if there exists a letter $y \in \mathsf{C}(\mathbf{v}_2)\backslash\{x\}$ with $y \notin \mathsf{C}(\mathbf{v}_1) \cup \mathsf{C}(\mathbf{v}_3)$, then y occurs in $\mathbf{u}_2$ by $\mathsf{ip}(\mathbf{u}) = \mathsf{ip}(\mathbf{v})$. It follows that $\mathsf{fp}(\mathbf{u}) = \cdots x \cdots y \cdots \neq \cdots y \cdots x \cdots = \mathsf{fp}(\mathbf{v})$. This contradicts the fact that $\mathsf{fp}(\mathbf{u}) = \mathsf{fp}(\mathbf{v})$.

Claim 2. $\mathsf{C}(\mathbf{v}_1) \cap \mathsf{C}(\mathbf{v}_3) \neq \emptyset$.

In fact, suppose that $\mathsf{C}(\mathbf{v}_1) \cap \mathsf{C}(\mathbf{v}_3) = \emptyset$. Then each letter except x in $\mathbf{v}_2$ must occur in just one of $\mathbf{v}_1$ and $\mathbf{v}_3$ by Claim 1. Since $\mathsf{ip}(\mathbf{u}) = \mathsf{ip}(\mathbf{u}_1)x \cdots$ and $\mathsf{ip}(\mathbf{v}) = \mathsf{ip}(\mathbf{v}_1)x \cdots$, it follows from $\mathsf{ip}(\mathbf{u}) = \mathsf{ip}(\mathbf{v})$ that $\mathsf{C}(\mathbf{u}_1) = \mathsf{C}(\mathbf{v}_1)$. Similarly, since $\mathsf{fp}(\mathbf{u}) = \cdots x \mathsf{fp}(\mathbf{u}_2)$ and $\mathsf{fp}(\mathbf{v}) = \cdots x \mathsf{fp}(\mathbf{v}_3)$, it follows from $\mathsf{fp}(\mathbf{u}) = \mathsf{fp}(\mathbf{v})$ that $\mathsf{C}(\mathbf{u}_2) = \mathsf{C}(\mathbf{v}_3)$. Thus $\mathsf{C}(\mathbf{u}_1) \cap \mathsf{C}(\mathbf{u}_2) = \emptyset$. Without loss of generality, we may assume that $\mathsf{C}(\mathbf{v}_2) \cap \mathsf{C}(\mathbf{v}_1) \neq \emptyset$. Define an assignment $\phi : \mathcal{X}^+ \to C$ by

$$t \mapsto \begin{cases} c, & \text{if } t \in \mathsf{C}(\mathbf{u}_1), \\ d, & \text{otherwise.} \end{cases}$$

Then $\mathbf{u}\phi = b$, $\mathbf{v}\phi = a$, and so $C \nvDash \mathbf{u} \approx \mathbf{v}$. It contradicts the fact that $\mathbf{S}_4' \vDash \mathbf{u} \approx \mathbf{v}$.

By Claim 1 there are two cases.

Case 1. $\mathsf{C}(\mathbf{v}_2)\backslash\{x\} \subseteq \mathsf{C}(\mathbf{v}_1) \cap \mathsf{C}(\mathbf{v}_3)$. Then $\mathbf{v}_2$ is a simple word by Remark 3.5. Since $\overleftrightarrow{\mathsf{m}}_x(y, \mathbf{u}) \equiv \overleftrightarrow{\mathsf{m}}_x(y, \mathbf{v}) \pmod 2$ and $\overleftrightarrow{\mathsf{m}}_x(y, \mathbf{u}) = 0$ for each $y \in \mathsf{C}(\mathbf{v}_2)\backslash\{x\}$, $\overleftrightarrow{\mathsf{m}}_x(y, \mathbf{v}) \equiv 0 \pmod 2$ for each $y \in \mathsf{C}(\mathbf{v}_2)\backslash\{x\}$. It contradicts the fact that $\mathbf{v}_2$ is simple.

Case 2. $y \in \mathsf{C}(\mathbf{v}_1)\backslash\mathsf{C}(\mathbf{v}_3)$ or $\mathsf{C}(\mathbf{v}_3)\backslash\mathsf{C}(\mathbf{v}_1)$ for some $y \in \mathsf{C}(\mathbf{v}_2)\backslash\{x\}$. Without loss of generality, we may assume that $y \in \mathsf{C}(\mathbf{v}_3)\backslash\mathsf{C}(\mathbf{v}_1)$ and y is the last letter in $\mathbf{v}_2$ which is the first occurrence in $\mathbf{v}$, that is,

$$\mathbf{v} = \mathbf{v}_1 x \mathbf{w}_1 y \mathbf{w}_2 x \mathbf{v}_3$$

such that $\mathbf{v}_2 = \mathbf{w}_1 y \mathbf{w}_2$ with $\mathbf{w}_1, \mathbf{w}_2 \in \mathcal{X}^*$, $y \in \mathsf{C}(\mathbf{v}_3)\backslash\mathsf{C}(\mathbf{v}_1\mathbf{w}_1)$ and

$$\mathsf{C}(\mathbf{w}_2) \subseteq \mathsf{C}(\mathbf{v}_1 x \mathbf{w}_1 y). \tag{3.4}$$

It follows from $\mathsf{ip}(\mathbf{u}) = \mathsf{ip}(\mathbf{v})$ and $\mathsf{ip}(\mathbf{v}) = \cdots x \cdots y \cdots$ that

$$y \in \mathsf{C}(\mathbf{u}_2)\backslash\mathsf{C}(\mathbf{u}_1). \tag{3.5}$$

Subcase 2.1. $\mathsf{C}(\mathbf{w}_2) \subseteq \mathsf{C}(x\mathbf{v}_3)$. Then $\mathsf{C}(\mathbf{w}_2)\backslash\{x, y\} \subseteq \mathsf{C}(\mathbf{v}_1 x \mathbf{w}_1) \cap \mathsf{C}(\mathbf{v}_3)$ by (3.4). (3.5) implies $\overleftrightarrow{\mathsf{m}}_y(x, \mathbf{u}) = 0$. It follows from $\overleftrightarrow{\mathsf{m}}_y(x, \mathbf{u}) \equiv \overleftrightarrow{\mathsf{m}}_y(x, \mathbf{v}) \pmod 2$ that $\overleftrightarrow{\mathsf{m}}_y(x, \mathbf{v}) \equiv 0 \pmod 2$. Hence $x \in \mathsf{C}(\mathbf{w}_2)$ since $x \notin \mathsf{C}(\mathbf{v}_3)$. On the other hand, since $\overleftrightarrow{\mathsf{m}}_x(y, \mathbf{u}) = 0$, $y \in \mathsf{C}(\mathbf{w}_2)$. Note that $y \in \mathsf{C}(\mathbf{v}_3)\backslash\mathsf{C}(\mathbf{v}_1 x \mathbf{w}_1)$ and $x \in \mathsf{C}(\mathbf{v}_1 x \mathbf{w}_1)\backslash\mathsf{C}(\mathbf{v}_3)$. It contradicts (CF4) by Claim 2.

Subcase 2.2. $z \notin \mathsf{C}(x\mathbf{v}_3)$ for some $z \in \mathsf{C}(\mathbf{w}_2)$. Then $z \in \mathsf{C}(\mathbf{v}_1)$ by Claim 1. It follows from $\mathsf{fp}(\mathbf{u}) = \mathsf{fp}(\mathbf{v})$ and $\mathsf{fp}(\mathbf{v}) = \cdots z \cdots x \cdots$ that

$$z \in \mathsf{C}(\mathbf{u}_1) \backslash \mathsf{C}(\mathbf{u}_2). \tag{3.6}$$

Without loss of generality, we may assume that z is the first letter in $\mathbf{w}_2$ which is the last occurrence in $\mathbf{v}$, that is,

$$\mathbf{v} = \mathbf{v}_1 x \mathbf{w}_1 y \mathbf{p}_1 z \mathbf{p}_2 x \mathbf{v}_3$$

such that $\mathbf{w}_2 = \mathbf{p}_1 z \mathbf{p}_2$ with $\mathbf{p}_1, \mathbf{p}_2 \in \mathcal{X}^*$, $z \notin \mathsf{C}(\mathbf{p}_2 x \mathbf{v}_3)$ and

$$\mathsf{C}(\mathbf{p}_1) \subseteq \mathsf{C}(z\mathbf{p}_2 x \mathbf{v}_3). \tag{3.7}$$

Then $\mathsf{C}(\mathbf{p}_1)\backslash\{y, z\} \subseteq \mathsf{C}(\mathbf{v}_1 x \mathbf{w}_1) \cap \mathsf{C}(\mathbf{p}_2 x \mathbf{v}_3)$ by (3.4) and (3.7). (3.5) together with (3.6) implies $\overleftrightarrow{\mathsf{m}}_y(z, \mathbf{u}) = 0$ and $\overleftrightarrow{\mathsf{m}}_z(y, \mathbf{u}) = 0$. It follows from $\overleftrightarrow{\mathsf{m}}_y(z, \mathbf{u}) \equiv \overleftrightarrow{\mathsf{m}}_y(z, \mathbf{v})$ (mod 2) that $\overleftrightarrow{\mathsf{m}}_y(z, \mathbf{v}) \equiv 0$ (mod 2). Hence

$$z \in \mathsf{C}(\mathbf{p}_1)$$

by $z \notin \mathsf{C}(\mathbf{p}_2 \mathbf{v}_3)$. A similar argument will show that $y \in \mathsf{C}(\mathbf{p}_1)$. Note that $y \in \mathsf{C}(\mathbf{p}_2 x \mathbf{v}_3)\backslash\mathsf{C}(\mathbf{v}_1 x \mathbf{w}_1)$ and $z \in \mathsf{C}(\mathbf{v}_1 x \mathbf{w}_1)\backslash\mathsf{C}(\mathbf{p}_2 x \mathbf{v}_3)$. It contradicts (CF4) by Claim 2.

Consequently, $\mathsf{a}(x, \mathbf{u}) = \mathsf{a}(x, \mathbf{v}) = 1$ or $\mathsf{a}(x, \mathbf{u}), \mathsf{a}(x, \mathbf{v}) \geq 2$ for each $x \in \mathsf{C}(\mathbf{u}) = \mathsf{C}(\mathbf{v})$, and so (1) holds. That (2) holds follows from (1) and (3.1). □

Lemma 3.8 *If* $\mathbf{S}_4' \models \mathbf{u} \approx \mathbf{v}$, *then* $\widehat{\mathbf{u}} = \widehat{\mathbf{v}}$.

Proof. Since A, B, C and A^* are subsemigroups of $\mathbf{S}_4'$, $\mathsf{C}(\mathbf{u}) = \mathsf{C}(\mathbf{v})$, $\mathsf{ip}(\mathbf{u}) = \mathsf{ip}(\mathbf{v})$, $\mathsf{fp}(\mathbf{u}) = \mathsf{fp}(\mathbf{v})$ and $\overleftrightarrow{\mathsf{m}}_y(x, \mathbf{u}) \equiv \overleftrightarrow{\mathsf{m}}_y(x, \mathbf{v})$ (mod 2) for all x and y. It follws form Lemma 3.7(1) that $\mathsf{m}(x, \widehat{\mathbf{u}}) = \mathsf{m}(x, \widehat{\mathbf{v}})$ for all $x \in \mathcal{X}$. Suppose that $\widehat{\mathbf{u}} \neq \widehat{\mathbf{v}}$. Note that $\mathsf{ip}(\widehat{\mathbf{u}}) = \mathsf{ip}(\mathbf{u}) = \mathsf{ip}(\mathbf{v}) = \mathsf{ip}(\widehat{\mathbf{v}})$ and $\mathsf{fp}(\widehat{\mathbf{u}}) = \mathsf{fp}(\mathbf{u}) = \mathsf{fp}(\mathbf{v}) = \mathsf{fp}(\widehat{\mathbf{v}})$. Then $\widehat{\mathbf{u}}(x, y) \neq \widehat{\mathbf{v}}(x, y)$ for some $x, y \in \mathsf{C}(\mathbf{u}) = \mathsf{C}(\mathbf{v})$. It follows from $\mathsf{ip}(\widehat{\mathbf{u}}) = \mathsf{ip}(\widehat{\mathbf{v}})$ and $\mathsf{fp}(\widehat{\mathbf{u}}) = \mathsf{fp}(\widehat{\mathbf{v}})$ that $\mathsf{a}(x, \widehat{\mathbf{u}}) = \mathsf{a}(x, \widehat{\mathbf{v}}) = \mathsf{a}(y, \widehat{\mathbf{u}}) = \mathsf{a}(y, \widehat{\mathbf{v}}) = 2$. Hence $\mathsf{a}(x, \mathbf{u}), \mathsf{a}(x, \mathbf{v}), \mathsf{a}(y, \mathbf{u}), \mathsf{a}(y, \mathbf{v}) \geq 2$ and

$$\mathbf{u} = \mathbf{u}_1 x \mathbf{u}_2 x \mathbf{u}_3 y \mathbf{u}_4 y \mathbf{u}_5, \qquad \mathbf{v} = \mathbf{v}_1 x \mathbf{v}_2 y \mathbf{v}_3 x \mathbf{v}_4 y \mathbf{v}_5,$$

for some $\mathbf{u}_1, \mathbf{v}_1, \mathbf{u}_3, \mathbf{u}_5, \mathbf{v}_5 \in (\mathcal{X}\backslash\{x, y\})^*$, $\mathbf{u}_2, \mathbf{v}_2 \in (\mathcal{X}\backslash\{y\})^*$, $\mathbf{u}_4, \mathbf{v}_4 \in (\mathcal{X}\backslash\{x\})^*$ and $\mathbf{v}_3 \in \mathcal{X}^*$ such that both $\mathsf{C}(\mathbf{u}_2)\backslash\{x\}$ and $\mathsf{C}(\mathbf{u}_4)\backslash\{y\}$ are nonempty. Since $\overleftrightarrow{\mathsf{m}}_x(y, \mathbf{u}) \equiv \overleftrightarrow{\mathsf{m}}_x(y, \mathbf{v})$ (mod 2) and $\overleftrightarrow{\mathsf{m}}_x(y, \mathbf{u}) = 0$, $\overleftrightarrow{\mathsf{m}}_x(y, \mathbf{v}) \equiv 0$ (mod 2), and hence $y \in \mathsf{C}(\mathbf{v}_3)$ by $y \notin \mathsf{C}(\mathbf{v}_2)$. Similarly, we may show that $x \in \mathsf{C}(\mathbf{v}_3)$. Therefore $x, y \in \mathsf{C}(\mathbf{v}_3)$. Furthermore, $\mathsf{C}(\mathbf{v}_3)\backslash\{x, y\}$ is nonempty by condition (CF4).

Claim 1. $\mathsf{C}(\mathbf{v}_3)\backslash\{x, y\} \subseteq \mathsf{C}(\mathbf{v}_1\mathbf{v}_2) \cup \mathsf{C}(\mathbf{v}_4\mathbf{v}_5)$.

In fact, suppose that $z \notin \mathsf{C}(\mathbf{v}_1\mathbf{v}_2) \cup \mathsf{C}(\mathbf{v}_4\mathbf{v}_5)$ for some $z \in \mathsf{C}(\mathbf{v}_3)\backslash\{x, y\}$. Then $\mathsf{ip}(\mathbf{v}) = \cdots y \cdots z \cdots$, and so $z \in \mathsf{C}(\mathbf{u}_4\mathbf{u}_5)$ by $\mathsf{ip}(\mathbf{u}) = \mathsf{ip}(\mathbf{v})$. It follows that $\mathsf{fp}(\mathbf{u}) = \cdots x \cdots z \cdots \neq \cdots z \cdots x \cdots = \mathsf{fp}(\mathbf{v})$. A contradiction.

Claim 2. $\mathsf{C}(\mathbf{v}_1\mathbf{v}_2) \cap \mathsf{C}(\mathbf{v}_4\mathbf{v}_5) \neq \emptyset$.

In fact, suppose $\mathsf{C}(\mathbf{v}_1\mathbf{v}_2) \cap \mathsf{C}(\mathbf{v}_4\mathbf{v}_5) = \emptyset$. Then $z \in \mathsf{C}(\mathbf{v}_1\mathbf{v}_2) \setminus \mathsf{C}(\mathbf{v}_4\mathbf{v}_5)$ or $z \in \mathsf{C}(\mathbf{v}_4\mathbf{v}_5) \setminus \mathsf{C}(\mathbf{v}_1\mathbf{v}_2)$ for each $z \in \mathsf{C}(\mathbf{v}_3) \setminus \{x, y\}$ by Claim 1. Now $\mathsf{ip}(\mathbf{u}) = \mathsf{ip}(\mathbf{v})$ together with

$$\mathsf{ip}(\mathbf{u}) = \mathsf{ip}(\mathbf{u}_1 x \mathbf{u}_2\mathbf{u}_3) y \cdots \quad \text{and} \quad \mathsf{ip}(\mathbf{v}) = \mathsf{ip}(\mathbf{v}_1) x \mathsf{ip}(\mathbf{v}_2) y \cdots$$

implies that $\mathsf{C}(\mathbf{v}_1\mathbf{v}_2) = \mathsf{C}(\mathbf{u}_1\mathbf{u}_2\mathbf{u}_3)$. Dually, we can show that $\mathsf{C}(\mathbf{v}_4\mathbf{v}_5) = \mathsf{C}(\mathbf{u}_3\mathbf{u}_4\mathbf{u}_5)$. Hence $\mathbf{u}_3$ is the empty word and $\mathsf{C}(\mathbf{u}_1\mathbf{u}_2) \cap \mathsf{C}(\mathbf{u}_4\mathbf{u}_5) = \emptyset$. By Claim 1, without loss of generality, we may assume that $\mathsf{C}(\mathbf{v}_3) \cap \mathsf{C}(\mathbf{v}_1\mathbf{v}_2) \neq \emptyset$. Define an assignment $\phi : \mathcal{X}^+ \to C$ by

$$t \mapsto \begin{cases} c, & \text{if } t \in \mathsf{C}(\mathbf{u}_1\mathbf{u}_2) = \mathsf{C}(\mathbf{v}_1\mathbf{v}_2), \\ d, & \text{otherwise.} \end{cases}$$

Then $\mathbf{u}\phi = b$, $\mathbf{v}\phi = a$, and so $C \nvDash \mathbf{u} \approx \mathbf{v}$. This is a contradiction.

Note that $y \in \mathsf{C}(\mathbf{v}_4 y \mathbf{v}_5) \setminus \mathsf{C}(\mathbf{v}_1 x \mathbf{v}_2)$ and $x \in \mathsf{C}(\mathbf{v}_1 x \mathbf{v}_2) \setminus \mathsf{C}(\mathbf{v}_4 y \mathbf{v}_5)$. It follows from Claim 2 and (CF4) that $\mathsf{C}(\mathbf{v}_3) \setminus \{x, y\} \nsubseteq \mathsf{C}(\mathbf{v}_1\mathbf{v}_2) \cap \mathsf{C}(\mathbf{v}_4\mathbf{v}_5)$, that is, $z \in \mathsf{C}(\mathbf{v}_1\mathbf{v}_2) \setminus \mathsf{C}(\mathbf{v}_4\mathbf{v}_5)$ or $z \in \mathsf{C}(\mathbf{v}_4\mathbf{v}_5) \setminus \mathsf{C}(\mathbf{v}_1\mathbf{v}_2)$ for some $z \in \mathsf{C}(\mathbf{v}_3) \setminus \{x, y\}$. Without loss of generality, we may assume that $z \in \mathsf{C}(\mathbf{v}_4\mathbf{v}_5) \setminus \mathsf{C}(\mathbf{v}_1\mathbf{v}_2)$ for some $z \in \mathsf{C}(\mathbf{v}_3) \setminus \{x, y\}$ and z is the last letter in $\mathbf{v}_3$ which is the first occurrence in $\mathbf{v}$, that is,

$$\mathbf{v} = \mathbf{v}_1 x \mathbf{v}_2 y \mathbf{w}_1 z \mathbf{w}_2 x \mathbf{v}_4 y \mathbf{v}_5$$

for $\mathbf{v}_3 = \mathbf{w}_1 z \mathbf{w}_2$ with $\mathbf{w}_1, \mathbf{w}_2 \in \mathcal{X}^*$, $z \in \mathsf{C}(\mathbf{v}_4\mathbf{v}_5) \setminus \mathsf{C}(\mathbf{v}_1\mathbf{v}_2\mathbf{w}_1)$ such that

$$\mathsf{C}(\mathbf{w}_2) \subseteq \mathsf{C}(\mathbf{v}_1 x \mathbf{v}_2 y \mathbf{w}_1 z). \tag{3.8}$$

It follows from $\mathsf{ip}(\mathbf{u}) = \mathsf{ip}(\mathbf{v})$ and $\mathsf{ip}(\mathbf{v}) = \cdots y \cdots z \cdots$ that

$$z \in \mathsf{C}(\mathbf{u}_4\mathbf{u}_5) \setminus \mathsf{C}(\mathbf{u}_1\mathbf{u}_2\mathbf{u}_3). \tag{3.9}$$

Hence $\overleftrightarrow{\mathrm{m}}_z(x, \mathbf{u}) = 0$ and so $\overleftrightarrow{\mathrm{m}}_z(x, \mathbf{v}) \equiv 0 \pmod 2$ by $\overleftrightarrow{\mathrm{m}}_z(x, \mathbf{u}) \equiv \overleftrightarrow{\mathrm{m}}_z(x, \mathbf{v}) \pmod 2$. Thus $x \in \mathsf{C}(\mathbf{w}_2)$ by $x \notin \mathsf{C}(\mathbf{v}_4\mathbf{v}_5)$. Similarly, we can show that $z \in \mathsf{C}(\mathbf{w}_2)$.

Note that $z \in \mathsf{C}(\mathbf{v}_4 y \mathbf{v}_5) \setminus \mathsf{C}(\mathbf{v}_1 x \mathbf{v}_2 y \mathbf{w}_1)$ and $x \in \mathsf{C}(\mathbf{v}_1 x \mathbf{v}_2 y \mathbf{w}_1) \setminus \mathsf{C}(\mathbf{v}_4 y \mathbf{v}_5)$. It follows from (CF4) that $\mathsf{C}(\mathbf{w}_2) \setminus \{x, z\} \nsubseteq \mathsf{C}(\mathbf{v}_1 x \mathbf{v}_2 y \mathbf{w}_1) \cap \mathsf{C}(\mathbf{v}_4 y \mathbf{v}_5)$. Which together with (3.8) implies that $\mathsf{C}(\mathbf{w}_2) \nsubseteq \mathsf{C}(\mathbf{v}_4 y \mathbf{v}_5)$, that is, $t \notin \mathsf{C}(x \mathbf{v}_4 y \mathbf{v}_5)$ for some $t \in \mathsf{C}(\mathbf{w}_2)$. Then $t \in \mathsf{C}(\mathbf{v}_1\mathbf{v}_2)$ by Claim 1. Thus $\mathsf{fp}(\mathbf{v}) = \cdots t \cdots x \cdots$. It follows from $\mathsf{fp}(\mathbf{u}) = \mathsf{fp}(\mathbf{v})$ that

$$t \in \mathsf{C}(\mathbf{u}_1\mathbf{u}_2) \setminus \mathsf{C}(\mathbf{u}_3\mathbf{u}_4\mathbf{u}_5). \tag{3.10}$$

Without loss of generality, we may assume that t is the first letter in $\mathbf{w}_2$ which is the last occurrence in $\mathbf{v}$, that is,

$$\mathbf{v} = \mathbf{v}_1 x \mathbf{v}_2 y \mathbf{w}_1 z \mathbf{p}_1 t \mathbf{p}_2 x \mathbf{v}_4 y \mathbf{v}_5$$

for $\mathbf{w}_2 = \mathbf{p}_1 t \mathbf{p}_2$ with $t \in \mathsf{C}(\mathbf{v}_1\mathbf{v}_2) \setminus \mathsf{C}(\mathbf{p}_2 x \mathbf{v}_4 y \mathbf{v}_5)$ such that

$$\mathsf{C}(\mathbf{p}_1) \subseteq \mathsf{C}(t \mathbf{p}_2 x \mathbf{v}_4 y \mathbf{v}_5). \tag{3.11}$$

Now $\mathsf{C}(\mathbf{p}_1)\backslash\{z,t\} \subseteq \mathsf{C}(\mathbf{v}_1x\mathbf{v}_2y\mathbf{w}_1) \cap \mathsf{C}(\mathbf{p}_2x\mathbf{v}_4y\mathbf{v}_5)$ by (3.8) and (3.11). On the other hand, $\overleftrightarrow{\mathsf{m}}_z(t,\mathbf{u}) = 0$ by (3.9) and (3.10). It follows from $\overleftrightarrow{\mathsf{m}}_z(t,\mathbf{u}) \equiv \overleftrightarrow{\mathsf{m}}_z(t,\mathbf{v})$ (mod 2) that $\overleftrightarrow{\mathsf{m}}_z(t,\mathbf{v}) \equiv 0$ (mod 2), and so $t \in \mathsf{C}(\mathbf{p}_1)$ since $t \notin \mathsf{C}(\mathbf{p}_2\mathbf{v}_4\mathbf{v}_5)$. Similarly, we may show that $z \in \mathsf{C}(\mathbf{p}_1)$. It follows from Claim 2 and (CF4) that $z \in \mathsf{C}(\mathbf{v}_1x\mathbf{v}_2y\mathbf{w}_1)\backslash\mathsf{C}(\mathbf{p}_2x\mathbf{v}_4y\mathbf{v}_5)$ and $t \in \mathsf{C}(\mathbf{p}_2x\mathbf{v}_4y\mathbf{v}_5)\backslash\mathsf{C}(\mathbf{v}_1x\mathbf{v}_2y\mathbf{w}_1)$. A contradiction.

Consequently, $\widehat{\mathbf{u}} = \widehat{\mathbf{v}}$. □

Lemma 3.9 *If* $\mathbf{S}_4' \models \mathbf{u} \approx \mathbf{v}$, *then* $e_i = f_i$ *for* $i = 1, \ldots, k$.

Proof. By Lemmas 3.7 and 3.8, we have

$$\mathbf{u} = x_1^{e_1}\mathbf{u}_1 \cdots x_{k-1}^{e_{k-1}}\mathbf{u}_{k-1}x_k^{e_k} \quad \text{and} \quad \mathbf{v} = x_1^{f_1}\mathbf{v}_1 \cdots x_{k-1}^{f_{k-1}}\mathbf{v}_{k-1}x_k^{f_k},$$

where $\widehat{\mathbf{u}} = x_1 \cdots x_k = \widehat{\mathbf{v}}$. Suppose that $e_i \neq f_i$ for some $i = 1, \ldots, k$ and let $x_i = x$. Without loss of generality, we may assume that x_i is the first occurrence of x in $\widehat{\mathbf{u}} = \widehat{\mathbf{v}}$. It follows from (CF1) and Lemmas 2.4 and 2.7 that $\mathsf{a}(x,\mathbf{u}), \mathsf{a}(x,\mathbf{v}) \geq 2$. By (CF2) we may assume that $e_i = 1$ and $f_i = 2$ and let

$$\mathbf{u} = \mathbf{u}_1x\mathbf{u}_2x\mathbf{u}_3, \qquad \mathbf{v} = \mathbf{v}_1x^2y\mathbf{v}_2x\mathbf{v}_3,$$

for some $\mathbf{u}_1, \mathbf{u}_3, \mathbf{v}_1, \mathbf{v}_3 \in (\mathcal{X}\backslash\{x\})^*$, $\mathbf{u}_2 \in \mathcal{X}^+$, $\mathbf{v}_2 \in \mathcal{X}^*$ with $\mathsf{h}(\mathbf{u}_2) \neq x$ and $y \neq x$. Then $y \notin \mathsf{C}(\mathbf{v}_1) \cap \mathsf{C}(\mathbf{v}_2\mathbf{v}_3)$ by (CF3), so y is its first or last occurrence in $\mathbf{v}$ and $\widehat{\mathbf{u}} = \widehat{\mathbf{v}} = \cdots xy \cdots x \cdots$. It follows that $y \in \mathsf{C}(\mathbf{u}_2)$ and $\mathbf{u}_2 = \mathbf{w}_1y\mathbf{w}_2$ with $\mathsf{C}(\mathbf{w}_1) \subseteq \mathsf{C}(\mathbf{u}_1) \cap \mathsf{C}(\mathbf{w}_2\mathbf{u}_3)$. Hence $\mathbf{w}_1$ is a simple word by Remark 3.5.

If $y \notin \mathsf{C}(\mathbf{v}_1)$, then $y \notin \mathsf{C}(\mathbf{u}_1)$. Suppose that $\mathbf{w}_1$ is not the empty word. Then $z = \mathsf{h}(\mathbf{w}_1) \neq x$ and

$$\overrightarrow{\mathsf{m}}_y(z,\mathbf{u}) = \overrightarrow{\mathsf{m}}_x(z,\mathbf{u}) + 1 \quad \text{and} \quad \overrightarrow{\mathsf{m}}_y(z,\mathbf{v}) = \overrightarrow{\mathsf{m}}_x(z,\mathbf{v}).$$

So $\overrightarrow{\mathsf{m}}_y(z,\mathbf{u}) \not\equiv \overrightarrow{\mathsf{m}}_y(z,\mathbf{v})$ (mod 2) since $\overrightarrow{\mathsf{m}}_x(z,\mathbf{u}) \equiv \overrightarrow{\mathsf{m}}_x(z,\mathbf{u})$ (mod 2). This contradicts Lemma 2.4 (3). Thus $\mathbf{w}_1$ is the empty word and $\overrightarrow{\mathsf{m}}_y(x,\mathbf{u}) = 1 \not\equiv \overrightarrow{\mathsf{m}}_y(x,\mathbf{v}) = 2$ (mod 2). A contradiction.

If $y \in \mathsf{C}(\mathbf{v}_1)\backslash\mathsf{C}(\mathbf{v}_2\mathbf{v}_3)$, then $y \in \mathsf{C}(\mathbf{u}_1)\backslash\mathsf{C}(\mathbf{w}_2\mathbf{u}_3)$. Suppose that $\mathbf{w}_1$ is not the empty word. Then $z = \mathsf{h}(\mathbf{w}_1) \neq x$ and

$$\overleftrightarrow{\mathsf{m}}_y(z,\mathbf{u}) = \overrightarrow{\mathsf{m}}_x(z,\mathbf{u}) - \overrightarrow{\mathsf{m}}_y(z,\mathbf{u}) + 1 \text{ and } \overleftrightarrow{\mathsf{m}}_y(z,\mathbf{v}) = \overrightarrow{\mathsf{m}}_x(z,\mathbf{v}) - \overrightarrow{\mathsf{m}}_y(z,\mathbf{v}).$$

By Lemmas 2.4 and 2.5, we have $\overrightarrow{\mathsf{m}}_x(z,\mathbf{u}) \equiv \overrightarrow{\mathsf{m}}_x(z,\mathbf{v})$ (mod 2) and $\overrightarrow{\mathsf{m}}_y(z,\mathbf{u}) \equiv \overrightarrow{\mathsf{m}}_y(z,\mathbf{v})$ (mod 2). Hence $\overleftrightarrow{\mathsf{m}}_y(z,\mathbf{u}) \not\equiv \overleftrightarrow{\mathsf{m}}_y(z,\mathbf{v})$ (mod 2), a contradiction. Thus $\mathbf{w}_1$ is the empty word and $\overleftrightarrow{\mathsf{m}}_y(x,\mathbf{u}) = 1 \not\equiv \overleftrightarrow{\mathsf{m}}_y(x,\mathbf{v}) = 2$ (mod 2), a contradiction.

Consequently, $e_i = f_i$ for $i = 1, \ldots, k$. □

Lemma 3.10 *If* $\mathbf{S}_4' \models \mathbf{u} \approx \mathbf{v}$, *then* $\mathbf{u}_i = \mathbf{v}_i$ *for* $i = 1, \ldots, k$.

Proof. By Lemmas 3.7, 3.8 and 3.9, we have

$$\mathbf{u} = x_1^{e_1}\mathbf{u}_1 \cdots x_{k-1}^{e_{k-1}}\mathbf{u}_{k-1}x_k^{e_k} \quad \text{and} \quad \mathbf{v} = x_1^{e_1}\mathbf{v}_1 \cdots x_{k-1}^{e_{k-1}}\mathbf{v}_{k-1}x_k^{e_k},$$

where $\widehat{\mathbf{u}} = x_1 \cdots x_k = \widehat{\mathbf{v}}$. It follows from (CF2) that both $\mathbf{u}_i$ and $\mathbf{v}_i$, $i = 1, \ldots, k$, are (possibly empty) simple words. To prove $\mathbf{u}_i = \mathbf{v}_i$ for $i = 1, \ldots, k$, it only need to prove $\mathsf{C}(\mathbf{u}_i) = \mathsf{C}(\mathbf{v}_i)$ for $i = 1, \ldots, k$ by (CF3) and Lemma 2.4 (1). Suppose that $y \in \mathsf{C}(\mathbf{u}_i) \setminus \mathsf{C}(\mathbf{v}_i)$ for some $i = 1, \ldots, k$ and $y \in \mathcal{X}$. Then $\mathsf{m}(y, \mathbf{u}_i) = 1$ since $\mathbf{u}_i$ is a simple word. Without loss of generality, we may assume that x_i is the first occurrence in $\widehat{\mathbf{u}} = \widehat{\mathbf{v}}$. There are two cases.

Case 1. x_{i+1} is the last occurrence in $\widehat{\mathbf{u}} = \widehat{\mathbf{v}}$. If $y \neq x_i, x_{i+1}$, then

$$\mathsf{m}(y, \mathbf{u}) = \overrightarrow{\mathsf{m}}_{x_i}(y, \mathbf{u}) + 1 + \overleftarrow{\mathsf{m}}_{x_{i+1}}(y, \mathbf{u}) \;\text{ and }\; \mathsf{m}(y, \mathbf{v}) = \overrightarrow{\mathsf{m}}_{x_i}(y, \mathbf{v}) + \overleftarrow{\mathsf{m}}_{x_{i+1}}(y, \mathbf{v}).$$

It follows from Lemmas 2.4 (3) and 2.5 (3) that $\mathsf{m}(y, \mathbf{u}) \not\equiv \mathsf{m}(y, \mathbf{v}) \pmod 2$. It contradicts Corollary 2.6. If $y = x_i$, then $x_{i+1} \neq x_i$ by (CF3) and

$$\mathsf{m}(x_i, \mathbf{u}) = e_i + 1 + \overleftarrow{\mathsf{m}}_{x_{i+1}}(x_i, \mathbf{u}) \;\text{ and }\; \mathsf{m}(x_i, \mathbf{v}) = e_i + \overleftarrow{\mathsf{m}}_{x_{i+1}}(x_i, \mathbf{v}).$$

It follows from Lemma 2.5 (3) that $\mathsf{m}(x_i, \mathbf{u}) \not\equiv \mathsf{m}(x_i, \mathbf{v}) \pmod 2$. It contradicts Corollary 2.6. If $y = x_{i+1}$, then $x_{i+1} \neq x_i$ by (CF3). A similar argument will yield a contradiction.

Case 2. x_{i+1} is the first occurrence in $\widehat{\mathbf{u}} = \widehat{\mathbf{v}}$. Then $y \neq x_{i+1}$. If $y \neq x_i$, then

$$\overrightarrow{\mathsf{m}}_{x_{i+1}}(y, \mathbf{u}) = \overrightarrow{\mathsf{m}}_{x_i}(y, \mathbf{u}) + 1 \;\text{ and }\; \overrightarrow{\mathsf{m}}_{x_{i+1}}(y, \mathbf{v}) = \overrightarrow{\mathsf{m}}_{x_i}(y, \mathbf{v}).$$

and so $\overrightarrow{\mathsf{m}}_{x_{i+1}}(y, \mathbf{u}) \not\equiv \overrightarrow{\mathsf{m}}_{x_{i+1}}(y, \mathbf{v}) \pmod 2$ by Lemma 2.4 (3). A contradiction. If $y = x_i$, then

$$\overrightarrow{\mathsf{m}}_{x_{i+1}}(x_i, \mathbf{u}) = e_i + 1 \;\text{ and }\; \overrightarrow{\mathsf{m}}_{x_{i+1}}(x_i, \mathbf{v}) = e_i$$

and so $\overrightarrow{\mathsf{m}}_{x_{i+1}}(y, \mathbf{u}) \not\equiv \overrightarrow{\mathsf{m}}_{x_{i+1}}(y, \mathbf{v}) \pmod 2$. It contradicts Lemma 2.4 (3).

Consequently, $\mathsf{C}(\mathbf{u}_i) = \mathsf{C}(\mathbf{v}_i)$ and so $\mathbf{u}_i = \mathbf{v}_i$ for $i \in \{1, \ldots, k\}$. □

Corollary 3.11 *Let* $\mathbf{u}$ *and* $\mathbf{v}$ *be words in canonical form. Then* $\mathbf{S}_4' \models \mathbf{u} \approx \mathbf{v}$ *if and only if* $\mathbf{u} = \mathbf{v}$.

Proof. It follows from Lemmas 3.7, 3.8, 3.9 and 3.10. □

Now we are ready to prove that $\Gamma^\star$ is an identity basis for the variety $\mathbf{S}_4'$.

Theorem 3.12 *The identity system* $\Gamma^\star$ *is an identity basis for the variety* $\mathbf{S}_4'$. *In particular,* $\mathbf{S}_4'$ *is finitely based.*

Proof. Up to isomorphism and anti-isomorphism, there are 10 non-permutative and non-idempotent semigroups of order four. Their multiplication tables of these 10 semigroups except A, B and C are given by (listed by lexicographical order) [4]:

```
aaaa aaaa aaaa aaaa aaaa aaaa aaaa
aaaa aaaa aaab aaab aaab abbd abcd
aacd abcd abca abcc cccc abbd acbd
dddd dddd aaad abcd aacd dddd dddd
```

It is routine to verify that the identities $\Gamma^\star$ hold in every of these 10 semigroups so that the variety $\mathbf{S}_4'$ satisfies the identities $\Gamma^\star$. It remains to show that any identity $\mathbf{u} \approx \mathbf{v}$ of the variety $\mathbf{S}_4'$ is a consequence of the identities $\Gamma^\star$. In the presence of Proposition 3.6, it suffices to assume that $\mathbf{u}$ and $\mathbf{v}$ are in canonical form. The identity $\mathbf{u} \approx \mathbf{v}$ is then trivial by Corollary 3.11 and so is vacuously a consequence of the identities $\Gamma^\star$. □

Corollary 3.13 $\mathbf{S}_4' = \mathbf{A} \vee \mathbf{B} \vee \mathbf{C} \vee \mathbf{A}^*$.

Proof. It follows from the proof of Lemmas 3.7, 3.8, 3.9 and 3.10 and Theorem 3.12. □

References

[1] J. Almeida, *Finite Semigroups and Universial Algebra*, World Scientific, Singapore, 1994.

[2] A. P. Birjukov, *Varieties of idempotent semigroups*, Algebra i Logika **9** (1970), 255–273 [Russian].

[3] C. C. Edmunds, *On certain finitely based varieties of semigroups*, Semigroup Forum **15** (1977), 21–39.

[4] C. C. Edmunds, *Varieties generated by semigroups of order four*, Semigroup Forum **21** (1980), 67–81.

[5] C. F. Fennemore, *All varieties of bands,* I, II, Math. Nachr. **48** (1971), 237–262.

[6] J. A. Gerhard, *The lattice of equational classes of idempotent semigroups*, J. Algebra **15** (1970), 195–224.

[7] J. A. Gerhard and M. Petrich, *All varieties of regular orthogroups*, Semigroup Forum **31** (1985), 311–351.

[8] M. Jackson, *Finte semigroups whose varieties have uncountably many subvarieties*, Journal of Algebra **228** (2000), 512–535.

[9] R. L. Kruse, *Identities satisfied by a finite ring,* J. Algebra **26** (1973), 298–318.

[10] E. W. H. Lee, *Identity bases for some non-exact varieties*, Semigroup Forum **68** (2004), 445–457.

[11] E. W. H. Lee, *Subvarieties of the variety generated by the five-element Brandt semigroup*, Internat. J. Algebra Comput. **16** (2006), 417–441.

[12] E. W. H. Lee, *Minimal semigroups generating varieties with complex subvariety lattices*, Internat. J. Algebra Comput. **17** (2007) 1553–1572.

[13] Y. F. Luo and W. T. Zhang, *On the variety generated by all semigroups of order three*, submited to J. Algebra.

[14] S. Oates and M. B. Powell, *Identical relations in finite groups*, J. Algebra **1** (1964), 11–39.

[15] P. Perkins, *Bases for equational theories of semigroups*, J. Algebra **11** (1969), 298–314.

[16] M. V. Sapir, *Problems of Burnside type and the finite basis property in varieties of semigroups,* Math. USSR-Izv. **30**(2) (1988) 295–314; translation of Izv. Akad. Nauk SSSR Ser. Mat. **51**(2) (1987), 319–340.

[17] L. N. Shevrin and E. V. Sukhanov, *Structural aspects of the theory of varieties of semigroups,* Soviet Math. (Iz. VUZ) **33**(6) (1989) 1–34; translation of Izv. Vyssh. Uchebn. Zaved. Mat. **6** (1989), 3–39.

[18] L. N. Shevrin and M. V. Volkov, *Identities of semigroups*, Soviet Math. (Iz. VUZ) **29** (11)(1985), 1–64; translation from Izv. Vyssh. Uchebn. Zaved. Mat. **11** (1985), 3–47.

[19] L. N. Shevrin, B. M. Vernikov and M. V. Volkov, *Lattices of Semigroup Varieties,* Russian Math. (Iz. VUZ) **53**(3) (2009) 1–28; translation of Izv. Vyssh. Uchebn. Zaved. Mat. **3** (2009), 3–36.

[20] A. V. Tishchenko, *The finiteness of a base of identities for five-element monoids,* Semigroup Forum **20** (1980), 171–186.

[21] A. N. Trahtman, *Finiteness of identity bases of five-element semigroups,* in: E. S. Lyapin (Ed.), Semigroups and Their Homomorphisms, Ross. Gos. Ped. Univ., Leningrad, 1991, 76–97 (in Russian).

[22] M. V. Volkov, *The finite basis problem for finite semigroups*, Mathematica Japonica **53** (2001), 171–199.

[23] W. T. Zhang and Y. F. Luo, *The subvariety lattice of the join of two semigroup varieties*, Acta Mathematica Sinica-English Series, **25**(3) (2009).

[24] W. T. Zhang and Y. F. Luo, *On varieties generated by minimal complex semigroups*, Order **25** (2008), 243-266.